Fourth Edition

Handbook of
ENVIRONMENTAL
HEALTH

Volume 1

*Biological, Chemical,
and Physical Agents of
Environmentally Related Disease*

Herman Koren • Michael Bisesi

NEHA

Co-published with the
National Environmental Health Association

CRC Press
Taylor & Francis Group
Boca Raton London New York

CRC Press is an imprint of the
Taylor & Francis Group, an **informa** business

CRC Press
Taylor & Francis Group
6000 Broken Sound Parkway NW, Suite 300
Boca Raton, FL 33487-2742

First issued in paperback 2017

© 2003 by Taylor & Francis Group, LLC

CRC Press is an imprint of Taylor & Francis Group, an Informa business

No claim to original U.S. Government works

ISBN-13: 978-1-56670-536-3 (hbk)
ISBN-13: 978-0-8153-7130-4 (pbk)

Library of Congress Card Number 2002016110

Library of Congress Cataloging-in-Publication Data

Koren, Herman.
    Handbook of environmental health / by Herman Koren and Michael Bisesi.
       p. cm.
    Includes bibliographical references and index.
    Contents: v. 1. Biological, chemical, and physical agents of environmental related disease — v. 2. Pollution interactions in air, water, and soil.
    ISBN 1-56670-536-3 (alk. paper)
    1. Environmental health — Handbooks, manuals, etc. 2. Environmental engineering — Handbooks, manuals, etc. I. Title: Environmental health. II. Bisesi, Michael S. III. Title.

RA565 .K67 2002
363.7—dc21
                                                    2002016110

Visit the Taylor & Francis Web site at
http://www.taylorandfrancis.com

and the CRC Press Web site at
http://www.crcpress.com

Dedication

To Donna Lee Koren and Christine Bisesi, our wives, dearest and very best friends, for all they have done to enhance our lives and encourage us to teach our students the true significance of improving the environment for all people.

Foreword

I have spent a career in this field. My years of experience have taught me that environmental health might best be described as a colorful, complex, and diverse spectrum of interrelated topics. These topics range from the individual to the compounded effects of pollution; to the impacts of contaminated air, water, land, food, and indoor and outdoor environments; and to biological, physical, chemical, and radiological hazards. Because this field is so broad, to protect both our environment and our people properly, it is important that one possesses a credible understanding of basic science, laws and regulations, governmental and private programs, disease and injury identification, and prevention and control. Moreover, because so many environmental concerns impact on other elements of the environment, it is important that even the specialists within our field appreciate and understand the implications of how their issues can impact other environmental concerns. A reading and comprehension of the material in these books can help greatly in building that overarching understanding that professionals in this field need to have.

The *Handbook of Environmental Health* has been, for the past 23 years, an excellent source for gaining that needed understanding of interrelated and current environmental topics. The presentations offered by this publication are particularly helpful inasmuch as they are comprehensive while concise and basic. The two volumes cover basic and applied chemistry relative to both toxicity to humans and the fate of natural and anthropogenic contaminants in the environment; basic and applied microbiology relative to both pathogenicity to humans and the fate of natural and anthropogenic contaminants in the environment; current status of each environmental problem area; discussion of the problem; potential for intervention; resources available for use, standards, practices, and techniques utilized to resolve the problem; surveillance and evaluation techniques; appropriate controls, laws, and regulations; future research needs; large number of current references; state-of-the-art graphics, and major environmental, including industrial hygiene, sampling and analytic instruments.

The fourth edition of Volume I, Chapters 1 and 2, provides a significant understanding of basic new environmental issues, energy, emerging infectious diseases, recent laws, emerging microorganisms, toxicology, epidemiology, human physiology, and the effects of the environment on humans. The remainder of the chapters discuss a variety of indoor environmental issues, including food safety, food technology, insect and rodent control, pesticides, indoor environment, institutional environment, recreational environment, occupational environment, and instrumentation. Some of the new and significantly expanded and updated sections include: new food codes and programs; emerging and reemerging insectborne disease; insecticide resistance; pesticides and water quality; indoor air pollution; asthma; monitoring environmental disease; homelessness and disease; emerging zoonoses; bacteria in hospitals and nursing homes; fungal and viral agents in laboratories; prions; guidelines for infection control; principles of biosafety; a variety of risk assessment techniques; and an updated overview of the occupational environment.

The fourth edition of Volume II discusses a variety of outdoor environmental issues including changes to the Clean Air Act; PM 2.5; toxic air pollutants; risk assessment and air, water, solid, and hazardous waste and the interrelationship between these areas; methyl bromide; air quality index; air, water, and solid waste programs; technology transfer; methyl tertiary butyl ether; toxic releases from waste; technical tracking systems; biological processes and solid and hazardous wastes; storm water runoff; ocean dumping; waste to energy; toxics release inventory; brownfields; contaminated governmental facilities; Superfund update; geographic information systems; pollution prevention programs; environmental justice; new laws; waterborne disease update; national mapping; EnviroMapper; safe drinking water standards; maximum contaminant level; crossconnections, backflow, and back-siphonage; sewage pretreatment technologies; leaching field chambers; constructed wetlands; drip and spray irrigation systems; peat bed filters; zone of initial dilution; new laws; wetlands; nonpoint source pollution; national water quality assessments; dredging waste; trihalomethanes; environmental studies; bioterrorism; and the Federal Emergency Management Agency. There are chapters on air quality management, solid and hazardous waste management, private and public water supplies, swimming areas, plumbing, private and public sewage disposal and soils, water pollution water quality controls, terrorism and environmental health emergencies, and instrumentation.

Two well-regarded environmental health professionals have written these two books. They have each conducted extensive research and are extremely knowledgeable about all areas of the environmental, industrial hygiene, and health related fields.

Dr. Herman Koren is a founding director of the environmental health science and internship program at Indiana State University and is a professor emeritus there. He has gained respect as a researcher, teacher, consultant, and practitioner in the environmental health, hospital, and medical care fields as well as in management areas related to these fields; nursing homes; water and wastewater treatment plants; and other environmental and safety industries for the past 47 years. Since his retirement from Indiana State University in 1995, he has continued his lifelong quest to gain and interpret the latest knowledge possible and to share it with students and other professionals. He also continues to give numerous presentations and workshops and has written and rewritten several books.

Dr. Michael Bisesi, professor and chairperson, department of public health, and associate dean, graduate allied health programs, Medical College of Ohio, School of Allied Health, has been in the environmental and occupational health fields for 20 years. He, too, is respected as a researcher, teacher, consultant, practitioner, and administrator. In addition to his environmental science and industrial hygiene accreditations, he is an expert in human exposure assessment and environmental toxicology. He also holds appointments in the School of Pharmacology and School of Medicine, has written numerous scientific and technical articles and chapters in scientific books, and is the author or co-author of several additional books.

The books are user-friendly to a variety of individuals including generalist professionals as well as specialists, industrial hygiene personnel, health and medical personnel, managers, and students. These publications can be used to look up specific information or to gain deeper knowledge about an existing problem area. The section on surveillance techniques helps the individual decide the extent and nature of a problem. The appropriate and applicable standards, rules, and regulations help the reader resolve a problem. Further information and assistance can be gained through the resource area and through the review of many of the updated bibliographical references. Except for Chapters 1, 2, 11, 12, in Volume I, and Chapters 8 and 9 in Volume II, all chapters follow the same format, thereby making the books relatively easy to use. The extensive index for both volumes in each book is also very useful.

Thank you, Professors Koren and Bisesi, for providing environmental health professionals, new or seasoned, generalist or specialist, with such a helpful resource.

Nelson Fabian
Executive Director
National Environmental Health Association
February 28, 2002

Preface

This handbook, in two volumes, is designed to provide a comprehensive but concise discussion of each of the important environmental health areas, including energy, ecology and people, environmental epidemiology, risk assessment and risk management, environmental law, air quality management, food protection, insect control, rodent control, pesticides, chemical environment, environmental economics, human disease and injury, occupational health and safety, noise, radiation, recreational environment, indoor environments, medical care institutions, schools and universities, prisons, solid and hazardous waste management, water supply, plumbing, swimming areas, sewage disposal, soils, water pollution control, environmental health emergencies, and nuisance complaints.

Sufficient background material is introduced throughout these texts to provide students, practitioners, and other interested readers with an understanding of the areas under discussion. Common problems and potential solutions are described; graphs, computerized drawings, inspection sheets, and flowcharts are utilized as needed to consolidate or clarify textual material. All facts and data come from the most recent federal government documents, many of which date from the late 1990s and early 2000s. Rules and regulations specified will continue to be in effect into the early 2000s. For rapidly changing areas in which the existing material used is likely to become dated, the reader is referred to the appropriate sources under resources and in the bibliography to update a given environmental health area or portion of an area as needed. This enhances the value of the text by providing basic and current materials that will always be needed and secondary sources that will enable the reader to keep up to date.

These books are neither engineering texts nor comprehensive texts in each area of study. Their purpose is to provide a solid working knowledge of each environmental health area with sufficient detail for practitioners and students. The text can be used in basic courses in environmental health, environmental pollution, ecology, and environment and people that are offered at all universities and colleges in the United States and abroad. These courses are generally taught in departments of life science, geology, science education, environmental health, and health and safety. For general areas of study, the instructor can omit specific details, such as resources, standards, practices and techniques, and modes of surveillance and evaluation. This same approach may be used by schools of medicine, nursing, and allied health sciences for their students. These texts are also suitable for basic introductory courses in schools of public health, environmental health, and sanitary science, as well as junior colleges offering 2-year degree programs in sanitary science and environmental science.

Practitioners in a variety of environmental health and occupational health and safety fields will find these volumes handy references for resolving current problems and for obtaining a better understanding of unfamiliar areas. Practitioners and administrators in other areas, such as food processing, water-quality control, occupational health and safety, and solid and hazardous waste management, will also find these reference books useful.

High school teachers often must introduce environmental health topics in their classes and yet have no specific background in this area. These books could serve as a text in graduate education courses for high school teachers as well as a reference source.

Public interest groups and users of high school and community libraries will obtain an overall view of environmental problems by reading Chapter 1; Chapter 2; and the sections in each chapter titled "Background and Status, Problems, Potential for Intervention, Resources, and Control." This volume also supplies a concise reference for administrators in developing nations because it explains tested controls and provides a better understanding of environmental problems; various standards, practices, and techniques; and a variety of available resources.

The material divides easily into two separate courses. Course I would correspond to the content of Volume I and would include Chapter 1, Environment and Humans; Chapter 2, Environmental Problems and Human Health; Chapter 3, Food Protection; Chapter 4, Food Technology; Chapter 5, Insect Control; Chapter 6, Rodent Control; Chapter 7, Pesticides; Chapter 8, Indoor Environment; Chapter 9, Institutional Environment; Chapter 10, Recreational Environment; Chapter 11, Occupational Environment; and Chapter 12, Major Instrumentation for Environmental Evaluation of Occupational, Residential, and Public Indoor Settings.

Course II, corresponding to the content of the Volume II, would include Chapter 1, Air Quality Management; Chapter 2, Solid and Hazardous Waste Management; Chapter 3, Private and Public Water Supplies; Chapter 4, Swimming Areas; Chapter 5, Plumbing; Chapter 6, Private and Public Sewage Disposal and Soils; Chapter 7, Water Pollution and Water Quality Controls; Chapter 8, Terrorism and Environmental Health Emergencies; and Chapter 9, Major Instrumentation for Environmental Evaluation of Ambient Air, Water, and Soil.

Because the problems of the environment are so interrelated, certain materials must be presented at given points to give clarity and cohesiveness to the subject matter. As a result, the reader may encounter some duplication of materials throughout the text.

With the exception of Volume I, Chapters 1, 2, 11, and 12, and Volume II, Chapters 8 and 9, all the chapters have a consistent style and organization, facilitating retrieval. The introductory nature of Volume I (Chapters 1 and 2) as well as the unusual nature of Volume II (Chapter 8) do not lend themselves to the standard format. Volume I (Chapter 12) and Volume II (Chapter 9) discuss instrumentation for the specific areas of each volume and therefore do not follow standard format.

In Volume I (Chapter 1), the reader is introduced to the underlying problems, basic concerns, and basic philosophy of environmental health. The ecological, economic, and energy bases provided help individuals understand their relationship to the ecosystem and to the real world of economic and energy concerns. It also provides an understanding of the role of government and the environmental health practitioner in helping to resolve environmental and ecological dilemmas created by humans. Chapter 2 on human health helps the reader understand the relationship between biological, physical, and chemical agents, and disease and injury causation.

In Volume II, Chapter 8, the many varied facets of terrorism and environmental emergencies, nuisances, and special problems are discussed. Students may refer to other chapters of the text to obtain a complete idea of each of the problems and the potential solutions.

The general format of Volume I, Chapters 3 to 11, and Volume II, Chapters 1 to 7, is as follows:

STANDARD CHAPTER OUTLINE

1. Background and status (brief)
2. Scientific, technological, and general information
3. Problem
   a. Types
   b. Sources of exposure
   c. Impact on other problems
   d. Disease potential
   e. Injury potential
   f. Other sources of exposure contributing to problems
   g. Economics
4. Potential for intervention
   a. General
   b. Specific
5. Resources
   a. Scientific and technical; industry, labor, university; research groups
   b. Civic
   c. Governmental
6. Standards, practices, and techniques
7. Modes of surveillance and evaluation
   a. Inspections and surveys
   b. Sampling and laboratory analysis
   c. Plans review
8. Control
   a. Scientific and technological
   b. Governmental programs
   c. Other programs
   d. Education
9. Summary
10. Research needs

- The background and status section of each chapter presents a brief introduction to, and the current status of, each problem area. An attempt has been made in each case to present the current status of the problem.
- The problem section is subdivided into several important areas to give the reader a better grasp of the total concerns. To avoid disruption in continuity of the standard outline, the precise subtitles listed may not be found in each chapter. However, the content of the subtitles will be present. The subtitle, impact on other problems, is given as a constant reminder that one impact on the environment may precipitate numerous other problems.
- The potential for intervention section is designed to succinctly illustrate whether a given problem can be controlled, the degree of control possible, and some techniques of control. The reader should refer to the controls section for additional information.

- Resources is a unique section providing a listing of scientific, technical, civic, and governmental resources available at all levels to assist the student and practitioner.
- The standards, practices, and techniques section is specifically geared to the reader who requires an understanding of some of the specifics related to surveys, environmental studies, operation, and control of a variety of program areas.
- The modes of surveillance and evaluation section explains many of the techniques available to determine the extent and significance of environmental problems.
- The control section presents existing scientific, technological, governmental, educational, legal, and civic controls. The reader may refer to the standards, practices, and techniques section in some instances to get a better understanding of controls.
- The summary presents the highlights of the chapter.
- Research needs is another unique section intended to increase reader awareness to the constantly changing nature of the environment and of the need for continued reading or in-service education on the future concerns of our society.
- The reference section is extensive and as current as possible. It appears as the last area in each volume and provides the reader with sources for further research and names of individuals and organizations involved in current research.

Acknowledgments

I extend thanks to Boris Osheroff, my friend, teacher, and colleague, for opening the numerous doors needed to obtain the most current information in the environmental health field, and for contributing his many fine suggestions and ideas before and after reading the manuscript; to Ed O'Rourke for intensively reviewing the manuscript and recommending revisions and improvements; to Karol Wisniewiski for giving his review and comments on the manuscript; to Dr. John Hanlon for offering his encouragement and for helping a young teacher realize his potential; to all the environmental health administrators, supervisors, practitioners, and students for sharing their experiences and problems with me and for giving me the opportunity to test many of the practical approaches used in the book; to the National Institute of Occupational Safety and Health, National Institutes of Health, National Institute of Environmental Health Science, U.S. Environmental Protection Agency, U.S. Food and Drug Administration, Cunningham Memorial Library, Indiana State University, Indiana University Library, Purdue University Library, and the many other libraries and resources for providing the material that was used in developing the manuscript; to my wife, Donna Koren, and my student, Evelyn Hutton, for typing substantial portions of the manuscript; to my daughter, Debbie Koren, for helping me organize the materials and for working along with me throughout the night at the time of deadlines to complete the work.

In the second edition, Kim Malone typed portions of the new manuscript, retyped the entire manuscript, and was of great value to me. Pat Ensor, librarian, Indiana State University, was of considerable value in helping gather large numbers of references in all areas of the book. A very special thanks goes to my sister-in-law, Betty Gardner, for typing a substantial portion of the new manuscript, despite recurring severe illness. Her cheerfulness during my low periods helped me complete my work. Finally, thanks go to my wife Donna for putting up with my thousands of hours of seclusion in the den, while I was working, and for encouraging me throughout the project and my life with her. She has truly been my best friend.

In the third edition, Alma Mary Anderson, C.S.C., and her assistants, Carlos Gonzalez and Brian Flynn, redid the existing illustrations and added new ones to enhance the manuscript. Professor Anderson directed the production of all the new artwork. In addition, I thank Bill Farms for his assistance with the original computer-assisted drawings for the chapters on instrumentation in both volumes. My wife Donna typed much new material, and the previously mentioned libraries, Centers for Disease Control, and the University of South Florida were most helpful in my research efforts. I would like to recognize Dr. Michael Bisesi, friend and colleague, who became my co-author.

In the fourth edition, Professor Anderson, advisor of the graphic design area in the department of art, Indiana State University, provided many excellent graphics and endless hours of work inserting new material and correcting previous material in the manuscripts. Without her help this new edition would not have been possible. Thanks also go to my daughter and son-in-law, Debbie and Kenny Hardas, who dragged me into the computer age by purchasing my first computer and by teaching me to use it.

Dr. Michael Bisesi acknowledges his appreciation of his wife Christine Bisesi, M.S., C.I.H., C.H.M.M.; his two sons Antonio (Nino) and Nicolas (Nico); and his parents Anthony (deceased) and Maria Bisesi for their love and support. In addition, he wants to acknowledge his mentors Rev. Francis Young; George Berkowitz, Ph.D.; Raymond Manganelli, Ph.D.; Barry Schlegel, M.S., C.I.H.; John Hochstrasser, Ph.D., C.I.H.; Richard Spear, H.S.D.; Christopher Bork, Ph.D.; Keith Schlender, Ph.D.; and Roy Hartenstein, Ph.D. for sharing their knowledge, wisdom, and encouragement at various phases of his academic and professional career.

About the Authors

Herman Koren, R.E.H.S., M.P.H., H.S.D., is professor emeritus and former director of the environmental health science program, and director of the supervision and management program I and II at Indiana State University at Terre Haute. He has been an outstanding researcher, teacher, consultant, and practitioner in the environmental health field, and in the occupational health, hospital, medical care, and safety fields, as well as in management areas of these areas and in nursing homes, water and wastewater treatment plants, and other environmental and safety industries for the past 47 years. In addition to numerous publications and presentations at national meetings, he is the author of six books, titled *Environmental Health and Safety*, Pergamon Press, 1974; *Handbook of Environmental Health and Safety*, Volumes I and II, Pergamon Press, 1980 (now published in updated and vastly expanded format by Lewis Publishers, CRC Press, as a fourth edition); *Basic Supervision and Basic Management*, Parts I and II, Kendall Hunt Publishing, 1987 (now published in updated and vastly expanded format as *Management and Supervision for Working Professionals*, Volumes I and II, third edition, by Lewis Publishers, CRC Press); *Illustrated Dictionary of Environmental Health and Occupational Safety*, Lewis Publishers, CRC Press, 1995, second edition due in 2004. He has served as a district environmental health practitioner and supervisor at the local and state level. He was an administrator at a 2000-bed hospital. Dr. Koren was on the editorial board of the *Journal of Environmental Health* and the former *Journal of Food Protection*. He is a founder diplomate of the Intersociety Academy for Certification of Sanitarians, a Fellow of the American Public Health Association, a 46-year member of the National Environmental Health Association, founder of the Student National Environmental Health Association, and the founder and advisor of the Indiana State University Student National Environmental Health Association (Alpha chapter). Dr. Koren developed the modern internship concept in environmental health science. He has been a consultant to the U.S. Environmental Protection Agency, the National Institute of Environmental Health Science, and numerous health departments and hospitals; and has served as the keynote speaker and major lecturer for the Canadian Institute of Public Health Inspectors. He is the recipient of the Blue Key Honor Society Award for outstanding teaching and the Alumni and Student plaque and citations for outstanding teaching, research, and service. The National Environmental Health Association has twice honored Dr. Koren with presidential citations for "Distinguished Services, Leadership and Devotion to the Environmental Health Field" and "Excellent Research and Publications."

Michael S. Bisesi, Ph.D., R.E.H.S., C.I.H., is an environmental and occupational health scientist and board certified industrial hygienist working full-time as professor and chairman of the department of public health in the School of Allied Health at the Medical College of Ohio (MCO). He also has a joint appointment in the department of pharmacology in the School of Medicine, serves as the associate dean of allied health programs, and is director of the Northwest Ohio Consortium for Public Health. At MCO, he is responsible for research, teaching, service, and administration. He teaches a variety of graduate level and continuing education courses, including toxicology, environmental health, monitoring and analytical methods, and hazardous materials and waste. His major laboratory and field interests are environmental  toxicology involving biotic and abiotic transformation of organics; fate of pathogenic agents in various matrices; and industrial hygiene evaluation of airborne biological and chemical agents relative to human exposure assessment. He also periodically provides applicable consulting services via Enviro-Health, Inc., Holland, OH.

Dr. Bisesi earned a B.S. and an M.S. in environmental science from Rutgers University and a Ph.D. in environmental science from the SUNY College of Environmental Science and Forestry in association with Syracuse University. He continues to complete additional postgraduate course work, including an MCO faculty development leave at Harvard and Tufts Universities in the summer of 1998. He also completed a fellowship at MCO in teaching and learning health and medical sciences and earned a graduate certificate.

Dr. Bisesi has published several scientific articles and chapters, including two chapters in the *Occupational Environment: Its Evaluation and Control* and three chapters in fourth and fifth editions of *Patty's Industrial Hygiene and Toxicology*. In addition, he is first author of the textbook *Industrial Hygiene Evaluation Methods* and second author of two other textbooks, the *Handbook of Environmental Health and Safety*, Volumes I and II. He is a member of the American Industrial Hygiene Association (fellow), American Conference of Governmental Industrial Hygienists, American Public Health Association, National Environmental Health Association, and Society of Environmental Toxicology and Chemistry.

Contents

Volume I

Contents

Volume II

Chapter 8   Terrorism and Environmental Health Emergencies

Chapter 9   Major Instrumentation for Environmental Evaluation of
                  Ambient Air, Water, and Soil

CHAPTER 1

Environment and Humans

Health and *safety* refer to the avoidance of human illness and injury through efficient use of the environment, a properly functioning society, and an inner sense of well-being. *Environmental health and safety* is the art and science of protecting human function; promoting aesthetic values; and preventing illness and injury through the control of positive environmental factors and the reduction of potential physical, biological, and chemical hazards.

To understand the relationship of the environment to humans and to understand how to protect humans from illness and injury, it will be necessary to discuss the ecosystem, ecosystem dynamics, and energy. Human impact on the environment and the various approaches, including risk assessment, epidemiological, economic, legal, and governmental aspects used to evaluate and resolve environmental problems, are discussed. To understand abnormal physiology, and the basis of human illness and injury, brief discussions on normal physiology, toxicology, and epidemic infectious disease are included. Finally, it is necessary to understand the role of professional environmental health practitioners, the skills that they need, and how they address the expanding scope of environmental problems.

ECOSYSTEM

Environments

Earth is divided into the lithosphere, or land masses, and the hydrosphere, or the oceans, lakes, streams, and underground waters. The hydrosphere includes the entire aquatic environment. Our world, both lithosphere and hydrosphere, is shaped by varying life forms. Permanent forms of life create organic matter and, in combination with inorganic materials, help establish soil. Plants cover the land and reduce the potential for soil erosion — the nature and rate of erosion affects the redistribution of materials on the surface of Earth. Organisms assimilate vast quantities of certain elements and molecules, such as carbon and oxygen. Animals, through respiration, release carbon dioxide into the atmosphere — carbon dioxide affects the heat transmission of the atmosphere. Organisms affect the environment and in turn are affected by it.

Two environments, biotic (living) and abiotic (nonliving), combine to form an ecosystem. An ecosystem can also be subdivided by more specific criteria into the following four categories: (1) abiotic, the nutrient minerals that are synthesized into living protoplasm; (2) autotrophic, the producer organisms (largely the green plants) that assimilate the nutrient minerals using energy and combine them into living organic substances; (3) heterotrophic, the consumers, usually the animals, that ingest or eat organic matter and release energy; and (4) heterotrophic reducers, the bacteria or fungi that return the complex organic compounds to their original abiotic state and release the remaining chemical energy. The biotic group in the ecosystem complex is essentially composed of the autotrophs, or producer organisms that synthesize organic substances, and the heterotrophs, or consumer or reducer organisms that decompose labile organic substances. The ecosystem is important when considering the food chain, which is in effect a transfer of energy from plants through a series of organisms that eat and, in turn, are eaten. Eventually, decay will start the process all over again.

The ecological niche is the combination of function and habitat of each of the approximately 1.5 million species of animals and 0.5 million species of plants on Earth. There are many interactions between species in the ecosystem, yet a balance is dictated by nature. The law of limiting factors states that a minimum quantity of essentials, such as nutrients, light, heat, moisture, and space, must be available within the ecosystem for survival of the organisms. In some instances where these limiting factors apply or where pesticides or other environmental elements are introduced into the ecosystem, the organism alters itself to exist within the new environment. This change is called mutation. Unfortunately, mutation becomes a serious concern in the area of pest control as well as in disease, because the new organism may be highly resistant to effective control and may therefore cause disease and physical destruction of plants and animals. The ecosystem is always in a dynamic instead of a static balance — changes in one part of the ecosystem cause changes in another.

Biosphere

The biosphere is that part of Earth — lithosphere and hydrosphere — in which life exists. However, this definition is not complete, because spores may commonly be found in areas that are too dry, too cold, or too hot to support organisms that metabolize. The biosphere contains the liquid water necessary for life; it receives an ample supply of energy from an external source, which is ultimately the Sun, and within it liquid, solid, and gaseous states of matter interface. All the actively metabolizing organisms operate within the biosphere. The operation of the biosphere depends on photosynthesis, during which carbon dioxide is reduced to form organic compounds and molecular oxygen. Oxygen, the by-product of photosynthesis, replenishes the atmosphere and most of the free water, which contains dissolved oxygen.

ECOSYSTEM DYNAMICS

Cycles

The ecosystem changes frequently. Several of the cycles that are important and that may be affected by humans include the hydrologic cycle, the carbon cycle, the nitrogen cycle, the phosphorous cycle, and energy flow. The hydrologic cycle is the movement of water from the atmosphere to Earth and back into the atmosphere. This is discussed more fully in Chapter 3 on water in Volume II.

The carbon cycle begins with the fixation of atmospheric carbon dioxide by means of photosynthesis performed by plants and certain algae. During this process carbon dioxide and water react to form carbohydrates, and free oxygen is simultaneously released into the atmosphere. Some of the carbohydrates are stored in the plant, and the rest are utilized by the plant as a source of energy. Some of the carbon that has been fixed by the plants is then consumed by animals, who respire and release carbon dioxide. The plants and animals die, decomposing by action of microorganisms and other catalysts in the soil, and the carbon in their tissues is then oxidized to carbon dioxide and returned to the atmosphere. The carbon dioxide is recycled through the plants and the process repeats itself (Figure 1.1).

The nitrogen cycle begins when atmospheric nitrogen is fixed or changed into more complex nitrogen compounds by specialized organisms, such as certain bacteria and blue-green algae. Some fixation may occur as a result of lightning, sunlight, or chemical processes; however, the most efficient nitrogen fixation is carried out by biological mechanisms. Other bacteria, fungi, and algae may also play an important role in nitrogen fixation. Basically the atmospheric nitrogen is changed into a nitrate that is absorbed by plants, eventually combining with other elements and becoming a plant protein. The plant protein decays when the plant dies, releases nitrogen as ammonia, and through bacterial oxidizing action becomes a nitrite; with further bacterial action the protein is reduced and released as atmospheric nitrogen. The plant protein may also be eaten by animals, may be broken down into molecules called amino acids, and eventually may be synthesized into an animal protein. Through decay of the dead animal or breakdown of excreted feces and urine, this protein is changed to ammonia. The ammonia returns to the nitrite stage through bacterial action, and again through bacterial action becomes atmospheric nitrogen (see Figure 1.1). A further description of the nitrogen cycle is found in Chapter 6 on sewage in Volume II.

In the phosphorus cycle, the element moves rapidly through similar stages, becoming locked in sediment or in biological forms such as teeth or bones. The primary sources of phosphorus for agriculture are phosphate rocks and living or dead organisms (see Figure 1.1).

Food Chain

The cycle of energy flow may also be described as the food web. The food web, or food chain (Figure 1.2), implies that an organism has consumed a smaller organism

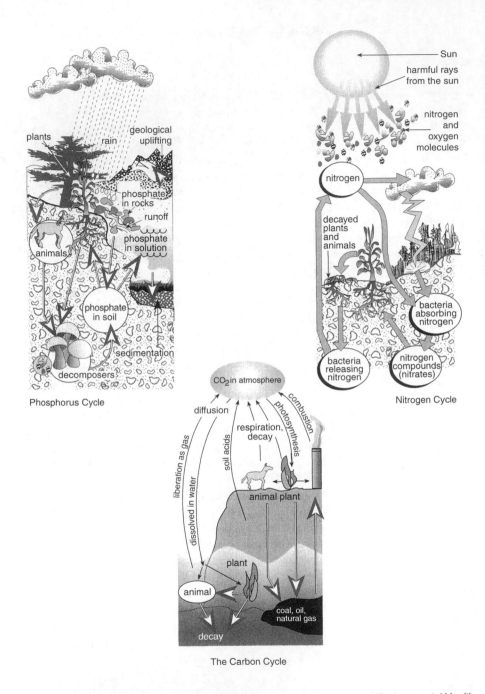

Figure 1.1   Ecosystem dynamics. (From Koren, H., *Illustrated Dictionary of Environmental Health and Occupational Safety,* CRC Press, Boca Raton, FL, 1995. With permission.)

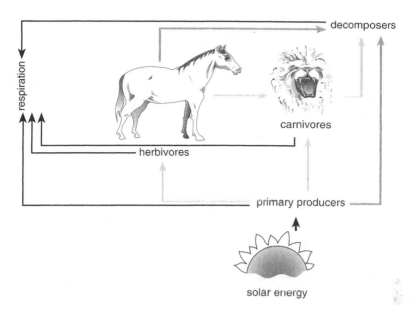

Figure 1.2 Food chain.

and is then consumed by a larger organism. Eventually, the microscopic plants and animals become the food supply for the small fish or animals that become the food supply for humans. The importance of the food chain is illustrated by biomagnification, in which the impurities found in water are concentrated in the lower forms of life and are reconcentrated substantially during the movement of the impurities through the food chain. For example, whereas people might only get 0.001 μg of mercury in drinking water, they might get 30 to 50 μg of mercury by consuming fish that have been bioconcentrating mercury.

Energy Cycles

Solar energy is absorbed by Earth and eventually reradiated into space as heat. The heat is distributed over the surface of Earth through circulation caused by the atmosphere and the oceans. Diurnal changes or changes occurring in a 24-hour period due to light, wind, temperature, and humidity are important near ground level and at high levels in the atmosphere. In certain localities land masses and sea breezes may affect the overall heat and weather patterns. However, total heat movement is not affected significantly by local conditions.

About 30% of the solar energy entering the atmosphere is deflected or scattered back toward outer space because of the atmosphere, the clouds, or Earth's surface. This portion of the solar energy is lost. About 50% of the incoming radiation reaches the ground or ocean, where it is absorbed as heat. The properties of the surface that receives the energy determine the thickness of the layer over which the available heat is distributed. In the oceans the surface wave motions effectively distribute the heat over a 300-ft layer of air. On land the energy transferred downward into the ground occurs very slowly through the process of molecular heat conduction. The penetration distance is very small. About 20% of the incoming solar radiation is absorbed as it goes through the atmosphere. In the upper atmosphere, oxygen and ozone molecules absorb an estimated 1 to 3% of the incoming radiation. This absorption occurs in the ultraviolet range and, therefore, limits the penetrating radiation to wavelengths longer than 300 nm. This absorption is very important, because it is the main source of energy for the movement of air in the upper atmosphere. This absorption also shields people on Earth from damaging effects of ultraviolet radiation. Most of the rest of the 20% is absorbed by water vapor, dust, and water droplets in the clouds. The energy that sustains all living systems is fixed in photosynthesis, as previously mentioned. Part of the energy fixed in plants and animals has been compressed over millions of years and is the source for stored energy, namely, coal, oil, and natural gas.

Current Ecosystem Problems

People and advanced technology affect the ecological niche by interfacing with the intricate function and habitat of various species of animals and plants. Discharges to air, water, and soil, including toxic and infectious chemical wastes, biological wastes, and radiological wastes create an environmental pressure that is detrimental to all life forms, including people. From the destruction of the rain forest of Brazil, because of the excessive clearing of trees, to the potential destruction of forests in the Northern Hemisphere due to acid precipitation, a huge number of ecosystems may be eliminated. Nature's diversity offers many opportunities for finding new pharmaceuticals, genetic mapping, genetic engineering, and potential power to improve crops. We are wasting our greatest natural resource on which we depend for food, oxygen, clean water, energy, building materials, clothing, medicine, socio-logical well-being, and countless other benefits.

Biological diversity includes two related concepts, genetic diversity and ecolog-ical diversity. Genetic diversity is the amount of genetic variability among individuals in a single species, whereas ecological diversity is the number of species in a community of organisms. An organism's ability to withstand the challenge of varying physical and chemical problems, parasites, and competition for resources is largely determined by its genetic makeup. The more successful organisms that survive pass on the genes to the next generation. An example of this would be in the case of climatic changes where plants have smaller, thicker leaves for losing less water in areas that are becoming arid. Genetic diversity is important in developing new crops to meet new conditions. This is how disease-resistant crops can be utilized to sustain life for people. Genetic variation among species is greater than within the species. It is difficult to determine which of these species are of extreme importance and will be essential to people in the future as new supplies of food, energy, industrial chemicals, and medicine. It is anticipated that the human population will increase by nearly 50% in the next 20 years, with much of this increase occurring in very poor nations. The resources needed to feed these individuals will be substantial, and many new forms of food will have to be developed to prevent starvation.

There appears to be adequate sources of fossil fuel for the immediate future. However, these fossil fuels, when burned, can potentially cause severe pollution. Eventually, fossil fuel sources will be depleted. It will then be necessary to utilize plants that can produce energy-rich materials, such as soybean oil. At present this process is very expensive.

Coastal waters contain several major natural systems. Because they directly abut the land, they are affected by activities on the land. Pollutants from coastal sewage treatment plants, industrial facilities, settling of air contaminants, and erosion of land, many hundreds of miles upstream, can cause considerable problems to the coastal waters. Oil slicks are extremely hazardous to ecosystems. Trace metals can be concentrated in the food chain and, thereby, become a hazard to people eating sea life. Toxic metals, such as mercury, lead, and cadmium, have been found in coastal waters.

Mollusks become indicators of environmental contamination and environmental quality. These shellfish, which live in the mud or sand bottoms of the aquatic ecosystems, accurately reflect some of the important characteristics of the water adjacent to the land These organisms may show a direct adverse effect when they accumulate chemical substances or microorganisms. They are good for systematic monitoring, because they are stationary in nature (except for the early egg and larva stages) and filter large quantities of water to feed themselves. Because they are of significant commercial importance, reliable quantitative data are available on harvesting and other aspects of the mollusk life cycle.

In the past, the Environmental Protection Agency (EPA) sponsored studies of the accumulations of chlorinated organic pesticides in shellfish. Residues of the pesticide dichlorodiphenyltrichloroethane (DDT) were found for years after the use of the chemical was banned in 1972. The Mussel Watch Program has shown elevated levels of polychlorinated biphenyls (PCBs). High-level concentrations of silver, cadmium, zinc, copper, and nickel have been detected in various bodies of water. Mussels have also shown the presence of radionuclei as a result of weapon-testing programs and also from effluent that has come from nuclear reactors. Records of commercial catches of clams and oysters indicate that the number of mollusks in coastal waters has been steadily declining over the years. This may result from overfishing, loss of habitat, and deterioration due to pollution and natural disasters. It has been shown that the building of a sewage treatment plant has the potential to adversely affect shellfish habitats by altering water salinity.

The continued loss of tropical forests is the single greatest threat to the preservation of biological diversity. Many of Central America's ecosystems may be affected by these problems.

Wetlands have exceptional value because of their location at the land and water interface. They provide ecological, biological, economic, scenic, and recreational resources. Many wetlands are among the most productive ecosystems on the planet, especially estuaries and mangrove swamps. At least half of the biological production of the oceans occurs in the coastal wetlands. From 60 to 80% of the world's commercially important marine fish either spend time in the estuaries or feed on the nutrients produced there. Coastal wetlands help protect inland areas from erosion, flood, waves, and hurricanes. The wetlands are an important economic resource, because they provide cash crops, such as timber, marsh hay, wild rice, cranberries, blueberries, fish, and shellfish.

Desertification is the process that occurs in semiarid lands where sterile sandy desertlike ecosystems are created from grasslands, brushlands, and sparse open forests due to a loss of organic carbon in the form of stable humus from the soils. This desertification can be stimulated beyond natural forces by overgrazing of animals and by misuse of the area. This can lead to mass starvation and death.

ENERGY

Problems

Energy is utilized in four primary areas, residential, commercial, industrial, and transportation. Residential use is approximately 11.1 quadrillion British thermal units (Btu). Industrial use is approximately 27.1 quadrillion Btu. Transportation use is approximately 24.7 quadrillion Btu. Where energy is used to produce electricity or heating, the primary consumption of the energy is significantly higher than its end use. The distinction between the end use and primary energy consumption is important because this indicates the level of efficiency of energy use. It also indicates the amount of unnecessary carbon emissions.

Residential demand is the largest energy-consuming sector in the United States. Households are responsible for 20% of all carbon emissions of which 63% is attributed to the fuels used to generate electricity. With continued growth in the economy and population, residential use will always be a major factor in the production of contaminants resulting from electrical usage. The energy is also used to provide heating, cooling, and ventilation, which accounts for most of the direct use of fossil fuels, and as a power source for refrigerators, freezers, dishwashers, clothes washers and dryers, and stoves, as well as lighting and other items. Further, there has been a steady increase in the use of electricity because people now have more appliances and computers.

Commercial demand consists of businesses and other organizations that provide services to stores, restaurants, hospitals, hotels, schools, jails, and other structures. This sector is the smallest of the energy users. The greatest savings in energy relate to the use of insulation and improvement in the efficiency of equipment.

Industrial demand includes the agriculture, mining, construction, and manufacturing industries. This sector uses energy to produce or process goods of all types, and also produces a wide variety of basic materials such as aluminum and steel that are used to produce goods. Coal is mainly used for a boiler fuel and for production of coke in the iron and steel industry.

Transportation demand includes the use of all types of vehicles including trains and planes. About 33% of all carbon emissions and about 78% of carbon emissions from petroleum consumption come from the transportation sector. People select their vehicles based on size, horsepower, price, and personal preferences. At present carbon emissions are continuing to grow because of the selection of heavier and less efficient vehicles, and because of the increase in vehicles present on the highways. The primary energy source is oil, with all its problems related to pollution. Alternative-fuel vehicle sales are dependent on vehicle price, cost of driving per mile, vehicle range, fuel availability, and commercial availability.

HISTORY

Energy problems were brought to the attention of the American public and the world in 1973 as a result of the Arab oil embargo and the actions of the Organization of Petroleum Exporting Countries (OPEC). Unfortunately, in the years preceding these dramatic events, the United States was buying an increasing amount of petroleum abroad and utilizing more natural gas than was readily available. As a result of these shortages, President Richard M. Nixon authorized the allocation of scarce fuels. By May 1973, the Office of Emergency Preparedness was reporting widespread problems of gasoline stations closing due to lack of fuel. Voluntary guidelines were announced and a variety of emergency actions taken including the lowering of the speed limit to 55 mi/hr. Project Independence was conceived with the goal of self-sufficiency in energy production by 1980. Congress established the Federal Energy Administration in December 1973, and the President created the Federal Energy Office to deal with the immediate crisis. Funding was provided for additional research and development in a variety of energy areas.

Unfortunately, by 1978 the amount of fuel being purchased abroad continued to escalate. The variety of programs at the federal level were so confusing, complex, and spread out over a variety of agencies that the administration in Washington, D.C. requested the establishment of a new federal Department of Energy with an administrator of cabinet rank. It was hoped that the new department would bring organization to the chaotic mass of confusing energy problems and move the country toward self-sufficiency.

Conservation

Conservation and development of new energy sources or expansion of existing sources were two major phases of energy programs. Energy conservation was of critical importance to the United States; yet despite predictions of severe shortages, conservation attempts had not been entirely successful and great amounts of energy were still wasted. Greater energy efficiency was needed in cars and in electrical and gas equipment. Better insulated homes and a general reduction in energy usage was essential. The United States had to become less dependent on oil and gas as sources of energy.

Oil and Gas

Oil and gas in 1972 accounted for nearly 78% of U.S. energy consumption. By 1980, oil and gas remained the greatest sources of energy used, despite the fact that the United States had about a 300-year supply of coal available. Although in 1979, Alaska was providing new oil, the country was more dependent than ever on foreign oil. However, it was understood that no matter how many sources of oil and gas were developed, eventually the supplies would be depleted; therefore, it was necessary to examine other kinds of energy sources. These sources included oil shale, nuclear energy, geothermal energy, solar energy, fusion energy, and coal. Oil shale was available, but at rather high costs. Additional research was necessary to determine how best to extract the oil from the shale both economically and without creating surface environmental problems.

Nuclear

Nuclear power could certainly provide a continuing source of available energy to the country. However, many permits had been delayed because of consumer groups concerned with the possibility of nuclear accidents and with the disposal of the radioactive wastes. The threat of terrorist groups obtaining wastes from the fusion process and utilizing these materials to terrorize the world was also of great concern. In 1973, the U.S. nuclear electrical generating capacity was 20,000 MW, or over 5% of the nation's total electrical capacity.

Geothermal

Geothermal energy, produced by tapping Earth's heat, was being used at the geyser sites in Oregon and California, where the Pacific Gas and Electric Company employed geothermal steam as a power source for a 400-MW electrical generating facility. Geothermal resources included dry steam, hot steam, and hot water. At that time, because of cost factors, only dry geothermal heat was used to drive electric turbines.

Solar

Solar energy was both economically and technologically feasible. The benefits of capturing the sun's energy had been long recognized. Funding for solar energy research had increased and considerable research was needed. The 1979 federal budget included $137 million for solar research and development. Techniques under investigation included direct thermal applications, solar electric applications, and fuels from biomass. Direct thermal application consisted of solar heating for cooling of buildings and hot water supply, and the use of solar heat for agriculture and industrial processes. Agricultural applications were the drying of food, crop drying, lumber drying, heating and cooling of greenhouses, and heating of animal shelters. The solar electrical applications included research on the generation of electricity from windmills, solar cells (photovoltaic), solar thermal electric systems, and ocean thermal energy conversion. Theoretically, these systems were well understood; however, their applications were not yet economically feasible. The fuels produced from biomass suggested large-scale use of organic materials, such as animal manure, field crops, crop waste, forest crops and waste, and marine plants and animals. The conversion process was technically feasible. During World War II, much of the liquid fuel of France consisted of methanol produced from wood. Processes requiring study included fermentation to produce methane and alcohol, chemical processes to produce methanol, and pyrolysis to convert organic waste material to low Btu gaseous fuels and oils.

Coal

Coal was the one major source of energy that could certainly be utilized to reduce U.S. reliance on other nations. The most serious deterrent was the many contaminants, particularly sulfur, found in coal. Sulfur could be removed by utilizing new technologies in coal mining and conversion. Because coal made up 85% of the U.S. fossil energy resource base, it was wise to utilize it as well and as quickly as possible. New coal technology under development promised two very important changes in the potential pollution areas: the reduction of sulfur oxide in particulate matter from direct burning of coal, and the provision of liquid and gaseous fuel substitutes for domestic oil and gas production. The emphasis in research had shifted from conventional combustion to direct fuel to fuel conversion, that is, coal gasification and coal liquefaction. Sulfur could be extracted even when coal was directly burned using fluidized beds in which crushed coal was injected into the boiler near its base. Coal was fluidized by blowing air uniformly through a grid plate. Sulfur dioxide was then removed by injecting the crushed coal with limestone particles less than ⅛ in. in diameter. The limestone particles were noncombustible and would bind the sulfur dioxide. This system also reduced the nitrogen oxides below emission standards for utility boilers of 0.7 lb of nitrogen dioxide per million Btu. The nitrogen dioxide could actually be reduced to 0.2 to 0.3 lb per million Btu. The gasification process involved the reaction of coal with air, oxygen, steam, or a mixture of these that yielded a combustion product containing carbon monoxide, hydrogen, methane, nitrogen, and little or no sulfur. The energy of the product ranged from 125 to 175 Btu per standard cubic foot for a low Btu gasification process to about 900 to 1000 Btu per standard cubic foot for a high Btu gasification process. The high Btu gasification process was comparable with the energy content of natural gas. The liquefaction transformed the coal into a liquid hydrocarbon fuel and simultaneously removed most of the ash and sulfur. Hydrogen was frequently added in many of the direct catalytic liquefaction techniques.

ENERGY DILEMMA

1980s

The trauma of the 1970s led to a more active role by government in the 1980s to try to resolve energy problems in the country. The Energy Security Act and the Crude Oil Windfall Act of 1980 gave the President additional power to encourage the development of oil and gas reserves and a synthetic fuels industry. Because coal was the greatest fuel resource, there was an attempt to substitute coal for oil in various industries. Energy was used more efficiently because of a variety of conservation programs. The federal conservation and solar programs were passed by Congress, and additional funds were allotted for weatherization assistance for low-income people and tax breaks for others who made their homes more efficient. Funds were provided by the Solar Energy and Conservation Bank to finance the use of solar energy systems. Solar wind and geothermal tax credits were made available for residential users. Funds were provided for research and development in the use of energy sources other than fossil fuels. The biomass program and the alcohol tax credits provided $1.6 billion through 1990 to promote the conversion of grain, farm residues, and other biomasses to alcohol fuels. Through 1990, $16.5 billion were allotted to improve transportation efficiency by enhancing public transportation programs. A fuel economy program was put in place and funds were provided for enforcement of the 55 mi/hr speed limit.

Auto efficiency standards were raised from 20 mi/gal for the 1980 model year vehicle to 27.5 mi/gal by 1985. A gas guzzler tax was added to new vehicles with mileage below 15 mi/gal. A building energy conservation code and an appliance efficiency standard code was established to improve fuel conservation and efficiency. As a result of all this effort, the growth of energy usage slowed substantially from the previous years.

By 1984, the national energy picture had continued to brighten substantially. Prices of gasoline had dropped and the price of crude oil had dropped significantly. This progress was due in part to the decontrol of the domestic petroleum markets and the removal of restrictions on energy use. Potential new sources of energy were under development in the oceans abutting the United States. This area is larger than the country's land mass and includes many of the deep water areas, as well as the outer continental shelf. Further, the OPEC nations were in disarray because of the long war between Iran and Iraq, and the propensity for cheating on the amount of oil that each country was permitted to produce to maintain their high income. Continued conservation of energy because of the many permanent changes made in energy use over the years helped keep the price of oil at lower levels.

In 1984, 1608 geothermal leases were in effect, in which 30 were capable of production and 12 actually were in production.

There was an estimated 2 trillion barrels of shale oil in the Green River formation in northwestern Colorado, northwestern Utah, and southwestern Wyoming. Of this amount, it was estimated that 731 billion bbl of oil could be extracted commercially. It was also estimated that in Utah, approximately 2 billion bbl of oil were recoverable with current technology. Approximately 100 to 200 million bbl could be recovered by surface mining. The question was not the availability of these energy resources, but the cost of developing them.

Since the Three Mile Island debacle of March 1979, numerous public protests occurred about the building and use of nuclear power as a source of energy. A need still exists for facilities for the permanent disposal of the high radioactive waste that is generated during the production of electricity in the nuclear power plants. Until very recently it was assumed that the nuclear waste was to be handled appropriately on-site. However, it has been disclosed that numerous leaks of nuclear waste from nuclear power plants have occurred over the decades. For example, serious problems in Ohio must be corrected to prevent disease. In 1990, the cleanup of these sites were estimated to cost $31.4 billion.

1990s

In 1988, the United States consumed more energy than in any previous year in its history. The demand for petroleum was about 17 million bbl/day, which was the highest point since 1980. However, the strong underlying energy efficiency trend continued, except for the use of energy in the transportation sector. In 1991, energy usage was 81.5 quad. A quad is equivalent to 1 quadrillion Btu. The U.S. domestic oil production, which was on the rise in the 1980s, started to decline again. The running aground of the *Exxon Valdez* in Prince William Sound, Alaska, caused an ecological debacle temporarily cutting the amount of oil coming from Alaska down to 20% of its normal flow. Although oil flow recovered and prices declined, one wonders about future catastrophes and their effect.

Because of lower oil prices during 1989, U.S. exploration decreased sharply. In 1989, there were only 740 drilling rigs, whereas in 1981 there were 4500 functioning drilling rigs. A need exists for a long lead time simply to get ready for further exploration for oil in the United States. Environmental and conservation groups attempted to limit further oil exploration in Alaska because of the incident that occurred in March 1989.

On August 2, 1990, Iraq invaded Kuwait. By the end of February 1991, Iraq had set on fire or damaged 749 oil facilities in Kuwait. The restoration of oil production from Kuwait was a long, complicated, and expensive process.

The demand for natural gas is expanding. This means gas production in excess of 20 trillion ft^3/year. Electricity, despite increased efficiency in usage, is projected to rise. Because no newly ordered nuclear power plants were on-line by the year 2000, fossil-fuel plants must continue to provide most of the generating capacity.

Despite the hope of the early 1970s, nuclear power has never achieved anywhere near the amount of electrical output that had been anticipated. This is in part due to the concern of citizens and a variety of lawsuits, and in a large part due to the Three Mile Island incident in 1979. The shoddy construction and need to obtain multiple permits, such as in Indiana, have further aggravated the situation. The U.S. Nuclear Regulatory Commission (NRC) has listed a series of unresolved safety issues. These include systems interactions, seismic design criteria, emergency containment, containment performance, station blackout, shutdown decay, heat-removal requirements, seismic qualification of equipment in operating plants, safety implications of control systems, hydrogen control measures and effects of hydrogen burns on safety equipment, high-level waste management, low-level waste management, and uranium recovery and mill tailings.

With the increasing problems related to oil and gas, as well as nuclear power, electric utilities and others increasingly have turned to coal and natural gas as sources of fuel and power. This occurs at a time when acid precipitation, in the form of acidic rain and snow, continues to be an issue. The damage caused by acid rain is increasing, as well as pressure from Canada to do something about it. Although new energy technologies related to coal, as well as other sources, have been put pretty much on a holding pattern, it seems that it may now be necessary to explore these further and to utilize other means of securing energy.

Although it is true that the Clean Air Act of 1990 put numerous additional constraints on the use of coal as a source of energy, at some time in the future it may be economically feasible to clean up the coal and use this abundant energy resource.

Current Concerns of Fossil Fuel

Oil

The World Resources Institute states that the world has already consumed between one-third and a little less than one-half of its ultimately recoverable oil reserves. Global production of oil is expected to start declining between 2007 and 2014. The economic impacts will depend largely on the price and availability of energy alternatives. The transportation sector, including motor vehicles, airplanes, and trains, will be most affected. Unfortunately, people in those sectors are not taking this information seriously. Oil demand will increase each year as the Third World countries start to recover from the economic neglect in which they have been involved for many years. By the 1970s, the average productivity of domestic wells began to decline, and domestic oil production leveled. Alaskan production at the end of the 1970s and through 1988 partially offset the declines in the lower 48 states. In 1989, Alaskan production declined. In 1995, the United States produced 6.5 million bbl/day with 79% of the total coming from onshore wells and 21% coming from offshore wells. By 1994, petroleum imports reached a 17-year high of 45% returning almost to the peak level of 47% in 1977. Domestic oil production will continue to decline increasingly, causing the country to import foreign oil.

Coal

In 1995, an estimated 1 trillion tons of coal were produced in the United States with approximately 47% coming from the West. Surface mining helped produce low-sulfur coal but also increased the potential for further contamination of the immediate environment. Coal production is expected to increase, with Western production growing to 50% of the total. Coal, the second leading source of carbon emissions behind petroleum products, is expected to produce 676 million tons of carbon emissions in 2020, or 34% of the total. Most of the increases in coal emissions result from electricity generation.

Gas

From 1990 through 1998, natural gas consumption in the United States increased by 14%. The greater use of natural gas as an industrial and electricity generating fuel is in part due to its clean-burning qualities in comparison with other fossil fuels. There has also been lower cost due to greater competition and deregulation in the gas industry as well as an expanding transmission and distribution network. However, based on current economic realities, these costs may vary widely.

Natural gas, when burned, emits less greenhouse gases and criteria pollutants per unit of energy produced than other fossil fuels because gas is more easily and fully combustible. The amount of carbon dioxide produced for an equivalent amount of heat is less for gas than oil or coal. However, because the major constituent of natural gas is methane, it also directly contributes to the greenhouse effect through venting or leaking of natural gas into the atmosphere.

Methane gas is 21 times as effective as carbon dioxide in trapping heat. Although methane emissions amount to only 0.5% of U.S. emissions of carbon dioxide, they cause about 10% of the greenhouse effect in the country. Natural gas consumption has increased considerably worldwide. Domestic natural gas production is expected to increase an average of 2% for the coming years.

Projected Increase in Carbon Emissions

The use of fossil fuels in the future will increase and along with it more carbon will be emitted. These emissions are projected to increase by an average of 1.3% a year to the year 2020, when they will be approximately 1.956 trillion tons. Increasing concentrations of carbon dioxide, methane, nitrous oxide, and other greenhouse gases may increase Earth's temperature and affect the climate.

Kyoto Protocol

From December 1 to 11, 1997, representatives from more than 160 countries met in Kyoto, Japan, to negotiate binding limits for greenhouse gas emissions for developed countries. The Kyoto Protocol sets emission targets for these countries relative to their emissions in 1990 to achieve an overall reduction of about 5.2% from current levels. These countries are known as Annex I countries. The reduction target is 7% for the United States and 6% for Canada and Japan, with European Union countries targeted at 8%. Non-Annex I countries have no targets under this protocol.

The greenhouse gases covered by the protocol are carbon dioxide, methane, nitrous oxide, hydrofluorocarbons, perfluorocarbons, and sulfur hexafluoride. Sources of emissions include energy combustion, fugitive emissions from fuels, industrial processes, solvents, agriculture, and waste management and disposal. The targets were to be achieved on average between 2008 to 2012. However, in 2001, President George W. Bush decided to remove the country from this agreement and to review the entire process further.

Cogeneration Systems

To meet the Kyoto Protocol, it will be necessary to change the way we produce electricity to make it more efficient. Currently we use more than 21 quadrillion Btu of energy from the combustion of coal, natural gas, and oil to produce the equivalent of only 7 quadrillion Btu of electricity available at the plant gate. Of the energy,

67% is lost as waste heat, which accounts for 346 million tons or about 24% of U.S. carbon emissions. Cogeneration systems simultaneously produce heat in the form of hot air or steam, and power in the form of electricity by a single thermodynamic process, usually using steam boilers or gas turbines; therefore, reductions of energy losses occur when process steam and electricity are produced separately. Advanced turbine systems will be needed to accomplish this.

Wood as an Energy Source

Wood has declined in use as a primary heating fuel in the residential area. However, as a secondary source of heating, it has increased. About 20% of households in the United States use woodstoves or fireplaces.

Nuclear Energy

Approximately 20% of the nation's electricity is generated by 109 operating nuclear reactors in 32 states. Nuclear power is used by 6 states for more than 50% of their electricity and by 13 additional states for 25 to 50% of their electricity. Nuclear power continues to be an important source of electricity worldwide, although its future is uncertain in some parts of the world. Nuclear power is used by 30 countries to produce almost 25% of their combined electricity generation. Worldwide new nuclear plants are being planned. However, in the United States there has been a campaign to shut down nuclear power plants where possible. Nuclear power is a good solution to the energy problem providing that it is engineered to avoid nuclear mishaps and will not contaminate the environment. The personnel need to be well-trained in the operation of the facility and must not attempt to take shortcuts, as was done in 1999 in Japan where people were subjected to radiation because of poor training and incompetence on the part of the personnel.

Renewable Energy

Renewable energy includes biomass, geothermal, hydrothermal, photovoltaics, solar thermal, and wind. Hydrogen is considered to be a viable sustainable energy carrier for transportation. At present approximately 7% of total energy used in the United States comes from renewable energy technologies. Facilitating technologies including energy storage, electricity transmission and distribution, and power electronics are essential to reduce the cost of these new technologies and to make them available for use in place of fossil fuels.

Biomass

Biomass is nature's storehouse of solar energy and chemical resources. Plant matter is a massive quantity of renewable energy, whether raised for that purpose or grown wild. Carbon dioxide from the atmosphere and water from Earth are combined in the photosynthetic process to produce carbohydrates that are the building blocks of biomass. The solar energy is stored in the chemical bonds of the structure and when burned efficiently the oxygen from the atmosphere combines with the carbon in the plants to produce carbon dioxide and water. This makes biomass a renewable resource for energy. Biomass production worldwide is about eight times the total annual world consumption of energy from all sources. Biomass residues have long been burned by forest industries and other industries to generate process steam electricity. Because biomass is stored solar energy, it can be used as needed to provide power when other renewable sources such as solar energy may not be available. Roughly 8000 MW of electricity capacity is currently available in the United States vs. <200 MW in 1979. Biofuels primarily from forest, agriculture, or municipal solid waste residues provide about 3.5% of U.S. primary energy. Because the physical and chemical composition of biomass feed stocks varies widely, there may be a need for some special handling or conversion technologies for specific biofuels. To generate electricity biofuels can be cofired with coal in conventional coal plants, burned in steam plants, gasified to power gas turbines, and used for fuel cells. Almost all biomass electric plants use steam turbines that are generally of a small scale, 30 MW. Biomass gasifies at lower temperatures and works more quickly than coal, reducing the gasification costs. After the biomass is gasified, the gas products are cleaned of particulate and other contaminants before being burned in a steam-injected gas turbine, or other turbine. Biomass systems are now cost competitive in many areas where a low-cost waste feedstock is available. Burning or gasifying biomass in a power plant generates much less sulfur oxides than coal but still does produce nitrogen oxides depending on combustion chamber temperatures.

The agricultural industry could produce large quantities of trees and grasses that would protect soils, improve water quality, and be converted to electricity, heat, liquid, or gaseous fuels. This could be a new cash crop for the agricultural industry.

Geothermal

Geothermal electricity generation systems extract heat from the ground to drive turbines. Geothermal energy is stored energy inside the Earth, either remaining from the original formation of the planet or generated by the decay of radioactive isotopes inside it. Only in certain active areas such as volcanic zones or along tectonic plates, is geothermal energy sufficiently concentrated and near enough to the surface to be used economically.

Geothermal energy comes in four forms:

1. Hydrothermal fluids
2. Geopressured brines
3. Hot dry rocks
4. Magma

Hydrothermal fluids are the only commercial geothermal resource at this time. They are hot water or steam in porous or fractured rock at depths of up to 4.5 km, and with temperatures ranging from 90 to 360°C. Geopressured brines are hot salty waters containing dissolved methane that are about 3 to 6 km below the Earth's surface. They are trapped under sediment layers at high pressures. Hot dry rock are regions where there is little or no water but considerable heat. Additional research is needed to turn this into a cost-effective technology. Magma is molten rock with temperatures of about 700 to 1200°C typically occurring at depths of 6 to 10 km.

Although magma energy is the largest supply of all geothermal resources, it is the most difficult to extract. Considerable research is needed to make this into an effective energy source. Geothermal energy is listed as a renewable resource, but it can be depleted if too much is withdrawn. This occurs when there is a more rapid removal of underground water and vapor resources than that which is replaced naturally.

Technologies used with geothermal resources for producing electricity include direct steam, single-flash, double-flash, and binary systems. The simplest technology is the piping of hydrothermal steam directly from underground reservoirs to drive turbines. These may be found in Yellowstone National Park. Single-flash units are similar to the steam units but use underground hot water instead. As it is pumped to the surface from the deep underground the pressures are reduced and steam is partially produced in a flash tank. A double-flash system uses a second flash tank that operates at pressures between the pressure of the first flash tank and air pressure. Binary systems pump hot water to the surface and then use a heat exchanger to transfer the heat to a working fluid. This fluid is vaporized by the heat and then drives the turbine. The environmental impact of geothermal power varies with the resource and the technology used. Direct steam and flash systems generally release gases to the atmosphere such as hydrogen sulfide. Small quantities of brines may also be released. Binary systems generally reinject all gases and brines into the reservoir.

There are about 170 geothermal power stations in 21 countries that produce about 5700 MW of electricity, The United States has the largest installed capacity.

Hydroelectricity

Hydroelectric generation systems use the energy in flowing water to turn a turbine. Currently, hydropower provides about 20% of the world's electricity supplies. In its conventional form with dam storage, hydropower can provide base, intermediate, or peak power. It is especially valuable in backing up intermittent power from solar and wind sources. U.S. hydropower resources are fairly well developed. More power, however, could potentially be obtained from existing facilities by upgrading equipment and installing equipment at dams that are not now used for power. Turbine efficiencies are typically in the 75 to 85% range, making hydroelectric power a fairly mature industry. Although hydroelectric power is reliable and cost-effective once constructed, and does not release carbon dioxide, a number of environmental concerns exist. These include inundating wildlife habitat; changing aquatic ecosystems; and changing water quality, especially temperature, dissolved oxygen and nitrogen, and sediment levels.

Photovoltaics

Photovoltaics, or solar cells, convert sunlight directly into electricity. Unlike wind turbines or solar thermal systems, photovoltaics have no moving parts. Instead they use solid-state electronics. Over the past three decades photovoltaic efficiencies and reliability have increased significantly.

A photovoltaic cell is made by depositing layers of various materials so as to create an intrinsic and permanent electric field inside. When light strikes the material, it can free an electron from weak bonds that bind it. Once freed, the electric field pushes the electron out of the photovoltaic cell and sends it through an external wire to become a source of power before returning to the cell completing the circuit. The higher the efficiency of the cell, the higher the cost is. A number of technologies are in use including thin-film plates, single crystal and polycrystalline flat plates, and concentrator systems (which are most costly). Thin film uses little material approximately 1/100 the size of a human hair on a low-cost substrate such as glass, metal, or plastic. Although a variety of toxic chemicals are used in the manufacture of the photovoltaic cells, the emission of these chemicals is minimal.

Solar Thermal

Solar thermal electric plants use mirrors to concentrate sunlight on a receiver holding a fluid or gas, thereby heating it and causing it to turn a turbine or to push a piston connected to an electrical generator. Solar thermal systems are typically categorized by the type of collector used such as the parabolic trough, central receiver, parabolic dish, and solar pond. Parabolic troughs systems account for more than 90% of the world solar electric capacity. The systems have long, 100 meters or more, trough-shaped mirrors with the tube at the focal line along the center. The trough tracks the sun's position in the sky. There is a clear glass with a black metal pipe carrying heat-absorbing fluid down the middle. To minimize heat loss from the black absorbing pipe back to the outside, it is covered by special coatings to reduce the amount of heat it radiates. Central receivers have a large field of mirrors known as heliostats, surrounding a fixed receiver mounted on a tower. Each of the heliostats independently tracks the sun and focuses the flight on the receiver where it heats a fluid or gas. The fluid or gas expands through a turbine. A parabolic dish uses a large dish or set of mirrors on a single frame with two-axis tracking to reflect sunlight on to the receiver mounted at the focus (Figure 1.3).

The solar resource varies by the hour, day, season, geography; and with the local climate. Sunlight at Earth's surface has two parts, direct or beam radiation coming directly from the sun and diffuse radiation that has been first scattered randomly by the atmosphere before reaching the ground. Together they are total or global radiation. Direct radiation is more sensitive to atmosphere conditions than diffuse radiation is. Heavy urban smog can reduce direct radiation by 40% and total radiation by only 20%.

Wind

Wind energy systems use the wind to turn their blades, which are connected to an electrical generator. Wind systems provide intermittent power based on the availability of the wind. Small wind systems are often backed up with battery storage. Large wind turbines can be either on an individual site or more commonly in "wind

Parabolic Trough

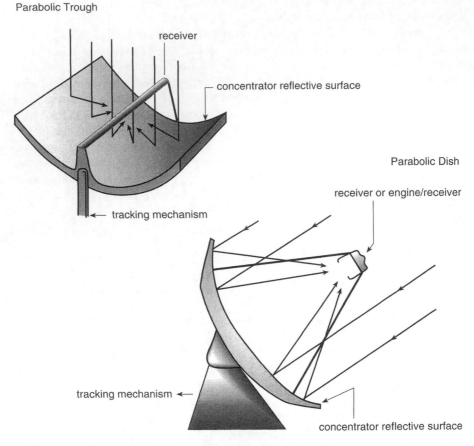

Figure 1.3   Solar thermal collectors. (From *Renewing Our Energy Resources*, Office of Technology Assessment, publication OTA-ETI-614, 1995.)

farms," and are connected to the electricity grid. Three key factors in wind energy
are the power of the wind and its speed, the variation in available wind speeds at
any given site over various time periods ranging from minutes to seasons, and the
variations in wind speed with different elevations above the ground. The power
available in the wind increases with the cube of the wind speed. Wind turbines must
handle a large range of power. This includes the speed at which the turbine reaches
its maximum power to the speed at which the turbine is stopped to prevent damage.
This has led to a variety of techniques for efficiently collecting power at low speeds
and for limiting and shedding excess wind power from the turbine blades at high
speeds. Because of the sensitivity of wind power to wind speed, it is necessary to
carefully choose a wind site. A 10% difference in wind speeds gives a 30% difference
in available wind power. Wind speeds can vary dramatically over the course of
seconds or minutes due to turbulence, hours due to diurnal variation, days due to
weather fronts, and months due to seasonal variations. The best locations are where
there are strong sustainable winds with little turbulence. The variation of winds with
distance above ground level is due to wind shear. Typically, winds at 50 meters are

about 25% faster and have twice as much power as winds at 10 meters above the ground. Wind shear places great stress on turbine blades.

There are two primary forms of wind turbines, the propeller style horizontal axis wind turbine and the less common vertical axis wind turbine. Turbine blades must manage the very high levels of stress while efficiently collecting energy over long periods of time and at a low cost. Large, modern wind turbines have become very quiet. At distances greater than 200 meters the swishing sound of the rotor blades is usually masked completely by wind noise in the leaves of trees or shrubs. Mechanical noise has virtually disappeared because of better engineering to avoid vibrations.

The United States produces more electricity from wind power than any other country in the world. The largest group of installations are located in three large wind development regions in California. New wind development programs are being tried in the Great Plains states because their wind capacity is several times greater than in California. Iowa and Minnesota are examples of this. Either in place of large wind farms or along with them organization of small clusters of wind developments is underway.

In Denmark, commercial-sized offshore wind turbine parks are becoming a reality. From 2000 to 2007, 750 MW of power will be built, with an additional 3300 MW planned to go on-line by the year 2030. Larger wind turbines, cheaper foundations, and new research on offshore wind conditions are increasing the confidence of the power companies, the Danish government, and the turbine manufacturers. There are increased wind speeds over the water.

Energy Used for Transportation

In the 1970s, the renewable fuel considered for transport in the United States was ethanol that came from corn. Converting corn to ethanol is expensive because of all the energy inputs needed to grow the corn and convert it to ethanol. Today, advances in biotechnology are allowing scientists to convert cellulose to sugars that can be fermented to ethanol by using cheaper feedstocks such as wood, grass, and corn stalks. Advances in gasification and catalysts are also lowering the cost of producing methanol and hydrogen from biomass.

The U.S. transportation system is essential to maintaining the economy of the country with highway transportation as a major factor. Highway transportation is dependent on internal combustion engine vehicles using almost exclusively petroleum. The United States consumes more than one third of the world's transport energy. Transportation uses about one fourth of total U.S. primary energy and nearly two thirds of the oil. The oil burning exacerbates local and global environmental problems. Motor vehicles account for 30 to 65% of all urban air pollution in the country and up to 30% of carbon dioxide emissions.

Methanol

Methanol is a liquid fuel that can be produced from natural gas, coal, or biomass. The major advantage to methanol is that it would require fewer changes in vehicle design. Flexible-fuel vehicles would be able to operate on methanol, ethanol, gasoline, or mixtures of these fuels. These types of vehicles would help in the transition from gasoline as a primary transportation fuel.

Ethanol

Ethanol, like methanol, is a liquid fuel that can be used in internal combustion engines. It can be produced from biomass. About one third of Brazil's automobile fleet runs on ethanol-produced sugars. The engines only require a minor modification to make this work. Ethanol and methanol produce less ozone and are less carcinogenic than gasoline, but ethanol produces formaldehyde and methanol produces acetaldehyde.

Hydrogen

Hydrogen is an extremely clean fuel that can be burned or electrochemically converted to generate electricity in fuel cells. Hydrogen can be produced from a variety of sources including:

1. Gas or coal
2. Biomass or municipal waste gasification
3. Direct photochemical conversion
4. Direct photobiological conversion
5. Thermal decomposition
6. Electrolysis of water

Producing hydrogen from gas or coal creates additional carbon dioxide and therefore is not considered to be the best process. Although there is an abundance of coal in the United States, new processes would have to be developed to make the new transportation fuel environmentally safe.

Biomass or municipal waste gasification is a direct method of producing hydrogen by extracting the hydrogen from fast growing grasses or trees. Biomass gasification relies on the technology that has been developed for other applications and therefore is one of the most advanced direct approaches to hydrogen production. It is the least expensive of the new methods of direct hydrogen production from renewable sources.

In direct photochemical conversion, sunlight strikes an electrolytic solution in which photosensitive semiconductors and catalysts are suspended. The solar energy is absorbed by the catalysts creating localized electrical fields that cause the electrolytic splitting of water into hydrogen and oxygen.

In direct photobiological conversion, various bacteria and algae have been isolated that produce hydrogen either from organic materials through digestion or directly from water and sunlight through photosynthesis. Both methods use entirely renewable resources including sunlight, biomass, and biological organisms to produce hydrogen.

In thermal decomposition concentrated solar thermal energy can achieve temperatures exceeding 5000°F. At these temperatures water thermally decomposes into hydrogen and oxygen. The problem facing engineers today is to capture the gases separately before they spontaneously recombine.

In electrolysis of water, an electrical current passes through electrodes immersed in an electrolytic solution divided into two chambers by a semipermeable membrane. The electrical energy splits the water molecule into hydrogen and oxygen gases, which are released and collected at the electrodes. The energy efficiency of electrolysis generally exceeds 80%. The problem is that unlike other renewable sources it takes electricity produced by other sources to carry out the process.

Transportation and storage of hydrogen gas is a key issue. A hydrogen infrastructure would have to be developed with safe storage and transportation of the fuel. In addition, there would be a need for a large onboard storage tank if the fuel was used directly in the engine. However, the fuel can be converted into electricity and the vehicle can be powered by a battery. A pipeline is the easiest form of hydrogen transmission, although the gas can be compressed or liquefied and transported in tankers as well. Advanced high-pressure tanks capable of containing pressures up to 10,000 lb/in.^{2} have been used in the space program. Most hydrogen is now stored at pressures up 3600 lb/in.^{2} which is common for natural gas and other compressed industrial gas storage systems. Hydrogen can be stored more easily than electricity until needed.

From the pure form of hydrogen, energy is released by burning or converted without combustion to electricity in fuel cells. A fuel cell is an electrochemical device without moving parts. This device uses catalysts, electrodes, semipermeable membranes, and electrolytic substances to control the combination of hydrogen and oxygen to produce energy and form water as a by-product. The energy released by the chemical reaction is captured electrochemically as electricity. Currently, there is a 50% efficiency rating. With emerging fuel cell technologies it is hoped that efficiency will rise beyond 60%.

In October 1996, the Hydrogen Future Act became law. The act mandates that government-sponsored hydrogen research, development, and demonstration project expenditures of over $100 million be conducted during the next 5 years in the United States. In addition, more than $20 million a year is available through the National Fuel Cells in Transportation Program. The Mercedes Corporation believes that fuel cells is the technology most likely to replace internal combustion engines in automobiles. Toyota Corporation, Chrysler Corporation, Daimler Benz, and other corporations are currently experimenting with fuel cell automobiles,

HEALTH IMPLICATIONS OF NEW ENERGY TECHNOLOGIES

Coal, which is our most abundant fossil fuel, presents a potential threat to human health and the environment. Coal miners excessively exposed to coal dust may suffer from lung diseases known as pneumoconioses, including black lung disease, silicosis, and emphysema. The fossil fuel by-products include two major groupings: coal tar, which comes from the combustion or distillation of coal, and polycyclic aromatic hydrocarbons. Polycyclic aromatic hydrocarbons (PAHs) are potential carcinogens found in mixtures within fossil fuels and their by-products. Coal tar products can cause skin and lung cancer through the cutaneous and respiratory routes, respectively.

Human exposure to complex mixtures of PAHs has been extensive, and carcinogenic effects following long-term exposure to them have been documented. Coke oven emissions and related substances, such as coal tar, have long been associated with excess disease. The first observation of occupational cancer, that is, scrotal cancer among London chimney sweeps, was made by Percival Pott in 1775. More recently, lung and genitourinary cancer mortality have been associated with coke oven emissions. Human tumorigenicity also has been associated with the exposure to creosote. Creosote is a generic term that refers to wood preservatives derived from coal tar creosote or coal tar oil, and includes extremely complex mixtures of liquid and solid aromatic hydrocarbons.

Bituminous coal is the starting point for coal liquefication and coal gasification. The mere process of breaking up the coal structure appears to release a large number of potentially carcinogenic compounds. Some of these compounds include the PAHs benz(a)anthracene, chrysene, and benzopyrene. In the coal gasification process, where coal is exposed to molecular oxygen and steam at temperatures of 900°C or higher, much of the hydrocarbon structure is destroyed; therefore, the levels of carcinogenic compounds can be greatly reduced. However, at these temperatures the hazards from carbon monoxide increase, as well as the potential hazards from other toxic agents such as hydrogen sulfide and other carbon–nitrogen products that may be released to the environment.

Coal liquefication occurs in the temperature range of 450 to 500°C. This is the most economical temperature range to produce pumpable liquid. During the initial conversion reaction, hydrogen is added to the coal, which produces a wide range of compounds containing condensed aromatic configurations. Many of these are carcinogenic. A potential also exists for the production of other toxic compounds, including carbon monoxide.

Oil shale is a generic name for the sedimentary rock that contains substantial organic materials known as kerogen. The kerogen content of oil shale may vary between 5 and 80 gal of oil per equivalent ton. The extraction of the oil from the oil shale occurs when the shale is heated to 350 to 550°C under an inert atmosphere. The products include oil vapor, hydrocarbon gases, and carbonaceous residue. Major concerns related to the retorting shale process include (1) the production of 1 ton of spent shale per barrel of oil; (2) the volume of the shale increasing by more than 50% during the crushing process and a considerable amount of alkali minerals contained in the oil shale possibly leaching into the groundwater supply; (3) the creation of a problem in areas where water is in short supply because of a large amount of water needed for controlling dust and reducing the alkalinity of spent shale.

During the processing of the oil shale, organic materials containing nitrogen, oxygen, and sulfur are produced in large quantities. Further, waste gases containing hydrocarbons, hydrogen sulfide, sulfur dioxide, and nitrogen oxides are formed.

Geothermal energy is a general term that refers to the release of the stored heat of the Earth that can be recovered through current or new technologies. The amount of contaminants found in the energy source varies dramatically from area to area.

In any case, geothermal fluids may contain arsenic, boron, selenium, lead, cadmium, and fluorides. These fluids also may contain hydrogen sulfide, mercury, ammonia, radon, carbon dioxide, and methane. Wastewater management from geothermal processes is an environmental problem if the spent liquids are put into surface water. They may contaminate the surface water and also cause an elliptical dish-shaped depression, such as occurred in New Zealand. If the liquids are injected under high pressure back into the geologic formation, they may enhance seismic activities. Another problem related to the production of geothermal energy is noise. Some noise levels reach 120 A-scale decibels (dBA). At this level, blowouts can occur and produce accidents with potential injuries. Airborne emissions and noise can affect human health. The most significant potential air emission problem is hydrogen sulfide gas. Hydrogen sulfide is a very toxic gas that produces immediate collapse with respiratory paralysis at concentrations above 1000 ppm by volume. Serious eye injuries can occur at 50 to 100 ppm by volume. Further, hydrogen sulfide has a very offensive odor (rotten eggs) that can be detected by the human nose at very low concentrations.

The photovoltaic solar cell is an efficient, direct energy-conversion device. Solar cells are used for powering most satellites. The major health hazards involved relate to the mining and the various steps needed in the production of the silicon cells. Other cells utilized include cadmium sulfide and gallium arsenide. The health hazards involved relate to the production and use of these chemicals in the occupational setting. If satellite solar power stations come on-line in the future, there will need to be an extensive research effort to determine if there will be potential biohazards due to the microwave beams being brought back to Earth from the solar space stations.

Biomass is composed of plant material and animal waste used as a source of fuel. Health hazards may be associated with the emissions from biomass combustion residuals, biomass gasification residuals, biomass liquefaction, and the related air and water pollutants. The burning of biomass creates the same primary air pollutants as in the burning of coal. Unburned hydrocarbons, sulfur oxides, and high-fugitive dust levels are created. Organic acids and minerals that affect water quality may be found in boiler water treatment chemicals, in leachates from ash residues, and in biomass storage piles. Residential wood burners, because of incomplete combustion, release carbon monoxide and unburned hydrocarbons that may include photochemically reactive chemicals as well as carcinogens. Emissions from biomass gasification may originate from the process stack, waste ponds, storage tanks, equipment leaks, and storage piles. These contaminants include oxides of nitrogen, hydrogen cyanide, hydrocarbons, ammonia, carbon monoxide, and particulates. The process water and condensates also may contain phenols and trace metals. The tars produced by the thermochemical decomposition of organic substances may contain PAHs, some of which are known carcinogens. Anaerobic digestion may lead to the production of odors associated with gaseous decay products such as hydrogen sulfide and ammonia gases. The effluent contains large quantities of biochemically oxygen-demanding organic materials, organic acids, and mineral salts.

Nuclear power plants have the potential for providing abundant supplies of electricity without contributing substantial amounts of pollutants to the environment

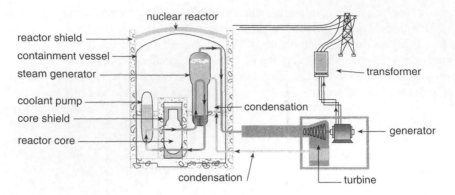

Figure 1.4   Nuclear energy site.

(Figure 1.4). The industry, however, has failed to deliver on the promise because the costs of making nuclear energy safe have spiraled out of control. The nuclear cycle starts with the mining of uranium and the resulting risk of lung cancer in miners attributable to the alpha radiation from the decay of radon-22 daughter products. Further, there are injuries associated with the mining and quarrying of uranium. During the milling of uranium ore, small airborne releases of radon and the residues or tailings still contain most of the radioactive species of the original ore and require careful disposal. Next, the uranium oxide is converted to uranium hexafluoride. This enrichment increases the percentage amount of fissionable uranium-235. Finally, fuel rods are fabricated during the processing of the uranium after it has been mined. (The potential amount of worker radiation exposure that comes from gaseous solid wastes is small.) The reactor then produces energy. Finally, after the fuel rods are spent, the irradiated rods are removed from the reactor cooler and stored for long periods of time to permit decay of the short-lived fission products before reprocessing. Radioactive water and waste must be contained.

Engineers can build reactors that are safer than those now in operation. The basic technology has been available for more than 25 years. This technology has been ignored in favor of water-cooled reactors, which have already been proved in nuclear submarines. However, these reactors are particularly susceptible to the rapid loss of coolant, which led to the accidents at Three Mile Island and Chernobyl. All nuclear reactors split large atoms into smaller pieces and thereby release heat. It is necessary to keep the core of nuclear fuel from overheating and melting into an uncontrollable mass that may breach the containment walls and release radioactive material. One way to prevent a meltdown is to make sure that the circulating coolant, which is water, will always be present in adequate quantities. To prevent mechanical failures from interrupting the transfer of heat, most reactors employ multiple backup systems. This technique is known as "defense in depth." The problem with this technique is that it can never be 100% safe against a meltdown.

The U.S. Department of Energy wanted to use a new strategy in Idaho Falls, ID. The agency wanted to build a series of four small-scale modular reactors that use fuel in such small quantities that their cores could not achieve meltdown temperatures under any circumstances. The fuel would be packed inside tiny heat-resistant ceramic spheres and cooled by inert helium gas. The whole apparatus would be buried below ground. The main problem is that the smaller units produce less electrical output and, therefore, are less economical initially. However, over time the units may become more efficient.

During the early morning hours on March 28, 1979, the most serious accident in the history of U.S. nuclear power took place. A series of highly improbable events involving both mechanical failure and human error led to the release of a considerable amount of radioactivity and the evacuation of preschool children and pregnant women within 5 mi of the Three Mile Island nuclear power plant, located near Middletown, PA. The accident occurred at Unit 2 of the complex. It was initially triggered by the failure of a valve and the subsequent shutting down of a pump supplying feed water to a steam generator. This in turn led to a "turbine trip" and a shutdown of the reactor, in itself was not the problem, because the backup systems that were used anticipate such failures. However, a pressure relief valve failed to close, which led to the loss of substantial amounts of reactor coolant, usually in quench tanks. Auxiliary feed water pumps were nonoperational because of closed valves, in violation of NRC regulations. The operator failed to respond promptly to the stuck relief valve. There were faulty readings on the control room instruments that led the operators to turn off the pumps for the emergency core cooling system. The operator shut down the cooling pumps more than an hour after the emergency began. Two days after the initial problems, a large bubble of radioactive gas, including potentially explosive hydrogen, was believed to have formed in the top of the reactor. Then President Jimmy Carter established a 12-member committee to review the accident and make recommendations. The major findings of the commission after a 6-month study concluded (1) the accident was initiated by mechanical malfunction and exacerbated by a series of human errors; (2) the accident revealed very serious shortcomings in the entire government and private sector systems used to regulate and manage nuclear power; (3) the NRC was so preoccupied with the licensing of new plants that it had not given adequate consideration to overall safety issues; (4) the training of power plant operators at Three Mile Island was extremely inadequate even though it conformed to NRC standards; (5) the utility owning Three Mile Island and the power plant failed to acquire enough information on safety to make good judgments; (6) an extremely poor level of coordination and a lack of urgency on the part of all levels of government existed after the accident occurred. The NRC had not made mandatory, state emergency or evacuation plans.

Other potential emergency situations have occurred in the past. On December 12, 1952, the accidental removal of four control rods at an experimental nuclear reactor at Chalk River, Canada, near Ottawa led to a near meltdown of the reactor uranium core. A million gallons of radioactive water accumulated inside. Fortunately, no accident-related injuries occurred. On October 7, 1957, a fire occurred in the reactor north of Liverpool, United Kingdom. Like the Chernobyl facility, the Windscale Tile 1 plutonium production plant used graphite to slow down neutrons emitted during nuclear fission. Contamination occurred on 200 square miles of countryside. Officials banned the sale of milk from the cows grazing the area for more than a month. It was estimated that at least 33 cancer deaths could be traced to the effects of the accident. On January 3, 1961, a worker's error in removing control rods from the core of an SL-1 military experimental reactor near Idaho Falls, ID, caused a fatal steam explosion, and three servicemen were killed. On March 22, 1975, a worker using a lighted candle to check for air leaks in the Brown Ferry reactor near Decatur, AL, touched off a fire that damaged electrical cables connected to the safety system. Although the reactor's coolant water dropped to dangerous levels, no radioactive material escaped into the atmosphere. On March 8, 1981, radioactive wastewater leaked for several hours from a problem-ridden nuclear power station in Tsuruga, Japan. Workers who mopped up the wastewater were exposed to radiation. The public did not become aware of the accident for 6 weeks until radioactive materials were detected. On January 4, 1986, one worker at the Kerr-McGee Corporation, a uranium processing plant in Gore, OK, died from exposure to a caustic chemical that formed when an improperly heated, overfilled container of nuclear materials burst. Some of the radiation flowed out of the plant. More than 100 people went to local hospitals.

April 26, 1986, the worst disaster in history relating to a nuclear plant occurred at Chernobyl in the former Soviet Union. A loss of water coolant seemed to trigger the accident. When the water circulation failed, the temperature in the reactor core soared over 5000°F, causing the uranium fuel to begin melting. This produced steam that reacted with the zirconium alloy of the fuel rod to produce explosive-type hydrogen gas. Apparently, a second reaction produced free hydrogen and carbon monoxide. When the hydrogen combined with the oxygen, it caused an explosion, which blew off the top of the building and ignited the graphite. Next, a dense cloud of radioactive fission products discharged into the air as a result of the burning graphite. In subsequent days, radioactive materials were found in Finland, Sweden, Norway, Denmark, Poland, Romania, and Austria. In Italy, freight cars loaded with cattle, sheep, and horses from Poland and Austria were turned back because of the concern due to abnormally high levels of radiation found in many of the animals. Britain cancelled a spring tour of the London Festival Ballet, which was to go to the former Soviet Union. In West Germany, officials insisted that children be kept out of sand boxes to avoid contamination. Slight amounts of radiation were found in Tokyo, in Canada, and in the United States. In the immediate vicinity of the reactor and up to 60 mi from the reactor, the topsoil may be contaminated for decades. The residents of Kiev, 80 mi from the disaster and the surrounding areas, were told to wash frequently and to keep their windows closed. They were warned against eating lettuce and swimming outdoors. Water trucks were used to wash down streets to wash away radioactive dust. It will be many years before an accurate determination is made about the number of people who will die as a result of this catastrophe. Approximately 100,000 people will have to be monitored for the rest of their lives for signs of cancer. In 1995, large populated areas surrounding the reactor site in the Ukraine and in nearby Belorussia remain contaminated with high levels of radioactivity. The cost of the cleanup could run as high as $358 billion.

On September 30, 1999, Japan had the worst nuclear accident in the history of its nuclear industry. Three workers were responsible for the accident when they used bucketlike containers to mix uranium. These men worked for the J. C. O. company, a private organization that ran the nuclear power plant in Tokaimura. The three workers were sent to the hospital with two of them suffering potentially lethal doses of radiation. Another 46 people were also exposed to the radiation. Apparently, J.C.O., a wholly owned subsidiary of Sumitomo Metal Mining Company, has admitted that it had for years deviated from government-approved procedures by having its own illegal manual. The company was not required to be prepared for possible atomic reactions because the uranium-processing plant was in principle not supposed to set off a reaction. The atomic reaction that occurred as the result of the accident is similar to that which happens in a nuclear reactor. Processing uranium, if done properly, does not cause an atomic reaction. By using the bucketlike containers instead of the proper equipment, the mixing could be shortened from 3 hr to 30 min. Firefighters called in to help the injured workers were never warned of a potential release of radioactivity and therefore did not have proper protective clothing. Nearly 2 hours elapsed between the accident and any notice to local residents that something had occurred.

In the United States, since 1980, each utility that owns a commercial nuclear power plant has been required to have both on-site and off-site emergency response plans as a precondition for obtaining and maintaining a license to operate the plant. The on-site emergency response plans are approved by the NRC, whereas the off-site plans are approved by the Federal Emergency Management Agency (FEMA). The NRC incident response operations has published over 500 reports on a broad range of operational experience since 1980. A congressional committee is currently investigating how to revitalize the nuclear power industry, because the energy produced does not include by-products that contribute to the amount of carbon dioxide or other contaminants released into the air.

It should be understood that the nation will have to make a myriad of difficult decisions concerning the conservation and use of fuel. Obviously the burning of carbonaceous fuels has significantly contributed to air pollution and accompanying health problems. The energy problem has affected our entire economy by the purchase of energy from abroad. An adequate program of conservation, development and use of energy sources within our own country will provide us with the needed energy and yet prevent the many potential hazards that may occur as a result of the burning of fossil fuels.

ENVIRONMENTAL PROBLEMS RELATED TO ENERGY

The major environmental problems related to energy are caused by pollutants created by fossil fuels; destruction of the natural environment by removal or spillage of fossil fuels; and effects of the continued rise in cost of fossil fuels on the economy, and therefore, on our way of life. The pollutants are detailed in Volume II, Chapter 1, on air pollution. With an increased need for fossil fuels comes an increase in the destruction of the natural environment. Improper fuel removal, fuel storage, or fuel transportation degrades the environment; destroys the aesthetic value of many of our most beautiful areas; and causes destruction of fish, wildlife, and plant life. The problems related to health, aesthetics, and economics are critical, because they are totally interrelated and therefore must be considered as a unit.

APPLICABLE CONCEPTS OF CHEMISTRY

Chemicals in general and those that contaminate the environment are classified as organic and inorganic. The fate of chemicals in environmental matrices of air, water, and soil relative to human exposure and toxicity are influenced by several fundamental properties.

Organic chemicals are based on carbon-12 ($^{12}_{6}C$) as the foundation element present in all molecules and compounds. The atoms that compose organic molecules are predominantly held together via intraatomic forces called covalent bonds. The covalent bonds involve sharing of electrons between the elements. Electronegativity is the measure of an atom attraction for orbital electrons. If the difference in electronegativity between two covalently bonded atoms is relatively low, then electrons are shared somewhat equally and the bond is considered nonpolar. When the difference in electronegativity between covalently bonded atoms is relatively high, then electrons between the bonded atoms are unequally shared and the bond is polar. Polar bonds are characterized as having a slightly positive end (pole) adjacent to the atom with lower electronegativity and a more negative end (pole) in the region adjacent to the more electronegative bonded atom.

Hydrocarbons are organic compounds composed of only carbon bonded to carbon and carbon bonded to hydrogen atoms. The covalent bonds formed between adjacent carbon atoms are nonpolar because no difference exists between electronegativity when carbon bonds to carbon, due to totally equal sharing of electrons. These bonds are called *coordinate covalent bonds*. The difference in electronegativities of carbon and hydrogen atoms is not substantial enough for a relatively positive pole and negative pole to exist. When only single carbon to carbon (C–C) bonds are present, the organic compound is classified as an alkane. Compounds containing double carbon to carbon (C=C) or triple carbon to carbon (C≡C) bonds are called alkenes and alkynes, respectively. These compounds are straight-chain aliphatics when they have distinct end groups or cyclic aliphatics when they form closed ring structures. Aromatic hydrocarbons consist of at least one or more benzene molecules, which are six-carbon ring structures with three alternating double bonds.

Functional groups are specific atoms or groups of atoms bonded to hydrocarbon structures. The presence and type of atom or group alters the chemical and physical properties of an organic molecule and, in turn, its function. According to the functional group, the organic molecules are classified as alcohols, aldehydes, acids, ketones, esters, thiols, and so forth. Examples of some common functional groups appear in Table 1.1.

Inorganic chemical compounds are not based on the element carbon as a foundation element and, accordingly, include all chemicals that are not classified as organic. Inorganic elements and, in turn, the composition of compounds include both metals and nonmetals. Some inorganic compounds, such as water, contain covalent bonds, but many compounds, such as salts, have ionic bonds. Because the ionic bonds are formed between atoms that have such a high difference in electronegativity, one or more electrons are lost by one atom and accepted or gained by an adjacent atom. As a result, the atom that loses the electron is oxidized and referred to as a *reducing agent* and the atom that gains the electrons is reduced and referred to as an *oxidizing*

Table 1.1 Examples of Some Organic Functional Groups

Hydrocarbons	R–H
Organohalides	R–X[a]
Alcohols	R–OH
Aldehydes	R–(C=O)–H
Acids	R–(C=O)–OH
Thiols	R–SH
Amines	R–NH$_2$
Amides	R–(C=O)NH
Nitro	R–NO$_2$
Ethers	R–O–R
Esters	R–O–(C=O)–R
Ketones	R–(C=O)–R
Peroxides	R–O–O–R
Nitriles	R–C≡N
Azo	R–N=N–R

Note: R represents an aliphatic or an aromatic group.

[a] Where X=Cl, Br, F, I.

agent. Thus, reducing agents lose or donate electrons which in turn causes their atomic charge to increase (oxidation) and the charge on atom that accepts or gains the electrons to decrease (reduction). Ionic bonds are polar because they form between a positively charged atom (cation) and a negatively charged atom (anion).

Biochemicals refer to compounds associated with living or once living organisms. The five kingdoms of organisms, Animalia, Plantae, Protozoa, Monera (bacteria) and Fungi, consist of biochemicals. The five major classes of large biochemicals, or biomacromolecules, are proteins (amino acid polymers), carbohydrates (simple monomeric- and oligomeric-, and complex polymeric-sugars), nucleic acids (nucleotide polymers such as deoxyribonucleic acid [DNA] and ribonucleic acid [RNA]), lipids (fats, waxes, steroids), and lignin (phenolic polymers). Only the plants contain lignin. Carbon (C), hydrogen (H), nitrogen (N), oxygen (O), phosphorus (P), and sulfur (S) account for >99% of the elements in living or once living organisms.

As implied earlier, relative to the environmental health sciences, the chemical and physical properties greatly influence the overall characteristics and fate of these molecules in the environment, when human exposure and impact are considered. Chemical properties, especially the elemental composition and arrangement in molecules and compounds, influence the likelihood that a chemical will be transformed biologically, chemically, or physically into various products and by-products. In relation, the products and by-products formed as a result of transformation reactions most commonly reflect elements present in the original matter or parent compound that underwent change. For example, natural organic matter, such as manures, biological sludges, and decaying vegetation, originated from and consists of living or once living matter. As a result, the matter contains various components including biomacromolecules that are elementally composed predominantly of combinations of C, H, N, O, P, and S in the form of polymers and oligomers. When transformation involves aerobic oxidation reactions, the ultimate products reflecting decomposition are carbon dioxide, water, nitrates, nitrites, phosphates, and sulfites and sulfates; or instead anaerobic reduction products methane, hydrogen, ammonia, and hydrogen sulfide.

Chemical and physical properties also influence transport (mobilization) and storage (immobilization) of contaminants in an environmental matrix or human system. Chemical functional groups can affect intermolecular bonding between contaminants and various structures found in soil, water, and living cells. For example, positively charged or cationic metals readily bond to and store at negatively charged oxygen atoms associated with silicon oxide structures of soil particles.

The physical properties of solubility of hydrophilic chemicals or insolubility of hydrophobic chemicals in water depends on the presence and number of nonpolar and polar bonds in the molecules, as well as their molecular shapes and three-dimensional structures (stereochemistry). For example, water molecules (H–O–H) have covalent bonds because the respective electrons are shared between one oxygen and each of the two hydrogen atoms. The bonds are actually polar covalent bonds because the electrons are unequally shared. Indeed, the entire water molecule is polar because of the polarity of the bonds and its stereochemistry. In general, relatively polar molecules are soluble or miscible in polar water (polar), whereas relatively nonpolar or lipophilic molecules are not soluble in water, but are soluble in fats or oils (lipids; nonpolar).

Density and specific gravity are related physical properties also important factors influencing the fate of contaminants in the environment and relevant to human exposure. The density of a solid, liquid, or gas is the mass of the substance for a given volume. For example, the density of water is 1 g/ml or 8.33 lb/gal. The specific gravity of a substance equals its density divided by the density of a reference substance such as water or air. For example, the specific gravity of water equals 1 because the density of water (8.33 lb/gal) is divided by a reference density, in this case water or 8.33 lb/gal. Relatively nonpolar and therefore insoluble contaminants in water that have a density greater than 8.33 lb/gal or specific gravity of >1 sink, and those with a density of <8.33 lb/gal or specific gravity of <1 float.

Air has a molecular weight of about 28.8 g/mol, based on a rounded approximation of 21% oxygen (O_2; mol wt = 32 g/mol) and 79% nitrogen (N_2; mol wt 28 g/mol). At normal temperature and pressure (NTP) or 25°C and 760 mmHg, 1 mol of air is equivalent to a volume of 24.45 l of gas (22.5 l/mol at STP or 0°C and 760 mmHg). The density of air at NTP, therefore, is 28.8 g/mol divided by 24.25 l/mol which equals approximately 1.18 g/l. Also, gaseous contaminants with a density greater than 1.18 g/l at NTP tend to sink in air and those with lower densities tend to rise in air.

Vapor pressure is the final physical property to be discussed here in this section. True gases, such as oxygen and nitrogen, are designated as such because they are totally in the gaseous state at standard temperature and pressure (STP) and NTP. Vapors are gases too, but are generated from volatile materials that are predominantly in a liquid and sometimes in a solid state at STP and NTP. Volatile liquids and solids can evaporate and generate gaseous vapors into the air as a result. Vapor pressure is a measure at a defined temperature of the pressure in the headspace of a closed container partially filled with a liquid, such as the organic solvent benzene. The measured pressure stabilizes at a point of equilibrium, meaning the number of molecules evaporating from the liquid state to the gaseous vapor state equals the number of molecules of the gaseous vapor state condensing back to the liquid state. Benzene solvent has a vapor pressure of 75 mmHg at 20°C and is very volatile. A direct relationship exists between increased temperature and increased vapor pressure. The higher the vapor pressure for a given liquid or solid, the more volatile it is or more readily it will evaporate.

Accordingly, chemical and physical properties, in combination with numerous other factors, influence the fate of chemicals. The fate relative to transport, storage, and alteration of chemicals is discussed briefly in the next section.

TRANSPORT AND ALTERATION OF CHEMICALS IN THE ENVIRONMENT

Humans are part of the flow of energy; in the biosphere we interact with thousands of plants and animals. Because of our power and productivity, we have the ability to alter the ecosystems of Earth, of which we are a part, in a beneficial or harmful manner. Technological advancement since the early 1800s has been revolutionary instead of evolutionary. In the great movement forward to improve our own lifestyle, we have destroyed many of the natural ecosystems and severely polluted much of the air, water, and soil. To better understand the nature of the environmental alterations, it is important to understand the role of chemicals in the environment and how they are transported and altered.

Much of the environment is made up of chemicals, the largest amount of which are occurring naturally and have few detrimental effects on humans. However, we change the form, distribution, and concentration of these chemicals and synthesize or produce, either as products or as by-products, chemicals that are naturally found in the environment, as well as chemical compounds that are not. It is necessary to understand the movement of those chemical compounds that are not natural and to understand the movement of all these chemicals through the environment, their concentrations, the degree of human exposure, and the potential hazards to human health, such as initiatives and promotion of cancer (Figure 1.5).

Once a contaminant is released into the environment, the chemicals are transported and transformed in a variety of complex ways. Chemicals are transported widely throughout the environment by active and passive movement through air, water, biota (biological life), and soil. In addition, they serve as vectors of transport. Chemicals are transformed by chemical or biochemical reactions, and diluted, diffused, or concentrated by physical or biological processes. Some substances diffuse upward, where they are degraded by ultraviolet light; or diffuse downward, where they are adsorbed onto the surfaces of suspended particulate matter. Other substances are dissolved in water droplets and returned to Earth by rainfall. Weather conditions may set up cycles of elimination of toxic substances from air onto land and then into water, and eventually reintroduction into the air. Where sources of pollution are in close proximity to sources of water usage, an extensive and potentially hazardous series of chemicals may be transported directly into the water source and may cause significant health problems following human exposure (Figure 1.6).

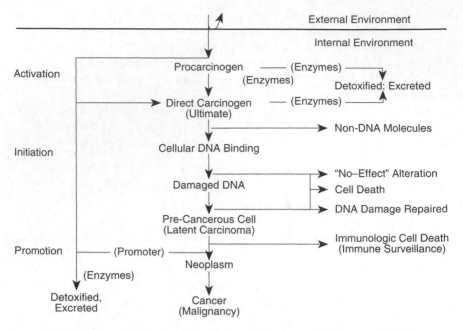

Figure 1.5   Chemicals in the environment. (From *Environmental Toxicology and Risk Assessment: An Introduction, Student Manual*, U.S. Environmental Protection Agency, Region V, Chicago, IL, Visual 3.8.)

Humans are exposed to chemical contamination through inhaled air, ingested water and food, absorption through the skin, and at times through a combination of these modes. We may be continuously, intermittently but repeatedly, or sporadically exposed. We may be exposed to chemicals present in substantial or in minute quantities. In some instances, chemicals present in minute quantities are more hazardous than those in large quantities. There is also direct intentional exposure through chemicals added to foods and cosmetics, and by those involved in the preparation of drugs. Good sense dictates limiting the level of exposure of any chemical that may cause adverse responses in an individual. Human health is endangered not only at the point of use or discharge of a specific chemical but also at varying points in the ecosystem. Unfortunately, many of the fundamental changes occurring in the environment are poorly understood. Reconcentration, specifically bioconcentration, is a process involving potentially harmful human exposure to chemicals (Figure 1.7). Plants and animals accumulate certain chemicals at levels higher than those of ambient environment. Chemicals also may be absorbed by air- or waterborne particles that are inhaled and concentrated in the lungs.

Dispersion of Contaminants

Contaminants tend to spread continuously within the environment. Indeed, when contaminants that are released into the environment accumulate to concentrations equal to or above an accepted threshold, then the ecosystem or environment becomes

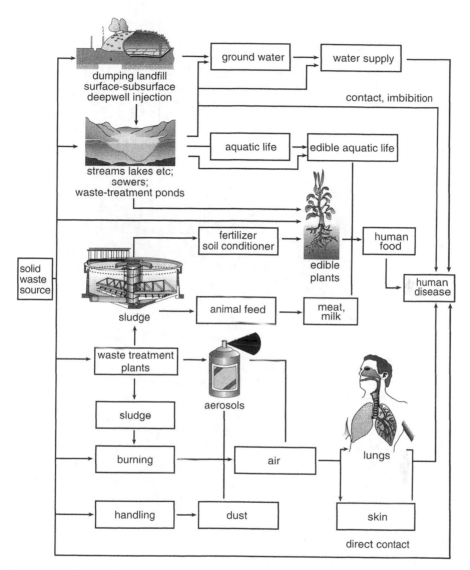

Figure 1.6 Chemical waste and human disease pathways (postulated).

polluted. The characteristics of the medium affect dispersion of the contaminants. Contaminants are actively dispersed via convective or turbulent manner. Convection is the circulatory motion that occurs in a fluid at a nonuniform temperature owing to the variation of its density and the action of gravity. Turbulence in the atmosphere is a complicated phenomenon that depends on the wind velocity; the direction of the wind; the altitude; the friction of the air over the surface of Earth; the temperature of the air; the pressure in the air; and the presence or absence of bodies of water, mountains, and flat areas. Dispersion from ground-level sources are seen almost immediately downwind from the source. Dispersion from tall smoke stacks or any

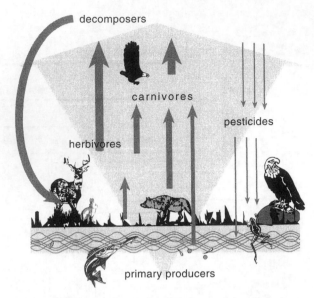

Figure 1.7   Bioconcentration. (Adapted from Koren, H., *Illustrated Dictionary of Environmental Health and Occupational Safety,* Lewis Publishers, CRC Press, Boca Raton, FL, 1995. With permission.)

elevated discharge may be found for considerable distances from the source. Contaminants also are passively dispersed through the atmosphere. The passive process is known as diffusion and implied movement of contaminants based on a concentration gradient. The contaminants diffuse through the air from areas of high to low concentration. This is most applicable to gases and vapors.

Dispersion in the hydrosphere is much more varied and complex than dispersion in the atmosphere. The dilution volumes and mixing characteristics, as well as the rate of transport of the contaminants, vary with rivers, lakes, estuaries, coastal waters, and oceans. The contaminants in the water may further contaminate the land, the groundwater supply, and the air.

The environment of the soil is very complex with regard to the dispersion of contaminants. The dispersion depends on the specific nature of the chemicals; the type, texture, and structure of soil; and other factors, such as moisture, pH, and temperature. Most chemical contaminants do not move readily through the soil once they enter it. Gases may easily diffuse through air pockets and pores in the soil. Soil water is the major means of moving chemicals downward through pores and channels to the lower horizons of the soil or upward to the point of evaporation. Water also may move the contaminants in a lateral manner or help in surface runoff depending on the slope of the soil. The more rapidly water percolates into and leaches through the soil, the more readily the chemicals may diffuse into the pores between the soil particles.

Alteration of Chemical Contaminants

Contaminants are removed from the atmosphere by a variety of means. The contaminants may either be in the original form or may have altered during the dispersion process. Particles are physically removed from the atmosphere by gravity, impacting on objects, and through rain, snow, or sleet. The larger the particle, the more rapidly it will settle out. The particles may attract gases, vapors, or viruses and bacteria. Gases and vapors also may be dissolved in moisture, or through chemical reactions in the atmosphere they may be chemically converted into other compounds. Photochemical interaction is the most significant catalyst for altering physical and chemical air contaminants. Ultraviolet energy from the sun can interact with air contaminants resulting in oxidation and transformation of chemical structures.

Chemicals found in water may be bioconcentrated in fish or plant life, deposited on surfaces, or converted into other products by the self-purification process that is related to chemical and biochemical oxidation. Physical self-purification may occur when the chemicals bond to suspended particles in water.

The persistence of chemicals in the soil is a complex function, related to physical, chemical, and biological factors. The major means of removal of contaminants in soil are by degradation due to biotic microbial action, abiotic chemical degradation, evaporation, volatilization from the surface, and uptake by vegetation.

To get some better understanding of transport and alteration of chemicals in the environment, seven chemicals have been chosen for discussion. All have been identified on the Comprehensive Environmental Response, Compensation, and Liability Act (CERCLA, Superfund) National Priorities List. They are part of a large group that pose a significant potential threat to human health, as determined by the Agency for Toxic Substances and Disease Registry. These chemicals are chrysene, chloroform, cyanide, lead, 2,3,7,8-tetrachlorodibenzo-p-dioxin (2,3,7,8-TCDD), tetrachloroethylene, and trichloroethylene.

Chrysene is found in the environment from natural sources, such as forest fires and volcanoes, and from people-made sources. Chrysene, like all PAHs, is formed during high-temperature pyrolytic processes. Combustion is the major source of environmental chrysene. Virtually all direct releases into the environment are to the air, although small amounts may be released to the water and land. Chrysene is removed from the atmosphere by photochemical oxidation and dry or wet deposition onto land or water. This chemical is very persistent in either soil or sediment. Biodegradation is a slow process. The half-life of chrysene in the soil is estimated to be 1000 days.

Incomplete combustion of carbonaceous materials is the major source of chrysene in the environment. Residential heating and open burning are the largest combustion sources. The inefficient combustion process creates uncontrolled emissions, with heating in the home as the greatest contributor to chrysene emissions. Hazardous waste sites can be concentrated sources of PAHs on a local scale and, therefore, may contain larger quantities of chrysene. Chrysene in the atmosphere is expected to be associated primarily with particulate matter, especially soot. Chrysene in aquatic systems is expected to be strongly bound to suspended particles or sediments. Chrysene is expected to be strongly sorbed to surface soils.

Chloroform is a colorless or water-white liquid used to make fluorocarbon-22. It is also used as a pesticide and as a solvent in the manufacture of pesticides and dyes. Chloroform is discharged into the environment from pulp and paper mills, pharmaceutical manufacturing plants; chemical manufacturing treatment plants, and from the chlorination of wastewater. Most of the chloroform released ends up in the air. It may be transported for long distances before becoming degraded by reacting with photochemically generated hydroxyl radicals. Significant amounts of chloroform from the air may be removed by precipitation. However, the chloroform may reenter the air by volatilization. When released to soil, chloroform either volatilizes rapidly from the surface or leaches readily through the soil to the groundwater, where it may persist for relatively long periods of time. It is estimated that its half-life in the atmosphere is 70 to 79 days. However, when chloroform is found in photochemical smog, its half-life is 260 days.

Volatilization is the primary mechanism for removal of chloroform from water. Its half-life in water may vary from 36 hr to 10 days. Although chloroform has been found to adsorb strongly to peat moss and less strongly to clay, typically it volatilizes rapidly in either dry or wet soil.

Although cyanides are naturally occurring substances found in a number of foods and plants and produced by certain bacteria, fungi, and algae, the greatest amount found in the environment comes from industrial processing. Hydrogen cyanide is used primarily in the production of organic chemicals. Cyanide salts are used primarily in electroplating and metal treatment. The major sources of cyanide released to water are reported to be discharges from metal finishing industries. The major source of the cyanide released to air is vehicle exhausts, and the major sources of cyanide released to soil appear to be disposal of cyanide waste in landfills and the use of cyanide-containing road salt. Cyanide released to air appears to be transported over long distances before reacting with photochemically generated hydroxyl radicals. The half-life of cyanide is 334 days. In water, cyanide may volatilize; or if in an alkali metal salt form, cyanide may readily disassociate into anions and cations. The resulting cyanide ion may then form hydrogen cyanide or react with metals to form metal cyanides. Insoluble metal cyanides are expected to adsorb to sediment and possibly bioaccumulate in aquatic organisms.

In soil, cyanides may occur in the form of hydrogen cyanide, alkali metal salts, or immobile metallocyanide complexes. At soil surface with a pH less than 9.2, hydrogen cyanide is expected to volatilize. In subsurface soil, where the cyanide present is at a low concentration, it is probably biodegradable. In soil with a pH less than 9.2, hydrogen cyanide is expected to be highly mobile, and where cyanide levels are toxic to microorganisms such as in landfills or in spills, the cyanide may reach the groundwater.

Lead is a naturally occurring element that may be found in Earth's crust and in all parts of the biosphere. It is released into the atmosphere by a variety of industrial processes and from leaded gasoline. The deposition of lead found in both soils and surface waters generally comes from the atmosphere. Lead is transferred continuously among air, water, and soil by natural chemical and physical processes, such as weathering, runoff, precipitation, dry deposition of dust, and stream or river flow. Long distance transport of up to about 600 mi may occur. Lead is extremely persistent in both water and soil. In the atmosphere, lead exists primarily in the particulate form. The chemistry of lead in the water is highly complex, because it may be found in a multitude of forms. The amount of lead that remains in solution depends on the pH of the water and the dissolved salt content. In most soils, the accumulation of lead is primarily a function of the rate of deposition from the atmosphere. Most lead is maintained strongly in the soil and very little is transported into surface water or groundwater.

The chemical TCDD, or simply dioxin (which is an inaccurate colloquial name except when it is used as a reference standard), is not intentionally manufactured by any industry. It is inadvertently produced in very small amounts as an impurity during the manufacture of certain herbicides and germicides. It is also produced during the incineration of municipal and industrial wastes. The environmental fate of 2,3,7,8-TCDD is not understood with certainty for air, water, and soil, but none-theless is considered a potent toxin.

Tetrachloroethylene is a nonflammable liquid solvent widely used for dry-clean-ing fabrics and textiles, and for metal-degreasing operations. It is used as a starter chemical for the production of other chemicals. It is also called perchloroethylene, perc, PCE, perclene, and perchlor. Most of the tetrachloroethylene used in the United States is released into the atmosphere by evaporative losses. The dry cleaning industry is the major source. When the chemical is released in water, it rapidly volatilizes and returns to the air. However, rainwater dissolves this chemical, and once again it falls to the land. The usual transformation process for tetrachloroeth-ylene in the atmosphere is the result of the reaction with photochemically produced hydroxyl radicals. The degradation products of this reaction include phosgene (a severe respiratory irritant) and chloroacetyl chlorides.

Trichloroethylene is used as a solvent for removing grease from metal parts. It is released into the atmosphere by evaporative losses. It may also be released into the environment through gaseous emissions from waste-disposal landfills, and it may leach into groundwater from waste-disposal landfills. It is dissolved in rainwater. The dominant transformation process for trichloroethylene in the atmosphere is its reaction with sunlight-produced hydroxyl radicals. The degradation products of this reaction include phosgene, dichloracetyl chloride, and formyl chloride.

ENVIRONMENTAL HEALTH PROBLEMS AND THE ECONOMY

The environment and the economy are not mutually exclusive; together they form an integral system. As the public demands a higher quality environment, the economy must adjust to meet this need. In the past, Americans have placed a high priority on convenience and consumer goods. There seemed to be no need for concern over the air, water, or the earth. The economic system was established such that incentives were given to individuals to better their position in our society without taking into account the problem of environmental degradation. The rich endowment of natural resources; the large amounts of investment, technology, and education; the influx of skilled immigrants; and the political and economic systems in the United States have combined to produce an economy with enormous material wealth. American business and industry, through its technological developments, produce huge amounts of goods and services. Consumers continually demand new products, and business and industry have accommodated consumers. Old products are dis-carded as quickly as new ones become available.

However, a new demand for better environmental quality and cleaner surroundings has emerged in the United States. The American public is demanding cleaner water, cleaner air, relief from noise and congestion, and more aesthetically pleasing environment. Concurrently, the supposedly unlimited supply of water and air is befouled with huge quantities of pollutants, creating severe health, economic, and aesthetic problems. Exotic chemicals have entered the environment and moved through the food chain to a point where they are becoming a potentially dangerous health hazard.

During the last 25 years, the production of synthetic organic chemicals, containers, packaging, and electronic equipment has increased sharply. As production has increased, consumer demand has increased. Unfortunately, although the products are available to the consumer at reasonable prices, these prices do not fully represent the cost to our society, because they do not include the cost of waste of our natural resources or the cost of environmental degradation. Due to the unfair economic incentive for use of raw materials vs. reuse of recycled materials, further destruction of our environment occurs. The dumping of wastes into the air, water, and land must end. The burden of the cost of the environmental cleanup and environmental improvement must be borne by all. Pollution costs include higher prices resulting from damage to crops, materials, and properties; and from disease and injury. The actual costs of control measures are difficult to assess, because one must take into account abatement programs; expenditures by business and industry; expenditures by government, research and development; monitoring devices; and assorted equipment and service purchases.

Dollars spent improving the environment are an investment in life and human society, although the benefits of these investments may not be readily seen. Imposing environmental controls results in the reallocation of resources, which in the short run may cause adverse economic effects, such as higher prices, temporary unemployment, and plant dislocations. The decrease in medical bills, the increase in recreational activities, the decrease in damage to materials, and the provision of a better society must be weighed against the temporary cost increases. Marginal firms that fail to meet these costs will be forced to close. The ultimate good to the greatest number of people should be the deciding factor in determining what should or should not be done to improve the environment. However, in any given area where changes are undertaken, adequate additional funds must be made available if the area is likely to suffer economic dislocation. In some cases where health is not endangered by pollution, it may be best to carefully and cautiously make temporary exceptions to certain standards in an effort to ease the strain on local economies. These exceptions should only be made with the greatest of care.

RISK–BENEFIT ANALYSES

In all areas, but particularly with regard to chemicals, it is necessary to compare the social benefits of a given environmental substance with the risk it causes to our society. Many pesticides are potentially extremely dangerous, and yet without the availability of pesticides the country and the world would be rampant with disease and would probably suffer from famines and starvation. It is therefore necessary to evaluate each of the environmental substances to determine the relative safety of the substance; the relative efficiency in production of the substance compared with the use of natural resources; and the relative impact on the environment at various stages of manufacture, use, or disposal. In all cases, risk–benefit analyses must be completely considered. The benefits include the value of the use of the substance to the consumer; its aesthetic value; the conservation of natural resources and energy resulting from the production of the substance; and the economic impact of employment, regional development, and balance of trade in the production of the substance. The risks include the adverse effect on health; the potential for death occurring as a result of use of the substance; the environmental damage to air, water, land, wildlife, vegetation, aesthetics, and property; and the misuse of natural resources and energy. Establishing risk–benefit analyses for each substance utilized within the environment is difficult. However, with analyses of these types, society can determine the necessary trade-offs between potential hazard and employment, health, gross national product, balance of trade, conservation of natural resources, and environmental quality.

Cost–Benefit

The four case studies to be discussed include the following: lead in gasoline, air toxics issues, incineration at sea, and municipal sewage sludge reuse and disposal.

Lead in Gasoline

In the case of the lead-in-gasoline rule, a careful quantitative analysis was used. Lead in gasoline has been regulated by the EPA for more than 15 years. There was a concern that leaded gasoline was being used in cars that required unleaded gasoline. This would affect the pollution control catalysts. A further concern was that the lead was a potential risk to children's health. Although previous rules had reduced lead in gasoline, in March 1984, the EPA set a very tight phase-down schedule. The final rule issued in March 1985 was more stringent than the original proposal. It required that lead in gasoline be reduced from 1.1 g per leaded gallon to 0.5 g per leaded gallon by July 1985, and to 0.1 g per leaded gallon by January 1986. Not all of the lead was removed from leaded gasoline, because of the potential effect it would have on older engines. Since the 1920s, refineries have added lead to gasoline as an inexpensive way of boosting octane. With the removal of lead, refineries had to do additional processing or use other additives, which tended to raise the cost of gasoline.

The EPA used a computer model of the refining industry that had been developed by the Department of Energy to estimate the cost of the rule. The computer model included various equations that showed how different inputs could be turned into different end products at varying costs and the constraints on the industry's capacity. To estimate the cost of the rule, the EPA first ran the model specifying the then current lead limit of 1.0 g per leaded gallon and computed the cost of meeting the demand for the refined products. It then ran the model specifying the lower lead limits and recomputed the overall cost. The difference between the original and anticipated lead limits was the estimated cost of the tighter standard. Based on that analysis, the EPA estimated that the rule would cost less than $100 million in the second half of 1985 to meet the 0.5 g per leaded gallon standard and approximately $600 million in 1986 to meet the 0.1 g per leaded gallon standard. They next ran extensive tests using more pessimistic assumptions, such as an unexpected high demand for high-octane unleaded gasoline, an increased downtime for equipment, and a reduced availability of alcohol additives. The 0.1 g per leaded gallon rule met virtually all conditions. It was extremely unlikely that a combination of these conditions would occur at the same time.

The benefits of the rule were estimated for the maintenance of good health in children and the improvement in educational efforts, compared with the efforts of children who were exposed to lead. Further, benefits were estimated based on the reduction in damage caused by excessive emissions of pollutants from vehicles using the wrong gasoline, the impacts on vehicle maintenance, and the impacts on fuel economy. The EPA also used existing data to study the relationship between levels of lead in the blood and blood pressure to make some estimate of the health effects that adults would have from this rule. In each category, the EPA first estimated the impact of reduced lead in physical terms. In the case of children's health effects, the agency used statistical studies relating to lead in gasoline and blood lead levels projected in children. The EPA also estimated the cost to people and society and the environment by allowing the lead levels to stay the same vs. the reduced lead levels. It was determined that the benefits of the lead rule exceeded the cost by a 3:1 ratio; and that if the potential benefits in reductions in blood pressure, illness, and death were to be counted, the ratio would rise to 10:1.

Air Toxics

Risk assessment techniques were used to determine the environmental hazards of toxic air emissions. A study was made to determine how much of the air toxics problems could be controlled by using existing EPA programs. A new toxic strategy was then developed and put into place. It had three main parts: (1) direct federal regulation of significant nationwide problems, (2) state and local control of significant pollutant problems that were national in nature, and (3) an increased study of geographic areas that were subjected to particularly high levels of air pollution. In the initial study, 42 air toxic compounds were evaluated, basically for their potential to cause cancer. A determination was made of how many people were exposed across the country. An estimate of the risk associated with each compound in this national exposure was determined. The estimate was of long-term cancer incidents (a 70-year time frame) associated with these compounds. The analysis suggested that the air toxics problem was complex, caused by many pollutants and sources, varied significantly from city to city, and even varied within a city. In the 1990s, the EPA reduced air toxic pollutants by nearly 1 million tons per year, which is almost a 10 times greater reduction than was achieved from 1970 to 1990. This success was related directly to the technology- and performance-based standards set by the EPA.

Apparently, preexisting EPA air toxic strategies that focused on regulating each pollutant as part of national emissions standards were too narrowly set up to be effective. A comparison of air quality data for 1970 and 1980, however, showed that significant reductions in national cancer incidents related to air pollution had occurred as a result of the reduction of air pollutants. Air toxic problems were examined in detail in Baltimore, Baton Rouge, Los Angeles, Philadelphia, and Phoenix. In the original study, 14 compounds were identified as causing the greatest concern for cancer risk among the general population, and how these compounds could be controlled was studied. Again, cancer was the indicator of health effects, because other health effect indicators were more difficult to determine. Emission sources were identified in each of the five cities for these 14 compounds. Human exposure modeling was used to determine the degree of risk experienced by the people of these selected areas and the cancer incidents expected from this risk. The assumption was made that there would be a full implementation of the criteria of the pollutant programs.

The greatest incidents of cancer are associated with vehicles or area sources, such as numerous small pollution sources, like wood stoves. Point-source pollution sources, which are large, relatively identifiable sources of pollution, such as utilities and steel plants, appear to account for little of the total incidents.

All these data suggest that the health problems of air toxics may require a targeted strategy to make the best use of available resources. The data also suggest that additional focus must be made on small and nontraditional sources to get the greatest benefit for the health of the individuals. Additional attention needs to be given to wood smoke, vehicles, waste oil burning, and gasoline stations. The risk assessment here has helped to highlight the important areas where the greatest amount of good results will occur from the smallest amount of dollar expenditures.

Incineration at Sea

Industry generates more than 70 billion gal of hazardous waste each year in the United States. This waste must be safely managed through treatment, storage, and disposal. Because the EPA is restricting the use of management practices through permitting, regulation, and enforcement programs, there is a need to determine how best to get rid of this hazardous waste. Incineration is a technique used to help destroy this hazardous waste. Public opinion, however, is strongly against the issuing of permits for new incinerators.

Between 1974 and 1982, the EPA issued permits for four series of burns conducted by the incinerator ship Vulcanus. Three of the burns were in the Gulf of Mexico and one in the Pacific Ocean. Public opposition to incineration at sea has intensified greatly. The EPA undertook a number of studies to determine whether incineration on land or in the oceans is more desirable.

The EPA developed a risk assessment study that compared the human and environmental exposure most likely to occur from releases of land-based incineration vs. ocean incineration. The study used existing information and added new analysis of the emissions, transport, fate, and alternate effects. Both systems used land transportation, transfer and storage operations, and final incineration. The ocean system also included an ocean transportation step. For the purposes of the study, the waste was assumed to be a combination of 35% PCBs by weight and 50% ethylene dichloride (EDC) by weight. These two chemicals were used to simplify the waste stream. The analysis was to determine the statistically expected amount of pollutant released from accidental spills and air emissions. The transportation and handling accounted for less than 15% of the expected releases, whereas the incineration accounted for about 85%.

During land transportation, there were two types of potential losses. They were vehicular accidents and spills from the containers in route. Vehicular accidents were expected to occur on the average of once every 4 to 5 years and container failure, once every 3 to 4 years. The analysis of transfer and storage of the waste was considered and three types of releases were anticipated: (1) spills when unloading the waste from tank trucks, (2) spills from equipment at the waste transfer and storage facilities, (3) and fugitive emissions in transfer and storage. Spills from transfer and storage components are infrequent events estimated to occur at about 0.50% per year, and from the tank trucks loading of waste to ship in the ocean system is approximately 0.002% per year. It was recognized that spills of this low quantity would likely be contained at the facility.

About 320 voyages of incineration ships have been made in the North Sea since 1972, and no casualties or spills have occurred. The spill rates from ships were based on worldwide historical data. It was estimated that the frequency of all spills for the Vulcanus would be about 1 per 12,000 operating years.

The study estimated and compared the possible human health and environmental effects due to incinerator releases and fugitive releases from the transfer and storage equipment.

The analysis of human health risk was based on the most exposed individual, who lives at the location, and the persons at the highest overall risk due to air concentrations from the incinerator stack and transfer storage facility. This risk is based on the cancer potential for an individual who has 70 years of continuous exposure. The risk on land-based incineration was 3 chances in 100,000, whereas the risk on ocean burning was in a range of 1 in 1,000,000 to 6 in 10,000,000. The relative risk on land for incineration of PCB waste is about 40 times more than on the ocean. For EDC waste, the ratio of land to ocean risk is about 30:1.

The conclusions of the incineration study were as follows: (1) incineration, whether on land or at sea, is a valuable and environmentally sound option; (2) no clear preference is made for ocean or land incineration as it relates to human health risk and the environment; (3) future demands will significantly exceed the capacity for disposal of hazardous waste; (4) continuing research is needed to improve current knowledge of combustion processes and effects; and (5) EPA needs to improve its public communications effort in the area of hazardous waste management.

Municipal Sludge Reuse and Disposal

Sludge management is a major part of municipal sewage disposal. Municipalities generate about 6.5 million dry tons of wastewater sludge a year. Municipal sewage sludge contains over 200 different substances, such as toxic metals, organic chemicals, and pathogenic and other organisms. The five major sludge use or disposal options currently available are land application, distribution and sale of sludge products, landfill, incineration, and ocean disposal that is used on a research or an emergency basis only. The EPA developed standards in February of 1993, for the use or disposal of sewage sludge. These standards are for land application, surface disposal, reduction of pathogens and vector attractions, and incineration. This Round One regulation also set limits for metals, and total hydrocarbons. These standards are in compliance with Section 405 of the Clean Water Act. They are titled, "Standards for the Use or Disposal of Sewage Sludge" (Code of Federal Regulations Title 40, Parts 257, 403, and 503, known as the Part 503 Sludge Rule). As the result of a citizen's suit, the EPA was required to propose Round Two regulations by December 15, 1999 and to take final actions by December 15, 2001, These changes were based on risk assessment results and involved only dioxin and dioxin-like compounds in biosolids. The proposed limit is 300 parts per trillion toxic equivalents of dioxins, above which biosolids may not be applied to the land. All facilities would be required to test the level of dioxins present in their biosolids before they could be land applied except where treatment plants treated less than 1 million gal/day of wastewater, and small businesses prepared less than 290 dry tons of sewage sludge a year.

Biosolids are treated sewage sludge. They are nutrient-rich organic materials resulting from the treatment of domestic sewage in a treatment facility. When treated and processed, these residuals can be recycled and applied as fertilizer to improve and maintain productive soils and stimulate plant growth.

Sludge disposal has always been a substantial portion of the cost of wastewater management. Over the last 20 years, restrictions have been placed on ocean dumping and landfill disposal, causing wastewater treatment facilities to consider agricultural use of sludge as a cost-effective alternative. About 36% of sludge is applied to the land for agriculture, production of turf grass, and reclamation of surface mining areas; 38% is land filled; 16% is incinerated; and the rest is discarded on the surface by other means.

The Midwest has a long history of using treated sludge on cropland, which is used to grow corn and small grains for cattle feed. In Madison, WI, the demand for sludge as a soil amendment exceeds the local supply. Coastal cities such as New York City and Boston send much of their sludge to other parts of the country, because they can no longer dispose the sludge at sea.

Pollution Prevention

Pollution prevention is a variety of techniques with a single objective — to achieve the most efficient use of resources to reduce or eliminate waste. Pollution prevention reduces or eliminates the generation of pollutants or waste, minimizes or eliminates the use of toxic materials in manufacturing, substitutes less harmful materials for toxic ones, reduces the chance of moving pollutants to a different media (e.g., air to water), and maximizes the efficient use of resources. Pollution prevention starts in the public sector with changes in a variety of public departments and their operations. It then includes the private sector and people at large. Better management of transportation saves fuel and reduces emissions. This is an example of an effective program. The substitution of calcium magnesium acetate for sodium chloride as a road salt reduces well contamination. Energy efficiency reduces the need for fuel to produce energy and reduces pollutants that come from burning hydrocarbons.

ENVIRONMENTAL HEALTH PROBLEMS AND THE LAW

The environmental health practitioner works for a unit of the local, state, or federal government; or, if working for industry, is constantly dealing with various levels of government. (For a detailed discussion on the functions of government, the reader is encouraged to refer to one of the many publications on local, state, and federal government.) For the purposes of this book, a brief discussion on federal–state relationships, legislative procedure, pressure groups, local government, and state and local finances is included.

The governmental system in a democratic society performs two distinct functions. It provides a foundation for debate on issues and a vehicle for the solution of problems. It also provides a service and regulatory function, because individuals in a complex society cannot provide for all their needs. These needs include adequate police, fire, and health protection. For the society to operate properly, the individual or group must adhere to rules and regulations formulated and enforced by the elected government.

According to the U.S. Constitution and amendments thereto, federal powers include the control of interstate and foreign commerce, conduct of foreign policy, and national defense. The federal government cannot levy direct taxes, other than income taxes, except in proportion to the population of the states. The federal government cannot abridge civil rights. The state is sovereign. The powers not delegated to the federal government or prohibited by the Constitution are reserved for the states or the people. The state powers include the administration of elections; establishment and operation of local government; education; intrastate commerce; creation of corporations; police force; and promotion of health, safety, and welfare. Although the federal government cannot dictate organization and administration of programs, or the establishment of policy, it has gained some control via its use of funds. For a state to receive federal funds, it must adhere to certain requirements established by the federal government for the use of these funds. These requirements include preparation and submission of plans, approval of plans by a central federal agency, establishment of necessary state agencies, provision for matching state funds, and supervision and auditing of state programs by the appropriate federal agencies.

Public policy is a result of the interaction of groups and individuals. The formulated policy is not necessarily ideal for the public. Major pressure groups — including business, industry, agriculture, labor, medicine, and religion — try to influence the legislature to pass laws beneficial to their own self-interest. These groups also pressure administrative agencies to make or change decisions on policies that are not in the best interest of the public. These groups compete with each other for public support. They use sizable sums of money to influence opinions on legislation. Citizens groups properly educated and stimulated by trained professional public health workers could use the techniques of the pressure groups to establish public policy that would be beneficial to the citizens and would improve the environment.

The state is divided into local government, including counties, townships, villages, cities, and boroughs. Local government functions may include all the functions of the state government, such as the powers of taxation; establishment of budgets; licensing; and administration of health, safety, and environmental programs. The local government is expected to deliver quality service and to protect the public. It should prevent the duplication of efforts, budgets, and facilities of the state government. Environmentalists must recognize that although concern for the environment is foremost in their minds, the county commissioners or township supervisors may be more concerned with other programs, such as road building, recreational facilities, and police and fire service. The environmentalist has to compete for budget monies with each of the other operating departments.

State and local governments obtain their funds for operation from taxes levied on sales, automobile licensing, gasoline, corporations, personal income, alcohol and tobacco, gross receipts, property, and establishment licensing. In many cases, tax rates become oppressive and yet income derived from taxes is inadequate to finance properly the necessary programs for the community. Then it becomes necessary for the federal government to supply funds; in this way, the federal government exercises control in some states even at the township level.

Law and Public Health

Public health policies are discussed briefly, because environmental health practitioners should have some understanding of the law to function in their capacity as either industry leaders or government officials. (For more detailed discussion, many books are available describing current laws relating to the environment.)

Law is the rule of civil conduct prescribed by the supreme power in a state commanding what is right and forbidding what is wrong. Law should represent the community desires or commands, apply to all members of the community, be backed by the full power of the government, and provide the administration of justice under the law for all people. The purpose of law is to protect by the regulation of human conduct of the individual from other individuals, groups, or the state, and vice versa. Statutory laws are basically legislative acts passed by a legislative body; common laws are established customs of the community; equity is a decision by a judge in a given situation where rules and regulations have been violated; and administrative laws are rules and regulations in a given area established by an agency authorized by the legislature. The value of administrative laws is that the rules and regulations established can be changed readily when new scientific data are available. If regulation were part of a statute, the legislature would have to go through the complex process of rewriting the act.

Public health law is that body of statutes, regulations, and precedents that protects and promotes individual and community health. Public health law is founded on the Preamble to the Constitution, which ordains that the government shall "... promote the general welfare"; and Section 8, Article 1 of the Constitution, which

"... provide(s) for the common defense and general welfare." The interpretation of these clauses and other clauses of the Constitution by the U.S. Supreme Court has established the legal basis for public health. The law of eminent domain empowers the state with the authority to seize, appropriate, or limit the use of property in the best interests of the community. This power is reflected in the establishment of zoning and land use regulations that are so essential to the preservation of a good environment.

Nuisance laws, originating in the Middle Ages, are used repeatedly in environmental health work. Basically they state that the use of private property is unrestricted only as long as it does not injure other persons or their property. If an injury occurs, a nuisance exists. Large numbers of public health officials use nuisance laws to eliminate problems caused by sewage, solid waste, air pollution, and insects and rodents.

The public health law owes its effectiveness to the police power of the state. In times of great stress, such as severe fires, floods, and outbreaks of diseases, the private property of an individual might be summarily appropriated, used, or even destroyed if the ultimate relief, protection, or safety of the community demands that such action be taken. The legislature can and does delegate police power to an administrative agency for use in the event of an emergency. If public officials fail to use the delegated police power, they are guilty of nonfeasance of office. The use of this police power is determined by the chief administrative office, usually the state secretary of health, in consultation with the governor of the state.

Administrative law is that series of rules, regulations, and standards needed to implement statutes. The development and use of sound rules, regulations, and standards based on strong scientific data put the burden of proof on the defendant, who has to show that the rules have not been fairly applied, rather than that they have limited value. Through judicial presumption, the judge assumes that the laws are correct, because they are based on sound scientific criteria or on the judgment of a group of experts. The judge only decides if the law is applied fairly.

Licensing is an important means of control and enforcement of environmental standards. If an individual or business does not meet the standards, a license is denied or revoked. An individual operating without a license is subjected to severe penalties.

Environmental health practitioners must be conscious of those actions that may occur in the enforcement of environmental health laws. These actions include: (1) misfeasance, which is the performance of a lawful action in an illegal or improper manner; (2) malfeasance, which is wrongful conduct by a public official; and (3) nonfeasance, which is an omission in doing what should be done in an official action.

Environmental Law

For historical purposes, the Rivers and Harbors Act of 1899 was passed by Congress to prevent the illegal discharge or deposit of refuse or sewage into navigable waters of the United States. This law was used before 1972 to penalize some corporations for polluting the water.

The National Environmental Policy Act of 1969 signed into law on January 1, 1970 (Public Law 91-190), amended by PL 94-52 and PL 94-83 in 1975, set forth the continuing responsibility of the federal government to:

1. Fulfill the responsibilities of each generation as trustee of the environment for succeeding generations
2. Assure safe, healthful, productive, aesthetically, and culturally pleasing surroundings for all Americans
3. Attain the widest range of beneficial uses of the environment without degradation, risk to health or safety, or other undesirable and unintended consequences
4. Preserve important historic, cultural, and natural aspects of our national heritage, and maintain, wherever possible, an environment that supports diversity and variety of individual choice
5. Achieve a balance between population and resource use that will permit high standards of living and a wide sharing of life's amenities
6. Enhance the quality of renewable resources and approach the maximum attainable recycling of depletable resources

The National Environmental Policy Act (NEPA) also states that all federal agencies shall:

1. Utilize a systematic, interdisciplinary approach ensuring the integrated use of the natural and social sciences and the environmental design arts in planning and in decision making when planning may be an impact on the human environment
2. Identify and develop methods and procedures, in consultation with the Council on Environmental Quality, ensuring that other environmental amenities and values be given appropriate consideration in decision making, along with economic and technical considerations
3. Include in every recommendation or report on proposals for legislation and other major federal actions significantly affecting the quality of the human environment, a detailed statement by the responsible official on:
   a. the environmental impact of the proposed action
   b. any adverse unavoidable environmental effects should the proposal be implemented
   c. alternatives to the proposed action
   d. relationship between local short-term uses of the environment and the maintenance and enhancement of long-term productivity
   e. any irreversible and irretrievable commitments of resources involved in the proposed action should it be implemented
4. Study, develop, and describe appropriate alternatives to recommended courses of action in any proposal involving unresolved conflicts, as well as alternative uses of available resources
5. Recognize the worldwide and long-range character of environmental problems and, where consistent with the foreign policy of the United States, lend appropriate support to initiatives, resolutions, and programs designed to maximize international cooperation in anticipating and preventing a decline in the quality of the world environment
6. Make available to states, counties, municipalities, institutions, and individuals advice and information useful in restoring, maintaining, and enhancing the quality of the environment
7. Initiate and utilize ecological information in the planning and development of resource-oriented projects

8. Establish the Council on Environmental Quality, whose functions are as follows:
    a. To assist and advise the President in the preparation of the environmental quality report, which will be submitted to the Congress starting in 1970 (this report is an excellent digest of various environmental health issues)
    b. To gather timely and authoritative information concerning the conditions and trends in the quality of the environment, currently and for the future
    c. To review and appraise the various programs and activities of the federal government
    d. To develop and recommend to the President, national policies to foster and promote the improvement of environmental quality
    e. To conduct investigations, studies, surveys, research, and analyses relating to ecological systems and environmental quality
    f. To document and define changes in the natural environment
    g. To report at least once every year to the President on the state and condition of the environment
    h. To make and furnish studies, reports, and recommendations with respect to matters related to policy and legislation

Since 1970 when the NEPA Act was passed and a Council on Environmental Quality was established, Congress has passed a large number of laws and amendments to the laws, related to a variety of environmental issues that potentially affect the health of people and affect the environment and ecosystems. These laws include, but are not limited to, the Asbestos School Hazard Detection and Control Act; Asbestos School Hazard Abatement Act; Asbestos Hazard Emergency Response Act; Chemical Safety Information Site Security and Fuel Regulatory Act; Clean Air Act (CAA); CERCLA; Consumer Product Safety Act (CPSA); Emergency Planning and Community Right-to-Know-Act; Energy Supply and Environmental Coordination Act; Energy Policy Act; Federal Insecticide, Fungicide, and Rodenticide Act (FIFRA); Food, Drug, and Cosmetic Act (FDCA) amendments; Food Quality Protection Act; Hazardous Materials Transportation Act; Lead Contamination Control Act; Medical Wastes Tracking Act; National Environmental Education Act; Noise Control Act; Nuclear Waste Policy Act; Occupational Safety and Health Act; Oil Pollution Act; Pollution Prevention Act; Radon Gas and Indoor Air Quality Research Act; Resource Conservation Recovery Act (RCRA); Surface Mining Control and Reclamation Act; Superfund Amendments and Reauthorization Act (SARA); Toxic Substances Control Act (TSCA); and Uranium Mill-Tailing Radiation Control Act.

A number of laws related to water and water pollution have been enacted since the original Rivers and Harbor Acts of 1899. These laws include clean water amendments of the Federal Water Pollution Control Act; Coastal Zone Management Act; Deep Water Port Act; Marine Protection, Research, and Sanctuaries Act; National Ocean Pollution Planning Act; Outer Continental Shelf Lands Act; Ocean Dumping Act; Ocean Dumping Ban Act; Port and Tanker Safety Act; Shore Protection Act; Safe Drinking Water Act (SDWA); Water Resources Planning Act; and Water Resources Research Act.

The major amendments to the CAA were added in 1977 and updated in 1983, with again a major amendment passed in 1990. There had been a substantial debate for the last 10 years concerning how to change the law. The congressional findings included: (1) that the predominant part of the nation's population was located in rapidly expanding metropolitan and other urban areas that generally cross the boundary lines of local jurisdictions and expand into multiple states; (2) that the growth and the amount of complexity of air pollution brought about by urbanization, industrial development, and increased use of motor vehicles has resulted in growing dangers to the public health and welfare, crops, property, and air and ground transportation; and (3) that the prevention and control of air pollution at its source was the primary responsibility of the state and local governments and that federal financial assistance and leadership was necessary for the improvement of air quality.

The law provides an elaborate federal–state scheme for controlling conventional pollutants, such as ozone and carbon monoxide. The 1990 amendments create tighter controls on tail pipe exhaust; reduction of acid rain, nitrogen oxides, and air toxics, which may be carcinogenic; and protection of the ozone layer by phasing out chlorofluorocarbons, carbon tetrachloride, methylchloroform, and hydrochlorofluorocarbons. Over the past 10 years, the EPA has added numerous regulations to further upgrade air quality.

The Chemical Safety Information, Site Security and Fuels Regulatory Relief Act of 1999 establishes new provisions for reporting and disseminating information under the provisions of the CAA. The law has two distinct parts that pertain to flammable fuels and public access to worst-case scenario data. Flammable fuels used as fuel or held for sale as fuel at a retail facility are removed from coverage under the risk management plan submitted to the EPA required by the CAA. Flammable fuels used as a feedstock or held for sale as fuel at a wholesale facility are still covered by the CAA. The law exempts worst-case scenario data from disclosure under the Freedom of Information Act and limits its public availability for at least 1 year. The federal government is to assess the risks of Internet posting of these data and to determine the benefits of public access to the data. These data are to be available to qualified researchers as long as they do not release it to the public.

CERCLA of 1980, also known as the Superfund, was designed to handle the problems of cleaning up the existing hazardous waste sites in the United States. The act, which was originally passed in 1980, was updated numerous times until 1988. The hazardous waste problems range from spills that need immediate attention to hazardous waste dumps that are leaking into the environment and posing long-term health and environmental hazards.

SARA, reauthorized and amended by CERCLA, expanded the federal government response and authority, clarified that federal facilities were subject to the same requirements as private industry. The CERCLA response effort is guided by the National Oil and Hazardous Substances Pollution Contingency Plan, better known as the National Contingency Plan. This plan describes the steps that responsible parties must follow in reporting and responding to situations in which hazardous substances are released into the environment.

The Emergency Planning and Community Right-to-Know Act of 1986 established state emergency response commissions, emergency planning districts, emergency planning committees, and comprehensive emergency plans. A list of extremely hazardous substances has been prepared and published.

This law was passed in response to concerns about the environmental and safety hazards potentially caused by the storage and handling of toxic chemicals. These concerns were triggered by the disaster in Bophal, India, in which more than 2000 people died or were seriously injured from the accidental release of methyl isocyanate. The various states and facilities are required to notify local emergency planning districts concerning materials stored at, or released from, those sites. The local community has to prepare plans that will deal with the emergencies relating to the hazardous substances and must inform local residents of the potential for serious problems.

The amendments to the Emergency Planning and Community Right-to-Know Act Sections 311 and 312 were passed in 1999. These final rules raise the gasoline and diesel fuel thresholds that trigger material safety data sheets reporting and chemical inventory reporting. It is now possible to store 75,000 gal of gasoline or 100,000 gal of diesel fuel entirely underground at retail gas stations without having to comply with the rules on underground storage tanks.

The Energy Reorganization Act of 1974 redirected federal energy efforts. Congress created the Nuclear Regulatory Commission and the Energy Research and Development Administration, which later became the Department of Energy in 1977. The Energy Reorganization Act also established the goal of efficient energy utilization while enhancing environmental protection.

The Energy Supply and Environmental Act of 1974 was updated in 1978. The purposes of this act were to provide for a way to assist in meeting the country's fuel requirements in a consistent, practical manner, and to protect and improve the environment. It allowed for coal conversion or coal derivatives to be used in place of oil in power plants. This, of course, can contribute to greater levels of air pollution.

The CPSA of 1970, updated in 1984, established the Consumer Product Safety Commission as an independent regulatory agency. The CPSA gives the commission the power to regulate consumer products and to oppose unreasonable risks of injury or illness. It also regulates consumer products, except for foods, drugs, pesticides, tobacco and tobacco products, motor vehicles, aircraft and aircraft equipment, and boats and boat accessories. The law authorized the commission to publish consumer product safety standards to reduce the level of unreasonable risks. The commission has recalled hair dryers containing asbestos because of potential hazards. It also may regulate carcinogens.

The Federal Facility Compliance Act amends Section 6001 of RCRA to specify that federal facilities are subject to "all civil and administrative penalties and fines, regardless of whether such penalties or fines are punitive or coercive in nature." Therefore, the federal government was made to adhere to the same legal framework as the private sector and to comply with all applicable environmental rules and regulations.

The FIFRA was originally passed by Congress in 1947 and was amended in 1996 by the Food Quality Protection Act. It provides for the registration of new pesticides; the review, cancellation, and suspension of registered pesticides; and the reregistration of pesticides. It is concerned with the production, storage, transportation, use, and disposal of pesticides. It also includes areas of research and monitoring.

A new safety standard, "reasonable certainty of no harm," must be applied to all pesticides used on foods.

The Food, Drug and Cosmetic Act, which is an amendment to the Food and Drug Act of 1906, and the federal FDCA of 1938, has been further amended numerous times. It was the first federal statute to regulate food safety. The amendment of 1938 established the general outlines for the authority of the Food and Drug Administration (FDA). Various parts of the law regulate food additives; food contaminants; naturally occurring parts of food or color additives to food, drugs, or cosmetics; and potential carcinogens. In 1958, the Food Additives amendment, better known as the Delaney clause, stated, "... that known additives shall be deemed to be unsafe if found to induce cancer when ingested by man and/or animal, or if it is found, after tests which are appropriate for the evaluation of food additives, to induce cancer in man or animal ...."

In 1996, the Congress replaced the outdated Delaney clause with a scientific-based data program used to determine food safety and pesticide availability. The new law, enacted on August 3, 1996, is the Food Quality Protection Act that establishes national uniform safe residue levels for pesticides and allows consideration of the benefits to nutrition and food supply of these pesticides. It also provides important incentives and better methods of registration for new chemicals, safer crop protection chemicals, and high-value minor crops. It establishes an extra margin of safety for residues on foods consumed in high amounts by infants and children. It requires consideration of chemical exposure from sources other than food, such as drinking water and home pesticide use. It requires consideration of common mechanisms of toxicity from other chemicals. The safety standard provided by the new law of reasonable certainty of no harm is a more flexible standard than the zero-risk Delaney clause. The extra margin of safety for infants and children would be imposed only when there were demonstrable health effects shown in the data in the registration of the pesticide. Risk assessments for food exposure would be based on the best available information. When data were absent or incomplete, better data would be obtained before regulatory decisions were made. Risk assessments would be based on actual, not theoretical health risks, when dealing with multiple sources of exposure. The implementation process for the new law would be open, transparent and follow established administrative procedures for federal rules and regulations. The registration of new crop protection products would be accelerated.

The Food Quality Protection Act of 1996 made changes in the FIFRA as well as in the federal FDCA, with the EPA establishing tolerances (maximum legally permissible levels) for pesticide residues in food.

The Hazardous Materials Transportation Act of 1975 as amended by the Hazardous Materials Transportation Uniform Safety Act of 1990 is the major transportation-related statute affecting the Department of Transportation. The objective of the law, according to Congress, is to improve the regulatory and enforcement authority of the Secretary of Transportation to protect the nation adequately against risk to life and property that are inherent in the transportation of hazardous materials and commerce. The 1990 act included provisions to encourage uniformity among different state and local highway routing regulations, to develop criteria for the issuance of federal permits to motor carriers of hazardous materials, and to regulate the transport of radioactive materials.

The Noise Control Act of 1972 was amended last in 1978. The findings of Congress were as follows: (1) that inadequately controlled noise presents a growing danger to the health and welfare of the population of the United States; (2) that the major sources of noise include transportation vehicles and equipment, machinery, appliances, and other products used in commerce; (3) that, although the primary responsibility for control of noise belongs to the state and local governments, federal action is needed to deal with major noise sources in commerce.

Because aircraft contribute considerable amounts of noise to the environment, two laws were passed and further amended to reduce these noise sources. The first law was titled an Act to Require Aircraft Noise Abatement Regulation. This law was signed in 1968 and amended several times until it became the Quiet Communities Act of 1978. Then Congress decided to control and abate aircraft noise and sonic booms. The second law was the Aviation Safety and Noise Abatement Act of 1979, enacted in 1980 and amended in 1982. The purpose of the act was to provide assistance to airport operators to prepare and carry out noise compatibility programs, to provide assistance to assure continued safety in aviation, and to serve other purposes. It helped establish a single, reliable system for measuring noise, and also a single system for determining the exposure of individuals to noise at airports.

In early 1981, the Director of the Office of Noise Abatement and Control at the EPA was informed that the White House Office of Management and Budget had decided to end funding for the noise control agency and that the matter was nonnegotiable. Of the 28 environmental health and safety statutes passed between 1958 and 1980, the Noise Control Act of 1972 is the only one stripped of budgetary support. Because Congress did not repeal the law, the EPA remains legally responsible for enforcing the regulations issued under this act, without any budgetary support legislated for that purpose. Noise continues to be a serious public health hazard.

The Occupational Safety and Health Act of 1970 established the Occupational Safety and Health Administration (OSHA), and the National Institute for Occupational Safety and Health (NIOSH). OSHA sets and enforces regulations to control occupational health and safety hazards, including exposure to carcinogens. NIOSH is responsible for research; for various evaluations, publications, and training; and for the regulation of carcinogens by supporting epidemiological research and recommending changes in health standards to OSHA.

The Occupational Health and Safety Act provides three mechanisms by law for setting standards to protect employees from hazardous substances. The act initially authorized OSHA to adopt the health and safety standards already established by federal agencies or adopted as national consensus standards. This authority was given for the first 2 years in 1972 and 1973. The act also authorizes OSHA to issue emergency temporary standards (ETSs) that require employers to take immediate steps to reduce workplace hazards. The ETSs may be issued by OSHA when the agency determines employees are exposed to grave danger and emergency standards are necessary. Although the public has not had a chance to comment on a standard, because of the nature of the potential hazard, it must be enforced. However, the final standard must be issued within 6 months of the emergency standard. The third way that OSHA sets standards is to issue new permanent exposure standards and to modify or revoke existing ones. However, the informal rule making that goes along with modifying or revoking existing standards is subject to court review. The OSHA approach to rule making can result in requirements in monitoring and medical surveillance, workplace procedures and practices, personal protective equipment, engineering controls, training, record keeping, and new or modified permissible exposure limits (PELs). The permissible exposure limits are the maximum concentration of toxic substances allowed in the workplace air.

NIOSH has published many occupational safety and health guidelines with technical information about chemical hazards for workers. In addition, the institute publishes the criteria documents (CDs), alerts, current intelligence bulletins (CIBs), health and safety guides (HSGs), symposium or conference proceedings, NIOSH administrative and management reports, scientific investigations, data compilations, and other worker-related booklets. The CDs are recommended occupational safety and health standards for the Department of Labor. Usually, a recommended exposure limit (REL) is part of the recommended standard. The CIBs relate important public health information and recommend protective measures to industry, labor, public interest groups, and academia. The HSGs provide basic information for employers and employees to help ensure a safe and healthful work environment.

OSHA in 1994 adopted standards related to air quality in indoor work environments. The standards were based on a preliminary determination that employees working in indoor environments faced a significant risk of impairment to their health due to poor indoor air quality, and that compliance with the provisions proposed in the notice would reduce that risk. The standards were proposed to apply to all indoor, but not industrial work environments with the environmental tobacco smoke provision also applying to industrial work environments. Employers were to implement controls for specific contaminants and their sources such as outdoor air contaminants, microbial contaminants, maintenance and cleaning chemicals, pesticides, and other hazardous chemicals within the work environment. The regulations were to include sick building syndrome, building-related illnesses, indoor air contaminants, microbial contaminants, environmental tobacco smoke, exposure to the sources, various health effects, risk assessments, and regulatory impact.

The Oil Pollution Act of 1990 streamlined and strengthened the EPA's ability to prevent and respond to catastrophic oil spills. A trust fund financed by a tax on oil is used to cleanup oil spills if the responsible party is unable or unwilling to do so. The law requires all oil storage facilities and vessels to submit plans detailing how they will respond to large discharges. EPA has published regulations for aboveground storage facilities whereas the Coast Guard has done so for oil tankers.

The Pollution Prevention Act of 1990 focused industry, government, and public attention on reducing the amount of pollution through cost-effective changes in production, operation, and raw materials use. Pollution prevention also includes increased efficiency in the use of energy, water, other natural resources; and the protection of our resource base through conservation, recycling, source reduction, and use of sustainable agriculture. Congress determined that the nation produced millions of tons of pollution each year and spent tens of billions of dollars controlling this pollution. There were significant opportunities for industry to reduce or prevent pollution at the source through cost-effective changes in production, operation, and raw material usage. Such changes offered industry substantial savings in reduced raw materials, pollution costs, and liability costs as well as protected the environment and reduced risks to workers' health and safety. A change was made in the focus on rules and regulations to emphasize multi-media management of pollution and source reduction instead of only treatment and disposal. Congress declared that it be the national policy that pollution should be prevented or reduced at the source whenever possible and that recycling where feasible be used in an environmentally safe manner.

The Radon Gas and Indoor Air Quality Research Act of 1986 included the following findings by Congress: (1) high levels of radon gas pose a serious health threat in structures in certain areas of the country; (2) certain scientific studies suggest that exposure to radon, including naturally occurring radon and indoor air pollutants, may cause a public health risk; (3) existing federal radon and indoor air pollution research programs are fragmented and underfunded; (4) need exists for adequate information concerning exposure to radon and indoor air pollutants; and (5) this need should be met by appropriate federal agencies. Additional radon legislation was passed in 1988 as part of TSCA.

RCRA of 1976 was updated through 1988. This law replaced the previous Solid Waste Disposal Act. The Used Oil Recycling Act of 1980 amended RCRA and then was incorporated into the main text of the act.

RCRA provides for regulating the treatment, transportation, and disposal of hazardous waste. Hazardous waste is defined as solid waste that may cause death or serious disease, or may present a substantial hazard to human health or the environment if it is improperly treated, stored, transported, or disposed. Solid waste includes solid, liquid, semisolid, or contained gaseous materials from a variety of industrial and commercial processes. This definition excludes solid or dissolved materials found in domestic sewage or related to irrigation, industrial discharges subject to CWA, or mining wastes. RCRA requires the EPA to develop and issue criteria for identifying the characteristics of hazardous wastes. Defining characteristics of hazardous wastes are:

1. It poses a present or potential hazard to human health and environment when it is improperly managed.
2. It can be measured by a quick, available, standardized test method; or it can be reasonably detected by generators of solid wastes through their knowledge of their wastes: ignitability, corrosivity, reactivity, and extraction procedure toxicity.

RCRA also includes the regulation of underground storage tanks that may be used for a variety of storage processes. Further, this act includes a program in medical waste tracking.

In 1996, the Land Disposal Program Flexibility Act was signed into law. It modifies the Hazardous and Solid Waste Amendments of 1984. It provides that special wastes are no longer prohibited from land disposal as long as they are not hazardous at the point of land disposal. Hazardous constituents of wastes are to be treated in such a way that they will be removed, destroyed or immobilized before the wastes go to permanent land disposal. Dilution is prohibited as a substitute for treatment of the wastes.

SARA of 1986 of CERCLA established an Alternative or Innovative Treatment Technology Research and Demonstration Program. Its function is to promote the development, demonstration, and use of new or innovative treatment technologies; and to demonstrate and evaluate new, innovative measurement and monitoring technologies. The SARA amendments require that an information dissemination program be established along with the research efforts. The three types of technologies include (1) available alternative technology, such as incineration; (2) innovative alternative technology, which is any fully developed technology for which cost or performance information is incomplete and the technology needs full-scale field testing; and (3) emerging alternative technologies that are in their early stage of development and that the research has not yet fully passed the laboratory or pilot testing phase.

TSCA was enacted in 1976 and amended through 1992. TSCA was enacted by Congress to allow for the regulation of chemicals in commerce, as well as before they even entered commerce. This policy includes (1) that chemical manufacturers and processors are responsible for developing data about the health and environmental effect of their chemicals; (2) that the government regulate chemical substances that pose an unreasonable risk of injury to health or the environment and act promptly on substances that pose imminent hazards; and (3) that regulatory efforts should not unduly hinder industrial innovation. TSCA is directed at hazardous substances, wherever they occur. The substances do not have to be within a special environment. As long as they may cause a danger to the public — including a significant risk of cancer, genetic mutations, birth defects, or other potential serious hazards — they may be restricted or even banned by the EPA. TSCA permits the EPA to regulate new or existing toxic substances by requiring the testing of new or existing chemicals and by requiring their restriction in production and use, or even outright banning of substances that pose an unreasonable risk to health or the environment. The Asbestos Emergency Response Program is part of TSCA.

Title 1 is the control of toxic substances. Title 2 is asbestos hazard emergency response. Title 3 is indoor radon abatement. Title 4 is lead exposure reduction. Control of toxic substances includes provisions for testing chemical substances and mixtures, manufacturing and processing notices, regulating hazardous chemical substances and mixtures, managing imminent hazards, and reporting and retaining information.

The Asbestos Hazard Emergency Response Act was enacted in 1986. Congress authorized the EPA to amend its TSCA regulations to impose more requirements on asbestos abatement in schools. In 1990, this law was amended by the Asbestos School Hazard Abatement Reauthorization Act, which required the accreditation of persons who inspect for asbestos-containing material in school, public, and commercial buildings. It also mandated the accreditation of persons who designed or conducted response actions with respect to friable asbestos-containing material in those buildings.

In 1988, the Indoor Radon Abatement Act was passed. The purpose of this law was to assist states in responding to the threat to human health posed by exposure to radon. The EPA was required to publish an updated citizens guide to radon health risk, and perform studies of the radon levels in schools and federal buildings.

In 1992, Congress passed the Lead Exposure Reduction Section. The purpose of this law was to reduce environmental lead contamination and prevent adverse health effects as a result of this exposure, especially in children. Provisions included identifying lead-based paint hazards, defining levels of lead allowed in various products, including paint and toys, and establishing state programs for the monitoring and abatement of lead exposure levels as well as training and certifying lead abatement workers.

There are many separate acts of Congress related to water quality and water pollution. These range from the CWA to proper management of the oceans near the continental United States and the improvement of water quality.

The Federal Water Pollution Control Act was amended by the CWA of 1977 and was amended numerous times through 1988. Since then Congress has continued to fund programs without passing a new law. In 1999, a bill was passed to authorize the Secretary of the Army to construct various projects to improve rivers and harbors. The initial act was passed in 1948 and was then referred to as the Federal Water Pollution Control Act. In 1972, Congress set the goals of achieving fishable, swimmable waters by 1983 and prohibiting the discharge of toxic pollutants in toxic amounts by 1985. In the 1977 amendments, Congress endorsed a new means for regulating toxic pollutants, which had been developed to settle a lawsuit between environmental organizations and the EPA. In 1987, Congress continued its emphasis on the control of toxic pollutants. Although the CWA Provisions are less directly related to human health than the provisions of the Safe Drinking Water Act, it still aims at regulating human exposure to carcinogens and other toxic materials. An important part of the CWA is the National Pollution Discharge Elimination System (NPDES), which creates permits for direct discharges into the waters. It is lawful to discharge a pollutant only if the discharge is in compliance with the NPDES permit, which has to be issued by the EPA or by states whose permit programs are approved by the EPA. To obtain a permit, a facility must have the following information submitted: (1) list of pollutants that must be regulated according to federal or state law, along with a permissible amount of each pollutant that may be discharged per unit of time; (2) monitoring requirements and schedules for implementing the pollution concentration requirements; and (3) special conditions concerning the pollutants that the agencies believe should be imposed on the polluters. These may include additional testing and procedures for spills of pollutants into the water.

The Coastal Zone Management Act of 1972 was updated through 1986. Congress determined that it was in the national interest to have effective management, beneficial use, protection, and development of the coastal zone. The coastal zone included the fish, shellfish, and other living marine resources and wildlife found in this ecologically fragile area. The act considered the increasing and competing demands of economic development, population growth, and harvesting of the fish and shellfish.

The Deep Water Port Act of 1974 was amended through 1984. Congress decided to authorize and regulate the location, ownership, construction, and operation of deep-water ports in waters beyond the territorial limits of the United States. The act provides for the protection of the marine and coastal environment beyond these territorial limits.

The Marine Mammal Protection Act of 1972 was used to protect and manage marine mammals and their products such as hides or meat. In 1994, Congress passed the Marine Mammal Protection Act Amendments. Their function was to reduce the incidental capture of marine mammals during commercial fishing operations, and to reduce harassment to marine mammals causing annoyance with the potential to injure. These amendments also allowed people to seek permits to photograph marine mammals.

The Marine Protection Research and Sanctuaries Act of 1972 was amended through 1988. Congress determined that unregulated dumping of material into the ocean waters endangered human health, welfare, and amenities; the marine environment; ecological systems; and economic potential. Congress decided that it was the policy of the United States to regulate the dumping of all types of materials into ocean waters and to prevent or strictly limit the dumping of any material into the ocean waters that would adversely affect health, welfare, and ecosystem. It regulated the transportation of the material and dumping at sea.

The National Ocean Pollution Planning Act of 1978 was updated through 1988. Congress determined that the activities of people in the marine environment can have a profound short-term and long-term impact on such an environment and can greatly affect ocean and coastal resources. It stated that a comprehensive federal plan should be developed for ocean pollution research and for the development and monitoring of the material that had been dumped, the fate of the material, and the effects of the pollutants on the marine environment.

The outer Continental Shelf Lands Act of 1953 was updated through 1988. The *outer continental shelf* means all the submerged lands lying seaward and outside of the areas of lands beneath navigable waters. This includes the subsoil and seabed attached to the United States. The outer continental shelf is of vital importance to the country, because it is a national resource reserve of a variety of potential minerals, oil, etc. Congress decided that since the exploration, development, and production of the minerals of the outer continental shelf will have significant impact on the coastal and noncoastal areas of the United States and on other affected states, it was in the national interest that this exploration be controlled by the federal government. The Secretary of the Interior, under this act, has the authority to authorize the leasing of the outer continental shelf; to enforce safety, environmental, and conservation laws and regulations; and to formulate and promulgate such regulations, as necessary, to prevent problems from occurring.

The Port and Tanker Safety Act of 1978 was updated through 1986. Congress decided that the navigation and vessel safety and protection of the marine environment were matters of major national importance. It said that the increased vessel traffic in the nation's ports and waterways created a substantial hazard to life, property, and marine environment, and that an increased need existed for supervision of the vessel and port operations. It was particularly interested in the handling of dangerous articles and substances on, or immediately adjacent to, the navigable waters of the country. It stated that advanced planning is critical in determining proper, adequate, and protective matters for the nation's ports and waterways, and the marine environment.

The Shore Protection Act was passed in 1988. Congress stated that a vessel may not transport municipal or commercial wastes in coastal waters without a permit from the Secretary of Transportation and without displaying a number or other marking on the vessel, as prescribed by the Secretary. The permit included the name, address, and telephone number of the vessel owner and operator, its transport capacity, and its history of cargo transportation during the previous year, including wastes. The Secretary of Transportation had the right to enforce regulations concerning loading, securing, offloading, and cleanup. The secretary can also refuse the permit.

The SDWA of 1974 was updated through 1996. It was originally passed to ensure a safe drinking water supply. The CWA was designed to control water pollution, but it did not provide authority to regulate polluted water discharged into nonnavigable waters, such as groundwater that often is a source of drinking water. The SDWA is used primarily to regulate water provided by public water systems, and the act contains several provisions that may be used to regulate hazardous substances, including carcinogens in drinking water. The SDWA is more directly concerned in protecting human health than the CWA. Under the SDWA, the EPA regulates contaminants that may have an adverse effect on the health of people. It then establishes the steps that the agency must go through, over time, to protect drinking water. The EPA published national interim drinking regulations in 1975. The regulations were to protect the health of the people to the extent feasible using current technology treatment techniques and other means that the administrator determined were generally available.

Congress also required that the EPA request a National Academy of Sciences study to determine the potential adverse health effects of contaminants in the water and to help establish recommended maximum contaminant levels (RMCLs). These RMCLs were the recommended maximum level to be set for contaminants to prevent the occurrence of any known or anticipated adverse effect. It had to include an adequate margin of safety, unless there was no safe threshold. In that case, the recommended maximum contaminant level could be set at zero. The RMCLs are not enforceable health goals but are used as guidelines for establishing enforceable drinking water standards. The maximum contaminant levels (MCLs) were to be as close to the RMCLs as feasible. The enforcement of the MCLs rested with the states, whereas EPA sets the MCLs (see 1986 SDWA amendments).

In 1996, SDWA amendments were passed by Congress. These amendments emphasize sound science and risk-based standard setting, small water supply system flexibility and technical assistance, community-empowered source water assessment and protection, public right to know, and water system infrastructure assistance through a multibillion dollar state revolving loan fund. The law provides for three drinking water infrastructure needs surveys, additional research on cancer risks from exposure to low levels of arsenic, proposed or final drinking water regulations including sulfates, guidance on special technologies for small systems, source water assessments, radon standards, and local source water assessments.

The Water Resources Planning Act of 1965 was updated through 1988. The act stated that to meet the rapidly expanding demand for water throughout the nation, it is declared to be the policy of the Congress to encourage the conservation, development, and utilization of water and related land resources of the United States. This was to be on a comprehensive and coordinated basis by the federal government, states, counties, and private enterprises with the cooperation of all the various agencies and governments, individuals, and businesses involved.

The Water Resources Act was passed in 1984. The Congress declared that the existence of an adequate supply of water of good quality for the production of materials and energy for the nation's needs and for the efficient use of the nation's energy and water resources is essential to the national economic stability and growth and to the well-being of the people. It also stated that the management of water resources is closely related to maintaining environmental quality and social well-being, and that the nation should make a continuing investment in water and related research in technology that was commensurate with growing national needs. This research and development of technology should include the development of a technology for the conversion of saline and other impaired waters to a quality usable for municipal, industrial, agricultural, recreational, and other beneficial uses.

How a Bill Becomes a Law

Although a passage of bills in the state legislatures and the federal government varies somewhat, all follow an approximate route from the introduction of the bill to the final passage by both houses of the state government or the federal government. The bills are then signed or vetoed by the governor of the particular state or by the President of the United States. To get a better understanding of the process of how a bill becomes a law, the procedure used in Indiana is now discussed. Recognize that this path varies somewhat with each of the kinds of government throughout the country. For a bill to become a law in Indiana, it must follow a prescribed path with carefully planned steps. The path has many detours, lost turns, stumbling points, barriers, and frustrations. The reason for all these problems is created by the unique system of checks and balances that we find in government in this country.

The process in Indiana is as follows: (1) a representative or senator decides to introduce legislation; (2) a legal specialist draws up a bill; (3) the bill is introduced and assigned to a committee; (4) the committee approves the bill or amends the bill and approves it or rejects it; (5) the house or senate votes to accept the committee recommendation; (6) the bill is printed; (7) the bill remains on each representative's desk or senator's desk for 24 hours and then placed on the house or senate calendar for a second reading; (8) the bill is ordered "engrossed" or reprinted to show amendments; (9) the bill passes a second reading (amendments may be added at this time); (10) the third reading of the bill occurs; (11) the speeches are made for or against the bill; (12) the vote is taken (in Indiana, the constitutional majority needed for either approval or rejection is 51 votes in the house or 26 votes in the senate); (13) the bill is then delivered to the opposite house of the state or federal government, where either a senator or representative has promised to sponsor it; (14) the bill goes through a similar process in the opposite house, where it may also be amended; (15) if the bill is amended, it must be approved by the house–senate conference committee; (16) the bill is then submitted to the house and senate for passage; (17) the bill when passed, is signed by the house speaker and the president of the senate; (18) the attorney general checks its constitutionality; (19) the bill is signed by the governor or vetoed and sent to the secretary of state; (20) the bill becomes an enrolled act if it is signed, and is printed and bound in the "Act of Indiana"; (21) the bill goes back to the house and senate if it is vetoed, with a constitutional majority of two-thirds vote needed to overrule the veto of the governor; and (22) the bill is either finally overruled or, if it has been enrolled, is an enrolled act, with it taking effect when distributed among the circuit court clerks of all the counties of the state (Figure 1.8).

CREATING FEDERAL LAWS AND REGULATIONS

At the federal level there are three major steps in creating a law and making it work. In Step 1, a member of Congress proposes a bill, which if approved will become a law. In Step 2, if both houses of Congress approve a bill, it goes to the President who has the option to either approve it or veto it. If approved, the new law is called an act, and the text of the act is known as a public statute. Examples of this would be the Clean Air Act, CWA, SDWA, etc. In Step 3, once an act is passed, the House of Representatives standardizes the text of the law and publishes it in the U.S. Code. The U.S. Code is the official record of all federal laws. Its database is available from the Government Printing Office (GPO), which is the sole agency authorized by the federal government to publish the U.S. Code. Because laws often do not include all the details, Congress authorizes certain governmental agencies, for example, the EPA, to create regulations. These regulations set specific rules about what is legal and what is not legal. For example, the EPA will issue a regulation to implement the CAA by stating what levels of sulfur dioxide are safe and how much industries can emit without being penalized.

To create a regulation the authorized agency such as the EPA will first decide if a regulation is needed; then the agency will research it, propose the proper wording, list it in the *Federal Register*, get public comments, revise the wording, and issue a final rule. At each stage in the process, the agency publishes a notice in the *Federal Register*. Twice a year, each agency publishes a comprehensive report that describes all the regulations on which it is working or has recently finished. These are published in the *Federal Register*, usually in April and October as the Unified Agenda of Federal and Regulatory and Deregulatory Actions. Once the regulation is completed and has been printed in the *Federal Register* as a final rule, it is codified by being published in the Code of Federal Regulations (CFR). The CFR is the official record of all regulations created by the federal government. It is divided into 50 volumes, called titles, each of which focuses on a particular area. Almost all environmental regulations appear in Title 40, which is revised every July 1. The laws are carried out by the regulatory agency by enforcing the various regulations through inspections and surveys; receiving periodic reports from industries and local and state government; following up on complaints; and performing additional research to determine if there are better ways to achieve the results sought. At any step in the regulatory

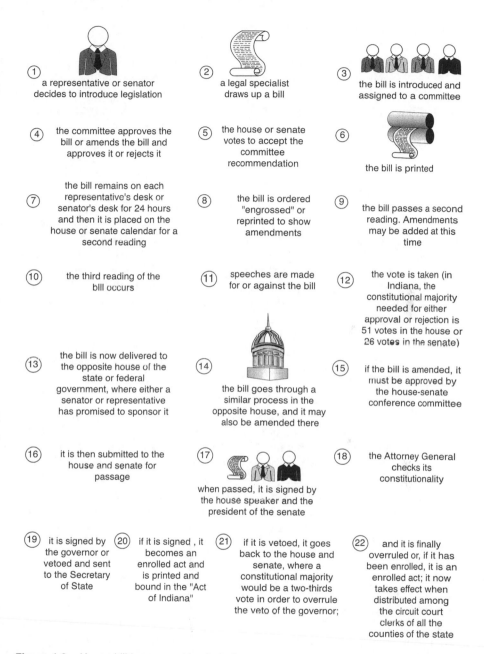

1. a representative or senator decides to introduce legislation

2. a legal specialist draws up a bill

3. the bill is introduced and assigned to a committee

4. the committee approves the bill or amends the bill and approves it or rejects it

5. the house or senate votes to accept the committee recommendation

6. the bill is printed

7. the bill remains on each representative's desk or senator's desk for 24 hours and then it is placed on the house or senate calendar for a second reading

8. the bill is ordered "engrossed" or reprinted to show amendments

9. the bill passes a second reading. Amendments may be added at this time

10. the third reading of the bill occurs

11. speeches are made for or against the bill

12. the vote is taken (in Indiana, the constitutional majority needed for either approval or rejection is 51 votes in the house or 26 votes in the senate)

13. the bill is now delivered to the opposite house of the state or federal government, where either a senator or representative has promised to sponsor it

14. the bill goes through a similar process in the opposite house, and it may also be amended there

15. if the bill is amended, it must be approved by the house-senate conference committee

16. it is then submitted to the house and senate for passage

17. when passed, it is signed by the house speaker and the president of the senate

18. the Attorney General checks its constitutionality

19. it is signed by the governor or vetoed and sent to the Secretary of State

20. if it is signed , it becomes an enrolled act and is printed and bound in the "Act of Indiana"

21. if it is vetoed, it goes back to the house and senate, where a constitutional majority would be a two-thirds vote in order to overrule the veto of the governor;

22. and it is finally overruled or, if it has been enrolled, it is an enrolled act; it now takes effect when distributed among the circuit court clerks of all the counties of the state

Figure 1.8 How a bill becomes a law in Indiana.

process any individual or group may file suit in a federal court for relief of a grievance against the regulation or proposed regulation and the regulatory agency involved. Many changes in regulations are due to these lawsuits.

ENVIRONMENTAL IMPACT STATEMENTS

NEPA, which established the Council on Environmental Quality (CEQ), requires each federal agency to prepare a statement of environmental impact in advance of each major action, recommendation, or report on legislation that may significantly affect the quality of the human environment. An environmental impact statement (EIS) is the heart of a federal administrative process designed to ensure achievement of national environmental goals. Each statement must access in detail the potential environmental impact of a proposed action and be continually conscious of environmental considerations. All federal agencies are required to prepare statements for matters under their jurisdiction.

The actions that the statement cover must be major and environmentally significant. The actions may be a precedent for much larger actions, which may have considerable environmental impact; or they may alter the future course of other environmental actions.

Each EIS must include (1) a detailed description of the proposed action, including information and technical data adequate to permit a careful assessment of environmental impact; (2) a discussion of the probable impact on the environment including any impact on ecological systems and any direct or indirect consequences that may result from the action; (3) any adverse environmental effects that cannot be avoided; (4) the alternatives to the proposed action that might avoid some or all the adverse environmental effects, including an analysis of costs and environmental impacts of these alternatives; (5) an assessment of the cumulative, long-term effects of the proposed action, including its relationship to short-term use of the environment vs. the environment's long-term productivity; (6) any irreversible or irretrievable commitment of the resources that might occur from the action or that would curtail beneficial use of the environment; and (7) a final impact statement that must include a discussion of problems and objections raised by other agencies (federal, state, and local), private organizations, and individuals during the draft statement's review process.

A draft statement must be prepared and circulated for comment at least 90 days before the proposed action. A final statement must be made public at least 30 days before the proposed action. Any agency unable to meet these requirements must consult with the CEQ.

NEPA requires each federal agency to consult with and obtain the comments of any other agency — state, local, as well as federal — that has jurisdiction by law or special expertise related to the environmental impact.

In its guidelines for preparing statements, CEQ lists the agencies that must be consulted in the following areas: air quality and air pollution control; weather modification; environmental aspects of energy generation and transmission; toxic materials; pesticides; transportation and handling of hazardous materials; coastal areas, including estuaries, waterfowl refuges, and beaches; historic and archaeological sites; floodplains and watersheds; mineral land reclamation; parks, forests, and outdoor recreational areas; noise control and abatement; chemical contamination of food products; food additives and food sanitation; microbiological contamination; radiation and radiological health; sanitation and waste systems; shellfish sanitation;

transportation and air quality; transportation and water quality; congestion in urban areas; housing and building displacement; environmental effects with special impacts on low-income neighborhoods; rodent control; urban planning; water quality and water pollution control; marine pollution; river and canal regulation and stream channelization; and wildlife. The guidelines also require that these statements be made available for public comment. Many individual agencies ask for such comments from interested parties and private organizations.

The impact statement procedure allows the public an opportunity to participate in federal decisions as they affect the human environment. The statements are announced in the *Federal Register*, and many agencies have other procedures to reach interested citizens.

A study of environmental impact statements for a 6-year period made by the CEQ indicated the following findings and recommendations: (1) the impact statement process can and should be more useful in agency planning and decision making; (2) the agencies with major EIS responsibilities should support high-level, well-staffed offices charged with implementing NEPA and the EIS process effectively; (3) the guidelines from CEQ are necessary to help agencies draw up EISs for broad federal programs or groups of projects; (4) the procedures that agencies use to notify other federal, state, and local agencies and the public of important EIS actions need improvement; (5) the agencies in consultation with CEQ should clearly define their expertise and jurisdiction for purposes of commenting on and improving EIS review; (6) the EIS process has contributed much to interagency coordination in environmental matters, but this coordination can be strengthened further; (7) the quality and content of EISs need continuing improvement; (8) several special issues arising from the EIS process require attention and remedy (among these are procedures for impact statements and federal permit actions); and (9) the adequacy of public participation in the EIS process needs thorough study and evaluation.

The environmental impact statements and their resulting conclusions can and have been challenged in the courts. Each year, numerous cases in the courts concern the statements. It is, therefore, necessary that the statements be completed in a thorough and comprehensive manner.

Cases have been filed in widely varying situations. *Calvert Cliff's Coordinating Committee vs. the NEC* was an early landmark case that set the direction of federal agency responsibilities under NEPA. The NEC was required to revise its activities to systematically analyze environmental impacts. Since then courts have handed down numerous decisions expanding federal agency responsibilities under NEPA. The court ruled in a case involving the Army Corps of Engineers that the findings by the CEQ indicated inadequate considerations were given to the safety and water quality effects of a project being carried out by the Army Corps of Engineers. The court required the corps to revise its impact statement and ordered the lower court to review this revised statement in light of the CEQ findings. Thus, for the first time, the CEQ findings as the principal overseer of the NEPA process were considered by a federal court.

Half of all the environmental impact statements have been written about road-building actions undertaken by the Federal Highway Administration. The statements have resulted in significant planning changes. The second largest number of statements have been prepared for watershed protection and flood-control projects. Statements have also been prepared in coal development and in strip mining concerning their effects on existing use of land and water. As a result of the evaluation and the review process of EISs, several federally sponsored projects have been suspended or modified in recent years. Many states have followed the federal approach and have passed environmental impact statement laws.

ENVIRONMENTAL HEALTH PROFESSION

The environmental health profession is composed of individuals whose efforts are directed toward controlling, preserving, or improving the environment so that people may have optimum health, safety, comfort, and well-being now and in future generations. Given the complexity of the environment, a variety of disciplines and career categories have developed, including engineering, composed of sanitary, environmental, mechanical, and chemical; education; and science, composed of generalists or specialists in biology, chemistry, medicine, etc. Another discipline is environmental health, composed of generalists or specialists in three basic types of categories: (1) sampling and analysis, routine inspections, and distribution of public information, (2) investigation consultation, planning, and education; and (3) supervision, administration planning, enforcement, and public relations.

Environmental Health Practitioner

The environmental health practitioner is an applied scientist and educator who uses the knowledge, comprehension, and skills of the natural, behavioral, and environmental sciences to prevent disease and to promote human well-being.

The environmental health practitioner makes inspections; conducts special studies; samples air, water, soil, and food; reviews plans; acts as an educator, public relations officer, and community organizer; plans programs; acts as a consultant to civic groups, business, industry, and individuals; and enforces or uses environmental and public health laws. The environmental health practitioner is involved in a multitude of program areas, including accident prevention, air pollution control, communicable diseases, environmental emergencies, engineering, food processing, and food handling, outbreaks of foodborne disease, hazardous substances control, housing, indoor environment, insect and rodent control, institutional environment, milk, noise control, nuisance abatement, occupational health, planning, product safety, radiation control, recreational sanitation, public and private sewage, solid and hazardous waste management, swimming pool sanitation, terrorism, water pollution, and public and private water supply.

The environmental specialist of today and the future must be a highly skilled, well-educated, and trained generalist possessing a multitude of competencies as an effective member of the health team. These competencies are divided into 6 general and 12 specific areas as follows:

General Competencies

General Science

1. Knowledge and comprehension of general inorganic and organic chemistry
2. Knowledge and comprehension of general biology
3. Knowledge and comprehension of general microbiology
4. Knowledge and comprehension of general college math, including algebra, trigonometry, and basic statistics
5. Knowledge and comprehension of physics (mechanics and fluids)
6. Knowledge and comprehension of epidemiological principles
7. Knowledge and comprehension of risk assessment techniques
8. Knowledge and comprehension of toxicological principles

Communications and Education

1. Knowledge and comprehension of various communications, verbal, written, and computerized
2. Knowledge and comprehension of how to work with people
3. Knowledge and comprehension of the use of audiovisual aids
4. Knowledge and comprehension of group dynamic techniques and group processes
5. Knowledge and comprehension of interviewing techniques
6. Knowledge and comprehension of teaching and learning principles
7. Understanding of the need for public information and proper relationships with the news media
8. Understanding of how to work with and motivate community organizations and industry groups including the use of small group dynamic techniques
9. Knowledge and comprehension of use of databases

Planning and Management

1. Knowledge and comprehension of techniques needed to establish a program in any of the environmental health areas
2. Knowledge and comprehension of computers, electronic data processing techniques and their application
3. Knowledge and comprehension of techniques used to establish priorities
4. Ability to design surveys and survey forms
5. Ability to use survey techniques to determine the extent of given environmental health problems
6. Ability to interpret survey findings
7. Ability to determine the advisability of legal action and when to initiate it

General Technical Skills

1. Knowledge and comprehension of the principles of learning and possession of skills in training, testing, evaluation, and use of aids in various areas of environmental health
2. Knowledge and comprehension of how to use inspectional and survey techniques to determine environmental health problems

3. Knowledge and comprehension of a variety of sampling techniques related to air, soil, water, food, hazardous chemicals, etc.
4. Ability to collect samples, complete record forms, and interpret results of laboratory samples accurately in light of surveys made
5. Ability to use survey and field sampling and test instruments

Administrative and Supervisory Skills

1. Knowledge and comprehension of environmental and public health laws, regulations, ordinances, codes, and their application
2. Knowledge and comprehension of supervisory techniques used in environmental health programs
3. Knowledge and comprehension of administrative techniques used in the management of environmental health programs
4. Knowledge and comprehension of the systems approach to the analysis of environmental health problems
5. Knowledge and comprehension of the vital role of continued maintenance in the permanent resolution of environmental control problems
6. Knowledge and comprehension of the relationship between health departments, other public agencies, voluntary agencies, business, and industry
7. Knowledge and comprehension of basic principles of economics and how they relate to the existence of environmental health problems and the potential for successful environmental health programs
8. Knowledge and comprehension of overall health problems and health priorities
9. Knowledge and comprehension of risk management techniques

Professional Attitudes

1. Desire to work with people and to use the environmental health sciences to resolve environmental health problems
2. Sense of obligation to fulfill the requirements of the job and to carry out assigned duties in a professional manner
3. Approach recipients of services with a cooperative attitude
4. Display courtesy in personal relationships with fellow employees
5. Accept constructive criticism from employees, peers, and the public
6. Sense of dedication to the environmental health profession reflected in participation in continuing education
7. Control emotions and perform in a mature manner during periods of stress
8. Desire to communicate public health principles

Specific Competencies

Environmental Chemical Agents

1. Knowledge and comprehension of potential chemical contaminants of food
2. Knowledge and comprehension of potential chemical contaminants of potable water supplies
3. Knowledge and comprehension of transport requirements for hazardous chemicals
4. Knowledge and comprehension of chemical weapons

5. Knowledge and comprehension of the techniques and procedures for identifying environmental chemicals
6. Knowledge and comprehension of the means of disposal for environmental chemicals
7. Understanding of decontamination of objects or substances that have been contaminated with environmental chemicals
8. Knowledge and comprehension of field tests used to determine the presence and concentration of environmental chemicals
9. Knowledge and comprehension of detergent and disinfectant chemistry
10. Ability to evaluate detergents in an "in-use" situation
11. Knowledge and comprehension of economic poisons and how they affect the human ecology of the region
12. Knowledge and comprehension of principles and practices in the formulation and application of economic poisons
13. Knowledge and comprehension of bait formulations used in pest control
14. Knowledge and comprehension of the safety features needed to prevent accidents with environmental chemicals
15. Knowledge and comprehension of disinfectant detergents and their use

Environmental Biological Agents

1. Knowledge and comprehension of the epidemiology of vectorborne diseases
2. Knowledge and comprehension of the natural habitat and control of common microorganisms and insects of public health and economic significance
3. Knowledge and comprehension of basic life cycles of microorganisms, insects, and rodents of public health significance
4. Ability to identify a variety of microorganisms and insects of public health or economic significance in the field
5. Knowledge and comprehension of environmental factors related to vector control
6. Ability to identify scope of field problems and to determine control activities required
7. Knowledge and comprehension of advantages and limitations of microbicides and insecticides and their effect on the ecology of the region
8. Knowledge and comprehension of the operation of sprayers and other pest control equipment
9. Knowledge and comprehension of the epidemiology of microbial, insect, and rodentborne diseases
10. Knowledge and comprehension of environmental procedures used in microbial, insect, and rodent control
11. Knowledge and comprehension of biological control of microbes, insects, and rodents
12. Knowledge and comprehension of bioweapons

Environmental Physical Agents

1. Knowledge and comprehension of public health and ecological effects of noise on the individual and community
2. Knowledge and comprehension of instrumentation and procedures involved in noise measurements
3. Knowledge and comprehension of existing laws pertaining to nuisances and noise abatement

4. Knowledge and comprehension of practical applications of control measures
5. Ability to implement surveys designed to define the extent of noise problem
6. Ability to evaluate results of surveys and to establish long-range and short-range goals for control
7. Knowledge and comprehension of work-related noise stress
8. Knowledge and comprehension of radiation theory and principles
9. Knowledge and comprehension of dangers of radiation
10. Knowledge and comprehension of use of radiation and radioisotopes
11. Knowledge and comprehension of effects of radiation
12. Knowledge and comprehension of radiological weapons
13. Knowledge and comprehension of safety precautions
14. Knowledge and comprehension of monitoring techniques and instrumentation used in radiation detection
15. Knowledge and comprehension of techniques of storage and disposal of radioactive materials
16. Knowledge and comprehension of techniques of transportation of radioactive materials
17. Knowledge and comprehension of techniques of decontamination
18. Knowledge and comprehension of legal requirements of transportation, use, storage, and disposal of radioactive materials

Air

1. Knowledge and comprehension of the different air pollutants and their sources
2. Knowledge and comprehension of the aerosolized weapons
3. Knowledge and comprehension of the relationship of weather conditions to air pollution
4. Knowledge and comprehension of effects of air pollutants on the biosphere
5. Understanding of the relationship of air pollution to topography
6. Knowledge and comprehension of microflow of air
7. Knowledge and comprehension of functional operation of air pollution control devices
8. Knowledge and comprehension of preventive measures in air pollution control
9. Knowledge and comprehension of corrective measures in air pollution control
10. Knowledge and comprehension of the practical applications of air pollution control procedures and techniques
11. Knowledge and comprehension of the principles of combustion engineering
12. Knowledge and comprehension of air sampling techniques and ability to conduct air sampling
13. Ability to implement surveys to clarify and identify the extent of problems
14. Ability to evaluate results of surveys in light of long-range and short-range problems and programs within the community
15. Ability to design and implement cost-benefit analysis of control programs
16. Knowledge and comprehension of air toxics

Water and Liquid Wastes

1. Knowledge and comprehension of water sources
2. Knowledge and comprehension of potable drinking water quality and standards (physical, chemical, biological, and radiological)

3. Knowledge and comprehension of waterborne diseases and how they are transmitted
4. Knowledge and comprehension of sampling and testing of all waters, including potable water
5. Interpretation of laboratory analysis of water samples
6. Knowledge and comprehension of legal aspects of water quality control
7. Knowledge and comprehension of different types of water usage
8. Understanding of the protection and selection of individual water supplies
9. Understanding principles of water treatment
10. Knowledge and comprehension of physical and biological composition of sewage, including common and exotic industrial wastes
11. Knowledge and comprehension of types of industrial wastes and their significance
12. Knowledge and comprehension of the effects of sewage discharge on water quality
13. Understanding of the epidemiology of sewage-associated diseases
14. Knowledge and comprehension of the technology and basic engineering principles related to water flow
15. Understanding the principles of individual sewage disposal
16. Knowledge and comprehension of principles of municipal sewage treatment
17. Knowledge and comprehension of small sewage treatment units
18. Knowledge and comprehension of the measurement of absorptive quality of soils
19. Knowledge and comprehension of the principles of nonwater sewage disposal
20. Knowledge and comprehension of the techniques used in problems of emergency situations related to water and sewage
21. Knowledge and comprehension of the techniques and potential hazards of sludge disposal
22. Knowledge and comprehension of the spread of bioweapons

Food

1. Knowledge and comprehension of food technology and its relationship to health
2. Knowledge and comprehension of principles of food manufacturing, processing, and preservation
3. Knowledge and comprehension of foodborne diseases and their control
4. Knowledge and comprehension of epidemiological techniques and procedures
5. Knowledge and comprehension of design, location, and construction of food establishments and their equipment
6. Knowledge and comprehension of principles of food establishment operations, housekeeping, and maintenance
7. Knowledge and comprehension of equipment design, operation, maintenance, and cleaning techniques
8. Knowledge and comprehension of methods of motivating industrial management to understand, accept, and carry out its responsibilities in the food environment, personnel training, and personal supervision
9. Knowledge and comprehension of legal requirements of food technology and food safety
10. Knowledge and comprehension of inspection, survey techniques, and significance of data
11. Knowledge and comprehension of the examination and licensure of food establishment managers
12. Knowledge and comprehension of techniques used by different cultural and ethnic groups in food growing and preparation

13. Knowledge and comprehension of institutional food handling practices
14. Ability to obtain public support for food programs
15. Knowledge and comprehension of characteristic and properties of milk
16. Knowledge and comprehension of dairy bacteriology
17. Knowledge and comprehension of milk production and processing
18. Knowledge and comprehension of legal standards of food and milk composition
19. Knowledge and comprehension of techniques used to investigate dairy farms
20. Knowledge and comprehension of milk processing operations and control
21. Ability to inspect pasteurization plants
22. Knowledge and comprehension of spread of bioweapons in food

Solid Wastes

1. Knowledge and comprehension of the types of solid waste generated in the community
2. Knowledge and comprehension of the types of waste generated by common industrial processes
3. Knowledge and comprehension of various methods of storage, collection, and disposal of solid waste
4. Knowledge and comprehension of public health and ecological aspects of solid wastes
5. Knowledge and comprehension of the use of systems analysis in waste disposal management
6. Knowledge and comprehension of economics of solid waste disposal
7. Ability to evaluate the results of solid waste surveys and to establish long-range and short-range goals
8. Ability to implement surveys to determine the extent of the solid waste problems
9. Ability to design, implement, and evaluate programs related to waste disposal vs. public health problems

Hazardous Waste

1. Knowledge and comprehension of the health and safety concerns related to hazardous waste sites including sites created by terrorism
2. Knowledge and comprehension of the effects of exposure to toxic chemicals in a hazardous waste site
3. Knowledge and comprehension of the route of entry to the body of hazardous chemicals, including inhalation, skin absorption, ingestion, and puncture wounds (injection)
4. Understanding of the potential health effects due to acute and chronic exposure to various chemicals at the hazardous waste site
5. Knowledge and comprehension of the symptoms of exposure to hazardous chemicals, such as burning, coughing, nausea, tearing eyes, rashes, unconsciousness, and death
6. Knowledge and comprehension of the potential chemical reactions that may produce explosion, fire, or heat
7. Understanding of the psychological effects of oxygen deficiency in humans related to an increase in specific chemicals in the immediate environment
8. Understanding of the health effects of ionizing radiation related to alpha radiation, beta radiation, gamma radiation, and x-rays
9. Knowledge and comprehension of techniques used to dispose of radioactive material

10. Understanding of the potential types of hospital and research facility waste that may cause biological hazards for the individual and may be spread through the environment
11. Knowledge and comprehension of the various safety hazards that may be found at hazardous waste sites
12. Knowledge and comprehension of the potential electrical hazards that may occur from overhead power lines, downed electrical wires, and buried cables that have been subjected to potential damage from hazardous waste situations

Population and Space Utilization

1. Knowledge and comprehension of increased population and its effect on the present and future needs of our society
2. Knowledge and comprehension of the health hazards related to congestion
3. Knowledge and comprehension of individual space needs
4. Knowledge and comprehension of the effects of different cultures on population control
5. Knowledge and comprehension of the use of community planning and zoning on space utilization
6. Knowledge and comprehension of establishment of priorities for the proper use of existing space

Indoor Environment

1. Knowledge and comprehension of cultural, economic, and sociological aspects of individual and multiple dwelling units
2. Knowledge and comprehension of housing conditions needed for health, comfort, and well-being
3. Knowledge and comprehension of impact of transportation on housing
4. Knowledge and comprehension of real estate laws and prevailing practices
5. Knowledge and comprehension of the various agencies involved in supervision and licensing of community shelters
6. Knowledge and comprehension of techniques used to evaluate individual and multiple dwelling units
7. Knowledge and comprehension of local, state, and federal housing programs
8. Knowledge and comprehension of zoning laws and their effect on the use of individual and multiple dwelling units
9. Knowledge and comprehension of the relationship of minority groups and poverty to housing use
10. Knowledge and comprehension of indoor air pollution problems
11. Knowledge and comprehension of spread of chemical/bioweapons in enclosed areas

Environmental Injuries

1. Knowledge and comprehension of public health and ecological aspects of environmental injury problems
2. Knowledge and comprehension of the instrumentation, material, and procedures involved in determining the causes of accidents
3. Knowledge and comprehension of epidemiological techniques used for studying accident problems

4. Ability to motivate voluntary corrective action on the part of the public
5. Ability to evaluate accidents and their causes

SUMMARY

Humans, who alone have the intellectual capacity to improve their life through science and engineering, share the Earth with numerous other biosystems, which might affect health. Humans contribute to the destruction of biosystems and to their own health problems through degradation of the environment by means of air, water, and land pollutants. The pollutants are transported biologically or physically through the environment and cause a variety of environmental problems that adversely affect much of the Earth. People use skill, good sense, and planning to avoid this destruction and provide for a wholesome, safe environment for future generations.

The problems of the environment are numerous and complex, and can only be resolved by determining the cause, means of prevention, and necessary controls of specific hazards and combined hazards through the use of epidemiological research, special study techniques, risk assessment, and risk management techniques. Not only the ecosystem, the food chain, the growth of population, and the energy cycle must be understood but also the use and abuse of energy and the impact of humans on their own environment. An economic determination as to the benefit–risk of each type of environmental impact must be made backed by reasonable decisions concerning these impacts. Further, the legal system and the trained environmental health practitioners, along with all levels of government, industry, and concerned citizenry, must be employed voluntarily to improve the environment of all living organisms.

It is certainly clear that individual or collective human decisions influence our environment to a considerable degree and that the quality of our life and the time and manner of our death are related to these decisions. It is not only important to eliminate specific environmental components that cause or contribute to disease or injury but also essential to develop a preventive strategy that will eliminate many of the environmental problems before they occur.

In this chapter, the general problems affecting the environment and some of the primary concerns have been discussed. In Chapter 2, environmental problems and human health are discussed. In the succeeding chapters, each environmental health area is treated separately. Also, the specific environmental problem is approached from background and status of the problem; necessary scientific, technological, and general information available; source, scope, and potential for disease related to the problem; potential for intervention; resources that may be utilized; standards, practices, and techniques available; modes of surveillance and evaluation available; specific controls; and research needs. (See Preface for a description of how best to use this book and Volume II.)

FUTURE

The environmental health practitioner, specialist, and scientist of the future must have appropriate knowledge of health and the environment and recognize the people-made pressures that have been placed on natural resources, both living and nonliving.

These pressures are far more severe than have ever been previously suspected. There is substantial evidence that potentially health-threatening groundwater contamination is a problem of increasing concern in the United States and that toxic chemicals in hazardous waste dumps and other ground storage may pose serious health and environmental threats and create public health problems. Toxic chemicals are present in the air, water, and workplace and are growing in quantity and complexity. Many examples of potential damage to the ecosystem, as well as to people, have been cited and are cited in the succeeding chapters.

Short-term research and short-term horizons have been the techniques used for trying to resolve long-term problems. Concern exists not only in government but also in the private sector. It is, therefore, necessary to have a better understanding of the interdisciplinary activities that will be needed to carry out the long-term research that will determine the kinds of risks that are occurring and the kinds of risk management that will be necessary in the future.

Research

New research techniques include molecular epidemiology, which is based on the measurement in the exposed individual of the interaction of a toxic chemical or its derivative with a tissue constituent or a tissue alteration resulting from exposure to the chemical. It provides an indirect measure of individual exposure. Recent research has determined means for detecting and measuring the interaction of a foreign chemical with easily accessible normal human constituents, such as chemical carcinogen interactions with DNA. It is now possible, at times, to detect a few altered DNA molecules out of millions of cells. Altered chromosomes can be determined, and important advances will be occurring in detecting and quantitating human exposure to foreign chemicals.

Susceptibility to chemical toxicants varies widely in the human host population. The host factors are genetic diversity, current and prior disease, sex, and age. There are approximately 2000 genetically identifiable human diseases. Genetic conditions are likely to enhance the risk to individuals of developing environmentally or occupationally associated adverse health effects. The extent of the risk is unknown. The extent of the risk of enhancement is unknown. Considerable study is needed in genetic diversity–susceptibility and biological mechanisms.

Exposure to humans, other animals, and the environment of pollutants or mixtures of substances is not clearly understood. No valid general rules exist for determining the presence of synergistic activity in the mixtures of chemicals to which these individuals or the environment is subjected.

The emerging disciplines of biotechnology and microelectronics are of considerable interest and potential concern. Biotechnology deals with genetic engineering, which hopefully will produce new medical, agricultural, chemical, and other products. Microelectronics involves the production of microelectronic shifts with the use of a variety of virtually unstudied chemicals, such as gallium arsenide, silicon, and halogenated hydrocarbon solvents.

Inadequate data are available for the physical, chemical, and ecological variations in freshwaters, oceans, and atmosphere over extended periods of time. It is, therefore, difficult to determine what is a natural change over a long period of time and what is caused by human activity. There is a need to understand a normal range of variations in ecosystems and how they can deal with the pollutants that are currently flowing into them and that will flow into them in the future.

As pollutants move from one environmental medium to another, the rate of transfer is not understood. Quantity assessments are necessary, and the results of these movements of pollutants need to be understood. At what rates are aerosols formed from chemically reactive organic pollutants and what effect do these aerosols have on the air–water interface, the precipitation of the aerosols, and the temperature and moisture content that may affect them? Study is needed to determine the behavior and biological effects of chemicals in various environmental media and what impact they may have on the overall global ecosystem.

The improvement in quantitative risk assessment is most likely to come from a better understanding of biological processes. Little is known about basic pharmaco-kinetic dynamics and the environmental mechanisms related to toxicity, other than cancer. Very little is known about actual exposure patterns, the potential of short-term biological screening to provide early prediction of effects, and a validation of risk assessments.

Environmental Changes and Their Consequences

To understand environmental changes and their consequences, it is necessary to know the movement from the point of exposure to the end point of disease. There are four rough stages in environmental toxicology. They include:

1. Exposure is where the ambient condition brings the organism into contact with the hazard, for example, a toxic material that may be found in the air or water.
2. The dose reaches the point where the internalized quantity of the hazard creates a specific body burden that may be toxic to an organ or cell.
3. Effects seen are markers of intermediate biological effects that are either a step in the toxicologic process or a parallel manifestation of effect. These indicators of exposure or dose are predictors of toxicity. They may include chromosome aberrations, mutations, and cell or cell killing enzyme activity.
4. The end point is the ultimate toxicological effect, which may be cancer, heart disease, or other diseases or injuries.

Perturbations are complex mixtures of substances or impacts, and equally complex patterns of human activities that influence exposure. Examples include hazard-ous waste dumps and related groundwater problems, as well as the dynamics of the ecosystems that may be affected by these. The perturbations, which also include the introduction of new technologies or changes in resource use patterns, affect ecosystems physically and biologically. The physical changes include nutrient and energy flow, physical structure of the system, and transport and transformation of substances that have been introduced. The biological changes include direct increases in populations due to secondary effects and resulting changes in linkages and functioning of the system as a whole. The ecosystems are related to exposures in humans or other animals and plants, of concern to humans, and finally end up with health effects that are either improper or proper. The improper health effects may be changes in target organs of the body or death to the total organism.

Recommendations for the Future

The research areas that have been previously discussed are summarized in some specific types of studies that need to be carried forward. They are:

1. Dynamics of stressed ecosystems. That is, changes occur in ecosystems that result from people-made stresses.
2. Ecologically significant end points. That is, the stressing of the ecosystems should become a basis for potential regulatory attention.
3. Ecological markers and sentinel events. That is, a study should be done to identify early indicators of potential ecological change.
4. Total toxicity of complex mixtures. That is, research should be devoted not only to the effects of individual substances but also to the complex of common mixtures, such as solvents, agriculture runoff, and wastewater sludges, as well as airborne chemical mixtures.
5. Movement of pollutants out of sinks. Sinks are ecological areas, such as wetlands or underwater sediment and landfills, that have been used for the disposal of toxic pollutants for many years. The pollutants may move across the media. Research is needed to identify under what conditions these materials or their conversion products will move out of the sinks and into ecosystems and human environments.
6. Human exposure. That is, data should be developed on human exposure from all environmental sources, ranging from inhalation in the ambient air, the workplace, and other indoor environments, to smoking and the ingestion of consumer products, as well as food and water, and the absorption of substances by skin contact.
7. Effects on multiple organs and physiological systems. That is, research should be expanded to study the neurobehavioral and immune systems.
8. Plant perturbations. That is, an analysis should be made of the effect of fossil fuel combustion and the potential for global temperature increases, changes in the carbon cycles, and other possible concerns.
9. Biotechnologies. That is, a better understanding is needed for the genetic manipulation of biological organisms to determine whether potentially serious consequences may occur.
10. Linkages among ecological systems. That is, an understanding needs to be found of the transport and fate of materials between heterogeneous ecosystem units.

CHAPTER 2

Environmental Problems and Human Health

INTRODUCTION

This chapter concentrates on problems related to human health, instead of the various other systems coexisting on Earth. The decay of the environment and the resultant effect on human health grows daily. The sun is shrouded by the smoke of industry and dwellings. Our eyes and respiratory system are irritated daily by air pollutants. Emphysema and lung cancer (as reported in March 2002, *Journal of the American Medical Association*) arc increasing sharply. Our noses are offended by acrid, noxious doses of chemicals, and overflowing sewage. Our auditory system is damaged by the noise of traffic, industry, construction, and jet aircraft.

Pesticides essential to food production and protection against the severe epidemics of the past, such as malaria, yellow fever, and plague, are misused and commonly abused. Too many pesticides have been applied carelessly by uninformed or indifferent individuals. Research on the ultimate effects of pesticides on humans and their environment has not kept pace with the sharp increase in usage in the last 25 years. Over 3 billion pounds of pesticides are applied annually in the United States alone.

We continue to use dangerous, misbranded, or adulterated food produced under unsanitary conditions despite several governmental enforcement acts, including the Pure Food and Drug Act of 1906; and the Food, Drug, and Cosmetic Act of 1938 and its numerous amendments in the 1950s, 1960s, 1970s, 1980s, 1993, and 1996. *Salmonella aureus* is present in most supplies of pooled raw milk. *Salmonella* organisms are found in 15 to 30% of raw dressed poultry and in many commercial egg products. *Clostridia perfringens* is present in more than half of the red meat sold. *Clostridia botulinum*, type E, is frequently found in raw fish. *Salmonella aureus* is present on the mucous membranes or skin of 30 to 50% of the population. Chemical residues, pesticides, and additives are components of food that may cause subtle problems. Although the overall incidence of foodborne disease is unknown, it is estimated from statistics of actual outbreaks that 76 million cases of gastroenteritis occur each year. *Salmonella* are the most frequent cause of disease. In smaller communities, foodservice programs are either inadequate or totally lacking. Major causes of foodborne disease includes: *Campylobacter, Escherichia coli* O157:H7; *Listeria, Salmonella, Shigella, Vibrio,* and *Yersinia.*

The residential environment for 6 million poor American families is the hazardous blighted slums of 19th century England. The healthful, pleasant, attractive, comfortable housing of the make-believe world of television is replaced in reality by tin and tarpaulin shacks; crowded, ramshackle houses; and tenements overflowing with garbage, flies, roaches, rodents, and sewage. Walls and floors deteriorate, lights fail, windows disintegrate, heating is poor, and ventilation is almost nonexistent. Lead poisoning, respiratory illness, and infectious diseases are endemic in this environment. Homelessness adds to the disease potential.

In the suburban areas, inadequate planning and poor land utilization have led to the subdivision nightmare, where children play in overflowing sewage and houses crack, settle, and are subject to flooding. Home accidents related to housing have sharply increased during the last 30 years, with resulting injuries and deaths.

Insects and rodents spread disease, cause annoyance, destroy crops, and ruin property. During the outbreaks of bubonic plague in Europe in the 14th and 15th centuries, an estimated 25 million people died of this ratborne disease. In the 1990s, bubonic plague was still present in southwestern United States. Yellow fever affected 23,000 out of a population of 37,000 people in Philadelphia in 1793, with 4000 people dying. In addition, malaria, encephalitis, spotted fever, tick fever, tularemia, and rickettsial pox have occurred regularly during the 19th and 20th centuries. In 1959, an epidemic of eastern encephalitis occurred in New Jersey; and in 1962, an epidemic of St. Louis encephalitis occurred in the St. Petersburg–Tampa Bay area of Florida. In 1975 and 1976, St. Louis encephalitis spread to many states. In 1986, Houston, TX, experienced its largest outbreak since 1980. It has also reoccurred in Long Beach, CA. Further, from 1984 to present, there has been a substantial amount of Lyme's disease from ticks. In 1993, hantavirus carried by deer mice caused a deadly disease called hantavirus pulmonary syndrome. In 1999, West Nile encephalitis was found in New York City. West Nile virus is a flavivirus commonly found in Africa, West Asia, and the Middle East. It is closely related to St. Louis encephalitis. Both diseases are transmitted by the bite of the *Culex* mosquito. Tens of thousands of rat bites occur each year, mostly to very young children.

Many individuals spend part of each day or most of their lives in a variety of institutions. Each institution, particularly if it provides sleeping accommodations, is in effect a small community, with all the environmental health problems discussed in this book, and the added problem of a rapidly shifting mobile population that is highly susceptible to disease and accidents.

Noise, or unwanted sound, causes temporary or permanent hearing loss, physical and mental disturbances, breakdowns in the reception of oral communications, reduced efficiency in performing work-related tasks, irritability, disruption of sleep and rest, and increased potential for accidents. Steady exposure to 90 dB of sound can cause eventual hearing loss. Heavy traffic can reach 90 dB, and the noise of jet planes, rock-and-roll bands, motorcycles, power mowers, auto horns, heavy construction, and farm equipment exceed this level.

Occupational hazards occur in all industries. In 1997, 6,026 workers died from occupational hazards and 3,300,000 were the victims of lost workday injuries. Billions of dollars are spent each year on worker's compensation cases for injuries and work-related illnesses. These problems are caused by accidents and various health hazards, such as toxic chemicals, including dusts, gases, fumes, mists, vapors; physical agents, including noise, pressure, ionizing radiation, and severe temperature variations; biological hazards, including insects, bacteria, fungi, and viruses; and other hazards, including unusual work-related posture, boredom, fatigue, repetitive motion, and monotony.

Low and high levels of radiation may adversely affect human health by causing genetic damage, burns, destruction of tissue, reduced life span, and cancer. Old x-ray and medical fluoroscopic machines may contain inadequate devices to attenuate unneeded or stray radiation. Improper use, storage, or disposal of radioactive material and wastes could lead to serious potential hazards.

Outdoor recreation is a complex and rapidly expanding area of the economy. Americans spend over $30 billion a year on recreation and over 100 million Americans vacation each year. The environmental health problems resulting from these activities are enormous. They include all areas discussed in these books, plus problems of mass migrations; inadequate housing, water, and sewage facilities; and high potential for accidents.

Soil is the single most important natural factor in choosing a site for construction of dwellings and for development of an effective, operating, on-site sewage disposal system. Unfortunately, lack of knowledge in this area, poorly designed tests, and inadequate supervision have led to severe settlement cracks in houses, flooded basements, flooded homes, slippage of hillsides and houses, soil erosion, contaminated or inadequate water supplies, and overflowing sewage. The disease and accident potential for humans is substantial.

Solid wastes generate enormous economic, aesthetic, social, and health problems. Our affluent society, improved technology, new packaging methods, and disposable items have increased the amount of solid waste per person per day from 2.7 lb in 1960 to 4.3 lb of waste in 1996, and growing. At this rate, our society will become inundated by waste unless changes in waste production and disposal are made immediately. Unsatisfactory storage, collection, and disposal of solid waste lead to insect and rodent problems, air pollution, offensive odors, accidents, fires, explosions, contamination of the water supply, degradation of the landscape, destruction of fish, and conversion of bodies of water into open sewers. Hazardous waste has complicated the problem enormously.

Water, our most precious resource, is used to sustain life, support the growth of food, develop business and industry, and provide recreation. Despite these essential human needs, individuals, municipalities, industries, commercial establishments, and agriculture continue to pollute our water supply. Historically many of our worst epidemics have been caused by contaminated water. Although in 2002 the epidemics are gone, waterborne outbreaks of disease still have occurred over the past 8 years and the potential for disease outbreaks continues. In Milwaukee, WI, a very large outbreak of cryptosporidosis occurred. Approximately 43,000 people became ill and 4400 were hospitalized.

Of the recent food and waterborne disease outbreaks, a few are now presented. On June 24, 1996, the Department of Health in Livingston County, NY, was notified of a cluster of diarrheal illness following a party on June 22. Of 189 people who had attended the party at a private residence and had eaten food that was catered by a local convenience store that sold gasoline, packaged goods, sandwiches, and pizza, about 30 became ill. People who became ill within 72 hours with diarrhea, as well as leftover food items and water samples, were tested. The individuals had either *Plesiomonas shigelloides* or *Salmonella serotype Hartford* or both isolated from stools. Interviewed were 52% or 98 people. Of the 60 people reporting illness, 56 or 57% of respondents met the case definition for the outbreak. Twenty food and beverage items were served at the party. Of these, three food items were associated with the illness: macaroni salad, potato salad, and baked ziti. The water source was an unprotected 10-ft deep dug well that was fed by shallow groundwater and may have received surface runoff from surrounding tilled and manured land and water from adjacent streams. A small poultry farm was located about 1600 ft upstream from the well. Farm field drainage systems discharged into the source water stream just above the well. A water sample collected at the store on June 27 showed no chlorine residual, indicating that the pellet chlorinator was not working. Well water used for food preparation and cleaning was probably contaminated as a result of rainfall on June 19 and June 20 that transported pathogens from the surrounding farmland.

On September 3, 1999, the New York State Department of Health received reports of at least ten children who were hospitalized with bloody diarrhea caused by *E. coli* O157:H7 infection from counties near Albany, NY. All the children had attended the Washington County Fair, which was held August 23 to 29, 1999. Approximately 108,000 people attended the fair that week. Subsequently, fair attendees also became infected with *Campylobacter jejuni*.

As of September 15, 1999, 921 people who had attended the fair reported diarrhea. Stool cultures from 116 people showed *E. coli* O157:H7, and 13 of these people were also coinfected with *C. jejuni*. Cases of diarrheal illness among fair attendees were reported from 14 New York counties and 4 states.

An environmental investigation of the fairgrounds on September 3 determined that although much of the fair was supplied with chlorinated water, a shallow well was used in one area to supply unchlorinated water to several vendors, who used this water to make beverages and ice. Cultures of the water from the well showed high levels of *E. coli*.

Chemical pollution has also created new hazards. Mercury has caused food poisoning in Japan and Sweden. Oil spills have caused the deaths of birds and fish, and have befouled once lovely beach areas. Insecticides have caused millions of fish to die and could well cause serious harm to people. Fertilizers are helping destroy our streams and lakes by transforming them into open sewers.

Potable water used for drinking and swimming may be unsafe because of physical, chemical, and bacteriologic hazards. Physical deficiencies exist because of inadequate groundwater sources, inadequate design of treatment plants, inadequate disinfection capacity, inadequate system capacity, improper design of swimming pools, and inadequate training of water plant and swimming pool operators. Poor, outmoded, unsafe plumbing can easily lead to contamination of potable water by nonpotable water through crossconnections and submerged inlets.

HUMAN SYSTEMS

Physiology

Physiology is the basic biomedical science dealing with the function of living organisms. To understand function it also is necessary to understand something about structure or anatomy. A major concern of physiology is how environmental factors influence the function of individuals. The intracellular processes in humans proceed properly if the fluid environment surrounding each cell is maintained in a nearly constant state. Temperature, oxygen supply, acidity, and nutrients must be at a nearly constant level. The external environment that the individual is exposed to, however, varies considerably. There are wide ranges of temperature, humidity, ionizing and nonionizing radiation, pressure, and assorted chemicals, as well as microorganisms, in this environment that can affect people. All these factors alone or in combination can elicit a biological response that can be harmful to the individual. To understand the adverse physiological reactions caused by chemical, physical, and biological agents, it is necessary to have some knowledge of the cell and a select group of organ systems.

Cell

The cell is the basic unit of all living organisms. The cell consists of the following components: cell membrane, endoplasmic reticulum (ER), Golgi apparatus, mitochondria, lysosomes, ribosomes, nucleus, and nucleoli (Figure 2.1). The cell membrane is the boundary of the cell between intracellular and extracellular water and

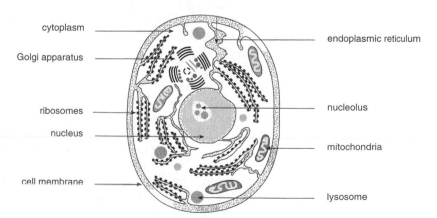

Figure 2.1 The human cell.

maintains cellular integrity by regulating influx and efflux of various substances. The membrane consists of a lipid bilayer, in which proteins are embedded. The endoplasmic reticulum serves as the cell circulatory system and is a site of protein and lipid synthesis. The Golgi apparatus synthesizes carbohydrates and then combines with protein to produce glycoprotein. The mitochondria function in cellular respiration and adenosine triphosphate (ATP) molecules synthesis. The lysosomes are the cell digestive system. The ribosomes synthesize proteins. The nucleus dictates how protein synthesis occurs. Therefore, the nucleus directs the other cell activities, such as active transport, metabolism, growth, and heredity. The nucleoli are essential in the formation of ribosomes. The nucleus stores, transcribes, and transmits genetic information. This information is stored in the deoxyribonucleic acid (DNA) molecules present in the nucleus. The sequence of the DNA base pairs is transcribed into the sequence of base pairs in messenger ribonucleic acid (mRNA) molecules. The mRNA then acts as a carrier of a coded message that is translated at the ribosome into the specific proteins they synthesize.

Metabolism consists of catabolism, a decomposition process in which relatively large food molecules are broken down to yield smaller molecules and energy; and anabolism, in which energy is used to synthesize relatively small molecules into larger molecules. Enzymes, hormones, and antibodies are all produced during the process of anabolism. Enzymes catalyze both catabolic and anabolic chemical reactions. Intracellular enzymes are synthesized within the cell and also function within the cell. Extracellular enzymes are synthesized in cells and function outside of the cells. Energy changes accompany metabolic reactions. Catabolism releases energy as heat or chemical energy. Chemical energy released by catabolism is first converted into, and stored briefly, as ATP molecules at the cellular level. ATP molecules, which are easily broken or hydrolyzed, then release the energy that can be utilized in anabolism.

Blood

The primary function of blood is the transport of various substances to and from the body cells and the exchange of materials, such as oxygen, nutrients, and waste materials, between the respiratory, digestive, and excretory organs. The secondary function of blood contributes to the homeostasis of fluid volume, pH, and temperature; blood is also necessary for cellular metabolism and defense against microorganisms. (Homeostasis is the maintaining of the internal environment of the body in a relatively uniform manner.) The blood consists of the erythrocytes or red blood cells, leukocytes or white blood cells, and thrombocytes or platelets. The red blood cells transport oxygen and carbon dioxide. The white blood cells are important in the immune response. The platelets initiate blood clotting that controls bleeding from wounds or incisions. The plasma is the liquid part of the blood minus its cells.

The immunologic response starts when bacteria or viruses invade the body. Scavengers, including blood cells called neutrophils, are among the first line of defense at the site of the infection. Neutrophils are produced in the bone marrow. They survive for a few days. Next, the complement system, which is a group of at least 20 proteins circulating in the blood, sticks to the microorganism, setting off a chain reaction that eventually will destroy it. Macrophages, which are long-lived scavengers, migrate through the body and engulf foreign matter, such as microorganisms, as well as cellular debris. They signal other cells in the immune system to confront and destroy the microorganisms. A macrophage, after ingesting the microbe, shows specific markers on its surface. These markers or antigens signal other immune system cells, called helper T cells, to come forth. T cells have their own set of receptors that can recognize specific antigens of a given microbe. The T cell grows and divides. The helper T cells start to reproduce when a protein, which is called interleukin-1, is released by a macrophage. The helper T cells then produce a variety of interleukins that activate other T cells and B cells. They also produce gamma interferon, which activates the macrophages.

The B cells that are stimulated by the helper T cells divide and mature into plasma cells. The plasma cells produce antibodies that are directed against the specific antigens. (Antibodies are proteins that recognize and bind to a specific microbe.) This either stops the microbe from going further or makes the microbe more vulnerable to macrophages and neutrophils. Antibodies also activate the complement system. It takes about 1 week for the immune system to be at its highest level of functioning. Killer T cells recognize and destroy virus-infected cells or cancer cells. They use lethal proteins that punch holes in the cell membrane and cause it to rupture. The suppressor T cells probably work by sending out chemical signals that either slow or stop the immune reaction after the organisms have been destroyed. This keeps the body from attacking itself. Memory cells are types of B and T cells that circulate through the body after an infection has occurred. The next time the antigens are released, the memory cells start the process of destroying the antigens. Cancer cells, whether they are created by chemicals, radiological agents, or possibly viruses, are surrounded by killer T cells because they are attracted by the surface antigens. These cells can release chemicals that break down the membrane of the cancer cells and cause the cells to die.

Lymphatic System

The lymphatic system is a specialized part of the circulatory system. It consists of a moving fluid, lymph. The lymph comes from the blood and tissue fluid, and returns to the blood. Lymph is a clear, watery fluid that is found in the lymphatic vessels. The interstitial (intercellular) fluid fills the spaces between the cells. The clear watery fluid, in fact, is complex and organized material. It makes up the internal environment of the body. Both lymph and interstitial fluids closely resemble blood plasma in composition. The main difference is that they contain a lower percentage of proteins than does the plasma. The lymphatic system, like the blood system, can carry contaminants throughout the body. The lymphatics return water and proteins from interstitial fluid to the blood from where they came. Lymph nodes have two major functions. They filter out injurious substances and phagocytose them. Unfortunately, sometimes the number of microorganisms entering the lymph nodes are greater than the phagocytes can destroy, and therefore the node becomes infected. Also, cancer cells that break away from a malignant tumor may enter the lymphatics, travel to the lymph nodes, and set up new sites of cancer growth. The lymphatic tissue of the lymph nodes forms lymphocytes and monocytes, which are the non-granular white blood cells and plasma cells. This process is called hemopoiesis.

Tissue Membranes

Tissue membranes are thin sheets of tissues covering or lining the various parts of the body. Four important membranes are mucous, serous, synovial, and cutaneous (skin). The mucous membranes line the cavities of passageways of the body that open to the exterior. This includes the lining of the gastrointestinal tract, respiratory tract, and genitourinary tract. Mucous membranes protect the underlying tissue, secrete mucus, and absorb water, salts, and other solutes. The serous and synovial membranes line the closed cavities of the body. The pleura is the serous membrane that lines the thoracic cavity. The mucous film of this lining consists of a superficial gel layer that traps inhaled particles in the respiratory system. The ciliated mucous membrane contains goblet cells. The cilia propel the mucus upward in the respiratory tract to the pharynx via mucociliary escalation, which causes a coughing reflex. Another serous membrane is the pericardium, the sac in which the heart lies. The synovial membrane has smooth, moist surfaces that protect against friction. It lines the joints, tendon sheaths, and bursae.

The cutaneous, or dermal, membrane (skin) protects the body against various microorganisms, sunlight, and chemicals. The skin helps to maintain the normal body temperature by regulating the amount of blood flowing through it and by sweat secretion. Skin consists of epidermis, dermis, and accessory organs of the skin, such as the hair, nails, and skin glands. The very top layer of the epidermis is composed of dead cells and is virtually waterproof. Because of the innumerable microscopic nerve endings throughout the skin, the body is kept informed of changes in the environment. Finally, beneath the dermis is the subcutaneous fatty tissue, which provides cushioning.

Nervous System

The nervous system is a major interface between humans and environment. It is involved in maintaining homeostasis. It controls posture and body movements. It is the center of the subjective experience, memory, language, and thought process that are peculiar to human activity. The fundamental unit of the nervous system is the neuron, which consists of cell body, dendrites, and axon, which may be several feet in length. The dendrites conduct impulses to the cell body of the neuron. Receptors that receive the stimuli that initiate conduction are the distal ends of dendrites of sensory neurons. (Distal is the farthest point away from the point of origin.) The neuron axon conducts impulses away from the cell body.

The connections between the nerve cells are called synapses. They play a key role in the transmission of impulses in the nervous system. A single cell may be connected to as many as 15,000 other cells by means of synapses. The transmission of impulses occur from the depolarization of the cell membrane brought about by chemical or mechanical events. The nerve impulse or action potential is a self-propagating impulse of electrical negativity that travels along the surface of the

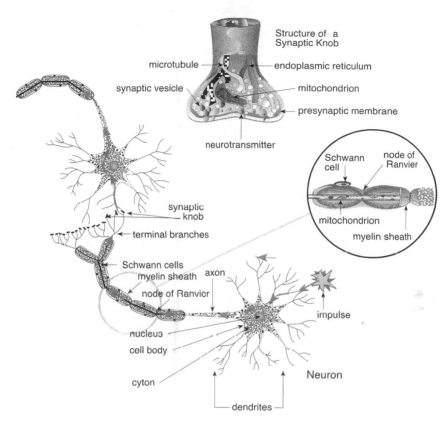

Figure 2.2 Neuron.

neuron cytoplasmic membrane. A stimulus to the neuron greatly increases the membrane permeability to sodium ions. Sodium ions rush into the cell at the point of stimulation. The membrane becomes depolarized at the point of stimulus. This occurs for an instant. As more sodium ions come into the cell, they produce an excess of positive ions inside the cell and leave an excess of negative ions outside the cell. The negatively charged point of the membrane sets up a local current, with the positive point adjacent to it. The local current acts as the stimulus. Within a fraction of a second, the adjacent point on the membrane becomes depolarized and its potential reverses from positive to negative. The cycle keeps repeating itself (polarization then depolarization) until the electrical current, in a wave motion, moves the full length of the neuron.

Many nerve impulses travel a route called the reflex arc consisting of two or more neurons arranged in a series that conduct impulses from the periphery or from the outside portion to the central nervous system (CNS, Figure 2.2). The receptors are stimulated, and the nerve impulse flows into the spinal cord and brain and back out to the effectors, which aid in the carrying out of the intended action. The synapse is the place where the nerve impulses are transmitted from one neuron to another neuron. For instance, if an environmental stimulus activates the receptor neuron, it then goes along the afferent nervous pathway to carry the action potentials to the central nervous system. Within the spinal cord or brain, there is an integrating center where the action potentials are sent out along different pathways from the CNS, the effector is activated, and the body responds.

The synapse is the place where the nerve impulses are transmitted from one neuron to another neuron. A synapse consists of a synaptic knob, a synaptic cleft, and the cytoplasmic membrane of the neuron dendrite or cell body. Because an action potential cannot cross the synaptic clefts, a chemical mechanism operates, where the chemical, which is called a neurotransmitter, is released from the synaptic knobs into the synaptic cleft. The neurotransmitter molecules diffuse rapidly across the microscopic width of the synaptic cleft and bind to specific protein molecules called neurotransmitter receptors. This leads to an opening of channels in the membrane through which sodium ions diffuse into and potassium ions diffuse out of the interior of the postsynaptic neuron. The message now moves forward.

Several different compounds serve as neurotransmitters. Acetylcholine is released at neuromuscular junctions with skeletal muscle cells. Acetylcholine is also released at neuromuscular junctions with smooth muscle cells, at cardiac muscle cells, and in neuroglandular junctions. Norepinephrine is released at other neuroglandular junctions and at some neuromuscular junctions with smooth muscle and cardiac muscle cells. The enzyme acetylcholinesterase rapidly halts the stimulation of muscle cells by acetylcholine. The enzyme hydrolyzes acetylcholine to acetate and choline.

Enzyme acetylcholinesterase (AChE)

↓

AChE bonds with acetylcholine

↓

Acetylcholine is hydrolyzed

Acetate + choline

↓

Acetylcholine deactivated

Chemical transmission of signal to muscle discontinues

Acetate + choline recycle back to originating neuron
and are stored as acetylcholine

The nervous system is composed of the CNS and the peripheral nervous system (Figure 2.3). The CNS consists of the brain and spinal cord. The cranial nerves, spinal nerves, and ganglia make up the peripheral nervous system. The nervous system is also classified by the effectors innervated, that is, the somatic nervous system and the autonomic nervous system. The somatic nervous system consists of the brain, spinal cord, cranial nerves, and spinal nerves. This system innervates the skeletal muscles. The autonomic nervous system is composed of the autonomic or

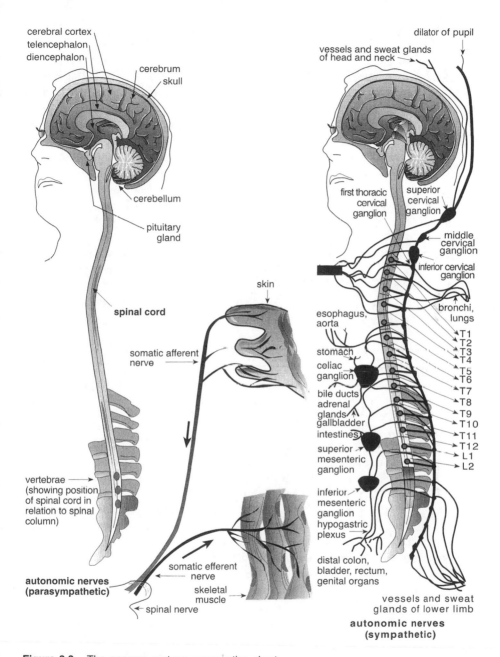

Figure 2.3 The nervous system: comparative chart.

visceral motorneurons, which innervate the cardiac muscle, smooth muscle, and glandular epithelial tissues.

The autonomic system is divided into two parts by function. The sympathetic part is an emergency system that greatly increases sympathetic impulses to most visceral effectors during stress and prepares the body for the use of maximum energy.

It increases the secretion of epinephrine by the adrenal medulla, which in turn increases and prolongs the effects of norepinephrine. The parasympathetic nervous system regulates many visceral effects under normal conditions. Its neurotransmitter, acetylcholine, stimulates the digestive juices, insulin secretion, and contraction of the smooth muscle of the digestive tract. The autonomic system does not function autonomously or independently of the CNS.

The CNS is intimately related to the endocrine system. The two systems perform the same general function for the body of communication, integration, and control. Whereas the nervous system sends nerve impulses conducted by neurons from one specific structure to another, the endocrine system sends tiny quantities of chemical messengers, known as hormones. Nerve impulses produce rapid, short-lasting responses, whereas hormones produce slower and longer lasting responses. The cerebral cortex sends impulses into the hypothalamus, which controls the production and secretion of six separate hormones of the interior pituitary. The hypothalamus secretes the antidiuretic hormone (ADH), which regulates the reabsorption of water into the kidney. The hypothalamus also controls the autonomic nervous system, which in turn controls the secretion of epinephrine.

Respiratory System

The function of the respiratory system is to distribute air and to act as a means for exchange of gases, so that oxygen may be supplied to body cells and carbon dioxide may be removed from them. The circulatory system is jointly responsible for meeting the respiratory needs of the body, because it provides a mechanism for oxygen transport to all the body cells. The nose, pharynx, larynx, trachea, bronchi, and bronchioles act as a conduit for air into the alveoli, where the actual gas exchange takes place with the blood capillaries. The nose also filters out impurities and warms, moistens, and chemically examines the air for substances that might become irritating to the mucous lining of the respiratory tract. This organ is also an organ of smell, and acts as an aid in phonation (the production of vocal sounds, especially speech). The pharynx is a corridor for the respiratory and digestive tracts. Both air and food pass through the structure before reaching the appropriate tubes. It is also important in phonation. The larynx or voice box is another part or the corridor to the lungs. It protects the airway from solids or liquids during swallowing. It is also the organ of voice production. The trachea is a continuation of the corridor through which air passes to the lungs. The bronchus is a continuation of the trachea and aids in the distribution of air to the interior of the lungs. The alveoli are surrounded by networks of capillaries. This is where the main function of the lungs occurs, that is, gas exchange between the air and the blood. The lungs contain the bronchus, bronchioles, alveoli, pleura (sac around the lungs), and alveolar sacs (Figure 2.4).

The conducting portion of the respiratory system provides little resistance to the movement of gases to the alveolar surface. Further, this area conditions the air and protects the lungs from the largest of infectious or potentially toxic particles in the air. The convoluted, moist, and richly vascular mucosa of the nose protects nose breathers from inhaling particles larger than 5 to 10 μm in diameter. Soluble gases

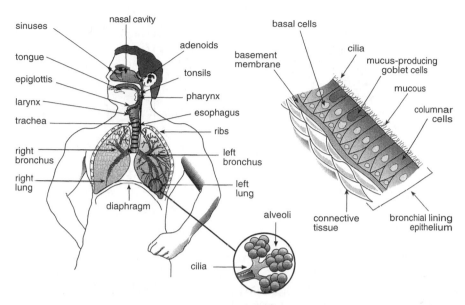

Figure 2.4 Conductive airway of the respiratory system, with details of the ciliated epithelial lining. (Adapted from *Environmental Toxicology and Risk Assessment: An Introduction, Student Manual,* U.S. Environmental Protection Agency, Region V, Chicago, IL, Visual 2.6.)

may be removed by absorption. The air is warmed and moistened, or cooled when the conditions are hot and dry. The tracheal–bronchial system is lined with cilia that constantly force mucus toward the larynx. The mucus carries with it the microorganisms and particles that have been trapped in it, as well as the macrophages that move out of the alveoli with material that has been scavenged from the alveolar surface. When the mucus reaches the pharynx, it is usually swallowed or may be spit out. These pulmonary mechanisms help prevent problems in the respiratory system due to the inhalation of dusts, fumes, and other materials that can cause disease or injury. The coughing mechanism helps in the removal of these materials.

Pulmonary ventilation is brought about by changes in the size of the thorax (chest). The contraction of the diaphragm and chest-elevating muscles leads to an expansion of lungs that decreases the alveolar pressure and creates an inspiration. Expiration is a passive process that begins when pressures change. This occurs when the muscles are relaxed, which leads to a decrease in size of the thorax, a decrease in the size of the lungs, and a pressure gradient in the alveoli that is greater than the atmosphere.

The exchange of gases in the lungs takes place between the alveolar air and venous blood that flows to the lung capillaries. The alveolar–capillary membranes allow gases to move back and forth. Oxygen enters the blood because the partial pressure of oxygen in the alveolar air is greater than in the venous blood. At the same time, the partial pressure of carbon dioxide in the venous blood is much higher than in the alveolar blood. This causes carbon dioxide to flow from the venous blood to the alveolar blood. Once the oxygen enters the blood, a small percentage is transported as a solute in the blood, whereas much more is transported as oxyhemoglobin in the red blood cells. The carbon dioxide is transported in the blood as a true solute in small quantities. A large amount of the carbon dioxide is transported as bicarbonate ions in the plasma and a moderate amount of the carbon dioxide is transported in the red blood cells as carbaminohemoglobin.

Carbon dioxide is the major regulator of respiration. An increase in blood carbon dioxide to a certain level may stimulate respiration, and a decrease may cause a decreased level of respiration. If the oxygen level of blood is low, this also may stimulate respiration to a certain point.

Gastrointestinal System

The digestive tract is largely a tube that is open at both ends. In a real sense, the contents of the gastrointestinal tract is exterior to the body (Figure 2.5). To get into the tissues or cells, the ingested materials have to go through an extremely acidic environment in the stomach, and have to be broken apart by enzymes. Some of the large molecules, such as cellulose, are unaltered and therefore are excreted in the feces. Almost all the digestion and absorption of food and water take place in the small intestine. The three basic food groups — the carbohydrates, proteins, and fats — are broken down in different ways. The carbohydrates, most of which are ingested in the form of starch, are split into disaccharides by the enzyme amylase, found in the saliva and in the pancreas. The disaccharides are then split into monosaccharides by enzymes in the small intestinal mucosa. The sugar molecules are then

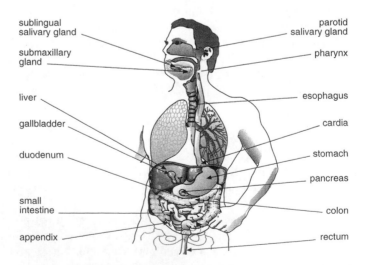

Figure 2.5   The gastrointestinal tract. (From *Environmental Toxicology and Risk Assessment: An Introduction, Student Manual,* U.S. Environmental Protection Agency, Region V, Chicago, IL.)

transported into the blood. The proteins are broken down further into peptides and into free amino acids, which are transported across the intestinal cells. Most of the fat digestion occurs in the small intestine from the combined actions of pancreatic lipase and bile salts secreted by the liver. The bile salts act chiefly as emulsifying agents. Fatty acids are resynthesized to triglycerides in the intestinal cells. They are then secreted into the lymphatics as small lipid droplets.

The function of the digestive system is to accept raw materials in the form of different foods, minerals, vitamins, and liquids, and to prepare them for absorption into the capillaries or lymphatics for distribution through the body. The main organs of the digestive system are the mouth, pharynx, esophagus, stomach, and intestine. The accessory organs that are open to the main organs are the salivary glands, liver, gallbladder, and pancreas.

The stomach serves as a food reservoir that expands and contracts depending on the amount of food present. The glands secrete most of the gastric juice, which is fluid made up of mucus, enzymes, and hydrochloric acid. The epithelial cells that form the surface of the gastric mucosa secrete mucus. The chief cells (zymogenic cells) secrete the enzymes of gastric juice. The parietal cells secrete hydrochloric acid and are also thought to produce a protein known as intrinsic factor. When the muscular coat of the stomach contracts, it churns up food and breaks it into small particles, which allow for mixing with the gastric juice. In time the muscular coat moves the gastric content into the duodenum. The stomach, in a limited quantity, absorbs water, alcohol, and certain drugs. It also produces the hormone gastrin in cells that are in the pyloric region.

The small intestine, which is about 6 meters in length, has three major divisions: the duodenum, the jejunum, and the ileum. The function of the small intestine is to complete the digestion of food, to absorb the end products of digestion into the blood and lymph, and to secrete hormones that help control the secretion of pancreatic juice, bile, and intestinal juice. The intestinal juice contains mucus from the intestine, digestive enzymes from the pancreas, and bile from the liver.

The lower part of the alimentary canal is called the large intestine. It is divided into the cecum, colon, and rectum. The main functions of the large intestine are the absorption of water, secretion of mucus, and elimination of the wastes of digestion. The peritoneum is a large, continuous sheet of serous membrane that lines the walls of the entire abdominal cavity and forms the serous outer coat of the organs.

The liver, which is the largest gland in the body, is one of the most vital organs. Liver cells detoxify a variety of substances. These cells secrete about a pint of bile a day. The liver carries on a number of important steps in the metabolism of proteins, fats, and carbohydrates. Liver cells store iron, vitamins A, B_{12}, and D. Poisonous substances that enter the blood from the intestine are circulated to the liver, where a series of chemical reactions occur that turn the substances into nontoxic compounds. The bile secreted by the liver is made up of bile salts, bile pigments, and cholesterol. The bile salts that are formed in the liver from the cholesterol are the most essential part of the bile that aids in the absorption of fats.

The gallbladder is a storage organ for the bile that is produced by the liver. When digestion is going on in the stomach and intestines, the gallbladder contracts, ejecting the concentrated bile into the duodenum. The pancreas is composed of two different types of glandular tissue, one exocrine and one endocrine. Embedded between the exocrine units of the pancreas are clusters of exocrine cells called islets of Langerhans or pancreatic islets. The pancreas secretes the digestive enzymes found in pancreatic juice. Its beta cells secrete insulin, and its alpha cells secrete glucagon.

The last step in the digestive process is defecation. The stomach is emptied in about 2 to 6 hours after a meal, depending on what is eaten. Chyme, which is a mixture of gastric juices and food that forms a milky white material, is ejected about every 20 sec into the duodenum. Defecation is brought about by a reflex that is caused by stimulation of the receptors in the rectal mucosa. Constipation occurs when the material in the lower colon in the rectum moves at a rate that is slower than normal. This causes extra water to be absorbed from the fecal mass and thereby hardens the stool. Diarrhea occurs when chyme moves too quickly through the small intestine, and therefore not enough water and electrolytes are absorbed. The loss of water, leading to dehydration and loss of electrolytes, makes diarrhea in infants a serious problem.

Urinary System

The urinary system consists of the kidneys, ureters, bladder, and urethra (Figure 2.6). The functions of the kidneys are to excrete urine, which eliminates various toxins and metabolic wastes. They also help maintain fluid, electrolytes, acid–base balance, and the proper level of blood pressure in the body. The ureters collect urine and drain it into the bladder. The bladder is a reservoir for storing urine and also acts as the organ that expels the urine through the urethra, which is the passageway for expulsion from the body.

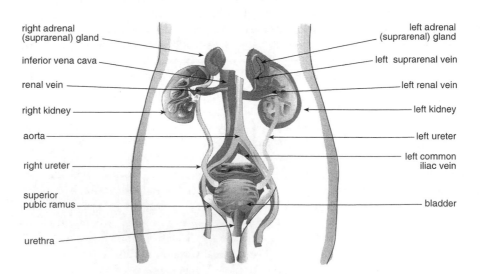

Figure 2.6   The urinary system.

TOXICOLOGICAL PRINCIPLES

Toxicology is the study of adverse human biochemical, morphological, and physiological effects resulting from exposure to harmful agents, such as chemicals and types of radiation, that are present in many environmental settings. The fate of these agents is dependent on the toxicokinetics or movement of toxic elements and molecules into, through, and out of the human body. The impact or toxicodynamics is dependent on factors related to the toxic agent, the exposure, and the individual exposed.

Toxicokinetics

Toxicokinetics refers to the fate or, more specifically, the *absorption, distribution, biotransformation*, and *elimination* of toxic agents or metabolites (Figure 2.7). Toxic agents enter and absorb into the body via three major modes and routes of entry: (1) inhalation into the respiratory system; (2) ingestion into the gastrointestinal system; and (3) contact with the dermal (skin) system. The toxic agent may deposit and interact at the site of initial contact causing local effects or may absorb across respiratory, gastrointestinal, and dermal cell membranes; and then the agent may enter the circulatory system, eventually causing systemic effects.

The lipid bilayer of cells that comprise the tissues of organ systems regulates the absorption of toxic agents into the body. Small water soluble or polar molecules, and both small and larger lipid soluble or nonpolar molecules most readily cross cell membranes. Indeed, based on the premise of solvent chemistry that "like dissolves like," nonpolar lipophilic agents can most readily cross the lipid bilayer that constitutes membranes. As a result, some of the most hazardous toxic chemicals are those considered fat soluble or lipophilic.

Inhalation brings chemicals into contact with the lungs. Most of the inhaled chemicals are gases or vapors generated from volatile liquids. Absorption in the lungs can be high, both because of the surface area and of the blood vessels in close proximity to the exposed surface of the respiratory system. Gases and vapors cross the lung by means of simple diffusion. The rate of absorption depends on the solubility of the toxic agent in the blood. Chemicals may also be inhaled in solid or liquid form as dusts or mists. Mists, if lipid soluble, may cross the cell membranes by passive diffusion. The absorption of solid particulates depends on the size and chemical nature of the particles. The rate of absorption of the particles from the alveoli is determined by the solubility of the chemical in lung fluids. Certain small insoluble particles may remain in the alveoli indefinitely. Particles of 2 to 5 μm are deposited into the tracheobronchiolar regions of the lungs. They are typically cleared by coughing and sneezing, or are swallowed and deposited in the gastrointestinal tract. Particles of 5 μm or larger are usually deposited in the nasopharyngeal region, where they are either expectorated or swallowed.

Ingestion brings the chemicals into contact with the tissues of the gastrointestinal system. The normal function is absorption of foods and fluids that are ingested. The gastrointestinal system is affected by absorbing toxic chemicals that are contained

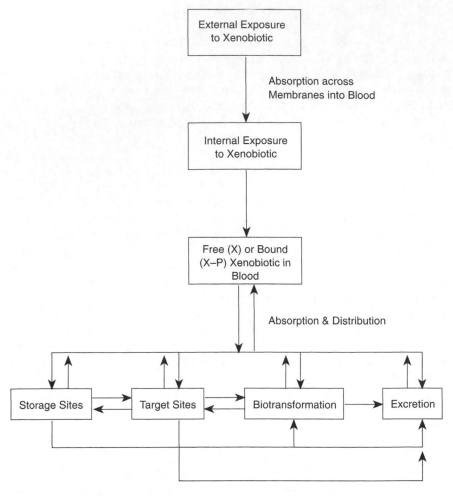

Figure 2.7   Possible fate of a chemical agent following external exposure.

in food or water. The degree of absorption usually depends on whether the chemical is easily soluble in water or easily soluble in organic solvents or fats. The lipophilic compounds, such as organic solvents, are usually well absorbed, because the chemicals can easily diffuse across the membranes of the cells that line the gastrointestinal tract. The hydrophilic compounds, such as metal ions, cannot cross the cell lining as easily as the lipophilic agents and have to be transported by systems in the cells. Many chemicals may bond to ingested food and, therefore, are not absorbed as efficiently as when ingested in water. Further, some chemicals may be altered by digestive enzymes or intestinal bacteria, and yield different chemicals with altered toxicological properties.

Absorption of toxic chemicals through the epidermal layer of the skin is hindered by the epidermal cells. These cells are densely packed layers of horny keratinized material. Absorption of the chemicals through the skin occurs more readily if the

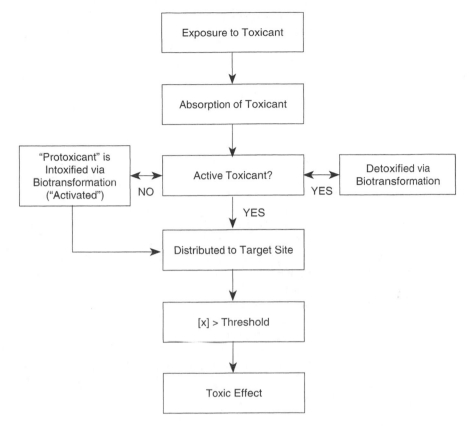

Figure 2.8 Protoxicant vs. toxicant.

skin is scratched or broken. In addition, the absorption of the chemicals by the skin
is increased if the agent is lipid soluble. A more rapid exposure to the body occurs
when the chemicals are accidentally or purposefully injected through the skin into
the body.

Once an agent absorbs into the blood, it may interact with blood cells or plasma
proteins. In addition or as an alternative, the agent can be distributed via the circulatory system and absorbed into other organ systems such as the liver, kidneys, and
brain. Some agents, such as lead and fluorides, may bond to and be stored temporarily
in bone tissue. Others, such as fat-soluble organic solvents, may distribute to and
be stored in adipose or fat tissue.

The originally absorbed agent, or parent compound, can be biotransformed into
a more toxic or less toxic metabolite prior to, during, or following distribution. A
parent compound that must be biotransformed or bioactivated into a toxic or active
metabolite can be called a protoxicant (Figure 2.8). In addition, nonpolar molecules
that absorb into the body most readily can be biotransformed into more polar
molecules that are excreted from the body more readily. The liver is the major site
of enzymes responsible for catalyzing biotransformation reactions, but numerous
other organ systems also are important sites for biotransformation reactions.

Toxic parent compounds or toxic metabolites are eliminated via biotransformation reactions that result in detoxification. In addition, toxic substances are eliminated via excretion, mainly in urine and feces. Toxic agents and metabolites also can be eliminated from the human body via mammary milk, perspiration, exhaled breath, hair, and nails.

Toxicodynamics

Toxicodynamics refers to the impact that a toxic agent may have on the human body. Several factors influence the potential interaction and related adverse effects that may be manifested following exposure. The dose–response relationship of a chemical and effect is the most fundamental concept in toxicology. The dose–response curve describes the relationship that exists between the degree of exposure to a certain dose and the magnitude of the effect, which is the response, by an exposed organism. No response is seen if the chemical is not present. As the amount of the chemical exposure increases, the response becomes more apparent and therefore increases. Many chemicals produce responses that show a threshold value. That is an exposure below which no response can be detected. The no observed adverse effect level (NOAEL) is the dose at which no adverse effects are observed. The lowest value where a significant adverse effect is first seen is the lowest observed adverse effect level (LOAEL). NOAEL and LOAEL values depend on the effect (end point) to be measured.

Some chemicals produce adverse effects that show a dose–response curve with no threshold. The reason for this is that the cells that are affected have little or no defense against the chemical and have little or no ability to repair or compensate for the damage that is done. It appears that no threshold exists for the effects of lead on the nervous system in infants and children. Chemicals that are carcinogenic are also considered to be part of the group that do not have thresholds. Therefore, any chemicals in these groups are considered to create a degree of risk at any exposure.

Dose–response curves also help show the toxic properties of the chemical and may be useful in comparing the toxicity of several chemicals. The midpoint of the dose–response curve is called the effective dose-fifty (ED_{50}). This is the dose that produces an effect in 50% of the test population. When a toxic effect is measured, the term used is TD_{50}. When a lethal effect is being measured the term used is LD_{50}.

The slope of a dose–response curve is an important variable in assessing the toxicity of the chemical. The steeper the dose–response curve, the more cautious the individual must be with exposure to a given chemical, because a small difference in the dose may produce a serious effect. To assess the toxicity of a chemical, it is important to determine the route of exposure, the length of exposure, the species and individual characteristics of the exposed organism, and the nature or end point of the toxic effect being measured. For a chemical to show a toxic effect, it must first gain access to the cells and tissues of the organism. In humans, the major routes of exposure or entry into the body are through ingestion, inhalation, and dermal adsorption.

In toxicity assessments using animals in comparison with human beings, it is important to understand that there is a difference in the toxic effect of a chemical in different species. The difference may relate to the absorption of the chemical, biotransformation or metabolism of the chemical, or the differences in anatomic function. The absorption of chemical across the skin, lungs, or gastrointestinal tract is determined by the properties of the cells at the surfaces of these tissues. Biotransformation of the chemical in the liver and kidney or other tissues is different in different organisms. The rate of metabolism of chemicals differs, as well as the kind of metabolic by-products.

The toxicity of many chemicals depends on the time of exposure and frequency of exposure. This time period is significant because some chemicals accumulate in the body over long periods of time and are not excreted. Toxic effects may depend on the duration of the exposure as it relates to the ability of these cells to repair themselves. Some adverse effects, such as some cancers and lung diseases, require an extended period of time to develop.

Individual characteristics within a given group, such as humans, may vary and cause different types of responses to chemicals. Some of these differences may be related to gender, race, or age. Nutritional status and dietary factors also may contribute to the differences in people and their reaction to different types of chemicals.

Adverse or toxic effects can involve a variety of possible target sites and end points such as the liver (hepatotoxicity); kidneys (nephrotoxicity); lung (pneumotoxicity); blood (hematotoxicity); nervous system (neurotoxicity); structural or functional abnormality usually to developing embryos or fetuses (teratogenicity); and growth of malignant tumors (carcinogenicity). Related observations include neurotoxicity, where damage occurs to reflexes, coordination, intelligence, memory, nerve impulses, and other behavioral types or concerns; carcinogenicity, where cancerous tumors are produced; hematological toxicity, where changes take place in hemoglobin levels, erythrocytes, leukocytes, platelets, and plasma components; hepatoxicity, where malfunctions occur in the liver, lipid metabolism, protein metabolism, carbohydrate metabolism, and metabolism of foreign compounds; inhalation toxicity, where a gross or a microscopic breakdown takes place in the anatomy or function of the lungs; mutagenicity, where alterations occur in chromosomes or DNA damage; renal toxicity, where damage happens in the kidneys; reproductive toxicity, where fertility levels are lower and more stillborns occur; and teratogenicity, where functional or physical defects occur in neonates.

All the major organ systems continue to be studied relative to their vulnerability following human exposure to various doses or concentrations of exogenous chemical and radiological agents. Among these organ systems is the endocrine system. Various organs are part of the endocrine system, including the pituitary gland, thyroid gland, pancreas, adrenal gland, and ovaries and testes. The endocrine system is associated with production and glandular secretion of hormones essential for maintaining normal developmental, growth, and physiological processes. Among these processes are metabolism, immune response, fertility, and fetal development. The endogenous hormones serve as chemical messengers that travel through the bloodstream in the human body and depart at specific cellular destinations. The hormones relay a message and initiate a response when they bind to specific receptors on the cells. Relaying of the hormonal messages and initiation of cellular responses are regulated by a feedback loop that returns messages back to the originating gland that secreted the hormone. Some examples of hormones include insulin, estrogen, testosterone, and adrenaline among others.

It has been known for decades that the endocrine system can be adversely affected in response to excessive exposure to some toxicants. However, attention has increased especially in the past decade concerning chemicals that are classified as endocrine disrupters. Research has suggested that various levels of organisms, including wildlife and humans, may be adversely affected by human-made or anthropogenic chemicals that either mimic or block hormones. It is believed by some that chemical hormone mimics actually bind to cellular receptors normally reserved for hormones and, in turn, initiate a normal response usually due to a specific hormone, such as estrogen. Alternatively, chemical hormone blockers competitively bind to cellular receptors preventing the hormones from attaching and blocking transfer of the hormonal message and subsequent initiation of normal response. A third descriptive category is the chemical hormone triggers. These chemicals bind to a cellular receptor and initiate an abnormal response not associated with the hormone that should have bonded to the receptor.

Much focus of the adverse effects studied to date includes decreased fertility and other reproductive effects observed in wildlife, including some mammals. For example, some specific observations have been decreased sperm count, abnormal development of genitals, and decreased population growth. Some of the chemicals cited as potential endocrine disrupters include chlorinated organics such as polychlorinated biphenyls (PCBs) and dichlorodiphenyltrichloroethane (DDT) metabolites. This area of toxicology is still under careful scientific scrutiny, with continued and expanded research warranted.

Carcinogenesis

Cancer is a general term for a group of related diseases that cause uncontrolled growth of certain types of cells and their related tissues. Chemical carcinogenesis generally means the induction by chemicals of malignant neoplasms or tumors not usually observed. The response to the chemical carcinogen varies with the species, strain, and sex of the experimental animal. Chemical carcinogens interact with other environmental agents that sometimes enhance and other times decrease their effect. Chemical carcinogens have a persistent and delayed biological effect, may be more effective in divided doses than in an individual large dose, and may have distinct interactions with host genetic elements. Carcinogens may either be direct initiators of the genetic change or act only as a promoter in combination with another initiator (Figure 2.9). At times, one carcinogen may be required to cause a second carcinogen to promote a response. They would then be called cocarcinogens. One of the major problems in determining the possibilities of carcinogenic activity are the long periods of latency, which may extend for many years. Typically, time periods for the ultimate production of the cancer may take from 15 to 40 years after the onset of the exposure and prior to disease manifestation.

The site of the development of a malignant tumor may vary considerably. It is based either on the portal of entry or on the site where biotransformation takes place. Asbestos causes cancer in the lung. Azo dyes produce tumors or cancer in the liver,

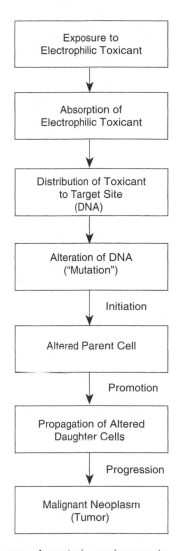

Figure 2.9 Schematic summary of genotoxic carcinogenesis.

which is the site of biotransformation. Radium produces cancer in the bone, where the radium is stored. The aromatic amines produce cancer in the bladder, the site at which excretion takes place.

The testing, assessing, and regulation of carcinogens is carried out by a variety of agencies, based on a large number of health, safety, and environmental laws that have been passed in the last 25 years. These agencies have issued guidelines and policies on how they intend to identify, evaluate, and regulate carcinogens. The Office of Technology Assessment (OTA) has been given the task of gathering together all the information and techniques used by these agencies in making their decisions. Many important issues are considered in assessing potentially carcinogenic chemicals. They are based on the interpretation of test data and the use of assumptions. These assumptions are derived from theories about how cancer is caused and decisions about what is considered appropriate public policy. The OTA has identified four important kinds of assumptions: (1) those used when the data are not available for a particular case; (2) those that are potentially testable but have not yet been tested; (3) assumptions that probably cannot be tested because certain experimental limitations exist; (4) those that cannot be tested because of ethical considerations.

Four kinds of evidence may be used for qualitatively identifying carcinogens. They are epidemiological studies, long-term animal bioassays, short-term tests, and structure–activity relationships.

Epidemiological studies are used to collect information about human exposures and diseases. Reports of individuals or clusters of cases help to generate hypotheses for later study. Many of the chemicals now known to be human carcinogens were first identified by physicians. Larger epidemiological studies were then devised and conducted. Descriptive epidemiological studies correlate risk factors, exposures, and diseases that are causes of death in specific groups. They are useful in establishing hypotheses for further study in providing clues about the hazards. Analytic epidemiological studies are used in comparing populations between a group that is exposed to the agent and a group that is not exposed. In case-control studies, the comparison is made between people with a given disease and those who do not have the disease.

Long-term animal bioassays are laboratory studies in which animals are exposed to suspected carcinogens for long periods of time (about 2 years). The animals are then examined for tumors. The tissues of animals of those who survive and those who die are compared. Short-term tests are used to examine genetic change in laboratory cultures of cells, in humans, or in other animals. These tests may take days or weeks to conduct.

Structure–activity relationships (SARs) are used to determine the chemical structures of substances and carcinogenicity. This is a comparison that helps to predict whether chemicals of a class closely related to carcinogens may also be carcinogenic.

The political considerations related to carcinogenesis are considerable. In regulatory proceedings, industry, labor, environmental groups, public interest organizations, and government may voice opinions that are frequently and substantially different. These groups place different values on the harm caused by the unnecessary regulation of a chemical that may later be determined safe. Even when accepting the value of animal data, many discussions take place on whether a particular animal study is reliable and whether the data apply appropriately to people.

The Food and Drug Administration (FDA) was the first agency to establish guidelines for toxicity. The FDA, under the 1958 Food Additives Amendment to the Food, Drug and Cosmetic Act, which included the Delaney clause, prohibited the intentional use of food and color additives that were determined to be carcinogenic in animals or humans. The Delaney clause does not apply to all food ingredients, because some were federally sanctioned prior to the 1958 amendment and some were considered to be generally safe. In the 1970s, the FDA began using quantitative assessments for certain environmental contaminants found in food. By the 1980s, the FDA began applying these techniques to food and color additives. Much discussion has centered around what is an approved analytic technique to determine if the food and color additives are contaminated with small amounts of carcinogenic impurities, or when the additive itself has been determined to be carcinogenic. In 1985, it was decided that an approved analytical technique could be defined as one that could detect residue contents as low as the level associated with the upper portion of human risk estimate of one cancer for every 1 million persons exposed. This technique requires a risk assessment to estimate what residue levels correspond to this risk level.

The Environmental Protection Agency (EPA) began to develop carcinogenic assessment guidelines during the regulatory proceedings on the suspension and cancellation of several pesticides. The attorneys for the EPA summarized the expert testimony and developed summaries that were called *cancer principles*. Considerable criticism of these cancer principles evolved. In 1977, the Environmental Defense Fund petitioned the EPA to establish a policy on classifying and regulating air pollutants that were carcinogenic. In response to a court order to assess the hazards and risks of a large group of substances related to the Clean Water Act (CWA), the EPA set a methodology for assessing human risk. In 1984, the EPA published a proposed revision of its carcinogen assessment guidelines, and also published proposed guidelines for assessing exposure to agents for mutagenicity and developmental toxicants. It also published proposed guidelines for risk assessments of chemical mixtures. The final version of these guidelines was published in 1986.

The carcinogen assessment guidelines published in the *Federal Register* described the general framework to be used in assessing carcinogenic risks and some of the principles to be used in evaluating the quality of the data and in making judgments concerning the nature and magnitude of the risk of cancer from suspected carcinogens. The various steps of risk assessment include hazard identification, dose–response assessment, exposure assessment, and risk characterization. The policy also presents a *weight-of-the-evidence* classification system. The EPA lists five groups:

1. Group A includes human carcinogens. This group is only used when sufficient evidence from epidemiological studies exist to support an association between exposure from agents and cancer.
2. Group B is the probable human carcinogens. In this group the weight of the evidence of human carcinogenicity as based on epidemiological studies is limited. However, the weight of evidence based on animal studies is sufficient.
3. Group C is the possible human carcinogens. In this group, the evidence of carcinogenicity in animals is limited and human data are lacking.
4. Group D is not classified as to human carcinogenicity. Human and animal evidence of carcinogenicity is inadequate.
5. In Group E, evidence of noncarcinogenicity for humans exists.

The Consumer Product Safety Commission (CPSC) published carcinogen guidelines in 1978. The guidelines were challenged in court and were thrown out. Subsequently, CPSC decided to use the guidelines adopted by the Interagency Regulator Liaison Group (IRLG).

The Occupational Safety and Health Administration (OSHA) and numerous other agencies have written regulations on carcinogens. A lot of arguments by various scientific groups, and numerous legal challenges of a variety of carcinogen guidelines have been made.

In 1993, the agency for Toxic Substances and Disease Registry developed a cancer policy framework based on an assessment of the current practice across programs within the agency. Most of the policies established by the various federal agencies declare that well-conducted positive epidemiological studies provide conclusive evidence for carcinogenicity. Some of the factors used in evaluating epidemiological studies include strength of association, level of statistical significance, information on dose–response relationship, biological plausibility, temporal relationships, accuracy of exposure and cause-of-death classification, adequacy of follow-up, and determination of the amount of time needed for latent effects to show up. A nonpositive study cannot be used to indicate an absence of a carcinogenic hazard. It can be used to establish upper limits of risk. Nonpositive studies were undertaken for arsenic, benzene, coke oven emissions, petroleum refinery emissions, and vinyl chloride. Subsequently, these chemicals were proved to be carcinogenic. Long-term animal bioassay data have been accepted as predictors for human beings. Substances shown to be carcinogenic in animals are presumed to present carcinogenic risk to humans. The reason for this conclusion is twofold. One, a number of chemicals that were first identified as animal carcinogens were subsequently confirmed as human carcinogens. Two, all chemicals accepted as human carcinogens, if properly studied in animals, have been shown to be carcinogenic in at least one species of animals. Although this cannot establish that all animal carcinogens also cause cancer in humans, in the absence of data on humans, it is biologically plausible. However, one must consider the amount of substance that is necessary to cause the potential carcinogenic effect in people and then determine if the chemical should be banned from use. Further, the route of administration in animal studies must be taken into account when determining the potential health effects in people. A big question is what do you do when you have conflicting animal data or conflicting animal and human data?

Quantitative risk assessment relates to dose–response determination, exposure estimation, and risk characterization (Figure 2.10).

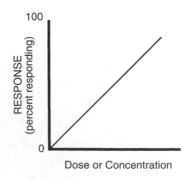

Figure 2.10   Dose–response curve.

Quantitative estimation is a serious problem because often a series of untestable assumptions must be made from an extrapolation of animal to human cases. The exposure levels are deliberately set at high quantities to maximize the probability of detecting a carcinogenic effect. Extrapolating from a high concentration to a low concentration, even when using human data, can be difficult. Therefore, it can be seen how difficult it must be to reach appropriate conclusions when using animal data. The relationship between the dose or exposure of an agent and the biological response of the human is one of the most fundamental in the fields of toxicology and epidemiology. The dose–response curve may have several different shapes, ranging from a straight line to different curves. The data from animals usually represent dose levels substantially higher than the range of human exposure. An important question is whether threshold limits exist for exposure to carcinogens. These limits would be set at a level where no possible effects result from the carcinogenic agent.

The development of cancer takes place in stages. These are typically called initiation, promotion, and progression. Initiation involves an alteration or mutation of the cell genetic material, or DNA. This mutation of DNA can remain latent (i.e., without apparent manifestation of disease) for years. Promotion involves the expression of mutated DNA and the transformation of latent initiated cells into tumors. Progression consists of the growth of the tumors and the development of metastasis, which is the transfer of cancerous cells to other tissues. Some chemicals are primarily initiators. Other chemicals act only as promoters. Some chemicals are both initiators and promoters and are called complete carcinogens. It is possible that the mechanism of promotion involves an alteration in body chemistry, cellular growth, and repair, and other processes.

Mutagenesis

Mutagenesis is brought about by a change in DNA transmitted during cell division (Figure 2.11). Electrophilic compounds or elements have an affinity for functional groups on endogenous molecules that are electron rich. Both DNA and various protein molecules contain electron-rich, or nucleophilic, function groups. Accordingly, toxic electrophiles may readily bind to DNA and protein, causing an alteration with the endogenous molecules. When an electrophile binds to nucleophilic DNA, an adduct is formed. If uncorrected, mutagenesis is initiated. Most carcinogenic events are initiated via a mutation. This change can cause deleterious effects to the individuals exposed or to their offspring. If the change occurs in either the sperm or egg cells, the mutagenesis may occur in future generations and is referred to as genetic mutation. The effect of the mutagen may either be long term or delayed; however, high doses may cause toxicity to occur rapidly. Mutagenesis has a number of characteristics in common with carcinogenicity. Indeed, most carcinogens are mutagens as previously discussed. Apparently, an absence of a threshold of a given chemical may cause the mutagenesis, as well as the carcinogenicity. The Environmental Mutagen Society in 1975 revealed that a variety of active compounds in all chemical classes could cause mutagenesis. A clear and present hazard is present for the human population now, as well as for succeeding generations.

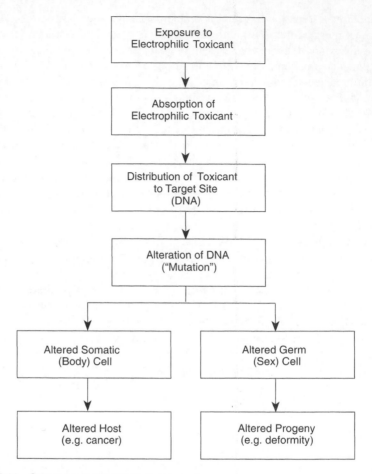

Figure 2.11   Schematic summary of mutagenesis.

A variety of test systems need to be used to determine whether any given chemical is mutagenic. Further monitoring studies of populations at great risk contribute significant information to determine what might be mutagenic and what might not. Genetic hazards to humans are typically predicted by bringing together the chemical and biological data generated from various tests and animal models. The degree of hazard or risk is usually a function of the perceived relevance of a test system that provides the positive responses and the total body of evidence that is available.

Reproductive Toxicity and Teratogenesis

A wide range of physical, chemical, and even biological agents can adversely affect reproductive outcomes. At least 50 chemicals, including heavy metals such as lead and cadmium, glycol ethers, organohalide pesticides, organic solvents, and chemical intermediates, have been shown to produce impairment of reproductive functions in animals. The methyl mercury poisoning in 1970 of the residents of

Minamata, Japan, demonstrated the teratogenic potential for chemical exposures. Occupational exposures can produce a wide range of adverse affects on reproduction. Paternal exposure before conception may be shown in reduced fertility, unsuccessful fertilization or implantation, or abnormal fetuses. Maternal exposure after conception may result in death of the fetus or structural and functional abnormalities in newborns. It is estimated that 560,000 infant deaths, spontaneous abortions, and stillbirths occur each year in the United States. It is also estimated that 200,000 live infants are born with some birth defect each year. The causes of most of the adverse outcomes are unknown. No recognized anatomic or physiological abnormalities are found in 6 to 30% of the infertile couples. The causes for 65 to 70% of the birth defects are not known.

The studies of occupational reproductive hazards have been related to exposures to lead, ethylene oxide, and anesthetic gases. Adverse effects have been found on the quality of semen in workers exposed to lead or ionizing radiation. The National Institute for Occupational Safety and Health (NIOSH) estimates that approximately 200,000 workers are potentially exposed to various glycol ethers, several of which show marked testicular toxicity in animals. An estimated 9 million workers are exposed to radiofrequency–microwave radiation, which has been shown to cause embryonic death and impaired fertility in animals. Personnel in hospital operating rooms are potentially exposed to waste anesthetic gases, and other hospital and industrial workers may be exposed to ethylene oxide. Both of these chemicals have been linked to an increased risk of spontaneous abortions in humans.

Teratology is the study of malformations induced during the development of an animal embryo or fetus from conception to birth. The teratogen causes a change to occur during early embryonic or fetal development, which results in anatomic defects, physiological defects, biochemical developmental errors, or behavioral effects. Teratogens also may cause the death of the embryo or fetus. Teratogenesis is usually considered to be a chronic effect, although the toxicity that caused it appeared over a short period of time during the developmental stage. The severity of the condition relates to the kind of exposure, the amount of the exposure, and when the exposure occurs during the developmental process.

Medical Surveillance

The adverse effects following toxic exposure may be biochemical, structural, or functional in nature. Although external environmental monitoring devices can be set up close to the individual to evaluate external exposure (see Chapter 12), they do not provide data concerning the internal exposure, absorbed dose, or potential signs of adverse impact. Fortunately, medical surveillance includes biological monitoring to detect and measure the parent chemical or its metabolites in tissues or excreta, or pathological effects of the toxin to the exposed individual. The individual's blood, urine, breath, hair, neurological response, and physical condition can be evaluated, however, to determine whether exposure to certain agents has occurred, and if so, at what levels. Biological monitoring data, in turn, can be compared with various recommended biological exposure indices.

Elevated liver enzymes may be due to exposure to and absorption of chemicals. Kidney function may be determined by abnormal elevations of normal blood constituents, such as microglobulins, urea, and creatinine. A standard blood count that evaluates the number and distribution of blood cell types and morphology helps access the effects of chemicals on the blood and blood-forming tissues. This test is relatively insensitive for most toxic effects. However, a bone marrow examination may be more informative, but it is very traumatic and expensive. A blood cholinesterase inhibition may be found in toxicity due to organophosphate exposure. Elevated carboxyhemoglobin in blood may be indicative of carbon monoxide poisoning.

Changes in skin texture, pigmentation, vascularity (changes in the channels in the body carrying fluids), hair, and nails, as well as the existence of lesions, can be detected with physical examinations. Neurological examinations can be used to detect disorders of intellect, memory, coordination, reflex changes, and motor and sensory problems. The electroencephalogram (EEG) records activity in the brain and the response to visual, auditory, or tactile sensations. This is particularly useful in determining any limiting of the function of the brain.

The respiratory system can be examined and analyzed by the use of pulmonary function tests. These tests can help differentiate between the normal lung volume and the restricted lung volume. The restrictions may be due to a chest–wall dysfunction or a disease such as pneumoconiosis. A reduction in the flow rate would be consistent with asthma or chronic bronchitis. By measuring the air remaining in the lung after a maximum expiration of air, it is possible to make some determination about emphysema being present. Chest x-rays are useful in measuring chronic pulmonary disease and cancer once the condition has occurred. Saliva may contain abnormal cells, which may be a means of determining whether asbestos bodies are present in the saliva. Although this is not a diagnostic procedure that confirms lung disease such as asbestosis, it still indicates asbestos exposure.

APPLICABLE CONCEPTS OF MICROBIOLOGY

Roles of Microorganisms

Microorganisms are important from several perspectives. For the most part, microorganisms serve extremely beneficial and essential purposes. Some microorganisms, such as some yeasts and bacteria, are used commercially in the manufacture of foods, beverages, and pharmaceuticals. Most commonly, microorganisms serve as the primary biotic (living) catalysts involved in transformation and stabilization of natural and anthropogenic matter generated within and released into the environment. These processes occur within the intestinal lumen of humans and other animals, as well as in ambient environmental matrices such as soil and surface water. Numerous microorganisms, however, are classified as pathogenic because they are associated with various infectious, allergenic, or intoxicating illnesses in humans and other animals. Pathogenic microbiological agents are also referred to as biological hazards or, more simply, biohazards.

Environmental health and safety focuses on: (1) microorganisms that play important roles in ecosystems by contributing to the transformation, decomposition, and humification of matter in air, water and soil; and (2) microorganisms that are deemed pathogenic to humans.

Microbes Involved in Transformation and Stabilization of Matter

The collective interaction of all microorganisms in the environment forms a heterogenous, complex web that can be simplified as a flow of energy through a series of food or trophic levels. Bacteria, which comprise the lowest trophic level, are ubiquitous in air, water, soil, and biological waste matter such as manures and sludges. Bacteria are unicellular prokaryotes; that is, they lack a membrane-bound nucleus and other membrane-bound organelles. They range in size from less than 1 μm to approximately 15 μm.

Both autotrophic and heterotrophic bacteria participate actively in aerobic and anaerobic transformation of matter. Heterotrophs, the most dominant, require organic compounds as a source of energy for growth and reproduction. Autotrophs utilize carbon dioxide (CO_2) as a source of carbon (C); and either light as their source of energy (photoautotrophs) or inorganic compounds (chemoautotrophs) such as sulfur (S), hydrogen sulfide (H_2S), ammonia (NH_3), nitrite (NO^{-2}), and iron (Fe^{+2}). The inorganic compounds serve as electron acceptors in oxidation–reduction (redox) reactions, through which energy is obtained.

Certain populations of bacteria grow on one substrate while simultaneously cooxidizing or cometabolizing another substrate not utilized to sustain their growth. The phenomenon of cometabolism, for example, accounts for the microbial transformation and detoxification of synthetic chlorinated pesticides, including those fractions found in soil. Many of these synthetics are normally classified as recalcitrant molecules because they are not regarded as a source of labile C-energy for any organism. Nonetheless, the residence time of synthetic organics is questionable due to their susceptibility to bacterial decomposition via cometabolism.

Actinomycetes are unicellular prokaryotes that possess characteristics of both bacteria and fungi. Some species form branched filaments or hyphae. The hyphae collectively constitute a mycelium, which is a characteristic feature of fungi; the actinomycetes, however, are more closely related phytogenetically to bacteria. Actinomycetes are implicated in accelerating the decomposition of a variety of organic compounds including the natural polyphenolic compounds lignin and humic acids.

Fungi are heterotrophic eukaryotic organisms. Unlike bacteria and actinomycetes, fungi possess a defined membrane-bound nucleus and organelles. Numerous species secrete polyphenoloxidase, an enzyme that catalyzes humification reactions.

Protozoans are unicellular microscopic eukaryotic organisms. They are able to locomote through the beating of numerous short hairlike processes as cilia or long hairlike processes as flagella, or through sol-gel transformations resulting in protoplasmic streaming. Protozoans are very active along with bacteria at the root–soil interface, called the rhizosphere. Through their role of consuming bacteria, fungi, actinomycetes, algae, nonliving particulates, and their own feces (coprophagy) for some species, protozoans play an important role in heterotrophic decomposition of matter by accelerating turnover of biotic and abiotic components.

Algae, some of which are monera and others are protista, are the only photosynthetic microbe in soil and water. They are located mainly in the upper 2 to 4 cm, where quanta of light energy may be secured for their photoautotrophic activity. Some algae, however, are heterotrophic and may be found, accordingly, in deeper layers of the soil, as well as superficially. Algae also grow on the surface of water and on some biological waste matter.

Nematodes are nonsegmented round worms. They range from 0.5 to 4 mm in length and 50 to 250 μm in width. Their preference of microhabitat in soil is similar to that of protista, because they commonly inhabit films of water bound to soil particles. They are also found in water. Nematodes consume soluble organic matter, bacteria, protista, fungi, and other nematodes. Like protista, their major contribution in heterotrophic decomposition is accelerated rate of turnover.

Pathogenic Microorganisms

Some bacteria, fungi, actinomycetes, protozoans, and helminths (e.g., tapeworms), as described earlier, may be pathogenic. In addition, viruses that are not true organisms, but instead are DNA or RNA bound to protein, also may be pathogenic.

Effective doses of microorganisms, otherwise referred to the number of microbes necessary to initiate an adverse response (illness), vary among humans. Indeed, unlike chemicals from a toxicological perspective, a paucity of dose–response data are available for microbial agents relative effective (adverse) responses in humans. The factors that most influence the probability of an adverse effect to be observed are the mode and route of exposure, the resistance of the exposed host, and the inherent virulence of the organism. Modes and routes of entry are primarily via inhalation into the respiratory system, ingestion into the gastrointestinal system, and entry through cuts or other openings in the dermal system.

Microorganisms may initiate disease via one or more general mechanisms of infection, allergic reaction, or intoxication:

1. *Infection* refers to the invasion of human systems by pathogenic microorganisms. The number of microbes present exceeds the hosts' immune defenses and results in manifestation of an infectious disease. Causative agents include bacteria, fungi, protozoa, chlamydia, rickettsia, viruses, and helminths.
2. *Allergy* refers to a hypersensitization reaction. Certain individuals can become overly sensitized to a foreign agent (antigen), such as proteinaceous components of nonliving microbes, following initial exposure. The initial exposure to an allergenic microbial antigen can result in production of antibodies against the specific antigen. Future exposure to that antigen not only may produce a secondary boosting of the antibody production response but also can cause tissue-damaging reactions. These tissue-damaging reactions are known as hypersensitivity reactions and the person is now considered hypersensitive or allergic to that antigen (Figure 2.12).
3. *Intoxication* refers to the interactions and reactions that can occur when a human is exposed to a chemical or biochemical agent that alters normal biochemical and or physiological activities and leads to illness or death.

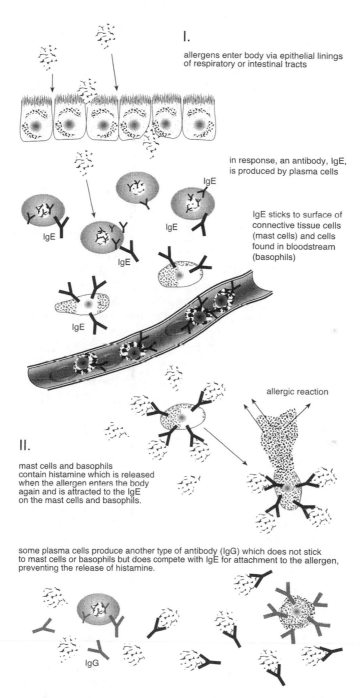

I.

allergens enter body via epithelial linings
of respiratory or intestinal tracts

in response, an antibody, IgE,
is produced by plasma cells

IgE

IgE sticks to surface of
connective tissue cells
(mast cells) and cells
found in bloodstream
(basophils)

IgE

IgE

IgE

allergic reaction

II.

mast cells and basophils
contain histamine which is released
when the allergen enters the body
again and is attracted to the IgE
on the mast cells and basophils.

some plasma cells produce another type of antibody (IgG) which does not stick
to mast cells or basophils but does compete with IgE for attachment to the allergen,
preventing the release of histamine.

IgG

Figure 2.12 Evolution of an allergic reaction.

Some microbes, such as the bacterium *Clostridium botulinum*, synthesize exotoxins that are secreted from their cells. In the case of botulism, the *C. botulinum* bacteria synthesize and secrete the neurotoxic exotoxin while growing in the food. Exotoxins are synthesized by both Gram-positive and Gram-negative bacteria and tend to consist of proteins. Many exotoxins are known by their mode of action on target cells and named accordingly as hemolysins, cytotoxins, neurotoxins, enterotoxins, and nephrotoxins. Exotoxins are important virulence factors for many microorganisms, enabling them to initiate and maintain an infection and possibly spread through a tissue. In addition, organisms such as *Staphylococcus aureus* produce an exotoxin while growing in the food; however, the toxin is described as an enterotoxin because of its action on the gastrointestinal tract. Exotoxins are important virulence factors for many microorganisms, enabling them to initiate and maintain an infection and possibly spread through a tissue.

Endotoxins are the lipopolysaccharide components of the cell wall of only Gram-negative bacteria. They are pyrogenic, producing fever in the body if released into the circulation following cell death of the bacteria. In addition to their pyrogenic effect, the lipopolysaccharide endotoxins also activate immune system complement, cause vasodilation, and activate a clotting cascade leading to shock and disseminated intravascular coagulation, with possible fatal results. Typically, microbial intoxication reactions are characterized by very short incubation periods — the time between the introduction of the toxin and the onset of symptoms — compared with infection where the organism has to grow in the body before its effects are demonstrated. Microbial intoxication causes an adverse effect to humans typically following introduction of preformed toxins and does not require growth of a microorganism in the body. However, some Gram-negative rod-shaped bacteria that produce *enterotoxins*, such as *Escherichia coli*, *Shigella* spp., and *Salmonella* spp., have to grow in the body before the toxin is released and symptoms are present.

Classification of Microorganisms

Five classes of biohazards have been established based on the collaborative effort among the U.S. Department of Agriculture (USDA), the U.S. Department of Health and Human Services (USDHHS), Centers for Disease Control and Prevention (CDC), and the National Institutes for Health (NIH). Classes 1 through 4 represent a ranking based on increasing hazard associated with increasing classification number. Class 5 represents animal pathogens excluded by law from the United States. It is possible for some organisms to fall under two classes. This depends on whether the organisms are present in tissues, in clinical specimens (e.g., Class 2), or in culture (Class 3). Outlined summaries and some examples follow:

Class 1: This class includes microorganisms not known to cause disease in healthy adult humans. Handling these organisms does not require special equipment or techniques normally required to deal with them. Bacterial, fungal, viral, rickettsial, chlamydial, and parasitic agents not listed in the higher classes are included; for example, nonpathogenic bacteria in the human body; bread molds; penicillin molds; various types of nonpathogenic protozoans and helminths; and nonpathogenic types of viruses, chlamydia, and rickettsia. More specifically, examples include bacteria *Bacillus subtilis* and *B. stearothermophilus*; fungi *Saccharomyces cerevisiae*; parasites *Naegleria gruberi*; viruses/rickettsia/chlamydia canine infectious hepatitis virus.

Class 2: This class includes microorganisms posing moderate risks. These organisms are indigenously present in the community and are associated with human diseases of varying severity, but not usually considered as transmissible by inhalation of aerosols. The organisms more commonly are thought to cause diseases by accidental inoculation, injection, or other means of cutaneous penetration. Class 2 agents typically can be controlled by using fundamental, proper laboratory techniques. Examples include bacteria such as members of the Enterobacteriaceae, *Vibrio* sp., and *Mycobacterium leprae*; fungi *Cryptococcus neoformans*; parasites *Toxoplasma gondii*, *Giardia lamblia*, and *Entamoeba* sp.; viruses, rickettsia, chlamydia such as *Chlamydia* sp., hepatitis A and B viruses, influenza virus, and herpes simplex virus. Although the parasite *Toxoplasma gondii* is assigned to Class 2, this parasite presents a special risk to the fetus of nonimmune females who might become pregnant.

Class 3: This class includes indigenous or exotic agents with potential for serious and lethal infections, especially by respiratory transmission. Class 3 agents are special hazards or include agents derived from outside of the United States requiring USDA permits for importation. Agents requiring special conditions for containment are also included. Handlers and distributors should have specialized training and possess a high degree of knowledge of microbiology. Control measures include separated and restricted access facilities and negative air pressure and filter systems. Examples include *Mycobacterium tuberculosis* and *B. anthracis*; fungi *Coccidioides immitis* and *Histoplasma capsulatum*; viruses, rickettsia, and chlamydia *Coxiella burnetii* and spotted fever group of rickettsia.

Class 4: This class includes exotic agents posing a high risk of life-threatening disease that are transmitted by the aerosol route, and for which no available vaccine or therapy exists. This class of agents requires the very strict containment because they are extremely hazardous to humans and have the potential to cause epidemics. Competency of handlers and distributors must be equal to that for the Class 3 agents. Containment requirements are same as for Class 3 plus the addition of requirements for wearing personal protective equipment and use of high-efficiency particulate air (HEPA) air filtration. Examples include Ebola virus, Lassa fever virus, and Marburg virus.

Class 5: This class includes nonindigenous pathogens of domestic livestock and poultry, the importation, possession or use of which is prohibited or restricted by law or by USDA regulations or administrative policy, and which may require special laboratory design beyond biosafety level 4 features. Foreign agents of bovine spongiform encephalopathy (mad cow disease), viral hemorrhagic disease of rabbits, and foot and mouth disease virus are included.

PRINCIPLES OF COMMUNICABLE DISEASE

Germ Theory of Disease

John Snow hypothesized that a specific microorganism caused cholera, yet this concept was not generally accepted during his time. The germ theory of disease was postulated on the basis of Louis Pasteur's research. Pasteur demonstrated that microorganisms caused fermentation and that these microorganisms were not spontaneously generated but could be found in the air. His work was followed by Robert Koch, who first isolated the organisms causing tuberculosis and Asiatic cholera. Koch introduced scientific rigor to the proof of the cause of disease. Koch's postulates were that the parasite had to be shown to be present in every case of the disease by isolation and pure culture; it could not be found in other diseases; once the parasite was isolated, it had to be capable of reproducing the disease in experimental animals; the parasite had to be recovered from the diseased animal. Unfortunately, Koch's postulates did not apply to viruses, because they could not be cultivated in pure culture and were host specific. His postulates also did not apply well to certain pathogenic organisms that infected a host but did not necessarily produce a recognizable disease.

Infectious Disease Causation

To understand the spread of infectious or communicable disease, it is necessary to understand the host–parasite relationship and the reservoir of disease. The reservoir of disease is defined as the living organisms or inanimate objects in which infectious agents live and multiply. A parasite is the microorganism causing a specific disease. A host may be either people or animals. The chain of infection of a given disease is usually illustrated as follows:

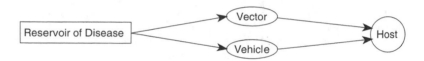

In the environmental health field, the practitioner attempts to halt the transmission of disease by eliminating the vector, which may be an arthropod, or controlling or purifying the vehicle, which may be an inanimate object — water, air, or solid waste. Generally, the reservoir of disease is difficult to control, with the exception of those diseases where hosts can be vaccinated. Eliminating the *Culex* mosquito, which in turn eliminates St. Louis encephalitis, is an example of breaking the chain of infection at the vector state. Purification and chlorination of water supplies, which destroys the microorganisms causing cholera, is an example of halting the spread of infection at the vehicle stage.

Humans are not only the recipients but often the carriers of disease. Carriers discharge microorganisms into the environment through feces and urine, and from the skin, nose, or pharynx. Because carriers may not exhibit symptoms of disease, they are dangerous in the transference of disease. Carriers may have a subclinical case of a disease, or may be in the incubation or in the convalescent period of disease. Some examples of diseases transmitted through carriers are polio, chicken pox, measles, hepatitis, diphtheria, salmonella, and staphylococcus infections. Disease may be transmitted directly from person to person through kissing, sexual intercourse, coughing, sneezing, direct contact of the skin, or indirectly through transmission by means of vehicles or vectors. Some examples of vehicles, also called fomites, include bedding, toys, surgical instruments, surgical dressings, contaminated food or water, and needles. Indirect transmission of disease may occur when dust and droplet nuclei enter the airstream through coughing, sneezing, talking, dusting, walking; or through various housekeeping or medical procedures. Q fever, histoplasmosis, tuberculosis, psittacosis, staphylococcus infections, and other diseases are transferred in this manner.

Disease may be spread from a common source or from secondary sources. Generally, the initial outbreak of an epidemic spreads from a common source; the secondary outbreak, which may occur days or weeks later, spreads from infected secondary sources, or the secondary environment, which became infected during the primary outbreak of disease. The scope of an outbreak of disease depends on the virulence of the organism, its pathogenicity, the number of unprotected individuals at risk, the various environmental factors, and the intervening factors used to control the disease. Virulence refers to the degree to which an infectious agent can produce serious illness. Pathogenicity refers to the ability of an agent to cause disease in an infected host.

The cause of the spread of infectious disease is generally complex, even when the primary cause, or agent, is known. The many contributing factors, or secondary causes, including the environmental factors, sharply influence the rate of disease occurrence. When the disease has run a specific course and new cases are not occurring, and if this is a spontaneous type of leveling of the disease process, it is recognized that an equilibrium has been established between the parasite and the host; and this equilibrium will last until new hosts or a more virulent form of the organism appears, or the environmental factors are changed.

Emerging Infectious Diseases

In 1994, the CDC developed a prevention strategy on emerging infectious disease threats for the United States. This strategy has four goals for dealing with infectious diseases that affect or threaten to affect the health of the public. These goals are:

1. Surveillance. Detecting, investigating, and monitoring emerging pathogens and disease
2. Applied research. Combining laboratory findings and epidemiological studies to enhance disease prevention
3. Prevention and control. Improving communication of public health information concerning emerging diseases and how to prevent and control them
4. Infrastructure. Strengthening support and personnel at local, state, and federal agencies and how they interact

Emerging infectious diseases are diseases of infectious origin whose incidence in humans has increased within the past 20 years, or a threat of increase exists. Examples of these emerging infectious diseases in the United States include:

1. *Escherichia coli* O157:H7 disease
2. Cryptosporidiosis
3. Coccidiodomycosis
4. Multidrug-resistant pneumococcal disease

5. Vancomycin-resistant enterococcal infections
6. Influenza A/Beijing/32/92
7. Hantavirus infections

Examples of these emerging infections outside of the United States include:

1. Cholera in Latin America
2. Yellow fever in Kenya
3. *Vibrio cholerae* 0139 in Asia
4. *E. coli* O157:H7 in South Africa and Swaziland
5. Rifvall fever in Egypt
6. Multidrug-resistant *Shigella dysenteriae* in Burundi
7. Dengue in Costa Rica
8. Diphtheria in Russia

Emerging infections are especially a problem in people with a limited immunity, such as those with human immunodeficiency virus (HIV) infection and those receiving medication for cancer or transplants. Because of widespread use and misuse of antimicrobial drugs, these drugs are beginning to lose effectiveness, and drug resistant organisms develop. Lack of appropriate surveillance techniques, laboratory tests, and knowledge by the public and physicians of these serious health problems have complicated the situation and increased the potential for disease.

Preventing Emerging Infectious Diseases

In 1998, the CDC presented its second phase of the nationwide effort to revitalize the national ability to protect the public from infectious diseases. It presented a plan for the next 5 years titled "Preventing Emerging Infectious Diseases: A Strategy for the 21st Century."

After World War II, with the advent of antibiotics and new vaccines, it was reasonable to believe that such bacterial diseases as tuberculosis and typhoid fever could be treated and such viral diseases as polio, whooping cough, and diphtheria could be eliminated. Improvement in a variety of environmental health areas including water quality and food helped lower the incidence of disease. The optimism was premature, because penicillin soon lost its ability to cure infections caused by *Staphylococcus aureus*. Soon new diseases appeared including different strains of influenza, acquired immunodeficiency syndrome (AIDS), multidrug-resistant strains of tuberculosis, Lyme disease, toxic shock syndrome, and Ebola hemorrhagic fever. By the early 1990s, health experts saw that infectious diseases in the United States and elsewhere were starting to increase. In 1992, the Institute of Medicine of the National Academy of Sciences emphasized the connection between U.S. health and international health. It described the major factors that contribute to the emergence of disease including:

1. Worldwide travel
2. International food supply
3. Centralizing food processing

4. Population growth, increased crowding and urbanization
5. Large population movements
6. Irrigation and changing the habitat of insects and animals
7. Risky human behavior
8. Increased use of antibiotics and pesticides
9. Increased human contact with tropical rain forests and other wilderness

CDC updated its original 1994 plan because of emerging threats, new scientific findings, new tools and technologies, changes in healthcare delivery, and new government commitments. The issues of emerging diseases included: antimicrobial resistance, foodborne and waterborne diseases, vectorborne and zoonotic diseases, diseases transmitted through blood, chronic diseases caused by infectious agents, and vaccine development and use.

There were six common bacteria in hospitals that were drug resistant. In recent years multistate foodborne outbreaks have occurred from *Shigella flexneri*, *Listeria monocytogenes*, *Salmonella enteriditis*, *Cyclospora cayetanensis*, *Escherichia coli* O157:H7, Norwalk virus, *S. infantis*, and hepatitis A. Human cases of rabies have increased significantly since 1990, with most of these cases associated with rabies strains found in bats. Since 1985, 21 episodes of red blood cells contaminated with *Yersinia enterocolitica* have been reported. The majority of peptic ulcers are caused by *Helicobacter pylori*. Chlamydia pneumonia may contribute to coronary heart disease. Infectious agents contribute to neoplastic diseases in humans. They are Epstein-Barr virus, *Helicobacter pylori*, hepatitis B virus, hepatitis C virus, human herpes virus-8, HIV, human papillomavirus, human T-cell leukemia virus, liver flukes, and *Shistosoma haematobium*. Vaccines have been developed and used for childhood diseases. Although new vaccines are currently being tested for Lyme disease, rotavirus gastroenteritis, and invasive pneumococcal disease, no effective vaccine exists for HIV, dengue, hepatitis C, and malaria.

New scientific findings indicate that many infectious microbes cause or contribute to the development of some chronic diseases. Human genes have been found that influence susceptibility to infection, severity of infection, and responsiveness to vaccination or treatment. The prion is a transmissible agent that appears to be responsible for certain neurological diseases including Creutzfeld-Jacob disease. New tools and technologies have extended electronic communications throughout most parts of the world including public health institutions. Innovations and biotechnology make it much easier to identify and track microbes.

The healthcare delivery systems of the United States have changed considerably. Because hospital stays are shorter, it is more difficult to determine patient outcomes including hospital-acquired infection. Home healthcare has grown at a considerable rate and is affecting the outcomes for disease for many patients, especially the elderly. A new emphasis is needed to measure the potential for infections in the home healthcare setting. The public is far more aware of the dangers of emerging infectious diseases. The federal government and the executive office have made elimination of these diseases a priority.

Special groups of people exist who are at greater risk than the normal population for emerging as well as existing diseases. They include: people with impaired defenses; pregnant women and newborns; and travelers, immigrants, and refugees. The ability to fight off disease can be impaired by illness, medical treatment, infectious or chronic diseases, age, nutrition, and environmental conditions. Pregnant women can transmit infections to the fetus that might result in prematurity, low birth weight, long-term disability, or death. The transmission may occur during pregnancy, delivery, or breast feeding. Exposure to adverse environmental conditions by the pregnant woman can also affect the fetus. People who move around a lot either nationally or internationally have an increased risk of contracting an infectious disease. Many Americans are visiting remote and tropical locations in the world, which exposes them to a variety of diseases that they would not normally contact.

The emergence of disease may be influenced by human activity such as development projects, agriculture, climate, and refugee movements. Climate changes can create shortages and famine that lead to disease. The beginning of the Zairian refugee crisis in 1994 where 1.3 million people fled resulted in 48,347 deaths predominantly from infectious diseases. This had occurred during other mass movements in the history of civilization. When moving into new areas, refugees may not be equipped immunologically against endemic diseases. Most refugee camps are overcrowded and have inadequate sanitation or medical care. Refugees often have severe shock or stress as well as poor nutrition that weakens their immune defenses. Diseases are readily transferred from animals to people in this type of environment.

The emergence of new pathogens may be a function of changes in host susceptibility as well as a change in the organism or transmission pathways. The factors influencing host susceptibility include increases in the number of immunocompromised patients; increased use of immunosuppressive agents, especially among people receiving cancer chemotherapy or undergoing organ transplants; aging of the population; and malnutrition. Organisms may become new pathogens because they have developed a new virulent gene or resistance to standard therapeutic methods. They also may have changed their transmission pathway to allow them to enter new, previously unexposed populations. Persons with AIDS show a clear increase in susceptibility to infection with salmonella species. *Cryptosporidium* causes an estimated 10 to 20% of AIDS-associated diarrhea cases. Cancer patients who have just undergone chemotherapy are far more susceptible to acquiring pseudomonas from raw produce. These patients appear to have an increased risk for *Salmonella septicemia*.

A cluster of disease is an occurrence where a greater than expected number of cases occur within a group of people, in a geographic area, or over a specific period of time. Some recent disease clusters include birth defects in the 1960s in children of mothers who took thalidomide during pregnancy, the outbreak of Legionnaire's disease in the 1970s from contaminated water in air-cooling units, and the initial cases of a rare type of pneumonia among homosexual men in the early 1980s that led to the identification of AIDS.

Cancer clusters may be suspected when people report that several family members' friends, neighbors, or co-workers have been diagnosed with cancer. Mesothelioma has been diagnosed in people who have been exposed to asbestos. This is a rare cancer of the lining of the chest and abdomen.

The CDC established four major goals for preventing emerging infectious diseases. They were:

1. Strengthening surveillance and response including better methods for gathering and evaluating data, using data to improve public health practice and medical treatment, and improving the ability to monitor and respond to emerging infectious diseases on a global scale
2. Integrating laboratory science and epidemiology to improve public health practice by developing new tools, identifying behaviors, environments, and those factors that put people at increased risk for infectious disease; and conducting research to develop and evaluate prevention and control strategies
3. Strengthening public health infrastructures and training by enhancing epidemiological and laboratory capacity, communicating electronically in an effective manner with other health jurisdictions, enhancing the country's ability to respond to diseases on a global scale and complex infectious diseases including bioterrorism, and providing training opportunities in the United States and worldwide
4. Implementing prevention and control strategies to help healthcare providers and other individuals change behaviors that increase the potential for disease transmission in this country and internationally

To carry out these goals and reduce the potential for emerging infectious diseases CDC has developed the following programs, committees, and publications:

1. Epidemiology and Laboratory Capacity Program
2. Emerging Infections Program
3. Provider-based sentinel networks
4. Foodborne Diseases Active Surveillance Network (FoodNet)
5. Emergency Department Sentinel Network for Emerging Infections
6. Infectious Diseases Society of America Emerging Infections Network
7. Sentinel network of travel medicine clinics
8. *Emerging Infectious Diseases Journal*
9. Hospital Infection Control Practices Advisory Committee
10. "The Cause: Careful Antibiotic Use to Prevent Resistance," a newsletter
11. Guidelines for the prevention of opportunistic infections in HIV-infected persons
12. Emerging Infectious Diseases Laboratory Fellowship Program

CDC is using new tools to improve surveillance. Molecular fingerprinting of microbes is similar to DNA fingerprinting used by forensic scientists. Fingerprinting techniques can distinguish among strains or isolates of bacteria, fungi, viruses, or parasites. It is possible to compare the genetic sequences of organisms as well as the sizes of nucleic acid fragments produced after digestion with special enzymes. For viruses, it is more common to sequence portions of the genome to compare different strains. Molecular epidemiology has been used to determine that cases of measles were imported most frequently from Western Europe and Japan. The strain of measles virus is isolated and sequenced to determine the country of origin. Molecular pathology has been used to diagnose diseases of unknown cause. In 1995, 13 people died of respiratory failure in Nicaragua. Hundreds more became ill. Although these symptoms were similar to dengue fever, all blood tests were negative. Eventually, a new test was developed for *Leptospira*, a bacterial pathogen found in animals and in humans. Although this organism does not typically involve the lungs, previous cases had been reported in Korea and China. Public health education programs were used to teach the public to get rid of rodents and to keep domestic animals out of their homes. They were warned about the spread of disease from urine found in mud or water. Pulsed-field gel electrophoresis is a method of molecular fingerprinting. It has been used to track outbreaks of foodborne disease. Finally, geographic data are being used to understand disease. Global positioning systems (GPSs) are handheld units that pinpoint locations by longitude, latitude, and altitude. Geographic information systems use a set of locations provided by GPS to create detailed computer maps from satellite-derived remotely sensed data. It is possible to link many occurrences such as outbreaks of mosquitoborne disease to mosquito breeding places.

Environmental Contaminants and Disease

Environmental contaminants interacting with genes can substantially affect a person's ability to develop cancer, which is a genetic disease involving one or more alterations in DNA or mutations in a gene. Cancer attributed to occupational or other environmental factors is higher among adult males because of a higher proportion of males employed in industry. More than half of the cancers observed in the third national cancer survey were caused or suspected to have been caused by an external agent. However, in evaluating cancer problems, it is important to recognize that cancer is also associated with socioeconomic status, cigarette smoking, alcohol, radiation, environmental chemicals, drugs, and biological agents.

A sharp increase has occurred in morbidity and mortality for bronchitis, for pulmonary emphysema, and to a lesser extent for asthma. These three conditions, which are collectively termed *chronic obstructive pulmonary disease* (COPD, Figure 2.13), are found more frequently among smokers than among nonsmokers. A higher mortality rate exists among the low-income urban population compared with the upper-income or rural population. Studies indicate that environmental conditions contribute to the increase in these diseases.

A sharp increase in cardiovascular conditions, particularly coronary heart disease, has been attributed to diet, cigarette smoking, and problems of the environment. Clinical experimental evidence suggests that exposure to carbon monoxide exacerbates symptoms of angina. A strong statistical correlation apparently exists with the season of the year, the yearly temperature change, and the mortality attributed to coronary heart disease. Although infectious agents contribute to this problem, the environment is also involved. Numerous, although controversial, publications concerning the relationship between water hardness and cardiovascular disease have appeared. Some association appears to exist between the level of trace metals found in water and the susceptibility to sudden death from arrhythmia. It is also suspected that carbon disulfide and freon gas contribute to arrhythmia in the occupational environment.

Many investigators are now studying the effect of environmental conditions on the outcome of pregnancy. During pregnancy the internal organism is more vulnerable to problems, because the detoxification mechanisms are altered and changes occur in protein, carbohydrate, and lipid metabolism. Concern exists today about teratogens, or substances producing abnormal variants, such as congenital malformations and

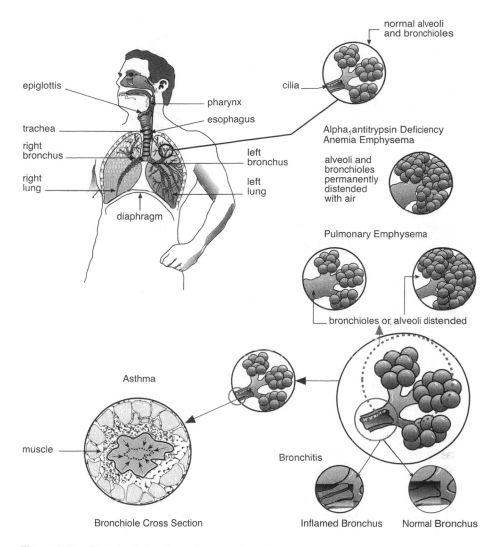

Figure 2.13 Chronic obstructive pulmonary disease.

spontaneous abortions. Although further research is needed in this area, certain chemicals are conclusively embryopathic, including methyl mercury, aminopterin, thalidomide, iodine deficiency, steroid hormones with androgenic activity, and carbon monoxide. Certain chemicals are suspected of affecting human prenatal development, including cortisone, vitamin A deficiency, and diethylstilbestrol.

Infants are susceptible to methemoglobinemia caused by the ingestion of nitrates or nitrites with drinking water. Chemicals may also affect the CNS during postnatal development, affecting particularly the cerebrum. Lead poisoning in children produces irreparable brain damage, permanent retardation, and eventual death. Long-term exposure to air pollutants by infants and children impairs respiratory health and aggravates chronic bronchitis, asthma, and other lung conditions. From 1971 to

1994, 356 waterborne outbreaks of disease were reported in the United States. The frequency of occurrence of waterborne disease is increasing. In addition to waterborne infections, an increasing number of deaths occurred due to chemical poisoning through the ingestion of water. In the reuse of water, problems are created by the presence of microorganisms and toxic minerals, which ordinarily are not found in the initial pure water supply.

EPIDEMIOLOGICAL PRINCIPLES

Epidemiology is the study of the factors that determine the occurrence and frequency of disease in a population. Epidemiology draws on the knowledge and skills of clinical medicine, microbiology, pathology, zoology, sociology, anthropology, toxicology, and environmental health sciences. Although originally the epidemiological technique was applied only to disease and disease causation, more recently it has been applied to accident causation and injuries and diseases caused by chemicals and environmental factors.

Two major categories encompass epidemiology. Descriptive epidemiology is the study of the amount and distribution of disease within a given population by person, place, and time; analytic epidemiology is the study of causes determining the relatively high or low frequency of disease in specific groups. Age, sex, ethnicity, race, social class, occupation, and marital status are the principal personal variables in epidemiological research.

Age is the most important determinant. Mortality and morbidity rates often show a relationship to age. Acute respiratory infections occur with highest frequency among the very young, and upper respiratory infections, in general, are of greatest concern to the very young and the very old. Sex, or gender, is another important epidemiological determinant. The mortality rate for emphysema and respiratory cancer is higher in males than in females, whereas the frequency of occurrence of diabetes is higher in females than in males. Ethnic groups and racial groups exhibit definite differences in morbidity and mortality; blacks have a substantially higher age-adjusted death rate from hypertensive heart disease, accidents, and tuberculosis. Cancer of the cervix is higher in blacks than in whites. Whites have a higher death rate from suicide and leukemia. Social class is an important determinant in the epidemiological study of a specific disease. The death rate for unskilled workers is considerably higher than for the professional classes; infant mortality rates follow the same pattern. Schizophrenia occurs more frequently in the lowest social classes of the urban community. Occupation influences susceptibility to given diseases. Pulmonary fibrosis is found among workers exposed to free silica; cancer is found among workers exposed to aniline dyes.

The physical and biological environment influences the occurrence of disease in a given area. Multiple sclerosis is known to occur most frequently in northern climates; malaria and yellow fever are most frequently found in southern climates; lung cancer is more prevalent in urban areas.

Time is another extremely important variable in the study of disease occurrence. If it is possible to determine the cycle of diseases, clues are available concerning the mutation of various organisms and also the potential effect of various chemicals on arthropods and on human beings. When the time element is determined in an outbreak of disease, it may be possible to quickly recognize the cause of the outbreak. Time is particularly important in outbreaks of foodborne disease. It is also important in determining degrees of exposure. Time, temperature, and concentration are the keys to many environmental health problems and often to the resolution of these problems or to the protection of the individual.

Descriptive epidemiology systematically summarizes the basic data available on health and the major causes of disease and death. These data permit the evaluation of trends in health and the comparison of these trends from subgroup to subgroup and from country to country. Descriptive epidemiology provides a basis for the planning and evaluation of health services and for the identification of health problems.

Analytic epidemiology seeks to discover the causes of problems by formulating hypotheses based on actual observations of existing diseases, conducting studies of outbreaks of disease, and evaluating the data to determine the accuracy of original hypotheses. The classic example of this method was the determination of the epidemiology of cholera by John Snow, the English physician, who realized through observation that cholera was spread through water from contaminated sources.

Epidemiological Investigations and Environmental Health

An epidemiological investigation is a systematic study of available current data, past records and data, and on-site investigations, when feasible. Epidemiological investigations vary with the type of study being conducted. Specific techniques are discussed in the chapters on water and food. A typical epidemiologic study consists of a preliminary review of material, further investigations, analysis of the data, a hypothesis concerning cause, and in some cases, the testing of the hypothesis through the review of new data. In the initial gathering of materials, standardized forms must be used to obtain consistent and complete information. The data are verified through diagnosis of the specific diseases, conditions, and injuries. The investigator determines the time, place, and individuals involved, and plots the cases by time and condition. The investigator asks these questions: how, why, what, when, where, and who.

A test hypothesis concerning the cause of the outbreak or the cause of the disease is theorized; additional cases are sought that either fit or do not fit the pattern. If the cases do fit the pattern, the hypothesis is strengthened. Comprehensive, accurate analysis of data is essential in determining the causes of diseases, conditions, or injuries. One of the principal problems facing the investigator is failure on the part of victims to report the disease to the proper authorities, which results in an incomplete picture of disease conditions. Determining the actual cause, or causes, of a given condition can be problematic; however, the trained investigator, in collaboration with the environmentalist and medical nursing team, uses available data in an attempt to control the problem. The work of an epidemiologist is similar to that of a detective: it is long and tedious, and frequently gives poor results. However, where the results are satisfactory, the population is saved from debilitating or death-causing diseases.

Epidemiological analysis of health data is made by age, race, sex, place, and time. Reports include incidence, prevalence, and case fatality, or morbidity of a given disease or condition. Variation by time and place reflects exposure and response to different environmental pollutants. Epidemiological analysis also includes variations in pulmonary function, cholinesterase levels, and potential toxic effects based on blood analysis of various chemicals.

Epidemiological methods are used to compare exposed and controlled groups. For example, the amount of lung disease in a specific group of workers is compared with that of the general population. The greatest problem in this type of study is controlling the significant variables, such as age, sex, economic status, ethnic group, and smoking habits for the two groups.

Another technique is a longitudinal study of a specific group over a period of time to determine the rate and degree of occurrence of a specific disease within varying time periods. This technique is particularly effective in determining the acute effects of exposure to given environmental pressures. The determination of community exposures, including not only the occupational but also other environmental settings, is difficult, because groups are usually poorly defined, vary within the geographic area, and receive different levels of exposure to substances.

In attempting to study the relationship between health and environmental exposure in a community, the subjects must be selected by geographic comparisons, by comparisons over time, or by dose–response relationships. In geographic comparison, the exposure is defined by a geographic area, for example, health problems experienced by individuals living in heavily air-polluted areas. A problem with this type of study is that a single measure or combinations of measures of exposure are used to represent the entire group exposure. In interpreting data from this type of study, it is important to account for possible difference in factors other than those under study. These differences include migration, age, and ethnic group.

In time comparison studies, a specific time period is compared with health effects over that time period. For instance, a community and its health problems are evaluated for a given period and then are reevaluated after a month, a year, or longer. If the time study period is less than a year, seasonal variation must be introduced. An attempt is made to understand differences in health effects in the population between the initial time period and the subsequent one. Occasionally geographic and time comparison methods are combined, producing a more reliable comparison than if the two were made separately.

In a surveillance study, available data are collected and analyzed, and departures from the norm are noted concerning the severity of environmental problems and health effects. This type of study is used to take corrective action as quickly as possible. In all data collection and epidemiological studies of environmental problems, recognition and control or elimination of variables contributing to illness are essential. Without this, data are of little or no value. In examining environmental influences on health, rare occurrences of illnesses, such as the sudden occurrence of liver cancer in the United States, are important indicators. The outbreak of liver cancer aided in the discovery of the problems related to occupational exposure to vinyl chloride.

Death certificates and morbidity data records from state compensation agencies provide information concerning occupational or other exposures to given environmental factors. This information is of value in developing the initial epidemiological study for a given environmental factor and its relationship to poor health. Other record sources include cancer registers, union records, social security records, and interviews with surviving employees or next of kin. The National Center for Health Statistics is another source of data, and is involved in epidemiological analysis by utilizing data available through various other components of the USDHHS, Social Security Administration, and other branches of government and industry. Although death certificates are required in all states, a National Death Index is needed that would provide the epidemiologist with information on a yearly basis. Great Britain and the states of California and Washington analyze death by occupation. This would be invaluable epidemiological data if provided on a national scale by the National Center for Health Statistics. The National Health Interview survey is a source of data on mortality, but until recently inadequate information on morbidity was provided. Better morbidity data are needed. A study that should provide better data is the National Health Examination Survey, developed to determine the possible role of water hardness on cardiovascular disease. The national census could be used to provide mortality and morbidity data for men and for women by occupation and social class. Each yearly census contributes an additional source of information that should be analyzed for environmental problems. The state health departments also gather and provide data on environmental health problems.

At present, data are available from the EPA, IOSH, and FDA. As can be seen, adequate data are available in some areas but are lacking in others. However, a tremendous need exists not only to provide data but also to coordinate data and to make them available to epidemiological, medical, and environmental support services.

Environmental Epidemiology

The establishment of contamination criteria or exposure limits can be done through epidemiological studies or toxicological studies. The epidemiological studies are concerned with the exposure of populations at their occupational settings or through the contamination of food, drinking water, or air. These studies provide a statistical association between the levels of contaminants and the reported effects. The toxicological studies are of groups of animals exposed intentionally in a variety of controlled laboratory experiments. In this situation, it is possible to define the doses, frequency of application, and metabolic pathways, areas of storage, and types and amounts of biological damage created by the chemical agent.

The advantage of using epidemiological data for establishing safe limits of human exposure to chemicals is that you do not have to have the compounding factor of interpreting animal data and extrapolating from this data. The epidemiological studies try to determine whether or not correlations exist between the frequency or prevalence of a disease or health condition in humans and some specific factor, such as the concentration of a toxic chemical in the environment. Another advantage to this type of determination is that typically large numbers of humans are involved in the study and the exposure levels are usually subclinical. Therefore, the data are directly relevant to the study and determination of the dosage. The disadvantages are (1) the human exposure to chemicals is fortunately very limited, (2) controlled studies of exposure in humans are neither feasible nor legal, and (3) the application of the available epidemiological data may be limited because of its quality or because many risk factors are involved in the study that cannot be truly limited or measured. These risk factors might relate to the ages of the individuals, their educational backgrounds, occupational histories, prior smoking habits, drugs or alcohol consumed, dietary practices, general health, and sexual and racial factors.

Another major problem is determining what the health endpoint is in a clear and concise manner. Death, of course, is a reliable and a definable statistic, but how do you measure the level of good health or poor health related to a combination of chemicals to which the individual has been exposed? Further, a group of chronic nonspecific diseases exist with nonspecific causations that may occur. Even if measurements are made of the appropriate chemicals, are they accurate and do they show the overall population exposure to the chemicals? The most effective types of epidemiological studies are of working populations where known groups of individuals are exposed to chemicals.

The two classes of experimental design in epidemiology are descriptive and analytic. The two types of descriptive studies include the case study and ecological study. The case study provides information for a single individual about the relationship of an exposure and disease. Many times case studies are the starting point for in-depth investigations. The ecological study is used to determine how a single factor is distributed in the population. The ecological study describes the disease in terms of prevalence, incidence, and mortality rates for the population. The ecological study compares the trends of disease in two or more populations or over time. It is used to generate hypotheses about the impact of an agent on the disease found in human populations. These hypotheses are then used to help design the analytic studies. The two types of analytic studies are the retrospective and prospective. In the retrospective study, the population under consideration is determined from death certificates, hospital records, and other sources of data indicating morbidity. A control group that is free of the disease is then chosen for comparison. The exposure history of the study group has been assessed and the causal relationship between exposure to the chemical or other substance and the level of disease is established.

In the prospective design, a cohort (a group) of disease-free patients are followed over time, and the development of disease is monitored. Of the pairs of individuals, one half is exposed to the test agent whereas the other half is not exposed. The development of disease is measured in both groups, and a determination is made of the causal relationship between the agent and disease. Other problems with the cohort study relate to diseases that rarely occur, sample size of the study, design management problems when the disease has a long latency period, and death of the subjects.

To get around some of the difficulties that have been mentioned, the Environmental Epidemiology Branch of the National Cancer Institute did comprehensive studies of 3056 counties in the United States for a 44-year period between 1950 and 1994. They were trying to do a human risk assessment for cancer. From the studies and maps that were drawn, they were able to identify cancer hot spots. They found Salem County, NJ, had the highest rate of bladder cancer mortality in the country. They then did ecological studies and generated maps of industries in that state. The maps included the total population of the county employed in a specific area, with 18 major industrial categories examined. They were able to use existing vital records and census data that eliminated extensive data collection.

In the comparison rates of mortality and morbidity, it was critical to adjust for confounding factors. (A confounding factor or confounder is a risk factor for the disease under study.) If you do not adjust for the confounders, then it is possible to get a mistaken assessment of the risk for disease as a result of specific agents. In cancer, age always is a confounder. Further, many cancers are sex or race specific. Smoking is a confounder in many cancers, including lung and bladder, when the individual is also exposed to environmental or industrial sources. Another problem is called bias. (Bias is a systematic error in the design, conduct, or analysis of a study that causes a mistaken estimation of the relationship of the study factor to the disease.) Some of the biases include a nonrespondent bias, a diagnostic suspicion bias, an exposure suspicion bias, and a family information bias.

Epidemiological investigations, when adjusted for bias and the confounding factors, can be of enormous importance because they provide information about humans under actual conditions to exposure to a specific agent. The surveillance systems used in epidemiology need to be set up in a very careful manner, because these systems vary widely in methodology, scope, objectives, and characteristics. The strength of the evaluation depends on the evaluator, how each of the items in the framework flow, and how they are established and assessed.

The public health significance of the health event must be established. In the concern about a few persons being exposed to highly toxic chemicals, a cluster of cases occurring, or a general outbreak of a specific disease, the health event may be measured by (1) the total number of cases, incidence, and prevalence; (2) the case to fatality ratio, which is an index of severity; (3) the mortality rate; (4) the index of loss productivity; (5) the index of the years of potential life loss; (6) the medical costs involved; and (7) the degree of preventability. Next, what are the objectives of the system? Describe the health event under surveillance and state a case definition for each health event. Draw a flowchart of the system. Describe the components and operation of the system, including (1) what the population under surveillance is, (2) what the period of time of data collection is, (3) what information is collected, (4) who provides the information, (5) how is it transferred, (6) how is it stored, (7) who analyzes the information, (8) how these data are analyzed and how often, (9) how often reports are disseminated and distributed and to whom they are sent, and (10) what the level of usefulness of the data from the surveillance system is. Each system has to be simple, flexible, acceptable, sensitive (collects considerable amounts of data), predictive of problems, representative, and timely.

Epidemiological studies that are adjusted for bias and confounding factors and meet the guidelines previously stated may be extremely beneficial in providing information about humans under actual conditions of exposure to specific agents. A well-designed, properly controlled study can provide competent results in health areas and the determination of risk assessment. The weight of evidence needed to determine whether an agent is potentially carcinogenic in humans or other kinds of health problems exist is to be based on a combination of epidemiological studies, long-term animal studies, pharmacokinetic studies, and relevant toxicological studies.

RISK ASSESSMENT AND RISK MANAGEMENT

Risk

Risk is the probability of illness or injury. It is the potential realization of unwanted consequences of an event. Both the probability of the event occurring and the magnitude of the consequences are involved in the term risk. Risk is a function of the hazard involved, the related dose–response data, and the magnitude and duration of human exposure.

Risk Assessment

Risk assessment is the determination of what the problems are. Risk management is the process of deciding what to do about the problems. Risk assessment is made up of four steps: hazard identification, dose–response assessment, exposure assessment, and risk characterization. Hazard identification involves the gathering of data on the substance, including information about the link between substance and adverse health effects. This step determines that such a link exists. Unfortunately, when the hazard identification is based on the extrapolation from animal data to humans, it is difficult at times to reasonably assert that the substance is a significant hazard. Once it is determined that a chemical is likely to cause a specific health effect in humans, it is necessary to establish the relationship between the amount of exposure (dose) and the effect that is produced. It is not entirely clear that safe levels or thresholds exist for toxic chemicals, because short exposures at low doses may cause damage to health in the future. This is a very complex area. It is easy enough to establish a health end point when the individual is subject to high doses of a chemical and an immediate result is seen in the body.

Exposure assessment is a determination of how much of the chemical is available to the individuals in a specific area. In addition, the duration of exposure is also considered. Its two components include an analysis of the path of transportation and an assessment of the impact of transformation, as well as the available dose and an analysis of the population characteristics. The best way of measuring human exposure is to do direct measurement or monitoring of the ambient conditions. The degree of exposure may vary from chemical to chemical. Human data typically are quite limited because of the types of monitoring, the expense, and the time required to gather information. Modeling often is used as a substitute. In this technique, data are fed into the computer on pollutant releases, release characteristics, meteorology, hydrology, geography, and other information. The exposed population is estimated through the use of census data, etc. Considerable uncertainty exists about these estimates. Risk characterization requires an evaluation of the information from the first three steps. For noncarcinogens, the margin of safety is estimated by dividing the experimental NOAEL by the estimated daily human dose. For carcinogens, the risk is estimated at the human dose by multiplying the actual human dose by the risk per unit of dose projected in dose–response modeling. It may be a range of risk produced by using different models and assumptions about dose–response curves and the relative susceptibility of humans and animals.

Hazard assessment has been based on the underlying principles that animal bioassays are indicative of probable human response and that no threshold of response to carcinogens exists. Technical guidelines have been established by various agencies in risk assessment as they relate to carcinogens.

Toxic substances can lead to adverse effects in the liver, kidneys, lung, and other body systems. These systems have certain thresholds at which they will respond to the chemicals. This further complicates the problem of risk assessment. Most risk assessments are for individual chemicals. However, in reality, people are subjected many times to chemical mixtures that would alter the risk assessment picture considerably. Are there carcinogenic or additive effects to the mixtures of the chemicals?

The exposure assessment is a combination of field monitoring, mathematical modeling, measurement of actual concentrations of chemicals in tissues, and laboratory modeling data. A complication in this exposure assessment is determining how many people are exposed to the chemicals through the air, soil, water, drinking water, or food. It is important to calculate the rate of uptake of the chemicals through breathing, eating, drinking, or absorption through the skin. Then it is essential to track the chemicals through the body and through the process of metabolism.

To summarize risk analysis it is necessary to determine: (1) the source of the chemical and how long it is released to the environment; (2) the pathways through which the chemical travels through the environment, such as the air, water, and food; (3) the behavior of the substances in the body and the pathways that the chemical takes during metabolism, as well as the toxicity of the metabolite products; (4) the estimated concentration or dose at the specific site in the organs and how long the organ is exposed to the chemical; (5) the persistence of the substance in the environment and how long it can be taken into a human being and then passed on to the specific organ; (6) the relationship between the dose the individual gets and the effect on the individual; and (7) the estimated risk to the exposed population.

Risk Management

Risk management helps set priorities. Many of the priorities set for the federal agencies, especially the EPA, have been decided by Congress. Because of potential emergencies, it is best that the agencies look at what the potential problems are and decide what is of the most pressing nature. Risk management produces a basis for balanced analysis and decision making. Because we are exposed to a complex, highly dilute mixture of chemicals that come through the air, water, land, and food, it is frequently difficult to determine which area must receive immediate attention. Rational decisions can be made on the basis of good scientific data. Risk management produces more efficient and consistent risk reduction policies. By doing complete evaluations, it is possible to determine how best to approach risk management. Risk management is the complex of judgment and analysis that uses the results of risk assessment to produce the decisions that are necessary to bring about an environ-mental action.

Inherent in risk management is the idea of comparability. The EPA has a variety of goals, some of which may be in conflict with others. For instance, deep ocean dumping of sewage sludge may reduce the human health risk in comparison with incineration or land spreading, but may have an adverse effect on marine ecosystems. Pollution control, and therefore the reduction of risk, is an incremental process. Consistency must be used in this instance because of the cost involved. At some point when increasing the effectiveness of the process, the cost needed to improve the process by a small increment could be beyond reasonable levels of expenditure. Risk management programs have been set up by the EPA for a variety of problem areas, such as hazardous waste, air pollution, and water pollution.

CHAPTER 3

Food Protection

BACKGROUND AND STATUS

Effective food protection is carried out on a daily basis by large numbers of personnel who attempt to make this essential component of human life safe, attractive, appetizing, nutritious, and free of disease or poison. The food and beverage industry is extremely large.

Many changes have occurred in the food industry since World War II. Today we use many worldwide food sources instead of depending primarily on home-grown food or food grown nearby. Our food is mass produced, and nationally and internationally distributed. Food establishments have increased in size, number, type, and complexity. The short-order food establishment is found everywhere, with many inexperienced people feeding large populations. Problems continue to increase in the new establishments, as well as in the problem establishments of the past, which still do exist, with management the key to the problem.

Food is contaminated by microorganisms from soil, water, air, surface, animals, insects, rodents, and people. Food is also contaminated by chemicals from soil, water, air, pesticides, herbicides, fertilizers, and radionuclides. Contamination may be introduced during processing, as demonstrated in outbreaks of botulism from vacuum-packed fish and tuna fish, or during any stage of production, transportation, storage, preparation, and serving. Raw foods contain disease-producing organisms. Salmonella is contained in raw dressed poultry. *Clostridium perfringens* is cultured from red meat in stores. *Salmonella aureus* is present in most pooled milk supplies and in cheddar cheeses. Salmonella is found in commercial egg products. *Vibro vulnificus* was identified in the bloodstream of persons with underlying liver disease who had fulminating infections after eating raw oysters or being exposed to seawater. *Cyclospora cayetanensis* foodborne outbreaks were traced to imported Guatemalan raspberries in 1996. *Listeria monocytogenes* is a known cause of meningitis and other invasive infections in immunocompromised hosts. Food is identified as the most common source of contamination. *Campylobacter jejuni*, an opportunistic bloodstream infection causing diarrheal disease, has been caused by poultry and raw milk. *Yersinia enterocolicita*, a rare disease in the United States but a common cause of diarrhea in Europe, had been frequently associated with undercooked pork. *Clostridium botulinum* type E has been found on raw fish coming from fresh- or saltwater. Shellfish have been contaminated with organisms causing infectious hepatitis and cholera. *Escherichia coli* O157:H7 has been associated with outbreaks of disease from hamburger and other foods. More than 10,000 different food items have been sold in stores. Each one has its own unique potential for causing the spread of foodborne disease. Besides acute foodborne disease, serious chronic sequelae can occur.

Many people carry pathogenic organisms that can be spread through food. *Salmonella aureus* may be found on the skin or mucous membranes of 3 to 50% of the population. Salmonella may be found in the feces of 0.2% of all people. *Clostridium perfringens* is carried by 80% of the population, and 6.4% of the fecal specimens from food handlers contain enteropathogenic *E. coli*.

Of total human exposure to most chemicals, including heavy metals, pesticides, and radionuclides, 80 to 90% comes from food consumption. Food is presently contaminated heavily by chemicals. These chemicals, such as pesticides, may be deliberately applied to crops and stored commodities to kill insects. Other chemicals, such as food additives, are employed to achieve some desirable effect, such as improved taste, color, and longevity. Drugs used in the care of animals and chemicals used in food-packaging materials may accumulate within food. Another group of chemicals enter by means of environmental pollutants or as fungal toxins. At present the most important chemicals and chemical contaminants of food for which measurements can be made include toxic metals, such as mercury, selenium, lead, arsenic, cadmium, and zinc; industrial chemicals, such as polychlorinated biphenyls, chlorinated dibenzo-*p*-dioxins, chlorinated dibenzofurans, and pesticides; fungal toxins, or mycotoxins, such as aflatoxin; and miscellaneous chemicals, such as the polynuclear aromatic hydrocarbons and nitrosamines. In addition, the level of foodborne disease due to microorganisms is increasing and will continue to increase because of the tremendous amount of food processed and sold in a variety of different types of operations.

This chapter is concerned with foodborne disease; food microbiology; plans review; physical facilities; storage, preparation, serving, and protection of food; housekeeping procedures; cleaning and sanitizing; solid waste disposal; insect and rodent control; personal health of employees; inspectional procedures; and public health laws.

SCIENTIFIC AND TECHNOLOGICAL BACKGROUND

Food Microbiology

Microorganisms are unicellular and microscopic, and perform positive or negative functions in our lives. Certain microorganisms are important in the production of vinegar (*Acetobacter*), sauerkraut (*Lactobacillus*), and bread (Saccharomyces). However, our major concerns are those organisms that can cause foodborne infections and food poisoning. Microorganisms are found in soil, water, and air; on

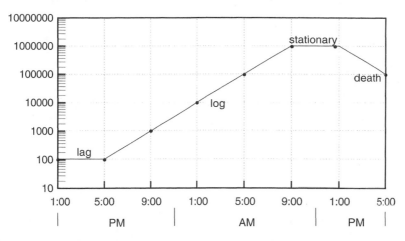

Figure 3.1   Bacterial growth curve. (From HACCP, *Regulatory Applications in Retail Food Establishments,* U.S. Food and Drug Administration, Rockville, MD, July 1991, p. 23.)

animals; in body openings; and on body surfaces. These organisms can be transmitted directly to food and water by hands or indirectly by fomites, insects, and rodents, and by aerosols due to coughing and sneezing. Bacteria can exist in the vegetative or spore form. The spore is resistant to heat and chemical destruction. Aerobic bacteria, such as *Pseudomonas*, grow in the presence of oxygen. Anerobic bacteria, such as *Clostridium*, grow in the absence of air. Facultative bacteria, such as *E. coli* and *Salmonella*, can grow under either condition.

Reproduction and Growth of Microorganisms

For bacteria to grow, they need an adequate temperature, proper water activity or water availability (aw), proper pH, presence or absence of oxygen (depending on the organisms), and nutrients. Bacterial cells reproduce by a single cell split to form two identical cells (transverse fission). This takes about 20 to 30 min for those bacteria that cause foodborne infections and poisonings. Bacterial growth follows four basic phases: the lag phase lasting about 2 hours; the logarithmic or growth phase, where growth is at a maximum rate; the stationary or resting phase, where bacteria die at the same rate as they are being produced (spores are produced here); the period of decline and death, or death phase, which may occur in 18 to 24 hours (Figure 3.1).

Infective Dose

The infective dose of a microorganism depends on the following:

1. Variables of parasite or microorganism
   - Variability of gene expression of multiple pathogenic mechanisms
   - Potential for damage or stress of microorganism
   - Interaction of the organism with food and environment
   - pH susceptibility of organism

- Immunologic uniqueness of organism
- Interactions with other organisms

2. Variables of the host
   - Age
   - General health
   - Pregnancy
   - Type of medications, either over-the-counter or prescription
   - Metabolic disorders
   - Alcoholism
   - Malignancy
   - Amount of food consumed
   - Variation in gastric acidity, as well as use of antacids
   - Genetic disturbances
   - Nutritional status
   - Immune response
   - Surgical history
   - Occupation

The complexity of all these factors contributes to the difficulty in trying to determine why some individuals become infected and others do not when they are exposed to the same infectious dose.

Environmental Effects on Bacteria

Temperature

The temperature of the medium in which bacteria live determines the rate of growth, multiplication, and death of the organism. Bacteria can generally survive very low temperatures, even though they are dormant; high temperatures will generally kill bacteria. The consistency of the medium, the number of organisms initially present, and the species also affects the rate of bacterial death. Generally the resistance of an organism to heat is greatest at a pH of 7.0, which is favorable for growth. As the acidity or alkalinity increases, the death of organisms due to heat increases. Psychrophiles are organisms found on uncultivated soil, in lakes and streams, on meats, and in ice cream, and live best at temperatures of 0 to 5°C. They frequently are the cause of spoilage in refrigerated foods. Mesophiles are parasites found on plants, or may be harmless parasites on animals and humans. They live best at temperatures between 20 and 45°C. Thermophiles are organisms causing off flavors in pasteurized milk. They live best at temperatures exceeding 55°C.

Radiation

Ultraviolet (UV) rays in sunlight are lethal to bacteria. UV wavelengths of 210 nm (1 nm = 1×10^{-9} meters) to 310 nm contain bactericidal power. The most effective UV wavelength against the majority of bacteria and some molds and viruses is 265 nm. Some of the shorter x-rays and gamma rays are more effective but are dangerous to people. However, under controlled conditions, these rays, along with high-energy electrons, are used for the irradiation of food. Radiation kills bacteria logarithmically and causes the remaining cells to form mutants susceptible to attack by bacteriophages.

Pressure

Bacteria are resistant to applied pressure.

Sound Waves

Supersonic vibrations rupture bacterial cell walls and destroy bacteria. Sound waves of low intensity but high frequency are most effective.

Moisture

Moisture is needed by microorganisms for growth and metabolism. Water activity or aw is the ratio of vapor pressure of food to that of pure water, which has a value of one. For growth and metabolism, most bacteria need an aw of 0.91, most yeast need an aw of 0.88, and most molds need an aw of 0.82. Because the osmotic pressure within the cell is usually greater than the surrounding media, water flows through the semipermeable cell membrane into the cell. By adding 10 to 15% sodium chloride or 50 to 70% sugar to the media, the water flows out of the cell, causing cell disruption and death of the bacteria. Drying the media stops the growth of bacteria and some die. As long as the product stays dry, spoilage or growth of disease-producing organisms will not occur.

Water molecules are loosely oriented in pure liquid water and can be easily rearranged. When other substances (solutes) are added to water, the water molecules orient themselves on the surface of the solute and the properties of the solution change. The microbial cell must compete with the solute molecules for free water molecules. Except for *Staphylococcus aureus*, bacteria are poor competitors, whereas molds are excellent competitors. A solution of pure water has a water activity of 1.00. The addition of solute decreases the water activity to less than 1.00. The water activity varies very little with temperature over the range of temperatures that support microbial growth. The water activity of a solution may dramatically affect the ability of heat to kill a bacterium at a given temperature.

Chemicals

A vast variety of chemicals will kill bacteria. These chemicals include antibiotics, alcohol, chlorine, bromine, iodine, iodophores, phenols, and quaternary ammonium chloride compounds. The effectiveness of the chemical is based on the organic load present, time, temperature, pH, specificity for organism, and mechanical action.

Detergents and Disinfectants

The detergent or disinfectant used in cleaning is determined by the soil load present, the type of soil, the type of water used, the cleaning operation, and the kind of disinfecting agent used for final sanitization. The degree of cleaning, which is a process of soil, tarnish, and stain removal from objects, is determined by the concentration of detergent, type of water, temperature, time, velocity, and type of cleaning procedure.

Soil varies in composition. It may contain water-soluble sugar, salts, and some organic acids, or it may contain water-insoluble mineral deposits or animal and vegetable fats, such as grease or carbon. Food particles may be bound to surfaces through absorption or by grease. Scale deposits may be formed by the interaction of alkaline cleaners and chemicals in the water. These chemicals include soluble iron and manganese salts, calcium and magnesium carbonate and bicarbonate, calcium sulfate, calcium chloride, magnesium sulfate, and magnesium chloride.

Adequate cleaning is enhanced through increasing the temperature of the cleaning liquid, because the rise in temperature decreases the bond between the soil and the surface, increases the solubility of materials, and increases the rate of the chemical reaction. Force or energy applied during cleaning helps to mechanically remove soil from surfaces. The cleaning procedure brings the detergent solution into intimate contact with the soil, removes the soil from the surface, disperses the soil into the solvent, and prevents redeposition of the soil on the clean surface.

Basic Detergent Mechanisms

The basic detergent mechanisms are as follows:

1. Establish intimate contact between detergent and soil through wetting and penetration. Wetting reduces the surface tension of the soil and allows the detergent solution to penetrate and spread out. Penetration is the action of a liquid entering into porous materials through cracks and pinholes.
2. Displace the soil from the surface through dissolving, peptizing, and saponification. Dissolving is a chemical action of detergents that liquefies water-soluble materials or soils. Peptizing is a chemical action that brings materials into a colloidal solution. Saponification is the chemical reaction between an alkali and animal or vegetable fat to produce soap.
3. Disperse the soil in the solvent through suspension, dispersion, or emulsification. Suspension is the action holding insoluble particles in solution. Dispersion is the action of breaking up clumps of particles and thereby suspending them. Emulsification is the action of breaking up fats and oils into small particles that are suspended in the cleaning solution.
4. Prevent redeposition of dispersed soil through rinsibility. Rinsibility is the condition of a solution, permitting it to be flushed easily and completely from a surface.
5. Soften water through precipitation, sequestration, and chelation. Precipitation is the physical settling of particles. Sequestration is the surrounding or tying up of magnesium and calcium compounds by sequestering agents. A sequestering agent is a substance that removes a metal ion from a solution system by forming a complex ion that does not have the chemical reactions of the ion that is removed. Chelation is the same action as sequestration, except with organic materials.

Detergent Formulation

All detergents differ and should be formulated to take care of specific types of soils and surfaces. The ideal detergent should act quickly, be completely soluble, have good wetting and penetrating action, dissolve food solids easily, emulsify fats, suspend particles in solution, rinse off easily, have water-softening properties, be noncorrosive to metal surfaces, and be economical in use dilution.

Types of Detergents

Manual immersion cleaners are used in hand dishwashing or equipment-washing procedures. These cleaners are usually a blend of anionic and nonionic organic liquids or powders containing special additives for corrosion inhibition and skin protection. Housekeeping cleaners are both liquid and powder and are used for cleaning a variety of surfaces, both by hand and machine.

Liquid all-purpose cleaners are composed of anionic and nonionic synthetic detergents that are low sudsing and will not damage or dull synthetic or wax floor finishes when used on floors. Heavy-duty cleaners or wax scrubbers are liquid products formulated to emulsify and soften wax and synthetic floor finishes so they may be easily removed prior to refinishing.

Low-sudsing cleaners are used in battery-operation automatic machines that both scrub and vacuum the floor. This operation is most useful on large floor areas. Wall cleaners may be either all-purpose cleaners or heavy-duty cleaners. Occasionally, where it is necessary to remove deteriorating paint residues, a special powdered wall cleaner including trisodium phosphate may be used.

Concrete cleaners are a powder type of general cleaner similar to the powdered wall cleaner. The concrete cleaner must be more highly alkaline and specifically formulated for cleaning soiled concrete surfaces.

Machine products are either powder or liquid in form and are specially designed for cleaning dishes and utensils. These products are highly alkaline, causing skin irritation if used in manual cleaning.

Acid detergents may be made of synthetic detergents and hydrochloric acid for use in toilet bowls and urinals. They effectively remove rust and other mineral stains from toilet bowls. These strong acid detergents should never be used in bathtubs and sinks, which are generally made of porcelain. Another milder type of organic acid detergent is composed of organic acids or phosphoric acids and nonionic detergent. These products are very useful for removing mineral films from stainless steel surfaces and also from the arms of dishwashers and the milk stone found in milk vats and containers.

Detergent Evaluation

The best type of detergent evaluation is an in-use test. The manager must determine if the product is accomplishing the cleaning job required. Obviously the foodservice manager must take into consideration the temperature of the cleaning solution, the time it is used, the concentration, the amount of friction used, and the initial amount of soil on the utensils or equipment.

Disinfection Evaluation Methods

The recommended disinfection evaluation method is the Association of Official Analytical Chemists (AOAC) Use Dilution Confirmation Test. The test is carried out by placing sterile, stainless-steel ring carriers in a broth culture of the test organisms and drying the rings for 20 to 60 min. The contaminated carriers are immersed in a tub of use solution of the germicide for 10 min at 20°C. The carriers are immersed in tubes of appropriate nutrient media and incubated for 48 hours. No growth must be present in at least 10 separate tubes from ring carriers. In practice, the U.S. Department of Agriculture (USDA) often requires perfect results in 60 or more ring carriers.

Definitions

Detergent disinfectants are disinfectants with cleaning ability. They are also called germicidal detergents, detergent germicides, disinfectant detergents, and germicidal cleaners.

A *disinfectant* is a product that kills all vegetative bacteria but does not kill spores. Germicide and bactericide are synonymous with disinfectant.

Sanitizer is a product that kills some, but not all, bacteria. It usually reduces bacterial counts to a generally acceptable level for public health standards.

Sterilization is a process that kills all living organisms, including spores. Sterilization may be achieved by the use of saturated steam at 121 to 123°C for 30 min, through dry heat at 160°C for 1 hour, or by the use of certain chemicals such as ethylene oxide gas for 2 hours.

A *bacteriostat* is a product that prevents bacteria from multiplying without actually killing the bacteria. Bacteriostats are ineffective on hard surfaces.

An *antiseptic* is an antibacterial agent used on the skin.

A *bactericide* is a substance that kills bacteria.

Sanitizing Agents

Sanitizing agents include various forms of chlorine, iodophores, quaternary ammonium compounds, and heat. Chlorine may be available as calcium hypochlorite at 70% available chlorine, or sodium hypochlorite at 2 to 15% available chlorine. Hypochlorites are effective sanitizers at a minimum concentration of 50 ppm of free available chlorine if applied for 1 min and if the surfaces to be sanitized are clean. Organic material, low temperatures, and high pH influence the bactericidal effectiveness. The temperature for the sanitizer should be at least 75°F, but not hot.

Chlorine may also be available as chloramine-T at 25% available chlorine, or dichlorocyanuric and tricyanuric acids at 70 to 90% available chlorine. These organic chlorine products having a slower bactericidal action are used at a minimum concentration of 200 ppm for 1 min. The temperature for the sanitizer should be at least 75°F, but not hot.

Iodophores are soluble complexes of iodine usually combined with nonionic surface-active agents. The iodophores have a rapid bactericidal action in an acid pH range in cold or hot water. They are less affected by organic matter than hypochlorites, are nontoxic in ordinary use, are noncorrosive, and are nonirritating to the skin. The yellow or amber color of the solution is proportional to its concentration. Iodophores are used at a minimum concentration of 12.5 ppm for 1 min. Quaternary ammonium compounds (QAC) are effective sanitizers at a minimum concentration of 200 ppm for 1 min. They are more stable in the presence of organic matter than chlorine or iodine compounds. They are noncorrosive and nonirritating to the skin. Hot water can be used as a sanitizer if the temperature in the final rinse at the entrance of the manifold is at least 170°F for 30 sec or at least 170°F for 30 sec when equipment is immersed in hot water.

Field Tests for Sanitizers

The concentration of chlorine in rinse water can be determined by use of indicator test paper impregnated with starch iodine or by use of an orthotolidine colorimetric comparison unit. Quaternary compounds can be tested by certain test papers that use color comparison. Iodophore concentrations can be determined by comparing the color of the solution with a color comparison kit. Hot water sanitizing can be determined by bacteriological sampling with Rodac contact plates or swab samples. An average plate count per utensil surface should not exceed 100 colonies.

PROBLEMS IN FOOD PROTECTION

Although our scientific technology is so advanced that we can put humans in space and give them protected life-support systems, including food, our food programs on Earth, in many cases, are the same as they were 60 years ago. The excellent programs are scientifically oriented and conducted by trained environmentalists; in our poor programs, investigators inspect food establishments to license them; in our nonexistent programs, the public is deprived of any type of protection from foodborne disease. Technological developments in food processing are occurring so rapidly that their public health implications are not being properly evaluated. New recently discovered production hazards include botulism and typhoid organisms in commercial products; and salmonella in liquid and powdered eggs and egg products, in instant nonfat dry milk, and in drugs and diagnostic reagents. Although the National Food Safety Initiative of 1997 has increased spending, coordination of agency efforts, and public awareness, governmental agencies still are unable to supervise industry adequately because of insufficient resources, inadequate workforce, preoccupation with legislative response, and inadequate channels of communication between the agency and industry. Public health agencies can no longer be satisfied with expending most of their efforts in supervising the foodservice industry; they must now concern themselves with the source of the raw food and the means of processing it. Because of widespread distribution of food, the potential for large outbreaks of disease has increased immensely. There is an increasing concern over long-term health effects of infectious agents, chemical residues, pesticides, additives, and radioactive fallout. Heavy metals, such as mercury, are found in the fish food chain consumed by humans. For most foodborne pathogens, no vaccines are available.

Source of Exposure

As of the year 2000, the Centers for Disease Control and Prevention (CDC) estimated that there were 76 million cases of acute gastroenteritis in the United States each year with an estimated cost of $6.5 to $35 billion per year. An estimated 325,000 people became seriously ill, and 5000 died from foodborne illness. Over 80% of the outbreaks of foodborne disease and over 60% of the cases reported are bacterial in nature. Salmonella are the most frequent agents. However, because less than 1% of the severe gastroenteritis cases are investigated thoroughly enough to determine the causative agents and routes of transmission, the true scope of this disease is not known. In fact, in known outbreaks of disease, about 50% of the causes are unknown.

Reporting is a major problem; because patients are either indifferent or reluctant to cause trouble, physicians usually only report fatal gastroenteritis, and industry is afraid to lose money. Investigations are hampered by nonreporting, delayed or inadequate reporting, inadequate professional staffs and laboratories, inability to obtain food samples rapidly or total absence of food samples, and lack of cooperation on the part of the institutions or industry. About 40% of the states report no outbreaks, and about 40% more report one or two outbreaks per year. California reports 30 to 50% of all foodborne outbreaks reported annually in the United States. Of all reported cases of disease due to water, milk, and food, 93% are caused by food products. Many outbreaks of disease are tracked back to comminuted food such as fish or meat products that have been chopped, flaked, ground, or minced. Improper holding temperatures compound the problem.

FoodNet

The Foodborne Diseases Active Surveillance Network (FoodNet) was established as a result of the imprecise and, in many cases, lack of data concerning foodborne diseases. Since 1996, this network has collected data to monitor the following nine foodborne diseases in the United States. They are campylobacteriosis, salmonellosis, shigellosis, cryptosporidiosis, E. coli O157:H7, yersiniosis, listeriosis, Vibrio infections, and cyclosporiasis. FoodNet provides a precise measure of the laboratory-diagnosed cases of specific foodborne illnesses and performs additional surveys and studies to interpret trends over time. The FoodNet area includes 30 million people. Inspections of meat and poultry processing plants in the United States mandated by the Pathogen Reduction and Hazard Analysis and Critical Control Points (HACCP) rule of the USDA seem to have affected the trends in a positive manner for salmonellosis and campylobacteriosis as shown in FoodNet studies. HACCP consists of production process controls, standard sanitation procedures, and microbial testing by both food processing plants and USDA to reduce foodborne illness by monitoring and decreasing microbial contamination in food processing plants.

The objectives of FoodNet are to describe the epidemiology of new and emerging bacterial, parasitic, and viral foodborne diseases of national importance; more precisely determine the frequency and severity of foodborne diseases in the United States; and determine the proportion of foodborne diseases caused by eating specific foods. To address these objectives, FoodNet conducts active surveillance and related studies with a population survey, a physician survey, and a case control study of *E. coli* O157:H7. Active surveillance is used to determine the actual number of laboratory-confirmed cases of illness caused by the targeted bacteria, viruses, and parasites. The laboratory survey is used to determine whether or not laboratories within the area are performing cultures for foodborne pathogens. The physician survey is used to determine whether or not doctors who see patients for diarrheal complaints are referring them for laboratory analysis. The population survey is used to determine the behavior of individuals within the area, both adults and children, concerning what foods are taken and what high-risk food consumption and preparation practices exist. The case control studies using *E. coli* O157:H7 help identify patient behavior, food preparation practices, and food consumption preferences that are strongly associated with the illness.

FoodNet surveillance has contributed to the detection and investigation of a large, multistate outbreak of *Listeria*. The current concern is not only for multistate outbreaks of disease but also for worldwide outbreaks of disease. Many of the organisms that cause diarrheal disease abroad, but are rare in United States today, may cause the severe foodborne outbreaks of tomorrow. Foods are shipped throughout the world for consumption by other people. People travel frequently and to many parts of the world where bacteria, viruses, and parasites are indigenous, but relatively unknown to U.S. citizens. These organisms can and will be transported quickly back to the United States and will cause serious future problems.

Foods that were traditionally thought to be safe such as shell eggs are now known to be contaminated by salmonella; and, therefore, many of the products that use these eggs can become sources of foodborne disease outbreaks. Fresh produce although still a rather limited problem, increasingly is becoming the source of pathogens that cause disease. The water may be contaminated, facilities for proper sewage disposal may be lacking, or the workers may carry the organisms. Low-level contamination of widely distributed commercial food products can be the source of outbreaks of disease. Surveillance and investigation of sporadic cases are needed, because they may be the initial cases for a large number of outbreaks from the same source. Existing surveillance systems provide a limited approach to determining multistate outbreaks. At the local level, the county or city health department responds to individual or group outbreaks, but may not necessarily share the information with other public health entities. Under the new FoodNet program this problem is corrected and the information is provided to a variety of health agencies.

Foodborne Infections

Foodborne infection is direct infection in which the organisms are ingested into the body through food or water and then continue to multiply. In some infections the organisms die off after the clinical disease is over. In other infections, such as the salmonella group, the individuals may continue to be carriers for varying periods of time. They pass off live organisms in their feces.

Streptococcus

Streptococcal infections are usually due to *Streptococcus fecalis* or *S. pyogenes*. The source of these organisms are the feces, nose, or throat. *Streptococcus pyogenes* may also come from infected wounds or dirty bandages. The reservoir of infection is people. It is communicable for 10 to 21 days.

The genus *Streptococcus* is made up of Gram-positive microaerophilic cocci, which are nonmotile and occur in chains or pairs. The genus is defined by a combination of antigenic, hemolytic, and physiological characteristics into Groups A, B, C, D, F, and G. Groups A and D can be transmitted to humans by food. Group A is one species with 40 antigenic types including *S. pyogenes*. Group D has five species including *S. faecalis*. Group A causes septic sore throat and scarlet fever as well as other pyogenic and septicemic infections. Sore and red throats, tonsillitis, high fever, headache, nausea, and vomiting occur within 1 to 3 days of the infectious dose (less than 1000 organisms) being consumed. Food sources include milk, ice cream, eggs, steamed lobster, ground ham, potato salad, egg salad, custard, rice pudding, and shrimp salad. In most cases, the food was allowed to be held at room temperatures for several hours between the preparation and consumption. The food could also be contaminated by sick food handlers, use of unpasteurized milk, and use of poor personal hygiene. Salad bars may be possible sources of infections.

Group D may produce a clinical syndrome similar to staphylococcal intoxication. Symptoms include abdominal cramps, nausea, vomiting, fever, chills, and dizziness in 2 to 36 hours following the ingestion of suspect foods, where the infectious dose is probably high. Food sources include sausage, evaporated milk, cheese, meat pies, pudding, raw milk, and pasteurized milk, with the food contaminated due to under-processing or unsanitary conditions. The diarrheal illness is self-limiting although acute in nature. Outbreaks are less common and are usually the result of preparing, storing, or handling food in an unsanitary manner.

Salmonella

Salmonellosis is a costly, highly communicable disease in the United States. An estimated 2 to 4 million human cases occur annually. Because the diseases are primarily food or waterborne, they are a potential threat to every person in the country. Although the true magnitude of the problem is not known, it has been determined that the highest incidents of reported cases occur among the very young and elderly. The interrelationships between people, animals, and fomites make salmonellosis a difficult problem to resolve.

Salmonella are rod-shaped, motile bacteria (nonmotile exceptions are *S. galli-narum* and *S. pullorum*), nonsporeforming, and Gram-negative. They are found in animals, especially poultry and pigs, as well as people. They are also found in or on water, soil, insects, rodents, factory surfaces, kitchen surfaces, animal feces, raw meats, raw poultry, raw seafood, eggs, milk and dairy products, sauces and salad dressings, peanut butter, cocoa, chocolate, etc. *S. typhi* and *S. paratyphi* normally cause septicemia and produce typhoid or typhoid-like fever in humans. Other salmonella usually produce milder symptoms. The acute symptoms are nausea, vomiting, abdominal cramps, diarrhea, fever, and headache. Chronic symptoms such as arthritis may follow in 3 to 4 weeks after the acute symptoms. The time of onset is 6 to 48 hours and the ineffective dose may be as few as 15 to 20 cells, depending on the age and health of the host and the strain of salmonella. Acute symptoms may last for 1 to 2 days or may be prolonged depending on the host, the dose, and the strain. The disease is caused by the organisms moving from the gut lumen into the epithelium out of the small intestine where the inflammation occurs. Evidence is available that an enterotoxin may be produced within the enterocyte. Several different types of salmonella had been isolated from the outside of egg shells. *S. enteritidis* may also be found inside the egg, in the yolk, which would indicate that the infected hen is depositing the organism in the yolk prior to shell formation. Foods other than eggs have also caused outbreaks.

Salmonella are divided into three groups based on their host preferences. Group 1 is primarily adapted to people and includes *S. typhi*, which causes typhoid fever, and *S. paratyphi* A, which causes paratyphoid fever. Group 2 is primarily adapted to animal hosts and includes several important pathogens of domestic animals, such as *S. cholerae-suis*, and serotypes of *S. enteritidis*. These organisms, especially serotype Dublin, cause gastroenteritis in humans. Children may be severely affected by these infections. Group 3 includes organisms adapted to either animals or humans showing no distinct preference. In any case humans may become carriers and transmit the organisms for years. Approximately 2400 serotypes exist that can cause illness in people.

In recent years, numerous outbreaks of *S. typhimurium, S. newport, S. montevideo, S. pullorum, S. derby, S. infantis,* and *S. enteritidis* have occurred. *Salmonella derby* was a particularly serious problem in many hospitals in the eastern United States during the 1950s and early 1960s. The problem became so severe that maternity wards and nurseries in certain hospitals were forced to close. An outbreak of *S. infantis* occurred in a major institution in the East, resulting in over 700 cases within a 30-day period. The organisms are usually spread through meat, poultry, eggs, and egg products contaminated directly or indirectly through contact with human or animal carriers. By the year 2000, outbreaks of salmonellosis accounted for 2 to 4 million infections and approximately 500 deaths per year.

In 1985, a salmonellosis outbreak involving 16,000 confirmed cases in 6 states was caused by low-fat and whole milk from one Chicago dairy. There the pasteurization equipment had been modified to facilitate the running-off of raw milk, resulting in the pasteurized milk, under certain conditions, becoming contaminated with raw milk input.

In August and September of 1985, *S. enteritidis* was isolated from employees and patrons of three restaurants of a chain in Maryland where at least 71 people became ill, resulting in 17 hospitalizations. Scrambled eggs from a breakfast bar were epidemiologically implicated.

From 1985 to 1992, there were 437 reported outbreaks of *S. enteritidis* with 15,162 cases of illness and 53 deaths. In 1993, egg salad from a salad bar was the cause of an outbreak. In other outbreaks, mayonnaise and unrefrigerated eggs were implicated. In 1994, in Minnesota, South Dakota, and Wisconsin, the outbreaks of disease were attributed to a variety of nationally distributed ice cream products. In 1995, outbreaks of *S. enteritidis* were associated with the consumption of raw shell eggs. In 1997, in Wales, three outbreaks were attributed to fish, eggs, and reconstituted mashed potatoes. More recently, salmonella outbreaks have been attributed to other serotypes. *Salmonella agona* was linked to toasted oats cereal in 1998. An increase in the cases of *S. agona*, with 209 people becoming sick, was reported in 11 states. In June 1999, in the United States and Canada, confirmed cases of *S. munchen* totaled 13,207 due to unpasteurized commercially prepared orange juice. Approximately 2400 different salmonella serotypes can cause human disease.

A cycle of infection includes feed and fertilizer, livestock, transportation, processing of food, processed foods, and animals or humans. Processed foods, in addition to those already mentioned, that are most commonly contaminated with salmonella include meat pies, pressed beef, sausage, cold cuts, smoked fish, reheated meats, gravies, dried egg products, frozen eggs, synthetic creams, custard, cream puffs, and prepared fish dishes.

Outbreaks of salmonellosis have also been traced to dirty cutting boards and other utensils and equipment. The reservoir of infection includes poultry, rodents, turtles, cats, dogs, and people. Its communicability varies from days to weeks.

Shigella

Shigella are Gram-negative, nonmotile, nonsporeforming, rod-shaped bacteria causing illness in people and occasionally in other primates called bacillary dysentery or shigellosis. The organism is found in a variety of salads, raw vegetables, milk and dairy products, and poultry. Contamination of these foods is usually through the oral–fecal route by means of unsanitary food handling by people and water contaminated with fecal material. The disease is caused by different members of the genus *Shigella*. The time of onset is 12 to 50 hours and the infective dose is as few as 10 cells depending on the age and the health of the host. The symptoms of the disease include abdominal pain, cramps, diarrhea, fever, vomiting, blood, pus, or mucus in the stools. The disease is caused when virulent *Shigella* organisms attach to, and penetrate epithelial cells resulting in tissue destruction. Some strains produce an enterotoxin and Shiga toxin very similar to the verotoxin of *E. coli* O157:H7. Infections are associated with mucosal ulceration, rectal bleeding, and severe dehydration. The fatality rate may be as high as 10 to 15%. Infants, the elderly, and the sick, especially those suffering from acquired immunodeficiency syndrome (AIDS), are most susceptible to this disease. Shigellosis is communicable for about 4 weeks. An estimated 300,000 cases occur each year in the United States. The number that comes from food is unknown, but thought to be substantial because of the low infectious dose. *Shigella sonnei* is a common cause of gastroenteritis, especially in young children who are in child care centers. *Shigella* have become resistant to some antimicrobial agents especially in other parts of the world. The most recent outbreaks of *S. sonnei* occurring in 1998 were reported in the United States and Canada from eating uncooked parsley.

Miscellaneous enterics

These rod-shaped bacteria are suspected of causing acute and chronic gastrointestinal disease. They may be found in forests, in freshwater, and in vegetables, where they are normal flora. They may be in the stools of healthy people. The individuals have two or more symptoms of vomiting, nausea, fever, chills, abdominal pain, or watery diarrhea, which occur 12 to 24 hours after ingesting food or water. Enterotoxins may cause the problems. Malnourished children may die. *Citrobacter freundii* has been suspected of causing outbreaks.

Cholera vibrio

Cholera is a severe gastrointestinal disturbance caused by the *C. vibrio*. Serogroup 01 causes symptoms of Asiatic cholera that varies from a mild watery diarrhea to an acute diarrhea with characteristic rice water stools. The onset of the illness is usually sudden, with the incubation periods ranging from 6 hours to 5 days. Abdominal cramps, nausea, vomiting, dehydration, and shock occur after severe fluid and electrolyte loss. Death may occur. The illness is caused by the ingestion of approximately 1 million organisms. In the United States, no major outbreaks of this disease have occurred since 1911. However, sporadic cases are usually associated with the consumption of raw shellfish, improperly cooked shellfish, or recontaminated shellfish after proper cooking. Most new cases are associated with poor countries, poor living conditions, and contaminated water. A person may be communicable for a few days, but may also become a carrier, which could last for several months. The reservoir of infection is people. In 1991, outbreaks of cholera in Peru became an epidemic and spread to other South American and Central American countries, including Mexico. Over 1 million cases of the disease and 10,000 deaths occurred. In Africa in 1994, thousands of people died from cholera when they were displaced because of a war.

Serogroup non-0 1 bacteria, which are related to *V. cholerae* Serogroup-0 1, infect only humans and other primates. The bacteria cause a disease less severe than cholera. Both the pathogenic and nonpathogenic strains of the organism normally inhabit marine and estuarine environments of the United States. The organism causes diarrhea, abdominal cramps, and fever with nausea and vomiting in about 25% of the cases. Diarrhea may be in some cases quite severe lasting from 6 to 7 days. An enterotoxin is suspected in the disease process, but this has not been proved. The effective dose is probably more than 1 million organisms and probably comes from the consumption of shellfish, especially raw oysters during the warmer months. Septicemia can occur, especially in individuals with cirrhosis of the liver or in those who are immunosuppressed.

Listeriosis

Listeriosis is caused by the organism *Listeria monocytogenes*. Listeria are Gram-positive bacteria that are motile because of flagella. Between 1 and 10% of humans may be intestinal carriers. They have also been found in at least 37 mammalian species. The ineffective dose is believed to vary with the strain and the susceptibility of the host. It is most frequently isolated from people, although the reservoir of infection can also be wild or domestic mammals and fowl. The disease is found at the extremes of age, during pregnancy, and in immunocompromised people. It may cause septicemia, meningoencephalitis, nausea, vomiting, delirium, coma, and death. The disease may be spread by people or food, or be a nosocomial infection. The incubation period is unknown, but is probably a few days to 3 weeks. Fetuses and newborn infants are highly susceptible, with the mother shedding the organism in urine for 7 to 10 days after delivery. An outbreak in 1985, in California, from contaminated soft cheese caused 90 to 100 deaths. From early August 1998 through January 6, 1999, at least 50 cases caused by a rare strain of the bacterium *L. monocytogenes*, serotype 4b, were reported to CDC by 11 different states. Six adults had died and two pregnant women had had spontaneous abortions. The vehicle of transmission of the disease apparently was hot dogs and possibly deli meats that were produced by one manufacturer under a variety of brand names.

Norwalk-Type Disease (Epidemic Viral Gastroenteritis)

Diarrhea caused by Norwalk virus, a 27-nm particle, which is a parvovirus-like agent, is a limited mild disease with symptoms of nausea, vomiting, abdominal pain, malaise, and low-grade fever. The occurrence of the disease is worldwide, with people the only reservoir of infection. Mode of transmission is probably the oral-fecal route through food and water. The incubation period is 24 to 48 hours, with communicability occurring during the acute stage of disease and shortly thereafter. Outbreaks of disease caused by Norwalk virus involve food such as fruits, vegetables, salads, eggs, clams, oysters, bakery items, and ice. Norwalk viruses and related caliciviruses are important causes of sporadic and epidemic gastrointestinal disease in the United States. An estimated 181,000 cases occur annually. An outbreak in 1996 was traced to sewage contaminating oyster beds from harvesting vessels. In 1998, U.S. Army trainees were hospitalized with vomiting, abdominal pain, diarrhea, and fever due to the ingestion of the microorganisms that were spread to cakes, pies, and rolls by an infected food handler.

Multiple Source Outbreaks

From October 1997 through October 1998, 16 outbreaks of gastrointestinal illness associated with eating burritos occurred in Florida, Georgia, Illinois, Indiana, Kansas, North Dakota, and Pennsylvania. All but one outbreak occurred in schools and most of the 1700 people affected were children. The median incubation period was approximately 15 min after eating lunch, and the median duration of the illness was approximately 4½ hours. The burritos produced by a commercial company contained beef, chicken, pinto beans, seasoning, textured vegetable protein, and tortillas. The symptoms included nausea, headache, abdominal cramps, vomiting, dizziness, and diarrhea. The short incubation period suggests that a preformed toxin or other short-acting agent was the cause of the illness. Many agents are suspect such as *Staphylococcus aureus* enterotoxin, *Bacillus cereus* toxin, mycotoxins, trace metals (e.g., fluorine, bromine, and iodine), plant toxins, pesticides, detergents, or unknown toxins.

Spiral Bacteria — the Gastric Helicobacters

Helicobacter pylori has become recognized as one of the most common human pathogens colonizing the gastric mucosa of almost all people who are exposed to poor sanitary conditions from childhood. However, it may also be found, although at a lower frequency, in groups at a higher socioeconomic status. The organism causes chronic active gastritis and is a major factor in the production of duodenal ulcers and, to a lesser extent, gastric ulcers. *Helicobacter pylori* is a Gram-negative bacterium with a curved, spiral, or gull-wing shape. In the United States and in other regions, direct contact or consumption of food or water contaminated by saliva, gastric juices, or feces may be major factors. It can be isolated from cats and therefore suggests that transmission from cats to humans is possible. *Helicobacter pylori* is considered to be a pathogen because it is always associated with chronic active gastritis, and the eradication of the bacterium always results in the resolution of the gastritis.

Milk from Rabid Cows

Rabies is a viral zoonosis that is usually transmitted by the bite of an infected animal. In Massachusetts from 1996 to 1998, 2 separate incidents involved 80 people drinking unpasteurized milk from cows that were rabid. Because of a potential contact with the saliva of the rabid animals, all 80 people were treated for possible rabies exposure. From 1990 to 1996, CDC received reports of 22 incidents of mass human exposure to rabid or presumed-rabid animals in the United States, with approximately 1900 people exposed. Since 1990, an annual average of 150 rabid cattle in the United States have been reported to the CDC. Transmission of the rabies virus in unpasteurized milk is theoretically possible.

Creutzfeldt–Jacob Disease

Creutzfeldt–Jacob disease (CJD) is a rapidly progressive fatal disease of the central nervous system (CNS) characterized by progressive dementia and myclonus. The disease usually lasts several years and is invariably fatal. The etiologic agent is thought to be an unconventional filterable agent. The exact mode of transmission in humans is not known. Since 1986, approximately 170,000 cases of bovine spongiform encephalopathy have occurred among about 1 million animals infected by contaminated feed in the United Kingdom. In 1995, two adolescents in United Kingdom died of CJD. A new variant strain of CJD has been reported in the United Kingdom (26 cases) and in France (1 case). Other mammals such as cats may also become infected. Apparently the infective agent is transmitted by the use of bovine offal in human food.

Viral Hepatitis

Hepatitis A virus (HAV) is classified with the enterovirus group of the *Picornaviridae* family. It has a single molecule of RNA surrounded by a small, 27-nm diameter protein capsid. Symptoms of the disease include jaundice, fatigue, abdominal pain, loss of appetite, intermittent nausea, and diarrhea. The infectious dose is unknown but believed to be 10 to 100 virus particles. HAV is spread through human feces to food, shellfish, and water, in turn consumed by people. The disease may also be transmitted by means of fomites. The incubation period is 10 to 50 days, usually 30 days. The reservoir of infection is people. Most cases are noncommunicable after the first week of jaundice. Viral hepatitis is endemic worldwide. An estimated 125,000 to 200,000 total infections with 100 deaths occur each year in the United States. The illness can be prolonged or relapsing, but no chronic infection persists. Food has been implicated in over 30 outbreaks of HAV since 1983. In 1987, in Louisville, KY, imported lettuce was the suspected food. In 1988, in Alaska, ice-slush beverage was suspected; in North Carolina, iced tea was suspected; and in Florida, raw oysters from an unapproved source were suspected. In 1989, in Washington, no food was identified. In 1990, in Georgia and Montana, frozen strawberries were suspected, while in Baltimore, MD, shellfish appeared to be the cause of the outbreak. In 1999, in North Carolina, along Interstate 95, travelers who dined at a restaurant in Smithfield were advised that they might have been exposed to HAV, because a worker at the restaurant had had the infection. Potentially 3000 diners could have been exposed.

The hepatitis B virus (HBV) infection is common in certain high-risk groups, such as intravenous drug abusers, homosexual men, people in hemodialysis centers, and certain health professions where routine exposure to blood or serous fluids occurs. The incubation period is usually 45 to 180 days, although symptoms may occur in as soon as 2 weeks. The reservoir of infection is people. People may be infective for years.

Hepatitis C (HCV) has symptoms including jaundice, fatigue, abdominal pain, loss of appetite, intermittent nausea, and vomiting. An estimated 35,000 to 180,000 infections with 8000 to 10,000 deaths occur each year in the United States. More than 85% of the infected people develop chronic infections and 70% develop chronic liver disease. Hepatitis C is transmitted by blood, through sexual contact, and from mother to infant. People at risk include injecting drug users, household contacts of infected persons, sexual contacts of infected persons, international travelers, and employees of, or those who attend, day care centers.

Hepatitis E virus (HEV) is often found in feces in patients who have the disease. It is clinically indistinguishable from hepatitis A disease. Symptoms include malaise, anorexia, abdominal pain, and fever. The infective dose is not known. HEV is transmitted by the oral–fecal route, by water, and from person to person. Spread by food is a potential cause. The incubation period varies from 2 to 9 weeks. The disease is usually mild and is over within 2 weeks. The fatality rate is 0.1 to 1% except in pregnant women where it approaches 20%. The disease is most frequently seen in the 15- to 40-year-old group.

Astrovirus

This unclassified virus contains a single positive strand of ribonucleic acid (RNA) of about 7.5 kilobase (kb), surrounded by a protein capsid of 28 to 30 nm in diameter. The virus may cause sporadic viral gastroenteritis in children under 4 years of age and about 4% of the hospitalized cases of diarrhea, with additional symptoms of nausea, vomiting, malaise, abdominal pain, and fever. The mode of transmission is through the oral–fecal route from person to person, or ingestion of contaminated food or water. The incubation period is 10 to 70 hours. The disease has been reported in England, Japan, and California. The reservoir of infection is people. Susceptibility is found in the very young and the elderly in an institutional setting.

Calicivirus

This virus is classified in the family Caliciviridae, containing a single strand of RNA surrounded by a protein capsid of 31 to 40 nm in diameter. The virus may cause viral gastroenteritis in children 6 to 24 months of age; and it accounts for about 3% of the hospitalized cases of diarrhea, with additional symptoms of nausea, vomiting, malaise, abdominal pain, and fever. The mode of transmission is the oral–fecal route from person to person, or the ingestion of contaminated food and water. The incubation period is 10 to 70 hours. The disease has been reported in the United Kingdom and Japan. The reservoir of infection is people. Susceptibility is found in the very young.

Enteric Adenovirus

This virus represents serotypes 40 and 41 of the family Adenoviridae containing a double-stranded dioxyribonucleic acid (DNA) surrounded by a distinctive protein capsid of about 70 nm in diameter. The virus may cause 5 to 20% of gastroenteritis in young children, with symptoms of diarrhea, vomiting, malaise, abdominal pain, and fever. The mode of transmissions are the oral–fecal route from person to person, ingestion of contaminated food or water, or possibly transportation by the respiratory route. The incubation period is 10 to 70 hours. The disease has been reported in the United Kingdom and Japan in children who are in hospitals or day care centers. The reservoir of infection is people. Susceptibility is found in young children.

Parvovirus

This virus belongs to the family Parvoviridae containing linear single-stranded DNA surrounded by a protein capsid of about 22 nm in diameter. The virus may cause viral gastroenteritis in all age groups, with symptoms of diarrhea, nausea, vomiting, malaise, abdominal pain, and fever. The mode of transmission is the oral–fecal route from person to person or ingestion of contaminated food, especially shellfish and water. The incubation period is 10 to 70 hours. The disease has been reported in the United Kingdom, Australia, and the United States. The reservoir of infection is people. Susceptibility is general.

Rotavirus

The virus belongs to the family Reoviridae and contains spherical RNA 65 to 75 nm in diameter, with a wheellike appearance. The virus may cause gastroenteritis with vomiting followed by diarrhea, malaise, fever, abdominal pain, and dehydration. Death may occur from dehydration or aspiration of vomitus. The mode of transmission is by person to person or may be through contaminated food. Identical or similar viruses are found in various animals. The disease has been reported in the United States and is the most common cause of diarrhea in infants and young children. The reservoir of infection is people and possibly animals. Susceptibility is general. Rotavirus A causes disease in the United States, and in China rotavirus B has caused millions of cases of disease that have been traced to drinking water supplies contaminated with sewage. Rotavirus C outbreaks have occurred in Japan and the United Kingdom.

Brucellosis

Brucellosis, also known as undulant fever or Bang's disease, is caused by *Brucella abortus*, *B. melitensis*, and *B. suis*. The organisms are transmitted through raw contaminated milk, dairy products made from raw milk, and contact with the tissues or discharges of infected animals. The incubation period is highly variable, usually 5 to 30 days. Reservoir of infection is animals. Communicability from person to person is very rare, with about 100 cases occurring in the United States each year. In 1992, an outbreak took place in the kill division of a pork processing plant in North Carolina.

Diphtheria

Diphtheria is caused by *Corynebacterium diphtheriae*. The organisms are transmitted from humans through milk or directly from human to human through discharges from the nose or throat. The incubation period is 2 to 5 days. The reservoir of infection is people. Communicability is typically 2 to 4 weeks.

Tuberculosis

Tuberculosis of the extra pulmonary type is caused by *Mycobacterium tuberculosis*. The organisms, both of the human and bovine type, are spread through raw contaminated milk or dairy products, and contact with infected animals and humans. The incubation period from infection to the primary lesion is about 4 to 12 weeks. The reservoir of infection is people and diseased cattle. Communicability lasts as long as infectious tubercle bacilli remain in the sputum.

Tularemia

Tularemia is caused by *Pasteurella tularensis*. The organisms are spread through the meat of wild rabbits, squirrels, other animals, and bites of infected ticks and deer flies. The incubation period is 2 to 10 days, usually 3 days. The reservoir of infection is wild animals and various hard ticks. The disease is not directly transmitted from person to person.

Diarrhea Caused by Escherichia coli

E. coli is the dominant species of bacteria found in animal, including human, feces when aerobic cultures are used. Normally *E. coli* is useful in the body, because it suppresses the growth of harmful bacterial species and it synthesizes large amounts of vitamins. A minority of *E. coli* strains are capable of causing human illnesses. At least four different types of enterovirulent or ectrodactyly-ecodermal dyplasia-clefting enteropathogenic (EEC) group of *E. coli* exist. They are enteroinvasive, enterotoxigenic (toxin producing), enteropathogenic, and enterohemorrhagic. The effective dose is unknown but from compiling data from various outbreaks including person-to-person spread in a day-care setting and nursing home setting, the dose, which is ten organisms, may be similar to that of *Shigella*. The invasive strains of *E. coli* cause a disease that is found in the colon and that produces fever and occasionally bloody diarrhea. The pathological changes are similar to those seen in shigellosis. The enterotoxigenic strains act more like *Vibrio cholerae* in producing profuse, watery diarrhea without blood or mucus, abdominal cramping, vomiting, and dehydration. The enteropathogenic strains belong to the group associated with outbreaks of acute diarrheal disease in newborn nurseries. The diseases are typically foodborne or waterborne, especially in areas with poor protection for the food and water, and endemic diarrhea is present. Traveler's diarrhea is most usually due to an enterotoxigenic *E. coli*. The disease is spread by the oral–fecal route. Contaminated hands, as well as improper protection of food and water, may lead to the outbreak of the disease. The incubation period is 12 to 72 hours. The reservoir of the infection is infected people. The disease is readily transmitted from person to person or from water, food, or fomites to people. An enterotoxigenic *E. coli* outbreak occurred in 1975 aboard two cruise ships due to contaminated crab meat. In 1980, an outbreak occurred in a Mexican restaurant and in a Texas hospital. Enteropathogenic *E. coli* has occurred for over 50 years in infant nurseries.

Hemorrhagic Colitis

This acute disease, caused by *E. coli* 015:H7 (called enterohemorrhagic strain), generally has symptoms of severe abdominal cramping, watery diarrhea (which may become grossly bloody), occasional vomiting, and possible low-grade fever. The incubation time is 2 to 9 days. The disease has been found in the Pacific Northwest, Canada, Michigan, Georgia, Missouri, Virginia, New York, Colorado, Wales, England, Europe, and Japan. The reservoir of infection is cattle, and the disease is transmitted by raw or improperly cooked ground beef or raw milk. It is not communicable from person to person. Susceptibility is general but the very young may develop renal failure and the very old may develop fever and neurological symptoms that can lead to death. This species of *E. coli* is controlled by proper refrigeration and cooking of ground beef, and with use of only pasteurized milk.

In the 1980s, outbreaks were associated primarily with hamburgers or ground beef. Approximately 30 outbreaks were recorded in the United States. Currently, the organism, which has 60 known serotypes, has been transmitted by a variety of foods including acidic ones. Apple cider, mayonnaise, sauces, alfalfa sprouts, drinking water, recreational water, and person-to-person contact have been associated with outbreaks of the disease. However, beef and beef products continue to be the primary cause. In intact cuts of beef such as steaks, roasts, briskets, and stew beef, the interior remains protected from the pathogens that may be present on the exterior of the beef, because it is unlikely for the pathogens to migrate below the surface. In nonintact beef products including beef that has been injected with solutions, mechanically tenderized, or reconstructed into formed entrees, the processing introduces pathogens beneath the surface and makes them much more difficult to eliminate.

In Japan, in 1996, an outbreak of *E. coli* O157:H7 resulted in 6000 cases of the disease and afterward 1000 secondary infections occurred in the families of the patients. Uncooked radish sprouts were the vehicle of the disease. The organisms were resistant to both acid and dryness. From 1990 to 1998 in Wales, continuing outbreaks of *E. coli* O157:H7 took place. The highest number of asymptomatic cases were among the 25- to 34-year age group, often the caretakers of the symptomatic patients. In Michigan and Virginia in 1997, simultaneous outbreaks of *E. coli* occurred, resulting from the consumption of alfalfa grown from the same seed lot. Since 1995, several outbreaks of salmonella infection have occurred from the consumption of contaminated alfalfa sprouts in United States. In 1997, in Colorado, an outbreak occurred as the result of eating frozen ground beef patties from a nationally distributed commercial brand. In 1999, at the Washington County fair in New York, a group of children were hospitalized with bloody diarrhea, with 921 people reporting diarrhea as of September 15. The outbreak was apparently due to the use of unchlorinated water to make beverages and ice. The water came from a shallow well that had high levels of coliform.

Campylobacter *Enteritis*

Diarrhea caused by *Campylobacter*, called campylobacteriosis, is an acute enteric disease of varying severity. It is caused by *C. jejuni* and *C. coli*. The organisms are an important cause of diarrheal disease in all parts of the world and in all age groups. *Campylobacter jejuni* is the most commonly reported bacterial cause of foodborne infection in the United States. It is responsible for more enteritis than either salmonella or shigella. *Campylobacter jejuni* is a Gram-negative slender, curved, and motile rod. It is a microaerophilic organism that requires reduced levels of oxygen, such as 3 to 5% oxygen and 2 to 10% carbon dioxide for optimal growth. It is relatively fragile and sensitive to environmental stresses such as drying, heating, disinfectants, and acidic conditions. The diarrhea it causes may be watery or sticky and contain blood and leukocytes. Other symptoms include fever, abdominal pain, nausea, headache and muscle pain. The infected dose is considered to be small, about 400 to 500 bacteria. The pathogenic mechanisms are not completely understood, but the organism produces a heat-labile toxin that may cause diarrhea. It may also be an invasive organism. Although complications are relatively rare, infections have been associated with reactive arthritis, septicemia, infections of nearly every organ, and Guillain–Barré syndrome (a demyelating disorder resulting in neuromuscular paralysis). *Campylobacter jejuni* infections increasingly are caused by antimicrobial-resistant strains.

In the United States, and an estimated 2.1 to 2.4 million cases of this disease occur each year. Most outbreaks have been associated with foods, unpasteurized milk, and unchlorinated water. These organisms are an important cause of traveler's diarrhea. The incubation period is 2 to 5 days, with a range of 1 to 10 days. The reservoirs of infection are farm animals, cats, dogs, rodents, and birds (including poultry). The disease may be transmitted for several days to several weeks. Chronic carrier states may occur in animals and poultry, with these possibly becoming the primary source of infection. Usually outbreaks of the disease are small, less than 50 people; however, in Bennington, VT, a large outbreak occurred (about 2000 people) when a temporary nonchlorinated water source was used as a water supply. Small outbreaks have occurred from drinking raw milk or from eating inadequately cooked or recontaminated chicken, meat, barbecued pork, and sausage. Other risk factors include traveling abroad, and contact with dogs and cats, especially juvenile pets or pets with diarrhea. The reservoir of infection includes migratory birds, seagulls, rodents, and insects that carry the organism on their exoskeletons. The retail chicken supply is especially infected with the microorganisms.

Yersiniosis

Yersiniosis enterocolitica is a small, rod-shaped, Gram-negative bacterium often isolated from clinical specimens such as wounds, feces, sputum, and mesenteric lymph nodes. It is not part of the normal human flora. *Yersiniosis pseudotuberculosis* has been isolated from the diseased appendix of humans. Both organisms have been isolated from wounds, feces, sputum, and mesenteric lymph nodes. They have also been isolated from pigs, birds, beavers, cats, and dogs. *Yersiniosis enterocolitica* has been found in ponds, lakes, meats, ice cream, and milk. The infective dose is unknown. The onset of the illness is within 24 to 48 hours after ingesting the contaminated food. The major complication is the performance of unnecessary appendectomies because one of the main symptoms of infections is abdominal pain of the lower right quadrant. The CDC estimates that about 17,000 cases occur each year in the United States. The most susceptible populations and possible complications for the disease are to the very young, the debilitated, the very old, and those people undergoing immunosuppressive therapy. The organism causes an acute enteric disease with acute watery diarrhea, enterocolitis, fever, headache, pharyngitis, anorexia, and vomiting. Numerous serotypes and biotypes are found worldwide, especially in infants and children. Transmission is by the oral–fecal route from contact with infected persons or animals, or from eating or drinking food or water contaminated with feces. The organism is communicable as long as symptoms last, which may be 2 to 3 months.

Plesiomonas shigelloides

Gastroenteritis caused by *P. shigelloides,* a facultative anaerobic, flagellated, Gram-negative, rod-shaped bacterium that has been isolated from freshwater, freshwater fish, shellfish, and many animals, is usually a mild, self-limiting disease with fever, chills, abdominal pain, nausea, diarrhea, or vomiting. In severe cases, diarrhea may be greenish-yellow, foamy, and blood tinged. Although this condition cannot yet be considered a definite cause of human disease, the organism has been isolated in the stools of patients with diarrhea, but also is sometimes isolated from healthy individuals. The disease occurs primarily in tropical or subtropical areas with rare infections reported in the United States or Europe. The mode of transmission is probably from contaminated water, drinking and recreational, or water used to rinse foods, especially shellfish that are uncooked or unheated. The mode of transmission is probably through the oral–fecal route. The incubation period is 20 to 24 hours. The reservoir of infection is not clear; it may be people or animals. A general susceptibility occurs especially with children under 15. People who are immuno-compromised or seriously ill with cancer and blood disorders are at greatest risk for severe symptoms. An outbreak occurred in New York in 1996, where *P. shigelloides* and *Salmonella Hartford* were involved. Macaroni salad, potato salad, and baked ziti were implicated. Unfiltered, untreated water contaminated the food.

Vibrio vulnificus

This bacterium causes gastroenteritis, primary septicemia, or wound infections. Generally symptoms include diarrhea and bullous skin lesions. Consumption of the microorganisms in raw seafood by people with chronic diseases, especially liver disease, may result in septic shock rapidly followed by death. Incubation time is 16 hours for gastroenteritis; the disease is sporadically found in the United States. The reservoir of infection is humans and other primates, with transmission through water, sediment, plankton, and consumption of raw shellfish. People with open wounds or lacerated by coral or fish can become contaminated in seawater. Bacteria are not communicable from person to person. The organism is controlled by avoiding seawater if one has open cuts and treating them properly if they do occur, by properly disposing of feces, and by cooking all shellfish thoroughly.

Bacterial Food Poisoning

Bacterial food poisoning or food intoxication is due to the consumption of food containing a toxin created by bacterial growth in the food.

Staphylococcus

Staphylococcus aureus is a spherical bacterium (coccus) that appears in pairs, short chains, or bunched, grapelike clusters. The organisms are Gram-positive with some strains capable of producing a highly heat-stable protein toxin that causes illness in humans. These enterotoxins need an infective dose of less than 1.0 µg to cause the intoxication. This toxin level is reached when the *S. aureus* population exceeds 100,000 per gram of contaminated food. Outbreaks of staphylococcus food poisoning, probably the most frequent kind of food poisoning, usually affect large groups of people at picnics, church suppers, hospitals, cafeterias, and other mass feeding operations. The organisms, which come from humans, are present in boils, carbuncles, pimples, hangnails, postnasal drip after colds, and wound infections. Humans then infect such foods as creams, hams, potato salad, chicken salad, ham salad, egg salad, meat and meat products, poultry, turkeys, custards, eggnogs, cream pies, eclairs, casseroles, and warmed-over foods. The organisms grow best at 50 to 120°F for a minimum of 5 hours. The bacterial growth does not alter the appearance or flavor of the food and does not produce off odors. The outbreak of the food poisoning occurs when individuals consume the enterotoxin produced by the staphylococcus. The enterotoxin causes inflammation and irritation of the stomach and intestine, resulting in vomiting and diarrhea. Because individuals vary in their susceptibility to the toxin, some become quite ill while others are not affected at all.

In July 1976, in Oakland County, MI, 87 senior citizens who were receiving hot meals through a special program fell ill with staphylococcus food poisoning. The beef base used in the preparation of the gravy was found to be the cause of the outbreak. Staphylococcal intoxication is due to exoenterotoxins from *S. aureus*, which may be present in the nose, on the skin, and in lesions of infected people and animals. The incubation period is 1 to 8 hours, usually 2 to 4 hours. The reservoir of infection is people or animals. The organisms are readily transmitted to foods to cause an intoxication.

An outbreak of staph food poisoning occurred in Texas when over 1300 children became ill from chicken salad improperly prepared, stored, and held at improper temperatures. In 1989, multiple outbreaks associated with eating canned mushrooms occurred.

On September 27, 1997, a community hospital in northeastern Florida notified the county health department about several people who had eaten a common meal and were under treatment in the emergency room for gastrointestinal illness. These individuals had attended a party where precooked ham had been served. The ham was sliced (16 lb) and placed in a plastic container ($14 \times 12 \times 3$ in.) that was covered with tinfoil and stored in a walk-in cooler for 6 hours. Those individuals who did not eat the ham stayed well, whereas the others who ate the ham became ill with nausea, vomiting, diarrhea, weakness, sweating, chills, and fatigue. The food preparer had cleaned the slicer in place instead of dismantling it.

Clostridium perfringens

Clostridium perfringens is an anaerobic, Gram-positive, spore-forming rod with toxin production in the digestive tract associated with sporulation. *Clostridium perfringens* food poisoning is caused by the consumption of large quantities (10^8) of the vegetative cells in food. Most reported outbreaks of this food poisoning have been associated with mass feeding, such as banquets at schools or hospitals and dining rooms of college residence halls.

The organisms are present everywhere, but come primarily from soil, human or animal intestinal tracts, fecal material, and sewage. Although the vegetative forms are killed by cooking the food, spore forms are not. The organisms, which are circulated through the air or by the hands of foodservice personnel, are found on meat (particularly large roast beefs and pork loins) and turkeys; in gravies and dressings; or in prepared dishes containing meat, poultry, fish, vegetables, or macaroni products. The organisms require 13 to 14 amino acids, 5 to 6 vitamins, temperatures of 60 to 125°F (the optimum temperature is 110 to 117°F), and anaerobic conditions for growth. Heating of foods such as gravies, using warmers, and slow cooling in ambient air of large pieces of meat cause the air to escape and provide excellent incubation temperatures for rapid growth of bacteria. The incubation period is 8 to 22 hours, usually 10 to 12 hours. The reservoirs of infection include the soil, the gastrointestinal tract of healthy people, and animals.

Clostridium welchii

Clostridium welchii type A food poisoning, similar to *C. perfringens* food poisoning, is usually caused by infected food handlers. Boiled, braised, steamed, stewed, or inadequately roasted meat allowed to cool slowly and then served warmed or cold the following day is the vehicle in the spread of this organism. (See *C. perfringens* for incubation period and reservoir of infection.)

Clostridium botulinum

Clostridium botulinum is an anaerobic, Gram-positive, spore-forming rod that produces a potent neurotoxin. The spores are heat resistant and can survive in foods that are incorrectly or minimally processed. Foodborne botulism is distinct from wound botulism and infant botulism. Although infant botulism and an undetermined category of botulism may have food as a source of spores, wound botulism does not. In wound botulism the spores infect a wound and produce toxins in the bloodstream that then reach other parts of the body. A few nanograms of toxin can cause the foodborne disease. *Clostridium botulinum* produces in food the most deadly toxin known to humans. A small amount of the pure toxin could kill thousands of people.

Between 10 and 34 outbreaks of botulism occur each year in the United States, mostly associated with home canned foods and occasionally commercially prepared foods. Sausages, meat products, canned vegetables, and seafood products have most frequently been involved. In 1987, eight cases of type E botulism occurred in New York and Israel from the consumption of an uneviscerated, dry-salted, air-dried whole whitefish called Kapchunka. One person died. A bottle of chopped garlic-in-oil mix caused three cases of botulism in Kingston, NY.

In 1994, a 47-year-old resident of Oklahoma was admitted to an Arkansas hospital with symptoms of progressive dizziness, blurred vision, slurred speech, difficulty in swallowing, nausea, facial paralysis, palatial weakness, and impaired gag reflex. He developed respiratory compromise and had to have mechanical ventilation. During the 24 hours before the onset of his illness he had eaten home-canned green beans and a stew containing roast beef and potatoes. The green beans were negative for botulism A toxin, but the stew was positive. The stew had been cooked, covered with a heavy lid, and left on the stove for 3 days before being eaten. It had not been reheated. An analysis of his stool culture yielded *C. botulinum* and detected type A toxin.

In 1998, the Thailand Ministry of Public Health was informed of six people who had a sudden onset of cranial nerve palsy suggestive of botulism. Apparently 13 people had been ill and 2 had died. These individuals had eaten home-canned bamboo shoots. Inadequate cooking of the bamboo shoots, anaerobic conditions in the can, and lack of an acidifier allowed the spores to germinate and produce the toxin.

Infant botulism, first recognized in 1976, affects infants under 12 months of age. It is thought to be caused by the ingestion of *C. botulinum* spores that colonize and produce toxin in the intestinal tract of infants. The only food source implicated is honey.

The organism is found throughout the world in the spore form in soil. The spore is resistant to heat, chemicals, and physical stress. Because home-canned, preserved, or processed foods may not be adequately cooked or processed, and the spore is so resistant to environmental changes, most outbreaks of botulism have been traced to home-cooked instead of commercially processed foods. However, outbreaks of botulism, as well as food containing the toxin, have been traced to commercially prepared foods. As the spores vegetate, bacteria grow and multiply, producing the toxin in anaerobic conditions during storage. Boiling the food for a few minutes destroys the toxin.

There are six known types of *C. botulinum*. Type A, associated with human illness, is the common cause of botulism in the United States. Type B, also associated with human illness, is most frequently found in soils in the world. Type C is associated with outbreaks of botulism in cattle, mink, waterfowl, and other animals. Type D, responsible for forage poisoning of cattle, is most commonly found in Africa. Type E, associated with human illness, is usually found in outbreaks associated with fish and fish products. Type F, associated with human illness, has only recently been isolated and is relatively rare. Type A and B toxins are found mainly in canned vegetables and fruits, although beef, pork, fish and fish products, milk and milk products, and condiments are also incriminated.

The key factors in the growth of *C. botulinum* and toxin production from the disrupted cells are pH, availability of oxygen, salt content, time, and temperature. A pH close to neutral favors growth, whereas a pH of 4.5, as is found in tomatoes, pears, and red cabbage, inhibits growth. Types A and B grow best at 95°F, but the temperature can range from 50 to 118°F. Type E grows best at 86°F, but can grow and produce toxin at 38 to 113°F. Although a food mass is aerobic, the conditions next to the bacterial cell may be anaerobic. Smoked fish can develop anaerobic conditions in the visceral cavity and under the skin. The interior of sausage can become anerobic.

Although some foods become foul and rancid due to growth of the organism, others show only minor changes in odor or appearance. In some foods no change occurs at all, despite the fact that the food is lethal. Vacuum-packed smoked fish is an example of a food causing serious outbreaks in recent years, yet showing no physical changes. The incubation period is 18 to 36 hours, when neurological symptoms usually occur, although it may last several days. The shorter the time, the more severe the disease and the higher the death rate. The reservoir of infection is soil marine sediment and the intestinal tract of animals and fish. Botulism is not communicable from people.

Vibrio parahaemolyticus

Vibrio parahaemolyticus is associated almost exclusively with seafood and is found in nearly all seafood products. The number of recorded cases in Japan range from 10,000 to 14,000 annually. Several outbreaks occurred in the United States when steamed crabs were eaten. The organisms can live in saltwater separate from the host. The incubation period is usually between 12 and 24 hours, but it can range from 4 to 96 hours. The reservoir of infection during the cold season is in marine areas, where the bacteria are found free in water or in fish and shellfish during the warm season.

Both pathogenic and nonpathogenic forms of the organism can be isolated from the marine and estuarine environments and from fish and shellfish found there. *Vibrio parahaemolyticus* is a Gram-negative bacteria. The symptoms of the disease include diarrhea, abdominal cramps, nausea, vomiting, headache, fever, and chills. The illness is usually mild to moderate although some cases may be hospitalized. An infective dose of >1 million organisms may cause the disease. All individuals who eat fish or shellfish are susceptible; however, the infection is not spread from person to person.

In 1997, raw oysters harvested from California, Oregon, Washington, and British Columbia and contaminated with *V. parahaemolyticus* caused illness in 209 people who ate them. The individuals reported diarrhea, abdominal cramps, nausea, vomiting, fever, and bloody diarrhea.

In 1998, an outbreak of *V. parahaemolyticus* was associated with eating raw oysters and clams from Long Island Sound. These individuals lived in Connecticut, New Jersey, and New York. Of the 23 sick people 22 had eaten or handled oysters, clams, or crustaceans. This was the fourth multistate outbreak of this infection in the United States since 1997.

Aeromonas hydrophila

Aeromonas hydrophila is a bacterium found in all freshwater environments and in brackish water. Some strains are capable of causing illness in fish and amphibians, as well as in humans who may acquire the infections through open wounds or by ingestion of sufficient numbers of the organisms in food or water. Some controversy exists as to whether the organism can cause human gastroenteritis. All people are susceptible to the disease although it is most frequently observed in very young children.

Bacillus cereus, *Food Poisoning*

Bacillus cereus food poisoning is a gastrointestinal disorder in which a sudden onset of nausea and vomiting, and in some cases diarrhea, can occur. The symptoms typically last about 24 hours. The infectious agent is *B. cereus,* which is a Gram-positive facultative aerobic spore former. The spore, when becoming a vegetated cell, produces two enterotoxins that are heat stable. These cause vomiting. One other enterotoxin is heat labile. This causes diarrhea. The incubation period is 1 to 6 hours where vomiting occurs, and 6 to 16 hours where diarrhea is most prominent. It is not communicable from person to person.

Chemical Poisoning

Chemicals accidentally introduced into food or leached into food from a variety of containers cause rapid illness and death at times. The incubation period is less than 30 min. The cause of the outbreak varies with the chemical and its origin.

Antimony

Antimony is leached from chipped gray enamelware by acidic foods. A typical antimony poisoning is caused by storage of a lemon punch in a large, chipped gray enamel pot.

Arsenic

Arsenic is found in ant, roach, or rodent baits; insecticidal fruit sprays; and herbicides. These poisons accidentally contaminate foods. Because arsenic is an accumulative poison, small doses over extended periods of time lead to poisoning. Arsenic is also a fairly frequent contaminant of drinking water.

Cadmium

Cadmium, present in plating materials of containers and trays, is leached into food by acidic substances, such as fruit juices. Cadmium might also seep from industrial operations into a water supply.

Chlorinated Hydrocarbons

Chlorinated hydrocarbons are synthetic chemical pesticides, such as dichlorodiphenyltrichloroethane (DDT), lindane, and endrin. Because these pesticides have been used extensively and are fat soluble, they can create long-term hazards. Excess quantities of these poisons cause nervous system disorders. The main modes of entry into the environment in an agricultural setting are through the air, soil, or water. This may occur through air spraying of chemicals, ground spraying, exposure from industrial sources, and spreading of sludge on soil. The chemicals may enter the groundwater supply, directly contaminate the food supply, or be bioconcentrated in the food chain.

Copper

Copper poisoning is caused by the leaching of copper from food contact surface, such as carbonated water-machine tubes, into food or drink.

Cyanide

Cyanide, found in silver polishes, can accidentally contaminate foods.

Lead

Lead is present in lead arsenate, a pesticide used on apples. Lead may also be leached by acidic beverages from improperly glazed pottery vessels and utensils. Automobile radiators used as condensers for illegally distilled whiskey add lead to the final product.

Two cases of pediatric lead poisoning were traced to the consumption of imported candy and foodstuffs. In 1997, family members revealed that a maternal aunt had returned from Mexico with tamarindo candy jam products packaged in ceramic jars, although this had been illegal since 1993. The two affected children had consumed these food products.

In 1997, in Connecticut, adult lead poisoning was traced to an Asian remedy for menstrual cramps. This was a previously unrecognized source of lead. However, it was known that adulterants including lead are present in Asian traditional or folk medicines. Folk remedies and cosmetics from East Indian, Pakistani, Chinese, and Latin American cultures have previously been shown to contain lead.

Mercury

Mercury, commonly found in many industrial processes, commercial products, and homes, is an insidious chronic poison, producing symptoms that resemble emotional and psychological disorders. Mercury poisoning also appears in an acute form. Methyl mercury, the principal form of mercury found in food, penetrates the placenta, causing birth defects. It is also found in mother's milk. Fungicides containing mercury are used on grain consumed by farm animals that are ultimately consumed by humans, leading to severe mercury poisoning and eventual death. Fish and shellfish that have consumed mercury and are later used as food also cause blindness, paralysis, and death in humans.

Organic Phosphate Compounds

Organic phosphates, such as dimethyldichlorovinyl phosphate (DDVP), parathion, and tetraethylpyrophosphate (TEPP), are used as insecticides, fungicides, and herbicides. Although those compounds in this group of poisons are most dangerous when inhaled, they also constitute a hazard when ingested with food.

Cholinesterase-inhibiting pesticides such as organic phosphates and carbamates, which are widely used in agriculture, can cause illness if they contaminate food or drinking water. Aldicarb is a regulated carbamate pesticide that is highly toxic. In 1998, 20 employees attended a company lunch prepared from homemade foods. Shortly after eating, several people developed neurological and gastrointestinal symptoms. Ten went to the hospital emergency room and two were hospitalized. Symptoms included abdominal cramps, nausea, diarrhea, dizziness, eye twitching, and vision problems. The lunch consisted of pork roast, boiled rice, cabbage salad, biscuits, and soft drinks. Only the cabbage salad was associated with the illness. It was prepared from two 1-lb bags of precut, prepackaged, cabbage in a bowl with vinegar and ground black pepper. The black pepper, on testing, revealed that it was contaminated with aldicarb. Aldicarb has also been implicated in the contamination of watermelons and cucumbers.

Polychlorinated Biphenyls

Polychlorinated biphenyls (PCBs) are found accumulated in fat tissues of wildlife in much of the world. PCBs are also found in fish; animal feed, which leads to contaminated meat, milk, and eggs; paints that have leached into feed; heat exchange fluids; and cardboard cartons from recycled paper. The potential for food poisoning increases with the increased availability of this family of over 200 chemicals resembling DDT that are stored in the fatty tissues. PCB poisoning has occurred in Japan. In 1968 in 1 reported outbreak, 1000 people became ill, resulting in 5 deaths from ingesting rice oil contaminated with PCBs.

Zinc

Zinc poisoning is caused by zinc leaching into acidic foods stored in galvanized containers.

Poisonous Plants and Fungi

Certain plants and fungi are poisonous. Castor beans contain a toxin called ricin; ergotism is caused by a parasitic fungus of rye (*Claviceps purpurea*); favism is due to the bean, *Vicia fava*; poisonous mushrooms are often fatal; shellfish poisoning is due to eating shellfish that have consumed a plankton called *Gonyaulax*; and cyanide is produced in green or sunburned potatoes and wild celery.

Aflatoxins

Aflatoxins are formed by molds on foods at harvest time, in storage, or in conditions of water damage. Aflatoxins are found in peanuts, brazil nuts, pecans, copra, corn, and cottonseed. Aflatoxin, known as a very potent carcinogen for certain animal species, can be produced whenever proper temperature, time, humidity, and proper strain of *Aspergillus flavus* is present. It is thought that some aflatoxins in Africa might be associated with liver cancer.

Aflatoxicosis is a poisoning that results from the ingestion of aflatoxins in contaminated food or feed. Aflatoxins produce acute necrosis, cirrhosis, and carcinoma of the liver in a number of animal species. In northwest India in the fall of 1974, 397 people became ill with aflatoxin poisoning and 108 people died. Contaminated corn was the major dietary constituent. The aflatoxin levels were between 0.25 and 15 mg/kg. The patients experienced high fever, rapid progressive jaundice, edema of the limbs, pain, vomiting, and swollen livers. Examinations showed extensive bile duct proliferation and periportal fibrosis of the liver together with gastrointestinal hemorrhage. In 1982, in Kenya, 20 people were hospitalized after consuming 38 μg/kg of body weight of aflatoxins. The relative frequency of aflatoxicosis in humans in the United States is unknown.

Fish Intoxications

Fish intoxications are caused by eating buffalo fish (Haff disease), tropical and subtropical marine fish (Ciguatera poisoning), tuna and mahi-mahi (scombroid poisoning), shellfish (paralytic, diarrheic, neurotoxic, and amnesia shellfish poisoning), and pufferfish (pufferfish, tetradon, fugu poisoning). Each of these diseases is caused by a toxin found in the seafood. The symptoms can vary from gastrointestinal to neurological to death.

Grayanotoxin

Grayanotoxin, or honey intoxication, is caused by the consumption of honey produced from the nectar of rhododendrons. The intoxication is rarely fatal and generally lasts for no more than 24 hours. The symptoms of the disease are dizziness, weakness, excessive perspiration, nausea, and vomiting. Other symptoms are low blood pressure or shock. The disease is rare in humans.

Gastrointestinal Basidiobolomycosis

Gastrointestinal basidiobolomycosis (GIB) is a new or emerging infection that may have been previously misdiagnosed as cancer or inflammatory bowel disease. Basidiobolomycosis of the gastrointestinal tract is very rare in people. As of August 1998, only six cases had been described in the world's medical literature, with two of them in the United States. In March 1999, the Arizona Department of Health Services notified the CDC about six cases of GIB, with three of them reported between January and March 1999, compared with three cases in the previous 5 years. The disease is caused by an invasive fungus. It usually begins with pain, and sometimes a mass that can be felt in the abdomen. The fungus is found mainly in the soil and on decaying vegetation. It has also been isolated from the river banks of tropical rivers in West Africa and has been found in association with some insects. The fungus is present in the gastrointestinal tract of reptiles, amphibians, and some bats. Surgical cultures of infected people have grown out *Basidiobolas ranarum*. There is a question whether eating contaminated fish or unwashed vegetables could be a means of transmission of the disease.

Fungi

Fungi may invade plants, causing them to undergo metabolic changes that produce toxic substances. The phytoalexins are examples of this. Genetic manipulation also produces new plant varieties having a stress mechanism that causes problems to the human being. It is known that compounds inhibiting the proteolytic activity of certain enzymes are found throughout the plant kingdom, particularly among the legumes. One of the better known inhibitor reactions is the inhibition of trypsin. Protease inhibitors may also be found in peanuts, oats, chickpeas, field beans, buckwheat, barley, sweet potatoes, rice, lentils, lima beans, navy beans, garden peas, white potatoes, wheat, and corn. Trypsin is essential for adequate digestion and utilization of food. The presence of the inhibitors decreases the digestion and absorption of the proteins, causing reduced growth and decreased efficiency in utilizing food. Because protein is the most costly part of the diet, the trypsin inhibitors create an economic as well as a public health effect.

Lectins

Certain plants contain substances that agglutinate red blood cells. These substances are called phytohemagglutinins or lectins. They are present in seeds, in red kidney beans, and in leaves, barks, and roots to a lesser extent. They produce important environmental toxins, such as ricin, from the castor bean. The acute disease is known as red kidney bean poisoning, kinkoti bean poisoning, etc. The time of onset is 1 to 3 hours after eating raw or undercooked kidney beans, with symptoms of extreme nausea, vomiting, and diarrhea. The toxic agent is at its highest concentration in red kidney beans that are uncooked. Outbreaks have occurred in the United Kingdom, whereas in the United States there are only anecdotal reports available. All persons regardless of age or gender are susceptible to the toxin.

Mushroom Poisoning

Mushroom poisoning is caused by the consumption of raw or cooked mushrooms or toadstools. Unfortunately, most individuals who are not experts in mushroom identification cannot recognize the differences between the poisonous and nonpoisonous species. The toxins involved in mushroom poisoning are produced naturally by the fungi themselves. Most mushrooms that cause human poisoning cannot be made nontoxic by cooking, canning, freezing, or any other means of processing. Mushroom poisonings are generally acute. The four categories of mushroom toxins are protoplasmic toxins that result in generalized destruction of cells followed by organ failure; neurotoxins that result in neurological symptoms such as coma, convulsions, hallucinations, excitement, and depression; gastrointestinal irritants that result in rapid, transient nausea, vomiting, abdominal cramps, and diarrhea; and disulfiram-like toxins that do not cause any symptoms unless the individual has consumed alcohol within 72 hours after eating them, and then a short-lived acute toxic syndrome occurs. Protoplasmic toxins include amatoxins (time of onset is 6 to 48 hours), which produce a poisoning with onset of symptoms of severe abdominal pain, persistent vomiting, watery diarrhea, extreme thirst, lack of urine production, loss of strength, pain-induced restlessness, and death due to irreversible liver, kidney, cardiac, and skeletal muscle damage; hydrazines (time of onset is 6 to 10 hours), which produce abdominal discomfort, severe headache, vomiting, diarrhea, liver damage, blood cell damage, and CNS damage; and orellanine (time of onset is 3 to 14 days), which produces an intense burning thirst, excessive urination, nausea, headache, muscular pains, chills, spasms, loss of consciousness, and kidney failure that may result in death. Neurotoxins include muscarine (time of onset is 15 to 30 min), which causes increased salivation, perspiration, lacrimation, abdominal pain, severe nausea, diarrhea, labored breathing, blurred vision, and rarely death resulting from cardiac or respiratory failure; ibotenic acid and muscimol poisoning (time of onset is 1 to 2 hours), which produces drowsiness, dizziness, hyperactivity, excitability, delusions, delirium, rarely death in adults but more often in children (muscimol is approximately five times more potent than ibotenic acid); and psilocybin toxin (time of onset is usually very rapid and the effects generally subside within 2 hours) produces a syndrome similar to alcohol intoxication sometimes accompanied by hallucinations in adults whereas in small children fever, convulsions, coma, and death may occur. The gastrointestinal irritants include the toxins produced by numerous mushrooms with symptoms of nausea, vomiting, diarrhea, and abdominal cramps. The time of onset is very rapid.

Plants

Certain substances, known as hepatotoxins, that are toxic to the liver are found in plants. These plants are poisonous to livestock and humans. The seeds of these plants, which are of the genus *Senecio*, contaminate wheat and corn harvested from the same land.

Pyrrolizidine Alkaloids Poisoning

The alkaloids are found in flour as well as in milk from cows that have been feeding on contaminated plants. The toxins affect the liver. Very few cases have been reported in the United States.

Goitrogens

Goitrogens, natural products found in plant foods eaten by humans and animals, cause hypothyroidism, which is an enlargement of the thyroid gland. Consumption of additional iodine controls this problem. The plants in this group of the genus *Brassica* include cabbage, turnip, mustard greens, radish, and horseradish.

Saponins

Saponins, comprising at least 400 species and 80 families of plants, produce the toxin known as glycoside that hemolyze red blood cells. Saponins are found in soybeans and alfalfa, and are of particular concern, because soybeans and alfalfa are highly nutritious plants that are essential to our society.

Mycotoxins

Mycotoxins, produced by fungi, produce acute and chronic effects. Many humans and animals suffer with these fungal metabolites. Cases that come to the attention of public health officials are those in which large amounts of the toxin have been consumed and result in serious illness or death. Mycotoxins are secondary metabolites of molds that may be toxic, carcinogenic, or hepatotoxic.

Species of *Aspergillus*, *Penicillium*, *Rhizopus*, and *Streptomyces* produce aflatoxins. Species of *Fusarium* produce trichothecenes. Species of *Penicillium* and *Aspergillus* produce penicillic acid.

Three major classes of molds invade agricultural products. They are field fungi, storage fungi, and advanced decay fungi. The physical factors that bring about mycotoxin production include moisture, temperature, mechanical injury, blending of grains, hot spots, and time. A variety of favorable conditions must be present for the mycotoxin for the mycotoxins to be produced. Aflatoxins can grow on peanuts, cotton seed, seed oil, corn, legumes, dried fruits, wines, and dairy products. Zearalenone is found almost totally on corn. Ochratoxin is found on food grains such as wheat and barley. Citrinin is a cocontaminate with ochratoxins. Patulin is produced on apple rot and is found in apple juice. Penicillic acid is found on tobacco and storage grain, such as wheat, corn, peanuts, and cotton seed. Trichothecenes are found on a variety of different foods. Aflatoxins not only cause toxic effects but also may be mutagenic and carcinogenic. When two or more toxins are combined, there may be a variety of toxic effects on experimental animals. Mycotoxins cause toxic effects in the liver, digestive tract, urinary system, skin, hematopoietic system, reproductive organs, and nervous systems.

Allergic Reactions

Individuals may have peculiar allergic reactions to any kind of food, ranging from simple headaches to unusual CNS disorders, including convulsions and death. Infants and children often complain of abdominal distress and sometimes of genitourinary tract problems. The cardiovascular system and skin can be involved in an allergic reaction.

Parasitic Infections

Most parasitic infections are due to the ingestion of parasites in food or water.

Amoebic Dysentery

Amoebic dysentery, or amebiasis, is caused by *Entamoeba histolytica*. It is a single-celled parasite, a protozoan, that infects predominantly humans and other primates. Dogs and cats can become infected but do not usually shed the cysts in their feces and therefore do not significantly contribute to the transmission of the disease. The infections may be accompanied by no symptoms, vague gastrointestinal distress, and dysentery with blood and mucous. Complications include ulcerative and abscess pain and, rarely, intestinal blockage. The infectious dose is theoretically the ingestion of one viable cyst. Water or food is contaminated with sewage or human feces containing the amoeba. Disease rates are higher in areas of poor sanitation, at mental institutions, and among homosexuals. It may cluster in households or institutions where sanitation is good. The incubation period varies from a few days to several months or years. Commonly it is 2 to 4 weeks. The reservoir of infection is a person who is usually a chronically ill or asymptomatic cyst passer. Communicability may continue for years. The most dramatic incident of an amoebic dysentery outbreak occurred in 1933 at the Chicago World Fair where 1000 people became infected and 56 died due to defective plumbing.

Cryptosporidiosis

Cryptosporidiosis is caused by *Cryptosporidium parvum*, a protozoa, which is an obligate intracellular parasite. It is currently thought that the form infecting humans is the same species that causes disease in young calves. The organism infects many herd animals. The infective stage of the organism, the oocyst, is 3 μm in diameter or about one half the size of a red blood cell. The symptom of the intestinal disease is severe watery diarrhea, but the individual may also be asymptomatic. In the pulmonary form of the disease the individual coughs and frequently has a low-grade fever accompanied by severe intestinal distress. The infectious dose is less than ten organisms. The organism could theoretically be transferred to any food that has been touched by a contaminated food handler. The incidence of the disease is higher in child day care centers. AIDS patients may have the disease for life, with the severe watery diarrhea contributing to death. The organism may also invade the pulmonary system and therefore be fatal.

During 1987, a waterborne outbreak of the disease caused illness in approximately 13,000 people who were exposed to the contaminated drinking water. On December 29, 1997, the Spokane, WA, Regional Health District received reports of acute gastroenteritis among people who attended a dinner catered by a local restaurant on December 18. There was a long (3 to 9 days) incubation period and diarrhea among those who became sick. The banquet buffet included 18 separate food and beverage items with 7 items containing uncooked produce. The food was prepared or served by 15 food workers, 2 of whom tested positive for the organism. Since 1993, three foodborne outbreaks of cryptosporidiosis in the United States have occurred. Outbreaks have been associated with drinking unpasteurized, fresh-pressed apple cider probably contaminated from a cow pasture, chicken salad contaminated by a food worker who operated a day care facility in her home, and apples contaminated by well water that had fecal material in it. The apples were used to make cider.

Cyclosporiasis

Cyclosporiasis is caused by the protozoan *Cyclospora cayetanenis*. To be infectious, the spherical, chlorine-resistant oocyst (8 to 10 mm) found in the feces of infected people must sporulate in the environment, a process that takes several days. The oocysts are variably acid fast. The incubation time is 1 week. Symptoms include watery diarrhea, frequent stools, loss of appetite, weight loss, nausea, vomiting, fatigue, and low-grade fever. Before 1996, only three outbreaks of this infection had occurred in the United States, whereas between May 1 and July 1996, almost 1000 laboratory-confirmed cases were reported to the CDC. Raspberries from Guatemala have been implicated in outbreaks including Canada in 1998. Cyclospora have also been found in the feces of chickens. The consumption of undercooked meat and exposure to contaminated drinking water are additional sources of this disease.

Trichinosis

Trichinosis, which is caused by *Trichinella spiralis*, has an incubation period of about 9 days, although it varies from 2 to 28 days. In heavy infections, the incubation period may be 24 hours. The live larvae are found in raw or improperly cooked pork, pork products, whale, seal, bear, or walrus meat.

From November 1998 through January 1999, in Germany, 11 different cities and districts reported 52 cases of trichinellosis. Symptoms included myalgia, fever, headache, facial edema, and diarrhea. The individuals had eaten raw or partially cooked ground meat including pork or raw pork sausage.

Tapeworms

Tapeworms are caused by the ingestion of live larvae in raw or insufficiently cooked beef, fish, or pork. The beef tapeworm is *Taeniasis saginata*; the fish tapeworm is *Diphyllobothrium latum*; and the pork tapeworm is *Taenia solium*. The beef tapeworm causes intestinal infections in people, whereas the pork tapeworm may cause intestinal infections and somatic infections by the larva. The larval disease, which is called cysticercus, is a tissue infection that may affect vital organs and therefore may cause fatality to occur. The beef tapeworm causes the intestinal infection, whereas the pork tapeworm may cause both diseases to occur. The fish tapeworm causes, in some individuals, a vitamin B_{12} deficiency anemia along with diarrhea. The incubation period for the pork tapeworm is 8 to 12 weeks. The incubation period for the fish tapeworm is 3 to 6 weeks. The reservoir of infection for beef and pork tapeworms is people. The intermediate host for the beef tapeworm is cattle, whereas the intermediate host for the pork tapeworm is pigs. The reservoir of infection for fish tapeworms is people and other hosts, including dogs, bears, and fish-eating mammals. The beef tapeworm may be communicable for as long as 30 years. The fish tapeworm is not communicable from person to person. An outbreak of *D. latum* occurred in Los Angles in 1980 when 39 people consumed sushi (a raw fish dish made of tuna, red snapper, and salmon). The salmon appeared to be the fish incriminated.

Eustrongylides

Larval *Eustrongylides* species are large, bright red roundworms, 25 to 150 mm long and 2 mm in diameter. They are found in freshwater fish, brackish water fish, and marine fish. The larva normally mature in wading birds such as herons, egrets, and flamingos. When the larvae are eaten in undercooked or raw fish, they can attach to the wall of the digestive tract and penetrate causing severe pain through the perforation of the gut and other organs. One live larva can cause an infection. The disease is very rare in the United States.

Roundworms

Roundworms may cause ascariasis and trichuriasis. The eggs of *Ascaris lumbricoides* and *Trichuris trichuria* cause these diseases when they are carried to the mouth by hands, other body parts, fomites, or foods. Infection with numerous *Ascaris* worms may result in a pneumonitis during the migratory phase when the larva that have hatched from the ingested eggs in the lumen of the small intestine penetrate into the tissues and by way of the lymph and blood systems reach the lungs. The larvae break out of the pulmonary capillaries into the air sacks, ascend into the throat, and descend to the small intestine where they grow. The *Trichuris* larvae do not migrate after hatching but molt and mature in the intestine. Although there are no major outbreaks, the occurrence of these diseases is quite common in individuals.

Anisakis

Anisakiasis is caused by *Anisakis simplex* and other roundworms through the consumption of raw or undercooked seafood. Symptoms may occur from 1 hour to 2 weeks after consumption of the food. The symptoms include a tingling or tickling sensation in the throat with coughing up of a nematode, acute abdominal pain similar to acute appendicitis, and nausea. Although small numbers of cases of the disease are reported, it is felt that many more people are becoming infected.

Gnathostomosis

Gnathostomosis is a foodborne zoonotic disease caused by several species of the nematode *Gnathostoma*. Between 1993 and 1997, in Acapulco, Mexico, 98 cases of this disease were clinically identified. The adult parasites are found in the stomachs of mammals such as dogs and cats. The feces containing the ova reach water, where the free-swimming, first-stage larvae are formed and then ingested by the minute copepod crustacean cyclops and the second stage larvae occur. Freshwater fish that eat cyclops are the second intermediate host. The larvae develops into the third state in the fish muscles. Consumption of this fish by cats, dogs, or other mammals results in development of the adults in the gut, thereby finishing the cycle. Humans acquire the infection by consuming raw or undercooked freshwater fish. The symptoms of the disease include swelling in the skin or in the organ affected by larvae. If the parasite migrates to a vital organ, it can cause severe illness or death.

Giardiasis

Giardiasis is a protozoan infection, usually of the upper small intestine, which is caused by *Giardia lamblia*. The symptoms of the disease, when they are there, include chronic diarrhea, abdominal cramps, fatigue, and weight loss. The disease is transmitted by contaminated water and less frequently by contaminated food. Person-to-person contact may also occur and may also contribute to the spread of the disease when the hands of a contaminated person transfer the cysts from the feces of the individual to the mouth of another individual. This may occur in a variety of institutions, including day care centers. The incubation period is 5 to 25 days or longer, with a possible 7- to 10-day period most frequent. The reservoir of infection is people. The period of communicability lasts during the entire period of the infection. One or more cysts may cause the disease.

Angiostrongyliasis

Angiostrongyliasis is a disease of the CNS that is caused by a nematode, *Angiostrongylus cantonensis*. The disease is spread through the ingestion of raw or improperly cooked snails, slugs, prawns, fish, and land crabs. The incubation period is usually 1 to 3 weeks, although it may be longer or shorter. The reservoir of infection is the rat. The disease is not transmitted from person to person.

Toxoplasmosis

Toxoplasmosis is a protozoal disease that may be spread from the mother to the child through the placenta, if she is infected, but may also be acquired by eating raw or undercooked infected pork or mutton. The disease may also be spread through water or dust contaminated by cat feces. The infectious agent is *Toxoplasma gondii*. The incubation period may be 10 to 23 days when consuming contaminated food. The reservoir of infection includes rats, pigs, cattle, sheep, goats, chickens, birds, and cats. The disease is not transmitted from person to person.

Scombroid Poisoning

Scombroid poisoning is caused by eating scrombroid fish or eating fish of the family mahi-mahi, which is also called dolphin fish. The poisoning is due to a histamine-like substance produced by several species of *Proteus* bacteria or other bacteria. The histamine is produced from histidine in the flesh of the fish. The individual person has headaches, dizziness, nausea, vomiting, peppery taste, burning throat, facial swelling and flushing, stomach pain, and itching of the skin. The disease may also be caused by tuna or mackerel. The cause of the disease is inadequate refrigeration of scromboid fish. The incubation time is 10 min to 1 hour. The disease is not transmitted from person to person, and the reservoirs of infection are the fish that have been mentioned.

CHRONIC AFTEREFFECTS OF FOODBORNE DISEASE

Several researchers had estimated that there are chronic aftereffects in 2 to 3% of foodborne disease cases, which result in increased morbidity and potential costs in the billions of dollars each year. Several bacteria, including salmonella, induce septic arthritis by spreading to the synovial space, thereby causing an inflammation. *Yersinia enterocolitica, Shigella flexneri, Campylobacter jejuni*, and *Escherichia coli* initiate aseptic or reactive arthritis, an acute, nonpurulent joint inflammation following infection elsewhere in the body.

Graves disease is an autoimmune thyroid disease mediated by autoantibodies to the thyrotropin receptor. The disease may be linked to an infection with *Y. enterocolitica* serotype 0:3. A suppressor cell dysfunction may be involved in Graves disease. Severe hypothyroidism may also result from chronic intestinal giardiasis due to an infection by *G. lamblia*.

Crohn disease and ulcerative colitis, inflammatory bowel diseases, have clinical characteristics of diarrhea, abdominal pain, fever, and weight loss. Although the cause of these diseases and the mechanisms for spontaneous exacerbations and remissions are unknown, much research has been done on transmissible agents, including foodborne pathogens. Recent immunocytochemical techniques demonstrated antigens to *Listeria monocytogenes, E. coli*, and *Streptococcus* species in Crohn disease tissues.

Superantigens interact with the T cell receptor by recognizing elements shared by a subset of T cells. Depending on the type of interaction or recognition there can be different consequences including proliferation and expansion, suppression, or alternation of the T cell receptor. Several foodborne bacteria including *staphylococcus, streptococcus, yersinia*, and *clostridium* have superantigens. Many of these are thought to be associated with several autoimmune disorders such as rheumatic heart disease, rheumatoid arthritis, multiple sclerosis, Graves disease, and Crohn disease.

Renal disease may follow colitis caused by *E. coli* O157:H7 and other enterohemorrhagic strains of *E. coli*. Hemolytic uremic syndrome can develop with some patients having acute renal failure, which is the leading cause of death in children; and thrombocytopenia, which is the leading cause of death in adults.

Guillain–Barré syndrome is a subacute, acquired, inflammatory demyelinating polyradiculoneuropathy that frequently occurs after acute gastrointestinal infection. The disease occurs worldwide and is the most common cause of neuromuscular paralysis.

Ciguatera poisoning is the most common foodborne disease related to the consumption of fin fish. The clinical syndrome is characterized by gastrointestinal, neurological, and sometimes cardiovascular problems. Amnesic shellfish poisoning first causes gastrointestinal symptoms followed by neurological dysfunction. Severe cases may be prolonged to become chronic with confusion, disorientation, lack of response to deep pain, dysfunction of the autonomic nervous system, seizures, loss of reflexes, and coma.

Toxoplasmosis causes congenital impairments including hearing loss, visual impairment, and slight to severe mental retardation. It may also cause encephalitis when the individual's immune system is impaired such as in AIDS.

Helminth parasites can cause serious disease in infected people. More than 1 billion people are infected with the largest intestinal nematode, *Ascaris lumbricoides*, which may cause an allergic response generated by the lung migratory phase resulting in pneumonia and the spontaneous development of idiopathic bronchial asthma. Viral agents induce autoimmune disorders potentially by molecular mimicry. Hepatitis A causes an acute hepatitis with jaundice in adults.

Metabolic activation is an important part of the toxic response to humans to mycotoxins. The secondary metabolites are highly toxic. They easily enter the human food supply because they are resistant to food processing and do not degrade at high temperatures. Ochratoxin A is nephrotoxic and carcinogenic. Aflatoxins have been implicated in both acute and chronic liver disease in humans. Other organs such as the kidney, spleen, and pancreas may also be affected.

Several foodborne pathogens have been either directly or indirectly associated with endocarditis and myocarditis, with heart damage that appears to be permanent. Persons with ankylosing spondylitis linked to enteric pathogens as the trigger show a high incidence of cardiac conduction abnormalities.

Enteric pathogen-induced diarrhea may lead to a variety of conditions including loss of fluids, anorexia, malabsorption of nutrients, and malnutrition. The diarrheal episodes may become chronic. Deaths due to diarrheal illness in the immunosuppressed and in persons with AIDS is nearly 80%.

Helicobacter pylori is the cause of chronic gastritis. The organism can survive in water, chilled foods, milk, and fresh vegetables for several days. It leads to acute gastritis and colonization of the stomach accompanied by chronic inflammation that may last for many years unless treated.

Evaluation of Risks Related to Microbiological Contamination by Food Workers

From 1975 to 1998, a total of 93% of the outbreaks involved food workers who were sick either prior to the outbreak or at the time of it. They were believed to be the source of infection.

It is believed that the cost of outbreaks due to foodservice workers can in some instances far exceed the costs associated with outbreaks due to person-to-person contact. These costs include the expenses involved in controlling the disease, medical treatment, business losses, losses in productivity for the initial cases, secondary cases, and potential exposure to many patrons.

Hand contact with food is a consistent mode of contamination and transmission of enteric disease, which may be either viral or bacterial in nature. Hepatitis A is an example of a disease spread by hand-contaminated foods that are not cooked or are improperly cooked. Ready-to-eat foods that are not washed, cooked, or receive additional preparation before consumption represent a serious threat to the consuming public. From 1988 to 1992, authorities reported 1435 outbreaks of disease included contributing factors. The most common reported practices contributing to the foodborne disease were improper holding temperatures of food (59%) and poor personal hygiene of food workers (36%). Specific food items implicated as vehicles for transmission were sandwiches, salads, and miscellaneous hot food items such as mashed potatoes and ham. Other foods included baked goods, beverages, fruit salads, and miscellaneous cold foods. Workers either wore gloves improperly or not at all, neglected hand washing, added parsley or other garnishments to dishes, and sliced meat or deboned chicken with bare hands.

Hand Washing

Hand washing is the removal of soil and microorganisms from the hands by means of friction and the use of detergents. Two types of microflora on the skin are resident and transient. The resident bacteria are normally found on the skin, in this case the hands. They may include *Staphylococcus aureus* as well as other organisms. They are not easily removed by mechanical friction because they are buried deep within the pores and are protected by sebaceous gland secretions. The transient organisms are of concern because they are readily transmitted by the hands unless removed by the mechanical friction of washing with soap and water or destroyed by the use of an antiseptic. These organisms are acquired from environmental sources and become attached to the outer epidermal skin. Hands as well as contaminated gloves serve as a means of transmission of disease, especially when they are wet.

SPECIAL FOOD PROBLEMS

The Vessel Sanitation Program was established in 1975, as part of the CDC, as a cooperative activity with the cruise ship industry. The goal of the program is to promote a level of food safety on passenger vessels that will lower the risk of gastrointestinal disease outbreaks and ensure a healthy environment for both passengers and crew. A comprehensive evaluation is made of the vessel twice a year and other reinspections are made as needed. The inspection consists of the ship's water supply; food protection during storage, preparation, and service; potential contamination of food and water; employee practices and personal hygiene; and general cleanliness, facility repair, and vector control. This program was adopted because of outbreaks of disease reported to the CDC from a variety of cruise ships. The original manual was based on the 1976 model food code. This has been updated numerous times and a draft titled "Vessel Sanitation Program Operations Manual," September 1999, is now in use.

If an outbreak of disease involves a foreign flag vessel, it is reported immediately to the Foreign Quarantine Section of the CDC. If an outbreak of disease occurs on an interstate conveyance such as a plane, bus, train, or vessel, it is the responsibility of the Food and Drug Administration (FDA). Interviews are conducted with the sick passenger, family members, or health professionals to determine the meal that was involved, the time of onset of symptoms, the history of eating suspect foods, and other potential exposures to the individual. The carrier is asked to reveal the names and phone numbers of other passengers so that the environmentalists can conduct a proper epidemiological investigation. The various health jurisdictions where the passengers reside need to be contacted and advised of the nature of the health problem.

Church dinners are usually prepared by volunteers in their homes and then transported at varying times and temperatures to the place of consumption of the food. Time, temperature, personal health, methods of preparation, and initial ingredients are uncontrolled. This has led to outbreaks of foodborne disease and intoxication.

Although fairs and special events are limited timewise, they provide an environment where tens of thousands of people congregate and eat. The food operations should be similar to a permanent food operation but usually are not. Problems include lack of hot and cold running water, inadequate refrigeration, unwholesome food, poor handling of food, flies and other insects, lack of screening, and improper waste disposal. These food operations must be licensed and inspected by the health department.

Vendors that sell food to the public from vehicles of any type must meet the same criteria as permanent food establishments. Problems related to vendors include lack of hot and cold running water, improper waste and liquid waste disposal, lack of refrigeration, improper cleaning and sanitizing, and inability of the health department to make inspections because of the high degree of mobility of the vendors. Sanitary conditions deteriorate very quickly in this type of operation. All vendors must be licensed.

Food stores must adhere to all the regulations of other preparation and serving establishments. Problems usually include unwholesome food or damaged and rusted cans, improper refrigeration, stacking above freezer line, inadequate cleaning of equipment and utensils, flies, roaches, rodents, and inadequate solid waste disposal.

The remnants of consumed food create solid waste problems. If garbage is improperly wrapped or stored, it becomes a fly and rodent attraction. If it is contaminated by highly infectious individuals, it becomes an infection problem. When garbage is removed to a landfill, it becomes part of the overall landfill solid waste disposal problem. When burned it creates air pollutants.

The individual infected with foodborne disease organisms may become a carrier and transmit the infection to others. Although the initial spread of disease may be due to a foodborne organism, the secondary occurrences may well be traced to other environmental factors. An example of this would be the serious and prolonged outbreaks of *Salmonella derby* that occurred in U.S. hospitals in the 1960s. The initial outbreak was thought to be caused by contaminated eggs. The secondary outbreaks were caused by the carrier spreading the organism in a variety of ways.

Individuals who become ill from foodborne disease require treatment either at home or in hospitals. The estimated cost to society of foodborne illness is between $6.5 and $35 billion.

POTENTIAL FOR INTERVENTION

The potential for intervention in food problems related to food protection and disease outbreaks consists of the techniques of isolation, substitution, shielding, treatment, and prevention. In the area of isolation, all prepared, stored, or served food must be kept from sources of hazardous materials, especially chemicals. The food also must be isolated from insects and rodents, and from contaminating sewage and water. In the area of substitution, mechanical processes should be substituted for hand operations in the preparation of food when possible. In the area of shielding, food on a buffet table must be protected from the customers. Treatment is used for food handlers with foodborne diseases. Prevention is the major technique utilized for intervention in disease processes. Prevention includes everything from refrigeration and freezing, adequate cooking temperatures, and proper storage, to removal of sick foodservice workers from food preparation and service.

The potential for intervention generally is very good; specifically, care should be given to the selection of food and special techniques used for the handling, processing, and serving of specific types of hazardous foods.

RESOURCES

Scientific and technological resources in the area of food include Institute of Food Technology; Canning Trade Inc.; International Association of Environmental, Milk, and Food Sanitarians; National Environmental Health Association; American Public Health Association; various land grant colleges such as the University of Iowa and Purdue University; and schools of public health such as University of Michigan, University of California, and University of Minnesota. Civic associations concerned with food are basically in the area of nutrition and provision of food for individuals who are starving or who have inadequate food supplies. The governmental organizations involved in food include the federal FDA, various local and state health departments, and many of the state departments of agriculture throughout the country. The USDA is another fine resource. County extension agencies are an immediate source of help. The U.S. Senate Agricultural Committee is an important source for new legislation. The Food Code of 1993, updated by the Food Codes of 1997 and 1999, recommendations of the U.S. Public Health Service Food and Drug Administration, are excellent. The Shellfish and Sanitation Branch, U.S. Food and Drug Administration, 200 C Street S.W., Washington, D.C., is an excellent source. Prevention, education, early detection, containment, and control information are obtainable from the FDA through the President's Food Safety Initiative that started in 1997.

STANDARDS, PRACTICES, AND TECHNIQUES

Food Quality Protection Act of 1996

The Food Quality Protection Act of 1996 amended the federal Food, Drug and Cosmetic Act (FDCA), as well as the Federal Insecticide, Fungicide and Rodenticide Act (FIFRA). The purpose of the Food Quality Protection Act was to repeal the Delaney clause for pesticides and establish a single safety standard for raw and processed food. It was also to protect children from pesticide exposure and allow the Environmental Protection Agency (EPA) to respond to changes in science as they occur. The major changes to the Federal FDCA are as follows:

1. Consider children's special sensitivity and exposure to pesticides and include an additional safety factor of up to ten times to account for the uncertainty in data as they relate to children and their tolerance for pesticide residues. The law requires an explicit determination that pesticide tolerances are safe for children.
2. Group together compounds with a common mechanism of action, such as one tolerance used for all organophosphates.
3. Consider the cumulative exposure of individuals to pesticides through contact with air, food, water, pets, household, and lawn and garden products.
4. Reevaluate all existing tolerances to pesticides to ensure that they meet the new standard.
5. Test pesticides for their potential as endocrine disrupters, which are compounds that mimic or block the effect of hormones such as estrogen. The chemical

manufacturers are required to provide data on their products, including potential endocrine defects.

The major changes in the FIFRA are as follows:

1. Establish a new category of minor use pesticides that includes those that were used on crops grown on less than 300,000 acres, and those that were not economically viable for the registrant, but safer than existing alternatives, or important for integrated pest management. These chemicals could not pose unreasonable risks.
2. Pesticide registrations were to be reviewed periodically on a proposed 15-year cycle.
3. The EPA was required to develop criteria for reduced-risk pesticides and to expedite their review for registration.

Enforcement by the FDA was enhanced by allowing the agency to impose civil penalties for tolerance violations. The various states have to adhere to the federal regulations. It is anticipated that tolerance levels for pesticides may be lowered. It is also anticipated that such factors as antimicrobial pesticides review and registration will be expedited and the regulatory overlap in liquid chemical sterilants will be ended.

FDA Modernization Act of 1997

The FDA Modernization Act of 1997 includes a section on food safety and labeling that eliminates the requirement of premarket FDA approval for most packaging and other substances that come in contact with food and migrate into it. Instead, the law establishes a process whereby the manufacturer can notify the agency about its intent to use certain food contact substances. Unless the FDA objects within 120 days, the manufacturer may proceed with the marketing of the new product.

Food Code of 1997

The U.S. Public Health Service has been providing model food codes since 1924, when it proposed the Grade A Pasteurized Milk Ordinance (its current name). The FDA's purpose in maintaining an updated model food code is to assist food control jurisdictions at all levels by providing them with a scientifically sound technical and legal basis for regulating the retail portion of the food industry. The model Food Code is neither federal law nor federal regulation and it does not preempt the laws and regulations of the various states. It represents the best advice of the FDA for a uniform system of regulations to ensure that food at the retail level is safe and properly protected. The revisions in this code reflect the input of many experts who have taught, studied, and used the earlier codes. Changes also reflect new federal laws and regulations administered by other federal agencies such as the EPA and USDA. The 1997 Recommendations of the U.S. Public Health Service and FDA provide the following:

1. Improved definitions that are more precise and more consistent with terminology and definitions found in related laws and regulations
2. Provisions modified to make them more consistent with national requirements and standards administered by other federal agencies and international bodies without compromising public health
3. Provisions rewritten to make their intent clearer and reduce confusion as well as the potential for inconsistent application
4. Improved user aids contained in the Annexes such as added references and updated public health reasons, model forms, guides, and lists
5. Expanded index with additional layman's terms to assist a broader base of users who find topics of interest

The code requires a preapproval of HACCP plans for food establishments that include flow diagrams, product formulations, training plans, and corrective action plan. (A copy of the 1997 or 1999 Model Food Codes can be obtained from the National Technical Information Service, at 703-605-6000.)

Food Code of 1999

The 1999 Food Code makes numerous editing changes in the 1997 document for internal consistency, for correcting some of the errors, and for clarification. In Chapter 3, it changes the cooking of pork from 155 to 145°F. It expands the section on juices to include fruit and vegetable juices and prohibits prepackaged unpasteurized juice. Further, it has provisions concerning raw seed sprouts and raw eggs. The rest of the chapters are somewhat modified as well as the seven Annexes.

Plans Review

A successful foodservice operation is dependent on good facilities, proper arrangement and utilization of equipment, good quality raw food, and well-trained personnel. State health departments and, where designated, local health departments are responsible for reviewing the plans of all new food establishments prior to construction, the conversion of an existing structure for food use, and the remodeling or change of a food establishment. Plans and specifications, including architectural, structural, mechanical, plumbing drawings, intended menu, and volume of food prepared, stored, served, and sold must be submitted to the health department, which gives preliminary approval for construction or remodeling and makes an inspection prior to licensing. The inspection includes evaluating material and grouting of floors; material, finish, and color of walls and ceilings; presence of insect- and rodent-proofing devices; quantity and quality of lighting in various areas; amount and type of ventilation; type of toilet facilities for males and females; type of water supply; hand washing facilities; construction of utensils and equipment; type of utensil washing and sanitizing; type of utensil storage; type of food storage areas; type and quality of food refrigeration; type of locker or dressing rooms; type of sewage disposal; and type of solid waste disposal. The foodservice operation is licensed and permitted to serve food after specifications set forth by the health department are met.

Foodservice Facilities

Physical Facilities

Floors in food preparation, serving, and storage areas should be composed of smooth, nonabsorbent, and easily cleaned materials to aid in cleaning and to prevent absorption of grease, organic material, and odors. Antislip materials or treatment may be used for safety purposes. When floor drains are used, they should be screened to prevent harborage of flies and roaches, and to enable easy cleaning. Nonrefrigerated, dry storage areas do not need to have nonabsorbent floors. Securely attached, tightly installed carpeting under coving or away from the wall may be used except in food preparation areas, storage areas, ware washing areas, hand washing areas, toilet room areas where urinals and toilets are located, and refuse storage rooms or areas (Figure 3.2).

Walls and ceilings should have light-colored, smooth, easily cleaned surfaces to facilitate good food preparation and foodservice habits and to provide more adequate distribution of light. A minimum of 70 fc of light are needed, whereas 50 fc are required on all food preparation surfaces; and 20 fc on all other surfaces to provide an environment conducive to the elimination of dirt, grease, insects, rodents, and accidents.

Plumbing and Water Supply

There must be an adequate quantity of bacteriologically and chemically safe hot and cold water under pressure from an approved source, because water can easily contaminate food directly or indirectly through equipment, utensils, and hands. Ice must be prepared from water from an approved source; and must be transported and stored in a sanitary manner, because ice is used in food and drink, and has been shown to contain numerous bacteriologic contaminants. Ice used for cooling must not then be used as a food.

Plumbing and drainage systems carrying human wastes have been incriminated in outbreaks of typhoid fever, paratyphoid fever, dysentery, and other gastrointestinal diseases. Transmission of these diseases usually occurs through cross-connections, overhead leakage, stoppage in drainage systems, and submerged inlets. Backflow to the potable water supply system may occur directly in coffee urns, dishwashing machines, and double-jacketed kettles. Back-siphonage is possible where submerged inlets occur as in drinking fountains, flushometer valves, flush tanks, garbage can washers, hose outlets, ice makers, lavatories, slop sinks, steam tables, and vegetable peelers. The backup of sewage is possible in enclosed equipment, such as ice makers, refrigerators, steam tables, and walk-in freezers. Sewage should be disposed in a public sewage system or in a properly operated on-site sewage disposal system to prevent the spread of enteric disease and mosquito breeding.

Separate sanitary toilet facilities with self-closing doors are needed to prevent the spread of enteric disease. The hand washing facilities must have hot and cold running water, soap, and disposable towels or air dryers to prevent the spread of disease through contaminated hands. (In-depth discussions on water supply, plumbing, and sewage disposal are found in Chapters 3, 5, and 6, Volume II.)

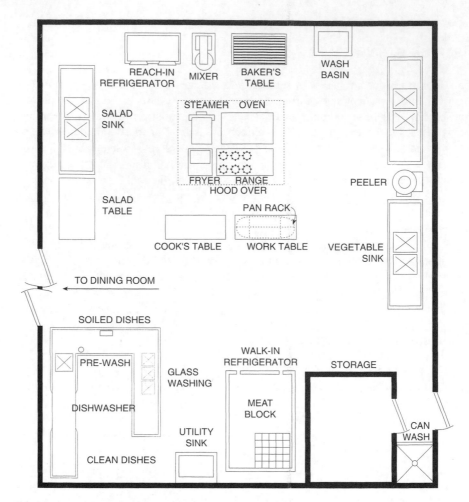

Figure 3.2   Layout of kitchen equipment.

Hot water has a corrosive effect on pipes. Hard water deposits are seven times greater at 180°F than at 140°F. Water softeners should be used where hardness exceeds 125 ppm. Because hot water is the sanitizing agent in machine ware washing, the force and water pressure, volume of water, and temperature are very important. A water pressure of 15 to 25 lb/in.² (which is 100 to 170 kPa) at the entrance to the rinse manifold is optimal. Water should be flowing at a rate of 9 gal/min in the rinse section. The gas or electric company should be called on to determine the size of the water heater necessary for any given establishment.

Ventilation Systems

Adequate ventilation must be provided for all rooms to reduce condensation, minimize soiling of walls and ceilings, and remove excessive heat and objectionable odors. All hoods should be equipped with noncombustible 2-in. commercial grease filters that are easy to install, remove, and clean. The surface of hoods should have sealed joints and seams, be accessible, and be easily cleaned. The function of the hood is to collect vapors, mists, particulate matter, fumes, smoke, steam, or heat. The velocity of the air capturing these items is called the capture velocity. The canopy hood is an overhead hood completely covering the equipment it is designed to serve. The filter in the hood should be a minimum of 2.5 ft above an exposed cooking flame, 4.5 ft above charcoal fires, and 3.5 ft above other exposed fires. Hoods should be so designed as not to interfere with normal combustion or exhaust of combustion products from commercial cooking equipment, other processes, and heat equipment.

To determine the total quantity of air exhausted from a hood, the following formula should be used:

$$Q = VPD$$

where
   Q = flow rate of air exhausted in cubic feet per minute
   V = capture velocity in feet per minute
   P = the perimeter of the open sides of the hood in feet
   D = distance between the cooking surface and the face of the hood in feet

Makeup air, composed of outside air equal to 100% of the quantity of the air exhausted, must be introduced into the building in such a way as to not interfere with the exhaust system.

Food Thermometers

Using a food thermometer is the only reliable way to ensure that food is safely processed. The food must be cooked to an internal temperature high enough to destroy any harmful bacteria. It is essential to use a thermometer when cooking meat and poultry to prevent undercooking and to prevent foodborne disease. An oven cord thermometer is a thermistor-type thermometer probe attached to a long metal cord that has a base unit with a digital screen on the other end. The probe is inserted into the food and a cord extends from the oven to the counter, to the stove top, or to oven door. Users program the thermometer for the desired temperature. When the food reaches the setting, the thermometer beeps. Users can monitor the internal temperature of the food they are cooking without opening the oven door. The thermometer–fork combination combines a cooking fork with a thermometer. A thermocouple probe is inserted in one of the tines of the fork, which helps it accurately measure the internal temperature of even the thinnest foods. The thermocouple can register a temperature within a few seconds. For a thermometer to work properly it must be placed in the food correctly, which means it should be inserted into the thick part or center of the food. When cooking beef, pork, or lamb roasts, insert the thermometer midway in the roast, avoiding the bone. If a food is irregularly shaped, place the thermometer in several different places. When cooking whole poultry insert the thermometer in the thick part of the thigh, where the temperature must reach 165°F. If stuffing is used, it must reach 165°F. When measuring the temperature of a thin food, such as a hamburger or pork chop, use a digital thermistor or thermocouple thermometer if possible. The temperature must reach 160°F.

There are four basic types of kitchen thermometers: liquid filled, bimetallic coil, thermistor, and thermocouple. The liquid filled contains a colored liquid, usually alcohol, inside the thermometer and expands and rises to indicate the temperature. This thermometer is not recommended because the glass may break and cause a hazard. The bimetallic-coil thermometer contains a coil in the probe made of two different metals with different rates of expansion. The coil, which is connected to the temperature indicator, expands when heated. The thermistor thermometer is a resistor (a ceramic semiconductor bonded in the tip with temperature-sensitive epoxy), which measures temperature. The thermocouple thermometer is the fastest reading. It shows a final temperature on a screen in seconds. Because it has a very small tip, it can accurately measure the temperature of very thin foods. Thermocouples measure temperature at the junction connected by fine wires located in the tips of the probes.

Storage of Food

Wholesome food can be readily contaminated if stored improperly. Food must be protected against unsanitary conditions, insects and rodents, poisons, cleaners, other chemicals, and bacterial growth. Temperature, time of storage, initial contamination, and moisture content are also vital factors to be considered in food storage.

Dry Storage

Unsanitary conditions result from dirt, dust, contamination by overhanging sewage pipes, sneezing, coughing, unclean utensils and work surfaces, and unnecessary handling.

Insects affecting food products can easily contaminate a large amount of grains, flour, and rice. These supplies should be stored in a clean, dry area, on racks at least 6 in. off the floor and 1 ft from walls, and should be rotated at least once every 2 weeks during the summer months and once every 3 to 4 weeks during the winter months. The keeping quality of the food is improved because air circulation is adequate and because it is possible to prevent insects and rodent harborage under or behind the food. Mice and rats contaminate large quantities of food through defecation and urination on the food and through nibbling. This hazard can be detected by using a blacklight to determine if urine is present and a flashlight to determine if fecal material is present.

All poisons, cleaning materials, and other nonfood items must always be clearly labeled, stored in their original containers, and kept apart from all foods. However, only those poisons and cleaning materials needed for the operation of the establishment can be kept here.

Refrigerated Food

Bacterial growth can be controlled by proper cooking and by storing food in shallow containers at or below 41°F or above 140°F. The temperature at which food is stored is probably one of the most important of all environmental conditions, because it affects the rate of decomposition and the growth of microorganisms. The time of storage is equally important in preventing disease and undesirable changes in taste, odor, and quality of the raw products. Therefore, it is essential that the initial contamination of raw food be minimized. Chemicals added to food limit food spoilage, but add unusual tastes and odors if the food is stored for long periods of time. The presence or absence of air determines the amount of food decomposition due to different types of bacteria. Moisture, which aids in the growth of bacteria, has a considerable effect on food spoilage.

Refrigeration of food has been practiced for many centuries. Food has been placed in cool caves, wells, springs, and running streams. Ice has been used for at least 200 years. Today, the modern refrigerator and freezer are a major means of storing and preserving food.

The rate of refrigeration is influenced by the heat transfer properties of the food, the volume of the food to be refrigerated, the kind of containers used, the heat conductivity of the containers, the agitation of the food, and the temperature difference between the food and the refrigerator unit. (See Table 3.1 for refrigerated storage temperatures and shelf life for cold perishable foods.)

Potentially hazardous ingredients for salads, sandwiches, filled pastry products, and reconstituted food need to be chilled to 41°F or below within 4 hours, or more rapidly if possible.

Frozen Food

The quality of frozen food is determined by the raw products, the method of preparation, the speed of freezing, the temperature and time involved in distribution and storage, and the method of packaging. The frozen food package must be strong, flexible, and prevent entrance or escape of liquid from the package. Desiccation of frozen foods is a physical change taking place when moisture escapes. Crystallization is also a physical change, usually occurring when frozen food is permitted to defrost and is then refrozen. Bacteria grow readily in the moisture film collected at the surface of the frozen foods when the temperature rises above 32°F. The general practice of defrosting frozen foods at room temperature is considered to be poor, because the outer surface becomes a fine bacterial medium while the inner core is still defrosting. Frozen foods should be defrosted in the refrigerator at temperatures not exceeding 41°F or should be cooked immediately. Frozen poultry, fish, shellfish, and frozen leftovers are particularly prone to contamination and bacterial growth, and may easily lead to outbreaks of food infection or food poisoning if not properly defrosted and adequately cooked. Frozen food temperatures should be below 0°F. Fishery products should be at −4°F or below for 7 days in a freezer, or −31°F or below for 15 hours in a blast freezer to destroy parasites.

Table 3.1   Refrigerated Storage Temperatures and Shelf Life

Product	Temperature (°F)	Maximum Storage Period
Meats		
Bacon	28–30	15 days
Beef (dried)	62–40	6 months
Brined meats	31–32	6 months
Beef (fresh)	30–32	1 day
Fish (frozen)	−4	6 months
Fish (iced)	30–32	10 days
Hams and loins	28–30	21 days
Lamb	28–30	14 days
Livers	20–22	6 months
Oysters (shell)	32–38	15 days
Oysters (tub)	32–38	10 days
Pork (fresh)	30–32	15 days
Pork (smoked)	28–30	15 days
Poultry (fresh)	28–30	5 days
Poultry (frozen)	0–5	10 months
Sausage (franks)	35–40	2 days
Sausage (fresh)	31–27	15 days
Sausage (smoked)	32–40	6 months
Veal	28–30	15 days
Miscellaneous		
Butter	35	6 months
Cheese (American)	32–34	15 months
Cheese (Swiss)	38–42	60 days
Cream (40%)	5–10	4 months
Eggs (frozen)	0–5	18 months
Eggs (fresh)	38–45	2 months
Milk	35–40	5 days
Oleo	34–36	90 days

Protecting Wholesomeness and Detecting Spoilage of Foods

All raw food products and processed food should be clean, wholesome, free from spoilage, free from chemical contaminants, and free from organisms that cause disease or intoxication. Meat, milk, shellfish, and similar products should come from approved sources. All leftovers should be discarded, having been handled extensively and probably having been kept at improper temperatures. All food processing and storage areas should be free of birds, animals, insects, and rodents. Frozen foods should be defrosted in such a way as to prevent rapid growth of bacteria. The best approach is to defrost under refrigeration at 41°F or below.

Visual Inspection of Foods

Decomposing meat may have an off-odor, be slimy, and have an off-color. Beef spoils from the surface inward; pork spoils from the bone outward. Decomposing fish have an off-odor, sunken eyes, and gray or greenish gills. An indentation remains on the flesh of the decomposing fish, which is easily pulled away from the bones.

Decomposing poultry is sticky or slimy, has off-odors, and exhibits darkening of wing tips. Canned foods can be swollen, dented, rusted, without labels, and have off-odors and off-color. With the exception of large pieces of beef, all of the previously mentioned foods should be discarded. Large pieces of beef can be trimmed by the butcher and the trimmings discarded. Ground hamburger and ground pork decompose rapidly; discoloration and off-odors result.

Thawing of Frozen Foods

To protect the wholesomeness of frozen foods and to prevent potential outbreaks of disease, the following thawing procedures should be followed: Cook frozen vegetables, small cuts of meat, chicken, fish, and prepared foods in the frozen state; thaw frozen fruits, juices, large cuts of meat, poultry, and shellfish in the refrigerator at or below 41°F.

Potentially Hazardous Foods

Fresh meats may contain *Clostridium perfringens*. Because growth of the bacteria and development of the toxin occur best at room temperatures, refrigerate meat immediately after cooking and until used.

Fecal streptococci and *Staphylococcus aureus* are also found on meat. Partially cooked hams may contain fecal streptococci and staphylococci. Smoked hams that are cooked to an internal temperature of 150°F are not sterile and may contain organisms. Because cold cuts, including hot dogs, are handled by personnel, there is a potential hazard of staphylococcal food poisoning or salmonella food infection.

Salmonella, the primary contaminant of poultry, causes many outbreaks of salmonellosis; *C. perfringens* and fecal streptococci may also be contaminants. Fish and shellfish coming from sewage-polluted waters may contain enteric organisms. *Clostridium botulinum* Type E is found on fish. Shellfish have been incriminated in outbreaks of infectious hepatitis and cholera. Canned foods, such as tuna fish and mushrooms, and prepared frozen foods, such as some pizza, have been shown to contain botulism toxin.

Chicken salad, tuna salad, ham salad, potato salad, custard-filled pastries, dairy products, smoked fishes, egg salad, and other mayonnaise or cream-based salads have led to substantial outbreaks of food poisoning and food infection. The lack of proper refrigeration, improper handling of the food, and saving of leftovers are the contributing factors. Food contaminated with water and smoke during fires, defrosted frozen foods, and unrefrigerated perishables are considered unwholesome and should be discarded.

Preparation and Serving of Food

Because food becomes contaminated easily during preparation or serving, and organisms already present in the food can cause disease and poisoning, the following rules of food protection should be followed:

1. Use only good quality, wholesome food.
2. Keep the food clean and free of insects and rodents.
3. Clean preparation and serving areas carefully.
4. Use only well-constructed, easily cleaned equipment.
5. Keep equipment very clean.
6. Handle food as little as possible.
7. Use good personal hygiene at all times.
8. Refrigerate perishables as quickly as possible.
9. Cook foods long enough to kill organisms.
10. Keep food below 41°F or above 140°F in serving areas.
11. Provide proper sneezeguards in cafeterias and smorgasbords.
12. Use indicating thermometers accurate to +2 or −2°F to determine the temperature of the refrigerator and steam table.

Be particularly careful of ground and chopped foods, foods made of several raw materials, rich or nutritious foods, foods with high moisture content, and foods that require considerable handling during preparation, such as salads and ground meats.

Utilize the proper precleaned equipment and tools for the job. Use forks, knives, tongs, spoons, scoops, or single-use gloves on hands. Remove small quantities of prechilled perishable ingredients from the refrigerator immediately prior to use. Wash all raw fruit and vegetables thoroughly before cooking or serving. Cook all raw animal foods such as eggs, fish, poultry, pork, and game to an internal temperature of at least 155°F for 15 sec. Cook all products that are stuffed, including pasta, to an internal temperature of at least 165°F for 15 sec. See Table 3.2 for roast beef and corned beef. Place all custard, cream filling, and puddings below 41°F.

When food is served, servers must be careful not to place their fingers on the eating surface of the dishware or silverware. Plates of food should not be stacked on each other, because the undersurface of plates could contaminate the food beneath it.

Raw animal food cooked in a microwave oven must be:

1. Rotated or stirred throughout or midway during cooking because of the uneven distribution of heat
2. Covered to retain surface moisture
3. Heated an additional 25°F above previously stated temperatures for raw animal food products, because of the shorter cooking times
4. Allowed to stand covered for 2 min after cooking to obtain temperature equilibrium

When food is served in smorgasbords, buffets, and cafeterias, it should be kept behind protective guards below the speaking level of people. The food-holding equipment should be made of nonabsorbent, smooth, easily cleaned, corrosion-resistant material. Glass used must have a safety edge. Because hot or steam tables and cold tables are not intended to alter the temperature of food, but instead maintain it at the proper temperature, the food has to be preheated to 140°F internal temperatures or precooled to 41°F internal temperature. Hot tables should be maintained at 160°F to achieve the 140°F needed in the food. Where ice is used for cooling, the ice should be ¾ in. in diameter or less, should be clean, and should come from

**Table 3.2   Oven Parameters Required for Destruction of Pathogens
on the Surface of Roasts of Beef and Corned Beef**

	Oven Temperature Roast Weight	
Oven Type	Less Than or Equal to 4.5 kg (10 lb)	Greater Than 4.5 kg (10 lb)
Still dry	177°C (350°F)	121°C (250°F)
Convection	163°C (325°F)	163°C (325°F)
High humidity[a]	Less than 121°C (250°F)	Less than 121°C (250°F)

**Minimum Holding Times Required at Specified
Temperatures for Cooking All Parts
of Roasts of Beef and Corned Beef**

Temperature °C (°F)	Time[b] (min)	Temperature °C (°F)	Time[b] (min)	Temperature °C (°F)	Time[b] (min)
54 (130)	121	58 (136)	32	61 (142)	8
56 (132)	77	59 (138)	19	62 (144)	5
57 (134)	47	60 (140)	12	63 (145)	3

[a] Relative humidity greater than 90% for at least 1 hr as measured in the cooking chamber or exit of the oven; or in a moisture-impermeable bag that provides 100% humidity.
[b] Holding time may include postoven heat rise.

Food Code 1993. U.S. Public Health Service, U.S. Department of Health and Human Services, Washington, D.C. 20204, p. 52.

an approved source. Food containers should be a maximum of 6 in. deep, and food should never be stacked above the ice level. Ice storage units need drains, which have an air gap, to prevent backflow of sewage. Tongs, forks, spoons, picks, spatulas, and scoops should be used by foodservice workers, if possible. If the customer uses these implements, the utensils should be changed frequently. Ice cream scoops, dippers, and spoons should be stored either in a cold running water well, which is frequently cleaned, or in a clean, dry manner. Sugar should be kept in closed containers or preferably in single-service packets. The remnants of all food served to the customer should be discarded.

Where food is transported in hot and cold food trucks to various parts of an institution, special care must be taken to preheat and precool the unit. The food containers and the truck must be carefully cleaned immediately on completion of the serving process. The set-up trays must be covered to avoid contamination by dust, aerosols, and microorganisms.

Single-service articles are an excellent method of serving food, because the articles are thrown away after each use and a premeasured portion of food can be set up at the processing plant or kitchen. Contamination by hands and improperly washed utensils is avoided. Caution must be used in the storage of single-service materials to prevent contamination by dust, dirt, sewage, insects, and rodents. Items should be discarded in lined containers with self-closing lids.

Washing Food

Washing or soaking meat or poultry can allow the bacteria to spread to all parts of the meat and to other ready-to-eat foods. Bacteria present on the surface of the meat or poultry will be killed by cooking to a temperature of 160°F. Bacteria in raw meat and poultry juices spread readily to other foods, utensils, and surfaces resulting in cross-contamination. Proper hand washing after handling raw meat or poultry is essential. Countertops and sinks must be scrubbed with hot soapy water, rinsed thoroughly, and then sanitized with liquid chlorine. The packaging materials from the meat or poultry can also cause cross-contamination. These materials should always be disposed in an appropriate manner. Eggs should not be washed before storing or using. The natural coating on just-laid eggs that helps prevent bacteria from permeating the shell is removed by washing during shell egg processing and is then replaced by a light coating of edible mineral oil that restores the protection. The extra handling of the eggs increases the risk of cross-contamination, especially if the shell becomes cracked. The major exception to the washing of raw food is produce. All fresh produce should be washed with cold running water to remove any possible dirt and reduce the bacteria that may be present. However, soap or detergent should not be used because it can be absorbed in the produce. Always cut away any damaged or bruised areas of fruits and vegetables because bacteria that may cause illness can grow there. Immediately refrigerate all prepared fresh-cut items such as salad or fruit to maintain quality and food safety. Unpasteurized fresh orange juice should not be used for drinking or in salads.

Microwave Oven Radiation Safety

Microwaves, a form of electromagnetic radiation, are waves of electrical and magnetic energy moving together through space. Microwaves fall in the middle of the electromagnetic radiation band. They have three characteristics that allow them to be used in cooking: they are reflected by metal; they pass through glass, paper, plastic, and similar materials; and they are absorbed by foods. Microwaves are produced inside the oven by an electron tube called a magnetron. The microwaves bounce back and forth within the metal interior until they are absorbed by the food. The microwaves cause the water molecules in the food to vibrate, thereby producing heat that cooks the food. The results are vegetables with much higher water content cooking more rapidly than other foods. The microwave energy is changed to heat as soon as it is absorbed by the food; therefore, the food cannot become contaminated. Although heat is produced directly in the food, microwave ovens do not cook food from the inside out. In thick foods, like roasts, the outer layers are heated and cooked primarily by microwaves whereas the inside is cooked mainly by the slower conduction of heat from the hot outer layers. Glass, paper, ceramic, or plastic containers allow the microwaves to pass through them and will not become hot from the microwaves, but instead from the food cooking inside.

The FDA recommends that microwave ovens or conventional ovens not be used in home canning, because both cannot produce or maintain temperatures high enough to kill the bacteria that cause disease. Microwave ovens must not have radiation leakage exceeding 5 mW/cm^2 of microwave radiation at approximately 2 in. from the oven surface. All ovens must have two interlock systems that stop the production of microwaves the moment the latch is released or the door is opened. Injuries that have occurred have been due to people being burned by hot food, splattering grease, or steam from the cooked food. The FDA does not require microwave ovens to carry warnings for people with pacemakers, because many other types of electronic products may also possibly interfere with certain electronic cardiac pacemakers. However, patients with pacemakers should consult their physicians before using microwave ovens. It is essential that all microwave ovens be kept thoroughly clean to help avoid leakage of the microwaves.

Strict enforcement of microwave oven exposure is necessary because many scientific questions concerning low-level radiation have not been answered. At high levels, microwaves may affect the human body by heating body tissue in the same way it heats food. Exposure to high levels of microwaves can cause painful burns. The lens of the eye is particularly sensitive to intense heat, and exposure to high levels of microwaves can cause cataracts. The testes are very sensitive to change in temperature, which can result in alteration or the killing of sperm, producing temporary sterility.

Design and Installation of Foodservice Equipment

The design of foodservice equipment should be based on good research, sound engineering, good environmental practices, and practical knowledge of experts from industry and the field of public health. The standard for equipment design is now basically set by the National Sanitation Foundation (NSF) in Ann Arbor, MI. This nonprofit organization develops standards based on the previously mentioned criteria. Foodservice equipment should contain the following design features:

1. Easily disassembled, cleaned, and maintained equipment with as few parts as possible
2. Smooth, nonabsorbent, nontoxic, odorless, and easily cleaned food contact surfaces
3. Food contact surfaces containing no toxic materials, such as cadmium, lead, or copper
4. Nontoxic, nonabsorbent, easily cleaned gaskets, packing, and sealing materials that are unaffected by food or cleaning products
5. Easily cleaned splash zone areas
6. Food product surfaces that will not chip, crack, or rust

There are use limitations for cast iron, ceramic, and crystal; utensils, galvanized metal, linens, napkins, and sponges; and pewter, solder, flux, and wood. Existing NSF standards or criteria should be reviewed at least every 3 years to ensure that the foodservice equipment keeps pace with the public health and industry advances. The competent foodservice manager should obtain copies of NSF standards and review them carefully before purchasing new equipment.

The installation of equipment is almost as important as its design and construction. If equipment is installed into the wall but is not flush with the wall, roaches may easily hide behind it. Overhead equipment such as hoods must be within easy reach, enabling foodservice personnel to remove the filters regularly and to clean the hoods and filters frequently. Large equipment should be mobile or installed on legs to facilitate cleaning behind and underneath. Drains should be located in areas in the floor where spillage is likely or where equipment is scrubbed. Hand washing facilities and service sinks should be conveniently located to ensure their use when needed. Clean metal storage shelves must be provided for the storage of cleaning equipment and utensils. See Figure 3.2 for a proper layout of kitchen equipment.

Housekeeping and Cleaning in Foodservice Facilities

In a satisfactory environment in a foodservice facility, soil is removed; bacteria are destroyed; and an aesthetically pleasing picture is provided through good physical facilities, proper color, adequate light, and well-managed cleaning practices. Soils are divided into four basic types: (1) fresh soil found immediately after use of equipment or facilities; (2) thin film of soil caused by ineffective cleaning within which microbes live; (3) built-up deposits of soil caused by consistent ineffective cleaning and composed of soap films, minerals, food materials, and grease; (4) soil caused by drying of heavy, crusty deposits or baking of organic material onto dishes or trays by improper ware washing.

Fresh soils and thin films of soil on floors and walls are removed by the use of a good detergent and spraying technique, and plenty of physical action. Built-up deposits are best removed by the use of detergent spraying techniques plus mechanical scrubbing machines and special heavy-duty detergents. Dried deposits on dishes are first soaked in a good detergent and hot water, and then scrubbed with plastic or metal pads. Dried deposits on baking trays are best removed by cleaning with live steam. The best approach to cleaning, however, is prevention of the accumulation of deposits and soil through good daily cleaning. (Live steam is dangerous. Be careful of its use.)

General Daily Cleaning

At the conclusion of each work day, an assigned foodservice worker should wash all kitchen, storage, and serving area floors in the following manner:

1. Starting at the far end of the room, spray an area of the floor measuring approximately 100 ft^2.
2. Spray an adjacent area of approximately the same size.
3. Pick up the detergent with a very clean mop and rinse in a clean pail of water.
4. Repeat steps 1, 2, and 3 until the entire floor is washed.
5. Clean floor drains.

All dry-food storage areas should be swept with a treated mop daily to decrease contamination by dust and dirt, and also to determine if an insect or rodent infestation exists. All outside areas, especially where solid waste is stored, must be checked frequently during the day and swept and washed daily. This prevents an unsightly mess and also helps destroy potential insect and rodent harborage.

Bathroom floors should be cleaned at least twice a day using the same procedure outlined for kitchen floors. In addition, the trash in the bathrooms should be emptied a minimum of twice daily, the sinks washed, and the soap and single-service towels refilled. Toilet bowls and urinals must be cleaned thoroughly. Deodorants are not a substitute for good cleaning procedures. Under no condition should food be stored in restrooms. Foodservice attendants must thoroughly scrub their hands for at least 20 sec with a cleaning compound and rinse thoroughly with clean water. Special attention must be paid to fingernails and between the fingers. After defecating or handling waste, it is recommended that hand washing take at least 2 min to be thorough. Finally, an alcohol-based, instant hand sanitizer solution should be used. The sanitizer should be the strength equivalent of 100 mg/l chlorine or above.

Periodic Cleaning

Walls and ceilings should be washed down with a good detergent periodically, but not less than once a month. Lighting fixtures should be taken apart and cleaned and light bulbs replaced at least once every 2 months or when the bulbs burn out.

Cleaning Schedules

The establishment of definite work assignments in writing, and the development of a cleaning manual and schedule are essential to good cleaning practices. A foodservice worker must be taught to disassemble and clean pieces of equipment and to utilize cleaning equipment and materials. A good foodservice manager supervises this part of the foodservice operation closely. Foodservice managers should inspect their own establishments on a weekly basis using a flashlight to see behind and under equipment. A self-inspection sheet can be obtained from the health department or industry, or may be found in Chapter 9 on the institutional environment.

Cleaning and Sanitizing Equipment and Utensils

Good cleaning consists of prescraping and prerinsing with cold or warm water as soon after usage of equipment and utensils as possible; soaking in hot detergent solution, without overloading or improper stacking; power scrubbing manually or mechanically; power rinsing; sanitizing; and air drying.

All equipment, including counters, tables, carts, display cases, steam tables, shelves, storage racks, salad and vegetable bins, drain boards, meat blocks, stoves, stove hoods, coffee urns, meat and vegetable choppers, meat tenderizers, food mixers, griddles, cutting boards, ice cream counter freezers, steam-jacketed kettles, vegetable peelers, and garbage cans, must be cleaned thoroughly each day. In all cases, empty the contents of the equipment and rinse and scrape thoroughly. For large, immovable equipment, spray all portions with a good QAC compound, scrub with clean brushes, and rinse with QAC. Scrape meat blocks and cutting boards clean and wash. Turn meat blocks regularly. Where food spillage occurs in the oven, sprinkle salt on food and heat the oven to 500°F. When the spillage is carbonized, cool oven and scrape food deposits. Ice cream counter freezers must be emptied each night and the contents discarded. Rinse the unit thoroughly with water while still assembled, add dishwashing detergent, operate for 1 min, rinse and drain all detergent, disassemble all parts, scrub thoroughly with a brush and detergent, rinse and reassemble, and use a final rinse of chlorine. Garbage cans should contain plastic liners, which are removed and replaced frequently. The cans should be scrubbed mechanically with a garbage can washer or with a brush at least once every 3 days.

Hand Ware Washing

Prescrape and prerinse all dishes and utensils, and place in the first compartment of a three-compartment sink containing detergent and hot water at 95°F or above. Scrub dishes well with plastic brushes. Do not use washcloths or sponges, because they spread microorganisms. Rinse dishes thoroughly in the second sink at a temperature of 110 to 120°F. In the third sink, use a chemical sanitizer or hot water at a minimum of 170°F for 1 min. Remove and allow to air dry. An accurate temperature measuring device must be kept in the sink or be readily available. In all cases, predetermine the amount of water to be used to ensure that sufficient detergent, sanitizer, or both are added. A test kit or other device needs to be provided to measure the concentration of the sanitizer solution accurately.

Machine Ware Washing

Prescrape and prerinse dishes and utensils. Prewash heavily soiled items carefully. Place objects in racks so that all parts are exposed to the spray. Wash at 150°F for 40 sec. Rinse at 180°F at a pressure of 15 to 25 lb/in.^{2} (100 to 170 kPa) for 30 sec. Some variations in time and temperature will occur based on the type of dishwashing machine used. Suitable thermometers and pressure gauges should be attached to each machine. The accuracy of these gauges should be checked during an inspection.

A good mechanical ware washing operation depends on the following: (1) selection of the proper size dishwashing machine; (2) proper-sized hot water boiler and booster heater; (3) proper layout of equipment and utilization of workforce; (4) adequate training for personnel; (5) proper supervisory control; (6) adequate storage of clean dishes and utensils; (7) discarding of chipped and cracked dishes, and rusted or corroded utensils; and (8) adequate ventilation of dishwashing area to reduce strain and maximize personnel efficiency.

Soiled dishes are caused by improper scraping, prerinsing or prewashing, inadequate detergent, improper wash-water temperature, inadequate time for wash and rinse, and improperly cleaned dishwasher. Check all operations carefully and make corrections where needed. Unclog wash and rinse nozzles. Films are caused by water hardness: use adequate amounts of detergent and water softeners. Greasy films are due to low pH, insufficient detergent, low water temperature, and dirty dishwasher: check alkalinity, detergent, temperature, and wash and rinse nozzles. Streaking is due to high pH and improperly cleaned dishwasher: reduce alkalinity and clean dishwasher. Cooked on egg or other proteins are caused by improper prerinsing and prewashing, and too high a washing temperature. Spotting is due to rinse water hardness, too high or low a rinse temperature, inadequate time between rinsing and storage; to correct, soften water, check temperatures, and allow adequate time for air drying.

Dishwashers should be cleaned at the end of each work day and descaled at least once a week. Daily cleaning consists of removing, soaking, and scrubbing scrap traps, the suction strainer, and wash and spray arms. Return to the machine and run the machine without using a load of dishes. Once a week, after disassembling and scrubbing, reassemble, add 7 fluid oz of phosphoric acid or a 2% solution of acetic acid to the wash water, and operate the machine for 1 hour; drain off the solution, add baking soda, or 2 cups of detergent, and run the machine for 15 min; drain the machine and rinse it several times.

Solid Waste, Recyclables, and Returnables

Solid wastes containing food are unsightly, odorous, and attract insects and rodents. Recyclables and returnables may also contain food. Unless a refuse can is in constant use in an establishment, it should have a tight-fitting lid. At night all receptacles should be tightly closed. The indoor storage area for refuse, recyclables, and returnables should be clean and effectively screened for flies. Self-closing mechanisms should be used for all exterior and bathroom doors. Garbage grinders should only be used when the liquid waste goes to a public sewer system.

Solid waste must be stored 12 in. off the ground inside or outside the premises in receptacles with tight-fitting lids. These containers should not absorb odors or leak water. The 10- or 20-gal cans should always have a plastic liner. Cans should be scrubbed at least every 3 days. All solid waste should be removed daily, or at the most every 3 days. Problems result from carelessness in dumping of waste, improper supervision by management, failure to empty cans at night, dirty cans, lack of a drain or clogged drain where cans are washed, failure to clean large metal storage boxes, failure to replace rusted or crushed cans and lids, and failure to use plastic bags. A complete discussion on solid waste disposal may be found in Volume II, Chapter 2, on solid and hazardous waste. Recyclables and returnables should be removed frequently enough to avoid unsightly conditions and odors, and to prevent insect and rodent problems.

Insect and Rodent Control

Flies, roaches, and rodents spread disease by contaminating food. Food product insects are a nuisance and a contaminant. Flies are present wherever there is highly organic material. They breed in floor drains or in other out-of-the-way areas. The best control for flies is removal of organic material, cleanliness, screening, and use of safe pesticides. Roaches enter the premises through cracks; along pipes; in packages; through doors, windows, and other openings; in clothing; or through broken sewer lines. Cleanliness, sealing of roach entrances, and a good, safe poisoning program is required. Mice and rats enter establishments through holes in the walls, open doors and windows, and ruptured sewer lines. Control occurs through removal of harborage and food, poisoning, and rodent proofing. Food product insects infest dried fruit, cereal, grain, flour, rice, candy, and nuts. Control occurs through initial fumigation, rotation of stock, and destruction of highly infested foods. Detailed discussions on insects, rodents, and pesticide use are included in Chapters 5 to 7.

Personal Health of Employees

An employee's health and cleanliness of person and clothing prevent the spread of many foodborne diseases. Hands are one of the greatest causes of the spread of disease. Hands must be scrupulously washed when starting work, after using the bathroom, after cleaning, and before preparing or serving food. Hands should be kept away from the head, mouth, nose, and other body parts. A 2-min hand and forearm scrub, including several soapings and rinsings, should be a routine practice for the worker. Foodservice attendants should never smoke while preparing or serving food, should have clean and covered hair, and should have clean clothing changes daily. A foodservice worker should not work if stricken with an illness due to *Salmonella typhi*, *Shigella* spp., *Escherichia coli* O157:H7, or HAV infection. Employees should be restricted from work if they exhibit symptoms of intestinal illness such as abdominal cramps or discomfort, diarrhea, fever, loss of appetite, vomiting or jaundice; boil or infected wound; past illness from an infectious agent that may create a potential carrier state; record of suspected cause or exposure to an outbreak of the previously mentioned organisms, or family members sick from these organisms. All these workers with infections or exposure to infections should be seen by a physician and a culture should be taken if needed. All individuals with diarrhea should have a stool specimen taken. If the specimen is positive for an enteric organism causing a foodborne illness, the worker should not be permitted to return to work until three consecutive negative samples are taken.

In keeping with the requirements of the 1993 Food Code as updated in 1997 and 1999, the Applicant and Employee Interview form (Figure 3.3), the Food Employee Reporting Agreement form (Figure 3.4), and the Applicant and Food Employee Medical Referral form (Figure 3.5), should be completed as needed. A properly protected and stored first-aid kit should be available for employee use.

MODES OF SURVEILLANCE AND EVALUATION

Flow Techniques and Inspectional Procedures

The department of health, or other designated departments at the state or local level, issues licenses for the operation of a foodservice establishment based on laws approved by the appropriate legislative bodies. The health department formulates rules and regulations to be followed by the establishments and conducts special and periodic evaluations on site to determine compliance and whether licenses should be issued or renewed. All new foodservice establishments must submit plans for approval by the health department prior to construction and must be evaluated on site prior to opening.

Form
1   **Applicant and Food Employee Interview**

Preventing Transmission of Diseases through Food by Infected Food
Employees with Emphasis on *Salmonella typhi, Shigella* **spp.,**
Escherichia coli O157:H7, and Hepatitis A Virus.

The purpose of this form is to assure that Applicants and Food Employees advise the
Person in Charge of past and current conditions described so that the Person In
Charge can take appropriate steps to preclude the transmission of foodborne illness.

Applicant or Employee name (print) _____
Address _____
_____
Telephone Daytime: _____ Evening: _____

TODAY:

Are you suffering from any of the following:

    1. Symptoms
          Abdominal cramps?   **YES/NO**
          Diarrhea?   **YES/NO**
          Fever?   **YES/NO**
          Prolonged loss of appetite (more than 3 days)? **YES/NO**
          Jaundice?   **YES/NO**
          Vomiting?   **YES/NO**
    2. Pustular lesions
          Pustular lesion on the hand, wrist or an exposed body part?
          (such as boils and infected wounds, however small)   **YES/NO**

PAST:

Have you ever been diagnosed as being ill with typhoid or paratyphoid fever
(Salmonella typhi), shigellosis *(Shigella* spp.), *Escherichia coli* O157:H7
infection, or hepatitis (hepatitis A virus)?   **YES/NO**
If you have, what was the date of the diagnosis _____

HIGH–RISK CONDITIONS

1. Do you have a household member attending or working in a setting where
   there is a confirmed outbreak of typhoid fever, shigellosis, *Escherichia coli*
   O157:H7 infections, or hepatitis A?   **YES/NO**
2. Do you live in the same household as a person diagnosed with typhoid fever,
   shigellosis, hepatitis A, or illness due to *E. coli* O157:H7?   **YES/NO**
3. Have you been exposed to or suspected of causing a confirmed outbreak of
   typhoid fever, shigellosis, *E. coli* O157:H7 infections, or hepatitis A? **YES/NO**
4. Have you traveled outside the United States within the last 50 days? **YES/NO**

Name, Address, and Telephone Number of your Doctor:
Name _____
Address _____
_____
Telephone Daytime _____ Evening _____

Signature of Applicant or Food Employee _____ Date _____

Signature of Permit Holder's Representative _____ Date _____

Figure 3.3   Food Code 1993. (From U.S. Public Health Service, U.S. Department of Health
and Human Services, Public Health Service, Washington, D.C., Annex 7, p. 1.)

Form
2   **Food Employee Reporting Agreement**

Preventing Transmission of Diseases through Food by Infected Food
Employees with Emphasis on *Salmonella typhi, Shigella* **spp.,**
Escherichia coli O157:H7, and Hepatitis A Virus.

The purpose of this form is to assure that Food Employees notify the Person in
Charge when they experience any of the conditions listed so that the Person In
Charge can take appropriate steps to preclude the transmission of foodborne illness.

I AGREE TO REPORT TO THE PERSON IN CHARGE:

FUTURE SYMPTOMS and PUSTULAR LESIONS:
1. Abdominal cramps
2. Diarrhea
3. Fever
4. Prolonged loss of appetitie (more than 3 days)
5. Jaundice
6. Vomiting
7. Pustular lesions:
    • Pustular lesion on the hand, wrist, or an exposed body part
    (such as boils and infected wounds, however small)

FUTURE MEDICAL DIAGNOSIS:

Whenever diagnosed as being ill with typhoid or paratyphoid fever *(Salmonella typhi),* shigellosis
(*Shigella* spp.), *Escherichia coli* O157:H7 infection, or hepatitis (hepatitis A virus).

FUTURE HIGH–RISK CONDITIONS:

1. A household member attending or working in a setting experiencing a confirmed outbreak of typhoid
   fever, shigellosis, *Escherichia coli* O157:H7 infections, or hepatitis A.
2. A household member diagnosed with typhoid fever, shigellosis, hepatitis A, or illness due to *E. coli*
   O157:H7.
3. Exposure to or suspicion of causing any confirmed outbreak of typhoid fever, shigellosis, *E. coli*
   O157:H7 infections, or hepatitis A.
4. Travel outside the United States within the last 50 days.

I have read (or had explained to me) and understand the requirements concerning my responsibilities
under the Food Code and this agreement to comply with:

    (1) Reporting requirements specified above involving symptoms, diagnoses, and high-risk
        conditions specified;
    (2) Work restrictions or exclusions that are imposed upon me; and
    (3) Good hygienic practices.

I understand that failure to comply with the terms of this agreement could lead to action by the food
establishment or the food regulatory authority that may jeopardize my employment and may involve legal
action against me.

Applicant or Food Employee Name (please print): _____

Signature of Applicant or Food Employee _____ Date _____

Signature of Permit Holder's Representative _____ Date _____

Figure 3.4   Food Code 1993. (From U.S. Public Health Service, U.S. Department of Health
and Human Services, Public Health Service, Washington, D.C., Annex 7, p. 2.)

Form

3   **Applicant and Food Employee Medical Referral**

Preventing Transmission of Diseases through Food by Infected Food
Employees with Emphasis on *Salmonella typhi, Shigella* **spp.,**
Escherichia coli O157:H7, and Hepatitis A Virus.

The Food Code specifies, under *Part 2-2 Employee Health Subpart 2-201 Disease or
Medical Condition,* that Applicants and Food employees obtain medical clearance from a
physician licensed to practice medicine whenever the individual:

1. Is chronically suffering from a symptom such as diarrhea; or
2. Meets one of the high-risk conditions specified in Paragraph 2-201.11(D) and is suffering
   from any symptoms specified in Subparagraph 2-201.11(B)(1).
3. Has a current illness involving *Salmonella typhi* (typhoid or paratyphoid fever), *Shigella* spp.
   (shigellosis), *Escherichia coli* O157:H7 infection, or hepatitis A virus (hepatitis), or
4. Reports past illness involving *Salmonella typhi* (typhoid or paratyphoid fever), *Shigella* spp.
   (shigellosis), *Escherichia coli* O157:H7 infection, or hepatitis A virus (hepatitis), if the
   establishment is serving a highly susceptible population such as preschool age children or
   immunocompromised, physically debilitated, or infirm elderly.

Name of Applicant or Food Employee being referred and population served:
(————————————(Name, please print)———————————). **Highly susceptible population   YES ☐ NO ☐**

REASON FOR MEDICAL REFERRAL:

The reason for this referral is checked below:
☐ Chronic diarrhea or other chronic symptom ——(specify)—— .
☐ Meets a high-risk condition specified in Paragraph 2-201.11(D) —(specify)— and suffers from a
   symptom specified in Subparagraph 2-201.11(B)(1) (specify)
   diagnosed or suspected typhoid or paratyphoid fever, shigellosis, *E.coli* O157:H7 infection, or
   hepatitis.
☐ Reported past illness from typhoid or paratyphoid fever, shigellosis, *E.coli* O157:H7 infection, or
   hepatitis.
☐ Other medical condition of concern per the following description:

_____

PHYSICAN'S CONCLUSION:

☐ Applicant or food employee is free of *Salmonella typhi* , *Shigella* spp., *Escherichia coli* O157:H7, or
   hepatitis A and may work as a food employee without restrictions.
☐ Applicant is an asymptomatic carrier of *Salmonella typhi* and is restricted from direct food handling
   duties in establishments not serving highly susceptible populations.
☐ Applicant or food employee is not ill but continues as an asymptomatic carrier of —(pathogen)— and
   should be excluded from working as a food employee in food establishments serving highly
   susceptible populations such as those who are preschool age, physically debilitated,
   immunocompromised, or infirm elderly.
☐ Applicant or food employee is suffering from *Salmonella typhi, Shigella* spp., *Escherichia coli*
   O157:H7 or hepatitis A virus and should be excluded from working as a food employee.

COMMENTS:

Signature of Physician _____ Date _____

Figure 3.5   Food Code 1993. (From U.S. Public Health Service, U.S. Department of Health
                 and Human Services, Public Health Service, Washington, D.C., Annex 7, p. 13.)

The function of the food inspection program is to protect people from contamination of food, which may cause disease. Several important facets of this inspection include temperature control, hygiene and personal practices, food handling practices, and evaluation of the premises and outside surroundings. Most outbreaks of disease are caused by improper personal practices by infected individuals and poor temperature control of food. Personal practices relate to the removal of individuals, who have infected wounds, skin infections, sores, open cuts, or wounds, from food handling. Foodservice handlers must wear protective clothing, hair covering, and gloves; and not wear jewelry or use tobacco or chewing gum in food-handling areas. Hand washing facilities with hot and cold running water, soap, and disposable towels or air dryers must be provided and used. Raw product processing must be separated from finished product processing, such as in shellfish production.

Temperature control is essential to avoid a buildup of harmful microorganisms. Food should not be kept in the danger zone between 41 and 140°F unless necessary and then only for short periods of time. Functional and accurate thermometers must be used in all food storage units. Temperatures must be monitored at all critical points in the process. Frozen foods must be thawed under refrigeration. A potable water, steam, and ice supply is mandatory. Chemicals, such as detergents, sanitizers, and pesticides, must be stored away from food. All equipment must conform to the standards set by the National Sanitation Foundation or other accredited bodies. Cleaning and sanitizing equipment and practices need to be evaluated. The outside surroundings are checked for sources of contamination, such as insects and rodents, bird harborage, drainage problems, solid waste, dust, and other contaminants. The manager of the food establishment must on a daily basis evaluate all potential sources of contamination and eliminate them. This self-evaluation process is reinforced by the periodic visits by the trained professional.

The environmental health practitioner inspects the foodservice establishment using the flow technique, whereby the flow of food from the point of delivery to the point of disposal is followed (Table 3.3) and problems are recorded. Simultaneously, cleanliness and the condition of floors, walls, ceilings, lighting, ventilation, hand washing facilities, and bathrooms are checked, as well as plumbing problems, water supply, sewage disposal, housekeeping procedures, and insects and rodents inside and outside the premises. The environmentalist should always carry a flashlight, clipboard, inspections sheets, pads, sanitizing testing equipment, light meter, and applicable rules and regulations of the health department.

If possible, the foodservice manager or supervisor should accompany the environmental health practitioner on the on-site evaluation; otherwise, at the conclusion of the on-site evaluation, the environmental health practitioner should ask the foodservice manager to go back through the establishment to see the major problems noted. Environmental health practitioners should complete the on-site evaluation form from their notes, and determine the length of time needed to correct each problem. If conditions are bad, another evaluation should be made within 1 week. If the health of the public is potentially endangered, conditions should be corrected at once. The environmental health practitioner should teach the foodservice manager to set up an adequate cleaning schedule and should make recommendations for carrying out of necessary corrections. It is important to discuss the public health

Table 3.3 Flowchart for Food Establishment by Type of Operation

Preparation Delivery	Storage	Preparation	Equipment	Serving	Cleaning and Storage	Waste Disposal
Raw foods	Refrigerator	Baking	Ovens	Milk containers	Removal of soiled utensils and dishes	Garbage disposal units
Finished products	Freezer	Frying	Stoves	Steam tables	Equipment cleaning	Garbage cans
	Dry storage	Cooking	Deep fryers	Cold tables	Dishwasher operation	Garbage rooms
		Salad making	Steam kettles	Counter freezers	Dishwashing techniques	Dempster dumpers
		Mixing	Slicers	Serving utensils	Use of cetergents	Garbage can cleaning
		Grinding	Grinders	Dishware	Sanitizing	Cleaning of exterior premises
		Chopping	Toasters	Silverware	Storage of clean utensils, pots, pans, and dishware	
		Separating	Coffee urns	Single service utensils	Storage of single service articles	
		Combining	Pots and pans	Ice cream dipeprs		
		Food handling	Cutting boards	Ice handling		
			Meat blocks	Food handling		
			Griddles	Automatic vending equipment		
			Can openers			

Note: The environmental health worker checks construction and cleanliness of equipment, temperatures, and time, where applicable, during inspection.

reasons for compliance of the establishment with certain rules and regulations. Additional on-site evaluations are made as needed. When the foodservice establishment will not comply, further action such as warning letters, administrative hearings, license suspension, license revocation, and court hearings are necessary. (See Figure 3.6 for an example of an inspection sheet.)

Foodborne Disease Investigations

When the health department is notified that a possible foodborne disease outbreak has occurred, it must move quickly to obtain the necessary information from persons who ingested the suspected food. It is also necessary to interview people who did not become sick and may have or may not have eaten the suspect food. Surveys of the food preparation, storage, and serving operations must be made, and samples must be taken and processed rapidly. Unfortunately, many outbreaks of foodborne disease occur at large gatherings, such as church dinners, reunions, and picnics. This type of food operation complicates the investigation procedures.

Epidemiological Study Techniques

Environmental health personnel and epidemiologists should complete detailed questionnaires for all persons who may have eaten the food or drinks suspected of having caused the foodborne disease. Once a pattern develops concerning the kinds of foods involved and the places in which the food was consumed, additional environmentalists should be detailed to make complete studies of all food operations. These studies must include a food history and the flow of preparation. The food, its source, method of preparation, storage, food temperatures, times, and other critical limits (such as the time needed at a certain temperature to prevent the growth of organisms that cause disease), and refrigeration must be determined. As part of this study, remnants of suspected foods should be collected aseptically in sterile containers or in their original containers and sent, refrigerated, immediately to the laboratory for analysis. Other samples collected include parent stocks of suspect foods; vomitus, stool samples, swabs of nose and throat, open sores, or wounds of food handlers (taken by a physician or nurse); insecticides; rodenticides and other poisons; suspect food containers; and food residues from slicing machines, cutting boards, etc. It is necessary to properly identify the origin of the sample, the date and time of collection, the name of the environmentalist, and a brief statement about the symptoms of the patients and the suspected organisms or chemicals.

Also, water samples should be taken. All sewage systems and plumbing should be evaluated. The presence of insects and rodents must be determined. It is essential as part of this study to determine if any food handlers were ill within a period of 6 to 8 weeks prior to the onset of the disease. If an individual food handler had diarrhea or vomiting, a fecal sample should be taken. Food handlers should also be checked for any boils, carbuncles, and respiratory infections that could cause an outbreak of staphylococcus food poisoning.

FOOD SERVICE
INSPECTION REPORT_____
HEALTH JURISDICTION

KEY: S=Satisfactory
U=Unsatisfactory
R=Repeat Violation
N=Not Apply

Name of Establishment   Owner or Operator's Name   Address

| Person Interviewed and Title | Date | Time | am pm | Current License | Yes ( ) No ( ) |

Rating

Facilities

1. Floors–clean, good repair
2. Walls–clean, painted, good repair
3. Ceiling–clean, painted, good repair
4. Lighting–fixtures clean, good repair, adequate
5. Ventilation–adequate, proper design
6. Filters–duct work clean, adequate
7. Bathrooms–convenient, self-closing doors, clean, free from odors, adequate supplies
0. I landwashing facilities–adequate, good repair, hot & cold running water, convenient, clean soap, single use towels or dryers
9. Plumbing–sewage & water supply, properly constructed, no cross-connections
10. Hot and cold running water
11. Screened doors & windows or other fly proofing devices

Food Protection

12. Ice–approved source
13. Approved source of food
14. Wholesomeness
15. Refrigeration of perishables below 41°F; Freezing below 0°F
16. Hazardous cooked foods above 140°F
17. Storage of foods
18. Protection of food from contamination
19. Preparation of food
20. Serving of food

Rating

Equipment and Utensils

21. Food Service Equipment–clean & good repair
22. Materials for cleaning & sanitizing
23. Equipment for cleaning–dishwashers, sinks, etc.
24. Cleaning Operation–temperatures, procedures, etc.
25. Sanitization–hot water or chemical
26. Cleanliness–equipment
27. Cleanliness–utensils, dishes, etc.
28. Storage of utensils, silverware, dishes, pots, pans, etc.
29. Construction of equipment, utensils, etc. (including thermometers)
30. Cleaning procedures for food contact surfaces
31. Cleanliness of food contact surfaces

Personal Health

32. Employee free from boils, diarrhea, etc.
33. Cleanliness of hands, garments, etc.

Solid Waste

34. Garbage storage & disposal
35. Rubbish storage & disposal
36. Outer premises

Insects and Rodents

37. Flies, roaches, rats, mice

Sampling Type

38. Sample number

Item   Recommendations   Date of Compliance

Figure 3.6   Foodservice inspection report.

Medical diagnosis of patients combined with good epidemiological data may determine the causative agent in the outbreak and assist the environmentalist in preventing further outbreaks of this type. If commercially prepared food is the cause

Investigation of a Food–Borne Outbreak Report

1. Where did the outbreak occur?	2. Date of outbreak: (Date of onset 1st case)
State_____ (1.2) City or Town_____County_____	_____(3-8)

3. Indicate actual (a) or estimated (e) numbers	4. History of Exposed Persons	5. Incubation period (hours):

3. Indicate actual (a) or estimated (e) numbers

Persons exposed _____ (9-11)
Persons ill _____ (12-14)
Hospitalized _____ (15-16)
Fatal cases _____ (17)

4. History of Exposed Persons

No. histories obtained _____ (18-20)
No. persons with symptoms _____ (21-23)
Nausea _____ (24-26) Diarrhea _____ (33-35)
Vomiting _____ (27-29) Fever _____ (36-38)
Cramps _____ (30-32) Other, specify _____ (39)

5. Incubation period (hours):
Shortest _____ (40-42) Longest _____ (43-45)
Approx. for majority _____ (46-48)

6. Duration of illness (hours)
Shortest _____ (49-51) Longest _____ (52-54)
Approx. for majority _____ (55-57)

7. Food-specific attack rates: (58)

Food Items Served	Number of persons who ATE specified food				Number of persons who did NOT eat specified food			
	Ill	Not ill	Total	Percent ill	Ill	Not ill	Total	Percent ill

8. Vehicle responsible (food item incriminated by epidemiological evidence)_____ (59-60)

9. Manner in which incriminated food was marketed:(Check all applicable)

(a) Food Industry   (6-1)
        Raw ....□ 1
        Processed ......□ 2
    Home Produced
        Raw ....□ 3
        Processed ......□ 4
                    (6-2)
(b) Vending Machine . □ 1

(c) Not wrapped ....□ 1 (6-1)
        Ordinary Wrapping ....□ 2
        Canned ....□ 3
        Canned–Vacuum Sealed. □ 4
        Other (specify) ....□ 5

(d) Room Temperature ...... □ 1 (6-2)
        Refrigerated .... □ 2
        Frozen .... □ 3
        Heated .... □ 4

If a commercial product, indicate brand name and lot number

10. Place of Preparation of
    Contaminated Item (65)
    Restaurant .... □ 1
    Delicatessen......... □ 2
    Cafeteria.............. □ 3
    Private Home........ □ 4
    Caterer................ □ 5
    Institution:
        School.............. □ 6
        Church............. □ 7
        Camp.............. □ 8
        Other, specify....... □ 9

11. Place where eaten   (66)
    Restaurant .... □ 1
    Delicatessen......... □ 2
    Cafeteria.............. □ 3
    Private Home........ □ 4
    Caterer................ □ 5
    Institution:
        School.............. □ 6
        Church............. □ 7
        Camp.............. □ 8
        Other, specify....... □ 9

DEPARTMENT OF HEALTH, EDUCATION AND WELFARE
PUBLIC HEALTH SERVICE
CENTERS FOR DISEASE CONTROL
BUREAU OF EPIDEMIOLOGY
ATLANTA, GEORGIA 30333

Figure 3.7 Investigation of a foodborne outbreak report.

of the foodborne disease, it is necessary to obtain data identifying the source of the food, when and where it was purchased, and any identifying codes or marks on the containers. This information should be immediately forwarded to the FDA. If the outbreak of disease is traced to a hospital or college cafeteria, intensive inspection, coupled with changes in procedures and good food handler training, will help prevent further outbreaks. If the outbreak of disease is traced to a church dinner or picnic, the environmentalist should provide information and suggest the presentation of a food-handling course to the group involved.

To gather data properly, an investigation of foodborne outbreak report (Figure 3.7), the application and food employee interview, the food employee reporting agreement, and the application and food employee medical referral should be analyzed. An HACCP study should be conducted. These forms are obtained from local or state health departments.

12. Food specimens examined: (67) Specify by "x" whether food examined was **original** (eaten at time of outbreak or **check up** (prepared in similar manner but not involved in outbreak)				
Item	Orig.	Check up	Findings Qualitative	Quantitative
Example: beef	x		C. perfringens, Hobbs type 10	2×10^6/gm

13. Environmental specimens examined: (68)

Item	Findings
Example: meat grinder	C. perfringens, Hobbs Type 10

14. Specimens from patients examined (stool, vomitus, etc.):   (69)

Item	No. Persons	Findings
Example: stool	11	C. perfringens, Hobbs Type 10

15. Specimens from food handlers (stools, lesions, etc.):   (70)

Item	Findings
Example: lesion	C. perfringens, Hobbs Type 10

16. Factors contributing to outbreak (check all applicable)

	Yes	No
1. Improper storage or holding temperature	☐1	☐2 (71)
2. Inadequate cooking	☐1	☐2 (72)
3. Contaminated equipment or working surfaces	☐1	☐2 (73)
4. Food obtained from unsafe sources	☐1	☐2 (74)
5. Poor personal hygiene of food handler	☐1	☐2 (75)
6. Other, specify	☐1	☐2 (76)

17. Etiology: (77, 78)
Pathogen _____   Suspected ....☐1   (79)
Chemical_____   Confirmed ....☐2
Other _____   Unknown ....☐3

18. Remarks: Briefly describe aspects of the investigation not covered above, such as unusual age or sex distribution; unusual circumstances leading to contamination of food, water; epidemic curve, etc. (Attach additional page if necessary)

Name of reporting agency: (80)	
Investigating official:	Date of investigation:

NOTE: Epidemic and Laboratory Assistance for the investigation of a foodborne outbreak is available upon request by the State Health Department to the Centers for Disease Control, Atlanta, Georgia 30333.

To improve national surveillance, please send a copy of this report to:
Centers for Disease Control
Attn: Enteric Diseases Section, Bacterial Diseases Branch
Bureau of Epidemiology
Atlanta, Georgia 30333

Submitted copies should include as much information as possible, but the completion of every item is not required.

Figure 3.7   Continued.

HAZARD ANALYSIS AND CRITICAL CONTROL POINT INSPECTION

An HACCP inspection is an in-depth inspection process used to resolve or prevent disease outbreaks by identifying the foods at greatest risk and the critical control points (Figure 3.8). These foods may be critical because they are naturally contaminated, or involve large volume preparation, multistep preparation, and

DEPARTMENT OF HEALTH AND HUMAN SERVICES
PUBLIC HEALTH SERVICE
FOOD AND DRUG ADMINISTRATION

HACCP INSPECTION DATA

EST. NAME: DATE:	PERMIT NO. TIME IN: :AM/PM	INSPECTOR TIME OUT: :AM/PM

Record all observations below–transfer violations to Inspection Report

FOOD TEMPERATURES/TIMES/OTHER CRITICAL LIMITS
Use Additional Forms If necessary

FOOD STEP	1.	CRITICAL LIMIT	2.	CRITICAL LIMIT	3.	CRITICAL LIMIT	4.	CRITICAL LIMIT
A. SOURCE								
B. STORAGE								
C. PREP BEFORE COOK								
D. COOK								
E. PREP AFTER COOK								
F. HOT/COLD HOLD								
G. DISPLAY/ SERVICE								
H. COOL								
I. REHEAT								

OTHER FOOD TEMPERATURES OBSERVED   Use steps from above for location

FOOD	TEMP. °C/°F	STEP	FOOD	TEMP. °C/°F	STEP	FOOD	TEMP. °C/°F	STEP

Figure 3.8   Food Code 1993. (From U.S. Public Health Service, U.S. Department of Health and Human Services, Washington, D.C., Annex 7, p. 5.)

MANAGEMENT/PERSONNEL OBSERVATIONS	

OTHER FOOD OBSERVATIONS	

EQUIPMENT, UTENSILS, AND LINEN OBSERVATIONS	

WATER, PLUMBING, AND WASTE OBSERVATIONS	

PHYSICAL FACILITIES	

Figure 3.8   Continued.

temperature changes in the foods. The critical control points are those few steps in a process that are most important to bacterial contamination, survival, or growth. Menu analysis and food flow helps identify critical control points. They also are an operation or part of an operation where actual or potential risks are usually found and where preventive or control measures that eliminate, prevent, or minimize a hazard that has occurred prior to this point can be exercised. The critical limit is the maximum or minimum value to which a physical, biological, or chemical parameter must be controlled at a critical point to limit the risk that the identified food safety hazard may occur. The HACCP analysis helps determine the critical control points to prevent disease.

Many of the HACCP principles have been in effect in the FDA-regulated low-acid canned food industry. HACCP was established for the seafood industry in an FDA rule that took effect in December 1997. The FDA has also incorporated HACCP in its Model Food Code. The USDA has established HACCP for meat and poultry processing plants, with most of these establishments required to start using this technique by January 1999. Very small plants did not have to comply until January 2000. The FDA has proposed requiring HACCP controls for fruit and vegetable juices and would establish these techniques throughout all areas of the food industry both for domestic and imported food. To help determine if such regulations are feasible, the FDA is conducting pilot programs with volunteer food companies processing cheese, frozen dough, breakfast cereals, salad dressing, bread, flour, and other products.

HACCP involves six principles:

1. Analyze potential hazards associated with food and the techniques used to control these hazards, including biological, chemical, or physical problems.
2. Identify critical control points in food production from the raw state, through processing and shipping, to consumption by the consumer, and in situations where the potential hazard can be controlled or eliminated.
3. Establish preventive measures with critical limits for each control point, such as setting minimum cooking temperatures and time to eliminate any harmful microbes.
4. Establish corrective action to be taken when monitoring shows that a critical limit has not been achieved.
5. Establish procedures to verify that the system is working properly, such as the use of time and temperature recording devices.
6. Establish effective record keeping to document the entire HACCP system.

New challenges to the U.S. food supply prompted the FDA to adopt a HACCP-based food safety system on a wide basis. Many of the emerging diseases that we now consider to be serious public health problems did not exist several years ago. These new diseases have made us much more conscious of the need for a scientifically based preventive program. The governmental agencies involved are more efficient and effective because the food establishments have to maintain proper records of their compliance with food safety laws over a period of time, This tends to place the responsibility for ensuring food safety on the food manufacturer, food distributor, or foodservice establishment.

FOOD PROTECTION CONTROLS

Until recently, the major controls used for food protection were control of temperature and time, and the proper preparation of food in a clean manner. Although these measures are still essential, there are new concerns due to the influx of chemicals into our food supply. Because specific controls do not exist at present, they are mentioned in a succeeding chapter under research needs.

Streptococcus infection control measures include pasteurization of milk and other dairy products, exclusion of persons with known streptococcal infections, and antibiotic treatment of known carriers or contacts.

Salmonella controls include pasteurization of milk and other dairy products, control and certification of shellfish areas, water treatment and chlorination, elimination of flies from human feces, and prohibition of food handling by individuals with a history of typhoid or paratyphoid fever.

Salmonellosis control measures include the use of good personal hygiene, removal of sick food handlers from duty, obtaining cultures from food handlers with diarrhea infections, and restricting them from work until three consecutive fecal samples are negative. In addition, storage of susceptible foods should be at temperatures below 41°F; the danger zone for salmonella growth, which occurs between 60 and 120°F (serve all hot foods above 140°F but precontrol to 165°F), should be avoided. Never stuff chicken or turkey before cooking, and cook in a separate pan; prepare small quantities of such dishes as egg salad, potato salad, chicken salad, or turkey salad; analyze feed for salmonella contamination and maintain disease-free animals; report immediately all suspected food infections to public health authorities and retain samples of food; and make complete epidemiological studies.

Salmonella enteritidis organisms apparently occur in 75% of cases associated with raw or inadequately cooked Grade A whole shell eggs. Scrambled eggs should be cooked at 250°F for 1 min and the egg temperature should be 165°F. Poached eggs should be cooked for 5 min in boiling water. Sunnyside fried eggs should be cooked at 250°F for 7 min uncovered and for 4 min covered. Eggs over easy should be cooked at 250°F for 3 min on one side and 2 min on the other. Boiled eggs should be submerged in boiling water for 7 min. Norwalk-type disease control measures are similar to salmonellosis control measures.

Shigellosis control measures include strict personal hygiene, elimination of carriers as food handlers, refrigeration of moist foods, cooking of foods at 165°F prior to serving, and elimination of flies. Yersiniosis is controlled by proper disposal of human, cat, and dog feces; protection of water supplies; proper hand washing; proper preparation of food; pasteurization of milk. Cholera control measures include filtration and chlorination of drinking water supply, growing of shellfish in certified sewage-free areas only, removal of known cases from food-handling operations, and vaccination when an outbreak of cholera occurs.

Hepatitis can be prevented by observing the same controls as stated under cholera. *Plasmodium shigelloides* can be controlled by purification of drinking water, protection of food from nonpotable water, and proper cooking. Campylobacterosis can be controlled by cooking meat thoroughly, pasteurizing milk, chilling foods rapidly in small quantities, and avoiding contaminating prepared food with raw food.

Brucellosis, diphtheria, and bovine tuberculosis control measures include pasteurization of milk and dairy products and isolation of carriers. In addition, for brucellosis and bovine tuberculosis, sick animals should be eliminated. Listeriosis can be controlled by cooking food thoroughly and pasteurizing milk. Tularemia control measures include the use of protective gloves when handling wild animals and the elimination of arthropods.

Staphylococcus food poisoning control measures include removing food handlers who have nasal discharges or skin infections, cooking foods thoroughly, refrigerating foods immediately, using custard pastries within 2 to 4 hours after preparation and discarding the remainder, and washing hands thoroughly and frequently. Bacillus cereus can be controlled by heating to 165°F for 5 min for the toxin.

Clostridium welchii or *C. perfringens* control measures include the elimination of known carriers from food handling, cooking meat thoroughly before consumption, cooling meat rapidly in the refrigerator immediately after cooking, using drippings from meat for gravies only on the day of cooking, and discarding all leftovers. Preparation of chicken, turkey, or beef pot pies should only be from freshly cooked pieces of meat.

Botulism can be prevented by proper cooking of foods. If home-processed foods are used, they should be boiled for a minimum of 15 min before eating. Discard any canned or bottled foods that show signs of spoilage including off-odors, off-tastes, gas or foams, and off-colors. If botulism is suspected, have a physician administer the proper antitoxin immediately.

Vibrio parahaemolyticus can be killed by thorough cooking. Where substantial mold damage appears, food should not be consumed because of the potential for mycotoxins.

Amoebic dysentery control measures include protection of water supplies from human excreta, elimination of carriers from food preparation, proper sewage disposal, and filtration and chlorination of water supplies. Giardiasis can be controlled by cooking food thoroughly, good personal hygiene, and control of sewage.

Beef and pork tapeworm control measures include proper inspections by trained veterinarians and thorough cooking of beef and pork. For fish tapeworms, cook fish thoroughly and avoid all raw smoked fish. Trichinosis control measures include cooking garbage fed to pigs to an internal temperature of 137°F, eliminating rats from pig farms, and cooking all pork to an internal temperature of 165°F.

Toxoplasmosis can be controlled by cooking foods thoroughly, freezing food at or below 5°F for 24 hours, and washing hands after handling raw foods. Angiostrongyliasis can be controlled by cooking foods thoroughly; freezing food at or below 5°F; and not eating raw freshwater prawns, raw land mollusks, and raw crab.

Scrombroid poisoning can be controlled by icing or refrigerating fish right after capture, eating fish quickly after proper cooking, and discarding sharp or peppery tasting fish. Enteric adenovirus can be controlled by proper cooking of food, good personal hygiene, and proper disposal of sewage.

Escherichia coli of all types can be controlled by chilling foods rapidly under 41°F in small quantities, cooking and heating to 165°F, practicing good personal hygiene, treating water, and properly disposing of sewage. Astrovirus, calcivirus, parvovirus, and rotavirus can be controlled by proper cooking, handling, and refrigeration of food if the virus is foodborne.

Chemical poisoning control measures include eliminating utensils and containers that may leach the chemical into solution and discontinuing the use of dangerous pesticides; protecting all food and food contact surfaces when using pesticides; and storing all chemicals in their original containers away from food storage, preparation, and serving areas. Intoxication from poisonous plants or animals can be prevented by not eating these plants or animals.

Foodservice Training Programs

The keys to a safe, sanitary food program are foodservice supervisors and personnel properly trained in evaluating menus for potential hazards; good techniques of purchasing, storing, preparing, transporting, and serving wholesome food; and preparation of an HACCP plan including flow diagrams, temperatures, times, and critical limits. Scheduling of cleaning operations, cleaning and sanitizing techniques, and proper use of equipment and materials must be understood and the necessary materials must be made available. Although all foodservice personnel are important, the supervisor is the main key. The supervisor schedules operations, teaches personnel, orders supplies, and inspects the establishment daily. Assistance is available from local and state health departments, which are happy to conduct necessary training programs. The supervisor should attend special training sessions held by the health department and should be tested and certified on an annual basis by the health department. Renewal of the establishment's license should be contingent on this certification.

An educational program directed at homemakers and individuals who prepare food for mass feeding operations, such as church dinners, is useful in controlling the numerous outbreaks of disease that may occur in the home or at special functions.

National Food Safety Initiative

In the past, local and state health departments and the CDC have worked together to try to solve outbreaks of disease that had occurred and on a limited basis to do surveillance of some foodborne pathogens. Recent changes in the food supply and in the consuming public have created a variety of new foodborne diseases as well as making existing ones more severe. In 1997, President Bill Clinton announced an administrative initiative to create a nationwide early warning system for foodborne diseases; enhanced food safety inspection; and expanded food safety research, training, and education. The National Food Safety Initiative includes rapid identification of foodborne hazards and development of an effective response; identification of large, diffuse outbreaks that have a low attack rate within one health jurisdiction that affect the health of many people within a larger geographic area; and evaluation of the nature and scope of food safety problems. The National Food Safety Initiative links CDC, state and local health departments, and other federal agencies within a powerful electronic network developed by CDC for rapid sharing of microbial subtyping information, including the digitalized images of subtyping patterns needed for molecular epidemiology; and enhances as well as expands the activities of FoodNet, the interagency foodborne disease portion of the emerging infectious disease active surveillance program.

PulseNet is the national molecular subtyping network for foodborne disease surveillance. This program helps in the rapid epidemiological investigation of foods that may have caused disease. Training is given to state public health laboratory workers to enhance their ability to find the causes of foodborne disease. New and reemerging diseases are scientifically investigated, and are written about in the *Emerging Infectious Disease Journal*. Research accomplishments include development of a quick test to identify *E. coli*; development of a DNA-based technique to detect *E. coli* O157:H7; development of an intervention technology to eliminate or inactivate microbial contaminants by the use of high hydrostatic pressures run at ambient or moderate temperatures; development of a rapid, sensitive, and reliable method capable of detecting low levels of Norwalk viruses in shellfish; and development of a method to improve detection of ochratoxins in grains. Finally, a national food safety education month has been established to reinforce food safety education and training among restaurant and foodservice workers and the general public.

Terminology

To better understand foodborne disease, it is necessary to have some knowledge of the following terminology:

- *Active immunity*. Antibodies produced in the person through contact with disease
- *Carrier*. An individual harboring specific infectious agents in the absence of discernible clinical disease who serves as a potential source or reservoir of infection for other humans
- *Chain of infection*. The spread of disease from the reservoir by means of a vehicle or vector to the host
- *Contamination*. The presence of a pathogenic organism in or on a body surface or an inanimate object
- *Communicable disease*. An illness due to an infectious agent or its toxic products transmitted directly or indirectly to a well person from an infected person or animal; or transmitted through an intermediate animal host, vector, or inanimate environment
- *Communicable period*. The time or times during which the etiologic agent may be transferred from an infected person or animal to humans
- *Endemic*. The regular occurrence of a fairly constant number of human cases of a disease within an area
- *Enterotoxin*. A toxin produced by *Staphylococcus aureus* that specifically affects cells of the intestinal mucosa causing vomiting and diarrhea and arising in the intestine
- *Epidemic*. The occurrence in a community or region of a group of illnesses of a similar nature, clearly in excess of normal expectancy, and derived from a common or propagated source
- *Epidemiology*. The study of the causes, transmission, and incidence of diseases in communities or other population groups
- *Etiological agent*. The pathogenic organism or chemical causing a specific disease in a living body
- *Exotoxin*. Toxic substance produced by bacteria found outside of the bacterial cell

- *Fomite.* An inanimate object not supporting bacterial growth but serving to transmit pathogenic organisms from human to human
- *Foodborne disease outbreak.* An incidence of disease occurring after the ingestion of a food shown to cause the disease
- *Hazard.* A biological, chemical, or physical property of food that may cause an unacceptable consumer risk
- *Host.* The living body, human or animal, that provides food and shelter for the disease organisms
- *Incidence.* The number of cases of disease occurring during a prescribed time period in relation to the unit of population in which they occur
- *Incubation period.* The time interval between the infection of a susceptible person or animal and the appearance of signs or symptoms of the disease
- *Infection.* The entry and development or multiplication of a particular pathogen in the body of humans or animals
- *Parasite.* An organism living on the tissues and waste products of the host within the body of the host; all disease-causing organisms classified as parasites
- *Passive immunity.* Antibodies produced in another host and then injected into the diseased person
- *Report of disease.* An official report, usually by a doctor, of the occurrence of a communicable or other disease of humans or animals to departments of health or agriculture or both, as locally required; reports to include those diseases requiring epidemiological investigation or initiation of special control measures
- *Reservoir of infection.* Humans, animals, plants, soil, or inanimate organic matter in which an infectious agent lives and multiplies and then is transmitted to humans, who themselves are the most frequent reservoir of infectious pathogenic agents
- *Resistance.* The sum total of body mechanisms that place barriers to the progress of invasion of pathogenic organisms
- *Vector.* A living insect or animal (not human) that transmits infections and diseases from one person or animal to another
- *Vehicle.* Water, food, milk, or any other substance or article serving as an intermediate means by which the pathogenic agent is transported from a reservoir and introduced into a susceptible host through ingestion, inhalation, inoculation, or by deposit on the skin or mucous membrane

SUMMARY

Food protection is an essential part of the environmental health program, considering the vast variety of food produced, the enormous potential for disease, and the short- and long-range problems of chemicals entering our food supply. Foodborne disease outbreaks probably account for millions of illnesses each year. The various long-range problems related to foodborne disease cannot even be estimated.

Food protection consists of proper plan review, the development and use of adequate facilities in a correct manner; the procurement and use of correct equipment; necessary cleanliness; proper storage, preparation, and serving of food; and adequate disposal of food remnants. The cleaning and sanitizing techniques utilized are an essential facet of food protection. Important factors also include the health and potential for spreading disease of humans. A variety of inspectional techniques are utilized on a regular and special basis, depending on the nature of the food establishment and the types of problems likely to occur. Proper techniques and foodborne disease investigations eliminate outbreaks and prevent future outbreaks of disease. All states and many counties and localities have food protection programs.

RESEARCH NEEDS

Research needs are discussed in Chapter 4 on food technology.

CHAPTER 4

Food Technology

BACKGROUND AND STATUS

Food is a perishable product consisting of proteins, carbohydrates, fats, vitamins, minerals, water, and fibers. It is used by an organism in sustaining growth, repairing tissues, maintaining vital processes, and furnishing energy. Great pleasure is derived from the proper preparation of food and from its consumption. Food is altered favorably or unfavorably by microorganisms, enzymes, insects, and environmental changes. Microbes help produce sauerkraut, bread, and cheese; ripen olives; and ferment milk. The action of the yeast *Saccharomyces cerevisiae* produces carbon dioxide in bread, making it porous and causing it to rise. The starches and proteins in bread are split by the action of the yeast, becoming more digestible, and characteristic flavors and aromas are produced. However, if certain spore-forming bacilli, principally *Bacillus mesentericus*, are introduced with the flour or yeast, discoloration and unpleasant odor results and the bread becomes unappetizing.

In this chapter, contamination, spoilage, and disease potential of certain types of foods are discussed, including milk and milk products, poultry, eggs, meats and meat products, fish and shellfish, and fresh produce (including juice). The nature of enzymes, microbiological spoilage, factors affecting spoilage, techniques of preservation of foods, chemical preservatives, additives, pesticides, fertilizers, and antibiotics are also discussed. The production, processing, testing, and environmental problems associated with each food group are addressed. Applied on-site evaluation techniques and inspection forms are introduced as needed. The foods were selected as the most frequent contributors to foodborne diseases, intoxication, or kinds of food spoilage that the environmental health practitioner is likely to encounter.

In the year 2000, malnourishment and death from starvation occur in many areas of the world. In poor countries, especially where population is growing rapidly, the problem has become critical. An estimated 2 billion people suffer from malnutrition and dietary deficiencies. More than 840 million people, disproportionately women and girl children, suffer from chronic malnourishment. The United Nations Food and Agricultural Organization (FAO) has identified 82 poor countries that are at a particular risk. From the 1960s until a few years ago, the world food supply kept pace with population growth, because of better seed varieties and irrigation. Between 1985 and 1995, however, 64 of the 105 countries studied by this organization had food production lagging behind population growth.

To combat hunger and death, and to improve the economy of many countries, more land is under cultivation and more environmental resources and people-made chemicals are utilized. Water, which is a prime resource for increasing food output, is plentiful in some areas and essentially lacking in others. The changing of water patterns to increase productivity has also brought disease. In Egypt, for example, the Aswan Dam stimulated a large outbreak of schistosomiasis. Fertilizers when used properly are excellent for increasing crop yield, but when improperly used they become contaminants and potential cancer-causing substances.

Plant disease, insects, pests, and weeds caused an estimated 30% loss in potential food production throughout the world, resulting in a sharply increased use of pesticides. Unfortunately, some pesticides are inherently dangerous or are dangerous when misused. In the Canete Valley of Peru, the cotton crop was heavily treated with dichlorodiphenyltrichloroethane (DDT) and other hydrocarbons. The sprays killed not only the pests but also their predators. The pests eventually developed resistance to DDT, whereas the predators did not, resulting in lost cotton crops. Numerous times pesticides contaminated raw food products and were concentrated as they were processed through the food or became a part of the runoff problem, resulting in varying levels of water pollution.

Additives are added to food to enhance flavor, taste, and keeping quality. Unfortunately, unintentional or intentional additives may be the cause of cancer or other related diseases. There is serious concern over the use of sodium in processed foods, because the sodium may produce adverse effects in hypertensive individuals. Other contaminants found in raw food before processing and remaining in the food until consumption include mercury, lead, cadmium, zinc, copper, manganese, selenium, and arsenic. In addition, a variety of pesticides, such as chlorinated hydrocarbons, polychlorinated biphenyls, and a vast group of chemicals are entering our food supply.

Animal feed may induce a toxic response in humans or animals. Unfortunately, there is a lack of research in this area, although scientists know the toxicants are found in animal tissue. Milk and dairy products are of particular importance, not only to the young but also to vegetarians. Milk and dairy products contain estrogens, nitrates, antibiotics, pesticides, radionuclides, and mycotoxins. Eggs cause disease, and may cause allergic disorders in children and adults due to a natural allergic response or to estrogens and antibiotics found in egg yolks. Aflatoxins have been found in corn.

A rapid increase has occurred in the use of frozen prepared foods and vacuum-packed foods. Improper processing of these foods has resulted in outbreaks of botulism, and other foodborne diseases.

Changing Consumer Lifestyles and Emerging or Reemerging Foodborne Pathogens

The consumer's desire for good health is having a great impact on the food processor. In the last 25 years the marketplace has changed and people are requiring

more perishable products, including fruits and vegetables, and more innovative packaging. Consumers are becoming averse to traditional chemical preservatives, which
has encouraged the food processors to find new processing and preservation technologies, which might not necessarily kill the microorganisms that cause disease. The
number of meals eaten away from home has increased dramatically, and the number
of home-delivered meals has also increased substantially. Each of these meals lends
to multiple handling and greater potential for foodborne disease. The local supermarket has become a source of ready-to-eat take-out foods on a very regular basis. Fast
food restaurants provide 48% of the takeout food, whereas restaurants provide 25%
and supermarkets provide 12%. There has been a proliferation in nontraditional food
preparation facilities. This has led to a large number of food handlers, who may be
improperly supervised and lack knowledge concerning the potential for spread of
disease through food. Consumers may lack adequate knowledge concerning the
spread of disease from different types of foods, especially fruits and vegetables.
Further they may contaminate areas where they are selecting foods.

SCIENTIFIC AND TECHNOLOGICAL BACKGROUND

Chemistry of Foods

Enzymes produce desirable and undesirable changes in foods. As an example,
enzymes may tenderize meat or cause off-colors, off-odors, and off-tastes. An
enzyme is an organic catalyst that speeds up a chemical reaction without altering
itself. It has the following characteristics: it acts most rapidly at body temperature,
it has a high degree of specificity for a given test, it is produced by living cells
including microorganisms, it continues to react after harvest of plants or slaughter
of animals, it is retarded by low temperatures, it is destroyed by boiling, and it may
react either within the cell (endozyme) in which it was produced or outside of the
cell (exozyme) in the tissue.

The inactivation of the enzyme phosphomonoesterase by heat is a reliable means
of measuring the efficiency of pasteurization of milk and ice cream in the phosphatase test. The enzyme is totally destroyed by proper pasteurization. However,
false positives can occur due to certain bacteria present in milk, including streptococcus and aerobacter.

Food Additives and Preservatives

A food additive is any substance or mixture of substances other than basic
foodstuffs added during the production, processing, storage, or packaging of foods.
Additives are used to preserve, emulsify, flavor, color, and increase nutritive value.
Intentional food additives include preservatives, antioxidants, sequestrants, surfac-
tants, stabilizers, bleaching and maturing agents, buffers, acids, alkalines, colors,
special sweeteners, nutrient supplements, flavoring compounds, and natural flavoring
material. Unintentional food additives or contaminants include radionuclides, insect
parts and excreta, insecticides and herbicides, fertilizer residue, other chemicals, dirt,
microorganisms, and any other unintentional materials.

Antioxidants preserve freshness in meats by preventing rancidity of fats and preserve appearance and taste of fruits by preventing discoloration. Mold inhibitors, such as calcium or sodium propionate, keep bread fresh for longer periods. Emulsifiers, such as gum arabic, are used to maintain the consistency of French dressing. Stabilizers and thickeners, such as lecithin, are used in ice cream. Monosodium glutamate, nonnutritive sweeteners, and other agents enhance the flavor of food. Additives that add nutritive value to foods include iodized salt to prevent goiter; vitamin D in milk to prevent rickets; thiamine, niacin, riboflavin, or iron in bread and cereals to provide individuals with sources of these essential vitamins and minerals; and vitamin A in margarine to help prevent malnutrition. The major restrictions on additives are that they do not cause short-term or chronic poisoning or cancer, do not make decomposed food appear fresh, or do not cause any other unfavorable side reactions. For example, in curing meats, nitrites must not exceed 120 ppm combined with an activator of ascorbates or erythorbates at 550 ppm.

In addition to the nitrates or nitrites that are purposely added to food, considerable amounts of these substances are introduced into the food chain through the use of nitrogen fertilizers. Certain vegetables accumulate nitrogen compounds based on the amount of nitrate and molybdenum in the soil, the light intensity, and the existing drought conditions. Further high concentrations exist from the accidental introduction of large amounts of nitrates or nitrites in the foods. When tests are conducted to meet the nitrate–nitrite standards, they should include not only levels added, but also levels that already exist. Nitrates may also be introduced by groundwater usage.

Sodium chloride is added to diets in varying degrees in individuals. A relationship is suspected between sodium chloride and human hypertension. The amount of salt added as an additive to food may well be a serious problem to the individual who has potential or existing hypertension.

Phosphates are introduced into food through the processing of poultry, in the soft drink industry, and in the production of modified starches. Phosphates may also be introduced through the use of various fertilizers. Excessive daily intake of phosphorus causes a premature cessation of bone growth in children, subsequently affecting final adult height.

Additional knowledge is needed concerning the effects of the intentional addition of food additives to our processed foods. Cyclamates have been removed from the market. It is difficult to predict which additives previously thought to be safe will be evaluated and found potentially hazardous to humans. Even the widely used saccharin has been under suspicion. The Food and Drug Administration (FDA) puts out a Food Additive Status List, which states the status of food additives that may be harmful to people.

Nitrates, Nitrites, and Nitrosamines

Nitrates and nitrites are used to preserve meats, including bacon, ham, hot dogs, pastrami, smoked fish, cured poultry, etc. Nitrite curing is more rapid than nitrate curing and, therefore, is used more often. Nitrites are added to the cured meat as a preservative. Nitrites are also effective microbial inhibitors, retard the development of rancidity, enhance the flavor of certain meats, and maintain the bright red color of meats. The nitrite in the cured meat does not actually add color, but fixes the color pigment, myoglobin. Nitrites are not themselves carcinogens but instead are precursors. Nitrosamines are produced from nitrites and amines, which may be found in various places, including the decomposition of proteins. Amines are found in food, drugs, and pesticides. Nitrosation is the reaction between nitrous acid and secondary or tertiary amines. Nitrous acid comes from the nitrite added to the meat, and amines come from the meat protein. Over a hundred nitrosamines have been isolated and identified, with approximately 75% of these carcinogens. Nitrosamines can cause cancer in animals after a single dose. Nitrosamines may also be found in imported and domestic beer.

Sulfites Used in Food

Sulfites have been used as preservatives in wine and other beverages, as well as in foods. Sulfiting agents retard the oxidation process of uncooked food, including vegetables, and extend the time of visual appeal. Fruits, green vegetables (especially lettuce), peeled or cut potatoes and apples, shrimp, seafood, and seafood salads are commonly sprayed or dipped into sulfiting agents. A sulfiting agent refers to sulfur dioxide, and several forms of inorganic sulfite that liberate sulfur dioxide under various conditions of use. Currently sulfur dioxide, potassium and sodium metabisulfite, potassium and sodium bisulfite, and sodium sulfite are considered to be safe for use in foods by the FDA. Sulfiting agents are added to foods because they control enzymatic and nonenzymatic browning, are antimicrobial, and act as an antioxidant and a reducing agent. Sulfiting agents are used in a wide variety of food products.

Unfortunately, it has been discovered that some individuals, especially asthmatics, may have a potentially severe adverse reaction to sulfites. The sulfites may induce an attack of asthma. Further, in some rare instances other types of hypersensitivity have occurred. Therefore, it is necessary that where sulfiting agents are permitted to be used, especially on salad bars, individuals be advised of this by the placement of a placard at the salad table.

Aspartame

Aspartame is marketed as either Equal® or Nutrasweet®. Aspartame is low in calories and has no odor or aftertaste. However, because of its sweetness, aspartame acts as a sugar substitute; therefore, it is useful in the diet of diabetics, heart patients, overweight patients, etc. There are some questions concerning the potential health effects of aspartame. Aspartame degrades into diketopiperazine at certain temperatures. The metabolism of aspartame, which is a peptide, proceeds in the same way as the metabolism of proteins. After metabolism, aspartame components become aspartic acid, phenylalanine, and methanol.

In some women of child-bearing age, the natural blood levels of phenylalanine fluctuate wildly. Therefore, small doses of aspartame might contribute to much higher levels of the phenylalanine of the individual. Some researchers have indicated that certain people had difficulty in metabolizing phenylalanine. High levels of phenylalanine in mothers apparently cause mental retardation in their babies. At lower levels, phenylalanine could affect the development of the fetus brain, thereby potentially decreasing intelligence quotient (IQ). Because aspartame is so useful in the diet of so many people, it is essential that further research be conducted on the potential health effects of aspartame on people. If these health effects are proved to occur, certain restrictions or advisories should be printed on the aspartame containers.

F D and C Dyes

F D and C dyes in themselves are not carcinogenic; however, when the dyes are metabolized, depending on which dye is utilized, a carcinogenic substance may be produced. As a result of this, at least 13 F D and C dyes have been banned from food use in the United States. The gastrointestinal tract is an important metabolic organ. It contains mostly anaerobic bacteria, which aid in metabolizing F D and C dyes with the possible production of azo reduction, which is the most important metabolic reaction of azo dyes that can contribute to carcinogenesis.

Anabolic Hormones

Anabolic hormones are used in cattle to increase weight. There is a potentially serious concern about the adverse affects to children who have been exposed to these hormones. In Puerto Rico, a study was done on young children that indicated premature development of the breasts, high incidence of ovarian cysts, very early development of vaginal bleeding, and uterus enlargement due to exposure to anabolic hormones.

Bioengineered Foods

Bioengineered foods require labeling if they contain a new protein with characteristics suggesting that it may be a food allergen for sensitive people. However, it is unlikely that most proteins introduced into food by bioengineering will be allergens. Only a fraction of the thousands of proteins are known allergens. Peanuts and soybeans account for 90% of problems.

PROBLEMS OF FOOD TECHNOLOGY

Contamination

Animal products contain microorganisms that are part of their normal flora, are present because of a disease process, or have been added at any point during the killing, processing, storage, preparation, and serving of the product. The organisms may be added by humans or from the environment. About 8% of fresh eggs contain microorganisms; dirty eggs are covered with organisms that easily penetrate the shell. The organism may be salmonella, shigella, and *Cholera vibrio*. Meat contains pseudomonas, proteus, *Escherichia coli*, clostridium, and many other organisms. Although the milk of a cow is sterile in a healthy udder, it frequently becomes contaminated by streptococci normally present in the milk ducts. Diseased cattle may be infected with pathogenic staphylococcus, streptococcus, tuberculosis, or brucellosis organisms. Plant products, such as lettuce, cabbage, carrots, fruits, and vegetables, are contaminated with microorganisms from the air and soil, and by workers handling food.

Those bacteria causing foodborne disease or foodborne intoxication are of particular concern. The spores of *Clostridium botulinum* require temperatures well above the boiling point for destruction. Although intoxication is infrequent, it is of such a serious nature that particular care must be taken in preparing high-risk foods. This is especially true in home canning. Some molds produce toxic materials known as mycotoxins. The best known of these are the aflatoxins, which have been found on peanuts, rye, wheat, millet, jellies, and jams.

Food Spoilage

The composition of food is important. Proteins are especially susceptible to spoilage by spore-forming Gram-negative rods, such as pseudomonas and proteus, and by molds. Carbohydrates are particularly affected by yeasts, molds, streptococcus, and micrococcus. Fats undergo hydrolytic decomposition and become rancid. The pH of acidic foods, such as fruits, is low enough to prevent most bacterial spoilage; however, yeasts and molds grow well on these products. Nonacidic foods are subject to bacterial spoilage. Foods with a moisture content exceeding 10% support the growth of microorganisms. Sugar and salt concentrations create an osmotic pressure that disrupts organisms; 5 to 15% of salt inhibits bacteria and 65 to 70% of sugar inhibits molds. The presence or absence of oxygen determines the kinds of microorganisms that may or may not grow.

Food spoilage is due to two principal causes, chemical (including the enzymes just discussed), and biological (caused by bacteria, molds, and yeasts). Spoilage in most fruits is caused by molds and yeasts. Damage to the surface of the fruit increases the rate of deterioration. Spoilage of vegetables is due to bacteria and molds. Spoilage of properly refrigerated meat is usually confined to the surface. This is the reason why in the aging of beef, the surface is cut away and the interior portion, which is now more tender, is used. However, putrefactive decomposition occurs rapidly in ground meat and ground fish, because the bacteria are distributed throughout the food mass. Fish fillets become slimy and proteolysis occurs within several days because of the enormous number of microorganisms present initially and the handling techniques used. Shucked shellfish contains large numbers of bacteria that can lead to putrefactive decomposition if the shellfish are not refrigerated immediately. Milk, because it is drawn from the cow at temperatures favorable to rapid bacterial growth, must be immediately cooled to prevent bacterial decomposition.

Canning Spoilage

Underprocessing of canned foods results in microbial spoilage in the presence of high heat-resistant spores. This spoilage by microorganisms is a potential problem, because the raw food products and the processed foods may be contaminated by air, water, hands, equipment, and added ingredients. Flat sour spoilage is characterized by the production of acid without the production of gas. A slightly disagreeable odor and change of color may be present in the canned food. A swollen can of food may be caused by gas-forming thermophilic organisms. Swollen cans of meat are most likely caused by organisms producing putrefaction. Anaerobic bacterial growth in canned foods is of particular concern, because *C. botulinum* may be causing the problem, instead of other members of the anaerobic family. Some characteristics of anaerobic activity are offensive odors, black sediment or residue, and reduction of oxygen. Spoilage in cans may also be due to a reaction between the food and the metal in the can, producing hydrogen gas; overfilling the can at too low a temperature; freezing the liquid portion of the food; inadequate removal of oxygen from the can before sealing; or sulfides present in the foods.

Reduced oxygen packaging is the reduction of the amount of oxygen in a package by mechanically evacuating the oxygen; displacing the oxygen with another gas or gases; or otherwise controlling the oxygen in a package to a level below 21% oxygen. The concern is the potential for supporting the growth of *C. botulinum*.

Sources of Exposure

Extraneous Materials

Insect parts and rodent hairs are unwanted components of raw agricultural materials and processed food. Although large amounts of these materials are unacceptable, little is known about the relationship of the quantity of material to microbiological contamination. More research is needed in this area.

Pesticides and Fertilizer

Raw food products are contaminated by pesticides, herbicides, and fertilizers. As an example, in 1989, the pesticide Alar made national news when it was found to be contaminating apples. After washing, a residue of the chemical may be present. The Pesticide Chemical Act of 1954, as updated by the Federal Insecticide, Fungicide, and Rodenticide Act (FIFRA) of 1988, prohibits interstate shipping of raw agricultural foodstuffs containing a residue of a pesticide unless it is safe or if the residue is within the tolerance level established by the FDA as safe. This law has been updated by the Food Quality Protection Act of 1998. A further discussion on pesticides appears later in this book.

Radioactive Fallout

Radioactive material mixed with soil and rock are spread over a large area when a nuclear weapon is exploded or when a nuclear accident occurs. The fallout causes harm to people, animals, crops, food, water, structures, and fields. Radioactive strontium and iodine are most dangerous. Radioactive strontium is chemically similar to calcium; it enters the bones and may cause cancer. It has a half-life of 28 years. Radioactive iodine, which has a half-life of 8 days, may cause cancer of the thyroid gland. The major danger to people is that strontium-90 may be taken in with grass by cows and then passed through milk to humans. Milk sheds are sampled regularly for levels of strontium-90.

Antibiotics in Food

Such antibiotics as penicillin, streptomycin, bacitracin, Chloromycetin, Terramycin, and Aureomycin are used in the preservation of fish and meat, primarily to retard spoilage. Antibiotics and sulfa drugs are commonly used in treatment of bovine mastitis in cows. Antibiotics have also been used as food additives in uncooked ground beef and pork products to preserve keeping quality. Until recently, the FDA permitted the addition of chlorotetracycline and oxytetracycline to poultry water to extend shelf life. The major question for consumers is whether the antibiotics are toxic if ingested over extended periods of time. Another serious question is whether substantial usage of antibiotics in foods causes the development of resistant organisms that may be harmful to humans. Other concerns include the potential for human allergic reactions to the antibiotics.

Heavy Metals and the Food Chain

The heavy metals are considered to be antimony, arsenic, cadmium, chromium, lead, mercury, nickel, silver, thallium, and uranium. All individuals are exposed to heavy metals from a variety of sources, including natural sources, land pollution sources, water pollution sources, and aerial sources. The natural sources are due to chemical weathering and volcanic activities. In freshwater systems, chemical weathering of igneous and metamorphic rocks, as well as soil, causes these trace metals to enter surface waters. The decomposition of plants and animals also adds trace metals to the surface waters. Dust from volcanic activities, smoke from forest fires, aerosols, and particulates on the surface of oceans add additional amounts of trace metals.

Mining operations contribute higher levels of heavy metals, because the material brought from below the surface of the ground is now exposed to the air. Additional heavy metals are added by the corrosion of water pipes and the manufacturing of consumer products. Wastewater treatment by means of the activated sludge process usually removes less than 50% of the heavy metals present in the liquid. Sewage sludge, when placed on top of the ground or sprayed on the ground, becomes a source of cadmium, lead, mercury, and iron. Industrial sludge also adds a significant amount of heavy metals. Storm water runoff from urban areas contribute lead, cadmium, chromium, and zinc. The salt used on streets may increase the mobilization of metal ions, such as cadmium, mercury, and lead. Leachate from sanitary landfills can increase the levels of lead and mercury. Agricultural runoff, which includes soil erosion as well as animal and plant residues, fertilizers, herbicides, fungicides, and other pesticides, adds to the heavy metal problem.

Aerial sources of heavy metals come from the smelting of metallic ores. Other aerial sources come from fossil fuels and the addition of metals to fuels as anti-knock agents. These enter the body via these routes:

1. Through respiratory surfaces
2. Through adsorption from water onto body surfaces
3. From ingested food particles or water through the digestive system

The two main routes of entry into plants are:

1. Through the plants surface above the ground
2. Through the root system

Terrestrial animals take up heavy metals through food and water. Further, a significant amount of absorption may occur through the lungs in some situations. Heavy metals can be bioconcentrated, because there is a progression upward through the food chain, and therefore, may be a biomagnification of the heavy metals in larger animals.

Use of Recycled Plastics in Food Packaging

Glass, steel, aluminum, and paper have been recycled for food contact use. Contamination has not been a major concern with glass and metals because they are generally impervious to contaminants and are readily cleaned at the temperatures used in their recycling. Pulp from reclaimed fiber in paper and paperboard may be used for food contact articles providing it meets the federal government criteria found in 21 Code of Federal Regulation (CFR)176.260. However, recycled plastic may be another problem. The manufacturers of food contact articles made from recycled plastic must assure that the recycled material, like virgin material, is of suitable purity for food contact use, and will meet all existing specifications for the virgin material. Plastic packaging can be recycled in several general ways and each one introduces distinct concerns about the contaminant residues that may be present in the material. Packaging material may be reused directly; undergo physical processing, such as grinding, melting, and reformation; or subjected to chemical treatment where its components are isolated and reprocessed for use in manufacture.

The Environmental Protection Agency (EPA) has adopted a widespread nomenclature that refers to physical reprocessing as secondary recycling (2 degree), chemical processing as tertiary recycling (3 degree), and primary recycling (1 degree). The EPA considers recycling to be the processing of waste to make new articles. Bottles intended for reuse are not considered recycling by the EPA, but instead considered a form of source reduction. The reuse of plastic bottles presents special concerns. Plastic bottles are more likely than glass to absorb contaminants that could be released back in the food when the bottles refill. Washing and sanitizing or sterilization in the bottles must be shown to be effective for removing contaminants to an acceptable level. Bottles must retain structural integrity and be functional after each cycle of washing and reuse. Secondary recycling does not alter the basic polymer during the process. Prior to melting and reforming the polymer, the ground, flaked, or pelletized resin is washed to remove contaminants. Recyclers must be able to demonstrate that the contaminant levels in preformed plastic have been reduced to sufficiently low levels to assure that the resulting packaging will not adulterate the food. In this process, some unique problems arise because the recycler has little or no control over the waste stream entering the recycling facility. This concern could be reduced by using sorting procedures that only allow certain types of waste to be utilized, such as soft drink bottles. Tertiary recycling consisting of chemical reprocessing may involve depolymerization of the used packing material with regeneration and purification of the monomers, which are then repolymerized. The use of a two-degree or three-degree recycled material as a nonfood contact layer of a multilayer food package is permitted providing the recycled resin was separated from the food by an effective barrier made from a regulated virgin resin or other appropriate material such as aluminum.

The Center for Food Safety and Applied Nutrition suggests that dietary exposure to contaminants from recycled food contact articles be 1 ppb or less in the daily diet. To achieve this dietary concentration of a contaminant below 1 ppb, assuming a 20 ml thick container, the maxim residue allowed for each type of polymer is as follows: PETE, with a density of 1.4 g/cc, has a maximum residue of 430 ppb; polystyrene with a density of 1.5 g/cc, has a maximum residue of 360 ppb; polyvinyl chloride (PVC) with a density of 1.58 g/cc, has a maximum residue of 180 ppb; polyolefins with a density of 0.965 g/cc, has a maximum residue of 96 ppb. Chemical testing should be done to ensure the safety of the packaging material.

Under no condition should containers that have been used for pesticides, or other chemicals, be used as recycled material for the production of any type of food packaging. The potential risk is enormous and the amount of money saved from this type of recycling is insignificant.

Environmental Problems in Milk Processing

Environmental problems on dairy farms vary with the size of the operation and the age of the structures and equipment. The older, smaller operations may have problems with poor ventilation, poor lighting, and improper construction of the physical facility. Common to all farms are the increased problems of waste handling, flies, excessive use of pesticides, improper storage of pesticides, inadequate and improper cleaning of pipelines, milking equipment, and bulk storage tanks. A safe water supply is essential. The basic problems in milk plants include cleaning of equipment and the constant use of properly operated, accurate instruments.

Environmental Problems in Poultry Processing

Basic problems in poultry plants include cleanliness of walls, floors, drains, equipment, conveyor belts, and interior and exterior surroundings. Blood is frequently splattered and mixed with dirt and feathers. This becomes very difficult to remove without constant, well-organized, efficient cleaning procedures. The amount of solid waste produced causes a serious removal problem. Sewage systems are frequently overloaded and the biological oxygen demand (BOD) is enormous. Rats frequently invade chicken-slaughtering plants because of the availability of food, blood, and other waste materials. Flies and roaches may become a serious problem. Water used in the chilling process is readily contaminated by birds, thereby continuing the cycle of contamination and bacterial growth. Improper supervision of cleaning, inadequate cleaning schedules, and poor understanding of cleaning techniques increase problems. Quaternary ammonium chloride compounds cannot be used as sanitizers in poultry drinking water.

Environmental Problems in Egg Processing

Environmental problems include all those previously related to poultry. Further, because of the nature of the egg, bacteria may be introduced and grow rapidly due to improper handling, unhealthy workers, unclean equipment, utensils, facilities, contaminated uniforms, insects, rodents, and general debris. The water supply may be unsafe, and there may be an inadequate quantity of hot and cold running water under pressure for use in all preparation areas. The solid waste problem is enormous because of the large quantity of shells and other materials that must be removed. A serious problem occurs when individuals, equipment, or air move from the raw eggs' storage and breaking areas to the pasteurizing and packaging egg-processing areas. Improper removal of detergents and use of chemicals readily contaminate eggs. The use and storage of insecticides must be carefully controlled, because the eggs may readily absorb odors, as well as a variety of chemical contaminants.

Organisms penetrate the eggshell and grow quickly within the egg itself. Because eggs are used raw in eggnog and partially cooked in scrambled eggs, the chance of spreading foodborne disease is high. In addition, eggs are frequently used in specialized therapeutic diets for high-risk populations, such as the very young, sick, chronically ill, and aged. The feed ingredients are of concern because they may be contaminated. Clean, fresh eggs properly handled after production are not a public health problem. However, where the eggs are dirty, cracked, or both, a potential condition arises in which pathogenic organisms could be introduced.

Environmental Problems in Meat Processing

The environmental problems relating to meat encompass all areas of stockyards, slaughterhouses, meat-processing plants, retail meat-cutting operations, meat storage, and sales. Concern exists about the removal of large quantities of solid waste, intense fly problems, and presence of rodents while the cattle are in stockyards. Disease may be spread from animal to animal and from human to animal.

At the slaughtering plant, organisms are added during many parts of the killing process. All equipment, appurtenances, and parts of the physical structure, such as walls, ceilings, and floors, contribute contaminants because of improper cleaning techniques. Adequate quantities of potable hot and cold running water may not be available for cleaning of equipment, physical structure, and hand washing. Cutting instruments and food contact surfaces may be improperly cleaned and sanitized. Meats may contain such pathogenic organisms as brucella, salmonella, streptococcus, mycobacterium, pseudomonas, and any other organisms that may inadvertently be introduced. Pork is hazardous because it may contain *Trichinella spiralis*. Refrigeration must be constantly monitored, because it is so essential to the control of bacterial growth. Employee health, personal cleanliness, and uniforms are also of concern.

At the retail market, all utensils, saws, and packaging equipment are readily contaminated and frequently improperly cleaned. Adequate quantities of hot and cold running water may not be available for equipment and hand washing. Employee health and personal hygiene practices, which are so essential because the meat is frequently handled, may be poor. The walls, ceilings, cutting tables, and other pieces of equipment are frequently improperly cleaned. Meat may be overstocked in the display cases and not rotated properly. Once again, refrigerator temperatures are of considerable concern. The number of microorganisms on meat varies from as few as 100 per gram of beef to several million per gram of beef, depending on slaughtering and processing techniques and the ultimate condition of the meat. Obviously, any chopped or ground meat may have bacterial counts in the millions.

Environmental Problems in Fish and Shellfish Processing

Fish and shellfish are potential vehicles for many organisms that cause disease in people. Fish acquire organisms from polluted water, harvesting equipment, processing equipment, or infected human beings. Shellfish are filter feeders and concentrate microorganisms, natural toxins, and chemical contaminants from water. Many shellfish, such as oysters, clams, and mussels, are consumed raw, posing a serious public-health hazard. Infectious hepatitis, salmonellosis, typhoid fever, and cholera have been spread by shellfish. In recent years, outbreaks of botulism type E were caused by fish. The heat processing of fish may not create a high enough temperature to kill organisms. Coagulase-positive staphylococci are frequently isolated from crab and shrimp meat. Fish meal is of particular concern, because it is produced in the United States or abroad, and it may be contaminated with innumerable organisms plus the feces from rodents and birds. Shigella organisms from imported shrimp have been involved in outbreaks of dysentery. Lately concern has increased about fish containing mercury and other chemicals found in polluted waters.

The major environmental problems in the harvesting of shellfish are due to the water in which the shellfish are cleaned, harvesting boats, lack of proper prewashing of the shellfish to remove mud, bacteriological quality of the washing water, and disposal of human excretion.

The major environmental problems at the shucking and packing plant are cross-contamination from the shucking room, which is a dirty room, to the packing room, which must be kept immaculately clean; the presence of flies, which spread enteric organisms; improper plumbing, with the resultant possible spread of disease-causing organisms; inadequate hand washing facilities; improper personal hygiene of workers, including inadequate hand cleaning; inadequate cleaning and bactericidal treatment; inadequate refrigeration of shell stock and shucked shellfish; and contamination of single-service containers. Because employees can easily cut their fingers or hands with sharp knives or shells, it is essential that anyone with an infection be removed from the shellfish-processing areas.

Seafood implicated in histamine poisoning are the families Scrombridae and Scromberesocidae, which includes tuna and mackerel. Nonscrombroid fish causing poisoning include mahi-mahi, bluefish, pink salmon, herring, anchovies, and sardines.

More than half of the U.S. seafood consumption comes from abroad and originates in some 172 different countries. It is difficult to determine where imported fish is harvested. The United States imports salmon from Switzerland and Panama, although both countries are not known for their salmon resources. The safety of seafood is critical. Until now, seafoodborne illnesses, in which the cause was known and was reported to the Centers for Disease Control and Prevention (CDC), were primarily due to ciguatoxin from a few reef fish species; scombrotoxin from tuna, mackerel, bluefish, and a few other species; and consumption of mollusks, mostly raw. The hazards from imported seafood will continue to grow because of lack of knowledge and appropriate technology for aquaculture, technology abuse, technology neglect, and lack of good public health practice. Hazards associated with the consumption of seafood are concerned with three areas, product safety, food hygiene involving dirty processing plants or unwholesome products, and mislabeling or economic fraud.

A common practice in many developing countries is the creation of numerous small fish pond impoundments. There may be adverse health effects because the aggregate shoreline of ponds is large, thereby causing higher densities of mosquito larvae and cercaria, which can increase the incidence and prevalence of diseases such as lymphatic filariasis and schistosomiasis. A lack of knowledge of the microbial profile of aquaculture products can affect human health, as shown in the transmission of streptococcal infections from tilapia to humans that has resulted in several meningitis cases in Canadian fish processors. A change in marketing strategies to sell live fish in small containers resulted in human vibrio infections from live tilapia in Israel in 1996. Increased hazards occurred with the use of untreated animal or human waste in aquaculture ponds to increase production. Food growers, for centuries, have grown various species of fish and vegetable crops in wastewater-fed ponds and wastewater sediment material. Animal pathogens can enter the aquaculture ponds from exotic animals who live in the vicinity of the ponds. The use of therapeutics and other chemicals in the ponds can alter the normal flora and lead to potential outbreaks of disease.

Produce

Over the last 10 years, outbreaks have increased, involving human disease associated with the consumption of raw vegetables and fruits or their unpasteurized juices. Changes have occurred in agronomic, harvesting, distribution, processing, and consumption patterns and practices that have probably contributed to this increase. *Listeria monocytogenes*, *C. botulinum*, and *Bacillus cereus* are naturally present in some soil, and therefore may be found on fresh produce. *Salmonella*, *E. coli* O157:H7, *Campylobacter jejuni*, *Vibrio cholerae*, parasites, and viruses may be found on fresh produce as a result of improperly composted manure, irrigation water containing untreated sewage, or contaminated wash water. Mammals, reptiles, birds, insects, and unpasteurized products of animal origin can cause contamination of produce. Human hands come in constant contact with whole or cut produce throughout the total system of growing, harvesting, packing, processing, shipping, and preparing produce for consumption. Contaminated surfaces do the same. Sick workers, healthy carriers of disease, lower sanitation standards, nationwide or worldwide distribution of produce from a single shipper, and year-round consumption of fresh produce, have facilitated widespread dissemination of potential pathogens and therefore potentially a sharp increase in foodborne disease.

Preharvest sources of contamination include feces, soil, irrigation water, water used to apply fungicides and insecticides, inadequately composted manure, dust in the air, wild and domestic animals including fowl and reptiles, insects and rodents, and human handling. Postharvest sources of contamination include feces; human handling by workers and customers; harvesting equipment; transport containers used for moving the produce from the field to the packing shed; wild and domestic animals including fowl and reptiles; insects and rodents; dust in the air; wash and rinse water; sorting, packing, cutting, and other processing equipment; ice; transport vehicles; improper storage including temperature and the physical environment; improper packaging and packaging materials; cross-contamination from other foods; improper display temperature; and improper handling after wholesale or retail purchase.

In Germany, 154 strains of *L. monocytogenes* were isolated from soil and plants, 16, from feces of deer and stag; 9, from moldy fodder and wildlife feeding grounds; and 8, from birds. Corn, wheat, oats, barley, and potato plants were analyzed with 10% of the corn plants and 13% of the grain plants infected with *L. monocytogenes*, which is a saprophyte that lives in a plant–soil environment and therefore could easily be contracted by humans and animals through many sources.

Irrigation and surface runoff waters containing raw sewage or improperly treated effluents from sewage treatment plants may include hepatitis A virus (HAV), Norwalk viruses, enteroviruses, and rotaviruses. Listeria, salmonella, and other bacteria have been found in the irrigation water. The application of sewage sludge or irrigation water to the soil is one way in which parasites can contaminate fruit and vegetables. *Ascaris ova* on tomatoes and lettuce have remained viable for up to 1 month, whereas *Endamoeba histolytica* have been viable for a few days. The *Escherichia coli* O157:H7 organisms can remain viable in bovine feces for up to 70 days. Cryptosporidium infection linked to consumption of unpasteurized apple juice was thought to have been caused by contamination of apples by calf feces.

Wild birds are known to disseminate *Campylobacter*, *Salmonella*, *V. cholerae*, *Listeria* species, and *E. coli* O157:H7. The pathogenic bacteria are apparently picked up as a result of the birds feeding on garbage, sewage, fish, or lands that are grazed with cattle or have had applications of fresh manure.

Impact of Food Processing on Other Environmental Problems

Food processing has an enormous impact on other environmental problem areas. Enormous quantities of solid waste are produced that must be processed in some manner, discharged to a receiving stream, burned, or removed to a landfill. The materials cause water pollution, air pollution, and soil pollution. For an in-depth discussion of these concerns, see Volume II.

Food Quality

Establishing standards or guidelines for food is a complex problem. Although microbiological standards have been established for water, milk, and dairy products, attempting to establish microbiological standards for solid foods is quite difficult. Virtually all foods, unless prepared in a sterile manner, contain bacteria. The poorer the practices, the higher the level of bacteria. However, numbers of bacteria, although indicative of possible poor handling practices, may not necessarily indicate exposure to microorganisms that may cause disease. Frequently the number of microorganisms exceed 10 million per 1 g of food. Numbers of microorganisms may be increased sharply if there are roaches, flies, ants, rats, or mice present in the establishment.

Imported Foods

Imported foods may be of considerable concern to the population. The foods may come from processing plants that are run in a very unsanitary manner. Water used in processing may be contaminated. The raw food stuff may be contaminated with microorganisms, pesticides, or other chemicals. Leaking jars, dented cans, moldy food, and insect parts have been found in areas where imported foods have been stored.

Economics of Food Processing

Food production is controlled in many cases by the amount of fertilizers available and the energy that can be utilized. High-cost energy tends to decrease the amount of food produced in various parts of the world.

Inflationary forces in the United States and in other industrial countries have caused the cost of food to rise sharply. The farmers in certain areas are facing difficulties in getting adequate prices for the food they produce. The short-range effect may be to drive some of the smaller farmers out of business. The long-range effect will probably be further increases in the cost of food. The larger farmers tend, especially in the cattle-raising business, to concentrate the yields of food and therefore concentrate the by-products, which become environmental contaminants.

POTENTIAL FOR INTERVENTION

Potential for intervention in the causation of disease or injury due to food processing varies with the type of processed food. With raw food products, a reduction in the use of pesticides under controlled conditions and the proper application of fertilizers and pesticides should help reduce the potential for disease due to these contaminants. A better understanding of the use of additives and preservatives, and the establishment of limitations based on this understanding, may be of further assistance in reducing disease potential. However, considerable research is needed in the entire area of food processing of raw foodstuffs to better understand the techniques for intervention in the disease process.

In the specific preparation of milk, meat, eggs, shellfish, and other highly perishable foods, the standards, practices, and techniques exist. Also, the potential for intervention and disease prevention is excellent.

RESOURCES

Scientific and technical resources include the food industry, the National Environmental Health Association, the International Association of Milk, Food, and Environmental Sanitarians; the American Public Health Association; the National Sanitation Foundation; the American Dietetic Association; the University of Iowa; the University of Michigan School of Public Health; the University of Minnesota School of Public Health; the University of California School of Public Health; and many others. Civic associations interested in food include the North American Vegetarian Society, the Nutrition Today Society, and the Society for Nutrition Education.

Governmental organizations include all state health departments and departments of agriculture, and many local health departments; EPA; Department of Energy; and various sections of the U.S. Department of Health and Human Services (USDHHS), including the CDC, National Cancer Institute, National Institute of Environmental Health Sciences, National Institute of Occupational Safety and Health, National Center for Toxicological Research, and FDA. In addition, the Department of Defense and the Consumer Product Safety Commission are concerned with food products. Departments of agriculture of the various states and the federal government are important resources in a variety of areas. The U.S. Senate Agricultural Committee is an important source of legislation and information. The U.S. Fish and Wildlife Service Department of the Interior is also an important resource.

The FDA, Consumer Affairs and Small Business Staff, Department of Health and Human Services, 5600 Fishers Lane, Room 13-55, Rockville, MD, is helpful to individuals trying to get additional information about food and food technology. The Food and Nutrition Service, Department of Agriculture, Room 512, 3101 Park Office Center Drive, Alexandria, VA, is useful for those concerned with nutrition aspects related to food. The Food Safety and Inspection Service, Room 1163, South Building, Department of Agriculture, Washington, D.C., is useful for all areas of food and food technology. The National Agricultural Chemicals Association, 1155 15th Street, N.W., Washington, D.C., has experts in the areas of agricultural chemicals, including pesticides and their effects on food. The USDA hot line for meat, poultry, and egg products food safety is 1-800-535-4555. The hot line for seafood safety is 1-800-332-4010. See Chapter 3 on food protection.

STANDARDS, PRACTICES, AND TECHNIQUES

Food Preservation Techniques

The shelf life of food can be extended by inactivating or destroying enzymes or by inhibiting or destroying microorganisms. Food preservation techniques include the use of drying, low and high temperatures, irradiation, salt and sugar, and chemical preservatives. Various chemicals are also added during processing to enhance flavor and to promote better color and odor.

In all food preservation, it is essential to start with as clean a raw material as possible and to prevent contamination from hands, soil, water, air, surfaces, processing, and storage equipment.

Drying

Drying, also called dehydration or desiccation, is one of the oldest methods of food preservation still in use today. Dates, figs, and raisins are air dried. Forced hot air is used in drying potatoes, fruits, and vegetables. Milk and eggs are sprayed into heated cylinders. Enzymes are inactivated prior to drying by blanching, which is the removal of color by boiling, or by use of sulfur dioxide or sulfites. Blanching removes sticky substances, checks browning, and makes vegetables easier to work with. Reconstitution of the dry food is aided by clustering the particles during drying, by adding a foaming agent such as nitrogen during concentration of the food prior to drying, or by causing the food to explode into particles. Drying will kill some organisms, but spores of bacteria, yeasts, and molds will survive and can later cause outbreaks of foodborne disease, or food spoilage.

Freeze Drying

Food is first frozen and water is then evaporated from the ice crystals without melting them. A small amount of heat is applied and pressure is reduced in this process, called sublimation.

Refrigeration and Freezing

Low temperatures retard food spoilage. The shelf life depends on the kind of food. Freezing is the removal of heat from a product. The formation of ice crystals in the food is due to free water or air being trapped in the package. During the freezing process, moisture from the warm food forms a vapor on the packaging material and freezes more rapidly than the food mass. Ice crystals break down cell structure and cause a physical deterioration in food quality. This can be avoided by rapid freezing of high-quality raw materials. Freezing does not improve the quality of food, and does not destroy many of the pathogenic organisms that may be present.

Heating

The rate of kill of microorganisms at high temperatures varies with the species of the organisms, presence of spores, quantity of organisms, consistency, pH and type of food medium, temperature, time, moisture, quantity of food, and ease of heat conduction and convection. To avoid overcooking a given food, which could affect food quality, the temperature and time of cooking should be adequate to kill pathogenic organisms, but not necessarily thermophilic or thermoduric organisms.

Pasteurization is a heat-treatment process well below the boiling point, which destroys all harmful microorganisms and improves the keeping quality of food. This process is used in the dairy industry and also in the preservation of dried fruits, syrups, honey, and juices.

Boiling is a common method of preparing and preserving food. Boiling does not kill all organisms, because at atmospheric pressure the temperature may not reach 212°F. Therefore, adequate time must be provided for a total bacterial kill. Although heating of food to boiling destroys botulinum toxin, the spores may survive boiling temperatures for a considerable period of time. Baking in an oven at a temperature of 350 to 400°F may not cause the food temperature to rise to 212°F because of moisture present in the food and because baked goods have low heat conductivity. Frying kills organisms on the surface of the food, but may not affect the organisms within the food mass.

Canning

Canning is the preservation of food by subjecting selected prepared foods to high temperatures in a permanently sealed container (Table 4.1). Containers may be made of glass and sealed with tin, aluminum, paperboard, or plastic caps, which are either automatically vacuum sealed or screwed on. Metal cans made from tin-coated steel or aluminum are also used. Lacquered cans are used for highly colored foods to

Table 4.1   Canning Flow Diagram

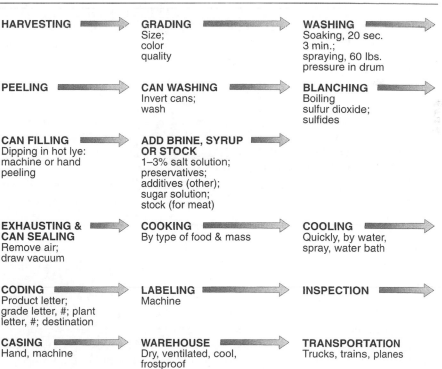

HARVESTING ➡ **GRADING** ➡ **WASHING** ➡
Size;
color
quality
Soaking, 20 sec.
3 min.;
spraying, 60 lbs.
pressure in drum

PEELING ➡ **CAN WASHING** ➡ **BLANCHING** ➡
Invert cans;
wash
Boiling
sulfur dioxide;
sulfides

CAN FILLING ➡ **ADD BRINE, SYRUP OR STOCK** ➡
Dipping in hot lye:
machine or hand
peeling
1–3% salt solution;
preservatives;
additives (other);
sugar solution;
stock (for meat)

EXHAUSTING & CAN SEALING ➡ **COOKING** ➡ **COOLING** ➡
Remove air;
draw vacuum
By type of food & mass
Quickly, by water,
spray, water bath

CODING ➡ **LABELING** ➡ **INSPECTION** ➡
Product letter;
grade letter, #; plant
letter, #; destination
Machine

CASING ➡ **WAREHOUSE** ➡ **TRANSPORTATION**
Hand, machine
Dry, ventilated, cool,
frostproof
Trucks, trains, planes

prevent color loss. Can sizes are as follows: no. 1, 10.94 oz; no. 2, 20.55 oz; no. 2½, 29.79 oz; no. 3, 35.08 oz; and no. 10, 109.43 oz.

The entire canning process is important. However, certain key items should be noted:

1. Use high-quality food from which gross dirt and microorganisms have been removed.
2. Heat-process foods as soon as the air has been exhausted.
3. Cool promptly to avoid overcooking and undesirable changes in texture and flavor and the germination and multiplication of spores of thermophiles.
4. Because food is commercially sterile (all bacteria, but not spores are killed), the resistant spores of *Clostridium botulinum* may not be destroyed in nonacid foods.

Testing

Canned products should be tested by incubating samples at 100°F to test for swells, incubating samples at 130°F for 2 weeks to test for flat sours, inserting thermocouples in center of food to check temperature, placing spore-forming test organisms in marked cans, and taste-sampling for quality.

Problems in Canning and Bottling

The following problems may occur in canning: microbial contamination, pinholes developing from chemical action of food on container, cans collapsing, glass containers breaking because of impact, thermal shock, internal pressure, and toxins developing.

Irradiation

Certain selected food, meat, poultry, fish products, vegetables, and spices have been preserved by irradiation. Organisms in food are destroyed by direct hits, causing ionization, fragmentation of DNA, and death. Some lethal effects are caused chemically by free radicals produced in the solvents. In irradiated water, hydroxyl ions, which are strong oxidizing agents, are formed. The hydrogen ions present are strong reducing agents. The oxidizing and reducing agents may produce off-flavors, and changes in taste, appearance, odor, and texture. On the beneficial side, radiation sterilizes very quickly; kills insects, parasites, helminths, pathogenic bacteria, molds, and yeasts effectively; keeps the temperature of the food from rising significantly; delays ripening; prevents sprouting, and extends the shelf-life of food by retarding spoilage.

Irradiated food is not radioactive, and does not contain hazardous radiolytic products. The radiolytic products include glucose, formic acid, acetaldehyde, and carbon dioxide that are naturally present in foods or formed by heat processing. The change in nutritional value of the food caused by irradiation depends on radiation dose, type of food, packaging, processing, and storage time. Other food processing technologies create similar problems. Vitamins A, E, C, K, and B_1 in foods are sensitive to radiation whereas riboflavin, niacin, and vitamin D are more stable. No abnormal chromosomes have been shown to occur from eating irradiated food. Irradiation at levels up to 10 kilogray (kGy) does not increase risk from botulism any more than other substerilizing techniques. Although irradiation suppresses microorganisms that cause spoilage, it cannot freshen up spoiled food. If toxins or viruses are already formed in foods, irradiation does not destroy them. No scientific evidence indicates any health hazard associated with irradiation of food containing pesticide residues and additives. Almost all commonly used food packing materials, including plastics, are unaffected by irradiated food.

The Joint Expert Committee on Food Irradiation concluded that irradiation to doses not exceeding 10 kGy is wholesome and safe for human consumption. The World Health Organization (WHO), in 1992, after a comprehensive review of data concluded that irradiation carried out under good manufacturing practices presented no toxicological hazard and caused no special or microbiological problems. All poultry and meat prepared from irradiation had to be clean, processed, and packaged in accordance with the Pathogen Reduction and Hazard Analysis and Critical Point (HACCP) Systems final rule of July 25, 1996.

As of May 1999, the FDA has issued approvals for irradiation of foods for the following categories: 1 kGy to control insects and other arthropods, and to inhibit maturation, ripening or sprouting, of fresh foods; 3 kGy to control foodborne pathogens in poultry; 4.5 kGy to control foodborne pathogens in refrigerated meat and 7.0 kGy in frozen meat; 10 kGy to control microorganisms in dry or dehydrated enzymes; and 30 kGy to control microorganisms in dry or dehydrated aromatic substances such as spices and seasonings. The FDA requires that the label includes the phrase "treated with radiation" or "treated by irradiation" plus an appropriate symbol.

Irradiation can cause chemical change in packaging, as well as in food, and this can affect the migration of the package components to the food. Sometimes, irradiation has been used in the manufacture or sterilization of packaging before the food is added. Irradiated packaging must comply with the appropriate FDA regulations and must not otherwise adulterate the food by releasing decomposition products that may render the food injurious. Irradiation of the food in a package is a special case, because a volatile decomposition product may enter the food from the package; therefore, the FDA requires that special testing be done to ensure the safety of the ultimate product. Irradiation pasteurization of solid foods with low doses of gamma rays, x-rays, and electrons effectively controls vegetative bacterial and parasitic foodborne pathogens. Because the residual risk for infection that remains after the proper harvest, production, processing, distribution, and preparation may cause unacceptable levels of illness or death, irradiation should be considered as a supplemental tool. Further, the admonition to cook food properly does not work if people are more interested in having undercooked and raw foods. Recent outbreaks of foodborne illness associated with undercooked meat and uncooked fresh produce, and the emergence of previously unrecognized foodborne hazards have stimulated interest in methods of pasteurizing solid foods without altering their raw appearance and characteristics.

Irradiation pasteurization is a well-established process with clearly documented safety and efficacy that can be put into widespread use as quickly as the facilities can be built to carry out this means of protection. Good sanitation practice guidelines and the use of HACCP principles can result in raw meat, poultry, and seafood having low levels of pathogens such as *Campylobacter, Cryptosporidium, E. coli, Listeria, Salmonella,* and *Toxoplasma. Vibrio* infections associated with the consumption of raw shellfish can be prevented with irradiation pasteurization, but the Norwalk-like viruses appear to be more radio resistant than vegetative bacterial pathogens. Higher levels of irradiation are also needed to inactivate HAV. To reduce the risk for these foodborne diseases it will be necessary to reduce the level of exposure of food to human feces. Irradiation pasteurization works well for meat, poultry, seafood, and soft fruit, but causes leafy vegetables and sprouts to wilt.

Because there was a potential concern for *Clostridium botulinum* growth and toxin production, the FDA has specified that the poultry irradiation dose be limited to 3 kGy and that air-permeable packaging be used for the product. It has not been shown that these restrictions are necessary or effective and discussions are going on considering making the regulations consistent with those for meat and meat food products. *Clostridium botulinum* is very rare in meat and poultry. Several studies have shown that the growth of this organism and other pathogens are inhibited by non-pathogenic lactic acid-producing bacteria, such as lactobacillus, that predominate in and on irradiated raw chilled meat, and grow well in anaerobic environments. The lactic acid producers do this through the production of both acid and bacteriocin, as well as by competing successfully with pathogens for nutrients.

Ionizing radiation can significantly reduce the levels of many pathogenic organisms found in meat food products. The radiation dose necessary to reduce the initial population of the pathogens by 90% is called the D-value, which ranges from 0.1 to 1 kGy for the following pathogens: *Campylobacter jejuni,* 0.18 kGy; *Clostridium perfringens,* 0.586 kGy; *E. coli* O157:H7, 0.25 kGy; *L. monocytogenes,* 0.4 to 0.64 kGy; *Salmonella* spp., 0.48 to 0.7 kGy; *Staphylococcus aureus,* 0.45 kGy; *Toxoplasma gondii,* 0.4 to 0.7 kGy; and *T. spiralis,* 0.3 to 0.6 kGy. Irradiation can also significantly extend the shelf life of meat food products through the reduction of spoilage-causing bacteria. As an example, beef can go from 3 to 9 days after irradiation, and lamb can go from 7 to 28 to 35 days after irradiation.

Milk

Milk is the fluid secretion of the mammary glands, practically free of colostrum, containing a minimum of 8.25% nonfat milk solids and 3.25% milk fat. Milk contains the following average constituents: water, 87.3%; fat, 3.7%; protein, consisting of casein, 2.9%; albumin, 0.5%; lactose, 4.9%; and ash, 0.7%, which includes salts of calcium, copper, iron, magnesium, manganese, potassium, and zinc. Milk contains lactose in true solution, casein in permanent suspension, and butter fat in temporary colloidal suspension. Milk is an excellent source of vitamins A and B_2, and a fair source of vitamins B_1 and E. It also contains vitamins D, K, niacin, pantothenic acid, B_6, B_{12}, folic acid, citrin, *p*-aminobenzoic acid, biotin, choline, and inositrol. It also contains enzymes, epithelial cells, leukocytes, yeasts and molds, extraneous foreign material, and bacteria (from within the udder from milking, people, processing equipment, and poor techniques). Microbial pathogens may include *E. coli* O157:H7, *Salmonella* spp., *L. monocytogenes*, and others.

Milk Production

Clean, safe, nutritious milk comes from healthy cows that live and are milked in an environment where clean, sanitized equipment, rapid cooling, and proper handling and storage techniques are utilized (Table 4.2 contains a diagram of the milking flow process).

Vat Pasteurization

The vat pasteurizer consists of a stainless steel jacketed vat with cover, agitator, inlet and outlet valves, airspace heater, and thermometers (indicating, recording, and airspace). It is essential that the milk is protected during the entire process, that pasteurization temperatures of 145°F be maintained for a minimum of 30 min, and that the milk be agitated.

The vat may be operated in the following manner:

1. Heat milk to pasteurization temperatures and maintain these temperatures by spraying hot water on the inside of the double-jacketed vat or by surrounding the inner jacket with heated coils.
2. Partially preheat milk by means of a heater and then bring milk up to pasteurization temperatures as in 1.
3. Preheat milk to pasteurization temperature and then maintain as in 1.
4. Remove milk from vat partially cooled either by turning off heating system or at pasteurization temperature.

The cover must have overlapping edges, have all its openings protected by raised edges, and have all condensate diverted from it.

The inlet and outlet of the vat must have leak-protector valves with the following design:

1. Seat in every closed position
2. Grooves at least 3/16 in. wide and 3/32 in. deep at center
3. Stops not reversible
4. Close coupled valve seats flush with inner wall of pasteurizer or so close to the wall that milk in the valve inlet is no more than 1°F colder than pasteurized milk

To ensure that all milk is pasteurized, the inlet and outlet lines must be properly sloped to allow drainage of all milk; the inlet valve must be closed during the holding and draining periods; and the outlet valves must be closed during filling, heating, and holding.

Constant agitation of the milk during pasteurization ensures that all the milk is uniformly heated and properly pasteurized. The airspace heater is used for the same purpose for the foam above the milk. The temperature of the air must be 5°F higher to ensure proper pasteurization.

Table 4.2   Milk Production Flow Diagram

MILKING

COW
Wash udder with warm water
and then with chlorine;
strip, examine foremilk;
rinse teat cups; sanitize
teat cups

MILKING MACHINE ➡️
Prewash, sanitize;
rinse in cold water;
sanitize; attach cups
to udder; set timer;
strip udder

STRAINING ➡️
Stainless steel
strainer with pad;
material on pad
indicates level of
sanitation in
milk house

**COOLING AND
STORAGE** ➡️
Mechanical milk cooler;
cool to below 41°F rapidly*
(within 2 hours)

TRANSPORTATION ➡️
Cans cooled below 41°F*;
bulk 41°F* or below in
stainless steel tank;
wash, sanitize
every 24 hours.

MILK PLANT

MILK PROCESSING

TRANSPORTATION ➡️
Raw milk–cans; bulk

**HOLDING TANK–
RAW MILK** ➡️
Storage 41°F or below*

BALANCE TANK ➡️
Constant level of
liquid

BOOSTER PUMP ➡️
Moves milk

REGENERATION ➡️
Hot pasteurized milk on
positive pressure side of
plate heats up cold, raw
milk on negative pressure
side

METERING PUMP
Opposite side
regenerator from
holding tank;
regulates amount
of milk passing

PASTEURIZATION

HOMOGENIZER ➡️
Suspends milk, milk
particles; diffuses
equally

Vat ➡️
145°F for 30 min.;
higher milk fat or
added sweeteners–
150°F for 30 min
**High temperature
short-time**
161° for 15 sec.;
higher milk fat or
added sweeteners–
166°F for 15 sec.
Ultra high temperature
by steam injection at
191°F for 1 sec.;
194°F for 0.5 sec. or
212°F for 0.01 sec.

REGENERATION ➡️
Cold, raw milk on
negative pressure side
cools hot pasteurized
milk

COOLER ➡️
Quickly lowers milk
temperature

PIPES ➡️
To storage containers;
to fillers

FILLERS ➡️
Bottles; cartons; cans;
bulk containers

TRANSPORTATION ➡️
Below 41°F

RETAIL STORES ➡️
41°F

CONSUMER

Note: The 1997 *Grade A Pasteurized Milk Ordinance*, Food and Drug Administration, recommends 45° or lower.

The indicating and airspace thermometers show the temperature of the milk at any given time. The recording thermometer, which is attached by wires to a chart-and-pen arm, records the temperature at which the milk is undergoing processing. The recording thermometer, which is checked daily against the indicating thermometer, must never read higher than the indicating thermometer. Charts should show the preheating time, pasteurization time, precooling time, airspace temperature, indicating thermometer temperature at a given time, record of unusual occurrences, name of milk plant, operator's name, location of recorder, and date.

High-Temperature Short-Time Pasteurization

High-temperature short-time (HTST) pasteurization (Table 4.3) consists of a system including:

1. Drawing of cold raw milk at 41°F or under from a constant-level tank
2. Raw milk heated as it flows on the negative pressure side of thin stainless steel plates in the regenerator, by hot pasteurized milk flowing in the opposite direction on the positive-pressure side of the plates
3. Raw milk still under negative pressure passing through a positive displacement timing pump that pushes it under positive pressure through the rest of the system
4. Heating raw milk to at least 161°F
5. Raw milk under pressure at 161°F flowing through holding tube for a minimum of 15 sec
6. Milk flowing past indicating thermometer and recorder controller at 161°F or higher and then through the forward-flow position of the flow-diversion valve (if the temperatures less than 161°F, the milk flows through the diverted-flow position into the diverted-flow line and back to the raw, constant-level tank)
7. Pasteurized milk flowing through the regenerator and cooling
8. Milk passing through cooler where temperature is dropped to 41°F or under and storing in a vat until packaging

The recorder chart must have a scale with a span greater than 30°F, including 12°F from the diversion temperature. The temperature must be accurate to within 1°F at diversion temperatures. The chart must not rotate more frequently than once every 12 hours. The thermometric response must be 4 sec to travel 12°F within a 19° range. The date, number of location, name and address of plant, reading of indicating thermometer at a given point, amount and name of product, product temperatures at beginning and end, time in forward-flow position, unusual occurrences, and operator's name must be recorded on the chart.

It is always important to maintain proper pressure throughout the system.

Ultrahigh Temperature Pasteurization

Ultrahigh temperature pasteurization is accomplished by steam injection or by use of a vacuum. In the steam process, automatically controlled injected steam is used to heat the milk to the following pasteurization temperatures: 191°F for 1 sec, 194°F for 0.5 sec, 201°F for 0.1 sec, 204°F for 0.05 sec, and 212°F for 0.01 sec. The

Table 4.3   High-Temperature Short-Time Flow

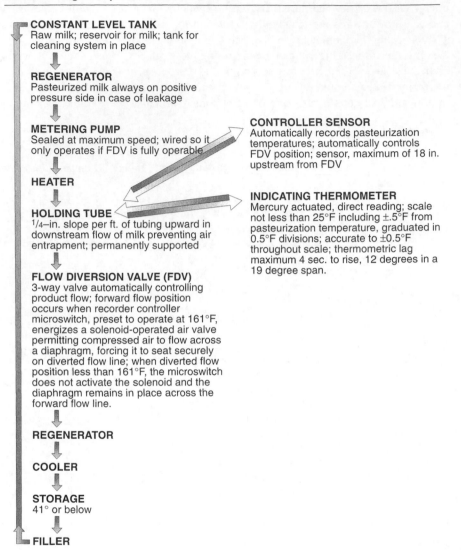

CONSTANT LEVEL TANK
Raw milk; reservoir for milk; tank for
cleaning system in place

REGENERATOR
Pasteurized milk always on positive
pressure side in case of leakage

METERING PUMP
Sealed at maximum speed; wired so it
only operates if FDV is fully operable

HEATER

HOLDING TUBE
1/4–in. slope per ft. of tubing upward in
downstream flow of milk preventing air
entrapment; permanently supported

FLOW DIVERSION VALVE (FDV)
3-way valve automatically controlling
product flow; forward flow position
occurs when recorder controller
microswitch, preset to operate at 161°F,
energizes a solenoid-operated air valve
permitting compressed air to flow across
a diaphragm, forcing it to seat securely
on diverted flow line; when diverted flow
position less than 161°F, the microswitch
does not activate the solenoid and the
diaphragm remains in place across the
forward flow line.

REGENERATOR

COOLER

STORAGE
41° or below

FILLER

CONTROLLER SENSOR
Automatically records pasteurization
temperatures; automatically controls
FDV position; sensor, maximum of 18 in.
upstream from FDV

INDICATING THERMOMETER
Mercury actuated, direct reading; scale
not less than 25°F including ±.5°F from
pasteurization temperature, graduated in
0.5°F divisions; accurate to ±0.5°F
throughout scale; thermometric lag
maximum 4 sec. to rise, 12 degrees in a
19 degree span.

product is then rapidly cooled and passed through the flow-diversion valve before storage. The reasons that the flow-diversion valve, which operates the same as in the HTST process, is located after the cooler are that a safety hazard, due to flashing steam, occurs if milk is diverted at 212°F; the milk would be diluted by steam if diverted at high temperatures; and the response time of the thermal limit controllers would be too slow to prevent the forward flow of raw milk.

In the vacuum process, the milk must be heated to at least 194°F. Constant temperatures must be maintained in the pasteurizing zone by a constant uninterrupted, adequate quantity of steam. The steam pressure must be 35 lb/in.^{2} and the vacuum drawn in the chamber must be 8 in. Ultrahigh temperature pasteurization is used to increase the shelf life of milk, inactivate enzymes, and improve flavor.

Dry Milk

Milk is dried by spraying in a vacuum chamber. Salmonella surviving in the dry milk are a potential hazard.

Cleaning

Proper cleaning is absolutely essential during all phases of milk production and milk processing. One technique involves disassembling all equipment, rinsing thoroughly with cold water, scrubbing all equipment with brushes using hot water and detergent, rinsing thoroughly, reassembling, and sanitizing with 200 ppm of chlorine. Sanitization may also be accomplished by the use of steam for 5 min or hot water at 180°F for 5 min. Acid cleaners should be used when necessary to aid in cleaning. Cleaning in place is a technique in which the circulating system is not disassembled. Instead, a cleaning solution is circulated through the system at a velocity of at least 5 ft/sec. The solution, which is alkaline and nondepositing, is heated to 120°F and circulated for at least 15 min. The solution is then drained and the system is thoroughly rinsed. Finally, chlorine or hot water is used for sanitization.

Ice Cream and Frozen Desserts

Frozen desserts are pure, clean, frozen, or semifrozen foods prepared by freezing while stirring a pasteurized mix composed of edible fats, nonfat milk solids, sugar sweeteners, flavors, and other ingredients such as eggs, fruits, and nuts. The pasteurization process is the same as used in milk, except that in the vats the pasteurization temperature must be 155°F for 30 min instead of 145°F for 30 min; and in the HTST system the pasteurization temperature is 175°F for 25 sec instead of 161°F for 15 sec. Contamination may enter the mix when fruits and nuts are blended into the mix after pasteurization and cooling, and before freezing. Poor techniques, careless handling, or contaminated raw products can raise the bacterial count of the mix.

Counter Freezers or Soft Ice Cream Freezers

Counter freezers or soft ice cream freezers are a continuation of the frozen dessert process. Handling of the mix after delivery to the stores and improper cleaning of the equipment contribute to extremely high bacterial counts. Typically, the mix is delivered in plastic bags, packed in boxes, or packed in cartons similar to large milk containers. The mix is poured into the dispenser and then added during the day as needed. At the end of the day, in many instances, the mix is drained into a bottle or jar and saved for the next day. The dispenser may or may not be rinsed or cleaned.

Proper technique consists of carrying out the following steps each evening: drain mix at end of day to waste; rinse dispenser thoroughly with cold water; wash inside

Table 4.4   Poultry Processing Flowchart

LIVE BIRD ➡	INSPECT AND WEIGH ➡	SHACKLE ➡
Healthy; vigorous		
BLEED ➡	SCALD ➡	DEFEATHER ➡
Thoroughly; quickly	135° to 140°F; helps remove feathers	
WASH ➡	EVISCERATE ➡	INSPECT ➡
Spray		
WASH ➡	CHILL ➡	PACKAGE ➡
Inside; outside	Quickly below 35°F; freeze	Cut up; whole; fresh, chilled, frozen; in prepared meal, either frozen, canned or fresh (chicken roll)

of machine with hot water and detergent for at least 5 min; rinse unit thoroughly; add to dispenser 200 ppm of chlorine in water and allow to stand for 2 min; drain thoroughly and air dry. Before refilling the following morning, the dispenser should be once again sanitized with 200 ppm chlorine for 2 min and allowed to drain.

Poultry

Poultry are domesticated birds, such as chickens, turkeys, ducks, geese, and pigeons, that are raised for food consumption. The widespread use of poultry and poultry products, the substantial amount of handling and processing, the kinds of bacteria indigenous to the birds (salmonella), and the long history of poultry-associated foodborne diseases makes poultry processing an important area of concern in public health.

Poultry are mass-produced in special housing with automated feed and water supplies. The feed ingredients may contain salmonella or other organisms transmitted to chicks and eventually to humans. Salmonella may also be introduced into the environment of the poultry by dust, air, domestic animals, rodents, fomites, human waste material, flies, pigeons, and other birds. (See Table 4.4 for a diagram of the poultry process.)

Eggs

Eggs are produced from domesticated chickens, ducks, geese, turkeys, and other domestic fowl for human consumption. Eggs are mass-produced in special units in which hens are automatically fed and provided water. Eggs have been implicated in outbreaks of disease. Intact-shell eggs have been implicated in outbreaks of *S. enteritidis*. (See Table 4.5 for a diagram of the egg process.) Egg pasteurization is usually performed in a high-temperature short-time pasteurizer and is similar to HTST milk pasteurization.

Americans consume an average of 234 eggs per person per year. Although only an estimated 1 in 20,000 eggs are contaminated with *S. enteritidis* in the United States,

Table 4.5   Egg Processing Flow Diagram

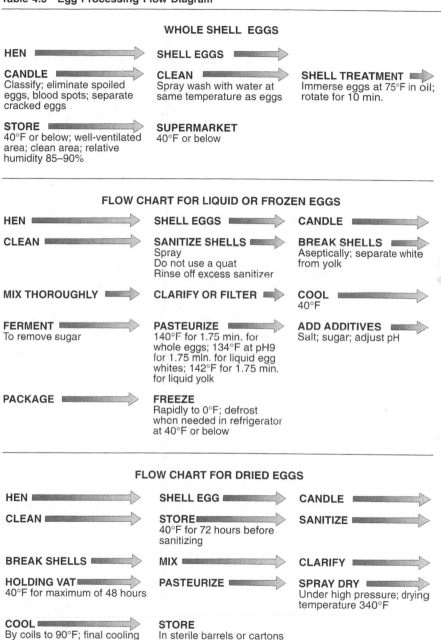

WHOLE SHELL EGGS

HEN ➡ SHELL EGGS ➡

CANDLE ➡
Classify; eliminate spoiled
eggs, blood spots; separate
cracked eggs

CLEAN ➡
Spray wash with water at
same temperature as eggs

SHELL TREATMENT ➡
Immerse eggs at 75°F in oil;
rotate for 10 min.

STORE ➡
40°F or below; well-ventilated
area; clean area; relative
humidity 85–90%

SUPERMARKET
40°F or below

FLOW CHART FOR LIQUID OR FROZEN EGGS

HEN ➡ SHELL EGGS ➡ CANDLE ➡

CLEAN ➡

SANITIZE SHELLS ➡
Spray
Do not use a quat
Rinse off excess sanitizer

BREAK SHELLS ➡
Aseptically; separate white
from yolk

MIX THOROUGHLY ➡ **CLARIFY OR FILTER** ➡

COOL ➡
40°F

FERMENT ➡
To remove sugar

PASTEURIZE ➡
140°F for 1.75 min. for
whole eggs; 134°F at pH9
for 1.75 min. for liquid egg
whites; 142°F for 1.75 min.
for liquid yolk

ADD ADDITIVES ➡
Salt; sugar; adjust pH

PACKAGE ➡

FREEZE
Rapidly to 0°F; defrost
when needed in refrigerator
at 40°F or below

FLOW CHART FOR DRIED EGGS

HEN ➡ SHELL EGG ➡ CANDLE ➡

CLEAN ➡

STORE ➡
40°F for 72 hours before
sanitizing

SANITIZE ➡

BREAK SHELLS ➡ **MIX** ➡ **CLARIFY** ➡

HOLDING VAT ➡
40°F for maximum of 48 hours

PASTEURIZE ➡

SPRAY DRY ➡
Under high pressure; drying
temperature 340°F

COOL ➡
By coils to 90°F; final cooling
to 40°F or below

STORE
In sterile barrels or cartons

this involves nearly 3.36 million eggs annually, thereby exposing a large number of
people to this disease. The CDC estimated in 1997, that there were 300,000 cases of
S. enteritidis in the United States, but in regions where quality assurance efforts were

most extensive, the amount of disease was less. Children, the elderly, and people with weakened immune systems are specially vulnerable to this infection. The costs associated with this disease are estimated to range from $150 million to $870 million annually.

In the unbroken fresh eggs, the *Salmonella* are usually in the yolk, but may also be found in the egg white. *Salmonella* of various serotypes are usually found in the digestive tracts of animals and frequently contaminate the environment. Originally, it was believed that the contamination of the shell eggs occurred primarily when organisms present on the egg passed through the shell into the yolk or white. However, recently it has been determined that there is transovarian salmonella contamination of the egg contents by *S. enteritidis*-infected laying hens.

CDC surveillance data show that the isolation of *S. enteritidis* from infected humans increased from 0.5 to 3.9 per 100,000 population throughout the United States from 1976 to 1994. From 1990 to 1994, a decrease of 8.9 to 7.0 per 100,000 occurred in the Northeast because of better quality assurance. In the Pacific region, there was a threefold increase in the isolates, with California reporting an increase from 11 to 38%.

Traditional egg-handling practices have been poor. Raw eggs were put into foods. Undercooked and nonrefrigerated eggs were considered to be safe. With the increase in *S. enteritidis* caused by the consumption of eggs, practices have been reevaluated. Temperature abuse (holding eggs and egg-containing foods at room temperature instead of under refrigeration), inadequate cooking, and preparing large volumes of egg-containing foods without proper temperature control have led to serious outbreaks of disease. Many of these outbreaks occurring between 1985 and 1998 were in commercial establishments, such as restaurants, hospitals, nursing homes, schools, and prisons. The 1996 to 1997 Food Consumption and Preparation Diary Survey showed that 27% of all egg dishes consumed were undercooked. On the average, each person in the survey consumed undercooked eggs 20 times a year.

Meat

Fresh, raw meats refer to the regular retail cuts of beef, veal, lamb, and pork. Processed meats are meats plus other ingredients, additives, and spices that have received special treatment, such as curing, smoking, or canning. Meats have been implicated in a number of outbreaks of disease including, most recently, ground beef-associated *E. coli* O157:H7.

To simplify the slaughtering process, only the slaughtering of cattle is discussed. (See Table 4.6 for a diagram of the slaughtering process.) In the slaughtering plant, the bones, feet, and condemned parts are sent to rendering plants. Unusable portions of the meat are used for pet food. However, pet food processing must be a totally separate operation from human food processing.

Refrigerated beef, after being cut into prime or retail cuts, may be wrapped and frozen quickly at temperatures of −10 to 0°F. Quick freezing and proper wrapping are essential to preserve the quality of the food. Freezing may cause slight changes in flavor or color.

Table 4.6   Meat Processing Flowchart

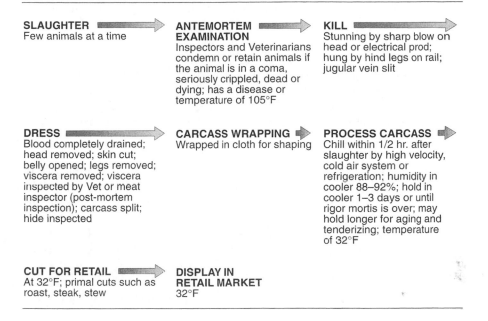

SLAUGHTER ⟶	**ANTEMORTEM** ⟶	**KILL** ⟶
Few animals at a time	**EXAMINATION**	Stunning by sharp blow on
	Inspectors and Veterinarians	head or electrical prod;
	condemn or retain animals if	hung by hind legs on rail;
	the animal is in a coma,	jugular vein slit
	seriously crippled, dead or	
	dying; has a disease or	
	temperature of 105°F	

DRESS ⟶	**CARCASS WRAPPING** ⟶	**PROCESS CARCASS** ⟶
Blood completely drained;	Wrapped in cloth for shaping	Chill within 1/2 hr. after
head removed; skin cut;		slaughter by high velocity,
belly opened; legs removed;		cold air system or
viscera removed; viscera		refrigeration; humidity in
inspected by Vet or meat		cooler 88–92%; hold in
inspector (post-mortem		cooler 1–3 days or until
inspection); carcass split;		rigor mortis is over; may
hide inspected		hold longer for aging and
		tenderizing; temperature
		of 32°F

CUT FOR RETAIL ⟶	**DISPLAY IN**
At 32°F; primal cuts such as	**RETAIL MARKET**
roast, steak, stew	32°F

Specialized Processing of Meats

Curing, the addition to meat of sodium chloride mixed with sugar and spices, was originally developed to preserve meats without the use of refrigeration. Sodium nitrites and nitrate are added to stabilize the color, because most people prefer a pink look to their meat. Sugar is added to counteract the hardening effect of salt and to improve flavor. These ingredients can be dissolved in water to form a brine solution or brine cure. The meat can be soaked in the brine, injected with brine, pumped with brine through the arteries, or rubbed with a salt and brine solution.

The smoking of meats is primarily for the addition of tastes, although some preservation takes place. Hardwood chips and sawdust are used in the smoke houses. Humidity and temperature are carefully regulated. Long-cured meats are smoked at an internal meat temperature of about 110°F. Short-cured meats are smoked at temperatures of 140 to 150°F and finally at temperatures of about 200°F. The length of smoking varies from a few hours to several months.

Cased meat products include sausages, frankfurters, and luncheon meats. Sausage is usually made from chopped meats, including beef, pork, veal, lamb, and mutton. It contains fillers, such as cereal, starch, soy flour, dry milk, and added water. Sausages may be cooked or smoked. Frankfurters, a type of smoked sausage, contain any of the previously mentioned products and may also contain chicken; they are always fully cooked before being sold. Luncheon meats contain a variety of meat products, other ingredients, spices, and water; they are usually precooked before being sold.

Canned meats are usually cooked after packing. The storage of meats in cans depends on the processing. In almost all cases, canned hams must be refrigerated.

Seafood

Fish must come from safe and clean fishing waters. Fish may be preserved by icing, salting, smoking, freezing, or canning. The fish are harvested from the ocean and may be processed on the boat. If so, they are gutted, scaled, cut into fillets, and iced. The fish may also be iced on the harvesting boat and then taken back to a fish-processing plant. Unfortunately, even artificially produced ice is not sterile because of handling, production, and storage, and may contaminate the fish.

In the smoking process, dehydration occurs, which helps in preservation. Smoking, however, is basically used for improving taste. The fish is first salted and then smoked with hardwood fumes in the same way that meat is smoked. Smoking may last a few hours or many days.

Fish preserved by canning are heat treated and commercially sterile. The keeping quality of the canned fish is based on the kind and amount of cooking. Fish should be flash frozen. One of the problems is that during defrosting, a considerable amount of water is lost from the fish and the flavor is impaired. When salting, fish must first be cleaned and cut, immersed in a 20% salt brine for 30 min, and placed in water-tight containers.

Fresh fish, when inspected, have the following characteristics: firm; no off-odor; no discoloration or slime; eyes bright, full, and moist; red gills, not gray or brown; and tight scales.

Shellfish Harvesting and Processing

Shellfish include all edible species of oysters, clams, and mussels. (See Table 4.7 for a diagram of the shellfish production process.)

MODES OF SURVEILLANCE AND EVALUATION

Inspections

Dairy farms are inspected based on the violations noted in the previous 12 months. Inspection can vary from once every 2 months to once a year (see Figure 4.1). Milk plants are inspected once every 3 months (see Figure 4.2).

Milk Testing

Tests are performed on raw milk as it enters the plant to determine odor, appearance, off-color, blood, viscosity, foreign matter, sediment, temperature, and drug residues.

Table 4.7   Shellfish Processing Flowchart

GROWING AREA ▰▰▰▰▰▷
Sanitary survey by industry and state authority under supervision of Food and Drug Administration (see NSSP program); evaluate sources of actual or potential pollution; bacteriological study of water and shellfish; chemical analysis of water and shellfish; graded as approved, conditional approval, restricted or prohibited

HARVEST ▰▰▰▰▰▷
From approved area only; from conditionally approved area only under supervision of health authorities, to ensure that shellfish are purified; from restricted areas only under controlled purification and supervision of health authorities; never from prohibited areas; human body waste must not be discharged from the harvest boat into the water

TRANSPORT ▰▰▰▰▰▷
Boats and truck, shellfish stored in thoroughly cleaned bins and away from bilge water; bins must be sanitized

WASH SHELLFISH ▰▰▰▷
Wash away mud; use potable water under pressure

SHUCK ▰▰▰▰▰▷
Separate from packing operation; not subject to flooding; fly-tight screening

PACK ▰▰▰▰▰▷
Clean room; surfaces washed, sanitized; impervious floors and walls; fly control measures; adequate safe, sanitary water supply; approved plumbing–avoid cross connections and submerged inlets; strict hand-washing procedures; use clean, sanitized single-service containers; label with packer's name, address, number, state, and date

REFRIGERATE ▰▰▰▰▰▷
Refrigerate quickly below 40°F; freeze quickly

CLEAN EQUIPMENT AND SANITIZE
Washing sinks, made of impervious, non-toxic materials; storage, shucking, and packing rooms cleaned within two hours after completing operations; sanitize with 100 ppm available chlorine

---

Laboratory tests conducted on raw milk include:

1. Direct microscopic count or breed count, which measures bacteria count, and identifies the type of bacteria and the source of contamination
2. Methylene blue test, which gives a rough estimate of the bacterial load; decolorization in half an hour, which indicates a high bacterial load; decolorization in 8 hours, which indicates a low bacterial load
3. Resazurin test, which is similar to the methylene blue test
4. Standard plate count
5. Thermoduric bacteria test, which indicates use of unclean equipment on the farm or in transportation
6. Coliform test, which indicates use of unclean equipment in production
7. Drug residue, which indicates usage in treatment of animals or in feed

Department of Health and Human Services
Public Health Service
Food and Drug Administration

DAIRY FARM INSPECTION REPORT

Inspecting Agency

Name and location of Plant:

POUNDS SOLD DAILY

Plant _____

Permit No. _____

Inspection of your plant today showed violations existing in the items checked below. You are further notified that this inspection sheet serves as notification of the intent to suspend your permit if the violations noted are not in compliance at the time of the next inspection. (See sections 3 and 5 of the Grade A Pasteurized Milk Ordinance.)

COWS

1. Abnormal Milk:
Cows secreting abnormal milk milked last or in separate equipment (a) _____
Abnormal milk properly handled and disposed of (b) _____
Proper care of abnormal milk handling equipment (c) _____

MILKING BARN, STABLE, OR PARLOR

2. Construction:
Floors, gutters, and feed troughs of concrete or equally impervious materials; in good repair (a) _____
Walls and ceilings smooth, painted or finished adequately, in good repair, ceiling dust-tight (b) _____
Separate stalls or pens for horses, calves, and bulls no overcrowding (c) _____
Adequate natural and/or artificial light; well distributed (d) _____
Properly ventilated (e) _____

3. Cleanliness:
Clean and free of litter (a) _____
No swine or fowl (b) _____

4. Cow Yard:
Graded to drain; no pooled water or wastes (a) _____
Cow yard clean; cattle housing areas & manure packs properly maintained (b) _____
No swine (c) _____
Manure stored inaccessible to cows (d) _____

5. Cleaning Facilities:
Two compartment wash and rinse vat of adequate size (a) _____
Suitable waterheating facilities (b) _____
Water under pressure piped to milkhouse (c) _____

6. Cleanliness:
Floors, walls, windows, tables, and similar non-product contact surfaces clean (a) _____
No trash, unnecessary articles, animals or fowl (b) _____

TOILET AND WATER SUPPLY

7. Toilet:
Provided; conveniently located (a) _____
Constructed and operated according to *Ordinance* (b) _____
No evidence of human waste about premises (c) _____
Toilet room in compliance with *Ordinance* (d) _____

8. Water Supply:
Constructed and operated according to *Ordinance* (a) _____
Complies with bacteriological standards (b) _____
No connection between safe and unsafe supplies; no improper submerged inlets (c) _____

UTENSILS AND EQUIPMENT

9. Construction:
Smooth, impervious, nonabsorbent, safe materials; easily cleanable, seamless hooded nails (a) _____
In good repair; accessible for inspection (b) _____

TRANSFER AND PROTECTION OF MILK

14. Protection From Contamination:
No overcrowding (a) _____
Produce and CIP circuits separated (b) _____
Improperly handled milk discarded (c) _____
Immediate removal of milk (d) _____
Milk and equipment properly protected (e) _____
Sanitized milk surfaces not exposed to contamination (f) _____
Air under pressure of proper quality (g) _____

15. Drug and Chemical Control:
Cleaners and sanitizers properly identified (a) _____
Drug administration equipment properly handled and stored (b) _____
Drugs properly labeled (name and address) and stored (c) _____
Drugs properly labeled (directing for use, cautionary statements, active ingredients) (d) _____
Drugs properly used and stored to preclude contamination of milk (e) _____

PERSONNEL

16. Hand-Washing Facilities:
Proper hand-washing facilities convenient to milking operations (a) _____
Wash and rinse vats not used as hand-washing facilities (b) _____

MILKHOUSE OR ROOM

5. Construction and Facilities:
Floors:
Smooth; concrete or other impervious material in good repair (a) ___
Graded to drain (b) ___
Drains trapped, if connected to sanitary system (c) ___

Walls and Ceilings
Approved material and finish (a) ___
Good repair (windows, doors, and hoseport included; (b) ___

Lighting and Ventilation
Adequate natural and/or artificial light; properly distributed (a) ___
Adequate ventilation (b) ___
Doors and windows closed during dusty weather (c) ___
Vents and lighting fixtures properly installed (d) ___

Miscellaneous Requirements
Used for milkhouse operations only; sufficient size (a) ___
No direct opening into living quarters or barn, except as permitted by Ordinance (b) ___
Liquid wastes properly disposed of (c) ___
Proper hoseport where required (d) ___
Acceptable surface under hoseport (e) ___
Suitable shelter for transport truck as required by this Ordinance (f) ___

Approved single-service articles, not reused (c) ___
Utensils and equipment of proper design (d) ___
Approved CIP milk pipeline system (e) ___

10. Cleaning:
Utensils and equipment clean (a) ___

11. Sanitization:
All multi-use containers and equipment subjected to approved sanitization process (See Ordinance) (a) ___

12. Storage:
All multi-use containers and equipment properly stored (b) ___
Stored to assure complete drainage, where applicable (a) ___
Single-service articles properly stored (b) ___

MILKING

13. Flanks, Udders, and Teats:
Milking done in barn, stable, or parlor (a) ___
Brushing completed before milking begun (b) ___
Flanks, bellies, udders, and tails of cows clean at time of milking; clipped when required (c) ___
Teats treated with sanitizing solution and dried, just prior to milking (d) ___
No wet hand milking (e) ___

17. Personnel Cleanliness:
Hands washed clean and dried before milking, or performing milk house functions; rewashed when contaminated (c) ___
Clean outer garments worn (d) ___ (e) ___
(a) ___
(b) ___

COOLING
18. Cooling:
Milk cooled to 45° F or less within 2 hours after milking, except as permitted by Ordinance (a) ___
Recirculated cooling water form safe source and properly protected; complies with bacteriological standards (a) ___ (b) ___

PEST CONTROL
19. Insect and Rodent Control:
Fly breeding minimized by approved manure disposal methods (See Ordinance) (a) ___
Manure packs properly maintained (b) ___
All milkhouse openings effectively screened or otherwise protected; doors tight and self-closing; screen doors open outward (c) ___
Milkhouse free of insects and rodents (a) ___
Approved pesticides; used properly (b) ___
Equipment and utensils not exposed to pesticide contamination (c) ___
Surroundings neat and clean; free of harborages and breeding areas (d) ___
Feed storage not attraction for birds, rodents or insects (e) ___
(c) ___
(d) ___
(e) ___
(f) ___
(g) ___
(h) ___

REMARKS:

DATE: SANITARIAN:

Note: Item numbers correspond to required sanitation items for Grade A raw milk for pasteurization in the Grade A Pasteurized Milk Ordinance.

Form FD 2359a (1/96)

Figure 4.1 Dairy farm inspection report. (From Grade A Pasteurized Milk Ordinance, Food and Drug Administration, 1997 Revision, Appendix M.)

Department of Health and Human Services
Public Health Service
Food and Drug Administration

MILK PLANT INSPECTION REPORT
(Includes Receiving Stations, Transfer Stations,
and Bulk Tank Cleaning Facilities)

Inspecting Agency _____

Name and Location of Plant: _____

POUNDS SOLD DAILY
Milk _____
Other Milk _____
Products _____
Total _____
Permit No. _____

Inspection of your plant today showed violations existing in the items checked below. You are further notified that this inspection sheet serves as notification of the intent to suspend your permit if the violations noted are not in compliance at the time of the next inspection. (See sections 3 and 5 of the Grade A Pasteurized Milk Ordinance.)

1. FLOORS:
Smooth; impervious; no pools; good repair; trapped drains (a) ____
2. WALLS AND CEILINGS:
Smooth; washable; light-colored; good repair (a) ____
3. DOORS AND WINDOWS:
All outer openings effectively protected against entry of flies and rodents (a) ____
Outer doors self-closing; screen doors open outward (b) ____
4. LIGHTING AND VENTILATION:
Adequate in all rooms (a) ____
Well ventilated to preclude odors and condensation; filtered air with pressure systems (b) ____
5. SEPARATE ROOMS:
Separate rooms as required; adequate size (a) ____
No direct opening to barn or living quarters (b) ____
Storage tanks properly vented (c) ____
6. TOILET FACILITIES:
Complies with local ordinances (a) ____
No direct opening to processing rooms; self-closing doors (b) ____
Clean; well-lighted and ventilated proper facilities (c) ____
Sewage and other liquid wastes disposed of in sanitary manner (d) ____
7. WATER SUPPLY:
Constructed and operated in accordance with Ordinance (a) ____
No direct or indirect connection between safe and unsafe water (b) ____
Condensing water and vacuum water in compliance with Ordinance requirements (c) ____
Complies with bacteriological standards (d) ____

Multi-use plastic containers in compliance (e) ____
Aseptic system sterilized (f) ____
13. STORAGE OF CLEANED CONTAINERS AND EQUIPMENT:
Stored to assure drainage and protected from contamination (a) ____
14. STORAGE OF SINGLE-SERVICE ARTICLES
Received, stored and handled in a sanitary manner; paperboard containers not reused except as permitted by the Ordinance (a) ____
15a. PROTECTION FROM CONTAMINATION:
Operations conducted and located so as to preclude contamination of milk, milk products, ingredients, containers, equipment, and utensils (a) ____
Air and steam used to process products in compliance with Ordinance (b) ____
Approved pesticides, safely used (c) ____
15b. CROSS CONNECTIONS:
No direct connections between pasteurized and raw milk or milk products (a) ____
Overflow, spilled and leaked products or ingredients discarded (b) ____
No direct connections between milk or milk products and cleaning and/or sanitizing solutions (c) ____
16a. PASTEURIZATION BATCH:
(1) INDICATING AND RECORDING THERMOMETERS:
Comply with Ordinance specifications (a) ____
(2) TIME AND TEMPERATURE CONTROLS
Adequate agitation throughout holding; agitator sufficiently submerged (a) ____
Each pasteurizer equipped with indicating and recording thermometer; bulb submerged (b) ____
Recording thermometer reads no higher than indicating thermometer (c) ____

(2) TIME AND TEMPERATURE CONTROLS:
Flow diversion device complies with Ordinance requirements (a) ____
Recorder controller complies with Ordinance requirements (b) ____
Holding tube complies with Ordinance requirements (c) ____
Flow promotion devices comply with Ordinance requirements (d) ____
(3) ADULTERATION CONTROLS:
Satisfactory means to prevent adulteration with added water (a) ____
16d. REGENERATIVE HEATING:
Pasteurized or aseptic product in regenerator automatically under greater pressure than raw product in regenerator at all times (a) ____
Accurate pressure gauges installed as required, booster pump properly identified and installed (b) ____
Regenerator pressures meet Ordinance requirements (c) ____
16e. TEMPERATURE RECORDING CHARTS:
Batch pasteurizer charts comply with applicable Ordinance requirements (a) ____
HTST pasteurizer charts comply with applicable Ordinance requirements (b) ____
Aseptic charts comply with applicable Ordinance requirements (c) ____
17. COOLING OF MILK:
Raw milk maintained at 45° F or less until processed (a) ____
Pasteurized milk and milk products, except those to be cultured, cooled immediately to 45° F or less in approved equipment; all milk and milk products stored thereat until delivered (b) ____

8. HAND-WASHING FACILITIES:
Located and equipped as required; clean and in
good repair; improper facilities not used (a)
9. MILK PLANT CLEANLINESS:
Neat, clear; no evidence of insects or rodents;
trash properly handled (a)
No unnecessary equipment (b)
10. SANITARY PIPING:
Smooth; impervious, corrosion-resistant, nontoxic,
easily cleanable materials; good repair; accessible
for inspection (a)
Clean-in-place lines meet *Ordinance*
specifications (b)
Pasteurized products conducted in sanitary piping,
except as permitted by Ordinance (c)
11. CONSTRUCTION AND REPAIR OF
CONTAINERS AND EQUIPMENT:
Smooth, impervious, corrosion-resistant, nontoxic,
easily cleanable materials; good repair; accessible
for inspection (a)
Self-draining; strainers of approved design (b)
Approved single-service articles; not reused (c)
12. CLEANING AND SANITIZING
OF CONTAINERS/EQUIPMENT:
Containers, utensils, and equipment effectively
cleaned (a)
Mechanical cleaning requirements of *Ordinance* in
compliance; records completer (b)
Approved sanitization process applied prior to use
of product-contact surfaces (c)
Required efficiency test in compliance (d)

Product held minimum pasteurization temperature
continuously for 30 minutes, plus filling time if product
preheated before entering vat, plus emptying time, if
cooling is begun after opening outlet (d)
No product added after holding begun (e)
Airspace above product maintained at not less than 5.0 F
higher than minimum required pasteurization temperature
during holding (f)
Approved airspace thermometer; bulb not less than
1 inch above product level (g)
Inlet and outlet valves and connections in compliance
with *Ordinance* (h)
16b. PASTEURIZATION-HIGH TEMPERATURE:
(1) INDICATING AND RECORDING THERMOMETERS:
Comply with *Ordinance* specifications (a)
(2) TIME AND TEMPERATURE CONTROLS:
Flow diversion device complies with *Ordinance*
requirements (a)
Recorder controller complies with *Ordinance*
requirements (b)
Holding tube complies with *Ordinance*
requirements (c)
Flow promoting devices comply with *Ordinance*
requirements (d)
(3) ADULTERATION CONTROLS:
Satisfactory means to prevent adulteration with added
water (a)
16c. ASEPTIC PROCESSING:
(1) INDICATING AND RECORDING THERMOMETERS:
Comply with *Ordinance* specifications (a)

Approved thermometer properly located in all
refrigeration rooms and storage tanks (c)
Recirculated cooling water from safe source and
properly protected; complies with bacteriological
standards (d)
18. BOTTLING AND PACKAGING:
Performed in a plant where contents finally
pasteurized (a)
Performed in a sanitary manner by approved
mechanical equipment (b)
Aseptic filling in compliance (c)
19. CAPPING:
Capping and/or closing performed in sanitary
manner by approved mechanical equipment (a)
Imperfectly capped/closed products properly
handled (b)
Caps and/or closures comply with *Ordinance* (c)
20. PERSONNEL CLEANLINESS:
Hands washed clean before performing plant functions;
rewashed when contaminated (a)
Clean outer garments and hair covering
worn (b)
No use of tobacco in processing areas (c)
21. VEHICLES:
Vehicles clean; constructed to protect milk (a)
No contaminating substances transported (b)
22. SURROUNDINGS:
Neat and clean; free of pooled water, harborages,
and breeding areas (a)
Tank unloading areas properly constructed (b)
Approved pesticides, used properly (c)

Remarks:

Date _____ Signature _____

1. A receiving station shall comply with items 1 to 15, inclusive, and 17, 20, and 22. Separation requirements of item 5 do not apply.
2. A transfer station shall comply with items 1, 4, 5, 7, 8, 9, 10, 11, 12, 13, 14, 15, 20, 22 and as climatic and operating conditions require, applicable provisions or items 2 and 3. In every case, overhead protections shall be required.
3. Facilities for the cleaning and sanitizing of bulk transport tanks shall comply with the same requirements for transfer sections

Figure 4.2 Milk plant inspection report. (From Grade A Pasteurized Milk Ordinance, Food and Drug Administration, 1997 Revision, Appendix M.)

Individual producer milk is not to exceed 100,000 organisms per milliliter in raw milk. Commingled milk is not to exceed 300,000 organisms per milliliter prior to pasteurization. After pasteurization the limit is 20,000 organisms per milliliter.

Laboratory tests conducted on pasteurized milk include:

1. Standard plate count, which determines the number of living bacteria and indicates the sanitary conditions under which milk was processed
2. Phosphatase test, which determines the efficiency of pasteurization
3. Coliform test, which determines the presence of coliform and indicates contamination after pasteurization
4. Residual bacterial count from milk processing and storage equipment and containers
5. Screening tests for abnormal milk
6. Disk assay for presence of antibiotics
7. Radioactive material testing

Vat Pasteurization Equipment Tests

All indicating thermometers, recording thermometers, and leak-protector valves used in vat pasteurization must be checked at installation and at least quarterly afterward to determine whether they are operating properly. The indicating thermometer, which must be accurate to +0.5 or –0.5°F within the pasteurization range or +0.5 or –0.5°F within the airspace range compared with a standard thermometer, is tested by inserting both thermometers into a well-agitated can of hot water at 145°F and making a direct reading. However, the indicating thermometers must read not less than the pasteurization temperature during the entire holding period. The error allowed in the thermometer accuracy is taken into account when determining the indicating temperatures.

All recording thermometers and pen arms must be checked at installation, quarterly, and when needed; and must be able to adjust to a wide range of fluctuating temperatures. The recorder thermometer is adjusted to read exactly the same as the previously tested indicating thermometer while immersed in a can of agitated water at 145°F. The recorder sensor bulb is then immersed in a can of boiling water for 5 min and into a can of ice water for 5 min. Finally, the recorder element is placed back into the can of agitated water at 145°F. The final temperature on the recorder chart must never read higher than the indicating thermometer.

The rotating time of the recorder chart is tested at installation and at least quarterly to make sure that the elapsed time is accurate. This is done by comparing the elapsed time on the recorder chart with an accurate watch. The recording-pen arm temperature notation is compared with the indicating thermometer while the product is in the vat. The recorder must not be higher than the indicating temperature.

Proper functioning of the leak-protector valve is determined by disconnecting the outlet pipe and carefully checking for leakage around the valve while product pressure is exerted on the valve.

High-Temperature Short-Time Pasteurization Equipment Tests

Indicating thermometers, including airspace thermometers, must be checked on installation and at least once every 3 months against a standard thermometer to make sure that they are within +0.5 and −0.5°F accuracy of the test thermometer. This is done by inserting the thermometer in an agitated 10-gal can of water heated to within −3°F of pasteurization temperatures and taking a direct reading.

Measure the thermometric response of the indicating thermometer on the pipeline by determining the time that it takes to raise the temperature of the indicating thermometer, which has been placed in a water bath 19°F higher than the lowest reading on the thermometer and above pasteurization temperatures. It should rise 12°F within 4 sec.

The recording thermometer should be checked against the indicating thermometer at installation and quarterly by an environmental health worker and daily by the plant operator. The recording thermometer should never read higher than the indicating thermometer. The accuracy of the recording thermometer temperature is tested in the same way that vat pasteurization temperature is tested, except that 161°F is used instead of 145°F. The time accuracy of the recorder controller is measured in the same way.

Cut-in temperatures of milk flow should be determined by raising the product temperature from 3°F below pasteurization temperatures to pasteurization at a rate of 1°F per 30 sec and determining the temperature at cut in. Cut-out temperature is determined by raising the pasteurized product flow slightly and then decreasing at a rate of 1°F per 30 sec. Determine the temperature at cut out. This ensures that milk is reaching full pasteurization temperatures.

The thermometric lag of the recorder controller must not exceed 5 sec. This is determined by inserting the recording thermometer in a can of hot water that is 7°F above pasteurization and determining the time it takes the mercury to rise from 12°F below pasteurization and the cut-in mechanism by the controller.

Check the flow-diversion valve for leakage. Measure the response time of the flow-diversion valve, which should not exceed 1 sec, by determining the elapsed time between the instant of activation of the cut-out mechanism and the fully diverted flow position.

Determine the time of product flow through the holding tube by electrodes and injection of 50 ml of saturated salt solution at the starting point. Determine the time it takes to reach the second electrode. This should be a minimum of 15 sec (see Figure 4.3).

Hazard Analysis and Critical Point-Based Inspections for Poultry and Meat

The 1993 outbreak of foodborne illness caused by the E. coli O157:H7 pathogen focused the attention of the public, the Congress, and the USDA on the fact that the system of meat and poultry inspection, based solely on visible detection, did not address the major causes of foodborne illness, which are microorganisms. On July 25, 1996, the Food Safety and Inspection Service of the USDA, which was responsible for ensuring the safety, wholesomeness, and accurate labeling of meat, poultry, and egg products, issued its landmark rule, pathogen reduction, HACCP systems. The rule discussed the serious problem of foodborne illness in the United States associated with meat and poultry products. It focused attention on the prevention and reduction of microbial pathogens on raw products, which may lead to disease. It clarified the

MILK PLANT EQUIPMENT TEST REPORT

Department of Health and Human Services
Public Health Service / Food and Drug Administration

TEST	TEST	TEST FREQUENCY	TESTED (X or NA)	RESULTS OF TEST (SEE REVERSE FOR WORKING)
1.	Indicating thermometers (including air space): Temperature accuracy	3 months		
2.	Recording thermometers: Temperature accuracy	3 months		
3.	Recording thermometers: Time accuracy	3 months		
4.	Recording thermometers: Checked against indicating thermometer	3 months		Daily by operator
5.	Flow Diversion device: Proper assembly and function (HTST and			
5.1	Leakage past valve seat(s)	3 months		
5.2	Operation of valve stem(s)	3 months		
5.3	Device assembly (micro-switch) single stem	3 months		
5.4	Device assembly (micro switches) dual stem	3 months		
5.5	Manual diversion - Parts (A, B, and C) (HTST only)	3 months		
5.6	Response Time	3 months		
5.7	Time delay interlock (dual stem devices) (Inspect)	3 months		
5.8	Time delay interlock (dual stem devices) (CIP)	3 months		
5.9	Leak Detect flush time delay (HTST only as applicable)	3 months		
6.	Leak-protect valves: Leakage (Vats only)	3 months		
7.	Indicating thermometers in pipelines: Thermometric response (HTST)	3 months		
8.	Recorder-Controller: Thermometric response (HTST only)	3 months		
9.	Regenerator Pressure Controls			
9.1	Pressure Switches (HTST only)	3 months		
9.2	Differential pressure controllers	3 months		
9.2.1	Calibration	3 months		
9.2.2	Interwiring Booster Pump (HTST only)	3 months		
9.2.3	Interwiring FDD (HHST and Aseptic)	3 months		
9.3	Additional Booster Pump interwiring (HTST only)			
9.3.1	With FDD	3 months		
9.3.2	With Metering Pump	3 months		

			3 months	Daily by operator (HTST)
10.	Milk-flow controls. Cut-in and cut-out temperatures (10.1, 10.2, or			
11.	Timing System Controls			
11.1	Holding time (HTST except magnetic flow meters)		6 months	Adjusted product holding time if
11.2	Magnetic Flow Meters (HTST only)		6. months	
11.2	Flow alarm (HTST, HHST, and Aseptic)		6 months	
11.2	Loss of signal alarm (HTST, HHST, and Aseptic)		6 months	
11.2	Flow cut in/cut out (HTST only)		6 months	
11.2	Time delay (After divert) (HTST only)		6 months	
11.3	HHST Indirect heating		6 months	
11.4	HHST Direct Injection Heating		6 months	
11.5	HHST Direct Infusion heating		6 months	
12.	Controller: Sequence logic (HHST and Aseptic) (12.1 or 12.2)		3 months	
13.	Product pressure-control switch setting (H-HST and Aseptic)		3 months	
14.	Injector differential pressure (HHST and Aseptic) (Injection)		3 months	
Remarks				

PLANT	IDENTITY OF EQUIPMENT	LOCATION	D	ATE	SANITARIAN

NOTE: This form is a supplement to the Milk Plant Inspection Reports FDA 2359, and these tests are in addition to the equipment requirements for which compliance is determined by inspection. See Appendix I, Grade A Pasteurized Milk Ordinance.

Figure 4.3 Milk plant equipment test report. (From Grade A Pasteurized Milk Ordinance, Food and Drug Administration, 1997 Revision, Appendix M.)

roles of government and industry in food safety, with industry accountable for producing safe food. Government was responsible for setting appropriate food safety standards, maintaining a strict inspection program to ensure the standards were met, and utilizing a strong enforcement program to deal with the plants that were not in compliance. The Pathogen Reduction and HACCP rule requires the following:

1. All meat and poultry plants must develop and implement a system of preventive controls to improve the safety of their products (HACCP).
2. All slaughter plants and plants producing raw ground products must meet the new pathogen reduction performance standards set by the agency for salmonella.
3. All meat and poultry plants must develop and implement written standard operating procedures for sanitation.
4. All meat and poultry plants must conduct microbial testing for generic E. coli to verify the adequacy of their process controls for the prevention of fecal contamination.

Under HACCP, a plant analyzes its processes to determine at what points hazards might exist that could affect the safety of its products. These points are called critical control points (such as chilling, cooking, filling, and sealing cans) and slaughter procedures (such as removal of internal organs). Once the critical control points are determined, the plant must establish critical limits, which are usually expressed as numbers related to time, temperature, humidity, water activity, pH, and salt concentration. The critical limits are based on scientific findings found in the scientific and technical literature or on recommendations made by experts. The plant establishes a monitoring system, takes corrective action, and keeps appropriate records of what has occurred. As of February 1, 2000, all plants small and large are now required to be in compliance with the rule.

Performance standards for *Salmonella* on raw products were established as a means of determining the degree of protection the HACCP system must achieve. However, the microbiological safety of the meat or poultry product at the point of consumption is based on several different factors. The organisms may be introduced at many points along the continuum from the live cattle to final preparation and serving of the food, and once in the product these organisms can multiply into larger numbers if the appropriate conditions exist. Small numbers of highly virulent organisms can cause disease. Susceptibility of the individuals to the organisms varies widely based on the type of population, such as the very young, the elderly, and those with compromised immune systems. Whereas *E. coli* O157:H7 organisms are so virulent that they can pose a significant hazard and therefore are not permitted in raw meat products, other organisms such as *Salmonella* need larger numbers to cause disease, and therefore are allowed in small numbers.

Performance standards for *Salmonella* are based on the class of the products, such as carcasses of cattle, swine, and chickens; and ground beef, ground chicken, and ground turkey. They are also based on the prevalence of *Salmonella*, as determined by the Food Safety and Inspection Service nationwide microbial baseline surveys. The standards are based on industry averages, instead of how much salmonella it takes to make someone sick. The agency does not have solid data on the relationship between rates of contamination and association with foodborne illness. The standards are to be adjusted in the future as new data becomes available. The standards are expressed in terms of the maximum number of positive samples that are allowed per sample set (samples sets are statistically determined and range from 51 samples for broilers to 82 samples for steers or heifers). In steers and heifers, only 1 of 82 samples is permitted to be positive, whereas in broilers, 12 of 51 samples are permitted to be positive. The samples are either positive or negative and are not expressed in terms of numbers of organisms. Test results are not used to condemn products. They are used to indicate whether the HACCP systems are working. The four major areas of concern in the existing system are adequacy of testing; potential for administrative delays, where potentially contaminated food would reach the public; inadequacy of enforcement because facilities can sell contaminated products for long periods of time; and database problems because of the potential inadequacy of sampling.

The Sanitation Standard Operating Procedures is a written plan used to ensure that sanitation problems are addressed and corrected for better compliance with existing regulations. It does not impose new requirements. The federal inspectors verify the plant plan.

Since January 1997, slaughter plants have been required to test carcasses for generic *E. coli* as an indicator of the adequacy of the plant's ability to control fecal contamination, which is the primary route of contamination for pathogenic organisms. The Food Safety and Inspection Service has adopted performance criteria for *E. coli* for each species of animal that reflect the frequency and levels of microorganisms on carcasses according to nationwide baseline surveys. The criteria are not guidelines and therefore are not regulatory standards. Results of the tests by themselves will not result in regulatory action.

Food safety hazards are not limited to microbial contamination. They also include natural toxins, chemical contamination, pesticides, drug residues, zoonotic diseases, decomposition, parasites, unapproved use of direct or indirect food or color additives, and physical hazards. The HACCP plan is concerned with hazard analysis. It uses a flowchart to describe the steps of each process in the establishment. In a meat or poultry plant it would include slaughter, raw product, type of heat treatment, temperatures, time in process and storage, and other potential contamination.

Poultry Testing

High bacterial counts indicate poor handling and processing techniques and improper refrigeration. Clean birds should be held at 32 to 34°F for a maximum of 4 days. If slime develops, the bacterial count per square centimeter of surface will probably be in the millions. A total count of organisms can be made using a swab technique or by pressing an agar plate against the surface of the bird. A rinse technique using sterile water may also be used.

The dye, resazurin, may be used to determine the condition of a bird, with swab samples in a solution noting the length of time the sample takes to decolorize to a fluorescent pink color. A fresh bird takes greater than 8 hours, a good bird takes greater than 5 hours but less than 8 hours, a fair bird takes greater than 3½ hours but less than 5 hours, and a poor bird takes less than 3½ hours.

Hazard Analysis and Critical Control Point-Based System for Shell Egg Processing

As in meat and poultry processing, a new system of HACCP inspections are used to improve egg processing and reduce the potential for foodborne disease, especially *S. enteritidis*. This system includes basic sanitation of premises and facilities; rodent and pest control; employee hygiene and health; safety of water; safety of food packing materials; environmental testing of chickens; use of SE-negative feed; use of chicks from SE-negative breeders; cleaning and disinfection of poultry houses and equipment; diverting of eggs to pasteurization if SE testing yields a positive; and washing, grading, sanitizing, packaging, cooling, and repackaging. In addition, requirements exist for safe handling statements on labels to warn customers about the risk of illness caused by *S. enteritidis*, and requirements for refrigeration of the shell eggs at 40°F or lower.

Egg Testing

The tests used with milk pasteurization equipment are also used with egg pasteurization equipment. In addition, total counts are taken from the whole eggs, egg white, egg yolk, salted yolk, sugared yolk, and blends. Tests are taken of liquid, frozen, or dried eggs. Yeasts and molds are also evaluated. Of particular concern are salmonella bacteria, especially *S. enteritidis*. Therefore, special tests are conducted to determine the kinds and quantities of salmonella present. Bacteria grow very rapidly in broken-out egg products when the temperature rises above 60°F. Although freezing and drying decreases the number of viable bacteria, these processes do not eliminate them; therefore, frozen and dried eggs can and have caused outbreaks of foodborne disease.

Meat Testing

Programs monitor pesticide residues, cleaning agents, antibiotics, salmonella, generic *E. coli,* and *E. coli* O157:H7 in meat tissue. Inspections are made at many stages during meat processing by USDA, FDA, and state and local health authorities. Labels are required to show all ingredients used in any type of processed meat. Total plate counts may be taken from fresh meats, cured meats, canned meats, and various meat products. Samples are usually taken of the surface and the interior of the meat. In addition, tests are conducted for yeasts and molds. Direct microscopic examinations are also used. The pH, salt, and nitrite levels are also determined.

Shellfish Testing

Shellfish are tested for coliform on a routine basis. However, where outbreaks of specific diseases, such as infectious hepatitis, cholera, typhoid fever, and bacillary dysentery occur, specific tests are conducted for these organisms. The bacterial content of shellfish can be sharply increased during any phase of harvesting, processing, and storage. Shellfish may also be tested for various chemicals.

Hazard Analysis and Critical Control Point-Shellfish Inspection System

A shellfish inspection system following the procedures established for HACCP in other foods is required for all fish, shellfish, and their products. Ensuring the safety of seafood creates special challenges for the industry as well as governmental agencies required to protect the public health. Seafood is unique in many ways. It comprises a variety of products covering hundreds of species that have little in common other than the fact that they have an aquatic origin. The range of habitats for edible species is diverse ranging from cold to warm water, bottom dwelling to surface feeding, deep sea to near the shore, and freshwater to saltwater. Fish are exposed to the bacteria and viruses that naturally occur in their environment as well as to those that enter the water through pollution. Chemicals, some of which are toxic to humans, can accumulate in fish as well. Fish can also accumulate natural toxins and parasites that are specific to marine animals. Fish are therefore subject to a wide range of hazards before harvest. Shellfish are particularly a concern because they gather bacteria and viruses as they feed. Worker skills, older processing plants, location of the plants in potential flood areas, inadequate sanitation, insects and rodents, improper processing, and improper storage temperatures are all critical points to be evaluated in the HACCP process. The firms that pasteurized their products, in many cases, did not have adequate controls to ensure that the process was properly carried out. In a survey conducted by the FDA, it was determined that 54% of the companies had not established the adequacy of the pasteurization process, 27% of the companies did not have temperature-indicating devices, 42% of the companies did not perform can seam evaluations, 43% of the companies did not perform cooling water sanitizer strength checks, 84% of the companies did not monitor the internal temperature of products during the various stages of processing or check the time, 14% of the companies did not have temperature-indicating devices on their finished product coolers, and 89% of the companies did not have temperature-recording devices.

HACCP inspections are particularly concerned with time and temperature at various stages of the process and therefore would cause the seafood process to be improved substantially. The HACCP inspection would help identify all biological, chemical, and physical hazards, as well as the critical control points in the process. Further, it would establish a means of verifying that the critical limits are adequate to control the hazards, and ensure that the HACCP plan was working properly and was thoroughly documented.

Fresh Produce Safety Evaluation System

To reduce the risk of foodborne disease, it is necessary to evaluate the entire system of production of the produce, from the use of the land, through the various steps of production and transportation, to sales to the consumer and final consumption of the product. Throughout this process the emphasis is on risk reduction instead of risky elimination, because current technologies cannot eliminate all potential food safety hazards associated with fresh produce that will be eaten raw. Microbial hazards, pesticide residues, and other chemical contaminants must be considered during the course of the evaluation. Critical limits need to be set for the major control points, which are as follows:

1. Water use including agricultural water, processing water, wash water, and cooling water
2. Manure and municipal biosolids production including agricultural practices for manure management, treatment to reduce pathogen levels, handing and application of untreated manure and treated manure, and other animal feces
3. Control of potential hazards due to worker health and personal hygiene
4. Use of sanitary facilities including toilet facilities, hand washing facilities, and sewage disposal
5. Control of potential hazards during general harvest and in equipment maintenance
6. Control of potential hazards in general packing and facility maintenance
7. Control of insects and rodents during all phases of the operation,
8. Control of potential hazards in transportation and storage facilities

Growers, packers, and shippers need to assume a proactive role in minimizing food safety hazards potentially associated with fresh produce. This can be accomplished through knowledge of the hazards and the imposition of a self-inspection program. Of necessity, governmental agencies need to provide accurate and reasonable information, and make inspections and surveys where appropriate.

FOOD QUALITY CONTROLS

The federal government protects all food shipped in interstate commerce through the enforcement of the following acts: Pure Food and Drug Act, passed in 1906 and updated to the present; the Wholesome Meat Act of 1967, amended in 1981; the Wholesome Poultry Products Act of 1968; and the Egg Products Inspection Act of 1970, revised in 1992; and the Food Quality Protection Act of 1996.

The Food Quality Protection Act of 1996 amended the FIFRA, and the Federal Food, Drug, and Cosmetic Act. These amendments fundamentally changed the way in which the EPA regulates pesticides. The requirements included a new safety standard of reasonable certainty of no harm that must be applied to all pesticides used on foods. On January 25, 1997, President Clinton announced the National Food Safety Initiative. The initiative included components for reducing the incidence of foodborne illness from farm to table. The key components included expansion of the federal food safety surveillance system; improved coordination between federal, state, and local health authorities; improved risk assessment capabilities; increased inspections; expanded research; consumer education; and strategic planning. The USDA and the USDHHS received additional funding to initiate changes to ensure the safety of a wider variety of food products from a broader range of hazards. The goals of the initiative were to enhance surveillance and investigation to improve outbreak response; strengthen coordination of agencies to improve efficiency; improve capability to estimate risks associated with foodborne hazards; expand inspection and compliance efforts; implement new preventive measures with a new emphasis on domestic and imported produce; facilitate the implementation of HACCP programs;

continue to build the national food safety education campaign; and accelerate food safety research. The objectives were to develop a science-based food safety system, which was needed in this changing world, and to provide adequate prevention, education, and verification procedures.

The National Food Safety Initiative focused on combating foodborne illness in two major ways. The FDA, USDA, CDC, and EPA have joined together to create a state-of-the-art science-based food safety system, which develops prevention strategies and programs to keep pathogens out of the food throughout the food chain from farm to table. This has involved education and inspection of producers, processors, distributors, foodservice workers, and consumers. The second technique involves the development of a strong outbreak response capability that provides for the early detection and containment of foodborne hazards on a national basis, before they become widespread. It also helps develop the knowledge to prevent future outbreaks. The development and use of the HACCP system has helped in determining where contamination occurs and in reducing subsequent risk. The development and adoption of model prevention programs to advance food safety was upgraded by the publishing of the 1999 Food Code. This is a model code of food safety guidelines for foodservice establishments. It promotes food safety by serving as a reference for more than 3000 state and local regulatory agencies that oversee food safety in restaurants, grocery stores, nursing homes, and other institutional and retail settings.

Fast and effective response to emerging pathogens in the food supply is essential to preventing widespread illness. The National Food Safety Initiative outbreak response program uses scientific research as the foundation for a multifaceted program that includes:

1. Scientists at the Foodborne Diseases Active Surveillance Network (FoodNet) sites located throughout the country are actively tracking laboratory-confirmed cases of foodborne disease in their nearby population centers. These findings are shared among these sites and with the CDC to gain a more accurate picture of the number of possible foodborne disease cases annually and to determine rapidly if the outbreak is occurring in a variety of locations in the country.
2. A computerized database, called PulseNet, has been created to match the DNA fingerprint of foodborne diseases, thereby accelerating the trace back process to the source of contamination.
3. The Epidemiology Rapid Assessment Team is available 24 hours a day to provide a rapid review of incoming reports of foodborne, disease and to coordinate trace back and containment efforts.
4. The Foodborne Outbreak Response Coordinating Group (FORCE-G) links federal, state, and local government agencies to enhance coordination and communication in responding to outbreaks. It uses resources efficiently and prepares for new and emerging threats to the food supply.

Because imported foods can be better monitored where they are produced, instead as they cross the U.S. border, new programs have been established to prevent contamination in countries that export to the United States. Foreign inspections of food establishments that produce food products at high risk for microbial contamination are now being conducted. When foodborne illness outbreaks associated with imported foods occur, follow-up investigations are conducted in the exporting countries.

Meat and Poultry Controls

Meat and poultry controls include the use of the HACCP plan and microbiological testing. The plan at a minimum includes a list of food safety hazards; critical control points designed to control these hazards; critical limits to be met at each of the critical control points; procedures to be used that will monitor each of the critical control points to ensure compliance with the critical limits; appropriate record-keeping system; verification procedures and frequency; and appropriate corrective actions. The establishment validates the adequacy of the HACCP plan in controlling the food safety hazards identified during the hazard analysis, and verifies that the plan is effectively implemented. The initial validation takes place on the completion of the hazard analysis and the development of the HACCP plan. During the plan validation period, the establishment repeatedly tests the adequacy of the critical control points, critical limits, monitoring and record-keeping procedures, and corrective actions as set forth in the plan. Ongoing verification activities include the calibration of process-monitoring instruments; direct observations of monitoring activities and corrective actions; and review of records generated and maintained in accordance with the rules and regulations of the governmental agency. Each year, every establishment reassesses the adequacy of the HACCP plan, and alters it, as needed. This also happens when changes in the process occur, such as different raw materials or sources of raw materials; different product formulations; different slaughter or processing methods or systems; change in production volume; change in personnel; and change in packaging; change in finished product.

Shell Egg Control

On August. 27, 1999, new federal regulations became effective, requiring that shell eggs packed in containers for consumer consumption be stored and transported under refrigeration at an ambient temperature not to exceed 45°F. Further, the packed shell eggs had to be labeled that "refrigeration was required." Consumers and food-services establishments were advised that bacteria, especially, *S. enteritidis*, could be found in the unbroken egg. Beyond refrigeration and labeling other controls included not eating raw eggs in milk shakes, Caesar salad, Hollandaise sauce, eggnog, or any other products requiring raw egg ingredients; purchasing clean eggs, which had been refrigerated; refrigerating eggs at home at or below 40°F; cooking eggs thoroughly; and using safe egg recipes. In addition, HACCP inspections were to be performed at the egg-producing businesses. *Salmonella enteritidis* environmental testing was to take place during egg production. With the great concern for *S. enteridis* infections the following steps were to be taken to protect the ultimate consumers:

1. Reducing the number of shell egg-containing eggs marketed to the consumer
2. Reducing exposure of consumers to shell egg-containing foods
3. Expanding and upgrading surveillance systems for human shell egg infections
4. Expanding surveillance and upgrading surveillance systems for poultry shell egg infections
5. Accelerating shell egg outbreak detection and initiating outbreak investigations, and improving completeness of outbreak investigations

6. Improving communication among federal, state, and local agencies involved in shell egg outbreaks and tracing back investigations
7. Ensuring adequate current information is available to make decisions about shell egg prevention, controls, surveillance, and education based on sound science
8. Educating individuals throughout the production to consumption continuum using science-based materials

Fish and Shellfish Control

Proper sanitation is the key to fish health management in modern aquaculture. A number of infectious diseases, especially external fungal infections, may be directly attributed to the accumulation of organic material in the tank or pond. Tanks or ponds may be sanitized between groups of fish by draining, drying, and in some cases using a chemical sterilant such as hydrated lime, which rapidly causes the pH in the treated areas to rise above 10, is lethal to parasites and bacteria, and helps in the elimination of ammonia tied up in the muds. All equipment must be thoroughly cleaned and disinfected and employees must wash their hands thoroughly with disinfecting soaps or detergents. Beside using the HACCP inspection system, the FDA promotes seafood safety through other means, including:

1. Setting standards for seafood contaminants especially for polychlorinated biphenyls (PCBs), pesticides, mercury, paralytic shellfish poisoning, and histamine in canned tuna
2. Assisting the Interstate Shellfish Sanitation Conference, an organization of federal and state agencies and members of the shellfish industry; the conference that develops uniform guidelines and procedures for state agencies that monitor shellfish safety
3. Entering into cooperative programs with states to provide training to state and local health officials who inspect fishing areas, seafood processing plants and warehouses, and restaurants and other retail establishments
4. Working with National Oceanic and Atmospheric Administration (NOAA) to close federal waters to fishing whenever oil spills, toxic blooms, or other phenomena threaten seafood safety
5. Sampling and analyzing fish and fishery products in agency laboratories for toxins, chemicals, and other hazards
6. Administering the National Shellfish Sanitation Program, which involves 23 shellfish-producing states, plus a few nonshellfish producing states, and 9 countries; the program that controls all environmental situations related to the growing, harvesting, shucking, packing, and interstate transportation of oysters, clams, and other molluscan shellfish

Each state supervises its local and imported fresh and fresh-frozen oysters, clams, and mussels, which must be certified by a certified dealer to have been grown, harvested, transported, processed, and shipped in accordance with the criteria. Plant inspections and record keeping are of primary importance. The FDA makes available for use the following: Sanitation of Shellfish Growing Areas — Part 1; Sanitation of the Harvesting, Processing, and Distribution of Shellfish — Part 2; and Interstate Certified Shellfish Shippers List.

Produce Control

The control of the quality of water coming into contact with the edible portion of produce is essential. It is necessary to identify the source and distribution of water that is used and to be aware of its relative potential as a source of pathogens. Agriculture water quality varies considerably, especially if it comes from surface waters, which may be subject to intermittent, temporary, or continuous wastewater discharges from upstream livestock operations. Groundwater may be infiltrated by surface water, such as in older wells with cracked casings, and may become contaminated. Agriculture water can be contaminated directly or indirectly by improperly managed human waste. Good agricultural practices may include protecting surface waters, wells, and pump areas from uncontrolled livestock or wildlife access to limit the extent of fecal contamination. Soil and water conservation practices such as grass or sod waterways, diversion berms, runoff control structures, and vegetative buffer areas may help prevent polluted runoff water from contaminating agricultural water sources and produce crops.

Processing water should be of such a quality that it will not contaminate produce. The water quality should be consistent with the U.S. EPA requirements for drinking water. Periodic water sampling and microbial testing should be performed. All water contact surfaces such as dump tanks, flumes, wash tanks, and hydrocoolers should be clean and sanitized as often as necessary to ensure the safety of the produce. Antimicrobial agents may help minimize the potential for microbial contamination.

Chlorine is commonly added to water at 50 to 200 ppm total chlorine, at a pH of 6.0 to 7.5 for postharvest treatments of fresh produce, with a contact time of 1 to 2 min. Ozone is also used to sanitize wash and flume water in packinghouse operations. UV radiation may be used to disinfect processing water. Washing fresh produce with potable water helps reduce the overall potential for microbial food safety hazards. Where produce is water sensitive, brushing, scraping, or blowing clean air can reduce potential microbiological hazards. Cooling water containing antimicrobial chemicals may reduce the potential for microbial contamination of produce.

Growers can reduce the potential for microbial hazards in produce by using good agricultural practices for handling animal manure. These include composting, treatment, proper manure storage, and application to the land to avoid crops.

Establishing training programs to teach all employees, including supervisors, full-time and part-time, about basic sanitation and good hygiene helps reduce disease. An important part of this training is teaching the importance of hand washing and providing adequate hand washing facilities.

All states have health departments and departments of agriculture that provide additional protection through inspection services. Counties, cities, and multiple county health departments provide for local inspection services to protect the food supply. When food processing is poor or unsanitary, necessary legal action is taken.

SUMMARY

The production of adequate quantities of food to satisfy the needs of the world continues to be an extremely serious problem because of the sharp increase in world population and the decrease in death rates brought about by better medical care. The production of food brings with it inherent difficulties, such as the use of scarce water supplies, energy supplies, and potentially hazardous fertilizers and pesticides. Additives to enhance color and taste and to preserve food are increasingly under suspicion of having the potential to cause a variety of diseases, including cancer. Current techniques of food technology have been incriminated in outbreaks of a variety of existing and emerging diseases because of a lack of adequate controls during food processing. As the quantity of prepared frozen foods and food shipped to varying parts of the country and the world increases sharply, the threat of disease increases rapidly. Existing controls must be reinforced and new controls must be established.

RESEARCH NEEDS

Toxicants that may be part of plant foods require further study. Long-term feeding studies need to be conducted in experimental animals to determine special problems, such as sensitization and allergies. Analytic methods should be developed to measure the important toxicants found in foods. These toxicants include the pressor amines, goitrogens, antinutritional factors, safrole-related compounds, and allergic hemagglutinins. Sampling techniques must be utilized both on raw agricultural products and prepared products. Studies of plant breeding to reduce the level of toxicants found in certain selected food plants are needed. New plant varieties should be investigated to determine if genetic manipulation has caused an increase in the production of toxicants. Research is required to determine if existing plants produce abnormal toxic metabolites under certain stress conditions. A catalog of toxic constituents of poisonous plants should be produced. A rapid, reliable, quantitative analytic method should be developed for paralytic shellfish poisoning. A complete study of the chemical nature of various poisons must be conducted on all paralytic shellfish poisoning. A legally binding action level for histamine in fish needs to be developed to protect people from scromboid poisoning. Chemical indicators need to be developed for detecting decomposed fish.

The mycotoxins must be studied in the laboratory and field situation to determine why they are produced. Methods must be developed for measuring the incidence of mycotoxin contamination in raw and finished foods. Studies are needed to determine the biotransformation of mycotoxins that occurs in food animals and to determine the potential for contamination of the food for human beings. Toxicological studies are needed to determine the amount of mycotoxin to which humans are exposed.

A complete study of the biological activities of nitrates, nitrites, and nitrosamines in foods and a study of the effect of these chemicals on humans are needed. Research is needed on the effects of high intakes of sodium chloride and phosphates. Special emphasis should be placed on the relationship of these chemicals to hypertension and bone growth. Because of the large numbers and quantities of compounds found in foods, ongoing studies should be made to determine the interaction of food additives with major dietary constituents, the alterations of these dietary items, and the possible resulting effects.

Considerable research is needed in understanding the movement of mercury through the food chain. An understanding is needed of the adverse effect of lead and on the range of adverse effects of cadmium, copper, zinc, magnesium, selenium, and arsenic found in food. Research is needed on pesticides to determine the level of food contamination and their adverse effects in humans. Additional research should be carried out on the consumption of food additives, the interaction of food additives and other residues present in food, the chemical nature of many of the additives, and the chemical changes brought about by cooking. This fundamental research is essential to determine which additives should be eliminated from food and which should remain.

Chemicals, including many of the coloring substances, are potential carcinogens. The use of risk assessment in determining potential hazards is an excellent technique. The first uses of risk assessment appear to have been for aflatoxins in 1978 and for PCBs in 1979. In 1982, risk-assessment techniques were used to determine carcinogenic impurities in color additives that were not covered by the Delaney clause (the Delaney clause is no longer in effect). Finally, risk assessment was extended to substances covered by the Delaney clause. Beside direct food additives, which may be potential carcinogens, indirect food additives may come from the packaging materials and food-processing equipment, color additives used for ingestion and for external purposes, potential contaminants of food or color additives, and unavoidable environmental contaminants in food and cosmetic ingredients.

In certain substances, such as saccharin, although the FDA has found the substance to be potentially carcinogenic, the benefit that the substance gives to the population at large is more important than the potential risk that has been shown in some studies with laboratory animals. The FDA also allows the continued use of methylene chloride to decaffeinate coffee, although it is banned for use in cosmetics. The use of this decaffeination agent is justified on the grounds that through assessment the risk is minimal to people. The indirect food additives, such as the packaging materials that have been mentioned may come in contact with food, as well as processing equipment. The FDA has banned outright two indirect food additives. They are flectol-H and NBOCA. The FDA has also banned certain uses of bottles made from acrylonitrile copolymers and PVC because of the concern of leaching of these chemicals into the food. Numerous other chemicals probably are part of packaging materials that should be studied as potentially hazardous to people through consumption of food.

Animal drug residues have been tested since 1962. The Congress in 1962 passed the diethylstilbestrol (DES) proviso that allows the use of carcinogenic drugs in animals providing that the residues cannot be detected in edible portions of tissue or foods derived from the living animals. These substances need to be studied further, because carcinogenic activity in humans may be a potential risk, over time, as was shown later with the drug DES. DES became the only animal drug successfully banned, because there was human evidence of potential cancer to humans. Other research concerns include the potential hazard to the food supply by groundwater or surface water contaminated by a variety of organic chemicals that may be toxic and hazardous. Additional study should be done on food crops to determine whether airborne pollutants that may be hazardous to people can be carried forward during the processing of food, therefore creating potential problems. The study of sludge-related contaminants put on land, and their potential effect in the food chain, need to be further researched.

The President's Council on Food Safety has set a goal of ensuring the development and use of a comprehensive scientific and technological food safety knowledge base to support prevention, inspection, surveillance, and education programs. The objectives for the future are:

1. Developing a national food safety research and technology infrastructure
2. Developing and improving data, methods, models, and measures to assess health effects, including a better understanding of the factors that affect sensitivity to foodborne illness, such as age and health status
3. Developing new, as well as improving existing, data, methods, models, and measures to assess exposure, including improved analytic and surveillance methods
4. Developing better, integrated uniform national and international risk assessment capability and conducting a wide variety of risk assessments
5. Developing and improving prevention and control methods and risk management practices through better integration of research
6. Coordinating and evaluating research on the highest priority food safety issues and efficiently leveraging federal agency research resources
7. Developing adequate technological support, including advanced modeling technology, for risk assessment and risk management

Risk assessment has been required by numerous federal laws, such as the Federal Crop Insurance Reform and Reorganization Act of 1994, and the Drinking Water Amendments of 1996. Sound risk assessments are important in various aspects of international trade, including the provisions of Codex Alimentarius and the World Trade Organization, the international bodies that govern standards for food safety, among other issues. All federal agencies with risk-management responsibilities for food safety established a joint consortium at which these agencies collectively advanced the science of microbial risk assessment. The goal of the consortium was to improve the quality of risk-assessment research by coordinating research priorities, eliminating redundancies, encouraging multidisciplinary research efforts, while working with outside experts from academia, industry, consumer groups, and private sources. The consortium's three primary functions were:

1. Developing an approach for setting methodological research priorities based on the value of information expected from each research activity
2. Serving as a clearinghouse for information about current and planned research projects pertinent to microbial risk-assessment techniques
3. Encouraging and, where possible, augmenting the research activities of the federal agencies to accelerate particularly critical research projects

The consortium plans to develop and validate exposure assessment models based on probabilistic methodology. Because risk assessment of foodborne illness is dependent on accurately estimating the probability that various quantities of a toxin or pathogen will be ingested by the consumer, it is necessary to develop models and simulations based on the probability of the occurrence of microbial pathogens and chemical hazards in food at all stages of the food chain. It is also necessary to determine how the commercial and home preparation operations are carried out; the food consumption patterns of the population, especially the sensitive subpopulations;

and the vehicles by which the sporadic and epidemic diseases spread. In the agricultural environment are there pathogen reservoirs, pathogens in the feed, or pathogens in the animal manure? What are the effects of the key processing steps on the levels of pathogens? It is also necessary to understand the following:

1. How to develop and validate dose–response–assessment models for use in risk assessment
2. Why certain organisms are becoming resistant to traditional preservation technologies, such as heat or cold, low pH, high salt, and disinfectants
3. Why certain pathogens in food-producing animals and their manures become resistant to antibiotics
4. How to develop prevention techniques including pathogen avoidance; reduction; and elimination from food animals and their manures, fresh fruits, fresh vegetables, and grains
5. How to reduce the stresses associated with the transportation of live animals and fresh produce when food production, processing, and consumption often occur thousands of miles apart

CHAPTER 5

Insect Control

BACKGROUND AND STATUS

Insects and rodents are responsible for numerous outbreaks of disease in humans and animals. These pests, or vectors, also cause extreme annoyance because of bites. They contaminate huge quantities of food, which must be destroyed because of the potential spread of disease. Contamination can be caused by vectors coming into contact with disease-causing microorganisms and carrying the organisms on their bodies or by vectors depositing feces or urine on food.

Vectorborne disease varies in prevalence from state to state and from year to year according to weather and climatic conditions. In determining the potential incidence of these diseases, it is necessary to identify and determine the prevalence of the vectors that act as primary transmitters. For instance, in an outbreak of viral encephalitis in St. Louis, it was determined that mosquitoes spread the disease to the south and through Illinois, Indiana, and other states in the country. Other types of data include the incidence of a given vectorborne disease in an area; the type of climate, including maximum and minimum temperatures, rainfall, prevailing winds, and their daily and monthly occurrence; life cycles of the vectors; seasonal fluctuation in vector populations and their requirements for life; any unusual conditions that may help spread the given vectorborne disease, such as floods, earthquakes, and other natural disasters.

Insects are extremely adaptable to the human environment. They are found in the air, on and under the soil, and in fresh or brackish water. They live on or in plants and animals, and compete fiercely with other species as parasites. Insects have caused enormous problems for many centuries. With numerous insecticides available, insects continue to plague humans, because they are capable of developing a resistance to insecticides.

Rat fleas have transmitted plague organisms to millions of people over the last 15 centuries. An estimated 25 million people or 25% of the population of Europe died during the great plagues of the 15th, 16th, and 17th centuries. Today, in the United States, humans still contract plague from rodents. Sylvatic plague, which is contracted from wild rodents, particularly groundhogs, in the western part of the country, is seen most frequently today. In recent years, plague occurred in some of the southwestern states, including New Mexico. The potential for serious outbreaks of plague always exists as long as rats and Indian rat fleas are present, and the organism can be transmitted from rodent to rodent and from rodent to human. Pneumonic plague is a highly contagious, pneumonia-like disease related to bubonic plague. Fortunately, pneumonic plague has not appeared recently in this country.

The majority of the important human vectorborne diseases cannot be prevented by vaccines or by chemotherapy. Their control is based on the ability to reduce the source of vectors and the contact between vector and humans. Although in the last 25 years there has been a sharp reduction in many of the insectborne diseases, they have not disappeared; and in some areas because of changes in environmental conditions or social conditions, the diseases are on the increase again. By 1972, malaria was eradicated from most of the world; yet in 1973 and 1974, during the period of conflict between Pakistan and India, malaria increased. Yellow fever was under control for many years, but there was a sharp increase in Nigeria in 1969. Recently malaria has occurred in New Jersey, New York, and Texas. Plague increased in South Vietnam from 1963 to 1973. Fatal human plague occurred in Arizona and Colorado in 1996.

Whenever wars have occurred in other areas, an increase in typhus and relapsing fevers have followed. In 1994, typhus fever again occurred in refugees from a war in Africa. As a result of movement from urban to rural areas, available sewage systems sharply decreased and overflowing sewage and mosquito problems increased. In California, people moving into the woodlands in the suburbs have become exposed to the vectors of California encephalitis and dog heartworm. As artificial lakes are created to enhance the aesthetic beauty of areas, mosquito problems increase. Schistosomiasis increased with the creation of major dams, especially the Aswan Dam in Egypt. Devastating epidemics of malaria occurred in Pakistan after the construction of the Sukkar Varrage. In India a severe epidemic of malaria occurred after the construction of the Mettur Dam. A detailed assessment of status is difficult to make, because limited statistics are available that can be organized into some uniform analysis. From 1982 to 1999, Lyme disease had become the leading tickborne illness in the United States, with over 125,000 reported cases.

A variety of diseases that have been typically found in other parts the world are now occurring in the United States. Ecological and economic factors, as well as changes in human behavior, have resulted in the emergence of new and the reemergence of existing but forgotten infectious diseases during the last 20 years. African tick-bite fever occurred among international travelers in 1998. A 34-person group traveled to Swaziland where 9 people contracted the disease. U.S. residents contracted dengue, which is a mosquito-transmitted acute disease, in either the Caribbean islands or Asia. In the United States, 179 people had symptoms of the disease in 1996 and submitted serum samples for testing to the Centers for Disease Control and Prevention (CDC). Ehrlichioses has emerged as a public health problem in the United States; the organism is transmitted by ticks. Fleaborne disease organisms could easily reemerge in epidemic form because of changes in vector–host ecology. The changing ecology of *Murine typhus* in southern California and Texas over the last 30 years is a good example of urban and suburban expansion affecting infectious disease outbreaks. In these areas, the classic rat–flea–rat cycle has been replaced by a peridomestic animal cycle involving cats, dogs, and opossums and their fleas. These animals are running wild. West Nile encephalitis is caused by viruses transmitted by mosquitoes. In late 1999, a West Nile-like virus was found in New York during an outbreak of St. Louis encephalitis; in 2001, it was found in Florida.

SCIENTIFIC, TECHNOLOGICAL, AND GENERAL INFORMATION

Description of Insects

To identify insects, one must understand something about insect biology (Figure 5.1). Insects have a hard outer skin called an exoskeleton. The exoskeleton serves as a means of attachment of muscles and also protects the internal organs from injury. In most insects, the outer parts of the body wall are hardened into place. These hardened areas, or sclerites, are joined together by flexible body wall segments called intersegmental segments. The scleritis may be covered with numerous small structures, such as hairs, scales, protuberances, and spines, which are useful for identification. Insect bodies are divided into three main regions: the head, thorax, and abdomen. The class arachnid, composed of such insects as ticks, mites, scorpions, and spiders, has one or two main segments instead of three (see Figure 5.4 in later section on insect development for adult arthropods of public health significance).

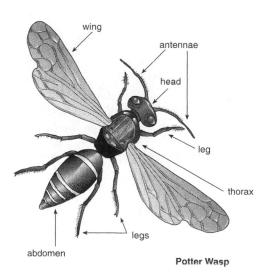

Figure 5.1   Typical insect.

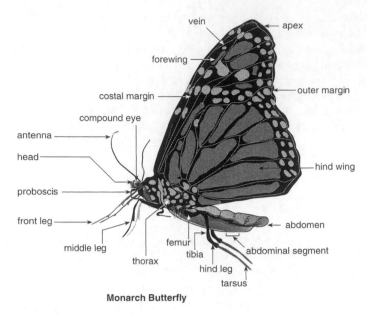

Figure 5.2   Insect: external structure.

External Structure

The following discussion of the external structure (Figure 5.2) is not meant to be a detailed description of all insects. For this type of information, it would be well to obtain a book on entomology. The first body region, or head, contains mouth parts, antennae, large compound eyes, and simple eyes (ocelli). The mouth parts are used in chewing, sponging, or piercing–sucking. Insects that have chewing mouth parts (e.g., roaches and silverfish) grind solid food. Insects that have sponging mouth parts (e.g., houseflies, blowflies, and flesh flies) suck up liquid or readily soluble foods. They eat such solids as sugar and all liquids; they regurgitate a drop of saliva to dissolve the sugar. Insects such as mosquitoes, deer flies, sucking lice, and fleas have piercing sucking mouth parts. They easily pierce the skin of animals and humans, and suck their blood. The number of pairs of antennae vary with the type of insect, as do the eyes.

The second part of the insect, the thorax, contains the three sets of legs and may contain two sets of wings attached to the last two segments of the three-part thorax. Insects are identified at times by observing the thorax. The wings, when present, contain a reinforcing structure called veins. The arrangement and number of wing veins aids in insect identification. The third body part, or abdomen, contains the spiracles, which are the external openings of the respiratory system and the external reproductive organs.

Internal Structure and Physiology

The digestive system that is part of the internal structure (Figure 5.3) of an insect consists of an alimentary canal. This canal runs from the mouth to the anus and is

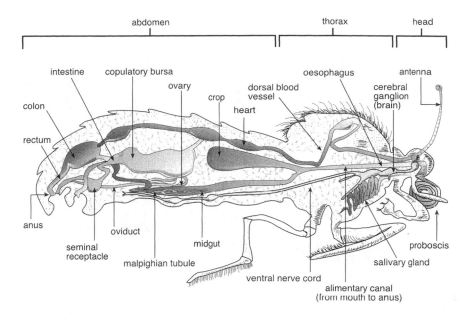

Figure 5.3   Insect: internal structure.

divided into several sections. Digestion takes place in the canal, and undigested food plus waste is removed as feces. The digestive system is an important part of the transmission of disease, because the insect can pick up an organism from one host and pass it on to another during feeding and defecation.

The circulatory system is not enclosed in blood vessels but circulates freely throughout the body cavity. In most insects, blood is colorless or greenish-yellow. Blood does not carry oxygen or expel carbon dioxide. Its major function is in the removal of waste products from body cells.

The nervous system of the insect contains a brain located in the head, a double nerve cord extending backward along the ventral surface of the body cavity, and a nerve center or ganglia. Air enters the insect through spiracles into large tracheal trunks to tracheae and tracheoles.

Most insects have two sexes that mate to produce eggs. Insects that lay eggs are oviparous, and insects that deposit larvae are larviparous. Some insects have only one sex, the female, and reproduce without fertilization. Ants and bees are examples of unisexual reproducing insects. This occurs part of the time to produce workers, which are then sexually sterile.

Insect Development

The life cycle of the insect starts with the fertilization of the egg and is completed when the adult stage is reached. Some insects go through an incomplete metamorphosis (Figure 5.4). This means that their life cycle consists of egg, nymph, and adult stages. Examples of insects with incomplete metamorphosis are roaches and body lice.

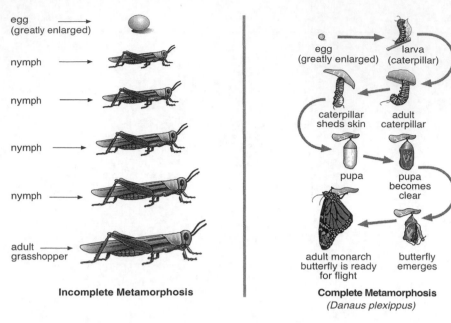

Figure 5.4   Insect life cycle: incomplete and complete metamorphosis.

Complete metamorphosis occurs in four stages, as egg, larvae, pupa, and adult. Examples of insects with complete metamorphosis are mosquitoes, flies, butterflies, and moths.

Insects molt or shed their protective exoskeleton while a new exoskeleton is being exposed. After the first day of molting, a new outer skin hardens and becomes colored.

Insect Senses and Behavior

Insects have a sense of touch, taste, smell, hearing, sight, balance, and possibly orientation. Because insect skin is hard, the sense of touch is conducted through antennae or feelers, as well as other parts of the body that are sensitive to contact and pressure. The sense of smell, located primarily in the antennae, is highly developed to aid in locating food, mates, and a place to deposit eggs. Insects vary in their degree of hearing ability. Sound waves are picked up by fine sensory hairs or special organs appearing on the side of the abdomen or the lower part of the front legs. They also hear through cuplike organs on the antennae. Visual power varies considerably among insects.

Although there is no proof that insects reason, they do have a series of instinctive, highly complex patterns that they follow. Certain stimuli cause them to behave in certain ways.

TYPES OF INSECTS OF PUBLIC HEALTH SIGNIFICANCE

The insects to be discussed include flies, fleas, lice, mites, mosquitoes, roaches, ticks, and food product insects. It is important to understand the biology, characteristics, breeding habits, and mechanisms of how each of these insects contribute to the spread of disease in order to institute adequate chemical, physical, biological, or mechanical controls (Figures 5.5 and 5.6).

Fleas

Fleas breed in large numbers where pets and livestock are housed. They spread readily through homes, buildings, and yards, attacking pets, livestock, poultry, and people. Fleas are the vectors of many important diseases, such as bubonic plague, murine typhus, tapeworms, salmonellosis, and tularemia.

Fleas obtain food and spread disease by biting humans and animals. Some individuals have a simple swelling as a result of the flea bite; others have severe generalized rashes. The species of fleas that commonly bite humans include the cat flea (*Ctenocephalides felis*), the dog flea (*C. canis*), the human flea (*Pulex irritans*), and the oriental rat flea (*Xenopsylla cheopis*).

Female fleas lay their eggs after mating on the hair of animals. The eggs drop to the floor and may become enmeshed in mats, webs, overstuffed furniture, or cellar floors. When the eggs grow to fleas, they seek a blood meal from animals or humans. In the case of rat fleas, the adult life is usually spent on the Norway or roof rat. Fleas may leave their animal host and bite humans (see Figure 5.7).

Flea Biology

Fleas are small, wingless insects that range in length from 1 to 8.5 mm. They feed through a siphon or tube after they have bitten a warm-blooded animal. The female must have a blood meal before producing eggs. Fleas cannot tolerate extremes in temperature and humidity. This is why they live close to the animals or humans that they infest. They may live on the individual's body or in the individual's shelter.

Fleas go through complete metamorphosis. Under favorable conditions, this may occur in 2 to 3 weeks. The eggs are laid in small batches over a long time period, which is interrupted by blood meals. The adults are usually able to feed within 1 day after emerging from the cocoon. The adult flea can live several weeks without food. This is the reason why people can take their dogs or cats along on vacation and return to still have a flea infestation.

Flies

Flies annoy humans; they bite, infest human flesh and the flesh of domestic animals, attack and destroy crops, and cause numerous, serious diseases. Although biting flies do not contain any type of toxin, they can kill the victim when attacking in the thousands. The victim may die of anaphylactic shock as the result of all the foreign protein injected into the body. Eye gnats do not bite, but they damage the delicate membrane of the eye.

Many different species of flies lay eggs or larvae in the flesh of humans and animals. The larvae invade the flesh and produce a condition known as myiasis.

Thousands of species of flies exist. The ones discussed in this chapter are those most common to humans and those that cause the most disease. These flies include

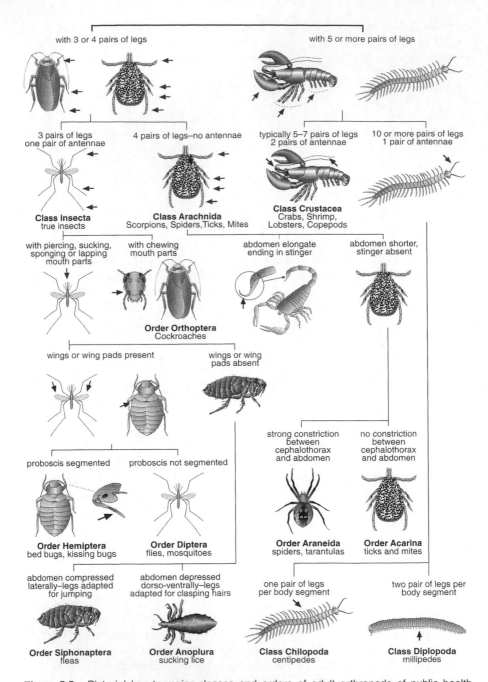

Figure 5.5   Pictorial key to major classes and orders of adult arthropods of public health importance. (From *Pictorial Keys to Some Arthropods and Mammals of Public Health Importance,* U.S. Department of Health, Education, and Welfare, Public Health Services, Washington, D.C., 1964, p. 3.)

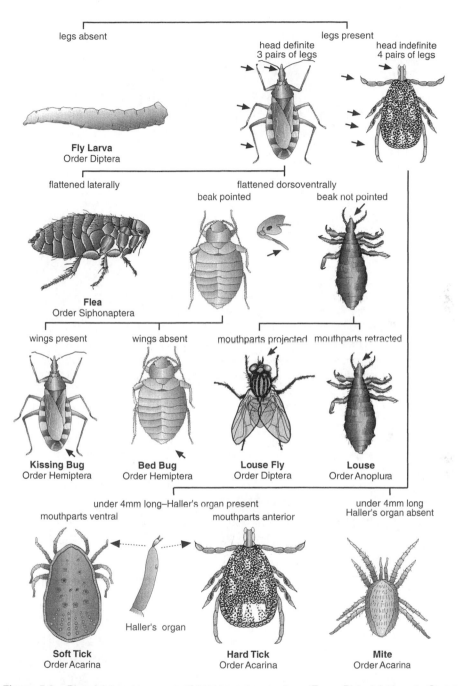

Figure 5.6 Pictorial key to groups of human ectoparasites. (From *Pictorial Keys to Some Arthropods and Mammals of Public Health Importance,* U.S. Department of Health, Education, and Welfare, Public Health Services, Washington, D.C., 1964, p. 4.)

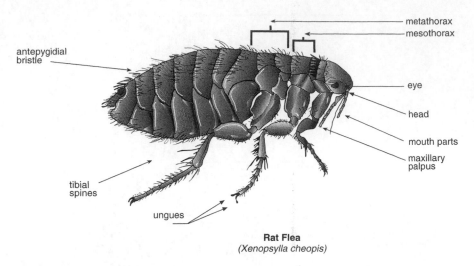

Rat Flea
(Xenopsylla cheopis)

Figure 5.7   Flea biology.

the common housefly (*Musca domestica*), the stable fly (*Stomoxya calcitrans*), the face fly (*Musca autumnalis*), the blowfly or bottle fly (*Calliphora cynomyopsis*, *C. phaenicia*, *C. lucillia*, and *C. phormica)*, the flesh flies (*Sarcophaga*), and the horse- or deerfly (*Tavanus chrysops*). The housefly is the most important vector of microorganisms causing foodborne and infectious diseases. These flies are most active during daylight hours, when they move from food source to food source to obtain a balanced diet. They utilize sugar and starches for energy and for extending their life span. They utilize proteins for the production of eggs. Flies feed about three times a day. Between feedings, they rest on floors, walls, interior surfaces, ground, grass, bushes, fences, and other surfaces. During feeding on solid food, the fly regurgitates fluids to dissolve the solids, because they can only take up liquid food. During this feeding process, they regurgitate fluids containing microorganisms that cause disease and they also defecate. Flies are inactive at night. They usually rest on upper surfaces of rooms (Figure 5.8).

Fly Biology

Flies belong to the order Diptera. This order has one or no pair of wings and halteres. Halteres are tiny knoblike structures located behind the wings and are considered to be the second pair of wings. The adult fly has three distinct body regions — the head, thorax, and abdomen. They have large compound eyes, one pair of antennae, sponging, and rasping or sucking mouth parts. The mesothorax, or middle section of the thorax, is much larger than the first and second sections. The size of the mesothorax is necessary for the powerful wing muscles used by the fly in flight.

Flies go through complete metamorphosis. The larvae feed differently and have a different habitat from the adult. The pupae are usually quiet and often are enclosed in a heavy puparium.

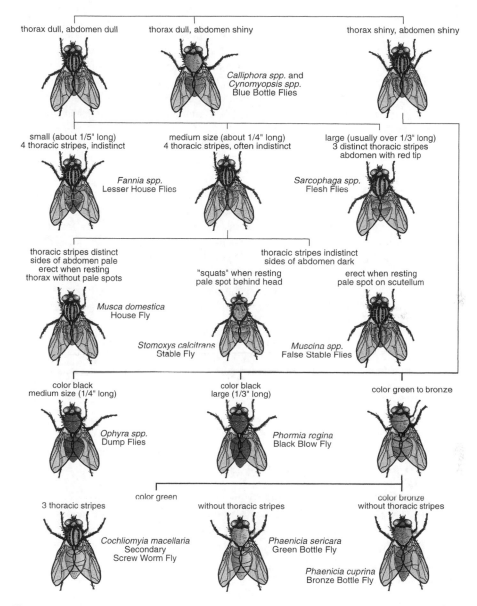

Figure 5.8 Pictorial key to common domestic flies. (From *Pictorial Keys to Some Arthropods and Mammals of Public Health Importance,* U.S. Department of Health, Education, and Welfare, Public Health Services, Washington, D.C., 1964, p. 37.)

The housefly lives in almost all parts of the world, except in the Arctic, in the Antarctic, and at extremely high altitudes. The normal life cycle runs from 8 to 20 days under average summer conditions. The female fly starts laying eggs within 4 to 20 days after it reaches adulthood. The eggs are deposited in batches of 75 to 150. The average female will lay five or six batches in cracks and crevices in the breeding medium away from direct light. The eggs hatch in about 12 to 24 hours during the summer months. The larvae grow quickly in the breeding medium. In warm weather the usual time for the larvae to develop through the three stages is 4 to 7 days. When the growth is completed, the larvae migrate from the growth medium and go into soil or under debris for the pupa stage. The pupa stage may be as short as 3 days or, at low temperatures, as long as several weeks. Following the pupa stage, the adult emerges. As an adult, the fly will mate and start the breeding process all over again. It is common to have two or more fly generations per month during warm weather. Adult flies can be kept alive for long periods of time at 50 to 60°F. This is why flies live through the winter and are ready to mate and start new fly populations during the warm months. Almost any kind of fresh, moist, organic matter is suitable for fly breeding. Flies can live in cereal, grain, animal manure, garbage, garbage-soaked soils, urine-soaked soils, or anywhere organic materials are present.

Houseflies move rapidly into new areas by flying. They travel as far as 6 mi within a 24-hour period, and possibly as far as 20 mi from their source of breeding. Flies are inactive below 45°F and are killed by temperatures slightly below 32°F. Their maximum activity occurs at 90°F. Activity declines rapidly at higher temperatures, and death occurs at 112°F. Humidity has a definite relationship to temperature. When the humidity is high and the temperature is low or very high, flies die rapidly. Above 60°F, flies live longest when the relative humidity is 42 to 55%. Flies tend to move toward light and are inactive at night. This is why the ordinary flytrap is successful. Flies are sensitive to strong air currents, and therefore are quiet on extremely windy days. During high winds, houseflies may be carried as far as 100 mi. The natural enemies of flies include fungi, bacteria, protozoa, roundworms, other arthropods, amphibians, reptiles, birds, and humans.

Face flies are barely distinguishable from houseflies in physical characteristics. The larvae develop in fresh animal feces and the pupa burrow into the soil. Adults hibernate in houses and barns. They suck blood and other exudates from the surfaces of mammals, although they cannot pierce the skin. They are particularly annoying because they are found around the eyes, nostrils, and lips.

Stable flies have a piercing proboscis, which sucks blood from humans or animals. They are vicious biters, usually found around stables and horses. Stable flies are suspected of transmitting disease, although no proof exists at this point.

Blowflies or bottle flies have varying degrees of iridescence in their basic coloring. Their eggs are deposited on carcasses of animals, on decaying animal matter, or in garbage containing animal matter. Eggs may also be deposited on fresh meat or decaying materials. On humans, the larvae of the flies can invade living tissue and cause a condition known as myiasis.

Flesh flies comprise a large number of species. They are usually light gray and have three, dark longitudinal stripes on the thorax. The abdomen has a checkered pattern. Most of the species breed in the flesh of animals or in their stools. They generally do not enter buildings, and therefore, are not very significant as vectors or nuisances.

Horseflies and deerflies are large flies that produce painful bites. They create a nuisance around swimming pools, streams, and sunny portions of damp woods.

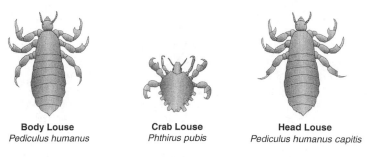

| **Body Louse** | **Crab Louse** | **Head Louse** |
| *Pediculus humanus* | *Phthirus pubis* | *Pediculus humanus capitis* |

Figure 5.9   The three basic kinds of human lice.

Lice

Lice bite severely, causing skin irritation and sometimes a generalized toxic reaction. They inject an irritating saliva into the skin during feeding. The itching for some people is extremely annoying. Lice have the capacity to act as vectors for numerous diseases. They are particularly a problem during times of war and disaster. The three basic kinds of human lice are the body louse (*Pediculus humanus*), the head louse (*Pediculus humanus capitis*), and the crab louse (*Phthirus pubis*) (see Figure 5.9). Today lice infestations occur rarely, but they still do occur. In 1974, in a school system in Indiana, a lice infestation occurred that spread throughout the school system.

Louse Biology

Lice go through an incomplete metamorphosis (see Figure 5.4). The stages are egg; first, second, and third nymph; and adult. The egg, also called a nit, attaches itself to the scalp, skin, or area around the pubic hair, as well as to underclothes. Body heat causes the eggs to hatch. The nymph then goes through its three molting stages until it becomes a sexually mature adult. The nymphal stages require 2 to 4 weeks if the clothing is removed from the body periodically and 8 to 9 days if the lice remain in contact with the body. The total life cycle may take about 18 days. Lice move quickly from person to person by means of clothing, bedding, or close contact. Infested brushes and combs are also excellent means of transmission of lice.

Mites

Mites are very small arthropods that are difficult to see with the naked eye. They do not have a distinct body segmentation. Their life cycle is very short, usually from 2 to 3 weeks. Mites increase in number very quickly under favorable conditions. They often infest food, stuffed furniture, and mattresses, and occur sometimes by the hundreds of thousands or millions. They enter premises with birds or rodents. The major problem is that they may cause several types of diseases, as well as annoyance or infestation of the body.

Mite Biology

Because mites differ so greatly from one another, it is not possible to give a general picture of them. The mite lays eggs that hatch into larvae and then pass through two or more nymphal stages to the adult stage. When trying to identify mites, use a good entomology textbook.

Mosquitoes

Mosquitoes are small, long-legged, two-winged insects belonging to the order Diptera and the family Culicidae. The adults differ from other flies by having an elongated proboscis and scales on their wing veins and wing margin. Of the 2600 species of mosquitoes, approximately 150 are found in the United States. The ones of most concern include *Anopheles, Aedes, Culex, Culiseta, Mansonia, and Psorophora*. The discussion that follows is concerned with the mosquitoes most hazardous or annoying to humans.

Female mosquitoes have piercing and blood-sucking organs; males do not suck blood. The four stages in the life cycle of the mosquito include the egg, larvae, pupa, and winged adult. The eggs are laid and hatched in quiet, standing water. The female lay eggs in batches of 50 to 200 or more, and may lay several batches of eggs. Some species glue the eggs together into a floating mass. Other species deposit eggs singly on the water, on the side of containers above the waterline, or in moist depressions. The incubation period is 3 days. It then takes 7 to 10 days for the larvae to mature. The larvae become adults in 4 to 10 days, during which time the larvae skin is shed four times. Mosquito larvae feed on minute plants and animals or fragments of organic debris.

It is difficult to determine the life span of adult mosquitoes in natural settings. However, in most of the southern species, the life span is probably only a few weeks during the summer months. In the North, the female of *Culex, Anopheles*, and other species hibernate. Mosquitoes inject saliva during feeding. The saliva causes the itchy feeling (Figure 5.10).

Mosquito Biology

Roughly equal numbers of male and female mosquitoes are produced. The male usually stays near the breeding place and mates with the female soon after reaching adulthood. Flight habits vary with the species of mosquito. Some stay within 100 yd of human habitation, and others travel as much as 10 to 20 mi. The preferred hosts, which include cattle, horses, other domestic animals, and humans, vary with the kind of mosquito. The female requires 2 days or more to digest a blood meal, lay eggs, and seek another blood meal. Mosquitoes may breed four or five times before they pass on disease to animals or humans. Because the spread of disease is intimately related to a given mosquito and its life cycle, each of the mosquitoes and the diseases that they transmit are discussed.

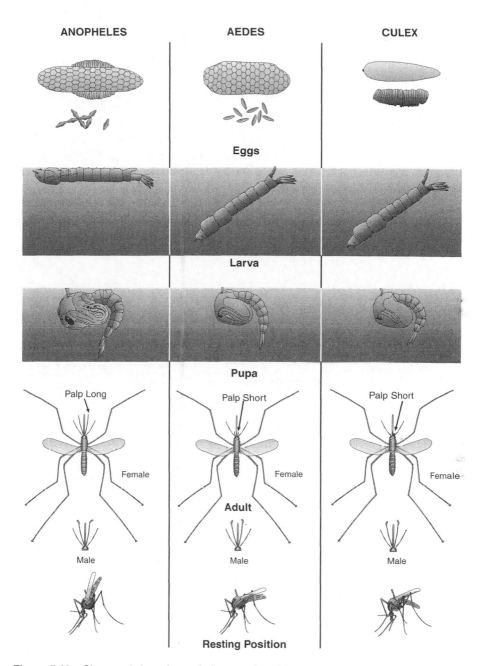

ANOPHELES **AEDES** **CULEX**

Eggs

Larva

Pupa

Palp Long Palp Short Palp Short

Female Female Female

Adult

Male Male Male

Resting Position

Figure 5.10 Characteristics of anophelines and culicines. (From *Pictorial Keys to Some Arthropods and Mammals of Public Health Importance,* U.S. Department of Health, Education, and Welfare, Public Health Services, Washington, D.C., 1964.)

Aedes

Of the 500 species of *Aedes* in the world, about 40 are common to the United States. They are of greatest concern in the northern areas. All *Aedes* lay their eggs singly, on the ground, at or above the waterline, in tree holes, or in a variety of containers, including tires. The eggs hatch after the area has been flooded. Some species can survive long dry periods. Breeding places for *Aedes* vary tremendously. They may breed in small, wet depressions left by rains or melting snows. Some breed in coastal salt marshes; others, in irrigation ditches. The *Aedes* are vicious biters and bloodsuckers. Some attack during the evening hours, some during the day, and some will bite at any time.

Aedes aegypti is a vector for yellow fever. It is found in the southeastern and southern United States. It is a semidomesticated mosquito that lives very well in artificial containers in and around human habitation. Mosquitoes may breed in anything that will hold water. The life cycle can vary from 10 days to 3 weeks or more. *Aedes aegypti* insects are susceptible to cold and cannot survive the winter unless they are in the southern United States. The adults prefer humans to other animals, therefore spreading disease from human to human. They usually bite around the ankles, under coat sleeves, or on the back of the neck. The adult lives about 4 months or more. Their flight range is normally from about 100 ft to 100 yd.

Anopheles

Anopheles mosquitoes are found throughout the United States. Most of the anophelines have spotted wings, whereas the culicines have clear wings. Anophelines rest in a position where their head, thorax, and abdomen are in a straight line, usually at an angle of 40 to 90°, whereas the culicines rest in a position almost parallel to the surface.

Anopheline eggs are laid singly on the water surface and float there. They are usually laid in batches of 100 or more, and hatch within 1 to 3 days.

The larvae are found either in fresh or brackish water. The larval stage takes from 4 to 5 days to several weeks, depending on the species of mosquito and the environmental conditions, especially the water temperature. The larvae eat microscopic plant life. Most adults are active only at night. They spend the daytime resting in dark, damp shelters. Their peak of activity occurs just after dark and just before daylight. Flight range is from less than 1 mi to several miles.

Anopheles quadrimaculatus is the most common species of *Anopheles* that causes malaria. It is frequently found in houses and is likely to attack humans. The mosquito is distributed throughout the southeastern United States and into the Midwest and Canada. It breeds primarily in permanent freshwater pools, ponds, and swamps that have aquatic vegetation or floating debris. It is most frequently found in shallow waters, in sunlit or densely shaded areas. The larvae can withstand temperatures below 50 to 55°F, although development will not be completed. Development starts to progress at temperatures of 65 to 70°F and is most active at temperatures of 85 to 90°F. It takes 8 to 14 days at the 85 to 90°F range for the larvae to complete their cycle. The female, after mating, lays eggs for 2 to 3 days after the first blood meal. A single mosquito may lay as many as 12 batches, totaling over 3000 eggs. The adults are inactive and rest during the daytime in cool, damp, dark shelters, such as buildings, caves, and under bridges. Feeding and other activities occur almost completely at night. They feed readily in houses on humans and on other warm-blooded animals. Usually the adults do not fly more than one-half mile from the breeding place.

Anopheles albimanus are found in southern Texas and the Florida Keys. The larvae live in freshwater ponds or brackish water. The adults enter dwellings during the night and bite people, and leave at dawn for forested areas. They have not been known to cause malaria in the United States.

Anopheles freeborni is the most important vector of malaria in the western United States. These mosquitoes enter homes and animal shelters and bite avidly at dusk and dawn. They breed in any permanent or semipermanent water that is at least partially exposed to sunlight. The larvae are also found in slightly brackish water. This species is particularly adapted to rice fields and sunny arid regions. The mosquitoes leave their hibernating places in February, obtain blood meals, and lay their eggs. As the season progresses and the area gets dryer, the mosquitoes travel longer distances, sometimes 10 to 12 mi to seek shelter. During the winter, they are in semihibernation.

Culex

The *Culex* include about 300 species, of which 26 are known to inhabit the United States. Only about 12 of these species are common or pests or vectors of disease. The *Culex* mosquitoes breed in quiet waters, artificial containers, or large bodies of permanent water, and often breed in areas where there are large quantities of sewage. The eggs are deposited in rafts of 100 or more. They float on the water surface until they hatch in 2 to 3 days. The adult females are usually inactive during the day, but bite during the night.

Roaches

Most of the 55 species of roaches known in the United States live outdoors and are not a problem to humans. Only five species are usually found indoors in the United States. The roaches mechanically carry dirt and organisms that may spread disease, and they destroy fabrics and book bindings. Their odors, which are quite offensive, ruin food. Roaches are broad, flattened, dark or light brown or black insects. They usually run at night to seek food and hide out during the daytime. When roaches are seen during the day, they are generally present in very large numbers. The females lay eggs in capsules called oothecae.

Roaches have been around for about 400 million years. Their fossil remains, which are similar to today's roaches, are abundant in the strata of certain types of soils. Roaches are present everywhere. It is important that moisture be present where they exist. They feed on cereals, baked goods, small amounts of grease, glue, starch, wallpaper binding, fecal materials, dead animals, etc. During the feeding process, they regurgitate a brown liquid that may contaminate the food. Roaches commonly rest in crevices, behind moldings of doors and window frames, in areas where pieces of equipment are joined together, under food preparation tables, in cabinets and closets, in the asbestos covering of hot water lines, in cardboard boxes, in and under debris, and in almost any other place where a tiny crack may exist. They are also found in the backs of radios, television sets, paper bags from the supermarket, and pockets or cuffs of pants or other places in clothing (Figure 5.11).

Roach Biology

The roach passes through incomplete metamorphosis. The stages include the egg, nymph, and adult. Roaches have an oval, flattened shape with the head hidden. They feed on almost anything. The eggs are enclosed in capsules called oothecae that are either deposited or carried at the end of the abdomen.

American Roach

The American roach, *Periplaneta americana*, is widely distributed throughout the world in mild climates, although it is a native of tropical and subtropical climates. The adult is reddish brown to dark brown. It usually forages for food on the first floor of buildings. Although it does not normally fly, it has the ability to do so. The female carries the egg capsule for a day or two, and then glues it to some object in a protected area. In 2 or 3 months, about 12 nymphs will hatch from each capsule. Adult females live for at least a year. During this period of time, they lay many capsules. The nymph, which is the same color as the roach, passes through 13 molts as it matures during the course of a year. It can grow very slowly where conditions are unfavorable. The length of the roach ranges from 1½ to 2 in.

Brown-Banded Roach

The brown-banded roach, *Supelle supellectilium*, is found throughout the world, although its main habitat is the tropical areas. The adult is about ½ in. in length, is light brown, and has mottled wings. The female is reddish brown and lighter. The eggs are laid and hatched in the same way as the American roach. The color of the nymph is the same as the adult. This roach matures in 4 to 6 months and lives in almost any part of buildings.

German Roach

The German roach, *Blattella germanica*, has a very wide range. It is most commonly found indoors and is the roach that readily becomes resistant to insecticides. The adult is light brown with longitudinal stripes on the back. The female carries oothecae for about a month and then drops it a day or so before the eggs are ready to hatch. Each capsule has about 30 nymphs. The adult is about 6/10 in. in size. The color of the nymph is the same as the adult. It undergoes six molts and matures in about 4 to 6 months, living in kitchens, bathrooms, and other areas.

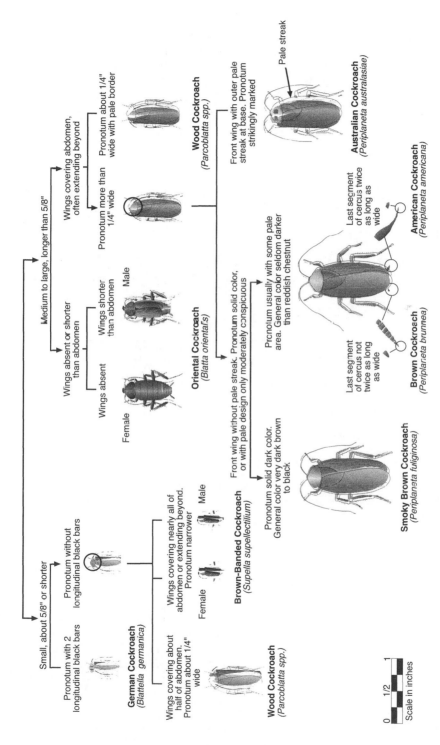

Figure 5.11 Pictorial key to some common adult cockroaches. (From *Pictorial Keys to Some Arthropods and Mammals of Public Health Importance*, U.S. Department of Health, Education, and Welfare, Public Health Services, Washington, D.C., 1964, p. 18.)

Oriental Roach

The oriental roach, *Blatta orientalis*, is spread throughout the world. The adult is black to dark brown, with vestigial wings in the female. It is quite sluggish. It usually forages in the basement or first floor of buildings. The eggs are laid in an egg capsule, which is glued to some object in a protected place. About 12 nymphs hatch from each capsule in about 2 to 3 months. The roach is 1 to 1¼ in. in length, with the female a little longer than the male. The color of the nymph is the same as the adult. It matures in about a year.

Ticks

Ticks belong to the class Arachnida. They are not true insects, but are important vectors of significant arthropodborne disease (Figure 5.12).

Tick Biology

Ticks belong to the same class as mites, spiders, and scorpions. They have three characteristics that distinguish them from insects: the head, thorax, and abdomen are

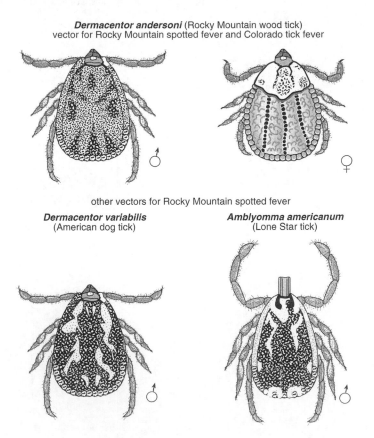

Figure 5.12   Tick vectors for Rocky Mountain spotted fever and Colorado tick fever.

all fused together in one body region; no antennae; and four pairs of legs in the nymph and adult stages instead of three. Ticks are divided into two groups, soft and hard.

The tick has a piercing organ that has barbs, allowing the tick to anchor itself to the host. It also has cutting organs to allow the piercing organ to enter the skin for a blood meal. The leathery body covering is capable of great distension with blood. The female tick is somewhat flattened and tapered toward the anterior end before it takes its blood meal. Ticks are adaptable for blood sucking because they have a powerful, pumping pharynx. The blood is pumped backward into the esophagus and then into the stomach.

Ticks metamorphose completely. The female hard tick feeds once and lays large batches of up to 18,000 or more eggs. Most of the soft ticks take several blood meals and then deposit 20 to 50 eggs in a batch. The eggs hatch in 2 weeks to several months, depending on the environmental conditions. The larvae have difficulty attaching themselves to a host, and therefore may go through long periods of starvation. After a blood meal, the engorged larvae usually drop to the soil, shed their skin, and enter the nymph stage. The nymph may wait for long periods for a suitable host. After it has its blood meal, the nymph drops from the host, molts, and becomes the adult. The life cycle may take from less than a year to 2 to 3 years or longer. Ticks survive if they can get another blood meal. During this period the sexual organs are formed, and both the female and male tick become bloodsuckers and feed several times before copulation. The male hard tick will have a blood meal, usually copulate with one or more females, and then dies. The female drops to the ground and starts to develop eggs over a period of several days. Once the eggs are deposited, the hard tick female dies. Unfed hard ticks can survive frigid winter weather if they are sheltered. Engorged ticks and eggs are less resistant. Ticks are most active during the spring and summer. They usually die quickly if the area is dry. Soft ticks can exist in much dryer situations.

Food Product Insects

Stored food product insects become a serious problem in the kitchen, warehouse, pantry, or any other place where dried foods are stored. These pests eat or contaminate the food, making it unfit for human consumption. They are also annoying, because they leave the infested food and crawl or fly around the area. These food product insects include the confused flour beetle, sawtooth grain beetle, flour grain beetle, larder beetle, black carpet beetle, cabinet beetle, grain weevil, rice weevil, Indian meal moth, Mediterranean flour moth, and bean and pea weevils. These insects may forage on dry foods, furs, skins, wool, silk, cheese, and almost any other product made from dried meats, cheese, or other organic materials.

ENVIRONMENTAL AND HEALTH PROBLEMS CREATED BY INSECTS

Destruction of Food and Material

Insects infest cattle, causing weight loss, annoyance, and eventually a reduction in the supply of acceptable food. Further, they infest stored foods and contaminate them with feces, body parts, or live insects. Common stored food product insects include grain and flour beetles; *Dermes pidae* beetles, which feed on a variety of plant and animal products, including birds, skins, live meat products, woolen and silk materials, cheese, and cereal grain products; grain weevils, which feed on stored whole grain; and bean and pea weevils, which infest dry beans and dry peas.

Transmission of Disease

Disease is transmitted from human to human, from human through vector to human, from human through vector to animal to human, or from human through vector through animal through vector to human. It is very important to understand these various types of disease transmission, because once the life cycle of the disease organism is known, it becomes much simpler to break the chain of infection and to stop the outbreak of the vectorborne disease.

Disease is transmitted mechanically; for example, flies pick up microorganisms of dysentery, typhoid fever, or cholera on their feet or body hair and transmit them to humans or their food; or biologically through various parts of the life cycle of the arthropod. In developmental biological transmission, the parasite grows or multiplies in the insect. For example, the viruses causing yellow fever or encephalitis increase in number within the body of the mosquito. In cyclic biological transmission, the parasite goes through an essential part of its life cycle in the arthropod, without increasing in number. For example, in filariasis, the mosquito sucks up a number of microfilaria into its gut and eventually these organisms enter the thoracic muscles, where they pass through several stages of their life cycle and then move to the proboscis of the mosquito. When the mosquito feeds, it infects the host. In cyclodevelopmental transmission, both an essential cycle occurs in the arthropod host and the parasites increase in number. An example of this would be the parasite *Plasmodium vivax*, which causes malaria.

In some cases, humans are accidentally infected with a disease that ordinarily would be an animal disease. This is called a zoonoses. An example of this is jungle yellow fever, where jungle mammals such as monkeys are normal hosts and wild mosquitoes transmit the disease from monkey to monkey. From time to time, the insect will bite a human, who then develops the disease. Another example of this is viral encephalitis transmitted by mosquitoes, such as St. Louis encephalitis, where the normal hosts would be wild birds.

Numerous tick- and miteborne diseases are spread by eggs from the infected parents to the next generation. This is congenital transmission of the pathogen. The eggs are laid either on humans or animals, and when they hatch and become insects they transmit the disease to the host. Examples of this would be tickborne typhus, tickborne encephalitis, tularemia, relapsing fever, and scrub typhus.

People can also be poisoned by the bites and stings of insects. The victim may suffer an insignificant blister, anaphylactic shock, or death. Anaphylactic shock can result from an individual reaction to the protein or a large amount of protein injected into the body by the arthropods. Some insects, such as the black widow spider, cause death by their extremely poisonous toxin.

Table 5.1   Some Arthropods Affecting Human Comfort

Mites		
Chigger	*Trombicula alfreddugesi*	Intense itching, dermatitis
Rat mite	*Ornithonyssus bacoti*	Intense itching, dermatitis
Grain itch mite	*Pyemotes ventricosus*	Dermatitis and fever
Scabies mite	*Sarcoptes scabiei*	Burrows in skin, causing dermatitis
Ticks		
Hard ticks	*Amblyomma* spp.	Painful bite
	Dermacentor variabilis, *D. andersoni* and *D. ixodes* spp.	Tick paralysis, usually fatal if ticks not removed
Soft ticks	*Ornithodoros* spp.	Some species very venomous
Spiders		
Black widow spider	*Latrodectus mactans*	Local swelling, intense pain and occasionally death
Brown spider	*Loxosceles reclusa*	Extensive loss of affected body tissue
Scorpions	Order: Scorpionda	Painful sting, sometimes death
Centipedes	Class: Chilopoda	Painful bite
Mayflies	Order: Ephemerida	Asthmatic symptoms from inhaling fragments
Caddis flies	Order: Trichopera	Asthmatic symptoms from inhaling hairs and scales
Lice	*Pediculus humanus* and *Phthirus pubis*	Intense irritation, reddish papules
Bugs		
Bed bugs	*Cimex lectularius*	Blood suckers, irritating to some
Kissing bugs	Family: Reduviidae	Painful bite, local inflammation
Beetles		
Blister beetles	Family: Meloidae	Severe blisters on skin from crushed beetles
Rove beetles	Family: Staphylinidae	Delayed blistering effect
Caterpillars	Order: Lepidoptera	Rash on contact with hairs or spines
Bees, wasps, and ants	Order: Hymenoptera	Painful sting, local swelling
Fleas	Order: Siphonaptera	Marked dermatitis frequent
Flies		
Punkies or biting midges	*Culicoides* spp.	Nodular swelling, inflamed
Blackflies	*Simulium* spp.	Bleeding punctures, pain, swelling
Sandflies	*Phlebotomus* spp.	Stinging bite, itching, whitish weal
Mosquitoes	Family: Culicidae	Swelling, itching
Horseflies	*Tabanus* spp.	Painful bite
Deerflies	*Chrysops* spp.	Painful bite
Stable flies	*Stomoxys calcitrans*	Painful bite

From *Introduction to Arthropods of Public Health Importance,* H.D. Pratt and K.S. Littig, DHEW Publication No. (HSM) 72-8139, 1972.

An infestation occurs when living insects burrow into the skin of the host. Lice or scabies burrow under the skin causing infestation. Table 5.1 lists some of the arthropods that affect human comfort.

Insectborne Diseases

African Tick-Bite Fever

African tick-bite fever is endemic in Africa and is transmitted by an infected *Amblyomma* tick carrying *Rickettsia africae*. This disease causes acute onset of fever, fatigue, chills, sweats, headache, myalgia, and arthralgias, which lasts about 4 days from onset. The erythematous annular skin lesions with dark centers may last for 2 months even with treatment. The number of U.S. residents traveling to Africa from 1986 to 1996 had increased by 70%. An estimated 19 million U.S. residents traveled overseas in 1996, including approximately 455,000 people going to Africa. The number of vectorborne diseases will continue to increase.

Bubonic Plague

Bubonic plague is a highly fatal infectious disease, with toxemia, high fever, reduced blood pressure, rapid and irregular pulse, mental confusion, and prostration. Bubonic plague can become pneumonic plague when the disease spreads to the lungs. The incubation period in bubonic plague is 2 to 6 days. The incubation period in pneumonic plague is 3 to 4 days. In 1994 an extremely serious outbreak of pneumonic plague occurred in India. The etiologic agent is *Yersinia pestis*. The insect involved is the oriental rat flea, *Xenopsylla cheopis*. The reservoir of infection is wild or domesticated rats, and possibly other rodents.

In September 1994, there was a reported epidemic of plague in India. This caused the CDC to enhance surveillance in the United States for imported pneumonic plague. From September 27 to October 27, the surveillance system identified 13 persons with suspected plague; however, no cases were confirmed.

In 1996, five cases of human plague, two of which were fatal, were reported in the United States. *Yersinia pestis* antigens were detected in the blood of dogs that were pets in the home. These dogs lived near prairie dog colonies that were infected. In the United States, most cases of human plague are reported from New Mexico, Arizona, Colorado, and California. From 1947 through 1966, a total of 390 cases of plague, resulting in 60 deaths, were reported.

Chagas Disease

Chagas disease, or American trypanosomiasis, is an infection caused by the parasite *Trypanosoma cruzi*. It is estimated that 16 to 18 million people are infected with the disease and that 50,000 will die each year. There are three stages of infection with Chagas disease, each one having different symptoms. In the acute phase that affects about 1% of the cases, characteristically a swelling of the eye on one side of the face occurs, usually at the bite wound or where feces were rubbed into the eye. Other symptoms are usually not specific for Chagas infection. These symptoms may include fatigue, fever, enlarged liver or spleen, swollen lymph glands, rash, loss of appetite, diarrhea, and vomiting. In infants and in very young children with acute Chagas disease, swelling of the brain can develop and cause death. In general,

symptoms last 4 to 8 weeks and then go away, even without treatment. The inde-
terminate stage begins 8 to 10 weeks after the infection, during which people do
not have symptoms. As many as 10 to 20 years after the infection, people may
develop the most serious symptoms of Chagas disease, which include: cardiac
problems, enlarged heart, altered heart rate or rhythm, heart failure, or cardiac arrest;
and enlargement of parts of the digestive tract, which results in severe constipation
or problems with swallowing.

In persons who are immunocompromised, Chagas disease can be very severe.
Not everyone develops the chronic symptoms. Although symptoms may occur within
a few days to weeks, most people do not have them until the chronic stage of the
infection. Chagas disease is carried by the Reduviid bug or kissing bug, which lives
in cracks and holes of substandard housing found in South and Central America.
The insects become infected after biting an animal or person who already has the
disease. The infection is spread to humans when an infected bug deposits feces on
a person's skin, when the person is sleeping at night. The person often accidentally
rubs the feces into the bite wound, an open cut, the eyes, or the mouth. The organisms
may also be transmitted from infected mothers to babies during pregnancy, at
delivery, or while breastfeeding. Transmission may also occur during blood transfusion and organ transplants. Eating uncooked food contaminated with infective
feces can cause the disease. Travelers are at risk.

Colorado Tick Fever

Colorado tick fever is an acute fever lasting 2 to 3 days, with occasional enceph-
alitis occurring. The incubation period is 4 to 5 days. The etiologic agent is a virus.
The arthropod involved is an infected tick, most frequently *Dermacentor andersoni*.
The reservoir of infection is small mammals.

Dengue Fever

Dengue fever causes sudden onset of fever, intense headache, joint and muscle
pain, rash, and prolonged fatigue, as well as depression. The incubation period is
typically 3 to 4 days, but most often 4 to 7 days. The etiologic agent is a virus of
the group B togaviruses. Various *Aedes*, including *Aedes aegypti* and *A. albopictus*
are responsible for causing this disease. The reservoir of infection is people, mosquitoes, and monkeys.

Dengue and dengue hemorrhagic fever have become a global health problem
including in the United States. These diseases are caused by one of four closely
related, but antigenically distinct virus serotypes (DEN-1, DEN-2, DEN-3, and
DEN-4), of the genus *Flavivirus*. Infection with one of the serotypes does not provide
cross-protective immunity. People living in the endemic areas can have all four
dengue infections during their lifetimes.

Dengue infections are primarily an urban disease of the tropics, and the viruses
that cause them are in a cycle that involves humans and *A. aegypti*, a domestic, day-
biting mosquito that prefers to feed on humans, and can be found in Asia, Africa,
and North America. After 35 years, epidemic dengue fever occurred in both Taiwan

and the People's Republic of China in the 1980s. In the Pacific, the viruses were reintroduced in the early 1970s after an absence of more than 25 years. In Africa, the viruses have increased dramatically since 1980. The *A. aegypti* eradication program, which was officially discontinued in the United States in 1970, eroded programs elsewhere; and the species began to reinfest countries from which it had been eradicated. With the increase in the mosquitoes found, there has been an increase in dengue fever. *Aedes* have been found in parts of Alabama, Arkansas, Florida, Georgia, Louisiana, Mississippi, North Carolina, South Carolina, Tennessee, and Texas. In 1995, the CDC reported 86 imported laboratory-diagnosed cases of dengue in the United States.

Ehrlichioses

Ehrlichiae are small, Gram-negative, obligately intracellular bacteria that reside within a phagosome. The first human ehrlichial infection was recognized in the United States in 1987. It was caused by a new species, *Ehrlichia chaffeensis*. In 1994, an ehrlichial pathogen was found within neutrophils, which would infect humans. It was closely related to the known veterinary pathogens *E. equi* and *E. phagocytophila*. Approximately 170 cases had been diagnosed, from 1994 to 1996, predominantly in the upper Midwest and northeastern states as well as in northern California. The disease caused by the bacteria ranges from subclinical to fatal. The median incubation period is 8 days. The clinical manifestations include fever, chills, malaise, myalgia, headaches, nausea, vomiting, cough, confusion, and rarely a rash. The median duration of illness including that for treated patients is 23 days. Involvement of the central nervous system, liver, respiratory system, renal system, and cardiovascular system may occur. Some patients have a life-threatening illness resembling toxic shock syndrome. Deaths have occurred in approximately 2 to 3% of patients including previously healthy children. The median age of patients is 44 years, and three quarters are male. The groups at greatest risk are golfers, campers, and others participating in the outdoor or rural environment, especially during the May to July period. The vector for the disease is the Lone Star tick and the American dog tick. The vertebrate host is a human or a deer for genogroup 1, and horses, cattle, sheep, or deer for genogroup 2. The organism may possibly be transmitted by a small rodent.

Encephalitis

Arthropodborne viral encephalitis includes eastern equine, western equine, California, St. Louis, and West Nile encephalitis. The symptoms of encephalitis include a short period inflammation, which involves parts of the brain, spinal cord, and meninges. Symptoms also include headaches, high fever, stupor, and disorientation. Fatality rates range from 5 to 60%. The incubation period is 5 to 15 days. The etiologic agents vary with the type of the disease. Eastern and western equine encephalitis are caused by a virus in group A togaviruses, whereas St. Louis encephalitis is caused by a virus in the group B togaviruses. Various mosquitoes, including *Culex tarsalis, Culiseta melanura,* and other mosquitoes, transmit the disease. The

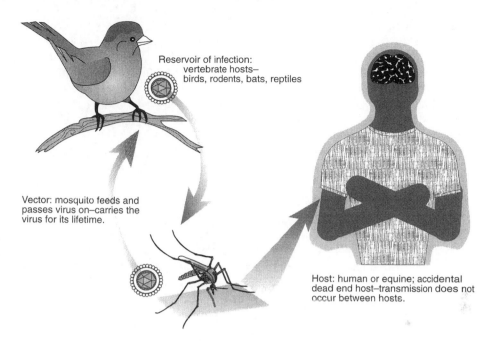

Reservoir of infection:
vertebrate hosts–
birds, rodents, bats, reptiles

Vector: mosquito feeds and
passes virus on–carries the
virus for its lifetime.

Host: human or equine; accidental
dead end host–transmission does not
occur between hosts.

Figure 5.13 Transmission of arboviral encephalitis.

reservoir of infection is suspected to be birds, rodents, bats, and reptiles. Surviving mosquito eggs or adults may keep the virus alive over the winter.

At risk groups for eastern equine and western equine encephalitis are residents of endemic areas, visitors, and persons who work outdoors or are involved in recreational activities. At risk groups for La Crosse virus–California encephalitis are children <16 years old, residents of woodland habitats, and those involved in outdoor activities. At risk groups for St. Louis encephalitis are all residents of areas where active cases have been identified. At risk groups for West Nile encephalitis are all residents of areas where active cases have been identified, but especially in those individuals >50 years of age. In October 1999, scientists from CDC investigating the outbreak of West Nile-like virus in New York thought that the virus had been present in the eastern United States for several months, possibly longer (Figure 5.13).

Lyme Disease

The early symptoms of Lyme disease include a rash that starts out flat or raised and may develop into blistering or scabbing in the center. The lesions may have a bluish discoloration. Other symptoms include fatigue, headache, neck stiffness, jaw discomfort, pain or stiffness in muscles or joints, slight fever, swelling of glands, or conjunctivitis. If the symptoms of the disease are untreated, later symptoms occur, which include complications of the heart, nervous system, or joints. Relapse and incomplete treatment responses occur. Complications of untreated early-stage disease include: 40 to 60% joint disease; 15 to 20% neurological disease; 8% carditis; and 10% or more are hospitalized, some with chronic debilitating conditions. The incubation period ranges from 3 to 32 days after a tick bite. Typically a red ringlike lesion develops at the site of the tick bite. The etiologic agent, which is a spirochete, is called *Borrelia burgdorferi*. A tiny tick found in northeastern United States, and as far south as Virginia, called *Ixodes dammini,* spreads the disease. In the South and Southwest, *Ixodes scaturlaria* spreads the disease. On the West Coast, *I. pacificus* spreads the disease. Dogs, cats, horses, cows, and goats are domestic animals that may carry the ticks and the disease. Also, wild animals may carry the ticks and the disease. These are the reservoirs of infection.

Lyme disease is the leading cause of vectorborne infectious illness in the United States with about 15,000 cases reported each year, although the disease is greatly underreported. Over 125,000 cases have been reported since 1982. Of these cases 90% were reported in the states of New York, Connecticut, Pennsylvania, New Jersey, Wisconsin, Rhode Island, Maryland, Massachusetts, Minnesota, and Delaware. The persons at risk are those in endemic areas who frequent sites where infected ticks are common, such as grassy or wooded locations favored by the white-tailed deer in the northeastern and upper Midwest states, and along the northern Pacific coast of California.

Malaria

The symptoms of malaria include fever, chills, sweating, headache, shock, renal failure, acute encephalitis, and coma. In nontreated cases, a fatality rate is greater than 10%. The incubation period varies with the type of plasmodium causing the disease. *Plasmodium falciparum* has an incubation period of 12 days. *Plasmodium vivax* has an incubation period of 14 days. *Plasmodium ovale* has an incubation period of 14 days. *Plasmodium malaria* has an incubation period of 30 days. In certain strains of *P. vivax*, the incubation period is 8 to 10 months. The etiologic agents have just been described. The *Anopheles* mosquito transmits the disease. The reservoir of disease is people.

Malaria is transmitted in large areas of Central and South America, the Indian subcontinent, the Middle East, Southeast Asia, and elsewhere. Worldwide it is estimated that 300 to 500 million clinical cases and 1.5 to 2.7 million deaths occur each year due to malaria. Although the disease was eradicated in the United States in the 1940s, about 1000 to 1400 cases of malaria are reported to CDC each year, with almost all of them acquired during international travel. Each year in the United States several cases, less than ten, are acquired stateside by congenital transmission, local mosquitoborne transmission, blood transfusion, or organ transplant. Today, environmental changes, spread of drug resistance, and increased air travel could lead to the reemergence of malaria as a serious public health problem. Anopheline mosquitoes are found in all 48 states of the contiguous United States; and they are capable of transmitting the malaria parasites, which are protozoa of the genus *Plasmodium*. The immature stages of the mosquito's life cycle (egg, larva, and pupa) are aquatic and develop in breeding sites, whereas the adult stage is terrestrial. Over the past years, mosquito control programs have been reduced, thereby creating a greater potential for more mosquitoes and possible malaria. In the 1990s, mosquitoborne malaria was found in California, New Jersey, Florida, and Houston Texas (Figure 5.14).

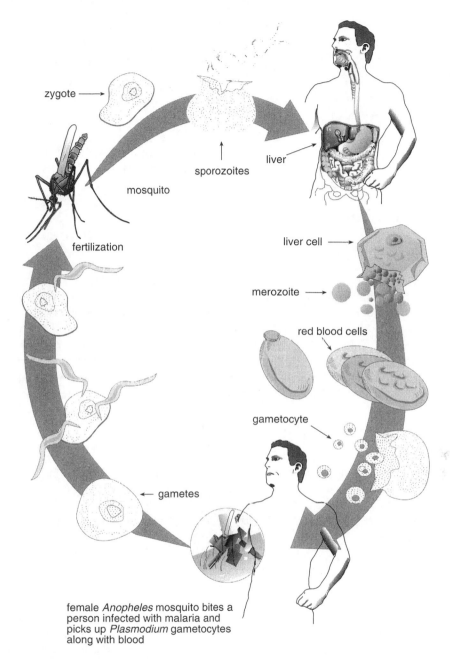

Figure 5.14 Mosquito (*Anopheles*) as a vector of infection (malaria).

Relapsing Fever

The symptoms of relapsing fever include periods of normal temperature and then elevated temperature. In untreated cases, the fatality rate is usually 2 to 10%, although it may exceed 50% in the epidemic louseborne disease form. The incubation period is 5 to 15 days but usually 8 days. The etiologic agent is *Borrelia recurrentis*. The disease is transmitted through either lice or ticks. The reservoir of infection for the louseborne disease is people. The reservoir of infection for the tickborne disease is wild rodents or through transovarian transmission.

Rickettsial Pox

The symptoms of the disease include skin lesions, chills, fever, rash, headache, muscular pain, and malaise. The incubation period is probably 10 to 24 days. The etiologic agent is *Rickettsia akari*. The rodent mite, *Liponyssoides sanguineus*, is the insect that spreads the disease through its bite. The reservoir of infection is the infected house mouse.

Rocky Mountain Spotted Fever

Rocky Mountain spotted fever is also known as tick fever or tickborne typhus fever. The symptoms of the disease include a sudden onset of a persistent fever, headache, and chills. A rash appears on extremities and spreads to most of the body. There is a 20% fatality rate in untreated cases. The incubation period is 3 to 10 days. The etiologic agent is *R. rickettsii*. The disease is spread by the bite of an infected tick or by contamination of the skin with the crushed tissues or feces of the tick. The reservoir of infection is the tick, which may transmit the disease through either transovarian or transstadial passage within the tick. The tick may also be transferred from animals to people.

Scabies

The symptom of the disease is an intense itching, especially at night. The incubation period ranges from several days to several weeks. The etiologic agent is *Sarcoptes scabiei*. The insect involved is the mite. The means of spread is through direct contact or by indirect contact by means of underclothing. The reservoir of infection is people.

Scrub Typhus

The symptoms are skin ulcers followed by fever, headache, profuse sweating, eruptions on various parts of the body, cough, and pneumonitis. The fatality rate ranges from 1 to 40%. The incubation period is from 6 to 21 days, but usually 10 to 12 days. The etiologic agent is *R. tsutsuganushi*. A mite causes the disease. The reservoir of infection is the infected larval stages of the mite.

Tularemia

Tularemia, which is also known as deerfly fever or rabbit fever, causes ulcers to form at the site of the inoculation by the insect. Headache, chills, and rapid rise of temperature occur. The fatality rate in untreated cases is about 5%. The incubation period is 1 to 10 days, but usually 3 days. The etiologic agent is *Francisella tularensis* (*Pasteurella tularensis*). Although the disease may be transmitted by handling or ingesting insufficiently cooked rabbit or hare meat, or by the drinking of contaminated water or inhalation of dust from contaminated soil, it is also spread by a variety of insects. They include the deerfly; mosquito, *Aedes*; wood tick, *D. andersoni*; dog tick, *D. variabilis*; and Lone Star tick, *Amblyomma americanum*. The reservoir of infection includes many wild animals, especially rabbits and beavers.

Typhus Fever — Epidemic (Louseborne)

Epidemic typhus fever causes symptoms of headaches, chills, prostration, fever, pain, and toxemia. In untreated cases, the rate of fatality varies from 10 to 40%. The incubation period is from 1 to 2 weeks, but usually 10 days. The etiologic agent is *R. prowazeki*. The most common insect spreading the disease is the body louse, *Pediculus humanus*. The reservoir of infection is people.

Typhus Fever — Fleaborne

Fleaborne typhus fever, also known as murine typhus or epidemic typhus fever, causes the same symptoms as the louseborne epidemic typhus fever but they are milder. If the individual is untreated, the death rate is about 2%. The incubation period is 1 to 2 weeks but usually 12 days. The etiologic agent is *R. typhi*. The insect involved is the flea, primarily the rat flea, which is *Xenopsylla cheopis*. Rats are the reservoir of infection, but cats and other wild or domestic animals may spread the disease.

Viral Hemorrhagic Fevers

The viral hemorrhagic fevers are a group of illnesses caused by several distinct families of viruses, some of which cause relatively mild illnesses whereas others cause severe, life-threatening disease. The viral hemorrhagic fevers are caused by viruses of four distinct families: arenaviruses, filoviruses, bunyaviruses, and flaviviruses. Each of these families share a number of features:

1. They are all ribonucleic acid (RNA) viruses, and all are covered in a lipid coating.
2. Their survival is dependent on the natural reservoir, which is an animal or an insect host.
3. The viruses are geographically restricted to the areas where their host species live; however, because some of the animals range over continents, such as the rodents that carry viruses which cause hantavirus pulmonary syndrome, the diseases can be widespread.
4. Humans are not the natural reservoir for any these viruses; however, the humans can become infected when they come into contact with the infected host. In some cases, the humans can then transmit the virus to others.
5. Human cases or outbreaks of hemorrhagic fevers caused by these viruses are sporadic and not easily predictable.

In rare cases other viral and bacterial infections can cause a hemorrhagic fever, such as scrub typhus. The viruses associated with most viral hemorrhagic fevers are zoonotic. These viruses naturally reside in an animal reservoir host or arthropod vector. The animals include the multimammate rat, cotton rat, deer mouse, house mouse, and other field rodents. Vectors include ticks and mosquitoes. Some of the hosts, especially for the Ebola and Marburg viruses, are unknown.

The viruses are initially transmitted to humans when they interact with the infected hosts or vectors. The viruses carried in rodent reservoirs are transmitted when humans come into contact with urine, fecal matter, saliva, or other body secretions from infected rodents. The viruses carried by the arthropod vectors are spread usually when the mosquito or tick bites a human or when the tick is crushed. With some viruses, the hemorrhagic fever can spread from one person to another directly through close contact with infected people or their body fluids. This is a secondary transmission.

The symptoms vary by the type of disease; however, initially they often include high fever, fatigue, dizziness, muscle aches, loss of strength, and exhaustion. People with severe cases often show signs of bleeding under the skin, in internal organs, or from body orifices (such as the mouth, eyes, or ears). Severely ill people may also show shock, nervous system malfunction, coma, delirium, seizures, and in some cases kidney failure. The incubation period varies with the type of disease. Everyone in contact with the initial reservoir or an infected patient is at risk.

Yellow Fever

The symptoms of the disease are fever, headache, backache, prostration, nausea, vomiting, and jaundice. The fatality rate in areas where yellow fever is endemic is less than 5%. In other areas, it may be as high as 50%. The incubation period is 3 to 6 days. The etiologic agent is a virus of the B togaviruses. The disease is spread by the bite of the infected *Aedes aegypti* mosquitoes.

Fleas

Many important diseases are transmitted by fleas beside bubonic plague. Murine typhus fever is a disease caused by *R. typhi*. The organism is transmitted by fleas to humans. The important vector in the United States is the oriental rat flea. The disease is spread from rodent to rodent and occasionally to humans. Infection occurs when the site of the flea bite is contaminated by the flea feces. Although murine typhus fever has largely disappeared, it still could be the cause of a serious epidemic.

Fleas spread tapeworm infestations. They act as an intermediate host for several tapeworms that usually infest dogs and cats or rodents, and occasionally infest humans. The dwarf tapeworm (*Hymenolepis mana*) frequently infests children.

Salmonella enteritidis, which often causes outbreaks of foodborne disease, may be transmitted from rats and mice to humans via fleas. The flea contaminates food with its feces or it contaminates humans directly.

Flies

Flies transmit disease mechanically or biologically. Because flies breed and feed in human feces, they are capable of carrying organisms mechanically to food or to people on their mouthpart, in their vomitus, on their hairs (which are on their body and legs), on the sticky pads of their feet, or in their feces. Mechanically transmitted diseases include typhoid fever, paratyphoid fever, cholera, bacterial dysentery, amoebic dysentery, pinworm, roundworm, whipworm, hookworm, tapeworm, and salmonella other than those that are the causative agents of typhoid and paratyphoid fever. Biologically transmitted diseases include African sleeping sickness, onchocerciasis (blinding filariasis), loiasis (African eyeworm disease), bartonellosis, and sandfly fever.

Lice

Lice are responsible for the spread of epidemic typhus fever, trench fever, and relapsing fever. Typhus fever has resulted in tens of thousands of deaths and hundreds of thousands of illnesses. Typhus fever is caused by a rickettsia. The rickettsia goes through part of its life cycle within the louse. It may change from the organism, causing murine typhus fever, which is rather mild, or to the mutant causing epidemic typhus fever, which is extremely serious. The rickettsia multiplies rapidly within the midgut of the louse after it has ingested blood with the organism *R. prowazeki*. The epithelial cells of the midgut become so loaded with rickettsiae that they rupture and are released into the contents of the gut and feces of the insect. The insect may die as the result of this. The disease is transmitted through bites or feces deposited on human skin.

Trench fever caused over a million cases of illness in soldiers on the western front in World War I. In World War II, the organism and disease once again appeared on the German–Russian front. The organism responsible for this disease is *R. quintana*.

Relapsing fevers are a group of closely related diseases in which the patient's temperature rises and falls at regular intervals. Louseborne relapsing fever caused epidemics in parts of Europe, Africa, and Asia. When the louse bites the patient with relapsing fever, it takes in about 1 mg of blood, which may contain many spirochetes. The spirochetes leave the gut of the louse and reappear in great numbers in the blood of the louse in about 6 days. The louse becomes infected for life. The disease is not transferred to human beings by bites or feces, but only by crushing the louse or damaging it in some way so that the blood contaminates the skin or mucous membrane of the human being.

Mites

Scabies or mangelike conditions are produced primarily by the mange or itch mites. Sometimes the mites cause only mild infections, but often they can cause serious infections, serious skin irritations, and severe allergic reactions. Outbreaks of scabies have occurred, especially during wartime and other emergency periods, when many people are together in close quarters. The mites burrow under the skin and leave open sores to become the source of the secondary infection. The victim becomes very pale and loses considerable sleep. The individual may have intense itching, redness, or rashes. The primary means of transmission of the scabies mite is through physical contact with an infected person. Itch and mange mites are also found on domestic animals.

Chiggers transmit the rickettsia causing scrub typhus, a disease that has afflicted Americans, especially during wartime while in the Orient. Scrub typhus occurs most frequently where there are tall grass fields and neglected coconut plantations. The rickettsia are transmitted from infected victims through the egg to the larvae, which then feeds on rodents and humans. In past outbreaks of scrub typhus, mortality rates varied from 3 to 50%, depending on the virulence of the strain of rickettsia. Chigger bites are thought to transmit hemorrhagic fever. The disease, which is probably caused by a virus, affected American troops in Korea. The disease is also found in Siberia and Manchuria. It is fatal in about 5% of the cases.

The house-mouse mite transmits *R. akari* from the house mouse to humans. This causes outbreaks of rickettsial pox. Another possible disease transmitted by mites is encephalitis, which is usually found in fowl or wild birds. A number of different mites, especially the chiggers and rodent and bird mites, may cause skin irritation or dermatitis. These mites are found in tropical or subtropical areas, including the subtropical areas of the United States. They are most abundant in wooded areas, swamps, roadsides, or any place inhabited by birds or wild rodents. Occasionally mites have been found to invade the respiratory tract of laboratory animals such as dogs, monkeys, and birds. Mites may serve as the intermediate host for tapeworms. The tapeworm is transmitted from one host to another. Of course, the possibility always exists that humans are the inadvertent host.

Mosquitoes — Aedes

Aedes aegypti is the principal mosquito vector of yellow fever. The virus lives in the victim's blood for 3 to 4 days at the onset of illness. During this time, a mosquito can bite the infected human and start the process of spreading yellow fever once again. The mosquito becomes infectious after a period of 10 to 14 days and can harbor the virus for life. There is no known means of infecting human beings other than by the mosquito bite, which transmits the filterable virus to the person. Destroying the breeding places of *A. aegypti* is the principal means of controlling the spread of disease. *Aedes aegypti*, which is a tropical nonnative vector of dengue fever and yellow fever, has been identified in several desert communities, including the city of Tucson, AZ, and the border towns of Douglas, Naco, and Nogales. Its distribution is probably spreading and therefore poses a public health risk in Arizona. It was first detected on the west side of Tucson in 1994.

Aedes albopictus (Asian tiger mosquito) was first found in the United States breeding in imported tires in Houston, TX, in August 1985. It was probably imported from northern Asia, more specifically, Japan. The mosquito is exceedingly adaptable. It is a container breeder where rainwater is stored and can be found in artificial containers including old tires, tree holes, and other artificial depressions. The mosquito also is a link between forest, rural, and urban areas, because it can live in both, especially in houses. The eggs survive cold and can live through a Midwestern winter. *Aedes albopictus* is an established vector of dengue fever and dengue hemorrhagic fever, and is two to three times more susceptible to infection with the dengue virus than *A. aegypti*. This insect transmits all four dengue serotypes transovarially and transstadially. In laboratory studies it has been shown to be a highly efficient vector of California encephalitis group viruses (La Crosse virus). This mosquito can also transmit yellow fever, Japanese encephalitis, eastern and western equine encephalitis, and possibly St. Louis encephalitis. It is a vicious biter and seeks out people to bite. *Aedes albopictus* is now widely distributed in the southeastern United States. It is a competent experimental vector of 7 alphaviruses including Chikungunya, eastern equine encephalitis, Mayaro, Ross River, western equine encephalitis, Venezuelan equine encephalitis, and Sindbis viruses. However, only eastern equine encephalitis has been isolated from this mosquito, when collected in nature, at this time. *Aedes albopictus* is also a competent experimental vector of the following flaviviruses: dengue serotypes 1, 2, 3, and 4; Japanese encephalitis; West Nile encephalitis, and yellow fever viruses. In the case of the St. Louis encephalitis virus, the amount of circulating viruses in the naturally infected avian hosts is generally insufficient to infect the mosquito. There is also now a concern about the potential for *A. albopictus* transmitting the La Crosse virus. In the late summer and fall of 1997, this *Aedes* species was found in the immediate area of chipmunks carrying the La Crosse virus in Peoria, IL, which is a long recognized focus of transmission of this virus.

Other problem-causing *Aedes* are discussed briefly in the following section. *Aedes atlanticus-tormentor-infirmatus* mosquitoes are vicious biters and have been known to drive cattle from woodlands. The virus of California encephalitis was isolated from these mosquitoes. *Aedes canavensis* species are serious pests in woodlands and also transmit the virus of California encephalitis. *Aedes cantator* insects live in salt marshes and migrate to shore towns and resorts, and can be a nuisance on the east coast. *Aedes cinerus* mosquitoes occasionally pester the northern states. *Aedes borsalis* species are severe pests for humans and cattle throughout the arid and semiarid parts of the western United States. These are vicious biters and attack during the day or night. *Aedes nigromis* inhabit the western plains from Minnesota to Washington and south to Texas and Mexico. These mosquitoes are serious pests that bite severely in the daytime. *Aedes punctor* and other related species are important pests in woodlands and recreational areas in the northeastern United States and mountains of the West. *Aedes sollicitans* insects live in salt marshes, are severe pests along the Atlantic and Gulf Coasts, and are also found in the Midwest. They are fierce biters that migrate just before dark. *Aedes seencerii* are found on the prairies of Minnesota, North Dakota, and Montana, and in several northwestern states. They are fierce biters that attack during the day. *Aedes sticticus* are found mostly in the northern states. They are usually quite abundant after a flood. They are severe biters during the evenings and also bite in cloudy or shady areas during the day. The *Aedes stimulans* are found in the northern states. They bite during the daytime. *Aedes taeniorhynchus* are found on coastal plains along the Pacific Coast. They are most abundant in salt marshes and are severe pests and fierce biters, primarily at night.

Aedes triseriatus are fierce biters found in tree holes, old tires, tin cans, and other artificial containers. *Aedes trivittatus* are fierce biters and annoying pests found in the northern United States. These mosquitoes usually rest in grass and vegetation during the day. *Aedes vexans* are found in rain pools or flood waters in the northern states from New England to the Pacific Coast. The adults travel 5 to 10 mi from breeding places. They viciously bite at dusk or after dark. Although there are many other species of *Aedes* too numerous to discuss in this volume, the principal species have been covered.

Anopheles

The major vectors of malaria include *Anopheles albimanus* in the Caribbean region, *A. freeborni* in the western part of the United States, and *A. quadrimaculatus* in most parts of the world and east of the Rocky Mountains.

Humans are the only important reservoir of human malaria. The disease is transmitted when the *Anopheles* ingest human blood containing the plasmodium in the gametocyte stage. The parasite develops into sporozoites in 8 to 35 days. The sporozoites concentrate in the salivary glands of the mosquitoes and are injected into humans as the mosquito takes its blood meals. The gametocyte appears in the blood of a host within 3 to 14 days after the onset of the symptoms. Malaria is also transmitted from person to person through blood transfusions or contaminated syringes.

The incubation period varies, depending on the organism, from 12 days to 10 months. The infection may be transmitted as long as the gametocytes are present in the patient's blood. The disease is carried by a patient for as little as 1 year or as long as a lifetime, depending on the organism. The *Plasmodium vivax* are usually carried from 1 to 3 years. *Plasmodium falciparum, P. ovale*, and *P. malariae* are also malaria parasites.

Other troublesome *Anopheles* include *A. punctipennis*, which are vicious biters out-of-doors and apparently do not enter homes readily. These insects breed in rain barrels, hog wallows, grassy bogs, swamps, and margins of streams. The *A. walkeri* breed in sunny marshes along the edges of lakes and in sawgrass, readily bite humans, and are good laboratory vectors of malaria.

Culex

Culex pipiens pipiens, the northern house mosquito, lives in the northern United States; *C. pipiens quinquexasciatus*, the southern house mosquito, lives in the southern United States. Both these mosquitoes are severe pests. They are extremely annoying, bite fiercely, and feed continuously on humans. Both these mosquitoes breed in huge quantities in rain barrels, tanks, tin cans, storm-sewer catch basins, poorly drained street gutters, polluted ground pools, cesspools, open septic tanks, effluent drains from sewage disposal plants, and any other highly unsanitary condition. The mosquitoes do not migrate far from their area of harborage. They are capable of transmitting innumerable diseases.

Culex tarsalis are naturally infected with the viruses of St. Louis and western encephalitis, and have the capability to transmit either of these diseases. The mosquito bites birds and then transmits the virus to other birds, horses, or humans. It is the most important vector of encephalitis in humans and horses in the western United States. The insect is active soon after dusk and enters buildings for its blood meal. It is widely distributed west of the Mississippi River, and also in southern Canada and northern Mexico. The larvae develop in a variety of aquatic settings. They develop in arid and semiarid regions, wherever water is trapped; in effluent from cesspools and other organic materials; and in artificial containers. The females deposit 100 to 150 eggs in a raft. Hatching occurs within 18 hours. Adults are active from dusk to dawn and remain at rest in secluded areas during the day. They are found on porches, on shaded sides of buildings, under bridges, or in other protected areas. The majority of these mosquitoes rest in grass and shrubs or along the banks of streams. They fly up to 11 mi, even though they normally stay within 1 mi of their breeding place.

Encephalitis, which is caused by an arthropodborne virus, is a group of acute, inflammatory diseases of short duration that involve parts of the brain, spinal cord, and meninges. The disease is not transmitted directly from human to human, but from mosquito to bird to mosquito to human.

Culiseta

There are ten species of *Culiseta* in the United States, of which five are fairly widespread. Although they are relatively unimportant as pests, two of the species are naturally infected with encephalitis viruses. It is not known, however, whether they spread encephalitis to humans. These two species are *Culiseta inornata*, which are frequently found in cold water, and *C. melanura*, which are found in most of the eastern United States from the Gulf states to Canada.

Mansonia

Of the three species of *Mansonia* found in the United States, one is widespread and common. *Mansonia perturbans* is a serious biter and pest found in the southern and eastern states, and also in the Great Plains, Rocky Mountain, and Pacific Coast states. In Georgia, this mosquito was found to be naturally infected with the virus of eastern encephalitis. It breeds in marshes, ponds, and lakes that have a thick growth of aquatic vegetation. The larvae develop slowly. The difficulty with this mosquito is that it cannot be killed in its larval stage by an ordinary surface larvacide, because its breathing equipment is inserted into plants through which it breathes. The females bite during the daytime in shady humid places, but they are most active in the evening and the early part of the night. They enter houses and bite viciously.

Psorophora

In the United States, 13 species of *Psorophora* are found, 10 of which are rather widely distributed in the southern and eastern states. Their breeding habits are similar to the *Aedes*. *Psorophora ciliata*, which breed in temporary pools, are vicious biters and serious nuisances in the South, Midwest, and Eastern United States. They attack during the day or evening. *Psorophora confinnis* are fierce biters that attack in the day or night. Their numbers are often so great that they kill livestock. They are principally found through the southern United States and in Nebraska, Iowa, New York, Massachusetts, and southern California. They breed in temporary rain pools, irrigation waters, or seepage pits. They are a great problem in rice fields. *Psorophora cyanescens* are severe pests that attack in the day or night. These are particularly abundant in Oklahoma, Arkansas, Alabama, Mississippi, Louisiana, and Texas. *Psorophora ferox* are vicious, painful biters found in wooded areas of the South and East.

Roaches

Although there are no records of outbreaks of disease caused by roaches, it is possible that they act as vectors. Roaches cross areas containing sewage or fecal material. They also cross food and could mechanically transmit enteric organisms from their breeding places to food. It is also possible that they transmit organisms through their feces and through the small amount of material regurgitated while they are feeding. Roaches have been shown to carry salmonella in their gut.

Roaches may contribute to asthma in children. Studies indicate that one third of children in inner city areas are allergic to roach allergen. Children who are sensitive to roach allergen and live in infested housing have 200% more hospitalizations than children who live in a good environment. Asthma is clearly one of the important reasons for the increased hospitalization.

Ticks

Ticks can transmit pathogenic organisms mechanically or biologically. Their mouthparts become contaminated when feeding on an infected animal and may in turn contaminate another animal or human while feeding. This is mechanical transmission. Further, ticks serve as reservoirs of viruses, rickettsiae, spirochetes, bacteria, and protozoa. These organisms are transmitted from the infected adults through the eggs to the larval, nymphal, and adult stages. This is called transovarial, or transstadial, transmission. This mechanism is of considerable concern because the tickborne disease can survive adverse weather conditions, because the organisms are being transmitted through the eggs.

Tularemia, or rabbit fever, is caused by the bacillus *Francisella tularensis*. This bacterial disease infects numerous hosts, including rabbits, rodents, and humans. Ticks transmit this disease organism through the genera *Dermacentor* and *Haenaphysalis*. In addition, the deerfly, which is also a blood-sucking arthropod, may be an important vector of the spread of tularemia. Ticks transmit the disease organisms to their own eggs, keeping the organism alive from generation to generation. Two peaks of tularemia occur in the United States, during the fall–winter season and also during the spring–summer season when the ticks and deerflies are most prevalent.

Protozoal diseases spread by ticks include cattle tick fever, Texas cattle fever, and anapromosis. The diseases are spread primarily from cattle to cattle. Rickettsial diseases include Rocky Mountain spotted fever and Q fever. Rocky Mountain spotted fever is caused by the organisms *R. rickettsia*. The wood tick, *D. andersoni*, and the rabbit tick, *H. leporistalustris,* spread the disease from animal to animal and from animal to human in the western part of the country. The American dog tick, *D. variabilis*, is the most important vector in the eastern part of the country. The Lone Star tick, *Amblyomma americanum*, is probably a vector of Rocky Mountain spotted fever in Texas, Oklahoma, and Arkansas. These *Dermacentor* ticks easily spread spotted fever, because the larvae and nymphs feed on infected animals and the adults attack humans or other large animals. Spotted fever has been reported in 46 of the 50 states in the United States. The western strain of the disease is much more virulent than the eastern strain.

Rocky Mountain spotted fever resembles typhus fever, but it is a more severe infection. The mortality rate may be as high as 25%. The rickettsiae invade the endothelial cells and smooth muscle of the arterioles, which may cause obstruction and destruction of the vessels. Although the disease is not transmitted directly from human to human, the tick remains infected for life and continues to transmit the disease to humans for periods of up to 18 months.

Q fever is a rickettsial disease caused by the organism *Coxiella burnetii*. Outbreaks of the disease often occur in stockyard workers. The wood tick and Lone Star tick, among others, spread rickettsiae to humans. The disease is also airborne or is passed through raw milk from infected cows. The tissues and feces of ticks become massively infected. It is suspected that the most common means of transmission occurs when the organisms are inhaled in dust or droplet form. About 1% of untreated victims die of Q fever. Relapsing fevers, which are tick or louseborne, are caused by spirochetes. The diseases occur on all continents, with the possible exception of Australia. The spirochetes causing relapsing fever most frequently in the United States belong to the genus *Borrelia*. The louseborne type of the disease is usually due to *B. recurrentis*. The tickborne disease is spread through soft ticks in the genus *Ornithodoros* found in 13 western states. The spirochetes are transmitted through the offspring of the ticks or through their rodent hosts, which then serve as reservoirs for the disease.

The viral disease spread by ticks is Colorado tick fever. New cases occur in the Rocky Mountain areas of the West. The wood tick, or *D. andersoni*, carries the disease. Although dog ticks are shown to carry the virus, they have never been incriminated in any outbreak of the disease. The reservoirs of infection include ground squirrels, porcupines, and chipmunks. Ticks, in addition to causing the previously mentioned diseases, may also cause a paralysis of considerable concern to both physicians and veterinarians.

Food Product Insects

Stored food product insects are a nuisance and contaminate food in large quantities. They must be destroyed and removed from the food supply. Their role as agents in the spread of foodborne diseases has not been proved.

Emerging and Reemerging Insectborne Disease

A complex combination of ecological, economic, and human behavior changes has caused a reemergence of previous diseases and an emergence of new infectious diseases during the past 20 years. Ecological changes have contributed to the emergence of both known, and at this point unknown, pathogens circulating in the complex host and vector systems of disturbed habitats. In the United States, Rocky Mountain spotted fever increased sharply in the late 1970s; Lyme disease, in the 1980s; and ehrlichioses, in the 1990s. *Murine typhus* in southern California and Texas has changed because of the changing ecology due to suburban expansion during the past 30 years. In suburban areas, vector fleas are most associated with human habitation through their natural hosts; the commensal rodents; and the peridomestic animals, such as cats, dogs, opossums, raccoons, and squirrels. Fleas are picked up by household pets and brought into the homes. Fleaborne disease organisms such as *Yersinia pestis*, *R. typhi*, *R. felis*, and *Bartonella henselae* are widely distributed throughout the world in endemic-disease foci, where components of the enzootic cycle are present. However, fleaborne diseases could reemerge in epidemic form because of changes in vector–host ecology due to environmental and human behavior modification. The rapid spread of pathogens to human populations is due to the frequent feeding behavior and extraordinary mobility of fleas.

Rickettsial pathogens are important causes of illness and death worldwide. They exist primarily in endemic and enzootic foci that occasionally give rise to sporadic or seasonal outbreaks. The arthropod vector is often more important than the host in the natural maintenance of the pathogen. Rickettsia use as vectors lice, fleas, ticks, and mites. All these insects live with humans in a variety of situations, especially where overcrowding occurs, and where people move into the suburban areas where many of these insects may be found. Depending on the type of rickettsia, thousands of human cases of the disease may occur (Figure 5.15).

Impact of Insects on Other Environmental and Health Problems

Pesticides are the major means of control of pests today. Unfortunately, the World Health Organization (WHO) estimates that approximately 500,000 cases worldwide of pesticide intoxication occur annually and that about 1% of the victims die as a direct result of contact with the pesticide. Pesticides also cause resistance in a disease-bearing vector, thereby exacerbating disease production.

Economics

It is difficult to make a good economic analysis of a pest control program. Public health specialists are unable to evaluate the associated costs and benefits of large or small programs; and economists fail to deal adequately with the benefits or values of health vs. illness, the value of human life, or even the value of a reduction in the number of pests. There also is uncertainty about the technical feasibility of various types of controls. Even the standard techniques used, such as source reduction,

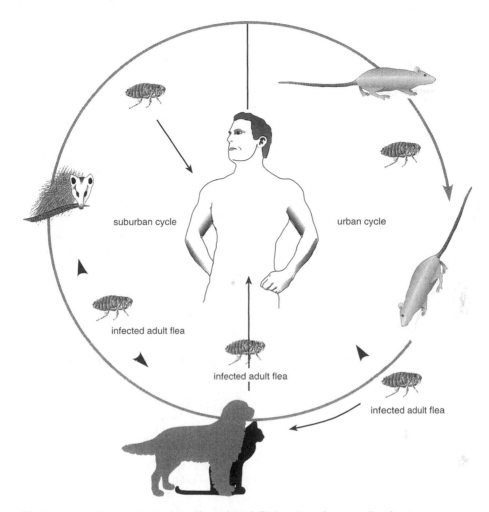

Figure 5.15 Urban and suburban life cycles of *Rickettsia* and mammalian hosts.

larvaciding, and adulticiding, vary tremendously with region and country. Nowhere has a major effort been made in specific data collection to justify or to establish cost-benefit ratios.

POTENTIAL FOR INTERVENTION

The potential for intervention is excellent in the general areas of insect control and specific pests. The techniques of intervention include removal of adult harborage, adulticiding, larvaciding, and biological control. (For additional information, see the section on controls.)

Intervention by isolation techniques is difficult to accomplish, because insects move rapidly through a given area. Substitution is another inadequate technique.

Shielding can be accomplished by use of proper-sized screens or blow-down air devices. Treatment is used for disease carriers but not effectively for the insects. Prevention is the strongest potential intervention technique in the spread of particular insects.

RESOURCES

Some scientific and technical resources available include Tulane University School of Public Health, Colorado State University, various land grant colleges such as Pennsylvania State University and Purdue University, Notre Dame University, American Public Health Association, National Environmental Health Association, and Entomological Society of America.

The civic and professional associations in pest control include American Mosquito Control Association, and International Association of Game-Fish and Conservation Commissioners. Pest control operator associations and chemical manufacturer associations are other important resources.

Government resources available include state health departments, many local health departments, the U.S. Department of Agriculture (USDA), National Bureau of Standards, Department of Defense, Environmental Protection Agency (EPA), Food and Drug Administration (FDA), National Science Foundation (NSF), U.S. Department of Health and Human Services (USDHHS), and county extension agents.

STANDARDS, PRACTICES, AND TECHNIQUES

Mosquitoes are by far the largest group of insects controlled by humans. The standards, practices, and techniques utilized vary tremendously from area to area. Generally, surveys are made to determine the extent and location of the mosquito populations and then extensive larvaciding or adulticiding programs are put into effect. In the event of an outbreak of disease, extensive larvaciding and adulticiding programs are again carried on in those areas where mosquitoes are most prevalent. In many areas there are unorganized mosquito control programs. These programs are put into effect when citizens demand action because of the annoyance due to mosquitoes. Generally, public works employees or contracted commercial firms do the actual control work. The work consists of the use of larvacides and fogging units, or mist units for adulticiding. Control may be done in a haphazard way or it may be carried out weekly on a regular basis where specific high-volume mosquito areas exist. Where organized mosquito programs are in existence, the community is taught first how to reduce mosquitoes through environmental controls and adequate prevention; then proper engineering approaches are used, as well as larvacides and adulticides. Although the individuals carrying out this work must be licensed by their states, and adhere to the principles of the EPA, a substantial need still exists to establish proper standards, practices, and techniques for insect control.

MODES OF SURVEILLANCE AND EVALUATION

Inspections and Surveys

Fly Surveys

The major value of a fly survey is to determine those areas in which flies are in greatest concentration, to determine the kinds of flies to be effectively controlled, and to determine the potential spread of disease. Several different types of fly surveys are available. They include flytraps, fly grills, and reconnaissance surveys. The selection of the sample area should be based on a clear understanding of what information is needed, the kinds of problems anticipated, the areas where the greatest number of flies are found, and the areas where the greatest refuse and other solid waste problems exist.

Flytraps consist of baited traps, flypaper, or fly cones. The flytraps have the advantage of capturing a cross section of the kinds of flies in a given area. They also provide an approximate count of the various species. The baited flytrap may be lined with animal feces, sugar, fish heads, or some other decaying matter. The flypaper strip gives a rapid count of flies, but this only represents a small portion of the fly population. Flies are drawn to fly cones by the attractant, and then as they attempt to escape they move upward toward a light that is in the fly cage.

Fly grills are the most widely used means of surveying for flies. The grill depends on the natural tendency of flies to rest on edges. The grill is placed over a natural attractant, such as garbage, manure, or other decaying materials. The number of flies landing on the grill during a 30-sec period is tabulated. It is possible to physically count the number of flies and make some judgment of the different kinds of flies on the grill. If there is a huge number of flies, the grill should be subdivided and a count made on the subdivided portion. A minimum of ten counts should be made in each subdivision sample, and the five highest counts should be recorded.

It is possible to make a rapid survey by driving through an area or walking through an area in which you would expect to find large numbers of flies. A rough estimate can be determined by checking the usual fly-breeding areas and making a physical count. In an attempt to determine if a fly survey or a fly control program is effective, it is wise to make a fly survey initially in the usual breeding areas and then return after the breeding areas have been eliminated and resurvey. Obviously, chemical control procedures should be used prior to the cleanup procedures so that the flies do not move on to other areas.

Mosquito Surveys

Surveys are necessary for the planning, operation, and evaluation of mosquito control programs. The basic survey is used to determine the species of mosquitoes, sources, breeding places, locations, quantities, flight distance, larval habits, adult resting places, and any other information that may be needed in the actual mosquito control program. It is also necessary to evaluate the ongoing program continually.

This is done by conducting surveys on a periodic basis during each season of the year, especially the mosquito-breeding season, to determine the effectiveness of chemicals and other controls. The surveys consist of making field evaluations of the quantity of mosquitoes; the types of mosquitoes; the kinds of mosquito breeding areas and potential mosquito breeding areas; the numbers and types of larvae; and the development of accurate, comprehensive maps that indicate where all the larval and adult problems may be found.

A larval mosquito study may be carried out by an environmentalist who goes to areas where there are artificial breeding places and standing bodies of water and removes water containing larvae by means of a white enamel dipper, which is about 4 in. in diameter. The dipper has an extended handle so that it may be lowered into the body of water. Some environmentalists prefer to use white enamel pans that are about 14 in. long, 9 in. wide, and 2 in. deep. The pan is used to sweep an area of water until it is half full. The water is then examined for larvae. Another technique is to examine cisterns, artificial containers, or other areas with a flashlight. A large bulb pipette or siphon is used to draw out water with larvae into a pan or onto a slide, and then observations are made. After observing the habits of the larvae in the water, environmentalists should always record the number of dips they make and the count of larvae. A rough estimate of larvae increases or decreases can be made over time by going to the same area and using the same dipping procedures.

Adult mosquito surveys can be carried out by setting up bait traps, window traps, carbon dioxide traps, insect nets, and light traps, or by checking the daytime resting places. Bait traps are put in areas where mosquitoes are expected to be found. The mosquitoes are attracted to a bait animal that is placed in a trap. When they enter the trap, they cannot bite the animal and they cannot be released. The mosquitoes are taken alive and evaluated. Bait animals have been such animals as horses, calves, mules, donkeys, and sheep.

Window traps use the same principle as the animal bait trap, except a human being is the attractant. The trap is set on the outside of the window. The mosquitoes are attracted to the building by the presence of the human being and then they enter the trap and cannot leave. Insect nets are used for collecting mosquitoes in grass and other vegetation. Light traps are traps that contain a light that acts as an attractant to the mosquitoes. Once the mosquitoes enter, they are incapable of leaving the trap area. An evaluation can be made of the adult resting places. The type of resting place depends on the kind of mosquito. This material has been previously covered.

Once all the information is gathered and a determination is made of the kinds of mosquitoes that are present in an area, comprehensive maps are drawn, and the control program is established and put into action.

Food Product Insect Surveys

The easiest means of determining the presence of food product insects is searching through dry food products and looking for live insects. The insects may also be present along the seam of the food bag. At times the insects can be seen crawling near the food or flying about. In any of these cases, a food product insect problem exists and preventive measures should be taken.

CONTROLS

Before dichlorodiphenyltrichloroethane (DDT) was used for mosquito problems, the principal control techniques consisted of identifying the species of mosquito and its breeding habits, and then designing a method that would eliminate the larva, either by means of engineering techniques or by means of chemical larvacides. Engineering techniques consist of improving drainage areas, adjusting the slope and depth of streams, and using methods and techniques for the removal of any obstructions hampering the flow of water; this in turn increases the potential for mosquito breeding. Biological control is the technique utilizing parasites, such as *Bacillus thuringiensis*, predators, and pathogens of insects to eliminate them. It also includes the use of sterilization techniques, genetic manipulation, management of the habitat, insect growth regulators, and biologically produced chemical compounds. In recent years there has been considerable interest in this area. Currently, mosquito biolarvicides have become an important part of integrated mosquito control management programs. Biological control is basically the fostering and manipulation of the natural enemies of insects. Fish are an excellent biological control for mosquito larvae. Over 1000 species of freshwater fish consume mosquito larva and pupa. However, the one that is most frequently used is the minnow *Gambusia affinis*. This minnow is a surface feeder and inhabits permanent waters. It has been used in the control of anopheline mosquito larvae to reduce malaria, and *Aedes aegypti* to reduce yellow fever. The technique of building insects out by removing harborage along with chemical control is now called integrated pest management.

Insect growth regulators are used as a means of control. Attractants lure the insect to trap or poison. Repellents are used on the skin and body as a means of control to prevent the insect from alighting and biting, or otherwise contaminating the individual. Because certain pesticides are banned as a result of new research, it is essential to reevaluate the chemicals in very frequent use. Read the labels for proper usage.

Integrated Pest Management

Integrated pest management promotes minimized pesticide use, enhanced environmental control, and environmental safety. The primary goal of an integrated pest management program is to prevent pest problems by managing the facility environment in such a way as to make it less conducive to pest infestation by integrating housekeeping, maintenance, and pest control services. Each program is tailored to the environment where it is applied. The various components of the program include:

1. Facility design, where the planning, design, and construction provides an opportunity to help incorporate features that exclude insects and rodents, minimize habitat, and promote proper sanitation
2. Monitoring, where traps, visual inspections, and special studies are used to identify areas and conditions that may promote insect and rodent activity
3. Sanitation and facility maintenance, where insect and rodent problems can be prevented or corrected by using proper cleaning, reducing insect and rodent habitat, and performing repairs that exclude pests

4. Communications, where a staff person meets with pest management personnel to assist in resolving facility issues that impact on insect and rodent problems (including necessary training of all personnel)
5. Record keeping, where a logbook is kept of insect and rodent activity and the conditions leading to the problems
6. Nonpesticide pest control, where such methods as trapping, exclusion, caulking, and sealing openings are used in conjunction with proper sanitation and structural repair
7. Pest control with pesticides, where the least toxic pesticides that are still effective are applied in the best and safest manner
8. Program evaluation and quality assurance, where evaluation is ongoing about all program activities and surveys of potential insect and rodent problems
9. Technical expertise, where a qualified, certified and licensed individual gives assistance to the individuals in the integrated pest management program of the facility
10. Safety, where the level of hazardous chemicals are reduced in usage and storage

Specific Scientific and Technical Controls

Flea Control

In flea control, infested animals and their habitats are treated. Pets can be treated with imidacloprid, fipronil, insect growth regulators, and other products prescribed by veterinarians. The animal quarters or the house that is infested with fleas should be treated with carbaryl, methoxychlor, or chlorpyrifos. Each of these chemicals is slightly toxic and has long residual periods. If there is concern about a rodent flea infestation, it is important to dust the rat run with any of the previously mentioned chemicals. The dust then gets onto the rats or mice and kills the fleas that are present. It is essential that the fleas be killed before a rat control program starts. Otherwise, the rats die and the fleas leave the dead animals and possibly attack humans. In addition, infested premises may be treated with 2% malathion or 0.5% diazinon. In barns and yards, 2.5% diazinon or 2.5% ronnel, malathion as a 4% dust, or carbaryl as a 5% dust may be used. Insect growth regulators, such as methoprene, are juvenile flea hormone analogues that disrupt egg development and sterilize adult cat fleas.

Fly Control

The four basic fly-control techniques include proper solid waste disposal, chemical control, mechanical and physical control, and biological control.

Because houseflies breed in garbage, decaying matter, feces, dead animals, and any kind of organic material, it is essential that this solid waste be removed and stored properly. The storage should take place in plastic bags within metal cans with tight-fitting lids. The refuse should also be stored, depending on the kind, in concrete containers (for animal manure), in large metal containers with tightly closed lids, or in refuse rooms where the doors and windows are fly proof. The maximum size of the can should be not greater than 30-gal capacity, because it becomes difficult to remove heavier cans. Where garbage, heavy refuse, or ashes are stored, the maximum size of the cans should be 20 gal. Collection should take place during the summer at least twice a week, because flies go from the eggs to the pupa stage within 5 days and then move into the ground. By removing the cans at least once every 4 days, the fly eggs, larvae, or pupa are removed and disposed. During the winter, the solid waste should be removed at least once a week. In food operations and in other situations where there are large quantities of organic waste material, this material should be removed on a daily basis. Solid waste disposal should occur in either an incinerator or a properly operated sanitary landfill. Garbage that is used for hog feeding must be cooked to an internal temperature of 137°F before it is served to the hogs. One other technique utilized for disposal is through the home garbage grinder.

Animal feed and feces become breeding areas for flies. Animal manure should be adequately stored, and dog and cat feces should be picked up and removed to prevent fly breeding. It is also essential to check the animal feed periodically to make sure flies are not breeding in it. In a home situation, dog food or cat food left outside becomes an attractant.

Large fly populations can breed in high weeds. It is therefore necessary that weeds be controlled either chemically or physically. The type of chemical chosen for control of flies is determined by the effect that is desired. Residual sprays are put on surfaces on which flies alight. These sprays should last for considerable periods of time. The chemicals include malathion mixed with sugar emulsion, dimethoate, and permethrin. Space sprays are used in areas where quick knockdown of flies is desired, and where the aim is to kill the adult fly in large quantities. The space spray should contain 0.1% synergized pyrethrum plus 1% malathion. Fly cords should be impregnated with 2.5% diazinon. The cord should be made of strong cotton, $3/32$ in. in diameter and strung at a rate of 30 linear ft of cord per 100 ft^2 of floor area. Cords are accepted because flies rest on them, absorbing the insecticide. A larvacidal treatment should consist of 2.5% diazinon emulsion, 1% malathion emulsion, or 2% dimethyldichlorovinyl phosphate (DDVP) emulsion. Dry baits can consist of diazinon, malathion, ronnel, or DDVP plus sugar. For wet bait add water.

It is important to realize that insects become quite tolerant and resistant to insecticides. As a result of this, the insecticides that are now recommended and probably the safest to use for humans at this time may not be those recommended for future use. Environmental health specialists should contact the local county extension agent if they have any questions concerning the problems of fly control and chemical safety.

It is known that flies not only are resistant to certain insecticides but also apparently develop an inherited resistance over several generations. As a result of this, flies are able to absorb a lower rate of insecticide, store the insecticide without being killed, excrete the insecticide, detoxify it, or use an alternate mode of accomplishing its body functions when a chemical blocks its normal route.

Several techniques are used in applying insecticides. They include space spraying for the adults; mist spraying through large blowers or insecticide bombs for adults; fog generators, which produce an aerosol or smoke for adults; and residual spraying with hand sprayers for adults and larvae. Flies can also be controlled chemically by treating the animal by the use of fly repellents or by the use of fly attractants to draw flies into traps or onto poison.

One of the most effective physical controls is fly screens, which must be 16 mesh in size. The screens have to fit tightly to the windows or door frames. Fly traps, which have been discussed under survey techniques, can be utilized for some fly control. Flies can be electrocuted if they cross an electrically charged field or alight on a screen that is electrically charged. Electronic fly traps, although effective in certain instances, can increase the number of respirable fly particles by 10 to 500 times over the background levels. Air shields can be installed so that air blows down and out from the doorway, thereby preventing flies from entering a building.

The biological control of flies is based on the use of sterile flies and on the dissemination of organisms that are pathogenic to the insect. These techniques are basically in experimental stages and are not generally used. Predatory animals consume a certain quantity of flies; however, it is necessary to be cautious that the predators do not themselves become pests. The most important means for control of flies is through the elimination of food and garbage. Therefore, it is extremely important that all exterior areas and all interior areas are thoroughly cleaned and all solid waste is removed.

Within the home, flies are best chemically controlled by the use of synergized pyrethrum plus malathion, diazinon, or baygon. Each of these chemicals must be used very carefully. DDVP strips may be used to kill flies. These resin strips last for about 3 months. One strip should be used per 1000 ft^3 of space. Care must be taken in the use of DDVP strips around food or food services and around individuals that have respiratory difficulties.

Louse Control

Body louse control is accomplished by washing the clothes of the individual in very hot water, drying, and then ironing. Wool clothing, which is especially good for louse habitation, should be thoroughly dry cleaned and then pressed. Clothing may also be put in a plastic bag and allowed to remain for a period of 30 days. This kills the existing lice as well as the eggs. The individual can be treated with a 1% malathion powder, which is effective against eggs and adult lice.

Head lice are controlled by cutting the hair very short and then washing it thoroughly. It is essential to kill both the lice and the eggs. The best technique for head louse control is to shampoo, dry the hair thoroughly, tilt the head back, cover the eyes, and apply a malathion or lindane lotion (which is worked into the hair and scalp, and combed through the hair). Shampoo the hair again after 24 hours; then dry, comb, and brush it to remove the dead lice and loosen the eggs.

In crab louse control, the pubic hair should be washed very thoroughly and the hair shaved off. A vaseline ointment containing pyrethrum or a 1% lindane cream or lotion may be used. It is essential to complete louse control by eliminating all the bedding and clothes used by the lice-infested individuals. These garments and materials should be washed very thoroughly or dry cleaned. It is also essential that hair is washed frequently and thoroughly.

Mite Control

Mites are controlled through the use of chemicals or proper cleanup. In trying to control mites within a house, or even within areas where animals live, pyrethrum bombs can be used as knockdown agents. Obviously the pyrethrum has to be synergized. As soon as the mites are knocked down, they must be vacuumed up and destroyed. For residual treatment use 1% malathion. This insecticide should be applied onto the top of foundations; around the plate and ends of joists, baseboards, edges of floor areas; and around windows and doors. On the outside, mites can be controlled with 0.5 to 1% malathion at a rate of 2.5 gal of 40% emulsifiable concentrate per acre.

Human infestation can be controlled by use of 68% benzyl benzoate or 12% benzocaine. The chemicals should be left on the body for 24 hours before washing. The second treatment is needed in 10 to 14 days. Scabies may also be treated with eurax, which is a salve containing 10% N-ethyl-O-crotontoluide in a vanishing cream. Two applications should be given at 24-hour intervals.

Mites are controlled through elimination of rats, house mice, birds, and so forth. With chigger mites, the environment should be modified to permit plenty of sunlight and air to circulate freely.

Larval and Adult Mosquito Control

Mosquito larvae are controlled mechanically, biologically, or chemically. Mechanical control involves emptying and removing all temporary containers, such as old tires and cans. If possible, depressions in the ground should be smoothed out. High weeds, which trap water, must be cut on a regular basis. It is also important to ensure that all sanitary landfills or any other type of landfill has a final slope of 0.1 to 0.5 ft/100 ft for drainage purposes. Clean, straighten, and drain all ditches so that water runs freely. Any growth in bodies of water, particularly around the shoreline, must be removed so that the water flows readily instead of sitting stagnant. Of particular importance is the closing or sealing of all seepage pits, septic tanks, and other areas where sewage may be trapped in a stagnant situation. Impounded waters can be a source of mosquito larvae breeding. This can be remedied by cleaning the major vegetation off the shoreline, filling, and removing the water from low places. The water level should be so controlled that a reduction in level of about $1/10$ ft per week occurs during the mosquito-breeding season.

Additional research should be done in the areas of mechanical mosquito larvae control, because huge numbers of mosquitoes inhabit marshy areas and areas where certain types of food, such as rice, are grown. It would be detrimental, at present, to change the ecology of these areas. On irrigated lands, the mosquito problems can be considerable. These insects generally find habitation in stored water, irrigation conveyance and distribution system, irrigated land, and drainage systems. The key to control is to use only the amount of water necessary and to have all systems freely moving so that stagnation of water cannot occur. The biological technique for eliminating larvae would be to use the toxins of *Bacillus thuringiensis*, to use insect growth regulators, or to stock the lakes and other impounded areas with top-feeding minnows of the gambusia variety. In residential areas, it is important to change the water in the birdbaths, stock garden ponds with goldfish, and remove all containers holding stagnant water.

Chemical control is accomplished by applying no. 2 fuel oil, no. 2 diesel oil, kerosene, or petroleum distillates. Petroleum distillates tend to be more environmentally friendly than the other petroleum compounds. The petroleum is toxic to the eggs, larvae, and pupa. Pyrethrum larvacides have been used for many years. Methoxychlor may act as a systemic poison and as a contact poison, which penetrates the body wall or respiratory tract. In addition, baytex and malathion have been used effectively in mosquito larvae control.

Bacillus sphaericus and *B. thuringiensis israelensis* are biological pesticides used for mosquito larvae control in water. When the larvae eat them, the pesticides release a protein that disrupts the feeding process of the mosquito larvae, causing them to starve to death.

At dark, mosquitoes are best controlled by use of screens and repellents, and by space spraying. Screens should be 16×16 or 14×14 mesh to the inch to keep mosquitoes out. In areas where very small mosquitoes exist, such as the *Aedes aegypti*, it is necessary to have a screen with a mesh of 16×20 or 16×23. Bed nets are useful in temporary camp areas where mosquitoes are quite prevalent. Bed nets are made of cotton or nylon cloth with 23 to 26 meshes per inch.

Several types of mosquito repellents are on the market today. They include bimethyl thalate, Indalone, and biethyl toluamide. The repellents are applied to the neck, face, hands, and arms. They prevent mosquito bites for a period of from 2 to 12 hours. This protection depends on the person, the species of mosquito attacking, and the abundance of mosquitoes available.

Space spraying for adult mosquitoes is accomplished by using aerosols, fogging, misting, dusting, or applying by airplane. The chemicals generally used include naled, sumithrin, resmethrin, and malathion. Care must be taken to avoid large concentrations of pesticides in any type of space spraying in residential areas. Where residential areas are space sprayed, personnel must ensure that the trucks have blinking red or orange lights as signals and that children are not riding their bicycles in and out of the mists. Serious accidents have been caused when automobile drivers were blinded by the fog or mist and drove into children on bicycles.

Residual spraying is another technique for the destruction of adult mosquitoes, where 2.5% malathion is used. However, the Asian tiger mosquito in Houston appears to be resistant to malathion. The chemicals are applied to surfaces, and they continue to kill mosquitoes for a period of 10 to as much as 32 weeks, depending on the type of insecticide and mosquito. Another valuable technique is the use of strips of DDVP in enclosed basins. These strips continue to give off vapona insecticide over a period of time and kill the mosquitoes within the catch basins.

It is important to realize that mosquitoborne outbreaks of disease have not been eliminated from this country or other countries. In the years to come, all health departments should develop adequate, comprehensive mosquito-control programs in conjunction with the general public and the federal agencies. Computerized programs can be utilized to aid in finding reoccurring mosquito breeding areas and determining level of control. Such an excellent program is used by the Marion County Health Department in Indianapolis, IN.

Roach Control

Roaches are controlled by keeping the premises extremely clean and by discarding paper bags, food cartons, and other materials from food stores and warehouses. Roaches are controlled chemically by the following insecticides: diazinon, 2% spray or 2% dust; malathion, 3% spray or 4% dust; baygon 1% spray or 2% dust. Borax is also used to control roaches by placing the powder in out of the way areas, such as wall openings, under cabinets, under bathtubs, in attics, and in other hard to get at places during construction. Roaches that repeatedly cross these areas will be killed by the chemical. German roaches may also be controlled by the use of an insect growth regulator called hydroprene, which augments the level of juvenile hormones. The insect develops a hormonal imbalance that neither allows it to remain an adolescent nor progress to an adult, causing the death of the insect. Roaches can also be killed by a desiccating dust such as diatomaceous earth. Sticky traps may be used with pheromones.

Tick Control

An individual is protected against ticks by keeping clothing buttoned and trouser legs tucked into socks, and by wearing the type of clothing that prevents ticks from penetrating and getting to exposed body parts. Individuals should always inspect their clothing and their bodies after they have been through a tick-infested area. The ticks should be removed immediately and destroyed. If ticks become attached, the simplest technique for removal is a slow straight pull that will not break off the mouth part and leave it in the wound. A drop of chloroform, ether, vaseline, or fingernail polish rubbed over the tick and the area help the removal. Tick repellents can also be used. However, no one repellent appears to be perfect for all ticks. Clothing may be treated with indalone, dimethyl tareate, or benzyl benzoate. Dogs and cats may be dusted with 0.75 to 1% rotenone or 3 to 5% malathion.

Ticks inside of buildings are controlled by the use of baygon or 0.5% diazinon. Sprays should contain 0.2 to 0.5% DDVP as a fumigant to drive the ticks from behind baseboards and from cracks in walls. Tick-infested areas on the outside may be treated with a 4% malathion dust or DDVP. The control program must go on for many months so that the infected eggs can be destroyed as they become adults. Another technique for the removal or destruction of ticks is to remove the host. Where dogs are infested with ticks in a certain area, the area and the dog should be carefully treated, and then the dog should be kept away from the usual sleeping quarters, which prevents the ticks from getting a blood meal.

Food Product Insect Control

Control of food product insects include the following: (1) locate the source of infestation by examining the foods and areas where infestation is most likely;

(2) clean all cabinets and storage areas thoroughly by vacuuming and washing with hot water; and (3) keep all shelving dry and spray with appropriate insecticides (these insecticides may include diazinon of 0.5% concentration or 1% baygon); (4) put only fresh food back into the area after drying and spraying; and (5) make sure that foods are rotated frequently. During the hot months, rotate the dry foods at least once every 2 weeks. During the cold months, the foods should be rotated at least once every month.

Insecticide Resistance and Vector Control

Insecticide resistance is a problem in all insect groups where they serve as vectors of emerging diseases. Each resistance problem is potentially unique and may involve a complex pattern of insect behavior and avoidance in the case of malaria vectors, or a biochemical resistance in the case of other vectors. The list of insecticide-resistant vector species includes 56 anopheline and 39 culicine mosquitoes, body lice, bedbugs, triatomids, 8 species of fleas, 9 species of ticks, flies and roaches. Resistance has developed to every chemical class of insecticide, including microbial drugs and insect growth regulators. Insecticide resistance is expected directly and profoundly to affect the reemergence of vectorborne diseases.

The two major forms of biochemical resistance are target-site resistance, which occurs when the insecticide no longer binds to its target; and detoxification enzyme-based resistance, which occurs when enhanced levels or modified activities of esterases, oxidases, or glutathione S-transferases interferes with detoxification.

Resistance to growth regulators, ivermectins, and other microbial agents has increased. The initial mechanisms that cause resistance to insect growth regulators were oxidase based. Resistance to ivermectins has resulted from a number of factors including oxidase, conjugation, and altered target-site mechanisms. Vectors have not yet demonstrated resistance to these compounds in the field. Microbial agents such as *B. sphaericus* and *B. thuringiensis* are considered insecticides because the principal agents are crystal toxins produced by the bacteria. The mechanisms of resistance to *B. sphaericus* are not yet defined, but more than one mechanism seems to be involved. The resistance to *B. thuringiensis* results from the binding of the toxin to the brush border in the lumen of the insect gut, or by enhanced digestion of toxin by gut proteases.

Pyrethroid resistance is emerging, despite early optimism that because of its rapid toxicological action, this newest class of insecticides would not produce resistance. Existing mechanisms of resistance are becoming enhanced and cross-resistance is occurring.

Bacterial Symbiosis in Arthropods and the Control of Disease Transmission

Bacterial symbionts may be used as vehicles for expressing foreign genes in arthropods. The selected genes can cause the arthropod to be incapable of transmitting a secondary microorganism that is pathogenic for humans. This is an alternative approach to the control of arthropodborne diseases. This approach is guided by the fact that certain arthropods use restricted food sources such as blood harboring bacterial symbionts; in some cases these symbionts can be cultured and genetically transformed to express a gene whose product kills a pathogen that the arthropod transmits; and normal arthropod symbionts can be replaced by genetically modified symbionts, resulting in a population of arthropod vectors that can no longer transmit the disease. This approach has worked in three species of Chagas disease vectors. Numerous questions need to be resolved concerning science, efficacy, safety, regulatory concerns in both laboratory and field research. These concerns include potential environmental and ecological hazards associated with release of genetically modified arthropod vectors of human disease; potential public health risks; an overall public perception, which can determine the future of programs.

Public Health Laws

A typical public health law gives the state board of health the power to develop rules and regulations to control nuisances that are dangerous to public health, to control fly and mosquito breeding places, and to control the spread of rodents. These provisions are typically found in the powers and duties of the state board of health. In addition, the food codes that come under the general public health laws have specific provisions that food establishments shall be protected by all reasonable means against flies, other insects, and rodents. Further, housing codes stipulate that dwelling units, school units, or any other type of dwelling or shelter must be free of insects and rodents. A health code may also contain a provision making it unlawful for any individual to maintain any lands, places, buildings, structures, vessels, or watercraft that are infested by insects and rodents. It may also stipulate that any person owning, leasing, occupying, possessing, or having charge of any land, place, building, structure, stacks, or quantities of materials (e.g., wood, hay, corn, wheat, and other grains), vessels, or watercraft must eliminate infestations of insects or rodents or be prosecuted under the pest eradication sections of the health code.

SUMMARY

Vectorborne disease continues to be a major deterrent to human settlement and agricultural development in certain areas of the country and the world. The insects discussed are of public health importance, including fleas, flies, mites, mosquitoes, roaches, ticks, and food product insects. These insects destroy food material and cause disease. The potential for intervention is excellent, providing that humans are willing to use a comprehensive program of habitat removal, biological control, chemical control, and good public health education. The pesticides used in chemical control, although effective, have caused potential hazards to humans, either through the short-range problem of intoxication or the long-range potential hazard of carcinogenesis or mutagenesis. The destruction of the vector should be supplemented by chemotherapy and vaccines to limit disease potential. In working toward a good pest control program, the potential hazards to the environment should be considered. The use of pesticides for all purposes should be coordinated to reduce the environmental degradation while increasing the health potential for society.

RESEARCH NEEDS

Additional study is needed to determine the best technique to be used for the control of arthropod vectors of disease or other public health pests, and the significance of this control on the ecology and behavior of the various ecosystems. A reporting system is needed where information from health departments of cities, counties, and states is sent to the CDC to help identify quickly the existing levels of insect or other pest problems. Additional study is needed to determine the problems related to environmental manipulation. Tolerance levels for biting nuisances should be established. It is necessary to have a combined study by agricultural experts, medical experts, and environmental experts to determine the problems of poison persistence and resistance as it relates to pesticide management.

CHAPTER 6

Rodent Control

BACKGROUND AND STATUS

Rodents are members of the order Rodentia, which includes squirrels, beavers, mice, rats, lemmings, porcupines, and chinchillas. These mammals have teeth and jaws adapted for gnawing. For the purpose of this chapter, four rodents — the Norway rat, the roof rat, the house mouse, and the deer mouse — are discussed (Figure 6.1).

The first three rodents, known as murine rodents (they are actually of the subfamily Murinae), are important because they are found everywhere that humans are found. Murine rodents live in buildings, destroy food and property, endanger health, and compete with humans for existence. They are found in all parts of the country and are dangerous whether they are out-of-doors or within the home.

In rural and semirural areas rodents pose the greatest problems by destroying and contaminating food. They inhabit outside areas, improperly managed solid waste disposal areas and attics, or other parts of the home, depending on the species. Further, the deer mouse (family Cricetidae) is a primary reservoir host for hantavirus infection.

Rodent infestations occur in urban areas and in luxurious suburbs where mice inhabit houses and rats inhabit lawns, undersides of doghouses, or any area where construction has taken place or sewers have been disrupted. The Norway rat typically lives in sewers or along creek or river banks and migrates to areas where food may be found.

Rat infestations become extremely significant within the inner city. To simply count rat bites or determine the amount of food loss or property damage is inappropriate. Human misery increases as the rat continues to spread. The rat is an important part of the inner city community. It is a symptom of the breakdown of the community, not only from an environmental standpoint, but also from a social and economic standpoint.

Rat control within the city is an extremely complex problem that must be resolved by involving not only the various municipal and state agencies but also the community itself. Behavioral patterns must be altered, lifestyles changed, social and economic factors improved, and serious concern with the rat problem shown if the rodent problem, including disease and community degradation, is to be eliminated.

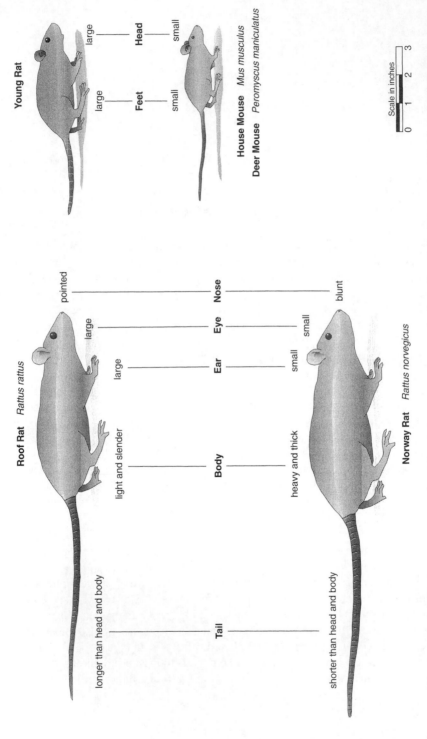

Figure 6.1 Field identification of domestic rodents. (From *Pictorial Keys to Some Arthropods and Mammals of Public Health Importance*, U.S. Department of Health, Education, and Welfare, Public Health Services, Atlanta, GA, 1964, p. 55, as adapted by H. Koren, 1994.)

More than 60 million people in the United States live in 300 cities where rats are a major environmental problem. The Urban Rat Control Program, funded by the federal government, was started in 1969 to eliminate the conditions that allowed rats to grow and multiply. The program emphasized the control and improvement of both the physical and social environments. As a result, intensive efforts in rodent control and solid waste removal reduced rat populations. Although the federal government by 1979 had spent many millions of dollars in attempting to remove or reduce rat populations, the results were insubstantial. The program was temporarily effective, but as soon as it was removed from an area, the rat problem generally returned. However, as of 2000, the city of Philadelphia, Marion county (Indianapolis), and some other communities were maintaining effective rodent control efforts. Rodents are developing immunity to anticoagulants in both the United States and Europe. As their resistance increases, so too will their numbers.

SCIENTIFIC, TECHNOLOGICAL, AND GENERAL INFORMATION

Characteristics and Behavior of Murine Rodents and Deer Mice

Murine rodents are capable of almost continual reproduction and produce large quantities of newborn mice and rats. Some of the young are killed or eaten. The newborn rodent is born blind, but gains sight in about 12 to 14 days. For about 21 days it depends on its mother for food. If forced to, it can start to live by itself within 3 to 4 weeks of birth. Sexual activity and fighting occurs in about 2 to 3 months. The young learn through experience and imitation. At 3 months, they are very active and completely independent (see Table 6.1 for detailed description of murine rodents).

Daily life patterns rarely vary, unless there is a sharp increase in the rodent population or a decrease in the food supply. When food is abundant, rodents are most active during the first half of the night; when food is in short supply, they are active at any time. Generally, however, if rodents are seen during the day, large numbers are present. Rodents tend to use regular paths and runways. They live as close to their food source as possible and travel to it with the least amount of exposure. This is why runways tend to be along the edges of walls or through covered areas.

Rodents have an avoidance instinct to alien objects that aids survival. Rodents sense and avoid unusual changes within the environment. They may avoid new food for several days. When poisoning, it is wise to prebait the area with unpoisoned bait for several days before poison is added. This is necessary when the poison kills swiftly; when the poison is an anticoagulant and death is slow, it is possible to start poison baiting immediately. In environments where there are constantly new strange objects, rodents are less prone to avoidance. Rodents also avoid areas containing several dead rodents, around poisoned bait, out of a sense of danger.

Rodents have tremendous versatility in climbing, jumping, reaching, and swimming. Mice and roof rats are excellent climbers; Norway rats vary from fair to good. Rodents can climb the vertical walls of most brick buildings as long as they can get

Table 6.1 Description of Murine Rodents and Deer Mouse

Characteristics	Norway Rat (*Rattus norvegicus*)	Roof Rat (*Rattus rattus*)	House Mouse (*Mus musculus*)	Deer Mouse (*Peromyscus maniculatus*)
Weight	16 oz	8–12 oz	1/2–3/4 oz	2/5–1 1/4 oz
Total length including the tail	12 3/4–18 in.	8 3/4–17 3/4 in.	6–7 1/2 in.	4 4/5–9 in.
Head and body	Blunt muzzle; heavy, thick body	Pointed muzzle; slender body	Small muzzle and body	Same
Tail	Shorter than head and body	Longer than head and body	Equal to or longer than head and body	Same
Ears	Small, close set	Large, prominent	Large, prominent	Same
Hind foot	1 1/2 in. or longer	Less than 1 1/2 in. in length	3/4 in. in length	
Teeth	Strong, well-developed; single pair of incisors in upper and lower jaw	Same	Same	
Color	Great range	Black, tawny, gray	Gray, brown, white	White feet; usually white underside; brownish upper surfaces
Distribution	United States and southern Canada	Southern Gulf states and Pacific Coast	United States and southern Canada	Throughout most of North America
Environmental distribution	Share human habitat	Less dependent on humans; live away from human habitation	Least dependent on humans; live anywhere	Same; usually forest and grasslands
Gestation	22 days	22 days	19 days	21–23 days
Mating	Female can mate again within 48 hours after giving birth	Same	Same	
Litter size	4–6 litters of 8–12 young each	4–6 litters of 6–8 young each	8 litters of 5–6 young each	2–4 litters or more, depending on weather; with 3–5 young each
Number weaned	20 per female annually	Same	30–35 per female annually	
Length of life	One year	One year	Same	2 years in wild, 5–8 years in captivity
Harborage	Outdoors — in buildings under foundations, in waste disposal areas; indoors — between floors	Outdoors — trees and dense vines; indoors — attics, between walls, enclosed spaces	Nest anywhere	Nest anywhere, but primarily out of doors; in Great Plains region they enter homes

Table 6.1 (continued)   Description of Murine Rodents and Deer Mouse

Characteristics	Norway Rat (*Rattus norvegicus*)	Roof Rat (*Rattus rattus*)	House Mouse (*Mus musculus*)	Deer Mouse (*Peromyscus maniculatus*)
	and walls, under solid waste, in any concealed area			and cabins
Range	100–150 ft	100–150 ft	10–30 ft	0.13–1.6/ha or greater
Droppings	To ¾ in.; capsule shaped	To ½ in.; spindle shaped	To ⅛ in.; rod shaped	
Food	Omnivorous — garbage, meat, fish, cereal	Omnivorous — usually vegetables, fruits, cereal	Omnivorous — usually cereals	Primarily seed eaters; also nuts, acorns, fruits, insects, larvae, fungi, and some green vegetation
Daily food requirements	¾–1 oz dry food; ½–1 oz water	½–1 oz dry food; 1 oz water	Nibbles; ¹/₁₀ oz dry food; ³/₁₀ oz water	

a toenail hold. They can also climb or cross wires easily. Rats can reach as much as 13 in. along smooth vertical walls. They can make a standing high jump of about 2 ft and with a running start, they can jump a little over 3 ft. Mice can high jump more than 2 ft with a running start. Rats can jump outward 8 ft horizontally when jumping or dropping at least 15 ft downward. With a running start, the distance is increased. Rats can swim as much as a half mile in open water. If they are placed in a tank of water, they repeatedly dive to the bottom to look for exit pipes. This is why they survive so well in sewer lines and use them as a transportation route.

As mentioned earlier, rodents live close to food sources if they can maintain their safety. Rats usually live within 100 to 150 ft of their food source, whereas the house mouse lives within 10 to 30 ft. Deer mice usually live within ⅓ to 4 acres. Rodents build their nests in any relatively quiet hiding place. Outside, they burrow into the ground. Norway rats burrow 5 to 6 ft; roof rats only burrow in the absence of Norway rats; house mice burrow extremely well if no other harborage is present. Rodents also nest in piled-up trash, in lumber piles, in discarded appliances, around stables, under animal houses, in old garages and sheds, and on water banks (streams, creeks, and rivers). Indoors, rodents nest between double walls, under floors, above ceilings, in attics and basements, and in any closed-in spaces behind counters, equipment, or stairwells. Their nests are usually bowl shaped, and are about 8 in. in diameter for rats and 5 in. in diameter for mice.

Rodent Senses

Rodents have the senses of touch, vision, smell, taste, and balance. In addition to the normal sense of touch, rats have highly sensitive whiskers with a complex nerve net at the base of each. Their bodies are covered by guard hairs, facilitating night travel.

Rodents are apparently color blind. Their sense of vision is underdeveloped, although they are able to detect objects as far away as 45 ft. Rodents have a keen sense of smell. They are attracted to the entrance of new bait boxes by the urine of other rodents, if the odors are strong enough to travel some distance. Human odor is commonplace, because rodents live with humans constantly.

The sense of taste is not as well developed in rodents as it is in humans. They probably cannot taste poisonous material in baits with the possible exception of red squill, which has a very bitter taste. Bait shyness is developed in rodents because the rodents become sick, not because they are able to taste the poison within the bait. In the case of anticoagulants, the poisoning is done so slowly, over a period of 7 to 10 days, that the rodent does not connect sickness with the anticoagulant.

Rodents have an excellent sense of balance. They usually land on their feet if tossed into the air. Rats and mice can sometimes fall two to three stories without injury. Mice have fallen as much as four stories without injury. This sense of balance enables rodents to cross from one building to another over wires and cables, or cross from a tree into the open window of a house.

Rodent Signs

Signs of rodent habitation apart from the sight of live or dead rodents include sounds, droppings, runways, tracks, rub marks, and burrows. The most frequently seen outdoor rodent sign is a burrow or gnawed material. The most frequently seen indoor rodent sign is rodent droppings. Obviously, the most positive proof of an infestation would be to sight live or dead rodents. Again, rodents travel at night; rodents seen during the day suggest a sizable infestation. A technique used in determining whether rodents are present in a building at night is going into the building and suddenly turning on the lights or flashing a large-beam flashlight. Either the rodents are seen or the noise of their flight is heard.

Sounds of rodents are very unique. You can hear them running, gnawing, or scratching within double walls or floors. You may also hear squeaking noises from young rodents in the nest or noises of adults fighting.

Droppings are a sure sign of the presence of rats and mice. The Norway rat droppings are ¾ in. long by ¼ in. in diameter. The roof rat droppings are generally smaller than those of the Norway rat. The house mouse droppings are usually about ⅛ in. long. The age of the dropping can be determined by color and texture. If the droppings are fresh, they are usually black, soft, glistening, and appear moist. If they are old, they are brittle, graying, and dry (see Figure 6.2).

Because rodents follow the same path frequently, they establish runways, tracks, and rub marks. The runways are easily identified because the area is smooth from the constant movement of the rodents. The area is also relatively free of dust, in contrast to surrounding areas. Tracks are detected by shining a light obliquely with the ground. The tracks are marks in dusty or muddy areas. Dark smears or rub marks are found along the walls, pipes, or rafters from the grease on the coats of the rodents. Generally the runways are along walls, under boards, behind stored objects, and in, around, or under accumulated solid waste. Tracing the runways, harborage, food

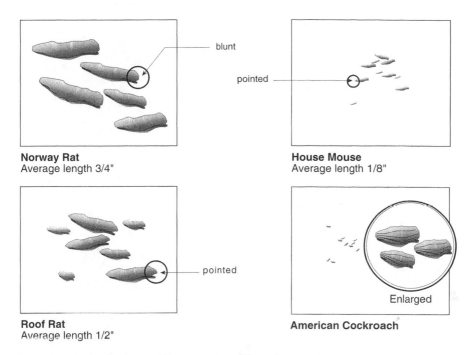

Norway Rat
Average length 3/4"

House Mouse
Average length 1/8"

Roof Rat
Average length 1/2"

American Cockroach

Figure 6.2 Recognizing rat and mouse signs. (From *Control of Domestic Rats and Mice,* U.S. Department of Health, Education, and Welfare, Atlanta, GA, 1972, p. 7.)

and water supplies, and methods of entry of the rodents into the building facilitates proper control measures (Figures 6.3 and 6.4).

Rodent Population Characteristics

A rodent population behaves like any other society. It goes through the characteristic stages of birth, growth, maturity, and death. Each population area has the capacity to support rodent populations according to the environment. This includes the amount of food, harborage, living space, and population in the area. The population size, affected by reproduction patterns, tends to increase in the spring and fall. Mortality rates counterbalance population increases. Female rats and mice live longer than males; however, accurate mortality data are difficult to obtain. Population density is also affected by movement. Rats and mice generally spend their lifetimes in a limited area. If within the home area there is a problem of sharply changing temperature or population increase, portions of the population move. Rodents frequently migrate to additional food sources. This behavior is important in a successful rodent control project, because if the food source or harborage is removed prior to poisoning, the rodents scatter and sharply increase the rodent population of other areas. The major food source for rodents is garbage, stored food products, crops, and grains. Harborage is a limited problem because rodents are so adaptable.

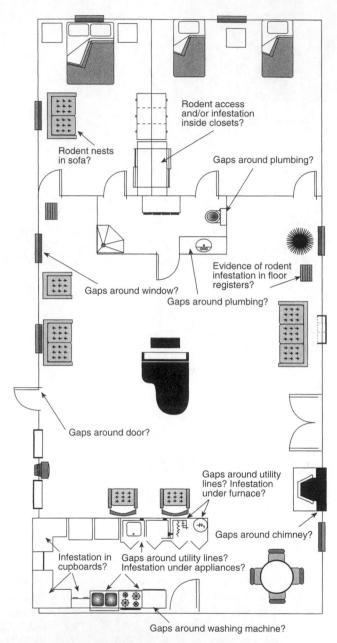

Figure 6.3   Internal rodent site check. (From Hoddenback, G., Johnson, J., and Disalvo, C., *Rodent Exclusion Techniques, A Training Guide for National Park Service Employees,* National Park Service, Washington, D.C., 1997, p. 7.)

Rodents are predators and will, in the absence of a food source, feed on each other; they are, of course, the victims of predators as well. Where the population of rodents is high, more are killed by predatory animals. Cats, with the exception of

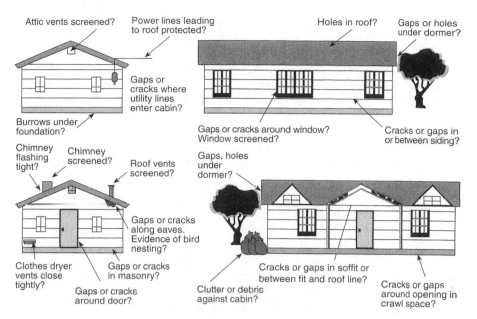

Attic vents screened?

Power lines leading to roof protected?

Holes in roof?

Gaps or holes under dormer?

Gaps or cracks where utility lines enter cabin?

Burrows under foundation?

Gaps or cracks around window? Window screened?

Cracks or gaps in or between siding?

Chimney flashing tight?

Chimney screened?

Roof vents screened?

Gaps, holes under dormer?

Gaps or cracks along eaves. Evidence of bird nesting?

Clothes dryer vents close tightly?

Gaps or cracks in masonry?

Gaps or cracks around door?

Clutter or debris against cabin?

Cracks or gaps in soffit or between fit and roof line?

Cracks or gaps around opening in crawl space?

Figure 6.4 External rodent site check. (From Hoddenback, G., Johnson, J., and Disalvo, C., *Rodent Exclusion Techniques, A Training Guide for National Park Service Employees,* National Park Service, Washington, D.C., 1997, p. 7.)

Siamese cats, are poor rodent predators. Although cats are good mouse catchers, they generally play with the mice rather than kill large quantities of them. Dogs, although good rat catchers, are not used for rodent control work because they are often mutilated by the rodents. The presence of domestic animals also provides a source of food for the rodents because owners leave pet food and water out for their pets, providing an excellent fresh food source for the rodents on a continuing basis. Other rodent predators are snakes and birds. Rodents compete fiercely for food and water, nesting sites, harborage, and females in heat. When Norway and roof rats compete, the fiercer Norway rats generally are successful. Norway rats compete with each other — many times over females. This competition among members of the same species is associated closely with their social organization. The aggressive animals live closest to the food source, whereas the least aggressive animals live farthest away from the food source and are in most danger of attack by predators.

RODENT PROBLEMS AND THEIR EFFECT ON HUMANS

Rodents cause psychological, sociological, economic problems, and human disease. It is estimated that each rat damages between $1 and $10 dollars worth of food and other materials each year by gnawing or feeding. They contaminate 5 to 10 times more food and materials than they utilize. Many fires of unknown origin are attributed to rats gnawing through the insulation of electrical wires. They also gnaw through walls, floors, and doors either to obtain food or to grind down their teeth, which grow at a rate of 5 in. a year. Americans lose an estimated $500 million to $1 billion annually from rodent damage.

It is difficult to properly determine the number of rat bites occurring each year, because many are unreported. Many years ago the author, while involved in a community rodent control program, conducted a survey of several city blocks within the inner city. At least 1500 rat bites had occurred during a year in the target rat control area. It was difficult to get good statistics, because many of the individuals interviewed failed to report the rat bites to the health department or other medical authorities. An estimated 45,000 individuals are bitten by rats each year. Apparently these figures are low. To get a proper figure, it would be necessary to conduct extensive surveys of many inner city areas throughout the country.

Rats bite helpless infants and defenseless aged or invalid adults. Rat bites are disfiguring; they may become infected and in rare cases lead to death. The psychological problem associated with rat bites is enormous. The individual is extremely frightened by the occurrence and may bear emotional scars for years.

During the course of one of the rat control projects participated in by the author, a mother came running out of the door screaming that her child was being bitten by a rat. When the health department extermination crew ran into the building, they found a large Norway rat clinging to a 6- or 7-year-old girl. The rat had to be beaten into unconsciousness before it would let go of the child. Rats that bite once will probably bite again. That is why it is so important to report rat bites and then to exterminate the premises where the bite has occurred.

An estimated 60 million Americans live in dilapidated and dirty inner city sections within the nation's larger cities. To these individuals the rat is the uninvited king that bites, destroys, stalks its prey, makes irritating noises, defecates, urinates, and spreads disease. It is a symbol of urban decay. Empty lots piled 3 and 4 ft high with garbage, trash, and debris; or houses with plaster falling down, electrical wires showing, floor boards ripped up, staircases collapsing, and overpowering odors are home to rats.

Today the sewer systems of our major cities are infested with rats. Sewers provide easy entrance and exit to areas, food, liquid, shelter, and constant temperature. The food found in sewers includes food wastes, undigested food in fecal material, ground garbage from garbage disposal units, and food that is washed or thrown into sewers. Rats enter sewers by burrowing into the ground, and then through a broken pipe into the sewer, or they swim through storm drains. Combined sewers have a far greater rate of rat infestation than the modern sanitary sewer, because the opportunity for entrance is increased. During the summer months, when the level of liquid decreases within the sewer, the rats may take to the exterior areas and search for food in yards and other areas. Rats generally do not leave the sewer area unless the competition for space and food becomes acute. Then they move to a new section of the sewer or leave the system altogether. Obviously, because rats are constantly in contact with all manner of infectious organisms, the opportunity for spreading disease is multiplied. When rodent problems occur in new neighborhoods, they are frequently traced to broken or improperly connected sewer lines.

Diseases

Rats and mice may be the reservoir of infection for many human diseases, including murine typhus fever (see Chapter 5), hantavirus pulmonary syndrome, plague (see Chapter 5), rat bite fever, rickettsial pox (see Chapter 5), salmonellosis (see Chapter 4), trichinosis (see Chapter 4), viral hemorrhagic fevers, and Weil's disease.

Hantavirus Pulmonary Syndrome

Hantavirus pulmonary syndrome is caused by *Hantavirus*, a genus in the family Bunyaviridae, which includes lipid-enveloped viruses with a negative-stranded ribonucleic acid (RNA) genome composed of three unique segments. Humans may be infected when they encounter and inhale aerosolized microscopic particles that contain dried rodent urine, feces, or saliva from infected deer mice. The disease might also be transmitted by rodent bites. It is not transmitted from person to person. The disease is extremely severe with a fatality rate of over 50%. The incubation period is about 1 to 2 weeks, but ranges from a few days to up to 6 weeks. The reservoir of infection is the deer mouse.

As of May 28, 1999, the Centers for Disease Control and Prevention (CDC) has confirmed 217 cases of hantavirus pulmonary syndrome in the United States. From January through May 1999, seven cases were confirmed in Colorado, New York, New Mexico, and Washington. An additional 11 suspected cases were reported in Arizona, California, Idaho, Iowa, Montana, New Mexico, and Washington. This is an increase over previous years. Hantavirus antibody prevalences in deer mouse populations surveyed during the spring of 1999 were 35 to 45% in some populations in New Mexico and up to 40% in Colorado. The increase in rodent populations and the prevalence of the virus have led to an increase in the disease in humans. The risk for human disease is proportional to the frequency with which people come into contact with infectious rodents, rodent population density, and prevalence of infection in rodents. Disturbing rodent droppings and sleeping near burrows where trash is present are contributing factors.

Rat Bite Fever

Rat bite fever or Haverhill fever is transmitted through the teeth and gums of a rat infected with the organism *Streptobacillus moniliformis*. This disease, which may also be spread by milk, is highly infectious and causes acute febrile symptoms. If untreated the fatality rate is 7 to 10%. The incubation period is 5 to 15 days, usually 8 days. The reservoir is infected rats.

Viral Hemorrhagic Fevers Caused by Arenaviridae

Arenaviridae is a family of viruses whose members are generally associated with rodent-transmitted diseases in humans. Each virus is usually associated with a particular rodent species. Arenaviruses cause infections that are relatively common in humans in some areas of the world and may lead to severe illnesses. The virus particles are spherical and have an average diameter of 110 to 130 nm. All are enveloped in a lipid membrane. In cross section, they show grainy particles that are ribosomes acquired from their host cells. Their genetic material is composed of RNA only. Their replication strategy is not completely understood, but new viral particles called virions are created by budding from the surface of their hosts' cells. The arenaviruses are divided into two groups: the New World or Tacaribe complex and the Old World or lymphocytic choriomeningitis (LCM) and Lassa complex. Viruses in these groups that cause illness in humans are as follows:

1. Lymphocytic choriomeningitis virus which causes lymphocytic choriomeningitis
2. Lassa virus which causes Lassa fever
3. Junin virus which causes Argentine hemorrhagic fever
4. Machupo virus which causes Bolivian hemorrhagic fever
5. Guanarito virus which causes Venezuelan hemorrhagic fever
6. Sabia virus which causes a disease found in Brazil that has not been named yet

These viruses are zoonotic, which means they are found in animals in nature. Each virus is associated with either one species or a few closely related rodents. Tacaribe complex viruses are usually associated with New World rats and mice of the family Muridae, and subfamily Sigmodontinae. The LCM/Lassa complex viruses are associated with Old World rats and mice of the family Muridae, and subfamily Murinae. These rodents are located in Europe, Asia, Africa, and the Americas. Only the Tacaribe virus found in Trinidad has also been isolated from a bat. The rodent hosts of the arenaviruses are chronically infected with the viruses, although these rodents do not appear to have obvious illnesses. Some of the Old World arenaviruses appear to be passed from mother to offspring during pregnancy. Some New World arenaviruses are transmitted among adult rodents through fighting and inflicting bites. The viruses get into the environment in the urine or droppings of the infected hosts.

Human infection is incidental to the natural cycle of the viruses and occurs when a person comes into contact with rodent excretions, materials contaminated by excretions, an infected rodent, contaminated food through ingestion, and tiny particles contaminated with rodent urine or saliva and which are inhaled. Some renaviruses, such as Lassa and Machupo viruses, are associated with secondary person-to-person and nosocomial transmission. This occurs when there is direct contact with the blood or other excretions containing virus particles of infected people. Airborne transmission has also been reported.

Lassa fever is an endemic disease in portions of West Africa. An estimated 100,000 to 300,000 cases with approximately 5000 deaths occur each year. The symptoms typically occur 1 to 3 weeks after the patient comes into contact with the virus. They include fever, retrosternal pain, sore throat, back pain, cough, abdominal pain, vomiting, diarrhea, conjunctivitis, facial swelling, proteinuria, and mucosal bleeding. Neurological symptoms include hearing loss, tremors, and encephalitis. Approximately 15 to 20% of patients hospitalized die from the illness.

Weil's Disease

Weil's disease or leptospirosis is caused by the organism *Leptospira incetro-haemorrahagiae*. The individual becomes infected by direct or indirect contact with infected rodent urine. The organism enters the body through the mucous membranes or through minute cuts or abrasions of the skin. The disease ranges from mild to severe, but is seldom fatal. The incubation period is about 10 days, with a range of 4 to 19 days. The reservoir of infection is rats or farm and pet animals.

In October 1996, the Illinois Department of Public Health was advised about five people who had an unknown febrile disease after returning from a white-water rafting trip on flooded rivers in Costa Rica. A total of 26 people from five states and the District of Columbia as well as from Costa Rica had participated in the event. Nine of the individuals became ill. Leptospirosis was the cause of the febrile illness. In 1995, in Nicaragua, 2000 people became ill after widespread flooding. In 1998, the Illinois Department Of Health, the Wisconsin Department of Health, the U.S. Department of Agriculture (USDA), and CDC in collaboration with other state and local health departments investigated an outbreak of acute febrile illness among athletes from 44 states and 7 countries who participated in triathlons in Springfield, IL, on June 21, 1998, and in Madison, WI, on July 5, 1998. A total of 90 athletes had illnesses that met the case definition of leptospirosis.

POTENTIAL FOR INTERVENTION

Isolation, substitution, shielding, treatment, and prevention techniques are all forms of intervention in controlling rat problems. The potential of each technique varies from excellent to poor, depending on the individual involved, the nature of the housing areas, the concern of citizens' organizations, and the government. Isolation is ineffective because it is almost impossible to isolate rodents. Substitution is another unacceptable procedure. Proper rodent-proofing techniques in all areas are used in the shielding technique. Treatment is the application of rodent-specific pesticides. Commercial applicators use the anticoagulants and other chemicals, whereas government programs use raticate and anticoagulants. Prevention is by far the most effective technique for reducing rodent problems. Prevention is the removal of all food and harborage, and proper maintenance of property to eliminate rodent entry.

RESOURCES

Scientific and technical or industry resources include, among others, National Pest Control Association, area schools of agriculture, National Environmental Health Association, and American Public Health Association.

Civic and professional resources consist of the community civic associations responsible for neighborhood beautification, neighborhood housing improvement, and specific rodent control programs. Pest control operators associations and chemical manufacturers associations are important resources.

Governmental resources include local and state health departments, the federal urban rat control program, and county extension agents. The city of Philadelphia Department of Public Health Division of Environmental Health is an important resource. This agency prepares a report annually of its activities, titled Philadelphia Environmental Improvement Program.

STANDARDS, PRACTICES, AND TECHNIQUES

Rodent Poisoning and Trapping

The safest rodent poisons include the anticoagulants, such as warfarin, pival, fumarin, and diphacinone. Somewhat less safe, but still effective, is the anticoagulant brodifacoum. Bromethalin is quite toxic to humans and pets, but is very effective. Although many other rodent poisons are used for rat and mouse control, they are too toxic to be used in the home. Some of the more toxic poisons include zinc phosphide and alpha naphthyl thiourea (ANTU). Rodent poisoning is discussed in detail in the chapter on pesticides.

Rodent trapping is only fairly successful for rats, because rats drag traps, if necessary chew off a leg caught in a trap, or even cause a trap to malfunction. Mouse trapping is successful if single mouse traps, instead of four-unit traps, are used if the trap is placed perpendicular to the wall and within the runway and if fresh bait is used on a routine basis. The bait should be changed each day or every other day. Mice are attracted by peanut butter, bacon, nuts, and other fresh foods. The bait should always be tied to the trigger so that the mouse will not steal it. Another technique of baiting is to take cornmeal or some other tasty food and make a tiny path up to the trigger and cover the trigger with the food.

Glue boards, which are similar to flypaper but sturdier, have been used but are not very successful. The rodent sticks while crossing the glue board and must then be killed and disposed of.

Food and Harborage Removal

The potential for rodent infestation is directly related to the amount of food and harborage present. Rodents find food in exposed garbage, pet food, or stored food, such as grains, fruits, meats, and vegetables. Rodents also find food sources in gardens, on farms, in horse manure, and in any other type of organic material. Although they prefer fresh food, rodents eat anything.

Most harborage is provided by poorly stored rubbish and trash, lumber, abandoned vehicles, discarded appliances, old shacks and sheds, high weeds, hay, double walls, false ceilings, drawers, attics of houses, and any other area where the rodents may be able to conceal themselves and nest. Frequently, rodents are found in cellars or closets within the home behind piles of dirty, discarded clothing.

An effective rat control program requires removal of available food supplies and shelter for the rodents. As many rats as possible must first be exterminated to avoid scattering during a comprehensive cleanup program. Reduced food supply decreases the rodent population because of competition for food and harborage. The cleanup technique is extremely important, not only for the reduction and final elimination of most of the rodents but also for the elimination of flies, roaches, mosquito-breeding areas, and for the reduction of potential disease.

All refuse should be stored in 30-gal metal cans with tight-fitting lids 12 in. off the ground. Only heavy-duty plastic is acceptable because rodents chew it and plastic cans tend to tip easily. Heavy-duty rubber cans may be used providing the lids screw on tightly. A good refuse container is rust resistant, watertight, tightly covered, easy to clean and handle, and of heavy-duty construction. Drums (50 gal), such as oil drums, should never be used as a refuse storage container, because the lids do not fit properly on the container and when full they are too heavy to lift.

Remove pet food immediately after mealtime, making certain to dispose of all spilled pet food as well. Pet food has been a vital source of rat problems in numerous instances.

All dry food products should be stacked on pallets 12 in. above the floor and 12 in. from walls. Rotate food at least once every 2 weeks during the summer and once a month during the winter. Clean behind and under dry food on a daily basis. All spilled foods should be removed immediately. It is important to use a bright light or flashlight when sweeping storage areas to locate all spilled food and to determine if rodent signs are present.

Materials such as lumber and pipes should be stored in columns 100 ft from the house and 12 in. above the ground. Because high weeds contribute not only to rodent problems but also to fly and mosquito breeding, it is essential that they be trimmed regularly and destroyed. Cut grass, brush, and dense shrubbery within 100 ft of the home.

Abandoned automobiles are rodent harborages and should be removed and disposed of promptly in keeping with air pollution or solid waste ordinances. Rickety old sheds are excellent rodent shelters. Materials usually found in these sheds should be removed and such sheds either refurbished or torn down. Numerous rodent problems are traced to old sheds, which provide excellent harborage for rats.

Solid Waste Disposal and Storage

Solid waste should be collected at least twice a week during the summer and at least once a week during the winter. All business areas should have solid waste collected on a daily basis. Many of the large metal containers found behind businesses are open invitations for rats to come, eat, and live. If these containers are improperly covered and maintained, food is spilled, units grow odorous, and attract flies, roaches, and rodents. It is essential that these containers are perfectly maintained, that the surrounding areas are policed several times a day, and that the lids are placed firmly on the units. Compactor trucks are best for collection of garbage and other types of solid waste. Compactors cannot be used for the collection of large items, such as refrigerators and lumber. Compactor trucks prevent blowing and spilling of materials, are leakproof, and are easy to load and unload. Solid waste taken to properly operated sanitary landfills sharply reduces potential rodent problems. Properly operated incinerators destroy solid waste materials; however, when improperly operated, partially burned organic material is removed from the incinerator and becomes an excellent rodent attractant.

Garbage may be ground up and sent through the sewers to sewage treatment plants. This technique is fine for the home but provides enormous quantities of fresh garbage for the rodents living in the sewers. Composting can be utilized; however, if the compost heap is improperly maintained, it also becomes a rodent attractant and supplies rodent food. Further information on solid waste disposal can be found in Volume II, Chapter 2, on solid and hazardous waste.

Safety Precautions for Hantavirus Infection

Residents of hantavirus-affected areas should follow these general household precautions to reduce the risk of exposure to this most serious disease:

- Eliminate potential insect parasites of rodents
- Eliminate rodents through poisoning and trapping with spring-loaded traps and peanut butter as an attractant
- Reduce rodent shelter and food sources within 100 ft of the home
- Before rodent elimination, ventilate buildings thoroughly for at least 30 min
- Wear rubber or plastic gloves when removing rodents or cleaning
- Before removing gloves, wash gloved hands in a general household disinfectant and then soap and water (3 tbsp. of bleach in 1 gal of water can be used). Then wash and disinfect hands thoroughly
- Leave several baited spring-loaded traps inside the house at all times
- Clean up rodent contaminated areas by spraying everything that has come in touch with the rodents with a general purpose household disinfectant, bag the materials or dead rodents, and bury in a hole 2 to 3 ft deep or incinerate. Then spray and mop floors, countertops, and surfaces; clean carpets; and launder clothes

In homes with hantavirus or heavy rodent infestation, wear disposable coveralls, rubber boots or disposable shoe coves, rubber or plastic gloves, protective goggles, half-mask air-purifying (or negative pressure) respirator with a high-efficiency particulate air (HEPA) filter or a powered air-purifying respirator with HEPA filters. Decontaminate on site, launder on site, or immerse launderable items in a disinfectant until laundry can be done.

All potentially infective waste material, including respirator filters that cannot be incinerated or buried deeply on site should be double bagged, marked as infectious waste, and disposed of in accordance with infectious hazardous waste regulations.

Workers in affected areas exposed to rodents should have baseline serum samples taken; understand symptoms, nature, and severity of the disease; and seek immediate attention if ill.

Campers and hikers in affected areas should avoid contact with rodents or rodent nests; avoid use of empty cabins or enclosed shelters unless thoroughly cleaned, disinfected, and ventilated; avoid sleeping on the bare ground or near rodent shelters, feces, or burrows; and use only chlorinated, iodinated, boiled, or bottled water.

Rodent Proofing

Rodent proofing is the elimination of all holes ½ in. or larger for rats and ¼ in. or larger for mice to prevent these rodents from entering an establishment. Rodent proof all doors, windows, gratings, vents, pipe openings, and foundation walls within easy access of rats. Techniques include cuffing and channeling, screening, use of metal guards, steel wool, concrete, and curtain walls. Channeling and cuffing is the application of metal bent at right angles to the bottom and sides of doors. Kickplates are also used on doors. The door must be lowered so that the opening beneath it is less than ½ in. from the floor. Vents and windows must be screened against rodents and flies. Existing fire control ordinances or rules and regulations should be checked to avoid violation. The screening should be 17-gauge galvanized hardware cloth for protection against rats, and 19-gauge (4 × 4½ in. mesh) for protection against mice. Where sheet metal is used, it must be 24-gauge galvanized sheet metal.

Metal guards are protection against rats entering a building by coming across wires and pipes. The metal guard is conical in shape, with the cone facing outward. The openings around pipes or conduits have to be covered with either sheet metal patches or with concrete bricking mortar. The difficulty with the use of concrete is that if it is in an area, especially around radiators, where expansion and contraction takes place, the concrete will expand and contract at a different rate than the wall; and may easily crack and crumble, thereby creating a new rat opening.

Curtain walls, which are L-shaped walls, are made of concrete. They are poured along the foundation walls to keep rats from burrowing under the foundation. The curtain has to be at least 2 ft below ground level, 1 ft in length, and 4 in. in height. (See Figure 6.5 for these types of rodent proofing.) Place metal roof flashing around the base of a wooden, earthen, or adobe dwelling to a height of 12 in. above the ground and 6 in. below the ground. Also, place 3 in. of gravel below the base of homes or under mobile homes to discourage burrowing.

Rodent proofing also calls for the elimination of dead spaces, such as double walls, double floors, or enclosed areas of stairways, whenever possible. If elimination is impossible, then these areas should be checked frequently for rodent infestation and measures taken as are needed for control. Once a building has been rodent proofed, adequate rat and mouse control work should take place within the building before it is occupied.

Continued Maintenance

Continued maintenance is a necessary part of any rodent control activity. Once food and harborage are removed, the last rodent is killed, and rodent proofing is complete, many assume the job is finished. It is not, because there are innumerable ways in which rodents may still invade a premises. New pipes may be put into a building. New construction may take place. Doors or windows may be left open. Walls may crumble. Screens may be broken. Garbage and trash may start to accumulate again, and new harborage may be created. It is essential that the individuals responsible, whether they be tenants, landlords, or homeowners, thoroughly survey the premises at least once every 2 weeks to determine if a rodent problem is

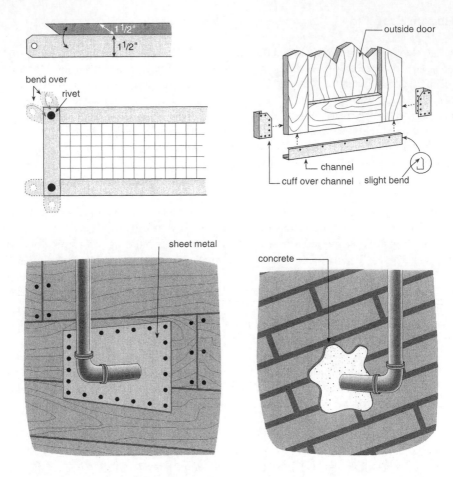

Figure 6.5   Rodent-proofing devices. (From *Control of Domestic Rats and Mice,* U.S. Department of Health, Education, and Welfare Public Health Service, Communicable Disease Center, Atlanta, GA, 1964, p. 22.)

redeveloping. Flashlights should always be used indoors to check for rodent signs. Continued maintenance is the key to any private or public rodent control program. Without it the conditions not only reoccur, but also the rat problem substantially increases, because it takes a period of time before a rat population can find some sense of equilibrium within its environment.

MODES OF SURVEILLANCE AND EVALUATION

Rodent surveys are conducted by professional personnel, technical personnel, members of the community, or businessmen who are trained to look for rat signs. There are four basic types of rodent surveys. The first one is a survey in a single building, business, or industry where rodents are suspected. The surveyor looks for previously discussed rodent signs, for food and water sources, for improper solid waste storage and removal, for rodent openings, and for any other potential rodent harborage. This information is noted on a survey form and the exact place where each of the problems occur is marked on a rough map on the back of the form. It is wise to use a system of symbols with explanations to avoid cluttering the map.

The second type of survey is one in which the environmental health practitioner or technician makes a study of an area and determines the types of problems in a given city block. This study includes such information as the address of a building, information concerning the building (i.e., residential, business, and vacant lot), potential rodent food and harborage, active rodent signs, potential rodent entries, and so forth. A map of the area is then drawn. It is also wise, at this time, to make a comprehensive housing survey to determine the rodent problem within houses and the types of housing problems needing correction.

A third type of survey is one in which a professional health educator or environmentalist goes from house to house with a questionnaire to determine from residents the types of rodent problems encountered. Again this information could be put on a check sheet. The information should be indicated as a questionnaire survey, in contrast to a physical survey. Together the two surveys are of considerable value in a rodent control program.

A fourth type of survey is one in which the environmentalist goes out to a vacant area, farmland, park, or other outdoors area away from homes to determine where rodent conditions exist. A detailed map is drawn of the area to indicate precisely where rodent problems are found and the extent of the problems. It is wise to utilize a simple complaint form to record rodent or other problems on an individual basis.

Ecological Studies of Rodent Reservoirs

In the past several years, emerging human diseases have been associated with small-mammal reservoirs. It is essential to understand the host and vector ecology to prevent and control disease. The rodentborne hemorrhagic fevers are among the most dramatic of the recently emerging infectious diseases. Hantavirus pulmonary syndrome is a good example of this. Recognition of these new rodentborne diseases has renewed interest in reservoir host ecology in the United States as well as in South America.

Once the primary risk or host has been identified, the following overlapping steps are used to understand the rodent reservoir ecology as it relates to human disease:

1. Determination of the geographic distribution of the host
2. Determination of the geographic range of the pathogen within the host range
3. Determination of the regional distribution of the host and pathogen and the habitats
4. Determination of the relative prevalence of infection among populations of the host, such as males, females, adults, and juveniles
5. Understanding of the host–pathogen dynamics
6. Development of a time and place predictive model

The geographic distribution of the host defines the maximum area in which the disease can be endemic, and therefore defines the area in which the human disease can be endemic. The studies of these populations help identify the levels of potential infection, the seasonal and year-to-year fluctuations in incidence and prevalence, and the environmental variables associated with changes in host density or rates of transmission. Predictive models may then be established, taking into consideration such factors as weather conditions, habitat, insect populations, rodent populations, and their potential for producing human disease.

RODENT CONTROL

Many of the techniques of rodent control include poisoning, trapping, removal of food sources, removal of harborage, storage of solid waste, alternate methods of disposal of solid waste, and continued maintenance. Additional control measures are used in large communities, single homes, sewers, and outside areas.

Community Rodent Control

Although rodent problems are not unique to any part of the urban environment, those that are of greatest concern exist within the inner city. Slum tenements and poorly kept houses, coupled with debris, litter, and other solid waste, contribute to the large rat populations. Where single-family, three-story, or four-story dwellings once existed, now dwellings have been divided into six- to eight-family apartments. The population density of a given dwelling can rise from 4 or 5 to 50 or 60. Obviously, bathroom accommodations are lacking and there is insufficient space for adequate living. Many of these dwellings have numerous rat entrances, usually bad sanitary conditions, and severe rat infestations. The human population is generally highly transient. Cohesiveness between members of the same building seems to be lacking.

Even where civic organizations and block councils exist, only a few residents may have participated in electing the officers and boards of directors. At times, individuals may simply appoint themselves as officials. Generally, these associations start out to improve their neighborhood by keeping taprooms out, obtaining stoplights or stop signs, improving recreational facilities and housing conditions, and possibly getting involved in some health matters. The typical single-block organization consists of a group of people who live near each other on a given street. These individual block organizations range from totally inactive to highly active. Usually within each block one finds one or more civic-minded people who are willing to work together on a project to improve the neighborhood. In many cases, the larger civic association does not truly represent the wishes of the individual block groups, because the more vocal and better-educated individuals tend to come to civic association meetings, whereas the less vocal tend to stay home and be concerned with individual problems. Unfortunately, if you only use these organizations, rat-control projects will be unsuccessful, because the people who need the most help are generally those who do not attend the civic association, home and school association, or health and welfare meetings. To stimulate this group of individuals to work together, it is important to go into each individual block and work on a house-to-house basis to attempt to set up small meetings in the homes.

A good approach is to knock on doors, introduce yourself officially as a member of the health department (make sure you always show your identification card), and explain the severity of the rat problems in the area, as well as in other areas. The environmentalist should also give some indication as to the nature of the programs that can be brought into the area to help improve it. These individuals within the blocks are then encouraged to come to a mass meeting at a local school to further discuss the rodent control problems and the kind of things that might be done to improve the environment. Each individual should be encouraged to bring friends, neighbors, and relatives to the meeting. Announcements should be distributed the day prior to the meeting and on the day of the meeting. Sound equipment should be used on a truck to announce the community meeting for rodent control at the local school. Explanations of the rodent problem given over the loudspeaker should be simple. The meeting place and time should be spelled out clearly.

It is essential at this mass meeting to have representatives from all the civic associations and the various official agencies to discuss informally the needs of the community as they see them, to listen to citizen problems, and then to explain some proposed techniques for conducting a community rodent control program. It is important that the individuals at this meeting vote to participate in each of the phases of the rat program. During such a meeting each of the citizen complaints should be recorded and submitted to the appropriate official agency for correction, where possible.

In one such meeting in which the author was involved, the community's major concern was the fact that the young people were constantly getting into trouble with the police. There was an apparent language barrier. When this problem was identified, the police department sent community relations people who spoke the language of the community into the area and worked with them to improve community–police relations. This helped not only to satisfy a need of the community, but also to indicate that the official agencies were interested in the public and that it would be a good idea to work together on community rodent control.

Another technique used to organize blocks is to pull together members of the community by the use of sound equipment at the time that rodent poison is placed by the health department. Here the members of any given block have an opportunity to meet face to face. This is an important time to hold an election of officers for that given block. Once officers are elected, the health educator in conjunction with the larger community organization should assist the block in developing its own block organization.

Another technique is to ask individuals, when you are going door to door prior to the first mass meeting, about the individual who they turn to when they have problems. Frequently, key people in a block are identified. These individuals are natural leaders and could be utilized as such in a block organization. Innumerable block organization techniques exist. The key is not necessarily which technique to use, but instead to help organize a block, stimulate the people to work together, and try to maintain the organization so that a ready supply of willing workforce exists for continued maintenance to keep the program from falling apart. These block organizations may be used later in helping bring satellite health clinics into an area to help increase immunizations, and so forth.

An important person in a block organization can be the health aide. This individual is a member of the community hired by the health department who receives specific health training. As a member of the community and therefore recognized by the community, and as a member of the health team, this individual can greatly assist in developing a comprehensive rodent control program, as well as other comprehensive health programs.

Presurveys

Three types of presurveys may be used. First is the horseback type of survey, where the environmentalist rides up and down alleyways or walks the city blocks to get some approximate idea of the kinds of existing housing conditions, the amount and type of solid waste, and the kind of rat signs readily visible. This survey gives a rough idea of the needs of the community.

The second kind of survey, the opinion attitude survey, should contain the following basic points: (1) what the extent of the community organization is, including size and scope of the group, potential leaders, and desire to organize; (2) what the needs and attitudes of the people are toward elimination of roaches, rats, and poor housing; (3) what the extent of the pest infestation is, including mice, rats, and roaches; (4) what action the citizen has taken against roaches, mice, and rats; (5) how many complaints the individual has made about pests, and to whom they have been made; (6) how many rat bites have occurred in the person's house in the last year, how many have been reported to the health department, and how many have been reported to others; and (7) whether the individual has heard about rat bites in the neighborhood within the last year, and if so, how many. The information in this opinion attitude survey, when compiled, gives the environmentalist some understanding of the community's self-concept and whether the individuals are anxious to help correct the rodent situation.

The technical survey consists of a professional evaluation of rodent signs inside and outside of the property, of potential rodent entrances, of availability of food and harborage, and of rat-proofing problem. The data should be compiled and combined with the information in the opinion attitude survey to make a good evaluation of the existing rodent problems within the community.

The survey data should be kept and used as a means of comparison after completion of the program. This can give the environmentalist a means of comparing the results of the actions taken and also the results of continued maintenance.

In addition, it is well to add the material from the files of the health department relating to rodent complaints, solid waste complaints, and rat bites that were reported to the health department within the areas under study. It is recommended that a comprehensive map with proper coding be drawn to indicate the degree of severity of rodent infestations and also the other environmental problems associated with rodent infestation.

Area Selection

The area selected for a community rodent control program should be based on the extent of community organization, the willingness of groups to form and work together, the number of complaints for a 3-year period relating to rats and solid waste, the number of rat bites for the past 7 years, and the various surveys that have been conducted. These proposed areas should be stated clearly on a map and in writing. The number of houses involved, the severity of infestation, and the extent of cleanup should then be presented to a committee composed of regulatory agencies and civic leaders. The ultimate decision of where to work should be made jointly by these representatives of the communities and the official agencies.

Planning

To plan an effective community rodent control program, it is necessary to establish a working committee structure. The steering committee can be composed of representatives of the health department, housing department, police sanitation unit, solid waste department, and community relations department. The functions of a steering committee are to plan details of operation, resolve day-to-day problems, supply general knowledge of existing community resources, determine on a departmental basis the availability of workforce, act as liaisons between the project and the individual official departments or agencies, and help evaluate and redirect the program when needed.

A second committee, called a citizens advisory committee, should be established and can be composed of representatives of the health and welfare council, civic associations, block councils, block leaders, unofficial agencies, local religious leader, local school principal, representative of the landlords, business person, and others as needed. The functions of this committee are to review the suggested programs developed jointly by official agencies and citizens and make recommendations to the health director or environmental health director who is the coordinator of the rodent control program concerning changes or additions. These recommendations may include selecting certain city blocks for intensive rat control activities, techniques to assist in stimulating community interest, and practical methods of carrying out effective community rodent control programs. The steering committee and citizens advisory committee should define the problem clearly in writing and establish goals, as well as develop the program and help in its implementation. The written definition of the problem is based on the previously mentioned surveys.

These goals should be of two types. The first is long range, to establish and maintain dwelling units, yard, building, and other areas that are free of rats; to help provide a community with a spirit of self-help and pride, leading to neighborhood betterment; to improve the total dwelling unit and neighborhood; and to stimulate interest to undertake similar programs in other infested areas. The second involves immediate goals, to remove garbage, trash, and other solid waste from yards, homes, buildings; to remove rodent harborage; to poison and otherwise destroy the rats in the neighborhood; to remove abandoned automobiles; to cut down weeds in vacant areas; to rodent-proof homes; to train community leaders in the various phases of rat control; and to encourage block organizations to promote successful programs and continued maintenance.

Program

The community program consists of a series of meetings and actual rat control activities. The first meetings, which are for stimulation and interest, have already been discussed. The major meeting called at a school or other similar facility has also been discussed. It might be added that, during the course of this major meeting, the responsibilities of government and each of its branches and the responsibility of each of the groups of tenants, landlords, businesses, and industry should be explained. The next meeting called should be a training meeting. Personal safety, safety of the child and family, removal of dead rodents, and covering up of rodent odors are taught at this time. It is important to note that only certified environmentalists or pest control operators can place rodent poisons. The training meeting is also used to present a total picture of how to rid an area of rats.

Therefore, further discussions, demonstrations, and visual aids in the removal and storage of solid waste material, in rodent control through rodent proofing, and in the necessity for constantly policing potential rodent problems and continually maintaining a rodent-free environment are conducted. At times, the general meeting may be combined with the training meeting. The individuals within the project decide which is better for their community.

During a discussion of cleanup techniques, it is important to mention the team approach, the use of teenagers, and the use of volunteers. Everyone can help, from a small child to an aged person. All that is necessary is to remove the solid waste to the curb in the best manner possible so that the solid waste agency workers can pick up the material and remove it. Additional meetings are held as needed during the course of the project to explain various aspects of the project. It is important to establish at the first meeting a timetable for the project and then to reinforce the dates of the timetable at each meeting.

In addition to meetings, other techniques used to stimulate community help are handouts, and especially mobile sound equipment, because the spoken message, especially in the early evening, is received by all; included are the working people within the family, who are then given the opportunity to participate in the rodent control program.

The second phase of the program consists of the actual implementation of activities. Implementation of activities are as follows:

1. For the first poisoning anticoagulant is used.
2. Clean up should be carried out by the community and any volunteers such as school children and Y-teen groups. All solid waste, including junk, garbage, refuse, and debris, have to be removed from basements, cellars, and vacant lots. The community supplies the workforce to bring this material to the curb. The city supplies the necessary personnel to pick up the material and truck it to the landfills, recycling center, or incinerators.

3. For additional poisonings use anticoagulants.
4. Total cleanup is carried out by the community on a special cleanup day.
5. The city should go into a direct enforcement program to get the individuals who do not comply with the wishes of the community to remove all their solid waste and destroy all their rats. Inspections should be made of all of the areas. Owners of houses in violation should be cited and if necessary brought to court. Vacant houses should be cleaned and sealed, and the cost should be assessed on the owners. Antilitter, antigarbage, and antirefuse storage ordinances should be enforced strictly. All vacant lots and alleyways should be cleaned by the community, and the waste should be removed by the solid waste agency.
6. The program should be evaluated by the environmentalist by making careful postsurveys of all areas. A comparison should be made between the presurvey and the postsurvey to determine the effectiveness of the program.
7. The community members, with the assistance of the environmentalist, should now stimulate their block organizations to work on continued maintenance and rodent proofing, where rodent proofing is possible. Individuals should attempt to rodent proof their own structures.

The schedule for the program is generally as follows: first week — first poisoning; second week — first cleanup; third week — second poisoning; fourth week — second cleanup; fifth week — additional poisonings; sixth week — survey; and seventh week forward — continued maintenance. The area-baiting concept is extremely important because rat pressures are reduced initially, and during cleanup rats do not scatter. Eventually, as more and more solid waste is removed, the rats die from lack of food or are poisoned by the additional rat poison available to them.

Individual Rodent Control

When an individual is faced with rodent problems, the premises should be surveyed to determine where the rats are coming from and appropriate measures of poisoning, cleanup, and rodent proofing should be taken. It is important that the individual contact the health department and inform it of the problem. In this way, the health department can determine whether this is more than a local situation. An individual should never use rodent poisons other than anticoagulants. If the situation is beyond the individual's control, a trained exterminator should be called to carry out the necessary rodent control functions.

In businesses, trained exterminators should be utilized for rodent control. However, owners should be alert to the use of highly toxic substances. Rat problems are the responsibility of the individual business owner.

Sewers

Rodents can be controlled in sewers by use of a poison, usually raticate or anticoagulant in paraffin blocks placed along the edges of the sewers. The rats consume the poison and usually die in the sewer.

Outside Areas

It is important to use adequate rodent poisoning. If the area is completely away from homes, carbon monoxide or chloropicrin can be pumped into the burrows to destroy the rats. Anticoagulants may be used.

Government Programs

Local and state governments have been involved in rodent control through the enforcement of nuisance laws, food laws, and other types of public health laws. The federal government became involved in rat control in 1967, when Congress authorized $40 million for a 2-year period for rat prevention. In the ensuing years, additional millions of dollars were allocated to the U.S. Department of Health and Human Services (USDHHS) for rat control projects in various communities. Since 1969, the Partnership for Health Amendment (PL 90-174) has initiated new rodent control programs and strengthened existing ones in cities with serious rat problems. Also since 1969, programs have been initiated in New York, Atlanta, Baltimore, Charlotte, Chicago, Cleveland, Milwaukee, Nashville, Norfolk, Philadelphia, Pittsburgh, St. Louis, and Washington, D.C. These rodent control programs generally consist of three phases: (1) preparatory phase, during which the concept was sold to community officials and mass media on how to control rats; (2) attack phase, consisting of surveys, education, cleanup, poisoning, and code enforcement; and (3) follow-up phase, which included continued maintenance. As part of these federal efforts, plans were drawn up for comprehensive rodent control programs. These included citizen participation, community information, and education; effective local administration; adequate municipal service, including garbage, trash, and junk pickup; enforcement of codes and ordinances; removal of dilapidated buildings; systematic poisoning; training and employment of local residents in rodent control efforts.

In 1979, the federally funded rat control projects were once again renewed. In the 1980s and 1990s, rodent control became solely a state and local effort. Philadelphia and New Orleans used some funds from Preventive Health Block Grants for rodent control. Ultimately, local citizens working with local government must solve the problem.

SUMMARY

Rodents live close to humans and can readily cause disease or other health-related problems. Although at present the level of disease caused by rodents is low, the potential is always there. Rodents not only destroy property and cause fires and disease but also bite helpless individuals and are the symbol of urban decay. Rodent control can be carried out if a given community is willing to work together to remove all the food and harborage, apply the necessary rodent poisons, and carry out proper rodent proofing. Rodents will continue to be a serious problem in our society because of a lack of adequate control programs.

RESEARCH NEEDS

It is necessary to develop a better understanding of the resistance that rats are starting to acquire toward anticoagulants. The mechanism of resistance must be understood in order to prevent the elimination of this relatively safe type of rodent poison. New rat control methods are needed. Research should be conducted for the development of single-dose chemosterilants and repellents. It is also necessary to determine how to motivate citizens to keep their homes and neighborhoods rat free.

CHAPTER 7

Pesticides

BACKGROUND AND STATUS

Pests cause a reduction in size, yield, storage, and market quality of crops and food and serve as vectors in the spread of diseases. To control pests, a series of chemicals, called pesticides, have been developed. Pesticides include acaricides or miticides, used against mites; algicides, used against algae; attractants, used to attract insects, birds, and other animals; chemosterilants, used to interfere with reproduction; defoliants, used to remove leaves from plants prior to harvest or to eliminate unwanted plants; fungicides, used against fungi; herbicides, used against weeds; insecticides, used against insects; molluscicides, used against slugs and snails; ovicides, used against insect eggs; repellents, used to drive animals or insects away; and rodenticides, used against rats, mice, and other rodents.

Pesticides are biologically active chemicals that kill or modify the behavior of problem insects, animals, microorganisms, weeds, and other pests. Pesticides are used as aerosols, sprays, and dust (in granular form) or as baits. They may be effective on contact, be taken up by the plant, enter the lungs or trachea of animals, or be eaten. The quantity of the pesticide used, the type of pesticide, and how it is used are very important in the control of troublesome pests.

Pesticides produce useful and harmful effects, depending on the type and quantity used and the method of application. About half of all pesticides are used in farming. Roughly 5% is used by governmental agencies and the balance, by residential and industrial users. Currently, we are using more than 1.1 billion lb of pesticides annually.

Pesticides are categorized by their lifetime of effectiveness as follows: nonpersistent, lasting several days to about 12 weeks; moderately persistent, lasting 1 to 18 months; persistent, which includes most of the chlorinated hydrocarbons such as dichlorodiphenyltrichloroethane (DDT), aldrin, and dieldrin, lasting many months to 20 years; and permanent, including mercury, lead, and arsenic, lasting indefinitely. Polychlorinated biphenyls (PCBs) used in asphalt, ink, and paper behave very much like the persistent pesticides and require close control to avoid contamination of the environment.

Pesticides that degrade or deteriorate rapidly are also of great concern because of their extreme toxicity and because of their nonselectivity in their action on animals, humans, and pests. Organophosphates would be an example of this type of pesticide.

A pesticide moves through an ecosystem in numerous ways. It is introduced by surface application, spraying, or other techniques and may stay in the air or be washed down by rain. The concentrations of the pesticide continue to increase in the soil over time; and where leaching occurs, the pesticide can move into surface or underground water supplies. Some pesticides become tightly bound to soil particles, polluting the surface waters when the surface particles are washed into them by the force of heavy rains. Some pesticides are ingested by minute, aquatic organisms, and scavengers and become concentrated as they move up through the food chain. It is known that oysters, for instance, will concentrate DDT in their tissues 70,000 times greater than amounts found in the surrounding waters. Fish also concentrate pesticides as part of the food chain. Eventually, the pesticides may reach humans and, at least in the case of DDT, are stored in the fatty tissues.

The major pesticide laws in effect in the United States were totally rewritten in 1972 and updated in 1978 and in 1988 as amendments to the 1947 Federal Insecticide, Fungicide, and Rodenticide Act (FIFRA). This law forbids anyone, including the federal government, from using a pesticide contrary to label instructions and gives the Environmental Protection Agency (EPA) the authority to restrict the use of pesticides to trained persons. The law applies to interstate and intrastate use and sale of the product. It provides screening procedures for new pesticides suspected of causing cancer, birth defects, or mutations. Based on this law, the EPA has taken action against the use of kepone, DDT, aldrin, dieldrin, heptachlor, chlordane, mirex, and mercury-based pesticides. At present, the only exception to these actions is made by the EPA if the agency believes that the benefits outweigh the potential adverse effects and no alternatives are available, if a significant health problem occurs without its use, or if an emergency exists.

In 1996, the Food Quality Protection Act became law. It made changes in FIFRA as well as the federal Food, Drug, and Cosmetic Act (FDCA), with the EPA establishing tolerances (maximum legally permissible levels) for pesticide residues in food. Tougher standards were set to protect infants and children from pesticide risks, which include an additional safety factor to account for developmental risks and incomplete data when considering the effect on infants and children, and any special sensitivity and exposure to pesticide chemicals that infants and children may have. As a result of this new law, the EPA announced in August 1999 the cancellation of the uses of the organophosphate pesticide methyl parathion and significant restrictions on the use of the organophosphate azinphos methyl on food typically eaten by children.

Under present EPA orders, all individuals, including public health workers, who are involved in the use of pesticides, must take comprehensive examinations and be registered in the use of pesticides by category of employment. Because the use of pesticides is increasing, the dangers may increase and the current status in succeeding years may deteriorate unless further action is taken.

Current Issues

The first pesticide act was enacted in 1947. When the USEPA was founded in 1970, the FIFRA authority was transferred from the U.S. Department of Agriculture (USDA) to the EPA. In 1972, Congress passed the Federal Environmental Pesticide Control Act as an amendment to the original pesticide act. It provided for direct controls on the use of pesticides, for classification of certain pesticides into a restricted category, for registration of the manufacturing plants, and for a national monitoring program for pesticide residues. Environmental effects and risks were added to the pesticide registration process.

The 1972, FIFRA amendments required a review of all the registered products in use. The review was to be completed by 1975. Unfortunately, because of the large amount of data to be collected and the large number of products to be assessed, the General Accounting Office (GAO) determined it would take until the year 2024 for this work to be done. As a result, the entire reregistration process simply broke down.

From the very beginning, the EPA had problems with the pesticide regulation program and with the process of implementing the 1972 amendments to FIFRA. However, this act still did not resolve the many problems related to pesticide regulation. The issues were registration, tolerances (which are standards) for pesticide residues, federal preemption of state tolerances, reregistration, inert ingredients of pesticide formulations, regulatory options, and pesticides in groundwater.

Under FIFRA the regulation of pesticides was done through the registration of the individual pesticide products. The products were not permitted to present unreasonable adverse effects to people or the environment if the pesticide was used on a food crop or animal feed. The regulations also required that a maximum acceptable level of pesticide residues remaining on a treated crop be determined by the EPA, and be monitored and enforced by the Food and Drug Administration (FDA). The data required for registration of pesticide products included health and environmental data, environmental fate, carcinogenicity, chemistry of the product, toxicity to fish life, and mutagenicity.

The EPA established tolerances for pesticide residues in foods. Tolerances determined the maximum amount of pesticide residue that could be permitted in food or animal feed so as not to be considered an adulteration of the product. Tolerance-setting procedures to protect human health include anticipated amount of pesticide residues found on food; toxic effects of these residues; estimates of the types and amounts of food that make up our diet; field trials of pesticide use and residues; toxicity studies; product chemistry data; and plant and animal metabolism studies including metabolites.

The EPA recognizes that the diet of infants and children may differ substantially from those of adults and that they may be exposed to pesticide residues in food at levels higher than adults receive. By using a computerized database known as the Dietary Risk Evaluation System (DRES), EPA combines survey information on food consumption and data on the pesticide residues to estimate dietary exposure. DRES breaks out a number of subgroups in the population, including infants, children and other age groups, several different ethnic groups, and regional populations. EPA appropriately identifies childhood or infant exposures for special consideration when looking at the risks for discrete periods of time. EPA also calculates a cumulative lifetime exposure that integrates the exposure rates experienced in infancy and childhood and takes the exposures experienced in adulthood. If risks are at an unacceptable level, then the EPA takes action to reduce those risks.

The federal government, under the 1988 law, preempted the tolerances established by state law. The problem was that the EPA standards were currently considered to be a floor and not a ceiling for standards for any given pesticide. Therefore, the states could have more stringent standards. However, if this occurred, it could have interfered or hindered the flow of products through interstate commerce and hindered the marketing of pesticide products, pest treatment services, and treated commodities. The states argued that the data that had been used to determine the allowable amount of pesticides in the food or feed product may not have been accurate, and over a long term it was the right of the states to protect their citizens.

The FIFRA amendments authorized the EPA to conduct a "generic" review of the safety of the active ingredient. The EPA had identified some 600 active ingredients considered to be commercially important among the 1500 active ingredients officially registered with them. An estimated 40,000 studies about these pesticides were in the EPA files. Only two pesticide active ingredients had been reregistered under this process until this point. Also under the 1988 law, all existing pesticides must be reviewed and reregistered, with industry required to provide the test data for the review. By 1993, the EPA had issued 31 reregistration documents.

The EPA also investigates inert ingredients. An *inert* ingredient is that part of the pesticide formulation that is not intended to have any pesticidal activity. It is used either to dilute the pesticide or to propel it or deliver it in some manner. Unfortunately, some of the inert ingredients have potentially adverse effects on people. Vinyl chloride gas is a human carcinogen that has been used as an aerosol propellant. Of the approximately 1200 compounds used as inert ingredients in pesticides, the EPA has determined that 55 are known toxics that may cause animal cancer or nerve damage; 51 compounds are structurally related to the known toxic compounds; and 900 compounds are of unknown toxicity.

In the past 3 years, the federal government, under the 1996 law, has registered 48 new safer pesticides that have a lower risk for infants and children than that of the organophosphates. Children are at a greater risk than adults because their internal organs are still developing and maturing; and their enzymatic, metabolic, and immune systems may provide less natural protection from chemicals, especially pesticides. At critical times in human development exposure to a toxin can permanently alter the way an individual's biological system operates. Children may also be exposed more to certain pesticides because they often eat different foods than adults. For example, children typically consume larger quantities of milk, applesauce, and orange juice per pound of body weight than adults. Children play on the floor or on the lawn where pesticides are commonly applied. They also put objects in their mouths, thereby increasing their chance of exposure to pesticides. The EPA is requiring hundreds of additional studies on pesticides to better understand their effect on children, especially developmental, acute, and subchronic neurotoxicity. All organophosphate residue limits have been reassessed. Other high-risk pesticides are receiving priority review including atrazine, aldicarb, and carbofuran. The EPA is requiring registrants of pesticides to provide additional data on approximately 140 pesticides. Many of these currently registered conventional food use pesticides have been observed to affect the nervous system in humans, laboratory animals, or both. Outstanding questions about these neurotoxic effects include:

1. Do these chemicals harm the nervous system following exposure during critical developmental stages before birth in the fetus, and after birth in infants and young children?
2. Are the effects in the young different from those observed in an adult?
3. If similar effects occur in both the young and the adults, are the young more or less sensitive than the adults to these effects?

The EPA uses the data collected in the studies in making decisions in the implementation of certain aspects of the Food Quality Protection Act tolerance-setting process, especially in making the "reasonably certain of no harm" finding and addressing the requirement that "in the case of threshold effects...an additional tenfold margin of safety for the pesticide chemical residue and other sources of exposure shall be applied for infants and children...." Additional studies will be done on the following compounds:

1. Cholinesterase-inhibiting carbamates including aldicarb, carbaryl, and carbofuran
2. Thio- and dithiocarbamates including mancozeb, maneb, and triallate
3. Pyrethrin and synthetic pyrethroids including deltamethrin, fenvalerate, and permethrin
4. Persistent organochlorines including dicofol, endosulfan, and lindane
5. Formamidines including amitraz, tridimefon, and tridimenol
6. Mectins including abamectin and emamectin
7. Phosphides including aluminum phosphide, magnesium phosphide, and zinc phosphide
8. Organotins including cyhexatin, fenbutatin oxide, and fentin hydroxide
9. Organoarsenicals including disodium methanearsonate, and cacodylic acid
10. Dipridyl compounds including diquat chloride, mepiquat chloride, paraquat bis-methyl sulfate and chloride
11. Other neurotoxic pesticides including carbon disulfide, imidachloprid, and nicotine

The Food Quality Protection Act not only established a new safety standard for pesticide residue limits in food but also for pesticide limits in feed (tolerances). To ensure that the new standard applies to all pesticides the EPA is required to reassess all 9721 tolerances and tolerance exemptions in effect when the law was passed in August 1996. The EPA surpassed the required 33% needed by August of 1999. By August 2002, 66% must be reassessed, and by August 2006, 100% must be reassessed. Two thirds of the reassessment completed are for pesticides in the highest priority group including the organophosphates, carbamates, carcinogens, and high hazard inert ingredients. This group consists of 228 pesticides and 5546 tolerances.

The reassessments at present have resulted in the revocation of over 1500 tolerances out of 3300 tolerances reassessment decisions. Approximately 500 of the revocations have been for organophosphates, 100 for carbamates, 1 for an organochlorine, and 220 for carcinogens.

The EPA can revoke the registration for a pesticide if the information indicates that the product presents an unreasonable risk to human health or the environment. This process is called *deregistration*. If the chemical is canceled, the stockholders, owners, and other individuals can demand payment of the EPA for their losses. In the last 23 years the EPA has canceled the registration of 36 potentially hazardous pesticides and has eliminated the use of 60 toxic inert ingredients in pesticide products.

The National Water Quality Assessment, Pesticide National Synthesis Project is conducted by the United States Geological Service. The first phase of intensive data collection was completed from 1993 through 1995 in 20 major hydrologic basins in the United States.The groundwater land-use studies are designed to sample recently (generally last 10 years) recharged groundwater beneath specific land-use and hydrogeologic settings. Pesticides were commonly detected in shallow groundwater in both agricultural and urban settings in the United States. Of the agricultural settings, 56.4% showed pesticides whereas of the urban settings, 46.6% showed pesticides. The maximum contaminant levels established by the EPA for drinking water were exceeded by only one pesticide, atrazine, which was greater than 3 µg/l at a single location. However, the relative infrequency with which pesticides exceeded drinking water criteria may not provide a complete assessment of the overall health and environmental risks associated with the presence of pesticides in shallow groundwater. Water quality criteria for protecting human health have only been established for 25 of the 46 pesticide compounds evaluated in the study. The drinking water criteria only consider the effects of individual compounds and do not consider additive or even synergistic toxic effects of exposure to multiple chemicals. Other pesticide compounds and their degradates exist that were not studied. Recent research also suggests that some pesticide compounds may cause harmful health effects at levels considered safe by current standards. Drinking water criteria do not provide for potential effects of pesticide compounds on aquatic systems, and the concentration of the compounds within aquatic life.

The contamination of major aquifers is largely controlled by hydrology and land use. Concentrations of nutrients and pesticides in 33 major aquifers were usually lower than those in the shallow groundwater underlying the agricultural and urban areas. Because the water that replenishes the major aquifers comes from a variety of different sources and land-use settings, higher quality water helps reduce the levels of contaminants in the aquifers. Deeper aquifers are usually more protected than shallow groundwater by impermeable layers. Fertilizers, manure, and pesticides have degraded shallow groundwater. Concentrations of nitrate exceeded the EPA drinking water standard of 10 mg/l as N in 15% of the samples collected in shallow groundwater beneath agricultural and urban land. Herbicides are also frequently found in the wells. In the groundwater and surface water, 58 pesticides were detected at least once at or above 0.01 µg/l.

Pesticides and Groundwater

Pesticides can reach the groundwater supply through misuse or mismanagement related to waste disposal, spills, leaching, etc. The four main issues related to pesticides in groundwater are:

1. Pesticides need to be detected in groundwater.
2. The EPA needs to determine at what level pesticide residues in groundwater should trigger action.
3. If a pesticide is detected in groundwater as a result of normal use and if the groundwater pesticide limits are exceeded, the EPA needs to decide upon an appropriate remedy.
4. The EPA needs to decide whether a pesticide should be immediately prohibited if it reaches substantial levels.

The National Pesticide Survey was conducted by the EPA in all 50 states between 1988 and 1990. Preliminary results indicate that 10% of the nation's community drinking water wells and about 4% of rural domestic wells have detectable residues of at least one pesticide. One or more pesticides exceeding health advisory or maximum containment levels are found in 1% of community wells and 0.8% of rural wells. More than 50% of the nation's wells contain nitrates, but fewer than 3% have concentrations of health concern.

Pesticides and Groundwater Strategy

"Pesticides and Groundwater Strategy" was released by the EPA in 1991. It includes the following six federal policies:

1. Encourage, where appropriate, less burdensome environmental agricultural practices concerning use of pesticides and fertilizers.
2. Determine appropriate regulatory approach to cancel chemicals that may threaten groundwater.
3. Confine legal sale and use of canceled pesticides to states that have EPA approved state plan.
4. Provide research and technical assistance supported by EPA to Department of Agriculture and U.S. Geological Survey to assess groundwater problems and vulnerable groundwater systems.
5. Provide clear instructions for the field use of pesticides by improving training and certification programs.
6. Promote and encourage companies to conduct more monitoring studies, develop safer pesticides, and prevent degradation of groundwater.

Risk–Benefit Balancing Under the Federal Insecticide, Fungicide, and Rodenticide Act

There are four steps in the EPA risk assessment process: hazard identification, dose–response assessment, exposure assessment, and risk characterization. The two steps in benefit assessment are biological analysis and economic analysis.

In hazard identification, the EPA evaluates the inherent toxicity of a pesticide, that is, the types and degrees of harmful effects a pesticide may cause by determining the effects on animals, especially cancer, in laboratory studies. In dose–response assessment, laboratory animals are exposed to various doses of the chemical during various time periods. The acute or chronic effects are determined and a no observed effect level (NOEL) is established where noncancer effects were found. The NOEL is divided by an uncertainty factor of 100 or more to determine the reference dose (RFD). At or below this level it is assumed that a lifetime exposure would not cause harmful effects. A negligible risk standard is used for cancer, where an individual's chance of cancer is 1 in 1 million if exposed for a lifetime.

Exposure assessment refers to the level, duration, and route of exposure of people to chemicals identified in laboratory tests as causing harmful effects. Risk characterization is the estimate of the risk from exposure to pesticides to people by integrating the preceding factors and extrapolating exposure in animals to humans. In benefit assessment, the EPA decides whether to cancel or approve a new pesticide by determining the effectiveness and economic value of a pesticide compared with alternative chemical and nonchemical controls.

Other Issues

Additional issues in the use of pesticides include food contamination, air pollution, potential indoor air pollution, preparation of professional pesticide applicators, and pesticides related to fish and wildlife. Pesticides in foods or on foods have become a major issue today. The scare related to the purposeful contamination of two grapes in 1990 brought an entire food supply to a total halt and created untold problems for the country providing the food, the agencies evaluating the food, and the public (who in some way ended up paying for the food through taxes or through the disposal of the rotted fruit). Some questions remain concerning the amount of pesticides that are introduced into the food chain and the concentration of these pesticides in the finished product because of the use of raw products that have been exposed to the chemicals. The consumer must be protected, but the food supply must not be destroyed.

Pesticide application can cause an air pollution problem. Air currents may carry pesticides to the wrong area or chemicals may be dumped on the wrong area and potentially affect people and animals.

A variety of pesticides are used on lawns and on pets. The individuals applying these pesticides are untrained in the storage, mixing, application, and disposal of the pesticides. Pesticides may be stored in garages or other enclosed areas, and therefore, constitute a potential indoor air pollution problem.

Professional pesticide applicators need to be certified and recertified by their state agencies. Many states have more stringent standards than the minimum standards established by the EPA. It is therefore necessary to determine the standards of a given state and then decide how best to train the applicators and test them.

Pesticides may have a long-range effect on fish and wildlife, depending on the type of accumulation that may occur. It has been known for years that DDT causes the thinning of egg shells, which in turn prevents the successful hatching of the chick in a variety of birds. Even a pesticide that is as valuable as diazinon has been found to be harmful to fish and wildlife and, therefore, can no longer be used on golf courses and sod farms.

SCIENTIFIC, TECHNOLOGICAL, AND GENERAL INFORMATION

Types of Pesticides of Public Health Importance

This section is basically concerned with the types of pesticides used to control insects, ticks, mites, spiders, rodents, and plants of public health significance. Pesticides should have certain qualities to be acceptable for use. They should be specifically toxic to harmful insects and so forth, harmless to humans, inexpensive and easily used, rapidly degradable to nontoxic substances, nonflammable, noncorrosive, nonexplosive, and nonstaining.

Insecticides may be used as stomach poisons, contact poisons that penetrate the body wall, fumigants that enter through the insect's breathing pores, desiccants that scratch or break the body wall or absorb into the waxy protective outer coating. Pesticides may also be listed as larvicides that kill larvae, ovicides that kill the insect's eggs, or adulticides that kill the adults.

Inorganic Insecticides and Petroleum Compounds

Prior to 1945, numerous inorganic pesticides were used widely. These included the arsenical, Paris Green, used against the potato beetle; hydrogen cyanide, used against red scale; lead arsenate, used against the gypsy moth; and sodium arsenite, an insecticide and a weed killer. All arsenicals are now banned. Compounds of copper, zinc, and chromium were also used as pesticides. Chlorine and sulfur made extremely toxic compounds and were used along with salts of arsenic, lead, mercury, and selenium. Unfortunately, many of these compounds were quite toxic to humans. Also, some of the insects developed resistance to certain inorganic pesticides.

Petroleum oils, such as kerosene, diesel oil, and no. 2 fuel oil, were used as mosquito larvicides. These oils, which are still in use, have certain toxic properties, because they penetrate the tracheae of larvae and pupa of mosquitoes and anesthetize them. A fraction of these oils mechanically interfere with the breathing process of insects, causing suffocation. Sulfur acts as a repellent against chiggers. Borax, or boric acid powder, is still used in buildings for roach and ant control with varying results.

Botanicals

Probably the earliest pesticides used were pyrethrum, which is extracted from the flowers of *Chrysanthemum cinerarifolium*; rotenone, which is derived from Peruvian cuve; and red squill, which is derived from the inner bulb of the plant *Urginea maritima* belonging to the lily family. These chemicals are highly specific to the pests and have a very low toxicity to humans. Today pyrethrum is used primarily in combination with other insecticides. However, permethrin synergized by piperonyl butoxide seems to be effective against mosquitoes that are organophosphate resistant. Piperonyl butoxide is derived from sesame. It is a synergist, because it enhances the effects of many botanicals. It inactivates enzymes on the bodies of insects and mammals that breakdown toxins. It reduces the amount of insecticide needed and increases the chance that insects will be killed instead of just temporarily paralyzed. Chronic exposure of humans to this synergist can cause damage to the nervous system. Further, sumithrin is effective against mosquitoes and mites. It is especially used to "disinsect" aircraft coming from foreign countries prior to landing. Methoprene is directly ovicidal to cat fleas and it also sterilizes the adult. Pyrethrum continues to be a quick insect knockdown agent. Pyrethrins are nerve poisons that cause immediate paralysis to most insects. Human allergic reactions are common, and cats are susceptible to pyrethrins.

Synthetic compounds similar to pyrethrum, called allethrin, resmethrin, sumithrin, and permethrin, have been developed and are utilized in the same manner as natural pyrethrum. Pyrethroids are axonic poisons (they poison the nerve fiber). They bind to a protein in nerves called the voltage-gated sodium channel. Pyrethroids bind to the gate and prevent it from closing properly, which results in continuous nerve stimulation. Rotenone is used to kill fish without leaving toxic by-products for human beings. Natives of some tropical countries crush and throw plants such as Derris and cuve into the water, and the chemicals present in the plants paralyze the fish. Rotenone is also utilized for killing of fleas and other ectoparasites on domestic pets. Red squill in its fortified state is used effectively in killing Norway rats. Because of the difficulty of obtaining these pesticides from abroad during World War II, and the military need for chemicals to kill disease-producing insects, the United States developed a series of organic pesticides in the early 1940s. In addition, such insecticides as DDT were recognized in Switzerland in 1939. Benzene hexachloride was recognized as an insecticide in 1940 in France and England.

Chlorinated Hydrocarbons

The chlorinated hydrocarbon insecticides are combinations of chlorine, hydrogen, and carbon, They act primarily as central nervous system poisons. The insect goes through a series of convulsions and finally dies. The first major chlorinated hydrocarbon was dichlorodiphenyltrichloroethane (DDT). DDT has been highly useful for the control of mosquitoes, flies, fleas, lice, ticks, and mites, reducing considerably the level of malaria, plague, typhus fever, yellow fever, encephalitis, and so forth.

Although DDT has been banned for general use in the United States, with the exception of a serious uncontrollable emergency, it is still an effective chemical for the control of mosquitoes, which may cause malaria or other diseases. DDT enters the ecosystem and is stored in animal fat. However, in public health a decision must be frequently made as to the relative importance of one hazard vs. another. Therefore, DDT is still used abroad in the interior of homes in areas where malaria is prevalent. The DDT, if applied carefully, leaves a residue on the structures that will last from

6 to 12 months. It also does not readily escape into the environment. The dosage in the residual spray should be 100 to 200 mg/ft^2.

Methoxychlor and dichlorodiphenyldichloroethane (DDD) are part of the DDT group. Methoxychlor, safer than DDT because it is less toxic to mammals, is utilized in many household sprays and aerosols, is readily metabolized and eliminated in the urine of vertebrates, and also does not remain within the environment for more than a short period of time. Methoxychlor is used to kill mosquito larvae and control flies and insects that attack livestock or occur in agricultural areas. As a larvicide, methoxychlor is applied at a rate of 0.05 to 0.20 lb/acre. DDD is now banned.

Benzene hexachloride (BHC), with a musty odor and a short residual life, was widely used in public health and in agricultural programs. BHC has now been canceled. The gamma isomer of BHC has significant insecticidal activity. BHC is currently used abroad as a residual spray at a dosage rate of 25 or 50 mg/ft^2 for the control of mosquitoes causing malaria or other diseases. It has a residual effect for about 3 months. Benzene hexachloride is a misnomer. It should be, technically, hexachlorocyclohexane.

Lindane, the pure gamma isomer of BHC, is highly effective as an ingested poison and as a residual contact insecticide; and has been used for control of lice, ticks, and other insects. Lindane vaporizers, used to control flies in food establishments, are dangerous and should never be used. Short-term exposure can cause high body temperature and pulmonary edema. Long-time exposure can lead to kidney and liver damage. Its use has been restricted. In Southeast Asia, lindane was used in irrigation waters to control rice stem borers. Unfortunately, the chemical killed the fish that were the protein sources for the local population. In overseas areas where the vector of plague, the flea *Xenopsylla cheopis,* still exists and where DDT is not effective, it is recommended that a 1% lindane solution be applied to ensure adequate control of this rodent flea. Lindane in a 1% emulsion may be used for treating infested household sites within the house and in the yard. To control body lice 1% lindane powders are used. The brown dog tick can be controlled with a 0.5% lindane spray, with spot treatments on baseboards, floors, wall crevices, and areas where the animal sleeps. Lindane can still be used for dogs. It is banned for any type of fumigation.

Chlordane, dissolvable in many solvents but not in water, is used to produce oil solutions, emulsifiable concentrates, wettable powders, and dusts. This chemical acts as a stomach poison, contact insecticide, and fumigant; is effective in spot control of ants, American roaches, silver fish; and has also been used extensively on soil insects, particularly termites. Chlordane had been extremely effective, although German roaches have built up a resistance to it. It is probably the least toxic of the chlordane series, which includes chlordane, heptachlor, aldrin, dieldrin, endrin, isodrin, and toxaphene. In July 1975, the EPA suspended the use of chlordane because it is suspected of causing cancer in animals and it readily contaminates the environment.

Heptachlor had been used effectively for mosquito larvicide control. Heptachlor production was also suspended by the EPA as a suspected link to cancer in animals and because it is highly toxic to humans. It remains for long periods of time in the environment and readily contaminates the water, soil, and air. Aldrin and dieldrin, considered to be effective chemicals for insect control, were found to be highly toxic when misused, because of killing fish, birds, mammals, and even human beings. These chemicals are also suspected of carcinogenic activity; and these too have been suspended from use since 1974. However, dieldrin still is used overseas at a rate of 25 or 50 mg/ft^2 as a residual spray in those areas where malaria or other mosquito-borne diseases are prevalent.

Endrin, one of the most poisonous of the chlorinated hydrocarbons, is highly toxic, persists in the environment, and is a hazard to animals and humans. This chemical should be used only when necessary and under strict supervision, is not recommended for general use, and is now banned for general use.

Isodrin and toxaphene are both highly toxic chemicals, persisting for long periods of time in the environment. They are not recommended for use as insecticides. They are now banned for general use.

Chlordecone, better known as kepone, is a very effective chlorinated hydrocarbon when used as insect bait. Kepone may last as long as 1 year without being altered. When used in proper dosages, that is, 0.125% peanut butter bait, kepone is effective against both roaches and ants, and produces a high kill; however, when used improperly or produced improperly, it can be very toxic to humans. Therefore, the production of the chemical should be closely monitored, the disposal of waste in the chemical process controlled, and the bait carefully handled. Kepone can be placed in a paraffin bait and still be very effective for the control of American roaches. Kepone has been banned.

Endosulfan is a chlorinated hydrocarbon insecticide and acaricide of the cyclodiene group that acts as a poison to a wide variety of insects and mites on contact. It is a hazardous, restricted-use pesticide.

Organophosphates

The organophosphates are derived from phosphoric acid and inhibit the enzyme cholinesterase. The poisoned synapse cannot stop the nerve impulse after it crosses the synapse. In many cases, these chemicals have replaced the chlorinated hydrocarbons, because they are effective against insects that have become resistant to the chlorinated hydrocarbons; they are biodegradable; they do not contaminate the environment for long periods; and they have fewer long-lasting effects on organisms that are not meant to be treated with these chemicals. However, organophosphates vary tremendously in toxicity. The organic phosphorous insecticides include tetraethylpyrophosphate (TEPP), chlorpyrifos, dichlorvos, phosdrin, and parathion, which are highly toxic; bayer 29493, baytex, dimethoate, fenthion, dimethyldichlorovinyl phosphate (DDVP), and diazinon, which are moderately toxic; and abate, gardona, dipterex, malathion, and ronnel, which are slightly toxic.

TEPP is used in greenhouses and on fruits and vegetables. It is highly toxic when mishandled and causes severe poisoning. It contaminates the environment for short periods of time. As of the year 2000, all eight tolerances have been revoked by the EPA.

Phosdrin and parathion are highly toxic insecticides that are fatal to humans if only one drop is placed in the eye. They should only be used by highly experienced, licensed, commercial operators. They are used as larvicides for mosquitoes at a rate of 0.1 lb/acre in rural areas away from children and animals. Phosdrin is now banned by the EPA. Most uses of parathion has been voluntarily canceled.

DDVP, dichlorvos, and vapona are all the same compound. This pesticide is useful as a fumigant because it is highly volatile; is highly toxic, but breaks down quickly; and is generally used in fly control as a spray or fog or in impregnated strips. In strip form DDVP is effective for mosquito control for 2.5 to 3.5 months if used at a rate of one strip per 1000 ft^3. It may also be used in catch basins with one strip suspended 12 in. below the catch-basin cover, per basin. Dichlorvos is used in a sugar solution as a bait for fly control. This chemical is mixed with water and used as an outdoor space spray for flies or mixed with water and used as a fly larvicide.

Dichlorvos, because it has only a short residual life, is most effective rapidly and presents a short residual hazard to the environment. However, vapona strips should not be suspended over food, because a drop of the chemical could collect at the bottom of the strip and fall into the food, creating a potential chemical food poisoning. It would not be wise to hang vapona strips in areas where individuals suffer from upper respiratory ailments, because the chemical is discharged slowly over a long period of time and could become either an irritation or a hazard to the individual. Dichlorvos should not be used around food. It is a restricted-use pesticide and may be used only by certified applicators.

Diazinon is utilized in fly and roach control and other insect control problems related to vegetables and fruits. It is a toxic chemical and should never be used in cases where there is potential contact with humans or pets. The residual period in the environment is fairly short, varying from 1 week to at most 2 months. Diazinon should not be used in poultry farms, because it is toxic to birds. In fly control, diazinon is mixed with petroleum compounds or water and used as a space spray, or mixed with water and used as a larvicide. In roach control, diazinon is effective in reducing or eliminating all roaches, including German roaches. At this time, only the German roach appears to be developing some resistance in some areas of the country to the compound. Solutions usually contain the following concentrations of diazinon: spray, 0.5%; dust, 1%. However, pest control operators are permitted to use 1% spray and 2 to 5% dust. Diazinon cannot be used on golf courses or sod farms. It is a restricted-use pesticide and may be used only by certified applicators.

Dipterex or trichlorfon is used in sugar and water as a bait for flies. In areas where garbage, organic materials, and manure are controlled, the fly bait works effectively and rapidly. Dipterex is also used as a bait with some success in roach control. It is a general use pesticide.

Abate, or temephos, is used at a rate of 0.05 to 0.1 lb/acre for larvicidal control of mosquitoes. It is a general use pesticide. Chlorpyrifos is used in mosquito control for ground-applied outdoor space, spraying at a rate of 0.0125 lb/acre; and also is used as a mosquito larvicide at a rate of 0.05 to 0.125 lb/acre. The chemical is effective in control of roaches at a 0.5% concentration. When painted on a surface over which German roaches crawl, chlorpyrifos has a strong residual effect, causing a kill of 90% of roaches for up to 1 year. The chemical persists in the environment from several days to as long as 1 year. Chlorpyrifos is a general use pesticide.

Dimethoate is used for outdoor space spraying for flies and also for larvicidal control outdoors for flies. Fenthion must be used by trained mosquito control personnel only, and is used in ground-applied outdoor space spraying for mosquitoes at a rate of 0.001 to 0.1 lb/acre. Fenthion is used as a larvicide at a level of 0.05 to 0.1 lb/acre, but must be carefully handled by trained personnel. Also, this chemical is effective in roach control, but must only be used by pest control operators as a spray at a rate of 2.0% concentration.

Gardona is a relatively safe, nonsystemic, broad-spectrum organophosphate used for fly and tick control. However, gardona is highly toxic to bees, but it persists in the environment for only short periods of time. Naled is moderately toxic to animals. It has a very short residual period in the environment, and is used as an outdoor ground-applied space spray at a rate of 0.02 to 0.1 mg/ft^2 for mosquito control. It is also used as an outdoor space spray for flies in liquid form, and as a bait in a sugar solution. Naled is corrosive to the eyes. Ronnel, or korlan, is used to control flies in an agricultural area. Its toxicity is slight and its residual period in the environment is very short. It is hazardous to livestock or dairy food. Ronnel has been canceled.

Malathion is a slightly toxic compound available for general use. It is a nonsystemic, wide-spectrum organophosphate insecticide used for control of flies, mosquitoes, household insects, ectoparasites, and head and body lice. It is available as an emulsifiable concentrate, a wettable powder, a dustable powder, and a liquid. Malathion may also be found in formulations with many other pesticides.

Carbamates

The carbamates are derived from carbonic acid. Most of the carbamates are contact insecticides. They inhibit the cholinesterase activity and act as nerve poisons, similar to the organic phosphorous compounds, but inhibition is reversible without antidotal treatment. Several of them produce a rapid knockdown, as produced by pyrethrum. The carbamates include sevin, which is also called carbaryl and dimetitan; baygon, which also called propoxur; and landrin.

Carbaryl is widely used in public health and agriculture. It is one of the safer insecticides for animals, but is highly toxic to bees. Carbaryl is formulated only as a solid, which is then used as a wettable powder, slurry, or dust. This chemical remains for a relatively short period of time in the environment. Carbaryl dusts are used in a 2 to 5% concentration to kill fleas on dogs and cats older than 4 weeks. It is also used in the United States to kill the oriental rat flea in murine typhus control programs. The sprays and dusts have been used in adult mosquito control. Carbaryl, when used as an outdoor space spray, is concentrated at a level of 0.2 to 1.0 lb/acre for the control of mosquitoes. It is a general use pesticide.

Dimetitan is highly toxic when ingested and moderately toxic when absorbed through the skin. It is impregnated into plastic bands and suspended near the ceilings of farm buildings for use in fly control.

Propoxur (baygon) acts as a stomach poison and contact poison in roach control and also in tick control. The spray has a long-lasting residual contact. This insecticide also differs from others in that it has a flushing or irritating action that forces insects out of hiding areas, and it has a rapid knockdown action. Baygon is used to control mosquitoes, flies, sandflies, ants, other insects, and the resistant brown dog tick. It has some toxicity for animals. Apparently, it lasts for short periods of time in the environment as a contaminant. Baygon is used in a dosage of 100 to 200 mg/ft² for residual spraying in mosquito control. For roach control, it is used either as a 1% spray or a 2% bait.

Landrin has been tested by the World Health Organization (WHO) as a residual spray in anopheles mosquito control.

Biolarvicides

Biolarvicides are naturally occurring crystalline delta-endotoxins produced by *Bacillus thuringiensis*, which is lethal to the larva of mosquitoes. The mosquito larvae ingest the delta-endotoxin, which reacts with gut secretions, causes gut paralysis and disruption of the ionic regulation capacity of the midgut epithelium, and results in death within minutes.

Insect Energy Inhibitors

Several chemicals that inhibit the production of energy are currently used as insecticides. Hydramethylnon belongs to the chemical class amidinohydrazone. This chemical binds to a protein called a cytochrome in the electron transport system of the mitochondrion. This binding blocks the production of adenosine triphosphate (ATP), thereby causing the insects to die while standing. Another insecticide currently available that inhibits energy production is sulfluramid. This chemical belongs to the halogenated alkyl sulfonamide class. The parent chemical is converted to toxic metabolites by enzymes in the body. Many new chemicals are under development for use as energy production inhibitors. Chemicals in the class pyrrole, thiourea, and quinazoline are showing great promise as pesticides that inhibit energy production.

Chitin Synthesis Inhibitors

Chitin synthesis inhibitors are often grouped with the insect growth regulators. The most notable chemical in use is benzoyphenyl urea. These chemicals inhibit the production of chitin, which is a major component of the insect exoskeleton. The insect is unable to synthesize new cuticle, thereby preventing it from molting successfully to the next stage. This class of insecticides include lufenuron, used for flea control on pets; diflubenzuron, used against fly larvae in manure; and hexaflumuron, used in termite bait stations.

Insect Growth Regulators

Insect growth regulators are chemicals that act on the endocrine or hormone systems of insects, causing them to remain in the immature state and preventing them from emerging as adults. Methoprene, hydroprene, pyriproxyfen, and fenoxy-carb, mimic the action of juvenile hormones and keep the insects in the immature state. Insects treated with these chemicals are unable to molt successfully to the adult stage and cannot reproduce normally. The ingestion of the insecticide by the larva maintains the high level of juvenile hormone and therefore the insect does not develop the physical features for adult emergence and it dies.

Pheromones

Insects send out chemical signals or pheromones that allow them to communicate with other members of their species. Usually the adult female produces these chemicals to attract males. Pheromones can be used in traps. However, the effectiveness of the trap can be lowered by rainfall, cool temperatures, and wind speed and direction. Each trap is specific for a different type of insect. Mating disruption with synthetic sex pheromones has been used for some fruit and forest trees. Further research is needed to see if this technique can be used with insects of public health importance.

Fumigants

Fumigants are gases that kill body cells and tissues after penetrating the body wall and respiratory tract of insects. They are purchased in either solid, liquid, or gaseous form. Fumigants are of limited use in public health work due to special hazards. They are flammable, toxic, highly reactive, and costly; tend to corrode metals or damage dyes in fabrics; and lack chemical stability. The fumigants include hydrogen cyanide, which is extremely hazardous to animals and humans; methyl bromide, which has little or no warning odor; carbon disulfide, which is highly flammable and explosive: chloropicrin (tear gas), which is highly irritating; ethylene dibromide, which is now banned and desorbs very slowly from certain products; ethylene oxide, which is highly flammable and explosive; phosphine, which may be a fire hazard; and sulfuryl fluoride, which is not recommended for food fumigation. In the year 2000, methyl bromide was phased out as an ozone-depleting substance.

The fumigants are important because they provide a means of destroying large quantities of insects that infest food and may also be utilized in homes where severe insect infestations exist. It is extremely important that all fumigants be handled very carefully and that they be applied only by trained, licensed pest control operators who understand the nature of the chemicals and their hazards.

Desiccants or Absorptive Dusts

Certain desiccants, which in effect damage the outer waterproof layer of the arthopod exoskeleton either by absorbing the fatty or waxy material or by abrasion, are used in insect control. These desiccants include finely powdered silica gels, silica aerosols, and diatomaceous earth. The desiccants affect the water balance, causing rapid water loss and death.

Other Types of Insecticides

Attractants are materials used to lure insects into traps or to make poison baits more inviting. Attractants include, for example, sugar, peanut butter, and fish. Sex hormones have also been tried. Some chemical attractants include methyl eugenol, ethyl acetate, and octyl butrate.

Repellents are substances that produce a reaction in insects that makes them avoid animals or humans. A good repellent works for several hours; is nontoxic, nonirritating, and nonallergenic; has a pleasant odor; is harmless to clothing and accessories; is effective against many insects; and is stable in sunlight. Some repellents include oil of citronella, sulfur, dimethyl phthalate, indalone, and *N,N*-diethyl-metatoluamide (DEET).

Piperonyl butoxide is a compound that acts as a synergist when added to insecticides. It is a low hazard compound in itself, is not known to create environmental problems, and is most effective when used in combination with insecticides that require a booster to do an effective killing job.

Red Squill

Red squill is an extremely effective rodenticide against Norway rats. It may be used fresh with live bait or water bait. Fortified red squill should be used, because the effectiveness of this natural chemical, which comes from the inner bulb scales *Urginea maritima*, varies with production techniques and the time of storage of the pesticide. Red squill kills very rapidly, and although a single rat may develop shyness to bait, rats in general continue to come back and feed on it for periods of time if the bait is mixed properly. In the past, it had always been recommended that red squill be mixed with fresh baits. This is an effective means of control of rats for a short period of time. However, if the red squill is mixed with cracked corn and rolled oats and bound with peanut oil, the peanut oil not only helps preserve the bait for weeks under all types of conditions but also acts as an attractant to the rats, who return and feed on the red squill bait. The poison is of considerable usefulness in public health, because it is one of the least hazardous rodenticides to humans and domestic animals. Red squill causes animals to regurgitate; because Norway rats cannot regurgitate, the poison produces cardiac arrest, convulsions, and respiratory failure. It is used at a ratio of 3.5 to 10.0% in baits. Red squill has been discontinued in the United States.

Cholecalciferol

Cholecalciferol is the activated form of vitamin D (vitamin D_3). Its toxic effect affects probably a combination of liver, kidney and possibly myocardium, with the last two toxicities due to hypercalcemia. This may cause death.

Anticoagulants

The anticoagulant poisons include fumarin; diphacinone; indandione; 2-isovalerylindane-1,3-dione (PMP), also known as valone; warfarin; warfarin plus, also known as sulfoquinoxalin; chlorophacinone; and brodifacoum. Each of these poisons can be used in liquid or dry bait, with the exception of brodifacoum, which is used in dry bait. They provide excellent control of Norway rats, roof rats, and also mice. Because the anticoagulant poison depends on an accumulative action, it is necessary for rodents to feed on poison bait for a period of several days. Generally, it takes from 1 to 2 weeks to get an effective kill of the rodents present. Brodifacoum can cause death in 4 or 5 days and present a lethal dose in one feeding. Some resistance has been noted to warfarin in Europe and also in some parts of the United States. However, the anticoagulants are still extremely useful and safe. It would take a large dose of anticoagulant poison bait to cause any harmful effects in humans or animals with the exception of brodifacoum, where smaller doses may be harmful. The chances of getting such large doses are apparently insignificant.

Fumarin is odorous, nonflammable, and soluble in water and oil. this anticoagulant is highly toxic to rats and mice and does not deteriorate in baits. It does present a slight problem of secondary poisoning to cats, dogs, and individuals applying the poison. Diphacinone is highly toxic to rats, cats, dogs, and rabbits. Again, the quantity of poison necessary to harm cats, dogs, and rabbits tends to protect them. Diphacinone is only hazardous to wildlife and fish if they feed on it continuously for a period of days. No deterioration is found in the bait. There is a slight possibility of hazard to the individual who is using the poison.

Indandione is an odorless compound that is also known as pival. It is soluble in water and oil, is moderately toxic to dogs, and is hazardous to fish and wildlife if eaten continuously over a period of time. This compound is slightly hazardous to the person applying the poison. PMP is odorless, is insoluble in water but soluble in oil, does not deteriorate in baits, has a slight chance of causing secondary poisoning to cats and dogs, and creates a slight problem for the applicator.

Warfarin is relatively insoluble in water and should not be added to baits that contain much vitamin K. It is highly toxic to cats, moderately toxic to dogs, and relatively nontoxic to humans, with a slight hazard to the applicator. No deterioration occurs in the baits. Warfarin plus is warfarin containing hydroxycoumarin and sulfaquinoxalin. This compound is available in ready-to-use baits. Sulfaquinoxalin inhibits vitamin K-producing bacteria in rodents and therefore increases the effectiveness of the warfarin. Smaller amounts are needed for control. Chlorophacinone in laboratory tests has shown good results against Norway rats. Difethialone is a newer anticoagulant. It is a single-feed anticoagulant that is effective on mice as well as on rats. Bromadiolone is another single-feed anticoagulant.

Brodifacoum is an effective anticoagulant that kills warfarin-resistant Norway rats and house mice, and can also be used for roof rats. It may be harmful or fatal if swallowed and must be kept away from humans, domestic animals, and pets. The product reduces the clotting ability of blood and causes hemorrhaging. The bait should be in permanent bait stations resistant to destruction and kept away from children, pets, domestic animals, and nontarget wildlife. It can be used around homes, industrial, commercial, agricultural, and public buildings, and in transportation facilities. It should not be used in sewers.

Warfarin and related compounds (coumarins and indandiones) are the most commonly ingested rodenticides in the Unites States, with 13,345 exposures reported in 1996. Care must be taken around these pesticides.

Other Rodenticides

Bromethalin is an effective rodenticide for rats and mice who are anticoagulant resistant. It may be harmful or fatal if swallowed and must be kept away from humans, domestic animals, and pets. It also causes eye irritation and should not come in contact with skin. The product causes an uncoupling of oxidative phosphoxlation in mitochondria with ATP depletion and leads to fluid buildup between the myelin sheath, pressure on the nerve axons, and increased spinal pressure. Acute poisoning has symptoms of headache, confusion, personality change, seizures, coma, and possible death. The bait should be in tamper-proof bait boxes or in areas inaccessible to children, pets, domestic animals, or wildlife. It can be used for control of Norway rats, roof rats, and house mice in and around homes, industrial and agricultural buildings, and similar structures. It may also be used in alleyways located in urban areas, and in and around port or terminal buildings, but should not be used in sewers.

Various other rodenticides are available. They each are discussed. Alpha naphthyl thiourea (ANTU), is a compound that causes death by inhibiting the clotting of blood and causing internal hemorrhaging. It is very toxic to Norway rats, but less effective on other species. A medium degree of hazard exists in its use, because no antidote is known. Tartar emetic is the best substance used when the compound is ingested accidentally. ANTU should not be used more than once a year, because the rat population will refuse the bait. The poison has also been used as a 20 to 25% tracking powder. ANTU is quite toxic to dogs, cats, and hogs, but is ineffective against roof rats.

Phosphorus bait is a fast-acting poison effective on Norway rats, roof rats, and roaches. It is highly toxic to humans, especially children. Phosphorus baits should only be used when absolutely necessary, in the absence of children, away from food, and under the strict supervision of pest control operators. Yellow phosphorus is no longer sold in the United States.

Zinc phosphide, usually used to kill rats and mice, is generally prepared as a 1% bait with meat or diced fruit. Tartar emetic can be added to this product to make it less hazardous to humans; however, it should never be placed in any area where children, dogs, or cats could consume it. This compound is an extremely hazardous poison and therefore should be used with the greatest of care. Its use is restricted.

Sodium fluoroacetate, or 1080, is an extremely effective poison against rats and mice. However, the poison is extremely hazardous to people, and therefore should only be used with the greatest of care by highly skilled, licensed pest control operators. this rodenticide causes death by paralyzing the heart and central nervous system. The degree of hazard of this chemical is so great that it is recommended that it never or rarely be used. The residue of the poisons must be destroyed by burning in an open field away from any possible human or animal activity. The operator must be extremely cautious in the destruction of this residue. This chemical has now been banned.

Fluoracetamide, or 1081, is very effective for rat control in sewers, in either dry or watered baits. This chemical should not be used for any other purpose because it is extremely toxic to both humans and animals. It should be applied only by trained pest control operators and should never or rarely be used. The remains of the poison and the dead rodents must be burned in an open field or buried so that they cannot be dug up. Extreme caution must be utilized if 1081 is used in rodent control. The sale of 1081 has been discontinued in the United States.

Norbormide, better known as raticate, is a dicarboximide. It is highly toxic for rats; slightly toxic for mice. It is extremely stable in all baits and environments. Thallium sulfate is used as a slow-acting rat and mouse poison. It produces a variety of neurological, circulatory, and gastrointestinal symptoms. Because of the danger in the use of thallium sulfate, it has been banned since 1972 by the EPA.

Arsenic trioxide, sodium arsenate, and sodium arsenite are odorless powders used for mouse and rat control. They are fast acting, but tolerance to the chemical can develop. The arsenic compounds are extremely toxic to humans. They should only be used by trained pest control operators where absolutely necessary, away from children and animals. Arsenicals are now banned.

Strychnine and strychnine sulfate are odorless compounds used in mouse baits. They may also be toxic to rats and are highly toxic to humans. It is extremely important that strychnine compounds are not used unless absolutely necessary and then away from humans, pets, and other animals; and used only by professionally trained pest control operators. These chemicals have now been banned.

Crimidine is a synthetic chlorinate pyrimidine that is very hazardous, causing violent convulsions similar to strychnine. Only specially trained personnel are allowed to use crimidine.

Zinc phosphide is effective against rats and mice. Because it is extremely toxic to all animals, including humans, it should only be used where absolutely essential and should be kept away from all animals and humans. It should be applied only by trained pest control operators.

As can be seen, many of the rodenticides are extremely dangerous to humans or other animals. It is essential that rodenticides be selected with great care and that they be used by trained professional people. It is preferable, whenever possible, to utilize anticoagulants or raticate, instead of the other types of rodenticides, because the chance of harming nontarget animals and humans is reduced.

Herbicides

Over 100 different chemicals act effectively as herbicides. They affect plants either through contact, as a systemic poison, or as a soil sterilant. Herbicides are important in public health work because they are used for the control of weeds and therefore reduce insect and rodent harborage, as well as decrease the amount of pollen present in the air. Contact herbicides kill plants through direct contact. These chemicals may be selective or nonselective and kill all plants. Systemic herbicides are also either selective or nonselective, therefore posing a problem to plants other than those that one would want to destroy. Soil sterilants unfortunately may remain in the environment for long periods of time, and therefore pose a problem to the environment. Inorganic herbicides are derived from the inorganic acids in which hydrogen is replaced by a metal. These herbicides produce a burning effect when coming into contact with the plants. Examples are calcium arsenate, sodium chlorate, and sodium borate. The metal organic compounds include those that have a metal ion complex combined with an organic portion of the molecule. The herbicides are usually used to control large areas of weeds, such as on railroad and highway right-of-ways. An example is disodium methane arsenate.

A third group of herbicides includes the carboxyl aromatic herbicides. This group has a carboxyl group and an aromatic group. These herbicides work as contact, systemic, and soil sterilants. The chemicals are categorized by five basic types:

1. Phenoxy herbicides, which are systemic in nature and usually last 30 to 60 days, are only slightly toxic to humans and animals. Examples are 2,4-dichlorophe-noxyacetic acid (2,4-D) (cautionary statement now added for grazing animals) and sesone.
2. Phenolactic acid is used for aquatic weed control and weed control in right-of-ways.
3. Benzoic acid compounds have a longer soil resistance and low toxicity to mammals. Examples are benzac and trysben.
4. Phthalic acid compounds act to prevent weed germination. They are persistent for about 30 days in the soil and are relatively nontoxic to mammals. Examples are dacthal and endothall.
5. Phthalamic acid compounds also prevent weed germination. They are relatively safe to humans and other warm-blooded animals. An example is alanap.

Other herbicides include aliphatic acid herbicides, which contain a carboxyl group and are temporary soil sterilants. Examples are dowpon and trichloracetic acid. Substituted phenol herbicides are used for contact killing and are applied by sprays in such areas as railroads and highway right-of-ways. These include dinoseb and pentachlorophenol. Their toxicity to mammals varies from moderate to very toxic. The nitrile herbicides are used for killing the seeds of broadleaf weeds. They are usually used in agricultural weed control.

The herbicides include chlorophenoxy compounds such as 2,4,-D and 2,4,5-trichlorphenoxyacetic acid (2,4,5-T). These chemicals are used for the control of weeds and unwanted plants. They affect the kidney, liver, central nervous system, and skin. They cause chloracne, and weakness or numbness of the arms and legs, resulting in long-term nerve damage. Dioxin, or 2,3,7,8-tetrachlorodibenzo-p-dioxin (2,3,7,8-TCDD), is an inadvertently produced trace contaminant in 2,4-D and 2,4,5-T and poses the most serious health risk of these chemicals. Dioxin is a colorless solid with no distinguishable odor. It does not occur naturally, and is not produced intentionally by any industry, except as a reference standard. Dioxin is inadvertently produced during the incineration of municipal and industrial waste containing chlorinated organic compounds. The compound is formed when accidental transformer-capacitor fires occur involving chlorinated benzenes and biphenyls. Dioxin may enter the body through contact with the skin and contaminated soils or materials. It may also enter through the ingestion of food that is contaminated. This food includes fish, cow's milk, and other foodstuffs. Inhalation of contaminated ambient air may contribute a small amount of total body intake; however, the inhalation of particulates such as fly ash may constitute a major source of exposure. In humans, 2,3,7,8-TCDD causes chloracne, which is a severe skin lesion that usually occurs on the head and upper body. It is disfiguring and may last for years after the initial exposure. Evidence suggests that 2,3,6,7-TCDD causes liver damage in humans as indicated by an increase in levels of certain enzymes in the blood. There is also suggestive evidence that the chemical causes loss of appetite, weight loss, and digestive disorders in humans. In rodents, the chemical, if administered during pregnancy, results in malformations of the offspring. 2,3,7,8-TCDD has been demonstrated to be a carcinogen in animals.

Biological Controls

Several techniques are being tried on an experimental basis to sterilize male insects and set them loose in areas where females are present. The aim is to reduce the fertility of the insects and therefore utilize a biological control. *Bacillus thuringiensis* is a naturally occurring soil bacterium that produces toxins causing disease in insects. This microbial insecticide is ideal for pest management because of its specificity to pests and its safety for humans and natural enemies of pests. To be effective, larvae must consume the bacteria during feeding. Predators have also been utilized to destroy insects. The only problem is that the predator may become a pest itself. In rodent control, biological control occurs when various environmental forces are utilized in the destruction of the rats by eliminating food supplies and harborage. The rat population density decreases. If a rat population is not tampered with by humans, it usually expands beyond the ability of the environment to support the number of rats. The overcrowding results in disease and competition, and therefore an increased mortality rate and decreased natality rate. Eventually, the population drops to a size that the environment can support.

Other Types of Pesticides

In addition to the pesticides mentioned, fungicides and bactericides are used to prevent plant disease caused by fungi and bacteria. The nematicides are used to control nematodes, which attack plants. Molluscicides are used to control mollusks, which affect fishing areas or plant areas. Piscicides (fish killers) are used to treat public waters. Their objective is to remove rough or trash fish from restocking lakes or game fish lakes. Avicides (bird killers) are used to control birds and pigeons in areas where they are troublesome and damage crops.

PROBLEMS CREATED BY PESTICIDES

Pesticides in the Environment

Pesticides may enter the environment by means of the air route, the food route, and the soil route. The environment is contaminated through the indiscriminate, uncontrolled, unmonitored, and excessive use of pesticides by all types of people, including the owners of households.

Air Route

Pesticides enter the air by means of aerial spraying, the use of mists and fog machines, and the application of pesticides by individuals using pressure containers. The pesticides, depending on the size of the particles and the volume discharged, the velocity of the air current, the temperature of the air, and other factors, may stay within a given area or may contaminate areas other than those intended. It is essential that great care be taken in the application of pesticides by the air route. The drift and weather conditions must be considered carefully. The human hazards are caused by inhalation, skin absorption, and ingestion of the pesticide. Pesticides may be transported over long distances if they attach to dust particles in the air. Further, they may be mixed in with other chemicals produced by a variety of air-polluting situations, causing secondary chemicals to form, which are in themselves very hazardous.

Water Route

Pesticides enter surface waters by being washed from the surface of the soil or from plants, houses, and agricultural areas. Some pesticides percolate down into underground water supplies through a water flow, providing access to pesticides injected into the soil purposely for the control of insects; or through rain or snow, washing the pesticides into the soil and slowly helping them percolate into the underground water (Figure 7.1). The use of pesticides must be carefully controlled and bodies of water should be regulated. Lakes and other bodies of water should be studied carefully before pesticides are used for either mosquito control or water weed control, because the pesticides may end up causing more harm than good. It should be recognized that pesticides and fertilizers are used extensively by home-owners and farmers. All these chemicals have a tendency to be washed into bodies of water through surface drainage and through storm sewer pipes.

Food Route

Pesticides, from time to time, have caused disastrous consequences when stored in the same vehicles transporting food. It is essential that this be forbidden, because any breakage or leakage would cause chemical food poisoning. Food treated with pesticides must contain the minimum quantity for effectiveness on insects and must not cause harm to humans.

Soil

The persistence of pesticides in the soil creates a situation in which not only is the soil contaminated but also the air may become contaminated by soil particles, or water may become contaminated by runoff. The chemicals used should not be

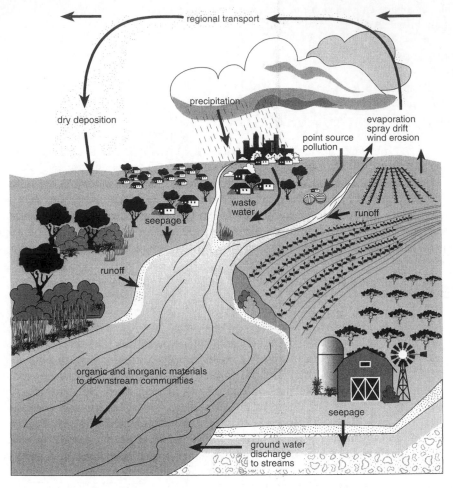

Figure 7.1   Pesticides in the hydrologic system. (Adapted from *Pesticides in Surface Waters,* U.S. Geological Survey Fact Sheet, FS-039-97.)

hazardous and should degrade rapidly within the soil. Those chemicals that are taken up by plants in the soil and are hazardous to humans should not be utilized.

The persistence of a pesticide in the soil depends partially on how it is transferred to the soil. Is it done through leaching, erosion, evaporation, or uptake of plants? The persistence also depends on how the pesticide is degraded. Erosion is still another factor. The algae, fungi, and bacteria found in the soil may use the organic chemicals present as a source of energy, and therefore may reduce some of the amounts of pesticides found in the soil. Chemical reactions may destroy some of the activity of pesticides, while enhancing the activity of other pesticides. Diazinon is broken down in acid conditions; however, the opposite is true for malathion (Figure 7.2).

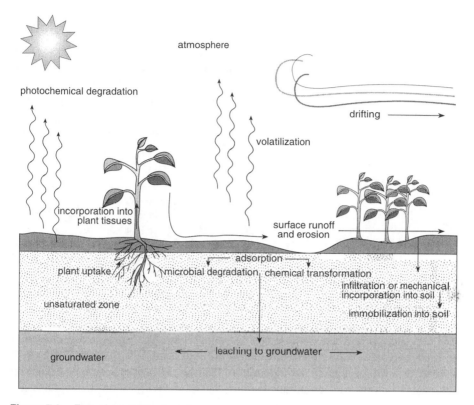

photochemical degradation

atmosphere

drifting ⟶

volatilization

incorporation into
plant tissues

surface runoff
and erosion

plant uptake

adsorption

microbial degradation, chemical transformation

infiltration or mechanical
incorporation into soil

unsaturated zone

immobilization into soil

groundwater

leaching to groundwater ⟶

Figure 7.2   Fate of pesticides in soil.

Household Use

Homeowners are often the ones most apt to contaminate their immediate environments or to provide opportunities for accidental poisoning of their children, themselves, and others around them. Many homeowners have no concept of the proper use of pesticides. They fail to read and understand the labels, use the pesticides under hazardous conditions, and generally store pesticides in areas where children play or can gain access, or where fires may occur. Pesticides are misused in gardens, in homes, and in the care of household plants.

Effects of Pesticides on Humans

It is difficult to evaluate fully the human risks of chronic pesticide problems, because very few studies have been made in this area and because of the complex nature of the problems involved. Further, variables such as age, sex, race, socioeconomic status, diet, state of health, length and state of exposure, and pesticide concentration level all profoundly affect the human response to pesticides. Although cases of acute pesticide poisoning occur, one should not extrapolate these results to chronic low-level exposures. The individual is not only exposed to pesticides in the environment but also to dusts and various environmental conditions that can alter the human response to any specific pesticide.

It is known that the organochlorine compounds, such as aldrin and dieldrin, can increase the excitability of the nervous system and may damage the liver. However, it is difficult to establish a correct diagnosis, because the symptoms vary. Certain compounds can penetrate the unbroken skin. Lindane is believed to cause hematologic disorders.

The organophosphate insecticides inhibit cholinesterase enzymes. Acute toxicity varies greatly from one compound to another. Organophosphates penetrate the skin easily. Carbamate pesticides also inhibit cholinesterase, but because the enzyme deactivates rapidly within the human body, it is difficult to measure the exposure based on this deactivation.

DDT is an organochlorine compound that becomes stored in the fat tissues of animals and humans. It also has considerable effects on fish and wildlife.

The difficulties involved in trying to gather adequate information on the human effects of pesticides contribute to the considerable uncertainty in determining which pesticides are safe and which are harmful. It is impossible to do a general epidemiological study, because all individuals have had some pesticide exposure; therefore, a control group cannot be selected. More often studies are made of acute poisoning due to pesticides among occupational groups, children, or other individuals who have accidentally poisoned themselves. From these data, we can generally determine the kinds of effects that humans have in the event of acute poisoning. However, even in acute poisoning, inadequate amounts of data are available; and it is difficult to make careful judgments about the absolute danger of the given pesticide, unless the pesticide is taken into the body in large amounts. Chronic poisoning becomes an even more complex and confusing issue.

The routes of entry of the pesticide include absorption through the intestines due to ingestion, absorption through the lungs due to inhalation of airborne pesticides, and penetration through the intact skin or absorption directly into the bloodstream through broken skin. The route of entry depends very much on the group of individuals studied and the use of the pesticide. Absorption through the intestine occurs when the residues that remain on food are ingested. This is probably the major route through which pesticides enter the body. In addition, accidental poisoning of children generally occurs through ingestion. Inhalation occurs when bug bombs or aerosol sprays are used to control roaches and other pests in the homes or when individuals inhale particles from fogs and mists that are used to control mosquitoes in exterior areas. Most skin contamination occurs in occupational environments. Pesticides may also be hazardous to humans because of the possibility of fires or explosions. This subject has been previously discussed. In 1998, in California, 34 farm workers became ill after exposure to carbofuran, abemectin, and mepiquat chloride (a growth regulator), following an aerial spraying that had occurred 2 hours earlier; 30 went to the hospital. From 1994 to 1997, in New York City, 25 people became ill when they used aldicarb as a rodenticide. From 1989 to 1997, as many as 16 cases of pesticide-related illness were attributed to occupational use of flea control products.

Considerable study is under way on the possibility of pesticides causing cancer. In the future, many of our existing pesticides may be banned because they are carcinogenic in animals. These studies have to be conducted with great care to ensure that false conclusions are not drawn. It is important not to ban pesticides that are valuable to humans and, at the same time, to protect humans from a new burden of additional carcinogenic agents.

Resistance to Pesticides

Pests, particularly insects, develop resistance to pesticides. Some insects are less susceptible to certain insecticides, and some are affected but not killed by insecticides. Over a period of time, resistant insects survive exposure to insecticides and reproduce new generations of increasingly resistant insects. Generally, two forms of resistance develop in insects: physiological and behavioral. In physiological resistance, the insect develops an immunity to the poison. The exoskeleton becomes less permeable to the insecticide, the insecticide is detoxified into less toxic chemicals, or the insecticide may be stored harmlessly in the body tissues or be excreted. Behavioral resistance is the ability of the insect to avoid lethal contact with the insecticide because it has developed protective habits or behavioral patterns. This includes such activities as mosquitoes changing resting places and flies avoiding baits. Resistance of insects to insecticides is increasing. In fact, cross-resistance has been known to occur. Studies have shown that insects resistant to certain types of chlorinated hydrocarbons may also develop resistance to the organophosphates. Examples of these insects would be the housefly, and the mosquitoes *Culex tarsalis* and *Aedes nigromaculis*. It is necessary from time to time to conduct surveys to determine whether insects are developing resistance to a given insecticide, and if so, the insecticide should be changed.

Economics

The major question in environmental economics as it relates to pesticides is whether the damage costs resulting from blighted crops, poor health, and higher death rates are greater or less than the potential benefits from increased crop yield, reduction of disease, and so forth. It is difficult to obtain a true picture of the cost-benefit ratio in the use of pesticides generally. Each pesticide needs to be studied and judged on an individual basis.

POTENTIAL FOR INTERVENTION

Intervention strategies include the use of the techniques of isolation, substitution, shielding, treatment, and prevention. The potential for intervention varies from poor to excellent, based on the understanding of the short-range and long-range problems associated with the particular pesticide; the preparation, storage, use, and disposal of the pesticide; and the training and ability of the pesticide applicator. Proper storage of pesticides to avoid fire and explosions, and contamination of individuals and their food are techniques of isolation. Substitution is the use of a less hazardous pesticide. Shielding is the use of safety glasses and protective clothing. Treatment is the technique used on an individual who becomes contaminated with a given pesticide. Prevention is the overall process of keeping pesticides out of the eyes, off the skin, away from the lungs, and away from clothing. Good housekeeping is part of the overall prevention technique.

RESOURCES

Scientific, technical, and industry resources include the Entomological Society of America, Pest Control Operators Association, American Public Health Association, National Environmental Health Association, various land grant colleges and universities, chemical manufacturers associations, and National Safety Council. Civic resources include the National Audubon Society, Wildlife Society, and Environmental Defense Fund.

Governmental resources include state and local health departments, USDA, National Bureau of Standards, Department of Defense, EPA (and its site for information on its cross-office multimedia: Persistent, Bioaccumulative and Toxic Chemical Initiative), Department of Health and Human Services (USDHHS), National Science Foundation, and so forth.

The area of pesticides continues to change so rapidly that what is used today may be banned tomorrow, and what does not exist today may exist tomorrow. Public health workers or environmental health workers should obtain a document from the EPA titled, "Suspended, Canceled and Restricted Pesticides." A second source is titled *Handbook of Pest Control* by Arnold Mallis. The third set of references would be the current rules, regulations, and laws of the EPA. A fourth resource is the Library of Congress, and a fifth resource is the U.S. Senate Agricultural Committee.

The National Pesticide Telecommunications Network may be reached by phone at 800-858-7378, which is a toll-free call. The Pesticide Producers' Association (1200 17th Street, N.W., Washington, D.C.) is another resource. The Synthetic Organic Chemical Manufacturers' Association (1330 Connecticut Avenue N.W., Washington, D.C.) is also another useful resource.

STANDARDS, PRACTICES, AND TECHNIQUES

Application of Pesticides

Pesticides are applied to standing water by the use of spray equipment. Oil solutions or emulsions may be utilized as well. The adult mosquito is controlled by the use of aerosol bombs, which are successful in the destruction of flying insects such as flies and mosquitoes. Fogging and misting techniques are used in large community programs as the major method of controlling the adult insect, especially the mosquito. Fog and mist are produced through specialized equipment mounted on driven vehicles and blown into the open air from the equipment. Dusting is used largely in agricultural areas. The dust is usually applied by airplanes (Figure 7.3). Residual spraying is carried out by applying the spray to surfaces upon which the

Figure 7.3   Introduction of pesticides into the environment by aerial spraying.

insects will alight or rest, usually from handheld equipment. Fumigation is the technique of releasing large quantities of aerosols quickly into an area. In catch basins, vapona strips are hung. This insecticide, a fumigant, is slowly released from the solid material.

The type of equipment utilized varies from the hand sprayer, which may be 1 to 3 gal in capacity, to the aerosol bomb; to the compressed air sprayer, which is usually 3 to 4 gal in capacity; to a variety of power equipment; to aircraft application.

Pesticide Labels and Names

All pesticides must be registered for use. The registration can be revoked by the EPA if the pesticide is determined to be a hazard to the community. Pesticide labels and names are very important. The label on a pesticide container specifies the name of the manufacturer; the name of the product; the active chemical ingredients and percentages of concentration; the type of chemical, whether it is an insecticide, rodenticide, and so forth; recommendations for specific uses; directions for use; precautions in storage; and precautions during use by personnel.

Because many chemicals can be sold under the same brand or trade name, it is essential to identify the actual, active chemical ingredients. The generic or common name of the ingredients must appear on the label. The label must state specifically what pests are controlled and what rate of application of insecticide is needed to control the pests affected by the particular pesticide. Because the application of the pesticide is extremely important, instructions should be listed clearly and carefully concerning correct and safe methods of application. All labels should specify the necessary precautions to be taken by personnel. Insecticide labels must read, *keep away from children*. Where highly toxic products are used, skull and crossbones are required on the label along with the word *poison* in red letters. The label should also contain the word *warning*, instructions for handling, the antidote, and the statement, *call a physician immediately*. Where the pesticide is not as toxic as stated earlier, the words *warning* or *caution* should be placed on the label; and it should be stated that a pesticide is a hazard and should be kept out of the hands of children. Common names of the pesticides should be used when possible.

Pesticide Formulations

To use pesticides properly, it is necessary to have some understanding of the form in which they come and also dilution specifications to achieve the proper formulation effective for a given pesticide in a given situation. Solvents dissolve insecticides, dispersing them evenly through the solution. The solvent is a carrier as well as a dilutant, depending on the amount used. Solvents may include volatile liquids, such as xylene, which evaporate after use and leave a residual deposit; or nonvolatile or semivolatile liquids, such as petroleum distillates, which leave the surface coated with the solution of the toxicant. The emulsifiers are surface-active agents that allow liquids to be mixed within liquids. The emulsifier forms a thin film around each minute droplet of oil, thereby keeping the oil from coalescing or separating into oil and water. Wetting agents or spreading agents allow the pesticide to penetrate a surface. Adhesives are used to improve the quality of the deposits of the pesticide. An example of such a material would be gelatin or gum. Perfumes and masking agents are utilized to cover the unpleasant odors of certain pesticides. An example would be oil of wintergreen.

Synergists are materials that help a pesticide work more effectively. Examples of synergists are piperonyl butoxide, sesamex, sulfoxide, and propylisome. Carriers and dilutants for dust are used to help deliver the insecticide more readily in an inexpensive manner. Examples of these carriers would be aptapulvite, bentonite, calcite, diatonmite, and talc.

An insecticide is used in various forms. The technical grade insecticide is the insecticide in concentrated form plus inert material. It is basically the purest chemical form. The insecticidal dusts are prepared by blending toxic ingredients, such as diazinon, into an inert carrier, such as talc. Wettable powders are the toxic ingredients plus the inert dust and a wetting agent such as sodium lauryl sulfate. Water is added and the material is agitated to form a suspension. The material must be continually agitated so that the chemical will not settle out. Wettable powders are used frequently in outbuildings.

In another form, the insecticidal solution may be purchased or prepared either as a concentrate or as a finished spray. The concentrates are diluted with oil to make the usable solution, which becomes a finished spray. The solution is not soluble in water. When a concentrate is mixed with water and an emulsifying agent is added, the insecticide spreads throughout the water. A finished emulsion is made by diluting a concentrate with sufficient water to form the concentration needed for use. Aerosol bombs typically contain 0.1 to 0.6% pyrethrums; 1 to 2% of malathion, or baygon; 10 to 12% petroleum distillates, and 85% propellant.

It is essential to have the correct dosage of pesticides for the proper type of pest control. Excess pesticides may cause unsightly residues, ruination of edible crops, toxicity to crops and animals, excess costs, and specific hazards to animals and humans. When preparing pesticides for use, it is necessary to determine the safe time intervals for application, the amount of residue acceptable by the FDA, the combination of pesticides that are safe or compatible, the treatment of products for animals, the toxicity of the pesticide to be used, the manner in which the equipment was cleaned prior to pesticide use, the rate of application of the pesticide, the size of the area, and the safety precautions to be taken by the operators. Always follow the instructions on pesticide containers to determine the dilution of the pesticide to the proper concentration.

Pesticide Application and Equipment

Pesticides may be applied as solids, liquids, or gases. Solids are usually applied as dusts by means of hand dusting, shaker cans, bellows, vault dusters, or power dusters. Solids may be applied as granules, pellets, gelatinous capsules, or poisonous baits. Liquids may be applied by pouring, painting, spraying, ejecting, or using aerosol bombs.

The method of pesticide application is based on which one effectively kills the pest, safety of the operation, expense, durability of the pesticide, and potential problems. Hand sprayers range from something as old fashioned as a flit gun, used for bedbug control, to various types of compressed air or other sprayers. The sprayers are regulated for the type of spray desired, the quantity desired, and the most desirable pressure for delivery of the spray. It is essential that all parts of the sprayer, including the hose and all fittings, are checked carefully to make sure that they are thoroughly clean and do not leak. A wand is a slender metal tube that is placed on the sprayer and connected to a nozzle that extends the length of the range of the sprayer. The nozzle is the most important part of the sprayer. It has to be fitted so that it will effectively deliver the kind of spray and the concentration desired. Ultralow volume (ULV) larviciding is very effective over land or water. It is necessary to have an air inversion with a 4 to 6°F difference between the hot upper air and the cooler lower air to have ULV work effectively. Late afternoon and early evening are the best times to larvicide with this technique.

Dusters are utilized in areas where rodent ectoparasites are found, where a fire hazard results from the use of oils, and where oil and water are not used advisably.

Powered sprayers or foggers are generally mounted on some type of mechanical equipment. The power unit distributes a large amount of spray at a high pressure. It is essential that this type of spraying be carefully controlled to avoid annoying residents of a home and also to avoid creating an accident hazard. Children have a habit of traveling in and out of fogging machine units, thereby creating accident hazards to the spraying crews, to automobiles, and to themselves. Wind direction and wind velocity must be taken into account whenever power fogging or spraying is utilized. The equipment should be operated to spray perpendicular to the wind and should be directed as close to the ground as possible. Mists and fogs are generally utilized to quickly knock down high concentrations of adult insects. They are economical and can be utilized to cover large residential areas in short periods of time. The mechanical fog generator breaks the insecticide into fine particles and then blows it into the air at high speed. Other types of fog generators are available that operate on either a pulsation principle, thermal principle, or steam principle.

Insecticides may also be injected into the soil to control a variety of pests, especially termites. Fumigation equipment is used to kill insects and rodents in large areas where people are not present. These would include railroad boxcars, hulls of ships, certain aircraft arriving from overseas, places where arthropodborne diseases are prevalent, food storage warehouses, and food-processing plants. Great care must be taken in fumigation.

Insecticides may also be applied by airplanes or helicopters (Figure 7.3). Care must be taken that the aircraft not be used as terrorist weapons. The insecticide must not drift into residential areas and affect animals or humans. Weather conditions, time of day, and so forth must be taken into account when aerial dispersion of insecticides is used.

Storage of Pesticides

Pesticides must be stored with great care, because they may be highly flammable, explosive, or toxic. Most insecticides, fungicides, and rodenticides can be stored in the same room. However, herbicides are quite volatile, and special precautions should be taken with their storage so that they do not contaminate the storage area or escape to the outside, thereby damaging plants.

All storage areas should maintain the following precautions: storage areas should be locked and located away from food, animal feed, plant seed, or water; pesticides should be stored in a dry, well-ventilated place, as directed on the labels; storage areas should be clearly marked as a pesticide storage area; signs should be posted listing the types of pesticides stored and the hazards therein; pesticides should be kept in their original containers; containers should be checked periodically for leaks, tears, and spills; inventory lists should be kept so that outdated materials are eliminated and shortages are clarified.

In the event of a fire, it is extremely important that the firefighters and the general public be protected from the fumes, residues, or washings due to the fire. Fires may originate because petroleum distillates are present; aerosol containers become overheated and explode; other flammable or explosive solvents are present; finely divided dust or powders explode; chlorates, which are flammable or explosive, are present; ammonium nitrate fertilizers are stored; or calcium hypochlorites are present, and may cause spontaneous ignition and explosion if contaminated by organic substances.

The hazards that exist are due to the presence of organophosphates, carbamates, and chlorinated hydrocarbons, which are highly toxic. Further, fumes from solvents, presence of gases, or any combination of substances may be toxic. Care must be taken that runoff water from firefighting, which may contain highly toxic pesticides in quantity, does not enter any of the environmental pathways.

In the event of a pesticide fire, a qualified physician should be available, hospitals should be alerted as to the type of potential hazard, and firefighters should be protected from poisoning by use of proper protective clothing and special self-contained breathing equipment. Great care must be given to the cleanup after the fire to make sure the firemen have removed all traces of the pesticide from their clothing and boots. All the areas where the pesticides are present must be cleaned with the utmost of care, and the resulting contaminated water must be trapped and treated.

Transportation of Pesticides

Transporting toxic chemicals can be very hazardous, with the possibility of accidents or leakage. It is essential that all regulations of the Interstate Commerce Commission and the Department of Transportation concerning proper identification of vehicles carrying hazardous chemicals be followed. The individuals operating these vehicles must be fully aware of procedures to follow in the event of an accident or emergency. Volatile pesticides should never be within the section of the vehicle holding the passengers or the driver.

The transporting vehicles should be properly built so that powders within paper bags are protected from rain and are not be punctured or torn. A vehicle should be able to be easily cleaned. Where pesticides are in liquid form, they must be in tightly closed original containers. Glass containers are not recommended. If they must be utilized, they should be packed and transported in such a way as to avoid breakage. Because pesticides are affected by high and low temperatures, they should be removed from trucks as soon as possible after delivery and stored in safe, locked facilities.

In the event of an accident involving the vehicle transporting pesticides, the drivers should immediately avail themselves of protective clothing and respirators, and should inform the fire department, police department, and health department of the accident. It is essential that fire, police, and health personnel understand the nature of the toxic material and the hazards therein. It is also extremely important that the public be kept as far away from the accident site as possible.

Disposal of Pesticides

Because pesticides in concentrated form or even in diluted form may constitute extremely serious hazards, unused pesticides and empty containers, must be disposed of in a safe manner. In all cases it is necessary to follow the laws, rules, and regulations stipulated by the states, federal government, and local legal bodies. Some types of disposal include ground disposal and incineration. Ground disposal may be very dangerous if the pesticide can contaminate either surface or groundwater supplies. Where this does not occur, it is possible to bury the pesticide in containers that do not deteriorate. This must be a deep burial so that the pesticide cannot be dug up inadvertently. Burial should not occur in normal landfills (see Volume II, Chapter 2 on solid and hazardous waste).

Many pesticides are destroyed during incineration. Incineration is acceptable if the resultant fumes or resultant waste escaping into the air is scrubbed out of the gases, concentrated, and either reincinerated or finally buried in containers that will not deteriorate. Pesticide containers should be handled separately and should be buried only in areas in which the water percolating through the ground cannot carry the pesticide into water supplies.

Ethylene dibromide (EDB), which has now been banned, is a perfect example of a chemical that is now considered to be very harmful. Leaking drums of EDB in Missouri showed how bad the disposal process had become.

The two classifications for wastes that affect the disposal of pesticide wastes are:

1. Hazardous waste
2. Solid waste

Pesticide waste consists of empty containers, excess mixture, rinse water, and material generated from cleanups of spills and leaks. If the pesticide or material is a solid waste, it can be disposed of with other solid wastes. If it is a hazardous waste, as determined by the label, it must follow the regulations of the Resource Conservation Recovery Act (RCRA). Farmers are exempt from complying with most of the RCRA, whereas commercial applicators are not. However, farmers still have to be concerned with local and state laws as well as the potential for contaminating the soil and groundwater.

Container rinsing is required by law and also helps save money. A typical 5 gal container will yield 0.5 oz of formulation, which saves between $8 and $10 per container. Immediate and proper rinsing removes more than 99% of the container residues. Properly rinsed pesticide containers pose a minimal risk for soil, surface water, or groundwater contamination.

Agricultural Pesticides and Water

Cleaning groundwater contaminated with one or more pesticides is complicated, time-consuming, expensive, and usually not feasible. The best solution is prevention. Good management practices include the following:

1. Determine the susceptibility of the soil to leaching by knowing the soil texture, organic matter content, soil moisture, and permeability affect on pesticide movement. Also determine the depth of the water table and if possible the geologic layers between the soil surface and groundwater.
2. Evaluate the pesticides to determine which one is less likely to leach into the groundwater.
3. Evaluate location of water sources such as wells, and make sure they are properly cased and grouted. Unprotected wells act as an immediate conduit for surface contamination to get into the aquifer. Mixing, storing, or disposing of pesticides should not occur within 100 ft of a well.
4. Read the label before you purchase, use, or dispose of a pesticide. Check for groundwater advisories or other water protection guidelines.
5. Consider the weather and irrigation. Delay pesticide applications if heavy or sustained rain is anticipated.
6. Measure pesticides carefully, calibrate the sprayer, mix and load pesticides carefully, and dispose of wastes properly.

Good management practices for surface waters follow the same items listed previously. In addition, the slope of the field, and the relative location of lakes, ponds, streams, canals, or wetlands need to be considered. It may be necessary to construct a berm or bank between the application site and surface bodies of water.

MODES OF SURVEILLANCE AND EVALUATION

Surveillance and evaluation consist of the epidemiological approach, the major incident approach, as well as analysis and monitoring. The epidemiological approach is used to study the general population and its exposure to various pesticides over long time periods. It is used on specific populations, such as workers exposed to pesticides during production, use, and storage, or individuals in communities exposed to large amounts of pesticides, especially individuals living in small farm communities. The epidemiological approach is a technique evaluating mortality and morbidity records of these individuals.

The major incident approach is used in situations such as the contamination of the city of Hopewell, VA, and its surroundings by the pesticide kepone. This particular incident caused widespread kepone poisoning. At least 70 victims were identified and 20 of these victims were hospitalized with untreatable ailments, including apparent brain and liver damage, sterility, slurred speech, loss of memory, and eye twitching. The National Cancer Institute reports that kepone causes cancer in test animals. Kepone was found in shellfish 60 mi down the James River from Hopewell. This type of incident is dangerous to a given community. It does provide, however, large quantities of data for further analysis and evaluation. It is hoped that these data can be utilized in preventing other types of major incidents.

Analysis and monitoring provide information on the methods by which chemicals escape into the environment, the levels at which they are harmful, the types of controls currently utilized, and effective control techniques. Analysis consists of identification of the chemical, substance, and chemical entities, and the quantitative measurement of amounts present. It also provides toxicological evaluation of the product, its isomers, by-products, secondary products, and unreacted intermediate products. During analysis, instruments capable of detecting concentrations in the range of 0.01 to 100 ppm should be used. For some chemicals it is necessary to detect even smaller quantities. Monitoring tracks specific chemicals through the environment. To monitor properly, it is necessary to consider a wide range of concentrations and potential toxicities and to understand the limit of detection of a given analytic method and the behavior of the chemical in the environment. How, when, and what to monitor, and when to stop monitoring are complex decisions that must be made by competent professionals.

Chemical structure, reactivity, basic physical and chemical properties, proper analytic methods and monitoring strategies, gathering of reliable analytic data, storage of the data, and retrieval and use of the data are essential to the proper techniques of surveillance and evaluation.

CONTROL OF PESTICIDES

Safety

Pesticides must be used with great care, because they are hazardous to humans. Most victims of pesticide poisoning are either workers in the occupational preparation or occupational use area, individuals who inadvertently have been affected by improper use of pesticides, or children who have eaten the pesticide. Individuals may become chronically ill from exposure for long periods of time or acutely ill from exposure to large quantities over a short period. Pesticides follow the respiratory route, are absorbed through the skin, or are ingested with food.

The solvents used for dilution may also be toxic. Poisoning by pesticides of children under 5 years of age is a serious public health hazard. It is essential that these materials be stored out of the reach of children. Pesticides should not be stored in pantries, under sinks, or in garages where children can reach them.

Safety rules include proper reading and understanding of labels; proper preparing and applying of the insecticide; proper storing of the insecticide in the original containers with the original labels; mixing of pesticides in well-ventilated areas; mixing and applying flammable pesticides in such a way that they are not near fires, defective wiring, smoking, or hot areas; avoiding eating, drinking, or smoking where pesticides are used; wearing of appropriate clothing and headgear; avoiding contamination of the food and water of humans and animals; avoiding inhalation of sprays and dusts; keeping equipment in good operating condition; avoiding the storage of partially used pesticides; proper disposing of pesticide containers; proper transporting of pesticides; proper storing of pesticides; and understanding of first aid measures where needed.

When accidental poisoning occurs, speed is the most essential concern. Proper treatment must be given at once and the individual taken to a hospital immediately. Poison information centers are available and usually can be reached by the telephone operator or local hospital. The poison information center or physician supplies information on immediate first aid, depending on the poison taken. It is essential that the label be read to determine the chemicals present and the antidotes and techniques used to counteract the poison. In no case should an individual be made to vomit if unconscious, in a coma, convulsing, or after consuming petroleum products and corrosive poisons. If eyes have been affected, they should be washed immediately with cold water for at least 5 min. Any delay may result in permanent injury. If the individual is in a poisonous atmosphere, remove this person as soon as possible to fresh air. It is best to reiterate that speed is absolutely essential in any type of pesticide poisoning. It is also essential to determine as soon as possible what chemicals are used and to get proper help from medical authorities.

Laws and Regulatory Agencies

The FIFRA of 1947 was superseded by the Federal Environmental Pesticide Control Act of 1972. The law forbids anyone, including the government, from using pesticides in a manner contrary to the label instructions and gives the EPA the authority to restrict use of certain pesticides to trained personnel in approved programs. It also extended control to all pesticides sold intrastate and interstate. In June 1975, regulations were published to establish a screening procedure. If a new pesticide was chemically suspected of causing cancer, heart defects, or mutations, it had to undergo testing before it could be declared safe. Any suspected pesticides that were already registered could retain their registration until a pending test proved that they were dangerous. These regulations also required that data be developed to determine when farm workers may reenter fields. On October 11, 1974, the EPA suspended the registration of aldrin and dieldrin because new evidence showed that they were imminent hazards. Animal experiments indicated that they were carcinogenic. The courts held that the burden of proof of the safety of a pesticide rested on the registrant and not with the government. The court also accepted animal test results to indicate the cancer risk to humans. In July, the EPA issued a notice of intent to suspend the use of heptachlor and chlordane based on animal experiments indicating that these pesticides may cause cancer. Pesticides containing myrex and phenyl mercury compounds had also been canceled. Although the EPA has the right to lift a ban in the event of a public health or other national emergency, generally the agency did not do so. The only one major exception involving the use of DDT occurred in 1974 in the forests of the northwest. DDT has been banned since 1972.

Under the 1972 act, all pesticides distributed, sold, offered for sale, held for sale, shipped, or delivered are required to be registered with the EPA. Registration consisted of filing a statement with the EPA administrator giving the name and address of the applicant or any other name appearing on the label; the name of the pesticide; a complete copy of the labeling of the pesticide, including directions for use; a full description of tests made of the complete formula on request; and a request for the type of classification, whether for general use, restricted use, or otherwise. The EPA administrator then approves the registration of the pesticide.

Another important part of the 1972 law states that the pesticide must be used only by certified applicators. Therefore, pest control operators have to obtain certification. Federal certification is handled by the EPA; state certification is submitted by the governor of each state to the EPA for approval. All operators after certification have to earn continuing certification hours credit to retain their licenses.

Registration of any pesticide is canceled if the registrant does not reapply within a 5-year period. In addition, the EPA can cancel the registration if the pesticide is found to be hazardous in any way. The EPA has the right to stop the sale and use, and to order removal and seizure of any pesticide that it deems hazardous or being used in a hazardous manner.

The 1972 law (section 171.3, categorization of commercial applicators of pesticides category number 8) stipulates that all state, federal, or other governmental employees using or supervising the use of restricted pesticides in public health programs have to be certified. Therefore, under the 1972 law, all pesticide operators and all public health workers involved in the control of pests of public health importance had to be certified by October 21, 1976, by the EPA and also by the state in which they were operating. The law was again updated in 1978 and in 1988.

The 1972 law was updated in 1975 to require impact statements and to require the EPA administrator to submit actions concerning health matters related to pesticides to a scientific advisory panel. The comments of the advisory committee and the administrator have been published in the *Federal Register* (see current issues for update on 1988 FIFRA amendments and 1996 Food Quality Protection Act in this chapter).

SUMMARY

Pesticides move through the various ecosystems in the environment in a variety of ways. When pesticides are ingested or otherwise carried by the target species, they stay in the environment. They may be recycled rapidly or further concentrated through bioconcentration as the pesticides move through the food chain. Most of the large volume of pesticides utilized do not reach their intended areas and therefore become contaminants. Pesticides are introduced into the environment by spraying or by surface application. The storage in body fat of a pesticide is based on its chemical nature, physical state, means of application, and atmospheric conditions. The persistence of pesticides in the air is influenced by both gravitational fallout and the washout caused by rain. The chemicals build up in the soil to concentrations that affect the various ecosystems, possibly contaminating soil, water, air, and various organisms. Pesticides can cause direct problems to humans through airborne contamination, drinking water, or food contamination. The pesticide problem is increasing, despite the fact that the Department of Health and Human Services, and the EPA, are working diligently to remove dangerous pesticides from the market. However, without pesticides our society would be in great trouble, because insects and rodents would consume valuable food supplies and cause a variety of serious diseases.

It is obvious from the preceding material and concerns under discussion in this chapter that the problem of various pests is unresolved and will probably continue as long as we exist. It is also clear that pesticides are hazardous and must be used with the greatest of care. Public health officials and pest control operators must be constantly on the alert to ensure that pesticide usage is proper. With new research, effective pesticides may be eliminated from the market because of their potential hazards as toxins, fire, and explosion hazards, and because they are potential carcinogens.

RESEARCH NEEDS

Research is needed on all common pesticides used for the control of pests of public health significance. It is necessary to determine whether the various chemicals can cause a resistance in the pest that might reduce or eliminate the effectiveness of large groups of pesticides. Chlorinated hydrocarbons must be studied to determine their reactions with food constituents and their potential harmful effect on humans. Recent epidemiological studies suggest that there might be a link between organochlorine compounds and breast cancer. Several pesticides have been shown to be altered by sunlight. It is important to study photochemical alterations to determine whether the efficiency of the pesticide is reduced in the natural environment. What are the issues relating pesticides binding to macromolecular components of plant and animal tissues used as food? Little is known about the chemistry of these bound residues. Organophosphate pesticides may be affecting commercial sprayers, because they have a significantly higher anxiety score on standardized tests than the control groups. This area also needs further study. Laboratory animals have difficulty in learning and memory because of exposure to pesticides. Chemical agents, such as the anticholinesterase pesticides in high doses, acutely affect the nervous system. Studies should be made over long periods of time at low-level doses to determine if they cause irreparable damage.

Under the federal Food, Drug, and Cosmetic Act (FDCA), the EPA, in cooperation with the USFDA, sets allowable limits for pesticide residues in food. These limits, which are called *tolerance levels,* are supposed to protect human health while allowing for the production of an adequate, wholesome, economical food supply. Tolerance limits need to be reevaluated for many existing chemicals

Techniques of genetic engineering, as they relate to the production of genetically engineered microorganisms under the FIFRA, need to be studied further. Although many pesticides have been approved in the past, they need to be further studied to determine if more stringent standards should be applied to their use. Studies should be conducted on dioxins, which are by-products of the production of pesticides. Studies should be conducted of the gas stream to determine the chemicals present and their potentials for causing serious health effects when pesticides are destroyed through incineration. Additional research is necessary in reducing the health risk of pesticides and also in reducing the risk to the environment. Evaluations of indoor air should be conducted to determine the level of pesticides present in the indoor air of the home and also the potential hazards. All new and previously registered pesticides should be screened for their potential to contaminate groundwater. Continuing studies are needed to determine the potential health effects of inert ingredients found in pesticide formulations.

CHAPTER 8

Indoor Environment

BACKGROUND AND STATUS

History

The indoor environment in its broadest sense includes housing, the immediate external environment of the structure, and other facilities where we may work.

Each of us understands to some extent the importance of our home environment. However, the physical environment for millions of less fortunate families is conducive to disease, behavioral problems, accidents, and social problems. Many times during the course of history, attempts have been made to create more attractive housing and to eliminate health-related environmental hazards. This chapter is concerned with the origins of housing problems; the types of health problems related to housing, neighborhoods and their effects on housing; the types of housing, the environment within dwellings; the structural soundness of dwellings; and the various techniques for improving housing conditions, indoor air pollution, and injury control.

In 1626, the Plymouth colony passed a law stating that new houses must be roofed with board. In 1648, wooden or plastered chimneys were prohibited on new houses in New Amsterdam, and chimneys on existing houses had to be inspected regularly. In Charlestown in 1740, it was declared that all buildings must be made of brooker stone because of a disastrous fire that had occurred. The reason that these communities began passing housing laws was the frequent fires that destroyed many lives, houses, possessions, and food. During the 1600s, some Pennsylvanians lived in caves along the water banks. This was declared illegal in 1687. Most of the early housing codes were primarily concerned with the fireproof nature or size of the structure and not with the basic indoor environment, which was very poor. Outdoor privies were used. In 1657, New Amsterdam passed a law that rubbish and filth could not be thrown into the streets or canals.

In the 1800s during the Industrial Revolution, millions of immigrants arrived in the United States. They were unable to support themselves, had to seek housing accommodations with relatives, or had to move into housing in poor condition. The cities, particularly New York, grew extremely overcrowded, and housing became the source of numerous outbreaks of disease. Only the fire department had any authority, until 1867, when the Tenement Housing Act of New York City was passed. This was the first comprehensive rule of its kind in the country. It stated that each room used for sleeping or occupied by an individual had to have adequate ventilation by means of a transom window, if windows to the outside were not available. The roof, stairs, and banisters had to be properly maintained. At least one water closet or privy had to be provided for every 20 persons. All houses had be cleaned to the satisfaction of the board of health, and all cases of infectious disease had to be reported. Authority was given to condemn houses unfit for human habitation. Additional laws were passed over the years to modify and strengthen the initial act.

In 1901, a new, improved Tenement House Act was passed, stipulating the amount of required space per person; the fireproofing of the structure; and the required number of bathrooms, room sizes, and so forth. Housing legislation was also enacted in Philadelphia, Chicago, and other cities. In 1892, the federal government passed a resolution that authorized an investigation of slum conditions in cities of more than 200,000 people. Very little money was allocated for the study, which limited the actual investigation. Although housing laws continued to be modified, enforcement of the codes was poor until the 1920s, when the state of housing in the United States grew extremely unsatisfactory, both within the cities and in the farm areas. By the 1930s, President Roosevelt reported to the country that one-third of the nation was ill-fed, ill-housed, and ill-clothed. The first federal housing law was passed in 1934, establishing a better system for residential mortgages through government insurance. The Federal Housing Administration (FHA) was created to carry out the objectives of the law. Many other governmental agencies became involved in the next few years in mortgage lending and in an attempt to clear slums and to increase urban renewal. However, it was not until after World War II that substantial housing acts were passed. The first one of these was the Federal Housing Act of 1949.

Industrialization, coupled with a substantial move from farms to cities, has helped to create the severe neighborhood and housing problems within the last 100 years. Industries established themselves wherever they were most profitable. If they contaminated the land, water, or air, it really was not of much importance. Obviously, because people had no access to the modern automobile, they had to live close to industry or to convenient public transportation. As new groups moved into areas to fill job needs created by industrial expansion, neighborhoods and houses were filled by a highly transient population. The nature of the population, the overcrowding that occurred, the nature of the industrial pollutants, and the cheaply constructed housing used as dwellings by workers were the factors contributing to the severe housing problems of the 20th century.

The 1940 census indicated for the first time the type of housing conditions found in cities. New Haven, CT, authorities assigned penalty scores to four items of the census. The resultant maps showed the concentration of substandard and presumably slum conditions. On the basis of these findings, a sample environmental survey of the dwelling units was conducted. This survey indicated the condition of the houses and the apparent condition of the neighborhoods. Points were given to such features as toilets, baths, means of exit, lighting, heat, rooms without windows, deterioration, number of people per room, and sleeping area in square feet per person. This survey indicated where slum clearance and housing improvement was needed. It was evident that any rebuilding program or slum clearance program had to be carried out in relation to the overall city plan. Revision and enforcement of housing regulations and control of blight in nonslum areas were necessary parts of the new thoughts on housing. Problems common to most areas included overcrowding due to large families, poor quality of dwellings available for minorities, and substandard properties in relation to the amount of rent paid.

Furthermore, the impact of decay during the years of World War II, the inability to build or improve housing from lack of materials, and the movement into the cities because of war production caused additional housing problems. At the end of the war, the birth rate expanded, marriages increased sharply, and the demand for new houses rose at an unusually rapid pace. Veterans moved to the suburbs with their large families, creating a new type of housing pressure in areas where people had previously lived on farms. The inadequacy of many of these new dwellings built immediately after the war, the inability of the soil to handle the enormous amount of on-site sewage, and the lack of proper water supplies contributed to a new wave of serious housing and environmental problems.

Poverty and poor housing are mutually inclusive; one feeds on the other. The poorer the individual, the more crowded the living quarters are. Poverty further reduces the chances of attaining a better education, occasional recreation, and occupational skills. These limitations create feelings of insecurity and helplessness and deprive the poor of alternatives. In many cases, the poor have great difficulty maintaining proper sanitary conditions in their dwellings. Either a serious problem exists before they arrive or they are unable to physically, mentally, financially, or emotionally cope with the upkeep of their dwellings. The situation tends to grow worse instead of better. Housing and poverty are not simple matters. In addition to poverty, the education background, lifestyle, health practices, outlook on life, age at marriage, and marital responsibilities must be understood. The housing of the poor leads to situations that intensify the potentially serious hazards that exist within the neighborhoods. Housing problems have been severely complicated by the presence of gangs and drug houses.

Urbanization and suburbanization have become the predominant population shifts in the United States. Lower income groups left the rural areas of Appalachia and the South to crowd into already overcrowded cities. This, coupled with increasing population growth, caused a severe housing problem, because the supply of new homes in these areas was minimal and the older properties continued to deteriorate at a rapid rate. Many of the people in the urban areas moved to the suburbs after attaining a higher socioeconomic status in the hope of finding a better way of life for themselves and their children. This has resulted in a huge suburban sprawl that has created innumerable environmental problems.

The inner city has, in many ways, become a jungle. People live not only with rodents and roaches and under the most severe conditions but also with the emergence of new hazard. The arsonist is creating a hazard with a price tag of $10 to

$15 billion annually. Residents are suffering injury and death because individuals either out of hate or for profit are setting fire to old structures. The destruction of property has increased the serious housing problem that already exists.

Increased housing costs have also placed a strain on housing for lower income families. The American dream of securing a good education and a good home is very difficult to realize. Due to this rise in cost, many individuals are unable to move to better living quarters and are forced to remain in overcrowded run-down dwellings. In existing homes, owners are having difficulty maintaining their property. Many homeowners have had to obtain second mortgages to simply repair or maintain their property. Many individuals become burdened with large debts and then have to move to less expensive housing to survive. In the coming months and years, the situation is not expected to improve but instead to deteriorate.

Current land-use patterns show a tendency toward further urbanization, which is consuming a significant amount of valuable land used for agriculture and by wildlife. Some communities have begun to renovate the older well-built structures found abandoned or underutilized in the inner cities; however, most communities are still ignoring this vital resource. Improper land use, improper use of existing structures, and improper growth are serious environmental problems in our society today.

In the last 50 years, the U.S. metropolitan population has more than doubled, with 8 out of 10 Americans living in an metropolitan area with populations of 1 million or more people. Despite the economic and social importance of cities, they have enormous challenges. All types of communities are concerned about crime, public health, affordable housing, barriers to residential mobility, fiscal stress, and the environment. The intense concentration of poor households and an eroding tax base have complicated the problems. Special problems have occurred because of the following:

1. Long-established patterns of economic activity and social organization have been disrupted because of global competition and technological innovation. Manufacturing employment has left urban cores for suburban, rural, and even foreign locations, and now less-skilled workers in inner-city communities are without jobs and incomes.
2. Widespread migration to the suburbs has compounded the urban crisis and left poor people in the big cities.
3. As upwardly mobile families left center cities for the suburbs, the poor became concentrated in inner city neighborhoods where education and job opportunities are severely limited.
4. A lack of cheap transportation has further increased the problem of joblessness.
5. More than half of all adults in these areas have less than a high school education.
6. Poor urban parents disproportionately report that groups supporting youth development such as scouts, organized sports, religious activities, and special classes are not readily available for their children.
7. The people living in neighborhoods of concentrated poverty include those who are not employed, single women heads of households, elderly persons, disabled people, immigrants, abandoned children, and homeless people. They are often

socially isolated and generally have more limited social networks to rely on for practical advice.

8. Cities with high poverty rates have high per capita expenditure for welfare, hospitals, police, fire, and education services.

9. The physical community structure as well as the housing tends to be very old and dilapidated.

10. Numerous old industrial sites now known as brownfields are contaminated, underused, or abandoned, posing serious problems for the cities and people because they degrade the environment, potentially cause disease or injury, represent lost opportunities for jobs, and contribute to a poor tax base.

11. Decent quality housing is widely available but remains unaffordable to many people because of low incomes and rising housing costs. From 1970 to 1994, the median income of renter households fell by 16%, whereas gross rents increased by 11%. The very low income earners pay 50% of their money for rent or live-in substandard housing. This has increased by almost 400,000 households to 5.3 million.

12. Despite the Fair Housing Act passed almost 40 years ago, discrimination still exists.

13. Finally, homelessness is one of the most serious housing-related problems in America. It is caused by crisis poverty and chronic and often untreated disabilities.

Indoor Air Pollution

During the last 30 years there has been a sharp increase and concern by the public and various levels of government about the quality of the air we breathe inside homes, schools, and workplaces. In the 1970s, energy conservation measures were widely supported by the federal government through income tax breaks, by state and local governments, and also by the public in general. The cost of energy had become extremely high. Energy conservation meant the development and use of new insulation materials and "tighter buildings." Formaldehyde became part of the insulation material in many buildings. In the mid-1960s, potential cancer risks due to asbestos were less understood. Asbestos has been found in a variety of buildings, especially in schools. Radon gas was observed in the 1970s. Particularly high levels of radon were found in the Reading Prong geologic formation that runs through New Jersey, Pennsylvania, and New York. Energy conservation also brought on the current problem of sick building syndrome or tight-building syndrome.

Indoor air pollutants were found in homes, offices, schools, hotels, restaurants, various buildings, buses, trains, and planes. In each of these environments, people were exposed to a wide variety of pollutants that came from smoking, building materials, rugs, furniture, appliances, pesticides, cleaning and deodorizing agents, etc. People were exposed to increased levels of carbon monoxide, nitrogen dioxide, metals, inorganic compounds, radioactive pollutants, and volatile organic compounds (VOCs). The best known inorganic compounds are mercury, chlorine, and sulfur. The particulates include the inorganic fiber asbestos and carbonaceous tobacco smoke. The VOCs include formaldehyde, benzene, and carbon tetrachloride. In recent years, increasing attention has been focused on indoor mold growth and associated gaseous by-products.

Good indoor air quality includes introduction and distribution of adequate ventilation air, control of airborne contaminants, and maintenance of acceptable temperature and relative humidity.

SCIENTIFIC, TECHNOLOGICAL, AND GENERAL INFORMATION

Neighborhoods and Their Effects on Housing

The neighborhood is that area comprising all the public facilities and conditions required by the average family for their comfort and existence. Residents of a neighborhood share services, recreational facilities, and generally an elementary school and shopping area. Some neighborhoods were artificially established in older cities, by a particular ethnic, religious, or racial group. Neighborhood community facilities include educational, social, cultural, recreational, and shopping centers; utilities and services include water, light, fuel, sewage, waste disposal, fire and police protection, and road maintenance.

The size of the neighborhood is difficult to assess, because it varies with the types of building structures, the density of the population, the identification of the population as a specific group, and the various services available. To establish limits for the area or population seems impractical. Small developments on the outskirts of a particular neighborhood should be included within the neighborhood if they utilize its services. Logical neighborhood boundaries also include such entities as rivers, topographical barriers, interstate highways, parkways, railroads, industrial areas, commercial districts, and parks.

Site Selection and the Neighborhood Environment

When developing a new neighborhood, it is essential to consider not only the previously mentioned services but also the land-use trends for the area; the presumed availability of transportation, public utilities, and schools; and the legal controls placed on the area by the local governing bodies. A new neighborhood should contain between 2000 and 8000 persons, with the most desirable size at about 5000 persons; it should be within a maximum walking distance of 0.5 mi from the local elementary school; it should not cover substantially more than 500 acres.

Site selection for the neighborhood is extremely important. The existing neighborhood may have to change, be razed, or be redeveloped to meet proper neighborhood and housing standards. The new neighborhood should avoid the problems of existing neighborhoods by locating on proper sites. In site selection, competent professional engineers should determine the geology of the land, the type of soil, and the type of weather conditions prevalent in the area; they also should consider whether disturbing conditions such as superhighways, heavy industry, or hog farms are nearby, and whether the site is in a flood plain.

Soil and subsoil conditions must be suitable for excavation, site preparation, location of utilities, weight-bearing capacity, and so forth. Test borings should be made to determine the types of soils and subsoil conditions. Soil with a high clay content shrinks when extremely dry, causing cracks in foundations and within houses. The type of groundwater and drainage conditions should also be evaluated. Drainage conditions after construction should be forecast. Areas in which basement flooding may occur, swamps or marshes are present, or the groundwater table is so high that it periodically floods the site should be avoided. It is also essential to plan a site that will not be flooded by surface waters from streams, lakes, tidal waters, or higher land. The land must not be too steep for proper grading and usage. Buildings should not be situated at an elevation above normal water pressure. The slope of the ground must also be evaluated. Adequate automobile and pedestrian traffic must be accessible to the community site. Sufficient land should be reserved for private yards, gardens, playgrounds, and neighborhood parks. The area should not be located in the vicinity of hazardous bluffs, precipices, open pits, or strip mines.

Water supply and sewage disposal are important keys to proper site selection. New sites in the community should not be approved unless public water and public or semipublic sewage are available. Today, due to poor planning or technical ignorance, millions of homes have overflowing on-site sewage disposal systems.

It is also necessary to determine the method of removal and disposal of solid waste. Because reasonably priced electricity is essential, the availability of electricity for the planned community must be determined before the site selection is approved. In addition, if gas is available, it should be included in these determinations. Telephone service must also be planned for the new site. The accessibility of firefighting crews and equipment, and the kind and amount of police protection have to be considered before the site can be definitely approved.

Major accident hazards have to be eliminated from the site and adjacent areas. Accident hazards include aircraft, cars, trains, and so forth. The site should not be located near the use of fire and explosion hazards, such as oil or gasoline storage, or firearms. Dumps, rubbish piles, and sanitary landfills should not be adjacent to the site. The site should also be a safe distance from unprotected bodies of water, such as strip mines and quarries. It is essential that railroad crossings are well marked and automatic signals are used. Excessive noise, odors, smoke, and dust should be avoided.

A site should be selected so that pollution from sewage plants, sewage outfalls, or farms do not contaminate any bodies of water running through the community. Adequate public transportation, and pedestrian and bicycle paths should be provided for proper recreation and access to community facilities out of the area of the new neighborhood. Other types of services, such as junior and senior high schools, large shopping centers, areas of potential employment, urban areas, outdoor recreation areas, public and private health services (including physicians, laboratories, and a hospital) must be accessible to residents. During the actual development of the neighborhood, proper grading techniques should be employed to ensure adequate surface drainage so that water does not stand and cause mosquito problems, or flow in such a way to cause erosion.

Property Transfer Assessments

A property transfer assessment is a determination of those areas of potential legal and economic liability that a buyer may assume by purchasing a business or industry that has actual or potential environmental problems. A nationwide trend in the federal and state legislatures is to hold individuals responsible for cleaning up environmental contamination on pieces of property that they own. This becomes a serious problem for the purchaser of any piece of property, because the environmental health hazards may be hidden or buried. These hazards may consist of spilled chemicals that are polluting the groundwater, asbestos insulation, or buried waste. The three major phases in determining the problem are as follows:

1. Examine available records to determine the prior land ownership and use. Contact appropriate agencies and review any hazardous materials handling practices that may have occurred at the site.
2. If necessary, conduct a more detailed investigation to determine how far local contaminants have already spread through the environment.
3. Develop and evaluate alternative site cleanup plans and costs. Only after the audit has been completed and the determination made concerning hazardous materials, should the buyer consider purchasing the property.

These properties are very much a part of the community and could be located near housing developments. If hazards are present, they may constitute an unacceptable risk for the community. At the time of sale of the property, the community has an excellent means of bringing about change in the buildings if needed.

Internal Housing Environment

The internal housing environment, which consists of heat, light, ventilation, plumbing, and so forth, affects the physical, emotional, and mental states of the occupants. An individual needs to be protected against the elements of heat, cold, disease, insects, and harmful chemicals. People need to know that when they arrive home they can leave the pressures of society behind them, can relax in safety and comfort, and then can face the challenges of society again.

Heat

The ideal thermal environment for a given individual varies with age, sex, conditions of health, and so forth. Most housing codes and air-conditioning engineers recommend a minimum indoor temperature of 68°F for winter and 75°F for summer; however, for some this temperature is too cold or too warm. The recommended temperature must be evaluated in the light of the types of individuals, the amount of air flow, and the amount of energy available. With increasing energy problems and rising energy costs, the entire concept of air-conditioning may well change in the future.

The most important aspect of the thermal environment, regardless of the selected temperature, is that the individual should not suffer from undue heat loss and that the heat within the building is not so excessive as to be oppressive. The temperatures within rooms vary tremendously from floor to ceiling depending on the window location, the type of heating devices, and the way in which the heating system is planned. Weatherproofing and insulating of buildings reduce heat loss and retain the cool air provided by air-conditioning during hot weather.

Light

The sources of lighting in a dwelling are both natural and artificial. Few dwellings have been built to maximize natural lighting from the sun's rays, which serve not only to illuminate, but can control temperature as well. Light is essential for cleanliness and avoidance of accidents, and it contributes to healthy mental attitudes.

Ventilation

Proper ventilation is necessary to maintain a thermal environment that allows adequate heat loss from the body, removes unnecessary chemicals, and permits proper aesthetic sensibility within the home environment. Heat loss is controlled by air temperature, relative humidity, air movement, and mean radiant temperature of the surrounding surfaces. Cool moving air is valuable for promoting restful sleep. Odors, which are not removed in a poorly ventilated area, affect the well-being of the individual.

Electrical Facilities

The electrical code, the housing code, and, if these are not available at a local level, the National Electrical Code should be consulted during installation of electrical wiring in the home. Electricity is the conversion of mechanical energy into electrical energy, often by means of a generator. Electrical voltage increases by several hundred thousand to more than a million volts when it passes through a transformer. The high voltage is necessary to increase the efficiency of the transmission of power over long distance. This high voltage is then reduced to the normal 115- or 230-V household current by a transformer located near the home Electricity is transmitted to the home by a series of wires called a service drop. For an electrical current to flow, it must travel from a higher to a lower potential voltage. The current flows between hot wires, which are colored black or red, at a higher potential to the ground or neutral wire, which is colored white or green, at a lower potential. The voltage measures the forces at which the electricity is delivered. Electrical current is measured in amperes, which is the quantity of flow of electricity. Wattage equals the number of volts times amperes.

Earth is an effective conductor. Near the house the usual ground connection is a water pipe of the city water system. Care must be taken that this ground is properly installed so that an individual is not shocked by touching both sides of the pipe if the water meter is removed.

Plumbing

Plumbing is defined as the practice, materials, and fixtures used in installing, maintaining, and altering all pipes, fixtures, appliances, and appurtenances that connect with sanitary or storm drainage, venting system, and public or individual water supply system. (An in-depth discussion of plumbing can be found in Volume II, Chapter 5.)

Structural Soundness

The structural soundness of a dwelling is related to the prevention of disease and injury, and the promotion of good health. The site selected should follow the requirements set forth under neighborhood site selection. The site must not be approved if a public sewage system is unavailable or the land is unable to absorb the fluid from on-site sewage systems.

The building must be placed on the site in such a way as to provide sufficient outdoor space, adequate air circulation, quiet, and safety. The location of the building may reflect certain conditions that cannot be corrected, such as the trajectory of the sun, the direction and velocity of winds, the general climate, and the nature of the water runoff from other properties or the potential for water problems created by heavy rains or snows. The ground slope is important, because this may well control the dryness and the safety of the structure. The site selection and the type of housing construction are dictated by local zoning, housing ordinances, and the amount of money available.

Daylight, direct sunlight, and heat from the sun play an important role in the placement of the structure on the site. The building should be located on the site to maximize year-round exposure to sunlight in each of the rooms. This means at least 1 hour of sunlight on a clear day in each room. Daylight should be available to all habitable rooms during the daytime hours to reduce the additional energy requirements and also to provide a more natural setting.

In site selection and housing construction, it is essential to consider the free circulation of air around and through the structure. This helps to improve the indoor thermal environment and to reduce the cost of air-conditioning. During the winter, proper insulation keeps the cold air out of the house. In summer thermal comfort is a combination of air temperature, air movement, and relative humidity. The air movement within the building is a combination of the amount of air that penetrates into the building and the wind velocity. It is generally most comfortable if the effective temperature within the structure in the evening is not more than 72 to 75°F.

Site selection must also be based on the avoidance of excessive noise due to exterior problems such as factories, railroads, and highways. By orienting bedrooms away from recreational areas within the house and noisy outdoor areas, individuals are able to sleep uninterrupted. The noise level should not exceed 50 db within the dwelling unit and 30 db within study or sleeping areas. In residential areas the site must be oriented so that the structure is placed an adequate distance from the road and other buildings.

Physical Structure

The basic physical structure of the house consists of the foundation, the framing, the roof, the exterior walls and trim, and roof covering. Because this chapter is not intended as a guidebook on building construction, many of the details that would ordinarily be covered are excluded.

Foundation refers to the construction below grade, such as the footings, cellar, and basement walls. It also refers to the composition of the earth on which the building rests and other supports, such as pilings and piers. The foundation bed may be composed of solid rock, sand, gravel, or clay. Sand and clay are the least desirable, because they may shift, in the case of sand, or swell and shrink, in the case of clay, leading to sliding and settling of the building. The footings should distribute the weight of the building over a large enough area of ground to ensure that the foundation walls stand properly. Footings usually are composed of concrete.

Foundation wall cracks occur for a variety of reasons, including the previously mentioned unfavorable subsoil, as well as improper construction or earthquake tremors. The foundation walls support the weight of the structure and transfer weight to the footings. These walls may be composed of stone, brick, or concrete. They should be moisture proof, which involves use of plastic sheeting joined with tar or asphalt and footing drains around the exterior of the walls to take moisture away from the property. The basement or cellar floor should have at least 6 in. of gravel beneath it and be composed of concrete. This protects the basement from rodents and flooding. The gravel distributes the groundwater moving under the concrete floor and reduces the potential of water penetrating the floor. Again a plastic shield should be laid before the concrete is poured. All basement doors and windows should be water tight and rodent proof.

HOUSING PROBLEMS

Health and the Housing Environment

There are approximately 5.3 million crowded substandard dwellings in the United States, with 4 million in such poor condition that they cannot be refurbished without major repairs. There are 92,200 deaths and millions of injuries due to unintentional events each year. The leading cause of death is motor vehicles, followed by fires, falls, drownings, and poisonings. The cost is over $500 billion a year.

As many as 25 million home accidents occur in the United States each year because of faulty appliances; electrical connections; poor lighting; broken furniture and equipment; and stairs, floors, and walls in need of repair. An estimated 45,000 cases of rat bite are reported annually. Because many incidents of rat bites are unreported, the problem is far greater than the actual statistics indicate. Each year 6000 cases of rat-transmitted diseases occur. Accidental poisoning due to inadequate storage facilities in the home and improper storage of hazardous materials and chemicals cause 8400 deaths and 1 million injuries. Over 400,000 children have unacceptably high levels of lead in their blood. Lead comes from paint chips, plaster chips, and windowsills and other woodwork in old houses painted with lead-based paint. Improperly constructed, installed, or maintained home-heating devices cause at least 1000 deaths and 5000 injuries from carbon monoxide poisoning annually. In 1998, accident-related deaths occurred in 28,200 homes and 6,800,000 disabling injuries occurred, costing billions of dollars.

In many areas, houses are abandoned because it is cheaper to leave them than to repair them or pay the property taxes. As a result, houses become dilapidated, vandalized, and eventually places where young children, teenagers, and adults can get into serious trouble. Not only are such buildings aesthetically unpleasant, but they also become dumping places for garbage, trash, and other junk; breeding places for roaches, mice, and rats; and meeting places for alcoholics, drug addicts, and criminals.

Human Illness

Environmentally related diseases are more prevalent in poor housing than in better housing. Although congestion within the house can lead to increased upper respiratory diseases because of the close contact with contaminated individuals, and the spread of diseases occurs often in kitchens where food is prepared along with other activities taking place (such as washing baby diapers), it is difficult to attribute any given disease to the housing problem. A number of additional variables should be considered. These include the basic personal hygiene practices and habit patterns of individuals, their susceptibility to certain diseases, and vast additional stresses. We know that within a defective housing structure various stresses are present, such as noise, improper lighting, inadequate space, improper ventilation, presence of myriad insects and rodents, and a variety of solid and hazardous waste. These wastes include aerosols, oven cleaners, drain openers, nail polishes, detergents, medicines, waxes and polishes, paints, pesticides, fertilizers, charcoal lighter fluids, gasolines, motor oils, and antifreezes. Surveys indicate that individuals living in substandard housing with the previously mentioned problems have higher infant mortality rates, greater level of disease, poorer health, more nutritional and dental problems, and numerous other health defects. Even though a specific disease may not be traced to a specific type of housing problem, with the exception of an outbreak of typhoid fever traced to a typhoid organism found in a defective plumbing system, it is still recognized that disease rates are higher among substandard housing dwellers.

Diseases caused by salmonella, staphylococcus, and streptococcus may be found in the housing environment. Other diseases can be caused by intestinal roundworms, hookworms, *Aspergillus*, *Blastocystis*, etc. Ascarids and hookworms are commonly found in dogs and cats and may easily be transmitted to humans especially children. *Aspergillus fumigatus* and *A. flavus* are transmitted by inhalation of the airborne fungi. The infection may be associated with dust exposure, especially during building renovation and construction. *Blastocytosis hominis* is found among people throughout the world. It is felt that the infection increases in areas where sanitation and personal hygiene are inadequate such as in poor housing.

Environmental stress not only reduces our ability to fight off infectious disease but also causes rather specific environmentally related diseases. For the chronically ill, environmental stress creates further problems, contributing to deterioration of their conditions. Noise is a specific environmental problem. Today humans live in shelters that have increasingly high levels of noise. The noise may be due to a lack of soundproofing in poorly or cheaply constructed buildings, and neighborhoods in or near commercially zoned areas and traffic routes. Noise causes nuisances, irritability, and loss of sleep. Eventually the noise may lead to actual, temporary, or permanent hearing loss. Some of the greatest sufferers from noise pollution are those who live within the vicinity of noisy industries or airports. The level of sound well exceeds that recommended for a home situation. A background noise of 35 dB is acceptable at night, with occasional sounds reaching 45 dB; 45 dB is acceptable during the day, with occasional sounds reaching 55 dB. Annoyance occurs in residential areas when the average continually exceeds 50 dB of noise in the background. Some data suggest that 60 dB may cause autonomic changes. Surveys indicate that individuals may get headaches or become nervous when the sound levels exceed 50 dB on a 24-hour average.

Ventilation is an important part of stress or the stress reduction factor within shelters. A minimum of 5 ft³/min per person of uncontaminated air should be provided within the residential environment. The use of mechanical ventilation today may be good or bad, and if fresh air is added instead of recirculating inside air, then home ventilation may be good. In recirculated air, as many as 100 identifiable contaminants have been found. These include pesticides, cleaners, bacterial contaminants, and viral contaminants.

Asthma is one of the most common and costly diseases in the United States. The health burden of asthma is increasing rapidly. The estimated number of asthma sufferers has more than doubled since 1980. Over 5 million children under the age of 18 suffer from the disease, with a disproportionate number occurring in African–American and Hispanic populations; especially severe cases happen in the urban inner cities. Over 5000 people die from asthma each year, and over 500,000 hospitalizations occur. The approximate cost is more than $15 billion a year.

The thermal environment helps reduce or increase stress, depending on the combinations of temperature and relative humidity present and the velocity of the air. A proper temperature within the home should range from 68 to 75°F with relative humidity ranging from 20 to 60%. Temperatures or relative humidities above these levels tend to cause discomfort, and in extreme cases may even lead to death if an individual is already subjected to serious diseases.

Hypothermia related deaths occur annually in various parts of the United States. Hypothermia is a medical emergency defined as the unintentional lowering of the core body temperature to less than or equal to 95°F. Environmental hypothermia results from a combination of heat loss from the body by convection due to the degree of wind exposure, conduction, and radiation to the surrounding ambient air. Mild hypothermia occurs when the body temperature is between 93 and 95°F, moderate between 86 and 93°F, and severe <86°F. The risk for death from hypothermia is related to age, preexisting disease, nutritional status, and alcohol and drug intoxication. Socioeconomic factors such as social isolation, homelessness, and chronic disease increase the risk. Neuroleptic drugs also predispose a person to hypothermia by inducing vasodilation and suppressing the shivering response. From 1979 to 1995 an annual average of 723 people died as a result of hypothermia.

Hyperthermia occurs when people are exposed to a sustained period of excessive heat. Heat-related mortality is more common in the very young and in the elderly, people with cardiovascular disease, social circumstances such as living alone, chronic health conditions such as respiratory diseases, and other factors that interfere with the ability of an individual to take care of oneself. Alcohol consumption that may cause dehydration, previous heat stroke, extreme exertion in hot environments, use of certain medications that interfere with the body's heat regulatory system may cause a variety of illnesses. These include heat stroke, heat exhaustion, heat syncope, and heat cramps. Heat stroke is a medical emergency with a rapid onset and increase within minutes of the core body temperature to greater than or equal to 105°F accompanied by lethargy, disorientation, delirium, and coma. From 1979 to 1996, an annual average of 381 deaths in the United States were attributed to excessive heat. Unfortunately, the elderly who live alone are highly susceptible to this type of condition and so are the homeless. As the planet warms up because of air pollutants and as the population ages, the poor, in particular, may be subjected to considerably more heat and may die from these conditions.

Illumination levels are essential in providing a good pattern of daily activity. When lights are too low or glaring, they cause fatigue and may lead to a variety of accidents or additional stress on an individual.

There are numerous cases where the external environment has contributed sharply to the problems within the home. In Texas, where a substantial Mexican–American population lives, many areas have silt and dirt filter through the cracks in the buildings into the kitchens. This contributes to improper handling of food and may lead to disease. In other areas, migrant workers have been periodically sprayed with pesticides used for the control of pests in the fields. These conditions are not unique. They may be found in every major or small city in the country.

Monitoring Environmental Disease

One of the national health objectives for Healthy People 2000 is to establish and monitor nonoccupational environmental diseases including asthma, heat stroke, hypothermia, heavy metal poisoning, pesticide poisoning, carbon monoxide poisoning, acute chemical poisoning, and methemoglobinemia. Surveys were done with all state environmental epidemiologists and territorial environmental epidemiologists. The only surveillance system that was fully operational was the childhood lead monitoring system. Surveillance systems at the local, state, and national levels are useful for assessing case investigations, implementing control activities, evaluating interventions, monitoring trends, and identifying risk factors.

Hazardous Waste Sites

About 25% of American children live within 4 mi of a hazardous waste site. These children often have greater exposures, greater potential for health problems, and less ability to avoid hazards. They may be contaminated by air, soil, or groundwater.

Homelessness and Disease

As many as 600,000 people are homeless on any given night. Millions have been homeless at one point or another in their lives. For the first time, women and children are occupying quarters reserved for skid row men. This is caused by crisis poverty that is transient or episodic. The housing problems are coupled with poor employment prospects because of poor schooling, obsolete job skills, domestic violence, poor parenting, and poor household management skills. Other individuals who are homeless may be those with chronic disabilities, alcoholism, drug abuse, severe mental illness, or other long-standing family difficulties. The people lack financial resources and have exhausted other types of support when they resort to living on the street. Lack of affordable housing is another contributor to homelessness. The homeless tend to be more susceptible to disease and therefore a potentially greater source of the spread of disease. Homeless children face significant barriers to receiving the same public education as the nonhomeless children. These young people also have serious emotional and developmental problems that can persist long after their families find permanent housing. Human immunodeficiency virus (HIV) and acquired immunodeficiency syndrome (AIDS) and tuberculosis are major afflictions of the homeless poor. *Bartonella quintana*, which caused epidemics of louseborne trench fever during the 20th century in distinct and diverse populations, such as 1 million troops in Europe in World War I, has emerged among homeless persons in North America and Europe. The organism was isolated from the blood specimens of ten patients in a single hospital in Seattle, WA, in a 6-month period. Homeless persons are also at risk for non-vectorborne infectious diseases such as meningococcal disease, pneumococcal disease, and diphtheria.

Unvented Residential Heating Appliances

Many heating appliances use the combustion of carbon-based fuels and therefore produce potential sources of health-threatening indoor air pollution. Although most of these devices are vented to the outside of buildings, some of them, such as kerosene- and propane-fueled space heaters, gas-fueled log sets, and cooking devices may be unvented. A recent 7-year study indicated that the percentage of adults using these devices is higher in the South, among low income groups, among blacks, and among rural residents. Unvented combustion space heaters were used by an estimated 13.7 million adults. Among the estimated 83 million adults who used a gas stove or oven for cooking, approximately 9.3% had used it for heating the premises. Unvented combustion heating appliances and gas stoves or ovens improperly used as heating devices often produce combustion by-products that exceed acceptable limits, create indoor air quality problems, and may cause exposure to toxic gases such as carbon monoxide. Unintentional, nonfire, nonautomobile poisonings from carbon monoxide exposure in permanent dwellings result in approximately 200 deaths and 5900 injuries a year.

Residential Fires and the Prevalence of Smoke Alarms

Children less than 5 years of age and adults over 65 most frequently die in fires. Although the number of deaths from residential fires has declined and meets the year 2000 objectives of 1.2 per 100,000 persons, too many deaths still occur. The seasonal use of heating devices, such as portable space heaters, and wood-burning stoves exacerbates this problem. The leading causes of residential fires are due to cooking and heating devices improperly placed or left unattended. Homes with smoke alarms have almost half as many fire-related deaths as those without them. Smoke alarm programs have been very effective; however, the maintenance portion is ignored in approximately 50% of the homes. Effective public health strategies to reduce residential fire-related injuries and death should include:

1. Proper smoke alarm installation
2. Monthly testing of smoke alarms
3. Reduction of residential fire hazards
4. Design and practice of fire escape plans
5. Fire safety education
6. Implementation of smoke alarm ordinances

Indoor Air Quality Problems

The serious concern over contaminants in the indoor air is due to the fact that they are not easily dispersed or diluted. The indoor air contaminant concentration is often many times higher than the outdoor concentration of these pollutants. At times indoor pollutants have exceeded the upper levels by 200 to 500%. Further, people spend as much as 90% of their time indoors, which subjects them to greater levels of risk. The degree of risk depends on how well the buildings are ventilated and the type, mixture, and amount of pollutants present in the structure. Improperly designed and operated ventilation systems add to these problems. Individuals complain of eye, nose, and throat irritations, fatigue, shortness of breath, sinus congestion, sneezing, skin irritation, dizziness, lethargy, headaches, nausea, irritability, and forgetfulness. Long-range health effects can include impairment of the nervous system as well as cancer. The harmful indoor air pollutants, in addition to the inorganic and organic compounds previously mentioned, include mold, fungi, viruses, and bacteria. In addition, there is a decrease in productivity, an increase in deterioration of furnishings and equipment, and a decrease in good morale and human relations.

Once chemical or microbial agents enter the indoor air environment, they may be difficult to remove. These substances are in constant motion, because of the air currents and people-type activities occurring within the building. The movement allows the substances to come in contact with other substances and either interact or become attached to other substances. Because there has been a reduction in ventilation rates from approximately 10 to 5 ft^3/min in these buildings, the amount of air turnover or potential removal of these substances has been decreased sharply.

The various contaminants, including microorganisms, may produce either acute effects if the contamination concentrations are high enough, or chronic effects if a prolonged exposure occurs to the contaminants at low concentrations over an extended period of time. *Sick building syndrome* is a general term applied to a group of illnesses where an acute exposure has occurred. This syndrome was rare before 1977, but by 1985 it was responsible for 13% of all the complaints received by National Institute for Occupational Safety and Health (NIOSH). In 1994, the Occupational Safety and Health Administration (OSHA) started publishing Indoor Air Quality Standards under Code of Federal Regulations (CFR) Title 29. As of the year 2000, this work is continuing because OSHA has declared that employees in indoor work environments face a significant risk of material impairment to their health due to poor indoor air quality. There are five different forms of this illness. The first form is an allergic response called humidifier fever. The symptoms are breathlessness, flu-like symptoms, and dry cough. The cause of the illness is the growth of microorganisms in the water that is used to cool the air-conditioning units, or microorganisms in the water of humidifiers. The second form includes common allergies. The symptoms are runny nose, sneezing, and irritated eyes. The causes are fungi, mites, pollens, and dust that have entered the indoor environment and are present in the air. The third type is infectious diseases. These diseases are caused by microorganisms that are found in cooling towers, ice machines, showerheads, and water condensers. The symptoms include coughing, fever, and pneumonia. Legionnaire's disease is an example of this. The fourth type is due to allergic reactions caused by synthetic material. The symptoms are rashes, irritation, itching, and eye problems. The causes are probably the release of chemicals within the buildings. The fifth type, the largest one of all, is due to unknown causes. The symptoms are increased mucus, irritation, sneezing, headache, sore eyes, and sore throat. Sometimes this is due to poor ventilation and low relative humidity. The symptoms appear when a person enters the building and disappear shortly after leaving the building. Three possible explanations for this fifth type include the release in very low concentrations of VOCs, psychological reactions by the employees because of dislike of their work, or psychological reactions to a traumatic situation. Building-related illness refers to illness brought on by a specific environmental agent such as the organism causing Legionnaire's disease.

Little information is available on the chronic effects of indoor air pollution. The problems of lung cancer caused by radon gas and the potential of cancer due to asbestos or health problems related to lead have been based more on what is occurring in industrial settings than that which is occurring in the pure indoor air environment of homes and businesses.

Age is a factor in determining how a person is affected by indoor air pollutants. The groups most affected are the very young from birth to 10 years of age, and the elderly 65 and up. The reasons for this include the amount of time they spend indoors, the condition of their immune system, their physical condition, medications that they may be taking, and their current health problems. In small children a relationship exists between the weight of the contaminant, especially if it is a chemical or inorganic material, and the weight of the child. In the elderly, many of the individuals already have some form of upper respiratory distress, and therefore, are more affected by indoor air pollutants. Some individuals are highly sensitive and are allergic to the particular contaminant. When an antigen enters the body, an immune response is elicited; therefore, the greater the dose an individual receives, the greater the response is, which in itself creates a harmful effect within the body.

The sources of indoor air contaminants include:

1. Outside sources
   a. Pollen, dust, spores, industrial pollutants
   b. General vehicle exhaust
   c. Odors from dumpsters
   d. Re-entrained (drawn back) air into building

    e. Radon from soil, construction materials, or water

    f. Leakage from underground storage tanks

    g. Chemicals from land disposal site

    h. Pesticides

2. Heating, ventilation, and air-conditioning equipment

    a. Dust or dirt in duct work

    b. Microbial growth in drip pans, humidifiers, coils, and duct work

    c. Improper use, storage, disposal of chemicals

    d. Leakage of refrigerants

    e. Emission from equipment such as volatile organic compounds, ozone, solvents, toners, ammonia

3. Human activities

    a. Smoking, cooking, body odors, cosmetics

    b. Cleaning materials, deodorizers, fragrances, trash, dust from sweeping

    c. Maintenance, airborne dust, volatile organic compounds, pesticides, microorganisms in cooling towers, animal dander, insect parts

4. Building components and furnishings

    a. Carpeting, textiles, deteriorated materials, asbestos, new furnishings

    b. Water damage, odors, molds, microorganisms

5. Accidental events

    a. Spills, fire damage, soot, odors

6. Special use areas

    a. Smoking rooms, laboratories, print shops, food preparation areas

The major route of entry for the indoor air pollutants is the respiratory system. The nasal passages are the major portals through which the air enters the body. Two filtering mechanisms are supposed to remove the particulate matter. The first is the nose, which includes its hairs. The second are mucous membranes that line the interior of the nasal passages. If these mechanisms do not work effectively, the particles may cause further harm. The oral route is the path through which the contaminants may go down into the gastrointestinal system and could eventually cause disease (such as cancer), depending on the type of chemical ingested. The eye is also a portal of entry for indoor air pollutants. Various chemicals can be dissolved in tears and therefore move into the bloodstream.

Residential Wood Combustion

The problems associated with wood combustion include incomplete combustion, type of wood, type of combustion chamber, inadequate ventilation, and burning of lumber that has been treated with chemicals. Carbon monoxide is probably the major pollutant emitted during the burning of wood. Nitrogen oxides are produced. They may be irritating to the eye and throat, and may cause coughing. Polycyclic aromatic hydrocarbons (PAHs) may be produced and may be carcinogenic. Further, a large portion of the particulates from wood combustion are in the respirable particle range, which is less than 3 μm. These particles can be deposited in the deep alveolar region of the lungs and therefore help contribute to pulmonary disease.

Environmental Tobacco Smoke

Passive smoking is an involuntary secondhand inhalation of tobacco smoke by nonsmokers. The nonsmokers may receive the exposure either through the workplace or through the home. Two groups of individuals are within the category of nonsmokers. The first one involves exsmokers who have quit direct smoking but continue to be chronic passive smokers, because they are still inhaling smoke from others. The second group includes true nonsmokers who are inhaling smoke from others. Of the two types of smoke entering the indoor environment the first is "mainstream" smoke, which is inhaled directly through the butt of the mouthpiece during the pumping process and then exhaled into the environment. The second is the "sidestream" smoke that is produced while the tobacco is smoldering. The four places in which nonsmokers are subjected to passive smoking problems are in private homes, public buildings, public transportation, and office buildings. Passive smoking can result in at least seven different types of potential health problems. They are:

1. Lung cancer — the risk of developing lung cancer is enhanced through passive as well as active smoking.
2. Chronic air flow limitation — passive smokers experience chronic airflow limitation when compared with a control group of individuals who have not been exposed to passive smoke.
3. Small-airway dysfunction — passive smoking increases the degree of small airway dysfunction.
4. Asthma symptoms — maternal smoking is a contributor to asthma development in the newborn child.
5. Acute cardiovascular response — nicotine is the primary constituent of tobacco smoke that affects the cardiovascular system. Acute cardiovascular alterations can result from the inhalation of smoke.
6. Pneumonia and bronchitis in infants — maternal smoking can increase the risk of infants developing pneumonia and bronchitis.
7. Carboxyhemoglobin affects on health — carbon monoxide is a gaseous constituent of tobacco smoke and it can combine with the hemoglobin in the blood to form carboxyhemoglobin.

Over 500 compounds are present in tobacco smoke. Proven animal carcinogens such as PAHs and N-nitrosamines are released from tobacco smoke. Carbon monoxide has been linked to arteriosclerosis and cardiovascular disease in people. Nitrogen oxides enhance the development of emphysema.

The potential dangers from the sidestream cigarette smoking to the nonsmoker are compounded by the amount of time of exposure to tobacco smoke, sex, age, ambient pollution exposure, size of family, occupation, drug consumption, alcohol consumption, and general health and immunologic response of the individual.

The Environmental Protection Agency (EPA) firmly states that the bulk of the scientific evidence demonstrates that secondhand smoke or environmental tobacco smoke causes lung cancer and other significant health threats to children and adults. Their reports have been peer-reviewed by 18 eminent, independent scientists who unanimously endorse methodology and conclusions. Although, a U.S. District Court had vacated several chapters of the EPA document, "Respiratory Health Effects of Passive Smoking: Lung Cancer and Other Disorders," that serve as the basis for the EPA classification of secondhand smoke as a group A carcinogen and the estimate that 3000 lung cancer deaths occur in nonsmokers each year from environmental tobacco smoke, the EPA is appealing this decision. The ruling was based largely on procedural grounds. None of the findings concerning the serious respiratory health effects of secondhand smoke in children was challenged. Infants and young children are especially sensitive to environmental tobacco smoke. In children, environment tobacco smoke exposure is causally associated with:

1. An increased risk of lower respiratory tract infections such as bronchitis and pneumonia.
2. An increased prevalence of fluid in the middle ear, symptoms of upper respiratory tract irritation, and small reductions in lung function.
3. Additional episodes and increased severity of symptoms in children with asthma.

Secondhand smoke is a combination of the smoke given off by the burning end of a cigarette, cigar, or pipe and the smoke exhaled from the lungs of smokers. This mixture contains more than 4000 substances, of which more than 40 are known to cause cancer in humans or animals and many of which are strong irritants.

Pharmacokinetics

Whether a chemical elicits toxicity depends not only on its inherent potency and site specificity but also on how the human system can metabolize and excrete that particular chemical. To produce health effects, the constituents of environmental tobacco smoke must be absorbed and must be present in appropriate concentrations at the sites of action. After absorption some contaminants are metabolized to less toxic substances, whereas others become more toxic or activate carcinogens. The extent of absorption, distribution, retention, and metabolism of these contaminants in the body depends on gender, race, age, and smoking habits of the exposed people.

Carbon Monoxide

In addition to cigarette smoke, which has been just discussed, carbon monoxide is also produced in the residential setting by improper combustion of fuel. On the average, the carbon monoxide levels appear to be higher in mobile homes than in regular housing. Improper ventilation in the tightly sealed mobile home and improperly functioning appliances are the two main reasons for higher carbon monoxide levels. One of the most common sources of emission is from the faulty furnace flues. The furnace is typically located very close to the living quarters, and therefore any escape of carbon monoxide may enter the breathing space of the individuals very quickly. In addition, gas stoves and space heaters used in mobile homes may be hazardous.

In the conventional house, carbon monoxide levels are increased by improper combustion and poor ventilation from gas stoves, gas water heaters, and gas furnaces.

The carbon monoxide is produced by the incomplete burning of the gas, and once produced may be distributed rapidly through the forced hot-air system. If there is an attached garage and heating or cooling vents are present in the garage, carbon monoxide produced from the automobile may seep through the vents into the remainder of the house. One other exposure in the house would be the improper burning of fuel in a fireplace.

Radon

Radon is a naturally occurring gas that results from the radioactive decay of radium. The primary sources of radon gas in the home are from the soil and rocks upon which the house is built (Figure 8.1). Radon gas may also be present in well water. Occasionally, radon gas may come from building materials. The radon enters the buildings through cracks in the foundation. When inhaled, radon can adhere to particles and then lodge deep in the lungs. This increases the potential risk of cancer.

The EPA estimates that radon may be responsible for 5000 to 20,000 lung cancer deaths per year. Radon gas is a particularly bad risk for smokers who already have a health risk of cancer ten times greater than nonsmokers. Radon-222 is a radioactive

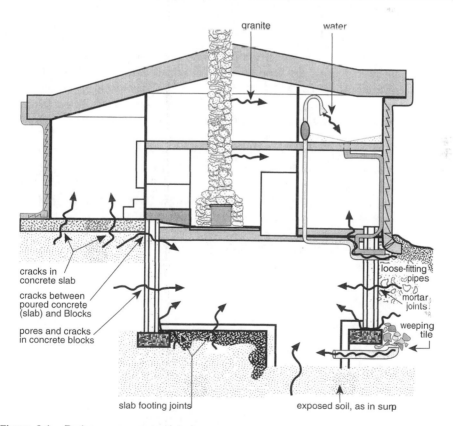

Figure 8.1 Radon access routes into house.

gas, with a half-life of 3.8 days. It is colorless, odorless, and chemically inert, but is fairly soluble in water. Radon gas is nine times heavier than air, and therefore remains close to the ground, diffusing into the stagnant air of wells or crawl spaces under houses. Radon goes through a series of transformations, emitting alpha, beta, and gamma radiation. The radioactive elements that result from the decay of radon are called radon progeny or radon daughters. The radon daughters formed during the decay process include polonium-218, lead-214, bismuth-214, polonium-214, lead 210, bismuth-210, polonium -210, and finally the stable isotope lead-206.

The amount of radon gas flowing into a home is primarily dependent on the radium content of the ground. Certain building materials, including concrete, brick, cement, granite, gypsum wallboard, and concrete blocks, have been found to contain radon. Radon daughters, unlike radon gas, are chemically active metals. These elements stick to anything with which they come in contact, such as particulates in the air, on furniture, and in the room. When radon daughters are attached to particulates and come in contact with the air passageways in the lungs, they may adhere to the surfaces. The biological effects of radon can be attributed mainly to its alpha-emitting products.

Radon concentrations in indoor air are influenced by four main groups of factors. They include the properties of the building material in the ground, building construction, meteorologic phenomena, and human activities. Construction materials, such as concrete, brick, and stone, can emit radon. Because radon originates from the bedrock, the levels of radon at Earth's surface is affected by the thickness of the unconsolidated materials, the type of unconsolidated materials, and the level of uranium concentration in the bedrock. The thicker the unconsolidated material and the farther away the bedrock is from the surface of the ground, the lower the radon level is. Coarse grain sizes of gravel and sand are highly permeable and therefore allow more radon to escape. Mud and wet clays have low permeability and therefore allow little or no radon to escape. If there is poor water percolation, then there is a low radon availability, unless the ground is the site of a high uranium yield area. The meteorologic phenomena relate to the amount of air that flows through the area and therefore is available for ventilating the buildings. The human activities include the rate of ventilation in the building, as well as the rate of smoking and burning of wood within the building.

Radon enters the building through dirt floors, cracks in concrete floors and walls, floor drains, sumps, joints, and tiny cracks or pores in hollow-block walls. Radon may also enter the house in water from private wells and may then be released into the home when the water is used in showers, in the kitchen, or in bathroom sinks.

Formaldehyde

Formaldehyde is a probable human carcinogen. It is responsible for a variety of acute health problems, such as eye and nose irritation and respiratory diseases. People who have lung diseases or an impaired immune system, children, and the elderly may be affected by formaldehyde. Formaldehyde is used in furniture, foam insulation, and pressed wood products, such as some plywood, particleboard, and fiberboard.

Formaldehyde is a colorless gas with a characteristic pungent odor. Each year 8 billion pounds of formaldehyde are produced in the United States. Formaldehyde is also found in cosmetics, deodorants, solvents, disinfectants, and fumigants. Urea formaldehyde (UF) is a resin that is produced from a mixture of urea, formaldehyde, and water. It is most commonly used as an adhesive in plywood, particleboard, and chipboard. The emissions from the UF boards and other materials are due to unreacted formaldehyde that remains in the product after its manufacture. It also occurs from the breakdown of the resin by its reaction with moisture and heat. UF foam has been used extensively as a thermal insulation in the walls of existing residential buildings. It is injected into the wall cavities through small holes that are sealed afterward. The insulation is accomplished by mixing the UF resin with a foaming agent and compressed air. The material then looks like shaving cream and can be pumped through a hose and forced into the wall cavity, where it will cure and harden.

Particleboard is made by saturating small wood shavings with UF resin and pressing the mixture at a high temperature into its final form. Particleboard can emit formaldehyde on a continuous basis for several months to several years. In buildings in which these products are used, either for partition walls or furniture, the formaldehyde can reach concentrations that cause eye and upper respiratory irritation. Where the air-exchange rates are low, the concentration can reach 1 ppm or more. In contrast, exposure to formaldehyde at a concentration as low as 0.05 ppm in sensitive people, under certain conditions of temperature and humidity, may cause burning of the eyes and irritation of the upper respiratory system. In other people, symptoms occur above 0.1 ppm. Concentrations higher than a few parts per million may produce a cough, constriction in the chest, and wheezing. Menstrual disorders are among the most frequent problems caused by exposure to formaldehyde.

The environmental factors mentioned earlier relating to temperature and humidity are important in determining the overall affect of formaldehyde on people. If the average indoor temperature is raised by 10°F, the levels of formaldehyde double. If the average indoor temperature is decreased by 10°F, an approximate 50% decrease occurs. An increase of relative humidity from 30 to 70% causes a 40% increase in the formaldehyde present. During very cold and windy days, if the house is not well insulated, the level of formaldehyde drops because of the infiltration of air and the mixing and dilution that occur. Generally, the greater the temperature difference, the more ventilated air is available for formaldehyde dilution. Formaldehyde levels indoors increase during the winter months and decrease during the summer months because of the amount of air that is available.

Other Volatile Organic Compounds

Other VOCs commonly found in the indoor environment include benzene from tobacco smoke, methylene chloride (dichloromethane) in paint strippers, and perchlorethylene emitted from dry-cleaned clothing; VOCs from varnishes, wax, and paints; and stored chemicals such as cosmetics, degreasing, and hobby products. VOCs can also be emitted from drinking water. Of the 20% of water supply systems having detectable amounts of VOCs, only 1% exceeds the 1996 Safe Drinking Water Act (SDWA) amendments standards for VOCs.

Pesticides may be used indoors or outdoors for termite proofing. Preservatives may be used for the treatment of wood. Even when pesticides and preservatives are used properly, they may release the VOCs. Wood preservatives, which may cause cancer, include creosote, coal tar, inorganic arsenicals, and pentachlorophenol. Household cleaners, solvents, and polishes of all types may release VOCs. VOCs also cause eye and respiratory tract irritation, headaches, dizziness, visual disorders, and memory impairment.

Asbestos

Asbestos fibers have been shown to cause lung cancer and other respiratory diseases. Asbestos has been used in the past in a variety of building materials, including many types of insulation and fireproofing materials, wallboards, ceiling tiles, floor tiles, asbestos wood and coal stove door gaskets, asbestos laboratory gloves and pads, asbestos stove pads and iron rests, and central hot-air furnace duct connectors containing asbestos. When buildings containing asbestos materials are demolished, tiny asbestos fibers are freed that then can be inhaled by the individuals. Even during the normal aging process of materials, the deterioration may be released as asbestos fibers, which can be inhaled and accumulated in the lungs.

Buildings constructed between 1945 and 1978 may contain asbestos material. These materials were applied by spraying or trowling onto ceiling or walls. Asbestos has been used as a thermal insulator to protect walls, floors, and people from intense heat produced by stoves and furnaces. Some door gaskets and furnaces, ovens, and wood and coal stoves may contain asbestos. Pipe insulation manufactured from 1920 to 1972 contained asbestos to prevent heat loss from the pipes and provide protection from being burned by hot pipes.

Asbestos was used because of its durability, flexibility, strength, and resistance to wear. It was also used because of its great insulating properties.

Asbestos fibers may remain airborne for periods of time depending on the fiber diameter, and to a lesser extent, the fiber length. An asbestos fiber settles in stale air in a 9-ft high room in approximately 4 hours if it is 5 μm long and 1 μm in diameter. The same length of fiber with a 0.1 μm diameter remains airborne up to 20 hours. If the air becomes turbulent, the fiber may stay airborne for long periods of time.

Asbestos has been found in office buildings where the practice was to fireproof those high-rise buildings with materials containing 10 to 30% asbestos. A major source of indoor-generated asbestos in those buildings is the recirculating of asbestos fibers by the air-conditioning system. A study of carpet vacuuming showed that overall airborne concentrations after vacuuming was four times as great as before vacuuming. In schools the problem of asbestos has become acute. The EPA estimates that approximately 15 million students and 1.4 million teachers and other school employees use buildings that contain asbestos. School corporations are now in the process of trying to correct this problem.

Asbestos may cause four major types of health effects. Lung cancer, which may take as long as 20 years to develop, is known to cause 20% of all deaths among asbestos workers. The total amount of lung cancer present in these workers depends on their length of exposure, age, and how long they worked with asbestos. Smoking increases the risk of dying from lung cancer by 92 times as much as nonsmoking. The level of lung cancer among other individuals exposed at far lower concentrations needs to be explored. Asbestosis is a disease that involves a permanent scarring of the lungs caused by the inhalation of asbestos fibers. Asbestosis lung scarring may continue after the exposure to the asbestos ends. The symptoms include coughing, shortness of breath, and broadening of the fingertips. Eventually heart failure may occur because of the lung obstruction.

Cancer can also develop in the digestive tract because of the consumption of asbestos in food, in beverages, and in the saliva that has been contaminated with cigarette smoke. Mesothelioma is a diffuse cancerous tumor that spreads very rapidly over the surface of the lung (pleura) and abdominal organs (peritoneum). This tumor also may be due to exposure to asbestos.

Nitrogen Dioxide

Nitrogen dioxide is produced from kerosene heaters, unvented gas stoves and heaters, and environmental tobacco smoke. It causes eye, nose, and throat irritation. It may cause impaired lung function and increased respiratory infections in young children. In homes with unvented appliances the indoor levels of nitrogen dioxide often exceed outdoor levels.

Biological Contaminants

Biological contaminants include bacteria, molds, mildew, viruses, animal dander, cat saliva, house dust mites, roaches, and pollen. Pollens originate from plants. Viruses are transmitted by people and animals. Bacteria are transmitted by people, animals, soil, and plant debris. Household pets are sources of saliva and animal dander. The protein in urine from rats and mice is a potent allergen. When it dries, it can become airborne. Contaminated central air-handling systems become breeding grounds for mold, mildew, and other sources of biological contaminants and can be distributed through the home. A relative humidity of 30 to 50% is generally recommended for homes to reduce these contaminants. Standing water, water-damaged materials, and wet surfaces also serve as a breeding grounds for molds, mildews, bacteria, and insects. House dust mites, which grow in damp, warm environments, are the source of one of the most powerful biological allergens. Some biological contaminants trigger allergic reactions including hypersensitivity pneumonitis, allergic rhinitis, and some types of asthma. Molds and mildews release disease-causing toxins. Symptoms of health problems include sneezing, watery eyes, coughing, shortness of breath, dizziness, lethargy, fever, and digestive problems. Allergic reactions typically occur after repeated exposure to specific biological allergens. Some diseases, like humidifier fever, are associated with exposure to toxins from micro-organisms that grow in ventilation systems and humidifiers. Children, elderly people, and people with breathing problems, allergies, and lung diseases are particularly susceptible to disease-causing biological agents in the indoor air.

More than 50 million Americans suffer from hay fever, asthma, and other allergic diseases. Many of these conditions are caused by exposure to allergens in indoor environments such as the home, workplace, and school, where we spend as much as 98% of our time. During an asthma attack, the airways get narrow, making it difficult to breathe. Asthma can even be fatal. About 17 million Americans have asthma. It is the leading cause of long-term illness in children.

Environmental Lead Hazards

Lead poisoning is one of the most common preventable public health problems related to children today. It affects as many as 1.7 million children, age 5 and under. Lead paint, the major high-dose source for children, is present on or in an estimated 30 to 40 million houses in the United States. This will continue to pose a hazard to children for many years to come. In addition to lead due to paint, lead may be found in house dust, garden or backyard soil, flaking paint of automobiles, and industrial emissions. Children are constantly exposed to lead hazards. Lead may also be found in water because old plumbing systems commonly had soldered lead joints. Lead can leach into the water from the joints or from lead pipes. Acidic or corrosive water can cause this leaching. Lead may be present in food such as canned fruits, vegetables, and juices. The lead may be taken up by crops or from the soil, or may be deposited on the crops from lead air contamination. Lead may also be available in pottery, fish sinkers, lead weights, jewelry making, antique ceramic doll making, other toys, metal sculpture soldering, special powders, and Chinese herbal medicines.

Workers may bring lead home to their children on their clothes. Children are more at risk for lead exposure than adults because they have more hand-to-mouth activity than adults, they absorb more lead than adults, and they are smaller than adults. Lead is a poison that effects virtually every system in the body. Whereas before the mid-1960s, a level above 60 µg/dl of blood was considered to be toxic, today above 15 µg/dl of blood is considered to be toxic. In fact, there may be adverse health effects above 10 µg/dl of blood. In children, very severe exposure, blood levels of 80 µg/dl, can cause coma, convulsions, and death. Blood lead levels as low as 10 µg/dL, which do not cause distinctive symptoms, are associated with decreased intelligence and impaired neurobehavioral development.

Other symptoms of lead poisoning include colic, abdominal pain, constipation, cramps, nausea, vomiting, anorexia, weight loss, hepatic effects, renal effects, interference with growth, and erratic behavior. Chronic lead exposure in adults can damage the cardiovascular, central nervous, renal, reproductive and hematologic systems. Of 20,000 adults tested in 1998, 4000 had excessive lead levels of 25 µg/dl or above (standard for adults).

Molds, Bacteria, and Other Biological Contaminants

Biological contaminants have been recognized as a substantial source of poor indoor air quality when present. These agents include microorganisms molds and other fungi, bacteria, viruses and protozoans, as well as pollens, animal hair and dander, and insect parts (e.g., dust mites, cockroaches). Some of these can cause infection, whereas some may contain or release harmful toxins, and most are potentially allergenic. It is thought that some of the health related generalized complaints or symptoms associated with exposure to certain biological agents in indoor environments include itchy and watery eyes, sneezing, runny nose, coughing, shortness of breath, dizziness, lethargy, fever, and even digestive problems.

Molds, such as *Aspergillus* spp. and *Stachybotrys chatarum*, and bacteria, such as *Legionella pneumophila*, have gained increasing attention as causative agents of human illnesses associated with the indoor environment. Indeed, infiltration of water into buildings due to leaks and condensation from air ducts and pipes contributes greatly to the growth and proliferation of these and other organisms. Excessively high relative humidity also is a contributing factor. The combination of excessive moisture and cellulosic building materials creates a somewhat ideal environment for microbes to flourish.

Stachybotrys chatarum synthesizes mycotoxins known as trichothecenes. The mycotoxins are present on the surface of its spores. This organism grows well on water-soaked cellulosic materials and is a problem in water-damaged buildings. The toxic effects from inhaling the spores are due to the presence of the toxins on the surface of the spores, not with growth of the fungus in the body. It is thought that infants' lungs are more susceptible and toxicity may lead to often fatal pulmonary hemorrhaging.

Reservoirs of water, such as fountains, cooling towers, humidifiers, and accumulations of standing water, may increase the potential for growth and proliferation of the bacterium *L. pneumophila*. When airborne, inhalation of water mists containing the bacteria can cause a pneumonia called Legionnaire's disease. The respiratory illness can be treated with antibiotics, but can be fatal.

Behavioral Problems

The density of individuals or the crowding factor within a dwelling can produce nervous irritation that is detrimental to mental and physical health. Crowding makes people more irritable, and causes interruptions and personal clashes that may end in deep-seated repressed bitterness or hatred. Crowding deprives individuals of personal space, which is a basic need of all animals. Privacy, another essential component for a healthy existence, is virtually impossible to obtain in crowded quarters where there is no place an individual can go to be alone. Sometimes it is necessary for people to "cool down" to avoid turning a misunderstanding into a disastrous argument. Even loving couples need time to be alone to assimilate what is occurring and to redirect their thoughts and their desires.

Overcrowding increases environmental stresses by increasing noise levels; decreasing ventilation, and increasing odors, poor housekeeping, and solid waste materials. Behavioral problems resulting from inadequate control in crowded dwellings may increase infectious diseases through lack of personal hygiene. This obviously amplifies the major oral–fecal, airborne, skin, and respiratory routes of transmission of disease.

Alcoholism is often a problem in overcrowded housing, not as a direct result of population density, but instead because alcoholic, unemployed, or chronically ill persons are frustrated by their inability to provide adequate facilities necessary for a healthy, happy family life. Where population density is high, children lack the privacy for proper study, which creates additional problems in attempting to raise their educational level. Generally, the cycle formed by poor nutrition leading to disease, and disease leading to poor nutrition, may also be applied to education and lack of ability to find adequate places to study or concentrate. Poor education may lead to poor housing and poor housing, in turn, may lead to poor education.

Types of Housing

Of the numerous types of dwellings, those under discussion in this section include the single-family home, multiple-family dwellings, trailers, transient housing, dormitories, and emergency housing.

Single-Family Dwellings

The single-family dwelling is defined as one unattached dwelling unit inhabited by an adult person plus one or more related persons. Single-family homes vary from small cabins to mansions. The significance of the single-family home is that it contains only family members or close friends, thereby reducing the potential for conflict and neglect of property. Single-family homes are subject to housing inspections, depending on the housing code, in like manner to other dwellings. These homes vary in upkeep from poor to excellent, and in environmental problems from great to none. Some single-family homes lack water, adequate sewage, proper waste disposal, adequate lighting, and plumbing. Others have excellent physical facilities, including sufficient space to accommodate family members or close friends.

Multiple-Family Dwellings

A multiple-family dwelling is any shelter containing two or more dwelling units, rooming units, or both. The multiple family dwelling varies from the duplex, which is a single structure with two complete and separate units, to apartment houses, which contain many separate dwelling units. Apartment houses vary from condominiums, where the occupants own their separate units, to reconverted houses, which are single-family dwellings converted to multiple-family dwellings. The latter group is of greatest concern to housing authorities. A subdivided single-family home once inhabited by four to seven people may be converted to a dwelling inhabited by six to eight different families with as many as 50 to 60 family members.

Further difficulty arises from the enormous number of people crammed into a structure originally built for smaller numbers. These structures tend to deteriorate rapidly. The windows break, the wallpaper and paint peels, the stairs decay, and solid waste may be found everywhere — in the hallways, in the basements, and in the backyards. Unfortunately, the tenants of such a structure are generally poor and lack the capabilities to move to a better housing situation. The types of housing problems discussed earlier in this chapter are all found in these overcrowded converted structures.

Apartment houses, unless well protected by guard systems, may be very dangerous for residents. Burglars, muggers, and attackers frequent poorly supervised apartment houses. Many high-rise buildings, built for the urban poor in the inner city on redeveloped land, have become places for crime, dirt, and destruction. The concept of multistory inner-city housing is a disastrous failure in reality. In St. Louis, a development that cost many millions of dollars had to be destroyed because the occupancy rate had fallen so sharply. It is impossible for the inner-city poor to live under these conditions in a satisfactory manner.

Travel Trailers and Mobile Homes

The major environmental health problems are the same as in other forms of housing. In addition, mobile homes are driven to out-of-the-way places and are used as permanent dwellings. Specific problems relate to access to adequate water supplies and on-site sewage disposal. Travel trailers are used as permanent residences in some areas. These units are neither large enough nor properly equipped to fill the needs and requirements of a permanent dwelling. In addition, dangerous storms frequently cause serious damage to this type of housing.

Transient Housing

Motels, hotels, and lodging houses are transient dwellings where occupants reside for a day, a week, and possibly longer. The problems emanating from this type of housing are similar to those found in hospitals, nursing homes, convalescent homes, and other short-term medical facilities. The major difference is the individuals within transient housing units are supposedly healthy individuals and are under no form of supervision. Many individuals fall ill when traveling, or may be carriers of a variety of diseases and spread the microorganisms throughout the transient housing environment.

The motel or hotel has the same basic problems as the medical facilities involving housekeeping, laundry, physical plant, food, solid waste disposal, insects and rodents, and so forth, but to a lesser degree.

Dormitories

College dormitories are basically similar to hotels, with the principal difference that individuals occupy dormitories for several months instead of several days. The one problem that frequently occurs in dormitories is the spread of contagious or infectious diseases. In the event of illness, a dormitory resident should be removed to the student health center, placed in isolation, if necessary, and properly treated.

Rehabilitation vs. Redevelopment

In the past, redevelopment was the technique most widely used for the elimination of slum areas and for the reuse of land. Run-down and neglected structures were repossessed and razed by the local government using the right of eminent domain. The land was then utilized for a variety of purposes, including the building of roads, parks, university buildings, and high-rise and low-rise housing for the poor. Several problems have developed as a result of redevelopment. The cost was extremely high. The families living in the area prior to redevelopment had to be relocated to neighboring areas, which in turn realized a devaluation in property and eventually were transformed into the same kinds of slums.

The high-rise, multiple-family dwellings, commonly referred to as *projects*, quickly deteriorated, becoming sources of crime and dirt. Unfortunately, the occupants of these high rises, many of whom were relocated due to redevelopment, were not trained in the proper use of these new buildings and facilities. Some of the better properties, which were vital resources and which could, in fact, have been rehabilitated, were destroyed.

ACCIDENTS AND INJURIES

Accidents are the fourth leading cause of death among the total population and the leading cause of death for the age group of 1 to 34. Accidents cost more than $500 billion dollars a year in cash, lost time, medical treatment, and insurance. The home is the most frequent place of injury. One third of all nonfatal accidental injuries and one fourth of all fatal injuries occur in the home and its surrounding environment. Over 15 million people injured annually in home accidents need medical attention. The two most susceptible or highest risk groups are the very old and the very young. Accidental deaths and injuries occur in all socioeconomic groups and in all areas. The home environment has been altered tremendously over the last few years because of many new kinds of labor-saving and recreational devices. People have not kept up with the safety precautions necessary to use these products effectively. There are 6.8 million disabling home injuries annually and 80,000 of these are permanently disabled. There are 28,000 deaths each year.

An accident may be defined as an event or incident resulting from the interaction of people with objects and the environment. A home accident occurs within the home or its immediate environment. Injury is some hurt or harm done to the body as a result of the accident. A hazard is any source of danger.

Accidents are due to glass, gas, tools, pesticides, toys and other consumer products, and so forth. The Consumer Product Safety Commission estimates that 28,000 people die and 300,000 are injured each year from consumer products. In 2000, the EPA estimated that 81,000 children were involved in pesticide poisonings or exposures, with 8400 deaths. It further estimates that 47% of all households with children under the age of 5 had at least 1 pesticide stored within reach of the child, and 75% of households without children under 5 had at least 1 pesticide stored within reach of a child. Of children's poisonings 13% occur at homes other than their own; of most people's exposures to pesticides 80% occur indoors. Measurable levels of up to a dozen pesticides can be found in the home. Accidents do not occur spontaneously; people are involved in one way or another. People must understand the potential hazards of the home environment and the precautions necessary to avoid accidents. They must understand the problems related to alcohol and drug use and their relationship to accidents. Small children must be supervised properly.

Dangerous Residential Areas

The two most dangerous areas in the entire house are the kitchen and bathroom. In the kitchen, fires may be caused because of thoughtless situations or inoperative equipment. People leave dish towels, paper, and other combustibles on or near stoves. Frying foods are left without proper supervision, and if the temperature of the oils becomes too high, they may ignite to cause a grease fire. People wear clothing with long, loose sleeves, resulting in food overturning or garments catching fire. Pot handles are turned outward and in reach of toddlers, who may easily upset hot liquids and suffer severe burns. Unfortunately, the kitchen becomes a storehouse for most aerosol-type cans. Although many of the labels clearly indicate that the cans should be stored away from direct sunlight and temperatures over 120°F, individuals do not read the labels, creating serious potential hazards. Because the aerosol can is under pressure, it can explode like a hand grenade at high temperatures and fragments, including flammable materials, may shoot out in all directions.

Houses using liquefied petroleum (LP) or any other gas must be checked carefully to ensure that there is no leakage. Gas leaks in stoves and ovens may be due to faulty or improperly cleaned equipment. Carbon monoxide poisoning is another danger that can occur when inadequate combustion of the gas takes place in the stove. If the flame is irregular, slow, and yellow, the burner is not working properly and carbon monoxide may be produced. Other incidents of carbon monoxide poisoning occur when heaters are improperly serviced or chimneys or flues are clogged. All equipment used for heating other than that serviced by the central heating system should be checked carefully, because the equipment may not be vented properly to the outside. As houses become better insulated, the incidence of carbon monoxide poisoning will increase.

Poisoning is a particularly serious hazard within the kitchen, bathroom, and garage. In these areas, children often have access to furniture polish, waxes, all types of household cleaners, bleach, medications, gasoline, and a variety of other materials stored for household use. Within the kitchen, falls occur from spilled grease, water, or other liquids. Individuals also create hazards by standing on rickety boxes or chairs to try to reach objects on cabinet shelves. The bathroom is a notorious area for accidents. Very serious falls occur because of wet and slippery floors, groggy individuals entering the bathroom at night, and narrow spaces in many bathrooms. Poisons are mistakenly swallowed by half-awake individuals when little or no illumination is available.

In homes, fireplaces are of special concern. The fireplace is a source of danger, especially to children. Individuals making fires must be certain that sparks do not ignite a room and that the children are kept at a safe distance. Handrails, stairs, and carpeting within halls and on stairwells should be checked for good condition to ensure that they are not torn, worn, or made slick. Adequate lighting is essential for these areas. Particular attention should be paid to the presence of children's toys on staircases.

Garage and other storage areas are extremely hazardous. Unfortunately, flammable liquids, including gasoline, paint, paint thinners, turpentine, and other such items, are kept in glass containers, in improper storage areas, and within reach of children. Other garage and storage area hazards include tool boxes and tools, pesticides, pesticide sprayers, garden equipment, lawn mowers, glue, poison, and other material that an individual saves, stores, or utilizes.

Electrical Hazards

Several common electrical violations are found during housing inspections:

1. The power supply is not grounded properly and does not have adequate capacity for major and minor appliances and lighting.
2. The panel box covers or doors are not sealed to prevent exposure to live wires.
3. The switch outlets and junction boxes are not covered to protect against electrical shock.
4. Frayed or bare wires are present as a result of drying out and cracking of the insulation around the wires, or constant friction and rough handling of the wires.
5. Electrical cords are run under rugs or other floor coverings, creating potential fire hazards.
6. Bathroom lights are not permanently installed in ceiling or wall fixtures and controlled by wall switches.
7. Inadequate light is used in stairwells and hallways and within habitable rooms.
8. Octopus outlets or several different appliances or lights are hooked into one outlet.
9. Excessive or faulty fuses are used.
10. Hanging cords or wires are used.
11. Long extension cords are used.
12. Temporary lighting is used in areas where the lighting comes from other types of electrical fixtures.

Heating Equipment

In the use of any of the fossil fuels it is important that carbon monoxide is not produced by faulty combustion. Coal also produces many excess volatile hydrocarbons and sulfur dioxide. Gas, which is colorless, may be detected by odors inherent or added to the gas. Gases form explosive mixtures when a leak occurs, causing additional hazards. Oil is as hazardous as the other two fuels, and in addition where leakage occurs, oil may be a fire hazard. Inadequate amounts of air added to any of the fuels creates hazardous by-products.

Hot water heaters may be especially hazardous, because gas can escape if the flame in the gas hot-water heater extinguishes, and if the pressure relief valve is not functioning properly.

Housekeeping

Areas where people live should be kept uncluttered and clean to promote good general emotional health and to prevent infestations of insects and rodents. A badly cluttered and dirty residence not only implies potential or actual infestation of roaches, mice, and rats but also gives some indication as to the lifestyle and emotional state of the individuals living within such premises. Solid waste and clutter, although in themselves environmental problems, are also symptomatic of other types of problems within our society, within the family, and with the individual. Because this book is not intended as a housekeeping manual, it is only necessary to state that keeping property free of solid waste, improperly stored foods, and clean, using adequate detergents and friction, can satisfy general housekeeping requirements.

Animal Bites and Other Animal-Caused Diseases

Animal Bites

Animal bites, especially among children, come from dogs, cats, rodents, and occasionally wild animals (especially raccoons). Animal bites, which are generally preventable, cause trauma, wound infection, psychological problems, potential for the hantavirus from the deer mouse, and potential for rabies. The trauma involves lacerations or puncture wounds typically in the extremities. However, in small children, the head, eyes, ears, and lips may be involved. Bite wounds may become infected with a variety of microorganisms. People at special risk include those who are immunocompromised or have a history of problems related to the spleen, diabetes, blood circulation, or other chronic conditions. Serious bites can cause profound psychological problems, especially in children. Care must always be taken in wild animal bites, as well as domestic animal bites, that rabies is not present. All bites should be reported to the local health department.

From 1979 through 1996, over 300 deaths occurred in the United States from dog attacks. Nonfatal dog bites in a single year, 1994, caused an estimated 4.7 million injuries to people. Of these individuals, approximately 800,000, sought medical care for the bite.

Echinococcosis

Echinococcosis, or alveolar hydatid disease, results from infection with the larval stage of *Echinococcus multilocularis*, a microscopic tapeworm found in foxes, coyotes, dogs, and cats. Although human cases are rare at this moment, infection in humans causes parasitic tumors to form in the liver; and less commonly, the lungs, brain, and other organs. If left untreated, the infection can be fatal.The infection can be found in various parts of the world including North America. Foxes, coyotes, and cats get infected when they consume the larvae while eating infected rodents, field mice, or voles. Once the animal becomes infected, the tapeworm matures in its intestine and lays eggs, and the infected animal passes eggs in the stools. Humans are exposed to these eggs either by directly ingesting food items contaminated with their stools, or by petting or handling household cats and dogs. Because we are now moving to areas where foxes and wolves have their habitat we are more often exposed to this disease.

Rabies

Rabies is a viral encephalomyelitis with a headache, fever, malaise, sense of apprehension, paralysis, muscle spasm, delirium, and convulsions after the bite of an infected animal. The incubation time is 2 to 8 weeks or as short as 10 days. It is caused by the rabies virus, which is a rhabdovirus that is found worldwide. Reservoirs of infections include wild and domestic dogs, foxes, coyotes, wolves, cats, raccoons, and other biting mammals including bats. Rabies is transmitted in the saliva from the bite of a rabid animal. The disease is communicable in dogs and cats for 3 to 5 days before the onset of clinical signs; during the course of the disease susceptibility is general. Rabies is controlled by vaccinations of all pets for rabies, 10-day detention for observation of dogs and cats who have bitten people, immediate evaluation of heads of animals for virus isolation in the brain when suspected of being rabid, and immediate immunization of an individual bitten by a rabid animal.

The epidemiology of rabies in the United States has changed substantially during the last half century, because the source of the disease has changed from domesticated animals to wildlife, raccoons, skunks, foxes, and bats. Human influence has contributed to these changes because people have moved into the area of these wild animals or have encountered them in recreational areas. An estimated 40,000 to 100,000 human deaths are caused by rabies each year worldwide. In addition, millions of people, especially in developing countries of the subtropical and tropical regions, are treated yearly. In the United States, because of a very expensive prevention and surveillance program running into the hundreds of millions of dollars, one to two deaths occur annually.

Emerging Zoonoses

In the past several years emerging diseases in the United States and throughout the world have increased. This pattern may continue to grow because of the changes in ecosystems brought about by populations ever on the move into new areas, much of them rural. Nearly all these emerging diseases involve either zoonotic or species-jumping infectious agents, for which prevention and control may not be well understood. An example of this was the emergence of the virus causing hantavirus pulmonary syndrome in the Southwest in 1993.

Many elements can contribute to the emergence of a new zoonotic disease, such as mutation, natural selection, evolutionary progression; individual host determinants including acquired immunity and physiological factors; host population determinants including behavioral characteristics, societal, transportation, and commercial factors; and environmental determinants including ecological and climatological factors. Global human and livestock animal populations have continued to grow, bringing increasingly large numbers of people and animals into close contact. Transportation has advanced so rapidly that people can travel around the world in less time than it takes to incubate most infectious agents. Bioterrorist activities supported by certain governments as well as individual groups are selecting zoonotic agents as their infectious agent of choice. Microorganisms and viruses may adapt to extremely diverse and changing conditions.

Histoplasmosis

Histoplasmosis is an infection caused by the inhalation of spores from the dimorphic fungus *Histoplasma capsulatum*, which may be found in and around old structures where birds and bats have been living. The mold grows well in soil rich with the bird or bat guano. Histoplasmosis is endemic in states in the Mississippi and Ohio River Valleys, including Kentucky, Illinois, Indiana, Missouri, Ohio, and Tennessee. Cases range from asymptomatic to mild to acute or chronic to disseminated. Mild cases are similar to a self-limited influenza. Acute or chronic pulmonary infections may occur. Disseminated disease is more likely to occur in the very young, the elderly, and the immunocompromised such as people under treatment for cancer with chemotherapy or people with AIDS. This can be life-threatening, with incubation periods ranging from 5 to 18 days.

Epidemiology of Injuries

Epidemiology, which is the fundamental science for studying the occurrence, causes, and prevention of disease, can also be used on a theoretical basis to study the problems related to injuries. To do this it would be necessary to determine the time, place, person injured, age, sex, nature of the injuries, factors involved in injury causation, and other environmental, and human factors.

Injuries, like diseases, do not occur at random. The elderly and persons age 15 to 24 have the highest fatality rates. The risk of fatal injury is two and one half times greater for males as it is for females. Males are also at greater risk for nonfatal injuries. Fracture rates are highest among older women who have osteoporosis or bone decalcification. Death rates, with the exception of homicides and suicides, are highest in rural areas, possibly because of differences in socioeconomic status, types of occupational and other exposures, and lower availability of rapid emergency care. Socioeconomic factors influence the incidence of homicide, assaultive injury, pedestrian fatalities, and fatal housefires. These occur at the highest rate among the very poor. Among the wealthy, the greatest number of injuries is caused by falls and home swimming pools.

Other factors causing injuries include high-risk jobs, poor housing, older cars, space heaters, and housing-related problems. These are especially true among the poor. The use and abuse of alcoholic beverages have a huge influence on all types of injuries, especially among young teenagers. About half the fatalities related to injured drivers or pedestrians are alcohol related. Alcohol is frequently detected in the blood of individuals who have been involved in a variety of different types of injuries. Usually the more severe the injury is, the higher the level of alcohol. Other types of behavioral factors are very difficult to determine, because they may be very transient.

The environment may contribute to the level of injuries that occur. Poor weather, as well as improperly maintained equipment, may contribute to injuries. The speed of emergency medical care is essential in controlling the severity of the injury that has occurred. This emergency medical care system helps reduce the numbers of fatalities among individuals.

OTHER ENVIRONMENTAL PROBLEMS

Noise

Noise is unwanted sound. Within the home environment, sleep or comfort is disturbed by loud television sets, grinding truck gears, fan motors, children, radios, dishwashers, garbage disposals, vacuum cleaners, washing machines, knocking pipes, and so forth. Noise within the home comes from a myriad of sources. Noise is transmitted directly through the walls or indirectly through walls, ceilings, and floors. Each person has a unique feeling about a comfortable sound level and the definition of noise. For this reason, special consideration should be made of neighbors and family members when creating loud sounds at varying times during the day and night.

Pests

Roaches, mice, flies, lice, ticks, rats, and mosquitoes are in the interiors and exteriors of housing units. The infestation may be brought to the unit by grocery bags, clothing, and equipment, or may be caused by migration of insects and rodents from their normal habitats.

Sewage

Sewage is overflowing in many housing developments in various areas of the country. Currently over 10 million homes have septic tank systems. Considerable research is needed to determine the proper techniques for sewage disposal for these properties. An in-depth discussion on individual sewage disposal can be found in Volume II, Chapter 6.

Drainage

Pools of stagnant water accumulate in poorly drained properties. These are excellent areas for mosquito breeding and may create health hazards. Electrical hazards must also be considered when standing water is encountered. Excessive dampness or water causes wood to rot and structures to deteriorate.

Solid Waste

The storage and disposal of solid waste in a proper manner is the responsibility of the occupant of the property. In many areas the local governmental unit provides adequate solid waste removal and disposal. In some areas the occupant must make private arrangements. In any case, where solid waste is not handled properly, insects and rodents, odor, and air pollution problems due to the burning of solid waste may occur. (Further discussion concerning this subject will be found in Volume II, Chapter 2 on solid and hazardous waste.)

External Housing Environment

The exterior premises in many ways are as important as the interior premises. Improper solid waste storage, poor pest and weed control, and accident hazards create housing problems. Exterior premises with piles of old lumber or other materials, dilapidated shacks or garages, abandoned automobiles, or other kinds of junk or waste are both hazardous and unsanitary. The exterior solid waste problem is often an indication of the interior housing environment.

Fires may be caused by improper storage or disposal of flammable liquids and by obstruction of electric power lines by trees, birds nests, and so forth. Other safety hazards on the exterior grounds include stepoffs, holes in the ground, rocks, broken glass, power equipment, and storage of LP gas.

Role of Government

The many policy inconsistencies that exist as a result of changes in federal programs and tax laws have adversely affected housing. For example, federal deductions are permitted for building depreciation. It might be wiser to allow a tax deduction for building maintenance to encourage investment in deteriorating buildings. The federal highway administration provides funds for the construction of radial highways that often cut through valuable urban neighborhoods. These plans might well be altered to avoid interference with neighborhoods, thereby reducing the potential for deterioration. In the past, federal housing policies have emphasized new building instead of revitalization of existing buildings and neighborhoods.

Housing Code Enforcement

There are many problems related to housing code enforcement. These include absentee landlords, properties belonging to estates, effective court procedures, tenants, political support, fiscal support, proper personnel, code enforcement, and relocation of families. Unfortunately, a high percentage of the difficult housing code violations involve absentee landlords or estates. The landlord may not correct the violations because of the cost involved, the destruction of previous corrections by the tenants, or the property tax write-off making it simpler to take a tax deduction than to spend capital on expensive repairs.

Tenants are also taken to court for violations of the housing code, because tenants may create as many problems as landlords. Where the tenant refuses or will not take care of the housing environment, the court needs to exercise its powers in the same way that it would exercise them over property owners.

It is one thing to notify the property owner of violations and quite another to secure compliance. Although many individuals correct violations because they are naturally law-abiding citizens, there are those who will attempt to avoid doing this. Unfortunately, politics may enter the situation. Political leaders generally are very careful about supporting vigorous code enforcement, because the code enforcement may in effect cause endless complaints from citizens and eventually result in the loss of votes.

Code enforcement at times cannot be carried out because the individuals in violation lack the money to make the necessary repairs. In the past, federal agencies, and in some cases state and local agencies, were able to provide funds for some corrections. These funds, however, are very indefinite and tend to vary from year to year based on the current political situation in Washington or in the various localities.

As a result of the uncertain political attitude toward code enforcement, inadequate funding is provided for the health department or housing department to carry out necessary code enforcement programs. This tends to deprive departments of trained competent individuals capable of performing necessary tasks and also to deprive the department of having adequate numbers of individuals to make a program workable.

One of the difficulties of code enforcement is that when a property is found to be unfit for human habitation, a family must be relocated. Necessary assistance must be provided to help these families relocate in a shelter that is acceptable for proper living. In some areas, the properties involved are so numerous that it becomes very difficult to remove the families from substandard housing and to relocate them to other areas. Either the new housing is unavailable or the family is unable to live its normal life in a new situation because of transportation and shopping difficulties, and difficulties in reaching friends or relatives.

Impact on Other Problems

Health statistics demonstrate a relationship between morbidity or mortality rates and physical surroundings. People living in economically depressed neighborhoods with poor facilities are less healthy than those living in the higher socioeconomic areas. Obviously inadequate nutrition, medical care, and stressful social conditions contribute to these health problems. A relationship of physical facilities to infectious disease and infant mortality is apparent. As the facilities and levels of cleanliness improve, the rate of disease occurrence seems to drop. The role of housing and neighborhood conditions on juvenile delinquency, psychiatric illness, mental retardation, and learning disabilities has been noted. Overcrowding and the effect that it may have on individuals is a concern. Substandard housing also contributes to air, water, and land pollution and to solid waste problems.

Economics

The economic costs of housing are related to high taxes for services, congestion, streams loaded with silt, air pollution, destruction of open space, and provision for a vast variety of services. The social costs are certainly difficult to evaluate but must be included. Other costs include the provision of schools, necessary streets and roads, and utilities. Noise, energy consumption, and erosion of land are added factors. Personal aspects must be considered in any cost analysis, including travel time to and from work, traffic accidents, crime, and cost involved in resolving emotional or psychiatric problems.

POTENTIAL FOR INTERVENTION

The potential for intervention is very difficult to assess in the area of housing. It is obvious that the technique of isolation can be utilized through the tearing down of unsafe structures. Substitution can be utilized by providing low-rise structures for high-rise ones in lower socioeconomic areas to eliminate problems that occur in the high-rise structures. Shielding is not an effective technique. Treatment includes renovation or improvement of the structures and the neighborhoods. Prevention is the most effective approach, because land planning coupled with adequate housing inspections and code enforcement, as well as good zoning, can help prevent problems from occurring. Prevention and control of indoor air pollutants and injuries can be accomplished through isolation, substitution, shielding, and treatment.

RESOURCES

Scientific and technical resources include the National Architectural Accrediting Boards Inc., Association of Collegiate Schools of Agriculture, various graduate professional schools of architecture, American Public Health Association, National Environmental Health Association, construction and general contractors, and so forth. Civic associations consist of a vast variety of homeowners associations, landowners associations, block councils, and community councils found in many cities or county areas.

The governmental resources include local and state health departments, local and state housing departments, National Bureau of Standards, Department of Defense, Department of Energy, General Services Administration, Department of Housing and Urban Development, Law Enforcement Assistance Administration, National Aeronautics and Space Administration, and National Science Foundation. A resource on safety is the Office of the Secretary, Consumer Product Safety Commission, Washington, D.C.

The American Public Health Association-Centers for Disease Control (CDC)-recommended minimum housing standards, as updated in 1996, may be obtained from the American Public Health Association (APHA), 1015 15th St., N.W., Washington, D.C. APHA also has an excellent new book on injury control.

The Marion County Health Department has an excellent housing program. Its ordinance is titled Housing and Environmental Standards Ordinance. Information may be obtained from the Marion County Health Department, 3838 Rural Street, Indianapolis, IN.

The U.S. Department of Housing and Urban Development is responsible for large numbers of programs related to community planning and development and housing. Copies may be obtained by writing to the U.S. Department of Housing and Urban Development, Washington, D.C.

Philadelphia has been at the forefront of injury prevention in homes. The Philadelphia Injury Control Program material may be obtained from the Philadelphia Department of Public Health, Division of Environmental Health, 500 South Broad Street, Philadelphia, PA.

Injury prevention is a serious concern that has been addressed by the Childhood Injury Prevention Resource Center, Harvard School of Public Health, Department of Maternal and Child Health, 677 Huntington Ave., Boston, MA. *Injury in America*, a major report of the Board of the National Research Council of the National Academy of Sciences, the National Academy of Engineering, and the Institute of Medicine, has been published by the National Academy Press, 101 Constitution Ave., N.W., Washington, D.C. The CDC, U.S. Department of Health and Human Services (USDHHS), Public Health Service, Atlanta, GA, has programs and published reports on injury control. The Consumer Product Safety Commission's Product Safety Hotline is 1-800-638-2772.

Indoor air pollution, radon gas reduction, lead contamination, and other associated programs and problems are under the control of the USEPA. Information may be obtained from them by contacting the USEPA, 401 M Street, S.W. Washington, D.C.; NIOSH, 1-800-356-4674; Indoor Air Quality Information Clearinghouse, 1-800-438-4318; Safe Drinking Water Hotline, 1-800-426-4791; Lead Clearinghouse, 1-800-424-5323; National Pesticide Telecommunications Network, 1-800-858-PEST; the National Radon Information line, 1-800-767-7236. For information on asbestos and formaldehyde, call 1-202-554-1404.

Additional resources for injury control include the Surgeon General's *1990 Injury Prevention Objectives for the Nation and Model Standards*, and *Workshop on Violence and Public Health Report*; *A Guide for Community Preventive Health Services*; *Developing Childhood Injury Preventions*; and *Administrative Guide for State, Maternal and Child Health Programs*; *Strategies for Injury Prevention*; and *Injury Prevention in New England*. (Contact the CDC, Atlanta, GA.) The National Center for Injury Prevention at CDC is an excellent resource.

STANDARDS, PRACTICES, AND TECHNIQUES

Rehabilitation

It is true that in some areas redevelopment has been necessary, but in other areas a better concept, known as rehabilitation, has been attempted. In rehabilitation, existing residential properties that are deteriorating are renovated with the aims of (1) ensuring improved housing that is livable, safe, and physically sound; (2) ensuring low-cost housing; (3) providing an acceptable minimum level for housing; (4) encouraging innovation and improved technology to reduce construction costs; and (5) utilizing instead of destroying resources that need upgrading. Obviously rehabilitative construction standards are not the same as new construction standards, because the work is done on existing buildings. However, all safety precautions do have to be taken into consideration and followed carefully.

All rehabilitated properties must have access to the property from the road by emergency vehicles, living units that exist independently of other living units each with its own entrance and exit to the building, and practical living units. Under rehabilitation, all highly dilapidated properties that cannot be rebuilt efficiently are torn down and the ground space is given to the community for use as playgrounds, parks, or parking lots, as decided by residents of the community. Adequate night lighting and open space are site criteria that must be provided; also, the land must be made free of flooding, sewage problems, and solid waste problems. The materials and products used for replacements or additions have to be of good quality, conforming to generally accepted practice. A building must be structurally sound. The exterior walls must be able to support the weight of the roof and also must be moisture free.

It is important in rehabilitation that individuals have sources for adequate food, hot water, baths, and proper sewage and solid waste systems. One must recognize that rehabilitated property may not look as good as new property. However, low-income families have the opportunity to live in good housing that provides all the essentials for prevention of disease, elimination of behavioral problems and accidents, and promotion of health.

Standards for Housing Utilities and Construction

Heat

The CDC housing code states that the temperature should be at least 68°F at a distance of 18 in. above the floor level. If a person needs a higher temperature because of age or physical condition, 70°F is required. Others believe that this temperature should be maintained at different points above floor level, such as at 5 ft. If carbonaceous fuel is used, the heating device or hot water heater must be vented to the outside and receive adequate air for proper combustion of fuel. In any case, the local housing code requirement is the one followed by health department personnel.

Light

In most communities, every habitable room must have at least one window or skylight facing outdoors, and the window area must be at least 8% of the total floor area of the room. Daylight should be used wherever possible in such a way that the amount, distribution, and quality of the light aids in, instead of detracts from, visual tasks.

Artificial light varies tremendously from area to area, and from room to room. Artificial light must always be provided in public halls and stairways, and should be adequate for all use within the house. The house that is properly illuminated enables the family to function more efficiently. Switches should be in convenient places, and the type of lighting should serve the various needs of the people during the day and night. Obviously, such areas as kitchens and bathrooms should be well lit to prevent accidents. Such places as furnace rooms, laundry rooms, and other areas where appliances are kept should have adequate lighting to make the task less demanding and to prevent accidents. Proper lighting is also very essential in preventing fatigue.

All electrical units, wiring, and appurtenances with electrical connections must be installed by electricians following electrical codes to avoid the occurrence of fires.

Plumbing

In the house, the piping for water service should be as short as possible, and elbows and bends should be reduced to maintain water pressure. The water line to the house should be at least 4 ft below the soil to prevent freezing. Valves are usually located outside the building so that the building supply may be turned off when it is necessary to service the building. Hot water heaters should be thoroughly evaluated and properly installed and ventilated. Usually a ¾-in. pipe is the minimum size required for water that must rise from the basement level to other parts of the house. The drainage system should have a main house drain a minimum of 6 in. in diameter. The house drain has to be sloped toward the sewer, usually at a level of ¼ in. fall per 1-ft length. Determination of the amount and size of drainage lines depends on the types of rooms within the house. Traps are used on sinks and toilets to prevent sewer gases from entering the property. All plumbing systems must be adequately ventilated directly through the roof. The most essential part of the plumbing is the proper size of the lines, adequate installation, proper ventilation, and proper drainage to avoid cross connections or other situations in which the potable water supply becomes contaminated or the sewage backs up through the toilets and various drains within the house.

Ventilation

The usual requirements for ventilation are that 45% of the minimum window area can be easily opened or an approved mechanical means of ventilation can be installed. Every bathroom and toilet room must comply with the various housing codes and rules on adequate ventilation. Ventilation must be to the outside. In many housing codes, two to six changes of air per hour are required.

Space Requirement

The APHA-CDC code is 150 ft^2 of total, habitable, room area per person. Where second-person occupancy occurs, the requirement is reduced to 100 ft^2. The code also requires that no more than two people are permitted in a habitable room. With the exception of rooms that have sloping ceilings, the habitable room should be a minimum of 7 ft in height. Where a room is less than 5 ft in height, it should not be used for living purposes. Individuals should not live in rooms, except in emergencies, that are totally below the ground surface. There should be some exposure to the outside. The exception to this rule is where construction is waterproof, window area is adequate, and other rules and regulations relating to light and ventilation are followed. Pipes, ducts, or other obstructions must not be less than 6 ft 8 in. from the floor. Where two people share a room, at least 70 ft^2 of floor space is required for the first person and 50 ft^2 of floor space, for the second. Living quarters must have easy access to a bathroom, and an individual must not have to pass through another sleeping room to get to a bathroom.

Water

All housing codes require that adequate quantities of running water be provided for a dwelling. This generally means 1 gal/min of hot and cold running water per each fixture in the house. Generally, 120°F is accepted as the maximum hot water temperature. However, the temperature may need to be lowered to prevent burns.

Kitchen Facilities

All kitchen facilities must contain sinks, cabinets or shelves, stoves, and refrigerators. The sink must be large enough for kitchen use and not a small hand-washing sink. Because the kitchen sink is used in the preparation of food and the cleaning of dishes, utensils, and equipment, it must have hot and cold running water under pressure and must be built so as to be utilized in an effective manner. The stoves must be adequately built, installed, and maintained to avoid electrical, fire, or carbon monoxide accidents. A refrigerator is needed to keep food under 41°F at all times. Freezers are highly desirable.

Physical Structure

Houses are framed in many different ways. However, all of them contain foundation walls and provide support for the outside walls of the building. The flooring system is made up of girders, joists, subflooring, and finished flooring, which may be composed of concrete, steel, wood, composition material, and so forth. Studs, which are usually 2 × 4 in. or 2 × 6 in., are used to provide a wall thick enough to allow the passage of waste pipes. Usually studs and joists are spaced 16 in. on center. All openings, such as windows or doors, must be framed with studs to provide support for that portion of the wall. The interior wall finish varies from plaster to wallboard, paneling, wood, and so forth. Interior stairways must be greater than 44 in. in width with a maximum rise of 8¼ in. and a minimum tread of 9 in. The handrail and all other parts of the stair enclosure should be no lower than 80 in.

The four basic types of windows are double-hung sash windows, which move up or down; casement windows, which are hinged at the side; awning windows, which are in panes on a horizontal axis; and sliding windows, which slide past each other. Many types of doors and door finishes are available.

The roof framing consists of the rafters, which support the roof and also create a place for roofing material. The exterior walls and trim may be made of a variety of materials. Their function is to enclose and weatherproof the building. Exterior walls also serve as weight-bearing walls. They are composed of wood, brick, stone, and so forth.

The roof covering must keep the house dry and intact. Roof covering composition includes asphalt shingles, asphalt built-up roofs, tar, slate, tile, and wood. Flashing is the metal that joins two portions of the roof where it forms a valley. This keeps water and air from getting into the building. Gutters are used to take water off the roof and away from the house. Insulation material is extremely important to keep the house cooler in the summer and warmer in the winter. Good insulation prevents the loss of vital energy.

Fire Safety and Personal Security

All dwellings should have at least two means of egress leading to a safe and open space at ground level. Individuals should not have to exit through someone else's dwelling unit. All entrance doors into a dwelling or dwelling unit should be equipped with a deadbolt or locking device. The entrance doors in a multiple dwelling should be equipped with a device that allows the occupants of the unit to see a person at the door without fully opening the door. All exterior windows and other means of egress should be equipped with locking hardware. Every dwelling unit should have at least one functioning smoke detector located on or near the ceiling in an area immediately adjacent to a sleeping area. These smoke devices should be tamper proof.

Travel Trailers and Mobile Homes

The travel trailer is a portable structure used as a temporary dwelling for travel, recreation, and vacation. The body width must not exceed 8 ft and the trailer must not exceed 4500 lb in weight and 9 ft in length. Travel trailers vary considerably in size, types of sleeping accommodations, and actual living facilities. The trailer may simply be built for sleeping or may have self-contained hot and cold running water, full kitchen, baths, gas tanks, air-conditioning, and lighting systems. The travel trailer has to be fitted for hooking into sanitary stations for removal of sewage. The trailer has toilets aboard and must also be fitted for taking on water.

The facilities used for travel trailer parking areas have to be carefully selected for good drainage, gentle sloping, and no obstructions. The roads around the parking areas must be curved to reduce the speed of the drivers. A separation of 15 ft between trailers is an absolute minimum. In trailer areas, provisions must also be made for service buildings where individuals can take showers, do laundry, and use toilet facilities. Solid waste has to be stored in tight-fitting containers that are waterproof, fly proof, and rodent proof. These containers must be removed daily to prevent pest infestation. When a travel trailer area plan is ready to be approved, it must indicate the area in site dimensions; the number, location, and dimension of the trailer spaces; the location and width of roads and walks; the location of the service building, sanitary station, and other facilities; the location of water and sewer lines; and the location of storm drains and catch basins.

The number of mobile homes has increased enormously within the last 15 years. Mobile homes may be on wheels or modular in nature. The mobile home is used in many cases as low-cost housing. Mobile homes may be placed on individual sites or in mobile home parks. The home placed on an individual site must follow all rules and regulations of the health department for proper water, sewage, site locations, and so forth. The site has to be zoned for this particular use. It should be well drained and free from topographical hindrances. It should not be located near swamps, marshes, and heavy industries; in flood plains; or on steep slopes. It should also be placed, if possible, in such a way that in the event of tornados, it will be protected as much as possible. Owners of these types of home either have to hook into an existing sewage system and obtain city water plus other utilities, or must dig a well and provide an adequate on-site sewage disposal system. (Well-drilling and on-site sewage disposal are discussed in further detail in Volume II, Chapters 3 and 6.)

Where a mobile home court is developed, a community development on the same order as any other type of community is required. All the roads have to be convenient and of adequate width for the parking and traffic loads. Streets should be approximately at right angles. Street intersections should be at least 150 ft apart, and the intersection of more than two streets should be avoided. The grades of the streets should not exceed 8% whenever possible, although for a short run as much as 12% is acceptable. The streets must be of all-weather construction and of hard, dense, easily drained material. Off-street parking is necessary. Walkways at least 3.5 ft wide must be provided within the area.

A mobile home should be on a minimum size lot of 2800 ft^2. A double-wide unit should be on a lot that is a minimum of 4500 ft^2. The mobile home should not occupy more than one third of a total lot area. Small lots contribute to overcrowding and create a poor aesthetic image of the area. If an area is developed for five or more mobile homes, a recreational area that is safe and free of traffic hazards should be made available for use by children and adults. A recreational area can be located next to recreation or service buildings. The minimum space should be ⅔ acre. Where swimming pools are constructed, they must comply with all the regulations of the various state and local health codes.

All buildings in a mobile home park must conform to the housing codes of the local and state areas. They must be kept in a sanitary manner, screened, properly ventilated, and properly lighted. Separate bathroom facilities have to be provided for men and women. If a laundry facility is present, it has to be adequate in size and properly installed, and the wastewater must go into adequate wastewater disposal units.

Before a mobile home development can be approved, the plans have to be submitted to the health department, zoning department, and other responsible authorities. Detailed construction plans, and the site and the location of all roads, buildings, and so forth have to be evaluated. It is essential, wherever possible, that the public water supply and public sewage disposal be used. If these are not available, a package treatment plant should be built. Access to firefighting equipment, police protection, and schools have to be determined before the site is approved.

In the mobile home park, a mobile home stand has to be provided for adequate foundation and anchoring of the mobile home. The stand must be properly graded and of such material that it can support the weight load, regardless of weather conditions. The anchors must be cast in place in concrete so that they will not be ripped out in the event of storms. Particular attention should be paid to this in those areas where tornados frequently occur. Because tornadoes knock over mobile homes or cause damage to them, all water and sewage hookups and electrical and gas hookups have to be installed in such a way as to avoid hazards and to provide adequate service. Telephone service is also needed. Further discussion concerning water and sewage are found in Chapters 3 and 6 in Volume II.

Storm water drainage systems have to be installed in such a way as to avoid flooding of the mobile home site. The drainage system must remove the heavy rain waters from the area. Solid waste containers, storage containers, and collection techniques have to meet the requirements of local and state codes. Further discussion concerning this subject can be found in Chapter 2, Volume II.

Insect and rodent control must also be considered in a mobile home park. Of particular importance is mosquito control, because many mobile home parks are in outlying areas where weeds or other breeding places, such as depressions in the ground, may be present. In addition, the areas must be free of cans, jars, buckets, old tires, and all other junk that serve as breeding places for mosquitoes. Fly- and rodent-control procedures are to be followed. It is essential that all garbage be properly stored and then discarded. Insect and rodent control are discussed in further detail in Chapters 5 and 6 of this book.

Electrical wiring, equipment, and appurtenances should be installed and maintained in accordance with local and state codes. Where these codes do not exist, it may be necessary to consult the National Electrical Code. If possible, distribution lines should be installed in underground conduits, which are placed at least 18 in. below the ground surface and at least 1 ft away from water, sewer, gas, or phone lines. If overhead power lines are installed, it must be done in such a way as to prevent accidents and should be available for service in emergencies. Grounding is necessary for all exposed, non-current-carrying metal parts of mobile homes and equipment. Adequate lighting has to be provided on exterior walkways, streets, and buildings.

The handling or storage of fuels, such as natural gas, LP gas, fuel oil, or other flammable liquids or gases, has to meet federal or state ordinances or the National Fire Protection Association Standards. It is best to have central storage and underground distribution of fuel. All fuel oil containers or other fuel containers on the lot where the mobile home stands are to be securely fastened to prevent accidents or fires.

It is essential that adequate fire protection be provided in mobile home parks. Applicable state and local codes and laws should be followed. If they do not exist, standards established by the National Fire Protection Association may apply. A water supply of adequate quantity must be available for fire hydrants within 500 ft of all mobile homes, service buildings, or other structures. A minimum of two 1.5-in. hoses must be available for hooking up to the water supply. At least 75 gal of water per minute per nozzle must be available at a pressure of 30 lb/in.^{2}. Where hydrants are not available, the types of equipment that can be utilized are limited to those that meet the National Fire Protection Association standards for adequate fire extinguishers. Class A extinguishers have to be provided for mobile homes.

Because carbon monoxide is a constant danger within the mobile home, it is essential that heating and cooking equipment be properly vented. If there is any question concerning this, the standards of the local or state authorities should be followed or the health department or fire department should make an evaluation of the equipment and the adequacy of the venting system.

MODES OF SURVEILLANCE AND EVALUATION

Housing Inspections and Neighborhood Surveys

Housing inspections and neighborhood surveys are used to determine the quality of housing and neighborhoods within a community. The American Public Health Association has done considerable work in this area. This association has developed tools of evaluation, including comprehensive forms, and has established point systems for determining into which category a neighborhood falls. (For more detailed information on housing, the reader is referred to the latest methods, techniques, and publications of the American Public Health Association, Philadelphia Department of Public Health, and Marion County, Indiana Public Health Department.)

Indoor Air Pollution Surveys

Indoor Air Quality Profile

When an indoor air pollution problem occurs or if you want to determine the current status of air quality in the building, follow this procedure:

1. Collect and review existing records, including heating, ventilation, and air-conditioning records, complaint records, design, construction and operating documents.
2. Conduct a walkthrough inspection of the building.
3. Collect detailed information on pollutant pathways, pollutant sources, and occupants.
4. Make an inventory of the heating, ventilation, and air-conditioning system repairs, adjustments, or replacements needed.
5. Obtain a set of Material Safety Data sheets (see Table 8.1) and an inventory of significant pollutant sources and their location.

The walkthrough inspection should include schedules and procedures on facility operation and maintenance, housekeeping, and pest control. Pressure relationships between special use areas and other rooms should be identified and any special circumstances related to areas where complaints have been received. Look for odors, dirty or unsanitary conditions, visible fungal growth, mold odors, drain pans, cooling towers, filters, staining and discoloration, smoke damage, presence of hazardous substances, improper storage of chemicals and cleaning substances, inadequate maintenance, occupant discomfort, overcrowding, blocked airflow, crowed equipment areas, new furniture, rugs, equipment, renovations, and painting. Also, collect air samples, bulk samples, and swab samples where appropriate.

Indoor air quality surveys almost always include initial characterization of carbon dioxide, temperature, and relative humidity as indicators of the quality of indoor ventilation. Carbon dioxide (CO_2) is a relatively innocuous gas, but it is a major product of exhaled air that can accumulate in rooms occupied by people or animals. Carbon dioxide can accumulate and exceed levels of 800 to 1000 ppm indicating that there is inadequate exchange of fresh and exhaust air from the indoor environment. Inadequate ventilation exchange rates can result in what is frequently described as stuffy, or stale, air. Carbon dioxide can be measured using direct-reading real-time monitoring instruments, including electronic meters and colorimetric detector tubes. Extremes of temperature or excessive air currents can result in an uncomfortable environment for occupants.

Also, extremes of relative humidity or moisture levels in air can contribute to discomfort. When air is too dry, or the relative humidity falls below 35%, occupants may complain about itchy or burning eyes and throat and dry skin. Levels above 55 to 60% relative humidity can result in a somewhat uncomfortable feeling of sticky or humid air. More importantly, excess humidity in indoor environments can foster

Table 8.1 Material Safety Data Sheet

Item	Possible Uses	Comments[a]
Substances covered	MSDSs may identify significant airborne contaminants	MSDSs may not be available on-site for many products Some components are listed as proprietary and are not disclosed MSDSs do not always highlight products most likely to be airborne Contaminant by-products inadvertently formed during manufacture are not always listed
Personal protection/ first aid	May suggest precautions for conducting source inspection	Usually relates only to high-level, worst-case exposures in general industry
Health effects	Generally presents types of health effects that may be expected primarily at high level (e.g., industrial) exposures	Symptoms listed may not occur at low-level concentrations found in indoor air MSDSs may not include more subtle IAQ aspects such as nuisance factors and sensitivity to mixtures
Physical data	Odor description may help identify sources Volatility may suggest which products are likely to be airborne Contaminants to expect in event of a fire or decomposition may be listed Reactivity data may suggest potential problems with storage or use	Reference material on how to use physical data information to predict IAQ impacts may be scarce
Control measures	Identifies proper storage and packaging procedures Identifies steps for cleanup of gross spills	Many office chemicals are kept in much smaller amounts than found in industrial settings Spill cleanup may not eliminate airborne contamination Does not specify routine emission controls

Note: Under OSHA regulations, responsible parties are required to document information on potentially hazardous products. These Material Safety Data Sheets (MSDSs) may be of limited help in identifying some products that may pose IAQ concerns. However, professional judgment and collection of additional information may be necessary in order to make full use of the MSDSs. The table summarizes some of the issues to keep in mind when deciding whether information from MSDSs is applicable to emission sources and exposures of concern in a building.

[a] A reasonable effort should be made to collect available MSDSs during IAQ profile development. Care should be taken to consider information that is relevant to IAQ concerns. Other important indicators of how a particular product may affect IAQ are available from direct odor and dust observations, a review of work practices, and interviews with operators and occupants. The manufacturer is a good source of follow-up information on a given product (phone number should be included on each MSDS).

Source: From *Building Air Quality — A Guide for Building Owners and Facility Managers,* U.S. Environmental Protection Agency, Indoor Air Division, and U.S. Department of Health and Human Services, National Institute for Occupational Safety and Health, Washington, D.C., 1991, p. 39.

unwanted, potentially pathogenic mold and bacterial growth on ceilings, walls, and floors. A device called a sling psychrometer is used to provide a direct-reading in real-time of the air temperature and relative humidity. Adequate mechanical ventilation described as general, or comfort, ventilation is usually necessary to regulate comfort variables of air turnover or exchange rate, temperature, and relative humidity.

The nature and type of other indoor air variables and contaminants vary depending on the use and condition of the rooms. Some other common indicators of indoor air quality include carbon monoxide (CO) gas, radon (Rn) gas, formaldehyde and other VOCs, and particulates including asbestos fibers, lead, and bioaerosols. Bioaerosols include bacteria, molds, and associated spores. All these substances are potentially either toxic or pathogenic to occupants and can be monitored using devices briefly mentioned here or in more detail in Chapter 12. Agent-specific, direct-reading, real-time monitoring devices are available to detect and measure almost instantaneously carbon monoxide gas, formaldehyde, and other VOCs. This is also commonly referred to as grab sampling. Longer term (e.g., 8- to 24-hour) continuous or integrated air sampling also can be conducted using air sampling pumps connected to specific sampling media such as filters for spherical (e.g., lead dust) and fibrous (e.g., asbestos fibers) particulates, solid sorbents for VOCs, solid or liquid sorbents for formaldehyde, and impactors plus agar-media for bioaerosols. Bulk sampling of gaseous air contaminants, such as VOCs, are often collected using evacuated rigid stainless steel containers or nonrigid plastic (e.g., Mylar, Tedlar) bags. Bulk sampling also includes collection of building materials (e.g., ceiling panels, floor tiles, and carpeting) to determine if contaminants such as lead dust, asbestos fibers, dust mites, or microbes are present. Surface wipe and swab sampling also are conducted when warranted.

The environmental lead survey is called for when a child's blood level exceeds 15 µg/dl. The survey is two part: an epidemiological study consisting of a field environmental investigation and an on-site analysis of paint samples. The two major routes of exposure for children to lead is through ingestion and inhalation. The environmental health specialist determines if the child:

1. Ingests, chews, or puts painted articles, paint chips, printed material, matches, tobacco items, dirt, etc. into the mouth
2. Consumes folk remedies that may be contaminated with lead
3. Plays or lives in, or has access to, any areas where cans of lead-based paints, pesticides, or other substances are used or stored
4. Eats from containers or utensils that are glazed or decorated with paint
5. Lives or sleeps (or both) in areas where there is crumbling paint or plaster
6. Lives near heavily traveled highways or lead-based industries
7. Lives near areas of old building renovation
8. Drinks water supplied by old plumbing

CONTROL OF HOUSING

Zoning and Land Use

Zoning is a means of ensuring that community land is used in the best possible way for the health and general safety of the community. To prevent disorderly patterns of growth, a community must plan. It would be very unfortunate if heavy industry and superhighways were placed in the center of residential areas. When a house is built where a zoning ordinance is in effect, the builders must comply with the zoning regulations, and they in turn are assured that the other houses in the area comply with the same regulations. A single-family dwelling, for example, cannot be converted into multiple-family units unless zoning laws permit this type of structure. It is important that public health officials, housing inspectors, developers, and planners have a good understanding of the plans of a community and the applicable zoning laws.

Zoning regulations have been in effect for several hundred years. They were originally passed to keep gunpowder mills and storehouses away from heavily populated areas of town. Later zoning was used to form fire districts and to prohibit certain types of highly flammable structures. Zoning has been used and misused in a variety of places. Proper usage would mean that zoning was tied to an overall community plan. Improper usage is to change the zoning by using variances, so that special interest groups or political groups can utilize structures or land as they see fit. The objectives of zoning are to regulate the height, bulk, and area of structures; to avoid undue levels of noise, vibration, and air pollution; to lessen street congestion through off-street parking and off-street loading requirements; to facilitate adequate provision of water, sewage, schools, parks, and playgrounds; to use areas that are not subject to flooding; and to conserve property values.

Zoning cannot correct existing overcrowding of substandard housing or change materials and methods of construction, because these are controlled by building codes. Zoning laws do not affect the cost of construction, because this is based on the economy. Zoning cannot be used to develop special regulations or covenants, and cannot design and lay out subdivisions, because this is controlled under other rulings. Zoning is concerned with lot size, usage, depth, and width; amount of open space; and types of buildings and alterations to buildings placed on the lot.

Role of Government

The four major problem areas in the role of government in land use and housing include the proper use of federal government program and federal policy, state government in action, local taking of land by right of eminent domain, and adequate municipal financing. A need exists for coordination and consistency between the federal, state, and local governments in land-use programs and policies. State governments should have greater input in land-use matters affecting large local areas. The regional metropolitan units of government should be concerned with land use and its effects on outlying, as well as immediate, areas. It is important that a sharp distinction be made between public interest and the rights of private ownership.

Because the major source of local government revenue is generally the property tax, it is important that land-use decisions be made in accordance with good, sound scientific and socioeconomic political decisions. The metropolitan area must be concerned not only with proper land use but also with misuse. Many unutilized buildings, if properly redesigned, could once again be used for shopping centers, office buildings, and so forth. Recycling of structures such as warehouses, garages, factories, and schools is not frequently considered. These facilities are therefore abandoned or razed. Recycling of these structures could possibly be done at a lesser expense than redevelopment, thereby aiding the rebuilding of the nuclear city, which is so necessary to the metropolitan area.

Because the Tenth Amendment to the Constitution grants the states the right to plan land use over nonfederal land, the states should devise uniform zoning laws and delegate their authority to proper local governments. They should also provide adequate supervision and evaluation. If this were done, instead of having an enormous number of local decisions made on zoning that affects land use, decisions might be made on a more rational basis and the decision makers would be responsible to some higher authority. Zoning would cease to be a political game and serve its proper function of providing adequate land use for the greatest number of people. Considerable legislative and judicial actions concerning the right of private property vs. the public interest have been taking place. It is essential that this question be viewed in detail and that firm long-range decisions be made to provide the greatest good for the greatest number of people and still protect private interests.

It is also wise to consider whether property tax revenue is an adequate base from which to obtain taxes to run the government. Should other major sources be utilized? If property does continue to be the principal source of taxation, then decisions must be made to provide proper land use and adequate taxing for a given area. Assessment practices vary from community to community and industries often receive tax breaks to attract them to a given area to provide jobs. All these factors must be considered in the evaluation of the proper use of land and the role of government in this use.

The federal government would do well to integrate all the various programs from housing through road construction to environmental controls in such a way that the states would have a better understanding of how land-use patterns could be properly determined. The federal government should coordinate land-use patterns with the states by the establishment of adequate interstate highway construction, pollution abatement facility construction, home mortgage guarantees, proper federal income tax deductions, and planning of areas surrounding federal lands.

The state governments should not only carefully plan and develop land-use and zoning laws but also establish standards for the use of floodplains, coastal areas, wetlands, agricultural lands, highways, and airports. The states should further set up a governing body to coordinate the innumerable local government agencies involved in land-use planning. It is essential that land use becomes not a single problem of one governmental unit but a coordinated approach by all agencies at all levels of government.

Housing Codes and Housing Laws

The state may regulate the use and enjoyment of individual property for the good of its citizens. This does not usually conflict with the constitutional protections given to the owner or occupant of a property. The power that the state has in establishing housing codes is based on the police power of the states, which has never been precisely defined. However, the general authority comes from the Constitution, which declares that the state has the power to protect the order, safety, health, morals, and general welfare of the society in which its citizens live. The courts have stated that municipalities may impose restrictions on individual property owners if these restrictions protect public health, safety, or welfare. In some court cases, it has been clearly held to be constitutional that housing codes may require windows and ventilation, screens, hot water, and other necessities of life. Local housing codes have been generally upheld by the courts.

For someone to conduct a housing survey, it has been generally agreed by the Supreme Court that an individual search warrant is not needed in a systematic area inspection unless the individual refuses entry to the environmentalist. It is not necessary for the health department to establish due cause, providing that a reasonable need exists to conduct periodic area-wide inspections based on passage of time, nature of the buildings, or condition of the entire area. However, under the Fourth Amendment to the Constitution, warrantless nonemergency inspections of residential and commercial premises cannot be carried out without the occupant's consent, unless a complaint has been lodged because of conditions supposedly existing within a given dwelling. In all cases where entry is refused to the environmental health practitioner, the county or city attorney should be consulted on the proper procedure used to gain entry to conduct the housing inspection.

It is difficult to determine a typical housing code. Although the American Public Health Association and the U.S. Public Health Service have issued, from time to time, a book titled *APHA-CDC Recommended Housing Maintenance and Occupancy Ordinance* (the 1986 edition, as updated in 1996, is called *Housing and Health, Recommended Minimum Housing Standards*), the local government can and does determine what it believes is a proper housing code. Therefore, codes differ from one area to another. As the housing codes are amended and changed by the local authorities, the tendency is toward decreasing stringent requirements to meet the needs of the governmental units. Where codes have been made stricter, it is because of federal requirements for obtaining urban renewal or other housing funds. It should be emphasized that housing requirements are minimum standards and that they affect the owners or landlords more than the tenants, although the tenants are minimally involved.

The entire establishment of the housing code and its subsequent use and enforcement cannot be based on any given political area, because it is necessary to look at the geographic instead of political boundaries. However, in reality, political boundaries still control the types of housing codes that are in effect. Although houses outside the city boundary are on a street adjacent to the city boundary, the housing code may be completely different. Most housing codes are municipally centered and therefore provide for the control of properties within the municipality and not outside it. Unfortunately, the contiguous areas become blighted because of lack of proper control.

Some states have attempted to impose statewide housing codes, including California, Maryland, Arkansas, and Georgia. Generally, these housing codes grant counties the authority to establish codes for their own area. A small number of counties have already done this. Unfortunately, the political nature of most counties and their varying subdivisions are such that conflicts arise and housing code enforcement becomes very difficult. An exception to this is Marion County, IN, with Indianapolis as its center city. The Marion County Health Department exercises the authority to regulate housing throughout the entire county. As a result, Marion County exercises greater control over housing code violations than many other areas of the country. Honolulu, HI, Denver, CO, and Dade County, FL, have city–county housing programs. Urban renewal, through the use of federal funds, has helped establish housing codes in a variety of jurisdictions. Thus, many urban areas have had an opportunity to raze large blighted areas and to reuse the land for other purposes; examples include Kansas City, Philadelphia, Baltimore, and St. Louis.

Effective court procedures may be needed to ensure compliance with housing codes. The procedures should be brief and court hearings should be held immediately following code violations. Conclusive evidence should be presented to enhance the procedure. Difficulties occur from long delays in hearing cases and because certain courts or judges are not particularly concerned with housing code violations.

A specific housing court would be able to exercise careful and quick judgment when needed. The court objective would be to correct the violation, not to punish the individual. A special court could also provide specific time blocks that could be utilized for better total enforcement. The court could use the punitive measures as a last resort and court appearances as a final attempt at an educational approach to correct the situation.

Personnel enforcing housing codes should be well-trained environmental health practitioners, who are assisted by environmental technicians. The practitioners should be responsible for the overall area in housing surveys. The technicians should be responsible for the house-to-house surveys. It is necessary for the technicians as well as the practitioners to understand the role of the housing code inspection in the overall problem of the environment. They need to understand how to work effectively with people to avoid the kinds of problems that lead to securing a warrant to enter a property. Good community and public relations on a one-to-one and one-to-community basis are absolutely essential to obtain good results. The function of environmental health technicians is to improve the housing environment, not to be a punitive police officer or to gather evidence to cause problems for a family. These technicians are there to evaluate the environmental health problems and to instruct the individual and family on corrections needed to improve their housing environment. Good, sound personal communications coupled with scientific educational approaches presented in a simple manner derive far better results than any series of court procedures. Once a court hearing occurs, partial defeat is suffered. Time that should have been used for helping people has been lost in courtrooms, and individuals forced to act against their will are not prone to maintain their property properly unless repeatedly penalized.

Community Programs in Housing

Housing programs work best when the community is involved and when agencies other than the health or housing department are also involved. The other agencies include planning, zoning, welfare, low-rent public housing, building inspection, and urban renewal.

Community groups have grown in numbers and importance in the last 50 years. They were formed in the cities for the purpose of neighborhood or community maintenance. Many of these groups are either ethnically or racially homogeneous. Individuals are often property owners and the groups are called *improvement associations*. With the change in neighborhoods, particularly in the inner city areas, community organizations have assumed different roles. Some of them are composed of individuals with a civic consciousness who want to help others. These individuals may or may not live within the specific neighborhood and may include business associations, clergy, educators, or other professionals. They may also include individuals who want to improve a neighborhood. Other neighborhood organizations form for specific interests, either because they are not getting something they want in the way of services or because they are seeking some means of providing a better place to live or a better social atmosphere.

Unfortunately, neighborhood associations are often difficult to maintain. They are funded by contributions and often where such organizations are most needed, residents are unable to contribute. There also appears to be a general situation of public apathy. If 5% of any given block or area actively participates in a neighborhood association, a considerable amount can be accomplished. The active group can carry out the necessary implementation and planning; and with good public relations techniques, other residents can become involved in special projects for the neighborhood. It is difficult to establish a metropolitan or regional organization that is of value to a given neighborhood, because these organizations are made up of individuals who are the presidents, chairpersons, or executive directors of large agencies. They have a solid understanding of their own agencies, but generally lack adequate understanding of given communities.

In attempting to maintain or build a community group, it is necessary to have trained public health educators working with the leaders or potential leaders in the planning process and throughout the entire process of the projects undertaken within the community. It is important that these individuals maintain a low profile. They should be facilitators instead of leaders, consultants instead of decision makers. They should help and guide the organization when guidance is needed or requested. In any community organization, it is important before any type of program is started, such as a housing program or rodent control program, that meetings of the community group be called to identify problems. Although the concern of the official agencies may be rodents or housing, the immediate concerns of the community may be police relations, recreational facilities, street repair, or better lighting. If the citizens receive an opportunity to voice their concerns and realize results, then they are more apt to work within their group to achieve the kinds of goals of interest to official agencies. The health educator should be involved in fact finding for and with the citizen groups, dissemination of information to the public, discussion of information among small groups, decision making on the basis of recommendations, testing decisions, establishing objectives, and helping to establish the program and the necessary follow-up and evaluation.

The community must identify its natural leadership and utilize this leadership in accomplishing ends established by the community and the official agencies. The community can identify those individuals who are most influential or powerful. These include church and business leaders, and professional, financial, governmental, educational, or labor leaders. These individuals can be utilized by the community in top coordinating positions. Other types of individuals, who actually get the work done, are the individuals to whom the community members turn when specific needs or problems arise. If these individuals exist in a community, they can be readily identified by other community members. When several individuals identify one person, a natural community leader has been found. These individuals are probably most effective in helping get a program across, because they live, work, and play within the community, and are respected by its members. These individuals can help determine the best approach to be used in establishing and implementing a program. It is wise to recognize and utilize such persons whenever and wherever possible.

INDOOR AIR CONTAMINANT CONTROLS

General Controls

To prevent or control indoor air contaminant problems, the following strategies may be used: source control, ventilation, air cleaning, exposure control. Source control would prohibit or limit smoking indoors, relocate contaminant-producing equipment, select products or raw materials to be used with lesser contamination potential, modify occupant activities, seal or cover the source, modify the environment, and remove the source (such as a chemical spill). Ventilation can be used to dilute the contaminants with outdoor air, and isolate or control contaminants by modifying air pressure relationships within areas. Air cleaning can be accomplished by using particulate filtration, electrostatic precipitation, negative ion generators, and gas sorption. Exposure control occurs when contaminant-producing activities are scheduled for off-hours and susceptible people are relocated.

Specific symptom patterns and suggested means of dealing with these problems can be found in Table 8.2. The three general types of air cleaners used to remove particles from the air are:

1. Mechanical filters installed in ducts to trap dust
2. Electronic air cleaners that use an electrical field to trap charged particles
3. Ion generators that use static charges to remove particles from the air

Specific Controls

Residential Wood Burning

The impurities or pollutants of residential wood burning can be reduced by modifying the wood burning stoves or fireplaces for greater efficiency; by providing adequate air for proper burning; by using dry wood; by using smaller pieces of wood; by avoiding burning of any rubbish, garbage, plastic, or colored newspapers; and by cleaning the furnace, chimney, and fireplace regularly.

Passive Smoking

Passive smoking can be prevented or controlled by adequate sources of natural or forced ventilation. Further control occurs if there is a physical separation of nonsmokers from smokers. Many public laws today require that smoking take place in only certain very limited designated areas.

Table 8.2   Symptom Patterns for Indoor Air Pollution Problems

Symptom Patterns	Suggestions
Thermal discomfort	Check HVAC condition and operation Measure indoor and outdoor temperature and humidity; see if extreme conditions exceed design capacity of HVAC equipment Check for drafts and stagnant areas Check for excessive radiant heat gain or loss

Common symptom groups
  Headache, lethargy, nausea, drowsiness, dizziness
	If onset was acute (sudden and/or severe), arrange for medical evaluation, as the problem may be carbon monoxide poisoning
	  Check combustion sources for uncontrolled emissions or spillage; check outdoor air intakes for nearby sources of combustion fumes
	  Consider evacuation/medical evaluation if problem is not corrected quickly
	  Consider other pollutant sources
	  Check overall ventilation; see if areas of poor ventilation coincide with complaints

Congestion; swelling, itching or irritation of eyes, nose, or throat; dry throat; may be accompanied by nonspecific symptoms (e.g., headache, fatigue, nausea)	May be allergic, if only small number affected; more likely to be irritational response if large numbers are affected
	  Urge medical attention for allergies
	  Check for dust or gross microbial contamination due to sanitation problems, water damage, or contaminated ventilation system
	  Check outdoor allergen levels (e.g., pollen counts)
	  Check closely for sources of irritating chemical such as formaldehyde or those found in some solvents

Cough; shortness of breath; fever, chills and/or fatigue after return to the building	May be hypersensitivity pneumonitis or humidifier fever; a medical evaluation can help identify possible causes Check for gross microbial contamination due to sanitation problems, water damage, or contaminated HVAC system
Diagnosed infection	May be Legionnaire's disease or histoplasmosis, related to bacteria or fungi found in the environment Contact your local or state health department for guidance
Suspected cluster of rare or serious health problems such as cancer, miscarriages	Contact your local or state health department for guidance
Other stressors Discomfort or health complaints that cannot be readily ascribed to air contaminants or thermal conditions	Check for problems with environmental, ergonomic, and job-related psychosocial stressors

Source: From *Building Air Quality — A Guide for Building Owners and Facility Managers,* U.S. Environmental Protection Agency, Indoor Air Division and U.S. Department of Health and Human Services, National Institute for Occupational Safety and Health, Washington, D.C., 1991, p. 56.

Carbon Monoxide

Carbon monoxide poisoning can be prevented in three specific ways:

1. Through proper ventilation, which provides adequate air for burning, if burning takes place, or dilutes the carbon monoxide present in the air
2. By use of proper equipment and methods to exhaust flue gases from the environment
3. By physical separation of all heating units and combustion appliances from people (oil and gas heating units placed outside the house or vented to the outside)

Radon Gas

Radon gas can be controlled through natural and forced ventilation. This is done by ventilating the lowest level of the house. Either fans are used to force outdoor air into the home or ventilation grills are put into areas to allow outside air to flow in by itself. A major disadvantage of this is the increased cost of heating and cooling. Another technique used to control radon gas is to seal the cracks and openings through which the gas passes from the soil into the house. This would mean to seal all openings around utility pipes; the tops of concrete block walls, chimneys, joints; and any other cracks or openings. It would be impossible to find every crack or dent. Another technique is to heat recovery ventilation. In this case, replace the radon-laden air with outdoor air. The indoor air is warmed or cooled by the radon-laden air exhaust. This procedure is similar to convection where heat energy flows from one gas to another gas. There is still a loss of indoor heating and cooling energy.

Another technique is through drain-tile suction. A drain tile is installed completely around the house and a fan is used to draw the radon from the drain tile. A subslab suction system may be used to reduce the radon from the slab. Techniques are also available for use in block-wall ventilation and the prevention of house depressurization. In house depressurization, it would be necessary to consider discontinuing the use of wood stoves and fireplaces, because they lower the air pressure in a house by consuming air and exhausting it to the outside. One other technique would be to have more pressure in the house by actually pressurizing the house and using fans blowing upstairs air into the basement.

Formaldehyde

There are seven different approaches that may be used in trying to alleviate the formaldehyde problem. They are as follows:

1. Interim measures are used, such as alternative housing or temporary housing, until the formaldehyde problem can be resolved. This is especially important for families with infants or young children, who are very sensitive to the irritating symptoms caused by formaldehyde. If the formaldehyde levels reach above 0.01 ppm, this measure should be considered.
2. The control of the indoor temperature and relative humidity can help significantly reduce formaldehyde levels. The recommended temperature levels for maximum formaldehyde reduction include lowering of the household temperature to 68°F during the heating season in northern states and the use of an air conditioner to maintain a constant indoor temperature of 70 to 72°F in the southern and southwestern states. The air conditioner also helps to dehumidify the air.
3. Source removal is the most effective control measure for formaldehyde. This would mean the removal of particleboard, subflooring, and urea formaldehyde foam insulation. This is very difficult and very costly. Other sources of formaldehyde emission include paneling, furniture, and cabinets. The removal of these sources is less costly and less difficult and can help reduce the amount of formaldehyde in the air.
4. Source treatment is an alternative to source removal but is generally less effective. This method consists of applying a coating material to the sources that emit

formaldehyde. The coating material acts as a barrier; prevents formaldehyde release; and also prevents moisture from entering the material, which helps promote the release of formaldehyde. Source treatment can be applied to unfinished wood products, particleboard subflooring, plywood paneling, particleboard shelving, cabinets, and countertop undersurfaces. The two primary coating materials used are polyurethane and a nitrocellulose-based varnish.

5. Ammonia fumigation, if done properly, can cause a reduction of 50 to 60% of the formaldehyde. With this approach the whole interior of the house is enclosed and exposed to ammonia gas over a 12-hour period at an indoor temperature of 80°F. It is essential that trained individuals do this, because ammonia is very toxic. At present this is the only method that is practical to reduce formaldehyde in environments that contain high concentrations. All surfaces must be vacuumed very carefully, because the ammonia combines with the formaldehyde to form hexamethylene tetramine, which is a fine dust.

6. Pure air purification systems remove formaldehyde from the air by absorption or adsorption. The effectiveness of the system is based on the chemical, the system design, the capacity to move air, and the levels of formaldehyde present. For adsorption, activated charcoal is relatively inefficient in reducing formaldehyde levels.

7. Forced ventilation systems push fresh air into the contaminated building and therefore reduce or dilute the level of formaldehyde present. The amount of reduction is based on the amount of airflow.

Other Volatile Organic Compounds

Other volatile organic compounds used in a house, such as solvents, paints, polishes, and household cleaners, should be only utilized in areas where ventilation is plentiful. The amount of material used should be strictly limited to that which is recommended on the container. VOCs are hazardous. They should be treated as such. Dispose of unused or little-used containers safely.

Asbestos and Lead

Techniques necessary for control of asbestos include:

1. Suppression
2. Removal

Suppression is the best way to avoid asbestos and lead dusts, because this controls resuspension of dusts into the air. Remediation which includes removal, a complicated task that must be carried out by individuals who are highly trained. The EPA, OSHA, and Department of Housing and Urban Development (HUD) have specific guidelines for the removal of asbestos and lead from buildings. In general, the asbestos- and lead-containing materials are prepared for removal. They are wetted with a water and surfactant mixture sprayed in a fine mist, allowing time between sprayings for complete penetration of the material. Alternatively, or in conjunction, high-efficiency particulate air (HEPA) filter vacuum cleaners are used to clean surfaces contaminated with friable materials. Once the material is totally wetted, it is then removed from the building and placed in thick plastic bags and sealed in such a way as to make the bags leak tight. They are then double bagged and placed in plastic-lined cardboard containers or plastic-lined metal containers and removed to the disposal site.

COMMUNITY PROGRAM AND INJURY CONTROL

Injury prevention is based on three general approaches, including:

1. Persuade the individuals at risk or injury to alter their behavior. An example of this would be the installation of smoke detectors.
2. Require an individual behavioral change by law or administrative rule. An example of this would be a law that smoke detectors must be installed in houses or apartments.
3. Provide automatic protection by changing the product or by correcting the design. An example of this would be the installation of built-in sprinkler systems.

The Philadelphia Injury Prevention Program is called the *Safe Block Intervention Program*. It is directed at reducing the occurrence, severity, and consequences of injuries in an economically disadvantaged community by creating safe blocks. Its goals are to conduct home modification programs, inspect all homes in the target area, and educate the residents about injury prevention methods. The safe block concept includes the use of a public health team that inspects residences for problems and then has the environmental modifiers to help implement the prevention program. After the initial visit, a second visit is made about 6 months later to determine whether the modifications remain intact, whether recommendations have been followed, and what the results are in injury control at this point. Obviously, data have to be gathered initially or prior to the intervention program on the level of injuries, types of injuries, etc.

An intervention program consists of modifications of the home in the following ways:

1. Place stickers for regional poison control centers and 911 on telephone.
2. Provide fire extinguishers for the kitchen.
3. Provide smoke detectors for homes without the detectors. It may also be necessary to inspect existing smoke detectors and replace batteries.
4. Provide bathtub nonslip strips.
5. Provide ipecac and instructions for safe use if a poisoning occurs.
6. Reduce the temperature of the hot water heater.
7. Provide light bulbs of higher intensity if lighting on the stairs is inadequate.
8. Rat proof the homes.
9. Install cabinet locks if young children are present.
10. Store poisons, medicines, and cleaning supplies out of the reach of children.
11. Install staircase gates at the top of the stairs and the bottom if children are present.
12. Inspect staircase hazards.
13. Inspect wiring or other hazards within the home.
14. Check space heaters and whether they are operational.

15. Inspect storage areas for newspapers, gasoline, and gasoline storage.
16. Provide education for the individuals in the home on how to avoid injuries from a variety of environmental sources.

This intervention program is considered to be operating at a good level.

Role of the Federal Government in Housing

Congress stated in the Housing Act of 1949 that the general welfare and security of the nation and the health and living standards of its people require housing production and related community development to correct the serious housing shortage, to eliminate substandard and other inadequate housing through the clearing of slums, and to provide decent homes and suitable living environments for all American families. To obtain this national housing objective, Congress provided for private enterprise to carry out the major portion of the program; government assistance to private enterprise to carry out the major portion of the program; government assistance to private enterprise to be given where needed; establishment of positive local programs; government assistance to eliminate substandard and other inadequate housing through clearance of slums and other related areas; and government assistance for decent, safe, and sanitary farm dwellings.

Federal government departments were given the responsibility of facilitating the steps leading to this national housing objective. They were to do so by encouraging and assisting in the production of housing of sound standards of design, construction, livability, and size; reducing the cost of housing without sacrificing standards; using new designs, materials, techniques, and methods of residential construction; developing well-planned integrated residential neighborhoods; and stabilizing the housing industry.

The 1965 Department of Housing and Urban Development Act stated that, as a matter of the general welfare and security of the nation and the health and living standards of the people, the national purpose was to have a sound development of the nation's communities and metropolitan areas. To carry out this goal and in recognition of the importance of housing and urban development in the national life, Congress established an executive department to handle the problems of housing and urban development.

The 1968 Housing and Urban Development Act reaffirmed the national goal set forth in 1949 of a decent home and suitable living environment for every American family. Congress recognized that the goal had not been fully realized and that many suffering low-income families still existed. It declared that special provisions should be made to assist families with incomes so low that they could not decently house themselves. In August of 1968, under Title Sixteen, Housing Goals and Annual Housing Report, Congress declared that the supply of housing was not moving ahead rapidly enough to meet the national housing goal of 1949. It stated that there was a need for construction or rehabilitation of 26 million housing units, 6 million of which were for low- and moderate-income families. It stipulated that the President had to set forth a plan to be carried out for a period of 10 years from June 30, 1968 to June 30, 1978 to eliminate all substandard housing by the latter date. The Department of Housing and Urban Development Act, Public Law 89-174, granted the

Department of Housing and Urban Development all the functions, powers, and duties of the Housing and Home Finance Agency, FHA, and Public Housing Administration. In addition, the National Mortgage Association and other departments relating to housing were placed under the control of HUD. This department was to have vast powers to improve housing in the United States.

The Housing and Community Development Act of 1974 stipulated that an allocation of housing funds be made to provide for local housing assistance plans. HUD, which received an extremely large budget from Congress, was responsible for the following areas: (1) Council for Urban Affairs; (2) environmental and consumer protection; (3) rent supplements; (4) disaster assistance; (5) establishment of advisory committees; (6) civil defense, as it related to vulnerability to attacks; (7) handicapped individuals and accessibility to buildings; (8) assignment of emergency preparedness functions; (9) Federal Council for Science and Technology; (10) coordination of federal urban programs; (11) National Institute of Building Sciences; (12) housing renovation and modernization under Title 1 of the revised July 31, 1975 basic laws and authorities on housing and community development; (13) mortgage insurance, Title 2; (14) war housing insurance, Title 6; (15) insurance for investment in rental housing Title 7; (16) armed services housing, Title 8; (17) national defense housing, Title 9; (18) mortgage insurance for land development, Title 10; (19) mortgage insurance for group practice facilities, Title 11; (20) FHA and VA interest rates; (21) Rehabilitation Act of 1973; (22) closing of military bases and mortgage defaults; (23) special assistance programs such as emergency homeowner relief, housing for the elderly, college housing rehabilitation loans, rent supplements, urban homesteading, training and technical assistance; (24) community development assistance programs such as block grants, urban renewal, public works planning, public facilities loans, public facilities grants, model cities, historic preservation, Lead-Based Paint Poisoning Prevention Act; and (25) rural and other nonhousing and urban development community development programs, such as the Rural Development Act, Headstart, regional action planning commissions, and Appalachian regional development.

In 1974, the Emergency Home Purchase Assistance Act was passed. The Emergency Housing Act of 1975 included Emergency Home Owner's Mortgage Relief Title I. In 1976, the Housing Authorization Act was passed by Congress.

In 1977, the Housing and Community Development Act was passed by Congress. It included special titles on Community Development; Housing Assistance and Related Programs; FHA; Mortgage Insurance and Related Programs; Lending Powers of Federal Savings and Loan Associations and Secondary Market Authorities; World Housing; National Urban Policy; Community Reinvestment; and miscellaneous provisions.

In 1978 to 1981, 1983, 1984, and 1987, additional housing laws were passed. However, in the 1980s, in reality, many housing programs were reduced in scope.

Programs of Housing and Urban Development — 1988 to 1989

HUD is responsible for a huge multitude of housing programs to help the citizens of the United States. Some 94 specialized programs are handled by HUD. They range from housing for the poor and the homeless, to the elderly, to nursing homes and hospitals, to public housing developments, to Indian housing. Extensive and extremely complex policy, development, research, and mortgage guarantee programs exist.

As can be seen, the federal government has made an enormous commitment to better housing for the American public. The ultimate goal has not yet been achieved. Considerable additional effort and money are needed. Beyond this, considerable research is required in the area of attitude change and human behavioral patterns toward the housing environment. The best housing may be destroyed readily unless the effect of environment on human behavior is thoroughly understood. It is strongly urged that environmental health practitioners, other public health officials, citizens bodies, and legislators work toward obtaining proper funds for behavioral analysis and training. It will then be necessary to institute the training to maintain the new environment in a proper manner for the individuals who will be living within it.

Housing and Urban Development Programs — 1990s into the 2000s

In 1990, the Cranston–Gonzalez National Affordable Housing Act emphasized homeownership and tenant-based assistance. Home housing block grants were issued. In 1992, the Federal Housing Enterprise Financial Safety and Soundness Act created the HUD Office of Federal Housing Enterprise Oversight to provide public oversight of FNMA and the Federal Home Loan Mortgage Corporation. The Empowerment Zone and Enterprise Community Program became law as part of the Omnibus Budget Reconciliation Act of 1993. In 1995, the blueprint for reinvention of HUD proposed sweeping changes in public housing reform and FHA consolidation of other programs into three block grants. By 1996, homeownership totaled 66.3 million households, the largest number ever.

A new philosophy by the federal government is changing the way in which housing programs and associated concerns are being handled. For years there was a proliferation of federal programs, complex regulations, and cumbersome bureaucratic procedures, which has limited the ability of communities to solve their own problems. Federal efforts to renovate housing, control crime and drugs, and increase employment have treated residents as passive clients and failed to involve them adequately in their problem-solving efforts. The failure to enlist local institutions and citizens, especially the women who play unique leadership roles in these communities, as partners in the decision-making process has left essential resources untapped, local priorities ignored, and has caused missed opportunities to strengthen the community home problem-solving capacities. This has changed.

The federal government is now promoting comprehensive planning in collaboration with many people at the local level through the Empowerment Zone and Enterprise Community Program. Distressed communities are receiving billions of dollars in tax incentives and additional funds in flexible grants to help bring capital back to the central city, create jobs within distressed neighborhoods, invest in education and training, and link residents to economic opportunity throughout their metropolitan region.

The federal government is also strengthening impact of community development corporations at the local level, while reducing onerous rules and regulations. Central cities are rich reservoirs of human and economic potential, and because many of the poor and much of the deteriorated housing surround the central city, the federal government is trying to use community-wide efforts to upgrade these areas. The Intermodal Surface Transportation Efficiency Act of 1991 authorizes $155 billion to develop an efficient, environmentally sound, national transportation system that provides a strong foundation for the country to compete in the global economy. Governments at all levels can pursue common policies to dismantle the barriers that separate the poor minority people from the rest of society through:

1. Developing fair-share affordable housing
2. Improving the portability of assistance from governmental agencies
3. Increasing intracommunity and intraregional mobility
4. Developing area-wide revenue sharing arrangements

Safe, healthy, and sustainable communities are created to improve life in the urban setting. The President's Council on Sustainable Development works within the federal government and with public and private partners to explore new means of achieving good community goals. As part of this, the Brownfields National Partnership brings together the resources of 15 federal agencies to clean up and redevelop contaminated land called brownfields. In the year 2000, approximately 100 communities were under evaluation for redevelopment planning and funding. A $1.6 billion incentive is used hopefully to leverage $6 billion in private sector cleanup over 11,000 brownfields sites in the next 3 years.

The National Homeownership Strategy called Partners in the American Dream is dedicated to help racial and ethnic minorities become homeowners. This would immediately improve housing and neighborhood communities. Individuals are provided with both the know-how and the necessary funding to purchase homes. This instills pride in oneself and therefore creates better homes and communities.

Emergency Housing

Emergency housing consists of the use of trailers provided by the HUD, gymnasiums, churches, or any other available buildings for use when normal housing has been destroyed, made unsafe, or inundated by floods. Emergency housing should be closely supervised by environmental health and other public health authorities to prevent the outbreak of contagious or infectious diseases. Adequate water, food, clothing, and blankets must be provided. Generally, the American Red Cross is most helpful in this type of situation.

ROLE OF THE FEDERAL GOVERNMENT IN INDOOR AIR QUALITY

The EPA has been very concerned with the problems of indoor air quality. The agency is currently operating under a variety of environmental laws, including the Toxic Substances Control Act; Federal Insecticide, Fungicide, and Rodenticide Act (FIFRA); the Safe Drinking Water Act (SDWA); the Resource Conservation and Recovery Act; Asbestos in School, Hazard Abatement Act of 1986; and Uranium Mill Tailings

Radiation Control Act. The EPA was given specific directions in establishing an indoor air quality program in the 1986 Superfund Amendments and Reauthorization Act.

Since 1982, the EPA has conducted research programs on indoor air quality. The research has been directed at increasing the understanding of personal exposure, emissions, health effects, and mitigation techniques. The Radon Gas and Indoor Air Quality Act of 1986 was directed at having EPA conduct research, implement a public information and technical assistance program, and coordinate federal activities on indoor air quality.

The EPA has taken regulatory action on asbestos, VOCs in drinking water, and certain pesticides. The agency requires schools to inspect for asbestos, to prepare management plans, and to take action when the schools find friable (easily crumbled) asbestos. The EPA has issued maximum contamination levels for the VOCs in water supplies that serve more than 25 people. The agency has also acted to control indoor exposure to pesticides, as well as requiring childproof packaging for certain pesticides. The EPA also prohibits indoor applications of the wood preservatives pentachlorophenol and creosote.

In the year 2000, indoor air pollution programs continue. The EPA is deeply involved in research and in teaching people how to improve the indoor environment and to reduce indoor air pollution.

Environmental Lead Hazard Abatement

Laws

The 1971 Lead-Based Paint Poisoning Prevention Act coordinated the efforts of health and housing people in dealing with the problem of lead-based paints in houses. It provided for the environmental management, control and prevention of lead-based paint illness. It also provided for screening and medical management of children who had ingested lead. In 1988, the Lead Contamination Control Act required the following:

1. Identification of water coolers that are not lead free
2. Repair or removal of water coolers with lead-lined tanks
3. Ban on the manufacture and sale of water coolers that are not lead free
4. Identification and resolution of lead problems in school drinking water
5. Authorization of additional funds for lead-screening programs for children

The reason for the enactment of this law is because of the very dangerous nature of children ingesting lead from water. The harmful health effects include serious damage of the brain and central nervous system, kidney, and liver. Obviously inhalation of lead dust is also a serious problem.

Abatement Procedures

Abatement procedures consist of emergency repair and permanent repairs. Emergency repairs consist of the following:

1. Scraping of all peeling, chipping, or flaking lead-based lead surfaces
2. Sanding with a rough grade of sandpaper on all lead-based paint surfaces
3. Applying inexpensive contact paper to lead-based paint surfaced with wallboard, plywood, etc.

Permanent repair methods include removing all lead in paint, scraping and sanding, using liquid paint removers, using heat to remove leaded paint, and covering all areas where lead may be on the walls. Lead in dust may be removed by the use of proper dustless cleaning techniques. The hazards associated with paint removal include open-flame torch or heating devices that may volatilize lead; machine sanding that produces lead dust, as does sandblasting; chemical strippers containing methylene chloride that are extremely toxic.

CDC Childhood Lead Poisoning Prevention Program

The Lead Contamination Control Act of 1988 authorized CDC to initiate programs to eliminate childhood lead poisoning in the United States. CDC created the Lead Poisoning Prevention Branch in 1990. Its responsibility was:

1. Developing programs and policies to prevent childhood lead poisoning
2. Educating the public and healthcare providers about childhood lead poisoning
3. Providing funding to state and local health departments to determine levels of childhood lead poisoning, to conduct medical and environmental follow-ups, and to develop neighborhood based programs
4. Supporting research at local, state, and federal levels

The CDC in its lead poisoning prevention effort has helped:

1. To initiate, develop, and improve lead poisoning prevention programs in 39 states and in more than 150 counties and cities
2. To identify more than 100,000 children with extremely high blood levels of 20 µg/dl or higher and expand the efforts to provide health education and medical follow-up
3. To improve the number and quality of inspections of lead hazards in houses leading to approximately 20,000 home inspections per year
4. To determine high-risk areas for lead poisoning, and subsequent programs to focus on prevention, screening, and hazard-remediation efforts
5. To develop comprehensive data management systems at state and local levels
6. To expand public laboratories to analyze blood and environmental samples

Local Childhood Lead-Prevention Program

The city of Philadelphia has developed a childhood lead prevention program that includes the screening of a large number of children in high-risk populations; a referral system that ensures a comprehensive diagnostic evaluation of every child with a positive screening test; and a program that ensures identification and elimination of the child's lead exposure source and a system that monitors the adequacy of the treatment of the child. Children from 6 months to 5 years of age are screened for undue lead absorption and or lead toxicity. These children come from high-risk areas. Medical management includes evaluating the medical aspects of the program; evaluation of the timeliness and appropriateness of referrals; evaluation of case records; consultation with physicians; and conduction of educational programs. All individuals in high-risk areas are followed up after the initial determination has been made concerning lead poisoning. Indeed, health departments throughout the country have implemented similar programs.

Environmental management is carried out through:

1. Investigations to determine whether lead-based paint or other sources of lead are present
2. Abatement by the removal of lead hazards, especially paint (difficult because of the total number of houses that have to be taken care of and the cost of doing this, with cost estimates for removing lead-based paint for an average three-bedroom house ranging from $8000 to $15,000 per house)

An education program is carried on by health professionals to teach parents the problems of lead, to make the general community aware of lead problems, and to provide an outreach program to assist individuals, hospitals, and physicians, as well as day care facilities, in understanding the risks of exposure to lead. The program that has been identified is operating in a successful manner. Other communities have also developed childhood lead programs.

SUMMARY

The indoor environment is a complex human environment. It encompasses facets of all other environmental problems. The indoor environment may include anything from an eight-family, three-story dwelling to a huge home with a swimming pool. The environment has an impact on the shelter, and the shelter has an impact on the environment. In addition, to fully understand the indoor environment, one must appreciate the problems of poverty, joblessness, welfare, socioeconomic status, racial and ethnic origins, and the belief in America that individuals' homes are their castles. A house in good condition provides a satisfactory environment for psychological, physical, and mental well-being. The house and neighborhood in poor order may furnish the impetus for crime, poor education, poor health, and fires; and creates a society that is helpless in its dealings with the rest of the world. Further, indoor air pollution has caused additional stress and potential for actual disease. Injuries are an overwhelming concern.

Many new theories abound on the relationship between housing and health-based or nonspecific responses, in which a given cause may create a variety of health patterns and a given syndrome of physical or emotional maladjustment may stem from a variety of causes. It is necessary to consider the causes and to carry out an analysis incorporating all of them. New housing developments, including new technology, materials, construction, and assorted consumer products, have added complexities to the housing environment. In some areas this has provided for the greater good of the individual. In other situations, this has created additional problems, because the new product, for example, when broken must be serviced; and the service cost in many cases exceeds the dollars that an individual has to take care of the problem.

Once again, it must be emphasized strongly that despite everything that the federal, state, and local governments, and various community groups may do to improve the indoor environment, the basic changes that occur must be brought about by the resident. It is recommended that the resident obtain a copy of "The Effect of the Man-Made Environment on Health and Behavior," by L.E. Hinkle, Jr., and W.C. Loring, CDC, Public Health Service, U.S. Department of Health, Education, and Welfare (USDHEW), Atlanta, GA, publication number (CDC) 77-8318, 1976. This publication may be obtained from the Government Printing Office, Washington, D.C., stock number 017-03-00110-8. In the future, conferences must be held between the specialists that examine the people-made environment and members of this environment for a fruitful exchange of ideas so that we may move forward to the goal proclaimed by Congress in the Housing Act of 1949 — the ultimate goal of a decent home and suitable living environment for every American family.

RESEARCH NEEDS

It is necessary to thoroughly evaluate the existing types of planning used in community development and redevelopment. From this evaluation, measures should be formulated to determine the best approaches to housing development and redevelopment. Research techniques are needed to determine the manner in which improved housing can reduce social problems. Behavioral studies of individuals living in poor housing should be conducted to determine the underlying sociobehavioral disturbances that occur in these environments and also to determine why certain individuals move into good housing and lower the physical structure to the level of their previous housing. Behavioral studies are needed to determine the best means to educate individuals to improve their housing environment and also to learn to live within an improved housing environment. Studies are needed to determine the critical variables in establishing new communities.

Indoor air has become a serious concern within the home as well as within institutions. The indoor air may contain a variety of contaminants created within the property or outdoor contaminants that have entered the property through cracks and openings, especially in older structures. Carbon monoxide, particles, sulfur dioxide, and oxidants enter buildings. A variety of odors created by external pollution sources or indoor pollution sources contribute to the air problem. The indoor contaminants consist of particulate matter from activities such as cooking; tobacco smoke; and vast assortments of cleaning materials, synthetic floor coatings, polishes, cosmetics, pesticides, paints, vapors, glues, and other contaminants that come from the workshop or other parts of the property.

It is necessary to conduct research and demonstration projects on how to evaluate the techniques that are used to reduce radon levels in new and existing homes. It is also necessary to develop new and more effective techniques for instantaneous detection and assessment if chemicals, radon gas, or microbes are present within the home. Additional research should be conducted on effective and inexpensive techniques for the removal of existing indoor air contaminants.

Surveys are needed to determine the patterns of polluting equipment and materials used in homes. Research priorities should then be developed to determine which areas need to be studied. Maximum exposure levels need to be determined and this information should be correlated with individuals' occupations to determine whether they are receiving additional stress from the home environment. Studies are needed of the best techniques to control indoor concentration of air contaminants by improving air-cleaning devices and developing inexpensive techniques for the removal of air pollutants at low air pressures.

Injury control in the indoor environment is essential. Further research is necessary on the causes of accidents and injuries in the indoor environment, their controls, and how best to limit the degree of the injury. Research emphasis must be placed on an economical approach to altering the environmental conditions leading to accidents and injuries, as well as the best techniques for product modification to enhance safety. New public health education tools need to be developed and evaluated to determine whether they are in fact reducing the levels of injuries in the indoor environment.

CHAPTER 9

Institutional Environment

BACKGROUND AND STATUS

The successful design and practice of an environmental health and safety program in an institutional setting is a comprehensive undertaking. It involves the coordination and integration of a broad range of people, both professional and nonprofessional, and their diverse and complimentary resources and skills. The institution is in effect a small community and therefore contains the numerous environmental, community health, social, emotional, and psychological problems attributed to or found in any small community. In addition, a considerable number of people congregate in a limited amount of space. The typical institution continues to grow in size and complexity. The budget of the institution and means of financing have become problems, because costs rise as new techniques and equipment are developed and as medical inflation continues to grow sharply.

People congregate in institutions. This congregation comprises a shifting population that may use the facilities for parts of a day, the entire day, or for an entire 24-hour period. As in any other situation where groups of people gather, the potential for the spread of disease is increased. Institutions contribute to air and water pollution, and solid waste problems. Further, each institution has its own unique set of occupational hazards and other types of safety hazards. Some concerns in institutions are limited to the basic environmental health and infection control problems, including water, air, solid waste, food, shelter, and personal spread of disease. Others include the many shops, or even the equivalent of comprehensive industries, that have all the occupational hazards found in these settings, including chemical, microbiological, and radiological hazards. A further complication is the fact that many institutions receive federal and state financing.

Institutions include primarily hospitals, nursing homes, convalescent homes, old age homes, schools, and prisons. (For a detailed presentation of the institutional setting, the reader is referred to *Environmental Health and Safety*, by this author, Pergamon Press, 1974, now out of print.)

Nursing and Convalescent Homes

Nursing homes exemplify in many ways a failure of public policy, public concern, and private enterprise. Although a good proportion of the patients are aged and have a variety of infirmities, including senility, many patients who still have active minds should be receiving the type of care to which they are entitled. It is much easier to medicate individuals to conform their behavior to the standards set by the nursing home than to allow them to maintain their dignity as human beings. Some nursing homes today are substandard. Overcrowding, unsanitary conditions, physical abuse and neglect, and potential firetraps exist. Each year a serious fire occurs in which unfortunate older people die or are injured. The basic causes behind this dilemma are the social, health, and economic trends of the past 90 years.

The United States became a rapidly growing industrial nation during the early part of the 20th century. People moved to the cities to find jobs and often to smaller houses with fewer rooms. They were unable to keep with them the grandparents or other family members who in the past had a secure place to live in their old age. Industrialization created more jobs that were filled by women and younger people. Today there are many two-earner families, with no one home to care for the elderly or children. Older persons were frequently left out of the job market. If they became disabled, their opportunity to get a job was almost totally lacking. This situation with the aged continues today and grows worse, because long-term inflation sharply reduces or totally eliminates the savings of a group of the aged, thereby creating additional problems.

Unfortunately, in the past, nursing homes did not evolve in an orderly manner but instead erupted, almost in the way an earthquake causes disruption within a community. To provide homes for the aged or sickly, many ill-equipped reconverted private houses were pressed into operation. Unfortunately, this contributed to the severity of the nursing home problem. Nursing homes were set up in run-down residential areas, in poor and inadequate facilities, and in difficult areas for family members to reach, unless they had access to private transportation. The nursing home movement received further impetus from the Social Security Act of 1935. This act provided for the elimination of public poorhouses and public assistance for residents of these institutions. The residents went to private institutions, because these were available.

Today the United States has an estimated 17,000 nursing homes, of which some 66% are proprietary or operating for profit. About 8% are government sponsored and about 25% are run by various nonprofit organizations, including religious groups. In some of these private nursing homes, individuals do not necessarily receive care commensurate with the amount of money that they pay. There are many fine proprietary nursing homes in the United States. However, others are interested primarily in making profit and not in providing adequate patient care. Congressional hearings in New York City demonstrated that, despite federal regulations for skilled nursing home facilities as required by Medicaid and Medicare, innumerable abuses were found. Obviously, because the average age of the people in institutions is approximately 80 years old, they seldom challenge their treatment and care.

Nursing homes, in effect, are medical facilities that should have the same type of personnel and programs as hospitals, but to a lesser degree. Obviously such things as operating rooms and emergency rooms do not exist. However, nursing homes should provide good patient care, medical and nursing supervision, physical therapy, occupational therapy, dental services, eye services, necessary drugs, and effective humane care. Nursing homes should be able to handle emergencies to save the lives of patients. The facilities should meet the requirements of skilled nursing home facilities under Title 19 of the Social Security Act. These requirements are updated from time to time by the U.S. Department of Health and Human Services (USDHHS). For current requirements, the Department of Health and Human Services should be contacted.

Old Age Homes

Old age homes encompass every type of shelter from a simple house where a few elderly people live to extensive beautiful facilities where the individuals have not only shelter and food but also adequate recreational facilities, medical care, and occupational and physical therapy. The old age home is an outgrowth of the same kinds of conditions that helped to create the nursing home.

Schools, Colleges, and Universities

Healthy school living requires considerable effort to provide good physical, emotional, and social well-being. The school or university administrator, along with the board of trustees and the various other administrative support and professional staffs, has as a goal the proper education of students at all levels. It is necessary to provide an environment that is wholesome and supportive of learning. This environment includes all the problems, programs, and services found in other institutional environments, but with some exceptions. In grade schools, the young age of the students increases the likelihood of accidents. In high schools, the students face potential accidents not only from various play areas but also from shops and laboratories. In colleges everything that applies to the high school continues to be a problem but is multiplied manyfold by the sheer numbers of students, by increasingly sophisticated and complex laboratory experiments and chemicals used, and by the fact that many of the students reside on the university grounds.

Prisons

Each correctional institution should provide a healthy place for a person to live. Because a healthy environment includes all the aspects of a good institutional environment and because custody means more than possession, the prisons of our country have a definite obligation to fulfill in taking care of prisoners. In the past there have been considerable problems in correctional institutions because of overcrowding, numerous existing health hazards, and poor environment.

The conditions in all types of institutions continue to deteriorate. Specific types of concerns are discussed under the section on problems. To avoid repetition in this chapter, most of the material discussed relates to the most complex type of institution, which is the hospital. In other institutions where problems, techniques, and so forth vary from the hospital setting, they too are discussed.

SCIENTIFIC, TECHNOLOGICAL, AND GENERAL INFORMATION

Hospitals

The hospital today is more than a place for patients who are terminally ill. The use of intensive care units, cobalt therapy, artificial kidneys, organ transplants, inhalation therapy, and so forth has made the hospital significantly different from its predecessor. It is a continually and rapidly changing environment in which diagnostic and treatment facilities are improving, staffs are better trained, and costs are soaring. The functions of the hospital are to give care to the sick and injured, provide necessary health education, conduct realistic research, and promote community health.

Hospitals, generally, are either governmental, proprietary (run for profit), or voluntary nonprofit. The voluntary nonprofit hospital accounts for about 54% of the nation's hospital beds. These hospitals are usually brought into operation by community groups, churches, or other organizations. Governmental hospitals, which have about 34% of the hospital beds, are operated under federal, state, or local budgets. Proprietary hospitals, which contain about 12% of the hospital beds, may either be individually owned or owned by a partnership or corporation. Hospitals may also be listed as general, special, short-term, or long-term.

Organization

The board of trustees or board of directors is the highest authority within the hospital. This board determines the direction that the institution may take and establishes policies that are consistent with its goals. It usually represents a cross section of businesspeople, educators, health leaders, and other distinguished individuals. They dedicate a significant part of the working time to the institution, usually without pay. Boards vary in size, although an adequate-sized group probably would be a maximum of seven to nine individuals. The board forms committees and so forth. The board evaluates and approves the hospital budget and also gives guidance to the chief executive officer. It is responsible for providing a bridge between the administration, the medical staff, and the public.

The chief executive officer, who may have a variety of titles from director to administrator, is the day-to-day chief administrator of the hospital. This administrator's job involves managing, coordinating, and directing the activities performed by the employees; and presenting the hospital problems to the board, implementing the board's decisions, and interpreting the board's actions to the hospital community. It is essential that the chief executive officer be a fine administrator, keep in close touch with all parts of the institution, and have a good working relationship with the many chiefs and supervisors of the various parts of the institution.

The medical staff directs patient care. It is responsible for giving professional care to the sick and injured, and for making careful and proper diagnosis. It is also responsible for maintaining its own efficiency, governing itself, and choosing new members. It participates in educational programs and gives advice to the chief executive officer and the board of trustees. The medical staff is divided into a series of committees that oversee the various operations of the hospital, including medical records, the surgical department, pediatrics, radiology, and so forth.

For the hospital to operate properly, it is essential that the three major groupings — the board of trustees, medical staff, and chief executive officer — cooperate and work together efficiently. The other professionals, such as the nursing staff, environmental practitioner, and safety officer, generally work through the chief administrative officer and therefore are extensions of this officer's administrative and supervisory control.

Electrical Facilities

Electrical energy is the major source of power on which almost every part of the hospital depends. Electrical supply must be adequate and always dependable. It is essential to have electrical supplies that switch on automatically when there is a loss of normal service. Electricity must be available on a 24-hour basis.

Heating, Air Conditioning, and Laminar Flow

A thermal environment should be maintained within the hospital to permit adequate heat loss but prevent excessive heat loss from the human body. At the same time, this environment must maintain an atmosphere that is reasonably free of chemical and bacterial impurities and odors, and one that is conducive to the best physiological and psychological comfort of the individual.

The function of the ventilation system, apart from providing adequate temperature and humidity control, is the control of the number of microorganisms in a given area. These microorganisms come from other areas into the target area, such as equipment and supplies, outside air, individuals within the immediate environment, infectious hazardous waste, other solid waste, blankets, or dirty linens. The function of mechanical ventilation in removing these organisms is carried out by literally blowing the organisms away from the areas that are to be kept free of contaminants. An example of this would be the wound or operating site on a patient. Several methods have been proposed to remove the air that may contain microorganisms from these critical areas. The first method directs a stream of air either horizontally or obliquely toward the operating room table to create turbulent airflow directly over the patient. The second method introduces air at the ceiling in such a way that only mild turbulence is created by the supply air mixing with the room air, and then it is exhausted. This creates a constant dilution of the air. The third method is the same as the second, except that it produces minimal turbulence when mixing with the room air and is then removed by downward displacement of the air to ports located on the walls near the floor. The most effective technique for removing microorganisms appears to be the third system. The turbulent system, although it removes a considerable amount of contaminated air, also brings more contaminated air across the wound site. Along with the air system, is the air filtration system that is used to remove particulate matter from the air.

Laminar flow of air provides the best total control of the air environment, which includes temperature, humidity, cleanliness of the air, and direction of flow. Laminar flow is simply the introduction of large volumes of clean air through a very large diffuser or perforated panel, which reduces the velocity of the incoming air, thereby preventing agitation and reintroduction of settled contaminants. It also decreases the contamination blown off personnel and equipment. At the opposite side of the room, the air is removed through a perforated area of the same size as the inlet diffuser. This is a single-pass technique in which the air comes from the wall or the ceiling; passes through ultrahigh efficiency particulate air filters; moves uniformly across the room and through the outlet to return ducts, which then go through the final filter and back through the room again. The amount of air flowing depends on the size and shape of the room and the height of the ceiling.

Heat and proper air control of temperature is important. Heat balance in the human body is affected by rate of metabolic heat produced in the body, rate of storage or change in body heat, rate of heat loss or heat gain by convection, rate of heat loss or heat gain by radiation, and rate of heat loss by evaporation of sweat. It is essential that the ventilation, rate of flow of air, and actual heating and cooling of the areas be done in such a way that the individual is comfortable. Comfort is difficult to define, because it is based not only on the previously mentioned items but also on the sex, age, and physical condition of the patient, and the sex, age, and type of activity of the worker. Generally speaking, within the hospital a temperature of 75°F would be considered a comfortable temperature for most patients. The room temperature, however, may have to be adjusted to the individual. Centers for Disease Control and Prevention (CDC) has publications available with recommendations for ventilation in hospitals.

Housekeeping

The function of the housekeeping service is to provide and maintain a clean, safe, and orderly environment that helps prevent disease and accidents, and promotes health. The individual, whether a hospital employee, visitor, or patient, is affected by aesthetic surroundings. It is important that all areas appear clean and in fact are clean to avoid hazards. Industry practices proper cleaning by using adequate procedures and techniques, proper task-oriented detergents, disinfectants, and proper equipment. It is important for other institutions to follow this example.

Maintenance

The maintenance department is responsible for the maintenance and operation of the physical plant of the hospital. The size of the department and its functions vary with the size and complexity of the hospital. Maintenance usually includes the operation of the power plant, electrical service, plumbing service, and heating; maintenance personnel include refrigeration mechanics, welders, carpenters, brick masons, roofers, gardeners, locksmiths, plasterers, painters, and air-conditioning specialists. The maintenance department is also responsible for all keys, especially the special keys that fit the narcotics cabinets, medicine cabinets, psychiatric areas, and prison areas. The proper operation of the maintenance department and the proper maintenance and repair of all equipment and parts of the physical structure are essential parts of a good environmental health program.

Plumbing

Hospitals probably have more piping systems than any other type of building. The common piping systems include cold potable water, chilled and recirculated drinking water, distilled water, fluid suction systems, vacuum cleaning systems, oxygen, fire sprinkling and stand pipes, lawn irrigation, air-conditioning, refrigeration systems, recirculated cooling water, drainage systems, soil and waste systems, vent systems, storm water systems, and building sewers.

Radioactive Materials

The use of radioactive isotopes in hospitals and other institutions has greatly increased in the last 50 years. Radioactive materials are found in dental clinics, nuclear clinics, pharmacies, diagnostic and therapeutic radiology laboratories, and patient rooms where radionuclides are used.

Water Supply

The water supply for hospitals and other institutions must be of high quality, physically, chemically, biologically, and radiologically. The potable water must meet the drinking water standards of the National Primary Drinking Water Regulations of 1989, as updated through 1991, and updated again by the Safe Water Drinking Act amendments of 1996, and subsequent regulations. Water must be provided in adequate quantities under proper pressure. It may then be further purified within the institution, if it is used in various processes. The water may be distilled, it may be turned into steam, or it may be sterilized. It is essential that the plumbing utilized in the distribution of water within the system be set up in such a way as to avoid any cross connections or submerged inlets.

Emergency Medical Services

The initial emergency medical service program began in 1966 when the Highway Safety Act provided federal grants for this purpose. Its primary objective was to establish self-supporting systems at the state and local levels. Each state provided a commission to receive the federal highway safety funds and to utilize them by establishing standards for ambulance services, their equipment, and personnel; and for institutions engaged in the training of ambulance personnel. The current laws, which were initially funded in 1975 and still are funded, also provide money for the certification and training of emergency medical technicians. Any person who serves in an ambulance and who administers emergency care to patients must be certified as an emergency medical technician by the commission. (See Chapter 8, Volume II, for more information.) The individual must obtain specific continuing education credits annually and be recertified on a yearly basis.

Fire Safety Programs

Fire prevention, protection, and suppression are an integral part of an institution's health and safety program. An effective fire prevention program includes early detection, prompt extinguishment, and evacuation of people. To accomplish this, the institution must define its needs in conjunction with skilled fire prevention specialists. It also must assign responsibilities for each part of the program to appropriate staff members; establish practical regulations; establish uniform procedures for fire prevention, fire protection, and fire suppression; give emergency instructions; provide appropriate training courses and educational programs; have adequate fire design control; make necessary inspections and corrections; and carry out fire drills.

Hand Washing and Hospital Environmental Control

Nosocomial infections (hospital-acquired infections) occur in approximately 5 to 6% of all patients admitted to acute care hospitals in the United States. They have contributed to more than 88,000 deaths. The cost of these infections runs into the billions of dollars. The CDC put out guidelines concerning hand washing and hospital environmental control in 1982 and then updated them in 1985. The guidelines contain six sections. They include hand washing; cleaning, disinfecting, and sterilizing patient care equipment; microbiological sampling; infectious waste; housekeeping; and laundry. In the year 2002, these guidelines are still in use. In addition, still in use are two supplements of the *Morbidity and Mortality Weekly Reports*, titled, "Recommendations for Prevention of HIV Transmission in Health Care Settings," published August 21, 1987; and the "Guideline for Prevention of Transmission of HIV and Hepatitis B Virus to Health Care and Public Safety Workers," published June 23, 1989. These two documents provide recommendations for appropriate decontamination of environmental surfaces and disinfection or sterilization of patient care equipment.

Direct contact is considered to be the primary means of transmission of nosocomial infections. Hand washing is considered to be the most important means of preventing these infections. The hand washing materials are extremely important, because they aid in the cleaning of the hands and the removal of not only soil but also microorganisms. The mechanical friction that is used in hand washing is also of considerable importance. The use of antimicrobial agents are of particular significance when personnel are dealing with newborns, between caring for patients in intensive care units, before caring for patients whose immune systems have been depressed, after caring for patients who have nosocomial infection, before labor and delivery, before an operation, and before any type of procedure where the body will be invaded in any way. Hand washing is the removal of soil and transient microorganisms from the hands, whereas hand antisepsis is the removal or destruction of transient microorganisms. Hygienic hand disinfection is a reduction of predominantly transient microorganisms with the use of germicidal agents or antiseptic detergents. Hygienic hand disinfection is used by both healthcare personnel and foodservice personnel to eliminate transient microorganisms.

The microflora of skin are of two types: resident and transient. Resident bacteria are those organisms usually found on the skin. Of these the only truly pathogenic organism is *Staphylococcus aureus*. Resident bacteria are not easily removed by mechanical friction, because they are buried deep within the pores and are protected by sebaceous gland secretions. Transient organisms are of concern because they are readily transmitted by hands unless removed by the mechanical friction of washing with soap and water or are destroyed by the use of an antiseptic solution. Transient organisms can be considered skin contaminants that are acquired from environmental sources and become attached to the outer epidermal skin layer.

Hands, as well as contaminated gloves, serve as vectors for transmission of transient microorganisms. These bacteria cause great concern to the foodservice and healthcare industries, because they are loosely attached to the surface of the skin and can easily contaminate food or patients. Hands, arms, and fingers become easily contaminated with fecal microorganisms after using the bathroom. These organisms include *Salmonella* spp., *Escherichia coli*, *S. aureus*, *Clostridium perfringes*, *Shigella*, and hepatitis A virus (HAV). Organisms from animal sources such as *Yersinia, Proteus, Campylobacter*, and *Klebsiella* can be transmitted from hands to food, equipment, and other people. Hand transfer can be a significant mode of transmission of bacteria and viruses from person to person, from person to surface to person, and from person to food. Bacterial transfer is enhanced by wet hands. Viruses are also readily transferred by hands, especially wet ones.

Rotaviruses are transmitted by the oral–fecal route and usually cause gastroenteritis in infants and small children as well as the elderly and immunocompromised individuals. Rotaviruses survive for long periods in contaminated water, on hard surfaces, and on hands. Rotaviruses can survive on hands for up to 4 hours. The virus can be readily transferred from hands to other surfaces and vice versa.

HAV survives well on environmental surfaces and on human hands for up to 7 hours, and can be easily transferred to and from hands and surfaces. It is resistant to many disinfectants. See Chapter 3 on food protection in this volume for specific hand washing procedures for food handlers.

The cleaning, disinfecting, and sterilizing of patient care equipment varies with the three categories of equipment in use. These categories include equipment for critical patients, semicritical patients, and noncritical patients. Disinfectants used in the healthcare setting include alcohol (to be used on patients only), chlorine and chlorine compounds, formaldehyde, glutaraldehyde, hydrogen peroxide, iodophors, phenolics, and quaternary ammonium compounds. These disinfectants are not interchangeable, and therefore the user must be aware of how the disinfectant works in different settings. Sterilization, which is the complete elimination or destruction of all forms of microbial life, is accomplished either by means of steam under pressure or by dry sterilization. Ethylene oxide, which has been used as a cold sterilization gas, is now listed under the Comprehensive Environmental Response, Compensation, and Liability Act (CERCLA) as a potential carcinogen. However, it still may be used if proper control techniques are put in place. The major control is to ventilate through a dedicated exhaust system that serves the sterilizer area only and routes the ethylene oxide directly to the outside air, away from people and building intake air systems.

Under the current CDC guidelines, the only microbiological sampling recommended is the placing of spore strips in sterilized materials to determine whether sterilization has occurred. In addition, the water used to prepare the dialysis fluids and the dialysate needs to be checked on a monthly basis. However, microbiological sampling could still be an effective technique if used when a massive outbreak of nosocomial infection occurs in an institution and no known cause for the outbreak is determined. It might be worthwhile to take air samples in the area where patients are becoming infected.

The area of infectious waste is discussed later. Typical nosocomial pathogens include *Pseudomonas aeruginosa*, *Enterobacter* sp., *Klebsiella* sp., Group B streptococci, and *Staphylococcus* sp. Other organisms include *Listeria monocytogenes*, *Corynebacterium*, *Propionibacterium*, and *Bacillus* sp., hepatitis B virus (HBV), hepatitis C virus (HCV), and *Mycobacterium tuberculosis*.

Housekeeping is an essential area for the reduction of nosocomial infections. The cleaning of hospital floors and other horizontal surfaces helps remove microbial contamination on these surfaces. Appropriate cleaning is another measure used to prevent the microorganisms that have landed on the surfaces from reintroduction into the air environment.

Hospital laundering significantly reduces numbers of microbial pathogens. The laundry then may be sterilized if used as drapes or other necessary items in specialized situations, such as the emergency room, operating rooms, delivery rooms, and other areas where sterile sheeting is required.

Human Immunodeficiency Virus, Hepatitis B Virus, and Hepatitis C Virus

The Health Omnibus Programs Extension Act of 1988 required that guidelines be developed for the prevention of transmission of human acquired immunodeficiency syndrome (AIDS) virus and hepatitis B virus (HBV) for healthcare and public safety workers. These guidelines were essential for reducing the risk in the workplace of becoming infected with the etiologic agent for AIDS and for determining the circumstances under which exposure to the etiologic agent may occur. The mode of transmission for HBV is similar to that of human immunodeficiency virus (HIV). The potential for HBV transmission in the occupational setting is greater than that for HIV. A larger body of experience relates to the control of the transmission of HBV in the workplace than that of the transmission of HIV. The general practices used to prevent the transmission of HBV also minimize the risk of transmission of HIV.

Other bloodborne transmissions of disease can be interrupted by adhering to the precautions used for protecting the workers against HBV and HIV. The workers involved not only include hospital personnel, but also fire service personnel, emergency medical technicians, paramedics, and law enforcement and correctional facility personnel.

HBV and HIV transmission in occupational settings occurs only by percutaneous inoculation, needle sticks, or cuts; or contact with an open wound, nonintact (chapped, abraded, weeping, or dermatitic) skin, or mucous membranes, blood-contaminated body fluids, or the concentrated virus. The probability of infection by the healthcare worker is related to the potential for the organism to be present in the patient's blood, etc. The risk of infection changes rapidly with the type of patient who is under care. The normal risk of HBV infection from the general population varies from 1 to 3 per 1000, whereas in the drug or homosexual population, the risk ranges from 60 to 300 per 1000 population.

As of December 1996, CDC had received reports of 52 documented cases and 111 possible cases of occupationally acquired HIV infection among healthcare workers in the United States. Of these, 45 were from needle sticks or cuts; 5 were from exposure to eye, nose, mouth, or skin; 1 was from an injury as well as exposure to a mucous membrane; 1 was due to unknown etiology; and 47 of the known cases involved contaminated blood.

Aerosols, which are particles less than 10 µm in diameter that float on air currents, have not been shown to transmit bloodborne pathogens. HBV in various studies has not been shown to be transmitted through the air. In situations where the concentration of HIV in blood is lower than the concentration of hepatitis B in blood this indicates that aerosols would not be a source of contamination for HBV.

From the 1990 to 1992 viral hepatitis surveillance program, the CDC reported the number of HBV infections in the United States to be 21,000 for 1990, 18,000 for 1991, and 16,000 for 1992, with approximately 25% of infected people developing acute hepatitis. In the year 2000, the latest report from the CDC showed about 13,600 cases. Approximately 6 to 10% may become HBV carriers, may be at risk of developing chronic active hepatitis and cirrhosis, and may spread the infections to others. From the viral hepatitis surveillance program of 1990 to 1992, the CDC reported that the infection rate in healthcare workers was 3.7/1000 in 1990, 3.5 per 1000 in 1991, and 3.9 per 1000 in 1992. In the year 2000, the latest report from the CDC indicated a rate of 3.1 per 1000. These workers' jobs entailed exposure to blood. The degree of risk not only to the healthcare workers but also to the emergency medical and public safety workers varies with the amount of exposure to blood that these individuals encounter.

Because of the serious concern from the federal government about the spread of HBV in the general population and among healthcare workers, several prevention programs were funded by Congress. The programs were developed in 1989 and the funding continues to the present. Areas of prevention include screening all pregnant women, infants, high-risk children and adolescents, and using vaccines and hepatitis B immunoglobulin where appropriate. Special precautions for healthcare workers were instituted by the CDC. These were updated in 1998 by the CDC for healthcare personnel. In 1999, the CDC issued new guidelines for prevention of surgical site infections.

HCV is most efficiently transmitted by repeated percutaneous exposures to blood, such as in the transfusion of blood and blood products from infectious people and the sharing of contaminated needles by drug users. Healthcare employees, especially those working with patients or in laboratory work, are at risk from HCV. Accidental needle sticks and cuts by sharp instruments appear to be the major route of infection. Nosocomial transmission is also possible if breaks in technique occur; disinfection procedures of contaminated equipment are improper; or patients share contaminated equipment. At least 85% of people with HCV infection become chronically infected and have chronic liver disease. In a 1998 report, the CDC estimated that an average 230,000 new infections of HCV occurred each year during the 1980s. During the 1990s, the annual number of cases had dropped to 36,000.

Tuberculosis

In 1998, over 18,000 cases of tuberculosis (TB) were reported in the United States, which was an 8% decrease from 1997 and a 31% decrease from 1992, when the height of the TB resurgence occurred. California, Florida, Illinois, New York, and Texas reported 54% of all cases. The decline in the overall numbers of reported TB cases is due to the apparent strengthening of control programs nationwide. The elimination of TB in the United States depends on eliminating TB cases among foreign-born persons. TB has been associated with immunodeficiency syndrome and has created problems in hospitals and other healthcare facilities for workers dealing with these individuals and the elderly.

Resistant Organisms

A resistant organism is one that is resistant to two or more unrelated antibiotics to which the organism is usually considered susceptible or to certain key drugs; an example is *S. aureus*, which is resistant to methicillin. Virulence of the microorganism is based on its ability to:

1. Attach to mucosal surfaces
2. Survive and penetrate the skin or mucous membrane
3. Multiply within the body
4. Inhibit or avoid host defenses
5. Produce disease or damage the host

The resistant organism then colonizes the area of the body site by growth and multiplication, which in fact may cause a carrier state. Colonization increases the reservoir of resistant organisms.

Resistance to antimicrobial agents has been going on for at least 40 to 45 years. Although it started in large teaching hospitals, it has spread readily to all hospitals, other healthcare facilities, and the community at large. Resistance can develop because of:

1. Antibiotic pressure due to prolonged or excessive exposure to certain antimicrobial agents
2. Disruption of the normal flora and colonization with more resistant organisms
3. Severity of illness and immunosuppression

Bacteria can produce antibiotic-inactivating enzymes, change cell wall permeability, alter target sites, change their susceptible metabolic pathways, or change cell wall binding sites to prevent antibiotic attachment. Although resistant organisms are not more virulent than nonresistant organisms, they cause more disease because they cannot be treated properly.

Resistant organisms are primarily spread by contact with hands of personnel, through droplets from coughing, sneezing, suctioning, ventilation equipment, and contaminated fomites.

PROBLEMS IN INSTITUTIONS

Hospitals

Because the institution is in effect a small community, all environmental health problems and necessary resulting services that would be found in any small community are found within the institution. These include air, food, housekeeping, insect and rodent control, laundry, lighting, noise, solid waste, water, and liquid waste. In addition, because of the uniqueness of the institution, other areas may develop specific environmental problems. These areas include the laboratories, surgical suites, and emergency rooms. Unsafe or careless practice of environmental protection in the institution not only endangers individuals within the institution but also may endanger the community at large.

Air

A significant difference between the normal population and the patient population is the increased susceptibility of the patient to infection and also the increased risks of spread of infection from the infected patient to others. This susceptibility to infection varies tremendously with the patient, the disease process, the chemical and radiological therapies, and the other techniques of treatment used. Pathogenic microorganisms, chemicals, and other contaminants are transmitted through the air. The transmission may be from humans or operating techniques to susceptible individuals. The organisms are found on clothing, bedding, and floors, and may readily become airborne. The rate at which the particles settle out is determined by their size. It has been observed that particles released from basements of buildings travel quickly trough the ventilation ducts to the upper stories of the buildings. This would indicate that microbiological or chemical contamination, which may occur anywhere, could easily be distributed to any or possibly all parts of the institution.

The types of air contaminants include dust, particles that range in size from 0.1 to 100 µg (approximate); fumes, solid particles that are formed by condensation of vapors from metals ranging in size from 0.001 to 2 µg; smoke, solid or liquid particulates that are produced by the incomplete combustion of fuel ranging in size from 0.1 to 30 µg; mists, small liquid droplets produced by atomizing, boiling, sneezing, and so forth, ranging in size from 0.5 to about 10 µg; vapors, gaseous phase of liquids that includes such combinations as gasoline; gases that are produced by chemicals found in the institution, ranging in size from 0.0001 to 0.001 µg; odors that are caused by a variety of processes or illnesses, ranging from 0.00001 µg to a much larger concentration; and microbial contaminants that range in size from 0.005 to 40 µg.

All these contaminants are produced by some form of personnel activity or operation of a variety of pieces of equipment. Further, these contaminants may readily be caused by carriers of specific organisms or shedders or dispersers of organisms. Some examples of patient equipment or other hospital equipment as the source of air contamination include microbial contaminants found in humidifier water, microbes found on cooling coils and other parts of the air-conditioning and ventilation system (*Pseudomonas aeruginosa*), nebulizing equipment, and inhalation equipment. Other air contaminants include skin scales, textile fibers, and dust particles.

The use of lasers may cause a potential hazard within the institutional environment. The laser vaporizes or fragments harmful viable or nonviable substances and may cause an airborne health hazard or a direct hazard to tissues. The principal indirect hazards and sources of air contamination are solvents used in the cleaning of equipment and gases, and fluids associated with the use of the equipment. Some of the toxic substances used in the laser facilities include benzene, carbon disulfide, carbon tetrachloride, cyclohexane, nitrobenzene, pyridine, toluene, xylene, mercury, and chlorine. The direct sources of air contamination vary with the type of laser. The carbon dioxide laser has a high potential for generating air contamination. The beam can readily melt, vaporize, and burn a wide variety of materials, depending on the power density and the ability of the material to absorb power. The hazards are very similar to those produced in welding. Some of these hazards include the production of nitrogen oxide and ozone.

Some additional potential environmental health air hazards involve the degree of filtration of exhaust air from vacuum cleaners; aerosols produced from solid-waste compactors; aerosols produced from bedpan flusher units; aerosols from centrifuges, blenders, and other laboratory devices; aerosols from hydrotherapy tanks; airflow around doorways of isolation and reverse isolation rooms; airborne contamination from housekeeping materials, including polishes, solvents, detergents, disinfectants, and pesticides; and airborne concentration from rehabilitation and occupational therapy areas, including glues, solvents, and materials such as fiberglass, dust from grinders, shavers, and sanders.

Equipment Design and Construction

Equipment can readily be contaminated by patients, professional staff, or hospital employees. The equipment may then become a source of infection for other individuals. Additional problems occur because of improper design or construction, making the equipment difficult to clean, disinfect, and sterilize. Faulty design of equipment may also contribute to occupational hazards or cause hazards to patients or visitors. Because of the great variety of equipment found in an institutional setting, it is impossible to discuss all the potential problems. Therefore, only a few pieces of equipment that may contribute to the spread of infection or cause other hazardous conditions are discussed.

Hemodialysis machines have been associated with the spread of HCV to patients and the potential spread of hepatitis to staff and employees. Worldwide, the average prevalence of antibody to HCV among hemodialysis patients is 20% with a range of 1 to 47%. Bronchoscopes, cystoscopes, and endoscopes are pieces of equipment that are not easily cleaned or decontaminated. Many types of electronic equipment also present a cleaning and decontamination problem.

Other problems related to electronic monitoring devices include the effects of the devices on patients and staff when they have been exposed for long periods of time. Plastic furniture and drapes are rapidly becoming introduced into the hospital environment. There is a poor understanding of what might occur if the drapes and furniture would start to smolder. Fumes from smoldering equipment mixed with gases utilized within the patients' rooms could present a toxic hazard to patients, staff, employees, and visitors. Hydrotherapy tanks are a constant source of potential infection problems. The tanks are difficult to clean because of the design and construction of the turbine pump and the agitator. The so-called bedpan sterilizer is actually a bedpan flusher unit that is inefficient, tends to become inoperative, is hard to clean, and becomes a serious potential source of infection. Bedpans, after they go through the flusher unit, may be more contaminated than before the cleaning operation.

Solid waste compactors push enormous quantities of contaminated waste together. They are difficult to clean and often leak, becoming a source of insect, rodent, and air contamination. Fiber-optic instruments are built in such a way that they are not readily sterilized. Inhalation therapy equipment and nebulizing, anesthesia, and infant resuscitation equipment have been known to cause pulmonary necrotizing infections.

Food

The dietary department is similar to a large commercial food operation. In addition, it has the problems of preparing diets for sick and debilitated individuals, and for preventing massive outbreaks of foodborne disease. (An in-depth discussion on food appears in Chapter 3.) This section is concerned with some of the problems that can be found in food operations within hospitals and other institutions. They are as follows: (1) inadequate supply of light, contributing to worker fatigue, inadequate removal of soil and wastes, and increases in insects and rodents; (2) poor housekeeping resulting from inadequate management of the large foodservice area that must operate on a least a 12-hour minimum basis and in some cases up to 24 hours a day; (3) accumulation of food particles, dirt, dust, and other soil on food equipment, such as cookers, stoves, ovens, deep-fat fryers, slicers, can openers, conveyors, sinks, cabinets, shelves, storage areas, pots, and pans; (4) coating of surfaces with greasy residues; (5) dirty and corroded refrigerators with worn gaskets on the doors; (6) poor ventilation in ware washing rooms; (7) corroded and clogged ware washers and spray arms within the ware washers; (8) inadequate steam pressure for washing, rinsing, and cleaning; (9) potential chemical food poisoning due to inadequate and improper storage of chemicals, insecticides, cleaning materials, and other poisonous materials; (10) inadequate and improperly placed hand washing facilities; (11) food handlers who are working although they have or have recently had diarrhea, vomiting, severe upper respiratory illness, or skin infections; (12) carry over of leftovers from one meal to another; (13) improper thawing of turkeys, meats, and other foods outside refrigeration; (14) improper cleaning of equipment, such as grinder heads and tenderizer blades; (15) reusing disposable water carafes and glasses; and (16) improper care and cleanliness of food carts that have refrigerator and heating units on them.

Other problems revolve around the preparation, storage, and use of infant formulas and tube feeding. Great care should be given to adequate cleaning, evaluation,
and supervision of all food areas to prevent an outbreak of foodborne disease.
Unfortunately, too many times in the past, thousands of patients have become ill
from avoidable outbreaks of foodborne disease.

Food-related illnesses and subsequent hospitalizations in the United States create
a potential source of a variety of nosocomial infections. It is estimated that foodborne
diseases cause approximately 76 million illnesses, 325,000 hospitalizations, and
5000 deaths in the United States each year. Any of the microorganisms discussed
in Chapters 3 and 4 of this volume can be the potential cause of these hospitalizations.
More than 200 known diseases are transmitted through food.

Hazardous Chemicals

Many thousands of raw materials are used in industry today. The toxicity of
these substances, the ways in which they are used, the chemicals to which they may
be bound, the degree of susceptibility of individuals, and the methods of control are
all intimately related to the possibility of toxic responses. The National Institute of
Occupational Safety and Health (NIOSH) lists over 8000 substances as toxic. It has
been noted that over 630 different chemicals are used in hospitals. Of this number,
about 300 are of unknown toxicity, about 30 are safe, and about 300 have hazardous
properties. These chemicals can be toxic or carcinogenic, and flammable or explosive. They include acids, alkalis, ammonia, assorted oxidizers, organic and inorganic
peroxides, oxides, permanganates, nitrates, nitrites, toluene, ether, lithium, potassium, sodium, and more.

Areas of the hospital that have the greatest usage of chemicals and the greatest
potential for hazardous situations include the departments of anatomic pathology,
clinical pathology, renal dialysis, and various storage areas. The laboratories contain
many solvents, fixants, and volatile hydrocarbons. In the patient rooms chemicals
are used for cleaning, degreasing, and decontamination. These may affect the patients
and the staff and personnel. The housekeeping department uses pesticides, polishes,
solvents, and detergent disinfectants. The physical plant stores gasoline, paints, paint
thinners, and assorted other materials used in the repair process.

The potential hazards of chemicals are enormous. It is necessary to determine
the kinds of chemical mixtures that are in the air. Will the mixtures be explosive,
flammable, toxic, or carcinogenic? For example, a very serious potential problem
relates to anesthetic gases. Anesthesiologists have an unusually high incidence of
headaches, fatigue, and irritability. Further, the incidence of spontaneous abortion
and a high incidence of abnormal pregnancies have occurred among females in the
operating room. These include embryotoxicity, mutagenesis, carcinogenesis, and
liver disease. Further, it has also been postulated that an exposed male can transmit
the defect to his unexposed wife. There is a lack of knowledge as to whether these
anesthetic gases can cause problems in patients receiving repeated doses of anesthesia who are also receiving immunosuppression therapy or coronary care. More
than 424,000 dentists, dental assistants, and dental hygienists are potentially exposed
to nitrous oxide used as an anesthetic. Occupational exposure to N_2O causes adverse

effects, such as reduced fertility, spontaneous abortions, and neurological, renal, and liver diseases.

Housekeeping

Housekeeping procedures are related to the direct or indirect transfer of pathogenic organisms from person to person. Many special problems arise within the institution. The housekeeping activities must be adapted and scheduled in such a way as to avoid the activities of medical staff, nurses, and other auxiliary personnel. Some rooms are overcrowded because of the use of essential equipment. Other rooms contain food spillage, dripping urine, broken glass, silver nitrate, chewing gum, infected or contaminated bandages, linens, and other materials. Housekeeping attendants, their equipment, and any other material that they may utilize can become a vehicle for the transmission of disease.

Because the institution is a vast conglomeration of buildings, floors, corridors, windows, and assorted rooms used for a variety of functions, the housekeeping task becomes even more difficult. The equipment and techniques utilized in bathrooms certainly must be modified for use in kitchens. X-ray areas, storehouses, laundries, garages, outside grounds, intensive care units, cardiac care units, and surgical units are all independent and have unique problems.

Insect and Rodent Control

Insects and rodents cannot be tolerated in an institutional setting. Not only are conditions favorable for their entrance and multiplication but also their normal propensity for transmitting disease is sharply increased by a substantial quantity of virulent organisms present in the environment. Insect and rodent control is an important part of the overall program of preventive medicine within the institution. These pests, which include German and American roaches, bedbugs, ants, flies, rats, mice, and fleas, may be found in foodservice areas, housekeeping areas, janitorial closets, nurses stations, clinical areas, patient areas, animal laboratories, conveying equipment and shafts, morgues and autopsy rooms, and laundries. Complete descriptions of the life cycles, problems, and controls of these insects and rodents appear in Chapters 5 and 6.

Structural problems are of concern, because wherever the maintenance department is involved in correcting plumbing or other types of repairs, holes may be left open in walls, floors, and ceilings. These openings provide a natural pathway for the invasion of rats, mice, or insects.

Food areas are a primary harborage for insects and rodents. These areas include preparation, serving, storage, dishwashing, and equipment. Food residues, crumbs, and other food materials may accumulate under sinks, around walls, on the bottoms of shelves, under tables, under equipment, and behind and under all other types of standing equipment within the preparation and serving area. In storage areas, spilled flour, sugar, beans, and other foods frequently accumulate. Storage areas provide an excellent source of food for insects and rodents, because these areas are generally warm and have sufficiently enclosed spaces for harborage. Insects and rodents breed in the residues from the evening meal, visiting-time snacking, encrusted step cans, uncovered wastebaskets, and baby formulas or foods dumped into sinks. Visitors frequently bring food into the institution and at times may even bring roaches in with their clothes or in food. The nurses stations, clinical areas, informal staff rooms, and other places where people congregate tend to be collection stations for food and become primary sources of insect and rodent problems. Animal laboratories are of special concern, because animal food is always available for the feeding of experimental animals. Roaches and rodents have a constant source of food, water, and harborage in these areas. Conveyors, tray carts, laundry carts, elevator shafts, plumbing, and electrical shafts are also sources of harborage and paths of travel for insects and rodents. Insects and rodents find powerful attractants in the laundry, because soiled and blood-encrusted linens are a food supply for insects and rodents. The solid waste disposal areas are frequently nothing more than open dumps and become sources of insect and rodent problems.

Laboratories and Biohazards

In addition to the fire hazards occurring in laboratories, biohazards also exist (Figure 9.1). Biohazards are more prevalent due to the considerable amount of experimental work conducted today. Two key points to recognize are that (1) it is necessary to identify and classify all biohazards, because any accident occurring in the presence of biohazard materials may result in infection; and (2) many biological agents have unknown or incompletely understood etiologies. The environment may impinge on laboratory animals or workers, on biohazardous agents or insects, on physical or chemical substances, and vice versa. Reported laboratory acquired infections only represent a fraction of those that have actually occurred. Many times unusual viruses are sent to laboratories in the United States for analysis. In 1994, at the Yale School of Medicine, a researcher was exposed to a rare and potentially lethal disease caused by the Sabia virus when a laboratory accident occurred. Other recent viruses are Marburg and Ebola viruses in Africa and the Junin, Machupo, and previously mentioned Sabia from South America. Virus "X" emerged from the rain forest in Southern Sudan, killed thousands, and disappeared.

Figure 9.1 Universal biological hazard symbol. Labels are usually fluorescent orange or orange-red, with lettering or symbols in a contrasting color. (Courtesy of the Centers for Disease Control.)

Bacterial Agents

Bacterial agents that cause laboratory-associated diseases are as follows:

1. *Bacillus anthracis* may be present in blood, skin lesion exudates, cerebrospinal fluid, pleural fluid, sputum, and rarely in urine and feces. Direct and indirect contact of the intact and broken skin may result in the infection.
2. *Bordetella pertussis*, a human respiratory pathogen, is the cause of whooping cough. People having common laboratory space with contaminated workers are at greatest risk.
3. *Brucella* (*abortus, canis, melitensis, suis*) have all caused illness in laboratory personnel. The illness may be due to hypersensitivity to the antigens. The agent may be present in blood, cerebrospinal fluid, semen, and occasionally urine. Sniffing the cultures as well as direct contact with the infectious agents causes the disease. This is the most commonly reported laboratory disease.
4. *Burkholderia* (*Pseudomonas*) *pseudomallei*, although rare, has caused disease through aerosol and skin exposure. The agent may be present in sputum, blood, wound exudates, and various tissues. It may also be present in soil and water samples from the endemic areas.
5. *Campylobacter* (*jejuni, coli*) is rarely a cause of laboratory-associated illnesses. However, some have been documented. Known reservoirs of infection include poultry, pets, farm animals, laboratory animals, and wild birds. The organisms may be present in fecal samples.
6. *Chlamydia* (*psittaci, pneumoniae, trachomatis*) were at one time among the most commonly reported laboratory-associated infections, with the highest case fatality rate. Infection is spread by exposure to infectious aerosols in the handling, care, or autopsy of naturally or experimentally infected birds. Some infections may be spread by infected mice and eggs.
7. *Clostridium botulinum* or toxin may be present in a variety of food products, serum, feces, soil, or surface water. The toxin may be absorbed after ingestion or following contact with the skin, eyes, or mucous membranes including the respiratory tract. Although only one laboratory-acquired case of botulism has been reported, the severity of the disease demands caution.
8. *Corynebacterium diphtheria* may be present in exudates or secretions of the nose, throat, pharynx, larynx, and wounds; in blood; and on the skin. Inhalation, and accidental parenteral inoculation, as well as ingestion are primary laboratory hazards.
9. *Escherichia coli*, cytotoxin-producing strains, may be found in a small proportion of patients with hematolytic uremic syndrome (especially children), and is responsible for most deaths associated with infections from these organisms. Domestic farm animals are significant reservoirs of the organisms. The enterohemorrhagic *E. coli* is usually isolated from feces, but also may be found in a wide variety of contaminated foods such as uncooked ground beef and unpasteurized dairy products. Ingestion is the primary laboratory hazard.
10. *Francisella tularensis*, which causes tularemia, has caused infections related to work with experimentally infected animals or their ectoparasites, The agent may be present in lesion exudates, respiratory secretions, cerebrospinal fluid, blood, urine, tissues from the infected animals, and fluids from the infected arthropods. Infection may be caused by direct contact with infectious material and the person's skin or mucous membranes, accidental inoculation, ingestion, and exposure to aerosols and infectious droplets.

11. *Helicobacter pylori* is an agent of gastritis. It may be present in gastric or oral secretions and stools. Although the means of transmission is not fully understood, it is believed to be by the fecal–oral or oral–oral route.

12. *Leptospira interrogans* has been associated with multiple infections and deaths in laboratories. Direct and indirect contact with fluids and tissues of experimentally or naturally infected mammals may cause the infection. In animals with chronic kidney infections, the agent is shed in the urine in large numbers for long periods of time. The agent may also be present in blood and tissues and may be spread by direct or indirect contact with the skin or mucous membranes.

13. *Listeria monocytogenes* had been isolated from soil, dust, human food, animals, and asymptomatic humans. Most cases of listeriosis have come from eating contaminated food products, mostly soft cheeses, raw meat, and unwashed raw vegetables. Naturally or experimentally infected animals are a source of exposure. Although ingestion is the most likely cause of exposure, the eyes and skin may be infected through direct contact.

14. *Legionella pneumophila* has caused a single documented laboratory-associated case of legionellosis, due to a presumed aerosol or droplet exposure during animal experimentation. The organism may be present in pleural fluid, tissue, sputum, and environmental sources.

15. *Mycobacterium* spp., other than *M. tuberculosis,* is thought to be related to accidents or incidents in the laboratory or autopsy room. Usually, these organisms are infectious but not contagious and may cause pulmonary diseases resembling tuberculosis and may also cause lymphadenitis and skin ulcers as well as soft tissue wound infections. Direct contact of the skin and mucous membranes with infectious materials and accidental inoculation are the primary laboratory hazards.

16. *Mycobacterium tuberculosis* and *M. bovis* may cause infections through exposure to aerosols. The incidence of tuberculosis in laboratory personnel working with *M. tuberculosis* is three times higher than those not working with this agent. The organism may be present in sputum, gastric lavage fluids, cerebrospinal fluid, urine, and in lesions from a variety of tissues.

17. *Neisseria gonorrhoea* have caused laboratory associated gonococcal infections. The agent may be present in conjunctival, urethral, and cervical exudates, synovial fluid, urine, feces, and cerebrospinal fluid. Accidental parenteral inoculation and direct or indirect contact of mucous membranes with infectious clinical materials are the known primary laboratory hazards.

18. *Salmonella*, including all serotypes except *S. typhi*, are present in a broad range of primary reservoir hosts (including birds, mammals, and reptiles), all of which may be a source of infection to laboratory personnel. The organism may be present in feces, blood, and urine; and in food, feed, and other materials. The worker is contaminated either through ingestion or through parenteral inoculation.

19. *Salmonella typhi* may be present in feces, blood, bile, and urine. Humans are the only known reservoir of infection through ingestion and parenteral inoculation, resulting in cases of typhoid fever.

20. *Shigella* spp. have caused outbreaks of shigellosis in laboratory personnel. Captive nonhuman primates are the only significant reservoir of infection beside humans. However, guinea pigs and other rodents may be the cause of infections. The organism is present in the feces and rarely in the blood.

21. *Treonema pallidium* is the organism causing syphilis. Humans are the only known natural reservoir of the disease. Syphilis has been transmitted to laboratory personnel

working with a concentrated suspension of the organism. It can be transmitted through the transfusion of fresh blood obtained from a patient with secondary syphilis.

22. *Vibrio cholerae* or *V. parahaemolyticus* organisms cause vibrionic enteritis, although rarely, in laboratory personnel. The disease comes from experimentally infected animals as well as naturally infected animals. The organisms may be found in feces. Ingestion and parenteral inoculation are the primary sources of infection.

23. *Yersinia pestis* is a proven although rare laboratory hazard. The agent may be present in blood, sputum, cerebrospinal fluid, feces, and urine from humans, depending on the clinical form and stage of the disease. Primary hazards to laboratory personnel include direct contact with cultures and infectious materials from humans or rodents, and infectious aerosols or droplets generated during the manipulation of cultures and infected tissues.

Fungal Agents

Fungal agents that cause laboratory-associated diseases are as follows:

1. *Blastomyces dermatitidis* has caused disease due to accidental parenteral inoculation with infected tissues or cultures containing the yeast forms of the organism. Pulmonary infections have occurred following the presumed inhalation of the organism. Soil and environmental samples may pose a hazard.

2. *Coccidioides immitis* is the agent causing coccidioidomycosis. The arthroconidia are between 2 and 5 μm and therefore are conducive to ready dispersion in air, and retention in the deep pulmonary spaces. The fungus may be present in clinical specimens and animal tissues, as well as in soil or other samples from natural sites (Figure 9.2).

3. *Histoplasma capsulatum* is the agent causing histoplasmosis. Pulmonary infections have resulted from handling mold form cultures. Local infection has resulted from skin puncture during autopsy of an infected human and from accidental needle inoculation of a viable culture. Collecting and processing soil samples from endemic areas has caused pulmonary infections.

4. *Sporothrix schencki* has caused a number of local skin or eye infections, due to accidents involving splashing of the culture material into the eye, scratching or injecting of infected material into the skin or biting by an experimentally infected animal.

Parasitic Agents

Parasitic agents that cause laboratory associated diseases are as follows:

1. *Plasmodium* spp., *Trypanosoma* spp., and *Leishmania* spp. have caused laboratory-associated disease through contact with lesion material from rodents, and contact with feces or blood of animals or insects. The infective stages may be present in the blood, feces, cerebrospinal fluid, bone marrow, or other biopsy tissue, lesion exudates, and infected arthropods. Depending on the parasite, the primary laboratory hazards are ingestion, skin penetration or microabrasions, accidental parenteral inoculation, and transmission by arthropod vectors.

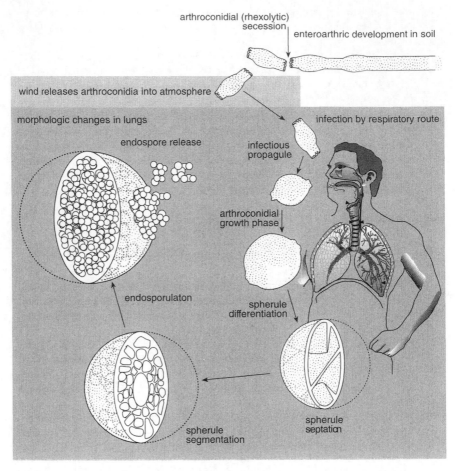

Figure 9.2   The dimorphic life cycle of *Coccidioides immitis.* (Adapted from Kirland, T. and Fierer, J., A reemerging infectious disease, *Emerging Infectious Diseases,* 2, 3, 1996. With permission.)

2. *Toxoplasma* spp., *Entamoeba* spp., *Isospora,* spp., *Giardia* spp., *Sarcocystis* spp., and *Crytosporidium.* spp., have all been reported to have caused laboratory-associated infections. Laboratory animal-associated infections provide a direct source of organisms for the infection of laboratory personnel who come in contact with feces of the experimentally or naturally infected animals. Other body fluids and tissues are potential sources of the parasites, with ingestion the most common method of contamination.

Prions

Prions that cause laboratory infections are as follows:

1. Scrapie prion from sheep and goats causes scrapie disease.
2. TME prion from mink causes transmissible mink encephalopathy.
3. CWD prion from mule deer and elk causes chronic wasting disease.

4. BSE prion from cattle causes bovine spongiform encephalopathy.
5. FSE prion from cats causes feline spongiform encephalopathy.
6. EUE prion from nyala and greater kudu causes exotic ungulate encephalopathy.
7. Kuru prion from humans causes kuru.
8. CJD prion from humans causes Creutzfeldt–Jakob disease.
9. GSS prion from humans causes Gerstmann–Straussler–Scheinker syndrome.
10. FFI prion from humans causes Gatal familial insomnia.

Prions are proteinaceous particles that lack nucleic acids. Prions are composed largely, if not entirely, of an abnormal isoform of a normal cellular protein. In mammals, prions are composed of an abnormal, pathogenic isoform of the prion protein (PrP), designated as PrP^{sc}. Unlike many viruses the properties of prions change dramatically when they move from one species to another. Surgical procedures on patients diagnosed with prion disease can lead to contamination of other individuals.

Rickettsial Agents

Rickettsial agents causing the greatest risk of laboratory infection are the *Coxiella burnetii*. The infectious dose of virulent phase one organisms in laboratory animals has been calculated to be as small as a single organism. Q fever, which is caused by this organism, is the second most commonly reported laboratory-associated infection involving 15 or more people. Exposure to infectious aerosols or parenteral inoculation are most likely sources of infection. All other rickettsia have been shown to cause disease in laboratory personnel. Accidental parenteral inoculation and exposure to infectious aerosols are the most likely sources of laboratory transmission.

Viral Agents

Viral agents that cause laboratory infections are as follows:

1. Hantavirus causes hantavirus pulmonary syndrome, which is a severe, often fatal disease. Most cases of human disease have resulted from exposures to naturally infected wild rodents. The laboratory transmission of the disease through the aerosol route is well documented. Exposures to rodent feces, fresh necropsy material, and animal bedding are associated with the risk. Other potential routes of laboratory infection include ingestion, contact of infectious materials with mucous membranes or broken skin, and animal bites.
2. HAV and HEV are hazards to animal handlers and others working with chimpanzees and other nonhuman primates, which are naturally or experimentally infected. HEV may cause severe or fatal disease in pregnant women. The organisms may be present in feces, saliva, and blood of infected humans and nonhuman primates. Ingestion of feces, stool suspensions, and other contaminated materials is the primary hazard to laboratory personnel.
3. HBV is one of the most frequently occurring laboratory-associated infections, with laboratory workers considered to be a high-risk group for acquiring the infection. These individuals are also at risk of infection with HDV. The organism may be present in blood and blood products of human origin, urine, semen, cerebrospinal fluid, and saliva. Parenteral inoculation, droplet exposure of mucous membranes,

and contact exposure of broken skin are the primary laboratory hazards. The virus may be stable in dried blood or blood components for several days.

4. HCV can cause infections in medical care workers, but not at the same rate as HBV. It has been detected primarily in blood and in serum, less frequently in saliva and rarely or not at all in urine or in semen. HBV appears to be relatively unstable to storage at room temperature, repeated freezing, and thawing.

5. Human herpes viruses are ubiquitous human pathogens and are commonly present in a variety of clinical materials submitted for virus isolation. They are especially a problem in immunocompromised people. *Herpes simplex* viruses 1 and 2 cause some risk through direct contact or aerosols.

6. Influenza virus, especially new strains, may cause human infection in the laboratory. The agent may be present in respiratory tissues or secretions of humans or most infected animals. Inhalation is the route of entry of the disease.

7. *Lymphocytic choriomeningitis* virus infection in laboratory workers occurs especially in areas where laboratory rodents have active or silent chronic infections. The agent may be present in blood, cerebrospinal fluid, urine, secretions of the nasopharynx, and feces and tissues of the infected animal hosts and possibly people.

8. HIV virus has been transmitted in the laboratory. The virus has been isolated from blood, semen, saliva, tears, urine, cerebrospinal fluid, amniotic fluid, breast milk, cervical secretion, and tissue from people.

Arboviruses and Related Zoonotic Viruses

Arboviruses and related zoonotic viruses have caused a variety of laboratory-associated diseases including vesicular stomatitis, Colorado tick fever, dengue fever, western equine encephalitis, and eastern equine encephalitis. As of 1998, there were 537 arboviruses registered with the American Committee on Arthropodborne Viruses.

Laundry

The laundry is a large service unit within the total hospital whose function is the collection and processing of soiled linens and the distribution of clean linens. Laundry and linens flowing to and from the laundry room can readily become a source of infection or a source of insect and rodent infestations. A well-run, reliable laundry service helps create an aesthetically clean environment. A poorly run service may become dangerous. Linens are processed in various ways. They may be cleaned within the institution or in commercial laundries. The institutional cleaning problems include the actual facility, which may be dusty and dirty. This facility may also have large quantities of lint on equipment, walls, and ceilings. The areas for sorting of soiled and blood-encrusted linens, if improperly kept, become a powerful attractant for insects and rodents. In many of these laundries, adequate screening is lacking.

The physical flow of linens is very important. Where laundry chutes are used, the exhaust air driven back into the nursing units by the piston effect of a falling laundry bag can contain staphylococci and other organisms. Trash chutes create the same type of problem. Where the linens are taken by cart to the laundry, the carts tend to become dirty and highly contaminated. In special areas where the linens are coming from rooms where the individuals may have infections, the linens need to be kept separate from the rest of the laundry, transported separately, and then washed in the bags before they are sorted and rewashed. Studies conducted in various hospital areas indicated that the blankets, mattresses, pillows, beds, and linens are highly contaminated by varying patient discharges including urine, blood, feces, and saliva. Bed-making and linen-handling procedures, when carried out in a vigorous manner, tend to circulate these organisms, creating a potential airborne hazard.

Commercial laundries may cause a link of infections from hospital to community to hospital to community and to other types of institutions. The commercial laundries have to have separate delivery and receiving trucks for dirty and clean linens to prevent contamination. The hospital linens must be washed separately from other commercially washed linens; otherwise there is an opportunity of infections.

Lighting

Illumination is a very complex subject of exceeding importance to any environment. It is well known that poor lighting results in fatigue, an increase in the number of accidents and injuries, poor housekeeping, and increased risk of infection. Much of our available light is not necessarily lost because of poor design, but instead because of poor maintenance of fixtures and lack of replacement of light bulbs. The color and shade of paint on walls, ceilings, and floors affects illumination. Dark colors absorb and decrease light; light colors reflect and utilize light properly.

Noise

Noise is unwanted sound. Environmental noise may cause temporary or permanent hearing loss, physical and psychological disorders, interference with voice communications, and disruption in job performance, rest, relaxation, and sleep. The increased use of therapeutic equipment and poor maintenance of electrical and mechanical equipment, coupled with lighter weight construction materials, tend to increase noise problems. Normal human activities increase the noise level. Sources of noise in hospitals or other institutions include autoclaves, communications systems, compactor units, dishwashing machines, electrical equipment, heaters, ventilators, air conditioners, and transportation systems. In addition, noise is created by comatose patients who are suffering from specific pains or who may be upset by the strange environment. On the exterior of the hospital, noises come from automobile traffic, compactor trucks, equipment, and emergency vehicles.

Patient Accidents

Many patients have accidents within institutions. The most frequent of these accidents are falls occurring within the immediate area of the patient's bed.

Plumbing

Although various plumbing systems are installed properly during hospital construction, changes are made frequently either as hospital space is utilized for different purposes or as problems occur within the institution. With the vast number of

Figure 9.3   Radura symbol.

activities undertaken within the hospital and the formidable kinds of microbiological and chemical contamination present, the hospital becomes a potentially hazardous area. Cross connections, submerged inlets, and back-siphonage problems occur. Plumbing fixtures become corroded, chipped, cracked, and blistered. As new hospital equipment is placed into operation, hospital plumbing surveys must be conducted to determine whether the equipment is putting such demands on the plumbing system that it creates improper pressure differentials and excess loads of waste that cannot be handled by the existing plumbing system. The potentially hazardous fixtures and equipment include bathtubs, laundry tubs, sinks in all areas — including the scrub rooms and operating rooms, hydrotherapy tanks, soup kettles, drinking fountains, laboratory sinks, urinals, laundry machines, and all tanks used for dilution. (An in-depth discussion of plumbing appears in Chapter 5, Volume II.)

Radioactive Material

There are numerous concerns about the use, storage, and disposal of radioactive materials found in body wastes. In the event of fires, special care must be taken to avoid contaminating firefighters and the general public. Other problems include improper shielding of medical x-ray and fluoroscopic machines, improper size of the x-ray beam, and improper beam filtration and alignment. Most of the newer dental x-ray machines have excellent built-in protection mechanisms. However, the older machines may still permit considerable stray radiation. Other sources of radiation are lasers, ultraviolet tubes, microwaves, and ultrasound (Figure 9.3).

Solid and Hazardous Wastes

Hospital wastes are discarded materials and by-products of normal hospital activities. The principal types of solid wastes are garbage, paper, trash, other dry combustibles, treatment room wastes, surgical room wastes, autopsy wastes, non-combustibles (such as cans and bottles), and various biological wastes (including wound dressings, sputums, placentas, organs, and amputated limbs). In the last

35 years the number of disposable items has increased from a few to hundreds. These items range from paper bedding to cardboard bedpans. This has caused a sharp increase in the volume and weight of solid wastes that must be disposed of on a daily basis. The average patient contributes an estimated 8.5 lb or 0.7 ft³ of solid waste daily. Containers and equipment into which this waste goes are unclean and are not easily cleaned. The waste is usually dumped into plastic bags and then removed to either trash rooms or outside to large metal containers. The containers tend to be dirty, filled with organic material, difficult to clean, and frequently left improperly cleaned. Unfortunately, hospital personnel leave open the doors to the metal units, creating open dumps, instead of a desirable means of temporary storage of solid waste.

Numerous additional problems are created by the presence of syringes, needles, examination gloves, catheters, enema basins, and petri dishes. This material is frequently contaminated or contains sharp items, causing safety and infection hazards. The solid waste of individuals with infectious or contagious diseases becomes considerably more dangerous to the other patients, the community, and the infected individual. Transportation and disposal of hospital solid waste and research animals to community incinerators and landfills deserves investigation. Many of the communities in the United States have unsatisfactory incineration systems or landfills.

Rubbish or trash chutes and trash rooms present additional hazards, because solid waste falling down a chute creates a piston effect, which forces air back into the patient areas. Further, as plastic bags hit the floor of the trash room they burst open and the organisms spew into the air, becoming readily capable of recirculating through the institution.

Hazardous waste that exhibits ignitability, corrosivity, reactivity, and EP toxicity may be found in the institution. (See Volume II, Chapter 2, on solid and hazardous waste.) Special precautions must be used for disposal of waste related to viral hemorrhagic fever, such as Ebola or Lassa Fever.

Surgical Suites

The surgical suite in the modern hospital is potentially more hazardous today than in former years, because of the number of new surgical techniques and the increased risk of infection due to modern drug therapy. Today many hospitals perform various organ transplants, such as kidneys, hearts, and parts of the blood vessels. It is critical that infection is prevented, patients and personnel are safeguarded from technological hazards, shock and trauma to the patients are minimized, physical comfort and emotional support are provided, and effective patient care is delivered by the staff in the surgical suite.

Some of the specific problems concerning the patients and the prevention of infection are elimination of cross traffic or back traffic of staff, patients, or soiled materials; separation of clean from dirty cases; preparation of operating rooms for handling general surgical procedures; establishment and observation of rules concerning aseptic techniques; immediate change of operating room garb by surgical staff; elimination of the possibility of contamination through air systems; provision of restricted flow zones; and proper cleaning and decontamination of operating rooms.

Hospital-Acquired Infections

Three basic kinds of infections are related to the hospital. The first is the infection of the patient present on admission to the institution. This infection can spread from patient to patient. The second type of infection appears during hospitalization and most frequently is hospital acquired. This may also spread from patient to patient. The third type of infection appears after the patient is discharged from the hospital. The incubation period for the infection occurs during the hospital stay. Individuals, once released, are in a position to spread the infection to members of their families, the community, and of course from the community back to the hospital. An example of the third type of infection would be some types of staph infections where the postpartum woman does not appear to have a breast abscess until she arrives home.

The hospital environment presents a special risk of infection to employees, patients, visitors, and community. The emergency room, admission areas, outpatient clinics, and surgical, medical, and obstetrical units are potential foci of infection. Patients, visitors, employees, supplies, equipment, and physical structure of the institution may aid in either harboring pathogenic organisms or introducing them to the environment. The organisms are transmitted by improper patient care, mishandling of contaminated materials and equipment, poor housekeeping, and inadequate physical facilities.

Hospital-acquired infections differ depending on hospital size, type, and patient usage; usage of antibiotics; employee supervision; time of year; and the quantity of virulent organisms that are introduced into the hospital environment. Some general types of infections found in institutions include Gram-negative bacteria from urinary catheter-associated infections and infections following prostatectomies, burn infections, and infections of the newborn; *Staphylococcus aureus* infections from nurseries, surgical suites, surgical recovery areas, and other areas of the institution; infant diarrhea; infections caused by salmonella and shigella or other enteric organisms; *Klebsiella* infections due to a Gram-negative bacteria found in thoracic surgery, in intensive care units, and on floors and other surfaces; *Serratia marcescens* infections due to organisms found in hexachlorophene dispensers; aerosols from ultrasonic cleaning devices; *Pseudomonas aeruginosa* found in the air and on surfaces; and *Bacillus anitratum* found in frozen plasma, water baths, and respiratory equipment, as well as in hospital food service, solid waste, and laundry operations.

Emerging and Reemerging Infection Problems

Gram-positive as well as Gram-negative bacteria may cause serious infections. *Corynebacterium* sp. and *B. antbrachis* can cause a significant disease in an immunocompromised person. *Listeria monocytogenes* can cause septicemia and meningitis. *Clostridium perfringens* can cause gas gangrene or necrotizing fasciitis in all types of patients.

Contaminated medical products in the United States have led to nosocomial infections. Between October 1970 and March 1, 1971, in 7 different states 8 hospitals experienced 150 bacteremias caused by *Enterobacter cloacae* or Gram-negative organisms of the Erwinia group. Nine deaths resulted. All cases were associated with intravenous (IV) fluids. The cause of the outbreaks was ultimately attributed to unexpected consequences of a new cap design for IV solution bottles. In 1999, a total of 22 respiratory cultures from 13 patients at a children's hospital in California were cultured positively for *Ralstonia pickettii*. Patients were mechanically ventilated with the exception of one who had a tracheostomy. The endotracheal suctioning protocol in this hospital included presuctioning with sterile saline. The sterile saline was found to be contaminated. Also, patients at two hospitals in Arizona had cultures positive for *Burkholderia cepacia*. Most of the isolates were from the respiratory tracks of patients in intensive care units (ICU). All the patients had been intubated and mechanically ventilated during their ICU stay. They had all received routine oral care that included swabbing with an alcohol- free mouthwash. The mouthwash was found to be contaminated.

Equipment such as catheters, bronchoscopes, and hemodialysis machines has been found to be the cause of outbreaks of nosocomial infection. The contaminated equipment was in direct contact with the person's body. The urinary tract is the most common site of nosocomial infection. More than 40% of the total number of infections reported by acute care hospitals are in the urinary tract. Approximately 600,000 patients are affected each year. Most of these infections, 66 to 86%, follow the use of instruments in the urinary tract, mainly urinary catheterization. The risk of acquiring a urinary tract infection depends on the method and duration of catheterization, quality of catheter care, and host susceptibility. These infections are caused by a variety of pathogens including *Escherichia coli, Klebsiella, Proteus, Enterococcus, Pseudomonas, Enterobacter, Serratia,* and *Candida.* Many of these microorganisms are part of the patients' endogenous bowel flora. However, they can also be acquired by cross contamination from other patients or hospital personnel or by exposure to contaminated solutions or nonsterile equipment.

Bronchoscopy is a useful diagnostic technique that can be performed safely by trained specialists in both inpatient and outpatient settings. Clusters of cases of disease may be found in various settings. *Mycobacterium tuberculosis, M. intracelluare,* and *P. aeruginosa,* have been found in these clusters. Most reported bronchoscopy-related outbreaks have been associated with inadequate cleaning and disinfection procedures.

Of the approximately 250,000 people receiving hemodialysis in the United States, approximately 183,000 received chronic hemodialysis. Those with chronic hemodialysis treatments are most apt to pick up bloodstream infections. Outbreaks of disease had been caused by inadequate disinfection of water treatment or distribution systems and reprocessed dialyzers. The bacterial bloodstream infection is a potentially severe complication and one of the most common causes of death among patients undergoing end-stage renal dialysis. Its costs exceed an estimated $1 billion per year.

Methicillin-resistant *Staphylococcus aureus* is becoming an emerging community-acquired pathogen among patients with established risk factors, such as, recent hospitalization, recent surgery, residents in a long-term care facility, or injecting-drug use. Recent deaths in Minnesota and North Dakota have highlighted this problem. The percentage of nosocomial *S. aureus* isolates that were methicillin resistant has increased from 2% in 1974 to approximately 50% in 1997.

Staphylococcus aureus, one of the most common causes of both hospital- and community-acquired infection worldwide, is now showing reduced susceptibility to vancomycin. This is extremely serious because vancomycin, at present, is the last uniformly effective antimicrobial available for treating serious *S. aureus* infections.

Antibiotic-resistant organisms have created a new challenge in infection control. Methicillin-resistant *S. aureus* is generally difficult to eradicate once it invades a hospital facility. The organism can remain viable in the environment for a long period of time. The organisms frequently cause bloodstream infections, pneumonia, osteomyelitis, and bacterial endocarditis. The infections typically occur in larger hospitals where overcrowding exists. Methods of control include isolation, bathing with antimicrobial agents such as chlorhexidine, and aggressive antibiotic treatment.

Enterobacter cloacae, E. agglomerans, and *E. aerogenes* can cause a wide variety of infections including bacteremia, meningitis, brain abscess, osteomyelitis, pneumonitis, and urinary tract infection. They become resistant during B-lactam antibiotic therapy. The increasing prevalence of nosocomial *Enterobacter* species infections may be related to the wide use of broad-spectrum antibiotics. These organisms are typically transmitted from patient to patient by hospital personnel.

Enterococcus faecium has become vancomycin resistant. It is also resistant to ampicillin. The major problem appears to be the large number of patients receiving broad-spectrum antibiotics. Several outbreaks among immunocompromised patients have occurred. Sepsis resulted in death. A resistant strain was involved in an outbreak spread by electronic thermometers.

New epidemics of cholera and dysentery have occurred worldwide. In 1994, outbreaks of intestinal disorders erupted on cruise ships. Virulent strains of *Streptococcus*-A-infections have caused many thousands of severe illnesses in the United States and Europe. Resistant TB has increased sizably. In Cincinnati an epidemic of whooping cough occurred in 1993.

Mycobacterium tuberculosis has reemerged as a multidrug-resistant organism, and has caused an increasing incidence of TB in some geographic areas. The prevalence of TB is not distributed evenly throughout all segments of the U.S. population. Some subgroups or people have a higher risk for TB than the general population. These individuals include contacts of persons who have active TB; foreign-born people from areas of the world with a high prevalence of TB such as Asia, Africa, the Caribbean, and Latin America; medically underserved populations such as some African-Americans, Hispanics, Asians and Pacific Islanders, American Indians, and Alaskan natives; homeless persons; current or former correctional-facility inmates; alcoholics and injecting-drug users; immunosuppressed individuals; and the elderly.

The TB organism is carried in airborne particles, or droplets nuclei, that can be generated when people who have pulmonary or laryngeal TB sneeze, cough, speak, or sing. The particles are an estimated 1 to 5 μm in size, and are spread through normal air currents in the room. Infection occurs when a susceptible person inhales the droplet nuclei containing the microorganisms. Usually, people who become infected have approximately 8 to 10% risk for developing active TB during their lifetime.

The transmission of *M. tuberculosis* is a recognized risk in healthcare facilities. Recently several outbreaks have occurred among HIV-infected people. The interval between diagnosis and death was brief, with a range of 4 to 16 weeks. Atypical mycobacteria are causing substantial illness and death throughout the world despite increased public health controls. Specific diseases caused by *M. ulcerans, M. marinum*, and *M. haemophilium* have increased in both healthy and immunocompromised patients in the last 10 years. These diseases have been reported in areas where they did not previously exist. All three of these organisms cause necrotizing skin lesions. The common reservoir of infection is stagnant or slow-flowing water.

Pseudomonas aeruginosa, an increasingly prevalent opportunistic human pathogen, is the most common Gram-negative bacterium found in nosocomial infections. It is responsible for 16% of nosocomial pneumonia cases, 12% of hospital acquired urinary tract infections, 8% of surgical wound infections, and 10% of bloodstream infections. Immunocompromised patients are at special risk with this organism. Pneumonia and septicemia with attributable deaths reach 30%. Outbreaks of this organism in burn units are associated with 60% death rates. The initial infection occurs when there is a break or breach of normal cutaneous or mucosal barriers such as in trauma, surgery, serious burns, or indwelling devices. The organism produces several extracellular products that after colonization can cause extensive tissue damage, bloodstream invasion, and dissemination.

Group A streptococcus (*S. pyogenes*), a common cause of pharyngitis and uncomplicated skin and soft tissue infections, can cause serious invasive infections including necrotizing fascitis, streptococcal toxic shock syndrome, and death. In 1997, in Maryland and California, 22 postoperative or postpartum patients had group A streptococcus infections. Strains of the organism isolated from patients with invasive disease had been predominantly M types 1 and 3 that produced pyrogenic exotoxin A or B or both. This organism is an old one that has become much more virulent. *Streptococcus pneumoniae* is a leading cause of illness and death in United States. It causes an estimated 3000 cases of meningitis, 50,000 cases of bacteremia, 500,000 cases of pneumonia, and more than 7 million cases of otitis media annually. The organism that had been uniformly susceptible to penicillin is now becoming drug resistant.

As can be seen by the previous information antimicrobial resistance is becoming a growing threat to public health and a growing source of concern to hospitals and other patient care facilities. Each year, nearly 2 million people in the United States get an infection as a result of receiving healthcare in a hospital. A large number of additional people get infections in nursing homes, outpatient centers, and other facilities. Bacteria, fungi, an even viruses can become resistant to drugs. Approximately 70% of the bacteria-causing infections are resistant to one of the drugs most commonly used to treat the infection.

Vancomycin-resistant *Staphylococcus aureus* is the pathogen of greatest concern. The incidence of infection and colonization with vancomycin-resistant bacteria has rapidly increased. Vancomycin-resistant enterococci have increased dramatically since 1989. Recent reports of outbreaks of enterococci have indicated that patient-to-patient transmission of the microorganisms can occur either through direct contact or through indirect contact by the hands of personnel, contaminated equipment, or surfaces. It is also a community problem, especially in Europe, where Avoparcin, a glycopeptide anti-microbial drug, has been used in subtherapeutic doses as a growth promoter in food-producing animals. Viral hemorrhagic fever, Ebola and Marburg, can be spread from person-to-person in the hospital setting. Although it is uncertain how the initial person becomes contaminated, the secondary contamination of other people is due to contact with infected body fluids, unsterilized syringes, needles, and other medical equipment. These are highly infectious diseases.

Modern travel and global commerce quickly spread known and unknown organisms. This, coupled with antibiotic resistance, creates serious new hazards for hospital employees and patients and may lead to sizable increases in hospital-acquired infections.

The extent of these infections and the susceptibility of the patients, whether they are burn patients, use special chemotherapy, or are newborn, elderly, or immuno-compromised, makes it necessary to understand the principles of epidemiology of the spread of these organisms, the means of determining potential hazards, and the institution of environmental, personnel, and patient controls.

Hospital-Wide Problems

1. Lack of training and inadequate supervision are often principal factors in the spread of infection.
2. Movement of personnel and supplies contributes to the dissemination of infectious organisms.
3. Failure to appropriately mark the chart of each infectious case can lead to contamination of unaware personnel.
4. Improper transportation and storage of sterile supplies can cause contamination.
5. Improper transportation and storage of infectious hazardous waste can cause contamination.
6. Failure to properly segregate clean and dirty linen, and failure to properly clean and disinfect bedding may increase the risk of infection.
7. Careless handling of any waste may introduce pathogenic organisms into the environment.
8. Because hospital food is often ground, creamed, and prepared in large quantities, the opportunities for contamination through improper handling and storage is increased.
9. Bedpans and bed urinals improperly removed, handled, stored, washed, and disinfected may cause a cross infection.
10. Supplies, instruments, and equipment may directly or indirectly contaminate patients.
11. Failure in any detail of isolation technique creates a risk of infection from employees, visitors, and other patients.
12. Failure in any detail of proper catheterization technique or follow-up care can cause urinary tract infections.
13. Other procedures, such as inhalation therapy and intravenous cut-downs, may be contributors to infection.
14. The flow of traffic in special areas is of unique importance, because the mere movement of traffic can lead to outbreaks of disease through the distribution of microorganisms.
15. Antibiotic-resistant organisms may colonize on the skin and therefore improper or ineffective treatment may quickly spread infection to others.

Routes of Transmission

Many factors influence the spread of infections within the institution. Therefore, it is frequently impossible to determine the actual causes surrounding the spread of a given disease. However, epidemiological studies should be conducted in all cases, because vital information can be gathered that is of value in efforts to reduce the numbers of organisms in the environment and in the patients. The routes of transmission of disease include droplets or droplet nuclei from person to person, enteric or cutaneous contact, fomites, food, water, insects, and rodents. The disease organisms may come from draining wounds, lesions, secretions, as well as healthy skin. They may come from patients, staff, visitors, or indirectly through the environment. It is known, for instance, that strep and staph infections come from nasal discharges. Staph infections also come from draining wounds, boils, infected decubitus ulcers, sores, hangnails, and infected pimples. Enteric organisms, which cause salmonellosis or shigellosis, are spread by fecal material. Most disease organisms within the institution are probably spread through improper or poor medical–nursing techniques. A smaller, but still significant proportion of organisms are spread through the environment.

Occupational Health and Safety

Hospitals have a real potential for occupational health and safety hazards, patient hazards, and hazards to visitors. The hospital contains not only all the problems and services of a small community but also the additional hazards brought to the institution by the need for special types of equipment, gases, chemicals, and operating procedures. Some specific problems include excessive heat from laundries, kitchens, boiler rooms, and furnace rooms; inadequate illumination in numerous areas of the institution; variety of chemicals, many of which are toxic, explosive, carcinogenic, or possible fire hazards; considerable noise from operational procedures in the physical environment; radioactive materials found in a variety of treatment and diagnostic areas; and research areas where animals contribute to disease and injuries. An in-depth discussion of particular safety hazards would take many pages. It is best to understand that the hazards exist and that they are found not only in the settings discussed but also in any setting within the institution (see Chapter 11 in this volume).

Fires may be caused by problems related to housekeeping, smoking, electrical appliances and their installation, flammable liquids, heating units, explosive materials, painting units, compressed gases, and so forth. Fires are caused by poor housekeeping and storage of solid waste, including combustible debris, oil- or paint-soaked rags, and wood scraps. Smoking is hazardous. Another serious hazard within the institution is the use and storage of flammable and explosive liquids and gases.

Nursing Homes

Special Environmental Problems

Sanitary conditions within nursing homes may change very rapidly as a result of inadequate budgets, insufficient numbers of employees, poor administration, or inadequate building structures and facilities. Personnel are often lacking proper in-service training. A need exists to improve supervision of personnel, establish standard operating procedures, and develop a better understanding of cleaning techniques and environmental controls. In some instances, the quality of food is poor, not only because of inadequate preparation but, more importantly, because of the purchase of cheap food that may come from areas where salvage operations took place. Salmonellosis, staph food poisoning, and other types of foodborne diseases appear regularly within the nursing home.

Contributing to the environmental health problem are the natures of patients and their mental and physical conditions. More than half the patients in nursing homes are disoriented at least part of the time. About 20% are in a state of confusion most of the time. The confused patient contributes not only to the accident problem but also to the spread of disease and infection. An estimated one third of the nursing home patients suffer from urine or feces incontinence or both. Obviously these wastes are excellent sources of contamination within the institutions. When incontinency combines with mental confusion and lack of adequate staff supervision, the chance of spreading environmentally related diseases increases sharply. Further, in many of the older homes, a problem with insect and rodent infestation exists. Even in the newer homes, infestations occur and can be spread because of lack of personnel and personnel training, and hoarding of food, providing a constant source for insects and rodents.

Special Safety Problems

The average age of the nursing home resident is 80 years. Residents generally have debilitating or crippling diseases, making them more prone to accidents. About 40% of the patients have cardiovascular conditions; about 75% have some heart or circulatory difficulties; 16% have the residuals of paralytic strokes; and about 10% are receiving care for fractures, primarily hip fractures. All these conditions may be coupled with arthritis, mental disorders, and diabetes. The physical conditions of the patients, plus their frequent intake of medications, make them prime subjects for accidents within the nursing home.

Other Special Problems

There is a perceived shortage of nursing personnel either because of lack of trained people or because of financial cutbacks. This causes unskilled staff to assume duties requiring training. The general quality of patient care is affected by this shortage. Unskilled labor is used within the nursing home to take care of many of the tasks of running the facility and to care for patients. The individuals who fill these positions receive low pay, are overworked and improperly trained, and unfortunately are often improperly screened to determine their temperament and emotional stability. A serious concern for potential abuse exists.

A lack of communication exists between hospitals and nursing homes. Records are not transferred quickly enough or in enough detail for the nursing home or the hospital emergency personnel to be able to care properly for the patient. Physicians, because of large workloads, make infrequent visits. When emergencies do occur, it is difficult to reach the necessary physicians to provide adequate patient care.

Problems arise with the patients, because they are prone to chronic illness and multiple impairments and disabilities. Many feel rejected by friends and family. Patients may believe they have come to the nursing home to die, and may lose the will to care for themselves and to recover and utilize their physical resources in the best way possible.

Enforcement is difficult because of small numbers of trained public health personnel, lengthy court procedures, frequent changes in federal regulations, budgetary problems, and general confusion concerning proper operation of the nursing homes. The high cost of nursing home care is alarming. The number of state and federal dollars supplied is substantial.

Old Age Homes

The types of problems found in the old age home vary with the number of people and the condition of the facility. They may be simple problems of inadequately prepared food, insects, and rodents, up to the many complicated problems found within the nursing home situation. It is often difficult to determine when an individual simply needs assistance with everyday life and when an individual needs nursing care. Because the population of the United States will increasingly fall into the 65-and-over category, proper evaluation of old age homes is essential to determine whether they are performing their necessary function in our society.

Prisons

Environmental health practitioners have recognized the problems in prisons for many years. Prison riots, such as those in Attica, NY, have further emphasized the problems and needs. The U.S. District Court for Alabama in the case *James vs. Wallace* held that the conditions of confinement within the Alabama penal system violated the judicial definition of "cruel and unusual punishment," and therefore improvements had to be made within the prisons. The improvements included reducing overcrowding; improving general living conditions; and improving food service, education, recreation, and rehabilitation programs. The court not only defined the problems but also stipulated that the prisons comply with the recommendations of the court or the court would order the prisons closed.

Several groups, including the American Public Health Association-National Environmental Health Association (APHA-NEHA) Jails and Prisons task force, the Federal Bureau of Prisons, the federal government, and the state of Kentucky, have conducted studies and made recommendations to improve penal institutions. A model code proposal included specific details on food, water, air, liquid waste, solid waste, vectors, recreational environment, housing, accidents and safety, industrial hygiene, public facilities, and emergency plans. It was obvious from the discussions and studies that an environmental health specialist was needed in penal and correctional systems to assist in changing the environment of the institution.

Recent outbreaks of TB have occurred in prison housing units for HIV-infected inmates. The organism spreads rapidly to other inmates. It may be readily transmitted to employees and visitors, and ultimately to the outside community.

Bedding

Frequently, bedding is unclean or in bad repair. Many institutions do not provide sheets or pillows. Linens and blankets, when available, are often dirty.

Food

Food problems are intimately related to the foodservice facilities and to the prisoners, who are the food handlers. Food is often contaminated from improper purchasing and storage, as well as inadequate training in food preparation, food hazards, and personal hygiene. The floors, walls, ceilings, and ventilation hoods are in poor repair in many institutions and contribute to the overall lack of cleanliness and the potential spread of foodborne disease.

Heat and Ventilation

Many buildings, because of their age and lack of repairs, provide inadequate heat and improper ventilation. Windows behind the bars have been smashed or are nonexistent. Proper heating and ventilation of buildings are necessary in providing a reasonable environment.

Insect and Rodent Control

Insects and rodents are found in about 25 to 35% of the institutions. Again this is due to the poor maintenance of buildings, to the age of the buildings, and to the nature of the institutions. In many cases, inadequate control procedures are used because of lack of proper understanding of insect and rodent control.

Laundry

Laundry service is not available in certain types of institutions unless friends or relatives bring clean clothing to the prisoners. Where laundries are present, they tend to be extremely hot and poorly ventilated.

Lighting

About one half of all the institutions surveyed in the state of Kentucky had inadequate illumination. About 45% of the light fixtures were in poor repair. Poor lighting was typically found in many institutions.

Plumbing

Plumbing problems abound within the penal institutions because of their age, overuse, and overcrowding. There is a serious lack of toilets and sinks. In many cases, the toilets, sinks, and shower facilities are in bad repair, dirty, or almost totally inoperative.

Safety

Many institutions lack adequate fire extinguishers or proper techniques available to take care of prisoners in the event of fires or accidents.

Occupational Health

Occupational health practices need improvement to conform with Occupational Safety and Health Administration (OSHA) regulations.

Schools

Environmental Health Problems

Many schools, ranging from child care centers to elementary through college, are substandard because of the existence of environmental health problems.

Food

The basic food protection problems found in the hospital apply also to the school. The additional concern that exists for children in substandard urban areas is that the school lunch may be the only nutritious meal received within the course of a day. For this reason, it is even more important that foodservice personnel provide a wholesome, safe meal in an attractive manner.

Within schools, mass feeding leads to leftovers, and leftovers and poor food-handling practices lead to serious outbreaks of foodborne disease. Bagged lunches become a particular hazard, because they are stored in school lockers until lunchtime. If food is perishable, the student may contract a foodborne disease. Often dust, dirt, roaches, and mice may get on the food bag or into the food. In universities, students have refrigerators in their rooms or take food back to the room for later snacking. This also leads to problems of potential foodborne disease and insect and rodent infestations.

Gymnasiums

Many accidents occur as a result of poor supervision, poor equipment, improperly trained individuals, poor lighting, horseplay, and so forth. Locker rooms present hazards, particularly through slipping, and sources for the spread of fungal and bacterial infections.

Laboratories

The problems of school laboratories are similar to those of hospitals, except that in junior and senior high schools fewer dangerous substances are used. The hazards are not only related to the types of substances used and the by-products of these substances, but more importantly to the inexperience of the students.

Plumbing

The special problems encountered in schools include the students stuffing the toilets with toilet paper, paper towels, and other objects, as well as wantonly destroying the facilities.

Roughly half the schools lack adequate hand washing facilities, soap, and single-service towels. About 10% lack adequate toilet paper. Because hands are probably the most common means of transmission of infectious diseases, it is important that these items be provided in the bathrooms.

Shops

Depending on the type of school, a variety of shops are used. These shops constitute a hazard for the students.

Ventilation, Heating, and Air Conditioning

Air heating, air cooling, humidity control, and air distribution are essential for the removal of body odors and for the comfort of the individual. Where air circulation is inadequate, temperature of the room is raised by body heat, humidity is raised by occupants' breath and perspiration, organic matter is released into the room causing odors, oxygen content is reduced, and carbon dioxide is increased. All these factors contribute to stuffiness and odors, and to an uncomfortable environment, contributing to drowsiness that may lead to accidents.

Water

In the school a specific problem occurs when water fountains are misused. All water fountains should be of an angular instead of a bubble type. Students should not have to put their mouths directly on the water source because this becomes a vehicle of contamination for other students.

Accidents

Accidents are the leading cause of death in all student age groups. About 16,000 students of school age lose their lives each year. Motor vehicle-related deaths account for 58% of these deaths, drownings, 15%, and fires and burns, 7%. Of these accidental deaths, 43% are associated with social life. About half of these occur in school buildings. The most hazardous areas are the gymnasium, where one third of the school-related deaths occur; the halls and stairs, where one fifth of the school-related deaths occur; shops and laboratories, where one fifth of the deaths occur; and other kinds of classrooms, where one seventh of the deaths occur. About 40% of the deaths occur on school grounds. Football is the most hazardous of sports.

Playground safety is a serious concern in the United States. Each year approximately 211,000 children receive emergency room care for injuries from playground equipment. This is the leading cause of injuries to children in school and to child care environments. Surveys conducted under the National Program for Playground Safety, at the University of Northern Iowa, indicate that inadequate supervision is the key contributor to playground injuries. Children who play on equipment inappropriate for their size, strength, and decision-making ability increase their injury risk. Of playground injuries, 70% involve falls to the ground, indicating that shock absorbent surfaces should be in place, such as sand, wood chips, or rubber. Poor equipment maintenance contributes to the problem.

Traumatic brain injury, which occurs primarily among children (especially adolescents and young adults) and people over 75, results in 50,000 deaths per year, 230,000 hospitalizations, and 1 million people treated and released from hospital emergency rooms. Children frequently fall on their heads. Traumatic spinal cord injuries occur in 10,000 people yearly. The cost of traumatic brain injuries is $37 billion a year, and the cost of spinal cord injuries is over $6 billion a year.

In a study of 1,900,000 students, it was determined that 24% of the injuries occurred within school buildings; 28%, on school grounds; 5%, going to and from school; 20%, at home; and 23%, in other types of situations.

The college is a complete, highly active community. It contains all the community safety problems, all types of industrial safety problems found in laboratories and industrial shop areas, and various types of home and recreational safety problems. In a study conducted in 22 colleges and universities serving 207,000 students, roughly 15,000 injuries were reported to the student health services. Of these injuries, 75% were incurred by men and 25% by women. Of the campus injuries 52% were either in the athletic or recreational facilities, 20% were in the residence halls, 15% were in the academic buildings, 11% were on the campus grounds, and 2% were in motor vehicles. It is obvious from these studies that schools, regardless of the level, can be dangerous places.

Bus Safety

Each year school buses are involved in accidents and in fatal injuries. This occurs because of the numbers of school buses traveling on the roads, poor roads and road conditions, bad weather, and in some cases inexperienced bus drivers. Many school buses are old, improperly maintained, and therefore, hazardous. Constant checks should be made of the equipment by utilizing systematic maintenance programs. All buses should be checked for levels of carbon monoxide at the back of the bus; carbon monoxide may seep in through the emergency door.

Hazardous Materials and Equipment

The storage and disposal of this material should be controlled by knowledgeable individuals. Laboratories present a special area of concern because of the lack of training of the student and the presence of potentially flammable liquids, explosive reactions, compressed gases, electrical units, and toxic substances.

Diseases

Many of the diseases found in hospitals and other healthcare settings as well as other institutional settings such as salmonellosis, shigellosis, and other oral–fecal route diseases may be found in child care settings as well as in schools. Typical childhood diseases, colds, flu, etc. may also be found. Recent outbreaks of invasive group A streptococcus have occurred in child care centers. Although this organism can cause common childhood diseases such as streptococcal pharyngitis and impetigo, it can also cause severe life-threatening invasive diseases including streptococcal toxic shock syndrome and necrotizing fascitis. In Boston, Toronto, and southern California, cases of these diseases have been documented.

Bacterial meningitis has occurred in child care settings. Meningitis, an inflammation of the membranes that carry the brain and spinal cord, may be caused by infection with either bacteria or viruses. Symptoms of the disease include sudden onset of fever, headache, neck pain or stiffness, vomiting, and irritability. In children the symptoms may rapidly progress to decreased consciousness, convulsions, and death. The common bacteria causing the disease are *Neisseria meningitis*, *S. pneumoniae*, or *Hemophilus influenzae* serotype b. These bacteria are carried in the upper back part of the throat (nasopharynx) of an infected person and are spread either through the air or by direct contact with secretions from the infected person.

Asthma, a chronic breathing disorder, is the most common chronic health problem among children. Children with asthma have attacks of coughing, wheezing, and shortness of breath, which may become very serious. Spasms of the air passages occur in the lungs due to swelling, inflammation, and increased mucus. Common cold viruses exacerbate this problem. Attacks may also be caused by:

1. Cigarette smoke
2. Stress
3. Strenuous exercise
4. Weather conditions, including cold, windy, or rainy days
5. Allergies to animals, pets, pollen, or mold
6. Indoor air pollutants, such as paint, cleaning materials, chemicals, or perfumes
7. Outdoor air pollutants, such as ozone

Pinworms are found in the child care setting. They are tiny parasitic worms that live in the large intestine. The female worms lay their eggs around the anus at night. Symptoms include anal itching, sleeplessness, and irritability. The condition may also be asymptomatic. Pinworms are readily spread within the child care setting. Head lice are spread by direct head-to-head contact. Brushes, combs, and linen are factors.

POTENTIAL FOR INTERVENTION

The potential for intervention is based on isolation, substitution, shielding, treatment, and prevention. All these techniques can be utilized to a greater or lesser extent in the institutional environment. The potential for intervention in hospitals is very good. In nursing homes and old age homes, the potential varies with the type of facility, the kind of staff, and the amount of administrative control. In schools the potential for intervention should be excellent. In prisons, it is poor.

Within all these areas isolation is a technique used to separate the contaminated or infected individual from the rest of the population. Isolation also refers to the separation of hazardous substances and gases, such as radiological material, hazardous chemicals, and explosive gases. Substitution is the use of a less hazardous chemical or technique for a more hazardous chemical or technique. Shielding is the utilization of special cabinets with special hood systems where highly dangerous or infectious materials are used. Shielding also refers to personal protective devices, such as gloves, gowns, masks, caps, and safety glasses. Treatment is utilized in all institutional settings for exposure to disease processes and chemical, physical, and radiological hazards. Prevention is the principal technique that should be used within the institutional setting.

RESOURCES

Scientific and technical resources include the national, local, and state chapters of American Medical Association; American Public Health Association; American College of Hospital Administrators; American Hospital Association; Association for Professionals in Infection Control, and Epidemiology, Inc. (202-296-2742); Catholic Hospital Association; Joint Commission of Accreditation of Hospitals; medical schools within all of the states; and assorted environmental health practitioners. In addition, other important sources include Foodborne Diseases Active Surveillance Network (Foodnet); National Notifiable Disease Surveillance System; Public Health Laboratory Information System; National Ambulatory Medical Care Survey; National Hospital Discharge Survey; WHONET, an information system developed to support the World Health Organization (WHO); BaCon Study (bacterial contamination of blood) sponsored by the Association of Blood Banks, American Red Cross, CDC, etc.; and SEARCH, a network of voluntary participants including hospitals, private industries, professional organizations, and state health departments to report the isolation of *S. aureus* with reduced susceptibility to vancomycin.

Civic resources consist of the various auxiliaries to the medical associations and to the hospitals, and a variety of foundations that provide funds for hospitals, such as the Ford Foundation and the Rockefeller Foundation.

Governmental resources include state health departments, National Bureau of Standards, Department of Defense, National Institutes of Health, Food and Drug Administration, National Science Foundation, Veterans Administration, USDHHS, and so forth. The CDC has issued these seven guidelines for the prevention and control of nosocomial infections:

1. *Guideline for Prevention of Catheter-Associated Urinary Tract Infections*
2. *Guideline for Hand Washing and Hospital Environmental Control*
3. *Guideline for Infection Control in Hospital Personnel*

4. *Guideline for Prevention of Intravascular Infections*
5. *Guideline for Isolation Precautions in Hospitals*
6. *Guideline for Prevention of Nosocomial Pneumonia*
7. *Guideline for Prevention of Surgical Wound Infections*

Numerous codes and standards should be utilized in the actual construction of an institutional facility or in the modernization of existing facilities. These codes and standards may be obtained from the following sources:

1. Superintendent of Documents, U.S. Government Printing Office, Washington, D.C. (for all government documents)
2. American National Standards Institute, 1340 Broadway, New York, NY
3. American Society for Testing and Materials, 1916 Race Street, Philadelphia, PA
4. American Society of Heating, Refrigerating, and Air Conditioning Engineers, United Engineering Center, 345 East 47th Street, New York, NY
5. Compressed Gas Association, 500 Fifth Avenue, New York, NY
6. International Conference of Building Officials, 50 South Los Robles, Pasadena, CA
7. National Association of Plumbing-Heating-Cooling Contractors, 1016 Twentieth Street N.W., Washington, D.C.
8. National Council on Radiation Protection, 4210 Connecticut Avenue N.W., Washington, D.C.
9. National Fire Protection Association, 60 Batterymarch Street, Boston, MA
10. Underwriters Laboratories, Inc., 207 East Ohio Street, Chicago, IL
11. National Bureau of Standards, U.S. Department of Commerce, Washington, D.C.
12. American Hospital Association, 840 North Lakeshore Drive, Chicago, IL

For an in-depth discussion of biohazard control, it is recommended that the reader obtain the book titled *Biosafety in Microbiological and Biomedical Laboratories*, 4th ed., U.S. Department of Health and Human Services, Centers for Disease Control and Prevention and National Institutes of Health, May 1999, U.S. Government Printing Office, Washington, D.C., 1999.

Very little has been written in the area of penal institution environmental health. It is recommended that the reader explore the publication *Corrections* by the National Advisory Commission on Criminal Justice, Standards, and Goals, Washington, D.C., published January 23, 1973, as a base of study. The most knowledgeable source on prisons is the Federal Bureau of Prisons, 320 1st Street, N.W., Washington, D.C.

STANDARDS, PRACTICES, AND TECHNIQUES

Hospitals

Physical Plant and Site Selection

The site of the medical facility must be well drained and free of obstructions or other natural hazards (see site selection for communities in Chapter 8 on the indoor environment). Its location should be accessible for fire apparatus, ambulances, and other types of service vehicles. Public transportation systems should be available to the patients and staff. Competent medical and surgical consultation from private practitioners or from medical schools should also be readily available. Paved roads need to be provided from the main entrance to the buildings and the emergency section. Paved walkways are necessary for pedestrian traffic. Off-street parking should be provided and be easily accessible to patients, staff, and visitors. Special curbing facilities and walkways are needed for wheelchair patients and the physically disabled. Parking spots located close to the entrance should be reserved for the handicapped. All public facilities, including bathroom facilities, should be designed for use by the handicapped. The medical facility should be accessible to major highways and conveniently located within reasonable distance of interstate highways to expedite the flow of patients or emergency vehicles to the institution. Heavily trafficked areas should be avoided. In planning the location, it is necessary to consider the path and proximity of existing air-polluting industries, future construction of industries, and possibility of flooding and natural disasters due to fault lines, underground mines, and other potential hazards.

According to the National Environmental Policy Act — Public Law 91-190, the site and the project of the hospital must be approved before construction can begin. An impact statement should be prepared, and any possible adverse environmental problems should be shown and techniques of control explained.

Interior Structural Requirements

The following items are requirements for the various parts of the typical hospital. The general nursing unit includes:

1. Patient rooms that have a maximum capacity of four patients.
2. Minimum floor area is 100 ft^2 for a single bedroom plus 80 ft^2 for each additional bed.
3. Each room should have a window that opens and is above the floor or grade level. The openings should be limited or screened to prevent accidents.
4. A nurses' call system should be provided in each room and in the bathrooms.
5. Each patient room should have its own lavatory or be attached to a bathroom so that the patient does not have to go into the central corridor to utilize the toilets or sinks.
6. Each room should be provided with wardrobes, lockers, or closets.
7. Provision should be made for visual privacy of the patients.
8. A series of service areas should be provided for the nursing unit to function properly, including nurses' station; nurses' office; storage and supply room; convenient hand washing facilities; charting facility; staff lounge and toilet facilities; staff lockers; conference rooms; patient examining rooms, with a minimum floor space of 120 ft^2; clean workrooms and holding rooms; dirty workrooms and holding rooms; drug distribution stations; clean linen storage closets; nourishment stations; equipment storage; parking areas for stretchers and wheelchairs away from the normal traffic flow; patient bathing facilities; and emergency equipment storage facility.

The nursing facility also needs isolation rooms that are completely and totally separated from all other patients and patient activity rooms. Isolation rooms must have private toilet facilities, sinks, and dressing areas for change of clothing and decontamination of soiled isolation gowns, masks, and so forth. Each hospital needs rooms for emotionally or mentally disturbed patients.

The ICU has critical space requirements, because it is used for seriously ill medical, surgical, or coronary patients. These patients are acutely aware of their surroundings and may be easily affected by them. As a result, it is necessary to eliminate all unnecessary noise, to provide individual privacy in keeping with the medical needs of the patient, and to provide adequate observation by the staff. Patients should have some outside view through a window. The window should not be higher that 5 ft above floor level.

The ICU includes the following:

1. Coronary patients should be housed in single bedrooms.
2. Medical and surgical patients should be housed in either single- or multiunit bedrooms.
3. All beds must be under direct visual observation by the nursing staff.
4. The clearance between beds in a multiunit bedroom must be a minimum of 7 ft.
5. Single bedrooms must have a minimum clear floor area of 120 ft^2 and at least 10 ft as a minimum width or length.
6. A single bedroom has to be provided for each patient for medical or surgical isolation.
7. Viewing panels, which can be covered by curtains for privacy, must be installed in the doors and walls to provide adequate visual coverage by the nursing staff.
8. IV solution supports must be provided for each patient so that they do not hang over the patient.
9. A hand washing sink is required in each patient room.
10. A nurses' calling system is needed and should be required.
11. Each coronary patient must have direct access to a toilet facility.
12. The location of the nurses' station should permit direct visual observation of the patient.
13. Hand washing facilities should be conveniently located at the nurses' station.
14. Charting facilities must be separated from the monitoring service.
15. The staff restrooms should contain toilets and adequate hand washing facilities.
16. Storage areas must be close to nursing personnel.
17. A clean workroom and a system of storage and distribution of clean and sterile supplies are mandatory.
18. A soiled or dirty workroom for proper storage of soiled equipment and supplies is necessary.
19. Bedpan flushing and sanitizing units are required.
20. A convenient 24-hour distribution system and station for medications is needed within the nursing area.
21. A clean linen storage closet is needed.
22. A nourishment station should be provided.
23. A storage room with easy access is needed for emergency equipment, such as inhalators and crash carts.

24. Patient lockers must be furnished for patients' personal effects.
25. A separate waiting room, including toilet accommodations, seating accommodations, and telephones, is needed for visitors.

Newborn nurseries can be a serious hazard to newborn infants if the design is poor and special precautions are not taken to protect the infants. The following criteria are essential in a newborn nursery:

1. One nursery should not open directly into another.
2. Nursery should be conveniently located to the postpartum nursing unit and obstetric facilities.
3. Nurseries should contain hand washing sinks at a ratio of 1 to 8 bassinets; emergency nurses' calling system; observation windows to permit viewing of the infants in public areas; charting facilities; full-term nurseries containing a maximum of 8 bassinets, or as many as 16 if the bassinets are isolets (minimum floor area for a bassinet should be 24 ft^2); special care nursery for high-risk infants and infants in distress (minimum of 40 ft^2 per bassinet); and workroom space.
4. Examination and treatment rooms or space for infants containing work counters, storage, and hand washing sinks.
5. Infant formula rooms should be provided if on-site formula preparation is conducted. However, it is recommended that for safety and also for economic reasons commercially prepared formulas should be utilized.
6. A closet for storage of all housekeeping supplies is needed.

The pediatric and adolescent unit is separated from the adult units to provide a quieter setting for the adults and to prevent the spread of childhood disease to the adult population and conversely. The pediatric nursery follows the same requirements as the other nurseries. The pediatric rooms conform to the same requirements as the other rooms within the general nursing facilities, with the additional need for an area used for dining, educational, or play purposes. Also, special storage closets and cabinets for storage of toys (that can be sanitized) are needed, and educational or recreational equipment should be installed.

The psychiatric unit should be constructed to provide a safe residence for patients and one in which patients are unable to hide, escape, injure themselves, or commit suicide. It must be flexible to care for the ambulatory inpatients and also to meet the needs of various types of psychiatric therapy. The unit should convey a noninstitutional atmosphere to facilitate recovery. The requirements for construction are as follows:

1. Maximum of four patients are allowed per room.
2. Minimum room area of clear floor space for single rooms should be 100 ft^2; multiple-patient rooms require 80 ft^2 per bed.
3. Each room should have a window with limited opening space and security screens.
4. Sinks must be provided in each patient's room.
5. Bathrooms must be provided in such a way that the patient does not have to go into the corridor to go to the bathroom. The bathroom door should unlock from the outside.

6. All other usual facilities found within a normal nursing unit are required, with the exception of all items that could be hazardous. These should be locked away and out of reach of the patients.
7. A special space for dining, recreation, and occupational therapy is needed, with a minimum floor space of 40 ft^2 per patient.

The number and types of operating rooms are based on the anticipated surgical load for the hospital. The surgical suite construction should be designed to prevent the possibility of contamination. The surgical suite must be located in a separate traffic area to exclude individuals unassociated with surgery. Surgical suites should meet the following requirements:

1. A minimum clear floor area of 360 ft^2, exclusive of all cabinets and shelves, per room, with the minimum dimension of 18 ft
2. An emergency communication system
3. Two x-ray film illuminators for each room
4. Special storage space for splints and traction equipment in orthopedic surgery rooms
5. A minimum clear floor area of 250 ft^2 for cystoscopic and other surgical endoscopic procedures
6. Recovery rooms arranged for visual control of patients, drug distribution, hand washing, and charting
7. Additional recovery space for outpatient surgical patients
8. Special service areas containing control stations, for visual surveillance of all traffic entering and leaving the operating suite; supervisors' office and station; sterilizing facilities with high-speed autoclaves, conveniently located near the operating rooms; drug distribution station; two scrub stations adjacent to and contiguous with the entrance to each operating room; dirty workrooms for all soiled equipment and materials; fluid waste disposal facilities convenient to the operating room; clean workrooms and supply rooms for clean and sterile supplies; anesthesia storage facilities, where flammable gases can be stored without causing a hazard; anesthesia workrooms for cleaning, testing, and storage of anesthesia equipment; medical gas storage areas; special equipment storage areas; emergency equipment storage areas; staff locker rooms, with adequate space for cleaning, changing, and resting; outpatient surgery change areas; preop holding areas that are under visual control of the nurses; stretcher storage areas out of the direct line of traffic; staff lounges; closets for storage of housekeeping supplies

Maximum capacity needs must be anticipated when constructing delivery and labor rooms to provide adequate facilities. Construction requirements for obstetric facilities include:

1. Delivery rooms should have a minimum clear floor area of 300 ft^2 with minimum dimension of 16 ft.
2. Rooms should have emergency communication systems.
3. Resuscitation facilities are required for newborn infants, including oxygen, suction, and compressed air.
4. Labor rooms should have a minimum floor area of 80 ft^2 per bed.
5. A minimum of two labor beds per delivery room are required.
6. Hand washing facilities should be provided within the labor room.

7. Adequate visual observation of the patients is needed.
8. Recovery rooms with a minimum of two beds and all ancillary facilities are required.
9. The usual service areas, as stipulated in the surgical suites should be provided.

The emergency suite is an extremely critical part of the entire hospital. When an emergency occurs, this is the place where a life is saved or lost. The emergency suite is a combination outpatient clinic, operating room for minor injuries, admissions area, and surgical suite. It is essential not only that the emergency room staff is properly trained and efficient in dealing with patients but also that the facilities are designed to facilitate emergency care. The emergency room facility should contain the following:

1. Well-marked entrance at grade level that is sheltered from the weather and easily accessible to ambulances and pedestrians
2. Reception and control area conveniently located near the entrance
3. Public waiting areas with bathroom facilities, telephones, and drinking fountains
4. Stretcher and wheelchair storage areas
5. Treatment rooms equipped with hand-washing facilities, containers, cabinets, and work counters
6. Treatment rooms containing medical suction, storage of emergency equipment such as emergency trays, defibrillators, cardiac monitors, and resuscitators
7. Staff work and charting areas
8. Clean supply storage areas
9. Sterile supply areas
10. Daily workroom areas for supplies and equipment
11. Patient restrooms
12. Communications equipment and an area where adequate communications between physicians at the hospital and ambulances can be provided
13. Adequate emergency surgical care and treatment in the event of a serious immediate surgical need when the individual cannot be taken to the operating room

The communications system is of vital importance, because the physician can provide guidance to the emergency medical technicians (EMTs) or paramedics on the care of a given patient. It is also useful in alerting the hospital to the type of arriving patient and the specific needs, so that the physician and nurses may be prepared to start treating the patient immediately.

The hospital also provides outpatient areas, clinics and other services, including diagnostic clinics, laboratories, physical therapy suites, occupational therapy suites, mortuary and autopsy rooms, pharmacies, administrative areas, public areas, medical records units, central supply, linen services or laundry, housekeeping, food areas, and solid waste disposal areas.

Electrical Equipment

All electrical equipment and facilities must adhere to the standards of the Illuminating Engineering Society, the Institute of Electrical and Electronic Engineering, Inc., and the codes and standards of the National Fire Protection Association. The hospital should have at least two separate sources of electricity, so that in the event that one fails the second one can immediately go into effect. The emergency system may be operated either as a separate service from the electric company or on portable generators or storage batteries. It is important that the emergency system be tested regularly to ensure working order should the regular system become inoperative. Because motors operate at a specific voltage, safety precautions should be provided in the event that voltage drops or increases sharply. The telephone and electric power conductors are not permitted in the same raceways, boxes, cabinets, and so forth. Any switchboards or panel boards should be protected to prevent shock. All hospital wiring must be placed in conduits to facilitate alterations and repairs. The conduits must be large enough to allow for expansion when needed. Receptacles should be installed in all required areas and at convenient heights. All receptacles should be grounded and explosion proof in locations where anesthetic gases are used. Patient rooms should have at least three double receptacles for a single bedroom with two outlets near the head of the bed. Multiunit bedrooms should also have two double receptacles at the head of each bed.

A variety of clocks are used in institutions; two types are wired and electronic. In the wired system individual clocks are controlled by a master clock; in the electronic system no control wiring connection is required. In locations where anesthetic or other hazardous gases are used, conductive flooring is required. The electrical resistance of these floors should be only moderately conductive to provide a conductive path for neutralizing static charges and to act as a resistor to limit current to the floor from the electrical system in the case of an electrical short. The possibility of electrostatic sparks causing ignition of the gases must be eliminated. As a further precaution, the electrical distribution system should be ungrounded in the operating and delivery rooms, with the exception of fixed nonadjustable lighting fixtures 8 ft or more above the floor and permanently installed x-ray tubes. All operating room floors and delivery room floors have to be monitored periodically for the buildup of static electricity. Generally, improper cleaning or removal of detergents and detergent residues contribute to an increase in electrostatic electricity.

Lighting fixtures must be installed to meet the comfort and work requirements of the particular workspace. Because these requirements vary, it is important to obtain from the Illuminating Engineering Society the current lists of lighting levels recommended for hospitals or other institutions. The voltage supplied to the x-ray unit should be nearly constant so that images and pictures can be uniform and consistently reproducible. It is recommended that an independent feeder with capacity sufficient to prevent a voltage drop greater than 2% be used. Elevators should be manufactured, inspected, and maintained in strict excellent working condition. Skilled elevator mechanics should be utilized for repairs. In the event of an emergency, the alternate source of electrical service in the elevator should automatically be shifted to elevator usage. Only one elevator should be so arranged.

The communications system within the hospital may include loudspeakers, chimes, coded bells, radio communications, telephones, buzzer systems, light signals, closed circuit television, and independently carried receivers. Because the communications system is so essential to the operation of the hospital, it is necessary that emergency electricity be provided for its operation in the event that the normal system goes out of order.

In addition, modern hospitals have remote dictation services. Fire alarm systems are necessary in every hospital. A manually operated system should be available and utilized, with the system tied in with the local fire department. Special electrical installations include the facilities of the electroencephalographic areas and medical electronic equipment areas.

Heating and Air Conditioning

Heating and air-conditioning systems should be designed to provide the following temperatures and humidities: operating room, 65 to 76°F at 45 to 60% relative humidity; delivery room, same as operating room; recovery room, 75 to 80°F at 50 to 60% relative humidity; nursery (observation for full term), 75 to 80°F at 50% relative humidity; nursery (premature), 75 to 80°F at 50 to 60% relative humidity; intensive care, 70 to 80°F at 30 to 60% relative humidity; and all other occupied areas at 75°F minimum temperature.

The ventilation system must be mechanically operated, with a minimum separation of 25 ft from air intake to air exhaust. The outdoor air intake should be a minimum of 8 ft above ground level, preferably on the roof. If on the roof, it should be 3 ft above roof level. The ventilation system must be designed to provide the general pressure relationships required in various areas of the hospital. For example, the air pressure should be positive within the operating room and negative within the corridors; and negative within the isolation room and positive in the corridors. In the operating room, the fresh air supply should be delivered from the ceiling and the air exhaust at floor level. Corridors should not be used to supply or exhaust air, with the exception of bathrooms or janitors' closets, which open directly onto the corridors. The ventilation systems in the operating and delivery rooms should have final filters located downstream from the main coil systems and a 99.7% efficiency rating in removing micron-sized particles. All other areas can have an efficiency rating of 80 to 90%. A manometer used in the central air systems will measure the flow of air through the system. All isolation rooms, laboratories, and other areas where potential environmental hazards exist must have ducts to the outside. Air should not be recirculated from these rooms to any other part of the hospital. In food preparation areas, the minimum exhaust rate is 100 ft^3/min/ft^2 of hood space area. The ventilation system in anesthesia and other rooms where hazardous gas is used must meet requirements of the National Fire Protection Association. Boiler rooms require sufficient outdoor air to maintain adequate combustion in the equipment.

Flammable and Explosive Liquids and Gases

In storing flammable and explosive materials, the following precautions should be used:

1. Gasoline, other flammable liquids, and gases should be handled and stored in specially designed containers and should be kept in well-ventilated areas. All flammable substances should be appropriately labeled.
2. Cleaning fluids should have high flash points.

3. Containers with flammable liquids should always be sealed when not in use, and caps should be replaced promptly after use.
4. All flammable liquid tanks should have vent pipes and accessible shut-off valves.
5. Flammable liquids and gases should not be kept in areas where individuals could trigger an explosion or fire.
6. All spills of flammable liquids should be cleaned immediately.
7. Approved fire extinguishers should be kept in areas where flammable liquids or gases are utilized.

Structural Features

The structural features of the hospital building must comply with established building, fire, and electrical codes. All footings must lie below the frost line, and the foundations must rest on solid earth or on pilings or piers that have been predetermined to be able to withstand the weight of the structure to avoid detrimental settling. The building materials must in all cases be fireproof. The minimum corridor widths must be 8 ft, 10 in., and the minimum door widths should be 3 ft, 10 in. All doors leading to patient rooms and bathrooms should be easily opened from the outside in the event of an emergency. Thresholds and expansion joint covers must be flush with the floor. Drinking fountains, telephone booths, vending machines, and other objects should not block corridors or exits. All parts of the building below the grade level must be waterproof and built to prevent the entrance of surface water. Storage rooms, patient bathrooms, and so forth must have a minimum width of 7 ft, 6 in. All patient rooms should have at least one window leading to the outside.

Boiler rooms, food preparation areas, mechanical equipment rooms, and laundries need to be insulated and ventilated to prevent the floor surfaces from exceeding normal floor temperatures. It is important that the boiler rooms not be located underneath patient or patient service areas. In the event of a fire or explosion, this could cause serious loss of life or injury.

Adequate cooling systems are needed for all areas of the hospital. Indirect cooling is most desirable in surgery suites, delivery areas, nurseries, emergency, and intensive care areas. The exhausting of air from various critical areas should be done in such a way that it does not interfere with the air movement patterns of the rest of the institution.

Where solid waste or linen chutes are used, they must comply with the applicable National Fire Protection Standards, which include approved glass service openings; chute openings located in rooms with a minimum of 1-hour fire-resistant construction; minimum diameter of the chute must be 2 ft; chutes must discharge directly into the refuse room, or in the case of linen into the linen room; and chutes must have sprinklers and provisions for washing the chutes and appropriate areas.

Housekeeping Services

One of the most effective techniques for organizing housekeeping services is to establish good work plans. The first step in planning work is to identify the type of surfaces to be cleaned and the amount of congestion, plus the potential hazards within the area. Once these have been identified and specified, time allotments are made for the amount of work to be produced by the average housekeeping person. The actual work scheduling is set up based on the competency of the individual, the kind of equipment available, and the type of problems encountered during cleaning. A daily cleaning schedule should be established for floor washing, dusting with treated cloths, and cleaning of bathrooms and other specialized rooms. This schedule should be followed on a daily basis. In the daily cleaning, a general-purpose, good detergent disinfectant should be utilized. Equipment used in a contaminated area should not be brought into other areas, because this is a direct means of spreading organisms from one area to another. Housekeeping personnel entering contaminated areas must take the same precautions in the use of caps, gowns, gloves, and masks as other personnel, to prevent the spread of infection.

Periodic cleaning or heavy-duty cleaning should be performed in areas where patients can be moved or in areas where patients are not present at all times. This includes cleaning of windows, walls, and stripping of wax and sealers from floors by the use of mechanical equipment. Electrical equipment should be utilized for this work, and the individuals should be specifically trained in the cleaning procedures and proper use of the equipment.

Special cleaning procedures should be carried out in isolation rooms or when terminal cleaning of the room is called for, after the patient with an infection or contagious disease has been transferred, discharged, or has died. Special cleaning techniques are also used in operating rooms, delivery rooms, nurseries, and rooms where reverse isolation has taken place. In special cleaning, it is essential that all equipment and materials be fresh, clean, and free of contaminating organisms. The cleaning equipment must not be transferred from special cleaning areas to other areas of the hospital. All disposable materials and equipment should be marked, and placed in special plastic bags to be disposed of separately. Equipment should be washed, cleaned, and sanitized before reuse when possible. All personnel should perform a 2-min medical scrub before entering and after leaving these areas.

Lighting

The desired levels of lighting vary tremendously from the areas where social activities and gatherings may take place with a minimum of 10 to 15 fc, to the cystoscopic tables, where a minimum of 2500 fc of light is required. It is suggested that the environmentalist obtain lighting standards from the various Illumination Engineering Society reports that are current for a given time period. Because innumerable places within the institution have different lighting requirements, it is difficult to keep an updated list in any text.

Plumbing and Water

All sinks used for patient care or in-service areas must have a water spout a minimum of 5 in. above the rim of the fixture. In surgical areas and in patient isolation areas, foot controls are necessary for the staff to cleanse hands before and after patient care or treatment. It is particularly important that water storage, such as distilling tanks, is tested frequently, because the water could become contaminated. Beside the normal water supply, institutions need a minimum pressure on the upper floors of 20 lb/in.^{2} during maximum demand periods. Additional water supply and water sources are needed for firefighting. These water supplies should be kept separate from the potable drinking water. (Further information concerning water, plumbing, sewage, cross-connections, and submerged inlets is found in Chapters 3, 5, and 6, Volume II.)

Emergency Medical Services

Typical ambulance specifications include front-disk power brakes, eight-cylinder engine, speed of at least 70 mi/hr when fully loaded, three forward transmission gears, and heavy-duty double-action shock absorbers. Another requirement is ambulance rescue equipment, including a number 10 ABC dry chemical fire extinguisher, primer wrecking bar, ropes, bolt cutters, and so forth. A third requirement is emergency care equipment, such as portable suction apparatus, bag mask ventilation unit, and oral pharyngeal airways for adults, children, and infants.

Training requirements include emergency medical technician courses successfully completed and passing grade on a written and practical examination, 24 hours of in-service instruction each year, and U.S. Department of Transportation emergency medical technician refresher training courses.

Fire Safety

The fire program should come under the control of a fire marshall or fire director trained by the fire department. The fire marshall should be a member of the overall safety committee and should be responsible for evaluating all fire hazards within the institution. The fire marshall, in conjunction with appropriate administrators, should determine the institutional needs and provide the following: supervision of fire prevention, protection, and suppression programs for the staff; administrative supervision and support for all subareas within the institution; coordination of all activities of all subareas; final authority on all matters related to fire prevention, protection, and suppression; assurance that periodic inspections are made of all parts of the building and that all equipment, facilities, procedures, and regulations are in keeping with good fire prevention policy; and leadership as administrative head should fires occur.

The subunit fire staffs, composed of individuals from various operating departments of the hospital, have responsibility for working directly in all program activities or emergencies under the fire director; conducting training of fire protection and suppression staffs; understanding the effects of extinguishing agents, such as cooling, smothering, anticatalytic, or inhibiting; scheduling and conducting fire-prevention inspections; reviewing and monitoring all activities that constitute fire hazards; holding regular conferences with all individuals involved when fire incidents occur; reviewing and approving plans and drawings for construction and alteration of buildings to ensure adequate fire protection; reviewing and approving storage plans for flammable materials; approving requests for open flame operations, such as welding; making technical investigations concerning fire incidents; preparing and publishing seasonal fire protection and prevention material; enforcing fire standards and rules in various work areas; helping develop, coordinate, and present plans for emergency evacuation of personnel, fire control, and salvage of property; conducting fire and evacuation drills; evaluating workers' knowledge of fire extinguishers, fire practice, and how to report fires; developing and posting operational instructions for extinguishers and fire hoses; posting appropriate posters and signs concerning fire problems; and working closely with the fire department.

In the event of a fire at an institution, a specific code should be used to alert the institutional personnel concerning the whereabouts and severity of the fire. At the same time, the fire department should automatically be notified of the existence of a fire and its location. Immediate action should be taken to remove individuals from the fire area and to extinguish the fire if possible. Generally, however, in the event of serious fires, fire department personnel should be the ones responsible for extinguishing the fire. Evacuation plans should be fully understood and personnel should be well trained in necessary evacuation procedures.

Fire equipment on the premises should be checked periodically to ensure that it is in proper operating condition. This should be done by institutional personnel and by the fire marshall's office.

Typical fire prevention rules in an institution include keeping stairwell doors closed at all times; keeping room and corridor doors closed; discarding all solid waste promptly; placing waste chemicals in special safety cans and removing them through specialized hazardous waste collection companies; providing adequate ashtrays; and keeping corridors free of obstructions.

Patient Safety

Patient safety rules include keeping the adjustable height beds in the low position, except during actual treatment; using safety or side rails on the bed; using safety vests that restrict but allow movement; properly indoctrinating patients to the hospital environment; using a good communications system, including signals within the bathrooms; properly positioning bedside tables and other furniture; properly using and instructing use of patients' aids such as walkers, wheelchairs, and canes; installing shower bars; providing stools within showers; providing adequate illumination during the day and night; giving special attention to patients most frequently prone to accidents, such as those on medication, the elderly, and those with diseases causing dizziness; properly medicating at designated times to avoid medication errors.

Radiation Safety

X-ray equipment should be monitored by competent individuals for stray radiation and improper functioning. Factors in radiation safety include the x-ray beam size, x-ray machine filtration, shielding, scatter measurements, tube-housing leakage, and fluoroscopic exposure rates. Surveys are conducted to determine the level of leaking radiation and to suggest recommendations for necessary corrections. The handling of radioactive material is of grave concern. The materials must be handled in such a way that the operator and the patient are not contaminated. Several specific controls include providing restricted areas for use of radioactive material; eliminating pipetting of radioisotopes by mouth; using impervious gloves; using proper cleaning techniques; and storing radioisotopes in containers that limit exposure to 100 mrem/hr maximum at a distance of 1 ft from the container. Hot radioactive materials must be transported in shatterproof containers and shielded, depending on the type of radiation. Radioactive wastes are stored in metal cans and removed to restricted areas, where they are disposed of using special techniques.

Animals containing radioactive materials are tagged, and all waste material, including their excreta, are collected and disposed of as all other radioactive material would be. The carcasses are wrapped in absorbent material and disposed of in a safe manner. Human sources of radioactive material include patients containing radionuclides that are used for treatment or diagnostic purposes. The internal beta–gamma emitters must be kept at the lowest practical level. Patients should not receive a total dose exceeding 10 Gy in any 12-month period. External beta–gamma emitters must not exceed the levels of the internal beta–gamma emitters. Because internal alpha emitters are particularly hazardous, special consideration must be given by a committee on radionuclides before they are used on humans. Nursing care for patients receiving radioactive isotopes must be carried out with the greatest caution, because the individuals or the nurses may have contaminated skin, may be contaminated by inhalation or ingestion of the radioactive materials, or may have their bodies irradiated by these materials. These problems are largely eliminated by good housekeeping, proper hand washing techniques, and clean work habits. In addition, equipment must be handled with great care.

Infectious Waste Management

An infectious waste management program provides protection to human health and environment from infectious waste hazards. An infectious waste management system should include the following:

1. Designation of infectious waste should be given.
2. Handling of infectious waste should be provided, including segregation, packaging, storage, transport and handling, treatment techniques, disposal of treated waste.
3. Contingency planning and staff training should be provided. The program needs to be designed for the particular situation. Such factors as the number of patients, the nature of the infectious waste, the quantity of infectious waste generated, the availability of equipment for treatment on-site and off-site, the regulatory constraints, and the cost are important in planning. The method of treatment varies with the waste type. The waste must be evaluated for its potential to cause disease. Such characteristics as the chemical content, density, water content, and microbiological content must be evaluated. Steam sterilization may be used for laboratory cultures, whereas incineration should be used for pathological waste. When possible, it is best to handle all infectious waste in the same way to cut down on

cost. In any planning program, it is necessary to determine the local standards for incinerators, regulations for water quality, the regulations and standards relating to chemical pollutants, thermal discharges, biological oxygen demand, total suspended solids, regulation of sharps, and any other environmental regulations at the local or state level.

4. A responsible person at the facility should prepare a comprehensive report that outlines the policies and procedures for the management of all infectious and hazardous waste in the facility. Reports should contain detailed procedures for all phases of the waste management program at the facility, including research, clinical laboratories, autopsy rooms, and other types of infectious waste from a variety of areas of the institution.

It is necessary to determine which of the wastes at the institution are to be considered infectious, and specific policies and instructions concerning this waste need to be presented in written material. The infectious waste should then be separated from the general waste stream to make sure that these wastes receive appropriate handling and treatment. The infectious waste should be segregated from the general waste stream at the point at which it is generated, either in the patient areas or in the laboratories. Infectious waste should be discarded directly into containers or plastic bags and marked with the universal biological hazard symbol, which is three circles with a gap overlaying a smaller circle. The symbol is fluorescent orange or orange red in color.

The infectious waste should be packaged to protect the waste handlers and the public from possible disease and injury. The packaging must be preserved through handling, storage, transportation, and treatment. In selecting the packaging container, it is necessary to determine the type of waste, the handling and transport of the packaged waste before treatment, the treatment technique, the special types of plastic bags, and the package identification. Liquid waste should be placed in capped or tightly stoppered bottles or flasks. Containment tanks can be used for large quantities of liquid waste. Solid or semisolid waste, such as pathological waste, animal carcasses, and laboratory waste, may be placed in plastic bags.

Sharps should be placed directly into impervious, rigid, and puncture-resistant containers to eliminate injury. The clipping of needles is not recommended, unless the clipping device contains the needle parts. Otherwise, an airborne hazard may occur. Sharps should be placed in such containers as glass, metal, rigid plastic, wood, and heavy cardboard. The sharps containers should be marked with the universal biohazard symbol. A single bag for containment of infectious waste is not adequate. Either a double plastic bag should be used or a single bag should be placed in a bucket, box, or carton that has a lid or a seal. Containers of sharps and liquids can be placed in other containers for transportation and storage.

When plastic bags of infectious waste are handled and transported, care must be taken to prevent tearing of the bags. Plastic bags containing infectious waste should not be placed in a trash chute or on a dumbwaiter. The proper practices for the handling of these plastic bags include loading by hand, transporting in such a way that minimal handling occurs, and putting the plastic bags in rigid or semirigid containers to prevent spills or breakage of the bags. Where recycled containers are used for transport in the treatment of bag waste, they should be disinfected between use. If the waste is to be incinerated, the containers must be combustible. If the waste is to be steam sterilized, the packaging materials must allow steam penetration and evacuation of air. Interference with the steam sterilization treatment may occur because of high-density plastics, which prevent effective treatment of the air, therefore not allowing the contents of the bag to reach the appropriate temperatures. Low-density plastics enhance steam penetration and allow air evacuation from the waste.

Several types of plastic bags are available for waste disposal. The quality of the plastic bag and its suitability to contain infectious waste is based on the raw materials used during manufacture; and the product specification suitability is based on the fitness and the durability of the plastic. It is most important that the plastic bag be tear resistant until the waste is treated. Tear resistance can be improved by not placing sharps, sharp items, or items with sharp corners in the bags or overloading the bag beyond its weight and volume capacity.

Although it is preferable to treat the infectious waste as soon as possible, it may not be practical to do so. Therefore, the waste has to be stored in an appropriate manner. Four important factors related to the storing of infectious waste are the packaging, storage temperature, duration of storage, and location and design of the storage area. The packaging must provide containment of the waste throughout the waste management process, and must deter rodents and insects from invading the packaging and thereby potentially becoming vectors of disease. As the temperature increases, the rate of microbial growth and putrefaction increases. This causes bad odors to be associated with the waste because of the retained organic matter present. Time is also a major consideration in treatment of infectious waste. It is suggested that the waste be stored under refrigeration in special units at 34 to 45°F for a maximum of 3 days. The storage areas should be extremely secure, have limited access, be free of rodents and insects, and display the universal biohazard label.

The infectious waste when transported on- or off-site should be in appropriate packaging, which has already been mentioned, and in containers that are leakproof, rigid, and appropriately marked. The regulation that describes the requirements for the packaging and shipping of biomedical material is found in 42 Code of Federal Regulation (CFR), Part 72, titled "Interstate Shipment of Etiologic Agents." This regulation is enforced by the CDC. The infectious waste is treated to reduce the hazard associated with the presence of infectious agents. This is discussed further in a later section on controls.

Once the infectious waste has been effectively treated, it is no longer biologically hazardous and may be mixed with and disposed of as ordinary waste, providing it does not pose any other hazards that are controlled by federal or state regulations. Treated liquid waste may be poured down the drain to the sewer system. Treated solid waste and incinerator ash may be disposed of in a sanitary landfill. Needles and syringes must be rendered nonusable before disposal. Treated sharps can be ground up, incinerated, or compacted. Body parts should be incinerated in special incinerators or buried, depending on religious law.

The institution should set up a contingency plan, which would provide for the following situations:

1. Spills of liquid infectious waste and cleanup procedures should be handled by personnel. Also, procedures should be available for the protection of the personnel and disposal of the spilled residue.
2. Rupture of plastic bags or other loss of contaminants and the necessary cleanup procedures, protection of personnel, and the repackaging of the waste should be provided.
3. Equipment failure and the alternate arrangements for waste storage and treatment should be planned.

Staff training is essential so that all technical, scientific, housekeeping, and maintenance personnel are protected. The training programs should be used for all new employees whenever infectious waste methods and practices change and when a new program is put into operation. Continuing education, including refresher courses, should be given periodically to protect the employees of the institution.

Medical Scrub

The medical scrub, which should take a minimum of 2 min, should be practiced by all personnel when hand washing procedures are indicated. The medical scrub consists of the following steps:

1. Remove all jewelry, rings, watches, etc.
2. Turn on the water (running water is essential).
3. Wet hands.
4. Soap hands with an appropriate hand washing and cleansing agent and wash them. Use considerable friction. Begin with hands, lead to wrists, then to elbows (be sure to clean well between fingers and around thumb).
5. Rinse (rinse hands, arms to elbow, letting water run off the elbow).
6. Clean nails under running water. Use orange stick and discard.
7. Repeat steps 4 to 6.
8. Be aware of skin irritation and the potential causes of it.
9. Dry hands. Wipe water from hands to elbow with one paper towel. Use second paper towel to finish drying.
10. Turn off faucet. Use second paper towel if knee-action control is absent. Use paper towels to turn off the faucets.

Essential Practices Related to Acquired Immunodeficiency Syndrome Virus and Hepatitis Virus

The essential practices related to the prevention of the transmission of the HIV and HBV viruses include the following:

1. Classification of the work activity
2. Development of standard operating procedures
3. Provision of training and education
4. Development of procedures to ensure and monitor compliance
5. Workplace redesign

Further, the worker needs to have available a program of medical management. This program should reduce the risk of infection by HBV and HIV. All workers involved in any exposure to blood or other body fluids should be vaccinated with hepatitis B vaccine.

Once an exposure has occurred to blood or other infectious body fluids and the person has not been adequately protected, a blood sample should be taken from the individual (with the individual's permission) to determine whether the surface antigen of hepatitis B or the antibody to HIV is present. If the individual has not previously been given the hepatitis B vaccine, a single dose of hepatitis B immune globulin is recommended within 7 days of exposure. The individual should also receive the vaccine series. If the person is possibly exposed to AIDS, it is recommended that the CDC in Atlanta, GA, be contacted and the most recent techniques approved by the CDC be followed.

Employees need to take precautions to prevent injuries from needles, scalpel blades, and other sharp instruments or devices that they use during various procedures and that they may handle during cleaning, disposal, or removal of these objects.

Schools

Physical Plant and Site Selection

Site selection is based on present and projected student population; distance from areas served; distance from sources of noise, air pollution, and other hazards; type of surface drainage; type and nature of industrial or highway construction; amount of acreage needed for athletic fields, parking, landscaping, and other recreational areas; access to good highways; suitability for a given neighborhood. The site is also determined by the type of soil; the soil conditions; and the recommendations of school architects, engineers, public health officials, and community groups.

The physical structure or structures should be designed and built to facilitate learning and to reduce the potential for hazards. The building construction must take into account all necessary lighting and acoustics, heating and ventilation, water supply and waste disposal, laboratories and gymnasiums, workshops, and classrooms. Before the plans are approved, the school architect submits them to an advisory committee of school personnel, and community groups, and finally to the state departments of education and health.

Housekeeping

A clean environment is conducive to a happy school day and therefore promotes good emotional health and the desire to learn. Many of the housekeeping tools and techniques described under the section on hospitals are appropriate for schools. (The book titled *Environmental Health and Safety* by this author mentioned earlier contains detailed housekeeping procedures.) Proper organization of the housekeeping or custodial staff, adequate scheduling, proper training of personnel, and use of proper cleaning agents and detergent disinfectants are essential to a good housekeeping program. The program should include, at the minimum, the following responsibilities for housekeeping personnel: maintenance of a clean, safe, and aesthetically satisfying environment; maintenance of grounds to prevent accidents and to promote aesthetic appeal; maintenance and orderliness of all apparatus and equipment; prevention of fires and promotion of fire safety; proper operation of heating, ventilation, and all other types of equipment; adequate storage of all materials utilized in construction; separate and proper storage of all hazardous equipment and materials, including chemicals and other potential fire hazards; and maintenance of proper records concerning housekeeping.

A variety of flooring materials are used within the school as within other kinds of institutions. These flooring materials include concrete, terrazzo, ceramic tile, rubber or vinyl tile, asphalt tile, magnesite, cork, and assorted woods. Concrete floors are quite common in school buildings. They are easy to clean, yet they sometimes pit or crack. Adequate maintenance is important. If the concrete is painted, it should be thoroughly cleaned with solvents to remove all oils and greases. Terrazzo floors are a mixture of cement and marble, or cement and granite chips. They are strong and durable, but acids, abrasives, and strong alkaline cleaners cannot be used because they will deface the terrazzo. It is important to put a sealer on this type of floor to prevent penetration. Ceramic tile floors, rubber or vinyl tile floors, and asphalt tile floors are not resistant to grease, oils, turpentine, or petroleum-based waxes. These floors look very attractive and must be given special care. Magnesite floors are somewhat like concrete and should be treated similarly. Cork floors may be cleaned by sanding lightly with steel wool and applying light water-emulsion wax. Wood floors consist of a variety of woods, ranging from very hard to soft. Although they are very pretty if maintained properly, many wood floors have difficulty withstanding the constant pressure of footsteps.

A special problem within schools are chalkboards. Black slate boards are difficult to clean and pose a problem for use in classroom instruction. It is preferable to use green or blue boards or other kinds of special materials. Erasers are best cleaned by using a special electric vacuum eraser cleaner. This should be carried out once or twice per week.

The following lighting standards are recommended: 10 fc of light for open corridors and storerooms; 15 fc for auditoriums, cafeterias, locker rooms, and stairways; 20 fc for reception rooms, gymnasiums, and swimming areas; 75 fc for classrooms, study halls, lecture rooms, libraries, shops, and laboratories; and 150 fc for drafting rooms, typing rooms, sewing rooms, and special learning rooms.

Maintenance is an essential part of lighting. A program should be established to include periodic cleaning of lamps and fixtures, replacement of burned out bulbs, and repainting of room surfaces where needed. Student seating should be arranged to eliminate shadows and glare. Window shades, drapes, or other types of window coverings should be provided to give balanced lighting and to eliminate glare problems.

Noise

Noise prevention starts during construction of the school. Site location, classroom design, and use of special noise-control materials are important. Noise must be suppressed, because it is a source of irritation, distraction, and emotional strain, and fosters inefficient performance in school children, teachers, and employees. Noise levels should not exceed the following: classrooms, 35 to 50 dB; cafeterias, 50 to 55 dB; outside noise, 70 dB; and other specialized rooms, 40 dB.

Space Utilization

Each elementary school should have a minimum of 12 acres of land for the first 200 students and 1 acre of land for each 100 additional students. In the kindergarten, 1050 ft^2 of space is needed for a maximum of 30 students. In grades one through three, 875 ft^2 are required for 30 students; and in grades four through six, 750 ft^2 per 30 students. In high schools, which include junior and senior high schools, 12 acres of ground are required for the first 300 students and 1 acre for each 100 additional students. Also, in high schools, 85% of the classrooms must have a minimum of 750 ft^2 for each group of 30 students. Other classroom sizes vary with the types of activities. All these requirements are minimum standards that can and should be exceeded by the school system for special types of study or where more space is desirable.

Plumbing

One sink should be provided for every 30 students in elementary schools; and one sink for every 40 students, in secondary schools. In elementary schools, a toilet is required for every 40 male students and one for every 30 female students. In secondary schools, one toilet is required for every 75 male students and one for every 45 female students. There should be at least 1 shower head for every 5 students and 1 drinking fountain for every 100 students.

The recommended classroom temperatures during the winter should be 68°F with slightly lower temperatures in gymnasiums. The temperature should be approximately 70°F in elementary schools. Fresh air should be added to the school at a rate of not less than 15 ft^3 of air per minute per person per room. Where the activity level is higher, the amount of room air should be increased to 20 or 30 ft^3 per minute per person. The air must be distributed within the room to reduce a chilling effect and to prevent discomfort. The velocity of air should not exceed 500 ft/min at the duct outlets. A recommended relative humidity would be 50%. The type of heating system varies with the school, its location, and the type of fuels available.

MODES OF SURVEILLANCE AND EVALUATION

Risk Assessment in Laboratories

Risk implies the probability that harm, injury, or disease will occur. In the microbiological and biomedical laboratories the assessment of risk is based primarily on the prevention of laboratory-associated infections. It helps to assign the biosafety levels for facilities, equipment, and practices to reduce the risk to the worker and the environment. In this sense, risk assessment can be qualitative or quantitative. In the case of certain chemicals, quantitative assessments can be done and level of risk determined. In the case of an unknown agent or unlabeled sample, the quantitative data may be incomplete or even absent. A few of the problems to be resolved include types, subtypes, and variants of infectious agents involving different or unusual vectors; difficulty of assays to measure an agent's amplification potential; and unique considerations of genetic recombinants. When performing a qualitative risk assessment, all the risk factors are first identified and explored. These factors are:

1. The pathogenicity of the infectious agent including disease incidence and severity.
2. The route of transmission — parenteral, airborne, or ingestive. The greater the aerosol potential the higher the risk
3. Agent stability involving the ability of the organism to survive over time in the environment. Taken into consideration are such factors as desiccation, exposure to sunlight or ultraviolet light, exposure to chemical disinfectants, and time.
4. The infectious dose of the agent and the complex nature of the interaction between the microorganisms and the host immune system
5. Concerning the concentration of the organism and the material in which the organism exists — there is a difference between the organism in solid tissue vs. blood or sputum or liquid. The volume of the concentrated material is important, as well as the number of times the material is handled.
6. The origin of the infectious material. Is it from a domestic or a foreign host? Is it from a human or an animal?
7. Use of animal studies when human data are not available
8. The availability of an effective prophylaxis or therapeutic intervention
9. Use of medical surveillance to monitor employee health
10. The experience and skill level of at-risk personnel in laboratories, maintenance, housekeeping, and animal care

Infections

Several techniques are used to determine the extent of possible hospital-acquired infections. The first technique is to make in-depth intensive microbiological studies of all critical areas and other areas of the hospital environment. These samples are taken through the air or by means of swabs and Rodac™ contact plates used on surfaces. As a general rule, the microbiological sampling technique is not effective and should only be utilized where a specific organism is sought and the cause of the disease outbreak cannot be determined. Microbiological sampling is also used on a research basis to determine the levels of microorganisms in given types of operations or where special study techniques have been used and the sources of the outbreak of disease cannot be traced.

A second technique involves the development of a surveillance and reporting program in which a nurse epidemiologist regularly carries out the following functions:

1. Reviews daily charts indicating that the patient has an infection or is suspected of having an infection
2. Holds conferences on a daily basis with nursing supervisors and nursing units to determine potential infection problems

3. Reviews all laboratory reports confirming infections and correlates the laboratory studies with the observations of the physicians and nurses
4. Reviews special reports submitted by physicians or supervising nurses where suspected infection problems are occurring
5. Lists and tries to determine the location and source of specific types of infections
6. Evaluates necropsy reports to determine if infections not previously reported are found at time of autopsy
7. Maintains a close liaison with the employee health center and evaluates employee infections as they may relate to hospital acquired infections
8. Prepares monthly statistical reports indicating levels of infection as determined by charts and laboratory tests and pinpoints where the higher levels of infection occur

The nurse epidemiologist may work for the hospital environmental health specialist, whose responsibility would encompass the overall environmental health, infection control, and occupational health and safety programs of the institution.

The third type of technique is a programmed approach for a team composed of the supervising nurse, the physician in charge of the unit, and the environmental health specialist, using specifically established study forms to determine the potential sources of infection within the institution. This team approach, systematically carried out, helps to establish the potential sources of infection, thereby assisting the individuals within their own unit and also in the hospital at large in determining the kinds of potential infections occurring within the institution. The programmed approach and the study form are found in Diagram 9.1 and Figure 9.4.

Accidents

The hospital building should be carefully evaluated and supervised to minimize accidents and injuries. It is important to evaluate all accidents and illnesses to determine whether any hazards were involved and to make the necessary recommendations to eliminate the hazardous situation or unsafe behavioral pattern. When investigating the incident, all possible facts should be collected. The equipment operation should be fully understood; processes, special conditions, fatigue factor, and other environmental factors related to the individual should also be known. The direct causes of an accident, such as faulty operating equipment or use of improper techniques, and the indirect causes, such as poor eyesight and lack of knowledge, should be determined. Additional contributory causes may be inadequate standards, inadequate training, lack of policy, lack of enforcement, faulty design, faulty maintenance, or failure of supervisors to perform their functions.

Reporting and recording of information notifies the institutional management of the incident, the results of the incident, and the contributory causes; in this way management can conduct proper evaluations and take necessary actions to prevent future injuries. Surveys, special studies, and investigations before accidents occur are of vital importance. A good survey should determine specific unsafe or unhealthy acts and conditions; determine the need for specific safeguards for people, equipment, and materials; detect inefficient, inadequate, or totally lacking safety programs; determine the condition of work areas, structural components, and service facilities; determine the condition of equipment; determine job procedures; identify the various types of

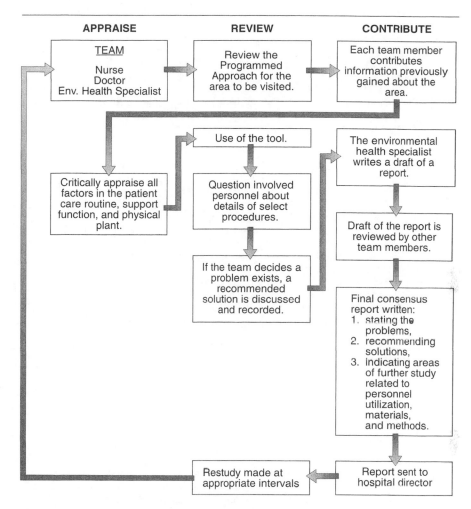

APPRAISE	REVIEW	CONTRIBUTE
TEAM Nurse Doctor Env. Health Specialist	Review the Programmed Approach for the area to be visited.	Each team member contributes information previously gained about the area.
Critically appraise all factors in the patient care routine, support function, and physical plant.	Use of the tool.	The environmental health specialist writes a draft of a report.
	Question involved personnel about details of select procedures.	Draft of the report is reviewed by other team members.
	If the team decides a problem exists, a recommended solution is discussed and recorded.	Final consensus report written: 1. stating the problems, 2. recommending solutions, 3. indicating areas of further study related to personnel utilization, materials, and methods.
	Restudy made at appropriate intervals	Report sent to hospital director

Diagram 9.1   Programmed approach to hospital infection control.

materials, chemicals, and other hazardous substances in use; identify the controls that should be employed to prevent accidents; determine whether these controls are actually operable; determine all the potential problems related to the environment, such as lighting, ventilation, heat, obstructions, slippery surfaces, and so forth; determine the kinds of existing hazards and the types of controls used for each type of operation.

Evaluation by the occupational health specialist includes routine surveys of chemical laboratories, storage, and handling areas; sampling and detection of airborne contaminants; evaluation of noise, lighting, temperature, ventilation, and so forth; regular testing of chemical fume hoods and biological safety cabinets to ensure proper functioning; evaluation and investigation of employee and student accidents; consultation services for special problems to make necessary recommendations for correction; routine surveys of shops, laboratories, and other specialized areas; and special studies concerning special hazards.

———————————————————————————————————— Hospital

PROGRAMMED APPROACH TO HOSPITAL INFECTION CONTROL

S=Satisfactory   I=Improvement   U=Unsatisfactory

Building Name and No.——————————— Floor————— Date ————Time ————

Names of Study Team————————————————————————————————————

ADMINISTRATIVE PROCEDURES

——— 1. Does any employee have a patient history of being a carrier of pyogenic infections?
——— 2. Do all employees report pyogenic infections?
——— 3. Is the employee who reports an infection immediately removed from patient care?
——— 4. Is there a restriction of traffic within the given patient area?
——— 5. Is there routine daily recording of infections on charts?
——— 6. Is indiscriminate antibiotic use discouraged?
——— 7. Are the interns given instruction in hospital infection control?
——— 8. Are the residents given instruction in hospital infection control?
——— 9. Are the nurses given instruction in hospital infection control?
———10. Are nonprofessional employees given instruction in hospital infection control?
———11. Are visitors restricted in time and number?
———12. Do visitors observe special isolation procedures when indicated?

STERILIZATION PROCEDURES

——— 1. Is each autoclave tested monthly for its ability to destroy bacterial spores?
——— 2. Are autoclave indicators or dates used?
——— 3. Are routine cultures made of utensils, trays, packs, instruments, catheters, gloves, sutures and parenteal solutions?
——— 4. Are records kept of all isolations of significant bacteria such as "staph"?
——— 5. Are all autoclaved items distinctly marked as sterile?
——— 6. Are autoclaved items stored separately from nonsterile items?
——— 7. Are autoclaved items protected from contamination while in storage?

LAUNDRY AND LINEN PROCEDURES

——— 1. Are linens washed in such a manner that all pathogenic bacteria are destroyed?
——— 2. Are bacteriological samples taken of the clean linen to insure that all pathogenic bacteria are destroyed?
——— 3. Are the clean, ironed linens exposed to the dirty linens in such a way that there may be a transfer of pathogenic bacteria?

Figure 9.4   Programmed approach study form. (From Koren, H., *Environmental Health and Safety,* Pergamon Press, Elmsford, NY, p. 140. With permission.)

_____ 4. Are clean linens protected during transportation to ward areas?
_____ 5. Are clean linens transported on the same truck as dirty linens?
_____ 6. Are all linens, blankets, and other bed covers gently removed from beds to prevent air dispersion of bacteria?
_____ 7. Are all linens from infected patients placed in a specially marked laundry bag and closed at once?
_____ 8. Do all linens from infected patients receive special handling?
_____ 9. Are laundry bags from isolation placed in clean outer bags?
_____10. Are all blankets sterilized before reuse?
_____11. Are mattresses washed with a germicide between patients?
_____12. Are pillows washed with a germicide between patients?
_____13. Are beds washed with a germicide between patients?
_____14. Are nonporous covers used for mattresses and pillows?
_____15. Are the laundry trucks and carts cleaned daily?
_____16. When not in use, are the trucks stored in an area which is free of gross contamination?

TRASH AND SEPTIC WASTE REMOVAL PROCEDURES

_____ 1. Are infectious materials either autoclaved or placed in sealed plastic bags before being discarded?
_____ 2. Is food waste from infected patients placed in sealed plastic bags?
_____ 3. Are paper bags used in isolation rooms for discarded tissues, dressings, tongue blades, and other disposable items?
_____ 4. Are they immediately closed and placed in a sealed plastic bag?
_____ 5. Are bags containing infectious materials marked "Danger, Infectious Materials Within"?
_____ 6. Do employees who remove infectious materials discard their gowns at the completion of the waste disposal and then follow proper hand washing procedures?
_____ 7. Do employees who remove infectious materials wear gowns?

FOOD HANDLING PROCEDURES FOR INFECTED PATIENTS

_____ 1. Are all clean dishes, trays, and eating utensils protected against contamination during transportation or storage?
_____ 2. Are disposable single service paper units discarded after each use?
_____ 3. Are all remnants of meals scraped into a plastic bag in the isolation room?
_____ 4. Are nondisposable utensils thoroughly and immediately disinfected after each use by infected patients?
_____ 5. Are all dishes, trays, and eating utensils washed and sanitized before reuse?
_____ 6. Is ice making equipment and ice storage equipment kept clean and sanitary?
_____ 7. Are there individual water containers and glasses for each patient?
_____ 8. Are these containers and glasses thoroughly washed and sanitized before reuse?
_____ 9. Are utensils for medications thoroughly washed and sanitized before reuse?

Figure 9.4 Continued.

BEDPAN HANDLING

_____ 1. Are all bedpans, urinals, enema containers and other equipment for collection of excreta sterilized before reuse by other patients?
_____ 2. Are all such contaminated utensils handled in an aseptic manner after use by patients?
_____ 3. Are bedpans stored off the floor in a cabinet?
_____ 4. Is toilet paper stored off the floor in a sanitary manner?

EQUIPMENT HANDLING AND SANITIZATION PROCEDURES

_____ 1. Is there an individual thermometer for each patient?
_____ 2. Are thermometers stored in an adequate concentration of germicide to kill all bacteria?
_____ 3. When equipment such as portable X-ray units, flashlights, otoscopes, opthalmoscopes, stethoscopes, ice bags, etc., are used for infectious cases are they cleansed thoroughly with a germicide solution before reuse?
_____ 4. Is there a Graduate Nurse assigned to the use of the dressing cart?

ISOLATION PROCEDURES

_____ 1. Are all patients with draining wounds isolated?
_____ 2. Is reverse isolation strictly enforced when indicated?
_____ 3. Are patients isolated in such a way and in such areas that proper procedure can be carried out?
_____ 4. Are isolated patients properly identified?
_____ 5. Is a suspected infected case reported the day it is noted?
_____ 6. Are clean gowns put on before entering isolation rooms?
_____ 7. Are the gowns discarded in containers in the isolation area?
_____ 8. Are hands washed with an acceptable technique before and after handling any contaminated material or patients?
_____ 9. Are all areas in which infectious cases have been treated thoroughly cleaned and sanitized before reuse?

CATHETERIZATION PROCEDURES

_____ 1. Are the genitalia thoroughly cleaned with an appropriate soap in preparation for the catheter?
_____ 2. Does the catheter come from a sterile package?
_____ 3. Does the syringe used for irrigation of the catheter come from a sterile package?
_____ 4. Are hands washed before donning sterile gloves?
_____ 5. Do the gloves used for sterilization come from a sterile package?
_____ 6. Does the urine drainage bottle overflow?
_____ 7. Is the urine bottle suspended above the floor?
_____ 8. Is the urine bottle autoclaved?
_____ 9. Is there a cotton plug in the air vent of the catheter?
_____10. Is an antiseptic present in the urine bottle?

DRESSING OF WOUNDS

_____ 1. Are dressings put on with gloved hands?

Figure 9.4 Continued.

_____ 2. Are hands washed before and after dressing patients' wounds?
_____ 3. Are dressing trays and carts arranged to prevent contamination of sterile and clean equipment?
_____ 4. Are dirty drapes or dirty dressings kept on dressing carts?
_____ 5. Does cold sterilization take place on the dressing cart?
_____ 6. Are wounds dressed at specifically scheduled times?

SURGICAL FACILITIES

_____ 1. Are the operating suites separate from each other and other patient areas?
_____ 2. Is cross traffic through these units from other hospital areas eliminated?
_____ 3. Are the dressing rooms for physicians and nurses accessible directly from the rest of the hospital and do they open directly into the operating suites?
_____ 4. Is there adequate separation between utility areas used for clean and those used for contaminated equipment, supplies, or waste?
_____ 5. Are all surgical wastes placed in plastic bags for disposal?
_____ 6. Are sub-sterilizers (rapid instrument sterilizers) accessible to operatiing rooms?
_____ 7. Are facilities for scrubbing directly accessible to operating rooms?
_____ 8. Are these facilities kept clean at all times?
_____ 9. Are locker rooms kept clean at all times?
_____ 10. Is there adequate space in clean areas for storage of sterilized items?
_____ 11. Are there facilities in the recovery room for isolation of infected cases?
_____ 12. Are there closed containers for all contaminated and soiled dressings and other contaminated material?
_____ 13. Are there adequate closed containers for disposal of gowns, masks, and booties in the locker room?

ANESTHESIA AND INHALATION THERAPY

_____ 1. Inhalation
a. Is the equipment cleaned and sterilized after use by patients?
_____ 2. Intravenous
a. Do the needles and syringes come in a sterile package?
b. Does the plastic tubing come in a sterile package?
_____ 3. Spinal
a. Is detergent used in cleaning needles?
b. Is all equipment sterilized?
c. Are solutions sterilized?
d. Are contents of ampoules sterile?
_____ 4. Conduction Anesthesia (local)
a. Are gowns, masks, gloves, worn when anesthesiologist does conduction anesthesia?
_____ 5. Is the anesthesia table stored outside of the O.R. within a traffic area?
_____ 6. Are endotracheal tubes cleaned and then sanitized in 70% alcohol plus Formalin? (1 part Formalin to 250 parts alcohol.)
_____ 7. Is the endotracheal tube then placed in a boiling sterilizer to boil off the Formalin?

Figure 9.4   Continued.

_____ 8. Is humidification maintained during inhalation therapy?
_____ 9. Is the equipment cleaned properly?
_____ 10. Are the nasal catheters sterilized?
_____ 11. Is the humidifying solution sterile?

SURGICAL PROCEDURES

_____ 1. Do all personnel change from street clothes before entering surgical suites?
_____ 2. Are booties put on at the O.R.?
_____ 3. Are surgical room clothing and shoes worn outside of the operating room suite?
_____ 4. Are surgical gowns put on and discarded properly?
_____ 5. Are adequate surgical gowns provided for the surgeons?
_____ 6. Are surgical masks discarded after each procedure?
_____ 7. Are surgical masks discarded when wet?
_____ 8. Are hands washed before donning surgical suit, mask, and cap?
_____ 9. Are patients brought to the O.R. on beds?
_____ 10. Are they returned to their beds?
_____ 11. Does the operating table leave the O.R. suite?
_____ 12. Does the operating team scrub its hands thoroughly in an acceptable manner with soap or liquid hexachlorophene detergent prior to an operation?
_____ 13. Is the patient "prepped" in an acceptable manner immediately prior to the operation and before the surgeon gowns?
_____ 14. Are the razors and razor blades for preoperative shaving cleaned and sterilized after each use?
_____ 15. Are disposable razors or razor blades used for preoperative shaving?
_____ 16. Is the nurse responsible for sterility of surgical procedures? (In unusual cases the surgeon makes this determination.)
_____ 17. Does the surgeon inform the nurse about a contaminated case?
_____ 18. Are precaution gowns provided for the circulating nurse and anesthesiologist?
_____ 19. Are contaminated patients scheduled for a separate operating room?
_____ 20. Are the contaminated cases scheduled last?
_____ 21. Are special precautions taken for T.B. cases in the O.R.?
_____ 22. Is the questionably contaminated case treated as a contaminated case?
_____ 23. On contaminated cases, are the gowns removed before the gloves?
_____ 24. Do personnel wash hands after a contaminated case and before going to the dressing room?
_____ 25. Is dropped equipment resterilized before reuse?
_____ 26. Is talking kept to a minimum during an operation?
_____ 27. Is the number of people in the O.R. kept to a minimum?
_____ 28. Is the suture package opened in an aseptic manner?
_____ 29. Is the adhesive tape stored in a satisfactory manner?
_____ 30. When operating in two different suites (grafting procedure), are separate instruments used?
_____ 31. When operating in two different suites, are different personnel used?
_____ 32. When operating in two different suites, do the personnel rescrub and regown?

Figure 9.4 Continued.

_____33. Are special precautions taken during an intestinal anastomosis? (Is a subset of instruments used?

_____34. Are the proper types of drapes used?

_____35. Is the proper thickness of drapes used?

_____36. Are dry drapes used?

_____37. Do they get wet during the procedure?

_____38. Are instruments retrieved from below the patient level and used?

_____39. Is sheet wadding sterilized when used on open procedures?

_____40. Are gloves changed when needles puncture them?

_____41. Are personnel with significant respiratory illnesses or with overt lesions excluded from the operating and delivery suites?

_____42. Are Kelly pads being used on the O.R. table?

_____43. Is the O.R. table cleaned properly?

_____44. Is the operating room equipment cleaned properly?

_____45. Is the recovery room clean?

_____46. Is the recovery room maintained with the same basic technique as the operating room?

_____47. Is the operating room thoroughly scrubbed and washed with a germicide solution and allowed to sit for at least 24 hours after an infected case?

_____48. Are the walls spot cleaned between all operations?

_____49. Are the floors and other horizontal surfaces cleaned between all operations?

_____50. Is a germicide solution used in the above mentioned procedures?

_____51. Is there routine bacteriological surveillance of aseptic techniques and sterilization In the O.R.?

_____52. Do surgeons wash their hands, put on masks, and don sterile gloves before performing intravenous cutdowns?

_____53. Are intravenous catheters inserted in an aseptic manner?

_____54. Are intravenous catheters removed promptly when not needed?

HOUSEKEEPING
See self-Inspection sheet for Housekeeping Program in Housekeeping Section.

_____

COMMENTS:

_____

Figure 9.4   Continued.

Foodservice Self-Inspection Concept

Consistent unbiased evaluation of items requiring correction is essential to the improvement of a hospital foodservice program. The best approach to be used by the foodservice managers is the self-inspection program. foodservice managers, on a weekly basis, should systematically inspect their own facilities; identify problem areas; determine whether improvements are carried out each week; and record the facts, which can be used for program changes or additional budget requests for training and workforce. For the foodservice managers to be able to carry out a good self-inspection of the facility, these individuals should be put through a training program, which would be conducted by the local health department, state health department, or in-house environmental health practitioner. The recommended program would have 19 sessions of 2 hours each. Each session would be conducted as a seminar, with the trainer getting necessary preliminary information, citing case histories and solutions, and then bringing the entire group of managers into the discussion concerning actual problems at the institution and potential solutions for them. The recommended training program is as follows:

- Session 1 — introduction by hospital administrative staff as well as introduction of the program itself
- Session 2 — review of general hospital problems related to infections, injuries, and hospital practices
- Session 3 — review of general hospital problems related to infections, injuries, and hospital practices (continued)
- Session 4 — Hazard Analysis and Critical Control Points (HAACP) principles and practices
- Session 5 — HAACP principles and practices (continued)
- Session 6 — the self-inspection form and technique
- Session 7 — field experience in use of the form
- Session 8 — additional field experience in use of the form
- Session 9 — review of foodservice department problems
- Session 10 — review of foodservice department problems and start using forms on a weekly basis
- Session 11 — general housekeeping principles and practices and discussion of results of inspections
- Session 12 — general housekeeping principles and practices and discussion of results of inspections (continued)
- Session 13 — general foodservice equipment and ware washing techniques, and discussion of results of inspections
- Session 14 — bacteriology and communicable disease control and discussion of results of inspections
- Session 15 — food preparation, storage, serving, and transportation and discussion of results of inspections
- Session 16 — insect and rodent control, and discussion of results of inspections
- Session 17 — supervisory techniques
- Session 18 — supervisory problems
- Session 19 — final exam

At the completion of the course, individuals should be given a certificate indicating completion or completion with honors (see self-inspection form in Figure 9.5), and HACCP Inspection Form (see Figure 3.8 in Chapter 3 of this volume).

Ventilation

To evaluate a ventilation system, it is necessary to evaluate the air patterns, air velocities, room air ventilation rates, and types of air supply and exhaust systems for moving the contaminated particles throughout the room. In this evaluation, the most important factor is the rapidity with which widespread contamination can be removed. The removal rate is determined by taking airborne samples of particles and microbial particles from various parts of the room during various operations.

Housekeeping

It is the function of the housekeeping department to determine the special housekeeping problems and, in consultation with the environmentalist, to determine those matters most critical to the environment. In-depth studies should be made of existing housekeeping practices, work plans, work scheduling, activities of housekeeping personnel, employee training, and work and time evaluations. Understanding is essential as to how organisms are spread, and what daily periodic and special cleaning techniques are required to fight the spread of these organisms.

Supervisors, after gaining familiarity with the total cleaning program, evaluate the work of the personnel and give them necessary additional in-service training when such training is warranted. The function of the supervisor is one of inspection, evaluation, training, encouragement, and disciplinary action, in the event that such action is needed.

Various housekeeping evaluation techniques have been proposed. The most effective appears to be the use of housekeeping self-inspection forms, where supervisors evaluate their own areas and the information is then tabulated by the environmentalist. This is supplemented by the use of a white-glove technique on surfaces to determine whether they are dirt free and the use of either contact or swab bacteriological samples or air samples. Because sampling is a very costly and an inaccurate means of determining the level of housekeeping in an institution, it is recommended that its use be limited to the evaluation of an actual outbreak of disease. Under ordinary circumstances, the use of self-inspection forms and a wet, white towel smeared across surfaces would be excellent indicators of the degree of cleanliness of a given area.

All hospital problems and problems of other institutions are evaluated by trained environmental health practitioners, who use the techniques and study forms discussed in other areas of this book.

The housekeeping self-inspection program is similar to the foodservice self-inspection program. The housekeeping training program for supervisors and managers would be similar to the one discussed under food service, but the specific topics would change from food to housekeeping. (See housekeeping self-inspection form, Figure 9.6.)

FOOD SERVICE INSPECTION REPORT

(Submit Weekly to Sanitation Officer)
Environmental Health Practitioner

_____ HOSPITAL
(name)
ENVIRONMENTAL SANITATION

KEY:
2–Satisfactory
1–Improvement Needed
0–Poor

BUILDING NAME AND NO.	FLOOR	DATE	TIME	TOTAL SCORE
POINTS				

ITEM
1. Perishable foods used within a 2-hour period of preparation or immediately refrigerated.
2. Perishable foods are not to be reused at any time.
3. All refrigerators kept below 41°F.
4. Ice making machines clean and operating properly.
5. Food carts washed and sanitized at ward kitchens.
6. Food protected from contamination: Preparation   Storage   Transportation
7. Hands thoroughly washed and cleaned after use of toilets and before preparation of food.
8. Clean outer garments worn at all times.
9. Soap, paper towels, and waste receptacles available at all times at handwashing sink.
10. Lavatory facilities clean and in good repair.
11. Dressing rooms, areas, and lockers kept clean.
12. All food handlers with diarrhea removed from duty until negative stool culture.
13. Food utensils, dishes, glasses, silverware, and equipment scraped, pre-flushed, and soaked.
14. Silverware, dishes, and glasses clean to sight and touch.
15. Silverware, dishes, and glasses sanitized.
16. Dishwasher wash and rinse temperatures checked daily   (insert temp. rinse)   (insert temp. wash)
17. Food utensils, such as pots, pans, knives, and cutting boards clean to sight and touch.
18. Grills and similar cooking devices cleaned daily.
19. Food-contact surfaces of equipment clean to sight and touch.
20. Tables and counters clean to touch.
21. Non-food-contact surfaces of equipment clean to sight and touch.
22. Washing and sanitizing water clean.
23. Single service food articles properly stored, dispensed, and handled.
24. Containers of food stored off floor on clean surfaces.

25. No wet storage of packaged food.

26. Sugar in closed dispensers or individual packages.

27. Poisonous and toxic materials properly identified, colored, stored, and used; poisonous polishes not present.

28. Clean wiping cloths used and properly restricted.

29. All late trays served after 8 p.m. scraped, rinsed, and stacked in previously cleaned sink.

30. Garbage and rubbish stored in metal containers with lids; containers adequate in number.

31. All garbage moved from area at end of working day.

32. Containers cleaned when empty.

33. When not in continuous use, garbage and rubbish containers covered with tight-fitting lids.

34. Garbage and rubbish storage areas adequate in size and clean.

35. All outer openings screened.

36. Absence of rodents.

37. Absence of flies.

38. Absence of roaches.

39. Floors kept clean.

40. Floors in good repair.

41. Floors and wall junctures properly cleaned.

42. Walls, ceilings, and attached equipment clean.

43. Walls and ceilings properly constructed and in good repair; coverings properly attached.

44. Walls of light color; washable to level of splash to shoulder height.

45. Lighting adequate.

46. Light fixtures clean.

47. Areas clean; no litter.

48. Ample quantities of cleaning supplies available.

REMARKS:

_____
Signature of Dietician or Nursing Supervisor

Figure 9.5   Foodservice inspection report. (From Koren, H., *Environmental Health and Safety*, Pergamon Press, Elmsford, NY, 1974, pp. 34–35. With permission.)

HOUSEKEEPING CHECKLIST
Prepare in Duplicate

This form based on a design by Herman Koren, R.S., M.P.H., Chief of Environmental Health and Safety at Philadelphia General Hospital, Philadelphia Dept. of Public Health. Printed and distributed by Economics Laboratory, Inc., 250 Park Ave., New York, NY 10017

RATINGS:
U–Unsatisfactory   (0)
S–Satisfactory   (1)
N–Not Apply

BUILDING NAME AND NO.	FLOOR & NAME	DATE	TIME	TOTAL SCORE

SUPERVISOR MAKING INSPECTION (Sign)

CUSTODIAN OR HOUSEKEEPING AIDE (Sign)

TOTAL SCORE: $\dfrac{S's}{S's + U's} \times 100 =$ %

Fill in all parts of the form. If the item is "Unsatisfactory" show a "U" under the **RATING** and the code under the **AREA CODE**. If an item is "Satisfactory" show an "S" under the **RATING** and nothing under the **AREA CODE**. For example, if the floors are found dirty in the Nurses' Station, Laboratory, and Supply Room, put a single "U" under **RATING** on line #2 (Floors-dirty). Then list N.R., Lab #3, and S.R. Use "N" if the item is not applicable. Under **TOTAL SCORE**, count up all the S's, then divide by the S's plus the U's to give you the percentage grade.

AREA CODES

W.–Ward–Give#	S.A.–Storage Area	A.R.–Animal Rooms
So.–Solarium	W.R.–Wait. Rm. & Lobby	K.–Kitchen
P.R.–Patient Room–Give#	Of.–Offices–Give#	S.R.–Supply Room
N.R.–Nurses' Station–Give#	Co.–Corridor	I.C.R.–Instr. Cleaning Rm.
B.R.–Bathrooms	Lab.–Laboratory–Give#	A.U.–Autoclave Room
U.T.–Utility Room	L.R.–Locker Rooms	M.R.–Miscellaneous Rm.
T.R.–Treatment Room		L.D.R.–Labor & Del.–Give#
E.R.–Examination Room		O.R.–Operating Room–Give#
J.S.C.–Janitor Supply Closet		
L.C.–Linen Closet		
S.S.C.–Sterile Supply Closet		

	AREA CODES	RATING		AREA CODES	RATING
FLOORS			**LAVATORIES**		
1. Dust			32. Toilet Paper		
2. Dirty			33. Wash Bowls		
3. Litter			34. Soap		
4. Spillage			35. Paper Towels		
5. Stains			**HOPPER ROOM**		
6. Warning signs			36. Dirty		
WALLS			**JANITORIAL SUPPLY CLOSET**		
7. Dust			37. Adequate supplies		
8. Splattering			38. Clean		
9. Cobwebs			39. Equip. stored properly		
10. Dirty			**PAILS**		
WINDOW SILLS, VENTILATORS			40. Dirty		
11. Dust			**MOP WATER**		
12. Spots			41. Dirty		
RADIATORS			**MOP HEADS**		
13. Dirty			42. Dirty		

DOOR LEDGES & FRAMES, TOPS OF CABINETS			43. Poor Condition	
14. Dirty			**REPAIRS**	
WARD SINKS			44. Repair reported below	
15. Wash bowl			**INSECT AND RODENT CONTROL**	
16. Wall			45. Mice	
17. Soap			46. Roaches	
18. Towels			47. Ants	
WASTE RECEPTACLES			48. Flies	
19. Need emptying			**LIGHTS**	
20. Dirty			49. Fixture dirty	
EQUIPMENT			50. Bulbs need replacement	
21. Dust			**WINDOWS AND SCREENS**	
FURNITURE			51. Dirty	
22. Dust			**STAIRWAYS**	
23. Spillage			52. Litter	
LAVATORIES			53. Dust	
24. Walls and windows			54. Spillage	
25. Partitions			55. Needs repair	
26. Floors			56. Needs bulbs	
27. Tubs			**ELEVATORS** (give number)	
28. Showers			57. Floor	
29. Urinals			58. Walls	
30. Toilet Seats			59. Tracks	
31. Toilet Bowls			60. Doors	

REPAIRS REPORTED	LOCATION	REPAIRS REPORTED	LOCATION
a. Wall needs repair		i. Stopped up floor drains	
b. Broken window		j. Hopper broken or stopped up	
c. Broken or missing window screens		k. Wash basin faucets leaking	
d. Ceiling needs repair		l. Wash basin stopped up	
e. Peeling paint			
f. Floor tile loose			
g. Floor tile broken			
h. Broken lighting fixtures or burned out bulbs			

Figure 9.6 Housekeeping self-inspection form. (From Koren, H., *Environmental Health and Safety*, Pergamon Press, Elmsford, NY, 1974, pp. 52–53. With permission.)

INSTITUTIONAL CONTROLS

The institutional environmental health specialist is the staff person most involved in carrying out comprehensive studies and presenting reports with combined recommendations for various committees and individuals. The environmental health professional has specific training in epidemiology; broad knowledge of biological, physical, and chemical hazards; extensive public health and environmental health training and experience; and intimate knowledge of the institutional environment, its risks, and problems. The function of this specialist is to promote better patient care and better care of all individuals in institutions by reducing the risk of infection and the opportunity for accidents, and by limiting the extent of injuries and improving the physical environment.

Environmental Control Measures

Environmental control measures include architectural considerations; proper installation of equipment; control of airborne contamination; control of surface contamination from linens, and solid and liquid wastes; and control of other environmental factors.

A good part of environmental control of infections in hospitals is based on initial architectural considerations. Basic problems of hospital design and construction always include traffic patterns; systems used for handling materials, equipment, and liquid and solid wastes; ventilation systems; special airflow control; and use of easily cleanable materials for surfaces. Traffic patterns are established in such a way as to keep the majority of individuals away from crucial areas, such as surgical suites, obstetric areas, isolation and reverse isolation areas, and nurseries. Further, the emergency room and its traffic patterns must prohibit movement of individuals to other critical areas of the institution, unless the patients are to be admitted to the institution. Even at that point, the emergency room patients, visitors, and families should never enter the critical areas that have been identified.

Materials handling, which is of great concern within the institution, must be organized to avoid the possibility of disease spread. Mechanical conveyors are designed for easy access for cleaning and inspecting to determine whether insect and rodent problems may exist. Certainly where such materials as bedpans and their contents are moved, specific consideration must be given to avoid taking them into clean rooms and contaminating other areas and personnel. The ventilation system should be designed so that potentially pathogenic or even questionable organisms, dust, lint, respiratory droplets, and dirt are not transmitted from dirty areas to clean areas. Laminar airflow is utilized in rooms where patients have or may have infections.

The materials used in construction should have good acoustical properties, durability, fire resistance, pleasant appearance, and be easily cleanable. Many other environmental controls are discussed in other sections of this book.

Control of infectious diseases for personnel and patients parallel each other. The reservoir of infection, that is, the individual who has the infection, or the insect or rodent carrying the organism, the secondary reservoirs of infection, must be treated, controlled, or removed from the presence of individuals who are susceptible to the infection. This is done using a surveillance program to determine the potential or actual carriers of disease; using medical aseptic precautions in the care of patients with communicable diseases, such as mumps or various staph infections; and controlling secondary reservoirs of infection through good housekeeping, removal of infected linens and solid waste, destruction of rodents and various insects, filtering and cleaning of ventilation systems, and control of other fomites that may cause infections to spread throughout the institution.

Personnel are exposed to infectious disease, because they treat diseased patients or they are involved in either removing waste products or cleaning the environment. All personnel are protected if adequate isolation techniques and procedures are utilized. Where masks are required, they must be clean, dry, and changed frequently. Masks are not an absolute barrier for droplets or aerosols. If gowns are donned and removed properly, and changed after leaving the patient's room, the opportunity for infection by organisms can be reduced. Hand washing is still probably one of the most effective techniques used for proper protection of personnel and patients.

All staff members, including all housekeeping and ancillary personnel, must perform a medical scrub before and after they leave the room where a patient is isolated. Gloves should be used when indicated and should be disposed of properly to avoid contaminating personnel or patients. Equipment and supplies used in patient care should stay within the isolation room until time for disposal. At that time, all equipment that cannot be properly sterilized or adequately sanitized should be disposed of in special double-bagged sacks marked *isolation*. Linens coming from isolation areas should also be stored in double bags and clearly marked isolation. Employees can further protect themselves by reporting to the hospital infirmary as soon as they have a suspected infection of any type, including diarrhea, vomiting, skin disorders, and so forth. Personnel should receive vaccinations for specific diseases in those areas where they may be subjected to the diseases. They should also be tested frequently in highly critical areas such as TB areas.

Control of Infectious Waste

Infectious waste may be controlled by steam sterilization (autoclaving), incineration, thermal inactivation, gas–vapor sterilization, chemical disinfection, and sterilization by irradiation. Incineration and steam sterilization are the most frequently used techniques for the control of infectious waste. Steam sterilization utilizes saturated steam within a pressure vessel (known as a steam sterilizer) at temperatures that will kill the infectious agents present. The two types of steam sterilizers are the gravity displacement type, in which the displaced air flows out the drain through a steam-activated exhaust valve; and the prevacuum type, in which a vacuum is pulled to remove the air before steam is introduced into the chamber. In both cases, the air is replaced with pressurized steam. The temperature of the treatment chamber continues to increase as the pressurized steam goes into the chamber. The treatment of the infectious waste by steam sterilization is based on time and temperature. The entire waste load has to be exposed to the proper temperature for a defined period of time. The decontamination of the waste occurs primarily because of the steam penetration. Heat conduction provides a secondary source of heat. The presence of residual air within the sterilizing chamber prevents effective sterilization, because the air will act as an insulator for the material.

Three factors that can cause incomplete displacement of the air include use of heat-resistant plastic bags; use of deep containers; and improper loading of the chamber. The type of waste, the packaging and containers, and the volume of the waste load are important to know when establishing how best to treat the waste during steam sterilization. Infectious waste with low density, such as plastic, is better for steam sterilization than high-density waste, such as large body parts and large quantities of animal bedding and fluids. Steam sterilization can be used for plastic bags of low-density, metal pans, bottles, and flasks. The infectious waste also includes other kinds of hazards, such as toxic material, radioactive material, or hazardous chemicals. Steam sterilization should not be utilized to treat these wastes.

The individuals who operate the steam sterilization process should be highly skilled in proper techniques to minimize exposure of people to the hazards due to the waste. Protective equipment, minimization of aerosol formation, and prevention of spillage of waste during loading of the autoclave are important factors to be considered. A recording thermometer should be used to make sure that the proper temperature is maintained through an appropriate period of time during the cycle. All steam sterilizers need to be routinely inspected and serviced. Steam sterilizers used for infectious waste must be in an area away from the central sterile supply and should never be used for sterilization of equipment or instruments.

Incineration is a process that converts combustible material into noncombustible residue or ash. The product gases are vented to the atmosphere through the incinerator stack. The residue may be disposed of in a sanitary landfill under proper conditions. The advantage of incineration is that it reduces the mass of the waste by as much as 95%, and therefore substantially reduces transport and disposal costs from the site of the incinerator. Incineration is especially good when used for pathological waste and contaminated sharps. If the incinerators are properly designed, maintained, and operated, the microorganisms present are killed in an effective manner. If the incinerator is operating poorly, then microorganisms may be released into the environment. When incinerating waste, it is necessary to consider the variation in waste composition, the rate of feeding of the waste into the incinerator, and the combustion temperature. The amount of moisture content and heating value of the waste affects its combustion. The rate at which the waste is fed into the incinerator affects the efficiency of the burn. The combustion temperature of 1600°F allows for a complete destruction of the microorganisms in the infectious waste and also allows for a considerable reduction of the material. The amount of air and fuel that is used has to be adjusted to maintain the combustion temperatures at the proper level, depending on the type of waste and volume.

If the infectious waste has special hazards attached to it, such as the inclusion of antineoplastic drugs, then the waste should be incinerated only in special incinerators that provide high enough temperatures and long enough time to completely destroy these compounds. The plastic content of the waste is also important, because many incinerators can be damaged by temperature surges caused by combustion of large quantities of plastic. Polyvinyl chloride and other chlorinated plastics are important, because one of the combustion products is hydrochloric acid, which is corrosive to the incinerator and may cause damage to the lining of the chamber and the stack.

Thermal inactivation includes treatment methods that utilize heat transfer to provide conditions that reduce the presence of infectious waste. The waste may be preheated by heat exchangers, or heat may be acquired by steam jackets. The amount of heat is predetermined, and the material is kept in a vessel for at least 24 hours. However, the temperature and holding time depends on the type of pathogens present in the waste. After the treatment cycle is complete, the contents of the tank are discharged.

Dry-heat treatment may be applied to solid infectious waste. The waste is heated in an oven that is operated by electricity. Because dry heat is less efficient than steam heat, it is necessary to have higher temperatures to accomplish the same type of kill. The typical cycle for dry-heat sterilization is treatment at 320 to 338°F for 2 to 4 hours.

Gas–vapor sterilization is an option that could be used for treating certain types of infectious waste. The two most commonly used chemicals are ethylene oxide and formaldehyde. However, considerable evidence suggests that both these chemicals are probable human carcinogens; therefore, considerable caution must be taken if the materials are used as gas–vapor sterilizing agents. In fact, it is best not to use ethylene oxide gas because of its toxicity and the fact that other options are available for treating the infectious waste. If formaldehyde gas is used to sterilize certain disposables, such as high-efficiency particulate air (HEPA) filters from biological safety cabinets, then only trained persons should do the work.

Chemical treatment is appropriate for liquid waste, but also may be used in treating solid infectious waste. To use chemicals effectively, it is necessary to consider the following: type of microorganism, degree of contamination, amount of protein materials present, type of disinfectant, concentration and quantity of disinfectant, contact time, temperature, pH, mixing requirements, and biology of the microorganism.

A new technology involving the use of ionizing radiation sterilization includes nominal electricity requirements, no steam, no residual heat in the treated waste, and performance of the system. The disadvantages include the high capital cost, the requirement for highly trained operating and support personnel, a large space requirement, and the problem of the ultimate disposal of the decayed radiation source. The ionizing radiation must be properly monitored.

Biological indicators are used to monitor the treatment process when some form of heat is utilized. These indicators are used in steam sterilization, incineration, and thermal inactivation. Spores of a resistant strain may be used as an indicator. An instantaneous reading may be gained by using a chemically induced color change when a particular temperature has been reached. The chemical indicators are not good for the sterilization process, because the temperature is only part of the sterilization effort.

Cleaning, Disinfecting, and Sterilizing Patient Care Equipment and Environmental Surfaces

Cleaning, the physical removal of organic material or soil from objects, is usually done by using water with detergents. Cleaning is used to remove microorganisms from equipment or surfaces. Sterilization is the destruction of all forms of microbial life. Disinfection is the process between cleaning and sterilization, which destroys substantial numbers but not all microorganisms. The degree of removal of microorganisms is based on the potential risk of infection involved in the use of the equipment or contact with the surface. The level of disinfection depends on amount of contact time, temperature, type and concentration of the active ingredients of the chemical germicide, and nature of the microbial contamination. In patient care areas, visibly soiled surfaces should be first cleaned and then chemically decontaminated. In the laboratory, large spills of cultured or concentrated infectious agents should be flooded with a liquid germicide, removed, and then decontaminated with a fresh germicide.

It is neither necessary nor possible to sterilize all patient care items. It is necessary to identify the items that may be cleaned, disinfected, or sterilized. For example, any microorganisms, including bacterial spores, that come in contact with normally sterile tissue including the vascular system of any patient, can cause infection, and therefore the equipment carrying them should be sterilized.

Isolation Procedures in Hospitals

The transmission of infection requires a source of infecting microorganisms, a susceptible host, and a means of transmission for the microorganisms. Human sources of the infecting microorganisms may be patients, personnel, or visitors, who have an acute disease, who are in the incubation period of disease, who are colonized by the infectious agent, or who are chronic carriers. Other sources of infecting microorganisms can be the patient's own endogenous flora, which may be difficult to control. Inanimate environmental objects that are contaminated including equipment and medications may lead to disease.

Resistance to pathogenic microorganisms varies greatly. Exposed individuals may be immune to infection, resist colonization, establish a commensal relationship and become an asymptomatic carrier, or may develop the clinical disease. Those factors such as age; underlying diseases; treatments with antimicrobials, corticosteroids, or other immunosuppressive agents; irradiation; and breaks in the first line of defense mechanisms caused by surgical operations, anesthesia, and indwelling catheters make patients more susceptible to infection.

Microorganisms are transmitted in hospitals by the following routes: contact transmission, either direct or indirect; droplet transmission generated by coughing, sneezing, talking, suctioning, etc.; airborne transmission by the dissemination of either airborne droplet nuclei of evaporated droplets containing the microorganisms, or dust particles containing the organism; common environmental transmission, including food, water, medications, equipment, and medical devices; vectorborne transmission, including mosquitoes, flies, and rats.

Isolation procedures include the following:

1. Hand washing and gloves should be used. Hand washing is the most important means of control for transmitting organisms from one person to another, and from a person to a surface to a person.
2. An individual secured in a single room where possible or in a special area where like infected patients are under treatment. The construction of the room and its

equipment is important. Air handling and ventilation must be such as to reduce the risk of transmission of microorganisms to others.

3. Transportation of infected patients should be at an absolute minimum. If it is necessary to move the patient, the following precautions should be taken: use of appropriate barriers such as masks and impervious dressings are worn by the patient; personnel in the area to which the patient is to be taken should be notified of the arrival as well as special precautions taken; patients are taught how to assist in the prevention of spread of their infectious disease.

4. Mouth and eyes are protected by masks, special respiratory protection, eye protection and face shields. Surgical masks may not be effective in preventing the inhalation of droplet nuclei. Disposable particulate respirators, certified by the CDC-NIOSH should be utilized.

5. Gowns and protective apparel are worn to provide barrier protection and to reduce opportunities for transmission of microorganisms.

6. Patient care equipment and articles need special handling and disposal as previously mentioned.

7. Laundry must be washed in an appropriate manner to remove microorganisms.

8. Dishes, glasses, cups, and eating utensils should generally be disposable. If they are not, they should go through a thorough cleaning procedure.

9. Routine and terminal cleaning of equipment and facilities needs to be thorough, frequent, with personnel using appropriate detergents and germicides.

Guideline for Infection Control in Healthcare Personnel

In 1998, the National Center for Infectious Diseases, National Immunization Program, and NIOSH prepared and published simultaneously in the *American Journal of Infection Control*, and the *Journal of Infection Control and Hospital Epidemiology*, an update of the guidelines. The diseases that healthcare personnel face have already been discussed in the area of laboratories in this chapter. A personal health program consists of coordination of all departments, immunization programs, medical evaluations, health and safety education, management of job-related illnesses and exposures to infectious diseases, counseling services for infected employees, and maintenance and confidentiality of personal health records.

Guideline for Prevention of Surgical Site Infection

The Guideline for Prevention of Surgical Site Infection was updated and published in 1999 in the *Journal of Infection Control and Hospital Epidemiology*. These guidelines apply to in hospital surgical procedures as well as same-day or outpatient operating rooms. An example of the classification of surgical site infections may be found in Figure 9.7. Some of the preoperative recommendations are as follows:

1. Use a preoperative shower or bath to decrease skin microbial colony counts
2. Do not do preoperative shaving of the surgical site the night before an operation, because it increases the risk of surgical site infections. Shaving should occur immediately before the operation.
3. Use antiseptic agents such as iodophors, alcohol-containing products, and chlorohexidine gluconate in skin preparation at the site of the incision.

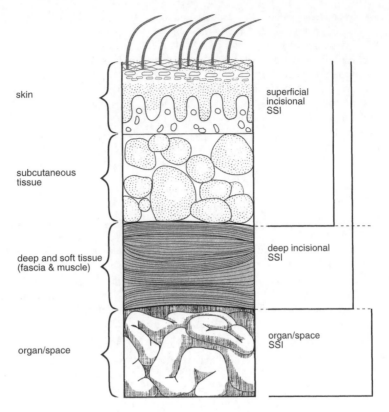

Figure 9.7   Cross section of abdominal wall depicting CDC classifications of surgical site infection. (From *Guideline for Prevention of Surgical Site Infection, J. Infect. Control Hosp. Epidemiol.,* 20(4).)

4. Surgical team members should perform surgical scrub immediately before putting on sterile gowns and gloves.
5. Remove all surgical personnel, who have active infections or are colonized with certain microorganisms, from the surgical team.
6. Use a surgical antimicrobial prophylaxis for a very brief time, shortly before an operation begins.

Special operating room concerns consist of:

1. Ventilation is needed because the operating room air may contain microbial-laden dust, lint, skin, or respiratory droplets. Operating rooms must be maintained at positive pressure in relation to hallways and adjacent areas. Laminar airflow and ultraviolet radiation in certain instances might help control the spread of infection.
2. Environmental surfaces including tables, floors, walls, ceilings, and lights are typically not a source of infection. However, it is important to clean the surfaces extremely well.
3. Surgical attire including scrub suits, caps or hoods, shoe covers, masks, gloves, and gowns as well as the drapes used on patients must be sterile.

Guidelines for Preventing the Spread of Vancomycin-Resistant Organisms

The CDC Hospital Infection Control Practices Advisory Committee recommends the following:

1. Prudent vancomycin use by physicians
2. Education of hospital staff concerning vancomycin resistance
3. Early detection and prompt reporting of vancomycin resistance in enterococci and other Gram-positive microorganisms by the hospital microbiology laboratory
4. Immediate implementation of appropriate infection control measures to prevent person-to-person transmission of vancomycin-resistant organisms
5. The development of institution specific plans

Control Measures for Transmission of Human Immunodeficiency Virus and Hepatitis B Virus

Control measures for transmission of HIV and HBV for healthcare workers and public safety workers include the general principles that have been discussed previously in the relationship to infectious body fluids and in the use of personal protective equipment. Precautions need to be taken with the individuals who are exposed to blood or body fluids through procedures related to cardiopulmonary resuscitation (CPR), IV insertion, trauma, and delivering babies.

Disposable gloves must be used by individuals in institutions, especially when coming in contact with patients where blood or body fluids are involved. This is particularly true in the emergency room. It is also true for fire and emergency medical service personnel. If public safety workers are involved in any type of body searches, they ought to also wear disposable gloves. The disposable gloves should provide for dexterity, durability, and fit in the task performed. Where large amounts of blood are likely to be encountered, the gloves must fit tightly at the wrist to prevent blood contamination of the hands around the cuff. Where multiple trauma victims are encountered, the gloves must be changed between patients. Contaminated gloves should be removed properly and placed in bags that are leakproof and can then be appropriately disposed of, as has been discussed earlier.

Masks, eyewear, and gowns should be utilized in institutions and in emergency vehicles as needed. The protective barriers are necessary to protect the workers from diseases that are borne by blood and other body fluids. The masks and eyewear (safety glasses) should be worn where blood or body fluids may be splashed. Changes in gowns or clothing should be available if necessary.

Although HBV and HIV infections have not been shown to occur during mouth-to-mouth resuscitation, a risk of transmission may still exist for such diseases as *Herpes simplex* and *Neisseria meningitis* during mouth-to-mouth resuscitation. Also theoretically a risk of HIV and HBV transmission exists during artificial ventilation of trauma victims. Disposable airway equipment for resuscitation bags should be used. Pocket mouth-to-mouth resuscitation masks designed to isolate emergency response personnel from contact with victims' blood and blood-contaminated saliva, as well as respiratory secretions and vomitus, should be utilized by emergency personnel.

Blood, saliva, and gingival fluid from all dental patients should be considered infectious during dental procedures. Gloves, masks, and protective eyewear should be worn during examination and treatment. Hand pieces should be sterilized after use with each patient. Blood and saliva should be cleaned from material used in the mouth and the material should be disposed of or sterilized. Dental equipment and surfaces that are difficult to disinfect, such as lights or x-ray unit heads should be wrapped in impervious-backed paper, aluminum foil, or clear plastic wrap.

Laboratories need to consider all blood and body fluids as infective and take the following precautions:

1. All samples must be put in a secure, nonleak container with a lid for transportation.
2. All technicians should wear gloves and masks as well as protective eyewear in case mucous membrane contact occurs.
3. Biological safety cabinets should be used if aerosols can be generated.
4. Mechanical pipetting only should be used.
5. Only use needles and syringes when needed.
6. Decontaminate laboratory work surfaces with an appropriate chemical germicide.
7. Decontaminate or dispose of infective waste properly.
8. Decontaminate scientific equipment contaminated with blood.
9. Use medical scrub before eating and leaving the laboratory.

Patient Control Measures

A major step in prevention of infection includes use of proper patient infection control procedures, the techniques utilized in employee control measures plus proper environmental controls. In addition, patients who are highly susceptible to infections should not be kept in the same areas with patients who have infections. If in doubt, remove the patient from the area. Patients should be taught good personal hygiene and body care to avoid recontamination by the discharged organisms. In hazardous situations, such as where blood bank blood is used, special techniques must be employed to avoid contaminating the patient. Whenever special procedures are used, such as hemodialysis treatments, tracheotomies, intravenous cutdowns, or change of surgical dressings, careful aseptic techniques must be employed.

Managing Patients with a Resistant Organism

Not all patients with resistant organisms show symptoms of such. Therefore, all suspected patients should be evaluated to determine whether they are colonized or infected with a resistant organism. If so, the normal infection control procedures may not work. Controls should include:

1. Very strict hand washing (medical scrub with an antimicrobial agent)
2. Using gloves when touching nonintact skin mucous membranes, moist body substances, or items contaminated by these items
3. Using proper glove removal, disposal, and hand washing between patients

4. Wearing a mask when working within 3 ft of a patient who has a respiratory infection involving a resistant microorganism
5. Wearing appropriate gowns and discarding same between patients
6. Notifying a facility when a patient with a resistant organism is admitted
7. Using appropriate antimicrobial agents
8. Giving additional infection control training to healthcare workers

Infection Control Committee

The infection control committee is a working committee of the institution. It is necessary that the committee consist of representatives of the medical staff, nursing staff, environmental health staff, infection control staff, and various members of the administrative staff. The function of the committee is to evaluate current and past levels of infection and to make necessary recommendations for changes in either medical–surgical techniques or other techniques used within the institution that may contribute to infections. The committee members also act as the supervisors and recipients of all special studies related to infections that are conducted within the institution. Once they have evaluated the study, they forward it, along with their recommendations, to the executive director of the institution for implementation.

Occupational Health and Safety

The occupational health and safety program consists of program planning; establishing program objectives to reduce frequency and severity of occupational hazards, increasing inspection coverage, charging supervisors with safety responsibilities, coordinating health and safety activities, planning and evaluating all safety activities, and planning and directing an adequate accident prevention program. It also includes safety and health training, education, and promotion; accident and illness investigation, reporting, and analysis; safety and health inspections; safety and health committees; safety and health engineering for hazard control; safety and health standards and regulations; fire prevention, protection, and suppression; motor vehicle safety; medical help and first aid; accident cost control; and program design, application, and evaluation.

Health and safety education is the process of transmitting knowledge and developing skills to produce safe practices. To accomplish this, the trainers must evaluate and the trainees must learn about hazardous conditions, safe and healthy practices, general rules of health and safety, special on-the-job problems, and job safety analysis.

The engineering approach is necessary in the initial layout and construction of the institution. Proper engineering eliminates many of the hazards already mentioned and provides not only a safe but also an aesthetically comfortable environment.

All institutions should provide preemployment and preplacement examinations. These examinations should determine not only the employee's medical history but also if the employee has specific problems with specific types of environmental conditions exist. It is well to determine whether the applicant can read and interpret written instructions and is able to react to posted instructions concerning safeguards.

In addition, periodic medical examinations should be given to employees to determine whether they are currently suffering from any health problems and need to be removed from their current environment and placed in a different type of job. An institutional occupational health program should include the diagnosis of occupational disease and injuries; the treatment of occupational disease and injuries; the treatment of nonoccupational illness and injuries; and the evaluation of the kinds of health problems, occurring over time, that may become serious health or safety problems. Special concern should be given to hearing, vision, employment of pregnant women in special hazardous situations, prevention and control of alcoholism, and establishment and use of emergency control centers in the event of toxicity.

Principles of Biosafety

Containment describes the safe methods for managing infectious materials in the laboratory environment where they are handled or maintained. The purpose of containment is to reduce or eliminate exposure of laboratory workers, other people, and outside environment to potentially hazardous agents. Good microbiological technique and use of appropriate safety equipment in well-designed facilities are essential. Safety equipment includes biological safety cabinets, closed containers, and other engineering controls designed to control or minimize exposures to hazardous biological materials. Safety equipment may also include gloves, coats, gowns, shoe covers, boots, respirators, face shields, safety glasses, or goggles. The four biosafety levels are:

1. Biosafety level 1, which is used when the practices, safety equipment, facility design, and construction are appropriate for secondary educational training and teaching laboratories, and for other laboratories in which work is done with microorganisms not known to consistently cause disease in healthy adults.
2. Biosafety level 2, which is used when the practices, equipment, and facility design and construction are appropriate for clinical, diagnostic, teaching, and other laboratories in which work is done with a broad spectrum of indigenous moderate-risk agents that are present in the community and associated with human disease of varying severity. HBV, HIV, salmonella, and *Toxoplasma* spp. are examples of these organisms. The potential for producing splashes or aerosols must be low. Primary hazards relate to accidental percutaneous or mucous membrane exposures, or ingestion of infectious materials. Extreme caution should be taken with contaminated needles or sharp instruments.
3. Biosafety level 3, which is used when the practices, safety equipment, facility design, and construction are appropriate for clinical, diagnostic, teaching, research, or production facilities in which work is done with indigenous or exotic agents with a potential for respiratory transmission that may cause serious and potentially lethal infection. *Mycobacterium tuberculosis*, St. Louis encephalitis virus, and *Coxiella burnettii* are examples of these organisms. More emphasis is placed on primary and secondary barriers to protect personnel and environment from exposure to potentially infectious aerosols. All laboratory manipulations are conducted in enclosed equipment. Primary hazards relate to autoinoculation, ingestion, and exposure to infectious aerosols.

4. Biosafety level 4, which is used when practices, safety equipment, facility design, and construction are appropriate for work with dangerous and exotic agents that pose a high individual risk of life-threatening disease, which may be transmitted through the aerosol route and for which vaccine or therapy is not available. Viruses such as Marburg or Congo-Crimean hemorrhagic fever are examples of these organisms. The primary hazards are respiratory exposure to infectious aerosols, mucous membrane and broken skin exposure to infectious droplets, and autoinoculation. All manipulations of potentially infectious diagnostic materials, isolates, and naturally or experimentally infected animals, pose a high risk of exposure and infection to laboratory personnel, community, and environment. Further, all personnel involved in treatment of infected individuals are at extreme risk. The laboratory worker must be completely isolated from the contaminated material. The biosafety level 4 facility is generally a separate building or a completely isolated zone with special ventilation requirements and waste management systems to prevent the release of viable agents into the environment.

Child Care Health and Safety Program

The Child Care Health and Safety Program is conducted by CDC with representatives of programs from various federal agencies, child advocacy programs in communities, national organizations, academic institutions, state and local government, and business community. CDC has developed an action plan organized into three main sections: public health information systems, epidemiological and evaluation research, and public health interventions.

Environmental Protection Agency and American Hospital Association Hospital Waste Reduction Program

The EPA has formed a voluntary partnership with the American Hospital Association and its member hospitals to:

1. Virtually eliminate mercury wastes generated by hospitals by the year 2005
2. Reduce overall hospital waste by humans by 33% by the year 2005, and 50% by the year 2010
3. Jointly identify additional substances to target for pollution prevention and waste reduction

Medical waste incinerators are the fourth largest known sources of mercury in the environment, accounting for approximately 10% of all emissions. The hospitals and association have reached a voluntary agreement with EPA to carry out the program.

Housekeeping

The selection of housekeeping personnel is important. Individuals should be selected for their desire to work, their intelligence, experience, age, physical makeup, character, and ability to read and understand housekeeping procedures.

Housekeeping training programs need specific formats. Individuals should be taught the various housekeeping techniques in the classroom and in the institution. They should also be taught the basic fundamentals of the spread of infection and the seriousness of the use, storage, and handling of dangerous and hazardous materials and chemicals. Instruction in isolation techniques from nursing personnel should also be included. By the completion of the course, new housekeeping employees should have a solid understanding of where they fit into the entire institutional process; what their specific functions are; and how they should carry out the tasks; the materials and chemicals to be used, how they can prevent accidents, spread of infectious diseases, and how to promote better health and welfare for all. After completing a week of this intensive training, individuals should receive 6 to 9 months of probationary work under close supervision. They should take a refresher course of at least 1 day every 6 months.

Supervisors and administrators generally reach their positions because of the good work done at a technical level. It is important that the supervisors and administrators receive training. They need to learn everything that their staff learns; in addition, they need a good understanding of how best to obtain an adequate quantity of quality work from the individuals working for them. It is one of the important functions of a supervisor and administrator to determine what the area looks like and how well the job is done. This could be accomplished by using, on a weekly basis, the housekeeping self-inspection sheet found in Figure 9.6.

A good housekeeping program integrates all housekeeping techniques with a self-inspection program, where the individual supervisor has an opportunity to evaluate the actual work being carried out in a given area. The supervisors and administrators may also want to earn continuing education credits while learning about basic supervision and administration by taking correspondence courses written by the author. If so, contact Continuing Education Department, Extended Services, Alumni Center, Indiana State University, Terre Haute, IN, or phone 1-800-234-1639, ext. 2522.

Financing and the Role of Government in Institutions

Nursing home care is paid for by a combination of funds from the individual, federal government, and private organizations. Nursing homes wishing to participate in federal government programs must adhere to the rules and regulations of the USDHHS. Regulations exist in the areas of Medicaid and also Medicare. These encompass all types of nursing home activities, including environmental health activities. For example, standards have been established for the maintenance of sanitary conditions in the storage, preparation, and distribution of food. Unfortunately, out of a booklet of some 49 pages on specific requirements for nursing home facilities under the Medicaid program, only one quarter of one page was given to the environment, and this portion concentrated on food and wastes. All other areas of the environment seemed to be ignored.

SUMMARY

It is obvious that the previous discussions of topics within the institutional environment are, of necessity, limited in scope due to the nature of the material presented. All these areas are evaluated by one or more public agencies, including the local and state health departments and the local and state welfare departments. In addition, various areas of the hospital environment are supervised and comply with the standards of the various professional societies that are specialists within the given area. Further, all hospitals must be accredited to operate. Accreditation comes from the Joint Commission on Accreditation of Hospitals, which evaluates physicians, facilities, techniques, and so forth. The Joint Commission, after duly studying a hospital facility, makes certain recommendations that must be carried out by the institution to maintain its accreditation. This accreditation is extremely important for the operation of the institution.

The institutional environment and its problems are so complex that it takes not only the environmental health specialist but also a true team effort on the part of medical, nursing, administrative, and other personnel to operate the institution properly and to provide a better, safer and disease-free environment for all.

RESEARCH NEEDS

There is need for continued assessment of potential sources and transmission of institutionally acquired infections in individuals and staff. New techniques should be developed to assess rapidly the modes of the spread of infection. New accident prevention techniques are needed. Studies to determine the impact of noise on relaxation and sleep are needed. Studies are also needed to determine the quantity and exposure to chemicals in institutions. These studies must include epidemiological, toxicological, monitoring, and laboratory analysis. The air levels of chemicals associated with housekeeping, laboratories, and other areas where solvents are used should be assessed.

CHAPTER 10

Recreational Environment

BACKGROUND AND STATUS

Recreation is a vital part of public health. Recreational activities provide necessary exercise for individuals to control weight problems, strengthen hearts, and provide an emotional release from the day-to-day cares faced in the home and at work. With increasing numbers of middle-class people, the development of the interstate highway system, and the overcrowding of many areas, individuals have sought to obtain their fun through recreational activities that take them into state parks, federal parks, various national forest reserves, and local park areas. Each of these areas, while providing the kind of necessary escape from tension and troubles, also provides an environment highly conducive to the spread of disease and injury. Individuals come from different areas, states, and even countries. They congregate in recreational areas and bring diseases from home. These diseases spread readily within the recreational areas because of existent environmental health problems. The public has increased sharply the number of visits to various recreational areas. In 1950, approximately 110 million people visited state parks. In the 1990s, over 500 million people visited state parks. The National Park Service in 1950 had approximately 25 million people visit the various park areas. In the 1990s, over 125 million people visited national parks. The U.S. Forest Service had approximately 30 million people visited forest service areas in 1950. In the 1990s, about 200 million people visited these areas. In the year 2002, these numbers continued to grow even in spite of proposals to limit the number of people entering National Parks to keep from destroying the environment. Facilities are being torn down and areas are being made off-limits. These figures do not include the millions of individuals who visit local and county park and recreational areas.

Water-oriented recreation has also increased at an unusually rapid rate. Since the end of World War II, there has been an enormous change in recreational activities in all categories. There has been a sharp increase in the number of individuals who wish to build homes on any type of water body. By the year 2002, the summertime participation in swimming increased 200% above 1965 levels; 80%, in fishing over 1965 levels; 200%, in boating over 1965 levels; and 360%, in water skiing over

1965 levels. Also camping has increased by 250%, picnicking, by 125%; and sight-seeing, by 160%. It is obvious from this smattering of data that water recreational areas have become of enormous value to our society from both the aesthetic and the financial points of view.

The American public makes an estimated 8 billion 1-day visits each year to a variety of recreation areas, including national parks; parks of other federal, state, county, and local agencies; and private sector facilities. This is an estimated increase of 40% over 1970. The tremendous number of visitors to these areas has intensified sharply the various pollution, land, wildlife, and historical site problems. The recreational environment is becoming an increasing source of concern to public health officials, ecologists, and private groups.

SCIENTIFIC, TECHNOLOGICAL, AND GENERAL INFORMATION

The scientific and technological background varies with the specific type of recreational environmental problem under discussion. See the appropriate chapters within this textbook and Volume II for each of the problem areas.

PROBLEMS IN RECREATIONAL AREAS

Because of sheer numbers of people, lack of adequate facilities, and subclinical carriers of a variety of diseases readily spread through the environment, it is essential that recreational environmental health be better understood. An example of the kinds of hazards encountered include improper drainage and soil permeability, inadequate sewage facilities, inadequate kitchen equipment and improper cleaning practices, improper refrigeration of food, contaminated water supplies, unsuitable solid waste disposal, vector problems, swimming pool hazards, other accident hazards, and inadequate training and physical fitness of individuals visiting recreational areas.

Recreational areas have peculiar problems that generally are not found in the average city or community. They include seasonal operation, problems of public behavior, vector and animal problems, noxious plants and weeds, remote locations, and protection of wildlife from humans and humans from wildlife.

Seasonal operation is one of the most difficult problems seen in a recreational area. As a result of this, equipment is outdated, inadequate, or broken. This equipment includes chlorinators, ware washing machines, swimming pool filters, food facilities, sewage treatment, and water facilities. Further, it is difficult to get experienced personnel, because they are not anxious to work (usually at minimum wage) for a 3- to 4-month period during the year. As a result of this, many of the individuals are either college students with no specific training in the environmental areas or job hoppers who have difficulty keeping jobs and do not do an adequate job within the recreational area. The maintenance of equipment, even where good equipment exists, generally is poor, because the problems have to be discovered by the operators of the equipment and frequently are not, and because it takes time for maintenance personnel to come from some central area to make necessary corrections within the recreational area. Another difficulty caused by seasonal operation is that of inspectional procedures. Because the environmental health personnel have many other functions apart from the recreational areas, by the time the studies have been completed and corrections made, the season may be over. It is necessary then to start again the following year with new management and new personnel.

Public behavior is a difficulty because some visitors are highly irresponsible. Some people dump garbage outside their cabins or trailers, leave solid waste scattered everywhere, and contribute to fly and mosquito problems. In addition, some visitors like to take souvenirs along to show where they spent their vacations. As a result of this, they cause damage that is expensive to repair. The money spent could have been used for improving environmental conditions.

In the recreational environment, the individual is exposed to animals, reptiles, and insects. Infection by rabid bats, ticks causing Rocky Mountain spotted fever and Lyme's disease, and mosquitoes causing encephalitis, as well as bees, hornets, and poisonous snakes become troublesome problems. In some areas, individuals have even been mauled by bears. The outdoor environment contains weeds causing hayfever and other allergic reactions, such as poison ivy and poison oak. Because in some areas electrical power is not available and roads are not accessible, it is difficult to obtain a proper water supply and to treat an injured person.

As has been mentioned, communicable disease is a very definite hazard, because the agent or reservoir of infection may be present or may have been brought into the environment. The vector or vehicle of disease is present in the form of insects, rodents, or inanimate objects. It is known that when large numbers of people gather together and associate closely, disease spreads rapidly. This, coupled with the lack of adequate environmental protection measures, contributes to the spread of communicable disease. When such a disease does occur, it should be immediately reported to the nearest health department and also to the park authorities.

Carbon monoxide poisoning can occur in the recreational setting. In 1999, a man and his three children and pet dog were found dead in a zipped up two-room 10- × 14-ft. tent at a campsite. The heating source was a propane gas stove, which was still found burning when investigators arrived. Then 2 weeks later, a 34-year-old man and his 7-year-old son were found inside their zipped up tent. A charcoal grill was the heating source. In both cases, there was a lack of proper ventilation. In the United States, the number of deaths per year from unintentional carbon monoxide poisoning ranges from 900 to 1500 cases.

Cruise ships have been implicated in outbreaks of a variety of diseases, especially those of an oral–fecal route nature. In addition, other diseases have been reported such as pneumonia and Legionnaire's disease.

Recreational waters have been frequently contaminated by industrial activity, farming, sewage effluent, effluent from the waste from ships, purposeful dumping at sea, accidents, and unauthorized discharges from industries with permits. These contaminants include fertilizers, sludge, effluent from primary and secondary treatment plants, runoff from farming and animal raising, acid deposition, chemicals and microbiological discharges.

Impact on Other Problems

The quality of the recreational experience is being eroded because of the vast number of individuals who are invading recreational areas. There are constant shortages of motels, campgrounds, stores, laundry facilities, gas stations, and space. The sewage systems and other utilities cannot service the increased numbers of people. Noise and physical damage have been caused by the use of light planes within the national parks. The alpine tundra close to access roads has been damaged by excessive use. Crime has increased sharply, and vandalism may eventually destroy our Indian cliff carvings and many of our natural wonders if it is not controlled. The national and state parks used to be free of pollution. This is no longer true. The automobile has brought air pollution to the park areas. Water pollution has increased sharply as a result of human interference with various watersheds, animal, and plant life. Congestion has become a part of the peak use periods of recreational sites. The increased numbers of people have created a transportation problem, because the existing systems cannot service them and a lack of adequate access to transportation exists. Problems have been created by haphazard private development of areas adjacent to recreation areas.

Economics

No real way is currently available for measuring the economic cost of misuse of the recreational environment. We do know that when conditions are poor, visitors do not return to the recreational areas, resulting in a tremendous loss of money to specific types of businesses. However, little understanding and information are available concerning the impact of economics as a problem of the recreational area.

POTENTIAL FOR INTERVENTION

The potential for intervention in the variety of environmental problems varies with the given problem and with the number of individuals who literally invade a recreational area. Human impact on the recreational environment can only be reduced by either increasing facilities and constraints, which may have tremendous economic or political costs, or by reducing the number of visits that can be made to recreational areas. The full potential for intervention can only be understood when enough research is conducted to determine what can and cannot be done within this unique environment.

RESOURCES

The resources for each of the environmental health problem areas are identified by problem areas in other parts of this book and in Volume II. See the appropriate problem area for the resources desired. In addition, many of the civic organizations, such as the Sierra Club, are concerned with recreational environmental problems. Governmental agencies include the Corps of Engineers, National Park Service, Department of the Interior, and Tennessee Valley Authority.

STANDARDS, PRACTICES, AND TECHNIQUES

In planning for a recreational site, it is essential to gather and analyze all pertinent data concerning the land, the environmental systems present, the environmental controls needed, and the use of the land. This type of preplanning concerns not only the recreational agencies and the health agencies but also the Department of Interior, the Department of Mines, and other agencies concerned with the health of individuals involved in the environment.

Although it is seldom possible to meet all desirable site criteria, a good many of these and, of course the most important ones, should be considered when developing a new campsite or altering a former recreational area. It is important that the area be well drained; gently sloping; free from topographical and other hindrances; accessible to sources of water supply, sewage disposal, and solid waste disposal; away from heavy traffic and noises; conveniently near major highways or helicopter landing strips to remove injured or seriously ill; away from swamps and marshes, where mosquitoes may breed and cause annoyance and disease; with adequate facilities for proper food storage and handling; and with adequate housing.

Water Quality

Water quality depends on watershed management, water supply, water usage, sewage, and plumbing. Watershed management is concerned with the supervision, regulation, maintenance, and use of water in the drainage basin to produce the maximum amount of water of desirable quality and to avoid erosion, pollution, and floods. The condition of the soil, the use of the land, and the construction of such amenities as roads all affect watershed management. To protect water at its source and to ensure its proper usage, it is necessary to preserve undeveloped areas and to prevent abuse of the watershed. In other areas, the quality of the watershed can be restored by controlling these problems. Preservation of existing quality or improvement of existing quality in effect means control of adjacent areas involved in construction, logging, raising of livestock and wild game, mining, land development, and waste disposal.

The construction activities that affect the quality of water are roads, railroads, power transmission lines, mines, and dams. These activities increase the amount of silt or the quantity of chemicals, oil, gasoline, and solid waste. Soil erosion occurs when construction is not carefully preplanned and too many trees are cut down.

It is essential that logging operations be carried out in such a way that erosion does not occur. Products of logging activities, decayed vegetation, pesticides, oil, and gasoline must not enter the water sources. All areas where construction roads are built must be carefully controlled to avoid erosion and watershed contamination.

Erosion is also caused by overgrazing of recreational areas by livestock and game. Grazing should be supervised by the proper authorities and wildlife controlled by hunting season programs.

Waste disposal of liquids and solids is a serious watershed problem. People who build cabins and do not install proper sewage systems cause sewage to seep into the watershed and watercourses. Garbage and solid waste are frequently carelessly discarded, becoming contaminants. Industrial operations must be controlled to avoid inadvertent contamination. If sanitary landfills are used, they should not leach into the watercourses.

Uncontrolled mining and processing of ores have destroyed many streams. The mines and the mine wastes not only spoil the land but also drain into bodies of water. Abandoned mines must be protected to avoid seepage into nearby bodies of water. In addition, roads, strip-mining operations, and stockpiling of various chemicals used in processing contribute to the water problem. Mining may cause the streams to be clogged with silt, colored matter, minerals, acids, and chemicals. Abandoned strip mines should be leveled and covered with the original soil or new top soil, and then seeded with an approved vegetative cover.

Pesticides, as explained in Chapter 7, are toxic to humans and animals. Depending on the pesticide and its use, it may become a contaminant of the watershed. It is important that the persistence and concentration of the pesticide, as well as its lethal effects, be thoroughly understood before the agent is used. Usage of pesticides in recreational areas is necessary, but must be handled carefully to prevent contamination of the watershed areas. Where recreational areas exist in watersheds, it is important for careful environmental controls to be followed to avoid human contamination.

Fires caused by carelessness, lightning, cooking, or bonfires must be carefully controlled to avoid damage to the forested areas and the watersheds. The greater the quantity of trees, the better the chances of avoiding erosion and maintaining the watershed. Conservation activities are carried out by a variety of agencies, including various park services, fire services, bureaus of land management, and the Corps of Engineers.

Water Supply

An adequate supply of pressurized water for drinking and other household purposes must be provided. The water supply must meet the bacteriological, chemical, physical, and radiological requirements established by laws and regulations updated periodically. In 1996, the Safe Drinking Water Act amendments were passed by Congress, emphasizing sound science and risk-based standard setting. These standards stipulate that the water supply system must include all the collection, treatment, storage, and distribution equipment needed for a proper, safe water supply. For firefighting purposes, water may be utilized from streams or creeks. To meet the requirements of proper water supplies, water should be taken from wells, springs, or infiltration galleries that are not subject to contamination. The water should be examined regularly to determine the level of safety. When water comes from underground or surface areas and it has been determined that low degrees of contamination are present, the water must be chlorinated. Where water contains impurities that require coagulation, sedimentation, and filtration for removal, the water should receive complete treatment. In inaccessible areas, water must be hauled to the places of use in tank trucks. In this case, the water must come from a safe source and the following precautions must be taken to prevent contamination:

1. The tank interior should be devoid of all defects. It must be thoroughly cleaned and disinfected before use.
2. The tank lining must be nontoxic.
3. The tank must be used for no other purposes.
4. Chlorine should be added to the water in the tank to ensure that a residual of at least 0.4 ppm of free chlorine is available after the chlorine has been in contact with the water for at least 20 min.
5. Only one hose that is utilized strictly for filling and unloading the tank should be available. The hose should be stored in a covered container in the truck when not in use and then thoroughly flushed with clean water before each use.

Further discussion of water supply and water treatment is found in Chapter 3 in Volume II.

In the event of emergencies, water suspected of contamination with bacteria may be used if any of the following procedures are followed:

1. Boil vigorously for 10 min.
2. Add five drops of 2% U.S.P. tincture of iodine to each quart of cold water and allow to stand for 30 min.
3. Add iodine tablets, one tablet for each quart of water and allow to stand for 30 min.
4. Add chlorine bleach at two drops to a quart of water and allow to stand for 30 min.
5. Add chlorine tablets at one tablet per quart of water.

It is essential that drinking fountains be constructed so that the individual's mouth does not come into direct contact with the water fountain itself.

Sewage

Safe disposal of human and domestic waste is essential in recreational areas. Improper sewage disposal contaminates the water, thereby ruining the water for its many recreational and drinking purposes. Sewage treatment systems must have adequate capacity and must meet all the necessary requirements set forth in federal and state water pollution control laws. Water carriage systems must be kept in good repair to prevent leakage into the underground water supplies. In many areas, water carriage systems are available, but sewage treatment systems are not. Septic tanks and subsurface disposal systems are utilized. In these cases, it is necessary to keep the on-site sewage disposal system away from wells, streams, or other sources of water supply. It is important that they do not overflow into the creeks or other water areas draining the campsite. The septic tanks and systems must not be permitted to flow downhill into water sources. (Additional discussions on this subject may be found in Chapter 6 in Volume II.) Nonwater carriage sewage disposal facilities are also necessary in recreational areas. These include such facilities as chemical and burn-out toilets and pit privies. Again it is essential that these units be so constructed that they do not contaminate the water supply or become a source for insect breeding.

It is important that privies be correctly built and be properly maintained. (Further information on this area may also be found in Chapter 6 in Volume II.)

Plumbing

Plumbing includes all the practices, materials, and fixtures used in installing, maintaining, extending, and altering any pipes, fixtures, appliances, and appurtenances. (Plumbing fixtures and materials, as well as an additional discussion on plumbing, may be found in Chapter 5 in Volume II.) It is known that back-siphonage and cross connections cause contamination of potable water supplies. Plumbing systems should be installed carefully by registered plumbers who adhere to the minimum requirements of the National Plumbing Code. The systems should be checked carefully to ensure their proper functioning. Air gaps, non-pressure-type vacuum breakers, or back flow preventors are utilized to prevent problems from occurring. Cross connections are most prevalent in the following recreational situations: direct connection of the water supply to cooling or condensor systems in refrigerators or air-conditioning systems, fish ponds with submerged inlets, fire hydrants with drains connected to sewers, frostproof water closets, kitchen and laundry fixtures with common waste and supply lines, underground water sprinkling systems that do not have vacuum breakers, and ice cube machine drainage lines connected to the sewer lines.

Comfort stations are mobile flush toilets used for the public. They are frequently used in recreational areas. It is necessary that comfort stations be constructed in such a way that any materials removed from them cannot contaminate water supplies or ground areas.

Shelter

Shelter consists of the site locations, structures to be utilized, and various campgrounds, picnic areas, and facilities.

Site Selection

The site selected must be away from swamps and mosquitoes, and must be off the main thoroughfares to reduce potential accident hazards. The access roads should be constructed with turns and loops so that vehicles have to reduce their speed. If feasible, one-way traffic patterns should be utilized. The roadways should be of all-weather construction where possible. The sites should be removed from sources of noise, air pollution, odors, or other kinds of contaminants. Recreational areas should be located on gently rolling well-drained land that is not subject to flooding and natural hazards. If possible, the camps should have an eastern exposure for the campers to receive the early morning sun and afternoon shade. Lightly wooded areas provide additional shade and help make the area more comfortable. If public utilities are available, they should be utilized.

Structures

The construction of buildings is important from a public health viewpoint, because people spend many hours within them. All units should meet the minimum requirements of the current proposed housing ordinance of the American Public Health Association. The following items are important for good and proper housing:

1. A habitable room should have at least one window or skylight facing outdoors.
2. Window or skylight should be readily opened.
3. Every bathroom or water closet should have adequate quantities of hot and cold running water.
4. Heating facilities should be available in those areas where temperatures are low at night. It is essential that the heating facility be well constructed, well engineered, and well ventilated.
5. All hallways and stairways should be properly lit.
6. Mosquito screening should be supplied in all housing units.
7. Foundations, floors, walls, ceilings, and roofs should be weather tight, watertight, and rodent tight.
8. Plumbing fixtures and water and waste pipes should be properly installed and in good working order.
9. At least 150 ft^2 of floor space should be provided for the first occupant and 100 ft^2 of floor space, for every additional occupant.
10. Cellars should not be used as rooms.
11. The buildings should be provided with or have easy access to adequate water supplies and sewage disposal.
12. Fire protection devices should be provided.

Additional material on housing can be found in Chapter 8 on the indoor environment.

Trailers

Trailer parking facilities must be so arranged that adequate water and sewage connections, and adequate shower and sanitary stations are provided. The sites must be well drained and free of obstructions. Further information on trailers also is provided in Chapter 8.

Campgrounds and Picnic Areas

Campgrounds and picnic areas are part of the area of shelter. These need to be selected in such a way that they are well drained, free of heavy undergrowth, and on solid and gently sloping ground. The layout should be such that cars have to travel slowly. Individuals should be able to walk comfortably to all areas of recreation, water supply, and sewage disposal and solid waste disposal. Adequate supervision should be provided within the park to maintain these areas in good working condition.

Food

The handling of food tends to be less stringent and the facilities less modern and well maintained than in other types of foodservice. In a recreational setting, food may be prepared for one individual, a family, several families, or a very large camp. It is essential that all the rules, regulations, codes, and good practices associated with food service be followed in the camp situation. At times it is difficult to refrigerate in this setting; however, because proper refrigeration of food is so essential to avoiding foodborne disease, this is an absolute must.

Food Service

To conduct an effective foodservice program in recreation areas, whether it is for a family or for a large group, it is necessary to start with good clean products, keep them well refrigerated, prepare them under proper conditions, and utilize them within adequate time spans. The whole area of food protection is discussed in detail in Chapter 3 on food protection. However, for the recreational area it is well to remember some of the most frequent food problems. Food must be prepared with clean utensils, and pots and pans of nontoxic materials. It is essential to maintain proper cleanliness. Hand washing is a must for the people who prepare and serve the food. Food that is left open should be disposed of instead of saved for other meals. Food should be served either on single-service plates, which are then thrown away, or on dishes and plates that are thoroughly washed and then sanitized. Solid waste from food should be stored in tightly closed containers to prevent breeding of flies and rodent problems. Because vending machines are used extensively in recreational areas, particular care should be given to these units. Vending machines must comply with the requirements of the vending of food and beverages codes. If foods are refrigerated, the temperatures must be adequately low. The machines should be cleaned carefully on a daily basis.

Milk

Milk, milk products, and frozen desserts must come from certified sources. Milk must be stored below 41°F at all times. Once the milk has been poured into pitchers, it is essential that it be utilized completely. It should not leave the original container and then be kept in a refrigerator for reuse. All milk, milk products, and frozen desserts should be inspected by the various county or state health departments.

Solid Waste

Solid waste consists of materials and entrails from fish cleaning, stables, eating, and life-associated processes. Solid waste handling is important in the prevention or reduction of flies, other insects, and rodents. People have a tendency, especially in recreation areas, to throw their solid waste on the ground or dispose of it in careless ways. Such carelessness causes a hazard and a nuisance and may lead to fires, odors, and unsightliness.

Fish Handling

Fish cleaning should take place in screen-enclosed facilities with nonporous floors. The washings from the fish should go into trapped floor drains. The traps should be cleaned thoroughly and the floor drains should be flushed. It is important that the fish cleaning is done on impervious, nonabsorbent sloping surfaces to ensure proper cleaning. Pressurized water is necessary for final cleanup. All the materials — entrails, fish heads, and so forth — should be put into garbage-grinding units or plastic bags that are fastened and then enclosed in metal cans with tight-fitting lids. The fish materials should be removed daily to an incinerator or a properly operated landfill. These materials should be buried as soon as they are deposited within the landfill. Fish-cleaning facilities must be scrubbed down very carefully to avoid attracting insects and rodents and becoming an odor nuisance.

Stables

Because horseback riding is popular in recreational areas, and cattle may also be found in certain recreational areas, it is important that the stables are properly maintained. Manure must be collected on a daily basis and put into concrete manure storage bins. These bins must be tightly closed to prevent fly breeding and odors. A second technique is to have the stables located on well-drained, gently sloping sites where the floors are made of concrete. The floors in the horse stalls should be paved with wooden blocks and sealed with asphalt or another impervious material. The rooms should be hosed down and the drainage either should go into a special holding tank to be pumped out later and removed to sewage treatment systems or should be put into waste stabilization ponds away from the recreational areas. It is essential that all areas of the stables be kept extremely clean at all times and effective insect and rodent control be practiced. Stables should be screened and rodent proofed.

Other Solid Wastes

The volume of garbage and other solid waste produced depends on the geographic location, the season, the number of people, the kinds of facilities, and so forth. It is essential that all solid waste be stored in watertight, rust-resistant, nonabsorbent, durable containers, which are covered with tight-fitting lids. The containers should be lined with plastic bags. The garbage, if not ground up in a garbage disposal unit, should be stored in the same way as other solid waste. It is important that the solid waste at campsites be removed on a daily basis to prevent rodents and other scavenger animals from knocking over the containers and scattering the solid waste. The material should be removed to either an incinerator or a sanitary landfill. Further information concerning solid waste disposal can be found in Chapter 2 in Volume II.

Swimming and Boating

Swimming takes place in artificial lakes, natural lakes, and swimming pools. An in-depth discussion on all types of swimming pools can be found in Chapter 4 in Volume II.

Swimming Pools

Swimming pools in recreational areas must be designed and constructed in such a way as to prevent accidents and the inflow of sewage or other types of contaminants. The water supply must be pure and feed into the pool through a system where an air gap exists. The sewer system must be separate from the swimming pool and in no way connected with it. Dressing rooms, toilets, and showers should be provided for the individuals using the pool. It is essential that all pools have lifeguards and that children or adults not be permitted to swim unless adequately trained lifeguards are present. It is also essential that the swimming pool water maintain a pH of 7.2 to 7.8 with a chlorine residual of 1 ppm. All pools should be clear at the deepest point and properly filtered. The types of records necessary, water quality, and so forth are discussed in Chapter 4, Volume II.

Outdoor bathing places, such as lakes, streams, rivers, and tidal waters, must be carefully checked before they are used for swimming. It is quite possible that the coliform aerogen group of bacteria from human feces may be present in the water. It is also possible that many organisms are present from runoff that comes from animal feces and from agricultural fields. It is important in outside bathing areas that all potential sources of harmful pollution are identified and these sources blocked or the bathing areas closed. These sources include sewage from boats, outlets from dwellings and other establishments, public sewer systems, leakage from improper dumping of solid waste or landfills, and runoff from land. In artificial areas, it is also essential to recognize that the bottom of the swimming area may be muddy, therefore making it difficult to view, or it may have step offs. Proper safety is an integral part of swimming area programs.

Swimming is the second most popular recreational activity in the United States with more than 350 million people participating each year, leading to outbreaks of disease. One of these is *Cryptosporidium parvum*, which has become a serious problem in recreational waters. The small size of the oocyst, which is 4 to 6 μm, and its resistance to many chemical disinfectants, for example, chlorine, create a challenge for standard filtration and disinfection procedures. The low dose required for infection and the prolonged excretion of high numbers of oocysts make *C. parvum* ideal for waterborne transmission. Chlorinated recreational water facilities, such as public swimming pools and water parks frequently used by large numbers of diapered children have been implicated in the spread of this disease.

Boating Areas

Boating has become one of the largest industries in the recreational field. More than 10 million pleasure boats are used for recreation in the United States. Many of these boats have galleys and toilet facilities. It is necessary to ensure that debris and garbage from the galley and the human fecal material from the various toilet facilities do not end up in the lake areas. These materials should be taken back to the marinas and emptied into holding tanks, where they are removed to sewage disposal plants for proper disposal. Other problems created by boats include waste oil, fuel, and any other refuse material that may be purposely or accidentally discharged into the waters.

In addition to the environmental problems involved in disposal, additional environmental problems occur in the obtainment of adequate quantities of safe water. It is important that the tanks holding pure water are not cross connected to sewage systems within the boat. These tanks must be properly washed, rinsed, and chlorinated. Water should be obtained from safe drinking sources. Food must be stored aboard the boats in refrigerated lockers until used.

In addition to the problems created by motorboats or boats driven by gasoline, many sailboats and rowboats are on the waters. All boat users must have a complete understanding of boating safety requirements and know what to do in the event of an emergency. It is essential that all water areas be clearly marked for hazardous situations.

MODES OF SURVEILLANCE AND EVALUATION

See the appropriate chapters for surveillance and evaluation by environmental health area.

CONTROLS

Recreational officials and agencies must work closely with public health agencies in an attempt to reduce potential and existing hazards. Individuals traveling into remote areas should have proper training in camping and hiking to prevent additional health and safety hazards. (See appropriate chapters for given environmental health areas in Volumes I and II.)

Safety

Recreational safety encompasses all forms of safety problems including problems related to site selection, types of buildings and equipment, fire and electrical wiring, heating systems, structural hazards, campgrounds, playgrounds, food service, swimming pools, bathing areas, and refuse disposal. In addition to these hazards found in all phases of life, it is essential to realize that the individuals who go out into the recreational environment are out there for fun. Many of them are not equipped to operate efficiently, effectively, and safely within this environment. They take risks, they act out of lack of understanding, and they lack proper equipment. Fire safety is a special concern, because many buildings are made of wood and it is difficult to get proper firefighting equipment into these areas. Individuals hike and climb without understanding proper techniques and get into trouble, resulting in lost lives or injuries. Many facilities in recreational areas have the kinds of structural hazards and poor electrical wiring that would never be tolerated within the home. The individual is on vacation and is unconcerned about these inadequacies. However, just these types of hazards lead to many serious accidents. Further, within the recreational environment, people hunt and many hunters get killed each year. People drown because they do not understand water safety.

Role of Federal Government

Since the creation of the Hot Springs Reservation in 1832, the federal government has been involved in the recreational environment. It manages hundreds of sites, and enforces all the environmental legislation that has been passed by Congress in the last several decades. For example, the Bureau of Land Management is responsible for managing 264 million acres of public lands and an additional 300 million acres of subsurface minerals. The National Park Service is responsible for conserving the natural and cultural resources of 375 plus sites, covering over 83 million acres. The Fish and Wildlife Service is charged with conserving, protecting, and enhancing fish, wildlife, and their habitats for the continuing benefit of the American people. The Bureau of Reclamation is charged with managing, developing, and protecting water and related resources in an environmentally and economically sound manner. The National Park Service was created on August 25, 1916 to manage areas reserved by Congress for the American people.

SUMMARY

To summarize the problems of the recreational environment and to introduce the kinds of programs that are necessary to resolve these problems properly, it is important that the individual read an entire book on the basic principles of environmental health and apply each section as it fits into a segment of the recreational environment. The difference between the recreational environment and other environments is that the former is more dangerous.

Humans have made increasing use of the environment for recreational purposes. These recreational uses by large groups of transient individuals cause destruction to the environment and intensify the potential for the spread of disease and injury. Limitations in the number of individuals using a given area, along with improvement of facilities and good educational approaches, are needed to enhance the recreational environment for humans and to protect it for succeeding generations.

RESEARCH NEEDS

Emerging Issues

There are at least ten major trends that will be redefining our society in the future. They will definitely have an effect on the recreational environment as well as all environmental issues. They are as follows:

1. There will be substantial demographic changes in our society. These include increasing diversity, aging of the population, population shifts, and change of family size and structure. It is anticipated that by the year 2050, racial and ethnic minorities will make up about half of all Americans. Also needed are new cultural experiences and a probable change in recreational and resource habits. In 1900,

only 4% of the population was over 65. By the year 2020, it is projected that 21% will be over 65. This will cause a change in recreation styles and use of facilities. The United States is becoming more urban, and therefore the individuals will seek relief in more country-style vacations. There are fewer traditional families of two parents and two children. It will be necessary to have the facilities available for different types of visitor groupings.

2. The information revolution has created access to large quantities of data concerning recreational areas. Individuals are more apt to select their own type of vacation instead of one that is planned by various facilities. This may lead to a sharp increase in accidents due to a lack of familiarity with hazardous conditions.

3. Individuals appear to have greater distrust of large, central institutions. Along with this, the individuals have a distrust of expert knowledge, leading to the potential for a greater number of accidents.

4. Individuals and communities are demanding involvement in the decisions, which leads to an alteration in what is being done, for good or for bad.

5. Our society has become highly litigious. This may lead to many lawsuits that could cause substantial financial problems.

6. People are much wealthier and therefore have more disposable income. People buy everything and expect to use it, such as all-terrain vehicles, watercraft, mountain bikes, and mountain skates. These items can lead to more contamination and a greater risk of accidents.

7. Our society has a decreasing reliance on, and identification with, primary production communities. This may lead to resentment on the part of rural and commodity-dependent communities.

8. Global integration will put new pressures for setting global standards for management and environmental quality in all areas, especially the recreational environment.

9. The demand for environmental quality will continue to grow, while the demands of people who degrade the environment will also continue to grow.

10. A chronic underinvestment in public lands and natural resource management exists. As the baby boomers retire, over the next 30 years, the ratio of workers to pensioners will shift, stretching the ability of the government to meet social needs, thereby intensifying the competition for federal dollars, and leading to less federal participation in the recreational environment.

CHAPTER 11

Occupational Environment

BACKGROUND AND STATUS

The occupational environment can be simply defined as any place, indoors or outdoors, where people work in return for financial or other remuneration. The prevention of occupationally related illnesses and injuries among workers is an important issue. Accordingly, occupational health and safety principles and practices are emphasized to identify, evaluate, and control sources and causes of hazards in the workplace environment, in turn decreasing risk of illness and injury to workers.

The hazards include chemical, physical, biological, ergonomic, mechanical, electrical, and psychological agents or factors that can potentially cause illness or injury to workers. The chemical stresses include toxic, flammable, reactive, or corrosive solids, liquids, and gases. These materials may become airborne in the form of dusts, fumes, mists, fibers, smoke, gases, and vapors. The physical hazards include ionizing radiation, nonionizing radiation, noise, vibration, and extreme temperatures and pressures. Biological hazards include insects, animals, bacteria, viruses, protozoans, fungi, and helminths. Ergonomic hazards include unusual body positions, repetitive motion, and poor equipment and workplace design. Mechanical and electrical hazards include instrumentation and conditions that may cause accidents and related injuries among workers. Psychological factors focus on emotional stressors and pressure encountered by workers, fatigue, monotony, and boredom.

Historical Perspective

Historically to the present, health and safety hazards have been associated with certain occupations. Effective action to protect the workers from related illnesses and injuries has ranged from either totally lacking or inadequate to extremely comprehensive in scope.

In the first century A.D., the hazard of working with sulfur and zinc was described and protective masks were used by workers. The Industrial Revolution in England and the resultant increased exposure to toxic materials and dangerous occupations were noted, but little was done to improve the industrial environment. Conditions

561

continued to deteriorate as the push for industrial expansion increased sharply. In the United States prior to 1911, employees had to sue their employers to collect damages for illness or injury due to their jobs. In 1911, the first state compensation laws were passed. More recently, workers have received increased protection through the following federal acts: Walsh-Healey Public Contracts Acts of 1936; Maritime Safety Amendments, Public Law 85–742 of 1958; Construction Safety Amendment, Public Law 91–54 of 1969; and other acts and administrative orders.

The Williams–Steiger Occupational Safety and Health Act, the most comprehensive and far-reaching federal legislation, was passed in 1970. The basic law was amended in 1974, 1978, 1982, and 1984. This legislation covers about 75% of the civilian labor force. Further special provisions of the act pertain to the 2.7 million federal government civilian employees.

The Occupational Safety and Health Act is a general law providing a broad spectrum of powers for use by the Secretary of Labor to reduce exposure to hazardous conditions in the occupational environment. The secretary has the authority to promulgate, modify, and revoke safety and health standards, make inspections, issue orders, and invoke penalties, require employers to keep adequate safety and health records, ask the courts to stop potentially dangerous situations, and approve or reject various state plans for Occupational Safety and Health Administration (OSHA) programs under the act. The Secretary of Labor works with the Secretary of Health and other agencies in a variety of ways. The Department of Labor conducts short-term training sessions for the federal personnel and for personnel from various states.

Under the Occupational Safety and Health Act, the U.S. Department of Health and Human Services (USDHHS) has primary responsibility for conducting health and safety research, evaluating hazards, determining and listing the toxicity of various substances, and developing and training workforce. The National Institute of Occupational Safety and Health (NIOSH) is the unit that is responsible for related research and education and training programs. The data from research help develop or modify criteria used for recommending health or safety standards. Surveillance and technical service programs, including hazard evaluation, also are conducted by NIOSH.

Over 117 million men and women are employed in the United States. All these individuals are exposed to occupational hazards to some degree. Although OSHA requires that occupational illnesses be reported, reporting has been poor. One reason is the inability or failure to identify and correlate potential occupational causes with adverse effects or illnesses. According to NIOSH, 125,000 new cases of occupational disease occur each year, but the figure may be an underestimate.

Economic Impact

The cost of occupationally related disease and accidents must include not only the loss of dollars due to death, disease, or injury but also the insurance, effect on society, and cost of loss of financial stability for the families involved. An additional cost involves the economic impact of OSHA standards and the corrections that must be made in equipment, facilities, and workers' procedures. OSHA must consider the technical feasibility and economic cost of each of its major proposed standards. The cost of implementation can be determined with some accuracy, but the resulting benefits cannot always be estimated. Overall, the loss of workers' time and production, as well as the securing of new equipment and facilities, has cost billions of dollars. Additionally, the cost of operating government is extremely high. It has also been estimated by the commission on federal paperwork that the government generates enough documents each year to fill 51 major-league baseball stadiums.

Occupational Health and Safety Professionals

Although many occupational health and safety principles and practices overlap, a distinction exists between the two areas. Whereas occupational health focuses mainly on illness prevention, safety emphasizes injury prevention. Occupational health professionals are traditionally called *industrial hygienists* and occupational safety professionals are commonly, but not exclusively, called *safety engineers*. The traditional industrial hygiene tetrad of anticipation, recognition, evaluation, and control of hazardous agents and factors that can cause illness and injury among workers is applicable to both occupational health and safety. Indeed, ideally both occupational health and safety issues must be simultaneously addressed to minimize risk of illness and injury among workers efficiently and effectively.

INTERRELATED PHASES OF OCCUPATIONAL HEALTH AND SAFETY

Anticipation and Recognition

Anticipation refers to a proactive estimation of health and safety concerns that are commonly, or at least potentially, associated with a given occupational setting or operation. Recognition closely parallels anticipation and involves actual identification of potential and actual hazards. Indeed, both phases can commence prior to entering a facility or site and continue throughout investigations, evaluations, and general operations as additional data are compiled.

Investigation and characterization involve gathering qualitative and quantitative data concerning a given facility, site, or operation. The qualitative and quantitative data include combinations of information, such as:

1. Identification and hazardous characteristics of physical, chemical, and biological agents and ergonomic, mechanical, and psychological factors
2. Types of unit processes and standard operating procedures
3. Number of operating personnel actually and potentially exposed
4. Proximity and distance of sources of exposure relative to workers, including layout of a facility or site
5. Review of records and documented health and safety programs

Knowledge of several unit processes in occupational settings ranging from office environments to the traditionally more hazardous manufacturing settings can provide insight about potential exposures. As much information as feasible is gathered during the anticipation and recognition phases and is reviewed and evaluated to design strategy necessary for preliminary on-premise survey and evaluation. In addition, the information is useful in developing preliminary health and safety plans to protect occupational health and safety personnel and other workers during specific operations or at a given location.

Evaluation

Anticipation and recognition continue during the evaluation phase, which involves a preliminary survey, including visual and instrumental monitoring, of a facility or site. Visual monitoring refers to the observation of processes or conditions at the facility or site, and subjectively and objectively identifying potential and actual hazards. Instrumental monitoring involves the use of monitoring equipment to record qualitative and quantitative environmental surveillance data. Chapter 12 provides a summary of major instrumentation used for identifying, detecting, and measuring various physical, chemical, and biological agents in occupational and nonoccupational environments.

Common instrumental monitoring focuses on atmospheric screening for combustible gases and vapors, molecular oxygen, toxic inorganic and organic gases and vapors, toxic inorganic and organic particulates, ionizing radiation, illumination, and sound levels. Meteorologic data, such as temperature, relative humidity, barometric pressure, wind speed, and wind direction, also are commonly collected. Mechanical and electrical hazards that may contribute to worker injury also are identified and evaluated. In addition, evaluation of control measures, including personal protective equipment and ventilation systems, is conducted to determine if the controls effectively reduce the potential for worker exposure. Occupational health and safety professionals also may be involved with the ergonomic evaluation of workers to determine if an efficient match exists between them and their physical workplace environment. Relatively recently, psychological factors, including stress encountered by workers, have received increased attention. Indeed, psychological issues are commonly addressed as behavioral aspects of ergonomics.

Qualitative and quantitative environmental surveillance data are augmented by medical surveillance data. Environmental surveillance can provide some data concerning external exposures or workplace conditions. Medical surveillance can possibly provide data about the impact or adverse effects due to external exposures to various agents or factors. The combination of environmental and medical surveillance data increases the probability of minimizing additional external and internal exposures, as well as establishing a correlation between cause and effect. The comprehensive environmental surveillance data, plus related medical surveillance data, are used to detect and measure hazardous agents and factors encountered by workers and to design strategy for control and remediation.

Data derived from evaluation are typically summarized in a technical report. The reports are important for documenting exposures and provide insight for future exposure prevention and remediation. The reporting of occupational illnesses, accidents, and incidents on appropriate forms to the safety supervisor, industrial hygienist, or other environmental health professional also is extremely important. Reporting of monitoring data and injury and illness statistics also is necessary to comply with various federal and state laws and to obtain worker's compensation, where necessary, for the victim of an accident or disease.

Control

Measures for controlling exposures to workers engaged in hazardous operations and activities involve implementation of administrative, engineering, and personal protective equipment controls. The implementation of administrative controls, such as development of a health and safety plan and standard operating procedures, should be a primary focus. Indeed, even if engineering and personal protective equipment controls are utilized, their full protective benefit is often compromised due to poor work practices. Engineering controls include appropriate process design, automation, guarding, shielding, grounding, and ventilation. When engineering controls are not adequate or feasible, however, personal protective equipment is frequently warranted. Most common personal protective equipment provides barriers between a worker and contaminated surroundings and sharp or heavy objects. A variety of equipment is available, including respirators, clothing and gloves, earplugs and muffs, glasses, goggles, shields, and rigid hard hats and shoes.

SCIENTIFIC, TECHNOLOGICAL, AND GENERAL INFORMATION

Overview

Occupational exposure limits have been established and recommended for many of the physical and chemical agents to limit the degree and duration of worker exposure. Fewer guidelines are available relative to biological agents. Monitoring and analysis are conducted to determine the levels of various physical, chemical, and biological agents present in the workplace, especially within the air matrix. Guidelines also have been set relative to ergonomic, electrical, and mechanical factors other than exposure to airborne physical, chemical, and biological agents that can cause adverse impact to human health.

The general characteristics or subcategories of the numerous agents and factors present in the occupational environment that can pose health and safety problems to workers exposed to elevated levels or unacceptable conditions were outlined earlier. A more detailed summary of three major categories of concern, namely, physical, chemical, and biological environmental agents and factors, follows.

Physical Agents

Physical agents include energy sources such as sound and vibration, nonionizing and ionizing radiation, and extremes of heat and cold stress. The hazard associated with these agents is typically due to the impact of a form of physical energy interacting with the human system.

Sound

Sound waves are disturbances in the uniform pressure of air molecules, detectable by the human ear, caused by vibrating objects. As objects vibrate back and forth, they alternately cause compressions and rarefractions (partial vacuums) of molecules in the air. Thus, sound is an energy form of vibration that may be conducted through various media including solids, liquids, or gases. The velocity of sound wave travel varies depending on the mass (density of the medium) and elastic reactions (pressure) of the molecules.

The amplitude or intensity of sound is related to sound pressure. Sound pressure is the deviation of air pressure from normal atmospheric pressure and is related to the amplitude of sound. The best value to use to determine sound pressure amplifications would be average pressure changes. One problem would be created if average pressure were used; average compression and rarefaction pressure change always equal zero or atmospheric pressure. In place of a simple average, a root mean square (RMS) of instantaneous pressures is calculated. RMS sound pressures are measured in dynes per square centimeter, newtons per square meter, microbars, and more commonly, micropascals. Sound power is another measure associated with the amplitude of a sound wave. It refers to the total sound energy radiated by the source. Sound power is typically expressed in the unit watts.

Sound wave amplitude may be defined either as the quantity of sound produced at a given location away from the source, or in terms of its overall ability to emit sound. Sound wave amplitude is the amount of pressure change from the average atmospheric pressure, which is generally described by sound pressure or sound intensity. The louder the noise, the higher the intensity. *Loudness* is a term that refers to an observer's perception or impression of loud vs. soft sounds.

Associated with the measurement of sound pressure and sound power is the logarithmic unit decibel (dB). By definition, the decibel is a dimensionless logarithmic unit relating a measured quantity such as sound pressure in micropascals ($\mu°$) or sound power in watts (W) to a reference quantity. Decibels are commonly used to describe levels of acoustic intensity, acoustic power, and hearing thresholds. Sound pressure levels are measured in decibels referenced to 20 $\mu°$. The reference value corresponds to the lowest sound pressure that produces an audible sound detectable by humans as follows:

$$SPL = 20 \log \frac{P_{meas}}{P_{ref}}$$

$$= 20 \log \frac{P_{meas}}{20 \ \mu P}$$

(11.1)

In addition to amplitude or intensity of sound waves, three other important characteristics of sound waves that are useful in understanding and evaluating sound in the occupational environment are frequency, wavelength, and speed (velocity). Frequency of sound waves is an important attribute that must be determined.

Frequency is the number of complete vibrations within a given period of time, is the total number of completed cycles (compressions and rarefactions) of a sound wave per second (cycles per second), and is measured in units of hertz (Hz). The A-scale refers to a frequency range of 20 to 20,000 Hz and is considered the frequency range for human hearing. Accordingly, sound measurements conducted for human exposure assessment are reported as decibels A-scale (dBA).

Wavelength refers to the distance required for one complete pressure cycle to be completed. This attribute is important when considering the effects of sound waves. For example, sound waves that are larger than surrounding objects diffract or bend around these obstacles. Small sound waves, on the other hand, refract or disperse and surrounding obstacles can function as effective sound barriers. Thus, wavelength is an important variable in noise control activities. The speed of sound or sound wave velocity is dependent on the ambient temperature of the medium and its environment and at about 22°C is approximately 344 meters/sec. The speed (c) of a sound wave is determined by the product of its wavelength (λ) and frequency (f):

$$c = f\lambda$$

$$344 \text{ m/sec} = f\lambda$$

(11.2)

Sound waves propagate in a spherical form from the source of generation. If sound measurements are taken at the same distance in any direction from a source under free-field conditions, the sound pressure levels should be the same. Free field refers to a location that is an open area; free of barriers or interferences from ceilings, floors, walls, adjacent equipment; and a wide variety of other items that can diffract, absorb, or transmit sound energy and either increase or decrease sound pressure levels. Under free-field conditions, the longer the distance is from the source the lower the sound pressure. Sound pressure changes abide by the inverse square law that states as the distance increases from the source, a corresponding decrease in intensity occurs based on sound level meter readings in decibels. If we are monitoring free-field sound, ideally there should be a 6 dB reduction when the distance doubles.

For example, if sound pressure readings are equal to 90 dB when taken 5 ft from a source, when the distance is doubled to 10 ft, the corresponding decreased sound pressure level should be 84 dB. The 6-dB reduction in sound pressure, however, is the expected pressure change under ideal conditions. Sound pressure measures failing to decrease by approximately 6 dB as the distance doubles from the source indicates that free-field conditions are not present.

Excessive or unwanted sound is referred to as *noise*. Sound energy travels as waves through air and is generated as a result of the oscillation of sound pressure above and below atmospheric pressure. The sound energy interacts with the human system by initially vibrating the tympanic membrane, or ear drum, which conducts the energy to the middle ear where the energy is conducted further via vibration of the ossicles. Ultimately, the energy enters the inner ear where pressure changes in the cochlea and movement of hair cells cause stimulation of the auditory nerve that transmits the signal to the brain (an example of ear–sound energy interaction may be found in Figure 11.1). Excessive levels of, and exposure to, sound at a given

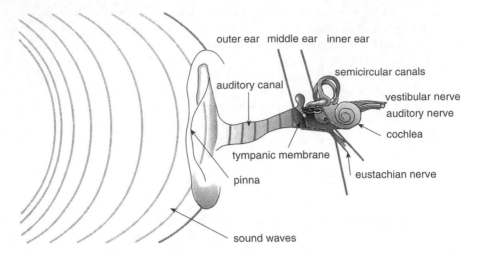

outer ear middle ear inner ear

semicircular canals

auditory canal

vestibular nerve

auditory nerve

cochlea

tympanic membrane

pinna

eustachian nerve

sound waves

Figure 11.1 Ear–sound energy interaction.

range of frequencies combined with a related excess in duration of exposure can result in damage to the auditory system and concurrent conductive or sensorineural hearing loss. Hearing loss associated only with aging is called presbycussis. Damage to the tympanic membrane (ear drum) or middle ear ossicles (bones) can result in conductive hearing loss. Sensorineural hearing loss results when the inner ear or auditory nerve is damaged. Noise also may cause physical and psychological disorders, interference with voice communications, disruption in job performance, and disruption of rest, relaxation, and sleep. Physical and psychological disorders are due to a change in the response of the nervous system. These effects are annoying and may lead to accidents, but basically are not due to physical changes in the body. Noise is responsible for increase in muscular activity; constriction in the peripheral blood vessels; acceleration of the heartbeat; changes in the secretion of saliva and gastric juices; increased incidence of cardiovascular disease; increase in ear, nose, and throat problems; and increase in equilibrium disorders. A noise level of 100 dB can interfere with speech. A level of 130 dB causes vibration of the viscera; 133 dB causes a loss of balance. The maximum 8-hour exposure is 90 dB. However, if this is maintained for a long period of time, loss of hearing can result. Any level above 130 dB is damaging. At 140 dB, the individual starts to experience pain. Annoyance due to noise is related to its loudness, frequency, and intermittency. As an example, if one were to sit in a quiet room and listen to water drip, it would eventually cause considerable annoyance and discomfort.

Vibration is an oscillating motion of a system ranging from simple harmonic motion to extremely complex motion. The system may be gaseous, liquid, or solid; periodic or continuous. Vibration, which in effect is the motion of a particle, is characterized by the displacement from the equilibrium position, velocity, or acceleration. Some common sources of vibration include tractor operation, pneumatic tools, heavy equipment vehicles, vibrating hand tools, chain saws, stamping equipment, sewing machines, and looms.

Overexposure to vibration while using hand tools may cause neuritis, decalcification of the carpal and metacarpal bones, fragmentation, and muscle atrophy. An extreme condition resulting from the use of power tools is called Raynaud's syndrome or dead fingers. The circulation in the hand is impaired because of extended periods of using vibrating tools. The fingers become white, devoid of sensation, and mildly frosted. The condition usually disappears when the fingers are warmed for some time, but in some cases the condition is so disabling that the individual must seek new work. At present, no generally accepted limits are set for a safe vibration level. This is an area where additional research is needed.

Nonionizing Radiation

Nonionizing radiation refers to waves of electromagnetic energy. In general and as shown earlier for sound, waves possess frequency, wavelength, and velocity. Frequency (f) is the number of vibrations per unit second, wavelength (λ) is one complete sinusoidal oscillation, and velocity (c) is the time rate change of displacement or the speed at which a wave travels a given distance. The three factors are related by velocity, often measured as the speed of light (3×10^8 meters/sec), which is equal to the frequency multiplied by the wavelength:

$$c = f\lambda$$
$$3 \times 10^8 \text{ m/sec} = f\lambda$$

(11.3)

Electromagnetic waves or radiation consist of oscillating electric and magnetic fields that are perpendicular to each other and each at 90° angles relative to the speed and direction of the traveling wave. Electromagnetic radiation energies make up a spectrum of categories that differ based on ranges of wavelengths (λ) and frequencies (f). In addition, the categories of the electromagnetic spectrum may be differentiated by quantity of photon energy. Photon energy is measured in units of joules (J), which refers to the work associated with a 1 N force moving an object a distance of 1 meter. The photon energy (E) is related to the product of Planck's constant (h; 6.624×10^{-34} J-sec) and wave electromagnetic wave frequency (f):

$$E = hf$$
$$= \left(6.624 \times 10^{-34} \text{ J sec}\right)f$$

(11.4)

The photon energy of electromagnetic radiation also can be expressed in units of electron volts (eV), which refer to energy gained by an electron when passing through a potential difference of 1 V; 1 eV is equivalent to 1.602×10^{-19} J. The power of electromagnetic radiation is related to energy over time and is measured in units of watts; 1 W represents power over time and is equivalent to 1 J/sec.

Examples of nonionizing electromagnetic radiation in order of increased wavelengths and inversely related decreased frequencies include ultraviolet (UV), visible

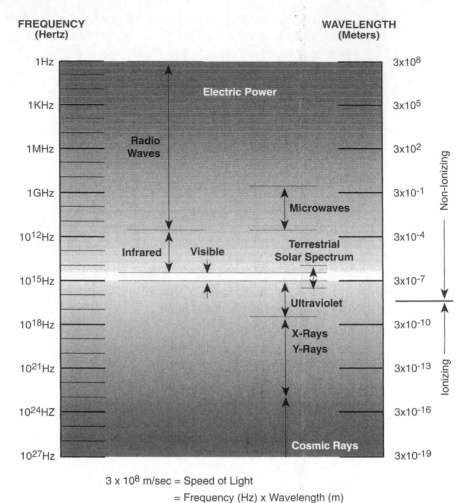

Figure 11.2   Levels of energy in the electromagnetic spectrum.

light, infrared (IR), microwaves (MW), radiowaves (RW), and extremely low frequency (ELF) electromagnetic field (EMF) energies (Figure 11.2).

The primary difference between ionizing radiation and nonionizing radiation is that ionizing radiation has sufficient energy (>10 eV) to dislodge subatomic orbital electrons on impact, thus causing ions to form. Nonionizing radiation has lower energies and longer wavelengths (>1.24 × 10⁻⁷ meters); therefore, it is incapable of dislodging orbital electrons, but may deposit thermal and vibrational energies leaving atoms in an excited state.

UV radiation is generated from natural and anthropogenic sources and is associated with potential damage to the eyes and skin. Indeed, excessive exposure to UV radiation from welding operations is associated with eye cataracts and exposure to sunlight can damage the eyes and increase the risk of developing dermal cancer.

Visible light is necessary for proper surface illumination and brightness. Too much or too little visible light can pose health and safety problems. Photons of visible electromagnetic light energy are transmitted linearly through air. The direction of light can be changed, however, depending on interaction with various media and incident light, it reflects from some surfaces. When light passes through different transparent media, it bends or refracts. In addition, when light passes through and exits a transparent medium it diffuses. Lighting is important in various environmental settings to increase comfort and safety. Several factors influence lighting, including the quantity and quality.

Illumination is related to the quantity of light or the light energy directed on a given area. Quality of light, however, refers to additional factors such as brightness, glare, and color. Glare causes discomfort and interferes with the successful completion of some work. The amount of glare is determined by the design of the lighting fixtures, the types of surfaces on which the light falls, the color of the surfaces and the environment, and the amount of outside light entering the industrial facility. Both the quantity and quality of illumination are evaluated in various environmental settings to determine whether lighting is adequate for specific and general tasks. In general, a minimum of 30 fc should be provided in all work areas, including receiving, opening, storing, and shipping areas. Footcandle levels may rise to as much as 2000 in certain examining areas. Poor qualitative and quantitative lighting conditions are associated with increased fatigue.

IR can occur from any surface that is at a higher temperature than the receiving surface. The transfer of radiant energy or heat occurs whenever the radiant heat emitted from one body is absorbed by another body. The organs of greatest concern associated with exposure to IR radiation are the eyes and skin. Radiant heat contributes to heat stress in exposed individuals. Skin erythema and burning, as well as lens and corneal cataracts, have resulted from exposure to specific wavelengths. Glass blowing, foundry, and boiler operations are common occupational sources of IR radiation.

Microwaves are electromagnetic nonionizing radiation with frequencies and wavelengths ranging from 1 MHz to 300 GHz and 0.1 cm to 10 meters, respectively. The energy is generated anthropogenically via devices that cause deceleration of electrons in an electric field; this, in turn, generates kinetic energy as microwaves. Depending on the frequency and wavelength, microwaves are transmitted for long distances through air. A major characteristic of microwave energy is that it can be readily absorbed by numerous materials, including human tissue, resulting in the generation of elevated temperatures almost uniformly within a given volume. Relative to human tissue, microwaves are more penetrating in relation to areas of higher water content, such as the eyes, skin, muscles, and visceral organs. In addition, higher frequencies and shorter wavelengths can increase the depth of absorption.

Microwaves cause an increase in heat in the whole body. Microwave absorption and heat generation are greatest in tissues having a high water content. Microwaves change polarity every half cycle, with the tissue molecules doing the same and thereby vibrating rapidly, which produces heat. They also cause damage to the lens of the eye and possible damage to the gonads.

Common uses and sources of microwave radiation include dielectric heaters, radio frequency sealers, radio and television broadcast stations, induction furnaces, communication systems (i.e., radar), and ovens. Microwave ovens are a common use and source of microwaves in both occupational and nonoccupational environments. Poorly aligned doors, accumulated food residues, and faulty door gaskets, hinges, and latches, however, can result in release of microwave radiation from inside to outside the microwave oven. Leaking microwave ovens can generate significant and dangerous amounts of microwave radiation into the external air. Accordingly, the ovens should be monitored periodically to determine whether microwave radiation is detected and measured outside of a closed and operating unit. Microwave oven performance standards specify less than 1 mW/cm^2 at any point 5 cm from the oven prior to purchase and less than 5 mW/cm^2 at any point 5 cm from the oven after purchase.

Specific ranges of wavelengths (0.1 cm to 300 meters) and frequencies (1 MHz to 300 GHz) of microwave and radiofrequency radiation can penetrate the human body and cause adverse effects to internal organs. Thermal impact including heat stress, teratogenesis, and ocular effects (cataracts and cornea damage) have been suggested as possible adverse effects. These thermal effects may include potentially damaging alterations in cells caused by localized increases in tissue temperature. The body's heat sensors, which are located in the skin, are not activated when the RF energy is absorbed deep within the body tissues.

Occupational and even nonoccupational exposures to very low and ELF EMFs have received increased attention. The nonionizing radiation energy consists of electric and magnetic fields. Electric fields are created by the presence of electric charges associated with operating and nonoperating electrical appliances and machinery connected to an electrical power source. Magnetic fields also are associated with electrical devices, but are generated in combination with electric fields when the devices are operating. EMFs are characterized by very low and extremely low frequencies and longer wavelengths. Although no specific demarcation exists, ELFs are below 10^3 Hz (1 kHz) and very low frequencies (VLFs) range between 10^3 to 10^7 Hz. Major sources of low frequency nonionizing radiation are electrical transmission lines (60 Hz) and numerous electrical devices present in the occupational environment. Occupations requiring installation and maintenance of electrical power lines and devices present some obvious potential sources of EMFs. Relatively recently, however, common office sources such as office computers and video display terminals (VDTs) have received more attention relative to VLF EMFs. Accordingly, monitoring is becoming more common to evaluate the levels of nonionizing EMF generated by this equipment in office settings.

ELF radiation has elicited increased concern, especially frequencies of 60 Hz. Sources of ELF include high voltage power lines, VDTs, and close proximity to electrical wiring such as electric appliances. Cancer, such as leukemia, and altered blood chemistry are some of the potential biological effects, according to preliminary research of ELF exposure. External exposures and the impact to health, however, remain in need of further investigation.

Light amplification by stimulated emission of radiation (laser) energy is an anthropogenic form of electomagnetic radiation consisting of intense light amplification and concentration in one wavelength and direction (Figure 11.3). Lasers have

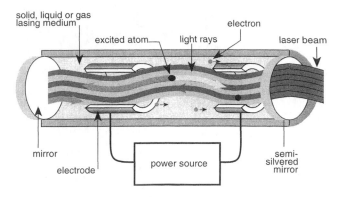

Figure 11.3   Laser.

been used in the construction industry for measuring distances and welding, and by some surgeons to repair torn retinas. The hazard to the individual is due to the point source of great brightness. The laser may also injure the eye, skin, or possibly internal organs. The laser also may oxidize chemicals in the ambient air.

Ionizing Radiation

Atoms are composed of three basic building blocks: the electron, which has an extremely small mass and is negative; the proton, which has a mass approximately 1800 times as large as the electron and is positive; and the neutron, which has about the mass of a proton and electron combined. Nonetheless, the mass of a proton and a neutron are considered approximately equal and the mass of an electron is essentially negligible. The protons and neutrons form the central core or nucleus of the atom. The electrons orbit the nucleus. The orbiting electrons do not have exact positions and, therefore, they are considered to be distributed as an electron cloud.

Elements differ based on the number of protons in their nuclei. Atoms of the same element have the same number of protons (the same atomic number), but may have different numbers of neutrons (different atomic weights). These atoms are called isotopes. For instance, the most common form of carbon (C-12) has six neutrons and six protons in the nucleus. However, an isotope of carbon (C-14) has eight neutrons and six protons. Carbon-14 is an unstable isotope of carbon and, therefore, is considered a radioisotope. Radioisotopes are atoms that have an unstable nucleus that emits energy in the form of ionizing radiation to become more stable. This often brings about a change from one element to another, which causes a change in the chemistry of the substance. For example, fission is the process occurring in nuclear reactors, and refers to the splitting of a uranium atom into two smaller atoms (Figure 11.4). The two smaller atoms are called fission products. These atoms have a greater number of neutrons compared with protons than the stable atom with the same number of protons. Therefore, for the atom to become stable, the neutron spontaneously ejects an electron and becomes a proton. The ejected electron is called a beta particle and travels at extreme velocities. The atom is said to have decayed. The atom has now become a new element.

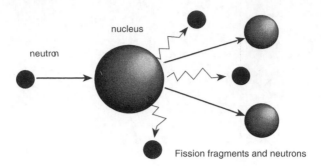

Figure 11.4   Nuclear fission.

Thus, chemical elements that are electrochemically unstable decay due to an imbalance of protons (p^+) or neutrons (n°) in the atomic nuclei and emanate energy in the form of ionizing radiation. The elements and compounds undergo natural reactions to achieve stability via emanation of atomic energy in the form of particulates and electromagnetic photons. The particles and photons impart excessive energy to matter with which they interact, resulting in ionization (Figure 11.5).

Ionizing particulate radiation consists of alpha particles ($2 n^\circ + 2 p^+$ in the form of a charged helium nucleus, $^4_2\text{He}^{+2}$); beta particles (negatron as e^- and positron as e^+); and neutron particles (n°). By emission of alpha particles, uranium becomes thorium and radium becomes radon. Ionizing electromagnetic radiation consists of photon energies as x-rays and gamma rays. Alpha or beta particles may be accompanied by gamma radiation, which is similar to x-ray. X-rays are of short wavelength, high frequency, and high energy. X-rays travel through space at the speed of light. They possess no real mass charge. X-rays are produced by bombarding a material with electrons that contain energy. Only about 1% of the electrons in the diagnostic region produce x-radiation. Most of the electrons give up their energy as they hit solid material and increase the heat of the material. The difference between the gamma ray and the x-ray is in their origin. A gamma ray is emitted from the nucleus of the atom when the nucleus goes at a higher or lower energy level.

The particulate and electromagnetic wave forms of ionizing radiation can interact via direct or indirect ionization of macromolecular or cellular components of the human body that, like those materials classified as toxic, may result in adverse biochemical and physiological changes manifested as abnormalities, illnesses, or premature deaths among those exposed or their offspring.

Ionization energies are measured in electron volts. The rate at which an element undergoes radioactive decay is referred to as activity and measured in units of curies (Ci) or becquerels (Bq) with 1 Ci equivalent to 3.7×10^{10} disintegrations per second (dps). The unit becquerel is an alternative to curies, where 1 Bq equals 1 dps. Half-life refers to the time required for 50% of a radioactive element to decay (disintegrate) and, accordingly, the activity to decrease. Each radionuclide decays with a specific half-life. During decay, it generally produces other radioactive substances. Radium-226 decays with a half-life of 1620 years. Iodine-131 decays with a half-life of 8 days. Sodium-24 decays with a half-life of 15 hours. The rate of decay is important to environmental health personnel.

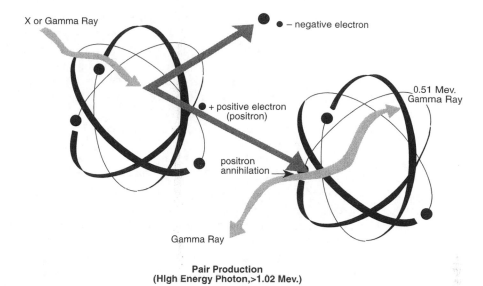

Pair Production
(High Energy Photon, >1.02 Mev.)

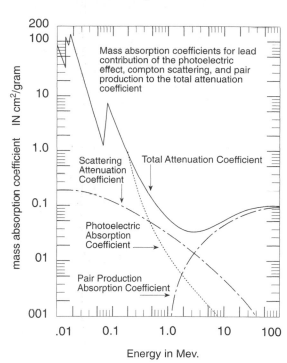

Figure 11.5 An example of ion production.

Activity does not specify the type of ionizing radiation and the degree of hazard. Exposure is the amount of ionization (electrical charge produced by x-rays or gamma rays) per mass or volume of air. Measured in units of roentgens (R = 2.58 C/kg =

2.083 × 10^9 ion pairs per cubic centimeter of air). Because 34 eV of energy is required to produce one ion pair in air, 1 R = 2.083 × 10^9 ion pairs per cubic centimeter of air × 34 eV per ion pair = 7.08 × 10^{10} eV/cc. In relation, 7.08 × 10^{10} eV/cc multiplied by 1.6 × 10^{-2} erg/eV = 0.113 erg/cc air. As a result, since 1 cc air has a density equal to 0.001293 g/cc, 1 R = 0.113 erg/0.001293 g air = 87.6 erg/g air. An erg, like the unit joule, is unit of measure for work or force through distance, with 1 J equivalent to 10^7 erg and 1 eV = 1.6 × 10^{-12} erg.

Exposure measured in units of roentgens has been replaced with the term *absorbed dose*. Absorbed dose is the amount of radiation energy (i.e., energy departed from any ionizing source) absorbed by the entire body or a specific body site. This is equivalent to the quantity of ionization energy deposited per unit mass of tissue (or other matter) and is measured in rads (r = 100 erg energy per gram of tissue). An alternative unit, *gray* is equivalent to 100 rads, which equals 1 J/kg. *Equivalent dose* refers to the quantity of absorbed dose multiplied by factors associated with a given form of ionizing radiation, such as the tissue weighting factor. The dose is measured in units of rem and, more recently, sieverts where 1 Sv = 100 rem.

Ionizing radiation is an important human hazard, because absorption of energy in the body tissues causes physiological injury by means of ionization or excitation. Ionization is a process involving removal of an electron from an atom or molecule leaving the system with a net positive charge. Excitation is the addition of energy to an atomic or molecular system, changing the system from a stable to an excited state. Radiation causes cell damage by upsetting the cell chemistry or physics, and by disrupting the ability of the cell to repair itself. Radiation causes somatic damage, which means that it affects the cells and organs during the lifetime of the human. It also causes genetic damage, which affects future generations. The amount of damage depends on the amount of radiation received, the extent of body exposure, the part of the body exposed, the amount of exposure, and the type of exposure, whether external or internal.

External radiation occurs when radioactive materials pass through the body. Internal radiation occurs when radioactive material is inhaled, swallowed, or absorbed through breaks in the skin. A large dose of acute radiation exposure causes nausea, fatigue, blood and intestinal disorders, temporary loss of hair, and potentially serious damage to the central nervous system. This might happen in the event of a nuclear war or a massive nuclear accident. Long-term effects include increased incidence of lung cancer suffered by miners of uranium and pitchblende; bone cancer, which occurred in luminous watch painters years ago who had swallowed small quantities of radium; and skin cancer and leukemia, which occurs in physicians and dentists from repeated exposure to radiation. Radiation also causes mutations in the genes of reproductive cells.

The extent of the hazard to humans is based on the dose, the body part, and the time of exposure. Large amounts of radiation to the whole body cause acute exposure. Large amounts of radiation to small parts of the body cause acute and chronic exposures. Small amounts of radiation to the whole body causes chronic exposure. Small amounts of radiation to limited portions of the body cause acute and chronic exposures.

Humans are exposed to radiation from the natural environment, including cosmic rays, Earth, and the atmosphere. Radiation from artificial sources include medical and dental sources, occupational sources, nuclear energy plants, radioactive fallout, radioactive dial watches, clocks, meters, radioactive static eliminators, television sets, isotope-tagged products, radioactive luminous markers, and certain household goods. Presently, the largest artificial source of radiation exposure is medical and dental x-ray machines. The possibility of acute radiation injury due to x-rays is minimal; however, long-term effects are not fully understood.

Industrial use of radiation produces additional sources of radioactive exposure. This includes the inspection of materials by means of radiographic inspection, fluoroscopic inspection, and continuous automatic inspection by x-rays to measure the thickness of material and the height at which something is filled. Personnel are checked for stolen articles, contraband, or weapons by whole-body fluoroscopes. The shoe-fitting fluoroscope has fortunately been eliminated in most states, because it was an important source of radiation, especially for small children. X-rays are also used to determine the diffraction pattern of crystalline materials.

Ionizing radiation has been used in industry for sterilizing foods and drugs, killing insects in seeds, toughening polyethylene materials, and activating chemical reactions in the petroleum process. One device used is the Van de Graaf apparatus, which is an electron beam generator. In the past, radium was used as a source for industrial radiography. It was also used in luminous paints and in the removal of static electricity generated by the movement of nonconducting materials at high speeds through machinery. The radium produced ionized air that removed the static charge. Today, other artificial radionuclides are used in industry, including cobalt-60, iridium-192, and cesium-137. Strontium-90 is used in some luminous paints and in the krypton switch lamp used by railroads in remote locations.

Atomic batteries are used to produce high-voltage, low-current sources in special instruments. Radionuclides are used to determine thickness, density, and depth of materials, and are used as tracers. Sources of radiation also include nuclear reactors, ventilation air, and cooling wastes. The latter come from nuclear power plants or nuclear ships and submarines. Thousands of mine workers are exposed to radiation in uranium mines and through milling.

Hot and Cold Temperature Extremes

Heat and cold stress are physiological responses to extremes of hot and cold temperatures, respectively. Accordingly, thermal agents must be considered as part of the evaluation of the occupational environment. The body generates metabolic heat via basal and activity processes. If the environment is too cold, the body loses heat faster than it can produce heat and a condition known as hypothermia can develop. Alternatively, if the environment is too hot and the body is not able to cool fast enough, heat stress can result. The potential for heat stress is exacerbated if an individual is also very active and simultaneously exposed to elevated temperatures.

The four factors that influence the interchange of heat between people and the environment include:

1   Air temperature
2. Air velocity
3. Moisture content of the air
4. Radiant temperature

The industrial heat problem is created when a combination of these factors produce a working environment that may be uncomfortable or hazardous to individuals because of an imbalance of metabolic heat production and heat loss.

Humans regulate internal body temperature within certain narrow limits (i.e., 97 to 99.5°F). Blood flows from the site of heat production in the muscles and deep tissues, which is the deep region or core, to the body surface, which is the superficial regions where it is cooled. The body produces sweat to help cool itself. Normally, the human body automatically controls the internal heat load. External heat load, which is produced by the environment, affects humans in different ways, based on physical fitness, work capacity, age, health, living habits, and degree of acclimation to heat.

Humans exchange heat with the environment through conduction and convection, radiation, and evaporation. Conduction is the transfer of heat through direct contact. If the substance in contact with humans, such as air, water, clothing, or other objects, is higher in temperature than the skin, the body gains heat; if the substance is lower in temperature than the skin, heat is lost. The rate of transfer is based mostly on the difference between the two temperatures. Air and water accelerate the transfer. Convection is the transmission of heat by the motion of air or water. Thermal or radiant heat is the transmission of energy by means of electromagnetic waves of IR radiation. Radiant energy, when absorbed, becomes thermal energy and results in an increase in the temperature of the absorbing body. Evaporation is based on air speed and the difference in vapor pressure between perspiration on the skin and the air. In hot, moist environments, evaporation may be limited or stopped. Cold is the absence of heat.

Thus, heat stress is influenced by several environmental factors, including air temperature measured in degrees Celsius or Fahrenheit, air movement measured in feet per minute, humidity measured in percentage of moisture, and radiant heat also measured in degrees Celsius or Fahrenheit. Elevated air temperature increases the heat burden on the human body. The adverse influence of air temperature is exacerbated by the presence of excessive water vapor in the air due to high humidity. Elevated humidity decreases the evaporation rate of perspiration from the skin. Perspiration and its subsequent evaporation is a natural defense mechanism to transport excessive heat from the body. Excessive water vapor in the air hinders the evaporation process and elimination of heat via evaporative cooling. The evaporative cooling is facilitated by increased air movement and, accordingly, also hindered by lack of sufficient wind or breeze.

Finally, radiant heat from various sources penetrates through the air and is absorbed by the body. In turn, the absorbed radiant heat increases the heat burden on the body. These thermal factors in combination with metabolic factors related to workers based on their physiology and degree of exertion influence the potential for individuals developing heat stress. In addition, the potential for heat-related disorders is increased by such factors as physical conditioning, age, weight, gender, and impermeable protective clothing.

The following are heat stress illnesses: heat stroke caused by considerable exertion in a hot environment, lack of physical fitness, obesity, dehydration, and so forth; heat exhaustion caused by extreme exertion and heat, lack of water, and loss of salt; heat cramps caused by heavy perspiration during hot work, large intake of water, loss of salt; heat syncope due to an unacclimated worker standing erect and immobile in the heat, causing blood to accumulate in the lower part of the body, which results in inadequate venous blood return to the heart; heat rash due to unevaporated sweat from the skin; heat fatigue caused by extreme heat for long periods of time; and interference with the heat-regulating center in the brain, causing body temperatures to rise to 108 to 112°F, which may result in brain damage or death. Certain safety problems are common in hot environments. Heat tends to promote accidents due to the slipperiness of sweaty palms, dizziness, or fogging of safety glasses. Also molten metal, hot surfaces, and steam create serious contact burns. Heat lowers an individual's mental alertness and physical performance. It also promotes irritability, anger, and overt emotional reactions to situations, which may lead to accidents.

Heat stroke symptoms include hot skin, usually dry, red, or spotted. The body temperature rises to 105°F or higher. The victim is mentally confused, delirious, possibly in convulsions, or unconscious. Unless the person receives quick and appropriate treatment, the individual will die. Individuals with these signs or symptoms of heat stroke need to be taken immediately to a hospital. Any emergency medical care prior to hospitalization should include removing the person to a cool area, soaking the clothing with water, and vigorously fanning the body to increase cooling.

Heat exhaustion includes several clinical disorders that have symptoms that may resemble the early symptoms of heat stroke. The person suffering from heat exhaustion still is able to sweat but experiences extreme weakness or fatigue, giddiness, nausea, or headache. In more serious cases the person may vomit or lose consciousness. The skin is clammy and moist, the complexion is pale or flushed, and the body temperature is normal or slightly elevated. In most cases, have the person rest in a cool place and drink plenty of liquids. If the case of heat exhaustion is mild, the individual can recover quickly. If the problem is severe take the individual to a hospital. Extreme caution must be taken concerning individuals who are on low sodium diets and who work in hot environments.

Heat cramps are painful spasms of the muscles that occur in individuals who sweat profusely because of heat exposure. They may be drinking large quantities of water but not replacing the body's salt loss. The water tends to dilute the body's fluids while the body is still losing salt. The low salt level in the muscles causes cramps, which may occur in the arms, legs, or abdomen. Again, be most cautious concerning individuals with heart problems or those on low sodium diets.

Heat rash, which is also known as prickly heat, occurs in hot, humid environments, where sweat is not easily removed from the surface of the skin by evaporation and the skin stays moist. The sweat ducts become plugged and a skin rash then appears. Extensive heat rashes may become complicated by infection. An individual can prevent the problem from occurring by resting in a cool place each day and by bathing regularly and drying the skin.

Transient heat fatigue is the temporary state of discomfort and mental or psychological strain coming from prolonged heat exposure. If persons are unaccustomed to heat, they are particularly susceptible to the problem. The individual suffers a decline in test performance, coordination, alertness, and vigilance, which may lead to accidents. Transient heat fatigue can be lessened or eliminated by using proper techniques of heat acclimatization.

Cold-related stress is initiated by a peripheral vasoconstriction, especially in the extremities, which results in a sharp drop in skin temperature. This helps reduce the body heat loss to the environment. However, this causes a chilling of the extremities, which results in the toes and fingers approaching freezing temperatures very rapidly. When the temperature of hands and fingers drop below 15°C, they become insensitive, malfunction, and the chance for accidents increases. The cooling stress is proportional to the total thermal gradient between the skin and the environment. The loss of heat through evaporation of perspiration is not significant at temperatures lower than 15 to 20°C. When vasoconstriction can no longer adequately maintain the body heat balance, shivering becomes an important mechanism for increasing the body temperature by causing the metabolic heat production to increase several times the resting rate. When proper clothing is used and the clothing does not become wet with water or sweating due to excessive work, then all parts that are protected by the clothing help maintain the body temperature. Exposed parts such as the face or fingers, may become excessively chilled and frostbite may occur.

Frostbite occurs when there is an actual freezing of the tissues with the mechanical disruption of the cell structure. The freezing point of the skin is −1°C. Increasing wind velocity increases heat loss and causes frostbite to occur more rapidly. When the velocity reaches 20 mi/hr, the exposed flesh freezes within about 1 min at −10°C. If the skin comes in contact with below-freezing-temperature objects, frostbite may occur, even though the environmental temperatures are warmer. The first symptom of frostbite is often a sharp, pricking sensation. Cold, however, produces numbness and anesthesia, which may allow serious freezing to develop without warning of acute discomfort. The injury from frostbite may range from a single superficial injury with redness of the skin, to deep cyanosis and gangrene.

Trench foot may be caused by long, continuous exposure to cold without freezing, combined with a persistent dampness or immersion in water. Edema, tingling, itching, and severe pain may occur. This is followed by blistering, superficial skin necrosis, and ulceration.

General hypothermia is an extremely acute problem that results from prolonged cold exposure and heat loss. If persons become fatigued during physical activity, they are more apt to have heat loss. As exhaustion approaches, the vasoconstrictor mechanism is overwhelmed and suddenly vasodilation occurs, with a rapid loss of heat and critical cooling. Alcohol consumption and sedative drugs increase the danger of hypothermia.

Chemical Agents

Chemical agents are classified as inorganic and organic and most notable for toxicity and flammability from an occupational health perspective. Aerosols, or particulates, are divided into dusts, mists, fumes, fibers, and smoke. Dusts are generated from mechanical actions such as crushing, grinding, or sawing solid materials. Mists are finely divided liquid droplets generated via agitation of liquids or certain operations such as spraying or dipping. Fumes result when a metal is heated to its sublimation point. The metal forms a vapor that rises into the air where it often reacts with molecular oxygen to form metal oxides. The vapor then condenses and forms a very small particle (<1 μm) called a fume.

Fibers are solids generated in much the same way as dusts. Fibers differ from dusts in their form. Dusts are generally regarded as spherical or round, whereas fibers have a defined length to width ratio (e.g., 3:1) making them appear more columnar. Smoke consists of small particles of carbon and other organic matter generated from incomplete combustion reactions. All aerosols, except fibers, are measured as weight per volume in units of milligrams per cubic meter of air (mg/m^3). Fibers are measured as counts per volume in units of fibers per cubic centimeter of air (f/cc).

Unlike gases and vapors, aerosols do not readily diffuse and uniformly distribute throughout their area of generation. In addition, particulates are classified as respirable and nonrespirable depending on particle size or diameter. Particulates with diameters <10 μm are considered respirable, that is, they can bypass the passages of the upper respiratory system. Particle diameters <5 μm can reach the lower regions and those <2 μm can travel to the terminal points of the bronchial system known as the alveolar region.

Gases and vapors are frequently generated in the occupational environment. True gases exist in the gaseous state at normal temperature (25°C) and pressure (760 mmHg), whereas sources of vapors exist as liquids or solids under these conditions. Gases and vapors form true solutions in the atmosphere, which means that they expand and mix completely with ambient air.

Gases and vapors are monodisperse in terms of molecular size. This means that all the molecules of a particular gas or vapor are the same size. Indeed, gaseous substances are not classified by size like aerosols. Unlike aerosols, gases and vapors are of molecular size and, accordingly, cannot be removed from the air via physical filtration methods. The gas or vapor must somehow be separated from the surrounding air molecules. Toxic gases and vapors are measured as volume per volume in units of parts per million. Flammable gases and vapors and oxygen gas are also measured as volume per volume, but in units of percent (%) gas or vapor in air. Note that 10,000 ppm equals 1%.

Toxic Chemicals

Toxic substances include metals, aqueous-based acids and alkalies, and petroleum-based solvents, among others. The human toxicity of chemical agents is related to several factors including the duration of exposure, concentration, and mode of contact. Inhalation and dermal contact are the primary and secondary modes of exposure, respectively, to chemical agents in the occupational environment. Accordingly, the occupational environment is commonly evaluated for airborne toxic chemical agents. Thus, exposure to chemical agents can occur directly via dermal contact with the material during handling and indirectly via respiratory and dermal contact with airborne agents. Materials also can be ingested via handling food with contaminated hands or as a result of mucociliary escalation of materials that were originally inhaled into the respiratory system and subsequently transported to the throat and mouth where they are swallowed.

Chemical agents are either organic or inorganic. The organic materials are either hydrocarbons or substituted hydrocarbons. The hydrocarbons are composed solely of carbon and hydrogen whereas substituted hydrocarbons are composed of carbon and hydrogen plus functional groups composed of elements such as chlorine, nitrogen, phosphorus, sulfur, or oxygen. The hazards associated with organic compounds and the fate of the compounds in the environment are mainly dependent on their chemical compositions and associated physical properties. Toxic inorganic materials are either metallic or nonmetallic elements, commonly in the form of salts, hydrides, and oxides.

Although distinctions are made between the subclasses of chemical agents, it should be noted that a single chemical component may exhibit a combination of characteristics. Indeed, in view of the fact that any chemical is toxic at an appropriate dose or concentration, then all flammable, corrosive, and reactive materials, as well as radiological and some biological agents are also toxic. The reverse, however, is not necessarily true. That is, all toxic materials do not exhibit other hazardous characteristics.

Toxic agents may induce biochemical and physiological changes in human systems following either systemic contact, via absorption into blood and tissues, or local contact. The changes may be ultimately manifested as adverse effects such as morphological and functional abnormalities, and illnesses and premature deaths among those exposed or their offspring (Figure 11.6). As suggested previously, toxicity is inherent in all compounds. The toxicity of a given agent, however, may be directly attributable to an original parent compound or indirectly attributable to an active metabolite formed via biotransformation in a human system. In addition, it should also be recognized that secondary toxicants can be generated by flammable, corrosive, and reactive chemicals in the form of toxic by-products released from reactions, fires, and explosions.

Toxic conditions also can be present when levels of atmospheric oxygen are not normal. Oxygen gas is a normal and obviously essential component in air at a concentration of approximately 20.8%. When levels decrease significantly below 20.8%, the atmosphere is designated *oxygen deficient* (e.g., <19.5% O_2) and when levels increase significantly, the atmosphere is *oxygen enriched* (e.g., >22% O_2). Deficient and excessively enriched atmospheres can result in toxic effects to individuals present due to hypoxia and hyperoxia, respectively. Enriched atmospheres also increase the potential for initiation of combustion and explosion reactions.

A summary of some representative toxic chemical agents encountered in the environment, including occupational settings, follows. *Crystalline silica* is a form of silicon oxide and is a natural component of the earth. Exposure is associated with underground mining, quartz-bearing rock, cutting of granite, manufacture of pottery, porcelain, abrasives, sand-blasting operations, stripping, mining, and steel and furnace making. The degree of hazard is based on the concentration of the dust, the percentage of free crystalline silica in the dust, and the duration of exposure. *Silicosis*

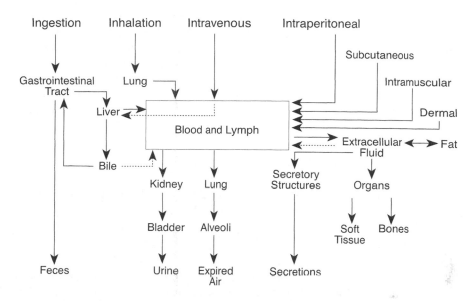

Figure 11.6 Routes of absorption, distribution, and excretion of toxicants in the body. (From *Environmental Toxicology and Risk Assessment: An Introduction, Student Manual*, U.S. Environmental Protection Agency, Region V, Chicago, IL, Visual 2.19.)

is a fibrogenic lung disease caused by the inhalation of crystalline silica. The disease has a relatively long latency period and frequently causes death. It also may be complicated by TB and acute lung infections.

Asbestos refers to a group of inorganic silicates occurring in nature as rocks. Asbestos particulates are characterized by a fibrous morphology instead of a common spheroid shape. Several types exist including chrysotile, crocidolite, amosite, anthophyllite, tremolite, and actinolite. Chrysotile represents about 90% of all asbestos processed in the United States. Asbestos has been used in the construction industry for roofing, plastics, insulation, floor tiling, and cement production. It also has been used in the automotive industry as friction material in brake and clutch linings and in the textile industry for fire-resistant clothing, safety equipment, and curtains. Inhalation of asbestos fibers has been associated with a fibrogenic lung disease called *asbestosis*. In addition, asbestos is associated with lung cancer and a cancer of the pleural lining called *mesothelioma*.

Metals are components of many inorganic compounds, commonly metallic oxides and salts, used in a variety of industrial and commercial areas. Metals bound to organic compounds are called organometallics. Organometallics, more lipid soluble than the inorganic metallic oxides and salts, can more easily be absorbed into the blood as a result. The blood, cardiopulmonary system, gastrointestinal system, kidney, liver, lungs, central nervous system, and skin are potential target sites for metals. The kidneys are the major target site and virtually all metals can accumulate and damage the renal system. Metals have a high affinity for indigenous protein groups, including enzymes. Many metals, including lead, arsenic, and mercury, bond to enzymes and inhibit essential catalytic activity. For example, lead can bond to and inhibit two different enzymes that catalyze reactions essential for formation of normal molecular components of red blood cells. If inhibited, levels of zinc proto-porphyrin (ZPP) and aminolevulinic acid (ALA) increase. Indeed, both ZPP and ALA are biomarkers of lead toxicity.

Arsenic, cadmium, chromium, nickel, and uranium are respiratory carcinogens. Beryllium and chromium are associated with ulceration of respiratory and dermal tissues. Beryllium also can cause a debilitating lung disease called berylliosis. Cobalt can damage the heart. Many metals, including arsenic, mercury, manganese, and lead are toxic to the nervous system causing adverse behavioral and neuromuscular effects. Metal oxide fumes generated from welding and foundry operations can cause an acute flu-like illness called *metal fume fever*.

Organic *solvents* are pure or substituted hydrocarbons derived from petroleum and are used to dissolve other organic materials. The organic solvents used in industry are divided into several chemical groups including hydrocarbons (aliphatic and aromatic), halogenated hydrocarbons, alcohols, ethers, glycol derivatives, esters, and ketones. Solvents are used in industry for extracting oils and fats, degreasing, dry cleaning, and printing inks. They are also present in a variety of products, including paints, varnishes, lacquers, paint removers, adhesive, plastics, textiles, polishes, waxes, thinners, chemical reagents, and drying agents. Direct dermal contact with liquid organic solvents can cause dermatitis.

In addition, nonpolar solvents can absorb through the skin and enter the blood. Solvents also pose an inhalation hazard because most volatilize and form airborne-solvent vapors. Virtually all organic solvents have some effect on the central nervous system and the skin. The narcotic effects of these solvents are extremely important because the worker becomes unable to use proper judgment and, as a result, may be involved in serious accidents. *Hydrocarbons* are powerful narcotic agents causing drowsiness and impaired cognitive and motor functions. In addition, many solvents cause damage to the blood, liver, kidneys, gastrointestinal system, eyes, and respiratory system. *Aromatic hydrocarbons*, which include benzene, ethyl benzene, toluene, and xylene, are commercial solvents and intermediates in the chemical and pharmaceutical industries. All the aromatic hydrocarbons cause a central nervous system depression marked by decreased alertness, headaches, sleepiness, and loss of consciousness. They cause a defatting, contact dermatitis. *Benzene* is extremely toxic to the blood-forming organs and causes anemia and leukemia. *Halogenated aliphatic hydrocarbons* include carbon tetrachloride, chloroform, ethyl bromide, ethyl chloride, ethylene dibromide, ethylene dichloride, methyl chloride, methyl chloroform, methylene chloride, tetrachloroethane, tetrachloroethylene (perchloro-ethylene), and trichloroethylene. These chemicals are commercial solvents and intermediates in organic synthesis. They affect the central nervous system, kidney, liver, and skin. All these chemicals cause a central nervous system depression, decreased alertness, headaches, sleepiness, and loss of consciousness. They also can damage the kidneys. These chemicals may cause anemia.

The halogenated aliphatic hydrocarbons cause liver changes that lead to fatigue, malaise, dark urine, liver enlargement, and jaundice. Tetrachloroethane, or *carbon tetrachloride*, causes serious damage to the liver and kidneys and is considered carcinogenic. *Alcohols* such as methyl or *n*-butyl are mildly narcotic, have low flash points, and therefore are highly flammable. Ethers are highly flammable and have strong narcotic properties. *Glycol derivatives*, such as ethylene glycol, are highly flammable and have a toxic effect on the nervous system and the blood. *Esters* are flammable and cause irritation to the eyes, nose, and upper respiratory tract. Ketones, such as acetone, are quite flammable. *Carbon disulfide* is a very hazardous solvent used in industry. It is highly flammable and highly toxic, acting mainly on the central and peripheral nervous systems.

Many inorganic and organic gases cause toxic effects. Oxides of nitrogen and sulfur, ozone, chlorine, formaldehyde, hydrogen chloride, hydrogen sulfide, and hydrogen cyanide are respiratory and ocular irritants. Hydrogen cyanide and hydrogen sulfide are chemical asphyxiants that can cause suffocation due to inhibition of a cellular enzyme causing inhibited utilization of inhaled molecular oxygen. Carbon monoxide, produced by the incomplete combustion of fuel, also is a chemical asphyxiant, but inhibits transport of inhaled oxygen by bonding to hemoglobin in red blood cells. Vinyl chloride is a flammable, compressed gas used as a raw material for the production of polyvinyl chloride and causes damage to the liver (including cancer), skin, vascular system, nervous system, and kidneys.

Insecticide manufacture and use causes potential hazards to the workers. The *organochlorine* insecticides are chlorinated ethanes, which include dichloro-diphenyltrichloroethane (DDT); and the cyclodienes, which include aldrin, chlordane, dieldrin, and endrin. They also include the chlorocyclohexanes such as lindane. They are used for pest control. They affect the kidney, liver, and central nervous system. The potential health effects include acute symptoms of apprehension, irritability, dizziness, disturbed equilibrium, tremor, and convulsions. The cyclodines may cause convulsions without any other initial symptoms. The chlorocyclohexanes can cause anemia. The cyclodienes and chlorocyclohexanes cause liver toxicity and can cause permanent kidney damage.

The *organophosphate* insecticides include diazinon, dichlorovos, dimethoate, trichlorfon, malathion, methyl parathion, and parathion. These chemicals are used for pest control. They affect the central nervous system, liver, and kidney. The major neurotoxic mechanism involves inhibition of the enzyme acetylcholine esterase. Depending on the amount of insecticide that has been absorbed by the individual, the acute symptoms range from headaches, fatigue, dizziness, increased salivation, crying, profuse sweating, nausea, vomiting, cramps, and diarrhea, to tightness in the chest, muscle twitching, and slowing of the heartbeat. Severe cases can bring about a rapid onset of unconsciousness and seizures, leading to death. A delayed effect may be weakness and numbness in the feet and hands, with long-term permanent nerve damage possible.

The *carbamate* insecticides include aldicarb, baygon, and zectran. They are used for pest control. They affect the central nervous system, liver, and kidney. Their potential neurotoxic effects involve inhibition of the enzyme acetylcholine esterase as discussed for the organophosphates.

Tobacco emissions consist of smoke particles and gaseous organic and inorganic molecules. Employees exposed in the workplace to toxic chemicals can receive additional exposure because of the presence of toxic chemicals in tobacco products. Cigarette smoking causes increased exposure to carbon monoxide. Workers are frequently exposed to carbon monoxide as part of their job. Therefore, the additional levels of carbon monoxide related to smoking can cause cardiovascular changes that are dangerous, especially to people who have coronary heart disease. Other chemicals found in tobacco that workers might be exposed to at their jobs include acetone, acrolein, aldehydes, arsenic, cadmium, hydrogen cyanide, hydrogen sulfide, ketone, lead, methyl nitrate, nicotine, nitrogen dioxide, phenol, and polycyclic aromatic compounds.

Tobacco products may become contaminated by chemicals used in the workplace and therefore increase the amount of toxic chemicals entering the workers' bodies. Smoking may contribute to an effect comparable to that which can result from exposure to toxic agents found in the workplace and therefore may cause an addictive biological effect. Combined worker exposure to chlorine and cigarette smoke can cause a more damaging biological effect than exposure to chlorine alone. Smoking may act synergistically with toxic agents found in the workplace and cause a more profound effect than the simple exposure to which the individual is subjected. Workers who were heavy smokers and installed asbestos had a higher level of cancer than those who were nonsmokers. Radon daughters also act synergistically with tobacco smoke. Smoking may contribute to accidents in the workplace. The accidents apparently are due to a lack of attention, preoccupation for the hand for smoking, irritation of the eyes, and coughing. Smoking can also contribute to fires and explosions where flammable or explosive chemicals are used or stored. Smokers have poor lung construction and a higher incidence of urinary abnormalities when they are exposed to cadmium. A chronic cough and expectoration occurs when individuals are exposed to both smoking and various ethers.

Flammable Chemicals

Flammable chemical agents include materials that serve as fuels (reducing agents) that can ignite and sustain a chain reaction when combined in a suitable ratio with oxygen (oxidizing agent) in the presence of an ignition source (heat or spark). Flammable agents are characterized as having a low flash point (i.e., <60°C), which is directly related to vapor pressure. In general, organic compounds such as organic solvents vaporize at relatively lower temperatures and, accordingly, are much more sensitive to heat than inorganic compounds. Flammable materials pose an obvious hazard from the standpoint of potential burns to human tissue. Indirectly during combustion, however, flammable materials can contribute to the formation of toxic atmospheres due to generation of by-products, such as strong irritants and chemical asphyxiants, and consumption of molecular oxygen during combustion.

Numerous organic gases (e.g., methane and propane) and solvent vapors act as combustible fuels if there is an adequate ratio of fuel to air. The combustible gas or vapor is a reducing agent and oxygen in air is an oxidizing agent. In the presence of an ignition source, an oxidation–reduction reaction involving the fuel and air can be initiated and result in the propagation of flame or an explosion. Most organic gases and vapors combust or explode within a range of ratios of fuel to air known as the *flammability range*. The extremes of the flammability range are referred to as the *lower flammability* or *explosive limit* (LFL or LEL) and *upper flammability* or *explosive limit* (UFL or UEL). The LFL or LEL is the minimum ratio of fuel to air that can support propagation of flame; ratios less than the LFL are considered too lean to support combustion. The UFL or UEL is the maximum ratio of fuel to air that supports propagation of flame; ratios greater than the UFL are considered too rich to support combustion.

Corrosive Chemicals

Corrosive chemical agents include those materials that on contact with human tissue can induce severe irritation and destruction of tissue due to accelerated dehydration reactions. Corrosive agents also include materials that can dissolve metal in a relatively short period of time. Typical examples of corrosives are organic and inorganic acids and bases. The strengths of acids and bases and the extremes of pH (i.e., <pH 2 and >pH 12.5, respectively) are directly correlated with the degree of corrosiveness.

Inorganic acids are compounds of hydrogen and one or more other elements that dissociate in water or other solvents to produce hydrogen ions. Inorganic acids include chromic, hydrochloric, nitric, and sulfuric. Inorganic acids, when mixed with other chemicals or combustible materials, cause fires or explosions. Organic acids include formic, acetic, propionic, chloracetic, and trifluoracetic. Acids are corrosive in high concentrations, destroy body tissue, and cause chemical burns to the skin and eyes. Inhalation exposure to acidic mists, vapors, and powders can irritate the respiratory system.

Alkalies are caustic substances that dissolve in water and accept hydrogen ions to form a solution with a pH higher than 7. These include ammonia, ammonium hydroxide, calcium hydroxide, potassium hydroxide, and sodium hydroxide. The alkalies in solid form or concentrated liquid solutions are more hazardous and destructive to tissues than most acids. They cause irritations of the eyes and respiratory tract and lesions in the nose. Potassium and sodium hydroxide, even in dilute solutions, soften the skin and dissolve skin fats.

Alkaline ammonia gas is an important part of many compounds containing nitrogen. It is used in making fertilizers, nitric acid, acetylene, urea, explosives, synthetic fibers, and synthetic resins. It is also used in producing paper products, photographic film, dyes, inks, glues, medicine, blueprints, and some cleaning solutions; and in ice-making machines and commercial refrigeration. Ammonia can be extremely irritating to the eyes, throat, and upper respiratory system. Large concentrations of ammonia can produce convulsive coughing by preventing breathing and can therefore cause suffocation in a short time. Ammonia also dissolves in the skin's moisture, causing a corrosive effect on the skin. Contact with anhydrous and aqueous ammonia causes first- and second-degree burns of the skin and eyes. Blindness can be the ultimate result of this contact. Ammonia gas in the presence of oil or other combustible materials causes fires.

Reactive Chemicals

Reactive chemical agents consist of chemically unstable materials that are typically characterized as either strong oxidizing or reducing agents. Chemical instability results in increased sensitivity to violent reactions that may result in extremely rapid generation of heat and gases; these, in turn, may culminate in ignition, explosion, or emission of toxic by-products. Some unstable compounds can react with air or water. Ethers can oxidize forming explosive peroxides that are sensitive to shock.

Biological Agents

Biological agents are commonly referred to as pathogenic organisms. Numerous identified pathogenic biological agents may be encountered in the environment. They consist of agents, which if introduced into the human body, may disrupt biochemical and physiological function via infectivity or induction of toxicity. The disruption can result in illness and death if the immune system is not able to destroy the biological agents. Infectivity is related to the virulence and the population density of organisms present at a given target site. Toxicity can be induced by microbiological agents that synthesize and release a chemical toxin. Examples of microbial pathogenic agents include bacteria, actinomycetes, rickettsia, fungi, protozoans, helminths, nematodes, and viruses. Unlike radiological and chemical substances, all biological agents, except viruses, are examples of biotic or living organisms. Viruses are abiotic or nonliving agents composed of biochemicals (i.e., proteins and nucleic acids) that may insert into and disrupt human cells. Microbiological agents are measured as counts per volume in units of colony-forming units per cubic meter of air (cfu/m^3).

Biological hazards vary with the type of occupation. The medical profession is exposed to staph and strep infections, viral infections, and many other types of organisms that their patients carry. Some healthcare workers and biomedical researchers are subjected to infectious hepatitis and other diseases, such as acquired immunodeficiency syndrome (AIDS). AIDS is caused by the human immunodeficiency virus (HIV) and has an insidious onset with an uncertain incubation period. People at greatest risk, beside those involved in sexual contact, purposeful needle contact, or necessary blood transfusions, are medical and other health-related personnel, as well as safety and security personnel. Individuals working with solid waste removal and disposal are also exposed to the same type of biological hazards. Research scientists are constantly working with bacteria, viruses, and other pathogenic microorganisms and are potentially exposed.

Laboratory exposures can result from the use of biological materials or laboratory work done on biological agents of unknown epidemiology or etiology. The worldwide literature reports an estimated 6000 cases of accidental infection have occurred in the laboratory. They include bacterial, fungal, parasitic, rickettsial, and viral infections. Among the causes of accidents that result in laboratory-acquired infection are oral aspiration through pipettes, accidental syringe inoculation, animal bites, spray from syringes, centrifuge accidents, cuts or scratches from contaminated glassware, cuts from instruments used during animal autopsy, and spilling or splattering of pathogenic cultures on floors and table tops. A large number of these infections have not been traced, and therefore whenever hazardous infectious material is used, extreme caution must be taken.

The zoonoses, which are diseases transmitted from animals to humans, are another biological hazard to which the worker is subjected. These zoonoses include anthrax, brucellosis, tetanus, encephalitis, leptospirosis, Q fever, rabies, salmonellosis, trichinosis, bovine tuberculosis, tularemia, ringworm, and other common fungus infections. Workers also acquire roundworm, hookworm, and many diseases transmitted by fleas, ticks, and flies. Many workers, in addition, are subjected to various dermatoses.

Examples of biological agents and related health effects that can be related to occupational exposures follow. *Cotton dust* is generated from the finely pulverized part of the cotton plant. The hazard is greatest in handling the raw cotton during the carding stage, but hazards also exist during other stages in the cotton manufacturing process. Workers who inhale this dust may acquire a respiratory disease called *byssinosis*.

The bacteria *Bacillus anthracis* causes an infectious disease called *anthrax*. The disease is an acute systemic disease involving the skin, gastrointestinal tract, or lungs. Inhaled anthrax is the most deadly form of the disease. In most cases the organism enters the body through a cut or an abrasion. The organisms are found in the carpet and leather-goods industries and have been produced for use as a weapon of bioterrorism. They are usually present in hair, wool, hides, skins, and soil. The incubation period is within 7 days, usually 2 to 5 days. Symptoms may vary with route of entry. Initial symptoms of inhalation anthrax may resemble a cold or the flu. Symptoms progress to severe breathing difficulties, shock, and death. Intestinal anthrax may follow the consumption of contaminated food. Initial symptoms are nausea, loss of appetite, vomiting, and fever, followed by abdominal pain, vomiting of blood, and severe diarrhea. Cutaneous anthrax begins with a reddish brown lesion that ulcerates and becomes a dark scab. Symptoms include internal hemorrhage, muscle pain, headache, fever, nausea, and vomiting.

Large reservoirs of *Salmonella* bacteria are present in chickens, human and animal foods, uncooked meat, vegetable products, and animal feeds. Ingestion of these organisms can cause serious gastrointestinal illness.

Tularemia is caused by the bacteria *Francisella tularensis* and occurs when an individual is scratched by a piece of bone from a wild rabbit. Other animals, such as muskrats, woodchucks, squirrels, and skunks, harbor and transmit this organism to humans. Tularemia is also spread from animals to humans by wood ticks, dog ticks, and the blood-sucking deerfly. The incubation period is 2 to 10 days, usually 3 days, and is related to the virulence of the infecting strain and the number of organisms.

Q fever is a rickettsial disease of rats, cattle, horses, sheep, goats, and dogs. Cattle barns and holding areas for goats and sheep become heavily contaminated with these organisms. The organism, *Coxiella burnetti*, is thought to be inhaled by the victim. Raw milk from infected cows has caused Q fever. The incubation period is 2 to 3 weeks, depending on the amount of organisms.

Aspergillosis is a fungus transmitted to humans through dust contaminated with *Aspergillus fumigatus*, *A. niger*, or *A. flavus*. It is a fungus disease of birds, ducks, chickens, and cattle. The incubation period is probably a few days to weeks.

Brucellosis comes from infected cattle and swine. The disease is transmitted by direct contact with the diseased animals and by drinking unpasteurized milk or milk products contaminated with *Brucella* sp. bacteria. The incubation period is highly variable, usually 5 to 30 days, occasionally several months.

Bacterial, fungal, or parasitic agents in the environment may cause various types of dermatitis. Fungus infections are found among kitchen workers, bakers, food handlers, fur, hide, wool handlers and sorters, barbers, and beauticians. Workers who handle grain or straw may also acquire parasites that cause grain itch and ground itch.

Viral hepatitis is insidious with anorexia, vague, abdominal discomfort, nausea, and vomiting, often progressing to jaundice. The occurrence is worldwide among drug abusers, homosexual men, and patients and employees of various medical care institutions. The reservoir is people and the transmission is through blood, semen, saliva, and vaginal fluids. The incubation period is usually 15 to 180 days with an average of 60 to 90 days.

Ergonomic and Psychological Factors

Ergonomic factors include human attributes, abilities, and limitations as applied to the living and occupational environments. Ideally, optimal conditions exist to maximize human health, comfort, and well-being while promoting performance efficiency and effectiveness through the appropriate design of tools, machines, tasks, and environments (Figure 11.7).

Specific human factors include psychological capabilities, physiological dimensions (anthropometrics) and capabilities (biomechanics), and psychosocial issues. Just a few of the many factors that can contribute to unfavorable conditions in the occupational setting include fatigue, boredom, occupational stress, vigilance, mental (memory and recall) abilities and limitations, circadian rhythms, sensory capabilities, and anthropometric and biomechanical attributes, as well as peer, labor-management, or organizational climate variables.

Some of the workplace factors of concern could include hazards associated with the machines, equipment, tools, and layout and design variables found in the occupational environment. Some of the many variables of concern when assessing the causes of occupational-related ergonomic problems involve confusing displays (gauges); controls that require excessive force to operate; hand tools requiring users to assume awkward body postures or positions; materials that are excessively heavy; or manual materials handling with unreasonable frequency, duration, pace, or transfer distance requirements. Light and vision adaptation when moving from dark to well-lighted environments could contribute to safety-related problems.

Work strain due to unusual postures or improper postural adjustment to the work results in back pain, headaches, nervousness, aching muscles, excessive perspiration, and depression. The most serious results of work stress are serious traumatic injuries, accumulative injuries, or death. Stress is due to excessive heat, improper chairs, poor work space, or any other type of situation in which individuals perform their work in an unusual situation or in a continuous manner. An example of this would be writer's cramp, caused by continued use of the same muscles for long periods of time. Other environmental factors, such as illumination, climate, and noise increase

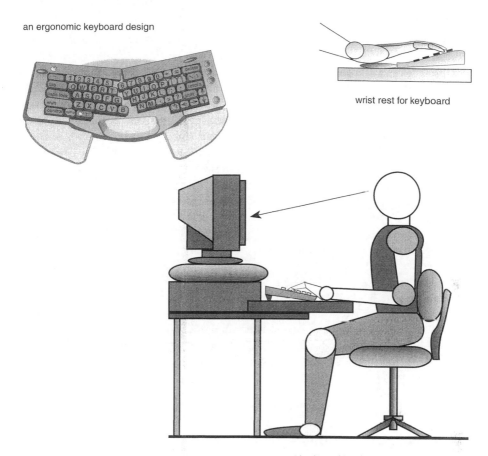

an ergonomic keyboard design

wrist rest for keyboard

correct body position for working at a computer

Figure 11.7 Some ergonomic considerations in computer usage.

stress. Restlessness and discomfort are often associated with sitting in certain hard-back chairs for extended periods.

Occupational stress can increase susceptibility to illness and injury. Stress is the nonspecific response of the body to any demand made on it. Stress may result in a fight-or-flight response to potentially dangerous situations. The body reacts through elevated heart rate and blood pressure, and a redistribution of blood flow to the brain and major muscle groups and away from the distal body parts. Subjective factors may play a much larger role in the experience of stress than objective factors. Job conditions that may result in stress include undue task demand, work conditions or work situations, exposures to chemical and physical hazards, pressure to accomplish tasks, and suppressed anger at inappropriate supervision and management decisions. Job stress may be a combination of conditions at work and at home that interact with the worker and result in acute disruption of psychological or behavioral homeo-stasis. If these reactions or disruptions are prolonged, it may lead to a variety of illnesses that are stress related. These illnesses include hypertension, coronary heart disease, alcoholism, drug use, and mental illness. Lower resistance to infectious diseases may also occur. The individual may be more apt to have accidents that may result in injuries.

The stressors (those factors that cause stress) may be sociocultural, such as racism, sexism, and economic variables; organizational, related to hiring policies, plant closing, and automation; work setting, such as the time and speed necessary to complete tasks, supervision and management problems, and ergonomics; interpersonal, such as marital problems, death, and illness; psychological, such as improper coping skills, poor self-image, poor communication, addictive behavior, and neurosis; biological, such as disease, poor sleep, poor appetite, chemical dependency, and biochemical imbalance; and physical or environmental, such as air problems, climate, noise, poor lighting, poor equipment design, exposure to radiation, and exposure to toxic substances. The distress cycle includes job stress, job dissatisfaction, organizational distress, stressful life changes, life and health risks, accident risks, and stress due to new technology.

Surveys suggest that 75 to 90% of all visits to primary care physicians are due to stress-related problems. These problems include backache, headache, insomnia, anxiety, depression, chest pain, hypertension, gastrointestinal problems, and dermatological problems. Stress also may lower resistance to a variety of diseases and contribute to a slow healing process. Uncertainty, doubt, lack of recognition, pressure related to time, and insecurity all are causes of health problems, and all may be related to an inordinate amount of stress created on the job.

OCCUPATIONAL HEALTH CONTROLS

General Measures for Controlling Workplace Agents and Factors

Substitution

Sometimes it is possible to substitute less hazardous materials, equipment, or even processes for more hazardous ones. Solvents such as dichloromethane and methylchloroform have been substituted for carbon tetrachloride. Toluene or xylene have been substituted for benzene. It is safer to dip an object into paint than to spray paint. This is an example of a process change.

Safety cans are much safer for storing flammable solvents than glass bottles. Another example of change in equipment would be the use of safety glass as opposed to regular glass in the fume hood and in glasses used by workers. The substitution of a wet method of operation for a dry method, where dust becomes a serious hazard, is another technique in good environmental control. An example of this would be the wetting of floors before sweeping up dust or the use of treated mops instead of regular dry mops in the hospital. The use of water drills by dentists has not only reduced the heat generated by the drills and the potential pain but also has reduced the number of particles that could fly into the dentists' eyes or injure their hands.

Isolation

Isolation is the physical separation of the worker from the hazard by use of a barrier, by increasing the distance from, or decreasing the time of exposure to, the hazardous material. Stored materials may be isolated physically from other materials and the worker or they may be stored in small units in specially ventilated rooms or under protective coverings. Radioactive materials may be stored in concrete vaults or in lead pigs. When the material to be mixed or processed is extremely hazardous and must be mixed by hand, either remote guidance controls should be used with mechanical hands within a specially protected container, or the individuals might use a glove box with their arms and hands in protective gloves and the rest of their bodies shielded from the area of material preparation.

Extremely hazardous equipment, such as that which is under high pressure, may be separated from the worker by reinforced concrete, steel, or armor plate. The work is performed by remote control and the worker observes the process through television cameras, mirrors, or periscopes. Remote-process control is used when the process itself is extremely hazardous. An example of this is automatic sampling analysis in petroleum processing plants, censors used on the production line, and processed equipment controlled through the use of a computer. Enclosures are another method of isolation. Enclosures are used in such operations as sandblasting, heat-treating, mixing, grinding, and screening. Enclosures also are useful tools in noise control.

Ventilation

Ventilation is used to provide a comfortable workplace environment and to keep dangerous materials from accumulating in the air. Ventilation is brought about when air is removed from a given area and fresh air replaces it. The exhausted air carries with it heat and possible odors that have been generated by people, processes, or equipment. The quantity and type of distribution of air by means of ventilation systems depends on the amount of heat produced, the kinds of contaminants getting into the air, the quantity of contaminants, and the ultimate air quality desired. Proper ventilation dilutes the concentration of a hazardous material and prevents fires and explosions. The various types include natural ventilation, forced general ventilation or dilution ventilation, general exhaust ventilation, and local exhaust ventilation (Figure 11.8).

Natural ventilation refers to airflow into the room through cracks, crevices, and open windows. Because it is difficult to control natural ventilation because of variations in temperature, pressure, wind velocity, and general weather conditions, this process is not very practical for an industrial environment.

Forced general ventilation or dilution ventilation is accomplished by blowing fresh air into a workroom. This may be used to control vapors given off by organic solvents at room temperature or to reduce the heat load within the working area. Exhaust ventilation is accomplished by mechanically withdrawing large quantities

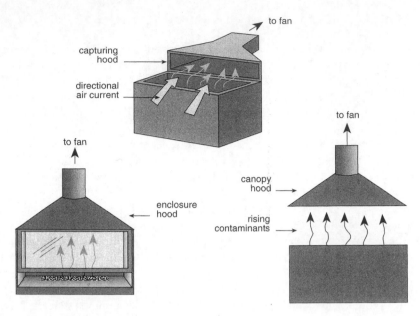

Figure 11.8 Local exhaust ventilation.

of air from the work area. The air is then replaced by fresh air sucked or blown into the work area. However, the movement of large quantities of air can cause drafts and may be uncomfortable for the worker. When this technique is used, it is preferable to have several low-velocity exhaust fans. Diluting the contaminant in the air, either by forcing in large quantities of air or by exhausting large quantities of air in the work area, does not reduce or eliminate the source of hazardous material introduced into the workroom. This is an important consideration when the hazardous material can cause a serious health problem for the worker.

General dilution ventilation is acceptable if the toxicity of the contaminant is low, the amount of material is produced at a low rate, the kind of material at low concentrations does not have serious effects on the individuals, and the material can be diluted adequately with a large volume of air. Local exhaust ventilation utilizes less makeup air, and therefore reduces costs due to heating or cooling. It helps control the generation and accumulation of particulates and vapors at the source. As a result, a lower if not negligible amount of contaminant is discharged into the air helping to reduce worker exposures and housekeeping problems. Because local exhaust ventilation controls the contaminant prior to discharge into the air, it is the preferred method for controlling worker exposure to toxic agents.

Personal Protective Equipment

Personal protective devices consist of respirators, protective clothing, gloves, shoes, helmets, eye and face shields and goggles, earmuffs, and plugs (Figure 11.9). Respiratory equipment is used to protect against inadequate oxygen, toxic gases, toxic vapors, toxic particulate matter, or any combination of these hazards. The type

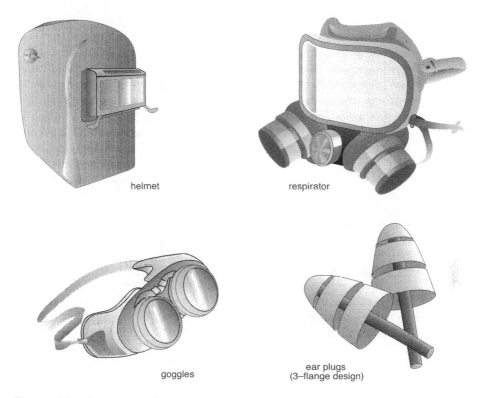

helmet

respirator

goggles

ear plugs
(3–flange design)

Figure 11.9 Some personal protective devices.

of respiratory equipment used is based on the kind of hazardous situation in which the individual works. It must be determined whether adequate oxygen is in the atmosphere, the types of contaminants in the air, the effects of the contaminant on the body, how quickly the effect can occur, the concentration of the material, and the length of time the worker is present within that environment.

The three basic types of respiratory equipment are:

1. Air-purifying respirators that contain special filters or sorbents for separating the contaminant from the inhaled air prior to entry into the respiratory system
2. Air-supplied equipment that has hoses and face pieces or hoods and is connected to a central system that provides breathing air via a stationary compressor or air cylinders
3 Self-contained equipment that provides breathing air from a cylinder carried by the individual

Protective clothing includes the use of gloves, aprons, boots, coveralls, and other articles made of impervious materials. Their function is to prevent chemicals from coming in contact with the skin. Special types of reflective aluminum clothing are used in areas where considerable radiant heat occurs. Helmets or hard hats are needed in any type of construction or in industries such as the steel industry to absorb impact from heavy materials.

The eyes and face are protected against contact with corrosive solids, liquids, and vapors; or impact with projectiles by using safety goggles, face shields, and glasses. It is important to protect the eyes against numerous hazards, such as glare, ultraviolet light, and other kinds of physical or chemical hazards. Such industries as grinding are extremely dangerous, because the particles may easily penetrate the eye or cornea, or cut the eyelid. Chemical burns in the eyes due to splash may cause temporary or permanent loss of vision.

Ear protection is provided by the use of earplugs or earmuffs. The devices attenuate sound levels so that a reduced level actually enters the auditory system. It is essential that the plugs and muffs are kept very clean to prevent ear infection, and the process should be evaluated, because the noise level may exceed the protection afforded by the muffs or plugs.

Housekeeping

The satisfactory environment is a complex interrelationship between the physical facilities, color, light, ventilation, aesthetics, dirt, chemicals, and microbiological hazards. The dark, shabby, dirty, littered environment is depressing and leads to accidents and poor occupational health practices.

The procedures and techniques of cleaning are extremely essential within the industrial environment. Attention must be paid to dustless removal, by the use of treated cloths, mops, and vacuum cleaners with filters; and inert or hazardous dusts that may fall on the floors, ledges, equipment, and partitions. Wet or damp sweeping is an important technique in the removal of this dust, provided it does not cause a slipping hazard or the chemicals present do not react in the presence of water.

It is essential to use proper detergents, disinfectants where needed, and equipment capable of performing a good housekeeping job. The personnel must have proper training in the use of the equipment. They must also have an understanding of the types of materials being removed from surfaces. It is extremely important that housekeeping personnel be well supervised so that a minimum of contaminants may be spread and a maximum of contaminants may be removed during the housekeeping process.

Where spills of toxic materials have occurred, previously established techniques for removal of the toxic material must be used. The workers must be trained to utilize these techniques and to know which techniques to use to protect themselves during the cleanup procedure. The housekeeping must be performed on a daily or shift schedule and also on special schedules depending on what must be done.

The types of equipment and floors to be cleaned vary with the industrial operation, making it difficult to list specific cleaning procedures. However, in all cases, materials used for cleaning must be evaluated so that they do not in some way combine with the hazardous or inert material to cause new hazards. Each industrial plant should be evaluated for its particular housekeeping problem by environmental or industrial hygiene consultants. Specialized programs should be established for each plant.

Education

The educational process is used in the occupational environment for training workers in job performance, retraining workers in their job assignments, or teaching new procedures; hopefully, this process can change attitudes toward the utilization of good personal practices. Although the machinery may be the best engineered, contain all available safeguards, and process and materials may not be causing a problem within the environment, lack of good attitudes on the part of the employees may lead to hazardous situations. Employees may improperly utilize equipment, remove guards, wear unsafe clothing, or allow their minds to wander during the work process. Good educational procedures should help workers develop a significant understanding of accident and health hazards. They should help workers to be flexible in their work habits. They should learn to work at a normal rate of speed, not intentionally violate existing rules and regulations, and carefully assess the potential problems of their unique environments.

Control of Physical Agents

Noise and Vibration

Noise is controlled by use of proper engineering, proper administrative techniques, and personal protective equipment. When replacing equipment, total noise levels for the area should be considered and pieces of equipment with lower noise levels should be utilized wherever possible. The proper maintenance of equipment is essential, because work or imbalanced parts, improper adjustments, inadequate lubrication, and improperly shaped tools contribute to the noise problem. Belt drives on machines should be substituted for gears wherever possible. Vibrations may be dampened, or decreased, by increasing the mass or stiffness of the equipment or material and by using rubber or plastic linings. The support of machines should be strengthened. Flexible mounts, hoses, or pipes should be used, if possible. Mufflers at either the intake or exhaust on internal combustion engines and compressors should be checked regularly and replaced when needed. The noise source and the operator should be isolated when possible.

Proper administrative controls reduce excessive noise exposure by adequate arranging of work schedules, removing workers to a low noise level when they have worked a high level during the day, dividing the work at high noise levels into several days or among several different people, running a noisy machine a small portion of the day, and running high-level noise machines when a minimum number of employees are present.

Personal protective equipment includes the use of plugs and earmuffs that reduce the sound level 25 to 40 dB. Further, a hearing conservation program should be implemented. The individuals subjected regularly or frequently to 90 dBA or above should be tested. A complete medical examination and previous work history should be completed for each worker and maintained on a regular basis. Individuals showing signs of physical damage or having spent long periods of time at high noise levels in the past should be assigned to a much quieter environment.

Vibration is reduced by isolating the disturbance from the radiating surface, reducing the response of the radiating surface, reducing the mechanical disturbance causing the vibration, using materials that deaden the vibration, reducing friction, and absorbing the energy created by the vibration. Proper design of equipment, adequate balancing of rotating machinery, and proper maintenance of machinery also act as effective controls.

Nonionizing Radiation

Individuals should be protected against ultraviolet rays by minimizing the time of exposure, keeping a maximum distance from the source, and using proper shielding. Glasses should be used when needed. Because ventilation does not control radiant heat, other techniques must be used. There techniques include installation of shields that reflect the radiant heat or absorb it without reradiating it to the individual and the use of reflective clothing, such as aluminized coats, pants, coveralls, and helmets. An individual should only be permitted to work within a hot area for short periods of time, wearing a breathing apparatus that prevents inhalation of hot vapors. Another means of protection is maintenance of proper distance between the source of heat and the worker.

The controls for microwave hazards include limiting the time and amount of exposure, and providing proper shielding and distance from the source. The controls for lasers include proper engineering to prevent exposure to the laser beam, proper procedures to avoid errors, and proper personnel equipment coupled with periodic eye examinations.

A pleasant visual environment is the result of adequate quantities and proper quality of illumination. Because the most common faults of lighting systems are low levels of illumination, glare, shadows, and poor brightness ratios, these factors must be carefully considered. The size, contrast, brightness, and time involved in visual activities must also be considered. Any glare present within the field of vision may cause visual discomfort. This glare may be either direct or indirect. Shadows should be eliminated, because they interfere with effective vision and are very annoying. Care must be taken to provide proper brightness ratios between the visual task and the immediate surrounding environment. For example, students should never sit in a darkened room with only a single source of light on their textbook. Whenever they look up their eyes must adjust from the brightness of the task to the darkened environment. This causes fatigue and reduces the students' ability to concentrate and absorb the material they are studying. The environment itself must be clean and pleasant. It is always better, if possible, to use pastel colors on walls and ceilings. Where it is impossible to eliminate a source of glare, the worker should be shielded from it and, if necessary, should wear protective goggles.

Ionizing Radiation

The type of controls used for ionizing radiation is based on the intensity and energy of the radiation source and whether it is an external or an internal hazard. X-rays and gamma rays are the most common types of external radiation hazard, for they can penetrate the entire body. Beta rays may or may not be an external hazard, depending on their energy. If the beta rays penetrate to the basal layer of the epidermis, they are hazardous. Neutrons, because of their high penetrating powers, are also external radiation hazards. Prevention of these hazards is accomplished through distance, proper shielding, and exposure time. The inverse square law applies to the reduction of radiation intensity due to point sources of x-ray, gamma ray, and neutron radiation. Good shielding includes proper enclosure of the radiation source and the use of diaphragms and cones to limit the beam to avoid stray radiation. Radiation scatters, so it is necessary to limit this beam and have proper absorptive shielding. The exposure time should be kept to an absolute minimum.

Other controls include using protective aprons and gloves, wearing of film badges, carrying of dosimeters, and evaluating machines to ensure that optimum voltage is used in the operation of the machine and the fastest film is utilized for best exposure. Because the radioactive material may enter the body through inhalation, it is important to wear proper masks in potentially dangerous areas and to contain the radioactive material in hoods with special exhaust systems. Food and water must be thoroughly monitored to ensure that individuals do not ingest radioactive material. The skin should be protected by the use of protective clothing. Absolute cleanliness must be practiced at all times and double containers should be used for all waste or cleaning materials.

Heat

Numerous controls are recommended for workers in hot environments, including acclimatization, proper work and rest periods, distribution of the work load, assigning younger and better equipped personnel to hot work, scheduling hot jobs for the coolest part of the day, regular breaks, frequent physical examinations, adequate quantities of drinking water, replacement of salt, personnel education concerning work problems related to heat and the effect of alcohol and drugs, protective clothing, shielding from the sun, adequate ventilation, and shielding from sources of radiant heat. Acclimatization involves a progressively increasing exposure to a given temperature based on a known activity over a period of typically at least a week.

Control of Chemical Agents

Exposure to chemical agents also is controlled by implementing administrative, engineering, and personal protective equipment controls. Toxic and flammable atmospheres are best controlled by minimizing generation by using less hazardous substitutes and controlling generation via process design and local and general dilution ventilation. When administrative and engineering controls are not feasible or they do not adequately control the hazard, use of personal protective equipment is warranted.

Control of Biological Agents

Special controls are needed when working with biological hazards. These controls include air pressure differentials; filtered supply air; filtered exhaust air; special airlock and pass-through autoclaves; ultraviolet barriers at through ways and in special laboratory areas; special treatment of contaminated sewage and waste; and isolation and separation of all the water, steam, natural gas, and other utilities. Additional controls are discussed in the section on controls of hazardous and toxic materials.

Control of Ergonomic and Psychological Factors

Tools and work situations must be selected to reduce as much work stress as possible. If the individual has difficulty reaching operating equipment, or suffers from poor posture, discomfort, or emotional strain, then physical illness or accidents may result. Proper design of the employee's work space is also essential. Care must be taken to eliminate fatigue. An evaluation must be made of individuals to determine whether they can effectively handle the physical and mental loads of the assignment. Workers should be trained for the task and then checked by a knowledgeable supervisor who has the ability to communicate with them.

An occupational stress management program should help the employees cope with the occupational stresses, as well as the personal and societal stresses. Many companies not only have developed stress management training programs but also have developed an employee assistance program. The function of these programs is to have individuals establish a mechanism to reduce stress, plan alternatives to problems, and assist individuals in planning techniques to deal with stress when the problems cannot be relieved. The actual program could consist of lectures, TV tapes, small group dynamics, and referrals to physicians when specific problems of a personal nature need to be determined and treated. All stress management programs should include techniques of relaxation therapy, good nutrition, preventive measures, and control of habits that may lead to stress. Above all else, a good supervision and management program to train the supervisors and managers would be extremely useful, because good management can help reduce stress levels among employees.

OCCUPATIONAL HEALTH AND SAFETY PROGRAM

General Aspects

The occupational health and safety program has as its major objectives protection of workers against health and safety hazards in the work environment; proper placement of workers according to their physical, mental, and emotional abilities; maintenance of a pleasant, healthy work environment; establishment of preplacement health examinations; establishment of regular, periodic health examinations; diagnosis and treatment of occupational injuries or diseases; consultation with the worker's personal physician, with the worker's consent, of other related health problems, such as heart disease; health education and counseling for employees; safety education for employees; establishment of research to identify hazardous situations or find means of preventing hazardous situations; and establishment of necessary surveys and studies of the industrial environment for the protection of workers, their families, and the community.

A good occupational health program not only benefits the worker but also is a value to the industry. It reduces the cost of worker's compensation insurance, reduces the cost of hospital insurance, reduces absenteeism and labor turnover, helps satisfy the legal requirements set forth in the OSHA laws and other federal and state laws, and creates a good working relationship between management and workers.

Hazard Communication Program

The Hazard Communication Program is required by the OSHA standards and by standards established for Federal Employee Occupational Safety and Health Programs. The program consists of three elements. They are hazard warning labels, material safety data sheets (MSDSs), and employee training programs. The standard that was issued in 1983 by OSHA requires chemical manufacturers and importers to make a comprehensive hazard determination for the chemical products they sell. The manufacturers, importers, and distributors must provide information concerning the health and physical hazards of these products. This is accomplished by means of warning labels. The manufacturing sector that uses the chemicals to turn out additional products must protect their employees by meeting the following standards:

1. Preparing a written hazard communication program, which includes an inventory of all hazardous chemicals used in their facilities
2. Obtaining an MSDS for each hazardous chemical they use
3. Displaying appropriate facility placards and warnings
4. Preparing a hazard communication training plan
5. Providing training to employees who are potentially exposed to hazardous chemicals in their facilities

The elements of the Hazard Communication Program should include preparation of a written hazard communication plan; identification and evaluation of chemical hazards in the workplace; preparation of a hazardous substance inventory; development of a file of MSDSs; provision for access to MSDSs for employees; assurance that incoming products have proper labels; development of a system of labeling within the facility; development of the training program; identification and training of employees who are potentially exposed to the hazardous chemicals; and evaluation of the programs and redirection where necessary. The definition of hazardous chemicals needs to be clearly stated to properly identify the potential problems within the workplace. The written plan must describe how complaints are to be handled, labeling, how MSDSs can be obtained and made available, and how information and training are to be provided. The inventory of all the toxic chemicals in the workplace have to be cross-referenced to the MSDSs. Provision must be made for employees to deal with accidental spills and leaks, and for making employees aware of the types of hazards that may occur as a result of the chemicals or the pipes carrying the chemicals.

A hazard assessment should be conducted by a trained environmental health professional such as an industrial hygienist. The survey involves identifying the chemicals, how they are used or produced, and what the quantities are. The environmental specialist should also determine whether warning labels are placed in readily observable places on drums or containers. Next, the specialist should set up a flowchart of each operation, starting at the place where the raw materials enter the building, are taken to storage, and then are introduced into the actual manufacturing operation. From this point, each step along the way in the manufacturing of the product should be shown and the potential problems listed under the particular step. The flowchart proceeds to the final step of production packaging and storage, and also to by-products, waste products, waste product storage, and disposal. By using the flowchart and diagramming the actual flow, it is easier to determine who may come in contact with the hazardous chemical, where it may occur, how it may happen, when it may happen, why it may happen, what the final results may be, and what kind of protective measures are in use by the employee to avoid contact with the chemical.

An air sampling survey should be conducted in those areas that appear to have potential chemical hazard problems. The air sampling survey should consist of a general air survey, specific air samples in high concentration areas, and monitoring of personnel with personal monitors at the breathing level.

An important part of the study includes ventilation controls related to local exhaust ventilation as well as general ventilation. Because the chemicals vary so, the requirement for respirators and protective equipment varies considerably.

Once the survey has been concluded, a hazardous substances inventory should be developed and a hazardous chemical MSDS file should be provided. When hazardous chemicals are utilized, they should be clearly identified by chemical name, trade name, and common name. All the hazardous ingredients in the substance should be listed. The physical and chemical characteristics of the substance, such as boiling and freezing points, density, vapor pressure, specific gravity, solubility, volatility, and general appearance and odor of the product, should be included. All physical hazards, such as carcinogenic, corrosive, toxic, and irritant, and other hazards, should be included. The route of entry should be clearly described. Special precautions and procedures should be established for spills, leaks, cleanup controls, emergencies and first aid. The program must provide access to the MSDSs for employees. All incoming labels and existent labels need to be checked.

Employee education and training are important requirements of the program. The employees need to clearly understand the operations of their work areas, the hazardous chemicals present, the means of prevention and control, the physical and health hazards associated with the chemicals, and the types of protective clothing or respirators that they should wear to avoid problems. Inherent to this is hazard recognition or the ability to recognize when conditions are not acceptable from health and safety perspectives. Finally, workers should be well aware of any symptoms of illness that may be attributed to the chemicals or other substances in the workplace.

SUMMARY

Occupational disease, injury, and death is of vital concern to our society because the costs and physical and emotional anguish are great. Occupational illnesses and injuries may be caused by physical, chemical, biological, ergonomic, and psychological factors. Although numerous techniques are available for hazard reduction and the potential for intervention is excellent in certain areas, it is extremely poor in others. Environmental health and clinical practitioners must work together for ongoing improvement.

The basic principles of occupational disease and injury control involve prevention primarily and treatment secondarily. To have proper prevention and treatment, it is necessary to study the individual employee and ensure that job work activities are carried out in an effective manner. It is also important to evaluate and control the level of anticipated and recognized hazards associated with the various agents in the workplace. Once the controls are established, both the environment and the worker must be constantly monitored to ensure that exposure levels to a given set of occupational hazards will not exceed the appropriate established standards.

CHAPTER 12

Major Instrumentation for Environmental Evaluation of Occupational, Residential, and Public Indoor Settings

BACKGROUND AND STATUS

Environmental evaluation includes the collection, detection, and measurement of representative samples from an environmental matrix such as air, water, soil, and even food and beverages. This chapter provides a summary of the major forms of major monitoring and analytic instrumentation that are used to evaluate several types of physical, chemical, and biological agents associated with occupational, residential, and public indoor settings. The air is a common matrix sampled and analyzed in these environmental settings because inhalation of contaminated air is considered a primary mode of toxic or pathogenic agent entry. In addition, the air serves as a major matrix for elevated sound levels, extremes of temperature and humidity, and transfer of radiation energies. Accordingly, this chapter only focuses on related instrumentation for detecting and measuring contaminants and energies in the air matrix. It must be noted, however, that although the majority of the instrumentation summarized in this chapter is used mainly to evaluate occupational and nonoccupational exposures in indoor environments, some also have application for evaluation of outdoor air contaminants and energies.

The data collected and analyzed are used to evaluate both actual and potential external exposure to chemical and some physical (e.g., noise) agents encountered by humans. In turn, the levels are compared with what are considered acceptable limits of exposure. These limits include the permissible exposure limits (PELs) enforced by the Occupational Safety and Health Administration (OSHA), recommended threshold limit values (TLVs) by the American Conference of Governmental Industrial Hygienists (ACGIH), and recommended exposure limits (RELs) by the National Institute for Occupational Safety and Health (NIOSH), among others. In the case of bioaerosols, exposure limits for airborne concentrations of microorganisms are lacking. Instead, for example, comparisons are made between concentrations indoors vs. outdoors, or one area vs. another area within a facility or outdoors. In other cases, a parameter may be monitored without concern necessarily for excessive exposure. For example, illumination is evaluated and compared with recommended guidelines to assure that there is an appropriate, neither inadequate nor excessive, quantity of lighting.

Environmental and occupational health and safety specialists also must evaluate engineering controls designed to reduce or eliminate airborne contaminants and to provide atmospheric comfort in indoor environments. Thus, both local, general, or dilution ventilation systems are evaluated to assure that the design and operation are efficient relative to the characteristics of the setting, including the type and levels of contaminants and the number of occupants present.

EVALUATION CATEGORIES

Monitoring and related analytic activities are divided into several categories to reflect the type of monitoring that is conducted. Categories are based on factors including time, location, and method of collection and analysis. Each serves a purpose in evaluating the occupational, residential, and public indoor environments to determine the degree of external exposure of occupants to various agents.

Instantaneous or Real-Time Monitoring

Instantaneous monitoring refers to the collection of a sample for a relatively short period ranging from seconds to typically less than 10 min. A major advantage of instantaneous monitoring is that both sample collection and analysis are provided immediately via direct readout from the monitoring device. The data represent the levels of an agent at the specific time of monitoring. Accordingly, instantaneous monitoring also is referred to as *direct reading* and *real-time* monitoring. Indeed, real-time monitoring is perhaps a more appropriate designation because some devices are already developed and some being designed for integrated or continuous monitoring that provide a direct readout or instantaneous result without need for laboratory analysis and associated delays. In addition, the main purpose of real-time monitoring is to reveal an agent's level is at an immediate point of time or during real-time.

The application of real-time monitoring varies. The strategy is used when preliminary information concerning the level of an agent is needed at a specific time and location. For example, real-time monitoring is commonly used for screening to identify agents and measure related levels during the initial stages of an indoor air quality investigation. This is important for developing follow-up monitoring strategy and to determine whether integrated monitoring (see the following section) is warranted. Real-time monitoring also is beneficial for determining levels of agents during short-term operations or specific isolated processes when peak levels are anticipated or suspected.

Integrated or Continuous Monitoring

Integrated monitoring refers to the collection of a sample continuously over a prolonged period ranging from more than 10 or 15 min to typically several hours.

Thus, integrated monitoring also is referred to as *continuous* monitoring reflective of the extended period of sample collection. Most work shifts are 8 hours, therefore, occupational exposure limits are most commonly based on an 8-hour exposure period. Accordingly, it is very common as well for sampling to cover the duration of the shift. Sampling for longer periods, however, is not uncommon.

Several strategies can be followed. For example, a sample run could be started immediately at the beginning of an 8-hour period and allowed to run until the end. Analysis of the sample would provide a single value representative of the level of a particular agent during the 8-hour period. The single value represents an integration of all the levels during the 8-hour period. The single value, however, does not provide information concerning fluctuations of levels that were higher, lower, or not detectable during shorter periods within the 8 hours. In addition, there is no indication as to the levels at specific times and locations during the 8 hours. As a result, an alternative strategy involves collection of several samples of shorter duration during the entire sampling period. In turn, analysis of the individual samples provides levels associated with specific times, tasks, and locations during the total (i.e., 8-hour) sampling period. Concentration (C) and corresponding sample time (T) data from one sample (C_1) or several individual samples (C_1 to C_n) can be time-weighted ($C \times T$) and averaged to provide a single overall time-weighted average (TWA) for the 8-hour period:

$$8 \text{ hr TWA} = \frac{C_1 T_1}{8 \text{ hr}}$$

or:

$$8 \text{ hr TWA} = \frac{(C_1 T_1) + (T_2 T_2) + \cdots + (C_n T_n)}{8 \text{ hr}} \tag{12.1}$$

The major advantage of integrated monitoring is that it provides a single value for the level of an agent over a prolonged period. The level of an agent can be determined during discrete times and locations within a work shift to assist in identifying factors that influence elevated values of exposure or external exposure. A major disadvantage associated with integrated monitoring is that in most cases samples must be submitted to a laboratory for analysis prior to knowing the measurement. This frequently results in a delay between sample collection and data reporting, a delay that potentially exposed occupants are not always willing to understand or accept.

Personal Monitoring and Area Monitoring

Personal monitoring involves direct connection of an integrated monitoring device to an individual. The device, in turn, collects a sample or records the intensity of an agent in the specific areas and during specific tasks conducted by an individual. Indeed, personal monitoring is frequently a form of mobile monitoring because the sampling device travels to the same areas at the same times as the individual that wears it.

If inhalation is the mode and the respiratory system is the route of entry of an agent, the monitoring device is positioned in the individual's breathing zone, that is, an area within a 1-ft radius of the individual's nose and mouth. Typically, an integrated monitoring device for personal sampling is attached near the individual's scapula or collar bone. Relative to evaluation and impact of sound levels, however, hearing is the major mode and the auditory system is the route of entry. When conducting personal monitoring for sound, therefore, the sampling device can be connected to the individual's hearing zone; a region within a 1-ft radius of the ear. This zone is ideally the ear itself, or more commonly, the trapezius region of the shoulder.

Instantaneous or real-time monitors also can be used to determine levels of agents in an individual's breathing and hearing zones. For example, a sound level meter can be held by the professional or technician conducting the monitoring in the auditory region of an individual whose exposure is under evaluation. This would provide an instantaneous assessment of sound levels in the individual's hearing zone at the specific time of monitoring.

Area monitoring is conducted to evaluate the levels of agents in a specific location, instead of evaluating levels encountered by a specific individual. Integrated monitoring devices are typically positioned in a stationary location. The data represent the level of an agent in the specific area for the sampled time. Instantaneous or real-time monitoring, however, involves area sampling in either a stationary or mobile mode. Instantaneous monitoring may be conducted while standing still in a given location or the monitor can be transported to various locations while intermittently checking the indicator or readout on the device.

EVALUATION OF AIRBORNE PARTICULATE
AND GASEOUS AGENTS

Integrated or Continuous Monitoring for Particulates

As defined in Chapter 11, several subclasses or forms of airborne particulates exist including dusts, fibers, and fumes. Exposure limits for airborne dusts, fibers, and fumes have been established for many individual elements and compounds based on their inherent toxicities. There are situations when it is important to measure and know the concentration of a specific particulate that is either highly, moderately, or slightly toxic. Other situations arise, however, when the concentration of a more general composition of particulates, including a combination of highly to only slightly toxic, is warranted. In addition, situations also exist in which no established exposure limits exist for a specific type of particulate air contaminant, but measurement of that material as total particulate provides important data during exposure assessment.

Total particulate refers to generic particulate mixtures consisting of various diameters, compositions, and inherent toxicities. These materials may be generically classified as nuisance dust or particulate not otherwise classified. The fraction of total particulate that consists of particles with diameters <10 μm is referred to as

Figure 12.1   Particulates or aerosols are separated and collected from contaminated air using a sampling filter. Polyvinyl chloride (PVC) and cellulose ester fiber (CEF) filter diameters are commonly 25 and 37 mm and specific pore sizes often range from 0.8 to 5.0 μm.

respirable particulate. The respirable fraction of particulates also can consist of mixtures of various compositions of particulates, but has a higher probability of entering the lower respiratory system because the smaller particle diameters (<10 μm) potentially permit the upper respiratory defense mechanisms to be bypassed.

Frequently, however, only knowing the concentration of more generic total or respirable particulates in the air is insufficient. For example, if a real or potential source of specific fibrous dusts exists, such as asbestos, it is important to know what the concentration of these specific particulates is due to their inherent toxicity. The morphology of fibers, unlike other classes of particles, is more columnar than spherical. As a result, the sampling and analysis involves methods that assure accurate collection and quantification of fibers, such as asbestos.

Specific metallic dusts and fumes are common particulates present in the environment for which many specific exposure limits for airborne levels have been established. If there is a real or potential source of metallic dusts or fumes, it is important to know the concentration of a specific metallic particulate, due to the associated toxicity. Indeed, most metallic dusts and fumes are considered highly toxic relative to nuisance dust or particulates not otherwise classified, and, accordingly, occupational exposure limits are much lower.

Specific sampling and analytic methods are available for dusts, fibers, and fumes. Sampling involves moving an airstream through a filter collection medium, selected in accordance to particle size (diameter) or composition of the particulates. Airborne particulates are electrostatically attracted to, intercepted by, and impacted on the filter medium, resulting in separation of the particulates from the air (Figure 12.1). The filter medium is contained in a plastic holder or cassette, which, in turn, is connected to an air-moving device or sampling pump via a short length of flexible tubing (Figure 12.2).

The two most common filter media are polyvinyl chloride (PVC) for collecting total and respirable dusts; and mixed-cellulose ester fiber (CEF) for collecting metallic dusts, fumes, and asbestos fibers. The common sample filter media range from 25- to 37-mm diameter and 0.8 to 5.0 μm pore sizes. The diameter of the filter provides a cross-sectional surface area for deposition and collection of particulate. The pore size causes particulates with diameters greater than the pore diameter to collect on the surface of the filter. Particulates with diameters less than the pore size, however, are collected within the pores due to an electrostatic attraction between the particles and the filter. In addition, the pores are tapered and allow interception and impaction of particulates within the medium. Indeed, collection efficiency increases as particulates accumulate on the filter.

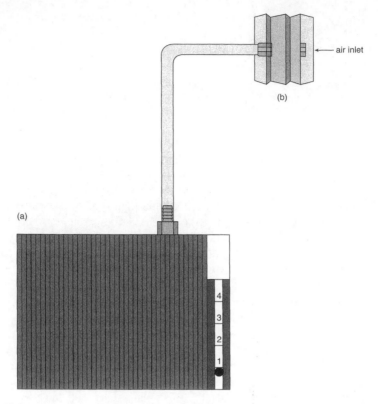

Figure 12.2 Airborne particulates such as total dust, metal dust, metal fumes, and asbestos fibers are sampled using (a) a high-flow or multiflow battery-operated air sampling pump connected via flexible tubing to (b) a membrane filter contained within a plastic cassette.

The "high-flow or multi-flow air sampling pumps used to collect air containing suspended particulates are typically operated at flow rates between 1.5 and 3 l/min. Vacuum pumps also are used, however, to collect very high volumes of air in a relatively short period and operate at flow rates reaching 1 ft^3/min (28.3 l/min). The total volume of air sampled is calculated by multiplying the flow rate of the air sampling pump (liters per minute) and sampling time (minutes) together. The volume collected is expressed in liters, which, in turn, is commonly converted to cubic meters or cubic centimeters. The volume actually represents how much air passed through the filter sampling medium, as shown in Equation 12.2:

$$\text{Volume } L = \text{Flow rate, } \frac{L}{\text{min}} \times \text{Time, min} \qquad (12.2)$$

The sample collection medium for *total dust* is prepared by preweighing a PVC filter for each field sample that is going to be collected. A filter is preweighed using a mechanical or electrobalance. The preweighed filter is subsequently positioned on

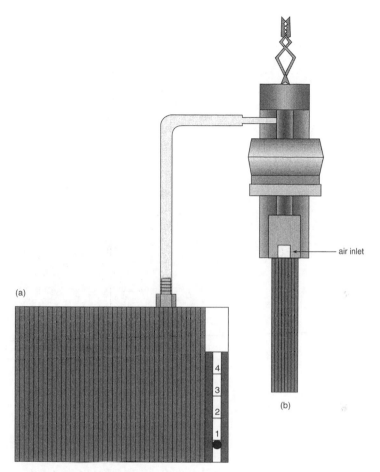

(a)

air inlet

(b)

Figure 12.3 Airborne respirable particulates are sampled using (a) a high-flow or multiflow battery-operated air sampling pump connected via flexible tubing to (b) a membrane filter contained within a plastic cassette positioned in a cyclone assembly.

a cellulose support pad in a three-stage plastic cassette and attached to an air sampling pump using a short length of flexible hose.

A PVC filter medium is prepared the same way for sampling *respirable dust*; however, the filter and a two-stage cassette are used in combination with a particle size-selective device or cyclone (Figure 12.3) to separate and collect the respirable fraction from total particulates. Only particles with diameters <10 μm flow through the upper cyclone, into the cassette, and onto the filter. Air enters the cyclone, which creates a vortex; and due to centrifugal force, the larger and denser particles flow downward and settle into the bottom of the device, called a grit cap. Simultaneously, particles with smaller diameters (<10 μm) flow upward in the opposite direction and enter the inlet port of a two-stage cassette, depositing and collecting on a preweighed PVC filter.

Gravimetric analysis involving a mechanical or electrobalance is used to measure the amount of either total or respirable particulate (Figure 12.4). Unlike specific

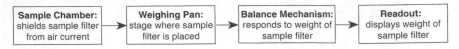

Figure 12.4   Major components of an electrobalance used for conducting gravimetric analyses of either sampled total or respirable particulates collected on a PVC filter.

particulate air contaminants such as asbestos fibers, metallic dusts, and fumes that require an analytic method that both identifies the specific analyte and measures the amount collected, gravimetric analysis of either total or respirable particulate only involves measurement of the amount (weight) collected on the filter. The gravimetric procedure involves comparison of postweight and preweight of filter. Theoretically, the preweight is only the weight of the filter and the postweight is the weight of the filter plus the collected particulate. The difference in weights (postweight minus preweight) is the mass of particulate collected. Gravimetric analysis of total and respirable particulate is the easiest of the major analytic methods. Both pre- and postweights of the PVC filters are determined as follows:

1. The filter is placed on a weighing pan inside the transparent glass sample enclosure with sliding door or sash to shield it from moving airstreams and related turbulence during weighing.
2. The mechanical or electrical balance device responds to the weight of the filter.
3. A digital readout is displayed in response to the mechanical or electrical signal from the balance mechanism and provides weight of filter in milligrams.

Dividing milligrams of particulate by cubic meters of air sampled provides results in milligram per cubic meter.

Sampling for asbestos fibers involves the same type of air sampling pump, but use of a different filter and cassette than those used to collect the more generic total or respirable particulate. A sample is collected using a 25-mm diameter CEF filter with 0.8-μm pore size positioned within a longer two-stage open-faced cassette. The open-faced, longer (50-mm extended cowl) cassette permits more uniform collection and distribution of fibers across the face of the filter.

The method of fiber analysis commonly involves *phase contrast microscopy*. The CEF filters are chemically digested and made transparent on a microscope slide so that individual fibers can be counted via microscopy. The method allows both identification of the particulate analyte as fibers and the count of the number of fibers collected.

Although transmission electron microscopy (TEM) is better for identifying and measuring specific types of fibers, phase contrast microscopy (PCM) still is the most common analytic method for counting fibers (Figure 12.5). Unlike TEM, PCM analysis is not generally used to identify the type of fiber. The number of fibers collected on the CEF filter are determined using PCM as follows:

1. A segment (ca. 25%) of the CEF sample filter is cut and mounted on a glass microscopy slide.
2. The filter section is digested and collapsed on a microscope slide via acetone vapor and, following mounting, is eventually positioned on the microscope stage for viewing.

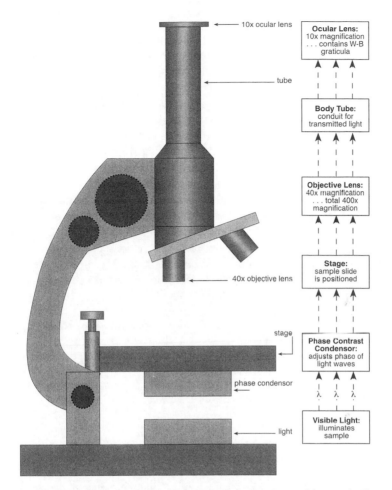

Figure 12.5 Major components of a phase contrast microscope used for conducting analysis of sampled fiber particulates collected on a CEF filter.

3. While looking through the ocular of the microscope, a Walton–Beckett (W-B) graticule positioned in the ocular appears as a grid superimposed over the microscope viewing field of the sample.
4. The calibrated graticule grid is used for measuring length, width, and number of fibers.

Any particle having a length to width ratio greater than 3:1 and a length of 5 μm or greater, is counted as a fiber. Samples are viewed at magnification power ×400 or ×450 and results are presented as the number of fibers per cubic centimeter (f/cc) of air collected.

It should be noted that phase contrast microscopy is a counting method and does not measure the specific properties of a substance. It is not inherently specific for asbestos. All particles meeting the criteria previously described (length-to-width ratio and length) are counted as fibers. Smaller fibers that may be present are not counted.

Metallic dusts and fumes also are separated from a contaminated airstream using filtration that involves electrostatic attraction, interception, and impaction of metal particles with a filter medium. The sample medium is a 37-mm diameter CEF or paper filter with 0.8-μm pore size. CEF filters are used because, relative to PVC filters, they hydrolyze (dissolve) easily in acid and ash; these characteristics are important during postsampling analysis of samples for metals. The smaller pore size (0.8-μm) filter is warranted because typically the variety of particle sizes is lower and the particle diameters are smaller for metal dusts and especially fumes relative to total dusts. The type of air sampling pump and the assembly of the CEF filter medium and cassette for metal dust and fume sampling involve the same procedure described for total particulate sampling, except for the differences in filters.

Metallic dusts and fumes are another example of specific particulate air contaminants that require an analytic method that both measures the amount of particulate collected and identifies the specific metallic analytes (e.g., Pb, Cd, Be). Analysis of the sample typically involves either atomic absorption spectroscopy (Figure 12.6) or inductively coupled plasma emission spectroscopy. The identity and amount of metal dust or fume collected on the CEF filter are determined using an atomic absorption spectrophotometer as follows:

1. Samples are prepared for analysis by digesting or ashing the CEF filter so that the metal air contaminant is present as a dissolved analyte.
2. The solution of dissolved analyte is injected or aspirated into the analytic instrument for qualitative and quantitative analysis.
3. The sample enters the chamber where extremely high temperatures fueled by gases such as oxygen and acetylene convert metal *ions* in the sample solution to ground state metal *atoms* via a process called *atomization*.
4. A cathode-ray lamp generates a specific wavelength of light directed at the sample chamber that is required for detection of specific metallic analyte.
5. A monochromator lens positioned between the sample chamber and detector is adjusted to wavelength absorbed by analyte.
6. The photodetector measures the intensity of light transmitted through the sample and monochromator lens, that is, the light not absorbed by the sample.
7. A direct readout receives an electronic signal from detector and displays data as absorbance.
8. The amount of metal detected and measured is determined based on comparison of measured absorbance to a standard curve of absorbance vs. known concentrations.

The concentration is expressed in units of milligrams per cubic meter based on the amount (mg) of metal detected and measured divided by the volume of air sampled (m^3).

Bioaerosols represent a group of viable or living particulates, such as bacteria, actinomycetes, fungi, and protozoa. Although specific exposure limits are not well established for airborne viable organisms, monitoring is often conducted to determine whether airborne organisms, such as bioaerosols, can be detected and measured for comparison with general recommended guidelines or background levels. The organisms are frequently combined with airborne mists, sprays, or dusts, with the particle size distribution classified as nonrespirable and respirable fractions.

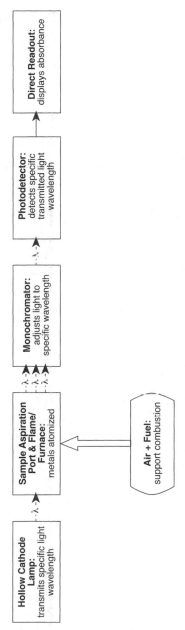

Figure 12.6 Major components of an atomic absorption spectrophotometer used for conducting analysis of sampled metallic particulates collected on a CEF filter.

The collection of airborne bioaerosols can be achieved through use of various methods, including filtration and impingement. A relatively accurate method, however, involves use of specialized sievelike devices called impactors connected with a short length of heavy gauge flexible hose to a high-flow vacuum air sampling pump (Figures 12.7 and 12.8). The sampling train is commonly operated at flow rate of 1 ft^3/min, which is equivalent to 28.3 l/min. To detect the presence of organisms, they are collected on media called nutrient agar, which enhances growth of colonies that are visible for counting. Dishes of agar are positioned within the impactor during monitoring so that the bioaerosols are initially separated based on particle sizes by the impactor and deposited on respective dishes of semisolid agar for growth, detection, and measurement. Common impactors consist of two and six stages.

Agar serves as a growth medium for numerous microorganisms, such as bacteria and fungi. The agar is prepared and sterilized using an autoclave prior to each use. The culture dishes also must be sterilized. The purpose of sterilizing the materials is to minimize the erroneous growth of microorganisms not associated with the actual air monitoring.

It is common to use a single-stage cascade impactor to measure total (i.e., sum of respirable plus nonrespirable) colony forming units. A two-stage cascade impactor can separate respirable from nonrespirable fractions. A six-stage cascade impactor, however, permits even more specific separation based on particle sizes. Prior to, and in between use, the cascade impactor should be disinfected using a chemical disinfectant or autoclave to minimize cross contamination of samples.

Bioaerosol analysis involves a preliminary incubation period to allow for growth of detectable and measurable colonies of microorganisms (Figure 12.9). The incubation time, temperature, humidity, and lighting varies depending on the type of organisms that were collected and are selected for analysis. Following a prescribed incubation, the culture dishes are observed for detectable and countable microbial colonies. Dishes can be positioned under a colony counter, which is basically an illuminated grid surface with a magnification lens, to facilitate detection and counting. The concentration of airborne bioaerosol is based on the number of colonies counted per volume of air sampled and is typically expressed as colony forming units per cubic meter (cfu/m^3) of air.

Instantaneous or Real-Time Monitoring of Particulates

Instrumentation and methods for conducting integrated or continuous personal and area monitoring and analysis of classes of airborne particulates have been summarized previously. Instrumentation and methods also are available, however, for conducting instantaneous or real-time area monitoring of generic total and respirable particulates as well as more specific airborne particulates such as fibers. The instrumentation typically involves battery-operated electronic meters that actively pump or blow air into the device where airborne particulates are detected and measured. The concentration of airborne particulates is indicated on a direct readout display (Figure 12.10).

Direct-reading instruments for monitoring airborne particulates are generally less accurate, but more convenient than integrated air monitoring techniques that require

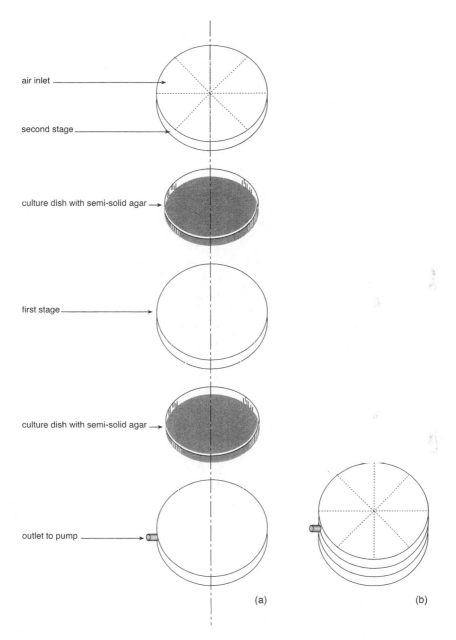

air inlet

second stage

culture dish with semi-solid agar →

first stage

culture dish with semi-solid agar →

outlet to pump →

(a)   (b)

Figure 12.7 An (a) exploded view and (b) assembled view of a two-stage cascade impactor containing two culture dishes filled with semisolid nutrient growth agar for sampling bioaerosols.

time for laboratory analysis. Their speed of operation allows rapidly changing conditions to be assessed at the specific time of monitoring.

Typically, the monitor is placed in a stationary position 4 to 6 ft above the working surface to collect area samples. Area samples also are collected using hand-held devices

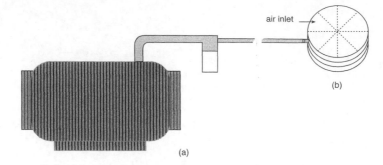

(b)

(a)

Figure 12.8   Airborne bioaerosols can be sampled using (a) a high-flow vacuum air sampling pump connected via flexible tubing to (b) a two-stage cascade impactor containing two culture dishes filled with semisolid nutrient growth agar.

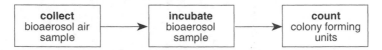

Figure 12.9   Following sample collection of bioaerosols onto culture dishes filled with nutrient growth agar, the dishes are incubated at a specific temperature for a specific duration prior to counting colony-forming units.

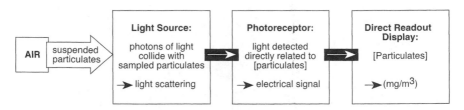

Figure 12.10   Various types of direct-reading aerosol and particulate analyzers can be used for instantaneous or real-time monitoring of airborne dusts and fibers. The schematic represents the mechanism of photometric detection and measurement of airborne particles.

that permit movement to various locations while simultaneously monitoring for airborne aerosols. Instantaneous area monitoring is important to determine the concentration of particulates within a given area, the sources of emission, the effectiveness of ventilation controls, the peak concentrations of particulates during specific operations or periods.

Several types of direct-reading aerosol meters are available. The devices differ based on application and method of detection and measurement. Two major principles of particulate detection and measurement are light scattering and electrical precipitation and oscillation. In addition, based on application, some devices differentiate between concentrations of spherical (mg/m^3) and fibrous (f/cc) particulates.

The x-ray fluorescence (XRF) meter uses a radioactive source to instantaneously detect and measure levels of metals. Applications are for evaluating painted surfaces and soils for metals mostly, but some developmental applications are for evaluating airborne metals. One major use is to evaluate homes and other dwellings for lead-based paint.The device works on the principle of x-ray fluorescence. It contains a small radio isotopic source. When the unit is placed against a surface, pressing the trigger opens a shutter that allows radiation to fall on the paint. If the paint contains lead, then the radiation stimulates the lead atoms to reemit characteristic x-rays. These are sensed by the detector in the unit. Solid-state electronic circuits convert the signals into the final reading.

Integrated or Continuous Monitoring of Gases and Vapors

Exposure limits and specific monitoring and analytic methods have been established for nonparticulate air contaminants known as *gases* and *vapors*. Airborne gases and vapors can be separated from a contaminated airstream via adsorption onto a solid medium or absorption into a liquid medium following contact. Accordingly, some integrated sampling methods for gases and vapors involve sorptive techniques using a solid medium such as activated carbon or silica gel as adsorbents. *Adsorption* (Figure 12.11) refers to the immediate bonding of a gas or vapor to the surface of a solid under normal conditions of temperature and pressure. Indeed, adsorption specifically refers to bonding at the surface interface where contact occurs between the respective surfaces of the solid adsorbent and the flowing gas or vapor molecules. Although adsorption of gases and vapors onto *solid adsorbent media* is preferred for conducting integrated monitoring, some compounds are not collected efficiently using these media. As a result, *liquid absorbent media* are used to collect relatively more soluble and reactive organic gases and vapors, as well as some inorganic gases and mists. Unlike adsorption where gas and vapor molecules bond to the surface of a solid medium, *absorption* (Figure 12.12) involves dissolution of molecules of gases and vapors into a liquid medium.

The solid adsorbents such as activated charcoal and silica gel are used most frequently to collect relatively water-insoluble and nonreactive gases or vapors. Solid

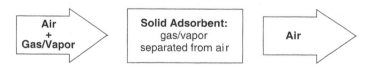

Figure 12.11   Organic gases and vapors are separated and collected from contaminated air via adsorption onto a solid adsorbent medium.

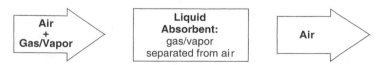

Figure 12.12   Organic gases and vapors are separated and collected from contaminated air via absorption into a liquid absorbent medium.

adsorbent media are characterized as small porous spheroid particles. This characteristic provides high specific surface area (i.e., square millimeter surface area per gram of media) for contact between airborne gases and vapors and the surface of the adsorbent. In turn, adsorption efficiency is directly related to the amount of specific surface area that is exposed. Other factors that affect collection efficiency include sample flow rate, concentration of the gas or vapor, temperature, and humidity. If flow rates are too high, not enough time is allowed for the contaminant to react with the adsorbent. The lowest flow rate that gives a sample of sufficient size for analysis should be utilized, but flow rates are rarely less than 10 cc/min. To compensate for high concentrations of gases or vapors, sampling time and or flow rates can be reduced, or a larger sorbent tube containing more adsorptive media can be utilized. Elevated temperature and humidity can increase the rate at which the surface of adsorbent media becomes saturated.

Activated charcoal is used for the collection of nonpolar organic vapors with boiling points greater than 100°C. Silica gel is generally used to collect polar substances that do not efficiently adsorb to activated carbon. Silica gel can be used to collect organics with a minimum of three carbons, compounds containing hydroxyl groups, halogenated hydrocarbons, and inorganic compounds. The silica medium is very hygroscopic and, accordingly, collection efficiency decreases under conditions of high relative humidity.

The adsorbent media are contained within flame-sealed glass tubes tapered at each end. Prior to sampling, both ends of the glass sampling tube are carefully snapped open for air to flow through during sample collection. Integrated monitoring for airborne gases and vapors involves a low-flow or multiflow pump set at a flow rate less than 1000 cc/min and commonly lower than 100 cc/min. The pump is connected to a short length of flexible hose and tube holder assembly, which in turn, contains the glass tube packed with the solid adsorbent (Figure 12.13). The sampling pump creates a vacuum that causes air to be pulled into the tube and through the front then the backup sections prior to eventual exhaust from an outlet port on the pump. In theory, if gases or vapors are present in the air as it passes though the column of solid adsorbent particles, they would be immediately removed from the airstream and collected via adsorption onto the solid medium. Subsequent to sample collection, the sampling tube is removed and plastic caps are placed on each end prior to laboratory analysis.

Gas chromatography (Figure 12.14) is the method of analysis most often utilized for contaminants collected on solid adsorbent tubes. Analysis consists of the following:

1. Solid adsorbent is removed from the glass sampling tube and placed into a vial to which a solvent such as carbon disulfide for charcoal and methanol for silica gel is added to desorb the gas or vapor contaminant from the solid adsorbent medium.
2. After a short incubation period for desorption, an aliquot of desorption solvent containing dissolved analytes is manually or automatically withdrawn from the sample vial and injected into a heated injection port of a gas chromatograph for vaporization.
3. Vaporized desorption solvent and analytes are transported (carried) to and through a heated column via an inert carrier gas (mobile phase).

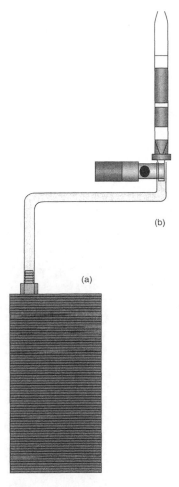

(b)

(a)

Figure 12.13 Organic gases and vapors are separated and collected from contaminated air using (a) a low-flow or multiflow battery-operated air sampling pump connected with flexible tubing to (b) a glass tube containing a solid adsorbent.

4. The desorption solvent and each analyte interact differently with the medium (stationary phase) coated or packed inside the column and, as a result, separate (partition) from each other.

5. The analytes elute off the column to a detector in the order of shortest to longest retention time on the column and are detected and plotted on a recorder as curves and peaks on a graph called a *chromatogram*; *retention time*, and *area under the curve* are indicated for each analyte.

The analytes are identified by comparing the retention time for the unknown sample analyte with the retention time for a known standard analyte. Analytes are quantified based on the area under the curve, or peak area (area ratio).

The alternative and typically less preferred sampling media, liquid absorbents, are contained in specialized sampling devices called *midget impingers*, or *midget*

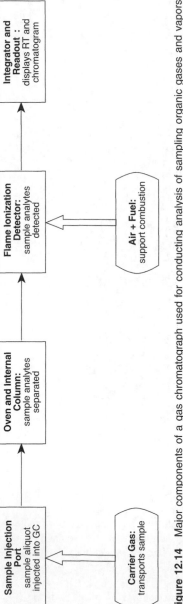

Figure 12.14 Major components of a gas chromatograph used for conducting analysis of sampling organic gases and vapors collected onto a solid adsorbent.

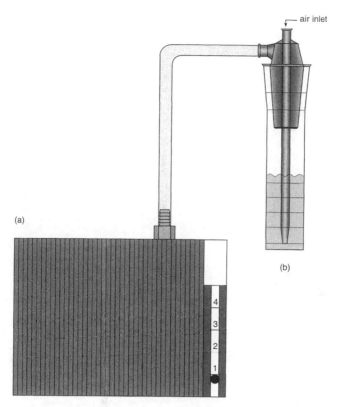

Figure 12.15 Organic gases and vapors are separated and collected from contaminated air using (a) a high-flow or multiflow battery-operated air sampling pump connected via flexible tubing to (b) a glass midget impinger containing a liquid absorbent.

bubblers, connected via flexible tubing to a high-flow sampling pump during monitoring (Figure 12.15). The devices enhance aeration of the liquid absorbent with the sampled air so that increased surface area and mixing is available. In turn, increased contact between the sampled gas and vapor molecules and the liquid absorbent enhances the efficiency of collection of the contaminant.

Midget impingers and bubblers are columnar glass devices that provide a reservoir for liquid absorption reagents. Inserted in the reservoir is a hollow glass inlet tube open at each end and tapered (impingers) or fritted (bubblers) at the bottom. The tube is attached and extends through the top of the reservoir. When the reservoir is filled with 10 to 20 ml of a liquid medium, the tapered or fritted end of the tube is immersed in the fluid. The tube serves as a conduit for the air to enter the impinger or bubbler and mix with, and dissolve into, the liquid absorbent.

Ultraviolet (UV) or *visible (Vis) spectrophotometry* is the method of analysis most often utilized for contaminants collected in liquid absorbents (Figure 12.16). The analysis proceeds as follows:

Figure 12.16 Major components of a UV and Vis spectrophotometer used for conducting analysis of sampled organic gases and vapors collected into a liquid absorbent.

1. Liquid absorbent plus dissolved contaminant is removed from the impinger or bubbler and mixed with reagents yielding a detectable analyte that maximally absorbs a wavelength of light in either UV or Vis spectrum.
2. A specific spectrum (UV or Vis) and wavelength of light is selected relative to the contaminant or analyte in the sample by adjusting the monochromator positioned between the light source and sample chamber.
3. A sample solution is placed in a special cuvette into the sample chamber where a specific wavelength of light is directed to the sample chamber where analyte molecules can absorb the light energy in direct relation (linearly) to its concentration (Beer's law).
4. Light not absorbed by the sample analyte exits the sample chamber to the detector that generates an electrical signal to the readout.
5. Intensity of the light detected is converted to display as absorbance (Abs) on a direct readout.

Absorbance reading for the analyzed sample (unknown concentration) compared with a standard curve consisting of absorbance readings corresponding to known concentrations of analyzed standards. The amount of analyte in the sample solution that was detected and measured by the UV or Vis spectrophotometer is extrapolated from the standard curve and divided by sampled volume of air to give concentration.

Instantaneous or Real-Time Monitoring of Gases and Vapors

Instantaneous, or real-time, direct-reading instruments also are used to detect and measure the concentrations of airborne gases and vapors. The gaseous agents detected and measured include combustible or toxic gases or vapors and oxygen gas.

The presence of a combustible or explosive atmosphere and either an oxygen-deficient or enriched atmosphere can pose immediate hazards to individuals. Monitoring devices are available that instantaneously detect and measure the level of airborne combustible gas or vapor and oxygen gas. The most common devices for detecting and measuring combustible and oxygen gases are used for area monitoring. Individual instruments are available to measure combustible gases or vapors and oxygen gas. It is more common, however, to use a combination unit that is designed to detect and measure both parameters. These combined devices are called combination combustible and oxygen gas meters or, more simply, combustible gas meters (Figure 12.17). The latter name, however, does not reflect the potential to monitor oxygen. Many of the combination meters also measure a toxic parameter such as carbon monoxide or hydrogen sulfide gases.

Monitoring of combustible gas or vapor and oxygen gas is instantaneous and represents real-time measurements of the analytes. The devices are typically electronic active-flow monitoring instruments that collect an air sample by automatically pumping or blowing air into the device. As the air flows through the instrument, combustible gas or vapor and oxygen gas interact with sensors or detectors for the respective analytes. The meters are generally calibrated for combustible gas or vapor using a known concentration of reference gas, such as pentane. Responses to other combustible gases or vapors actually detected and measured during monitoring,

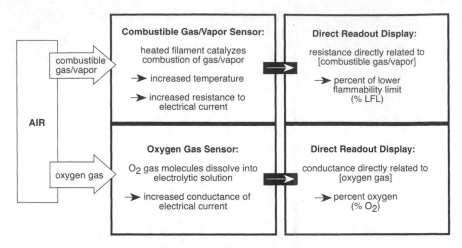

Figure 12.17   Various types of direct-reading combined combustible gas and vapor and oxygen analyzers can be used for instantaneous or real-time monitoring of combustible and oxygen-enriched or oxygen-deficient atmospheres. The schematic represents two respective mechanisms of detection and measurement of airborne combustible gases and vapors and oxygen gas.

however, must be adjusted via conversion factors to determine the exact concentration of a specific gas. For oxygen, some meters are calibrated using a known concentration of molecular oxygen gas.

Common combustible gas meters contain a heated platinum filament that internally ignites the combustible gas or vapor. The ignition of a combustible gas or vapor generates a detectable and measurable quantity of heat. This heat of combustion changes the resistance of the filament in proportion to the concentration of the combustible gas or vapor present. The readout of the meter indicates the concentration of combustible gases or vapors as percentage of the lower explosive limit (LEL). This actually represents a percentage of the lower flammability limit (LFL) or LEL of the calibration gas. Various makes and models are available.

Combustible gas meters are intended for use only in normal atmospheres, not atmospheres that are oxygen enriched or deficient. Oxygen concentrations significantly less than or greater than normal (ca. 20.8%) may cause erroneous readings. Accordingly, the level of airborne oxygen gas is usually measured simultaneously. Leaded gasoline vapors, halogens, and sulfur compounds interfere with the filament and decrease its sensitivity. Silicone-containing compounds can damage the filament.

Oxygen gas meters typically involve an oxygen-sensing device and a meter readout. In some units, air is drawn to the oxygen detector with an aspirator bulb or pump, whereas in other units, equalization due to diffusion is allowed to occur between the ambient air and the sensor. An electrochemical sensor is used to determine the oxygen concentration in air. Components of a typical sensor include a counting and sensing electrode; a housing containing a basic electrolytic solution; and a semipermeable Teflon membrane. Oxygen molecules diffuse through the membrane into the electrolytic solution. A minute electric current is produced by the reaction between the oxygen and the electrodes. This current is directly proportional to the oxygen content of the sensor. The current passes through the electronic circuit, resulting in a needle deflection on the meter. The meter is calibrated to read 0 to 10%, 0 to 25%, or 0 to 100% oxygen. When evaluating an unknown environment, the oxygen indicator should be used first to determine whether an oxygen-deficient or an oxygen-enriched atmosphere exists.

It also is frequently useful and necessary to instantaneously detect and measure the concentration of an airborne toxic inorganic or organic gas or vapor. Devices are available that consist of a manual air sampling pump with detector tube media to serve this purpose. An air sample can be collected, and within a minute or two, the presence of an individual contaminant or a class of contaminants can be determined qualitatively and quantitatively. Various detector tubes are available for the numerous contaminants of concern.

Detector tubes are very useful for initial screening of areas and developing strategies for more comprehensive integrated or continuous monitoring techniques for specific gaseous contaminants of concern. Detector tubes for real-time monitoring also are used concurrently during integrated sampling to trace sources of exposure and to track variations in exposure levels throughout the sampling area and period.

There are two classes of detector tubes, length of stain and colorimetric. In both cases, a sealed glass tube is filled with a chemically coated solid sorbent that has been impregnated with an appropriate chemical reagent to react specifically with the contaminant of interest. Prior to sampling, both ends of a glass detector tube are carefully snapped open for air to flow through during sample collection. The detector tube is inserted into the inlet orifice of the air sampling pump. Two major types of air sampling pumps are available for use with detector tubes. Both types are manual and are referred to as *piston pumps* and *bellows pumps* (Figure 12.18).

As contaminated air is drawn through a detector tube using a manual pump, molecules of the gaseous contaminant contact the solid medium and react. The reaction results in either a dark stain or color development, depending on the type of tube (i.e., length of stain or colorimetric). The length of stain or the intensity and shade of color development correspond with the concentration of the gas or vapor contaminant in the sampled air. Length of stain tubes either have the concentration scale printed on them, or the ratio of the length of stain to the total media length is compared against a chart of corresponding concentrations. When contaminated air is drawn through a colorimetric tube, the reagent-impregnated sorbent yields a progressive change in color intensity. At the completion of sampling, the sample tube is compared with a chart of tinted colors that correlate to the concentration of the gas or vapor present in the atmosphere.

Detector tubes are one of the easiest monitoring devices to use, but data must be interpreted with caution because the tubes are known to be relatively inaccurate. Accuracy is only ± 25% in many cases and for some contaminants error is greater. Therefore, a sample detector tube may reveal a concentration of a contaminant equal to 200 ppm, for example, but the actual concentration may be 150 to 250 ppm (200 ppm ± 25% or 200 ± 50 ppm).

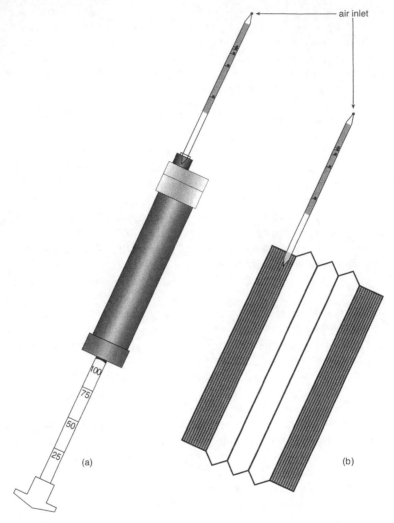

Figure 12.18   A direct-reading (a) piston pump or (b) bellows pump with an attached detector
tube can be used for instantaneous or real-time monitoring of inorganic and
organic gases and vapors.

Electronic devices also are available for instantaneous or real-time area monitoring of organic gases and vapors. These devices operate based on different principles, but also share several similar characteristics. The devices consist of air
sampling pumps that cause active flow of air and airborne contaminants into a
chamber. The air and related organic gas and vapor molecules flow across a detector
that qualitatively detects and identifies the contaminant as organic and quantitatively
measures the concentration in parts per million on a direct readout display
(Figure 12.19). In most cases, the identity of the specific components of the airborne
organic gas and vapor may be unknown or uncertain. Accordingly, data are commonly reported as total organic vapor (TOV).

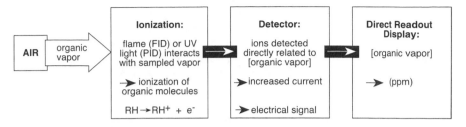

Figure 12.19 Various types of direct-reading organic gas and vapor analyzers can be used for instantaneous or real-time monitoring of toxic gases and vapors. The schematic represents the mechanism of flame ionization detection and measurement of airborne organic vapors.

Meters with flame ionization detectors (FID) measure the concentration of total ionizable organic gases and vapors; and, if the identity of the gas or vapor is known, they also measure the concentration of specific organics in the atmosphere. Common components of a meter with an FID are battery-operated air-sampling pump, FID, cylinder of compressed hydrogen gas, readout, and sampling probe.

Response varies depending on the number of carbon atoms in the molecules. A gas or vapor is pumped into the instrument and passed through a hydrogen flame at the detector that ionizes the organic vapors. When most organic vapors burn, positively charged carbon-containing ions are produced, which are collected by a negatively charged collecting electrode in the chamber. An electric field exists between the conductors surrounding the flame and a collecting electrode. As the positive ions are collected, a current proportional to the hydrocarbon concentration is generated on the input electrode. This current is measured with a preamplifier that has an output signal proportional to the ionization.

Meters with a photoionization detector (PID) also measure the concentration of total ionizable organic gases and vapors. Meters with a PID consist of a readout unit comprised of an analog meter, a rechargeable battery, an amplifier, a UV lamp, and a sensor ionization chamber.

An electrical pump pulls the gas or vapor sample past a UV source. Constituents of a sample are ionized, producing an instrument response, if their ionization potential is equal to or less than the ionizing energy supplied by the instrument UV lamp utilized. The radiation produces an ion pair for each molecule of contaminant ionized. The free electrons produce a current directly proportional to the number of ions produced. The current is amplified, detected, and displayed on the meter. By varying the electron volts of UV light, a wide range of organic compounds can be detected and quantified.

Infrared spectrophotometric meters can detect and measure concentrations of specific organic gases and vapors. The instruments have direct reading scales, recorder output, and a built-in sampling pump. Compounds are detected by an infrared spectrophotometer-based absorption of a discrete wavelength of infrared light. In turn, absorption is proportional to the concentration of gas or vapor.

Portable gas chromatographs are versatile instruments for instantaneous monitoring. The principle of detection and measurement is similar to bench top laboratory gas chromatographs described earlier. The instruments separate compounds and permit detection and quantification of specific organic compounds. Sampled gas or vapor is passed through a column packed with a liquid phase on solid support or a solid-phase support. Components of the sampled airstream separate as they pass through the column based on their inherent chemical characteristics. The individual components (analytes) are identified based on their retention time on the column and detected via a variety of commercially available detectors, typically an FID.

EVALUATION OF AIRBORNE SOUND LEVELS

Integrated or Continuous Monitoring of Sound

Integrated or continuous personal monitoring using audio dosimeters is frequently conducted to detect and measure sound levels when evaluating the occupational environment. Following monitoring, data are calculated to determine whether levels are within standard or recommended criterion levels. A criterion level is the continuous A-weighted sound level that constitutes 100% of an allowable exposure. For example, a standard criterion level is 90 dBA for an 8-hour day. If noise exposures greater than 90 dBA are based on a 5 dB "doubling exchange rate," then workers are permitted to be exposed to 95 dBA for 4 hours, 100 dBA for 2 hours, 105 dBA for 1 hour, and so on. Allowable time (T) can be calculated for exposure to a given sound pressure level (L_a) based on a criterion level of 90 dBA and a doubling exchange rate of 5 as follows:

$$T = \frac{480 \text{ min}}{2^{(L_a - 90)/5}} \tag{12.3}$$

To calculate noise exposure in locations or job classifications where sound pressure levels vary, the total sound pressure exposure or dose must be determined. Dose (D) is calculated based on the actual exposure time at a given sound pressure level (C) divided by the allowable exposure time at that level:

$$D\% = \left(\frac{C_1}{T_1} + \frac{C_2}{T_2} + \cdots + \frac{C_n}{T_n} \right) \times 100 \tag{12.4}$$

If the total dose exceeds 100% or unity, then individuals exposed to the various sound sources at the measured levels would be excessively exposed. To determine the equivalent sound pressure level (L_{eq}) in A-scale decibels for an 8-hour day based on dose (D) values, the following formula is used:

$$L_{eq} = 90 + 16.61 \log \frac{D\%}{100} \tag{12.5}$$

To determine the equivalent sound pressure level (L_{eq}) in A-scale decibels for dose (D) exposure durations not equal to an 8-hour shift, the following formula is used where T is the actual time in hours:

$$L_{eq} = 90 + 16.61 \log \frac{D\%}{12.5\,T} \tag{12.6}$$

Audio dosimeters are special sound level meters (SLMs) that integrate sound pressure levels over time and are used to measure personal noise exposures. Audio dosimeters (Figure 12.20) are composed of a microphone attached to a meter that

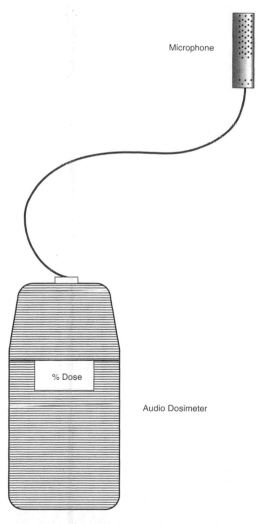

Figure 12.20   Audiodosimeter for integrated or continuous monitoring of sound levels.

is similar to a sound level meter (see below). The detected sound levels, however, are integrated into memory and measurements are indicated via direct readout. Internal and external computerized readout devices are available. The audio dosimeter microphone is typically clipped to an individual's clavicle (collar) or trapezius (upper shoulder) region provided it is within 12 to 18 in. of the ear (i.e., hearing zone). Ideally, the microphone of the device should be connected to the worker's ear, but this is not always practical or desired. The meter is either placed in the shirt pocket or attached to the belt. Audio dosimeters integrate the A-scale sound pressures that the worker is exposed to over an entire shift (or monitoring period), and total exposure or dose is then determined.

By calculating the integrated dose obtained with the audio dosimeter and inserting this value into one of the equivalent sound pressure level formulas presented earlier, it is possible to determine level of exposure. Many dosimeter models may be downloaded to a computer and printouts can be generated recording a wide variety of measures obtained throughout the monitoring activity.

Instantaneous or Real-Time Monitoring of Sound

The SLM and octave band analyzer (OBA) are two generic types of sound monitoring instruments typically used when assessing noise problems in various environmental settings. The SLM is an instrument used to monitor air pressure changes resulting from environmental sound and noise. The OBA is a special SLM for measuring pressure levels of sound for octave frequency bands and is referred to as a sound level meter octave band analyzer (Figure 12.21). The octave band analyzer is actually an electronic filter attached to or incorporated with an SLM.

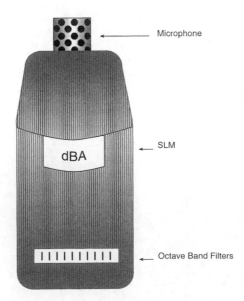

Figure 12.21 Combination sound level meter and octave band analyzer for instantaneous or real-time monitoring of sound levels at specific frequency ranges.

There are a wide variety of SLMs and SLM/OBAs commercially available. These instruments are used in the occupational environment to assist in detecting and measuring sound levels and determining whether a noise problem exists. SLMs consist of three major generic components; a microphone, a meter, and a connection for an octave band analyzer.

Although each sound level meter is slightly different, each may have one or more sound measurement scales corresponding to a range of frequencies. Three more common scales are labeled A, B, and C. The A-scale consists of frequencies (20 to 20,000 Hz) closely associated with human hearing and is the mandated scale for monitoring activities involving human exposure assessment. The B-scale is used in sound research activities, and the C-scale is used for engineering and maintenance or trouble-shooting activities.

There are four types of approved SLMs commercially available that differ by their certified accuracy. Type 0 has an accuracy of ±0 dB and is the most accurate SLM available. It is predominantly used for laboratory research applications. Type 1 SLMs are instruments that have an accuracy of ±1 dB. These are precision units intended for extremely accurate field and laboratory measurements. Type 2 SLMs are general-purpose sound level meters with an accuracy of ±2 dB. Type 2 SLMs better set at A-scale are mandated for use in evaluating the occupational environment for noise. The last type of SLM is the type S, or special purpose, sound level meter. It is equivalent to the old type 3 unit with design tolerances similar to those listed for the type 1 unit. The Type S SLM differs from the type 1 because these instruments are not required to have all the functions of the units previously listed.

Although the SLM without octave band electronic filter can be set at a scale (A, B, or C) that represents a relatively broad range of frequencies, the SLM/OBA can be adjusted to specific frequencies that represent relatively narrow octave band frequencies. The specific center band frequencies are 31.5, 63, 125, 250, 500, 1000, 2000, 4000, 8000, and 16,000 Hz. Each center band frequency represents an individual range or band of frequencies.

EVALUATION OF AIR TEMPERATURE FACTORS

Instantaneous or Real-Time Monitoring of Elevated Temperature Factors

Several devices and procedures are available for evaluating the occupational environment for conditions that may contribute to heat stress. The most common method, however, involves determination of the Wet-Bulb Globe Temperature (WBGT) Index. The method involves collection of two to three major measurements that reflect the contribution of environmental factors to heat stress.

The measurements include the dry-bulb (DB) temperature, the natural wet-bulb (WB) temperature, and the globe temperature. The DB temperature measures the ambient temperature. The natural WB temperature records the ambient temperature, but reflects the influence of humidity and air movement. High humidity or low air movement result in an elevated WB temperature. The globe temperature reflects the contribution of radiant heat. An integration of all three measurements yields a WBGT Index. For indoor and outdoor settings where there is no solar load, only two measurements are needed, natural WB and globe temperature (GT):

$$WBGT_{in} = 0.7WB + 0.3GT \qquad (12.7)$$

For outdoors in sunshine, the air temperature based on the DB temperature must also be measured:

$$WBGT_{out} = 0.7WB + 0.2GT + 0.1DB \qquad (12.8)$$

Although the WBGT is easy and simple to use, it does not in itself include a factor for the activity or exertion of an individual. It is not possible to determine an allowable exposure time directly from the WBGT. Guidelines have been developed, however, which eliminate these difficulties. The work load for an individual is determined by estimating the worker's metabolic rate. A chart has been developed to aid in this process. An occupational exposure limit has been presented based on the amount of time a worker is involved in continuous work and the level of work being performed.

Although measurement of the ambient air temperature by itself is insufficient to determine heat stress, it is an integral part of heat stress determinations. Various methods are available to assess ambient air temperature. The most common method of measurement is the DB thermometer. A DB thermometer consists of a hollow glass tube with bottom reservoir of mercury. The range of the thermometer should be −5 to +50°C and accurate to ±0.5°C. Temperatures beyond the range of a DB thermometer can break the thermometer. When the measurements are taken, the DB thermometer must be shielded from radiant heat sources so that only the temperature of the ambient air can be detected. Thermoelectric thermometers, or thermocouples, are also used to measure the ambient air temperature.

The black globe thermometer, or Vernon globe, is the standard method for measuring radiant heat. A black globe thermometer consists of a 6-in. diameter hollow copper sphere that is painted matte black. A thermometer is inserted so that its bulb is centered inside the globe. The range of the thermometer should be −5° to +100°C and accurate to ±0.5°C. The black globe absorbs radiant heat. The thermometer inside the globe is allowed to reach equilibrium. The energy measured is generally termed the mean radiant temperature, because some heat is lost from the globe via convection.

As the temperature of air rises, it holds more water vapor. When the air at a given temperature holds all the water vapor that it can, it is said to be saturated. The water in the air exerts a vapor pressure that is measured in millimeters of mercury. Humidity in the air is usually measured with a WB thermometer or a sling psychrometer. The WB thermometer consists of a standard DB thermometer covered with a cotton sleeve or wick moistened with distilled or deionized water. The covered thermometer bulb is partially immersed in water to keep the wick wet. As the wick evaporates water, a certain amount of heat energy is dissipated through evaporative cooling. The bulb is then cooled by the heat absorbed during evaporation of the water, and an equilibrium is reached between the evaporation rate and the water vapor pressure in the air. The temperature is indicated on the thermometer. At full saturation, the WB thermometer temperature equals the DB temperature because no evaporative cooling can be experienced.

The thermometers and accessories needed to measure WB temperature, globe temperature, and DB temperature can be assembled and positioned in a ring stand (Figure 12.22). Commercially available electronic WBGT meters are also available (Figure 12.23). The electronic devices integrate the temperature measurements and provide a direct readout for individual parameters and the combined WBGT.

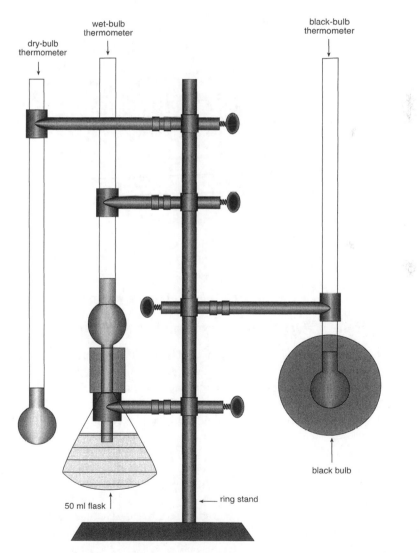

Figure 12.22 Assembly for measuring wet-bulb globe temperature (WBGT) indices for heat stress.

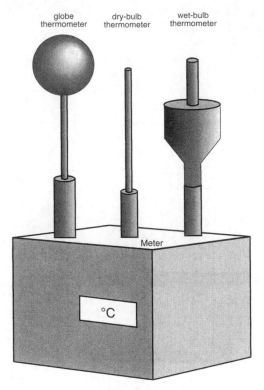

globe
thermometer

dry-bulb
thermometer

wet-bulb
thermometer

Meter

°C

Figure 12.23   Electronic meter for measuring wet-bulb globe temperature (WBGT) indices for heat stress.

EVALUATION OF ILLUMINATION

Instantaneous or Real-Time Monitoring of Illumination

Visible light is generated from both natural and anthropogenic sources. Luminous flux refers to the total energy from a source of light that is emitted and can influence sight. The luminous flux is measured in standard units called lumens, which refer to the total radiant power or flux shining on a 1-ft^2 surface. Illumination is the density of the luminous flux on a given area of a surface. Thus, illumination equals lumens per area. Relative to an area of 1 ft^2, illumination is measured in units of footcandles or lux.

Light detection and measurement is called *photometry*. An electronic illumination meter is used to measure lighting. The devices consist of a relatively flat photosensitive sensor with a color attenuator or filter connected to a detector (Figure 12.24). Light energy focused directly on the sensor or photocell causes emission of electrons and related generation of increased current. The current generated is directly related to the intensity of light. Measurement of illumination, in foot candles or Lux, is indicated on a direct readout display.

Illumination levels are measured and recorded at various points within a given area in relation to the activity that is conducted at the location. The distances are

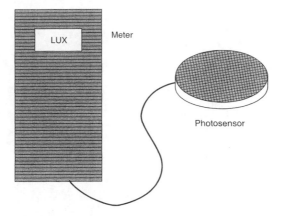

Figure 12.24 Illumination meter for instantaneous or real-time detection and measurement of light.

measured between the light source and the work surfaces, such as assembly lines and desk tops. Illumination is then monitored at the specific work surfaces and general areas of a given area.

EVALUATION OF MICROWAVE RADIATION

Instantaneous or Real-Time Monitoring of Microwave Radiation

Several types of microwave detectors are available, including thermal devices that respond to increased temperatures from absorbed radiation by reflecting a change in resistance or voltage. Electronic microwave meters that convert microwave frequency into direct current are commonly used for instantaneous area monitoring of nonionizing radiation from microwave ovens. The meters consist of an antenna, or a probe, that is connected to a detector and amplifier (Figure 12.25). A spacer may be present on the probe so that a uniform distance is maintained between the source and the antenna. When evaluating a microwave oven, a 5-cm distance is required between the oven door and the probe during monitoring.

A thorough inspection of the microwave oven is conducted as part of the evaluation to check the electrical wiring and plug, door alignment, door seal condition and cleanliness, and door hinge and latch condition. Following inspection, materials used for the standard loading of the unit must be placed in the microwave oven. Operation of ovens with no load may cause damage to the unit and must be avoided. Accordingly, a standard load typically consists of a 500 ml nonmetal (i.e., plastic, glass) beaker filled with approximately 275 ml of tap water. Monitoring is performed while the door is closed and the oven is operating by keeping the probe 5 cm from the surface of the unit and scanning the door area. Following the evaluation, a thermometer is immersed in the beaker of water to verify an increase in the temperature of the fluid due to microwave energy.

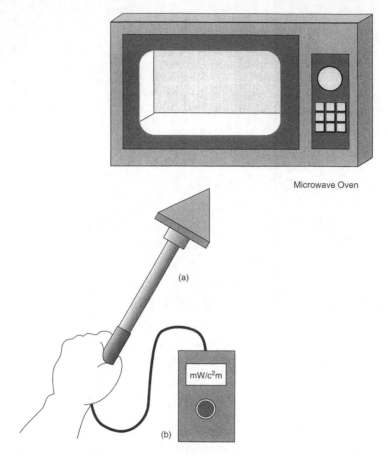

Figure 12.25 Evaluation of a microwave oven for instantaneous or real-time detection and
measurement of microwave radiation using (a) a survey probe attached to (b)
a direct-reading meter.

EVALUATION OF SOURCES OF ELECTRIC AND MAGNETIC FIELDS

Instantaneous or Real-Time Monitoring of Electric and Magnetic Fields

Monitoring requires use of a specialized electronic meter with antenna that can detect and measure low frequency electric and magnetic fields (Figure 12.26). The antenna is typically a diode–dipole design and serves as an electronic sensor of electric and magnetic field energies. The root mean square (RMS) of the detected signal is computed and the measurement is displayed on a direct readout meter. Magnetic fields are commonly measured in units of milligauss (mG) and electric fields in units volt/meter (V/m).

Distances and heights relative to the sources of electromagnetic radiation must be considered during monitoring. Measurements are recorded at various vertical and

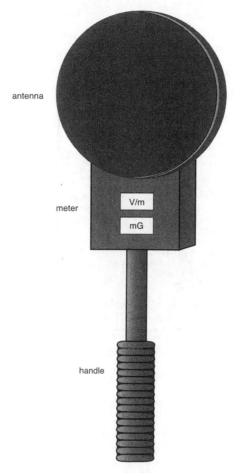

Figure 12.26 Combined electric and magnetic field meter for detection and measurement of nonionizing electric and magnetic fields.

horizontal distances from sources such as electrical appliances and devices. In addition, it is useful to conduct measurements when the suspected sources are turned "off" and then when turned "on." The contribution of electric and magnetic energies to the area then can be attributed to the operating source.

EVALUATION OF IONIZING RADIATION

Integrated or Continuous Monitoring of Ionizing Radiation

Film badges and pocket dosimeters are two common devices used for conducting personal integrated monitoring of ionizing radiation (Figure 12.27). Film badges consist of photographic film coated with a thin layer of a special emulsion containing silver halide (i.e., silver bromide). The film is positioned within a special badge that

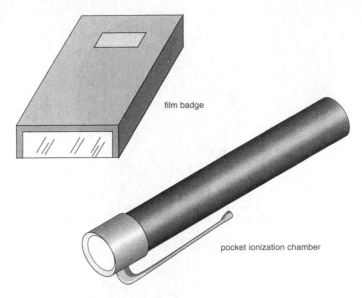

Figure 12.27 Integrated or continuous personal sampling (a) film badge and (b) pocket dosimeter for detection and measurement of ionizing radiation.

can be modified depending on the use. Interaction with ionizing radiation causes the silver halide to ionize forming silver cations. The silver cations are attracted to negatively charged regions of the film yielding free-silver compounds. After use, the film is removed from the badge and submitted for processing. Development of the film rinses away silver halide and silver ions. Free silver, however, remains and its presence causes the film to exhibit increased opacity in proportion to the amount of ionizing radiation exposure. The measurement of exposure is determined using a densitometer that measures the degree of film opacity.

Pocket dosimeters, like the film badge, are typically clipped to an individual's lapel or shirt pocket. They are similar to ionization chamber technology. The devices reveal accumulated exposure based on a direct readout visible in the dosimeter. A similar type of penlike dosimeter, however, requires a separate readout device to determine the measurement of absorbed dose.

Another common personal monitoring device for ionizing radiation is the thermoluminescence dosimeter (TLD). Energized electrons resulting from ionizing radiation interact and are trapped within a phosphor crystal. Thermal analysis of the dosimeter causes generation of detectable light. The measured light emission reflects the level of absorbed dose.

Radon Gas

Radon gas sampling can be accomplished by using charcoal canisters or alpha-track detectors. A trained technician may also test for radon gas levels within the home. Charcoal canisters and alpha-track detectors require exposure to the air in the homes for a specified period of time and then are sent to the laboratory for analysis. For charcoal canisters, the test period is 3 to 7 days. For alpha track detectors the test period is 2 to 4 weeks. Technicians can do a direct radon measurement at the time of analysis. These include real-time radon measurement, continuous radon readers, and integrating readers. The biggest problem with measuring radon is that the levels of radon fluctuate considerably during different time periods.

The appropriate technique for using radon detectors is to first do a screening measurement. This gives you an idea of the highest radon level in the home. It also advises you if it is necessary to continue with radon testing. The screening measurement should be made in the least livable area of the home, probably the basement if one exists. All the windows and doors should be closed for at least 12 hours prior to the start of the test and kept closed as much as possible throughout the testing period. This helps keep the radon level relatively constant throughout the testing period. It is recommended that the short-term radon measurements be done during the cool months of the year. Next, determine the need for further measurements. In most cases the screening measurement is not a reliable measure of the average radon level to which the family is exposed. The screening measurement only indicates the potential for a radon problem, because radon levels vary from season to season and from room to room, based on the level of ventilation that is present. If in the screening measurement the following results occur, then these results act as a guide to further measurements:

1. If the screening measurement result is greater than about 1.0 WL (working levels) or greater than about 200 pCi/l, it is necessary to perform follow-up measurements as soon as possible. Expose the detectors for no more than 1 week. Consider taking immediate action to reduce radon levels in the home.
2. If the screening measurement result is about 0.1 WL to about 1.0 WL or about 20 pCi/l to about 200 pCi/l perform follow-up measurements. Expose the detectors for no more than 3 months.
3. If the screening measurement result is about 0.02 WL to about 0.1 WL or about 4 pCi/l to about 20 pCi/l, perform follow-up measurements. Expose detectors for about 1 year or make measurements of no more than 1 week in each of the four seasons.
4. If the screening measurement result is less than about 0.02 WL or less than about 4 pCi/l, follow-up measurements are probably not required.

Instantaneous or Real-Time Monitoring of Ionizing Radiation

Instruments also are available for conducting instantaneous or real-time area monitoring of ionizing radiation. Levels of ionizing radiation that are detected and measured vary depending on several factors, including the source and the distance from it. Instantaneous area monitoring devices commonly consist of a probe connected to a meter with a direct readout display (Figure 12.28). Selection of monitoring devices depends on the type and level of ionizing radiation that is present.

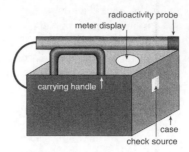

Figure 12.28   Geiger–Müller meter for instantaneous or real-time detection and measurement of ionizing radiation.

An ionization chamber is a real-time electronic monitoring device that consists of a chamber filled with a known volume of gas. Within the chamber is a positively charged anode and negatively charged cathode. When airborne ionizing radiation enters the chamber, it interacts with the gas molecules and forms primary ion pairs. The ions are attracted to the respective electrodes which, in turn, creates a change in voltage. The change in voltage and accompanying current resulting from the ions is proportional to the amount of ionizing radiation detected. The measurement is displayed on a direct readout. Ionization chambers are useful for area measurements of relatively high levels of ionizing radiation. The meters do not, however, differentiate between or identify the types of ionizing radiation detected.

Proportional counters are electronic meters that function very similarly as ionization chambers and are used for real-time area measurements of ionizing radiation. Indeed, they are a special type of ionization chamber. The major difference, however, is that these meters cause a magnification of ionization when ionizing radiation interacts with the gas molecules in the chamber. A higher voltage is generated and produces secondary ions. As a result of secondary ion formation, the current is amplified and corresponds to increased sensitivity of the instrument. Accordingly, proportional counters respond to lower levels of ionizing radiation than standard ionization chambers. The instrument is especially useful for detecting, measuring, and identifying alpha and beta particles.

Geiger–Mueller (G-M) meters also operate based on the principle of ionization as discussed for the preceding two devices. The instrument, however, exhibits even greater secondary ionization than described for the proportional counter. Accordingly, the sensitivity to detect low-level ionizing radiation is even greater. Unfortunately, the electrical current that is generated as a result of the interaction between the detected ionizing radiation and the gas molecules within the chamber and the secondary ions formed and attracted to the electrodes is not typically proportional to the ionization energy present. The G-M meter, therefore, is most useful as a screening device to detect low levels of ionizing radiation, but is not as effective for measuring accurate doses. High-level radiation can cause the meter to block, resulting in no response when levels of ionizing radiation are actually present. Major application is for area measurements of low levels of beta and gamma radiation.

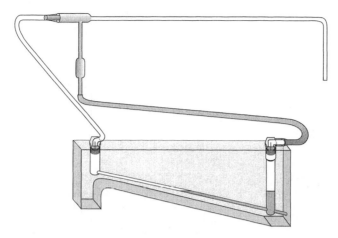

Figure 12.29   Pitot tube and inclined manometer for measuring velocity pressures at designated cross-sectional traverse points inside a duct of a local or general ventilation system.

EVALUATION OF AIR PRESSURE, VELOCITY, AND FLOW RATE

Instantaneous or Real-Time Monitoring within Ventilation Systems

Inside measurements are collected within ducts of local and general ventilation systems to determine pressures such as velocity pressure. In turn, air transport velocities and flow within the systems can be calculated.

To measure air velocity pressures as well as static and total pressures, the most common devices used are the Pitot tube connected to an inclined fluid manometer (Figure 12.29). A Pitot tube is a concentric tube that is designed to simultaneously account for total and static pressures and via differential pressure to measure velocity pressure. The inclined manometer contains a fluid such as water or oil. The Pitot tube is connected to the manometer and inserted into a duct. Air pressures within the duct cause the fluid in the manometer to deflect. The deflection is measured in inches of water ($''H_2O$). Several velocity pressure measurements are taken horizontally and vertically across the area of the duct lumen. In turn, the velocity pressures (*VP*) are converted to velocity (*V*) and averaged to yield average transport velocity (*V*) in feet per minute (ft/min). Average velocity multiplied by the cross-sectional area (*A*) of the duct in square feet (ft²) yields volumetric flow rate in cubic feet per minute (ft³/min), as follows:

$$Q \frac{\text{ft}^3}{\text{min}} = V \frac{\text{ft}}{\text{min}} \times A \text{ ft}^2 \tag{12.9}$$

Instantaneous or Real-Time Monitoring Outside Ventilation Systems

Measurements outside the system include face and capture velocity measurements. Face velocity refers to the movement of air in feet per minute at the immediate

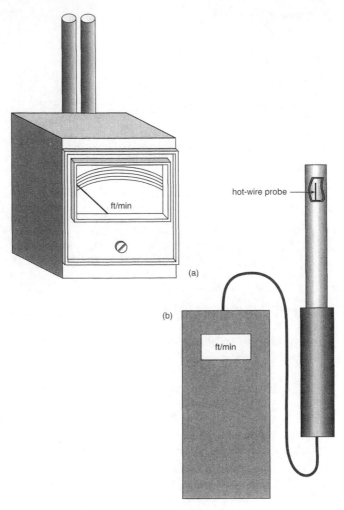

Figure 12.30   (a) Swinging-vane anemometer and (b) thermal or hot wire velometer for measuring face and capture velocities at designated points outside of a local or
general ventilation hood, vent, or diffuser.

opening of a hood, vent, or diffuser. Capture velocity is a measurement of velocity
at a specific distance from the source of contaminant generation to the face of the
hood.

Measurements of face and capture velocities outside of a ventilation system are
commonly determined using a swinging vane anemometer or a thermal anemometer
(Figure 12.30). The swinging vane anemometer is a mechanical device that involves
a pendulum mechanism that deflects in response to moving air. The degree of
deflection is related to the velocity of the air. A readout indicates the quantitative
velocity. Thermal velometers are battery operated and consist of a probe containing
a thin wire that is heated to a stable temperature. Movement of air across the wire

cools it. The degree of cooling is related to the air velocity shown quantitatively on the meter readout.

Velocity measurements are taken at several points at the face of a hood, vent, or diffuser and averaged to give average face velocity in feet per minute (fpm). For capture velocity, several measurements are taken at a distance from the source to the exhaust hood opening and averaged to give average capture velocity also in units of fpm. In addition to these instruments used for collecting quantitative data, a qualitative estimate of airflow can be made using smoke tubes containing titanium tetrachloride to determine whether air contaminants, based on the movement of the visible cloud of generated smoke, are captured by an exhaust system.

Bibliography

CHAPTER 1

Adams, J., Clinical relevance of experimental behavioral teratology, *Neurotoxicology,* 7(2):44–90, 1980.

American Gas Association, *Total Energy Resource Analysis*, Arlington, VA, January 1988.

Archer, V.E., *Health Implications of New Energy Technologies,* Ann Arbor Science, Ann Arbor, MI, 1980.

Billings, C.H., Health risk assessment, *Public Works,* 117:107, 1986.

Carlo, G.L. and Hearn, S., Practical aspects of conducting reproductive epidemiology studies within industry, *Progr. Clin. Biol. Res.,* 160:49–65, 1984.

Chen, R., Detection and investigation of subtle epidemics, *Prog. Clin. Biol. Res.,* 163B:39–43, 1985.

Committee on the Institutional Means for Assessment of Risks to Public Health, Commission of Life Sciences, National Research Council, *Risk Assessment in the Federal Government: Managing the Process*, National Academy Press, Washington, D.C., 1983.

Darmstadter, J., *Energy in America's Future: The Choices Before Us,* Johns Hopkins Press, Baltimore, MD, 1980.

Department of the Environment, *Heavy Metals in the Environment,* Her Majesty's Stationery Office, London, 1983.

Evans, D.W., Effects of ionizing radiation, *Can. Med. Assoc. J.,* 129:10–11, 1983.

Evans, W.E., *Applied Pharmacokinetics,* Applied Therapeutics, San Francisco, 1980.

Fawcett, H.H., *Hazardous and Toxic Materials,* John Wiley & Sons, New York, 1984.

Gorbaty, M.L., *Coal Science,* Academic Press, New York, 1982.

International Atomic Energy Agency, *Nuclear Power Reactors in the World,* Vienna, Austria, April 1997.

Larson, T.E., *Cleaning Our Environment: A Chemical Perspective,* American Chemical Society, Washington, D.C., 1983.

Pochin, E.E., *Nuclear Radiation: Risks and Benefits,* Clarendon Press, Oxford, 1983.

Porter, C.R., *Coal Processing Technology,* American Institute of Chemical Engineers, 1981.

SCEP, Study of critical environmental problems, in *Man's Impact on the Global Environment,* MIT Press, Cambridge, MA, 1983.

Sorensen, H., *Energy Conversion Systems*, John Wiley & Sons, New York, 1983.

U.S. Congress, Office of Technology Assessment, *Studies of the Environmental Costs of Electricity,* OTA/ETI-134, U.S. Government Printing Office, Washington, D.C., September 1994.

U.S. Department of Energy, Energy technologies and the environment, *Environmental Information Handbook,* DOE/EH-007, October 1988.

U.S. Energy Information Administration, *International Energy Annual 1996*, DOE/EIA-0219 (96), Washington, D.C., February 1998.

U.S. Energy Information Administration, *International Energy Outlook 1996*. DOE/EIA-0484 (98), Washington, D.C., April 1998.

U.S. Energy Information Administration, *Natural Gas Monthly*, August 1997. DOE/EIA-0130 (97/08), Washington, D.C., August 1997.

U.S. Energy Information Administration, *Performance Profiles of Major Energy Producers 1996*, DOE/EIA-0206 (96), Washington, D.C., February 1998.

U.S. Energy Information Administration, *Annual Energy Outlook 1998*, DOE/EIA-0383 (98), Washington, D.C., December 1998.

U.S. Energy Information Administration, *Annual Energy Outlook 1997*, DOE/EIA-0383 (97), Washington, D.C., December 1997.

U.S. Environmental Protection Agency, *Environmental Outlook 1980*, Washington, D.C., 1980.

U.S. Environmental Protection Agency, Office of Policy and Resource Management, *Environmental Mathematical Pollutant Fate Modeling Handbook/Catalogue* (draft), Contract no. 68-01-5146, Washington, D.C., 1982.

Wagner, P., *Coal Processing Technology*, American Institute of Chemical Engineers, 1980.

Washington Policy and Analysis, *Natural Gas: A Strategic Resource for the Future*, Washington, D.C., 1988.

Whitmore, R.W., *Methodology for Characterization of Uncertainty in Exposure Assessments*, Research Triangle Institute, Research Triangle Park, NC, 1985.

Zitko, V., Graphical display of environmental quality criteria, *Sci. Total Environ.*, 72:217–220, 1988.

CHAPTER 2

Alter, M.J., Mast, E.E., The epidemiology of viral hepatitis in the United States, *Gastrointestinal Clin. N. Am.*, 23:437–455, 1994.

Ames, B.N., Identifying environmental chemicals causing mutation and cancer, *Science*, 204:587–592, 1979.

Angulo, F.J., Swerdlow, D.L., Bacterial enteric infections in persons infected with human immunodeficiency virus, *J. Clin. Infect. Dis.*, 21:S8493, 1995.

Bates, J., Jordens, J.Z., Griffiths, D.T., Farm animals as a putative reservoir for vancomycin-resistant enterococcal infection in man, *J. Antimicrobial Chemotherapy*, 34:507–516, 1994.

Centers for Disease Control and Prevention, Status report on the childhood immunization initiative: reported cases of selected vaccine-preventable diseases — United States, 1996, *MMWR*, 46:665–671, 1997.

Centers for Disease Control and Prevention, Reduced susceptibility of *Staphylococcus aureus* to vancomycin — Japan, 1996, *MMWR*, 46:624–626, 1997.

Centers for Disease Control and Prevention, *Staphylococcus aureus* with reduced susceptibility to vancomycin — United States, 1997, *MMWR*, 46:765–766, 1997.

Centers for Disease Control and Prevention, Preventing emerging infectious diseases, a strategy for the 21st century, Atlanta, GA, 1998. Centers for Disease Control and Prevention, Guidelines for Evaluating Surveillance Systems, *MMWR*, 37(S-5), 1988.

Centers for Disease Control and Prevention, Hepatitis A associated with consumption of frozen strawberries — Michigan, March 1997, *MMWR*, 46:288, 1997.

Council for Agricultural Science and Technology, Foodborne pathogens: risks and consequences, Task Force Report No. 122, September 1994.

Doll, R., Relevance of epidemiology to policies for the prevention of cancer, *Human Toxicol.,* 4(1):81–96, 1985.

Doyle, D.K., Teratology: a primer, *Neonatal Network,* 4:24–29, 1986.

Fein, G.G. et al., Environmental toxins and behavioral development, *Am. Psychol.,* 38(11):1188–1197, 1983.

Fein, G., *Intrauterine Exposure of Humans to PCB's: Newborn Effects,* U.S. EPA, Duluth, MN, 1984.

Fochtman, E.G., *Biodegredation and Carbon Adsorption of Carcinogenic and Hazardous Organic Compounds,* U.S. EPA, Cincinnati, OH, 1981.

Frederick, G.L., Assessment of teratogenic potential of therapeutic agents, *Clin. Invest. Med.,* 8(4):323–327, 1985.

Goldsmith, J.R., Epidemiology monitoring in environmental health, introduction, and overview, *Sci. Total Environ.,* 32:211–218, 1984.

Green, G., Ionizing radiation and its biological effects, *Occupat. Health,* 38:354–357, 1986.

Gruenewald, R., Blum, S., Chan, J., Relationship betweeen human immunodeficiency virus infection and salmonellosis in 20–59-year-old residents of New York City, *J. Clin. Infec. Dis.,* 18:358–363, 1994.

Heinonen, O.P., Slone, D., Shapiro, S., *Birth Defects and Drugs in Pregnancy,* John Wright-PSG, Boston, MA, 1982.

Hendee, W.R., Real and perceived risks of radiation exposure, *West. J. Med.,* 138:380–386, 1983.

Hinman, A., 1889 to 1989: a century of health and disease, *Public Health Rep.,* 105:374–380, 1990.

Hodgson, E., *Introduction to Biochemical Toxicology,* Elsevier, New York, 1980.

Hospital Infections Program, NCID, CDC, PHS, DHHS, national nosocomial infections surveillance report, data summary from October 1986–April 1996, issued May 1996, *Am. J. Inf. Control,* 24:380–388, 1996.

Inskip, H., Beral, V., Fraser, P., Haskey, J., Epidemiological monitoring: methods for analyzing routinely-collected data, *Sci. Total Environ.,* 32:219–232, 1984.

Institute of Medicine, *Antimicrobial Resistance: Issues and Options,* National Academy Press, Washington, D.C., 1998.

Institute of Medicine, *Emerging Infections: Microbial Threats to Health in the United States,* National Academy Press, Washington, D.C., 1994.

Jefrey, A.M., DNA modification by chemical carcinogens, *Pharmacol. Terminol.,* 28:237–241, 1985.

Khoury, D.L. *Coal Cleaning Technology,* Noyes Data Corp., Parkridge, NJ, 1981.

Khoury, M.J., Holtzman, N.A., On the ability of birth defects monitoring to detect New teratogens, *Am. J. Epidemiol.* 126(1):136–143, 1987.

Lederberg, J., Shope, R.E., Oaks, S.C., Jr., Eds., *Microbial Threats to Health in the United States,* National Academy Press, Washington, D.C., 1992.

Legator, M.S., Rosenberg, M., Zenick, H., *Environmental Influences on Fertility, Pregnancy, and Development,* Alan R. Liss, New York, 1984.

Lippmann, M., Schlesinger, R.B., *Chemical Contamination in the Human Environment,* Oxford University Press, New York, 1980.

MacDonald, S.C., Pertowski, C.A., Jackson, R.J., Environmental public health surveillance, *J. Public Health Manage. Pract.,* 2:45–49, 1996.

Marcus, J., Hans, S.L., Neurobehavorial toxicology and teratology, Department of Psychiatry, University of Chicago, 1982.

Mattila, K.J., Valtonen, V.V., Nieminen, M.S., Asikainen, S., Role of infection as a risk factor for atherosclerosis, myocardial infarction, and stroke, *J. Clin. Infec. Dis.,* 26:719–734, 1998.

Meriweather, R.A., Blueprint for a national public health surveillance system for the 21st century, *J. Public Health Manage. Pract.,* 2:16–23, 1996.

Meyers, V.K., Chemicals which cause birth defects — teratogens a special concern of research chemists, *Sci. Total Environ.,* 32:1–12, 1983.

Monath, T., Arthropod-borne viruses, in *Emerging Viruses,* Morse, S., Ed., Oxford University Press, New York, 1993, p. 144.

Morgenstern, H., Uses of ecologic analysis in epidemiological research, *Am. J. Public Health,* 72:1336–1344, 1982.

Nau, H., Species differences in pharmacokinetics and drug teratogenesis, *Environ. Health Perspec.,* 70:113–129, 1986.

Neubert, D. et al., Principles and problems in assessing prenatal toxicity, *Arch. Toxicol.,* 60:238–245, 1987.

Office of Science and Technology Policy fact sheet: Addressing the threat of emerging infectious diseases, The White House, Washington, D.C., June 12, 1996.

Oftedal, P., Brogger, A., *Risk and Reason — Risk Assessment in Relation to Environmental Mutagens and Carcinogens,* Alan R. Liss, New York, 1986.

Porter, S.B., Sande, M.A., Toxoplasmosis of the central nervous system in the acquired immunodeficiency syndrome, *N. Engl. J. Med.,* 327:1643–1648, 1992.

Rosenberg, M.J., Halperin, W.E., The role of surveillance in monitoring reproductive health, *Teratogen. Carcinogen. Mutagen.,* 4(1):15–24, 1984.

Rowland, M., Tozer, T., *Clinical Pharmacokinetics: Concepts and Applications,* Lea & Febiger, Philadelphia, 1980.

Schottenfeld, D., Fraumeni, J., Jr., *Cancer Epidemiology and Prevention,* W.B. Saunders, Philadelphia, 1982.

Slutsker, L., Ries, A.A., Greene, K.D. et al., *Escherichia coli* O157:H7 in the United States: clinical and epidemiological features, *Annu. Intern. Med.,* 126:505–513, 1997.

Smith, A.W., Skilling, D.E., Cherry, N., Mead, J.H., Matson, D.O., Calicivirus emergence from ocean reservoirs: zoonotic and interspecies movements, *J. Emerg. Infect. Dis.,* 4:13–20, 1998.

Sorvillo, F.J., Lieb, L.E., Waterman, S.H., Incidents of Campylobacteriosis among patients with AIDS in Los Angeles County, *J. Acquired Immune Deficiency Syndro. Hum. Retrovirol.,* 4:598–602. 1991.

Swaab, D.F., Mirmiran, M., Functional teratogenic effects of chemicals on the developing brain, *Monogr. Neural Sci.,* 12:45–57, 1986.

Tappero, J., Schuchat, A., Deaver, K.A., Mascola, L., Wenger, J.D., Reduction in Listeriosis: effectiveness of prevention efforts? *J. Am. Med. Assoc.,* 273:1118–1122, 1995.

Thacker, S.B., Stroup, D.F., Future directions for comprehensive public health surveillance and health information systems in the United States, *Am. J. Epistem.,* 14:383–397, 1994.

Thacker, S.B., Stroup, D.F., Buehler, J.W., Public health surveillance, in *Oxford Textbook of Public Health,* 3rd ed., Detels, R., Holland, W.W., McEwan, J., Omenn, G.S., Eds., Oxford University Press, New York, 1997, pp. 35–50.

Ujino, Y., Epidemiological studies on disturbances of human fetal development, *Arch. Environ. Health,* 40:181–184, 1985.

Ulsamer, A.G., White, P.D., Preuss, P.W., Evaluation of carcinogens: perspective of the consumer product safety commission, in *Handbook of Carcinogen Testing,* Milman, H.A., Weisburger, E.K., Eds., Noyes Publications, Park Ridge, NJ, 1985.

United Nations Program on HIV/AIDS and the World Health Organization, AIDS epidemic update: December 1998, World Health Organization, Geneva, Switzerland, 1999.

U.S. Congress, Office of Technology Assessment, Impacts of antibiotic resistant bacteria (OTA-H-29), U.S. Government Printing Office, Washington, D.C., 1995.

U.S. Department of Agriculture, Pathogen reduction; hazard analysis and critical control point systems; proposed rule, *Fed. Regist.*, 60:6774–6889, 1995.

U.S. Department of Health and Human Services, National Toxicology Program, Explanation of levels of evidence of carcinogenic activity, 1986.

U.S. Department of Health and Human Services, National Toxicology Program, Review of current DHHS, DOE, and EPA research related to toxicology, fiscal year 1986, Publ. No. NTP-86-087, Research Triangle Park, NC, 1987.

U.S. Environmental Protection Agency, Guidelines for carcinogen risk assessment, *Fed. Regist.*, 51:33992, 1986.

U.S. Environmental Protection Agency, Guidelines for mutagenicity risk assessment, *Fed. Regist.*, 51:34006, 1986.

U.S. Environmental Protection Agency, Risk assessment and management: framework for decision making, Washington, D.C., 1984.

U.S. Environmental Protection Agency, U.S. Department of Agriculture, Food safety from farm to table: a national food-safety initiative, Government Printing Office, Washington, D.C., May 1997.

U.S. Public Health Service, Healthy people 2000 — a mid-course review and 1995 revisions. U.S. Department of Health and Human Services, Public Health Service, Washington, D.C., 1995.

Vainio, H., Current trends in the biological monitoring of exposure to carcinogens, *Scand. J. Environ. Health*, 11:1–3, 1985.

Wartak, J., *Clinical Pharmacokinetics*, University Park Press, Baltimore, 1983.

Wilson, M.E., Travel and the emergence of infectious diseases, *J. Emerg. Infect. Dis.*, 1:39–46, 1995.

Winter, M.E., *Basic Clinical Pharmacokinetics*, Applied Therapeutics, San Francisco, CA, 1980.

Working Group on Emerging and Re-emerging Infectious Diseases, Committee on International Science, Engineering, and Technology, National Science and Technology Council, Infectious disease — a global threat, U.S. Government Printing Office, Washington, D.C., 1995.

Yip, R., Sharp, T.W., Acute malnutrition and high childhood mortality related to diarrhea, *J. Am. Med. Assoc.*, 270:587–590, 1993.

CHAPTER 3

Abbot, J., Analysis of the response of *Saccharomyces cerevisiae* cells to *Kluyveromyces lactis* toxin, *J. Gen. Microbiol.*, 137(1757): 15–18, 1991.

Abdul-Raouf, U.M., Beuchat, L.R., Ammar, M.S., Survival and growth of *Escherichia coli* O157:H7 on salad vegetables, *Appl. Environ. Microbiol.*, 59:1999–2006, 1993.

Akinboade, O.A., Hassan, J.O., Adejinmi, A., Public health importance of market meat exposed to refuse flies and air-borne microorganisms, *Int. J. Zoonoses*, 11(1):111–114, 1984.

Allen, J.E., Maizels, J.M., Immunology of human helminth infection, *Int. Arch. Allergy Immunol.*, 109:3–10, 1996.

Baddour, L.M., Gaia, S.M., Griffin, R., Hudson, R., A hospital cafeteria-related food-borne outbreak due to *Bacillus cereus*: unique features, *Infect. Control*, 7(9):4652–4655, 1986.

Bakka, R.L., Making the right choice — cleaners, Ecolab, Inc., Food and Beverage Division, St. Paul, MN, 1995.

Barrett, T.J., Lior, H., Green, J.H. et al., Laboratory investigation of a multistate foodborne outbreak of *Escherichia coli* O157:H7 by using pulsed-field gel electrophoresis and phage typing, *J. Clin. Microbiol.,* 32:3013–3017, 1994.

Bean, N.H., Goulding, J.S., Lao, C., Angulo, F.J., Surveillance of foodborne disease outbreaks — United States, 1988–1992, *MMWR,* 45(SS-5):1–55, 1996.

Behar, S.M., Porcelli, S.A., Mechanisms of autoimmune disease induction: the role of the immune response to microbial pathogens, *Arthritis Rheumatol.,* 38:458–476, 1995.

Bettelheim, K.A., Evangelidis, H., Pearce, J.L., Sowes, E., Strockbine, N.A., Isolation of *Citrobacter freundii* strain which carries the *Escherichia coli* O157 antigen, *J. Clin. Microbiol.,* 31:760–761, 1993.

Birkhead, G.S., Morse, D.L., Levine, W.C., Fudala, J.K., Kondracki, S.F., Chang, H. et al., Typhoid fever at a resort hotel in New York: a large outbreak with an unusual vehicle, *J. Infect. Dis.,* 167:1228–1232, 1993.

Bitzan, M., Ludwig, K., Klemt, M., Konig, H., Buren, J., Muller-Wiefel, D.E., The role of *Escherichia coli* O157 infections in the classical (enteropathic) haemolytic uraemic syndrome: results of a Central European, multicenter study, *Epidemiol. Infect.* 110:183–196, 1993.

Blumenthal, D., Food irradiation toxic to bacteria, safe for humans, *FDA Consumer,* 11–15, November 1990.

Bockemuhl, J., Aleksic, S., Karch, H., Serological and biochemical properties of Shiga-like toxin (Verocytotoxin)-producing strains of *Escherichia coli,* other than O-group 157, from patients in Germany, *Zentralbl. Bakteriol.,* 276:189–195, 1992.

Bolton, C.F., The changing concepts of Guillain-Barré syndrome, *N. Engl. J. Med.,* 333:1415–1417, 1995.

Boren, T., Falk, P., Roth, K.A., Larson, G., Normark, S., Attachment of *Helicobacter pylori* to human gastric epithelium mediated by blood group antigens, *Science,* 262:1892–1895, 1993.

Boufford, T., Making the right choice–sanitizers, Ecolab, Inc., Food and Beverage Division, St. Paul, MN, 1996.

Boyce, T.G., Pemberton, A.G., Addiss, D.G., *Cryptosporidium* testing practices among clinical laboratories in the United States, *Pediatr. Infect. Dis. J.,* 15:87–88, 1996.

Brewster, D.H., Browne, M.I., Robertson, D., Houghton, G.L., Bimson, J., Sharp, J.C.M., An outbreak of *Escherichia coli* O157:H7 associated with a children's paddling pool, *Epidemiol. Infect.,* 112:441–447, 1994.

Briley, R.T., Teel, J.H., Fowler, J.P., Investigation and control of a *Shigella sonnei* outbreak in a day care center, *J. Environ. Health,* 56(6):23–25, 1994.

Brockman, R.A., Lenaway, D.D., Humphrey, C.D., Norwalk-like viral gastroenteritis: a large outbreak on a university campus, *J. Environ. Health,* 57(10):19–22, 1995.

Brown, P., Kidd, D., Riordan,T., Barrell,A., An outbreak of foodborne *Campylobacter jejuni* infection and the possible role of cross-contamination, *J. Infect.,* 17(2):171–176, 1988.

Bruckner, D.A., Amebiasis, *Clin. Rev. Microbiol.,* 5:356–369, 1992.

Bull, P., *Prev. Food-Borne Dis.,* 22(2):207–208, 1988.

Bunning, V.K., Immunopathogenic aspects of foodborne microbial disease, *Food Microbiol.,* 11:89–95, 1994.

Bunning, V.K., Lindsay, J.A., Archer, D.L., Chronic health effects of foodborne microbial disease, *World Health Stat. Q.,* 50:51–56, 1997.

Burslem, C.D., Kelly, M.J., Preston, F.S., Food poisoning — a major threat to airline operations, *J. Soc. Occup. Med.*, 40:97–100, 1990.

Bush, M.F., The symptomless *Salmonella* excretor working in the food industry, *Community Med.*, 7(2):133–135, 1985.

Buzby, J.C., Roberts, T., Economic costs and trade impacts of microbial foodborne illness, *World Health Stat. Q.*, 50(1/2):57–66, 1997.

Campbell, M.E., Gardner, C.E., Dwyer, J.J., Isaacs, S.M., Krueger, P.D., Ying, J.Y., Effectiveness of public health interventions in food safety: a systematic review, *Can. J. Public Health*, 89(3):197–202, 1998.

Campbell, M., Sahai, V., Vettoretti, I., Northan, A., *Shigella sonnei* outbreak in Minde-moya, Ontario, *PHERO*, February 26:40–44, 1993.

Cebula, T.A., Payne, W.L., Feng, P., Simultaneous identification of *Escherichia coli* of the O157:H7 serotype and their Shiga-like toxin type by MAMA/multiplex PCR, *J. Clin. Microbiol.*, 33:248–250, 1995.

CDC, Foodborne hepatitis A — Alaska, Florida, North Carolina, Washington, *MMWR*, 39(14):228–232, 1990.

CDC, Foodborne hepatitis A — Missouri, Wisconsin, and Alaska, 1990–1992, *MMWR*, 42(27):526–529, 1993.

CDC, Foodborne outbreak of diarrheal illness associated with *Cryptosporidium parvum* — Minnesota, *MMWR*, 44:783–784, 1996.

CDC, Outbreaks of *Escherichia coli* O157:H7 infection and cryptosporidiosis associated with drinking unpasteurized apple cider — Connecticut and New York, October 1996, *MMWR*, 46:4–8, 1997.

CDC, *Vibrio vulnificus* infections associated with raw oyster consumption — Florida, 1981–1992, *MMWR*, 42:405–407, 1993.

Center for Science in the Public Interest, Dine at your own risk: the failure of local agencies to adopt and enforce national food safety standards for restaurants, The Center, Washington, D.C., 1996.

Centers for Disease Control and Prevention, *Escherichia coli* O157:H7 infections associated with eating a nationally distributed commercial brand of frozen ground beef patties and burgers — Colorado, 1997. *MMWR*, 46:777–778, 1997.

Centers for Disease Control and Prevention, Foodborne outbreak of diarrheal illness associated with *Cryptosporidium parvum* — Minnesota, 1995, *MMWR*, 45(36):783–784, 1996.

Centers for Disease Control and Prevention, Outbreaks of *Salmonella enteritidis* gastroenteritis — California, 1993.

Centers for Disease Control and Prevention, Outbreak of *Vibrio parahaemolyticus* infections associated with eating raw oysters — Pacific Northwest, 1997, *MMWR*, 47:457–462, 1998.

Centers for Disease Control and Prevention, *Salmonella* surveillance: annual tabulation summary, 1997, U.S. Department of Health and Human Services, CDC, Atlanta, GA, 1998.

Centers for Disease Control and Prevention, Surveillance for foodborne disease outbreaks, United States, 1988–1992, *MMWR, CDC Surveill. Summ.* 45:1–66, 1996.

Chapman, P.A., Siddon, C.A., Zadik, P.M., Jewes, L., An improved selective medium for the isolation of *Escherichia coli* O157:H7, *J. Med. Microbiol.*, 35:107–110, 1991.

Chappell, C.L., Matson, C.C., *Giardia* antigen detection in patients with chronic gastrointestinal disturbances, *J. Fam. Pract.*, 35:49–53, 1992.

Chart, H., Cheasty, T., Cope, D., Gross, R.J., Rowe, B., The serological relationship between *Yersinia enterocolitica* O9 and *Escherichia coli* O157 using sera from patients with yersiniosis and haemolytic uraemic syndrome, *Epidemiol. Infect.*, 107:349–356, 1991.

Chen, Z., Wang, B., Xu, H. et al., Spiral shaped bacteria in the human gastric biopsy, *Hua-Hsi I Ko Ta Hsueh Hsueh Pao*, 24:392–394, 1993.

Cody, S.H., Glynn, M.K., Farrar, J.A. et al., An outbreak of *Escherichia coli* O157:H7 infection from unpasteurized commercial apple juice, *Ann. Intern. Med.*, 130:202–209, 1999.

Cohen, M.L., Tauxe, R.V., Drug-resistant *Salmonella* in the United States: an epidemiologic perspective, *Science*, 234(4779):964–969, 1986.

Cook, K.A., Swerdlow, D., Dobbs, T. et al., Fresh-squeezed *Salmonella*: an outbreak of *Salmonella hartford* associated with pasteurized orange juice — Florida {Abstract}, *EIS Conference Abstract*, 38–39, 1996.

Conrad, B., Widmann, E., Trucco, G., Rudert, W.A., Behboo, R., Ricordi, C. et al., Evidence for superantigen involvement in insulin dependent diabetes mellitis aetiology, *Nature* (London), 371:351–355, 1994.

Cook, K., Boyce, T., Langkop, C. et al., A multistate outbreak of *Shigella flexneri* 6 traced to imported green onions, presented at the 35th Interscience Conference on Antimicrobial Agents and Chemotherapy, San Francisco, CA, September 1995.

Corbe, S., Barton, P., Nair, R.C., Dulbert, C., Evaluation of the effect of frequency of inspection on the sanitary conditions of eating establishments, *Can. J. Public Health*, 75(6):434–438, 1984.

Coulombe, R.A., Biological actions of mycotoxins, *J. Dairy Sci.*, 76:880–891, 1993.

Cullen, D.J.E., Collins, B.J., Christiansen, B.J. et al., When is *Helicobacter pylori* infection acquired?, *Gut*, 34:1681–1682, 1993.

Derouin, F., Pathology and immunological control of toxoplasmosis, *Braz. J. Med. Biol. Res.*, 25:1163–1169, 1992.

de Wit, J.C., Kampelmacher, E.H., Some aspects of bacterial contamination of hands of workers in food service establishments, *Zentralbl. Bakteriol. Mikrobiol. Hyg. B.*, 186(1):45–54, 1988.

Dowell, S.F., Groves, C., Kirkland, K.B. et al., A multistate outbreak of oyster-associated gastroenteritis: implications for interstate tracing of contaminated shellfish, *J. Infec. Dis.*, 171:1497–1503, 1995.

Doyle, M.P., Schoeni, J.L., Survival and growth characteristics of *Escherichia coli* associated with hemorrhagic colitis, *Appl. Environ. Microbiol.*, 48:855–856, 1984.

Drumm, B., Perez-Perez, G.I., Blaser, M.J., Sherman, P.M., Intrafamilial clustering of *Helicobacter pylori* infection, *N. Engl. J. Med.*, 322:359–363, 1990.

E. coli in salami possibly linked to illness outbreaks, *Food Chem. News*, 36:19, 1994.

Estes, M.K., Atmar, R.L., Hardy, M.E., *Norwalk and Related Diarrhea Viruses in Clinical Virology*, Churchill Livingstone, New York, 1997, 1073–1095.

Farkas, J., *Irradiation of Dry Food Ingredients*, CRC Press, Boca Raton, FL, 1988.

Farmer, J.J., Davis, B.R., H7 antiserum-sorbitol fermentation medium: a single-tube screening medium for detecting *Escherichia coli* O157:H7 associated with hemorrhagic colitis, *J. Clin. Microbiol.*, 22:620–625, 1985.

Farr, S.B., Kogoma, Y., Oxidative stress response in *Escherichia coli* and *Salmonella typhimurium*, *Microbiol. Rev.*, 55:561–585, 1991.

Fauchere, J., Blaser, M.J., Adherence of *Helicobacter pylori* cells and their surface components to HeLa cell membranes, *Microb. Pathol.*, 9:427–439, 1990.

Feng, P., Identification of *Escherichia coli* O157:H7 by DNA probe specific for an allele of uidA gene, *Mol. Cell. Probes*, 7:151–154, 1993.

Feng, P., Hartman, P.A., Fluorogenic assays for immediate confirmation of *Escherichia coli*, *Appl. Environ. Microbiol.*, 43:1320–1329, 1982.

Ferguson, D.A., Li, C., Patel, N.R., Mayberry, W.R., Chi, D.S., Thomas, E., Isolation of*Helicobacter pylori* from saliva, *J. Clin. Microbiol.,* 31:2802–2804, 1993.

Flegr, J., Hrdy, I., Influence of chronic toxoplasmosis on some human personality factors, *Folia. Parasitol (Praha),* 41:122–126, 1994.

Flegr, J., Zitkova, S., Kodym, P., Frynta, D., Induction of changes in human behavior by the parasitic protozoan *Toxoplasma gondii, Parasitology,* 113:49–54, 1996.

Foegeding, P.M., Foodborne pathogens: risks and consequences, Council for Agricultural Science and Technology, Task Force Report, 1221, 1994.

Food and Drug Administration, Food code 1997, U.S. Department of Health and Human Services, Public Health Service, Food and Drug Administration, Rockville, MD, 1997.

Food and Drug Administration, Hazard analysis and critical control point; procedures for the safe and sanitary processing and importing of juice; food labeling; warning notice statements; labeling of juice products; proposed rules, *Fed. Regist.,* 63:20449–20486, 1998.

Food Safety and Inspection Service, FSIS/CDC/FDA sentinel site study: the establishment and implementation of an active surveillance system for bacterial foodborne diseases in the United States, USDA Report to Congress, Washington, D.C., 1997.

Foodborne and Diarrheal Diseases Branch, *Shigella* surveillance: annual tabulations summary, 1996, U.S. Department of Health and Human Services, Public Health Service, CDC, National Center for Infectious Diseases, Division of Bacterial and Mycotic Diseases, Foodborne and Diarrheal Diseases Branch, Atlanta, GA, 1997.

Fox, J.G., Non-human reservoirs of *Helicobacter pylori, Aliment. Pharmacol. Ther.,* 9 (suppl. 2):93–103, 1995.

Fox, J.G., Yan, L.L., Dewhirst, F.E. et al., *Helicobacter bilis* sp. nov., a novel *Helicobacter* species isolated from bile, livers, and intestines of aged, inbred mice., *J. Clin. Microbiol.,* 33:45–54, 1995.

Fratamico, P.M., Buchanan, R.L., Cook, P.H., Virulence of an *Escherichia coli* O157:H7 sorbitol-positive mutant, *Appl. Environ. Microbiol.,* 59:245–252, 1993.

FSIS pathogen reduction/HACCP, *Fed. Regist.,* Washington, D.C., July 6, 1996.

Gay, C.T., Marks, W.A., Riley, H.D., Bodensteiner, J.B., Hamza, M., Noorani, P.A. Bobele, G.B., Infantile botulism, *South. Med. J.,* 81(4):457–460, 1988.

Gelderblom, W.C.A., Effects of temperature and incubation period on production of *Fumonisin* B1 by *Fusarium moniliform, Appl. Environ. Microbiol.,* 56(1733):75–78, 1990.

Glass, K.A., Loeffelholz, J.M., Ford, J.P., Doyle, M.P., Fate of *Escherichia coli* O157:H7 as affected by pH or sodium chloride and in fermented, dry sausage, *Appl., Environ. Microbiol.,* 58:2513, 1992.

Goodacre, J.A., Brownlee, C.E.D., Ross, D.A., Bacterial superantigens in autoimmune arthritis, *Br. J. Rheumatol.,* 33:413–419, 1994.

Gunzer, F., Bohm, H., Russman, H., Bitzan, M., Aleksic, S., Karch, H., Molecular detection of sorbitol-fermenting *Escherichia coli* O157:H7 in patients with hemolytic-uremic syndrome, *J. Clin. Microbiol.,* 30:1807–1810, 1992.

Handt, L.K., Fox, J.O., Dewhirst, F.E. et al., *Helicobacter pylori* isolated from the domestic cat: public health implications, *Infect. Immun.,* 62:2367–2374, 1994.

Hayashi, S., Chan, C.C., Gazzinelli, R., Roberge, F.G., Contribution of nitric oxide to the host parasite equilibrium in toxoplasmosis, *J. Immunol.,* 156:1476–1481, 1996.

Hedberg, C.W., Levine, W.C., White, K.E., Carlson, R.H., Winsor, D.K., Cameron, D.N. et al., An international foodborne outbreak of shigellosis associated with a commercial airline., *JAMA,* 268(22):3208–3212, 1992.

Hedberg, C.W., White, K.E., Johnson, J.A., Edmonson, L.M., Soler, J.T., Korlath, J.A. et al., An outbreak of *Salmonella enteritidis* infection at a fast-food restaurant: implications for food handler-associated transmission, *J. Infect. Dis.,* 164:1135–1140, 1991.

Heilmann, K.L., Borchard, F., Gastritis due to spiral shaped bacteria other than *Helicobacter pylori:* clinical, histological and ultrastructural findings, *Gut,* 32:137–140, 1991.

Hemalatha, S.G., Drumm, B., Sherman, P.J., Adherence of *Helicobacter pylori* to human gastric epithelial cells *in vitro, Med. Microbiol. Immunol.,* 35:197–202, 1991.

Hepatitis A in restaurant clientele and staff — Quebec, *Can. Comm. Dis. Rep.,* April 1998.

Herwaldt, B.L., Ackers, M.L., The Cyclospora Working Group, An outbreak in 1996 of cyclosporiasis associated with imported raspberries, *N. Engl. J. Med.,* 336:1548–1556, 1997.

Herwaldt, B.L., Lew, J.F., Moe, C.L., Lewis, D.C., Humphrey, C.D., Monroe, S.S., Characterization of a variant strain of Norwalk virus from a foodborne outbreak of gastroenteritis on a cruise ship in Hawaii, *J. Clin. Microbiol.,* 32(4):861–866, 1994.

Hill, G.W., Levels of bacteria, fungi, and endotoxin in bulk and aerosolized corn silage, *Appl. Environ. Microbiol.,* 55(1039):200–210, 1989.

Hlady, W.G., Klontz, K.C., the epidemiology of *Vibrio* infections in Florida, 1981–1993, *J. Infect. Dis.,* 173:1176–1183, 1996.

Isaacson, P.G., Spencer, J., Is gastric lymphoma an infectious disease?, *Hum. Pathol.,* 24:569–570, 1993.

Jackson, L.S., De Vries, J.W., Bullerman, L.B., Eds., *Fumonisins in Food,* Plenum Press, New York, 1996.

Jeppesen, C., Media for *Aeromonas* spp., *Plesiomonas shigelloides* and *Pseudomonas* spp. food and environment, *Int. J. Food Microbiol.,* 26:25–41, 1995.

Johnson, H.M., Torres, B.A., Soos, J.M., Superantigens: structure and relevance to human disease, *Proc. Soc. Exp. Biol. Med.,* 212:99–109, 1996.

Kay, R.A., The potential role of superantigens in inflammatory bowel disease, *Clin. Exp. Immunol.* 100:4–6, 1995.

Kapasi, K., Inman, R.D., HLA-B27 expression modulates Gram-negative bacterial invasion into transfected L-cells, *J. Immunol.,* 148:3554-3559, 1992.

Keene, W.E., McAnulty, J.M., Hoesly, F.C. et al., A swimming-associated outbreak of hemorrhagic colitis caused by *Escherichia coli* O157:H7 and *Shigella sonnei, N. Engl. J. Med.,* 331:579–584, 1994.

Keusch, G.T., Hamer, D., Joe, A., Kelley, M., Griffiths, J., Ward, H., *Cryptosporidia* — who is at risk, *Schweiz. Med. Wochenschr.,* 125:899–908, 1995.

Khalil, S.S., Rashwan, E.A., Tumor necrosis factor-alpha (TNF-alpha) in human toxoplasmosis, *J. Egypt. Soc. Parisitol.,* 26:53–61, 1996.

Klein, P.D., Graham, D.Y., Gaillour, A., Opekun, A.R., Smith, E.O., Gastrointestinal Physiology Working Group, Water source as risk factor for *Helicobacter pylori* infection in Peruvian children, *Lancet,* 337:1503–1506, 1991.

Knight, K.G., Effect of temperature on aflatoxin production in Mucuna pruriens seeds, *Appl. Environ. Microbiol.,* 55(532):70–75, 1989.

Kotb, M., Bacterial pyrogenic exotoxins as superantigens, *Clin. Microbiol. Rev.,* 8:41–126, 1995.

Kotb, M., Infection and autoimmunity: a story of the host, the pathogen and the copathogen, *Clin. Immunol. Immunopathol.* 74:10–22, 1995.

Kuipers, E.J., Uyterlinde, A.M., Pena, A.S., Roosendaal, R., Pals, G., Nelis, G.F. et al., Long term sequelae of *Helicobacter pylori* gastritis, *Lancet,* 345:1525–1528, 1995.

Lange, W.R., *Ciguatera* fish poisoning, *Am. Fam. Physician,* 50:579–584, 1994.

Latman, N.S., Walls, R., Personality and stress: an exploratory comparison of rheumatoid arthritis and osteoarthritis, *Arch. Phys. Med. Rehabil.*, 77:796–800, 1996.

Lee, R., Peppe, J., George, H., Pulsed-field gel electrophoresis of genomic digests demonstrates linkages among food, food handlers, and patrons in a foodborne *Salmonella javiana* outbreak in Massachusetts, *J. Clin. Microbiol.* 36(1):284–285, 1998.

Leung, D.Y., Meissner, H.C., Fulton, D.R., Murray, D.L., Kotzin, B.L., Schlievert, P.M., Toxic shock syndrome toxin secreting *Staphylococcus aureus* in Kawasaki syndrome, *Lancet*, 342:1385–1388, 1993.

Lew, J.F., Swerdlow, D.L., Dance, M.E., Griffin, P.M., Bopp, C.A., Gillenwater, M.J. et al., An outbreak of shigellosis aboard a cruise ship caused by a multiple-antibiotic-resistant strain of *Shigella flexneri*, *Am. J. Epidemiol.*, 134(4):413–420, 1991.

Liu, Y., van Kruiningen, H.J., West, A.B., Cartun, R.W., Cortot, A., Colombel, J.F., Immunocytochemical evidence of *Listeria, Escherichia coli* and *Streptococcus* antigens in Crohn's disease, *Gastroenterology*, 108:1396–1401, 1995.

Lo, S.V., Connolly, A.M., Palmer, S.R., Wright, D., Thomas, P.D., Joynson, D., The role of the presymptomatic food handler in a common source outbreak of foodborne SRSV gastroenteritis in a group of hospitals, *Epidemiol. Infect.* 113:513–521, 1994.

Lossos, I.S., Felenstein, I., Breuer, R., Engelhard, D., Foodborne outbreak of Group A-hemolytic streptococcal pharyngitis, *Ach. Intern. Med.*, 152:853–855, 1992.

Luo, G., Seetharamaiah, G.S., Niesel, D.W., Zhang, H., Peterson, J.W., Prabhakar, B.S., Klimpel, G.R., Purification and characterization of *Yersinia enterocolitica* envelope proteins which induce antibodies that react with human thyrotropin receptor, *J. Immunol.*, 152:2555–2561, 1994.

MacKenzie, W.R., Schell, W.L., Blair, K.A. et al., Massive outbreak of waterborne *Cryptosporidium* infection in Milwaukee, Wisconsin: recurrence of illness and risk of secondary transmission, *Clin. Infect. Dis.* 21:57–62, 1995.

March, S.B., Ratnam, S., Sorbitol-MacConkey medium for detection of *Escherichia coli* O157:H7 associated with hemorrhagic colitis, *J. Clin. Microbiol.*, 23:869–872, 1986.

Mariani-Kurkdijian, P., Denamur, E., Milon, A., Picard, B., Cave, H., Lambert-Zechovsky, N. et al., Identification of *Escherichia coli* O103:H2 as a potential agent of hemolytic-uremic syndrome in France, *J. Clin. Microbiol.*, 31:296–301, 1993.

Marriott, N.G., *Principles of Food Sanitation*, Chapman & Hall, New York, 1994, pp. 85–113; 114–166.

Martin, D.L., Gustafson, T.L., Pelosi, J.W., Suarez, L., Pierce, G.V., Contaminated produce — a common source for two outbreaks of *Shigella* gastroenteritis, *Am. J. Epidemiol.*, 124(2):299–305, 1986.

Martin, S.M., Bean, N.H., Data management issues for emerging diseases and new tools for managing surveillance and laboratory data, *Emerg. Infect. Dis.*, 1:124–128, 1995.

Mascola, L., Lieb, L., Chiu, J., Fannin, S.L., Linnan, M.J., *Listeriosis*: an uncommon opportunistic infection in patients with acquired immunodeficiency syndrome. A report of five cases and a review of the Literature, *Am. J. Med.*, 84(1):162–164, 1988.

Mazzuchelli, L., Wilder-Smith, C.H., Ruchti, C., Meyer-Wyss, B., Merki, H.S., *Gastrospirillum hominis* in asymptomatic, healthy individuals, *Dig. Dis. Sci.*, 38:2087–2089, 1993.

Meehan, P.J., Atkeson, T., Kepner, D.E., Melton, M., A foodborne outbreak of gastroenteritis involving two different pathogens, *Am. J. Epidemiol.*, 136:611–616, 1992.

Millar, D., Ford, J., Sanderson, J., Withey, S., Tizard M., Doran, T., Hermon-Taylor, J., IS900 PCR to detect *Mycobacterium paratuberculosis* in retail supplies of whole pasteurized cow's milk in England and Wales, *Appl. Environ. Microbiol.*, 62:3446–3452, 1996.

Millard, P.S., Gensheimer, K.F., Addiss, D.G. et al., An outbreak of cryptosporidiosis from fresh-pressed apple cider, *JAMA,* 272:1592–1596, 1994.

Miller, F.W., Genetics of autoimmune diseases, *Exp. Clin. Immunogenet.,* 12:182–190, 1995.

Mohle-Boetani, J.C., Stapleton, M., Finger, R. et al., Community wide Shigellosis: control of an outbreak and risk factors in child daycare centers, *Am. J. Pub. Health,* 85:812–816, 1995.

Morgan, D., Newman, C.P., Hutchinson, D.N., Walker, A.M., Rowe, B., Majid, F., Verotoxin producing *Escherichia coli* O157:H7 infections associated with the consumption of yogurt, *Epidemiol. Infect.,* 111:181–187, 1993.

Morris, A., Ali, M.R., Thomsen, L., Hollis, B., Tightly spiral shaped bacteria in the human stomach: another cause of active chronic gastritis?, *Gut,* 31:134–138, 1990.

Mouzin, E., Mascola, L., Tormey, M.P., Dassey, D.E., Prevention of *Vibrio vulnificus* infections: assessment of regulatory educational strategies, *JAMA,* 278:576–578, 1997.

Moyer, L., Warwick, M., Mahoney, F.J., Prevention of Hepatitis A virus infection, *Am. Fam. Physician,* 54:107–114, 1996.

Nastasi, A., Mammina, C., Villafrate, M.R., Scarlata, G., Massenti, M.F., Diquattro, M., Multiple Typing of *Salmonella typhimurium* isolates: an epidemiological study, *Microbiologica,* 11(3):173–178, 1988.

National Restaurant Association, in *1997 Restaurant Industry Pocket Factbook,* The Association, Washington, D.C., 1997.

Neal, G.E., Genetic implication in the metabolism and toxicity of mycotoxins, *Toxicol. Lett.,* 82–83:861–867, 1995.

Nelson, M., Wright, T.L., Case, M.A., Martin, D.R., Glass, R.I., Sangal, S.P., A protracted outbreak of foodborne viral gastroenteritis caused by Norwalk or Norwalk-like agent, *J. Environ. Health,* 54(5):50–55, 1992.

Nelson, P., Mycotoxicology: introduction to the mucology, plant pathology, chemistry, toxicology, and pathology of naturally occurring mycotoxicoses in animals and man, *Am. Scientist,* 78(471):130–133, 1990.

Olle, P., Bessieres, M.H., Malecaze, F., Seguela, J.P., The evolution of ocular toxoplasmosis in anti-interferon gamma treated mice, *Curr. Eye Res.,* 15:701–707, 1996.

Ollinger-Snyder, P., Matthews, M.E., Food safety issues: press reports heighten consumer awareness of microbiological safety, *Dairy, Food Environ. Sanitation,* 14:580, 1994.

Oomes, P.G., Jacobs, B.C., Hazenberg, M.P., Banffer, J.R., van der Meche, F.G., Anti-GM1 IgG antibodies and *Campylobacter* bacteria in Guillain-Barré syndrome: evidence of molecular mimicry, *Ann. Neurol.* 38:170–175, 1995.

Owen, R.J., Bacteriology of *Helicobacter pylori, Bailliere's Clin. Microbiol.,* 9:415–446, 1995.

Padhye, N.V., Doyle, M.P., *Escherichia coli* O157:H7: epidemiology, pathogenesis, and methods for detection in food, *J. Food Prot.,* 55:555–565, 1992.

Padhye, N.V., Doyle, M.P., Rapid procedure for detecting enterohemorrhagic *Escherichia coli* O157:H7 in food, *Appl. Environ. Microbiol.,* 57:2693–2698, 1991.

Parashar, U.D., Dow, L., Frankhauser, R.L., Humphrey, C.D., Miller, J., Ando, T. et al., An outbreak of viral gastroenteritis associated with consumption of sandwiches: implications for the control of transmission by food handlers, *Epidemiol. Infect.,* 121:615–621, 1998.

Parrish, M.E., Public health and nonpasteurized fruit juices, *Crit. Rev. Microbiol.,* 23:109–119, 1997.

Patterson, T., Hutchings, P., Palmer, S., Outbreak of SRSV gastroenteritis at an international conference traced to food handled by a post-symptomatic caterer, *Epidemiol. Infect.,* 111:157–162, 1993.

Perez-Perez, G.I., Dworkin, B.M., Chodos, J.E., Blaser, M.J., *Campylobacter pylori* antibodies in humans, *Ann. Intern. Med.,* 109:11–17, 1988.

Pestka, J.J., Bondy, G.S., Immunologic effects of mycotoxins, in *Mycotoxins in Grain,* Miller, J.D., Trenholm, H.L, Eds., Eagan Press, St. Paul, MN, 1994, pp. 339–358.

Peterson, M.C., Clinical aspects of *Campylobacter jejuni* infections in adults, *West. J. Med.* 161:148–152, 1994.

Pudymaitis, A., Armstrong, G., Lingwood, C.A., Verotoxin-resistant cell clones are deficient in the glycolipid globotriosylceramide: differential basis of phenotype, *Arch. Biochem. Biophys.,* 286:448–452, 1991.

Rahyaman, S.M., Chira, P., Koff, R.S., Idiopathic autoimmune chronic hepatitis triggered by Hepatitis A., *Am. J. Gastroenterol.* 89:106–108, 1994.

Rampling, A., *Salmonella enteritidis* five years on, *Lancet,* 342:317–318, 1993.

Rice, E.W., Sowers, E.G., Johnson, C.H., Dunnigan, M.E., Strockbine, N.A., Edberg, S.C., Serological cross-reaction between *Escherichia coli* O157 and other species of the genus *Escherichia, J. Clin. Microbiol.,* 30:1315–1316, 1992.

Robens, J.F., Richard, J.L., Aflatoxins in animal and human health, *Rev. Environ. Contam. Toxicol.,* 127:69–94, 1992.

St. Louis, M.E., Botulism, in *Bacterial Infections of Humans: Epidemiology and Control,* 2nd ed., Evans, A.S., Brachman, P.S., Eds., Plenum Medical, New York, 1991, pp. 115–126.

Samadpour, M., Grimm, L.M., Desai, B., Alfi, D., Ongerth, J.F., Tarr, P.I., Molecular epidemiology of *Escherichia coli* O157:H7 strains by bacteriophage lambda restriction fragment length polymorphism analysis: application to a multistate foodborne outbreak and a daycare center cluster, *J. Clin. Microbiol,* 31:3179–3183, 1993.

San Joaquin, V.H., *Aeromonas, Yersinia,* and miscellaneous bacterial enteropathogens, *Pediatr. Ann.,* 23:544–548, 1994.

Sattar, S.A., Springthorpe, V.S., Ansari, S.A., Rotavirus, in *Foodborne Disease Handbook: Diseases Caused by Viruses, Parasites, and Fungi,* Vol. 2, Hui, Y.H., Gorham, J.R., Murrell, K.D., Cliver, D.O., Eds., Marcel Dekker, New York, 1994, pp. 81–111.

Schmidt, H., Montag, M., Bockemuhl, J., Heesemann, J., Karch, H., Shiga-like toxin II-related cytotoxins in *Citrobacter freundii* strains from humans and beef samples, *Infect. Immun.* 61:534–543, 1993.

Schoenfeld, Y., George, J., Peter, J.B., Guillain-Barré as an autoimmune disease, *Int. Arch. Allergy Immunol.,* 109:318–326, 1996.

Schoeni, J.L., Doyle, M.P., Variable colonization of chickens perorally inoculated with *Escherichia coli* O157:H7 and subsequent contamination of eggs, *Appl. Environ. Microbiol.,* 58:2513, 1992.

Schofield, G.M., Emerging foodborne pathogens and their significance in chilled foods, *J. Appl. Bacteriol.,* 72:267–273, 1992.

Schwartz, H.J., Latex: a potential hidden "food" allergen in fast food restaurants, *J. Allergy Clin. Immunol.,* 95:139–140, 1995.

Scott, E., Bloomfield, S., The survival and transfer of microbial contamination via cloths, hands and utensils, *J. Appl. Bacteriol.,* 68:271–277, 1990.

Seppel, T., Rose, F., Schlaghecke, R., Chronic intestinal giardiasis with isolated levothyroxine malabsorption as reason for severe hypothyroidism: implications for localization of thyroid hormone absorption in the gut, *Exp. Clin. Endocrinol. Diabetes,* 104:180–182, 1996.

Shapiro, R.L., Hatheway, C., Becher, J., Swerdlow, D.L., Botulism surveillance and emergency response: a public health strategy for a global challenge, *JAMA,* 278:433–435, 1997.

Shapiro, R.L., Hatheway, C., Swerdlow, D.L., Botulism in the United States: a clinical and epidemiologic review, *Ann. Intern. Med.,* 129:221–228, 1998.

Shrittematter, C., Lang, W., Wiestler, O.D., Kleihues, P., The changing pattern of HIV associated cerebral toxoplasmosis, *Acta. Neuropathol. (Berl.),* 8:475–481, 1992.

Sieper, J., Braun, J., Pathogenesis of spondylarthropathies, *Arthritis Rheumatol.,* 38:1547–1554, 1995.

Sipponen, P., Kekki, M., Seppala, K., Siurala, M., The relationship between chronic gastritis and gastric secretion, *Aliment. Pharmacol. Ther.,* 10 (Suppl. 1):103–118, 1996.

Smith, J.L., *Cryptosporidium* and *Giardia* as agents of foodborne disease, *J. Food Protect.,* 56:451–461, 1994.

Smith, J.L., Foodborne toxoplasmosis, *J. Food Safety,* 12:17–57, 1991.

Smoot, D.T., Mobley, H.L.T., Chippendaele, G.R., Lewison, J.F., Resau, J.H., *Helicobacter pylori* urease activity is toxic to human gastric epithelial cells, *Infect. Immun.,* 59:1992–1994, 1991.

Snyder, O.P., A "safe hands" hand wash program for retail food operations, Hospitality Institute of Technology and Management, 1997.

Solnick, J.V., O'Rourke, J., Lee, A., Tompkins, L.S., Molecular analysis of urease genes from a newly identified uncultured species of *Helicobacter, J. Infect. Dis.,* 168:379–383, 1993.

Soweid, A.M., Clarkston, W.K., *Plesiomonas shigelloides:* an unusual cause of diarrhea, *Am. J. Gastroenterol.,* 9:2235–2236, 1995.

Stehr-Green, J.K., McCaig, L., Remsen, H.M., Rains, C.S., Fox, M., Juranek, D.D., Shedding of oocysts in immunocompetent individuals infected with *Cryptosporidium, Am. J. Trop. Med. Hyg.,* 36:338–342, 1987.

Stolle, A., Sperner, B., Viral infections transmitted by food of animal origin: the present situation in the European Union, *Arch. Virol.,* 13 (Suppl):219–228, 1997.

Swerdlow, D.L., Woodruff, B.A., Brady, R.C. et al., A waterborne outbreak in Missouri of *Escherichia coli* O157:H7 associated with bloody diarrhea and death, *Ann. Intern. Med.,* 117:812–819, 1992.

Swift, A.E., Swift, T.R., Ciguatera, *J. Toxicol. Clin. Toxicol.,* 31:1–29, 1993.

Szabo, R.A., Todd, E.C.D., Jean, A., Method to isolate *Escherichia coli* O157:H7 from food, *J. Food Prot.,* 49:768–772, 1986.

Tarr, P.I., *Escherichia coli* O157:H7: clinical, diagnostic, and epidemiological aspects of human infection, *Clin. Infect. Dis.,* 20:1–10, 1995.

Taylor, D.N., Blaser, M.J., The epidemiology of *Helicobacter pylori* infection, *Epidemiol. Rev.,* 13:42–59, 1991.

Tebutt, G.M., Development of standardized inspections in restaurants using visual assessments and microbiological sampling to quantify the risks, *Epidemiol. Infect.,* 107:393–404, 1991.

Tesh, V.L., O'Brien, A.D., The pathogenic mechanisms of Shiga toxin and the Shiga-like toxins, *Mol. Microbiol.,* 5:1817–1822, 1991.

Thomas, J.E., Gibson, C.R., Darboe, M.K., Dale, A., Weaver, L.T., Isolation of *H. pylori* from human feces, *Lancet,* 340:1194–1195, 1992.

Todd, E.C.D., Domic acid and amnesic shellfish poisoning: a review, *J. Food Prot.,* 56:69–83, 1993.

Torok, T.J., Tauze, R.V., Wise, R.P., Livengood, J.R., Sokolow, R., Mauvais, S. et al., A large community outbreak of *Salmonellosis* caused by intentional contamination of restaurant salad bars, *JAMA,* 278(5):389–395, 1997.

Tozzi, A.E., Niccolini, A., Caprioli, A. et al., A community outbreak of haemolytic-uraemic syndrome in children occurring in a large area of Northern Italy over a period of several months. *Epidemiol. Infect.,* 113:209–219, 1994.

United States Department of Agriculture, United States Department of Health and Human Services, United States Environmental Protection Agency, Food safety from farm to table: a new strategy for the 21st century, Discussion Draft, United States Department of Agriculture, Washington, D.C., 1997.

United States General Accounting Office, Food safety: information on foodborne illnesses, Report to Congressional Committees, Washington, D.C., GAO/RCED-96-96, 1996.

United States Public Health Service, Food and Drug Administration, Food Code, 1999, National Technology Information Service, Springfield, VA, chap. 3, p. 43.

Unusual *E. coli* strain causes food borne illness in Montana, *Food Chem. News,* 36:37, 1994.

Villar, R.G., Shapiro, R.L., Busto, S. et al., Outbreak of type A botulism and development of a botulism surveillance and antitoxin release system in Argentina, *JAMA,* 281:1334–1338, 1340, 1999.

von Herrath, M.G., Oldstone, M.B.A., Virus-induced autoimmune disease, *Curr. Opin. Immunol.,* 8:878–885, 1996.

Wainwright, R.B., Heyward, W.L., Middaugh, J.P., Hatheway, C.L., Harpster, A.P., Bender, T.R., Foodborne botulism in Alaska, 1947–1985: epidemiology and clinical findings, *J. Infect. Dis.,* 157(6):1158–1162, 1988.

Warburton, A.R.E., Wreghitt, T.G., Rampling, A., Buttery, R., Ward, K.R., Perry, K.R. et al., Hepatitis A outbreak involving bread, *Epidemiol. Infect.,* 106:199–202, 1991.

Weagant, S.D., Bryant, J.L, Bark, D.H., Survival of *Escherichia coli* O157:H7 in mayonnaise-based sauces at room and refrigerated temperatures, *J. Food Prot.,* 57:629–631, 1994.

Weltman, A.C., Bennett, N.M , Ackman, D.A., Misage, J.H., Campana, J.J., Fine, L.S. et al., An outbreak of Hepatitis A associated with a bakery, New York, 1994: the 1968 "West Branch, Michigan" outbreak repeated, *Epidemiol. Infect.,* 117:33–41, 1996.

Willshaw, G.A., Smith, H.R., Roberts, D., Thirlwell, J., Cheasty, T., Rowe, B., Examination of raw beef products for the presence of Vero cytotoxin producing *Escherichia coli,* particularly those of serogroup O157, *J. Appl. Bacteriol.,* 75:420–426, 1993.

Wolf, M.W., Misaki, T., Bech, K., Tvede, M., Silva, J.E., Ingbar, S.H., Immunoglobulins of patients recovering from *Yersinia enterocolitica* infections exhibit Graves' disease-like activity in human thyroid membranes, *Thyroid,* 1:315–320, 1991.

Wolfe, M.S., Giardiasis, *Clin. Rev. Microbiol.,* 5:93–100, 1992.

Wong, H.D., Chang, M.H., Fan, J.Y., Incidence and characterization of *Bacillus cereus* isolates contaminating dairy products, *Appl. Environ. Microbiol.,* 54(3):669–702, 1988.

Wright, D.J., Chapman, P.A., Siddon, C.A., Immunomagnetic separation as a sensitive method for isolating *Escherichia coli* O157 from food samples, *Epidemiol. Infect.,* 113:31–39, 1994.

Yu, D.T., Thompson, G.T., Clinical, epidemiological and pathogenic aspects of reactive arthritis, *Food Microbiol.,* 11:97–108, 1994.

Zhao, T., Doyle, M.P., Besser, R.E., Fate of enterohemorrhagic *Escherichia coli* O157:H7 in apple cider with and without preservatives, *Appl. Environ. Microbiol.,* 59:2526–2530, 1993.

Zhao, T., Doyle, M.P., Besser, R.E., Fate of enterohemorrhagic *Escherichia coli* O157:H7 in commercial mayonnaise, *J. Food Prot.,* 57:780-783, 1994.

CHAPTER 4

Adams, C., Use of HACCP in meat and poultry inspection, *Food Technol.,* May 1990.

Adams, C., Sachs, S., Government's role in communicating food safety information to the public, *Food Technol.,* May 1991.

Adams, M.R., Hartley, A.D., Cox, L.J., Factors affecting the efficiency of washing procedures used in the production of prepared salads, *Food Microbiol.,* 6:69–77, 1989.

American Meat Institute, Putting the food handling issue on the table: the pressing need for food safety education, American Meat Institute and Food Marketing Institute, Washington, D.C., 1996.

Andrews, L.S. et al., Food preservation using ionizing radiation, *Rev. Environ. Contaminant Toxicol.,* 154, 1998.

Austin, B., Environmental issues in the control of bacterial diseases of farmed fish, in *Proceedings of the 31st International Center for Living Aquatic Resources Management Conference, 1993, Manila, Philippines,* Pullin, R.V.C., Rosenthal, J., Maclean, J.L, Eds., The Center, Manila, 1993, pp. 237–235.

Barmore, C.R., Chlorine: are there alternatives?, *Cutting Edge,* Spring, 4–5, 1995.

Bates, J., Jordens, J.Z., Griffiths, D.T., Farm animals as a putative reservoir for vancomycin-resistant enterococcal infection in man, *J. Antimicrob. Chemother.,* 34:507–514, 1994.

Bauman, H., HACCP: concept, development, and Application, *Food Technol.,* May 1990.

Bean, N.H., Goulding, J.S., Lao, C., Angulo, F.J., Surveillance for foodborne disease outbreaks — United States 1988–1992, *MMWR,* 45(no. SS-5), 1996.

Besser, R.E., Lett, S.M., Weber, J.T., Doyle, M.P., Barrett, T.J., Wells, J.G., Griffin, P.M., An outbreak of diarrhea and hemolytic uremic syndrome from *Escherichia coli* O157:H7 in fresh pressed apple cider, *JAMA,* 269:2217–2220, 1993.

Beuchat, L.R., *Listeria monocytogenes:* incidence in vegetables, *Food Control,* 7:223–228, 1996.

Beuchat, L.R., Pathogenic microorganisms associated with fresh produce, *J. Food Prot.,* 59:204–206, 1996.

Bohach, C.H., Personal communication regarding survival of *E. coli* in sheep manure, *J. Appl. Environm. Microbiol.,* December 1, 1997.

Brackett, R.E., Antimicrobial effect of chlorine on *Listeria monocytogenes, J. Food Prot.,* 50:999–1003, 1987

Branen, A.L, ed., *Antimicrobials in Foods,* 2nd ed., Marcel Dekker, New York, 1993, pp. 469–537.

Bruhn, C.M., Consumer attitudes and market response to irradiated food., *J. Food Prot.,* 58:175–781, 1995.

Bruhn, C.M., Consumer concerns: motivating to action, *Emerg. Infect. Dis.,* 3:511–515, 1997.

Buras, N., Microbial safety of produce from wastewater fed aquaculture, Environment and Aquaculture in Developing Countries, in *Proceedings of the 31st International Center for Living Aquatic Resources Management Conference, 1993, Manila, Philippines,* Pullin, R.V.C., Rosenthal, J., Maclean, J.L, Eds., The Center, Manila, 1993, pp. 285–295.

Bureau of the Census, Economics and Statistics Administration, U.S. Department of Commerce, Population Estimates, October 1998.

CDC, The foodborne diseases active surveillance network, 1996, *MMWR,* 46:258–261, 1996.

CDC, Foodborne disease outbreaks, 1983–1987, *MMWR, CDC Surveill. Summ.,* 39(SS-1):1–45, 1990.

CDC, Outbreak of *Vibrio parahaemolyticus* infections associated with eating raw oysters — Pacific Northwest, 1997, *MMWR,* 47:457–462, 1998.

CDC, Update: outbreaks of cyclosporiasis–United States, 1997, *MMWR,* 46:461–462, 1997.

Centers for Disease Control and Prevention, Invasive infection with *Streptococcus iniae*-Ontario, 1995–1996, *MMWR,* 45(30):650–653, 1996.

Centers for Disease Control and Prevention, Outbreaks of *Salmonella* serotype *enteritidis* infection associated with consumption of raw shell eggs — United States, 1994–1995, *Morbidity Mortality Wkly. Rep.,* 45(34), August 30, 1996.

Cook, K., Boyce, T., Puhr, N., Tauxe, R., Mintz, E., Increasing antimicrobial-resistant *Shigella* infections in the United States, presented at the 36th Interscience Conference on Antimicrobial Agents and Chemotherapy, New Orleans, LA, September 1996.

Domestic septage regulatory guidance — a guide to the EPA rule, EPA 832-B-92-005, September 1993.

Eckert, J.W., Ogawa, J.M., The chemical control of postharvest diseases: deciduous fruits, berries, vegetables, and root/tuber crops, *Ann. Rev. Cytopathol,* 26:433–463, 1988.

Environmental regulation and technology control of pathogens and vector attraction reduction, EPA 1625/1-92/013, December 1992.

Ercolani, G.L., Occurrence and persistence of cultural clostridial spores on the leaves of horticultural plants, *J. Appl. Microbiol.,* 82:137–140, 1997.

Foegeding, P.M., Roberts, T., Foodborne pathogens: risks and consequences, Task Force Report No. 122, Council for Agricultural Science and Technology, Ames, IA, 1994.

Foerster, S.B., Kizer, K.W., DiSogra, L.K., Bal, D.G., Krieg, B.F., Bunch, K.L., California's "5 a Day for Better Health" campaign: an innovative population-based effort to effect large-scale dietary changes, *Am. J. Prevent Med.,* 11:124–131, 1995.

Food and Drug Administration, HHS, Safe handling label statement and refrigeration at retail requirements for shell eggs, *Fed. Regist.,* 63(166), Final Rule, August 27, 1998.

Food Marketing Institute, Backgrounder: foodborne illness, The Institute, Washington, D.C., 1996.

Food Marketing Institute, Food safety: a qualitative analysis, The Institute, Washington, D.C., 1996.

Food Marketing Institute, Trends in the United States: consumer attitudes and the supermarket, The Institute, Washington, D.C., 1996.

Food Safety and Inspection Service, HACCP implementation: first year *Salmonella* test results — January 26, 1998 to January 25, 1999, United States Department of Agriculture, Food Safety and Inspection Service, Office of Public Health and Science Publications, Washington, D.C., 1999.

Food Safety and Inspection Service, USDA, Refrigeration and labeling requirements for shell eggs, *Fed. Regist.,* 63(166), Final Rule, August 27, 1998.

Food Safety and Inspection Service, USDA, *Salmonella enteritidis* risk assessment for shell eggs and egg products, final report, June 12, 1998.

Food Safety and Inspection Service, USDA and Food and Drug Administration, HHS, *Salmonella enteritidis* in eggs, *Fed. Regist.,* 63(96), Proposed Rules, May 19, 1998.

Frost, J.A., McEvoy, M.B., Bentley, C.A., Andersson, Y., An outbreak of *Shigella sonnei* infection associated with consumption of iceberg lettuce, *Emerg. Infect. Dis.,* 1:26–29, 1995.

Garrett, E.S., Role of diseases in marine fisheries management, transactions of the 50th North American Wildlife and Natural Resources NOAA Conference 1986, Tech. Memo. NMS, F/NWR-16, National Marine Fisheries Service, Washington, D.C., 1986.

Garrett, E.S., Lima dos Santos, C., Jahncke, M.L., Public, animal, and environmental health implications of aquaculture, *Emerg. Infect. Dis.,* 3:453–457, 1997.

Garrett, S., Hudak-Roos, M., Use of HACCP for seafood surveillance and certification, *Food Technol.,* May 1990.

Giese, J.H., Sanitation: the key to food safety and public health, *Food Technol.,* December 1991.

Gill, C.O., Delacey, K.M., Growth of *Escherichia coli* and *Salmonella typhimurium* on high-pH beef packaged under vacuum or carbon dioxide, *Int. J. Food Microbiol.,* 13:21–30, 1991.

Gold, H.S., Moellering, R.C., Antimicrobial-drug resistance, *N. Engl. J. Med.,* 335:1445–1453, 1996.

Groth, E., Communicating with consumers about food safety and risk issues, *Food Technol.,* May 1991.

Haug, R.T., *The Practical Handbook of Compost Engineering,* Technomics Publishing, Lancaster, PA, 1993.

Hecht, A., Willis, J., Sulfites: preservatives that can go wrong?, *FDA Consumer,* 11, 1983.

Hlady, W.G., Klontz, K.C., The epidemiology of *Vibrio* infections in Florida, 1981–1993, *J. Infect. Dis.,* 173:1176–1183, 1996.

Hollingsworth, P., The changing role of fast food, *Food Technol.,* 51:24, 1997.

Infectious Disease Surveillance Center, Verotoxin-producing *Escherichia coli* (enterohemorrhagic *E. coli*) infections, Japan, 1996–June 1997, *IASR,* (Tokyo, Japan), 18:153–154, 1997.

International Consultative Group on Food Irradiation, Irradiation of red meat: a compilation of technical data for its authorization and control, August 1996.

Jahncke, J.L., The application of the HACCP concept to control exotic shrimp viruses, *Proceedings of the NMFS Workshop on Exotic Shrimp Viruses,* New Orleans, LA, June 1996.

Janssen, P.J., van-der-Heijden, C.A., Aspartame: review of recent experimental and observational data, *Toxicology,* 50(1):1–26, 1988.

Jaquette, C.B., Beuchat, L.R., Mahon, B.E., Efficacy of chlorine and heat treatment in killing *Salmonella stanley* inoculated onto alfalfa seeds and growth and survival of the pathogen during sprouting and storage, *Appl. Environ. Microbiol.,* 62:2212–2215, 1996.

Jaykus, L.A., Epidemiology and detection as options for control of viral and parasitic foodborne disease, *Emerg. Infect. Dis.,* 3:529–539, 1997.

Kalish, F., Extending the HACCP concept to product distribution, *Food Technol.,* June 1991.

Kapperud, G., Rorvik, L.M., Hasseltvedt, V. et al., Outbreak of *Shigella sonnei* infection traced to imported iceberg lettuce, *J. Clin. Microbiol.,* 33:609–614, 1995.

Kimura, A., Reddy, S., Marcus, R. et al., Chicken, a newly identified risk factor for sporadic *Salmonella* serotype *Enteritidis* infections in the United States: a case-control study in FoodNet sites {Abstract}, in *Program and Abstracts of the Infectious Diseases Society of America 36th Annual Meeting,* Denver, CO, November 12–15, 1998, 178.

Krebs-Smith, S.M., Report on per capita consumption and intake of fishery products [letter], in *Natl. Rep.,* National Academy Press, Washington, D.C., 1991.

Lagunas-Solar, M.C., Radiation processing of foods: an overview of scientific principles and current status, *J. Food Prot.,* 58:186–192, 1995.

Lambert, A.D., Smith, J.P., Dodds, K.L., Combined effect of modified atmosphere packaging and low-dose irradiation on toxin production by *Clostridium botulinum* in fresh pork, *J. Food Prot.,* 54:94–101, 1991.

Lambert, A.D., Smith, J.P., Dodds, K.L., Shelf life extension and microbiological safety of fresh meat: a review, *Food Microbiol.,* 8:267–297, 1991.

Levine, W.C., Griffin, P.M., and the Gulf Coast *Vibrio* Working Group, *Vibrio* infections on the gulf coast: results of first year of regional surveillance. *J. Infect. Dis.,* 167:479–483, 1993.

Lin, C.-T., Morales, R.A., Ralston, K., Raw and undercooked eggs: a danger of Salmonellosis, *Food Rev.,* 20:27–32, 1997.

Lund, B.M., Bacterial spoilage, in *Post-Harvest Pathology of Fruits and Vegetables,* Dennis, C., Ed., Academic Press, London, 1983, p. 219.

Mahaffy, K.R., Heavy metal exposure from foods, *Environ. Health Perspect.,* 12:63, 1983.

Mahon, B.E., Ponka, A., Hall, W.N. et al., An outbreak of *Salmonella* infections caused by alfalfa sprouts grown from contaminated seeds, *J. Infect. Dis.,* 175:876–882, 1997.

Management Summary Update: Critical Strategic Issues 1996–2001, Technomics, Chicago, 1997.

Matilla-Sandholm, T., Skytta, E., The effect of spoilage flora on the growth of food pathogens in minced meat stored at chilled temperature, *Lebensm-Wiss Technol.,* 24:116–120, 1991.

Mazollier, J., Ivè gamme, Lavage-desinfection des salades, *Infros-Crifl,* 41:19, 1988.

Microbiological safety evaluations and recommendations on fresh produce, report by the National Advisory Committee on Microbiological Criteria of Food, March 5, 198. Copies are available from Dr. Richard Ellis, 6913 Franklin Court, 1400 Independence Ave., SW, Washington, D.C.

Millard, P.S., Gensheimer, K.F., Addiss, D.G., Sosin, D.M., Beckett, G.A., Houk-Jankowski, A., Hudson, A., An outbreak of Cryptosporidiosis from fresh-pressed apple cider, *JAMA,* 272:1592–1596, 1994.

Minnesota Department of Health, Foodborne and Waterborne Outbreak Summary, 1995, Minneapolis, MN.

Molbak, K., Neimann, J., Outbreak in Denmark of *Shigella sonnei* infections related to uncooked "baby maize" imported from Thailand, *Eurosurveillance Wkly.,* 2:980813, 1998. Available at <http://www.outbreak.org.uk/1998/980813.html> accessed April 2, 1999.

Monk, J.D., Beuchat, L.R., Doyle, M.P., Irradiation inactivation of foodborne microorganisms, *J. Food Protect.,* 58:197–208, 1995.

Mott, K.E., A proposed national plan of action for schistosomiasis control in the United Republic of Cameroon (unpublished document WHO/SCHISTO/86,88), World Health Organization, Geneva, 1996.

National Academy of Sciences, The Effects of Human Health of Subtherapeutic Use of Antimicrobials in Animal Feeds, U.S. Government Printing Office, Washington, D.C., 1980.

National Advisory Committee on Microbiological Criteria for Foods, Hazard analysis and critical control system, *Int. J. Food Microbiol.,* 16:1–23, 1992.

National Agricultural Statistics Service, USDA, Layers and Egg Production 1998 Summary, January 1999.

National Marine Fisheries Service, The draft report of the model seafood surveillance project; a report to Congress, Office of Trade and Industry Services, National Seafood Inspection Laboratory, Pascagoula, MS, 1995.

National Marine Fisheries Service, Fisheries of the United States 1995, current fisheries statistics No. 9500, U.S. Department of Commerce, Washington, D.C., 1996.

Neff, J., Will home meal replacement replace packaged foods?, *Food Process.,* 57:35, 1996.

Nguyen-The, C., Carlin, F., The microbiology of minimally processed fresh fruits and vegetables, *Crit. Rev. Food Sci. Nutr.,* 34:371–401, 1994.

Osterholm, M.T., Potter, M.E., Irradiation pasteurization of solid foods: taking food safety to the next level, *Emerg. Infect. Dis.,* 3:575–576, 1997.

Paul, P., Chawla, S.P., Thomas, P., Kesavan, P.C., Effect of high hydrostatic pressure, gamma-irradiation and combination treatments on the microbiological quality of lamb meat during chilled storage, *J. Food Saf.,* 16:263–271, 1997.

Pierson, M.D., Corlett, D.A., Jr., Eds, *HACCP Principles and Applications,* Van Nostrand Reinhold, New York, 1992.

Pullin, R.S.V., Discussion and recommendations on aquaculture and the environment in developing countries, in *Environment and Aquaculture in Developing Countries, Proceedings of the 31st ICLARM Conference,* Pullin, R.S.V., Rosenthal, H., MacLean, J.L., Eds., 1993, pp. 312–338.

Radomyski, T., Murano, E.A., Olson, D.G., Murano, P.S., Elimination of pathogens of significance in food by low-dose irradiation: a review, *J. Food Prot.,* 57:73–86, 1994.

Ratafia, M., Aquaculture today: a worldwide status report, *Aquaculture News,* 3:12–13, 18–19, 1994.

Reiners, S., Rangarajan, A., Pritts, M., Pedersen, L., Shelton, A., Prevention of foodborne illness begins on the farm, Cornell Cooperative Extension, Cornell University, Ithaca, NY.

Resurreccion, A.V.A., Galvez, F.C.F., Fletcher, S.M., Misra, S.K., Consumer attitudes toward irradiated food: results of a new study, *J. Food Prot.,* 58:193–196, 1995.

Rhodes, R.J., Status of world aquaculture, *Aquaculture Mag.,* 17th Annual Buyers Guide, 4–18, 1987.

St. Louis, M.E., Morse, D.L., Potter, M.E. et al., The emergence of grade A eggs as a major source of *Salmonella enteritidis* infection–new implications for control of Salmonellosis, *JAMA,* 259:2103–2107, 1988.

Schlosser, W.D., Henzler, D.J., Mason, J. et al., The *Salmonella enterica* serovar *Enteritidis* pilot project, in *Salmonella enterica Serovar Enteritidis in Humans and Animals: Epidemiology, Pathogenesis, and Control,* Saeed, A.M., Ed., Iowa State University Press, Ames, IA, 1999, pp. 353–365.

Shigeshi, T., Ohmori, T., Saito, A., Taji, S., Hayashi, R., Effect of high hydrostatic pressure on characteristics of pork slurries and inactivation of microorganisms associated with meat and meat products, *Int. J. Food Microbiol.* 12:207–216, 1991.

Son, T.Q., Hoi, V.S., Dan, T.V., Nga, C., Toan, T.Q., Chau, L.V. et al., Application of hazard analysis critical control point (HACCP) as a possible control measure against *Clonorchis sinensis* in cultured silver carp *Hypophthalmichthys molitrix,* paper presented at 2nd Seminar on Foodborne Zoonoses: Current Problems, Epidemiology and Food Safety, Khon Kaen, Thailand, December 6–9, 1995.

Steginck, L.D., The aspartame story: a model for the clinical testing of a food additive, *Am. J. Clin. Nutr.,* 46 (Suppl. 1):204–215, 1987.

Stephenson, J., New approaches for detecting and curtailing foodborne microbial infections, *JAMA,* 277:1337–1340, 1997.

Stevenson, K.E., Implementing HACCP in the food industry, *Food Technol.,* May 1990.

Tauxe, R.V., Emerging foodborne disease: an evolving public health challenge, *Emerg. Infect. Dis.,* 3:425–434, 1997.

Thayer, D.W., Extending shelf life of poultry and red meat by irradiation processing, *J. Food Protect.,* 56, 1993.

Thayer, D.W., Boyd, G., Elimination of *Escherichia coli* O157:H7 in meats by gamma radiation, *Appl. Environ. Microbiol.,* 59, 1993.

Thayer, D.W., Boyd, G., Radiation sensitivity of *Listeria monocytogenes* on beef as affected by temperature, *J. Food Sci.,* 60, 1995.

Thayer, D.W., Josephson, E.S., Brynjolfsson, A., Giddings, G.G., Radiation pasteurization of food, Issue Paper No. 7, Council for Agricultural Science and Technology, Ames, IA, 1996.

Turner, G.E., Ed., Codes of practice and manual of procedures for consideration of introductions and transfers of marine and freshwater organisms, EIFAC Occasional Paper 23, FAO, Rome, 1988.

Uno, Y., The toxicology of mycotoxins, *CRC Crit. Rev. Toxicol.,* 14(2):99–132, 1985.

U.S. Department of Agriculture, Food Safety and Inspection Service, Take out foods–handle with care, *The Food Saf. Educator,* 1:2, 1996.

U.S. Department of Agriculture, Vegetables and specialties/VGS-269/July; Fruits and tree nuts/FTS-278/October, USDA Economic Research Service, Washington, D.C., 1996, p. 22, 81.

U.S. EPA, Domestic septage regulatory guidance, a guide to the EPA 503 rule, EPA, Office of Water Regulations and Standards, 832-B-92-005, September, 1993.

U.S. EPA, A plain English guide to the EPA part 503 biosolids rule, EPA 1832-R-93-003, Washington, D.C., 1994.

U.S. General Accounting Office, Food safety: U.S. lacks a consistent farm-to-table approach to egg safety, GAO/RCED-99-184, July 1999.

U.S. Public Health Service, FDA, 1997 Food Code, U.S. Department of Health and Human Services, Food and Drug Administration, Washington, D.C.

Van Beneden, C.A., Keene, W.E., Werker, D.H. et al., A health food fights back: an international outbreak of *Salmonella newport* infections due to alfalfa sprouts, in *Program and Abstracts of the 36th Interscience Conference on Antimicrobial Agents and Chemotherapy,* American Society of Microbiology, Washington, D.C., 1996.

Van de Bogaard, A.E., Jensen, L.B., Stobberingh, E.E., Vancomycin-resistant enterococci in turkeys and farmers [letter]. *N. Engl. J. Med.,* 337:1558–1559, 1997.

Virus re-emerges at shrimp farm, *Aquaculture News,* 4(11):1–15, 1996.

Wallace, J.S., Cheasty, T., Jones, K., Isolation of verocytotoxin-producing *Escherichia coli* O157:H7 from wild birds, *J. Appl. Microbiol.,* 82:399–404, 1997.

Wang, W., Zhao, T., Doyle, M.P., Fate of enterohemorrhagic *Escherichia coli* O157:H7 in bovine feces, *J. Appl. Environ. Microbiol.,* 62(7), 1996.

Watson, D.H., Toxic fungal metabolites in food, *CRC Crit. Rev. Food Sci. Nutr.,* 22(3):177–198, 1985.

Wei, C.I., Huang, T.S., Kim, J.M., Lin, W.F., Tamplin, M.L., Bartz, J.A., Growth and survival of *Salmonella montevideo* on tomatoes and disinfection with chlorinated water, *J. Food Prot.* 58:829–836, 1995.

Weltman, A.C., Bennett, N.M., Ackman, D.A. et al., An outbreak of Hepatitis A associated with a bakery, New York, 1994: the "West Branch, Michigan" outbreak repeated, *Epidemiol. Infect.,* 117:333–341, 1996.

World Aquaculture Society, Abstracts of the annual International Conference and Exposition of the World Aquaculture Society and The National Aquaculture Conference and Exposition of National Aquaculture Association, Seattle, WA, February 19–23, 1997.

World Health Organization, Chemistry of food irradiation, in *Safety and Nutritional Adequacy of Irradiated Food,* 1994.

Wu, F.M., Doyle, M.P., Beuchat, L.R., Mintz, E., Swaninathan, B., Factors influencing survival and growth of *Shigella sonnei* on parsley, presented at the sixth annual meeting of the Center for Food Safety and Quality Enhancement, Atlanta, GA, March 1999.

Zepp, G., Kuchler, F., Lucier, G., Food safety and fresh fruits and vegetables: is there a difference between imported and domestically produced products?, *Vegetables and Specialities, Situation and Outlook Report,* ERS/USDA, VGS-274:23–28, April 1998.

Zhang, S., Farber, J.M., The effects of various disinfectants against *Listeria monocytogenes* on fresh-cut vegetables, *Food Microbiol.,* 13:311–321, 1996.

Zhuang, R.-Y., Beuchat, L.R., Angulo, F.J., Fate of *Salmonella montevideo* on and in raw tomatoes as affected by temperature and treatment with chlorine, *Appl. Environ. Microbiol.,* 61:2127–2131, 1995.

CHAPTER 5

Anon., The dengue situation in Singapore, *Epidemiol. Bull.,* 20:31–33, 1994.

Azad, A.F., Radulovic, S., Higgins, J.A., Noden, B.H., Troyer, M.J, Flea-borne Rickettsioses: some ecological considerations, *Emerg. Infect. Dis.,* 3:319–328, 1997.

Azad, A.F., Sacci, J.B. Jr., Nelson, W.M., Dasch, G.A., Schmidtman, E.T. Carl, M., Genetic characterization and transovarial transmission of a novel Typhus-Like Rickettsia found in cat fleas, *Proc. Natl. Acad. Sci. U.S.A.,* 89:43–46, 1992.

Bakken, J.S., Dumler, J.S., Chen, S.M., Eckman, M.R., Van Etta, L.L., Walker, D.H., Human granulocytic ehrlichiosis in the upper midwest United States, *JAMA.* 272:212–218, 1994.

Bakken, J.S., Krueth, J., Wilson-Nordskog, C., Tilden, R.L., Asanovich, K., Dumler, J.S., Human granulocytic ehrlichiosis (HGE: clinical and laboratory characteristics of 41 patients from Minnesota and Wisconsin, *JAMA,* 275:199–225, 1995.

Biosafety in Microbiology and Biomedical Laboratories. 3rd ed. Atlanta (GA): Centers for Disease Control and Prevention, 1993, Report No. 93:8395.

Bouma, M.J., Sondorp, H.E., van der Kaay, H.J., Climate change and periodic epidemic malaria, *Lancet,* 343:1440, 1994.

Brandling-Bennett, A.D., Pinheiro, F., Infectious diseases in Latin America and the Caribbean: are they really emerging and increasing?, *Emerg. Infect. Dis.,* 2:59–61, 1996.

Briseno-Garcia, B., Gomez-Dantes, H., Argott-Ramirez, E., Montesano, R., Vazquez-Martinez, A.-L., Ibanez-Bernal, S. et al., Potential risk for dengue hemorrhagic fever: the isolation of serotype dengue-3 in Mexico, *Emerg. Infect. Dis.,* 2:133–135, 1996.

Brook, J.H., Genese, C.A., Bloland, P.B., Zucker, J.R., Spitalny, K.C., Malaria probably locally acquired in New Jersey, *N. Engl. J. Med.,* 331:22–23, 1994.

Brouqui, P., Dumler, J.S., Lienhard, R., Brossard, M., Raoult, D., Human granulocytic ehrlichiosis in Europe, *Lancet,* 346:782–783, 1995.

Brouqui, P., Harle, J.R., Delmont, J., Frances, C., Weiller, P.J., Raoult, D., African Tick-Bite Fever: an imported spotless Rickettsiosis, *Arch. Intern. Med.,* 157:119–124, 1997.

Calisher, C.H., Medically important arboviruses of the United States and Canada, *Clin. Microbiol. Rev.,* 7:89–116, 1994.

Campbell, G.L., Hughes, J.M., Plague in India: a new warning from an old nemesis, *Ann. Intern. Med.,* 122:151–153, 1995.

Castelli, F., Caligaris, S., Matteelli, A., Chiodera, A., Carosi, G., Fausti, G., Baggage malaria in Italy: cryptic malaria explained?, *Trans. R. Soc. Trop. Med. Hyg.,* 87:394, 1993.

CDC, Dengue fever at the U.S. — Mexico border, 1995–1996, *MMWR,* 45:841–844, 1996.

CDC, Imported dengue — United States, 1995, *MMWR,* 45:988–991, 1996.

Centers for Disease Control and Prevention, *Addressing Emerging Infectious Disease Threats: a Prevention Strategy for the United States,* U.S. Department of Health and Human Services, Public Health Service, Atlanta, GA, 1994.

Centers for Disease Control and Prevention, Dengue-type infection — Nicaragua and Panama October–November 1994, *MMWR,* 44:21–24, 1995.

Centers for Disease Control and Prevention, Human granulocytic ehrlichiosis — New York, 1995, *MMWR,* 44:593–595, 1995.

Centers for Disease Control and Prevention, *Lyme Dis. Surveillance Summary,* 8:2, 1997.

Centers for Disease Control, Mosquito-transmitted malaria — California and Florida, 1990, *MMWR,* 40:106–108, 1991.

Centers for Disease Control and Prevention, Outbreak of ebola viral hemorrhagic fever — Zaire, 1995, *MMWR,* 44:381–382, 1995.

Centers for Disease Control and Prevention, Transmission of *Plasmodium vivax* malaria — San Diego County, California, 1988 and 1989, *MMWR,* 39:91–94, 1990.

Centers for Disease Control and Prevention, Update: outbreak of hantavirus infection — Southwestern United States, *MMWR,* 42:441–443, 1993.

Consumers unhappy with food regulations, *Nature* (London), 389:657, 1997.

Cully, J.F., Jr., Street, T.G., Heard, P.B., Transmission of La Crosse Virus by four strains of *Aedes albopictus* to and from the Eastern Chipmunk (*Tamias striatus*), *J. Am. Mosq. Control Assoc.,* 8:237–240, 1992.

Daniels, T.J., Falco, R.C., Schwartz, I., Varde, S., Robbins, R.G., Deer ticks (*Ixodes scapularis*) and the agents of Lyme disease and human granulocytic ehrlichiosis in a New York City park, *Emerg. Infect. Dis.,* 3:353–355, 1997.

Dawson, J.E., Ewing, S.A., Susceptibility of dogs to infection with *Ehrlichia chaffeensis,* causative agent of human ehrlichiosis, *Am. J. Vet. Res.,* 53:1322–1327, 1992.

Dawson, J.E., Warner, C.K., Baker, V., Ewing, S.A., Stallknecht, D.E., Davidson, W.R. et al., *Ehrlichia*-like 16S rDNA sequence from white-tailed deer (*Odocoileus virginianus*), *J. Parasitol.,* 82:52–58, 1996.

Dean, A.G., Dean, J.A., Coulombier, D., Brendel, K.A., Smith, D.C., Burton, A.H. et al., Epi Info, Version 6: a word processing, database, and statistics program for epidemiology on microcomputers, Centers for Disease Control and Prevention, Atlanta, GA, 1994.

Delmont, J., Brouqui, P., Poullin, P., Bourgeade, A., Harbour-acquired *Plasmodium falciparum* malaria, *Lancet,* 344:330–331, 1994.

Dohm, D.J., Logan, T.M., Barth, J.F., Turell, M.J., Laboratory transmission of sindbis virus by *Aedes albopictus, A. aegypti,* and *Culex pipiens* (Diptera: Culicidae), *J. Med. Entomol.,* 32:818–821, 1995.

Dumler, J.S., Asanovich, K.M., Bakken, J.S., Richter, P., Kimsey, R., Madigan, J.E., Serologic cross-reactions among *Ehrlichia equi, Ehrlichia phagocytophila,* and human granulocytic erhlichia, *J. Clin. Microbiol.,* 33:1098–1113, 1995.

Dumler, J.S., Bakken, J.S., Ehrlichial diseases of humans: emerging tick-borne infections, *Clin. Infect. Dis.,* 20:1102–1110, 1995.

Dupont, H.T., Brouqui, P., Faugere, B., Raoult, D., Prevalence of antibodies to *Coxiella burnetti, Rickettsia conorii,* and *Rickettsia typhi* in seven African countries, *Clin. Infect. Dis.,* 21:1126–1133, 1995.

Estrada-Franco, J.G., Craig, G.B., Jr., Biology, disease relationships, and control of *Aedes albopictus,* Pan American Health Organization Technical Paper No. 42, 1995.

Francy, D.B., Karabatsoss, N., Wesson, D.M., Moore, C.G., Jr., Lazuick, J.S., Niebylski, M.L. et al., A new arbovirus from *Aedes albopictus,* an Asian mosquito established in the United States, *Science,* 250:1738–1744,1990.

Grimstad, P.R., Kobayashi, J.F., Zhang, M., Craig, G.B., Jr., Recently introduced *Aedes albopictus* in the United States: potential vector of La Crosse Virus (Bunyaviridae: California Serogroup), *J. Am. Mosq. Control Assoc.,* 5:422–427, 1989.

Gubler, D.J., Dengue hemorrhagic fever: a global update [Editorial], *Virus Inf. Exchange News,* 8:2–3, 1991.

Gubler, D.J., Dengue and dengue hemorrhagic fever in the Americas, in WHO, Regional Office for SE Asia, New Delhi, Monograph, *SEARO,* 22:9–22, 1993.

Gubler, D.J., Clark, G.G., Community-based integrated control of *Aedes aegypti:* a brief overview of current programs, *Am. J. Trop. Med. Hyg.,* 50:50–60, 1994.

Gubler, D.J., Clark, G.G., Dengue/dengue hemorrhagic fever: the emergence of a global health problem, *Emerg. Infect. Dis.,* 1:55–57, 1995.

Gubler, D.J., Trent, D.W., Emergence of epidemic dengue/dengue hemorrhagic fever as a public health problem in the Americas, *Infect. Agents Dis.,* 383–393, 1994.

Hackstadt, T., The biology of Rickettsiae, *Infect. Agents Dis.,* 5:127–143, 1996.

Hagar, J.M., Rahimtoola, S.H., Chagas' heart disease, *Curr. Probl. Cardiol.,* 20:825–924, 1995.

Haines, A., Epstein, P.R., McMichael, A.J., Global health watch: monitoring impacts of environmental change, *Lancet,* 342:1464–1469, 1993.

Halstead, S.B., The XXth century dengue pandemic: need for surveillance and research, *Rapp. Trimest. Statist. Sanit. Mondo.,* 45:292–298, 1992.

Hanson, S.M., Craig, G.B., Jr., *Aedes albopictus* (Diptera: Culicidae) eggs: field survivorship during Northern Indiana winters, *J. Med. Entomol.,* 32:599–604, 1995.

Hardalo, C.J., Quagliarello, V., Dumler, J.S., Human granulocytic ehrlichiosis in Connecticut: report of a fatal case, *Clin. Infect. Dis.,* 21:910–914, 1995.

Harrison, B.A., Mitchell, C.J., Apperson, C.S., Smith, G.C., Karabatsos, N., Engber, B.R. et al., Isolation of Potosi virus from *Aedes albopictus* in North Carolina, *J. Am. Mosq. Control Assoc.,* 11:225–229, 1995.

Hawley, W.A., The biology of *Aedes albopictus, J. Am. Mosq. Control Assoc.,* 4 (Suppl. 1):1–40, 1988.

Henderson, D.A., Surveillance systems and intergovernmental cooperation, in *Emerging Viruses,* Morse, S.S., Ed., Oxford University Press, New York, 1993, pp. 283–289.

Herwaldt, B.L., Juranek, D.D., Laboratory-acquired malaria, leishmaniasis, trypanosomiasis, and toxoplasmosis, *Am. J. Trop. Hyg.,* 48:313–323, 1993.

Higgins, J.A., Radulovic, S., Schriefer, M.E., Azad, A.F., *Rickettsia felis:* a new species of pathogenic Rickettsia isolated from cat fleas, *J. Clin. Microbiol.,* 34:671–674, 1996.

Higgins, J.A., Sacci, J.B., Jr., Schriefer, M.E., Endris, R.G., Azad, A.F., Molecular identification of Rickettsia-like microorganisms associated with colonized cat fleas (*Ctenocephalides felis*), *Insect Mol. Biol.,* 3:27–33, 1994.

Ho, B.C., Ewert, A., Chew, L.M., Interspecific competition among *Aedes aegypti, A. albopictus* and *A. triseriatus* (Diptera: Culcidae): larval development in mixed cultures., *J. Med. Entomol.,* 26:615–623, 1989.

Hollander, A.K., Environmental impacts of genetically engineered microbial and viral biocontrol agents, in *Biotechnology for Biological Control of Pests and Vectors,* Maramorosch, K., Ed., CRC Press, Boca Raton, FL, 1991, pp. 251–266.

Hoy, M.A., Impact of risk analyses on pest-management programs employing transgenic arthropods, *Parasitol. Today,* 11:229–332, 1995.

Hoy, M.A., Laboratory containment of transgenic arthropods, *Am. Entomol.,* 43:206–256, 1997.

Jackson, L.A., Spach, D.H., Emergence of *Bartonella quintana* infection among homeless persons, *Emerg. Infect. Dis.,* 2:141–143, 1996.

Karp, B.E., Dengue fever: a risk to travelers, *Maryland Med. J.,* 46:299–302, 1997.

Kelly, P.J., Beati, L., Matthewman, L.A., Mason, P.R., Dasch, G.A., Raoult, D., A new pathogenic spotted fever group Rickettsia from Africa, *J. Trop. Med. Hyg.,* 97:129–137, 1994.

Kirchoff, L.V., American trypanosomiasis (Chagas' disease), *Gastroenterol. Clin. N. Am.,* 25:517–532, 1996.

Kirchoff, L.V., American trypanosomiasis (Chagas' disease) — a tropical disease now in the United States, *N. Engl. J. Med.,* 329:639–644, 1993.

Kitron, U., Kazmierczak, J.J., Spatial analysis of the distribution of Lyme disease in Wisconsin, *Am. J. Epidemiol.,* 145:558–566, 1997.

Kitron, U., Michael, J., Swanson, J., Haramis, L., Spatial analysis of the distribution of La Crosse encephalitis in Illinois, using a geographic information system and local and global spatial statistics, *Am. J. Trop. Med. Hyg.,* 57:469–475, 1997.

Krogstad, D.J., Malaria as a reemerging disease, *Epidemiol. Rev.,* 18:77–89, 1996.

Kumar, S., Tamura, K., Nei, M., *MEGA: Molecular Evolutionary Genetics Analysis,* Pennsylvania State University, University Park, PA, 1993, p. 16802.

Lanciotti, R.S., Lewis, J.G., Gubler, D.J., Trent, D.W., Molecular evolution and epidemiology of dengue-3 viruses, *J. Gen. Virol.,* 75:65–75, 1994.

Larsen, H.J.S., Overnes, G., Waldhland, H., Johansen, G.M., Immunosuppression in sheep experimentally infected with *Ehrlichia phagocytophila, Res. Vet. Sci.,* 56:216–224, 1994.

Layton, M., Parise, M.E., Campbell, C.C., Advani, R., Sexton, J.D., Bosler, E.M., Zucker, J.R., Malaria transmission in New York City, 1993, *Lancet,* 346:729–731, 1995.

Lederberg, J., Shope, R.E., Oaks, S.C., Jr., Eds., *Emerging Infections: Microbial Threats to Health in the United States,* Institute of Medicine, National Academy Press, Washington, D.C., 1992.

Local transmission of *Plasmodium vivax* Malaria — Houston, Texas, 1994, *MMWR,* 44:295–303 1995.

Lockheart, J.M., Davidson, W.R., Dawson, J.E., Stallknecht, D.E., Temporal association of *Amblyomma americanum* with the presence of *Ehrlichia chaffeensis* reactive antibodies in white-tailed deer, *J. Wildl. Dis.,* 31:119–124, 1995.

Loevinsohn, M.E., Climatic warming and increased malaria incidence in Rwanda, *Lancet,* 343:714–717, 1994.

Magnarelli, L.A., Anderson, J.F., Fish, D., Transovarial transmission of *Borrelia burgdorferi* in *Ixodes dammini* (Acari: Ixodidae), *J. Infect. Dis.,* 234–236, 1987.

Magnarelli, L.A., Denicola, A., Stafford, K.C., Anderson, J.F., *Borrelia burgdorferi* in an urban environment: white-tailed deer with infected ticks and antibodies, *J. Clin. Microbiol.,* 33:541–544, 1995.

Magnarelli, L.A., Stafford, K.C., III, Mather, T.N., Yeh, M.-T., Horn, K.D., Dumler, J.S., Hemocytic Rickettsia-like organisms in ticks: serologic reactivity with antisera to *Ehrlichiae* and detection of DNA of agent of human granulocytic ehrlichiosis by PCR, *J. Clin. Microbiol.,* 33:2710–2714, 1995.

Mandl, C.W., Holzmann, H., Kunz, C., Heinz, F.X., Complete genomic sequence of Powassan virus: evaluation of genetic elements in tick-borne versus mosquito-borne flaviviruses, *Virology,* 194:173–184, 1993.

Mantel, C.F., Klose, C., Sheurer, S., Vogel, R., Wesirow, A.L., Bienzle, U., *Plasmodium falciparum* malaria acquired in Berlin, Germany, *Lancet,* 346:320–321, 1995.

McLean, R.G., Ubico, S.R., Cooksey, L.M., Experimental infection of the Eastern chipmunk (*Tamias striatus*) with the Lyme Disease spirochete (*Borrelia burgdorferi*), *J. Wildl. Dis.,* 29:527–532, 1993.

Meslin, F.-X., Global aspects of emerging and potential zoonoses: a WHO perspective, *Emerg. Infect. Dis.,* 3:223–228, 1997.

Mitchell, C.J., Geographic spread of *Aedes albopictus* and potential for involvement in arbovirus cycles in the Mediterranean Basin, *J. Vector Ecol.,* 20:44–58, 1995.

Mitchell, C.J., The role of *Aedes albopictus* as an arbovirus vector, Proceedings of the Workshop on the Geographic Spread of *Aedes albopictus* in Europe and the concern among public health authorities, December 19–20, 1990, Rome, Italy, *Parassitologia,* 37:109–113, 1995.

Mitchell, C.J., Vector competence of North and South American strains of *Aedes albopictus* for certain arboviruses, *J. Am. Mosq. Control Assoc.,* 7:446–451, 1991.

Mitchell, C.J., Morris, C.D., Smith, G.C., Karabotsos, N., Vanlandingham, D., Cody, E., Arboviruses associated with mosquitoes from nine Florida counties during 1993, *J. Am. Mosq. Control Assoc.,* 12:255–262, 1996.

Mitchell, C.J., Smith, G.C., Miller, B.R., Vector competence of *Aedes albopictus* for a newly recognized *Bunyavirus* from mosquitoes collected in Potosi, Missouri, *J. Am. Mosq. Control Assoc.*, 6:523–527, 1990.

Mitchell, C.J., Smith, G.C., Karabatsos, N., Moore, C.G., Francy, D.B., Nasci, R.S., Isolation of Potosi virus from mosquitoes collected in the United States, 1989–1994, *J. Am. Mosq. Control Assoc.*, 12:1–7, 1996.

Monroe, M.C., Morzunov, S.P., Johnson, A.M., Bowen, M.D., Artsob, H., Yates, T. et al., Genetic diversity and distribution of *Peromyscus*-borne hantaviruses in North America, *Emerg. Infect. Dis.*, 5:75–86, 1999.

Moore, C.G., McLean, R.G., Mitchell, C.J., Nasci, R.S., Tsai, T.F., Calisher, C.H. et al., Guidelines for arbovirus surveillance programs in the United States, Fort Collins, CO, U.S. Department of Health and Human Services, 1993.

Moore, C.G., Mitchell, C.J., *Aedes albopictus* in the United States: ten year presence and public health implications, *Emerg. Infect. Dis.*, 3:329–334, 1997.

Munderloh, U.G., Kurtti, T.J., Cellular and molecular interrelationships between ticks and prokaryotic tick-borne pathogens, *Annu. Rev. Entomol.*, 40:221–243, 1995.

Munderloh, U.G., Madigan, J.E., Dumler, J.S., Goodman, J.L., Hayes, S.F., Barlough, J.E. et al., Isolation of the equine granulocytic ehrlichiosis agent, *Ehrlichia equi*, in tick cell culture, *J. Clin. Microbiol.*, 34:664–670, 1996.

Novak, M.G., Higley, L.G., Christianssen, C.A., Rowley, W.A., Evaluating larval competition between *Aedes albopictus* and *A. triseriatus* (Diptera:Culcidae) through replacement series experiments, *Environ. Entomol.*, 22:311–318, 1993.

Pan American Health Organization, *Guidelines for the Prevention and Control of Dengue and Dengue Hemorrhagic Fever in the Americas*, Pan American Health Organization, Washington, D.C., 1994.

Pancholi, P., Kolbert, C.P., Mitchell, P.D., Reed, K.D., Dumler, J.S., Bakken, J.S. et al., *Ixodes dammini* as a potential vector of human granulocytic ehrlichiosis, *J. Infect. Dis.*, 172:1007–1012, 1995.

Pinheiro, F.P., Corber, S.J., Global situation of dengue and dengue haemorrhagic fever, and its emergence in the Americas, *World Health Stat.*, 50:161–169, 1997.

Qiu, F.X., Gubler, D.J., Liu, J.C., Chen, Q.Q., Dengue in China: a clinical review, *Bull. WHO*, 71:349–359, 1993.

Rai, K.S., *Aedes albopictus* in the Americas, *Annu. Rev. Entomol.*, 36:459-84, 1991.

Raoult, D., Roux, V., Nidihokubwayo, J.B., Bise, G., Baudon, D., Martet, G. et al., Jail fever (epidemic typhus) outbreak in Burundi, *Emerg. Infect. Dis.*, 3:357–360, 1997.

Richter, P.J., Kimsey, R.B., Madigan, J.E., Barlough, J.E., Dumler, J.S., Brooks, D.L., *Ixodes pacificus* as a vector of *Ehrlichia equi*, *J. Med. Entomol.*, 33:1–5, 1996.

Rigau-Perez, J.G., Gubler, D.J., Vorndam, A.V., Clark, G.G., Dengue in travelers from the United States, 1986–1994, *J. Travel Med.*, 4:65–71, 1997.

Rigau-Perez, J.G., Gubler, D.J., Vorndam, A.V., Clark, G.G., Dengue surveillance — United States, 1986–1992, *MMWR*, 43:7–19, 1994.

Rikihisa, Y., The tribe *Ehrlichia* and ehrlichial diseases, *Clin. Microbiol. Rev.*, 4:286–308, 1991.

Roberts, D.R., Laughlin, L.L., Hsheih, P., Legters, L.J., DDT, global strategies, and a malaria control crisis in South America, *Emerg. Infect. Dis.*, 3:295–302, 1997.

Sanders, E.J., Borus, P., Ademba, G., Kuria, G., Tukei, P.M., LeDuc, J.W., Sentinel surveillance for Yellow Fever in Kenya, 1993 to 1995, *Emerg. Infect. Dis.*, 2:236–238, 1996.

Savage, H.M., Smith, G.C., Mitchell, C.J., McLean, R.G., Meisch, M.V., Vector competence of *Aedes albopictus* from Pine Bluff, Arkansas, for a St. Louis encephalitis virus strain isolated during the 1991 epidemic, *J. Am. Mosq. Control Assoc.*, 10:501–506, 1994.

Schriefer, M.E., Sacci, J.B., Jr., Dumler, J.S., Bullen, M.G., Azad, A.F., Identification of a novel rickettsial infection in a patient diagnosed with murine typhus, *J. Clin. Microbiol.*, 32:949–954, 1994.

Schriefer, M.E., Sacci, J.B., Jr., Higgins, J.A., Taylor, J.P., Azad, A.F., Murine typhus: updated role of multiple urban components and a second typhus-like Rickettsiae, *J. Med. Entomol.*, 31:681–685, 1994.

Schwartz, I., Fish, D., Daniels, T.J., Prevalence of the Rickettsial sgent of human granulocytic ehrlichiosis in ticks from a hyperendemic focus of Lyme disease, *N. Engl. J. Med.*, 337:49–50, 1997.

Schwartz, I., Varde, S., Nadelman, R.B., Wormser, G.P., Fish, D., Inhibition of efficient polymerase chain reaction amplification of *Borrelia burgdorferi* DNA in blood-fed ticks, *Am. J. Trop. Med. Hyg.*, 56:339–342, 1997.

Serufo, J.C., Montes de Oca, H., Tavares, V.A., Souza, A.M., Rosa, R.V., Jamal, M.C. et al., Isolation of dengue virus type 1 from larvae of *Aedes albopictus* in Campos Altos City, state of Minas Gerais, Brazil, *Mem. Inst. Oswaldo Cruz.*, 88:503–504, 1993.

Sexton, D.J., Rollin, P.E., Breitschwerdt, E.B., Corey, G.R., Myers, S.A., Hegarty, B.C. et al., Brief report: severe multi-organ failure following Cache Valley virus infection, *N. Engl. J. Med.*, 336:547, 1997.

Shantharam, S., Foudlin, A.S., Federal regulation of biotechnology: jurisdiction of the U.S. Department of Agriculture, in *Biotechnology for Biological Control of Pests and Vectors*, Maramorosch, K., Ed., CRC Press, Boca Raton, FL, 1991, pp. 239–250.

Simonson, L., Levin, B.R., Evaluating the risk of releasing genetically engineered organisms, *Tree*, 3:27, 1988.

Smoak, B.L., McClain, J.B., Brundage, J.F., Broadhurst, L., Kelly, D.J., Dasch, G.A. et al. An outbreak of spotted fever rickettsiosis in U.S. Army troops deployed to Botswana, *Emerg. Infect. Dis.*, 2:217–221, 1996.

Sorvillo, F.J., Gondo, B., Emmons, R., Ryan, P., Waterman, S.H., Tilzer, A. et al., A suburban focus of endemic typhus in Los Angeles County: association with seropositive domestic cats and opossums, *Am. J. Trop. Med. Hyg.*, 48:269–273, 1993.

Tassniyom, S., Vasanawathana, S., Chirawatkul, A., Rojanasuphot, S., Failure of high-dose methylprednisolone in established dengue shock syndrome: a placebo-controlled, double-blind study, *Pediatrics*, 92:111–115, 1993.

Telford, S.R., III, Armstrong, P.M., Katavolos, P., Foppa, I., Olmeda Garcia, A.S., Wilson, M.L. et al., A new tick-borne encephalitis-like virus infecting New England deer ticks, *Ixodes dammini*, *Emerg. Infect. Dis.*, 3:165–170, 1997.

Telford, S.R., III, Dawson, J.E., Katabolos, P., Warner, C.K., Kolbert, C.P., Persin, D.H., Perpetuation of the agent of human granulocytic ehrlichiosis in a deer tick–rodent cycle, *Proc. Natl. Acad. Sci. U.S.A.*, 93:629–614, 1996.

Telford, S.R., III, Lepore, T.J., Snow, P., Warner, C.K., Dawson, J.E., Human granulocytic ehrlichiosis in Massachusetts, *Ann. Intern. Med.*, 123:277–279, 1995.

Teutsch, S.M., Considerations in planning a surveillance system, in *Principles and Practice of Public Health Surveillance*, Tuetsch, S.M., Churchill, R.E., Eds., Oxford University Press, New York, 1994, pp. 18–30.

Thompson, D.F., Malone, J.B., Harb, M., Faris, R., Huh, O.K., Buck, A.A. et al., Bancroftian Filariasis distribution and diurnal temperature differences in the southern Nile Delta, *Emerg. Infect. Dis.*, 2:234–235, 1996.

UK inquiry into genetic engineering of crops, *Nature* (London), 392:534, 1998.

Vaughan, J.A., Azad, A.F., Acquisition of murine typhus rickettsiae by fleas, *Ann. N.Y. Acad. Sci.*, 590:70–75, 1990.

Wadman, M., EPA to be sued over gene-modified crops, *Nature* (London), 389:317, 1997.

Walker, D.H., Barbour, A., Oliver, J.H., Sumler, J.S., Dennis, D.T., Azad, A.F. et al., Emerging bacterial zoonotic and vector-borne diseases: prospects for the effects on the public health, *JAMA,* 275:463–469, 1996.

Walker, D.H., Dumler, J.S., Emergence of erlichiosis as human health problem, *Emerg. Infect. Dis.,* 2:18–29, 1996.

Walker, D.H., Raoult, D., *Rickettsia rickettsii* and other spotted fever group *Rickettsia* (Rocky Mountain spotted fever and other spotted fevers), in *Principles and Practices of Infectious Diseases,* 4th ed., Mandell, G.L., Bennett, J.E., Dolin, R., Eds., Churchill Livingstone, New York, 1721–1726, 1995.

Werren, J.H., Hurst, G.D.D., Zhang, W., Breeuwer, J.A.J., Stouthamer, R., Majerus, M.E.N., Rickettsial relative associated with male killing in the ladybird beetle (*Adalia bipunctata*), *J. Bacteriol.,* 176:388–394, 1994.

White, D.J., Chang, H.G., Benach, J.L., Bosler, E.M., Meldrum, S.C., Means, R.G. et al., The geographic spread and temporal increase of the Lyme disease epidemic, *JAMA,* 266:1230–1236, 1991.

World Health Organization, Dengue and dengue haemorrhagic fever, India, *Wkly. Epidemiol. Rec.,* 71:335, 1996.

World Health Organization, Vector resistance to pesticides, 15th Report of the WHO Expert Committee on Vector Biology and Control, *WHO Tech. Rep. Ser.,* 818:62, 1992.

Zanotto, P.M., Gao, G.F., Gritsun, T., Marin, M.S., Jiang, W.R., Venugopal, K. et al., An arbovirus cline across the Northern Hemisphere, *Virology,* 210:152–159, 1995.

Zanotto, P.M., Gould, E.A., Gao, G.F., Harvey, P.H., Holmes, E.C., Population dynamics of flaviviruses revealed by molecular phylogenies [see comments], *Proc. Natl. Acad. Sci. U.S.A.,* 93:548–553, 1996.

Zucker, J.R., Barber, A.M., Paxton, L.A., Schultz, L.J., Lobel, H.O., Roberts, J.M. et al., Malaria surveillance — United States, 1992, *MMWR,* 44 (SS-5):1–17, 1992.

CHAPTER 6

Benenson, A.S., Ed., *Control of Communicable Diseases in Man,* 16th ed., American Public Health Association, Washington, D.C., 1995:391–392.

Carballal, G., Videla, C.M., Merani, M.S., Epidemiology of Argentine hemorrhagic fever, *Eur. J. Epidemiol.,* 4:259–274, 1988.

CDC, Outbreak of acute febrile illness and pulmonary hemorrhage–Nicaragua, 1995, *MMRW,* 44:841–843, 1995.

CDC, Outbreak of acute febrile illness among athletes participating in triathlons — Wisconsin and Illinois, 1998, *MMWR,* 47:585–588, 1998.

Cheek, J., Bryan, R., Glass, G., Geographic distribution of high-risk, HPS areas in the U.S. Southwest, in *Proceedings of the 4th International Conference on HFRS and Hantaviruses,* March 5–7, 1998, Centers for Disease Control and Prevention, Atlanta, GA, 1998.

Childs, J.E., Korch, G.W., Glass, G.E., LeDuc, J.W., Shah, K.V., Epizootiology of hantavirus infections in Baltimore: isolation of a virus from Norway rats, and characteristics of infected rat populations, *Am. J. Epidemiol.,* 126:55–68, 1987.

Drebot, M.A., Artsob, H., Dick, D., Lindsay, R., Surgeoner, G., Johnson, A. et al. Hantavirus surveillance in Canadian rodents, *Am. J. Trop. Med. Hyg.,* 57:143, 1997.

Faine, S., Leptospirosis, in *Bacterial Infections, Topley and Wilson's Microbiology and Microbial Infections,* 9th ed. Vol. 3, Hausler, W.J., Jr., Sussman, M., Collier, L., Balows, A., Eds., Arnold, London, 1998, pp. 849–869.

Farr, R.W., Leptospirosis, *Clin. Infect. Dis.,* 21:1-8, 1995.

Farrar, W.E., Leptospira species (Leptospirosis), in *Principles and Practice of Infectious Diseases,* 4th ed., Mandell, G.L., Bennett, J.E., Dolin, R., Eds., Churchill Livingstone, New York, 1995, pp. 2137–2141.

Friedland, J.S., Warrell, D.A., The Jarisch-Herxheimer reaction in leptospirosis: possible pathogenesis and review, *Rev. Infect. Dis.,* 12:207–210, 1991.

Fulhorst, C.F., Bowen, M.D., Ksiazek, T.G., Rollin, P.E., Nichol, S.T., Kosoy, M.Y. et al. Isolation and characterization of Whitewater Arroyo virus, a novel North American arenavirus, *Virology,* 224:114–120, 1996.

Glass, G.E., Childs, J.E., Korch, G.W., LeDuc, J.W., Association of intraspecific wounding with hantaviral infection in wild rats *(Rattus norvegicus), Epidemiol. Infect.,* 101:459–472, 1988.

Glass, G.E., Watson, A.J., LeDuc, J.W., Childs, J.E., Domestic cases of hemorrhagic fever with renal syndrome in the United States, *Nephron.,* 68:48–51, 1994.

Goodman, R.A., Thacker, S.B., Solomon, S.L., Osterholm, M.T., Hughes, J.M., Infectious diseases in competitive sports, *JAMA,* 271:862–867, 1994.

Gravekamp, C., Van de Kemp, H., Franzen, M. et al., Detection of seven species of pathogenic leptospires by PCR using two sets of primers, *J. Gen. Microb.,* 139:1691–1700, 1993.

Hafner, M.S., Gannon, W.L., Salazar-Bravo, J., Alvarez-Castaneda, S.T., Mammal collections in the western hemisphere, American Society of Mammalogists, Lawrence, KS, 1997.

Hutchinson, K.L., Rollin, P.E., Peters, C.J., Pathogenesis of North American hantavirus, Black Creek canal virus, in experimentally infected *Sigmodon hispidus, Am. J. Trop. Med. Hyg.,* 59:58–65, 1998.

Jackson, L.A., Kaufmann, A.F., Adams, W.G. et al., Outbreak of leptospirosis associated with swimming, *Pediatr. Infect. Dis. J.,* 12:48–54, 1993.

Kaufmann, A.F., Weyant, R.S., Leptospiraceae, in *Manual of Clinical Microbiology,* 6th ed., Murray, P.R., Baron, E.J., Pfaller, M.A., Tenover, F.C., Yolken, R.H., Eds., American Society for Microbiology, Washington, D.C., 1995, pp. 621–625.

Leirs, H., Mills, J.N., Krebs. J.W., Childs, J.E., Akaibe, D., Woollen, N. et al., Search for the ebola reservoir in Kikwit: reflections on the vertebrate collection, *J. Infect. Dis.,* 1998.

Mills, J.N., Childs, J.E., Ksiazek, T.G., Peters, C.J., Velleca, W.M., *Methods for Trapping and Sampling Small Mammals for Virologic Testing,* U.S. Department of Health and Human Services, Atlanta, GA, 1995, p. 61.

Mills, J.N., Ellis, B.A., McKee, K.T., Calderón, G.E., Maiztegui, J.I., Nelson, G.O. et al. A longitudinal study of Junin virus activity in the rodent reservoir of Argentine hemorrhagic fever, *Am. J. Trop. Med. Hyg.,* 47:749–763, 1992.

Mills, J.N., Ellis, B.A., McKee, K.T., Ksiazek, T.G., Oro, J.G., Maiztegui, J.I. et al., Junin virus activity in rodents from endemic and nonendemic loci in Central Argentina, *Am. J. Trop. Med. Hyg.,* 44:589–597, 1991.

Mills, J.N., Ellis, B.A., McKee, K.T., Jr., Maiztegui, J.I., Childs, J.E., Reproductive characteristics of rodent assemblages in cultivated regions of central Argentina, *J. Mammal..* 73:515–526, 1992.

Mills, J.N., Johnson, J.M., Ksiazek, T.G., Ellis, B.A., Rollin, P.E., Yates, T.L. et al., A survey of hantavirus antibody in small mammal populations in selected U.S. national parks, *Am J. Trop. Med. Hyg.,* 58:525–532, 1998.

Mills, J.N., Ksiazek, T.G., Ellis, B.A., Rollin, P.E., Nichol, S.T., Yates, T.L. et al., Patterns of association with host and habitat: antibody reactive with Sin Nombre Virus in small mammals in the major biotic communities of the southwestern United States, *Am. J. Trop. Med. Hyg.,* 56:273–284, 1997.

Mills, J.N., Yates, T.L., Childs, J.E., Parmenter, R.R., Ksiazek, T.G., Rollin, P.E. et al. Guidelines for working with rodents potentially infected with hantavirus, *J. Mammol.,* 76:716–722, 1995.

Niklasson, B., Hornfeldt, B., Lundkvist, A., Bjorsten, S., LeDuc, J., Temporal dynamics of Puumala virus antibody prevalence in voles and of Nephropathia Epidemica incidence in humans, *Am. J. Trop. Med. Hyg.,* 53:134–140, 1995.

Rupp, M.E., *Streptobacillus moniliformis* endocarditis: case report and review, *Clin. Infect. Dis.,* 14:769–772, 1992.

Schmidt, K., Ksiazek, T.G., Mills, J.N., Ecology and biologic characteristics of Argentine rodents with antibody to hantavirus, in *Proceedings of the 4th International Conference on HFRS and Hantaviruses,* March 5–7, 1998, Centers for Disease Control and Prevention, Atlanta, GA, 1998.

U. S. Department of Health and Human Services, Center for Disease Control, *Hantavirus* infection — southwestern United States: interim recommendations for risk reduction, *MMWR* 42(RR-11), July 30, 1993.

U. S. Department of Health and Human Services, Center for Disease Control, Laboratory management of agents associated with hantavirus pulmonary syndrome: interim biosafety guidelines, *MMWR* 43(RR-7), May 13, 1994.

Washburn, R.G., *Streptobacillus moniliformis* (Rat-bite fever), in *Principles and Practice of Infectious Diseases,* Vol. 2, Mandell, G.L., Bennett, J.E., Dolin, R., Eds., Churchill Livingstone, New York, 1995, pp. 2084–2086.

Yahnke, C.J., Meserve, P.L., Ksiazek, T.G., Mills, J.N., Prevalence of hantavirus antibody in wild populations of *Calomys laucha* in the central Paraguayan Chaco, in *Proceedings of the 4th International Conference on HFRS and Hantaviruses,* March 5–7, 1998, Centers for Disease Control and Prevention, Atlanta, GA, 1998.

Yates, T.L., Tissues, cell suspensions, and chromosomes, in *Measuring and Monitoring Biological Diversity: Standard Methods for Mammals,* Wilson, D.E., Cole, F.R., Nichols, J.D., Rudran, R., Foster, M.S., Eds., Smithsonian Institution Press, Washington, D.C., 1996, pp. 275–278.

Yates, T.L., Jones, C., Cook, J.A., Preservation of voucher specimens, in *Measuring and Monitoring Biological Diversity: Standard Methods for Mammals,* Wilson, D.E., Cole, F.R., Nichols, J.D., Rudran, R., Foster, M.S., Eds., Smithsonian Institution Press, Washington, D.C., 1996, pp. 265–274.

Zaki, S.R., Shieh, W.-J., Epidemic Working Group at the Ministry of Health in Nicaragua, Leptospirosis associated with outbreak of acute febrile illness and pulmonary haemorrhage, Nicaragua, 1995 {Letter}, *Lancet,* 347:535–536, 1996.

CHAPTER 7

Baker, D.B., Wallrabenstein, L.K., Richards, R.P., in *New Directions in Pesticide Research, Development, Management, and Policy: Proceedings of the Fourth National Conference on Pesticides,* Virginia Polytechnic Institute and State University, Blacksburg, VA, 1994, pp. 470–494.

Barbash, J.E., Resek, E.A. *Pesticides in Ground Water: Distribution, Trends, and Governing Factors,* Ann Arbor Press, Chelsea, MI; *Pesticides in the Hydrologic System* series, Vol. 2, 590 p.

Bardin, P.G., Van Eeden, S.F., Moolman, J.A. et al., Organophosphate and carbamate poisoning, *Arch. Intern. Med.,* 154:1433–1441, 1994.

Baron, R.L., Carbamate insecticides, in *Handbook of Pesticide Toxicology,* Vol. 3, Hayes, W.J., Jr., Laws, E.R., Jr., Eds., Academic Press, San Diego, CA, 1991.

Barron, M.G., Bioconcentration, *Environ. Sci. Technol.,* 24(11):1612–1618, 1990.

Battaglin, W.A., Goolsby, D.A., *J. Hydrology,* 196:1–25, 1997.

Belluck, D., Benjamin, S., Pesticides and human health, *J. Environ. Health,* 53(7):11–12, 1990.

Campt, D., Pesticide regulation is sound, protective, and steady, *Chem. Eng. News,* 69:44, 1991.

Chapra, S.C., Boyer, J.M., Fate of environmental pollutants, *Water Environ. Res.,* 64(4):581–593, 1992.

Christensen, E., Pesticide regulation and international trade, *Environment,* 32:2–3, 1990.

Coats, J.R., in *Pesticide Transformation Products: Fate and Significance in the Environment,* Somasundaram, L., Coats, J.R., Eds. ACS Symposium Series 459, American Chemical Society, Washington, D.C., 1991, pp. 10–31.

Connolly, J.P., Application of a food chain model to polychlorinated biphenyl contamination of the lobster and winter flounder food chains in New Bedford Harbor, *Environ. Sci. Technol.,* 25(4):760–770, 1991.

DeBleeker, J., Willems, J., Van Den Neucker, K. et al., Prolonged toxicity with intermediate syndrome after combined parathion and methyl parathion poisoning, *Clin. Toxicol.,* 30:333–345, 1992.

DeBleeker, J., Willems, J., Van Den Neucker, K., Colardyn, F., Intermediate syndrome in organophosphorus poisoning: a prospective study. *Crit. Care Med.,* 21:1706–1711, 1993.

Gallo, M.A., Lawryk, N.J., Organic phosphorus pesticides, in *Handbook of Pesticide Toxicology,* Vol. 2, *Classes of Pesticides,* Haves, W.J., Laws, E.R., Eds., Academic Press, San Diego, CA, 1991.

Garcia-Repetto, R., Martinez, D., Repetto, M., Coefficient of distribution of some organophosphorus pesticides in rat tissue, *Vet. Hum. Toxicol.,* 37:226–229, 1995.

Gianessi, L.P., Anderson, J.E., *Pesticide Use in U.S. Crop Production: National Data Report,* National Center for Food and Agricultural Policy, Washington, D.C., February 1995 (Revised April 1996).

Gilliom, R.J., Thelin, G.P., U.S. Geological Survey Circular No. 11331, 1996.

Goswamy, R., Chaudhuri, A., Mahashur, A.A., Study of respiratory failure in organosphosphate and carbamate poisoning, *Heart Lung,* 23:466–472, 1994.

Howard, P.H., *Handbook of Environmental Fate and Exposure Data for Organic Chemicals,* Vol. 1–3, Lewis Publishers, Boca Raton, FL, 1989.

Jamal, J.A., Neurological syndromes of organophosphorus compounds, *Adverse Drug React. Toxicol. Rev.,* 16(3):133–170, 1997.

Katona, B., Wason, S., Superwarfarin poisoning, *J. Emerg. Med.,* 7:627–631, 1989.

Kruse, J.A., Carlson, R.W., Fatal rodenticide poisoning with brodifacoum, *Ann. Emerg. Med.,* 21:331–336, 1992.

Larson, S.J., Capel, P.D., Majewski, M.S., *Pesticides in Surface Waters: Distribution, Trends, and Governing Factors,* Ann Arbor Press, Chelsea, MI, 1997, 373 p.

Lifshitz, M., Shahak, E., Bolotin, A., Sofer, S., Carbamate poisoning in early childhood and in adults, *Clin. Toxicol.,* 35:25–27, 1997.

Lima, J.S., Reis, C.A., Poisoning due to illegal use of carbamates as a rodenticide in Rio De Janeiro, *Clin. Toxicol.,* 35:687–690, 1995.

Litovitz, T.L., Smilkstein, M., Felberg, L., Klein-Schwartz, W., Berlin, R., Morgan, J.L. 1996 report of the American Association of Poison Control Centers Toxic Exposure Surveillance System, *Am. J. Emerg. Med.,* 15:447–500, 1997.

Mack, R.B., Not all rats have four legs: superwarfarin poisoning, *N.C. Med. J.,* 55:554–556, 1994.

Majewski, M.S., Capel, P.D., *Pesticides in the Atmosphere: Distribution, Trends, and Governing Factors,* Ann Arbor Press, Chelsea, MI, 1995, 214 p.

Manz, A., Berger J., Cancer mortality among workers in chemical plant contaminated with dioxin, *Lancet,* 338(8773):959–964, 1991.

Meggs, W.J., Hoffman, R.S., Shih, R.D. et al., Thallium poisoning from maliciously contaminated food, *J. Toxicol. Clin. Toxicol.,* 32:723–730, 1994.

Mersie, W., Seybold, C. J., Agric. Food Chem., 44(7):1925-1929, 1996.

Miller, C.S., Mitzel, H.C., Chemical sensitivity attributed to pesticide exposure versus remodeling, *Arch. Environ. Health.,* 50:119–129, 1995.

Montgomery, J.H., *Agrochemicals Desk Reference: Environmental Data,* Lewis Publishers, Chelsea, MI, 1993.

Pogoda, J.M., Preston-Martin, S., Household pesticides and risk of pediatric brain tumors, *Environ. Health Perspect.,* 105:1214–1220, 1997.

Risk Assessment of Pesticides, *Chem. Eng. News,* 69(1):27–55, 1991.

Rosenstock, L., Keifer, M., Daniell, W. et al., Chronic central nervous system effects of acute organophosphate pesticide intoxication, *Lancet,* 338:223–227, 1991.

Scott, J.C., Water-Resou. Invest. Rep. U.S. Geol. Surv No. 90-4101, 1990.

Sericano, J.L., Wade, T.L., Atlas, E.L., Historical perspective on the environmental bioavailability of DDT and its derivatives to Gulf of Mexico oysters, *Environ. Sci. Technol.,* 24:1541–1548, October 1990.

Squillace, P.J., Zogorski, J.S., Wilver, W.G., Price, C.V., *Environ. Sci. Technol.,* 30(5):1721–1730, 1996.

Steenland, K., Jenkins, B., Ames, R.G. et al., Chronic neurological sequelae to organophosphate poisoning, *Am. J. Public Health.,* 84:731–736, 1994.

Sullivan, J.B., Blose, J., Organophosphate and carbamate insecticides, in *Hazardous Materials Toxicology,* Sullivan, J.B., Krieger, G.R., Eds., Williams & Wilkins, Baltimore, MD, 1992, pp. 1015–1026.

U.S. Environmental Protection Agency, Drinking Water Regulations and Health Advisories, U.S. Environmental Protection Agency, Office of Water. U.S. Government Printing Office, Washington, D.C., 1996.

Walls, D., Smith, P.G., Mansell, M.G., *Int. J. Environ. Health Res.,* 6:55–62, 1996.

Worthy, W., Household pesticide hazards examined, *Chem. Eng. News,* 69(18):50, 1991.

Worthy, W., Pesticides, nitrates found in U.S. wells, *Chem. Eng. News,* 69(18):46–48, 1991.

Zilberman, D., The economics of pesticide use and regulation, *Science,* 253(5019):518–522, 1991.

CHAPTER 8

Axelson, O., Cancer risks from exposure to radon in homes, *Environ. Health Perspect.,* 103:37–43, 1995.

Barnes, P.F., El-Hajj, H., Preston-Martin, S. et al., Transmission of tuberculosis among the urban homeless, *JAMA.* 275:305–307, 1996.

Berkelman, R.L., Stroup, D.F., Buehler, J.W., Public health surveillance., in *Oxford Textbook of Public Health,* 3rd ed., Detels, R., Holland, W.W., McEwan, J., Omenn, G.S., Eds., Oxford University Press, New York, 1997, pp. 735–750.

Bourhy, H., Kissi, B., Lafon M., Sacramento, D., Tordo, N., Antigenic and molecular characterization of bat rabies virus in Europe, *J. Clin. Microbiol.,* 30:2419–2426, 1992.

Breisch, S.L., Asbestos: an environmental liability, *Saf. Health,* 141(5):60–62, 1991.

Brody, D.J., Pirkle, J.L., Kramer, R.A. et al., Blood lead levels in the U.S. population: Phase 1 of the Third National Health and Nutrition Examination Survey (NHANES III, 1988 to 1991), *JAMA*, 272:277–283, 1994.

Bureau of the Census, Economic and Statistics Administration, U.S. Department of Commerce, Population estimates, Accessed March, 1999.

Burt, C.W., Injury-related visits to hospital emergency departments: United States, 1992. Advance Data, Number 261, National Center for Health Statistics, Hyattsville, MD, February 1, 1995.

Burton, J.D., Checklists and questionnaires aid investigation of indoor air quality, *Occup. Health Saf.*, 60(5):82–85, 1991.

Burton, J.D., Physical, psychological complaints can be the result of indoor air quality, *Occup. Health Saf.,*. 60(3):53, 1991.

Canadian Center for Occupational Health and Safety, Asbestos, *Cheminfo*, 12001-29-5, 1990.

Canadian Center for Occupational Health and Safety, Carbon monoxide, *Cheminfo*, 630-08-0, 1990.

Canadian Center for Occupational Health and Safety, Formaldehyde, *Cheminfo*, 50-00-0, 1991.

Canadian Center for Occupational Health and Safety, Formic acid, *Cheminfo*, 64-18-6, 1991.

Canadian Center for Occupational Health and Safety, Vinyl chloride, *Cheminfo*, 75-01-4, 1991.

Castell, J., Mismatch between risk and money spent often due to public opinion, *Saf. Health*, 141(2):34–36, 1990.

Castell, J., Reproductive health hazards in the workplace, *Saf. Health*, 141(6):70–72, 1990.

CDC, Adult blood lead epidemiology and surveillance — United States, first quarter, 1998 and annually 1994–1997, *MMWR*, 47:907–911, 1998.

CDC, Adult blood lead epidemiology and surveillance — United States, second quarter, 1997, *MMWR*, 46:1000–1002, 1997.

CDC, Adult blood lead epidemiology and surveillance — United States, third quarter, 1997, *MMWR*, 47:77–80, 1998.

CDC, Compendium of animal rabies control, 1997: National Association of State Public Health Veterinarians, *MMWR*, 46(RR-4), 1997.

CDC, Guidelines for evaluating surveillance systems, *MMWR*, 37(S-5), 1988.

CDC, Human Rabies — Montana and Washington, 1997, *MMWR*, 46:770-4, 1997.

CDC, Human Rabies — Texas and New Jersey, 1997, *MMWR*, 47:1–5, 1998.

CDC, Human rabies prevention — United States, 1999: recommendations of the Advisory Committee on Immunization Practices (ACIP), *MMWR*, 48(RR-1), 1999.

CDC, Rabies prevention — United States, 1991: recommendations of the Immunization Practices Advisory Committee (ACIP), *MMWR*, 40(RR-3), 1991.

CDC, Hypothermia-related deaths — Cook County, IL, November 1992–March 1993, *MMWR*, 42:917–919, 1993.

Centers for Disease Control and Prevention, Compendium of animal rabies control, 1995, *MMWR*, 44:(RR-2):1–9, 1995.

Centers for Disease Control and Prevention, Human rabies — Alabama, Tennessee, and Texas, 1994, *MMWR*, 44:269–272, 1995.

Centers for Disease Control and Prevention, Human rabies — Texas, Arkansas, and Georgia, 1991, *MMWR*, 40:765–769, 1991.

Centers for Disease Control and Prevention, Human rabies — Washington State, 1995, *MMWR*, 44:625–627, 1995.

Centers for Disease Control and Prevention, Raccoon rabies epizootic: United States, 1993, *MMWR*, 43:269–273, 1994.

Centers for Disease Control and Prevention, Translocation of coyote rabies — Florida, 1994. *MMWR,* 44:580-1-7, 1995.

Cole, G.E., Survey aids in assessing the impact of IAQ on worker health, *Occup. Health Saf.,* 60(5):38–51, 1991.

Crawford, R., Campbell, D., Ross, J., Carbon monoxide poisoning in the home: recognition and treatment, *Br. Med. J.,* 301:977–979, 1990.

Dalton, M.J., Robinson, L.E., Cooper, J., Regnery, R.L., Olson, J.G., Childs, J.E., Use of *Bartonella* antigens for serologic diagnosis of cat-scratch disease at a national referral center, *Arch. Intern. Med.,* 155:1670–1676, 1995.

Debbie, J.G., Rabies control of terrestrial wildlife by population reduction, in *The Natural History of Rabies,* 2nd ed., Baer, G.M., Ed., CRC Press, Boca Raton, FL, 1991, pp. 477–484.

DeWerth, D.W., Borgeson, R.A., Aronov, M.A., Development of sizing guidelines for vent-free supplemental heating products — topical report, Gas Appliance Manufacturers Association and Gas Research Institute, Arlington, VA, 1996.

Donoghue, E.R., Graham, M.A., Jentzen, J.M., Lifschultz, B.D., Luke, J.L., Mirchandani, H.G., National Association of Medical Examiners Ad Hoc Committee on the Definition of Heat-Related Fatalities, Criteria for the diagnosis of heat-related deaths. National Association of Medical Examiners, *Am. J. Forensic Med. Pathol.,* 18:11–14, 1997.

Drancourt, M., Mainardi, J.L., Brouqui, P. et al., *Bartonella (Rochalimaea) quintana* endocarditis in three homeless men, *N. Engl. J. Med.,* 332:424–489, 1995.

Elgart, M.L., Pediculosis, *Dermatol. Clin.,* 8:219–228, 1990.

Evans, P., An integrated approach to indoor air quality, *Consult. Specify. Eng.,* 10(1):48–54, 1991.

Feder, J.M., Jr., Nelson, R., Reiher, H.W., Bat bite? {Letter}, *Lancet,* 350:1300, 1997.

Fishbein, D.B., Robinson, L.E., Rabies, *N. Engl. J. Med.,* 329:1632–1638, 1993.

Ford, E.S., Kelly, A.E., Teutsch, S.M., Thacker, S.B., Garbe, P.L., Radon and lung cancer: a cost-effectiveness analysis, *Am. J. Public Health,* 89:351–357, 1999.

Freeman, R., Perks, D., Carboxyhemoglobin, *Anesthesia,* 57:45–51, 1990.

Gershman, K.A., Sacks, J.J., Wright, J.C., Which dogs bite? A case-control study of risk factors, *Pediatrics,* 93:913–917, 1994.

Gesser, H D., Removal of aldehydes and acidic pollutants from indoor air, *Environ. Sci. Technol.,* 24(4):495–497, 1990.

Gilbert, E.S., Dagle, G.E., Cross, F.T., Analysis of lung tumor risks in rats exposed to radon, *Radiat. Res.,* 145:350–360, 1996.

Grossman, D., Jr., Ambient levels of formaldehyde and formic acid, *Environ. Sci. Technol.,* 25(4):710–714, 1991.

Hall, J.R., *The U.S. Fire Problem and Overview Report: Leading Causes and Other Patterns and Trends.* National Fire Protection Association, Fire Analysis and Research Division, Quincy, MA, 1998.

Hanlon, C.A., Buchanan, J.R., Nelson, E. et al., A vaccinia-vectored rabies vaccine field trial: ante- and post-mortem biomarkers, *Rev. Sci. Technol.,* 99–107, 1993.

Hanlon, C.A., Trimarchi, C., Harris-Valente, K., Debbie, J.G., Raccoon rabies in New York state: epizootiology, economics, and control, Presented at the 5th Annual International Meeting of Rabies in the Americas, Niagara Falls, Ontario, Canada, 1994, Abstract, p. 16.

Harty, S., Governmental risk management: risk managers need to assess air quality problems, *Bus. Insurance,* 25(21):12–14, 1991.

Heckerling, P., Leikin, J., Mturen, A., Terzian, C., Segarra, D., Screening hospital admissions from the emergency department for occult carbon monoxide poisoning, *Am. J. Emerg. Med.,* 8:301–304, July 1990.

Hertz-Picciotto, I., Toward a coordinated system for the surveillance of environmental health hazards {Comment}, *Am. J. Public Health,* 86:638–641, 1996.

Huston, C.S., Heat Cramps, in *The Merck Manual,* Berkow, R., Ed., 7th ed., Merck & Co., Rahway, NJ, 1992, p. 2511.

Ilano, A.L., Raffin, T.A., Management of carbon monoxide poisoning, *Chest,* 97:165–169, 1990.

Jackson, L.A., Spach, D.H., Kippen, D.A., Sugg, N.K., Regnery, R.L., Sayers, M.H., Stamm, W.E., Seroprevalence to *Bartonella quintana* among patients at a community clinic in downtown Seattle, *J. Infect. Dis.,* 173:1023–1026, 1996.

Karter, M.J., Jr., Fire loss in the United States during 1995, National Fire Protection Association, Fire Analysis and Research Division, Quincy, MA, 1996.

Katzel, J., A common sense approach to controlling indoor air quality, *Plant Eng.,* 45(8):32–38, 1991.

Kennedy, T., Six steps for improving IAQ, *J. Prop. Manage.,* 56(2):44–46, 1991.

Kilbourne, E.M., Heat waves and hot environments, in *The Public Health Consequences of Disasters,* Noji, E.K., Ed., Oxford University Press, New York, 1997, pp. 245–269.

Kilbourne, E.M., Choi, K., Jones, T.S., Thacker, S.B., Field investigation team, risk factors for heat stroke: a case-control study, *JAMA,* 247:3332–3336, 1982.

King, A.A., Meredith, C.D., Thomson, G.R., The Biology of Southern Africa Lyssavirus Variants, in *Lyssaviruses,* Rupprecht, C.E., Dietzschold, B., Koprowski, H., Ed., Springer-Verlag, New York, 1994, pp. 267–296.

Kjrgaard, S., Human reactions to a mixture of indoor air volatile organic compounds, *Atmos. Environ.,* 25(8):1417–1426, 1991.

Koehler, J.E., Quinn, F.D., Berger, T.G., LeBoit, P.E., Tappero, J.W., Isolation of *Rochalimaea* species from cutaneous and osseous lesions of bacillary angiomatosis, *N. Engl. J. Med.,* 327:1625–1631, 1992.

Krebs, J.W., Smith, J.S., Rupprecht, C.E., Childs, J.E., Rabies surveillance in the United States during 1996, *J. Am. Vet. Med. Assoc.,* 211:1525–1539, 1997.

Krebs, J.W., Smith, J.S., Rupprecht, C.E., Childs, J.E., Rabies surveillance in the United States during 1997, *J. Am. Vet. Med. Assoc.,* 213:1713–1728, 1998.

Krebs, J.W., Strine, T.W., Smith, J.S., Rupprecht, C.E., Childs, J.E., Rabies surveillance in the United States during 1993, *J. Am. Vet. Med. Assoc.,* 205:1695–1709, 1994.

Kriess, K., The sick building syndrome: where is the epidemiological basis?, *Am. J. Pub. Health,* 80(10):1171–1173, 1990.

Lee, D.H., Seventy-five years of searching for a heat index, *Environ. Res.,* 22:331–356, 1980.

Levesque, B., Dewailly, E., Lavoie, R., Carbon monoxide in indoor ice skating rinks: evaluation of absorption by adult hockey players, *Am. J. Pub. Health Assoc.,* 80:594–597, 1990.

Lubin, J.H., Boice, J.D., Jr., Lung cancer risk from residential radon: meta-analysis of eight epidemiologic studies, *J. Natl. Cancer. Inst.,* 89:49–57, 1997.

Lubin, J.H., Steindorf, K., Cigarette use and the estimation of lung cancer attributable to radon in the United States, *Radiat. Res.,* 141:79–85, 1995.

MacDonald, S.C., Pertowski, C.A., Jackson, R.J., Environmental public health surveillance, *J. Public Health Manage. Pract.,* 2:45–49, 1996.

Mallonee, S., Istre, G.R., Rosenberg, M. et al., Surveillance and prevention of residential-fire injuries, *N. Engl. J. Med.,* 335:27–31, 1996.

Malven, F., Threat based definition of health, safety, and welfare, in design: an overview, *J. Int. Des. Edu. Res.* 16(2):5–15, 1990.

Mansfield, G., Air quality monitors, *Saf. Health,* 141(5):50–51, 1991.

Maurin, M., Roux, V., Stein, A., Ferrier, F., Viraben, R., Raoult, D., Isolation and characterization by immunofluorescence, sodium dodecyl sulfate-polyacrylamide gel electrophoresis, western blot, restriction fragment length polymorphism-PCR, 16S rRNA gene sequencing, and pulsed-field gel electrophoresis of *Rochalimaea quintana* from a patient with bacillary angiomatosis, *J. Clin. Microbiol.,* 32:1166–1171, 1994.

Mendez, D., Warner, K.E., Courant, P.N., Effects of radon mitigation vs. smoking cessation in reducing radon-related risk of lung cancer, *Am. J. Public Health,* 88:811–812, 1998.

Meriweather, R.A., Blueprint for a national public health surveillance system for the 21st century, *J. Public Health Manage. Pract.,* 2:16–23, 1996.

Meslin, F.X., Fishbein, D.B., Matter, H.C., Rationale and prospects for rabies elimination in developing countries, in *Lyssaviruses,* Rupprecht, C.E., Dietzschold, B., Koprowski, J., Eds., Springer-Verlag, New York, 1994, pp. 1–26.

Moseley, C., Indoor air quality problems, *J. Environ. Health,* 53(3):19–21, 1990.

National Center for Health Statistics, Compressed Mortality File, U.S. Department of Health and Human Services, Public Health Service, CDC, Atlanta, GA, 1999.

National Center for Health Statistics, National summary of injury mortality data, 1995, U.S. Department of Health and Human Services, CDC, National Center for Health Statistics, Hyattsville, MD, 1997.

National Center for Health Statistics, Plan and operation of the Third National Health and Nutrition Examination Survey, 1988–1994, DHHS Publication No. (PHS)94-1308. (Vital and Health Statistics; Series 1, No. 32), U.S. Department of Health and Human Services, Public Health Service, CDC, Hyattsville, MD, 1994.

National Electronic Injury Surveillance System, National Consumer Product Safety Commission, Washington, D.C., 1995.

NCIPC, National summary of injury mortality data, 1988–1994, Centers for Disease Control and Prevention, Atlanta, GA, 1996.

New systems monitor air pollution exposure, *J. Environ. Health,* 52(4):338, 1990.

New York State Energy Research and Development Authority, Critique of guidelines on use of unvented gas space heaters, New York State Energy Research and Development Authority, Albany, NY, 1997.

Noah, D.L., Drenzek, C.L., Smith, J.S. et al., Epidemiology of human rabies in the United States, 1980 to 1996, *Ann. Intern. Med.,* 128:922–930, 1998.

Nowak, R.M., *Walker's Mammals of the World,* 5th ed., Johns Hopkins University Press, Baltimore, 1991, p. 198.

Peters, K.D., Kochanek, K.D., Murphy, S.L., Deaths: final data for 1996, National Vital Statistics Reports. Vol. 47, No. 9., National Center for Health Statistics, Hyattsville, MD, 1998.

Petty, K.J., Hypothermia, in *Harrison's Principles of Internal Medicine,* 14th ed., Vol 1, Fauci, A.S., Braunwald, E., Isselbacher, K.J. et al., Eds., McGraw-Hill, New York, 1998, pp. 97–99.

Pirkle, J.L., Brody, D.J., Gunter, E.W. et al., The decline in blood lead levels in the United States: the National Health and Nutrition Examination Surveys (NHANES), *JAMA,* 272:284–291, 1994.

Public Health Service, Healthy people 2000: national health promotion and disease prevention objectives — full report, with commentary, DHHS Publication No. (PHS)91-50212, U.S. Department of Health and Human Services, Public Health Service, Washington, D.C., 1990.

Public Health Service, Healthy people 2000 — midcourse review and 1995 revisions, U.S. Department of Health and Human Services, Public Health Service, Washington, D.C., 1995.

Rademaker, K., Common sense solutions to poor indoor air, *Occup. Hazards,* 53(10):41–44, 1991.

Reactive, volatile organic compounds major contaminants in IAQ, *Occup. Health Saf.,* 59:106, 1990.

Residential formaldehyde, *J. Environ. Health.,* 55:34–37, 1990.

Robbins, A.H., Niezgoda, M., Levine, S. et al., Oral rabies vaccination of raccoons (*Procyon lotor*) on the Cape Cod isthmus, Massachusetts, Presented at the 5th Annual International Meeting of Rabies in the Americas, Niagara Falls, Ontario, Canada, 1994, Abstract, p. 29.

Roscoe, D.E., Holste, W., Niezgoda, M., Rupprecht, C.E., Efficacy of the V-RG oral rabies vaccine in blocking epizootic raccoon rabies, Presented at the 5th Annual International Meeting of Rabies in the Americas, Niagara Falls, Ontario, Canada, 1994, Abstract, p. 33.

Rupprecht, C.E., Hanlon, C.A., Niezgoda, M., Buchanan, J.R., Diehl, D., Koprowski, H., Recombinant rabies vaccines: efficacy assessment in free-ranging animals, *Onderstepoort J. Vet. Res.,* 60:463–468, 1993.

Rupprecht, C.E., Smith, J.S., Raccoon rabies — the re-emergence of an epizootic in a densely populated area, *Semin. Virol.,* 5:155–164, 1994.

Sacks, J.J., Kresnow, M., Houston, B., Dog bites: how big a problem?, *Injury Prev.,* 2:52–54, 1996.

Sacks, J.J., Lockwood, R., Hornreich, J., Sattin, R.W., Fatal dog attacks, 1989–1994, *Pediatrics,* 97:891–895, 1996.

Sacks, J.J., Sattin, R.W., Bonzo, S.E., Dog bite-related fatalities from 1979 through 1988, *JAMA,* 262:1489–1492, 1989.

Samet, J.M., Spengler, J.D., *Indoor Air Pollution: A Health Perspective,* Johns Hopkins University Press, Baltimore, MD, 1991.

Sasaki, D.M., Middleton, C.R., Sawa, T.R., Christensen, C.C., Kobayashi, G.Y., Rabid bat diagnosed in Hawaii, *Hawaii Med. J.,* 51:181–185, 1992.

Sattin, R.W., Falls among older persons: a public health perspective, *Annu. Rev. Public Health,* 13:489–508, 1992.

Semenza, J.C., Rubin, C.H., Falter, K.H. et al., Risk factors for heat-related mortality during the July 1995 heat wave in Chicago, *N. Engl. J. Med.,* 35:84–90, 1996.

Sheps, D., Herbst, M., Hinderliter, A., Adams, K., Production of arrhythmias by elevated carboxyhemoglobin in patients with coronary artery disease, *Ann. Intern. Med.,* 113:343–351, September 1990.

Singh, G.K., Kochanek, K.D., MacDorman, M.F., Advance report of final mortality statistics reports, 1994, Monthly Vital Statistics Report, Vol. 45, No. 3, Supplement, National Center for Health Statistics, Hyattsville, MD, February 1, 1996.

Slater, L.N., Welch, D.F., *Rochalimaea* species (recently named *Bartonella*), in *Principles and Practice of Infectious Diseases,* 4th ed., Mandell, G.L., Bennett, J.E., Dolin, R., Eds., Churchill Livingstone, New York, 1995, pp. 1741–1747.

Smith, S., Controlling sick building syndrome, *J. Environ. Health,* 53(3):22–23, 1990.

Spach, D.H., Callis, K.P., Paauw, D.S. et al., Endocarditis caused by *Rochalimaea quintana* in a patient infected with human immunodeficiency virus, *J. Clin. Microbiol.,* 31:692–694, 1993.

Spach, D.H., Kanter, A.S., Dougherty, M.J. et al., *Bartonella (Rochalimaea) quintana* bacteremia in inner-city patients with chronic alcoholism, *N. Engl. J. Med.,* 332:424–428, 1995.

Stein, A., Raoult, D., Return of trench fever [Letter], *Lancet,* 345:450–451, 1995.

Swanepoel, R., Barnard, B.J.H., Meredith, C.D. et al., Rabies in Southern Africa, *Onderste-poort J. Vet. Res.,* 60:325–346, 1993.

Tappero, J.W., Mohle-Boetani, J., Koehler, J.E. et al., The epidemiology of bacillary angiomatosis and bacillary peliosis, *JAMA,* 269:770–775, 1993.

Thacker, S.B., Stroup, D.F., Future directions for comprehensive public health surveillance and health information systems in the United States, *Am. J. Epidemiol.,* 140:383–397, 1994.

Thacker, S.B., Stroup, D.F., Parrish, R.G., Anderson, H.A., Surveillance in environmental public health: issues, systems, and sources, *Am. J. Public Health,* 86:633–638, 1996.

Thomas, D., Pogoda, J., Langholz, B., Mack, W., Temporal modifiers of the radon-smoking interaction, *Health Phys.,* 66:257–262, 1994.

Thompson, B., Indoor air quality, *J. Environ. Health,* 53(3):38–40, 1990.

Track radon in the workplace, *Saf. Health,* 143:42–45, 1991.

Traffic safety facts 1995: alcohol, National Highway Traffic Safety Administration, National Center for Statistics and Analysis, Washington, D.C., 1996.

Tsongas, G., Harge, W.D., Field monitoring of elevated carbon monoxide production from residential gas ovens, in *Proceedings of the American Society of Heating, Refrigerating, and Air-Conditioning Engineers, Inc., Indoor Air Quality 1994 conference,* Atlanta, GA, 1994.

U.S. Department of Health and Human Services, Healthy people 2000 review, 1997, U.S. Department of Health and Human Services, Hyattsville, MD, 1997.

U.S. Environmental Protection Agency and U.S. Consumer Product Safety Commission, The inside story: a guide to indoor air quality, EPA Document No. 402-K-93-007, Office of Radiation and Indoor Air, April 1995.

Wadden, R.A., Schaff., P.A., *Indoor Air Pollution,* John Wiley & Sons, New York, 1983.

Waldvogel, K., Regnery, R.L., Anderson, B.A., Caduff, R., Caduff, J., Nadal, D., Disseminated cat-scratch disease: detection of *Rochalimaea henselae* in affected tissues. *Eur. J. Pediatr.,* 153:23–27, 1994.

Wallance, L., Surprising results from a new way of measuring pollutants, *EPA J.,* 13:15–16, 1987.

Walsh, P.J. *Indoor Air Quality,* CRC Press, Boca Raton, FL, 1984.

Wandeler, A., Nadin-Davis, S.A., Tinline, R.R., Rupprecht, C.E., Rabies epizootiology: an ecological and evolutionary perspective, in *Lyssaviruses,* Rupprecht, C.E., Dietzschold, B., Koprowski, J., Eds., Springer-Verlag, New York, 1994, pp. 297–324.

Welch, D.F., Pickett, D.A., Slater, L.M., Steigerwalt, A.G., Brenner, D.J., *Rochalimaea henselae* sp. nov., a cause of septicemia, bacillary angiomatosis, and parenchymal bacillary peliosis, *J. Clin. Microbiol.,* 30:275–280, 1992.

West, R., Hack, S., Effects of cigarettes on memory search and subjective ratings, *Pharmacol. Biochem. Behav.,* 38:281–286, 1991.

White, F., Indoor air quality: what managers can do, *Employ. Relat. Today,* 17(2):93–101, 1990.

Woolf, A., Fish, S., Azzara, C., Dean, D., Serious poisonings among older adults: a study of hospitalization and mortality rates in Massachusetts 1983–1985, *Am. J. Public Health.,* 80:867–869, July 1990.

WHO, Guidelines for treatment of cystic and alveolar echinococcosis in humans, *Bull. WHO,* 74:231–242, 1996.

World Health Organization, Oral immunization of foxes in Europe in 1994, *Wkly. Epidemiol. Rec.,* 70:89–91, 1995.

World Health Organization, World survey of rabies 28 for the year 1992. World Health Organization, Geneva, 1994.

Yoshida, M., Adachi, J., Watabiki, T., Tatsuno, Y., Ishida, N., A study on house fire victims: age, carboxyhemoglobin, hydrogen cyanide, and hemolysis, *Foren. Sci. Int.,* 52:13–20, 1991.

Zangwill, K.M., Hamilton, D.H., Perkins, B.A. et al., Cat scratch disease in Connecticut: epidemiology, risk factors, and evaluation of a new diagnostic test, *N. Engl. J. Med.,* 329:8–13, 1993.

Zirschy, J., Witherell, L., Cleanup of mercury contamination of thermometer workers' homes, *Am. Ind. Hyd. Assoc. J.,* 48(1):81–84, 1987.

CHAPTER 9

Aarestrup, F.M., Occurrence of glycopeptide resistance among *Enterococcus faecium* isolates from conventional and ecological poultry farms, *Microb. Drug Resist.,* 1:255–257, 1995.

Aarestrup, F.M., Ahrens, P., Madsen, M., Pallesen, L.V., Poulsen, R.L., Westh, H., Glycopeptide susceptibility among Danish *Enterococcus faecium* and *Enterococcus faecalis* isolates of animal and human origin, *Antimicrob. Agents Chemother.,* 40:1938–1940, 1996.

Acheson, D.W.K., Breuker, S.D., Donohue-Rolfe, A., Kozak, K., Yi, A., Keusch, G.T., Development of a clinically useful diagnostic enzyme immunoassay for enterohemorrhagic *Escherichia coli* infection, in *Recent Advances in Verocytotoxin-Producing "Escherichia coli" Infections,* Karmali, M.A., Goglio, A.G., Eds., Elsevier Science, Amsterdam, 1994. pp. 109–112.

Adcock, P.M., Pastor, P., Medley, F. et al., Methicillin-resistant *Staphylococcus aureus* in two child-care centers, *J. Infect. Dis.,* 78:577–580, 1998.

Agerton, T., Valway, S., Gore, B. et al., Transmission of a highly drug-resistant strain (Strain W1) of *Mycobacterium tuberculosis, JAMA,* 278:1073–1077, 1997.

Albert, M.J., Siddique, A.K., Islam, M.S. et al., Large outbreak of clinical cholera due to *Vibrio cholerae* non-O1 in Bangladesh, *Lancet,* 3341:704, 1993.

Alter, M.J., Epidemiology of hepatitis C in the West, *Semin. Liver Dis.,* 15:5–14, 1995.

American Academy of Pediatrics, *1997 Red Book: Report of the Committee on Infectious Diseases,* 24th ed., Peter, G., Ed., American Academy of Pediatrics, Elk Grove Village, IL, 1997, p. 415.

American Conference of Governmental Industrial Hygienists, Threshold limit values and biological exposure indices for 1991–1992, American Conference of Governmental Industrial Hygienists, Cincinnati, OH, 1991.

American Hospital Association, OSHA's final bloodborne pathogens standard: a special briefing, item no. 155904, 1992.

American Institute of Architects, Committee on Architecture for Health, General hospital, in *Guidelines for Construction of Equipment of Hospital and Medical Facilities,* The American Institute of Architects Press, Washington, D.C., 1993.

American Institute of Architects Committee, *Guidelines for Design and Construction of Hospital and Health Care Facilities,* American Institute of Architects Press, Washington, D.C., 1996.

American Public Health Association/American Academy of Pediatrics, Caring for our children: national health and safety performance standards: guidelines for out-of-home child care programs, American Public Health Association/American Academy of Pediatrics, Ann Arbor, MI, 1992.

American Public Health Association, Benenson, A.S., Ed., *Control of Communicable Diseases Manual,* 16th ed., American Public Health Association, Washington, D.C., 1995.

American Society of Heating, Refrigerating, and Air Conditioning Engineers, Health facilities, in *1991 Application Handbook,* American Society of Heating, Refrigerating, and Air Conditioning Engineers, Atlanta, GA, 1991.

American Society of Heating, Refrigerating, and Air-Conditioning Engineers, Laboratories, in *ASHRAE Handbook, Heating, Ventilation, and Air-Conditioning Applications,* 1999, chap. 13.

Archibald, L., Phillips, L., Monnet, D., McGowen, J.E., Tenover, F., Gaynes, R., Antimicrobial resistance in isolates from inpatients and outpatients in the United States: increasing importance of the intensive care unit, *Clin. Infect. Dis.,* 24:211–215, 1997.

Arthur, A., Molinas, C., Depardieu, F., Courvalin, P., Characterization of Tn1546, Tn3-related transposon conferring glycopeptide resistance by synthesis of depsipeptide peptidoglycan precursors in *Enterococcus faecium* BM4147, *J. Bacteriol.,* 175:117–127, 1993.

Association for the Advancement of Medical Instrumentation, Flash sterilization: steam sterilization of patient care items for immediate use (ANSI/AAMI ST37-1996), Association for the Advancement of Medical Instrumentation, Arlington, VA, 1996.

Association for the Advancement of Medical Instrumentation, Selection of surgical gowns and drapes in health care facilities (AAMI TIR No. 11–1994), Association for the Advancement of Medical Instrumentation, Arlington, VA, 1994.

Bates, J., Jordens, J.Z., Selkon, J.B., Evidence for an animal origin of vancomycin-resistant enterococci [Letter], *Lancet,* 342:490–491, 1993.

Bates, J., Jordens, J.Z., Griffiths, D.T., Farm animals as a putative reservoir for vancomycin-resistant enterococcal infection in man, *J. Antimicrob. Chemother.,* 34:507–516, 1994.

Bean, N.H., Martin, S.M., Bradford, H., PHLIS: an electronic system for reporting public health data from remote sites, *Am. J. Public Health,* 82:1273–1276, 1992.

Beck-Sague, C.M., Dooley, S.W., Hutton, M.D. et al., Outbreak of multidrug-resistant *Mycobacterium* tuberculosis infections in a hospital: transmission to patients with HIV infection and staff, *JAMA,* 268:1280–1286, 1992.

Beck-Sague, C.M., Sinkowitz, R.L., Chinn, R.Y., Vargo, J., Kaler, W., Jarvis, W.R., Risk factors for ventilator-associated pneumonia in surgical intensive care unit patients, *Infect. Control Hospital Epidemiol.,* 17:374–376, 1996.

Beigel, Y., Ostfeld, I., Schoenfeld, N. Clinical problem solving: a leading question, *N. Engl. J. Med.,* 339:827–830, 1998.

Benenson, A.S., Ed., *Control of Communicable Diseases Manual,* 16 ed. American Public Health Association, Washington, D.C., 1995.

Bennett, S.N., Peterson, D.E., Johnson, D.R. et al., Bronchoscopy-associated *Mycobacterium xenopi* pseudoinfections, *JAMA,* 278:1073–1077, 1997.

Bergen, G.A., Shelhamer, J.H., Pulmonary infiltrates in the cancer patient, *Infect. Dis. Clin. N. Am.,* 10:297-326, 1996.

Birnbaum, D., Schulzer, M., Mathias, R.G., Kelly, M., Chow, A.W., Handwashing versus gloving, [Letter], *Infect. Control Hosp. Epidemiol.,* 12:140, 1991.

Bisno, A.L., *Streptococcus pyogenes,* in *Principles and Practices of Infectious Diseases,* 4th ed., Mandell, G.L., Bennett, J.E., Dolin, R., Eds., Churchill Livingstone, New York, 1995, pp. 1786–1799.

Bloom, B.R., Murray, C.J.L., Tuberculosis: commentary on a reemergent killer, *Science,* 257:1055–1064, 1992.

Bogaard, A., London, N., Driessen, C., Stobberingh, E., Prevalence of resistant fecal bacteria in turkeys, turkey farmers and turkey slaughterers [Abstract], in *Program and Abstracts of the 36th Interscience Conference on Antimicrobial Agents and Chemotherapy, New Orleans,* American Society for Microbiology, Washington, D.C., 1996, p. 86.

Bokete, T.N., O'Callahan, C.M., Clausen, C.R., Tang, N.M., Tran, N., Moseley, S.L. et al., Shiga-like toxin-producing *Escherichia coli* in Seattle children: a prospective study, *Gastroenterology,* 105:1724–1731, 1993.

Bonten, M.J.M., Hayden, M.K., Nathan, C., Van Voorhis, J., Matushek, M., Slaughter, S. et al., Epidemiology of colonization of patients and environment with vancomycin-resistant enterococci, *Lancet,* 348:1615–1619, 1996.

Boyce, J.M., Opal, S.M., Chow, J.W. et al., Outbreak of multi-drug resistant *Enterococcus faecium* with transferable VanB class vancomycin resistance, *J. Clin. Microbiol.,* 32:1448–1453, 1994.

Boyle, J.F., Soumakis, S.A., Rendo, A. et al., Epidemiologic analysis and genotypic characterization of a nosocomial outbreak of vancomycin-resistant enterococci, *J. Clin. Microbiol.,* 31:1380–1385, 1993.

Breiman, R.F., Butler, J.C., Tenover, F.C., Elliott, J.A., Facklam, R.R., Emergence of drug-resistant pneumococcal infections in the United States, *JAMA,* 271:1831–1835, 1994.

Brewer, S.C., Wunderink, R.G., Jones, C.B., Leeper, K.V.J., Ventilator-associated pneumonia due to *Pseudomonas aeruginosa, Chest,* 109:1019–1029, 1996.

Briss, P.A., Sacks, J.J., Addiss, D.G., Kresnow, M., O'Neil, J., A nationwide study of the risk of injury associated with day care center attendance, *Pediatrics,* 93:364–368, 1994.

Brogan, T.V., Nizet, V., Waldhausen, J.H., Rubens, C.E., Clarke, W.R., Group A streptococcal necrotizing fasciitis complicating primary *Varicella*: a series of fourteen patients, *Pediatr. Infect. Dis. J.,* 14:588–594, 1995.

Bruning, L.M., The bloodborne pathogens final rule, *AORN. J.,* 57:439–461, 1993.

Burke, J.P., Infections of cardiac and vascular prostheses, in *Hospital Infections,* 4th ed., Bennett, J.V., Brachman, P.S., Eds., Lippincott-Raven, Philadelphia, 1998, pp. 599–612.

Burwen, D.R., Bloch, A.B., Griffin, L.D., Ciesielski, C.A., Stern, H.A., Onorato, I.M., National trends in the concurrence of tuberculosis and Acquired Immunodeficiency Syndrome, *Arch. Intern. Med.,* 155:1281–1286, 1995.

California Department of Corrections, Public Health Infectious Disease Advisory Committee, tuberculosis protocols for human immunodeficiency virus infected inmates, California Department of Corrections, Sacramento, CA, 1998.

California Department of Health Services, California Tuberculosis Controllers Association, and the California Conference of Local Health Officers, Guidelines for coordination of TB prevention and control by local and state health departments and California Department of Corrections, California Department of Health Services, Berkeley, CA, 1998.

Carlton, J.T., Geller, J.B., Ecological roulette: the global transport of non-indigenous marine organisms, *Science,* 261:78–82, 1993.

CDC, 1997 USPHS/IDSA guidelines for the prevention of opportunistic infections in persons infected with Human Immunodeficiency Virus, *MMWR,* 46(RR-12), 1997.

CDC, Active Bacterial Core Surveillance Report, Emerging infections program network, Group A streptococcus, 1998, Available at <http://www.cdc.gov/ncidod/dbmd/abcs/gas98.pdf>, accessed August 5, 1999.

CDC, Anergy skin testing and preventive therapy for HIV-infected persons: revised recommendations, *MMWR,* 46(RR-15), 1997.

CDC, Draft guidelines for preventing the transmission of tuberculosis in healthcare facilities, 1994, *MMWR,* 43(RR-13):1-132, 1994; *Fed. Regist.,* 59(208):54242–54303, 1994.

CDC, Draft guidelines for preventing the transmission of tuberculosis in health-care facilities, second edition, *Fed. Regist.,* 58:52810–52854, 1993.

CDC, Guidelines for preventing the transmission of *Mycobacterium tuberculosis* in health care facilities, 1994, *MMWR,* 43(RR-13), 1994.

CDC, Guidelines for preventing the transmission of tuberculosis in health-care settings, with special focus on HIV-related issues, *MMWR,* 39(RR-17):1–29, 1990.

CDC, Hepatitis A associated with consumption of frozen strawberries, Michigan — March, 1997, *MMWR,* 46:288, 1997.

CDC, Hospital Infection Control Practices Advisory Committee, Guideline for prevention of nosocomial pneumonia, *Infect. Control Hosp. Epidemiol.,* 15:587–627, 1994.

CDC, Interim guideline for prevention and control of staphylococcal infection associated with reduced susceptibility to vancomycin, *MMWR,* 46:626–628, 635–636, 1997.

CDC, Initial therapy for tuberculosis in the era of multidrug resistance: recommendations of the Advisory Council for the Elimination of Tuberculosis, *MMWR,* 42(RR-7):1–8, 1993.

CDC, Lead poisoning associated with use of traditional ethnic remedies — California, 1991–1992, *MMWR,* 42:521–524, 1993.

CDC, Lead poisoning following ingestion of homemade beverage stored in a ceramic jug — New York, *MMWR,* 38:379–380, 1989.

CDC, Management of patients with suspected viral hemorrhagic fever, *MMWR,* 37(3S):1–16, 1988.

CDC, National Action Plan to Combat Multidrug-Resistant Tuberculosis, U.S. Department of Health and Human Services, Public Health Service, CDC, Atlanta, GA, 1992.

CDC, Nosocomial enterococci resistant to vancomycin — United States, 1989–1993, *MMWR,* 42:597–599, 1993.

CDC, Nosocomial group A streptococcal infections associated with asymptomatic health care workers — Maryland and California, 1997, *MMWR,* 48:163–166, 1999.

CDC, Nosocomial infection and pseudoinfection from contaminated endoscopes and bronchoscopes — Wisconsin and Missouri, *MMWR,* 40:675–678, 1991.

CDC, Nosocomial transmission of multidrug-resistant tuberculosis among HIV-infected persons — Florida and New York, *MMWR,* 40:585–591, 1991.

CDC, Nosocomial transmission of multidrug-resistant tuberculosis to healthcare workers and HIV-infected patients in an urban hospital — Florida, *MMWR,* 39:718–722, 1990.

CDC, Preventing lead poisoning in young children: a statement by the Centers for Disease Control, October 1991, U.S. Department of Health and Human Services, Public Health Service, Atlanta, GA, 1991.

CDC, Protection against viral hepatitis: recommendations of the Advisory Committee on Immunization Practices (ACIP), *MMWR,* 39(RR-2):1–27, 1990.

CDC, Prevention and treatment of tuberculosis among patients infected with human immunodeficiency virus: principles of therapy and revised recommendations, *MMWR,* 47(RR-20), 1998.

CDC, Prevention and control of tuberculosis in correctional facilities: recommendations of the Advisory Council for the Elimination of Tuberculosis, *MMWR,* 45(RR-8), 1996.

CDC, Prevention of perinatal group B streptococcal disease: a public health perspective, *MMWR,* 45(RR-7), 1996.

CDC, Prevention of *Varicella*: recommendations of the Advisory Committee on Immunization Practices (ACIP), *MMWR,* 454(RR-11), 1996.

CDC, Recommendations for prevention and control of tuberculosis among foreign-born persons: report of the Working Group on Tuberculosis among Foreign-Born Persons, *MMWR,* 47(RR-16), 1998.

CDC, Recommendations for preventing the spread of vancomycin resistance: recommendations of the hospital infection control practices advisory committee (HICPAC), *MMWR,* (RR-12), 1995.

CDC, Reduced susceptibility of *Staphylococcus aureus* to vancomycin — Japan, 1996, *MMWR,* 46:624, 1997.

CDC, Risks associated with human parvovirus B19 infection, *MMWR,* 38:81–88, 1989.

CDC, *Staphylococcus aureus* with reduced susceptibility to vancomycin — United States, 1997, *MMWR,* 46:765–766, 1997.

CDC, Surveillance for waterborne-disease outbreaks — United States, 1993-1994, *MMWR,* 45(SS-1):1–33, 1996.

CDC, Tuberculosis elimination revisited: obstacles, opportunities, and a renewed commitment., Advisory Council for the Elimination of Tuberculosis (ACET), *MMRW,* 48(RR-9), 1999.

CDC, Update on adult immunization: recommendations of the Immunization Practices Advisory Committee (ACIP), *MMWR,* 40(RR-12):1–94, 1991.

CDC, Update: universal precautions for prevention of transmission of human immunodeficiency virus, hepatitis B virus, and other bloodborne pathogens in health-care settings, *MMWR,* 37:377–382, 387–388, 1988.

CDC, Vital and health statistics: ambulatory and inpatient procedures in the United States, 1996, DHHS Publication No. 99-1710, U.S. Department of Health and Human Services, CDC, National Center for Health Statistics, Hyattsville, MD, 1998.

Centers for Disease Control and Prevention, 1998 FoodNet Surveillance Results, Preliminary Report, Atlanta, GA, 1999.

Centers for Disease Control and Prevention, Addressing emerging infectious disease threats: a prevention strategy for the United States, U.S. Department of Health and Human Services, Atlanta, GA, 1994.

Centers for Disease Control and Prevention, Assessing the public health threat associated with waterborne cryptosporidiosis: report of a workshop, *MMWR,* 44(RR-6):19, 1995.

Centers for Disease Control and Prevention, Drug-resistant *Streptococcus pneumoniae* — Kentucky and Tennessee, 1993, *MMWR,* 43:23–25, 31, 1993.

Centers for Disease Control and Prevention, Exposure of passengers and flight crew to *Mycobacterium tuberculosis* on commercial aircraft, 1992–1995, *MMWR,* 44:137–140, 1995.

Centers for Disease Control and Prevention, Hospital Infections Program, National Nosocomial infections surveillance (NNIS) report, data summary from October 1986–April 1996, issued May 1996: a report from the NNIS system, *Am. J. Infect. Control,* 24:380–388, 1996.

Centers for Disease Control and Prevention, Immunization of health-care workers: recommendations of the Advisory Committee on Immunization Practices (ACIP) and the Hospital Infection Control Practices Advisory Committee (HICPAC), *MMWR,* 46(RR-18):1–42, 1997.

Centers for Disease Control and Prevention, Laboratory management of agents associated with hantavirus pulmonary syndrome: interim biosafety guidelines, *MMWR,* 43(RR-7), 1994.

Centers for Disease Control and Prevention, Nosocomial enterococci resistant to vancomycin — United States, 1989–1993, *MMWR,* 42:597–599, 1993.

Centers for Disease Control and Prevention, Prevalence of penicillin-resistant *Streptococcal pneumoniae* — Connecticut 1992–1993, *MMWR,* 43:216–217, 223, 1994.

Centers for Disease Control and Prevention, National Institutes for Health, *Biosafety in Microbiological and Biomedical Laboratories,* 3rd ed., U.S. Department Health and Human Services, Public Health Service, Atlanta, GA, 1993.

Centers for Disease Control and Prevention, Public Health Service (PHS) guidelines for the management of health care worker exposures to HIV and recommendations for postexposure prophylaxis, *MMWR,* 1998.

Chow, J.W., Kuritza, A., Shlaes, D.M., Green, M., Sahm, D.F., Zervos, M.J., Clonal spread of vancomycin-resistant *Enterococcus faecium* between patients in three hospitals in two states, *J. Clin. Microbiol.,* 31:1609–1611, 1993.

Classen D.C., Evans, R.S., Pestotnik, S.L., Horn, S.D., Menlove, R.L., Burke, J.P., The timing of prophylactic administration of antibiotics and the risk of surgical wound infection, *N. Engl. J. Med.,* 326:281–286, 1992.

Committee of Infectious Diseases, American Academy of Pediatrics, Recommendations for the use of live attenuated *Varicella* vaccine, *Pediatrics,* 95:791–796, 1995.

Committee on Obstetric Practice, American College of Obstetricians and Gynecologists, Prevention of early-onset group B streptococcal disease in newborns, American College of Obstetricians and Gynecologists, Washington, D.C., 1996, ACOG Committee Opinion No. 173.

Coque, T.M., Tomayko, J.F., Ricke, S.C., Okhyusen, P.C., Murray, B.E. et al., Vancomycin-resistant enterococci from nosocomial, community, and animal sources in the United States, *Antimicrob. Agents Chemother.,* 40:2605–2609, 1996.

Council for Agricultural Science and Technology (CAST), Foodborne pathogens: risks and consequences, CAST, Task Force Report No. 122, September 1994.

Danila R., Besser, J., Rainbow, J. et al., Population based active surveillance for invasive group A streptococcal disease: comparison of pulsed-field gel electrophoresis testing and *emm* gene typing [Abstract P-4.17], in Program and Abstracts of the International Conference on Emerging Infectious Diseases, Atlanta, GA, March 8–11, 1998.

Davies, H.D., McGeer, A., Schwartz, B. et al., Invasive group A streptococcal infections in Ontario, Canada: Ontario group A streptococcal study group, *N. Engl. J. Med.,* 335:547–554, 1996.

de Bentzmann, S., Roger, P., Bajolet-Laudinat, O., Fuchey, C., Plotkowski, M.C., Puchell, E., Asialo GM1 is a receptor for *Pseudomonas aeruginosa* adherence to regenerating repiratory epithelial cells, *Infect. Immunol.,* 64:1582–1588, 1996.

Della-Porta, A.J., Murray, P.K., Management of biosafety, in *Anthology of Biosafety I: Perspectives on Laboratory Design,* Richmond, J.Y., Ed., American Biological Safety Association, Mundelein, IL, 1999.

DePaola, A., Capers, G.M., Moters, M.L. et al., Isolation of Latin American epidemic strain of *Vibrio cholerae* O1 from U.S. Gulf Coast, *Lancet,* 339:624, 1992.

Department of Health and Human Service, Department of Labor, Respiratory protective devices: final rules and notice, *Fed. Regist.,* 60(110):30336–30402, 1995.

Department of Labor, Occupational Safety and Health Administration, Occupational Exposure to bloodborne pathogens: final rule, *Fed. Regist.,* 56(235):64175–64182, 1991.

Doctor, A., Harper, M.B., Fleisher, G.R., Group A beta-hemolytic streptococcal bacteremia: historical overview, changing incidence, and recent association with *Varicella, Pediatrics,* 96:428–433, 1995.

Donnelly, J.P., Voss, A., Witte, W., Murray, B., Does the use in animals of antimicrobial agents, including glycopeptide antibiotics, influence the efficacy of antimicrobial therapy in humans?, [Letter], *Antimicrob. Chemother.,* 37:389–390, 1996.

Dooley, S.W., Jarvis, W.R., Martone, W.J., Snider, D.E., Jr., Multidrug-resistant tuberculosis [Editorial], *Ann. Intern. Med.,* 117:257–258, 1992.

Driver, D.R., Valway, S.E., Morgan, M., Onorato, I.M., Castro, K.G., Transmission of *Mycobacterium tuberculosis* associated with air travel, *JAMA,* 272:10311–10335, 1994.

Dryden, M.S., Keyworth, N., Gabb, R., Stein, K., Asymptomatic foodhandlers as the source of nosocomial Salmonellosis, *J. Hosp. Infect.,* 28:195–208, 1994.

Dunn, M., Wunderink, R.G., Ventilator-associated pneumonia caused by *Pseudomonas* infection [Review], *Clin. Chest Med.,* 16:94–109, 1995.

Edmond, M.B., Ober, J.F., Weinbaum, J.L., Pfaller, M.A., Hwang, T., Sanford, M.D. et al., Vancomycin-resistant *Enterococcus faecium* bacteremia: risk factors for infection, *Clin. Infect.Dis.,* 20:1126–1133, 1995.

Embil, J., Ramotar, K., Romance, L. et al., Methicillin-resistant *Staphylococcus aureus* in tertiary care institutions on the Canadian prairies, 1990–1992, *Infect. Control Hosp. Epidemiol.,* 15:646–651, 1994.

Engelgau, M.M., Woernle, C.H., Schwartz, B., Vance, N.J., Horan, J.M., Invasive group A streptococcus carriage in a child care center after a fatal case, *Arch. Dis. Childhood,* 71:318–322, 1994.

Facklam, R., Beall, B., Efstratiou, A. et al., *Emm* typing and validation of provisional M types for group A streptococci, *Emerg. Infect. Dis.,* 5:247–253, 1999.

Fahey, B.J., Koziol, D.E., Banks, S.M., Henderson, D.K., Frequency of nonparenteral occupational exposures to blood and body fluids before and after universal precautions training, *Am. J. Med.,* 90:145–152, 1991.

Falck G., Kjellander, J., Outbreak of group A streptococcal infection in a daycare center, *Pediatr. Infect. Dis. J.,* 11:914–919, 1992.

Falkinham, J.O., Epidemiology of infection by nontuberculosis mycobacteria, *Clin. Microbiol.,* 9:177–215, 1996.

Farley, D., Dangers of lead still linger, *FDA Consumer,* 16–21, 1998.

Farley, T.A., Wilson, S.A., Mahoney, F., Kelso, K.Y., Johnson, D.R., Kaplan, E.L., Direct inoculation of food as the cause of an outbreak of group A streptococcal pharyngitis, *JID,* 167:1232–1235, 1993.

Favero, M.S., Bond, W.W., Sterilization, disinfection, and antisepsis in the hospital, in *Manual of Clinical Microbiology,* 5th ed., Barlows, A. et al., Eds., American Society for Microbiology, Washington, D.C., 1991, pp. 183–200.

Feldman, M., Bryan, R., Rajan, S., Scheffler, L., Brunnert, S., Tan, H. et al., Role of flagella in pathogensis of *Pseudomonas aeruginosa* pulmonary infection, *Infect. Immun.,* 66:43–51, 1998.

Fergie, J.E., Shema, S.J., Lott, L., Crawford, R., Patrick, C.C., *Pseudomonas aeruginosa* bacteremia in immunocompromised children: analysis of factors associated with a poor outcome, *Clin. Infect. Dis.,* 18:390–394, 1994.

Food and Drug Administration, Increased surveillance of jellied fruit candy in ceramic vessels and of all products packed in ceramic vessels from Mexico due to lead, U.S. Department of Health and Human Services, Food and Drug Administration, Rockville, Maryland, August 1998, Import Bulletin no. 21-B12.

Food safety from farm to table: a national food-safety initiative, U.S. Environmental Protection Agency, U.S. Department of Health and Human Services, U.S. Department of Agriculture, Government Printing Office, May 1997.

Foodborne disease outbreaks, 5-year summary, 1983–1987, *MMWR,* 39(SS-1):15–57, 1992.

Fraser, V.J., Jones, M., Murray, P.R. et al., Contamination of flexible fiberoptic bronchoscopes with *Mycobacterium chelonae* linked to an automated bronchoscope disinfection machine, *Am. Rev. Resp. Dis.,* 145:8583–8585, 1992.

Fridkin, S.K., Welbel, S.F., Weinstein, R.A., Magnitude and prevention of nosocomial infections in the intensive care unit, *Infect. Dis. Clin. North. Am.,* 11:479–496, 1997.

Frieden, T.R., Munsiff, S.S., Low, D.E. et al., Emergence of vancomycin-resistant enterococci in New York City, *Lancet,* 342:76–79, 1993.

Garner, J.S., The CDC hospital infection control practices advisory committee, *Am. J. Infect. Control.,* 21:160–162 1993,

Garner, J.S., Hospital Infection Control Practices Advisory Committee, guideline for isolation precautions in hospitals, *Infect. Control Hosp. Epidemiol.,* 17:53–80, 1996.

Garrett, D.O., Jarvis, W.R., The expanding role of health care epidemiology — home and long-term care, *Infect. Control Hosp. Epidemiol.,* 17:714–717, 1996.

Gaynes, R.P., Horan, T.C., Surveillance of nosocomial infections, in *Hospital Epidemiology and Infection Control,* Mayhall, C.G., Ed., Williams & Wilkins, Baltimore, 1996, pp. 1017–1031.

Gilboa-Garber, N., Towards anti-*Pseudomonas aeruginosa* adhesion therapy, in *Toward Anti-Adhesion Therapy for Microbial Diseases,* Kahane, O., Ed., Plenum Press, New York, 1996, p. 39–50.

Goldmann, D.A., The role of barrier precautions in infection control, *J. Hosp. Infect.,* 18:515–523, 1991.

Goldmann, D.A., Platt, R., Hopkins, C., Control of hospital-acquired infections, in *Infectious Diseases,* Gorbach, S.L., Bartlett, J.G., Blacklow, N.R., Eds., W.B. Saunders, Philadelphia, 45:378–390, 1992.

Goldmann, D.A., Weinstein, R.A., Wenzel, R.P., Tablan, O.C., Duma, R.J., Gaynes, R.P. et al., Strategies to prevent and control the emergence and spread of antimicrobial-resistant microorganisms in hospitals. A challenge to hospital leadership, *JAMA,* 275:234–240, 1996.

Govan, J.R., Deretic, V., Microbial pathogenesis in cystic fibrosis: mucoid *Pseudomonas aeruginosa* and *Burkolderia cepacia, Microbiol. Rev.,* 60:539–574, 1996.

Gordon, S.M., Serkey, J.M., Keys, T.F., Ryan, T., Fatica, C.A., Schmitt, S.K. et al., Secular trends in nosocomial bloodstream infections in a 55-bed cardiothoracic intensive care unit, *Ann. Thorac. Surg.,* 65:95–100, 1998.

Gordts, B., Claeys, K., Jannes, H., Van Landuyt, H.W., Are vancomycin-resistant enterococci (VRE) normal inhabitants of the GI tract of hospitalized patients?, [Abstract], in *Program and Abstracts of the 34th Interscience Conference on Antimicrobial Agents and Chemotherapy, Orlando,* American Society for Microbiology, Washington, D.C., 1994, p. 145.

Graves, E.J., Gillium, B.S., Detailed diagnoses and procedures, national hospital discharge survey, 1995, National Center for Health Statistics. *Vital Health Stat.,* 13, 1997.

Gubler, J.G., Salfinger, M., von Graevenitz, A., Pseudoepidemic of nontuberculous mycobacteria due to a contaminated bronchoscope cleaning machine: report of an outbreak and review of the literature, *Chest,* 101:1245–1249, 1992.

Guideline for preventing the transmission of *Mycobacterium tuberculosis* in health-care facilities, *MMWR,* 43(RR-13), 1994.

Guideline for prevention of nosocomial pneumonia, *Respiratory Care,* 39:1191-1236, 1994, or *MMWR* 44(RR-12), 1995.

Gurevich, I., Body substance isolation, [Letter], *Infect. Control. Hosp. Epidemiol.,* 13:191, 1992.

Handwerger, S., Raucher, B., Altarac, D. et al., Nosocomial outbreak due to *Enterococcus faecium* highly resistant to vancomycin, penicillin, and gentamicin, *Clin. Infect. Dis.,* 16:750–755, 1993.

Handwerger, S., Skoble, J., Identification of a chromosomal mobile element conferring high-level vancomycin resistance in *Enterococcus faecium, Antimicrob. Agents Chemother.,* 39:2446–2453, 1995.

Hayden, M.K., Koenig, G.I., Trenholme, G.M., Bactericidal activities of antibiotics against vancomycin-resistant *Enterococcus faecium* blood isolates and synergistic activities of combinations, *Antimicrob. Agents Chemother.,* 38:1225–1229, 1994.

Health Care Financing Administration, ESRD facility survey data, 1996, U.S. Department of Health and Human Services, Health Care Financing Administration, Washington, D.C., 1997.

Hedberg, C.W., Savarino, S.J., Besser, J.M., Paulus, C.J., Thelen, V.M., Myers, L.J. et al., An outbreak of foodborne illness caused by *Escherichia coli* O39:NM, an agent not fitting into the existing scheme for classifying diarrheogenic *E. coli, J. Infect. Dis.,* 176:1625–1628, 1997.

Helmick, C.G., Griffin, P.M., Addiss, D.G., Tauxe, R.V., Juranek, D.D., Infectious diarrheas, in *Digestive Diseases in the United States: Epidemiology and Impact,* Everhart, J.E., Ed., U.S. Department of Health and Human Service, National Institutes of Health, National Institute of Diabetes and Digestive Diseases, U.S. Government Printing Office, Washington, D.C., 1994, pp. 85–123.

Hernandez, M.E., Bruguera, M., Puyuelo, T., Barrera, J.M., Sanchez-Tapias, J.M., Rodes, J., Risk of needle-stick injuries in the transmission of hepatitis C virus in hospital personnel, *J. Hepatol.,* 16:56–58, 1992.

Herold, B.C., Immergluck, L.C., Maranan, M.C. et al., Community-acquired Methicillin-resistant *Staphylococcus aureus* in children with no identified predisposing risk, *JAMA,* 279:593–598, 1998.

Herwaldt, L.A., Pottinger, J., Cofin, S.A., Nosocomial infections associated with anesthesia, in *Hospital Epidemiology and Infection Control,* Mayhall, C.G., Ed., Williams & Wilkins, Baltimore, 1996, pp. 655–675.

Hill, A.V., Genetics of infectious disease resistance, *Curr. Opin. Genet. Dev.,* 6:348–353, 1996.

Hill, G.J., Hill, S., Lead poisoning due to Hai Gen Fen, *JAMA,* 273:24–25, 1995.

Hiramatsu, K., Aritaka, N., Hanaki, H., Kawasaki, S., Hosada, Y., Hori, S. et al., Dissemination in Japanese hospitals of strains of *Staphylococcus aureus* heterogeneously resistant to vancomycin, *Lancet,* 350:1670–1673, 1997.

Hochberg, J., Murray, G.E., Principles of operative surgery: antisepsis, technique, sutures, and drains, in *Textbook of Surgery: The Biological Basis of Modern Surgical Practice,* 15th ed., Sabiston, D.C., Jr., Ed., W.B. Saunders, Philadelphia, 1997, pp. 253–263.

Hoge, C.W., Schwartz, B., Talkington, D.F. et al., The changing epidemiology of invasive group A streptococcal infections and the emergence of streptococcal toxic shock-like syndrome, *JAMA,* 269:384–389, 1993.

Horsburgh, C.R., Jr., Epidemiology of diseases caused by nontuberculosis mycobacteria, *Respir. Infect.,* 11:244–254, 1996.

Horsburgh, C.R., Jr., *Mycobacterium avium* complex infection in the acquired immunodeficiency syndrome, *N. Engl. J. Med.,* 324:1332–1338, 1996.

Hospital Infection Control Practices Advisory Committee, Recommendations for preventing the spread of vancomycin resistance, *MMWR,* 44(RR-12), 1995.

Institute of Medicine, *Antimicrobial resistance: issues and options,* National Academy Press, Washington, D.C., 1998.

Institute of Medicine, *Emerging Infections: Microbial Threats to Health in the United States,* 1st ed., National Academy Press, Washington, D.C., 1992.

Jackson, M.M., Lynch, P., An attempt to make an issue less murky: a comparison of four systems for infection precautions, *Infect. Control Hosp. Epidemiol.,* 12:448–450, 1991.

Jackson, M.M., Lynch, P., Body substance isolation, [Letter], *Infect. Control. Hosp. Epidemiol.,* 13:191–192, 1992.

Jarvis, W.R., Bolyard, E.A., Bozzi, C.J. et al., Respirators, recommendations, and regulations: the controversy surrounding protection of health care worker from tuberculosis, *Ann. Intern. Med.,* 122:142–146, 1995.

Johnson, R., Clark, R., Wilson, J., Read, S., Rhan, K., Renwick, S. et al., Growing concerns and recent outbreaks involving non-O157:H7 serotypes of verocytoxigenic *Escherichia coli, J. Food Protect.,* 59:1112–1122, 1996.

Jordens, J.Z., Bates, J., Griffiths, D.T., Faecal carriage and nosocomial spread of vancomycin-resistant *Enterococcus faecium, J. Antimicrob. Chemother.,* 34:515–528, 1994.

Kakosy, T., Hudak, A., Naray, M., Lead intoxication epidemic caused by ingestion of contaminated ground paprika, *J. Toxicol. Clin. Toxicol.,* 34:507–511, 1996.

Karanfil, L.V., Murphy, M., Josephson, A. et al., A cluster of vancomycin-resistant *Enterococcus faecium* in an intensive care unit, *Infect. Control Hosp. Epidemiol.,* 13:195–200, 1992.

Kessler, M., Hoen, B., Mayeux, D., Hestin, D., Fontenaille, C., Bacteremia in patients on chronic hemodialysis: a multicenter prospective survey, *Nephron,* 64:95–100, 1993.

Klare, I., Heier, H., Claus, H., Witte, W., Environmental strains of *Enterococcus faecium* with inducible high-level resistance to glycopeptides., *FEMS Microbiol. Lett.,* 106:23–30, 1993.

Klare, I., Heier, H., Claus, H., Witte, W., *Van* A-mediated high-level glycopeptide resistance in *Enterococcus faecium* from animal husbandry, *FEMS Microbiol. Lett.,* 125:165–172, 1995.

Klare, I., Heier, H., Claus, H., Böhme, G., Marin, S., Seltmann, G. et al., *Enterococcus faecium* strains with *van* A-mediated high-level glycopeptide resistance isolated from animal foodstuffs and fecal samples of humans in the community, *Microb. Drug. Resist.,* 1:265–273, 1995.

Klein, R.S., Universal precautions for preventing occupational exposures to human immunodeficiency virus type 1, *Am. J. Med.,* 90:141–153, 1991.

Kluytmans, J., Surgical infections including burns, in *Prevention and Control of Nosocomial Infections,* 3rd ed., Wenzel, R.P., Ed., Williams & Wilkins, Baltimore, 1997, pp. 841–865.

Knudsen, R.C., Risk assessment for biological agents in the laboratory, in *Rational Basis for Biocontainment: Proceedings of the Fifth National Symposium on Biosafety,* Richmond, J.Y., Ed., American Biological Safety Association, Mundelein, IL, 1998.

Kolmos, H.J., Svendsen, R.N., Nielsen,S.V., The surgical team as a source of postoperative wound infections caused by *Streptococcus pyogenes, J. Hospital Infect.,* 35:207–214, 1997.

Kulshrestha, M.K., Lead poisoning diagnosed by abdominal X-Rays, *J. Toxicol. Clin. Toxicol.,* 34:107–108, 1996.

Laboratory confirmed Salmonella surveillance. Annual summary, 1997, Centers for Disease Control and Prevention, Atlanta, GA, 1999.

Lanphear, B.P., Linnemann, C.C., Cannon, C.G. et al., Hepatitis C virus infection in health care workers: risk of exposure and infection, *Infect. Control Hosp. Epidemiol.,* 15:745–750, 1994.

Larson, E.L. and the 1992, 1993, and 1994 APIC Guidelines Committee, APIC guideline for handwashing and hand antisepsis in health care settings, *Am. J. Infect. Control,* 23:251–269, 1995.

Lederberg, J., Shope, R.E., Oaks, S.C., Jr., Eds., *Emerging Infections: Microbial Threats to Health in the United States,* National Academy Press, Washington, D.C., 1992.

Lee, J.T., Surgical wound infections: surveillance for quality improvement, in *Surgical Infections,* Fry, D.E., Ed., Little, Brown, Boston, 1995, pp. 145–159.

Leggiadro, R.J., Barrett, F.F., Chesney, P.J., Davis, Y., Tenover, F.C., Invasive pneumococci with high level penicillin and cephalosporin resistance at a mid-south children's hospital, *Pediatr. Infect. Dis. J.,* 13:320–322, 1994.

Levine, W., Griffin, P., Gulf Coast *Vibrio* Working group, *Vibrio* infections on the Gulf Coast: results of first year regional surveillance, *J. Infect. Dis.,* 167:479–483, 1993.

Levins, R. Awerbuch, T., Brinkmann, U. et al., The emergence of new diseases, *Am. Scientist,* 82:52–60, 1994.

Lew, D.P., Waldvogel, F.A., Infections of skeletal prostheses, in *Hospital Infections,* 4th ed., Bennett, J.V., Brachman, P.S., Eds., Lippincott-Raven, Philadelphia, 1998, pp. 613–620,

Linden, P.K., Pasculle, A.W., Manez, R., Kramer, D.J., Fung, J.J., Pinna, A.D. et al. Differences in outcomes for patients with bacteremia due to vancomycin-resistant *Enterococcus faecium* or vancomycin-susceptible *Enterococcus faecium, Clin. Infect. Dis.,* 22:663–670, 1996.

Lowy, F., *Staphylococcus aureus* infections, *N. Engl. J. Med.,* 339:520–532, 1998.

Mack, M.G., Hudson, S., Thompson, D., A descriptive analysis of children's playground injuries in the United States, 1990–1994, *Inj. Prev.,* 3:100–103, 1997.

Mack, M.G., Thompson, D., Hudson, S., Playground injuries in the 90's, *Parks Recreation,* 33:88–95, 1998.

Maguire, G.P., Arthur, A.D., Boustead, P.J., Dwyer, B., Currie, B.J., Clinical experience and outcomes of community-acquired and nosocomial methicillin-resistant *Staphylococcus aureus* in a Northern Australian hospital, *J. Hosp. Infect.,* 38:273–281, 1998.

Maloney, S., Welbel, S., Daves, B. et al., *Mycobacterium abscessus* pseudoinfection traced to an automated endoscope washer: utility of epidemiologic and laboratory investigation, *J. Infect. Dis.,* 169:1166–1169, 1994.

Martin, M.A., Reichelderfer, M., APIC guidelines for infection prevention and control in flexible endoscopy, *Am. J. Infect. Control,* 22:19–38, 1994.

Mastro, T.D., Farley, T.A., Elliot, J.A. et al., An outbreak of surgical-wound infections due to a group A *Streptococcus* carried on the scalp, *N. Engl. J. Med.,* 323:968–972, 1990.

McCaig, L.F., McLemore, T., Plan and Operation of the National Hospital Ambulatory Medical Care Survey, National Center for Health Statistics, Hyattsville, MD, 1994.

McCaig, L.F., Stussman, B.J., National hospital ambulatory medical care survey: 1996 emergency department summary, advance data from *Vital and Health Statistics: No. 293,* National Center for Health Statistics, Hyattsville, MD, 1997.

McCarthy, S.A., Effect of sanitizers on *Listeria monocytogenes* attached to latex gloves, *J. Food Saf.,* 16:231–237, 1996.

McCarthy, S.A., McPhearson, R.M., Guarino, A.M., Toxigenic *Vibrio cholerae* O1 and cargo ships entering Gulf of Mexico, *Lancet,* 339:624–625, 1992.

McKenna, M.T., McCray, E., Jones, J.L., Onorato, I.M., Castro, K.G., The fall after the rise: tuberculosis in the United States, 1991 through 1994, *Am. J. Public Health,* 88:1059–1063, 1998.

Meier, P.A., Infection control issues in same-day surgery, in *Prevention and Control of Nosocomial Infections,* 3rd ed., Wenzel, R.P., Ed., Williams & Wilkins, Baltimore, 1997, pp. 261–282.

Mendelson, M.H., Gurtman, A., Szabo, S., Neibart, E., Meyers, B.R., Policar, M. et al., *Pseudomonas aeruginosa* bacteremia in patients with AIDS [Review], *Clin. Infect. Dis.,* 18:886–895, 1994.

Mermin, J., Townes, J., Gerber, M., Dolan, N., Mintz, E., Tauxe, R., Typhoid fever in the United States, 1985–1994, *Arch. Intern. Med.,* 158:633–638, 1998.

Miller, M.L., A field study evaluating the effectiveness of different hand soaps and sanitizers, *Dairy Food Environ. Sanit.,* 14:155–160, 1994.

Mitsui, T., Iwano, K., Masuko, K. et al., Hepatitis C virus infection in medical personnel after needlestick accident, *Hepatology,* 16:1109–1114, 1992.

Mobarakai, N., Landman, D., Quale, J.M., *In vitro* activity of trospectomycin, a new aminocyclitol antibiotic against multidrug-resistant *Enterococcus faecium, J. Antimicrob. Chemother.,* 33:319–321, 1994.

Moellering, E.C., Jr., The Garrod Lecture: The enterococcus — a classic example of the impact of antimicrobial resistance on therapeutic options, *J. Antimicrob. Chemother.*, 28:1–12, 1991.

Mohle-Boetani, J., Schuchat, A., Plikaytis, B.D., Smith, D., Broome, C.V., Comparison of prevention strategies for neonatal group B streptococcal infection: a population based economic analysis, *JAMA*, 270:1442–1448, 1993.

Montecalvo, M.A., Horowitz, H., Gedris, C., Carbonaro, C., Tenover, F.C., Issah, A. et al., Outbreak of vancomycin, ampicillin and aminoglycoside-resistant *Enterococcus faecium* bacteremia in an adult oncology unit, *Antimicrob. Agents Chemother.*, 38:1363–1367, 1994.

Moore, M., McCray, E., Onorato, I.M., Cross-matching TB and AIDS registries: TB patients with HIV co-infection, United States, 1993–1994, *Public Health Rep.*, 114:269–277, 1999.

Moore, M., McCray, E., Onorato, I.M., Trends in TB in the United States, 1993–1997 [Abstract], presented at the 29th World Conference of the International Union Against Tuberculosis and Lung Disease, Bangkok, Thailand, November 23–26, 1998.

Morris, J.G., Shay, D.K., Hevden, J.N., McCarter, R.J., Perdue, B.E., Jarvis, W. et al., Enterococci resistant to multiple antimicrobial agents, including vancomycin. Establishment of endemicity in a university medical center, *Ann. Intern. Med.*, 123:250–259, 1995.

Muller, B.A., Steelman, V.M., Hartley, P.G., Casale, T.B., An approach to managing latex allergy in the health care worker, *Environ. Health*, July/August 8–16, 1998.

Mulroy, M.T., Filchak, K., Gaudio, M., *What You Should Know about Lead Poisoning: a Resource Manual for Childcare Providers*, Connecticut Department of Public Health, Hartford, CT, 1–10, 1997.

Murray, B.E., The life and times of the enterococcus, *Clin. Microbiol. Rev.*, 3:46-65, 1990.

Nafziger, D.A., Saravolatz, L.D., Infection in implantable prosthetic devices, in *Prevention and Control of Nosocomial Infections*, 3rd ed., Wenzel, R.P., Ed., Williams & Wilkins, Baltimore, 1997, pp. 889–923.

Namura, S., Nishjima, S., Asada, Y., An evaluation of the residual activity of antiseptic handrub lotions: an 'in use' setting study, *J. Dermatol.*, 21:481–485, 1994.

Namura, S., Nishjima, S., Mitsuya, K., Asada, Y., Study of the efficacy of antiseptic handrub lotions: an 'in use' setting study, *J. Dermatol.*, 21:405–410, 1994.

Nardell, E.A., Dodging droplet nuclei: reducing the probability of nosocomial tuberculosis transmission in the AIDS era, *Am. Rev. Respir. Dis.*, 142:501–503, 1990.

Nataro, J.P., Kaper, J.B., Diarrheogenic *Escherichia coli*, *Clin. Microbiol. Rev.*, 11:1–60, 1998.

Nicoletti, G., Boghossian, V., Borland, R., Hygienic hand disinfection: a comparative study with chlorhexidine detergents and soap, *J. Hosp. Infect.*, 15:323–337, 1990.

Office of Science and Technology Policy, The White House, Fact sheet: addressing the threat of emerging infectious diseases, The White House, Washington, D.C., June 12, 1996.

Olsen, R.J., Lynch, P.L., Coyle, M.B., Cummings, J. et al., Examination gloves as barriers to hand contamination in clinical practice, *JAMA*, 270(3):350–353, 1993.

Pablos-Mendez, A., Raviglione, M.C., Laszlo, A. et al., Global surveillance for antituberculosis-drug resistance, *N. Engl. J. Med.*, 338:1641–1649, 1998.

Panlilio, A.L., Culver, D.H., Gaynes, R.P. et al., Methicillin-resistant *Staphylococcus aureus* in U.S. Hospitals, 1975–1991, *Infect. Cont. Hosp. Epidemiol.*, 13:582–586, 1992.

Papanicolaou, G.A., Meyers, B.R., Meyers, J., Mendelson, M.H., Lou, W., Emre, S. et al., Nosocomial infections and mortality, *Clin. Infect. Dis.*, 23:760–766, 1996.

Park, C.H., Gates, K.M., Vandel, N.M., Hixon, D.L., Isolation of Shiga-like toxin producing *Escherichia coli* (O157 and non-O157) in a community hospital, *Diagn. Microbiol. Infect. Dis.*, 26:69–72, 1996.

Passador, L., Iglewski, B.H., Quorum sensing and virulence gene regulation in *Pseudomonas aeruginosa*, in *Virulence Mechanisms of Bacterial Pathogens*, 2nd ed., Roth, J.A., Ed., American Society for Microbiology, Washington, D.C., 1995, pp. 65–78.

Paulson, D.S., A comparative evaluation of different hand cleansers, *Dairy Food Environ. Sanit.*, 14:524–528, 1994.

Paulson, D.S., Evaluation of three handwash modalities commonly employed in the food processing industry, *Dairy, Food Environ. Sanit.* 12:615–618, 1992.

Paulson, D.S., Foodborne disease: controlling the problem, *Environ. Health*, May:15–19, 1997.

Paulson, D.S., Get a handle on contamination, *Food Qual.*, April:42–26, 1996.

Paulson, D.S., To glove or to wash: a current controversy, *Food Qual.*, June/July:60–64, 1996.

Paulson, D.S., Variability evaluation of two handwash modalities employed in the food processing industry, *Dairy Food Environ. Sanit.*, 13:332–335, 1993.

Payment, P., Siemiatycki, J., Richardson, L., Renaud, G., Franco, E., Prevost, M., A prospective epidemiological study of gastrointestinal health effects due to the consumption of drinking water, *Int. J. Environ. Health Res.*, 7:5–31, 1997.

Pearson, M.L., Jereb, J.A., Frieden, T.R. et al., Nosocomial transmission of multidrug-resistant *Mycobacterium* tuberculosis: a risk to patients and health care workers, *Ann. Intern. Med.*, 117:191–196, 1992.

Petrosilla, N., Puro, V., Ippolito, G., and Italian Study Group on Blood-Borne Occupational Risk in Dialysis, Prevalence of Hepatitis C antibodies in healthcare workers, *Lancet*, 344:339–340, 1994.

Phillips, I., Eykyn, S., Laker, M., Outbreak of hospital infection caused by contaminated autoclaved fluids, *Lancet*, 1:1258–1260, 1992.

Platt, R., Guidelines for perioperative antibiotic prophylaxis, in *Saunders Infection Control Reference Service*, Abrutyn, E., Goldmann, D.A., Scheckler, W.E., Eds. W.B. Saunders, Philadelphia, 1997, pp. 229–234.

Polish, L.B., Tong, M.J., Co, R.L. et al., Risk factors for hepatitis C virus infection among health care personnel in a community hospital, *Am. J. Infect. Control.*, 21:196–100, 1993.

Pollack, M., *Pseudomonas aeruginosa*, in *Principles and Practice of Infectious Diseases*, 4th ed., Mandell, G.L., Benett, J.E., Dolin, R., Eds., Churchill Livingstone, New York, 1995, pp. 1980–2003.

Prpic-Majic, D., Pizent, A., Jurasovic, J., Pongracic, J., Restek-Samarzija, N., Lead poisoning associated with the use of ayurvedic metal-mineral tonics, *J. Toxicol. Clin. Toxicol.*, 34:417–423, 1996.

Prusiner, S.B., Prion diseases and the BSE crisis, *Science*, 278:245–251, 1997.

Prusiner, S.B., Prions, in *Virology*, 3rd ed., Fields, B.N., Knipe, D.M., Howley, P.M. et al., Eds., Lippincott-Raven, Philadelphia, 1996.

Pugliese, G., Lynch, P., Jackson, M.M., *Universal Precautions: Policies, Procedures, and Resources*, American Hospital Association, Chicago, IL, 1991, pp. 7–87.

Ramamurthy, T., Garg, S., Sharma, R. et al., Emergence of novel strain of *Vibrio cholerae* with epidemic potential in Southern and Eastern India, *Lancet*, 341:703–704, 1993.

Reichler, M.R., Allphin, A.A., Breiman, R.F. et al., The spread of multiple-resistant *Streptococcus pneumoniae* at a day care center in Ohio, *J. Infect. Dis.*, 166:1346–1353, 1992.

Rhame, F.S., The inanimate environment, in *Hospital Infections*, 3rd ed., Bennett, J.V., Brachman, P.S., Eds., Little, Brown, Boston, MA, 1992, pp. 299–333.

Rhinehart, E., Smith, N., Wennersten, C. et al., Rapid dissemination of beta-lactamase-producing aminoglycoside-resistant *Enterococcus faecalis* among patients and staff on an infant and toddler surgical ward, *N. Engl. J. Med.*, 323:1814–1818, 1990.

Richard, P., Le, F.R., Chamoux, C., Pannier, M., Espaze, E., Richet, H., *Pseudomonas aeruginosa* outbreak in a burn unit: role of antimicrobials in the emergence of multiple resistant strains, *J. Infect. Dis.,* 170:377–383, 1994.

Riley, R.L., *Principles of UV Air Disinfection,* Johns Hopkins University, School of Hygiene and Public Health, Baltimore, MD, 1991.

Rowe, P.C., Orrbine, E., Lior, H., Wells, G.A., McLaine, P.N., A prospective study of exposure to verotoxin-producing *Escherichia coli* among Canadian children with haemolytic uraemic syndrome. The CPFDRC co-investigators, *Epidemiol. Infect.,* 110:1–7, 1993.

Roy, M.C., Surgical site infections after coronary artery bypass graft surgery: discriminating site-specific risk factors to improve prevention efforts, *Infect. Control Hosp. Epidemiol.,* 19:229–233, 1998.

Roy, M.C., Perl, T.M., Basics of surgical-site infection surveillance, *Infect. Control Hosp. Epidemiol.,* 18:659–668, 1997.

Rubin, L.G., Tucci, V., Cercenado, E., Eliopoulos, G., Isenberg, H.D., Vancomycin-resistant *Enterococcus faecium* in hospitalized children, *Infect. Control Hosp. Epidemiol.,* 13:700–705, 1992.

Rudnick, J.R., Kroc, K., Manangan, L., Banerjee, S., Pugliese, G., Jarvis, W., Are U.S. hospitals prepared to control nosocomial transmission of tuberculosis?, Epidemic Intelligence Service Annual Conference, [Abstract], 1992, p. 60.

Rutala, W.A., APIC guideline for selection and use of disinfectants, *Am. J. Infect. Control.,* 24:313–342, 1996.

Rutula, W.A., Disinfection, sterilization, and waste disposal, in *Prevention and Control of Nosocomial Infections,* 2nd ed., Wenzel, R.P., Ed., Williams & Wilkins, Baltimore, MD, 1993, pp. 460–495.

Rutula, W.A., Mayhall, C.G., The Society for Hospital Epidemiology of America Position Paper: Medical waste, *Infect. Control Hosp. Epidemiol.,* 13:38–48, 1992.

Sader, H.S., Pfaller, M.A., Tenover, F.C., Hollis, R.J., Jones, R.N., Evaluation and characterization of multiresistant *Enterococcus faecium* from 12 U.S. medical centers, *J. Clin. Microbiol.,* 32:2840–2842, 1994.

Sahm, D.F., Olsen, L., *In vitro* detection of enterococcal vancomycin resistance, *Antimicrob. Agents Chemother.,* 34:1846–1848, 1990.

Sartori, M., La Terra, G., Aglietta, M. et al., Transmission of hepatitis C via blood splash into conjunctiva, *Scand. J. Infect. Dis.,* 25:270–271, 1993.

Scheckler, W.E., Brimhall, D., Buck, A.S., Farr, B.M., Friedman, C., Garibaldi, R.A. et al., Requirements for infrastructure and essential activities of infection control and epidemiology in hospitals: a consensus panel report, *Infect. Control Hosp. Epidemiol.,* 19:114–124, 1998.

Schwartz, B., Elliott, J.A., Butler, J.C. et al., Clusters of invasive group A streptococcal infections in family, hospital, and nursing home settings, *Clin. Infect. Dis.,* 15:277–284, 1992.

Sepkowitz, K.A., Occupationally acquired infections in health care workers, Part I, *Ann. Intern. Med.,* 125:826–834, 1996.

Sepkowitz, K.A., Occupationally acquired infections in health care workers, Part II., *Ann. Intern. Med.,* 125:917–928, 1996.

Shay, D.K., Maloney, S.A., Monteclavo, M., Banerjee, S., Wormser, G.P., Arduino, M.J. et al., Epidemiology and mortality risk of vancomycin-resistant enterococcal bloodstream infections, *J. Infect. Dis.,* 172:993–1000, 1995.

Slaughter, S., Haden, M.K., Nathan, C., Hu, T.C. Rice, T., Van Voorhis, J. et al., A comparison of the effect of universal use of gloves and gowns with that of glove use alone on acquisition of vancomycin-resistant enterococci in a medical intensive care unit, *Ann. Intern. Med.* 125:448–456, 1996.

Schuchat, A., Lizano, C., Broome, C., Swaminathan, B., Kim, C., Winn, K., Outbreak of neonatal listeriosis associated with mineral oil, *Pediatr. Infect. Dis. J.,* 10:183–189, 1991.

Slutsker, L., Schuchat, A., Listeriosis in humans, in *Listeria, Listeriosis, and Food Safety,* Ryser, E., Marth, E., Eds., Marcel Dekker, New York, 1999, pp. 75–96.

Stevens, D.L., Streptococcal toxic shock syndrome: spectrum of disease, pathogenesis, and new concepts in treatment, *Emerg. Infect. Dis.,* 1:69–78, 1995.

Summary of notifiable diseases, United States, 1997, *MMWR,* 46(54), 1997.

Tappero, J., Schuchat, A., Deaver, K., Mascola, L., Wenger, J., Reduction in the incidence of human listeriosis in the United States. Effectiveness of prevention efforts, *JAMA,* 273:1118–1122, 1995.

Tarr, P.I., Neill, M.A. Perspective: the problem of non-O157:H7 Shiga toxin (verocytotoxin)-producing *Escherichia coli* [Comment], *J. Infect. Dis.,* 174:1136–1139, 1996.

Tenover, F.C., Arbeit, R.D., Goering, R.V., How to select and interpret molecular strain typing methods for epidemiological studies of bacterial infections: a review for healthcare epidemiologists, *Infect. Control Hosp. Epidemiol.,* 18:426–439, 1997.

Tenover, F.C., Arbeit, R.D., Goering, R.V. et al., Interpreting chromosomal DNA restriction patterns produced by pulsed-field gel electrophoresis: criteria for bacterial strain typing, *J. Clin. Microbiol.,* 33:2233–2239, 1995.

Tenover, F.C., Tokars, J., Swenson, J., Paul, S., Spitalny, K., Jarvis, W., Ability of clinical laboratories to detect antimicrobial agent-resistant enterococci, *J. Clin. Microbiol.,* 31:1695–1699, 1993.

Thompson, D., Hudson, S., National action plan for the prevention of playground injuries, National Program for Playground Safety, Cedar Falls, IA, 1996.

Travis, J., Invader threatens Black, Azov Seas, *Science,* 262:1366–1367, 1993.

U.S. Congress, Office of Technology Assessment, Impacts of antibiotic-resistant bacteria (OTA-H-629), U.S. Government Printing Office, Washington, D.C., 1995.

U.S. Congress, Office of Technology Assessment, Risks to students in school, U.S. Government Printing Office, Washington, D.C., 1995.

U.S. Consumer Product Safety Commission, Handbook for public playground safety, Washington, D.C., 1997.

U.S. Department of Health and Human Service. 42 CFR Part 84: Respiratory protective devices; proposed rule, *Fed. Regist.,* 59:26849–26889, 1994.

U.S. Department of Labor, Occupational Safety and Health Administration, Criteria for recording on OSHA form 200, OSHA instruction 1993; standard 1904, U.S. Department of Labor, Washington, D.C., 1993.

U.S. Department of Labor, Occupational Safety and Health Administration, Occupational exposure to bloodborne pathogens: final rule, CFR part 1910.1030, *Fed. Regist.,* 56:64004–64182, 1991.

U.S. Department of Labor, Occupational Safety and Health Administration, Record keeping guidelines for occupational injuries and illnesses: the Occupational Safety and Health Act of 1970 and 29 CFR 1904.OMB no. 120-0029, Washington, D.C., 1986.

Vugia, D.J., Peterson, C.L., Meyers, H.B. et al., Invasive group A streptococcal infections in children with *Varicella* in southern California, *Pediatr. Infect. Dis. J.,* 15:146–150, 1996.

Webster, R.G., Influenza: an emerging disease, *Emerg. Infect. Dis.,* 4(3), 1998.

Weinstein, R.A., Epidemiology and control of nosocomial infections in adult intensive care units, *Am. J. Med.,* 91:179–184, 1991.

Weinstein, R.A., SHEA consensus panel report: a smooth takeoff, *Infect. Control Hosp. Epidemiol.,* 19:91–93, 1998.

Wheeler, M.C., Roe, M.H., Kaplan, E.L., Schlievfert, P.M., Todd, J.K., Outbreak of group A streptococcus septicemia in children: clinical, epidemiologic, and microbiological correlates, *JAMA,* 266:533–537, 1991.

Whitney, C.G., Plikaytis, B.D., Gozansky, W.S., Wenger, J.D., Schuchat, A., Prevention practices for perinatal group B streptococcal disease: a multistate surveillance analysis, *Obstet. Gynecol.,* 89:28–32, 1997.

Wiblin, R.T., Nosocomial pneumonia in *Prevention and Control of Nosocomial Infections,* 3rd ed., Wenzel, R.P., Ed., Williams & Wilkins, Baltimore, 1997, pp. 807–819.

Wilson, M.E., Levins, R., Spielman, A., *Disease in Evolution: Global Changes and Emergence of Infectious Diseases,* New York Academy of Sciences, New York, 1994, p. 740.

Wilson, M.E., Disease in evolution: introduction, in *Disease in Evolution: Global Changes and Emergence of Infectious Diseases,* Wilson, M.E., Levins, R., Spielman, A., Eds., New York Academy of Sciences, New York, 1994, 740, pp. 1–12.

Wilson, M.E., *A World Guide to Infections: Diseases, Distribution, Diagnosis,* Oxford University Press, New York, 1991.

Wong, E.S., Surgical site infections, in *Hospital Epidemiology and Infection Control,* Mayhall, C.G., Ed., Williams & Wilkins, Baltimore, 1996, pp. 154–174.

Wong, E.S., Stotka, J.L., Chinchilli, V.M., Williams, D.S., Stuart, C.G., Markowitz, S.M., Are universal precautions effective in reducing the number of occupational exposures among health care workers?, *JAMA,* 265:1123–1128, 1991.

Woodwell, D.A., National ambulatory medical care survey: 1996 summary, advance data from *Vital and Health Statistics; No. 295,* National Center for Health Statistics, Hyattsville, MD, 1997.

Working Group on Emerging and Reemerging Infectious Diseases, Committee on International Science, Engineering, and Technology Council, Infectious disease — a global health threat, U.S. Government Printing Office, Washington, D.C., 1995.

Working Group on Prevention of Invasive Group A Streptococcal Infections, Prevention of invasive Group A Streptococcal Disease among household contacts of case-patients: is prophylaxis warranted?, *JAMA,* 279:1206–1210, 1998.

World Health Organization, Cholera in the Americas, *Wkly. Epidemiol. Rec.,* 67:33–39, 1992.

Zabransky, R.J., Dinuzzo, A.R., Huber, M.B., Woods, G.L., Detection of vancomycin resistance in enterococci by the Vitek AMS System, *Diagn. Microbiol. Infect. Dis.,* 20:113–116, 1994.

Zuckerman, J., Clewley, G., Griffiths, P., Cockcroft, A., Prevalence of hepatitis C antibodies in clinical healthcare workers, *Lancet,* 343:1618–1620, 1994.

Zurawski, C.A., Bardsley, M., Beall, B. et al., Invasive group A streptococcal disease in metropolitan Atlanta: a population-based assessment, *Clin. Infect. Dis.,* 27:150–157, 1998.

CHAPTER 10

ANSI/NSPI 1, Standards for Public Swimming Pools, in *Pool and Spa Water Chemistry,* Taylor Technologies, Sparks, MD. 1994, p. 40.

Ault, K., Estimates of non-fire carbon monoxide poisonings and injuries, U.S. Consumer Product Safety Commission, Washington, D.C., 1997.

CDC, Carbon monoxide poisonings associated with snow-obstructed vehicle exhaust systems–Philadelphia and New York City, January 1996, *MMWR,* 45:1–3, 1996.

CDC, Carbon monoxide poisoning at an indoor ice arena and bingo hall — Seattle, 1996, *MMWR,* 45:265–267, 1996.

CDC, Unintentional carbon monoxide poisonings in residential settings — Connecticut, November 1993–March 1994, *MMWR*, 44:765–767, 1995.

Cobb, N., Etzel, R.A., Unintentional carbon monoxide-related deaths in the United States, 1979 through 1988, *JAMA*, 266:659–663, 1991.

Fayer, R., Speer, C.A., Dubey, J.P., The general biology of *Cryptosporidium*, in *Cryptosporidium and Cryptosporidiosis*, Fayer, R., Ed., CRC Press, Boca Raton, FL, 1997, pp. 1–41.

Gyurek, L.L., Finch, G.R., Belosevic, M., Modeling chlorine inactivation requirements of *Cryptosporidium parvum* oocysts, *J. Environ. Engl.*, 123:865–875, 1997.

Joce, R.E., Bruce, J., Kiely, D., Noah, N.D., Dempster, W.B., Stalker, R. et al., An outbreak of cryptosporidiosis associated with a swimming pool, *Epidemiol. Infect.*, 107:497–508, 1991.

Kebabjian, R.S., Disinfection of public pools and management of fecal accidents, *J. Environ. Health*, 58:8–12, 1995.

Kilani, R.T., Sekla, L., Purification of *Cryptosporidium* oocysts and sporozoites by cesium chloride and percoll gradients, *Am. J. Trop. Med. Hyg.*, 36:505–508, 1997.

Korich, D.G., Mead, J.R., Madore, M.S., Sinclair, N.A., Sterling, C.A., Effects of ozone, chlorine dioxide, chlorine, and monochloramine on *Cryptosporidium parvum* oocyst viability, *Appl. Environ. Mirobiol.* 56:1423–1427, 1990.

Medema, G.J., Schets, F.M., Yeunis, P.F.M., Havelaar, A.H., Sedimentation of free and attached *Cryptosporidium* oocysts and *Giardia* cysts in water, *Appl. Environ. Microbiol.*, 64:4460–4466, 1998.

Parker, J.F.W., Smith, H.V., Destruction of oocysts of *Cryptosporidium parvum* by sand and chlorine, *Water Res.*, 27:729–731, 1993.

Sorvillo, F.J., Fujioka, K., Nahlen, B., Tormey, M.P., Kebabjian, R.S., Mascola, L., Swimming-associated cryptosporidiosis, *Am. J. Public Health*, 82:742–744, 1992.

U.S. Bureau of the Census, *Statistical Abstract of the United States: 1995*, 115th ed., The Bureau, Washington, D.C., 1995. p. 260.

Venczel, L.V., Arrowood, M., Hurd, M., Sobsey, M.D., Inactivation of *Cryptosporidium parvum* oocysts and *Clostridium perfringens* spores by a mixed-oxidant disinfectant and by free chlorine, *Appl. Environ. Microbiol.*, 63:1598–1601, 1997.

CHAPTER 11

Bisesi, M.S., Kohn, J.P., *Industrial Hygiene Evaluation Methods*, Lewis Publishers/CRC Press, Boca Rotan, FL, 1995.

DiNardi, S., Ed., *The Occupational Environment — Its Evaluation and Control*, American Industrial Hygiene Association Publications, Fairfax, VA, 1997.

Plog, B.A., Niland, J., Quinlan, P.J., Eds., *Fundamentals of Industrial Hygiene*, 4th ed., National Safety Council, Itasco, IL, 1996.

Talty, J.T., *Industrial Hygiene Engineering: Recognition, Measurement, Evaluation, and Control*, Noyes Data Corp., Park Ridge, NJ, 1988.

CHAPTER 12

Abramowitz, M., *Contrast Methods in Microscopy: Transmitted Light*, Vol. 2, Olympus Corp., Lake Success, NY, 1987.

Abramowitz, M., *Microscope Basics and Beyond*, Vol. 1, Olympus Corp., Lake Success, NY, 1985.

American Conference of Governmental Industrial Hygienists, *Air Sampling Instruments*, 9th ed., ACGIH, Cincinnati, OH, 2001.

American Industrial Hygiene Association, *Fundamentals of Analytical Procedures in Industrial Hygiene*, AIHA, Akron, OH, 1987.

Berger, E.H., Ward, W.D., Morrill, J.C., Royster, L.H., Eds., *Noise and Hearing Conservation Manual*, American Industrial Hygiene Association, Akron, OH, 1986.

Bisesi, M.S., Kohn, J.P., *Industrial Hygiene Evaluation Methods*, Lewis Publishers/CRC Press, Boca Raton, FL, 1995.

Chou, J., *Hazardous Gas Monitors: A Practical Guide to Selection, Operation and Applications*, McGraw-Hill, New York, 2000.

DiNardi, S., Ed., *The Occupational Environment — Its Evaluation and Control*, American Industrial Hygiene Association Publications, Fairfax, VA, 1997.

Gollnick, D.A., *Basic Radiation Protection Technology*, 3rd ed., Pacific Radiation Corp., Altadena, CA, 1994.

Kenkel, J., *Analytical Chemistry Refresher Manual*, Lewis Publishers, Chelsea, MI, 1992.

Malansky, C.J., Malansky, S.P., *Air Monitoring Instrumentation: A Manual for Emergency, Investigatory, and Remedial Responders*, Van Nostrand Reinhold, New York, 1993.

Plog, B.A., Niland, J., Quinlan, P.J., Eds., *Fundamentals of Industrial Hygiene*, 4th ed., National Safety Council, Itasco, IL, 1996.

Wight, G.D., *Fundamentals of Air Sampling*, Lewis Publishers, Chelsea, MI, 1994.

Index

VOLUME I

Index

VOLUME II

AA, *see* Atomic absorption spectroscopy
Abandoned wells, 289, *see also* Wells
Abiotic synthesis, 415
Abrasion, 56
ABS, *see* Alkyl benzene sulfonate
Absolute pressure, 13, 14, 375
Absolute temperature, 13, 15, *see also* Gases
Absolute zero pressure, 14
Absorbents, 71, 697
Absorber, 211
Absorption, 26, 212, 215, 478
Absorption ponds, 604
Absorption trenches, 471, 501–502, 504–505,
      see Fig. 6.27
Accidents
    hazardous waste sites, 169, 222
    swimming pool areas, 347, 371
Acello filters, 430, *see* Fig. 6.8
Achromobacter, spp., 562
Acid, 538
    mine drainage, 624
    pulp process, 153
Acid cation exchanger, 624
Acid digestion, 437, *see also* Sludge
Acid (Kraft) pulp process, 153
Acid rain, 47, 49
Acid shock, 49
Acid/alkalinity, 266, 538
Acids, deposition, 47–49, Fig. 1.9
    acid rain, 47, 49
    acid shock, 49
    acid snow, 49
Acoustic barrier particulate separation, 69,
      see also Air pollution
Actinomycetes, 572

Activated aeration, 432, *see also* Sludge
Activated carbon, 159
    advanced treatment of sewage, 445–446
    air sample collection for gases and vapors, 697
    secondary treatment plants, 491–493
    wastewater treatment, 582, 598, 603
    water treatment, 271
        chemical wastes, 603
Activated charcoal, *see* Activated carbon
Activated sludge process, *see also* Sludge
    hazardous waste, 218
    microorganism-contaminated aerosols,
        582–583
    phosphate removal from sewage, 448–450
    problems with, 482–483
    secondary treatment of sewage, 430, 432–433,
        see Fig. 6.9
Active waste, 88
Acts of terrorism, types, 652
Acute hazardous waste, 104, *see also* Hazardous
      waste
Additional air, 134, *see also* Incineration
Adenoviruses, 331, 480, 573, 574
Adiabatic lapse rate, 9
Adsorbers, 71
Adsorption, 212, 445–446, 467–468, 575
Advanced water treatment (sewage), 438,
        see Fig. 6.14, *see also* Primary
        treatment; Secondary treatment;
        Sewage treatment
    adsorption, 445–446
    distillation, 447
    electrodialysis, 447
    foam separation, 446–447
    freezing, 447–448

THE POWER OF FIVE: BOOK FOUR

NECROPOLIS

ANTHONY HOROWITZ

WALKER
BOOKS

For Nicholas

Based on an idea first published in 1989 as *The Day of the Dragon*

First published 2008 by Walker Books Ltd
87 Vauxhall Walk, London SE11 5HJ

This edition published 2009

2 4 6 8 10 9 7 5 3 1

Text © 1989, 2008 Anthony Horowitz
Cover illustration and design © 2009 Walker Books Ltd
Power of 5 logo™ © 2008 Walker Books Ltd

This book has been typeset in Frutiger Light and Kosmik Bold

Printed and bound in Great Britain by Clays Ltd, St Ives plc

British Library Cataloguing in Publication Data:
a catalogue record for this book is available from the British Library

ISBN 978-1-4063-2108-1

www.walker.co.uk

"A supernatural adventure of the most chilling kind... The tension explodes into the sort of chase scenes that make this author a favourite."

The Times

"Amongst the best work the author has produced to date: vivid, exciting and intelligent."

INIS

"The supernatural excitement will leave readers breathless."

Pittsburgh Post-Gazette

"The Horowitz formula is up to strength ... fuel for the nightmares of Horowitz fans."

The Daily Telegraph

"High-octane fantasy story-telling ... this race-along novel will be ferociously devoured by the author's many teenage fans."

The Bookseller

"Horowitz fans are in for a treat ... so taut and scary that at times it had me looking over my shoulder when I was reading it."

The Children's Buyer's Guide

"This fab book from Anthony Horowitz's supernatural sequence The Power of Five is unlike any book of his you've ever read!"

Kraze Club

"Horowitz excels at getting into the minds of his teenage characters and testing them to the limits. This is a supernatural thriller all boys will be desperate to get their hands on."

School Library Association

"If Harry Potter and a new Power of Five were coming out on the same day, it would be really hard to decide which to read first."

The Sunday Times

ANTHONY HOROWITZ is one of the most popular children's writers working today. Both The Power of Five and Alex Rider are No.1 bestselling series and have been enjoyed by millions of readers worldwide. Anthony is particularly excited by *Necropolis*, which he sees as a major step in a new direction. For a start, it's his first book with a full-blooded female at the heart of the action. It also develops the themes that began with *Raven's Gate* and sets up the epic finale which he plans to begin soon. Anthony was married in Hong Kong and went back there to research the book. Everything you read is inspired by what he saw.

The hugely successful Alex Rider series, which has spurred a trend of junior spy books, has achieved great critical acclaim and Anthony has won numerous awards including the Booksellers Association/Nielsen Author of the Year Award, the Children's Book of the Year Award (at the British Book Awards), and the Red House Children's Book Award. The first adventure, *Stormbreaker*, was made into a blockbuster movie, starring Alex Pettyfer, Ewan McGregor, Bill Nighy and Robbie Coltrane.

Anthony's other titles for Walker Books include the Diamond Brothers mysteries; *Groosham Grange* and its sequel, *Return to Groosham Grange*; *The Devil and His Boy*; *Granny* and *The Switch*. Anthony also writes extensively for TV, with programmes including *Midsomer Murders*, *Poirot* and the drama series *Foyle's War*, which won the Lew Grade Audience Award. He is married to television producer Jill Green and lives, reluctantly, in London with his two part-time sons, Nicholas and Cassian and their dog, Lupus.

You can find out more about Anthony and his books at:
www.anthonyhorowitz.com
www.powerof5.co.uk
www.alexrider.com

Other titles by Anthony Horowitz

The Power of Five (Book One): *Raven's Gate*
The Power of Five (Book Two): *Evil Star*
The Power of Five (Book Three): *Nightrise*

The Devil and His Boy
Granny
Groosham Grange
Return to Groosham Grange
The Switch

The Alex Rider series:

Stormbreaker
Point Blanc
Skeleton Key
Eagle Strike
Scorpia
Ark Angel
Snakehead

The Diamond Brothers books:

The Falcon's Malteser
Public Enemy Number Two
South By South East
The French Confection
I Know What You Did Last Wednesday
The Blurred Man
The Greek Who Stole Christmas

CONTENTS

NIGHTRISE CORPORATION

STRICTLY CONFIDENTIAL

Memorandum from the Chairman's Office: October 15

We are about to take power.

Four months ago, on June 25, the gate built
into the Nazca Desert opened and the Old Ones finally
returned to the world that once they ruled. They are
with us now, waiting for the command to reveal themselves and to begin a war which, this time, they cannot
lose.

Why has that command not been given?

The triumph of Nazca was tainted by the presence of
two children, teenaged boys. One has already become
familiar to us … indeed, we had been watching him for
much of his life. His name, or the name by which he
is now known, is Matthew Freeman. He is fifteen and
English. The other was a Peruvian street urchin who
calls himself Pedro and who grew up in the slums of
Lima. Between them, they were responsible for the
death of our friend and colleague, Diego Salamanda.
Incredibly, too, they wounded the King of the Old
Ones even at the moment of his victory.

These are not ordinary children. They are two of the
so-called Gatekeepers who were part of the great

battle, more than ten thousand years ago, when the Old Ones were defeated and banished. It is absolutely crucial this time, if our plans are to succeed and a new world is to be created, that we understand the nature of the Five.

1. Ten thousand years ago, five children led the last survivors of humanity against the Old Ones. The battle took place in Great Britain, but at a time when the country was not yet an island, before the ice sheets had melted in the north.

2. By cunning, by a trick, the children won and the Old Ones were banished. Two gates were constructed to keep them out: one in Yorkshire, in the north of England, the other in Peru. One gate held. The other we managed to smash.

3. The Five existed then. The Five are here now. It is as if they have been reborn on the other side of time … but it is not quite as simple as that. They are the *same* children, somehow living in two different ages.

4. Kill one of the children now and he or she will be "replaced" by one of the children from the past. This is the single, crucial fact that makes them so dangerous an enemy. Killing them is almost no use at all. If we want to control them, they have to be taken alive.

5. Alone, these children are weak and can be beaten. Their powers are unpredictable and not fully in

their control. But when they come together, they become stronger. This is the great danger for us. If all five of them join forces at any time, anywhere in the world, they may be able to create a third gate and everything we have worked for will be lost.

The fifth of the Five

So far only four of the children have been identified. The English boy and the Peruvian boy have now been joined by American twin brothers, Scott and Jamie Tyler, who were revealed to us by our Psi project. At the time, they were working in a theatre in Nevada.

Note that Scott Tyler was thoroughly programmed whilst held captive by our agents in Nevada, USA. Although he was subsequently reunited with his twin brother, it is still possible that he can be turned against his friends. A psychological report (appendix 1) is attached.

We know very little about the fifth of the Five. She is a girl. Like the others, she will be fifteen years old. We expect her to be of Chinese heritage, quite probably still living in the East. In the old world, her name was Scar. It is certain that the other four will be searching for her and we have to face the fact that they may succeed.

We must therefore find her first.

We have agents in every country searching for her. Many politicians and police forces are now working actively for us. The Psi project continues throughout

Europe and Asia and we are still investigating teen-agers with possible psychic/paranormal abilities. There is every chance that the girl will reveal herself to us. It is likely that she still has no idea who and what she really is.

Once we have the girl in our hands, we can use her to draw the rest of them into a trap. One at a time, we will bring them to The Necropolis. And once we have all five of them, we can hold them separately, imprison them, torture them and keep them alive until the end of time.

Everything is now set. The Gatekeepers have no idea how strong we are, how far we have advanced. Our eyes are everywhere, all around the world, and very soon the battle will begin.

We just have to find the girl.

Ia sakkath. Iak sakkakh. Ia sha xul.

ROAD SENSE

The girl didn't look before crossing the road.

That was what the driver said later. She didn't look left or right. She'd seen a friend on the opposite pavement and she simply walked across to join him, not noticing that the lights had turned green, forgetting that this was always a busy junction and that this was four o'clock in the afternoon when people were trying to get their work finished, hurrying on their way home. The girl just set off without thinking. She didn't so much as glimpse the white van heading towards her at fifty miles an hour.

But that was typical of Scarlett Adams. She always was a bit of a dreamer, the sort of person who'd act first and then think about what she'd done only when it was far too late. The hockey ball that she had tried to thwack over the school roof, but which had instead gone straight through the head-mistress's window. The groundsman she had pushed, fully clothed, into the swimming pool. It might have been a good idea to check first that he could swim. The twenty-metre tree she'd climbed up, only to realize that there was no possible way back down.

Fortunately, her school made allowances. It helped that

Scarlett was generally popular, was liked by most of the teachers and even if she was never top of the class, managed to be never too near the bottom. Where she really excelled was at sports. She was captain of the hockey team (despite the occasional misfires), a strong tennis player and an all-round winner when it came to summer athletics. No school will give too much trouble to someone who brings home the trophies and Scarlett was responsible for a whole clutch of them.

The school was called St Genevieve's and from the outside it could have been a stately home or perhaps a private hospital for the very rich. It stood in its own grounds, set back from the road, with ivy growing up the walls, sash windows and a bell tower perched on top of the roof. The uniform, it was generally agreed, was the most hideous in England: a mauve dress, a yellow jersey and, in summer months, a straw hat. Everyone hated the straw hats. In fact it was a tradition for every girl to set the wretched thing on fire on their last day.

St Genevieve's was a private school, one of many that were clustered together in the centre of Dulwich, in South London. It was a strange part of the world and everyone who lived there knew it. To the west there was Streatham and to the east Sydenham, both areas with high-rise flats, drugs and knife crime. But in Dulwich, everything was green. There were old-fashioned tea shops, the sort that spelled themselves "shoppes", and flower baskets hanging off the lampposts. Most of the cars seemed to be four-by-fours and the mothers who drove them were all on first-name terms. Dulwich College, Dulwich Preparatory School, Alleyn's, St Genevieve's ... they were only a stone's throw away from each other, but of course nobody threw stones at each other. Not in this part of town.

It was obvious from her appearance that Scarlett hadn't been born in England. Her parents might be Mr and Mrs Typical-Dulwich – her mother tall, blonde and elegant, her father looking like the lawyer he always had been, with greying hair, a round face and glasses – but she looked nothing like them. Scarlett had long black hair, strange hazel-green eyes and the soft brown skin of a girl born in China, Hong Kong or some other part of Central Asia. She was slim and small with a dazzling smile that had got her out of trouble on many occasions. She wasn't their real daughter. Everyone knew that. She had known it herself from the earliest age.

She had been adopted. Paul and Vanessa Adams were unable to have children of their own and they had found her in an orphanage in Jakarta. Nobody knew how she had got there. The identity of her birth mother was a mystery. Scarlett tried not to think about her past, where she had come from, but she often wondered what would have happened if the couple who had come all the way from London had chosen the baby in cot seven or nine rather than cot eight. Might she have ended up planting rice somewhere in Indonesia or sewing Nike trainers in some city sweatshop? It was enough to make her shudder … the thought alone.

Instead of which, she found herself living with her parents in a quiet street, just round the corner from North Dulwich station which was in turn about a fifteen-minute walk from her school. Her father, Paul Adams, specialized in international business law. Her mother, Vanessa, ran a holiday company that put together packages in China and the Far East. The two of them were so busy that they seldom had time for Scarlett – or indeed, for each other. From the time Scarlett had been five, they had employed a full-time housekeeper to look after all of them. Christina Murdoch was short, dark-haired and seemed to have no sense of humour at all. She had come to London from Glasgow and her father was a vicar. Apart from that, Scarlett knew little about her. The two of them got on well enough, but they had both agreed without actually saying it that they were never going to be friends.

One of the good things about living in Dulwich was that Scarlett did have plenty of friends and they all lived very nearby. There were two girls from her class in the same street and there was also a boy – Aidan Ravitch – just five minutes away. It was Aidan who had prompted her to cross the road.

Aidan was in his second year at The Hall, yet another local private school, and had come to London from Los Angeles. He was tall for his age and good-looking in a relaxed, awkward sort of way, with shaggy hair and slightly crumpled features. There was no uniform at his school and he wore the same hoodie, jeans and trainers day in day out. Aidan didn't understand the English. He claimed to be completely mystified by such things as football, tea and *Dr Who*. English policemen in particular baffled him. "Why do they have to wear those stupid hats?" He was Scarlett's closest friend, although both of them knew that Aidan's father worked for an American bank and could be transferred back home any day. Meanwhile, they spent as much time together as they could.

The accident happened on a warm, summer afternoon. Scarlett was thirteen at the time.

It was a little after four and Scarlett was on her way home from school. The very fact that she was allowed to walk home on her own meant a lot to her. It was only on her last birthday that her parents had finally relented … until then, they had insisted that Mrs Murdoch should meet her at the school gates every day, even though there were far younger girls who were allowed to face the perils of Dulwich High Street without an armed escort. She had never been quite sure what they were so worried about. There was no chance of her getting lost. Her route took her past a flower shop, an organic grocer's and a pub – The Crown and Greyhound – where she might spot a few old men, sitting in the sun with their lemonade shandies. There were no drug dealers, no child snatchers or crazed killers in the immediate area. And she was hardly on her own anyway. From half past three onwards, the streets were crowded with boys and girls streaming in every direction, on their way home.

She had reached the traffic lights on the other side of the village – where five roads met with shops on one side, a primary school on the other – when she noticed him. Aidan was on his own, listening to music. She could see the familiar white wires trailing down from his ears. He saw her, smiled, and called out her name. Without thinking, she began to walk towards him.

The van was being driven by a twenty-five-year-old delivery man called Michael Logue. He would have to give all his details to the police later on. He was delivering spare parts to a sewing machine factory in Bickley and, thanks to the London traffic, he was late. He was almost certainly speeding as he approached the junction. But on the other hand, the lights were definitely green.

Scarlett was about half-way across when she saw him and by that time it was far too late. She saw Aidan's eyes widen in shock and that made her turn her head, wanting to know what it was that he had seen. She froze. The van was almost on top of her. She could see the driver, staring at her from behind the wheel, his face filled with horror, knowing what was about to happen, unable to do anything about it. The van seemed to be getting bigger and bigger as it drew closer. Even as she watched, it completely filled her vision.

And then everything happened at once.

Aidan shouted out. The driver frantically spun the wheel. The van tilted. And Scarlett found herself being thrown forward, out of the way, as something – or someone – smashed into her back with incredible force. She wanted to cry out but her breath caught in her throat and her knees buckled underneath her. Somewhere in her mind she was aware that a passer-by had leapt off the pavement and that he was trying to save her. His arm was around her waist, his shoulder and head pressed into the small of her back. But how had he managed to get to her so fast? Even if he had seen the van coming and sprinted towards her immediately, he surely wouldn't have reached her in time. He seemed to know what was going to happen almost before it did.

The van shot past, missing her by inches. She actually felt the warm breeze slap her face and smelled the petrol fumes. There had been two books in her hand: a French dictionary and a maths exercise book ... an hour and a half's homework for the evening ahead. As she was carried forward, her hand and arm jerked, out of control, and the books were hurled into the air, landing on the road and sliding across the tarmac as if she had deliberately thrown them away. Scarlett followed them. With the man still grabbing hold of her, she came crashing down. There was a moment of sharp pain as she hit the ground and all the skin was taken off one knee. Behind her, there was the screech of tyres, a blast of a horn and then the ominous sound of metal hitting metal. A car alarm went off. Scarlett lay still.

For what felt like a whole minute, nobody did anything. It was as if someone had taken a photograph and framed it with a sign reading ACCIDENT IN DULWICH. Then Scarlett sat up and twisted round. The man who had saved her was lying stretched out in the road and she was only aware that he was Chinese, in his twenties, with black hair, and that he was wearing jeans and a loose-fitting jacket. She looked past him. The white van had swerved round a traffic island, mounted the pavement and smashed into a car parked in front of the primary school. It was this car's alarm that had gone off. The driver of the van was slumped over the wheel, his head covered in broken glass.

She turned back. A crowd had already formed – perhaps it had been there from the start – and people were hurrying towards her, rushing past Aidan, who seemed to be rooted to the spot. He was shaking his head as if denying that he had been to blame. There were twenty or thirty school kids, some of them already taking photographs with their mobile phones. A policeman had appeared so quickly that he could have popped out of a trapdoor in the pavement. He was the first to reach Scarlett.

"Are you all right? Don't try to move…"

Scarlett ignored him. She put out a hand for support and

eased herself back onto her feet. Her knee was on fire and her shoulder felt as if it had been beaten with an iron club, but she was already fairly sure that she hadn't been seriously hurt.

She looked at Aidan, then at the white van. A few people were already helping the driver out, laying him on the pavement. Steam was rising out of the crumpled bonnet. Next to her, the policeman was speaking urgently into his shoulder mike, doing all the stuff with Delta Bravo Oscar Charlie, summoning help.

Finally, Aidan made it over to her. "Scarl...?" That was his name for her. "Are you OK?"

She nodded, suddenly tearful without knowing why. Maybe it was just the shock, the knowledge of what could have been. She wiped her face with the back of her hand, noticing that her nails were grimy and all her knuckles were grazed. Her dress was torn. She realized she must look a wreck.

"You were nearly killed...!" Why was Aidan telling her that? She had more or less worked it out for herself.

Even so, his words reminded her of the man who had saved her. She looked down and was surprised to see that he was no longer there. For a moment she thought that it was a conjuring trick, that he had simply vanished into thin air. Then she saw him, already on the far side of the road – the side that she had been heading towards – hurrying past the shops. He reached a hair salon on the corner, where a woman with hair that was too blonde to be true had just come out. He pushed past her and then he was gone.

Why? He hadn't even stayed long enough to be thanked.

After that, things unravelled more slowly. An ambulance

arrived and although Scarlett didn't need it, the van driver had to be put on a stretcher and carried away. Scarlett herself was examined but nothing was broken and in the end she was allowed to go home. Aidan went with her. A WPC accompanied them both. Scarlett wondered how that would go down with Mrs Murdoch. Somehow she knew it wasn't going to mean laughter and back-slapping at bedtime.

In fact, the accident had several consequences.

Paul and Vanessa Adams were told what had happened when they got home that night and as soon as they had got over the shock, the knowledge of how close they had come to losing their only child, they began to argue about whose fault it was: their own for allowing Scarlett too much freedom, Aidan for distracting her, or Scarlett for showing so little road sense, even at the age of thirteen. In the end, they decided that in future Mrs Murdoch would take up her old position at the school gates. It would be another nine months before Scarlett was allowed to walk home on her own again.

The identity of the man who had saved her remained a mystery. Where had he come from? How had he seen what was about to happen? Why had he been in such a hurry to get away? Mrs Murdoch decided that he must be an illegal immigrant, that he had taken off at the sight of the approaching policeman. For her part, Scarlett was just sorry that she hadn't been able to thank him. And if he was in some sort of trouble, she would have liked to have helped him.

That was the night she had her first dream.

Scarlett had never been one for vivid dreams. Normally she got home, ate, did her homework, spent forty minutes on her

PlayStation 3 and then plunged into a deep, empty sleep that would be ended all too quickly by Mrs Murdoch, shaking her awake for the start of another school day. But this dream was more than vivid. It was so realistic, so detailed that it was almost like being inside a film. And there was something else that was strange about it. As far as she could see, it had no connection to her life or to anything that had happened during the day.

She dreamed that she was in a grey-lit world that might be another planet ... the moon perhaps. In the distance, she could see a vast ocean stretching out to the horizon and beyond – but there were no waves. The surface of the water could have been a single sheet of metal. Everything was dead. She was surrounded by sand-dunes – at least, that was what she thought they were, but they were actually made of dust. They had somehow blown there and – like the dust on the moon – it would stay the same forever. She walked forward. But she left no footprints.

There were four boys standing together, a short distance away.

The boys were searching for her. If she listened carefully, she could actually hear them calling her name. She tried to call back, but although there was no wind, not even a breeze, something snatched the words away.

The boys weren't real. They couldn't be... Scarlett had never seen them before. And yet somehow she was sure that she knew their names.

Scott. Jamie. Pedro. And Matt.

She knew them from somewhere. They had met before.

That was the first time, but over the next two years she had the same dream again and again. And gradually, it began to change. It seemed to her that every time she saw the boys, they were a little further away until finally she had to get used to the fact that she was completely on her own. Every time she went to sleep, she found herself hoping she would see them. More than that. She needed to meet them.

She never spoke about her dreams, not even to Aidan. But somewhere in the back of her mind she knew that finding the four boys had become the single most important thing in her life.

THE DOOR

Two years later, Scarlett had turned fifteen – and she had become an orphan for a second time.

Paul and Vanessa Adams hadn't died but their marriage had, one inch at a time. In a way, it was amazing they had stayed together so long. Scarlett's father had just started a new job, working for a multi-national corporation based in Hong Kong. Meanwhile, her mother was spending more and more time with her own business, looking after customers who seemed to demand her attention twenty-four hours a day. They were seeing less and less of each other and suddenly realized that they preferred it that way. They didn't argue or shout at each other. They just decided they would be happier apart.

They told Scarlett the news at the end of the summer holidays and for her part she wasn't quite sure what to feel. But the truth was that in the short term it would make little difference to her life. Most of the time she was on her own with Mrs Murdoch anyway and although she'd always been glad to see her parents, she'd got used to the fact that they were seldom, if ever, around. The three of them had one last meeting in the kitchen, the two adults sitting with grim faces and large glasses of wine.

"Your mother is going to set up a company in Melbourne, in Australia," Paul said. "She has to go where the market is and Melbourne is a wonderful opportunity." He glanced at Vanessa and in that moment Scarlett knew that he wasn't telling the whole truth. Maybe the Australians were desperate for exotic holidays. But the fact was that she had chosen somewhere as far away as possible. Maybe she had met someone else. Whatever the reason, she wanted to carve herself a whole new life. "As for me, Nightrise have asked me to move to the Hong Kong office…"

The Nightrise Corporation. That was the company that employed her dad.

"I know this is very difficult for you, Scarly," he went on. "Two such huge changes. But we both want to look after you. You can come with either of us."

In fact, it wasn't difficult for Scarlett. She had already thought about it and made up her mind. "Why can't I stay here?" she asked.

"On your own?"

"Mrs Murdoch will look after me. You're not going to sell the house, are you? This is my home! Anyway, I don't want to leave St Genevieve's. And all my friends are here…"

Of course, both her parents protested. They wanted Scarlett to come with them. How could she possibly manage without them? But all of them knew that it was actually the best, the easiest solution. Mrs Murdoch had been with the family for ten years and probably knew Scarlett as well as anyone. In a way, they couldn't have been happier if they had suggested it themselves. It might not be conventional but it was clearly for the best.

27

And so it was agreed. A few weeks later, Vanessa left, hugging Scarlett and promising that the two of them would see each other again very soon. And yet, somehow, Scarlett wondered just how likely that would be. She had always tried to be close to Vanessa, recognizing at the same time that they had almost nothing in common. They weren't a real mother and daughter and so – as far as Scarlett was concerned – this wasn't a real divorce.

Paul Adams left for Hong Kong shortly afterwards and suddenly Scarlett found herself in a new phase of life, virtually on her own. But, as she had expected, it wasn't so very different from what she had always been used to. Mrs Murdoch was still there, cooking, cleaning and making sure she was ready for school. Her father telephoned her regularly to check up on her. Vanessa sent long e-mails. Her teachers – who had been warned what had happened – kept a close eye on her. She was surprised how quickly she got used to things.

She was happy. She had plenty of friends and Aidan was still around. The two of them saw more of each other than ever, going shopping together, listening to music, taking Aidan's dog – a black retriever – out on Dulwich Common. She was allowed to walk home from school on her own again. In fact, as if to recognize her new status, she found herself being given a whole lot more freedom. At weekends, she went into town to the cinema. She stayed overnight with other girls from her class. She had been given a big part in the Christmas play, which meant late afternoon rehearsals and hours in the evening learning her lines. It all helped to fill the time and to make her think that her life wasn't so very unusual after all.

Everything changed one day in November. That was when Miss Chaplin announced her great Blitz project – a visit to London's East End.

Joan Chaplin was the art teacher at St Genevieve's and she was famous for being younger, friendlier and more easygoing than any of the dinosaurs in the staff room. She was always finding new ways to interest the girls, organizing coach trips to exhibitions and events all over London. One class had gone to see the giant crack built into the floor of the Tate Modern. For another it had been a shark suspended in a tank, an installation by the artist, Damien Hirst. Weeks later, they had still been arguing whether it was serious art or just a dead fish.

As part of their GCSE history coursework, a lot of the girls were studying the Blitz, the bombing of London by the Germans during the Second World War. Miss Chaplin had decided that they should take an artistic as well as a historical interest in what had happened.

"I want you to capture the spirit of the Blitz," she explained. "What's the point of studying it if you don't feel it too?" She paused as if waiting for someone to argue, then went on. "You can use photography, painting, collage or even clay modelling if you like. But I want you to give me an idea of what it might have been like to live in London during the winter of 1940."

There was a mutter of agreement around the class. Walking around London had to be more fun than reading about it in books. Scarlett was particularly pleased. History and art had become two of her favourite subjects and she saw that here was an opportunity to do them both at the same time.

"Next Monday, we're going to Shoreditch," Miss Chaplin

went on. "It was an area of London that was very heavily bombed. We'll visit many of the streets, trying to imagine what it was like and we'll look at some of the buildings that survived."

She glanced outside. The art room was on the ground floor, at the back of the school, with a view over the garden, sloping down with flower-beds at the bottom and three tennis courts beyond. It was Friday and it was raining. The rain was sheeting down and the grass was sodden. It had been like that for three days.

"Of course," she went on. "The trip won't be possible if the weather doesn't cheer up – and I have to warn you that the forecast hasn't been too promising. But maybe we'll be lucky. Either way, remember to bring a permission slip from your parents." Then she had a sudden thought and smiled. "What do you think, Scarlett?"

It had become a sort of joke at St Genevieve's.

Scarlett Adams always seemed to know what the weather was going to do. Nobody could remember when it had first started but everyone agreed – you could tell how the day was going to be simply by the way Scarlett dressed. If she forgot her scarf, it would be warm. If she brought in an umbrella, it would rain. After a bit, people began to ask her opinion. If there was an important tennis match or a picnic planned by the river, have a word with Scarlett. If there was any chance of a cross-country run being called off, she would know.

Of course, she wasn't always right. But it seemed she could be relied upon about ninety per cent of the time.

Now she looked out of the window. It was horrible outside.

30

The clouds, grey and unbroken, were smothering the sky. She could see raindrops chasing each other across the glass. "It'll be fine," she said. "It'll clear up after the weekend."

Miss Chaplin nodded. "I do hope you're right."

She was. It rained all day Sunday and it was still drizzling on Sunday night. But Monday morning, when Scarlett woke up, the sky was blue. Even Mrs Murdoch was whistling as she put together the packed lunch requested by the school. It was as if a last burst of summer had decided to put in a surprise appearance.

The coach came to the school at midday. The lesson – combining art and history – was actually going to take place over two periods plus lunch and, allowing for the traffic, the girls wouldn't be back until the end of school. As they pulled out of St Genevieve's, Miss Chaplin talked over the intercom, explaining what they were going to do.

"We'll be stopping for lunch at St Paul's Cathedral," she said. "It was very much part of the spirit of the Blitz because, despite all the bombing, it was not destroyed. The coach will then take us to Shoreditch and we're going to walk around the area. It's still a bit wet underfoot so I want us to go indoors and the place I've chosen is St Meredith's, in Moore Street. It's one of the oldest churches in London. In fact there was a chapel there as long ago as the thirteenth century."

"Why are we visiting a church?" one of the girls asked.

"Because it also played an important part in the war. A lot of local people used to hide there during the bombing. They actually believed it had the power to protect them ... that they'd be safe there."

She paused. The coach had reached the River Thames, crossing over Blackfriars Bridge. Scarlett looked out of the window. The water was flowing very quickly after all the rain. In the distance, she could just make out part of the London Eye, the silver framework glinting in the sunlight. The sight of it made her sad. She had ridden on it with her parents, at the end of the summer. It had been one of the last things the three of them had done while they were still a family.

"...actually took a direct hit on October 2, 1940." Miss Chaplin was still talking about St Meredith's. Scarlett had allowed her thoughts to wander and she'd missed half of what the teacher had said. "It wasn't destroyed, but it was badly damaged. Bring your sketch books with you and we can work in there. We have permission and you can go anywhere you like. See if you can feel the atmosphere. Imagine what it was like, being there with the bombs going off all around."

Miss Chaplin flicked off the microphone and sat down again, next to the driver.

Scarlett was a few rows behind her, sitting next to a girl called Amanda, who was one of her closest friends and who lived in the same road as her. She noticed that Amanda was frowning.

What is it?" she asked.

"St Meredith's," Amanda said.

"What about it?"

It took Amanda a few moments to remember. "There was a murder there. About six months ago."

"You're not being serious."

"I am."

If it had been anyone else, Scarlett might not have believed them. But she knew that Amanda had a special interest in murder. She loved reading Agatha Christie and she was always watching whodunnits on TV. "So who got murdered?" she asked.

"I can't remember," Amanda said. "It was some guy. A librarian, I think. He was stabbed."

Scarlett wasn't sure it sounded very likely and when the coach stopped off at St Paul's, she went over to Miss Chaplin. To her surprise, the teacher didn't even hesitate. "Oh yes," she said cheerfully. "There was an incident there this summer. A man was attacked by a down-and-out. I'm not sure the police ever caught anyone, but it all happened a long time ago. It doesn't bother you, does it, Scarlett?"

"No," Scarlett said. "Of course not."

But that wasn't quite true. It did secretly worry her, even if she wasn't sure why. She had a sense of foreboding which only grew worse the closer they got to the church.

The art teacher had chosen this part of London for a reason. It was a patchwork of old and new, with great gaps where whole buildings and perhaps even streets had been taken out by the Germans. Most of the shops were shabby and depressing, with plastic signs and dirty windows full of products which people might need but which they couldn't possibly want: vacuum cleaners, dog food, one hundred items at less than a pound. There was an ugly car park towering high over the buildings, but it was hard to imagine anyone stopping here. The traffic rumbled past in four lanes, anxious to be on its way.

But even so there were a few clues as to what the area might once have been like. A cobbled alleyway, a gas lamp, a red tele-phone box, a house with pillars and iron railings. The London of seventy years ago. That was what Miss Chaplin had brought them all to find.

They turned into Moore Street. It was a dead end, narrow and full of puddles and pot-holes. A pub stood on one side, opposite a launderette that had shut down. St Meredith's was at the bottom, a solid, red-brick church that looked far too big to have been built in this part of town. The war damage was obvious at once. The steeple had been added quite recently. It wasn't even the same colour as the rest of the building and didn't quite match the huge oak doors or the windows with their heavy stone frames.

Scarlett felt even more uneasy once they were inside. She jumped as the door boomed shut behind her, cutting out the London traffic, much of the light – indeed, any sense that they were in a modern city at all. The interior of the church stretched into the distance to the silver cross, high up on the altar, caught in a single shaft of dusty light. Otherwise, the stained glass win-dows held the sun back, the different colours blurring together. Hundreds of candles flickered uselessly in iron holders. She could make out little side-chapels, built into the walls. Even without remembering the murder that had happened there, St Meredith's didn't strike her as a particularly holy place. It was simply creepy.

But nobody else seemed to share her feelings. The other girls had taken out their sketch books and were sitting in the pews, chatting to each other and drawing what they had seen outside. Miss Chaplin was examining the pulpit – a carving of an eagle. Presumably, most Londoners chose not to pray at two o'clock in the afternoon. They had the place to themselves.

Scarlett looked for Amanda, but her friend was talking to another girl on the other side of the transept so she sat down on her own and opened her pad. She needed to put the murder out of her mind. Instead, she thought about the men and women who had sheltered here during the Blitz. Had they really believed that St Meredith's had some sort of magical power to avoid being hit, that they would be safer here than in a cellar or a Tube station? She thought about them sitting there with their fingers crossed while the Luftwaffe roared overhead. Maybe that was what she would draw.

She shivered. She was wearing a coat but it was very cold inside the church. In fact it felt colder inside than out. A movement caught her eye. A line of candles had flickered, all the flames bending together, caught in a sudden breeze. Had someone just come in? No. The door was still shut. Nobody could have opened or closed it without being heard.

A boy walked past. At first, Scarlett barely registered him. He was in the shadows at the side of the church, between the columns and the side-chapels, moving towards the altar. He made absolutely no sound. Even his feet against the marble floor were silent. He could have been floating. She turned to follow him as he went and just for a second his face was illuminated by a naked bulb, hanging on a wire.

She knew him.

For a moment, she was confused as she tried to think where she had seen him before. And then suddenly she remembered.

It was crazy. It couldn't be possible. But at the same time there could be no doubt.

It was the boy from her dreams, one of the four she had seen walking together in that grey desert. She even knew his name.

It was Matt.

In a normal dream, Scarlett wouldn't see people's faces – or if she did, she would forget them when she woke up. But she had experienced this dream again and again over a period of two years. She'd learned to recognize Matt and the others almost as soon as she was asleep and that was why she knew him now. Short, dark hair. Broad shoulders. Pale skin and eyes that were an intense blue. He was about her age although there was something about him that seemed older. Maybe it was just the way he walked, the sense of purpose. He walked like someone in trouble.

What was he doing here? How had he even got in? Scarlett turned to a girl who was sitting close to her, drawing a major explosion from the look of the scribble on her pad.

"Did you see him?" she asked.

"Who?"

"That boy who just went past."

The other girl looked around her. "What boy?"

Scarlett turned back. The boy had disappeared from sight. For a moment, she was thrown. Had she imagined him? But then she saw him again, some distance away. He had stopped in front of a door. He seemed to hesitate, then turned the handle and went through. The door closed behind him.

She followed him. She had made the decision without even

thinking about it. She just put down her sketch book, got up and went after him. It was when she reached the door that she asked herself what she was doing, chasing after someone she had never met, someone who might not even exist. Suppose she ran into him? What was she going to say? "Hi, I'm Scarlett and I've been dreaming about you. Fancy a Big Mac?" He'd think she was mad.

The door he had passed through was in the outer wall underneath a stained glass window that was so dark and grimy that the picture was lost. Scarlett guessed it must lead out into the street, perhaps into the cemetery if the church had one. There was something strange about it. The door was very small, out of proportion with the rest of St Meredith's. There was a symbol carved into the wooden surface: a five-pointed star.

She hesitated. The girls weren't supposed to leave the church. On the other hand, she wouldn't exactly be going far. If there was no sign of the boy on the other side, she could simply come back in again. The door had an iron ring for a handle. She turned it and went through.

To her surprise, she didn't find herself outside in the street. Instead, she was standing in a wide, brightly lit corridor. There were flaming torches slanting out of iron brackets set in the walls, the fire leaping up towards the ceiling which was high and vaulted. The corridor had no decoration of any kind and it seemed both old and new at the same time, the plasterwork crumbling to reveal the brickwork underneath. It had to be some sort of cloister – somewhere the priests went to be on their own. But the corridor was nothing like the rest of St Meredith's. It was a different colour. It was the wrong size and shape.

It was also very cold. The temperature seemed to have fallen dramatically. As she breathed out, Scarlett saw white mist in front of her face. It was as if she were standing inside a fridge. She had to remind herself that this was the first week of November. It felt like the middle of winter. She rubbed her arms, fighting off the biting cold.

There was a man, sitting in a wooden chair opposite her, facing the door. She hadn't noticed him at first because he was in shadow, between two of the torches. He was dressed like a monk with a long, dirty brown habit that went all the way down to his bare feet. He was wearing sandals, and a hood over his head. He was slumped forward with his face towards the floor. Scarlett had already decided to turn round and go back the way she had come, but before she could move, he suddenly looked up. The hood fell back. She gasped.

He was one of the ugliest men she had ever seen. He was completely bald, the skin stretched over a skull that was utterly white and dead. His head was the wrong shape – narrow, with part of it caved in on one side, like an egg that has been hit with a spoon. His eyes were black and sunken and he had horrible teeth which revealed themselves as he smiled at her, his thin lips sliding back like a knife wound. What had he been doing, sitting there? She looked left and right but they were on their own. The boy called Matt – if it had even been him – had gone.

The man spoke. The words cracked in his throat and Scarlett didn't understand any of them. He could have been speaking Russian or Polish … whatever it was, it wasn't English. She backed away towards the door.

"I'm very sorry," she said. "I think I've come the wrong way."

38

She turned round and scrambled for the handle. But she never made it. The monk had moved very quickly. She felt his hands grab hold of her shoulders and drag her backwards, away from the door. He was very strong. His fingers dug into her like steel pincers.

"Let go!" she shouted.

His arm sneaked over her shoulder and around her throat. He was holding her with incredible force. She could feel the bone, cutting into her windpipe, blocking the air supply. And he was screaming out more words that she couldn't understand, his voice high-pitched and animal. Another monk appeared at the end of the corridor. Scarlett didn't really see him. She was just aware of him rushing towards them, the long robes flapping.

Still she fought back. She reached with both hands, clawing for the monk's eyes. She kicked back with one foot, then tried to elbow him in his stomach. But she couldn't reach him. And then the second monk threw himself onto her.

The next thing she knew, she was on her back, her arms stretched out above her head. Her legs had been knocked out from underneath her. The two men had grabbed hold of her and there was nothing she could do. She twisted and writhed, her hair falling over her face. The monks just laughed.

Scarlett felt her heels bumping over the stone cold floor as the two men dragged her away.

FATHER GREGORY

The cell was tiny – less than ten metres square – and there was nothing in it at all, not even a chair or a bench. The walls were brick with a few traces of flaking paint, suggesting they might have been decorated at some time. The door had been fashioned out of three slabs of wood, fastened together with metal bands. There was a single window, barred and set high up so that even for someone taller than Scarlett, there wouldn't be any chance of a view. From where she was sitting, slumped miserably on the stone floor, she could just make out a narrow strip of sky. But even that was enough to send a shiver down her spine.

It was dark. Not quite night, but very nearly. She realized that it would be pitch black in the cell in just a couple of hours as they hadn't left her a candle or an electric light. But how was that possible? It had been around two o'clock when she had entered St Meredith's and the sun had been shining. Suddenly it was early evening. So what had happened to the time in between?

Scarlett was shivering – and not just because of the shock of what she had been through. It was freezing in the cell. There was no glass in the window and no heating. The bare brick-work only made it worse. Fortunately she had been wearing her winter coat when she set off on the school trip and she drew it around her, trying to bury herself in its folds. She had never been so cold. She could actually feel the bones in her arms and legs. They were so hard and brittle that she thought they might shatter at any time.

Desperately, she tried to work out what had happened. For no reason that she could even begin to imagine, a man she had never met had grabbed her and thrown her into a cell. Could she have strayed into a secret wing of St Meredith's, some-where that no one was meant to go? The single strip of sky told her otherwise. That and the freezing weather. She remembered that the monk had spoken in a foreign language.

She was no longer in London.

It seemed crazy but she had to accept it. Maybe she had blacked out at the moment she had been seized. Maybe they had drugged her and she had been unconscious without even knowing it. Everything told her that this wasn't England. Somehow she had been spirited away.

With a spurt of anger, she scrambled to her feet and went over to the door. She wasn't just going to sit here and wait for them to come back. Suppose they never did come back? She might die in this place. But she quickly saw that there was no way through the door – not unless it was unlocked from the other side. It was massive and solid, with a single keyhole built for an antique key. She tried to squint through it but there was nothing to see. She straightened up, then hammered her fists against the wood.

"Hey! Come back! Let me out of here!"

But nobody came. She wasn't even sure if her voice could be heard outside the cell.

That left the window. Could she possibly climb up, using the rough edges of the brickwork to support herself? Scarlett tried but her fingertips couldn't get enough grip, and anyway the bars at the top were too close to squeeze through, even assuming she could drop down on the other side. No. She was in a solid box with no trapdoors, no secret passages, no magic way out. She would just have to stay here until somebody came.

She sank back into a corner, trying to preserve what little body warmth she had left by curling herself into a ball. The strange thing was, she should have been terrified. She was completely helpless, a prisoner. This was an evil place. But she still couldn't accept the reality of what had happened to her and because of that it was difficult to feel scared. This was all like some bad dream. Once she had worked out how she had got here, then maybe she could start worrying about what was going to happen next.

An hour passed, or maybe two. Finally there was a rattle of a key in the lock, the door swung open again and two monks came into the cell. Scarlett couldn't say if they were the ones who had grabbed her in the corridor as all these people were dressed the same way. Their hoods were up and they were skeleton-thin. Even if you stood them up against a wall, it would have been difficult to tell them apart.

One of them barked out a command in the strange, harsh language she had heard before and when he saw that she didn't understand, made a rough gesture, telling her to stand up. Scarlett did as she was instructed. Her face gave nothing away but she was already thinking. If they took her out of here, maybe she would be able to break away. She would run back down the corridor and find the nearest exit. Whatever country she was in, there would have to be a policeman or someone else around. She would make herself understood, somehow find her way home.

But right now, the two monks were watching her too closely. They led her out with one standing next to her and the other directly behind, so close that she could actually smell them. Neither man had washed, not for a long time. As they reached the corridor, Scarlett hesitated and felt a hand pushing her roughly forward. She turned left. The three of them set off together.

Where was she? The place had the feel of an old palace or a monastery, but one thing was certain – it had been abandoned long ago. Everything about it was broken down and neglected, from the peeling walls to the paved floor, which was slanting and uneven with some sort of mould growing through the cracks. Naked light bulbs hung on single wires (so at least there was electricity) but they were dull and flickering, barely able to light the way. The air was damp and there was a faint smell of sewage.

Scarlett noticed an oil painting in a gilt frame. It showed a crucifixion scene, but the colours were faded, the canvas torn. An antique cabinet with two iron candlesticks stood beneath it, one door open and papers scattered on the floor. The three of them turned a corner and for the first time she was able to see outside. A series of arches led onto a terrace with a garden beyond. Scarlett stopped dead. Her worst fears had been realized. She knew now that she definitely wasn't in England.

The garden was covered in snow. There were trees with no leaves, their branches heavy with the stuff. The ground was also buried and, in the distance, barely visible in the darkness, she could see white-topped mountains. There were no other buildings, no lights showing anywhere. The monastery was in some sort of wilderness – but how had she got here? Had she been knocked out and put on a plane? Scarlett searched back in her memory but there was nothing there … nothing to indicate a journey, leaving England or arriving anywhere else. Then one of the monks jabbed her in the back and she was forced to start moving again.

They came to a hallway, lit by a huge chandelier, not electric but jammed with rows of candles, at least a hundred of them, the wax dripping slowly down and congealing into a series of growths that reminded Scarlett of the sort of shapes she had once seen in a cave. Some of it had splattered onto a round table beneath. An empty bottle lay on its side along with dirty plates and glasses, mouldering pieces of bread. There had been a dinner here – days, maybe weeks before. There were no rats or cockroaches. It was too cold.

Several doors led out of the hallway. The two monks led her to the nearest of them. One of them opened it. The other pushed her inside. He had hurt her and Scarlett spun round and swore at him. The monk just smiled and backed away. The other man went with him. The door closed.

She turned back and examined her new surroundings. This was the only half-way comfortable room she had seen so far.

It was furnished with a rug on the floor, two armchairs, book-shelves and a desk. It was warmer too. A coal fire was burning in a grate and although the flames were low she could feel the heat it was giving out and smell it in the air. More paintings hung on the walls, also with religious subjects. There was a window, but it had become too dark to see outside.

A man was sitting behind the desk. He also wore a habit, but his was black. So far he had said nothing but his eyes were fixed on Scarlett and, with an uneasy feeling, she walked over to him. He was the oldest man she had seen – at least twenty years older than the others, with the same bald head and sunken eyes. There were tufts of white hair around his ears and he had thick white eyebrows that could have been glued in place. His nose was long and too thin for his face. His fingers, spread out across the surface of the desk, were the same. He was watching Scarlett intensely, and as she drew closer she saw that there was a growth – a sty – sitting on one of his eyes. The whole socket was red and dripping. It was as if, like the rest of the building, he was rotting away. Scarlett shuddered and felt sick.

The man still hadn't spoken. Scarlett drew level with him so that the desk was between them. Despite everything, she had decided that she wasn't going to let him intimidate her. "Who are you?" she demanded. "Where am I? Why have you brought me here?"

His eyes widened in surprise. At least, one of them did. The diseased eye had long since lost any movement. "You are English?" he said.

Scarlett was taken by surprise. She hadn't expected him to speak her language. "Yes," she said.

"Please. Sit down..." He gestured at one of the chairs. "Would you like a hot drink? Some tea should be arriving soon."

Scarlett shook her head. "I don't want any tea," she snapped. "I want to go back where I came from. Why are you keeping me here?"

"I asked you to sit down," the monk said. "I would suggest that you do as you are told."

He hadn't raised his voice. He didn't even sound threatening. But somehow Scarlett knew it would be a mistake to disobey him. She could see it in his eyes. The pupils were black and dead and slightly unfocused. They were the sort of eyes that might belong to someone who was mad.

She sat down.

"That's better," he said. "Now, let's introduce ourselves. What is your name?"

"I'm Scarlett Adams."

"Scarlett Adams." He repeated it with a sort of satisfaction, as if that was what he had expected to hear. "Where are you from?"

"I live in Dulwich. In London. Please, will you tell me where I am?"

He lifted a single finger. The nail was yellow and bent out of shape. "I will tell you everything you wish to know," he said. His English was perfect although it was obvious that it wasn't his first language. He had an accent that Scarlett couldn't place and he strung his words together very carefully, like a craftsman making a necklace. "But first tell me this," he went on. "You really have no idea how you came here?"

46

"No." Scarlett shook her head. "I was in a church."

"In London?"

"Yes. I went through a door. One of the people here grabbed hold of me. That's all I can remember."

He nodded slowly. His eyes had never left her and Scarlett felt a terrible urge to look away, as if somehow he was going to swallow her up.

"You are in Ukraine," the man said, suddenly.

"Ukraine?" Everything seemed to spin for a minute. "But that's…"

It was somewhere in Russia. It was on the other side of the world.

"This is the Monastery of the Cry for Mercy. I am Father Gregory." He looked at his guest a little sadly, as if he was disappointed that she didn't understand. "Your coming here is a great miracle," he said. "We have been waiting for you for almost twenty years."

"That's not possible. What do you mean? I haven't been alive for twenty years." Scarlett was getting tired of this. She was feeling sick with exhaustion, with confusion. "How come you speak English?" she asked. She knew it was a stupid question but she needed a simple answer. She wanted to hear something that actually made sense.

"I have travelled all over the world," Father Gregory replied. "I spent six years in your country, in a seminary near the city of Bath."

"Why did you say you've been waiting for me? What do you mean?"

The door suddenly opened and one of the monks came in,

carrying a bronze tray with two glasses of tea. Scarlett guessed that Father Gregory must have ordered it before she was brought in because there was no obvious method of communication in the office, no telephone or computer, nothing modern apart from a desk lamp throwing out a pool of yellow light. The monk set down the tray and left.

"Help yourself," Father Gregory said.

Scarlett did as she was told. The liquid was boiling hot and burned her fingers as she lifted the glass. She took a sip. The tea tasted herbal and it was heavily sugared, so sweet that it stuck to her lips. She set it down again.

"I will tell you my story because it pleases me to do so," Father Gregory said. "Because I sometimes wondered if this day would ever come. That you are sitting here now, in this place, is more than a miracle. My whole life has been leading to this moment. It is perhaps the very reason why I was meant to live."

Scarlett didn't interrupt him. The more he talked, the more passionate he became. She could see the coal fire reflected in his eyes, but even if the fire hadn't been there, there might still have been the same glow.

"I was born sixty-two years ago in Moscow, which was then the capital of the Soviet Union. My father was a politician, but from my earliest age, I knew that I wanted to enter the Church. Why? I did not like the world into which I had been born. Even when I was at school, I found the other children spiteful and stupid. I was small for my age and often bullied. I never found it easy to make friends. I did not much like my parents either. They didn't understand me. They didn't even try.

"I was nineteen when I told my father that I wanted to take holy orders. He was horrified. I was his only son and he had always assumed that I would go into politics, like him. He tried to talk me out of it. He arranged for me to travel around the world, hoping that if I saw all the riches that the West had to offer, it would change my mind.

"In fact, it did the exact opposite. Everything I saw in Europe and America disgusted me. Wealthy families with huge homes and expensive cars, living just a mile away from children who were dying because they could not afford medicine. Countries at war, the people killing and maiming each other because of politicians too stupid to find another way. The noise of modern life; the planes and the cars, the concrete smothering the land. The pollution and the garbage. The people, in their millions, scurrying on their way to jobs they hated..."

Scarlett shrugged. "So you weren't happy," she said. "What's that got to do with me?"

"It has everything to do with you and if you interrupt me again I will have you whipped until the skin peels off your back."

Father Gregory paused. Scarlett was completely shocked but didn't want to show it. She said nothing.

"I entered a seminary in England," he continued, "and trained to become a monk. I spent six years there, then another three in Tuscany before finally I came here. That was thirty years ago. This was a very beautiful and very restful place when I first arrived, a refuge from the rest of the world. The weather was harsh and, in the winter, the days were short. But the way of life suited me. Prayer six times a day, simple meals and silence while we ate. We cultivated all our food ourselves. I have spent many hundreds of hours hacking at the barren soil that surrounds us. When I wasn't in the fields, I was helping in the local villages, tending to the poor and the sick.

"A holy life, Scarlett. And so it might have remained. But then everything changed. And all because of a door in a wall."

Father Gregory hadn't touched his tea, but suddenly he picked up his glass between his finger and thumb and tipped the scalding liquid back. Scarlett saw his throat bulge. It was like watching a sick man take his medicine.

"It puzzled me from the start. A door that seemed to belong to a different building with a strange device – a five-pointed star – that had nothing to do with this place. A door that went nowhere." He lifted a hand to stop her interrupting. "It went nowhere, child. Believe me. There was a brief corridor on the other side and then a blank wall.

"The monastery was then run by an abbot who was much older than me. His name was Father Janek. And one day, walking in the cloisters, I asked him about it.

"He wouldn't tell me. A simple lie might have ended my curiosity, but Father Janek was too good a man to lie. Instead, he told me not to ask any more questions. He quickened his pace and as he walked away, I saw that he was afraid.

"From that day on, I became fascinated by the door. We had an extensive library here, Scarlett, with more than ten thousand books – although most of them have now mouldered away. Some of them were centuries old. I searched through them. It took me many years. But slowly – a sentence here, a fragment there – a story began to emerge. But in the end, it was one book, a secret copy of a diary written by a Spanish monk in

50

1532 that told me everything I wanted to know."

He stopped and ran his eyes over the girl as if she were the most precious thing he had ever seen. Scarlett was revolted and didn't try to hide it. The eyes underneath the white eyebrows were devouring her. She could see saliva on the old man's lips.

"The Old Ones," he whispered, and although Scarlett had never heard those words before, they meant something to her; some memory from the far distant past. "The diary told me about the great battle that had taken place ten thousand years ago when the Old Ones ruled the world and mankind were their slaves. Pure evil. The Bible talks of devils ... of Lucifer and Satan. But that's just story-telling. The Old Ones were real. They were here. And the one who ruled over them, Chaos, was more powerful than anything in the universe."

"So what happened to them?" Scarlett asked. Her voice had almost dropped to a whisper. Apart from the flames, twisting in the hearth, everything in the room was still.

"They were defeated and cast out. There were five children..." he spoke the word with contempt. "They came to be known as the Gatekeepers. Four boys and a girl." He levelled his eyes on Scarlett and she knew what he was going to say next. "You are the girl."

Scarlett shook her head. "You're wrong. That's insane. I'm not anything. I'm just a schoolgirl. I go to school in London..."

"How do you think you got here?" The monk pointed in the direction of the corridor with a single trembling finger. Some sort of liquid was leaking out of his damaged eye, a single tear. "You have seen the monastery and the snow. You know you are not in London now."

"You drugged me."

"You came through the door! It was all there in the diary. There were twenty-five doorways built all around the world. They were there for the Gatekeepers so that when the time came, they would be able to travel great distances in seconds. Only the Gatekeepers could use them. Nobody else. When I pass through the door, I find myself in a corridor, a dead end. But it's not the same for you. It brought you here."

Scarlett shook her head. Nothing she had heard made any sense at all. She didn't even know where to begin. "I'm not ten thousand years old," she said. "Look at me! You can see for yourself. I'm fifteen!"

"You have lived twice, at two different times." He laughed delightedly. "It's beyond belief," he said. "Finally to meet one of the Gatekeepers after all these years and to find that she has no idea who or what she is."

"You mentioned there was an abbot here," Scarlett said. "I want to talk to him."

"Father Janek is dead." He sighed. "I haven't told you the rest of my story. Maybe then you will understand." He nodded at her glass. "You haven't drunk your tea."

"I don't want it."

"I would take what you are given while you still can, child. There is much pain for you ahead."

Scarlett's tea was right in front of her. Briefly, she thought about picking it up and flinging it in his face. But it wouldn't do much good. It was probably lukewarm by now.

"The discovery of the diary, along with all the other fragments, changed my life," the monk continued. "I began to think about the reasons why I had come to the monastery in the first place. Did I really think that religion – prayer and fasting – would help me change the world? Or was I just using religion to hide from it? Suddenly I knew what had brought me here. Hatred. I hated the world. I hated mankind. And praying to God to save us was ridiculous. God isn't interested! If He was, don't you think He'd have done something centuries ago?

"My whole life had been devoted to an illusion. All those prayers, the same words repeated again and again. Did they really make any sense? Of course not! The cries for mercy that would never come. Kneeling and making signs, singing hymns while, outside in the street, people were killing each other and trying to make as much money as they could to spend on themselves and to hell with everyone else. Do you never read the papers? What do you see in them except for murder and lust and greed, all day, every day? Do you not see the nature of the world in which you live?

"There is no God, Scarlett. I know that now. But there *are* the Old Ones. They are our natural masters. They deserve to rule the world because the world is evil and so are they."

He paused for breath. Scarlett looked at him with a mixture of pity and disgust. She had already decided that this wasn't about God or about religion. It was about a man who had nothing inside him. The years had hollowed out Father Gregory until there was nothing left.

"I will finish my story and then you must be taken back to your cell," he said. "You will not be staying with us very long, Scarlett. You have a long journey to make. You will not return."

Scarlett said nothing. She knew that he was trying to frighten her. She also knew that he was succeeding. A long journey … where? And how would they take her there? Would they force her through another door?

Father Gregory closed his eyes for a few seconds, then continued.

"When I came here, there were twenty-four brothers at the Monastery of the Cry for Mercy," he explained. "Some of them, I knew, felt the same as me. They were disillusioned. Their life was hard. There were no rewards. The local people, the ones they were helping, weren't even grateful. Gradually, I began to sound them out. I shared with them the knowledge I had discovered. How many of them would abandon their religion and turn instead to the Old Ones? In the end, there were seven of us. Seven out of twenty-four. Ready to begin a new adventure.

"We could of course have left. But I already knew that was out of the question. We were here for a reason, and that reason was the door. It had been here long before the monastery existed. Indeed, why was the monastery built in this place at all? It was because the architects knew that the door was in some way magical even if they had forgotten what its true purpose was. Do you see, child? The monastery was built *around* the door just as you will find holy places connected to the other doors all over the world: churches, temples, burial sites, caves.

"The seven of us agreed that we would stay here and serve the Old Ones. We would guard the door and should a child ever pass through it, we would know that we had found one of the Gatekeepers and we would seize hold of them just as we have taken hold of you…"

54

"What happened to Father Janek and the other monks?" Scarlett asked, although she wasn't sure she wanted to know.

"I killed Father Janek," the monk replied. "I crept into his room while he slept and cut his throat. Then we continued around the monastery and did the same to all the others. Seventeen men died that night and in the morning the corridors were awash with blood. But don't mourn for them. They would have died happily. They would think they were going to Heaven, into the embrace of their God.

"We have been here ever since. Of course, with so few of us, the monastery has fallen into disrepair. Once, the villagers brought us food because they revered us. Now they give it to us because they are afraid. We have survived a very long time, always waiting, always watching the door. Because we knew that you would come. And recently we realized that our time had come. We were expecting you."

"How?"

"Because the Old Ones have returned to the world. Even now, they are gathering strength, waiting to take back what was always theirs. Their agents have contacted us. Very soon, we will hand you over to them. And then we will have our reward."

"What will happen to me?"

"The Old Ones will not kill you. You don't need to be afraid. But they will need to keep you close to them and you still must pay for what you did to them so many years ago."

"I didn't do anything. I don't know what you're talking about..."

He nodded his head sadly. "A great pity," he murmured.

"I had expected more of you. A warrior or a great magician. But you really are nothing. A little girl, as you said, from school. Maybe the Old Ones will let me torture you for a while before you go. I would like that very much. To pay you for the disappointment. We will see…"

He stood up and went over to the door. He walked with a limp and it occurred to Scarlett that as well as the diseased eye, he might have a withered leg. It took him a while even to cross the room and she briefly wondered if she might be able to overpower him. But it wouldn't have done any good. When he opened the door, the two monks who had brought her there were waiting on the other side.

"They will take you back to your cell," he said. "They will also bring you food and water. I imagine you will be with us a few days."

Scarlett stood up and walked past him. There was nothing else she could do. For a brief moment, the two of them stood shoulder to shoulder in the doorway. Father Gregory reached out and stroked her hair. Scarlett shuddered. She didn't even try to hide her revulsion.

"Goodbye, Scarlett," he said. "You have no idea how glad I am that we have met."

Scarlett let the two monks walk her away. She didn't look back.

DRAGON'S BREATH

They took Scarlett back to the same cell she had occupied – but they had been busy while she was away. Someone had carried in a bed, although the moment she saw it she knew she wasn't going to be allowed the privilege of a comfortable sleep. It was little more than a cot with sagging springs and a metal frame and she wouldn't even be able to stretch out without her feet going over the end. There were just two coarse blankets to protect her from the chill of the night and no pillow.

They had also supplied her with a table, a chair and a bucket which she guessed she would be expected to use as a toilet, although she didn't even want to think about that. A candle in a glass lantern now lit the room and they had provided her with a meagre dinner. A bowl of thin, vegetable soup, a hunk of bread and a mug were waiting on the table. There was a spoon to eat with – and if Scarlett had any thought of using it as a weapon, her hopes were soon dashed. It was flimsy, made of tin. They hadn't bothered with a knife or a fork.

She didn't feel like eating yet. If anything, the sight of the starvation rations brought home the full horror of her situation. These people were utterly merciless. They wanted her to live but they didn't care how miserable or painful her life became – they had made that much clear. Scarlett sat down on the bed and sank her head into her hands. She thought she was going to cry, but the tears didn't come. The Old Ones. The Gatekeepers. The twenty-five doors around the world. Everything that Father Gregory had said seemed to spin round and round her, sucking her ever further into a tunnel of misery and despair. How could this have happened to her? Could any of it really be true?

Somehow, she forced herself to go over it, to unpick the words. Much of what Father Gregory had said sounded completely insane. But at the same time, she had to admit that a lot of it was strangely familiar. There were echoes. There had been strange incidents in her life and they had taken place long before she walked through the church door.

The dreams, for one. Father Gregory had mentioned five children – four boys and a girl. Scarlett had been dreaming exactly the same thing for almost two years. And how had this all started? She had actually seen Matt, in St Meredith's. He had been the one who had led her through the door, although now she wondered if he had really been there at all. He had been silent, ghost-like. It wasn't that she had imagined him. But perhaps what she had experienced was some sort of vision. If he had really gone through the door, wouldn't he be here now?

And then there was the door itself. Scarlett had tried to persuade herself that she had been drugged and kidnapped, but the more she thought about it, the more she accepted that it hadn't happened that way. Father Gregory had told her the truth. She had gone through a door in London and ended up in Ukraine. There had been no flight, no drugs. And if she accepted that, what choice did she have but to accept the rest?

She went over to the table and examined the food. It looked far from appetizing, but she made herself swallow it, the soup cold and greasy, the bread several days old. It was all she was going to get and she needed her strength. The candle in the lamp was only an inch tall and she wondered how long it would last. When it went out, she would be left in total blackness. The thought made her shudder. There was already so much to be afraid of but being on her own, locked up in the dark was somehow worse than any of it.

It would be better if she could sleep. She didn't undress. It was far too cold to even think of taking off her coat. She climbed onto the bed and pulled the two blankets over her, burrowing into them like an animal in a cave. She lay like that for a long time and when sleep did finally come she didn't even notice it. She only knew that she was no longer awake when she realized that she had begun to dream.

She was back in the strange, airless world that she had visited many times. She recognized it and she was glad to be there. She was desperate to see Matt and the other three boys. If anyone could help her, they could. At least they might show her a way to break out.

But there was no sign of them. While part of her slept, alone in her cell, the other part was stranded here, alone on the edge of a grim and lifeless sea.

Something in the dreamworld had changed. Scarlett became aware of it very slowly, not seeing anything but sensing it, a sort of throbbing in the air that was coming from very far away, from the other side of the horizon. She heard a faint rumble of thunder and saw a tiny streak of lightning, like a hairline crack in the fabric of the world. Her head was pounding. She noticed the water, the surface of the ocean, begin to shiver. A gust of wind tugged at her hair. The sand, or the grey dust, or whatever it was, spun in eddies around her feet, then leapt up, half blinding her and stinging her cheeks. She backed away, knowing that she needed to hide. She still didn't know what she was hiding from.

And then, in a single moment, the ocean split open. It was as if it had been sliced in half by some vast, invisible knife – and the black water rushed in, millions of gallons pouring from left and right into the chasm – a mile long – that had been formed. At the same time, something rose up, twisting towards the surface. At first, she thought it was a snake, some sort of monstrous sea serpent that had been resting for centuries on the ocean bed and had only now woken up. She smelled its breath – how was that possible ... how could you smell anything in a dream? – and cried out as it rushed towards her, its eyes blazing, flames exploding around its mouth. It was a dragon! Straight out of ancient folklore. And yet it was horribly real, howling so loudly that she thought her head would burst.

SIGNAL ONE

The two words had appeared in front of her. They were written in neon: huge red letters hanging from some sort of frame, the light so intense that they burned her eyes. Where had they come from? They must have risen out of the ground because only a moment before the landscape had been empty.

The neon buzzed and flickered as some sort of electric power coursed through it. Scarlett looked down at her hands and saw that they were blood red, reflecting the light. It was as if she were on fire.

SIGNAL ONE... SIGNAL ONE...

It flashed on and off. The dragon was there one minute, then gone the next, lost in the darkness, reappearing in the light. But each time she saw it, it was a little closer. The wind was blasting her. If it got any stronger, it would throw her off her feet. She tried to run but she couldn't move. The dragon opened its mouth, showing teeth like kitchen knives.

And that was when she woke up and found herself still lying on top of the bed and covered by the two blankets, but with the first, dreary light of the morning creeping in through the window and ice cold all around.

Scarlett sat up. She was already beginning to shiver. What had that all been about? Signal One? She had never seen the two words written down before. She had no idea what they meant, even if she was certain that they must be important. They had been shown to her for a reason.

She looked up at the window and guessed that it must be about five or six o'clock in the morning. It was difficult to say without her watch. Presumably the monks would bring her some sort of breakfast. They had made it clear that they needed to keep her alive. Could she somehow overpower them when they came in, fight her way through the door and make a run for it? She doubted it. The monks were thin and malnourished but they were still a lot stronger than her. If only she had a weapon! That would make all the difference.

Sitting on the edge of the bed, she searched through her pockets. All she had was a blunt pencil, left over from art class, a comb and an Oyster card. The sight of it made her sad. It was so ordinary, a reminder of everything she had left behind. How many thousands of miles was she now from London buses and Tube trains?

There was nothing she could use. She considered taking off her coat, throwing it in the face of whoever carried in her food. But it was a stupid plan. She still didn't know there was going to be any food and anyway, it wouldn't work. They would just laugh at her before they took her away and whipped her or whatever else they planned to do.

There had to be a way out of the cell. Scarlett got up and examined the door a second time, running her hands over the hasps, pressing against it with all her weight. It was so solid it might as well have been cemented into the wall. That just left the window. There were three bars and no glass. The cell had been built to house a man, not a child – and certainly not a girl. Might it be possible to squeeze through after all?

She hadn't been able to reach the window before but maybe these monks, as clever as they might be, had made a mistake. They had supplied her with a table and a chair. Quickly, she dragged the table over to the window, put the chair on top and climbed up.

For the first time, she was able to look outside. There was a view down a hill, the ground steep and rugged, thick patches of snow piled up against black rocks. A few buildings stood in the near distance, scattered around. They looked like barns and abandoned farm houses which might belong to the monastery but which were more likely part of a village, just out of sight. A series of icicles hung above her, suspended from a guttering that ran the full length of the building. She had forgotten how cold it was but she was quickly reminded by a sudden snow flurry, blowing in off the roof. Her lips and cheeks were already numb. It had to be less than zero out there.

There was no way down. The bars were too close together and even if she had managed to slip through, she was at least twenty metres above the ground. Try to jump from this height and she would break both her legs.

She was still in the cell two hours later when the door opened and they finally brought her something to eat.

Breakfast was a bowl of cold porridge and a tin mug of water, carried in by a monk she hadn't yet met – for his face certainly wasn't one that she would have forgotten. It was horribly burned. One whole side of it was dead and disfigured as if he had fallen asleep with his head resting on an oven. Scarlett turned her eyes away from him. Was there anyone at Cry for Mercy who hadn't rotted over the past twenty years? A second monk stood with him, guarding the door.

"You ... eat ... little ... girl." Burnt Face was proud of his English but his accent was so thick she could barely make out the words.

He set the tray down, and Scarlett moved towards him. Her hands were clasped behind her back and she was clearly on the edge of tears. "Please," she said. "Please let me out..." Her voice was trembling.

The sight of the girl, pale and bleary-eyed after the long night, seemed to amuse him. "Out?" He sneered at her. "No out..."

"But you don't understand..." She was closer to him now and as he straightened up she brought her hands round and lashed out.

She was holding an icicle.

She had broken it off the guttering and she was holding it like a knife. The point was needle sharp. Using all her strength, she drove it into the flesh between his shoulder and his neck. The monk screamed. Blood gushed out. He fell to his knees, as if in prayer.

Scarlett was already moving. She knew that she had to take advantage of the surprise, that speed was all she had on her side. The second monk had frozen, completely shocked by what had just happened. Before he could react, she threw herself at him, head and shoulders down, like a bull. She hit him hard in the stomach and heard the breath explode out of him. His hands grabbed for her but then he was down, writhing on the floor. She pulled away and began to run.

According to Father Gregory, there were just seven monks in the Monastery of the Cry for Mercy and she had just taken out two of them. How long would it be before the ones that remained set off after her? Scarlett had to find the door that had brought her here. She knew where it was – a short way down the corridor, only a minute from the cell. With a bit of luck, she would be gone before they knew what had happened.

It was only when she had taken twenty paces that she knew she had gone wrong. Somehow she had managed to get lost. She was in another long corridor and it was one that she didn't recognize. There was a picture of some holy person hanging crookedly on the wall. An ornate wooden chest. Another passageway with a flight of stone steps leading down. For a moment they looked tempting. They might lead her out of the monastery. But at the same time, she knew they would take her further away from the door. The door was the fast way back to St Meredith's. She had to find it.

In the distance, a bell began to ring. Not a call to prayers. An alarm. She heard shouting. The second of the two monks – the one she had hit – must have recovered. Forcing herself not to panic, she continued forward even though she knew she was heading in the wrong direction and that the further she went, the more lost she would become. She heard flapping ahead of her, the sound of sandals hitting the stone floor and a moment later another monk appeared. He saw her and cried out. There was an opening to one side. She took it, passing between wood-panelled walls and a great tapestry, hanging in shreds, the fabric mouldering away.

The passage emerged in a second corridor and with a surge of relief she realized that she knew where she was. Somehow she had found her way back. There was the table with the candlesticks, the painting of the crucifixion. The door was just beyond. There was nobody in the way.

The noise of the sandals. If the monk had been barefooted, Scarlett might not have heard him. But even without looking round, she knew that someone had caught up with her, that he was running towards her even now. In a single movement she reached out, grabbed a heavy, iron candlestick and swung it round. She'd timed it exactly right. The end of the candlestick smashed into the side of the monk's bald head, knocking him out. Scarlett hit him a second time, just to be sure, then dropped the candlestick and made for the door.

Someone appeared at the far end of the corridor.

It was Father Gregory. He saw Scarlett and screamed something – maybe in English, maybe in his own language. The words were trapped in his throat. The door was now between the two of them, exactly half-way. Scarlett wondered if she could reach it. Father Gregory was dancing on his feet as if he had just been electrocuted. His good eye was wide and staring, making the other one look all the more diseased. Scarlett was about thirty metres away, panting, gathering all her strength for one last effort.

The two of them set off at the same moment.

In a way it was weird. Scarlett wasn't running away. She was actually hurtling towards the one man she most wanted to avoid. But she had to reach the door before he did. She had made her decision. It was the only way home.

Father Gregory was surprisingly fast. His limp had disappeared and he moved with incredible speed, his fury propelling him forward. Scarlett didn't dare look at him. She was aware of him getting closer and closer but her eyes were fixed on the door. There it was in front of her. She lunged forward and grabbed hold of the handle, but at the same moment his hands fell on her, seizing hold of the top of her coat, his fingers against her neck. She heard him cry out in triumph. His breath was against her skin.

She didn't let go of the door. She wasn't going to let him drag her back. Instead, she dropped down, twisting her shoulders so that the coat was pulled over her head. She had already undone the buttons and she felt it come loose, falling away.

Father Gregory lost his balance and, still holding the coat, fell backwards. Scarlett was free. She jerked the door open and threw herself forward. For a few seconds her vision was blurred. The doorway seemed to rush past. She heard Gregory screaming at her, suddenly a long way away.

The door slammed shut behind her.

She was lying, sobbing and shaking on the floor of St Meredith's. And there was a man standing in front of her, a young policeman, dressed in blue, staring at her with a look of complete bewilderment.

"Who are you?" he demanded.

"I'm ... Scarlett Adams." She could barely get the words out.

"Where have you been? What have you been doing?" The policeman shook his head in disbelief. "You'd better come with me!"

FRONT PAGE NEWS

Scarlett had only been missing for eighteen hours but she was a fifteen-year-old student on a school trip in the middle of London, and her disappearance had been enough to trigger a major panic with newspaper headlines, TV bulletins and a nationwide search. Both her parents had been informed at once and Paul Adams was already on a plane, on his way back from Hong Kong. He was actually in mid-air when Scarlett was found.

Scarlett had begun to realize that she was in trouble almost from the moment she found herself back in St Meredith's, sitting opposite the policeman who had immediately launched into a series of questions.

"Where have you been?" he began.

Scarlett was still in shock, thinking about her narrow escape from Father Gregory. She pointed at the door with a trembling finger. "There…"

"What do you mean?" The policeman was young and out of his depth. He had already radioed for backup and an ambulance was on the way. Even so, he was the first on the scene. There might even be a promotion in this. He took out a notebook and prepared to write down anything Scarlett said.

"The monastery." Scarlett muttered. "I was in the monastery."

"And what monastery was that?"

"On the other side of the door."

The policeman walked over to the door and opened it before Scarlett realized what he was going to do. At the last minute, she screamed at him, a single word.

"Don't!"

She had visions of Father Gregory flying in, dragging her back to her cell. She was sure the nightmare was about to begin all over again. But the policeman was just standing there, scratching his head. There was no monastery on the other side of the door, no monks – just an alleyway, a brick wall, a line of rubbish bins. It was drizzling – grey, London weather. Scarlett looked past him. She couldn't quite believe what she was seeing.

And that was when she knew that she was going to have to start lying. How could she explain where she had been and what had really happened to her? Magic doors? Psycho monks in Ukraine? People would think she was mad. Worse than that, they might decide that the whole thing had been a schoolgirl prank. She would be expelled from St Genevieve's. Her father would kill her. She had to come up with an answer that made sense.

The next forty-eight hours were a nightmare almost as bad as the one she had left behind. More policemen and paramedics arrived and suddenly the church was crowded with people all asking questions and arguing amongst themselves. Scarlett didn't seem to be hurt but even so she was wrapped in a blanket and whisked off to hospital. Somehow, the press had already found out that she was back. The street was jammed with photographers and journalists threatening to mob her as she was bundled into the ambulance and there were more of them waiting when she was helped out on the other side. All Scarlett could do was keep her head down, ignore the flashes of the cameras and wish that this whole thing would be over soon.

Mrs Murdoch had been called to the hospital and stayed with Scarlett as she was examined by a doctor and a nurse. The housekeeper was looking shell-shocked. It was obvious that nothing like this had ever happened to her before. The doctor took Scarlett's pulse and heart rate and then asked her to strip down to her underwear.

"Where did you get these?" He had noted a series of scratches running down her back.

"I don't know…" Scarlett guessed that she had been hurt in her final confrontation with Father Gregory but she wasn't going to talk about that now. She was pretending that she was too dazed to explain anything.

"How about this, Scarlett?" The nurse had found blood on her school jersey. "Is this your blood?"

"I don't think so."

The jersey was placed in a bag to be handed over to the police for forensic examination. It occurred to Scarlett that they would be unable to find a match for it … not unless their database extended all the way to Ukraine.

Finally, Scarlett was allowed to take a shower and was given new clothes to wear. Two policewomen had arrived to interview her. Mrs Murdoch stayed with her and just for once Scarlett was glad to have her around. She wouldn't have wanted to go through all this on her own.

"Do you remember what happened to you from the time of your disappearance? Perhaps you'd like to start when you arrived at the church…"

The policewomen were both in their thirties, kind but severe. The rumour was already circulating that Scarlett had never been in any danger at all and that this whole thing was a colossal waste of police time. By now, Scarlett had worked out what she was going to say. She knew that it would sound pretty lame. But it would just have to do.

"I don't remember anything," she said. "I wasn't feeling well in the church. I was dizzy. So I went outside to get some fresh air – and after that, everything is blank. I think I fell over. I don't know…"

"You fainted?"

"I think so. I want to help you. But I just don't know…"

The two policewomen looked doubtful. They had been on the force long enough to know when someone was lying and it was obvious to them that Scarlett was hiding something. But there wasn't much they could do. They asked her the same questions over and over again and received exactly the same answers. She had fallen ill. She had fainted. She couldn't remember anything else. And what other explanation could there be?

The interview ended when Paul Adams appeared. A taxi had brought him straight from Heathrow Airport and he burst into the room, his suit crumpled, his face a mixture of anxiety, relief and irritation, all three of them compounded by a generous dose of jet lag.

"Scarly!" He went over and hugged his daughter.

"Hello, Dad."

"I can't believe they've found you. Are you hurt? Where have you been?" The two policewomen exchanged a glance. Paul Adams turned to them. "If you don't mind, I'd like to take my daughter home. Mrs Murdoch…"

They left the hospital by a back exit, avoiding the press pack who were still camped out at the front. By now, Scarlett was exhausted. She had been found mid-morning, but it was early evening before she was released. She was desperate to go to bed and once she got there, she slept through the entire night. Maybe that was just as well. She would need all her strength for the headlines that were waiting for her the next day.

MISSING SCHOOLGIRL FOUND AFTER JUST ONE DAY

POLICE ASK – WAS THIS A PRANK?

Mystery still surrounds the return of fifteen-year-old schoolgirl, Scarlett Adams, who was discovered by police, just one day after she went missing on a school trip. Scarlett was feared abducted after she vanished during a visit to St Meredith's church in East London, prompting a national search. She was later found unhurt inside the church itself.

Although she received hospital treatment for minor scratches, there was no indication that she had been assaulted or kept against her will.

So far, the girl – described as "bright and sensible" by the teachers at the £15,000-a-year private school that she attends in Dulwich – has been unable to offer any explanation, claiming that she is suffering from memory loss. Her father, Paul Adams, a corporate lawyer, angrily dismissed claims that the whole incident might have been a schoolgirl prank. "Scarlett has obviously suffered a traumatic experience and I'm just glad to have her back," he said.

Meanwhile, the police seem anxious to close the file. "What matters is that Scarlett is safe," Detective Chris Kloet said, speaking from New Scotland Yard. "We may never know what happened to her in the eighteen hours she was gone but we are satisfied that no crime seems to have been committed."

The report had been sent ten thousand miles by fax. It was being examined by a boy in a room in Nazca, Peru. The boy got up and went over to a desk. He held the sheet of paper under a light. There was a picture of Scarlett next to the text. She had been photographed holding a hockey stick with two more girls, one on either side. A team photo. The boy examined her carefully. She was quite good-looking, he thought. Asian, he would have said. Almost certainly the same age as him.

"When did this arrive?" he asked.

"Half an hour ago," came the reply.

The boy's name was Matthew Freeman. He was the first of the Gatekeepers and, without quite knowing how, he had become their unelected leader. Four months ago, he had faced the Old Ones in the Nazca Desert and had tried to close the barrier, the huge gate, that for centuries had kept them at bay.

He had failed. The King of the Old Ones had cut him down where he stood, leaving him for dead. The last thing he had seen was the armies of the Old Ones, spreading out and disappearing into the night.

It had taken him six weeks to recover from his injuries and since then he had been resting, trying to work out what to do next. He was staying in a Peruvian farmhouse, a *hacienda* just outside the town of Nazca itself. Richard Cole, the journalist who had travelled with him from England was still with him. Richard was his closest friend. It was he who had just come into the room.

"It's got to be her," Matt said.

Richard nodded. "She was in St Meredith's. She must have gone through the same door that you went through. God knows what happened to her. She was missing for eighteen hours."

"Her name is Scarlett."

"Scar." Richard nodded again.

Matt thought for a moment, still clutching the article. He had spent the past four months searching for Scarlett in the only way that he could – through his dreams. Night after night he had visited the strange dream world that had become so familiar to him. It had helped him in the past. He was certain that she had to be there somewhere. Perhaps it would lead him to her, helping him again.

And now, quite unexpectedly, she had turned up in the real world. There could be no doubt that this was her, the fifth of the Five. And she was in England, in London! A student at an expensive private school.

"We have to go to her," Matt said. "We must leave at once."

"I'm checking out tickets now."

Matt turned the photograph round in the light, tilting it towards himself. "Scar," he muttered. "Now we know where she is."

"That's right," Richard said. He looked grave. "But the Old Ones will know it too."

MATT'S DIARY (1)

I never asked for any of this. I never wanted to be part of it. And even now, I don't understand exactly what is happening or why it had to be me.

I hoped that writing this diary might help. It was Richard's idea, to put it all down on paper. But it hasn't worked out the way I hoped. The more I think about my life, the more I write about it, the more confused it all becomes.

Sometimes I try to go back to where it all began but I'm not sure any more where that was. Was it the day my parents died? Or did it start in Ipswich, the evening I decided to break into a warehouse with my best friend ... who was actually anything but? Maybe the decision had already been made the day I was born. Matthew Freeman. You will not go to school like other kids. You won't play football and take your A-levels and have a career. You are here for another reason. You can argue if you like, but that's just the way it's got to be.

I think a lot about my parents even though sometimes it's hard to see their faces, and their voices have long since faded out. My dad was a doctor, a GP with a practice round the corner from the house. I can just about remember a man with a beard and gold-rimmed glasses. He was very political. We were recycling stuff long before it was fashionable and he used to get annoyed about the National Health Service – too many managers, too much red tape. At the same time, he used to laugh a lot. He read to me at night... Roald Dahl... *The Twits* was one of his favourites. And there was a comedy show on TV that he never missed. It was on Sunday night but I've forgotten its name.

My mum was a lot smaller than him. She was always on a diet, although I don't think she really needed to lose weight. I suppose it didn't help that she was a great cook. She used to make her own bread and cakes and around September she'd set up a production line for Christmas puddings which she'd flog off for charity. Sometimes she talked about going back to work, but she liked to be there when I got back from school. That was one of her rules. She wouldn't let me come home to an empty house.

I was only eight years old when they died and there's so much about them I never knew. I guess they were happy together. Whenever I think back, the sun always seems to be shining which must mean something. I can still see our house and our garden with a big rose bush sprawling over the lawn. Sometimes I can even smell the flowers.

Mark and Kate Freeman. Those were their names. They died in a car accident on their way to a wedding and the thing is, I knew it was going to happen. I dreamed that their car was going to come off a bridge and into a river and I woke up knowing that they were both going to die. But I didn't tell them. I knew my dad would never have believed me. So I pretended I was sick. I cried and kicked my heels. I let them go but I made them leave me behind.

I could have saved them. I tell myself that over and over again. Maybe my dad wouldn't have believed me. Maybe he would have insisted on going, no matter what I said. But I could have poured paint over the car or something. I could even have set fire to it. There were all sorts of ways that I could have made it impossible for them to leave the house.

But I was too scared. I had a power and I knew that it made me different from everyone else and that was the last thing I wanted to be. Freakshow Matt ... not me, thanks. So I said nothing. I stayed back and watched them go and since then I've seen the car pull away a thousand times and I've yelled at my eight-year-old self to do something and I've hated myself for being so stupid. If I could go back in time, that's where I would start because that's where it all went wrong.

After that, things happened very quickly. I was fostered by a woman called Gwenda Davis who was related in some way to my mother – her half-sister or something. For the next six years, I lived with her and her partner, Brian, in a terraced house in Ipswich. I hated both of them. Gwenda was shallow and self-centred but Brian was worse. They had what I think is called an abusive relationship which means that he used to beat her around. He hit me too. I was scared of him – I admit it. Sometimes I would see him looking at me in the same way and I would make sure my bedroom door was locked at night.

And yet, here's something strange. I might as well admit it. In a way, I was almost happy in Ipswich. Sometimes I thought of it as a punishment for what I'd done – or hadn't done – and part of me figured that I deserved it. I was resigned to my life there. I knew it was never going to get any better and at least I was able to create an identity for myself. I could be anyone I wanted to be.

I bunked off school. I was never going to pass any exams so what did I care? I stole stuff from local shops. I started smoking when I was twelve. My friend, Kelvin, bought me my first packet of Marlboro Lights – although of course he made me pay him back twice what they'd cost. I never took drugs. But if I'd stayed with him much longer I probably would have. I'd have ended up like one of those kids you read about in the newspapers, dead from an overdose, a body next to a railway line. Nobody would have cared, not even me. That was just the way it would have been.

But then along came Jayne Deverill and suddenly everything changed because it turned out she was a witch. I know how crazy that sounds. I can't believe I just wrote it. But she wasn't a witch like in a pantomime. I mean, she didn't have a long nose and a pointy hat or anything like that. She was the real thing: evil, cruel and just a little bit mad. She and her friends had been watching me, waiting for me to fall into their hands because they needed me to help them unlock a mysterious gate hidden in a wood in Yorkshire. And it seemed that, after all, I wasn't just some loser with a criminal record who'd got his parents killed. I was one of the Five. A Gatekeeper. The hero of a story that had begun ten thousand years before I was born.

How did I feel about that? How do I feel about it now?

I have no choice. I am trapped in this and will have to stick with it until the bitter end. And I do think the end will be a hard one. The forces we're up against – the Old Ones and their allies around the world – are too huge. They are like a nightmare plague, spreading everywhere, killing everything they touch. I have powers. I've accepted that now and recently I've learned how to use them. But I am still only fifteen years old – I had my birthday out here in Nazca – and when I think about the things that are being asked of me, I am scared.

I can't run away. There's nowhere for me to hide. If I don't fight back, the Old Ones will find me. They will destroy me more surely and more painfully than even those cigarettes would have managed. After I was arrested, I never smoked again, by the way. That was one of the ways that I changed. I think I have accepted my place in all this. First of all, I have to survive. But that's not enough. I also have to win.

At least I'm no longer alone.

When this all began, I knew that I was one of five children, all the same age as me, and that one day we would meet. I knew this because I had seen them in my dreams.

Pedro was the first one I came across in real life. He has no surname. He lost it – along with his home, his possessions and his entire family when the village in Peru where he lived was hit by a flood. He was six years old. After that, he moved to the slums of Lima and managed to scratch a living there. The first time I saw him, he was begging on the street. We met when I was unconscious and he was trying to rob me. But that was the way he was brought up. For him, there was never any right or wrong – it was just a question of finding the next meal. He couldn't read. He knew nothing about the world outside the crumbling shanty town where he lived. And of course he could hardly speak a word of English.

I don't think I'd ever met anyone quite so alien to me ... and by that I mean he could have come from another planet. For a start (and there's no pleasant way to put this) he stank. He hadn't washed or had a bath in years and the clothes he wore had been worn by at least ten people before him. Even after everything I'd been through, I was rich compared to him. At least I'd grown up with fresh tap water. I'd never starved.

Almost from the very start we became friends. It probably helped that Pedro decided to save my life when the police chief, a man called Rodriguez, was cheerfully beating me up. But it was more than that. Think about the odds of our ever finding each other, me living in a provincial town in England and him, a street urchin surviving in a city ten thousand miles away. We were drawn together because that was how it was meant to be. We were two of the Five.

Pedro is pure Inca: a descendant of the people who first lived in Peru. More than that, he's somehow connected with Manco Capac, one of the sun gods. The Incas showed me a picture of Manco – it was actually on a disc made of solid gold – and the two of them looked exactly the same. I'm not sure I completely understand what's going on here. Is Pedro some sort of ancient god? If so, what does that make me?

Like me, Pedro has a special power. His is the ability to heal. The only reason I'm able to walk today is because of him. We were both injured in the Nazca Desert. He broke his leg, but I was cut down and left for dead ... and I would have died if he hadn't come back and stayed with me for a couple of weeks. It's called radiesthesia, which is probably the longest word I know.

I've only managed to spell it right because I've looked it up in the dictionary. It's something to do with the transfer of energy. Basically, it means that I got better thanks to him. And as a result, Pedro is more than a friend. He's almost like a long-lost brother – and if that sounds corny, too bad. That's how I feel.

And then came Scott and Jamie Tyler.

They really were brothers ... twins, in fact. Formerly the telepathic twins, performing with The Circus of the Mind at The Reno Playhouse in Nevada. While Pedro and I had been fighting (and losing) in the Nazca Desert, they'd been having adventures of their own, chased across America by an organization called the Nightrise Corporation. They'd also managed to get tangled up in the American election and were there when one of the candidates was almost assassinated.

Scott and Jamie are more or less identical. They're thin to the point of being skinny and you can tell straight away that they have Native American blood – they were descended from the Washoe tribe. They have long, dark hair, dark eyes and a sort of watchful quality. Physically, I would have said that Jamie was the younger of the two, but when they finally reached us – they travelled through a doorway that took them from Lake Tahoe in Nevada to a temple in Cuzco, Peru – he was very much in charge. His brother had been taken prisoner and tortured. We're still not sure what they did to him and Pedro has spent long hours alone with him, trying to repair the damage. But Scott is still suffering. He's withdrawn. He doesn't talk very much. I sometimes wonder if we'll be able to rely on him when the time comes.

It's been more than four months since I faced the Old Ones

in the Nazca Desert and I still haven't recovered from my own injuries. I'm in pain a lot of the time. There are no scars but I can feel something wrong inside me. Sometimes I wake up at night and it's as if I've just been stabbed. Even Pedro still has a limp. So between the four of us, I certainly wouldn't bet any money on our taking on unimaginable forces of darkness and saving the world. I'm sorry, but that's how it is.

Jamie is very bright. He seems to see things more clearly than any of us, mainly because he was there at the very start. It's too complicated to explain right now, but somehow he travelled back in time and met us … before we were us. Yes. There was a Matt ten thousand years ago who looked like me and sounded like me and who may even have been me. Jamie says that we've all lived twice. I just hope it was more fun the first time.

Four months!

We've all been hanging out in this house near the coast, to the south of Lima. It belongs to a professor called Joanna Chambers who's an expert on pretty much anything to do with Peru. The house is wooden and painted white, constructed a bit like a hacienda, which is a Spanish farmhouse. There's a large central room which opens onto a veranda during the day and a wide staircase that connects the two floors. Everything is very old-fashioned. There are scatter rugs and a big open fireplace and fans turn slowly beneath the ceiling, circulating the air.

We've passed the time reading, watching TV (the house has satellite and we've also shipped in a supply of DVDs) and surfing the net, looking out for any news of the Old Ones. The professor insists that we do three or four hours of lessons, although it's been ages since any of us went to school and Pedro never *stepped into one in his life. We've played football in the gar-den, passing the ball around the llamas that wander onto the grass, and we've gone for hikes in the desert. And, I suppose, we've been gathering strength, slowly recovering from every-thing we've been through.*

But even so, there have been times when it all seems unreal, sitting here, doing nothing in the full knowledge that some-where in the world the Old Ones must be spreading their power base, preparing to strike at humanity. They'll be making friends in all the right places... As far as we know, they could be all over Europe. Their aim is to start a total war, to kill as many people as possible and then to toy with the rest, maiming and torturing until there's nobody left. Why do they want to do this? There is no why. The Old Ones feed on pain in the same way that can-cer will attack a healthy organism. It's their nature.

Sometimes, in the evening, the six of us will play Perudo, which is a Peruvian game, a bit like liar dice. Me, Richard, Pedro, Scott, Jamie and the professor. We'll sit there, throwing dice and behaving as if nothing is happening, as if we're just a bunch of friends on an extended holiday. And secretly I want to get up and punch the wall. We're safe and comfortable in Nazca. But every moment we're here, we're losing. Our enemy is gaining the upper hand.

What else can we do? The Old Ones have disappeared. And even if we knew what they were doing, we're not yet strong enough to take them on. Only four of the Gatekeepers have come together. There have to be five.

And now there are. At last we've found Scar.

It's hard to believe that today I actually held a picture of her in my hands. Now she has a name – Scarlett Adams. We know where she lives. We can actually reach out to her and tell her the truth about who she is – or was.

Ten thousand years ago, she was in charge of her own private army. Jamie actually met her and fought with her at the final battle when the King of the Old Ones was banished and the first great gate was constructed. She must have a power – we all do. But he never found out what it was. When he met her, he said she was brave and resourceful. She could ride a horse, fight with a sword, lead an army of men who were at least twice her age. But she never did anything that looked like magic ... at least, not anything that he noticed.

Very soon, we will leave Nazca. I really want to see England again.

And now I'm going to bed.

Richard is worried that Scar turning up is the start of a new phase. The Old Ones have left us alone but now they'll have been alerted. If they were planning a move against us, this is the time when they'll make it.

But I don't care. There are five of us and that means that soon this whole thing will be over. We'll get together and do whatever it takes to bring it all to an end. After that, I'll go back to school. I'll take my GCSEs and my A-levels. I'll have an ordinary life.

That's all I want. I can hardly wait.

LAST NIGHT IN NAZCA

Twenty-four hours after the fax had arrived, Professor Chambers organized a dinner. It was her way of saying goodbye. The following day, Matt, Richard, Pedro, Jamie and Scott would be leaving for England – the professor had arranged passports for all of them – and at last she would have the house to herself.

Joanna Chambers had spent most of her life in Peru, studying the Incas, the ancient Moche and Chimu tribes and, of course, the Nazca Lines. She was an expert on a dozen different subjects, a qualified pilot, a good shot with a rifle or a handgun and a terrible cook. Fortunately, the meal had been prepared by a local help: creole soup, followed by *lomo saltado* – a dish made with grilled beef, onions and rice. There were two jugs of Pisco Sour, a frothing, white drink made from grape brandy, lemon and egg white – it tasted much better than it looked.

Richard Cole was sitting at the head of the table. He had changed in the past few months. His hair had been bleached by the sun and he had grown it so that it fell in long strands over his collar. He had a permanent, desert tan and although he didn't quite have a beard, he looked rough and unshaven.

Tonight, he had changed into jeans and a white, linen shirt. Normally he slouched around in shorts and sandals and if the house had been nearer the sea, he might easily have been mistaken for a surfer. He started every morning with a five-mile run. He was keeping himself in shape.

Scott and Jamie Tyler were sitting on one side of the table, together as usual. Matt and Pedro were on the other. There was one empty seat and someone had placed the article with the picture of Scarlett Adams on the table in front of it, as if she were there in spirit.

All six of them were in a good mood. The food had been excellent and the drink had helped. Upstairs, their cases were packed and ready in the various rooms. Professor Chambers waited until the food had been cleared away, then tapped a fork against her glass and rose to her feet. Matt had never seen her wearing a dress and tonight was no exception. She had put on a crumpled safari suit and there was a small bunch of flowers in her buttonhole.

"We ought to go to bed," she began. "You have a long journey to make tomorrow – but I just want to wish you *bon voyage*. I can't say I'm too sorry that you're finally on your way…" There were protests around the table and she held up a hand for silence. "It's been impossible to get any work done with all your infernal noise, football games out on the front lawn, four boys clumping up and down the stairs, and all the rest of it.

"But I will miss you. I've enjoyed having you here. That's the truth of it. And although it's wonderful that Scar has finally turned up, I can't help wondering what lies ahead of you." She stopped for a moment. "I feel a bit like a mother sending my sons off to war. I can only hope that one day I'll see you again. I can only hope that you'll come back safe."

She lifted her glass.

"Anyway, here's a toast to all five of you. The Five, I should say. Look after yourselves. Beat the Old Ones. Do what you have to do. And now let's get some hot chocolate and have a final game of Perudo. You have an early start."

Later that night, Richard and Matt found themselves standing on the veranda outside the main room. It was a beautiful night with a full moon, an inky sky and stars everywhere. Matt could hear classical music coming from inside the house. Professor Chambers had an old-fashioned radio that she liked to listen to while she worked. Scott and Jamie were sharing a room on the first floor. Pedro was probably watching TV.

"I can't believe we're going home," Matt said.

"England." Richard gazed into the darkness as if he could see it on the horizon. "Do you have any idea what happens when we get there?"

Matt shook his head. "I don't know. I've thought about it. I've tried to work out some sort of plan. Maybe it would be easier if we knew what the Old Ones have been doing all this time." He thought for a moment. "Maybe we'll know when the five of us get together. Maybe it will all make sense."

Matt stared into the darkness. The nights in Nazca were always huge. Even without seeing it, he could feel the desert stretching out to the mountains. There seemed to be five times more stars in the southern hemisphere than he'd ever seen in Europe. The sky was bursting with them.

"What you said yesterday..." He turned to Richard. "About the Old Ones..."

"They were looking for kids with special powers," Richard said. "That's how they found Scott and Jamie. If Scarlett went through the door at St Meredith's, they'll know about it. They'll have read the article too."

"You think they'll be waiting for us?"

"Scarlett's being watched by the Nexus. Her father's with her. She took a couple of days off school. So far everything seems OK. She doesn't seem to be in any danger."

Richard had been in constant touch with the Nexus, the strange collection of millionaires, politicians, psychics and churchmen who knew about the Old Ones and had come together in a sort of secret society to fight them. It had to be secret because they were afraid of being ridiculed. How could they admit that they believed in devils and demons? The Nexus had made it their job to look after Matt and the other children. At one stage, they had paid for him to go to a private school. They were still paying for everything while the four of them were out here.

And they were also protecting Scarlett Adams. They had moved in the moment she had been identified in the national press, hiring a team of private detectives to watch over her night and day. They were lucky that she lived in England. That made things easier. One of the Nexus members was a senior police officer called Tarrant and he had arranged for all her calls to be monitored. Meanwhile, Scarlett had gone back to school. Her father was still with her in London and there was a Scottish helper living in the house. By now, Richard knew a great deal about her. She was in the school play. She had a boyfriend called Aidan and she regularly beat him at tennis. She seemed to have a happy life.

Richard and Matt were about to rip all that up. Somewhere inside him, Matt felt guilty about that – but he knew he couldn't avoid it. She had been born for a purpose. His job was to tell her what that purpose was.

Somewhere, an owl hooted in the darkness. The house was on the outskirts of Nazca but the two of them could make out the lights of the town, twinkling in the distance. Everything was very peaceful but they knew that it was an illusion. Soon the whole world would change.

"I'm not sure you should go," Richard said, suddenly.

"What do you mean?" Matt was surprised. Everything was ready. The tickets had been bought.

"I've been thinking about it … this trip to England. You and Pedro and Scott and Jamie … all on the same plane. Suppose the Old Ones have got control of American air space. They could smash you into the side of a mountain. Or a building."

"They don't want to kill us," Matt said. He was fairly sure about that. "If they kill us, we'll all be replaced by our past selves. That's how it works. And what good will that do them? They'll only have to start searching for us all over again. It's easier for them to keep us alive."

Richard shook his head. "They could still force the plane down somewhere and capture you."

Matt considered the possibility. The trouble was that the Old Ones had been silent for months. They seemed to have slipped into the shadows, as if they had never existed at all. Richard had been scouring the Internet, waiting to hear of a news event, some horror happening somewhere in the world that might suggest that the Old Ones were involved. There were plenty of stories. The war in Afghanistan. Ethnic cleansing in Darfur. Misery and starvation in Zimbabwe. But that was just everyday news. That would happen even without the Old Ones. He had been looking for something worse.

"What do you think they're doing?" Matt asked. "Why do you think they haven't shown themselves?"

Richard shrugged. "I guess they've been waiting," he said.

"Waiting for what?"

"Waiting for Scar."

There was a movement on the veranda and Matt tensed for a moment, then relaxed. He could tell it was Professor Chambers, even without turning round. The smell of her cigar had given her away and sure enough, there it was. She was clutching it in one hand with a glass of Peruvian brandy in the other.

"Are you two going in?" she asked. "I'm putting on the alarms."

The house was completely surrounded by a security system that had been installed shortly after Richard, Pedro and Matt had arrived. There were no fences or uniformed guards – the professor had said she couldn't live like that. The system was invisible. But there was a series of infra-red beams at the perimeter, and the garden itself had pressure pads concealed in different places under the lawn. Most sophisticated of all was the radar dish mounted on the roof, sweeping the entire area. It could pick up any movement a hundred metres away. That was how they had been living. It might look as if they were free, but they had all been aware that they were actually in a state of siege.

"We were just talking about tomorrow," Richard said.

"It'll be here soon." Chambers blew smoke. "It's after ten. Shouldn't you be in bed?"

Richard tapped Matt on the shoulder. "After you."

The three of them went inside. Matt said goodnight to Richard and climbed the stairs to the small room which he had chosen at the back of the house. He liked it there. When he was lying in bed, his head was directly underneath a slanting roof with a skylight so, lying on his back, he could look up at the stars. His case – a small canvas bag – was already packed and sitting on the floor. He wasn't taking much with him. If he needed anything in London, he could always buy it there.

Matt undressed quickly, washed and slipped between the sheets. For the last few months, he had been searching for Scar in the only way that he could – in his dreams. Time and again he had visited the dreamworld. He had been there so often that he knew the landscape well: the shoreline stretching along a great sea with everything dead and grey, the island where he had once found himself trapped.

The dreamworld baffled him. Was it a dream or was it a real world? That was the first question. And was it there to help him or to throw him off balance? On the one hand it was a frightening place, conjuring up strange, violent images that he couldn't understand: giant swans, walking statues, guns and knives. But at the same time, Matt didn't think he was in any danger there. The more he visited it, the more he felt it was on his side. He wondered if anyone actually lived there – or was it simply there for the Gatekeepers, its only inhabitants?

At any event, he had gone back there almost every night, floating out of the bed, out of the room, out of himself. Then he had begun to travel, searching for Scar. Sometimes he would see a flicker of lightning, an approaching storm. Once, he found footprints. Another time he came upon a grove of trees, which at least proved that the place wasn't entirely dead, that things could grow there.

But there had never been any sign of Scar.

There was no point in searching for her tonight. In just twenty-four hours he would be meeting her anyway. But even so – maybe it was just habit – he found himself back in the dreamworld almost at once. As usual, he was on his own. He was climbing a steep hill, but it took no more effort than if he had been walking on level ground. Far behind him, the wilderness stretched out, wide and empty.

And then he noticed something strange. The ground underneath his feet had changed. He knelt down and examined it, brushing aside the grey dust that covered everything. It was true. He was standing on a path fashioned out of paving stones that had been brought here and laid in place. He could see the joins, the cement gluing everything together. Even though he was asleep, Matt felt a surge of excitement. A man-made path! This was completely new and confirmed what he had always thought. The dreamworld was inhabited. There might be buildings, even whole cities there.

He looked up. The path had to lead somewhere. There could be something on the other side of the hill.

But he wasn't going to find out – not then. Suddenly he was

awake. Someone was shaking him, calling his name. The lights were on in his room. He opened his eyes. It was Richard.

"Wake up, Matt," he was saying. "There's someone here."

THE MAN FROM LIMA

Matt heaved himself out of bed, threw on some shorts and a T-shirt and ran downstairs barefoot. The whole house was awake. There were lights on everywhere and the alarm system was buzzing, warning them that somebody was approaching.

It had already occurred to him that this sudden interruption must be connected to the fact that Scarlett had been found. If all five of the Gatekeepers were now out there and known to each other, that made them a greater danger to the Old Ones, and it was no surprise that they'd want to take action. It was exactly what he and Richard had been worrying about. On the other hand, it could be a false alarm. Over the past four months, there had been plenty enough of those. Sometimes the children came out from the town, looking for food or something to steal. Professor Chambers kept llamas for their wool, and one of them might have broken loose. The system was sensitive. Even a bat or a large moth might have been enough to set it off.

Matt hurried into the main room. There was a computer standing on a table in the corner and it had already activated itself, automatically connecting to the radar on the roof. It showed a single blip moving slowly and purposefully towards the front door. It was half past eleven at night. A bit late for a visitor.

Jamie and Scott had come downstairs, fully dressed. Pedro followed them – barefoot like Matt, but then he often preferred to walk without shoes. When the two boys had first met, he had been wearing sandals made out of old car tyres and he still mistrusted proper trainers. He was yawning and pulling on a sweater. Joanna Chambers had arrived ahead of everyone. She was wearing an old dressing gown. Matt watched her open the gun cabinet and take out a rifle. So far, nobody had spoken.

"What's happening?" Jamie asked.

"A single figure moving through the garden." She nodded at the computer. "It looks like there's only one of them, but we can't be sure."

Richard went over and examined the screen. "I'd say he's trying not to be seen," he muttered. "Why don't we take a look at him?"

He leaned over and pressed a switch. This was another part of the security system. The entire garden was instantly lit up by a series of arc lamps so bright that it was as if he had set off a magnesium flare. Matt blinked. It was quite shocking to see the brilliant colours, the wide green lawn, so late at night.

There was a single figure, a man, trapped in the middle of the lawn. He was dressed in a linen jacket, jeans and a polo shirt, buttoned up to the neck. There was a canvas bag across his shoulder. As the lights had come on, he had frozen and stood there with his hands half-covering his eyes, momentarily blinded. He seemed to be on his own. He certainly wasn't carrying any visible weapons. Richard opened the French windows.

Professor Chambers stepped outside.

"Stay where you are!" she shouted. "I have a gun pointing at you."

"There is no need for that!" the man shouted back in heavily accented English. "I am a friend."

"What do you want?"

"I want to speak to the boy. Matthew Freeman. Is he here?"

Richard glanced at Matt who moved forward, stepping through the French windows. He was careful not to go too far. Professor Chambers lifted the gun, covering him. "What's your name?" he called out.

"Ramon." The man cupped his hand over his eyes, shielding them, trying to make him out.

"Where have you come from?"

"From Lima." The man hesitated, unsure what to do, whether to move forward or not. He seemed to be pinned there by the light. "Please ... are you Matthew? I am here because I want to help you."

Pedro had come over to the window. He was standing next to Matt. "Why does he come, like a thief, in the middle of the night?" he muttered. Matt nodded. He knew that Pedro was the most suspicious of them all. Maybe it was something to do with the life he'd once led.

Richard agreed. "We can ask him to come back in the morning," he muttered.

But Matt wasn't so sure. "What do you want?" he shouted.

The man hadn't moved. "I will show you when I am inside," he said. He looked around him. "Please ... it is not safe for me out here."

Matt knew he had to make a decision. It was something he was finding more and more. Although he was in the professor's house and she and Richard were far older than him, he always seemed to be the one in charge.

Quickly, he turned over the options. They were all supposed to be leaving the house at ten o'clock the next morning, driving up to Lima to catch the flight that would take them to London. This was no time to be meeting with complete strangers. On the other hand, there were six of them and one of him. Professor Chambers had a weapon. And the man seemed genuine enough.

"All right!" Matt called out. "Come in…"

The man began to walk towards the house. At the same time, Richard went over to the cabinet and reached inside. There was another gun there. He wasn't taking any chances.

The man came into the main room, Professor Chambers following him with the rifle. Now that he was inside, Matt could see that he was a few years older than Richard, with the dark hair and olive skin of a native Peruvian. He had obviously been on the road for a while. He was dusty and unshaven and his clothes were crumpled with sweat patches under the arms. There was a haunted look in his eyes. From the look of him, he didn't seem to be a threat.

The first thing he did was to take a pair of spectacles out of his top pocket and put them on. Now he looked like a school teacher or perhaps an accountant working in a small, local office. He had a cheap watch on his wrist and his shoes were scuffed and down-at-heel. He looked straight at Matt. "Are you Matthew Freeman?" He blinked. "I did not think I would find you here."

"Sit down," Richard said.

The man sat on the sofa with his back to the French windows. Richard pressed the button that turned off the garden lights and everything outside the room disappeared into blackness again. It had clouded over during the night. The moon and the stars had disappeared. Richard came back over to the sofa and sat down on one of the arms. He hadn't reset the security system. But then the visitor wouldn't be staying very long. Scott and Jamie perched on the edge of the coffee table. Professor Chambers sat in a chair with the rifle between her knees.

"So what do you want?" she demanded.

"I will tell you everything you want to know," Ramon said. "But can I first ask you for a drink? I have been travelling all day and I had to wait until night before coming here. Believe me, if I had been seen I would have been killed."

"I'll get it," Pedro said. He got up and went into the kitchen, returning a moment later with a glass of water. The man took it in both hands and gulped greedily.

"How do you know about me?" Matt asked.

"I know a great deal about you, Matthew. May I call you that? I know how you came to Peru and I think I know what you have been doing since you arrived here. I was present, also, the night you came to the *hacienda* at Ica, although perhaps you did not see me. I was there because I was hired to work for Diego Salamanda."

Ramon must have known the effect the name would have on everyone in the room. Salamanda had been the chairman and owner of a huge news corporation in South America. Deliberately deformed as a child – his head had been grotesquely stretched – he had used his power and wealth to bring back the Old Ones. Matt and Pedro had gone to his *hacienda* searching for Richard, and later on Matt and Salamanda had confronted each other in the Nazca Desert. Matt had killed him, turning back the bullets fired from his own gun.

"Please ... do not think of me as your enemy," Ramon continued, hastily. "I swear to you that I was not part of his plans." He paused. Beads of sweat were standing out on his forehead. "I am not even in business. I am a lecturer at Lima University and *Señor* Salamanda paid me to help him with a special project. I should explain that my speciality is Ancient History." He bowed in the direction of Professor Chambers. "I have heard you speak many times, *Señora*. I was there, for example, last April when you gave the presentation at the Museo Nacional de Antropologia. I thought it was a brilliant talk."

Professor Chambers thought for a moment. "It's true that I was there," she said. "But anyone could know that."

"*Señor* Salamanda told me that he was in possession of a diary which he wanted me to interpret on his behalf," Ramon went on. "The diary had been written in the sixteenth century by a man called Joseph de Cordoba. This man travelled here to Peru with the Spanish conquistadors. Salamanda told me that he bought the diary from a bookseller in London, a man called William Morton."

"He didn't buy it," Matt said. "He stole it. He killed William Morton to get it." Matt knew because he had been there at the time. Morton had been demanding two million pounds but all he had got was a knife in the back.

"I did not know these things," Ramon exclaimed. "I was

innocent. My job was to work only on the text, to unlock its secrets and I spent many, many hours in his office and also at his home in Ica. The diary was never allowed to leave his side. He made it clear to me from the start that it was the most precious thing to him in the world. And as I read it, as I began to study it, I realized why. It told this extraordinary history ... the Old Ones, a battle many thousands of years ago and a gate that could be unlocked by the stars."

He lowered his head.

"I know that I am responsible for what happened last June. I did the work that I was paid to do and I helped Salamanda to open the gate. I have allowed a terrible thing to happen and it has been on my conscience ever since." He twisted on the sofa, urging them to believe him. "I am not a bad man. I am a Catholic. I go to Church. I believe in heaven and hell. And I have been thinking ... what can I do to make amends for what I have done? What can I do to undo the damage that I have caused? And I knew, finally, that I must find you. So I came."

"How did you know where we were?" Jamie asked.

"*Señor* Salamanda often mentioned the name of Professor Chambers. I guessed that you would be with her and I have brought you something. You will not shoot me if I reach into my bag?"

He glanced at the professor, then reached beside him. He took out an old, leather-bound book and laid it on the table. Nobody in the room said anything. But they all knew what it was. It was hard to believe that it was actually there, in front of them. The cover was dark brown with a few faint tracings of gold, tied with a cord. The edges of the pages were rough and uneven. Matt recognized it at once. It contained everything they needed to know about the Old Ones. It might even describe how they could be defeated.

"It is the diary of the mad monk," Ramon said.

And it was. The small, square book sitting there in the middle of the table was, supposedly, the only copy in the world. There was no limit to how many secrets it might contain, how valuable it might be.

"How did you get it?" Richard demanded.

"I stole it!" Ramon took out a handkerchief and wiped it across his forehead. "I thought it would be impossible but in fact it was easy. You see, I still had my electronic pass-key to the office of Salamanda News International in Lima. And I had this crazy idea. Maybe the key had not been cancelled. *Señor* Salamanda was dead but surely they had forgotten about me. Two days ago I returned to the office. Nobody saw me, although by now they will know that it is gone. I took it from his desk and hurried away into the night. It is possible that the cameras will have identified me and that they will be searching for me even now."

Richard was still suspicious. "What do you want from us?" he asked. "Do you want us to pay you?"

Ramon shook his head. "Can you not understand me?" he exclaimed. He clasped his hands in front of him. "I am twenty-eight years old. Next year I hope to be married. When I was given this work by *Señor* Salamanda, I knew nothing. It was just, for me, a job.

He pushed the diary away.

"Here! You can have it without payment. It is yours. I brought it to you only because I thought you might make use of it in this great..." He searched for the word in English. "...*lucha*. Struggle. I want nothing from you. I am sorry that I came."

There was a pause. Matt knew that he had just been given a fantastic prize. The diary might explain the dreamworld. It might tell them the history of the twenty-five doorways that stood in so many different countries. Who had built them, and when? It might even help them work out what they were supposed to do when the five of them finally met in London. Ramon was right. Salamanda had been prepared to kill to get his hands on the diary and now it had just been handed to them, out of the blue.

Jamie leaned forward and picked it up. He unwound the cord and the diary opened in his hands. He examined the page in front of him. It was covered in handwriting which would have been almost unreadable even if it hadn't been in Spanish. There were tiny diagrams in the margins. Suddenly his eyes lit up. He pointed to a single word.

"Sapling," he said. "That was my name when I went back in time. Sapling was killed and I took his place."

The diary was real. Matt had no doubt of it. But what about the man who had brought it to them? He looked genuine, but Richard had been expecting some sort of trap and this could well be it. Suddenly Matt had an idea. There was an easy way to find out. "Jamie," he said. "Ask him if he's telling the truth."

Jamie understood at once. But before he could act, Scott stood up. "I'll do it," he said.

Scott walked forward and stopped in front of the visitor.

He looked Ramon straight in the eyes. "Are you telling the truth?" he demanded.

"On my mother's grave," Ramon replied, crossing himself and then kissing his thumb. "I'm only here because it is the right thing to do. Because I want to help."

Scott concentrated. This was his power, the ability that had kept audiences entertained for the many months when he was performing in Reno. They had thought it was a trick but in fact it was real. He could read minds.

Unfortunately, it wasn't quite as easy as that sounded. It wasn't like throwing a switch. Scott and Jamie had a connection with each other. When they were in the same room or even a short distance away, they could communicate with each other just by thinking. But when it came to other people, strangers like Ramon, what they saw was confused, chaotic. Nothing was ever black and white.

Perhaps a minute passed. Then Scott nodded. "He's telling the truth," he said.

"I promise you..." Ramon knew that he had been tested in some way. The words came pouring out. "I don't care if you don't trust me. I'll leave you with the diary. I'll go. I have no other reason to be here."

"You said it wasn't safe for you outside," Richard said. "Were you followed?"

Ramon shook his head and swallowed nervously. "I don't think so. After I had taken the diary, I hid in Lima. I wanted to see if the police would come. Then, when nothing happened, I took a tourist bus to Paracas. I thought it was less likely that I would be noticed that way. By now they will know that the diary is missing. They will know that I have taken it. And although Salamanda is gone, there are people in his organization who will still wish to continue what he began."

"So where will you go now?" Professor Chambers asked. "Do you have somewhere to hide?"

"I was hoping…" Ramon began. There was a strange sound, a whistling that came through the air, then the tearing of fabric. He looked down. There was something sticking out of his shirt. Puzzled, he reached down and touched it, then tried to pull it free. It wouldn't move and when he released it, his hand was wet with blood.

They had all heard it but hadn't realized what it was. A fence post. It had been thrown with impossible force from out of the darkness. It must have travelled more than fifty metres before the pointed end smashed into the back of the sofa penetrating through the leather and padding before impaling the man who was sitting there. Ramon's eyes widened. He tried to speak. Then he slumped forward, pinned into place, unable even to fall.

The alarms hadn't gone off. The radar screen was empty. Professor Chambers sprang to her feet and pressed the button to turn on the outside lights. Nothing happened.

Something was moving in the garden. There were figures, edging forward, dressed in filthy, tattered clothes that hung off them as if they were rotting away. Matt could just make them out in the light spilling from the room. It was suddenly very cold and he knew at once that dark forces were at work and whatever they were, coming towards him, they weren't human.

They had come for the diary.

NIGHT ATTACK

Slowly, determinedly, they closed in on the house.

There were more than a dozen of them: nightmare figures, shuffling across the lawn. Where had they come from? Matt could imagine them climbing out of the local cemetery. There was something corpse-like about them. A gleam of light from the living room caught one of their faces and he saw glistening bone, one empty eye socket, dried blood streaking down the side of the cheek and neck. At that moment he was sure of it. These creatures couldn't be killed. They were already dead.

As if to prove him wrong, Professor Chambers stepped forward and fired a shot at the nearest of them. Matt saw a great gout of blood explode out of the back of its head. It fell face down and lay, shuddering in the grass. So at least they could be stopped! She fired again, hitting another of them in the shoulder. The creature twitched as if shrugging off the bullet. Blood spread across what was left of its shirt. But it kept on coming. It didn't seem to feel pain.

Richard was already on his feet, loading the revolver that he had taken from the gun cabinet. A few weeks before, Matt had smiled when he had stumbled across him, shooting tin cans in the desert. Now he was glad that Richard had decided to practise.

When the attack had begun, Scott and Jamie had snatched up a couple of makeshift weapons – anything they could get their hands on. Jamie had a baseball bat. Scott had found a kitchen knife which he was holding in front of him, the blade slanting up. Pedro had backed away to the other side of the room. He was standing with his back to a full-length window, his eyes darting left and right, waiting for the first attack.

He hadn't looked behind him.

"Pedro...! Watch out!" Richard shouted the warning.

One of the creatures was looming out of the shadows on the other side of the glass. Pedro spun round just in time to see a dead, white face, staring eyes, grey lips, hands stretching towards him. The creature didn't stop. It walked straight through the window, smashing the glass which cascaded all around it, and came into the room with blood streaming down its face. Shards of broken glass were sticking out of its flesh, but it didn't seem to notice. Richard lifted his revolver and shot it twice in the head. It crumpled and fell at Pedro's feet. At the same time, Richard twisted round and fired again. Another of the creatures had reached the open French windows and was about to step inside. It threw up its hands and fell back with a bullet between its eyes.

But there were still many more of them moving slowly across the lawn, unafraid of dying, determined only to reach the house. Perhaps Ramon had been a diversion after all. While he had been talking, the night attackers had completely surrounded the house. Matt heard the sound of wood splintering upstairs and knew that some of them must have climbed up to the balcony and broken in that way. Jamie stepped forward and grabbed the diary, which he threw to Scott in a single movement. Scott caught it without even looking and slipped it into his jacket. Neither of them had spoken and Matt knew that the two of them must have communicated telepathically. He had seen them do it often enough. Each one of them knew instantly what the other was going to do. They were almost like reflections of each other.

Richard was reloading. Joanna Chambers fired again. She pulled some more bullets out of her dressing gown pocket but even as she fumbled with the loading mechanism, one of the creatures launched itself at her, grabbing hold of her with one hand, lifting an ancient-looking knife with the other. The blade was black with a broken, serrated edge. It stabbed down.

Matt stopped him.

Six months ago, he wouldn't have been able to do it. But then he had been alone. Now four of the Gatekeepers had come together and Scott, Jamie and Pedro had added their power to his. All he had to do was think about it and the blade snapped in half. The creature screamed in pain and a wisp of smoke rose from the palm of its hand as the hilt of the knife burned into it. By now, Chambers had loaded her rifle. She fired a single shot at point blank range, putting it out of its misery.

"We can't control them!" Jamie shouted.

If these creatures had been fully human, he might have been able to make them turn round and leave the house. He and Scott didn't just read minds. They were also able to control them. All their lives, the two brothers had recognized that they were living under a curse. Always, they had to be careful what they said. One unguarded thought, one word spoken in anger, could turn them into murderers. Once, Scott had almost killed a boy at school. And later, when their foster father committed suicide, Scott had known that he was secretly to blame.

But this time it wasn't going to work. Their attackers didn't seem to have minds that could be controlled. It was as if they had already been programmed to kill with no thoughts of their own. And there were too many of them. Matt glanced into the garden. It was still very dark outside but he could make out a whole crowd of them, moving relentlessly across the lawn. There were more at the back of the house and yet more of them upstairs.

Matt heard a horrible gargling sound and turned just as a man – or the remains of one - stumbled over the sofa and launched himself at him. The man was naked to the waist, sweat and slime dripping off his chest. Matt nodded and the man was flung backwards, crashing into the wall. He slid to the floor and lay still.

"They're on the stairs."

It was Scott who had seen them. The creatures from the balcony were making their way down, their movements slow, almost robotic. Jamie ran forward with the baseball bat and swung it into the face of the first man that he reached. There was a crunch of breaking bone. The man crumpled.

Matt looked all around him, wondering where the next attack was going to come from. At the same time, he smelled something. His eyes had begun to water and he was aware that it was getting more difficult to breathe. The temperature had risen too. Richard fired again, hitting one creature, then used the revolver as a club, smashing it into a second. "The house is on fire!" he yelled.

Matt didn't need to be told. Smoke was pouring down the staircase, sucked into the ground floor by the turning fans. He could already hear the crackle of burning wood. Stretched out in the hot Nazca sun – it almost never rained in this part of Peru – the professor's house would be bone dry. There were fire extinguishers in all the rooms, but they weren't going to be given a chance to use them. Left to itself, the fire would consume the whole building in minutes.

Richard fired two more shots but then the gun clicked uselessly in his hand. He rummaged in his pocket, searching for more ammunition. Professor Chambers blasted off another round, but she too had only a few bullets left. And the creatures kept on coming. Kill one and another two or three would take its place. There seemed to be no end to them. Matt saw another one appear on the stairs, holding an iron post similar to the one that had killed Ramon. It had been torn free from the garden fence. He watched as the creature lifted it up to its shoulder, realized too late what it was about to do.

The creature flung the rod like a spear, aiming straight at Pedro. Matt shouted a warning. Pedro twisted round. The missile turned once in the air and then struck him a glancing blow on the side of the head. He cried out and fell to the floor, dazed and bleeding. Another creature – dressed bizarrely in the rags of a dinner suit – closed in on him. Matt couldn't reach him. He was too far away. But Scott was there. He still had the kitchen knife. He was standing between Pedro and his attacker. Matt waited for him to move.

Scott did nothing. He stood where he was, frozen to the spot. He wasn't even blinking. Matt could see his chest heaving and his hands seemed to be locked in place, the fingers bent. His whole body was rigid.

Matt knew what was happening. He had seen it before. Scott wasn't afraid. He wasn't a coward. But he had spent weeks with Nightrise, with the woman called Susan Mortlake and in that time they had got into his mind. It was hard to imagine how much pain they had put him through, trying to turn him against his friends. This was the result. In moments of stress, he simply shut down. Even Pedro had so far been unable to help him. The wounds were too deep.

Pedro was lying still. There was a gash on the side of his head. Jamie was lashing out with the baseball bat, using it like a club or a sword. Matt looked for a weapon but couldn't see one. The man in the dinner jacket had reached Pedro and was standing over him. He had produced a second weapon, an axe which he was holding in both hands. Desperately, Matt searched across the room, saw a jagged piece of broken glass on the floor and – using his power – swept it through the air and into the creature's throat. The creature screamed horribly and fell back in a fountain of its own blood.

"We have to get out of here!" Richard shouted.

The air was full of smoke. It was getting harder to breathe inside, but running out into the fresh air would be suicide. Nobody would be able to see anything in the darkness – and if these creatures had night vision they would be in total command. Matt stood there, cursing himself. There were tears streaming down his cheeks. He knew that this was happening because of Scarlett. He had been expecting it. So why hadn't he been better prepared?

At any event, he knew that Richard was right. They had to get out of the house before they suffocated. The smoke didn't seem to have any effect on the attackers. It was as if their lungs had rotted away and they didn't need to breathe. Jamie threw the baseball bat at one of the creatures on the stairs, then ran over to his brother. Matt reached Pedro and helped him to his feet. At least he didn't seem to be too badly hurt. Professor Chambers blasted away with the rifle, clearing a way to the French windows.

"Look out!"

It was Richard who had shouted the warning. Matt looked up just in time to see part of the ceiling come crashing down in a chaos of orange fire and black smoke. The flames were leaping up at the night. It seemed that most of the roof and part of the second floor had gone. Taking Pedro with him, he threw himself to one side and the falling debris missed him by inches, crashing down onto the sofa where Ramon, the man who had started all this, was sitting. The iron rod that had killed him was slanting out of his chest. He was watching it all like a disinterested spectator.

The six of them staggered out into the garden leaving the burning house and the remaining creatures – nine or ten of them – behind. Professor Chambers fired one last shot. "No more ammo!" she called out to Richard but there was a strain in her voice and Matt wondered if she had been hurt. He looked at her in alarm. There was a patch of red spreading across the front of her dressing gown. A dark gash showed in the material. But she wasn't going to let the pain slow her down. "How about you?" she demanded.

"Two more bullets…" Richard replied.

Two more bullets and the attackers were everywhere. Matt could see them clearly in the light of the flames, their eyes glowing red, their hands clutching knives, axes, chains and lengths of barbed wire which they flailed like whips. Pedro was leaning against him, blood running down the side of his face. Scott and Jamie were standing together, catching their breath. They had made it outside but they had nowhere to run. Another creature lumbered towards Professor Chambers, who stood where she was, clutching her wound. Richard shot it twice.

Matt was almost ready to give up. He couldn't believe that it was going to end this way, surprised and surrounded in a garden in Nazca. Was this what the fight had all been about? He was a Gatekeeper. He had returned to the world after ten thousand years. Was he really going to allow himself to be beaten so easily?

And then the night exploded a second time, with lights bursting out all around them, slanting in from every direction. Matt and Pedro stood where they were, swaying on their feet. Jamie moved towards his brother. Richard and Professor Chambers swung round with their now useless guns. They were trapped, huddled together in a group with the blazing house behind them, the lights in front, surrounded on all sides. Matt tried to see who it was that had arrived at this late stage. Did he have the power to send them back? He bowed his head, drawing on the last of his strength.

Then, as if from nowhere, a volley of arrows was fired in

his direction. But not at him. They had been aimed deliberately over his head. Some of the creatures on the edge of the house cried out and fell back as they were hit. Another volley followed, taking out more of them. The lights were coming from the headlamps of four or five cars that had driven to the edge of the garden and parked in a semicircle. There were men running across the lawn. There were several gun shots. One of the men stopped and reached out for Professor Chambers who more or less collapsed in his arms. The others continued into the house, blasting away with hand guns, searching for any remaining attackers and setting to work, fighting the fire.

And suddenly Matt knew who they were.

They had helped Pedro and him when they had first come to Peru, spiriting them out of Cuzco through a network of underground tunnels. The two boys had stayed with them in their hidden city, Vilcabamba, high up in the mountains. They were Incas, the tribe that had once ruled Peru, but which had been reduced to little more than a handful of survivors, living in secret. They had promised to look after Matt and the other Gatekeepers while they were in Peru. And they had come, true to their word.

They were armed with guns as well as their own traditional weapons and they made short work of the attackers. Machetes swung through the darkness, slicing into rags and flesh. Bullets hammered through the night. It was over very quickly. Matt, Pedro, Scott and Jamie waited on the lawn while the last of the creatures was finished off. Richard was now helping to support Joanna Chambers. All the colour had left her face. She was barely able to stand.

One of the Incas came over to them. He was short with broad shoulders and a dark, serious face. "Are you OK?" he asked.

"We're all right," Richard said. "But Professor Chambers has been hurt."

"I am Tiso. We came when we heard the first alarm. I am sorry. We arrived too late."

"We're just glad you're here," Richard said. "Can we go back into the house? We need to get her inside..."

But it was another half hour before the Incas had put out the flames and they could get back in. The roof and part of the first floor had gone, but there were still two bedrooms that were habitable and, once the debris and the dead bodies had been cleared, the six of them would be able to camp out on the ground floor.

The house would never be the same again. Matt looked at the charred wood and the soiled carpets, the broken windows and debris, and felt a mounting sadness. It had been such a beautiful place. Professor Chambers had lived there for much of her life but then he and the others had come along and ruined it for her. In a few hours, they were supposed to be departing – on their way to London. And this was the mess that they were leaving behind.

Tiso and some of the other Incas helped carry Professor Chambers into her study. Richard went with her and Pedro followed too. His healing powers were going to be needed more than ever, although it looked as if the professor might be too badly injured even for him. She needed medical help, and sure enough a doctor arrived a few minutes later, urgently summoned from the nearby town. Matt, Scott and Jamie stayed outside while she was examined. None of them said anything. They were exhausted. Just a few hours before they had been laughing together, having dinner and playing dice games. And now this!

Matt glanced at Scott. "Where's the diary?" he asked. At that moment he almost wished they didn't have it. It didn't matter how valuable it was. It had so far brought them nothing but trouble.

Scott took it out of his jacket pocket and handed it over. "I'm sorry," he said. His voice was low. "I didn't help you, back there. I didn't help Pedro. I wanted to. But..." His voice trailed off.

"It doesn't matter," Matt said. "Everything happened so quickly. Anyway, Pedro's going to be OK."

"What are they doing in there?" Jamie stared at the closed study door. His voice was angry. He kicked out at the sofa were Ramon had been sitting. The dead man had been carried outside but there was still a great gash in the leather to remind them of what had happened. He turned to his brother. "You got it wrong," he said. "You said he was telling the truth."

Scott blushed – with embarrassment or perhaps with anger. "I thought he was telling the truth," he said.

"You may have been right," Matt interrupted. The two brothers seldom argued and he was surprised to see them starting now. "We can't be sure that Ramon was responsible for what happened tonight. He told us he was in danger and he was certainly right about that. They killed him. So maybe the rest of his story was true."

116

"Can we use it?" Scott asked.

Matt opened the diary. There was a page covered with diagrams. One of them looked a bit like a motor car, though as if drawn by a child, and he remembered that Joseph of Cordoba – the mad monk – was supposed to have been able to predict the future. He flicked through it. Some of the pages had been marked with a modern pen. Someone had scribbled down words and figures, underlining certain areas of the text. Diego Salamanda? The diary had belonged to him and he could have spent weeks deciphering it. It seemed that he had left some of his handiwork behind.

Matt tried to make sense of some of the words but the monk had written in ancient Spanish and anyway his handwriting was almost illegible. "I can't read this language," he said. "And although Pedro can speak it, he can't read…"

"Maybe the professor will be able to work it out," Jamie suggested.

Professor Chambers. Matt remembered how Richard had looked when he had helped carry her in. The two of them had been inside for a long time.

And then the door of the study opened. Pedro came out. He shook his head briefly and sat down, looking miserable. The doctor followed him. He muttered a few words to Richard, then left the house, doing his best to avoid eye contact. That was when Matt knew that it wasn't going to be good news.

"Matt…" Richard called him over to the door. "She wants to see you," he said. His voice was hoarse. "She wants to say goodbye."

"Is she…?" Matt realized what he'd just been told. "She

can't be dying," he said. "What about Pedro? Can't he help her?"

"It's too late for Pedro. There's nothing he can do." Richard sighed. "We've called an ambulance for her and it's on its way now. But she's not going to make it. I'm sorry, Matt. I don't know how it happened but she was stabbed. There's been a lot of internal bleeding and…" He stopped and took a deep breath. "She's not in any pain. The doctor's seen to that. But there's nothing more we can do for her. Do you want me to come in with you?"

"No…" Matt went into the study.

Joanna Chambers was lying on the day-bed that she liked to use as a place to think when she was working. As usual, her desk was completely covered in papers along with a bottle of brandy and a box of her favourite cigars. The old-fashioned radio that she liked to listen to was next to her computer but it was turned off, silent, and somehow that made Matt sadder than anything else, the thought that she would never listen to it again.

She was still in her dressing gown but someone had drawn a blanket over her legs and chest. There was only one light on and it was burning low, casting a soft glow across the room.

He thought she was asleep but as he closed the door, she looked up. "Matt…?"

He went over to her. "The ambulance is on its way," he muttered. "The doctor says…"

"Don't tell me any stuff and nonsense," she cut in, and just for a moment she sounded exactly like her old self. "There's nothing they can do for me and anyway I'm not going into any local hospital. Dreadful place." She tried to shift her position but she didn't have the strength. "Come and sit next to me," she said.

Matt did as he was told. His eyes were stinging and there was an ache in his throat. Why did it have to happen like this? Why couldn't she be all right? He remembered Professor Chambers as he had first seen her, piloting her own plane. She had worked out the secret of the Nazca Lines and she had been with him, in the middle of the desert, when they were attacked by the condors. He knew that without her, he would never have located the second gate. And since then, she had looked after them, never once complaining as her house was invaded and her work interrupted.

Matt had used his power to protect himself. Why hadn't he been able to do the same for her?

"Now you listen to me," she said. She found his hand and clasped it. "You mustn't be upset about me. You have a very great responsibility, Matt. I don't think you have any idea yet what is going to be asked of you. And how old are you? Fifteen! It's not fair…"

She closed her eyes for a few seconds, fighting for breath.

"The Old Ones will be beaten," she said. "Ever since the world began, there's always been good and evil, but somehow we've managed to muddle through. You'll see. It may not be easy. What happened today … silly, really. We should have known they would come."

She let go of his hand. She couldn't manage very much more.

"That's what I wanted to tell you," she said. Her voice was

fading away. "I'm so glad I met you, really. I'm glad we had our time here. I've always loved this place, always been happy here…"

She pointed at the door with one finger, telling him to leave her. Matt did as she said. Richard was waiting for him outside.

The ambulance arrived ten minutes later. But it was too late. Professor Chambers was already dead.

COUNCIL OF WAR

Matt woke up with the smell of burnt wood in his nostrils and the taste of it in his mouth. He had slept for about six hours but he might as well not have bothered. Even before he got out of bed, he knew that he was as tired as he had been when he got into it shortly after two o'clock the night before.

He'd had to share with Pedro. His own room had been destroyed by the fire, along with everything inside it – and it was only as he opened his eyes the following morning that he realized exactly what that meant. He no longer had a passport. He wasn't going to be travelling anywhere today, certainly not on a commercial flight – and that must have been just what the attack had set out to achieve. The Old Ones didn't want him arriving in London. They didn't want him anywhere near Scarlett Adams. And although there were policemen and private detectives looking out for her, she was completely isolated. One in England. Four in Peru. It certainly didn't add up to the Five.

Pedro was sitting, cross-legged on his bed, wearing only a pair of shorts. There was a plaster on the side of his head. Matt guessed that he had been awake for a while. Pedro was always

121

the first to get up, but then, of course, in his old life he would have been begging on the streets of Lima, waiting for the commuter traffic long before dawn. The two boys had been lying next to each other in twin beds.

"So what do we do now?" Pedro asked.

"I don't know, Pedro." Matt got out of bed and pulled on a fresh T-shirt. "We'll have to meet and decide."

"Will we still go to England?"

"Yes."

Pedro hadn't spoken very much about the journey and Matt suspected that he was finding it difficult to get his head around it. He had never been out of Peru in his life. Even the notion of getting on a plane was completely alien to him. He had only flown once and that had been in a helicopter which had crashed. The thought of spending fifteen hours in the air and landing in a completely different world unnerved him.

"I am sad that the professor is dead," he said. "She was very kind."

"I know." Matt wondered if he could have saved her. Was her death his fault? It seemed to him now that she had been doomed from the moment they had arrived, although he knew she would never have seen it that way. Even so... It had been two days since they had received the fax with the news about Scar. He wished now that they had all left at once.

There were now just five of them remaining: Matt, Pedro, Scott, Jamie and Richard. They met outside, sitting at a wooden table in the shade of a silk-cotton tree – a kapok, as it was also known. Professor Chambers had liked taking the boys around her garden, showing them all the different plants and talking about them. This one had somehow found its way out of the rainforest, she had said, and she couldn't understand how it was growing here at all. The table had been set up in the shade, the umbrella-shaped canopy and creamy white flowers of the kapok shielding them from the sun.

They might have been safer in the house but they could hardly bear to look at it, the ruin that it had become. Somehow, it didn't seem likely that the Old Ones would return ... not in the daylight. And anyway, the Incas were somewhere close. There was no danger of a second attack. Richard had brought out a tray of iced lemonade and a plate of *empanadas*, the little cheese pastries that they had often devoured. But nobody was hungry. They were exhausted and unhappy. Nobody knew what they were supposed to do.

One thing was sure. They couldn't stay here much longer. The house still had water and electricity and they might even be able to repair the roof. But there was no alarm system. The Incas couldn't protect them indefinitely. And – more to the point – none of them wanted to be here. The moment Professor Chambers had been taken from the *hacienda*, all of its life seemed to have gone with her.

"OK..." It was Richard who was the first to speak and Matt was grateful to him for breaking the silence, for taking control. He was wearing a clean polo shirt and jeans, but he looked completely worn out, as if he hadn't slept at all.

"This is a council of war," he said. "Because it looks as if the war has finally arrived. We have to talk about last night. We have to deal with it and put it behind us. And I might as well start by saying that it was mainly my fault." He held up a hand before anyone could interrupt him. "When Ramon came to the house, I turned off the security system. But I never put it back on again. Not the radar, anyway. Maybe that was the idea. Maybe that was why he was sent to us. A diversion…"

"It was my fault too," Scott cut in. "Matt wanted me to look into his mind and I did. But somehow he managed to fool me. I thought he was telling the truth."

"Maybe he *was* telling the truth," Matt said. "He brought us the diary … and do you really think he would have just sat there and allowed himself to be killed? Maybe they followed him from Lima. The whole point of last night could have been simply that they wanted the diary back."

"The question we've got to ask ourselves is – what are we going to do next?" Richard said. "It's been more than forty-eight hours since Scarlett Adams appeared in the newspapers. The Nexus are still watching her but we can't leave her on her own much longer. On the other hand…" He nodded at Matt. "Matt has lost his passport so he's not flying anywhere."

"We can use the door," Jamie said. "The same one that Scott and me came through. All we have to do is get to the Temple of Coricancha in Cuzco. We walk in … we walk out in London. We don't need a plane."

It seemed obvious. It was exactly the reason why the doors had been built in the first place. But Richard shook his head. "We can't use the doors," he said. "Think about it, Jamie. Salamanda had the diary and he obviously studied it carefully. If the Old Ones are looking for us – and it seems pretty likely that they are – that's exactly how they'll expect us to travel."

"Maybe they never saw the diary," Pedro said. "It was in

the office of *Señor* Salamanda. He could never have shown it to them."

Richard was still unhappy. "It's too much of a risk. Anyway, they know about the door in St Meredith's. Scarlett went through it. That's probably what started all this. They could be waiting for us there. I know it's boring, but I reckon we're much safer taking planes."

"But Matt doesn't have a passport," Scott said.

"The Nexus can get us into America," Richard replied. "I spoke to Nathalie Johnson this morning and she's sending a private plane. It's already on its way. And she's been in touch with John Trelawney. The two of them have enough clout to get us through immigration. They can also get Matt a new passport. After all, they didn't have any difficulty getting Pedro his. It'll take a couple of days but we could be in England by Tuesday."

Scott and Jamie had met Nathalie Johnson before they came to Peru. She was an American businesswoman who had made a fortune out of computers before she had been drawn into the Nexus. John Trelawney was the senator who had been fighting in the presidential election. The result was going to be announced in just one day and he was still the favourite to win. The two of them were powerful friends.

Jamie considered what Richard had said. "All right, then." He shrugged. "Let's go."

"Not all of us," Matt said.

There was a sudden silence around the table. All eyes were turned on him.

"I think we should separate," he said.

"Are you crazy...?" Scott began.

"Why?"

"What do you mean, Matt?"

Everyone was talking at once. Matt wasn't surprised. Even as he had decided what he was going to do, he had known that the rest of them would be against it. They were supposed to stick together. Finding each other, coming together … it was what their lives were all about. Five Gatekeepers. So far, against all the odds, four of them had managed to do exactly that. They were hours away from finding the fifth. It seemed completely mad to split up now.

"We've just got to be careful," Matt explained. "Richard and I were talking about it last night, before we were attacked. If all four of us get onto one plane and the Old Ones somehow manage to get control of it, they'll have us at their mercy. They'll be able to do anything with us. All four of us at once."

"So what are you saying?" Jamie asked.

"We can't stay here," Pedro added.

"I'm going to London with Richard," Matt said. "We'll meet the Nexus as soon as we can and we'll meet Scarlett as soon as we know it's safe." He turned to Jamie. "I'd like you to come with us."

Jamie opened his mouth but said nothing. He understood the implications of what Matt had just suggested.

"You're leaving me behind," Scott muttered. His voice was low and sullen.

"It's just for a few days. A week, no longer."

"Is this because I screwed up last night?"

"You didn't screw up." Matt had to choose his words carefully. In a way, Scott was right. He might not be to blame, but he still couldn't be completely trusted. Matt looked at him, slumped back from the table with his hands in his pockets, and saw the cold anger in his face. And there was something else. A sort of cruelty. When Scott had lived ten thousand years ago, his name had been Flint and it suited him. Sitting in the garden, his eyes were as hard as stone.

"Scott and I don't like being apart," Jamie said.

"I know that and I'm sorry," Matt said. "It's true that we're stronger together. That's why I want to stay in pairs. Two and two. If anything goes wrong in London, I'll need someone to back us up."

"So why not take Pedro?"

"Because Pedro doesn't know London. He's never been to England."

"Nor have I."

Matt sighed. "Jamie … if you really don't like the idea, I'll go on my own. I don't mind doing that. I just don't think we should all go. That's all. I'm trying to do what's best for everyone."

"And since when did you get to tell everyone what they should do?" Scott demanded. "I thought we were meant to be equal. Who put you in charge?"

There was another long pause. Richard opened his mouth as if to say something, then changed his mind. The day was getting warmer as the sun climbed over the mountains, but the atmosphere right then was anything but. Matt looked across the lawn to the track that led back to the town of Nazca. He had been there a couple of days ago, kicking a football, waiting for Professor Chambers to get back from the shops. Now she was dead, her house was in ruins and the four of them were at each other's throats. How could things have gone wrong so quickly?

"Scott, I don't think…" Jamie began.

"Are you on his side?" Scott directed his anger at his brother.

"We're all on the same side," Matt cut in. "And if we turn against each other, we might as well give up."

"You've never been on my side, Matt. You've never trusted me, not from the day I arrived here. Well, you go without me. You can all go without me. I don't care."

Scott got up angrily, knocking his chair over behind him. He didn't even notice. He walked away in the direction of the house and disappeared through the front door. Nobody spoke. Then Jamie stood up. "I'm sorry, Matt," he said. "I'll go and talk to him. He'll be all right."

Jamie followed his brother. That just left Richard, Pedro and Matt. Richard poured out a glass of the lemonade. He offered it to Matt who shook his head. Richard drank it himself.

"Where do you want me to go?" Pedro asked. "I do not think it is good for us to stay here."

Matt sighed. "I thought you'd go back to Vilcabamba with Tiso and the other Incas," he said. "I was hoping you could spend a bit more time with Scott…" Pedro understood. Scott still needed help after his experiences as a prisoner of Nightrise.

"I do what I can," he said. "But Scott has a lot of pain. There are things happening here…" He tapped the side of his head. "I do not understand."

"You were nearly killed last night. He didn't help you."

"Yes. But he and Jamie are very close. Twins. Maybe it is not

128

such a great idea to split them up."

There didn't seem anything more to say. Pedro collected the jug and the glasses and carried them in. Richard and Matt were left on their own.

"That went well," Matt said, gloomily.

Richard finished his lemonade and set the glass down. "Don't be too hard on yourself," he said. "We're all feeling bad about last night, the death of Joanna. Jamie will talk to Scott. He knows you're doing the best you can. They'll work it out."

"I hope so."

"In just a week, you'll be in Vilcabamba. All of you. You've got the diary now. And despite what happened last night, you all came out of it OK. None of you was badly hurt. I'm sure you've made the right decision, Matt. It's all going to work out."

But Matt wasn't so sure. He twisted round and looked at the house, at the scorched wood, what was left of the roof, and suddenly he was aware that something was wrong, that it didn't quite add up.

If Ramon had been able to find them so easily, why had it taken the Old Ones so long? And if they had wanted the diary back so badly, why hadn't they sent a larger force? Matt had seen the sort of creatures the Old Ones had at their disposal. They had crawled out of the floor of the Nazca Desert ... the armed soldiers, the giant animals, the hoards of shape-changers. But they hadn't been there last night.

Was he making the right decision, splitting them up? Or was this what he was meant to do? Was he reacting to decisions that had already been made?

Later that afternoon, two cars came to the house. One would take Pedro and Scott to Arequipa, the famous "White City" in the south of Peru. They would have to stay there overnight before flying to Cuzco. Because of the thin air high up in the Andes, planes were only able to take off and land in the morning. Two of the Incas would go with them and then escort them up through the cloud forest to Vilcabamba.

Jamie, Richard and Matt had a shorter drive to Nazca airport where a private plane was already waiting to fly them up to Miami. They would wait in Miami until Matt's new passport arrived and then they would cross the Atlantic to England. If things went well, they would only be apart for a few days.

Matt took one last look at the professor's house. The town children would probably raid it in the next few days, stripping it of anything of value. He had been there for a long time. He had almost begun to think of it as his home but now it was nothing. Burned out. Broken. Empty.

Richard loaded their bags into the boot.

"Vilcabamba," Matt said.

"Vilcabamba," Pedro agreed.

The two of them shook hands. Scott and Jamie said nothing – but Matt knew that they were communicating even so.

It was all over very quickly. The four boys climbed into their different cars and went their separate ways.

THE HAPPY GARDEN

In London, Scarlett Adams was trying to get back to her old life.

The doctors had decided there was nothing wrong with her. The police had asked more questions but had finally given up. Maybe she had suffered from amnesia. Maybe the whole thing about her disappearance had been a schoolgirl prank – but either way they had better things to do. Even the press had decided to leave her alone. A new president, a man called Charles Baker, had just been elected in the USA, and according to all the reports, there had been something strange about the way the votes had been counted. It was turning into a huge scandal and that left no room in the papers for a girl who had been missing for less than a day.

Just forty-eight hours after he had flown all the way to England, Paul Adams went back to Hong Kong.

Scarlett understood why he couldn't stay with her. He had only recently started his new job, working in the legal department of a huge company involved, amongst other things, in the manufacture of computer equipment and software. It hadn't made a good impression, shooting off to London at such short notice. He had to get back again.

Back to Nightrise.

Paul Adams took Scarlett out to dinner on his last night at home. The two of them went to a little Italian restaurant that he liked in Dulwich. He ordered half a bottle of wine for himself and a lemonade for her and the two of them sat facing each other trying to think of things to say. Paul was wearing expensive jeans and a jersey that didn't really suit him. The truth was that he was only really comfortable in a jacket and tie. It was like a second skin to him. Maybe it was his age. He was forty-nine years old and he had been a lawyer for more than half that time, devoting his life to contracts, complicated reports and charts. It was hard to imagine what he had been like as a teenager.

"Are you going to be all right, Scarly?" he asked.

"Yes." Scarlett nodded.

Neither of them had spoken very much about St Meredith's. Paul Adams seemed to have accepted her story. She had fallen ill. She had forgotten whatever had happened. Scarlett wondered why she hadn't confided in him. He had always been kind to her. Why was she lying to him now?

"I'm sorry, Dad," she said.

"There's nothing to be sorry about." Paul Adams paused and sipped his wine. "Do you really have no idea what happened to you?"

"I wish I did."

"You could tell me, you know. I wouldn't be cross with you. I mean, if there's some sort of secret or something you're afraid of..."

Scarlett shook her head. "I told the police everything."

Paul Adams nodded. Then the waiter arrived with spaghetti carbonara for him, a pizza for Scarlett. There was the usual business with the oversized pepper grinder, the sprinkle of parmesan cheese. At last they were on their own again.

"How's the job going?" Scarlett asked. She had deliberately changed the subject.

"Oh. It's not too bad." Paul Adams twirled his fork in the spaghetti. "Do you want to come to Hong Kong for the Christmas holidays? I've spoken to your mother and she's happy for me to have you this year. I'll get a few days off and we can travel together."

"I'd like that," Scarlett said, although she wondered what it would be like, travelling, just the two of them. They seemed to have grown apart so quickly.

They ate in silence. Paul Adams didn't seem to be enjoying his food. He left half of it, then took off his glasses and began to rub them with his napkin. Looking at him just then, Scarlett thought how old he had become. It wasn't just his hair that was going grey. It was all of him.

"I'm sorry, Scarly," he said. "I'm afraid I've rather let you down, haven't I? If I'd known that Vanessa and I weren't going to stay together … maybe we should have thought twice about adopting a child, although of course I'm glad we did. I think the world of you. But it hasn't been fair. Leaving you on your own with Mrs Murdoch."

"It was my decision," Scarlett reminded him.

"Well, yes. I suppose it was."

"Why do you have to work in Hong Kong?" Scarlett asked.

"It's a wonderful opportunity. Not just the money. Nightrise has offices all over the world and if I can work my way up the ladder..." His voice trailed off. "I'll only be there a couple of years. I've told them already. Then I want them to transfer me to the London office and we'll be together again."

"Don't worry about me, Dad. I'll be all right."

"Will you, Scarly? I hope so."

He left on the early flight the next day.

Scarlett had already gone back to school – and that hadn't been easy either. The headmistress, a grey-haired woman who looked more severe than she actually was, had made a speech in assembly, telling everyone to leave her alone, but of course they had been all over her, bombarding her with questions, desperate to know where she had really been. Scarlett had been on TV. She was a minor celebrity. Some of the younger girls had even asked her for an autograph. On the other hand, some of the teachers had been less than happy to see her – Joan Chaplin in particular. The art teacher had taken some of the responsibility for Scarlett's disappearance and she in turn blamed Scarlett for that.

The next couple of days passed with the usual routine of lessons and games. There were piles of homework and rehearsals for the Christmas play. Everything had returned to normal – at least, that was what Scarlett told herself. But in her heart, she knew that nothing was really normal at all. Maybe it never would be again.

She had already decided that there was only one person she could talk to and tell the truth about her disappearance. Not her father. Not Mrs Murdoch. It had to be Aidan. He was her closest friend. He wouldn't laugh at her. She had already texted him and the two of them met after school and walked home together, taking their time, allowing the other school kids to stream ahead.

She told him everything: the door, the monastery, Father Gregory, the escape. She was still talking as they turned into Dulwich Park, opposite the art gallery, taking the long way round past the playground and across the grass.

"Do you think I'm mad?" she asked, when she had finished. There had been times when she had begun to wonder herself. Could it be that the official version of events was actually true? Had she somehow hit her head against a wall and dreamed the whole thing?

"I always thought you were pretty strange," Aidan said.

"But to dream something like that..."

"You don't make it sound like a dream." His eyes brightened. "Hey – maybe we could go back to the church. We could go through the door a second time and see what happened."

Scarlett shuddered. "I couldn't do that."

"Why not? If you went with me, at least it would prove it was true."

"I couldn't go back. They might be waiting for me. They'd grab me and the whole thing would just start again."

"I'd protect you!"

"They'd kill you. They'd kill both of us."

They had reached the other side of the park and were coming out of the Court Lane Gate on the north side. From here the road cut down to the lights where, two years before, Scarlett had almost been killed.

Scarlett had just turned the corner when she saw the car.

It was a silver Mercedes with tinted windows so that although she could make out two people inside it, she couldn't see their faces. It was parked on the opposite side of the road and she might not even have noticed it ... except that it was the fourth time she had seen it. It had been in the street that morning, parked outside The Crown and Greyhound when she was on her way to school. Once again, there had been two people sitting inside. It had overtaken her when she was walking to the Italian restaurant with her father. And she had seen it from her bedroom, cruising down the street where she lived. She had made a note of the registration number. It contained the letters GEN which just happened to be the first three letters of St Genevieve's. That was why she remembered it now.

She stopped.

"What is it?" Aidan asked.

"Those two men." She pointed at the car. "They're watching me."

"Scarl..."

"I mean it. I've seen them before."

Aidan looked in their direction. "Maybe they're journalists," he said. "You're still a mystery. They could be after an interview."

"They've been following me."

"I'll ask them, if you like."

They must have seen him coming or guessed what he had in mind. As Aidan stepped off the pavement, the driver started the engine up and tore away, disappearing round the corner with a screech of tyres.

Scarlett didn't see the Mercedes again but that wasn't the

end of it. Quite the opposite. It told her something that she had been feeling all along.

She was being watched. She was sure of it. It had crept up on her over the past few days, before Paul Adams had left, a sense that she was trapped, like a specimen in a laboratory glass slide. She had found herself gazing at complete strangers in the street, convinced that they were spying on her. When she walked past a security camera outside a shop or an office it almost seemed to swivel round, its single, glass eye focusing on her – and she could imagine someone in a secret room far away, staring at her on a television monitor, picking her out from the crowd.

Even when she was on her own in her room she had got the sense of someone eavesdropping, and after a while, just the flapping of a curtain would be enough to unnerve her. When she made phone calls – it didn't matter if it was her mobile or a landline – she was sure she could hear something in the background. Breathing. A faint echo. Someone listening.

She wasn't imagining it. It was there.

Scarlett had tried to tell herself that none of this was possible. She knew that there was a word for what she was experiencing. Paranoia. Why would anyone bother to watch her? Nobody was watching her. She was just freaked out by what had happened before.

"There were five children. They came to be known as the Gatekeepers. Four boys and a girl. You are the girl."

It was when she saw the Mercedes with Aidan that Scarlett understood that what had started at St Meredith's wasn't over yet. It had only just begun.

The next day – Friday – was miserable. Scarlett hadn't slept properly. She was snappy with Mrs Murdoch and managed to make a spectacular mess of a maths test at school. She didn't want to be in class. She just wanted to go back to her room and close the door – to shut "them" out, even though she didn't have any idea who "they" might be.

That evening, she got a phone call. It was Aidan.

"Hi, Scarlett," he said. "I was wondering … do you want to come to a movie tomorrow?"

Just that one sentence and she knew that something was wrong. Scarlett didn't reply immediately. She cradled her mobile in the palm of her hand, playing back what she had just heard. First of all, Aidan never called her Scarlett. He called her Scarl. And there had been something weird about his tone of voice. He hadn't asked her out as if he really meant it. He sounded fake, as if he was reading from a script.

As if he knew he was being overheard.

She lifted the mobile again. "What do you want to see?"

"I don't know. The new Batman or something. We can go into the West End…"

And that was odd too. Why travel all the way into town? Dulwich had a perfectly good cinema.

"OK," she said. "What time do you want to meet?"

"Twelve?"

"I'll see you here…"

Aidan arrived at exactly midday, dressed in his trademark hoodie and jeans. As they walked over to the Tube station together, Scarlett wondered if she hadn't read too much into the conversation the night before. He seemed completely relaxed and cheerful. The two of them chatted about school, football, fast food and the American election which was still in the news. Aidan was interested in politics even if it left Scarlett completely cold.

"Charles Baker is a creep," he said. "I can't believe anyone voted for him as President. The other guy, Trelawney, should have walked it."

"So why didn't he?"

"I don't know. Some people are saying they screwed up the voting slips. But I'm telling you, Scarl, the wrong guy won."

They reached the cinema, the Empire in Leicester Square, but as they approached the box office, Aidan suddenly grabbed Scarlett and dragged her to one side. In an instant, his whole mood had changed. He made sure there was no one else around, then hurriedly began to speak.

"Scarl, I've got to tell you. Something really weird has happened."

"What is it?" Scarlett was completely thrown.

"I didn't know whether to tell you or not. But yesterday, when I called you, I was told to do it! This guy came up to me when I was coming out of school."

"What guy?"

"I'd never seen him before. At first I thought he was trying to sell me something. He was Chinese. A young guy. He asked me to get a message to you."

"Why didn't he tell me himself?"

"I'm only telling you what he told me." Aidan ran a hand through his long, shaggy hair. There was still no one in this part of the foyer. A short distance away, a family of four was just going in to the film. "He just came up to me and asked if he could talk to me. He knew my name. And he knew I was your friend."

"What did he want?"

"Listen … I don't want to freak you out but he told me that he couldn't approach you himself because your phone was bugged and you were being watched. He said you were in danger." Aidan paused. "Has this got something to do with what happened in the church?"

"I don't know, Aidan," Scarlett said. All her fears had just been confirmed. She didn't know if that made her feel better or worse. She looked around her. "So where is this mysterious Chinese man? Are we meeting him here?"

"No. He's round the corner … in a restaurant. The Happy Garden. It's in Wardour Street, about five minutes away."

"So why are we here?"

"That was my idea. I had to tell you what was going on, but I couldn't do it on the Tube in case someone was listening. I'm sorry, Scarl. I didn't want to lie to you but this guy sounded really serious. And it was only yesterday we saw that car at the park." Aidan drew a breath. "You don't have to go," he said. "Maybe you shouldn't go."

"Why not?"

"Maybe you should go to the police."

Scarlett had to admit that he had a point. Everyone knew that when a strange adult approached a kid outside school, it was time to dial 999. But she had already made up her mind. If she didn't go to this restaurant, she might never find out who the man was or what he wanted.

"The Happy Garden," she muttered. "What sort of name is that?"

"It's a Chinese restaurant," Aidan said.

"Oh yes," she nodded. "I suppose it would be." She thought for a moment. "Did the man say anything else?"

"Yes. He said that the two of you had met before. On Dulwich Grove, two years ago. He must have been talking about the accident…"

If Scarlett had had any doubts, that decided it. The man who had saved her, who hadn't waited to be thanked, had been Chinese. It had to be the same person. But what was he doing back in her life?

"What time am I meant to be there?" she asked.

"Half past one."

She looked at her watch. It was just after one o'clock. "We're going to be early."

"So you're going?"

"I've got to, Aidan. I don't think anything too bad can happen in the middle of a Chinese restaurant. And anyway, you'll be with me." She paused. "Won't you?"

"Sure." Aidan nodded. "I wouldn't leave you on your own. Anyway, I can't wait to find out what this is all about."

They left the cinema the way they'd come, slipping quietly into the crowds in Leicester Square. It was unlikely that anyone had followed them all the way from Dulwich but Scarlett wasn't taking any chances. They turned up an alleyway that led into Chinatown, an area that was packed with Chinese restaurants and supermarkets. From here, they crossed over Shaftesbury Avenue, heading for the address that Aidan had been given.

The afternoon was surprisingly warm. It was lunch-time, there were lots of people around. The smell of fried noodles hung in the air.

The explosion happened just as they were about to turn the corner into Wardour Street. They didn't just hear it. They felt it too. The pavement actually shuddered under their feet and a gust of warm air punched into them, carrying with it a cloud of dust and soot. If they had been just ten seconds earlier, they might have been hit by the full impact. A bomb had gone off. A large one. It had happened somewhere near.

"Stop…!" Aidan began.

He was too late. Scarlett had already run forward and turned the corner.

A scene of devastation greeted her on the other side. A building about half-way up the road had been blown to pieces. It was as if someone had punched a giant fist into it. There was glass and debris all over the pavement, and tongues of flame were licking out of the shattered brickwork. A taxi must have been passing at the moment the bomb went off. All its windows were broken and the driver had tumbled out, blind, blood pouring down his face. A woman was standing nearby, screaming and screaming, her clothes in tatters, covered in blood and broken glass. There was smoke everywhere but Scarlett could make out several bodies, lying still, some of them in rags. She had seen images like this on TV, in Baghdad and Jerusalem. But this was Soho, the centre of London. And she knew that she'd almost been part of it. It might have been Aidan and her, lying in the rubble.

Aidan had caught up with her. "We should go," he said.

"But the restaurant..."

"That *is* the restaurant."

Scarlett couldn't move. She stared at the gaping hole, the smoke billowing out, the smashed furniture and the bodies. It was a restaurant. He was right.

"Come on...!" Aidan pleaded.

Scarlett could already hear the sirens of the police cars and ambulances moving in from some other part of the city. It was amazing how quickly they had been alerted. She allowed Aidan to lead her away. She didn't want to be found there. Part of her even wondered if she might somehow have been to blame.

It was the first story on the news that night. A restaurant called The Happy Garden had been the target of a lethal attack. Three people had been killed and a dozen more injured by a bomb that had been concealed under one of the tables. According to the police, this wasn't a terrorist incident. They put the blame on Chinese gangs which had been operating in the West End.

"Police today are speculating that the attack is the result of rising tension within the Chinese community," the newscaster said.

Scarlett watched the broadcast with Mrs Murdoch. The housekeeper was knitting. "Weren't you in Soho today, Scarlett?" she asked.

"No," Scarlett lied. "I was on the other side of the town. I was nowhere near."

"This is the most serious attack so far," the report went on. "It follows other incidents involving gangs in Peckham and Mile End. Any witnesses are urged to come forward and

Scotland Yard has set up a special phone line for anyone with any information that might help."

Scarlett texted Aidan that night before she went to bed and he texted back. They both agreed that it was just a coincidence. Despite what they had thought earlier, it would be absurd to suggest that a restaurant in the middle of London had been blown up just to stop them meeting someone there.

But as she turned out the lights and tried to get to sleep, Scarlett knew that it wasn't. The newscaster had been lying. The police were lying. There were no gangs ... just an enemy who was still playing with her and who wouldn't stop until she was completely in their control.

MATT'S DIARY (2)

<u>Sunday</u>

A bomb has gone off in London. I've just been watching it on the television news and I wonder if it might have something to do with Scarlett. Richard thinks it's unlikely. According to the reports, the bomb had been hidden in a restaurant in Chinatown. It was something to do with Chinese gang warfare. Three people have been killed.

I saw the images on the big plasma screen TV in my hotel room. Dead people, ambulances, screaming relatives, smoke and broken glass ... it was hard to believe that it was all happening in the middle of Soho. You just don't expect it there. It made me feel even further away than I actually was.

Miami. I've never been here before and I certainly never dreamed that I'd wind up in a five-star hotel overlooking the beach, surrounded by Cadillacs, Cuban music and palm trees. The Nexus has certainly put us up in style while we wait for my new passport to arrive. The only trouble is, it's taking longer than we had hoped. We're now booked onto a flight leaving on Monday evening and we'll have to kick our heels until then. Scarlett will just have to manage without us for a couple more days. We'll be with her soon enough.

It feels strange, being back in a big city after spending so much time in a backwater like Nazca. Miami is full of rich people and expensive houses. It's too cold to swim at this time of the year, but a lot of life still seems to be happening in the street. We didn't do much today. I bought myself some new clothes, replacing the stuff that got lost in the fire. We walked. And tonight we ate on Ocean Drive, a long strip of fancy cafés and bars with bright pink neon lights, cocktails and live bands. It was good to be able to enjoy ourselves, sitting there, watching the crowds go past.

Nobody noticed us. For a few hours we could pretend we were normal.

Monday afternoon

This morning, the passport finally arrived, delivered in a brown, sealed envelope by a motorbike rider who didn't say a word. Terrible photograph. The Nexus have sent Jamie a new passport too, and they've decided that we should both travel under false names, for extra security. So now I'm Martin Hopkins. He is Nicholas Helsey. Richard is going to stay as himself but then, as far as we know, nobody is trying to kill him.

We have economy tickets. The Nexus could have flown us first class but they didn't want us to stand out.

We had our final meal on Ocean Drive. A huge plate of nachos and two Cokes. Richard had a beer. I wondered what the waiter must have made of us: Richard in a gaudy, Hawaiian shirt, sitting between two teenagers, the two of us wearing sunglasses even though there wasn't a lot of sun. We'd bought

146

them the day before and hadn't got round to taking them off. We liked them because they kept us anonymous. If anyone had asked, we were going to say that he was a teacher and that we were on a school exchange. It was a pretty unlikely story – but nothing compared to the truth.

I've spoken to Pedro via satellite phone a couple of times while we've been here. He and Scott reached Vilcabamba without any problem. We've agreed to contact each other every day while we are apart. If there's silence, we'll know something is wrong. Pedro told me that Scott was OK. But Scott didn't come on the line.

Jamie asked me something today. It took me by surprise. "Why did you really leave Scott behind? You didn't think you could rely on him, did you?"

"I never said that."

"But you thought it." He lowered his voice. "You have no idea what he went through with Mrs Mortlake. It was worse than anything you can imagine."

"Has he talked to you about it?"

Jamie shook his head. "He's put up barriers. He won't go there. He's not the same any more. I know that. But you have no idea how he looked after me all those years. When Uncle Don was beating me around or when I was in trouble at school, Scott was always there for me. The only reason he got caught was that he was helping me get away." He suddenly took off his sunglasses and laid them on the table. "Don't underestimate him, Matt. I know he's not himself right now, but he'll never let you down."

I hope Jamie is right. But I'm not sure.

I looked across the road. There were some little kids throwing a ball on a lawn beside the beach. A couple of rollerbladers swung by. A pale green convertible drove past with music blaring. And just a few metres away, we were talking about torture and thinking about a war that we might not be able to win. Two different worlds. I know which one I'd have preferred to be in.

We finished eating and went back to the hotel. Our car was already there. The concierge carried out our cases and then it was a twenty-minute drive across the causeway. The water, stretching out on both sides, looked blue and inviting. We reached Miami International Airport and went in, joining the crowds at the check-in desks. Thousands of people travelling all over the world. And this is what I was thinking…

Suppose the Old Ones are already here. Suppose they control this airport. We are allowing ourselves to be swallowed up by a system … tickets, passports, security. How do we know we can trust it, that it will take us where we want to go, or even let us out again?

We got to the baggage check. Richard took one look at the X-ray machines and stopped. "I'm an idiot," he said.

"What is it?"

He was carrying a backpack on his shoulder, cradling it under one arm. He'd had it with him at the restaurant too and I knew that among other things, the monk's diary was inside. But now he was watching as people took out their computers and removed their belts and I could see that he was furious with himself. "The tumi," he said. "I meant to transfer it to my main luggage. They'll never let it through."

The tumi *is a sacrificial knife. It was given to him by the prince of the Inca tribe just before we left Vilcabamba. I could understand Richard wanting to keep it close to him. It was made of solid gold, with semi-precious stones in the hilt, and it must have been worth a small fortune. But this was a mistake. He might try to argue that the* tumi *was an antique, an ornament or just a souvenir, but given that the airlines wouldn't even allow you to carry a teaspoon unless it was made of plastic, there was no way it was going to be allowed on the plane.*

It was too late to do anything now. There was a long line of people behind us and we wouldn't have been allowed to turn back. Richard dumped the bag on the moving belt and grimaced as it disappeared inside the X-ray machine. I suppose he was hoping that the security people might glance away at the right moment and miss it. But that wasn't going to happen. The bag came out again. It was grabbed by an unsmiling woman with her name – Monica Smith – on a badge on her blue, short-sleeved shirt.

"Is this yours?" she asked.

"Yes." Richard prepared for the worst.

"Can you unzip this, please?"

"I can explain…" Richard began.

"Just open it, please."

The tumi *was right on the top. I could see the golden figure of the Inca god that squatted above the blade. I watched as the woman, wearing latex gloves, began to rifle through Richard's clothes. Briefly, she picked up the diary, then put it back again. She examined a magnifying glass that Richard had bought in Miami, trying to decipher the monk's handwriting. But she*

149

didn't even seem to notice that the tumi was there. She closed the bag again.

"Thank you," she said.

Richard looked at me. Neither of us said anything. We snatched up our belongings and hurried forward. It was only afterwards that we understood what had happened.

The tumi has another name. It's also known as the invisible blade. When the prince of the Incas gave it to Richard, he said that no one would ever find it, that he would be able to carry it with him at any time. He also warned Richard that one day he would regret having it – something neither of us really like to think about.

But now we both realized what we had just seen. It was a bit of ancient magic. And it was all the more amazing because it happened in the setting of a modern, international airport.

<u>Monday night</u>

We took off exactly on time and once the seat belt signs had been turned off, I sat back in my seat and began to write this. In the seat next to me, Jamie had plugged himself straight into the TV console, watching a film. Richard was across the aisle, working with a Spanish dictionary, trying to unravel the diary.

A bit later, I fell asleep.

And that was when I went back. I had wanted to visit the dreamworld again, ever since I had discovered the path set into the side of the hill. Was it really possible that a civilization of some sort had once lived there? Might they be living there still? The dreamworld was a sort of in-between place, connecting where we were now with the world that Jamie had visited and

150

where he had fought his battle, ten thousand years ago. It was there to help us. The more we knew about it, the better prepared we would be.

I was right where I wanted to be, back on the hillside, half-way up the path. But that was how the dreamworld worked. Every time I fell asleep, I picked up exactly where I had left off. So if I woke up throwing a stone into the air, when I went back to sleep I would immediately catch it again. And I was wearing the same clothes that I had on the plane. That was how it worked too.

The hill became steeper and the path turned into a series of steps. They had definitely been made by human hand. As I continued climbing up, they became ever more defined and when I finally reached the summit I found myself on a square platform with some sort of design – it looked like a series of Arabic letters – cut into it. The letters made no sense to me, but then I lifted my head and what I saw was so amazing that I'm surprised I didn't wake up at once and find myself back on the plane.

I was looking at a city, sprawling out in all directions, as far as the eye could see. More than that. From where I was standing, high up on the hill, I could see thousands of rooftops stretching all the way to the horizon, perhaps ten miles away, but I got the impression that if I managed to walk all the way to the other side, it would continue to the next horizon and maybe to the one after that.

It was impossible to say if the city was ancient or modern. It somehow managed to be both at the same time. Some of the buildings were huge, cathedral-like with arched windows and domes covered in tiles that could have been silver or zinc. Others were steel and glass structures that reminded me of an airport terminal and then I realized that there were actually dozens of them and they were all identical, radiating out of central courtyards like the spokes of a wheel. Towers rose up at intervals, again with silver turrets. Everything was connected, either by spiral staircases or covered walkways.

There were no parks and no trees. There weren't any cars or people. In fact, I wasn't looking at a city at all. This vast construction was one single building: a massive cathedral, a massive museum, a massive … something. It was a mishmash of styles, some parts must have been added hundreds or even thousands of years after others – but it was all locked together. It was one. I couldn't work out where the centre was. I couldn't see where it had originally begun. Nor could I imagine how it had come into being. It was as if someone had taken a single seed – one brick – and dropped it into a bubbling swamp. And this, after thousands of years of growth, was the result.

Leaving the platform behind me, I walked down the other side of the hill and made my way towards the outer wall. I was now following a road with a marble-like surface and it was taking me directly towards a great big arch and, on the other side of it, an open door. The air was very still. I could actually hear my heart beating as I approached. I didn't think I was in danger, but there was something so weird about this place, so far removed from my experience, that I admit I was afraid. I didn't hesitate though. I passed through the arch and suddenly I was inside, in a long corridor with a tiled, very polished floor and a high, vaulted ceiling held up by stone pillars: not quite a church, not quite a museum, but something similar to both.

"Can I help you?"

Another shock. I wasn't on my own. And the question was so normal, so polite that it just didn't seem to belong to this extraordinary place.

There was a man standing behind a lectern, the sort of things lecturers have in front of them when they talk. He was quite small, a couple of inches shorter than me, and he had one of those faces ... I won't say it was carved out of stone (it was too warm and human for that) but it somehow seemed ages old, gnarled by time and experience.

From the look of him, I would have said he was an Arab, a desert tribesman, but without any of the trappings such as a headdress, white robes or a dagger. Instead, he was dressed in a long, silk jacket – faded mauve and silver – with a large pocket on each side and baggy, white trousers. A beard would have suited him but he didn't have one. His hair was steel grey. His eyes were the same colour. They were regarding me with polite amusement.

"What is this place?" I asked.

"This place?" The man seemed surprised that I had asked. "This is the great library. And it's very good to see you again."

A library. I remembered something Jamie had told me. When he met Scarlett at Scathack Hill, she had mentioned visiting a library to him.

"We've never met."

"I think we have." The man smiled at me. I wasn't sure what

language he was speaking. In the dreamworld, all languages are one and the same and people can understand each other no matter where they've come from. "You're Matthew Freeman. At least, that's the name you call yourself. You're one of the Gatekeepers. The first of them, in fact."

"Do you have a name?"

"No. I'm just the Librarian."

"I'm looking for Scarlett," I said. "Scarlett Adams. Has she been here?"

"Scarlett Adams? Scarlett Adams? You mean ... Scar! Yes, she most certainly has been here. But not for a long time. And she's not here now."

"Do you know where I can find her?"

"I'm afraid not."

We were walking down the corridor together, which was strange because I couldn't remember starting. And we had passed into a second room, part of the library ... it was obvious now. I had never seen so many books. There were books on both sides of me, standing like soldiers, shoulder to shoulder, packed into wooden shelves that stretched on and on into the distance, finally – a trick of perspective – seeming to come together at a point. The shelves began at floor level and rose all the way to the ceiling, maybe a hundred rows in each block. The air was dry and smelled of paper. There must have been a million books in this room alone and each one of them was as thick as an encyclopaedia.

"You must like reading," I said.

"I never have time to read the books. I'm too busy looking after them."

"How many of you are there?"

"Just me."

"Who built the library?"

"I couldn't tell you, Matt. It was already here before I arrived."

"So what are these books? Do you have a crime section? And romance?"

"No, Matt." The Librarian smiled at the thought. "Although you will find plenty of crime, and plenty of romance for that matter, among their pages. But all the books in the library are biographies."

"Who of?"

"Of all the people who have ever lived and quite a few who are still to be born. We keep their entire lives here. Their beginnings, their marriages, their good days and their bad days, their deaths – of course. Everything they ever did."

We stopped in front of a door. There was a sign on it, delicately carved into the wood. A five-pointed star.

"I know this," I said.

"Of course you do."

"Where does this door go?"

"It goes anywhere you want it to."

"It's like the door at St Meredith's!" I said.

"It works the same way ... but there you have only twenty-four possible destinations. In your world, there are twenty-five doors, all connecting with each other – although none of them will bring you back here. This library, on the other hand, has a door in every room and I have absolutely no idea how many rooms there are and wouldn't even know how to count them."

The Librarian gestured with one hand. "After you."

"Where are we going?"

"Well, since you're here, why don't we have a look at your life? Aren't you curious?"

"Not really."

"Let's see..."

We went through the door and for all I knew at that moment we crossed twenty miles to the other side of the city. We found ourselves in a chamber that was certainly very different from the one we had left, with plate-glass windows all around us, held in place by a lattice-work of steel supports. Maybe this was one of the airport terminals I had seen. The books here were on metal shelves, each one with a narrow walkway and a circular platform that moved up and down like a lift but with no cables, no pistons, no obvious means of support.

We went up six levels and shuffled along the ledge with a railing on one side, the books on the other.

"Matt Freeman ... Matt Freeman..." The Librarian muttered my name as we went.

"Are they in alphabetical order?" I asked. All the volumes looked the same except that some were thicker than others. I couldn't see any names or titles.

"No. It's more complicated than that."

I looked back at the door that we'd come through. It was now below and behind us. "How do the doors work?" I asked.

"How do you mean?"

"How do you know where they'll take you?"

He stopped and turned to look at me. "If you just wander

through them, they'll take you anywhere," he said. "But if you know exactly where you want to go, that's where they'll take you."

"Can anyone use them?"

"The doors in your world were built just for the five of you."

"What about Richard?"

"You can each take a companion with you, if you're so minded. Just remember to decide where you're going before you step through or you could end up scattered all over the planet."

We continued on our way but after another couple of minutes, the Librarian suddenly stopped, reached up and took out a book. "Here you are," he said. "This is you."

I looked at the book suspiciously. Like all the others it was oversized, bound in some grey fabric, old but perhaps never read. It looked more like a school book than a novel or a biography. I noticed that it had fewer pages than many of the others.

"Is that it?" I asked.

"Absolutely." The Librarian seemed disappointed that I wasn't more impressed.

"That's my whole life?"

"Yes."

"My whole life up to now…"

"Up to now and all the way to the end."

The thought of that made my head swim. "Does it say when I die?"

"The book is all about you, Matt," the Librarian explained patiently. "Inside its pages you will find everything you have ever done and everything you will do. Do you want to know when

157

you next meet the Old Ones? You can read it here. And yes, it will tell you exactly when you will die and in what manner."

"Are you telling me that someone has written down everything that happens to me before it happens?" I know that was exactly what he had just said but I had to get my head around it.

"Yes." He nodded.

"Then that means that I've got no choice. Everything I do has already been decided."

"Yes, Matt. But you have to remember, it was decided by you."

"But my decisions don't mean anything!" I pointed at the book and suddenly I was beginning to hate the sight of it. "Whatever I do in my life, the end is still going to be the same. It's already been written."

"Do you want to read it?" the Librarian asked.

"No!" I shook my head. "Put it away. I don't want to see it."

"That's your choice," the Librarian said with a sly smile. He slid the book back into the space it had come from. But I had one last question.

"Who wrote the book?" I asked.

"There is no author listed. All the books in the library are anonymous. That's one of the reasons why it makes them so hard to catalogue."

I was beginning to feel miserable. The dreamworld seemed to exist to help us, but every time we came here it was simply confusing. Jamie and Pedro had both found this too. "You call yourself a librarian," I snapped at the man. "So why can't you be more helpful? Why don't you have any answers?"

He tapped the spine of the book. "All the answers are here," he said. "But you just refused to look at them."

"Then answer me this one question. Am I going to win or lose?"

"Win or lose?"

"Against the Old Ones." I swallowed. "Am I going to get killed?"

"We are experiencing some turbulence…"

The Librarian was still looking at me, but he hadn't spoken those words. With a sense of frustration, I felt myself being sucked away. There was someone leaning over me. A member of the cabin crew.

"I'm sorry I've had to wake you up," she said. "The captain has put on the seat belt sign."

I looked at my watch. We still had four more hours in the air. Richard and Jamie were asleep but I knew I wouldn't be able to join them. I took out my notepad and started writing again.

Four hours until London.

Soon we will be home.

CROSSING PATHS

Scarlett thought she'd be safe, back at school. She'd slip back into the crowd and nobody would notice her. After all, nothing exciting ever happened at school. Wasn't that the whole point? So, for the first time in her life, she found herself looking forward to the next Monday morning. There would be no bombs, no strange men in cars, no cryptic messages. She would be swallowed up by double maths and physics and everything would be all right.

But it didn't happen that way.

Shortly before lunch, she was called into the headmistress's office. There was no explanation, just a brief: "Mrs Ridgewell would like to see you at twelve fifteen." Scarlett was nervous as she climbed the stairs. In a way, she'd been expecting trouble ever since the trip to St Meredith's. She had been the centre of attention for far too long and for all the wrong reasons. Her work had gone rapidly downhill. She'd been told off twice for daydreaming in class. And then there had been that terrible maths test. The teachers had already decided that all the publicity had gone to her head and Scarlett fully expected Mrs Ridgewell to read her the riot act. Get your head down.

Pull your socks up. That sort of thing.

But what the headmistress said came right out of the blue.

"Scarlett, I'm afraid you're going to be leaving us for a few weeks. I've just had a phone call from your father. It seems that some sort of crisis has arisen…"

"What crisis?" Scarlett asked.

"He didn't say. He was very mysterious, if you want the truth. But he wants you to join him immediately in Hong Kong. In fact, he's already arranged the flight."

There was a moment's silence while Scarlett took this in. There were all sorts of questions that she wanted to ask, but she began with the most obvious. "Has this got something to do with what happened to me?"

"I don't think so."

"Then what?"

"He didn't say." Mrs Ridgewell sighed. She had taught at St Genevieve's for more than twenty years and it showed. Her office was cluttered and a little shabby, with antique furniture and books everywhere. A Siamese cat – it was called Chaucer – lay asleep in a basket in a corner. "You haven't had a very good term, have you Scarlett?"

"No." Scarlett shook her head miserably. "I'm sorry, Mrs Ridgewell. I don't know what's going on, really. Everything seems to have gone wrong."

"Well, maybe we should look on the bright side. A complete break for a few weeks might do you good. I'll ask your teachers to prepare some work for while you're out there – and, of course, we're going to have to recast the Christmas play. I have to say that it is all very inconvenient."

"Didn't he say anything?"

"I've told you everything I know, I'm afraid. I thought he would have discussed it with you."

"No. I haven't heard from him."

"Well, I'm sure there's nothing to worry about. He told me he'd ring you tonight. So you've just got time to say goodbye to your friends."

"When am I leaving?"

"Your flight is tomorrow."

Tomorrow! Scarlett couldn't believe what she was hearing. Tomorrow was only a few hours away. How could her dad have done this to her? He hadn't mentioned anything when they were in the Italian restaurant. What crisis could possibly have arisen in less than a week?

Scarlett spent the rest of the day in a complete daze. Her friends were equally surprised, although the truth was that she was beginning to get a bit of a reputation. She was weird. First the church and now this. She didn't even get to see Aidan. She looked for him on the way home and tried texting him, but he didn't reply. Mrs Murdoch had already heard the news. She had started packing by the time Scarlett got home. And she didn't seem pleased.

"Not a word of warning," she muttered. "And no explanation. What do you suppose I'm meant to do, sitting here on my own?"

Paul Adams rang that night as he had promised, but he didn't tell Scarlett anything she wanted to know.

"I'm really sorry, Scarly..." His voice on the line was thin and very distant. "I didn't want to do this to you. But things have happened … I don't want to explain until I see you."

"But you've got to tell me!" Scarlett protested. "Is Mum all right? Is it you?"

"We're both fine. There's nothing for you to worry about. It's just that there are times when a family has to be together and this is one of them."

"How long am I staying with you?"

"A couple of weeks. Maybe longer."

"Why?" There was silence at the other end of the line. "Can't you tell me anything?" Scarlett went on. "It's not fair. It's the middle of term and I'm going to miss the school play and all the parties and everything!"

"Look, I'm just going to have to ask you to trust me. You'll be here in twenty-four hours and I want to explain everything to you face to face, not over the phone. Can you do that for me, Scarly? Just wait until you get out here … and try not to think too badly of me until you arrive."

"All right." What else could she say?

"I've booked you into business class, so at least you'll be comfortable. Make sure you bring lots of books It's a long flight."

He rang off. Scarlett stood there, holding the receiver. She was feeling resentful and she couldn't stop herself. This wasn't fair. She was being bundled onto a plane and flown to Hong Kong as if she were a parcel being sent by Fed-Ex. She was fifteen years old. Surely she should have some control over her own life?

The taxi came at midday. Scarlett's flight was leaving Heathrow at half past three. Mrs Murdoch helped carry the cases out and load them into the back and the two of them got in together. The housekeeper was coming with her as far as the airport and would then return to the house alone. It was a grey, overcast day and the weather reflected Scarlett's mood. She twisted round as they pulled away and watched the house disappear behind her. She knew she was only going to be abroad for a couple of weeks but even so she couldn't escape a strange feeling. She wondered if she would ever see it again.

They reached the bottom of the street and were turning left into Half Moon Lane. And that was when it happened. A car crash. Scarlett only saw part of it and it was only later that she was able to piece together what had happened. A car had been driving towards them – it had just come from the main road – and a second car, a BMW, had suddenly pulled out in front of it. Scarlett heard the screech of tyres and the smash of impact and looked up in time to see the two cars ricocheting off each other, out of control. One of them had been forced off the road and was sliding down a private driveway. She could make out at least three people inside.

"London traffic!" The taxi driver sniffed. He completed the turn and they picked up speed.

Scarlett twisted round and looked out of the back – at the crumpled bonnet of one of the cars, steam rising into the air, glass scattered on the road. A bus had been forced to stop and the driver was climbing down, perhaps to see if he could help. The accident was already disappearing into the distance behind them and she supposed it was just a coincidence. It couldn't mean anything.

But even so it made her uneasy. It reminded her of the

moment – two years ago, and just a short distance away – when she had almost been killed. And that made her think of the man who had contacted Aidan, wanting to meet her at the restaurant that had been blown to pieces before she could arrive. Scarlett sank back into her seat, feeling anxious, unable to control what was happening to her. Mrs Murdoch gazed out of the window with no expression on her face.

They parted company at the airport. Scarlett was flying as an unaccompanied minor – what the airline called a Skyflyer Solo. She had to suffer the indignity of a plastic label around her neck before she was led away. She said goodbye to Mrs Murdoch, hugging her awkwardly. Then she picked up her hand luggage and headed for the departure gate.

It had been so close. None of them would ever believe just how close it had actually been.

Matt Freeman had landed at the same airport earlier that morning. There had been a uniformed chauffeur waiting for him and the others, and soon they were sitting in the air-conditioned comfort of a new Jaguar, being driven to their hotel. Richard was dozing in the front seat. He had spent much of the flight working on the diary and had barely slept at all. Jamie was looking out for his first sight of the city. Matt could see that so far he was disappointed. They were driving through a wasteland of blank, modern warehouses and unwelcoming hotels – the sort of places that always surround airports – and Matt wanted to tell him that this wasn't London at all.

But then, twenty minutes later, they turned off the motorway and suddenly they were in the city itself, passing the Natural History Museum in Kensington – it was still closed for repairs following Matt's last visit there – then the Victoria and Albert Museum, Harrods and Hyde Park Corner. Jamie stared, open-mouthed. He had spent much of his life in the desert landscape of Nevada and he wasn't used to seeing anything that was actually old. For him, London with its monuments and palaces was another world. He saw red buses, pigeons, policemen in blue uniforms, taxis... It was like falling into a pile of picture postcards. His one disappointment was that Scott wasn't with him. The two brothers had never been so far apart.

The driver took them to a hotel in Farringdon, a quiet part of London with narrow streets and a meat market that had been around when the animals were driven there in herds rather than delivered from Europe, pre-packed in boxes. The Tannery, as it was called, was small and anonymous – Richard and Matt had stayed there before. It was just a few minutes away from the private house where the Nexus met. By the time they arrived, it was eleven o'clock. A meeting had been arranged for half past seven that evening, giving them the rest of the day to relax and unwind from the long flight.

They made their way into a reception area which was like the front room of someone's house, with thick carpets, flowers and the comforting tick of a grandfather clock. The receptionist was a tight-lipped woman who took care not to give too much away. She glanced disapprovingly at Richard – still in his Hawaiian shirt, looking more like a beach bum than ever – and the two boys who were with him, then asked for their passports and slid forward some forms for them to sign.

"How many nights?" she asked.

"We're not sure," Richard said.

"Two rooms. I see they've been prepaid..."

The telephone rang. The receptionist plucked the receiver as if it were an overripe fruit and held it to her ear. "The Tannery Hotel," she said. A moment's silence. Her eyes fluttered and she handed the phone to Richard. "It's for you, Mr Cole."

Richard took the phone. Whatever he was hearing, it wasn't good news. He muttered a few words, then put the phone down.

"What is it?" Matt asked.

"Scarlett Adams... She's leaving London."

"What?" Matt couldn't believe what he had just said. "Where's she going?"

"We can still catch her." Richard looked at his watch. "She's going to Hong Kong. She's booked on the three thirty flight..."

"Not back to Heathrow!" Jamie groaned.

"No." Richard weighed up the options. He was finding it hard to concentrate. He needed a shave more than ever and his eyes were red with jet lag. "We can't intercept her at Heathrow," he said. "It's too public. She's never met us. She might not even want to talk to us. But her taxi isn't collecting her until midday. We can reach her before she leaves."

The decision had been made. The three of them dumped their luggage with the receptionist, turned round and walked out again. Fortunately, the driver was still waiting. Richard went up to him and told him where they wanted to go. The driver didn't argue. Matt and Jamie got back in again.

They hadn't even seen their rooms. The next moment they were off again, threading their way through Farringdon and down to Blackfriars Bridge. But it was now approaching the lunch hour and London had changed. Although they had made good progress from the airport, the traffic had snarled up. Every traffic light was red. It felt as if the entire city had turned against them.

"Who was it on the phone?" Matt asked.

"Susan Ashwood. She's already in London."

Miss Ashwood was a medium who also happened to be blind. Matt had first met her in Yorkshire and it had been she who had introduced him to the Nexus.

"How did she know?" Matt asked.

"The Nexus are still bugging Scarlett's phone. They had two people following her too…"

It didn't look as if they were going to make it. The whole of South London had become one long traffic jam. The car crossed Tower Bridge – giving Jamie a quick glimpse of the River Thames and St Paul's – but after that, the city just felt drab and overcrowded with an endless stretch of cheap shops and restaurants punctuated by new office developments that would have looked out-of-date the moment they were built. Bermondsey, Walworth, Camberwell … they crawled from one district to the next without ever noticing where one ended and the next began and all the time they were aware of time ticking away. Half past eleven, twenty to twelve … they didn't seem to be getting any nearer.

"This is hopeless," Richard said. "Maybe we'd better go to Heathrow after all."

The driver shook his head. "We're nearly there," he said.

They dropped down a steep hill – Dog Kennel Hill, it was called – and, looking out of the window, Matt began to feel something very strange. He had never visited this part of London ... he was sure of it. And yet, at the same time, he knew where he was. He glimpsed a radio mast in the distance, a road sign pointing to King's College Hospital. They meant something to him. He had been here before.

And then it hit him. Of course he knew this part of the city. He had lived here – from the time when he was a baby to when he had been about eight years old.

He should have remembered it. It hadn't been that long ago. But perhaps he had blocked it out. It wouldn't have been surprising after everything he had been through. Now it all came flooding back. The mast belonged to Crystal Palace. He had often played football there. He had gone into the hospital on his seventh birthday with suspected food poisoning. He remembered sitting miserably in reception – short trousers – with a plastic bowl balanced on his knees. They drove past a very ordinary house but Matt knew at once who lived there. It was a boy called Graham Fleming who had been his best friend at school. The two of them had always thought they would be inseparable. Matt wondered if he was still living there. What would he say if the two of them met now?

And there was something else he remembered. If he went past Graham's house, turned the corner and walked past the old scout hut, he would come to a small, terraced house in a leafy street where all the houses were small and terraced. Number 32. It would have a green door and – unless they'd finally mended it – a cracked front step. It was his home. That was where he had once lived.

"How much further?" Richard asked.

The driver glanced at his sat nav. "We're a minute away," he said.

They went through a traffic light at a busy junction, then drove up towards North Dulwich station, turning into Half Moon Lane which was just opposite. Matt felt dazed. It was extraordinary to think that for half their lives, he and Scarlett had almost been neighbours. They might have passed each other a dozen times without even knowing it. She lived in Ardbeg Road, which was the next on the left, and just for a moment the way ahead was clear. The driver accelerated, glad finally to be able to use the Jaguar's power.

"Look out!" Richard shouted out the warning too late.

A car shot out from a private drive and smashed right into them.

Matt saw everything. He heard the roar of an engine and that made him turn his head. The car was coming straight at them. The driver was staring at them, his hands clenched on the wheel, not even trying to avoid them. He was middle-aged, clean-shaven – and there was no emotion in his face. He should have been scared. He should have been showing some sort of reaction, knowing what was about to happen. But there was nothing at all.

Half a second later, there was a huge crash of metal against metal as he smashed into them.

The other car was a four-by-four, a BMW, and it was like

being hit by a tank. The Jaguar was swept off the road, the world tilting away as it was hurled towards a wide, modern house with a short driveway sloping steeply down to the front door. There was a second collision as it hit the door, more crumpling metal. The house alarms went off. Jamie cried out as he was thrown sideways, his head hitting Matt's shoulder. Matt tasted blood and realized that he had bitten his tongue. The Jaguar was lying at an angle, almost underneath the front wheels of the BMW which was still on the road above them. Both the windows on the driver's side had shattered. The engine had cut out.

For a moment, nobody moved. Then Richard swore – which at least meant he was alive. He twisted round in the front seat. "Are you two all right?" he asked.

"What happened?" Jamie groaned.

"An accident…" Richard said. "Idiot … wasn't looking where he was going."

He was wrong. Matt knew that already. He had seen what had happened. The BMW driver had been waiting for them, knowing they would come this way. Why else would he have shot out like that, slamming straight into them? Matt had seen him, gripping the wheel. He had known exactly what he was doing.

Richard was already out of the car.

"Wait…" Matt said.

But Richard hadn't heard. He staggered up onto the road, only now becoming aware that he was in pain. There were no cuts or bruises but, like all of them, he had suffered from whiplash.

171

"What the hell do you think you were doing?" he demanded.

The driver of the BMW had got out and was standing in the road. He was a middle-aged man, well-built, wearing a long, black coat and leather gloves. His mouth was soft and flabby, with small teeth, like a child. His skin was very pink. He had curly hair. His head was almost perfectly round, like a football.

"I'm so terribly sorry," he said. "I didn't see you. I was in a hurry. I hope none of you are hurt."

Richard was still angry but he suddenly knew something was wrong. "You did it on purpose," he said. His voice had faltered. "You tried to kill us."

"Not at all. I just pulled out without looking. I can't tell you how sorry I am. Thank goodness you don't seem to be seriously hurt."

By now, Matt and Jamie had joined him. There was nothing they could do for their driver and they left him, unconscious in the front seat. Jamie stared at the man and the colour drained out of his face. He knew at once what he was looking at. It was the last thing he had expected to find here.

"Matt..." he whispered. "He's a shape-changer."

Matt didn't doubt him. Jamie had met shape-changers when he had gone back in time. Shape-changers were able to take on human form but it didn't suit them. It didn't quite fit. One of them, an old man who had suddenly become a giant scorpion, had almost killed him at the fortress at Scathack Hill. He knew what he was talking about. And Matt could see it for himself. Everything about the BMW driver was fake, even the way he stood there, stiff and unnatural, like a dummy in a shop

172

window. The words he was saying could have been written out for him, on a script.

"I'm insured," he continued. "There's absolutely nothing to worry about. It was my fault. No doubt about it."

Richard stared. None of them knew quite what to do. Barely a minute had passed since the collision but already other people were arriving on the scene. A bus, on its way to Brixton, had pulled up and the driver was climbing out of his cabin, coming over to help. Two more cars had stopped further up the road. Matt had seen a taxi pull out of Ardbeg Road and thought it might be coming their way, but it had already turned off and driven away.

They couldn't risk a fight. They were in the middle of a suburban, South London street. If they challenged the shape-changer, if he decided to drop his human form, all hell would break loose. And already the police had arrived. A squad car turned the corner and pulled over. Two officers got out.

"Good afternoon, officers." The BMW driver was pretending that he was pleased to see them. "Glad you're here. We're in a bit of a pickle."

His language was as fake as the rest of him and for just a few seconds, Matt was tempted to take him on, to show the entire crowd what was really happening here. He could use his own power. Without so much as moving, he could tear a strip of metal off the shattered car and send it flying into him. There were a dozen witnesses on the scene. How would they react when the blushing, curly-haired BMW driver turned into a half-snake or a half-crocodile and bled green blood? Maybe it was time to show the world the war that was about to engulf it.

It was Richard who stopped him.

"No, Matt."

He must have seen what Matt was thinking because he muttered the two words under his breath, never taking his eyes off the man who was standing in front of them. Matt understood. For some reason, the shape-changer was playing with them. It was pretending that this was just an ordinary accident. If he took it on, if he began a fight here in the street, innocent people might get hurt. And he was in England with a fake passport and a false name. This was the wrong time to be answering questions. Right now he had everything to lose.

"I'm so very sorry," the shape-changer said.

"I saw what happened!" the bus driver exclaimed. He nodded at the BMW driver, his face filled with outrage. "He pulled out at fifty miles an hour. He didn't look. He didn't signal. It was all his fault."

"Is anyone hurt?" one of the policeman asked.

"Our driver," Richard said.

The right-hand side of the Jaguar had taken the full force of the impact and it looked as if the driver had suffered a broken arm. He was only semi-conscious and in pain. One of the policemen helped him out and laid him on the pavement and they waited about fifteen minutes for an ambulance to arrive. Meanwhile the other officer began questioning the BMW driver – "Mr Smith". He had no ID.

"I was on my way to Chislehurst. I'm a piano teacher. I pulled out without looking. I can't tell you how dreadful I feel…"

Matt watched as they breathalysed him and it almost made him smile, seeing him blow into the machine. His breath wasn't human and if he'd drunk a crate of whisky it was unlikely that it would register. Meanwhile, their driver was loaded into an ambulance and driven off to hospital. Thirty minutes or more had gone by and Richard was desperate to be on his way, but the police weren't having any of it. They would have to take a statement. Will you come with us to the station, sir? There was no way out. Richard, Matt and Jamie were driven away.

It was almost four o'clock by the time the police finished with them. Even if they had wanted to go to Heathrow, it would have been too late. Scarlett would already be in the air, on her way to Hong Kong.

They left the police station and dropped into a local café but Matt refused the offer of a drink. He was angry and depressed. The Old Ones were out-manoeuvring him at every turn. They seemed to know exactly what he was going to do and the trap they had set had been childishly simple. He didn't mention the taxi that he had seen pulling out of Ardbeg Road, but it had already occurred to him that Scarlett might well have been inside it. Their paths had finally crossed ... but seconds too late.

"Let's go to her house," Matt suggested.

"Why?" Richard didn't even look up from his tea.

"I don't know. She could still be there. But even if she isn't, now that we've come this far..."

Neither Richard nor Jamie spoke.

"I'd just like to see where she lives," Matt said.

The three of them walked back to Ardbeg Road. It reminded Matt a little of the street where he had once lived. All the

175

houses were terraced with bay windows, neat front gardens and shrubs to hide the wheelie bins. Scarlett's was about half-way down.

They rang the bell, not expecting it to be answered, but after about half a minute the door opened and they found themselves being examined by a short, stern-looking woman with black hair tied back and eyes that seemed to be expecting trouble.

"Yes?" she said. She had a Scottish accent.

"We're looking for Scarlett Adams," Matt said.

"I'm afraid you've missed her. She left this morning."

Richard moved forward. "Do you live here?" he asked.

"Yes. I'm the housekeeper. Are you friends of Scarlett's?"

"Not exactly," Matt said. "We've just arrived from America. We were hoping to see her."

"That's not going to be possible. She's going to be out of the country for a while."

"Do you know when she'll be back?"

"It could be a week or two. I'm very sorry, if you'd been here just a few hours ago, you'd have caught her. Do you want to leave a message?"

"No, thank you."

"Right."

The woman closed the door.

And that was it. There was nothing more to be done. For a moment, nobody spoke. Then Richard sighed. "Anyone fancy a trip to Hong Kong?" he said.

PUERTO FRAGRANTE

Originally, there had been twelve members of the Nexus – the organization that existed only to fight the Old Ones. Professor Sanjay Dravid had been the first to be killed, stabbed at the Natural History Museum the same night that he had met Matt. Later on, a man called Fabian had also died. That just left ten – powerful people who lived in America, Australia, Europe ... all over the world.

They had all flown in to meet Matt and Jamie and at half past seven that evening they came together in the secluded, wood-panelled room which was their London base.

The building, which the Nexus owned, stood between two shops and there was nothing, no name or other marking, to suggest that it was anything but a private house. The room itself, up on the first floor, was equally plain. It could have been the meeting place of some small business, perhaps a firm of expensive solicitors. There didn't seem to be much there – just a long table with thirteen antique chairs, a handful of telephones and a computer and a lot of clocks showing the time all over the world. But the glass door that slid open automatically and then hissed shut, sealing itself as the ten men and women came in, suggested that there might be more to the place than met the eye. A sophisticated camera blinked quietly in the corridor. And the Nexus arrived one at a time, each one entering a different six-digit code before they were allowed in.

Matt wasn't looking forward to seeing them again. He knew that they were supposed to be on his side, but even so he felt a certain dread entering the room. It was like facing ten head teachers at the same time, knowing he was about to be expelled. There were only two people there who he felt he knew. He had met Susan Ashwood, the medium, at her home near Manchester, and although he had thought she was completely mad, at least he was fairly sure that her heart was in the right place. And he had got to know Nathalie Johnson in the past few months. She was the American computer billionaire who had helped Scott and Jamie and she had travelled down to Nazca a couple of times to make sure they were all right.

But that still left eight strangers. There was an Australian, broad and bullish with a round face and close-cropped hair. His name was Harry Foster and he owned a newspaper empire. Next to him, there was a bishop who dressed like a bishop and talked like a bishop but who hadn't actually told Matt his name. He was about sixty years old. Tarrant, the senior policeman who had helped put taps on Scarlett's phone, was at the head of the table, dressed in a smart blue and silver uniform.

Among the others, Matt had noted a Frenchman in an expensive suit, a small Chinese man who was continually rubbing his hands, a German who was something big in politics and two others who had made no impression on him at all.

They might all be world leaders. But tonight they just looked tired and scared.

Richard, Jamie and Matt had taken their places at the table, bunched together at one end. The three of them were in a gloomy mood. Scarlett Adams, the fifth Gatekeeper, had turned up in London and they had flown thousands of miles to see her. But she had slipped through their fingers and even as they sat there, she would be thirty thousand feet up in the air. Every word that they spoke, every second that passed, only carried her further away.

"We made a mistake." Nathalie Johnson came straight to the point. "We knew who she was. We knew where she lived. We should have approached her ourselves."

"It was my fault," Susan Ashwood said. "I didn't want to frighten her. I thought it would be easier for her if she heard it all from you." She turned to Matt. "I hoped you'd be here sooner. I didn't realize we'd have to wait for the new passports."

"I thought you had people watching her," Matt cut in. "Weren't there two private detectives or something?"

"They were ex-policemen," Tarrant said. "Duncan and McKnight. Good men, both of them. I'd worked with them before." He paused. "Scarlett may have caught sight of them. They were parked in a car outside a park in Dulwich and they had to be more careful after that. They kept their distance. But they were still on top of the case. Until last night..."

"What happened" Richard asked.

"They've both disappeared. Vanished without a trace. I've tried to contact them but with no luck. I have a feeling they may have been killed."

There was a brief silence while the rest of the room took this in. It was obvious to all of them that they had underestimated the Old Ones. From the moment Scarlett had been identified, they had been running rings around the Nexus.

"So why has she gone to Hong Kong?" Matt asked.

"Her father is there," Tarrant replied. "He's a lawyer. He works for the Nightrise Corporation."

"Nightrise?" Jamie spoke for the first time. Jet lag had hit him badly and he was exhausted. He'd only managed to keep himself awake with a black coffee and a can of Red Bull. "They're the people who came after Scott and me. Are you saying her dad is one of them?"

"Nightrise is a legitimate business," Nathalie Johnson reminded him. "They have offices all over the world. They employ hundreds of people. The vast majority of them probably have no idea who – or what they're working for."

"Even so..."

"We don't know, Jamie. His name is Paul Adams. He's divorced. He and his wife adopted Scarlett fifteen years ago and as far as we can tell, he doesn't know anything about the Old Ones."

"So what do we do now?" Richard asked. "Scott and Pedro are still in Peru. Matt and Jamie are here. And Scarlett will soon be in Hong Kong. The one thing we know is that we have to get the five Gatekeepers together. How are we going to do that?"

"You may have to follow her there."

It was the bishop who had spoken and the other members of the Nexus nodded. But for his part, Matt wasn't so sure. He knew nothing about the city except that some of the toys he'd played with when he was younger had been manufactured there. MADE IN HONG KONG: it had always been a sign that they would probably break five minutes after they came out of the packaging. Certainly, he had no desire to go there. He had flown enough for one week.

"If I may..." The Chinese man had a soft, very cultivated voice. He hadn't spoken until now. He was small, with heavy, plastic glasses and an off-the-peg suit. Perhaps he adopted this sort of appearance on purpose. It was as if he didn't want to be noticed. "My name is Mr Lee," he said, bowing his head towards Matt. "If you are thinking of making the journey to Hong Kong, I may be able to help you. I have connections throughout Asia and especially in that area. However, I would like to make one observation if I may."

He waited for someone to speak against him, as if he was nervous that there might be someone at the table who didn't want to hear what he had to say. When nobody protested, he went on.

"There is something very strange happening in Hong Kong," he began. "I know the place well. In fact, I was there – passing through – just a week ago. On the face of it, there is nothing I can put my finger on. Life continues as normal. Business is done. Tourists arrive and leave. But there is something in the city that makes no sense. How can I put it? There is an atmosphere there that is not pleasant. Friends of mine who live there, people I have known for many years, seem to be in a hurry to leave and when I ask them why, they are afraid to say. And those who remain are nervous."

"The Old Ones are there," Susan Ashwood said, as if she had known all along. She worked as a medium, talking to ghosts. Matt wondered if they had told her.

"That is what I believe, Miss Ashwood," Mr Lee agreed. "It is hardly a coincidence. Nightrise is based in Hong Kong. It is quite possible that much of the city is now in the control of the Old Ones. And if that is the case, then the moment this girl, Scarlett Adams, arrives there, it will be as if she is in prison and none of us will be able to reach her."

"We have to reach her," Richard said. "If we don't, we might as well all pack up. There have to be five Gatekeepers."

"Then we have to get her out of there – and that means following her. We have failed here in London. Maybe Matthew and Jamie will have more success over there."

"You want to send the two of them to Hong Kong?"

"They have certain powers, Mr Cole, which may be of use to them." Mr Lee nodded. "Yes. In my opinion they must find a way to enter the city, but without the Old Ones knowing they are on their way."

"The two of them travelled here with false names and false ID," Tarrant said. He sounded disapproving. "They can use them again."

"Absolutely." The Australian, Harry Foster banged a fist on the table. "They could be on the next flight out of here. There must be fifty thousand people a day flying in and out of Hong Kong. Who's going to notice a couple of kids in a crowd like that?"

"I don't agree." Susan Ashwood shook her head. "If Mr Lee is correct and the Old Ones are there, it would be complete madness to attempt to go in by air. Matt and Jamie would be seized the moment they stepped off the plane ... and I don't care how many people there are at the airport."

"I have an office in Hong Kong," Harry Foster said. "I could look in there on my way back to Australia. Why don't you let me try and find her? I can explain what's going on and she – and her father, for that matter – can leave with me. I'll take them down to Sydney and you can pick them up there."

"I think it's too dangerous," Mr Lee said.

"Well, at least I can get a message to her. Let her know the score." The Australian took out a pad and scribbled a note to himself "A letter to warn her that she's in danger. I can get someone in my Hong Kong office to deliver it by hand."

"I think we have to be very careful," Susan Ashwood said. "We all know what happened today. The Old Ones were waiting outside her house in Dulwich. They knew Matt was on his way and they were determined to stop him." She glanced at Tarrant. "You had two men watching Scarlett and now you say they may have been killed. How many more mistakes do we have to make before we realize what we're up against?"

"Then maybe it's time to use one of the doors," Richard said.

He had the diary and he slid it onto the table in front of him. All ten members of the Nexus stared at it. Only a few months before, they had been prepared to spend two million pounds to get their hands on it and here it was, right in front of them. They wanted to reach out and touch it. And yet at the same time they were afraid of it, as if it was a snake that might bite.

"I've been trying to work this out ever since Ramon brought

it to us," Richard went on. "I've read bits of it, though I won't pretend I've understood very much … even with a Spanish dictionary and a magnifying glass. But there is one thing we do know. Twenty-five doors were built around the world for the Gatekeepers to use. They all connect with each other and they can all be found in sacred places. One of them is in St Meredith's. When Matt went through it, it took him directly to the Abbey of San Galgano in Tuscany."

"Scott and me found one of the doors in a cave at Lake Tahoe," Jamie added. "It took us to the Temple of Coricancha in Cuzco, Peru."

"That's four of them," Richard said. "But there are twenty-one more and our friend, the mad monk, may have helped us. He's made a list…"

He unfastened the diary and opened it, laying it flat so that everyone could see. Everyone leaned forward. There was a very detailed map covering two pages, drawn in different colours of ink. It was just about recognizable as the world, although a world seen by a child with only a basic knowledge of geography. America was the wrong shape and it was too close to Europe. Australia was upside-down.

Joseph of Cordoba had used more care decorating his work. He had sketched in little ships, crossing the various oceans with their sails unfurled. Insect-sized animals poked out of the different land masses, helping to identify them. There was a tiger in India, a dragon in China and, at the North Pole, what could have been a polar bear.

"I don't know how much you know about old maps," Richard said, "but for what it's worth, I studied them a bit at university.

I did politics and geography. This one is fairly typical of the six-teenth century. That was a time when maps were becoming more important. Henry VIII was one of the first monarchs to realize how much they could give away about a country's defences. And everyone was using them to steal everyone else's trade routes. You see these little bags here?" He took out a pencil and pointed. "They're probably bags of spice. Joseph may have drawn them to represent the Spice Islands because that was what everyone wanted."

"There are stars," Jamie said.

They were scattered all over the pages; the five-pointed stars that he and Matt knew so well.

"That's right. There are twenty-five of them – one for each door. The only trouble is, like a lot of the maps being drawn at the time, this one isn't very accurate. As far as I can make out, there seem to be doors in London, Cairo, Istanbul, Delhi, Mecca, Buenos Aires and somewhere in the outback of Southern Australia. There's one here, close to the South Pole. But the world's changed quite a lot in five hundred years and trying to identify the exact locations isn't going to be easy."

"You mentioned a list," Tarrant said.

"Yes…" Richard turned a page and sure enough there was a long row of names, all of them in tiny handwriting. "The prob-lem we've got here is that the names don't quite match up with the modern places and half of them are in Spanish. Here's one, for example. Muerto de Maria. It took me half the night to work that one out."

"The death of Mary," the bishop translated.

"Or Mary's death," Richard said. "Do you get it? Marydeath.

Or the church of St Meredith in London. It's like a crossword clue although I don't suppose Joseph was doing it on purpose to confuse us. Coricancha isn't named at all. It's just represented by a flaming sun – but then, of course, the sun was sacred to the Incas."

"Is there a door in Hong Kong?" Matt asked.

"There's certainly a door somewhere nearby," Richard said. He turned the page back to the map. "You can see it here ... and if you look at the list, there's a reference to a place called Puerto Fragrante and a little dragon symbol. But that could be anywhere."

"May I see?" Mr Lee reached out and took the diary in both hands, holding it as if he was afraid it was about to crumble away. He looked at the map, then the list, then turned another page. "Someone has written in pencil," he said. "The words 'Tai Shan'." He glanced at Richard. "Was that you?"

Richard shook his head. "That must have been Ramon," he said. "He made notes all over it when he was trying to decipher it for Salamanda, but as far as I can see, he didn't have time to work out too much. Anyway, he was mainly focusing on the Nazca Lines."

"There is a door in Hong Kong!" Mr Lee exclaimed. "I can tell you that for certain. And I can even tell you exactly where it is." He laid the diary down. "Puerto Fragrante – the Spanish for Fragrant Harbour, I think – is another clue," he said. "In Cantonese, Fragrant Harbour translates as Heung Gong. Or in other words, Hong Kong. The city was originally given that name because of the smell of sandalwood that drifted across the sea. Whoever studied the diary has been good enough to confirm it for us. *Tai Shan* means 'the mountain of the East'. It is where the sun begins its daily journey. It is also the place where human souls go when they die. There is a very old and very sacred temple with that name in Hong Kong, in a part of the city called Wan Chai..."

There was a sense of relief in the room. It was as if they had all made their minds up. Even Susan Ashwood nodded her head in agreement and seemed to relax. Only Matt didn't look so sure.

"You could leave tonight," Harry Foster said. "If things went your way, you could actually be there to meet her at the airport. You could pull her out before the Old Ones even knew you'd arrived."

"Wait a minute," Matt said. "We flew here from Miami because we didn't think the doors were safe. Why has anything changed?" Nobody answered so he went on. "Salamanda had the diary. He'll have found out about the temple..."

"Not necessarily," Foster insisted. "This guy, Ramon, was working on it. But he may not have passed on everything he knew. Anyway, Salamanda's dead."

"Maybe there is an element of risk..." Susan Ashwood began.

"It's more than a risk. It's a trap."

Matt hadn't sat down and worked it out. It was just that all the doubts that had been in his mind had somehow come together and he could suddenly see everything very clearly.

"The whole thing is a trap," he said. "And it always has been, right from the start. Why were we attacked in Nazca? Why was Professor Chambers killed? It's because the Old Ones wanted to get us on the move. They wanted us to do exactly what we've done.

"Think about it. Scarlett Adams goes through the door at St Meredith's and suddenly the whole world knows about her. She's in all the newspapers and the Old Ones find out who she is. And then, the very next day, a university lecturer called Ramon turns up in Nazca. Somehow he's managed to track us down. He tells us that he's managed to steal the one thing we most want and he hands it across without even asking for money. Why? Because he goes to Church! Because he's planning to get married! His whole story was ridiculous. And it wasn't true. The Old Ones *wanted* us to have the diary."

"They killed him to get it back," Nathalie said.

"Did they? I think Ramon was as surprised to get that fence post through his chest as we were to see it happen. He must have been programmed – either drugged or hypnotized – to stop Scott and Jamie seeing into his mind. And then they killed him to make us believe that he had been telling the truth. Otherwise, it would have all seemed too easy."

Matt took a breath. Normally, he didn't like being the centre of attention but this time he knew he was right.

"All along, there was something that bothered me about that night in Nazca," he went on. "If they really wanted the diary back so badly, why did they send such a small force? What happened to the giant spider, the fly-soldiers, the shape-changers, the death-riders?" He turned to Jamie. "You've seen them. You've fought them. Nazca was peanuts compared to what you went through."

Jamie nodded but said nothing.

"They want me to come to Hong Kong. That's what this has all been about." Matt was getting tired. He had no idea what time it was according to his body clock. He just wanted to crawl into bed and forget everything for ten hours. "First of all they got us out of Nazca. They managed to split us up. And now they've given us a nice invitation to walk straight into their hands. The moment I go through that door, I'll be finished. They're using Scarlett to get at me. I hurt them. I wounded their leader, Chaos – the King of the Old Ones, or whatever he calls himself. They want to make me pay."

There was a long silence.

"What do you want to do, Matt?" Susan Ashwood asked. And that made a change. Normally the Nexus told him what they wanted him to do.

"I still have to go to Hong Kong," Matt said.

"Matt..." Richard began.

Matt stopped him. "What Miss Ashwood said was right. They're not going to let any of you get anywhere near Scarlett. It has to be the two of us, Jamie and me. And you too, Richard, if you want to come. But maybe we can use this situation to our advantage. The Old Ones expect us to turn up in the Temple of Tai Shan. That's how they've arranged the trap. But suppose we arrive another way? We could still take them by surprise."

"You could go in by sea," Foster said. "There are cruise ships going in and out of Hong Kong all the time."

"May I suggest something?" Mr Lee interrupted, asking permission again. "The best way to enter Hong Kong might be through Macau. It is part of China, a small stub of land on the South China Sea – and like Hong Kong it is a Special

Administrative Region, which is to say, it is – at least in part – independent. You can fly from one to the other in a very short time. Helicopters make the journey several times a day."

"And how do we get to Macau?" Richard asked.

"You cannot fly there direct. I believe you will have to go via Singapore. But it is, if you like, a back door into Hong Kong – and one that the Old Ones may have overlooked." He took out a handkerchief and polished the lenses of his glasses. "More than that, I have a connection in Macau who may agree to help you. He has many resources. In fact, if anyone knows the truth about what is going on in that part of the world, it will be him."

"Wait a minute..." Richard was worried and he didn't try to hide it. He was wishing he'd never mentioned the diary in the first place. "Matt ... are you really sure you have to go there?" he asked. "You've already said that it's you that they want. You say it's a trap. Now you're walking straight into it."

"We need Scarlett," Matt replied, simply. "They have her. We can't win without her." He looked round. "Jamie, will you come with me?"

Jamie shrugged. "I've always wanted to see Hong Kong."

"Then it's agreed." Matt turned back to Mr Lee. "How quickly can you get in touch with your friend?" he asked.

"His name is Han Shan-tung," Lee replied. "He is a man with great influence. He has many friends inside Hong Kong. But it may not be easy to find him. He travels a great deal. You may have to wait."

"We can't wait."

"It will just be a few days. But trust me. It would be foolish to enter the city without his support."

A few days. More waiting. Matt thought about Scarlett. In a few hours' time she would be landing in Hong Kong. What would she find when she got there? How would she manage on her own?

But there was no other way. Somehow she would have to survive until he got there. He just hoped it wouldn't be too long.

WISDOM COURT

The nightmare started almost from the moment Scarlett arrived at Hong Kong Airport.

She was still a Skyflyer Solo and the airline had arranged for an escort to meet her at the plane and to take her through immigration and customs. His name was Justin and he was dark-haired, in his early twenties, dressed like a member of the cabin crew.

"Did you have a good flight?" He spoke with an Australian accent and seemed friendly enough.

"It was OK."

"You must be tired. Never mind. I'll see you through to the other side. Is this your first time in Hong Kong?"

"Yes."

"You're going to love it here!"

He prattled on as Scarlett followed him to passport control. It would have been easy to find her own way – there were signs written in English as well as Chinese – but she was glad to have company after eleven hours sitting on her own in what had felt like outer space. The worst thing about the flight hadn't been the length or the boredom. It had been the sense of disconnection.

She was going somewhere she didn't want to go, not even know-ing why she was going there. What could be so urgent that her father had made her travel all this way? And why hadn't he been able to tell her on the phone?

The airport was surprisingly quiet, but then it was only six o'clock in the morning and perhaps there hadn't been that many international flights. Even so, Scarlett felt uneasy. She examined the people around her as they stood on the travelator, being carried down the wide, silver and grey corridors. The other passengers looked more dead than alive, bleary-eyed and pale. Nobody was talking. Nobody seemed happy to be there.

And there was something else that struck her. Everyone was heading the same way. They were all pouring into the main building. People might be arriving in Hong Kong but, this morn-ing at any rate, no one seemed to be leaving.

They arrived at immigration, joining a queue that snaked back and forth up to a line of low, glass booths with officials in black and silver uniforms, seated on low stools. They all looked very much the same to Scarlett – small, with brown eyes and black, spiky hair. She put the thought out of her head. She was probably being racist.

And then it was her turn. The official who took her passport and arrivals card was young, polite. He opened the passport and examined her details and as he did so, she noticed a surveillance camera just above him swivel round to examine her too. It was quite unnerving, the way it moved, without making any sound, somehow picking her out from the rest of the crowd.

"Scarlett Adams." The official spoke her name and smiled. He wasn't asking her to confirm it. He was just reading it off the page as if he didn't quite understand what it meant. Then he reached out for his stamp, inked it and brought it down on the passport with a bang.

And at that exact moment, he changed. Did it really happen or was her mind playing tricks with her after the long flight? It was his eyes. As the stamp hit the page, they seemed to flicker as if someone had blown smoke over them and suddenly they were yellow. The pupils, which had been brown a second ago were now black and diamond-shaped. The passport official glanced up at her and smiled and right then she was afraid that he was going to leap out of his booth and tear into her. His eyes were no longer human. They were more like a crocodile's eyes.

Scarlett gasped out loud. She couldn't help herself. She was paralysed, staring at the thing in front of her. The escort, standing next to her, hadn't noticed anything wrong. Nobody else had reacted. There was a stamp as another visa was issued in the booth next door and Scarlett glanced in that direction as a student with a backpack was allowed through. When she looked back, it was over. The official was normal again. He was holding out her passport, waiting for her to take it. She hesitated, then snatched it from him, not wanting to come into contact even with the tips of his fingers as if she was half expecting them to turn into claws.

"We need to pick up your bags," Justin said.

"Right..."

He looked at her curiously. "Is something the matter, Scarlett?"

"No." She shook her head. "Everything's fine."

The cases took about ten minutes to arrive. Scarlett's was

one of the first off the plane. Justin picked it up for her and the two of them went through the customs area, which was empty. Presumably nobody bothered smuggling anything into Hong Kong. The arrivals gate was directly ahead of them and Scarlett hurried forward. Despite everything, she was looking forward to seeing her father again.

He wasn't there.

There were about a hundred people waiting on the other side of the barriers, quite a few of them dressed in chauffeur uniforms, some of them holding names on placards. She saw her own name almost at once. It was being held by a black man in a suit. He was tall and bald with a face that could have been carved – it showed no emotion. Somehow, he didn't seem to belong in Hong Kong. It wasn't just his colour. It was his size. He towered over everyone else, staring over the crowd with empty eyes as if he didn't want to be there.

There was a woman standing next to him and Scarlett took a dislike to her at first sight. Was she even a woman? She was certainly dressed in women's clothes, with a grey dress, anorak and fur-lined boots that came up to her knees. But she had the face and the physique of a man. Her shoulders were broad and square. Her neck was thick-set. She wore no make-up although she was badly in need of it. She had skin like very old leather. She was Chinese and half the height of the chauffeur, with black hair hanging lifelessly down and thick, plastic glasses that wouldn't have flattered her face even if there had been some-thing to flatter. She reminded Scarlett of a prison warden. It was impossible to guess her age. Forty? Fifty? She didn't look as if she had ever been young.

Scarlett went over to her.

"Good morning, Scarlett," the woman said. "Welcome to Hong Kong. I hope you had a good flight."

"Who are you?" Scarlett asked. She wasn't in any mood to be polite.

The woman didn't take offence. "My name is Mrs Cheng," she said. "But you can call me Audrey. This is Karl." The man in the suit lowered his head briefly. "Shall we go to the car?"

"Where's my dad?" Scarlett asked.

"I'm afraid he couldn't come."

"Where is he?"

"I will explain in the car."

The escort – Justin – had listened to all this with growing concern. It was his job to hand Scarlett over to the right person and that clearly didn't seem to be the case. "Excuse me a minute," he interrupted. He turned to Scarlett. "Do you know these people?"

"No," Scarlett said.

"Well, I'm not sure you should go with them." He turned back to the woman. "Forgive me, Mrs Cheng," he went on. "I was told I was delivering this girl to her father. And I'm not sure…"

"You're being ridiculous," Mrs Cheng interrupted. "You can see quite clearly that we were waiting for her. We are both employed by the Nightrise Corporation and were sent here by her father."

"I'm sorry. She doesn't know you and right now I'm responsible for her. I think you'd better come over to the desk and talk to my supervisor."

Scarlett was beginning to feel embarrassed with two adults quarrelling over her, especially in the middle of such a public place. But Justin and Mrs Cheng had reached an impasse. The Chinese woman was breathing heavily and two dark spots had appeared in her cheeks. She was struggling to keep her temper. Suddenly she snapped out a command, her voice so low that it could barely be heard. The chauffeur, Karl, lumbered forward.

"Now hold on a minute..." Justin began.

It looked as if Karl was going to punch him. But instead he simply reached out and laid a hand on Justin's shoulder, his long, black fingers curving around the escort's neck. There was no violence at all. Then he leant down so that his eyes were level with the other man.

And Justin caved in.

"You're making a fuss about nothing," Mrs Cheng said.

"Yes..." He could barely get the word out.

"Why don't you phone the Nightrise offices when they open? They'll tell you everything you want to know."

"There's no need. Of course the girl can go with you."

"Let him go, Karl."

Karl released him. Justin swayed on his feet, then abruptly walked away. It was as if he had forgotten about Scarlett. He wanted nothing more to do with her.

"Let's be on our way, Scarlett. We've wasted enough time here."

Scarlett picked up her case and followed Karl and Mrs Cheng down an escalator. A sliding door led to a private road with a number of smart executive saloons and limousines waiting for their pick-ups. Karl took the case and hoisted it into the boot.

Meanwhile, Mrs Cheng had opened the door, ushering Scarlett into the back.

"Where are we going?" Scarlett asked.

"We will take you to your father's apartment."

"Is he there?"

"No." Audrey Cheng spoke English like many Chinese people, cutting the words short as if she were attacking them with a pair of scissors. "Your father had to go away on business."

"But that's not possible. He just got me out of school. He made me come all this way."

"He has written a note for you. It will explain."

They had left the airport. Karl drove them across a bridge that looked brand new with steel cables sweeping down like tendrils in a web. The airport had been built on an island, one of several that surrounded Hong Kong. Everything here was cut into by the sea.

They reached the outskirts of the city and Scarlett saw the first tower blocks, five of them in a row. They warned her just how different this world was going to be, how alien to everything she knew. All five tower blocks were exactly the same. They had almost no character. And they were huge. Each one of them must have had a thousand windows, stacked up forty or fifty floors in straight lines, one on top of another. From the road, the windows looked the size of postage stamps and anyone looking out of them would have been no bigger than the Queen's head in the corner. It was impossible to say how many people lived there or what it would be like, coming home at night to your identical flat in your identical tower, identified only by a number on the door. This was a city that was far bigger than the people who lived in it. Hong Kong would treat its inhabitants in the same way that an ant hill looks after its ants.

The motorway had turned into an ugly, concrete flyover that twisted through more office and apartment blocks. It was only seven o'clock in the morning but already the traffic was building up. Soon it would start to jam. Looking down, Scarlett saw what looked suspiciously like a London bus, trundling along with far too many passengers crammed on board. But it was painted the wrong colours, with Chinese symbols covering one side. Hong Kong had once belonged to the British, of course. It had been handed back at the end of the Nineties and although it was now owned by China, it more or less looked after itself.

They passed a market where the stalls were still being set up and made their way down a narrow street with dozens of advertisements, all in Chinese, hanging overhead. Finally, they turned into a driveway that curved up to a set of glass doors in a smaller tower block. Scarlett saw a sign: WISDOM COURT. The car stopped. They had arrived.

Wisdom Court stood to the east of the city in what had to be an expensive area, as it had the one thing that mattered in a place like this: open space. The building was old-fashioned – with brickwork rather than steel or glass. It was only fifteen storeys high and stood in its own grounds. There was a fore-court with half a dozen neat flower-beds and a white, marble fountain, water trickling out of a lion's head. There were two more lions with gaping mouths, one on each side of the door. Inside, the reception area could have belonged to a smart hotel. There were palm trees in pots and a man in a uniform sitting behind a marble counter. Two lifts stood side by side at the end of the corridor.

They went up to the twelfth floor, Karl carrying the luggage. Audrey Cheng had barely looked at Scarlett since they had left the airport, but now she fished in her handbag and took out a key which she dangled in front of her as if to demonstrate that she really did have a right to be here. They reached a door marked 1213. Mrs Cheng turned the key in the lock and they went in.

Was this really where her father lived? The flat was clean and modern, with a long living room, floor-to-ceiling windows and three steps down to a sunken kitchen and dining room. There were two bedrooms, each with their own bathroom. But at first sight there was nothing that connected it with him. The paintings on the wall were abstract blobs of colour that could have hung in any hotel. The furniture looked new … a glass table, leather chairs, pale wooden cupboards. Had Paul Adams really gone out and chosen it or had it been there when he arrived? Everything was very tidy, not a bit like the warm and cosy clutter of their home in Dulwich.

But looking around, Scarlett did find a few clues that told her he had been there. There were some books about the Second World War on the shelves. He always had been interested in history. The fridge had some of his favourite food – a packet of smoked salmon, Greek yoghurt, his usual brand of butter – and there was a bottle of malt whisky, the one he always drank, on the counter. Some of his clothes were hanging in the wardrobe in the main bedroom and there was a bottle of his aftershave beside the bath.

And there was the note.

It was printed, not written, in an envelope addressed to Scarlett and it wasn't signed. Scarlett wondered if he had asked his secretary to type it. He only used two fingers and usually made lots of mistakes. The note was very short.

Dear Scarly,
Really sorry to do this to you but something came up and I've got to be out of Hong Kong for a few days. I'll try to call but if not, enjoy yourself and I'll see you soon. No need to worry about anything. I'll explain all when we meet.
Dad

Scarlett lowered the note. "It doesn't say when he'll be back," she said.

"Maybe your father doesn't know."

"But he's the only reason I'm here!"

Mrs Cheng spread her hands as if to apologize but there was no sign of any regret in her face. "This afternoon I will take you into the place where your father works," she promised. "We will go to Nightrise and you will see the chairman. He will tell you more."

Karl had carried Scarlett's suitcase into the spare bedroom. So far he hadn't said a word. He was waiting at the front door.

"I'm sure you're tired," Mrs Cheng said. "Why don't you have a rest and we can explore the city later. Maybe you would like to do some shopping? We have many shops."

Scarlett didn't want to go shopping with Audrey Cheng.

It seemed that the two of them were going to be together until Paul Adams returned. It wasn't fair. Had she really swapped Mrs Murdoch for her? But she was certainly tired. She had barely slept on the plane. Right now, in London, it would be about midnight.

"I would like a rest," she said.

"That's a good idea. I will be here. Call if there is anything you need."

Scarlett went into her room. She undressed and had a shower, then lay on the bed. She fell asleep instantly, darkness coming down like a falling shutter.

And once again she returned to the dreamworld, to the desert and the sea. She could sense the water behind her but she was careful not to turn round. She remembered the creature that had begun to emerge – the dragon or whatever it was – and didn't want to see it again.

Everything was very still. Her head was throbbing. There was something strange in the air. She looked for the four boys that she had once known so well and was disappointed to find that they were nowhere near.

Something glowed red.

She looked up and saw the sign, the neon letters hanging in their steel frame. They were flashing on and off, casting a glow across the sand around them. But the words were different. The last time she had seen them, they had read: SIGNAL ONE. She was sure of it.

Now they had changed. SIGNAL THREE. That was what they read. And the symbol beside them, the letter T, had swung upside-down.

SIGNAL THREE
SIGNAL THREE

What did it mean? Scarlett didn't know. But behind her, far away in the sea, the dragon saw it and understood. She heard it howling and knew that once again it was rushing towards her, getting closer and closer, but still she refused to turn round.

And then it fell on her. It was huge, as big as the entire world. Scarlett screamed and after that she remembered nothing more.

THE CHAIRMAN

The view was amazing. Scarlett had to admit it despite herself. She had never seen anything quite like it.

It was the middle of the afternoon, her first day in Hong Kong, and she was standing in front of a huge, plate glass window, sixty-six floors up in the headquarters of the Nightrise Corporation. The building was called The Nail and looked like one too, a silver shaft that could have been hammered into its position on Queen Street. She was in the chairman's office, a room so big that she could have played hockey in it, although the ball would probably have got lost in the thick-pile carpet. Paintings by Picasso and Van Gogh hung on the wall. They were almost certainly original.

From her vantage point, Scarlett could see that the city was divided in two. She was staying on Hong Kong Island, surrounded by the most expensive shops and hotels. But she was looking across the harbour to Kowloon, the grubbier, more down-at-heels neighbour. The two parts were separated by what had to be one of the busiest stretches of water in the world, with ships of every shape and size somehow criss-crossing around each other without colliding. There were cruise ships, big enough to hold a small army, tied up at the jetty with little *sampans*, Chinese rowing boats, darting around them. Tugs, cargo boats and container ships moved slowly left and right while nimbler passenger ferries cut in front of them, carrying passengers over to the other side and back. There were even a couple of junks, old Chinese sailing ships that seemed to have floated in from another age.

The Hong Kong skyscrapers were in a world of their own, each one competing to be the tallest, the sleekest, the most spectacular, the most bizarre. And there was something extraordinary about the way they were packed together, so many billions of tons of steel and glass, so many people living and working on top of one another... It had already reminded Scarlett of an ant nest but now she saw it was for the richest ants in the world. There weren't many pavements in Hong Kong. An intricate maze of covered walkways connected the different buildings, going from shopping centre to shopping centre, through whole cities of Armani and Gucci and Prada and Cartier and every other million-dollar designer name.

There was very little colour anywhere. If there were any trees or parks, they had been swallowed up in the spread of the city and even the water was like slate. Although it was late in the day, the light hadn't changed much since the morning. Everything was wrapped in a strange, silver mist that made the offices in Kowloon look distant and out of focus.

While she was being driven there, Scarlett had noticed quite a few people in the street had covered their mouths and noses with a square of white material, like surgeons, so that only their eyes showed. Was the air really that bad? She sniffed a couple

205

of times but could detect nothing wrong. On the other hand, the air in the car was almost certainly being filtered. The same was true of the office. The windows here were several centimetres thick, cutting out all the noise and the smells of outside.

"It's quite a sight, isn't it?"

Scarlett turned round. A man had crept up on her without making any sound. He was a European, about sixty, with white hair and thin, silver glasses and although he was smiling, trying to be friendly, she found herself recoiling from him … as if he were a spider or a poisonous snake. There was something very unnatural about the man. He had clearly had a lot of work done to his face – Botox or plastic surgery – but there was a dead quality to his flesh. His eyes were a very pale blue, so pale that they had almost no colour at all.

This was the chairman of the Nightrise Corporation. It had to be. He was wearing an expensive suit, white shirt and red tie. Very successful people have a way of walking, pushing forward as if they expect the world to get out of the way, and that was how he was walking now. He had a deep, throaty voice – he could have been a heavy smoker – and spoke with a faint American accent. There was a silver band on the middle finger of his left hand. Not the wedding finger. Scarlett somehow doubted that he would be married. Who in their right mind would choose to live with such a man?

"It's all right," Scarlett said.

The chairman seemed disappointed by her reaction. "There is no greater city on the planet," he muttered. He pointed out of the window. "That's Kowloon. Some people say that the best reason to go there is to admire the views back again but there are many museums and temples to enjoy too. You can take the Star Ferry over the water. The crossing is quite an experience, although it is one I have never enjoyed."

"Do you get seasick?"

"No." He shook his head. "When I was twelve years old, a fortune-teller predicted that I would be killed in an incident involving a boat. I'm sure you will think me foolish, but I am very superstitious. It is something I have in common with the Chinese. They believe in luck as a force, almost like a spirit. This building, for example, had to be built in a certain way, with the main door slanting at an angle and mirrors placed at crucial points, according to the principals of *feng shui*. Otherwise, it would be considered unlucky. And you see over there?" He pointed to a factory complex on the other side of the water, in Kowloon. "How many chimneys does it have?" he asked.

Scarlett counted. "Five."

"It has four real chimneys. The extra one is fake. It is there because "four" is the Chinese word for death but on the other hand they believe that five brings good luck. Do you see? They take these things very seriously and so do I. As a result I have never been close to the water and I have certainly never stepped on a boat."

He gestured at a low, leather sofa opposite his desk. "Please. Come and sit down."

Scarlett did as she was told. He came over and joined her.

"It's a great pleasure to meet you, Scarlett," he said. "Your father told me a lot about you."

"Where is my father?"

"I'm afraid I owe you an apology. I'm sure you were

207

disappointed that he wasn't here to meet you. The fact is that we had a sudden crisis in Nanjing."

"Is that in China?"

"Yes. There was a legal problem that needed our immediate attention. Obviously, we didn't want to send him. But your father is very good at his job and there was no one else."

"When will he be back?"

"It shouldn't be more than a week."

"A week?" Scarlett was shocked. "Can I talk to him?" she asked.

The chairman sighed. "That may not be very easy. There are some parts of China that have very bad communications. The landlines are down because of recent flooding and there are whole areas where there's no reception for mobile phones. I'm sure he will try to call you. But it may take some time."

"So what am I meant to do?" Scarlett asked. She didn't even try to keep the annoyance out of her voice.

"I want you to enjoy yourself," the chairman said. "Mrs Cheng will be staying with you until your father returns and Karl will drive you wherever you want to go. There are plenty of things to do in Hong Kong. Shopping, of course. Mrs Cheng has the necessary funds. There's a Disneyland out on Lantau. We have all sorts of fascinating markets for you to explore. And you must go up to The Peak. Also, I have something for you."

He went over to the desk and opened a drawer. When he came back, he was holding a white cardboard box. "It's a small gift for you," he explained. "By way of an apology."

He handed the box over and she opened it. Inside, on a bed of cotton wool, lay a pendant made out of some green stone, shaped like a disc and threaded with a leather cord. Looking more closely, Scarlett saw that there was a small animal carved into the centre; a locust or a lizard or a cross between the two, lying on its side with its legs drawn up, as if in the womb. It was very intricate. If the work hadn't been so finely done, it might have been ugly.

"It's jade," he explained. "And it's quite old. Yuan Dynasty. That's thirteenth century. Can I put it on you?"

He reached forward and lifted it out of the box. Compared to the delicacy of the piece, his fingers looked thick and clumsy. Scarlett allowed him to lower it over her head although she didn't like having his hands so close to her throat.

"It looks beautiful on you, Scarlett," he said. "I hope you'll look after it. It's very valuable, so you don't want to leave it lying around." He got to his feet. "But now I'm afraid I will have to abandon you. I have a board meeting. I'd much rather not go. But even though I'm the chairman, they still won't accept my cry for mercy. So I'll have to say goodbye, Scarlett. It was a pleasure meeting you."

My cry for mercy...

Why had he said that? Cry for Mercy was the name of the monastery where Scarlett had been kept prisoner, on the other side of the door. Of course, he couldn't possibly have known that but nonetheless he had chosen the words quite deliberately. Was he taunting her? The chairman was already moving back to the desk, but even as he had turned Scarlett thought she had detected something in his eyes, behind his silver-framed glasses. Was she imagining it? He had just given her an expensive gift. And yet, for all his seeming kindness and concern, she could have sworn she had seen something else. A brief flash of cruelty.

Scarlett spent the rest of the afternoon shopping – or window shopping anyway. She didn't actually buy anything, which was unlike her. Back in England, Aidan had often teased her that she'd lash out money on a diving suit if it had the right designer label. But she wasn't in the mood. She wondered if she'd caught a cold. It was still very damp, with a thin drizzle that hung suspended in the air without ever hitting the ground. She was also more aware of the silver-grey mist that stretched across the entire city, even following her into the arcades. The skyscrapers disappeared into it, the top floors fading out like a badly developed photograph. There was no sense of distance in Hong Kong. The mist enclosed everything so that roads went nowhere and people and cars seemed to appear as if out of nothing.

She asked Audrey Cheng about it.

"It's pollution," she replied, in a matter-of-fact sort of voice. "Its not ours. It blows in from mainland China. There's nothing we can do." She looked at her watch. "It's time for supper, Scarlett. Would you like to go home?"

Scarlett nodded.

And then a man appeared, a little way ahead of them. Scarlett noticed him because he had stopped, forcing the crowd to separate and pass by him on both sides. They were in Queen Street, one of the busiest stretches in Hong Kong, surrounded by glimmering shop windows filled with furs, gold watches, fancy cameras and diamond rings. The man was young, Chinese, dressed in a suit with a white shirt and a striped tie. He was holding an envelope.

"Scarlett…" he began.

He disappeared. The moment he spoke her name, the crowd closed in on him. It was one of the most extraordinary things Scarlett had ever seen. One moment, the people had been moving along the pavement – hundreds of them, complete strangers. But it was as if someone, somewhere had thrown a switch and suddenly they were acting as one. Scarlett tried to look past the seething mass but it was impossible. She thought she heard a scream. Then the crowd parted. The man had gone.

Only the envelope remained. It was crumpled, lying on the pavement. Scarlett moved forward to pick it up but someone got there ahead of her … a pedestrian walking past. It was just a man going home. She didn't even get a chance to look at his face. He snatched up the envelope and took it with him, continuing on his way.

"What was that?" Scarlett demanded.

"What?" Audrey Cheng looked at her with empty eyes.

"That man…"

"What man?"

"He called out my name. Then everyone closed in on him." She still couldn't take in what she had just seen. "He had a letter. He wanted to give it to me."

"I didn't see him," Mrs Cheng said.

"But I did. He was right there."

"You still have jet lag." Audrey Cheng signalled and Karl drew up in the car. "It's easy to imagine things when you're tired."

211

Scarlett was glad to get back to Wisdom Court even though she wished her father had been there to greet her. She was going to sleep in his room. Audrey Cheng had taken the guest bedroom. Karl, it seemed, would spend the night elsewhere. She had been completely shaken by what she had seen. How could a whole crowd behave like that? She remembered the way they had suddenly turned. They could have been controlled by some inner voice that she alone had been unable to hear.

She ate dinner, said goodnight to Mrs Cheng and went to her room. She hadn't finished unpacking and it was as she took out the last of her clothes that she made a discovery. Someone had placed a guidebook for Hong Kong at the bottom of her suitcase. She assumed it must have been Mrs Murdoch and if so, it was a kind gesture – although it was odd that she hadn't mentioned it. She flicked through it. "The World Traveller's Guide to Hong Kong and Macau. Fully illustrated with thirty colour plates and comprehensive maps." It was new.

But that wasn't the only thing she found that night.

Scarlett had brought a little jewellery with her – a couple of necklaces and a bracelet Aidan had given her on her last birthday. She decided to keep them safe by putting them into one of the drawers in the dressing table. As she pulled, the drawer stuck. That was probably why nobody had noticed that it wasn't completely empty. She pulled harder and it came free.

There was a small, red document at the very back. It took Scarlett a few seconds to recognize what it was, but then she took it out and opened it.

It was her father's passport.

Paul Edward Adams. There was his photograph. Blank face,

glasses, neat hair. It was full of stamps from all over the world and it hadn't yet expired.

The chairman had lied to her.

If her father had left his passport in the flat, he couldn't possibly have travelled to China. And now that she thought about it, there had been something strange about the note he had left her. Why had he typed it? It hadn't even been signed. It could have been written by anyone.

It was eleven o'clock in Hong Kong. Four in the afternoon in England. Scarlett got into bed but she couldn't sleep. She lay there for a long time, thinking of the passport, the passport official with the crocodile eyes, the chairman joking about the cry for mercy, the man who had tried to give her a letter.

She had only been in Hong Kong for one day. Already she was wishing she hadn't come.

CONTACT

Over the next few days, Scarlett tried to forget what had happened and put all her energies into being a tourist. There had to be another explanation for her father's passport. He might have a second copy. Or maybe his company had been able to arrange other travel documents for his visit to China. It was, after all, just the other side of the border. She made a conscious decision not to think about it. He would be back soon – and until then she would treat this as an extended holiday. Surely it had to be better than being at school!

So she took the Star Ferry to Kowloon and back again and had tea at the old-fashioned Peninsula Hotel – tiny sandwiches and palm trees and a string quartet in black tie playing classical music. She went to Disneyland which was small and didn't have enough fast rides, but which was otherwise all right if you didn't mind hearing Mickey Mouse talking in Cantonese. She went up to The Peak, a mountain standing behind the city which offered panoramic views as if from a low-flying plane. There had been a time when you could see all the way to China from there, but pollution had put an end to that.

She visited temples and markets and went shopping and did

everything she could to persuade herself that she was having a good time. But it didn't work. She was miserable. She wanted to go home.

For a start, she was missing her friends at school, particularly Aidan. She had tried texting him but the atmosphere seemed to be interfering with the signal and she got nothing back. She tried to call her mother in Australia but Vanessa Adams was away on a trip. Her secretary said that she would call Scarlett back but she never did.

And it was worse than that. Scarlett didn't like to admit it. It was so unlike her. But she was scared.

It was hard to put her finger on what exactly was wrong, but her sense of unease, the fear that something was going to jump out at her from around the next corner, just grew and grew. It was like walking through a haunted house. You don't see anything. Nothing actually happens. But you're nervous anyway because you know the house is haunted. That was how it was for Scarlett. But in her case it wasn't a house – it was a whole city.

First of all, there were the crowds, the people in the street. Scarlett knew that everyone was in a hurry – to get to work, to get to meetings, to get home again. In that respect, all cities were the same. But the people in Hong Kong looked completely dead. Nobody showed any expression. They walked like robots, all of them moving at the same pace, avoiding each other's eyes. She realized now that what she had seen in Queen Street hadn't been an isolated incident. It was as if the city somehow controlled them. How long would it be, Scarlett wondered, before it began to control her too?

The strange, grey mist was still everywhere. Worse than

215

that, it seemed to be getting thicker, darker, changing colour. Mrs Cheng had said it was pollution but it seemed to have a life of its own, lingering around the corners, hanging over everything. It drained the colour from the streets and even transformed the skyscrapers: the higher storeys looked dark and threatening and it was easy to imagine that they were citadels from a thousand years ago. They didn't seem to belong to the modern world.

And then there was Wisdom Court. From the moment she had arrived there, Scarlett had been aware that something was wrong. It was just too quiet. But after two days there, going up and down in the elevator, in and out of the front door, she suddenly realized. *She hadn't seen anybody*. There were no sounds coming from the other flats, no doors slamming or babies crying. No cars ever pulled up. No smells of cooking or cleaning ever wafted up from the other floors. Apart from Mrs Cheng, she seemed to be living there entirely on her own.

Of course, there was the receptionist. She had barely registered him to begin with. He was always sitting in the same place, in front of a telephone that never rang, staring at a front door that hardly ever opened. He wore a black jacket and a white shirt. His face was pale. And he never changed. Nobody ever replaced him.

How was that possible? Scarlett found herself examining him more closely. The same man in the same place, morning, noon and night. Didn't he ever eat? Didn't he need toilet breaks? It could have been a corpse sitting there and once that thought had entered her head, she found herself hurrying through the reception area, doing her best to avoid him. Not that it would have made any difference. He never spoke to her once.

On the third evening, after their visit to Disneyland, she challenged Mrs Cheng. The Chinese woman was making dinner, tossing prawns and bean shoots in a wok.

"Where is everybody?"

"What do you mean, Scarlett?"

"We're on our own, aren't we? There's nobody else in this building."

"Of course there are other people here." Mrs Cheng turned up the flame. "They're just busy. People in Hong Kong have very busy lives."

"But I haven't seen anybody. There's nobody else on this floor."

"Some of the flats are being redecorated."

Scarlett gave up. She knew when she was being lied to. It was just another mystery to add to all the others.

The next day, Mrs Cheng took her to a market in an area known as Wan Chai. As usual, Karl drove them. By now, Scarlett had got used to the fact that he accompanied them everywhere and never spoke. She even wondered if he was able to. His role seemed to be to act as a bodyguard. He was always just a few paces behind.

Scarlett had always liked markets and in Hong Kong there was a vibrant street life, sitting side by side with the expensive Western shops and soaring offices. She had been keen to explore the Chinese streets, the stalls piled high with strange herbs and vegetables, soup noodles bubbling away in the open air and the signs and advertisements, all in Chinese, filling the sky like the flags and banners of an invading army.

And yet these markets were full of horrible things. She saw dozens of live chickens trapped in tiny cages and – next to them – dead ones, beaten utterly flat and piled up like deformed pancakes. On the stand next door there was an eel cut into two pieces, surrounded by a puddle of blood. A goat's head hung on a hook, its eyes staring lifelessly, severed arteries spilling out of its neck. It was surrounded by the other pieces of what had once been its body. And finally there was a whole fish, split lengthways, the two bloody halves lying side by side. That was in many ways the most disgusting sight of all. The wretched creature was still alive. She could see its internal organs beating.

Mrs Cheng took one look at it and smiled. "Fresh!" she said.

Scarlett wondered how long she could stay in Hong Kong without becoming a vegetarian.

They continued on their way, walking past a row of meat shops. Mrs Cheng was going to cook in the flat again that night and she was looking for ingredients. As they paused for a moment, Scarlett noticed one of the butchers staring at her. He was completely bald with a large, round head and a strange, childlike face. He seemed fascinated by her, as if she were a film star or visiting royalty. And he wasn't concentrating on what he was doing.

He was chopping up a joint of meat with a small axe. Scarlett watched the blade come down once, twice…

On the third blow, the butcher missed the meat and hit his own left hand. She actually saw the metal cut diagonally into the flesh, almost completely severing his thumb. Blood spouted. But that wasn't the real horror.

The butcher didn't notice.

He raised the axe again, unaware that his hand was lying flat on the chopping board, the thumb twitching, the pool of blood widening. He was so interested in Scarlett that he hadn't noticed what he'd done. Scarlett stared at him in total shock and that must have warned him because then he looked down and backed away immediately, cradling the injured hand, disappearing into the dark interior of the shop.

What sort of man could just about cut off his own hand without any sort of reaction? On the chopping board, human blood mingled with animal blood. It was no longer possible to tell which was which.

Scarlett didn't eat meat that night. And as soon as she had finished dinner, she went back to her room. The flat had cable TV and she watched a rerun of an old British comedy. It didn't make her laugh but at least it reminded her of home. She was thinking more and more about leaving. If her father didn't arrive soon, she would insist on it. How could this have happened to her? How had she found herself on the wrong side of the world, on her own?

She went over to the window and looked out.

Hong Kong by night was even more stunning than it was by day. The windows were ablaze – thousands of them – and all the skyscrapers used light in different ways. Some seemed to be cut into strange shapes by great slices of white neon. Others changed colour, going from green to blue to mauve as if by some sort of electronic magic. And quite a few of them carried television screens so huge that they could be read all the way across the harbour, advertisements and weather information glowing in the night, reflecting in the dark water below.

One such building was directly opposite her. As she gazed out, thinking about the butcher, thinking about the still-living fish that had been cut in half, she found herself being drawn almost hypnotically towards it. It must have belonged to some sort of bank or financial centre – the screen was displaying the performance of stocks and shares. But even as Scarlett watched, the long lists of numbers were wiped from left to right and replaced by four letters in burning gold.

It was her own name, or at least half of it. She smiled, wondering what the letters actually stood for. South China Associated Railways? Steamed Chicken And Rice? But then four more letters appeared, tracking from the other side.

And that was no abbreviation. It *was* her. Scarlett. The two blocks had formed her name and now they were flashing at her as if trying to attract her attention. She stood at the window, not quite believing what she was seeing. Was someone really trying to send her a message, using an electric sign on the side of a building to get it across?

A few seconds later, the screen changed. Now it had turned white and the message it was displaying read:

Scarlett was taken aback. Maybe she was mistaken after all. What did it mean? PG Tips was a type of tea, wasn't it? It was also a type of certificate, before a film. And what about the figure 70?

Scarlett waited, hoping that the sign would change a third time and tell her something more – but nothing happened. It seemed to have frozen. Then, abruptly, it went black, as if someone had deliberately turned it off. At the same moment she heard police sirens, a lot of them, racing through the streets on the other side of the harbour in Kowloon.

There was a knock at the door.

Scarlett went over to the bed and sat down, then quickly picked up a magazine and opened it. Although she wasn't quite sure why, she had decided that she didn't want to be found at the window. "Come in," she called.

The door opened and Audrey Cheng came in. She was wearing a tight jersey that showed off the shape of her body – round and lumpy. Her black hair was tied back in a bun. Her eyes, magnified by the cheap spectacles, were full of suspicion. "I just wanted to check you were all right, Scarlett," she said.

"I'm fine, thank you very much," Scarlett replied.

"Are you going to bed?"

"In a few minutes."

"Sleep well." She seemed pleasant enough, but Scarlett saw her eyes slide over to the window and knew exactly why she had come in. It was the message. She wanted to know if Scarlett had seen it.

And it *was* a message. She was sure of it now. Someone was trying to reach her and had decided that this was the only way.

There was some sort of sense in that. A man had tried to hand her an envelope and he had been dragged off the pavement. Mrs Cheng and Karl were watching her all the time. Perhaps this was the only way.

But what did it mean? Scarlett had never been any good at puzzles. Aidan had always laughed at her attempts to do a crossword and she had come bottom in the school quiz. PG 70. It obviously had nothing to do with tea. Could it be an address, a map reference, the registration of a car? She went back over to the window and looked out again but the screen was still dark. Somehow, she doubted it would come back on again.

Eventually she stopped thinking about it and tried to go to sleep – and that was when the answer suddenly arrived. Maybe not thinking about it had helped. PG. Wasn't that an abbreviation for page? Could it be that someone was trying to make her look at page seventy? But in what? There were about forty or fifty books in the bedroom, most of them old history books that could have nothing to do with Hong Kong.

She got out of bed and picked one off the shelf at random. Sure enough, page seventy took her to a fascinating description of the way Paris had been laid out in the nineteenth century. She tried a dictionary that had been lying on the table. Page seventy began with "Bandicoot … a type of rat" and continued with a whole lot of words beginning with B. How about a page in the telephone book? That would make sense if someone was trying to get in touch.

And then she remembered. There had been one book which she hadn't packed but which had turned up mysteriously in her luggage. A guide to Hong Kong and Macau.

She went back to her case. She hadn't even taken it out – but then she hadn't needed a guide, not with Karl and Mrs Cheng ferrying her every step of the way. She carried it over to the light, flicking through to page seventy and found herself reading a description of somewhere called Yau Ma Tei – "a very interesting area in Kowloon," the text said. "Yau Ma Tei means 'hemp oil ground' in Cantonese, although you are unlikely to see any around now." There was a photograph opposite of a market selling jade, which reminded her of the amulet that the chairman had given her. She was wearing it now and wondered if he had bought it there.

She was about to throw the book down – another false lead – when she noticed something. There was a pencil line against the text. It was so faint that she had almost missed it – but perhaps that was deliberate. The line drew her attention to a single paragraph.

> **Tin Hau Temple.** You shouldn't miss this fascinating temple in a quiet square just north of the jade market. Tin Hau is the goddess of the sea, but the temple is also dedicated to Shing Wong, the city god and Tou Tei, the earth god. Admission is free. And watch out for the fortune-tellers who practise their trade in the streets outside. If you're superstitious, you can have your palm read or your future foretold by a "bird of fortune".

And at the very end of the paragraph, also in pencil, was a message: *5.00 p.m.*

Scarlett didn't get very much sleep that night. Someone was trying to reach her – and the risk was so great that they'd had to take huge precautions. First, they'd slipped a book into her case. Maybe they'd bribed someone at the airport. Then they'd somehow taken over a whole office block to draw her attention to it. The message had been clever too. PG 70. Anyone whose first language was Chinese would have had difficulty working out what it meant. It had taken her long enough herself.

She had to visit the temple and she had to be there at five o'clock. Maybe someone who knew her father would be there. Maybe they'd be able to tell her where he really was.

There was a fire in Hong Kong that night. The office building with the giant screen burned to the ground and when Scarlett woke up, the air was even greyer and hazier than ever, the smoke mixing in with the pollution. She looked out of the window but she couldn't see the other side of Victoria Harbour. The whole of Kowloon was covered in fog.

Mrs Cheng was more chatty than usual at breakfast. She mentioned that nine people had been killed and insisted on turning on the television to see what had happened. And, sure enough, there it was on a local news channel. The image was a little grainy and the announcer was speaking in Chinese but Scarlett recognized the building, directly opposite Wisdom Court, right on the harbour front. The images had been taken the night before and there were flames exploding all around it, the reflections dancing in the black water. Half a dozen fire engines had been called to the scene.

But the firemen weren't doing anything. The camera panned

over them. None of them moved. None of them even unwound their hoses.

They just stood there and let the building burn.

BIRDS OF FORTUNE

The Tin Hau Temple was a low, narrow building, crouching behind a wall and surrounded by trees, almost as if it didn't want to be found. There were tower blocks on every side, the dirty brick walls crowding out the sky, but in the middle of it all there was a space, a wide square with trees that seemed to sprout out of the very concrete itself. Some benches and tables had been set out and there were groups of old men playing a Chinese version of chess. A few tourists were milling around, taking photographs of each other against the green, sloping roofs of the temple. The air smelled faintly of incense.

It hadn't been easy getting Mrs Cheng to bring her here.

From the very start, Scarlett knew she had to be careful. Mrs Cheng had shown her the news report for a reason. She hadn't been fooled by Scarlett's act of the night before and she was letting her know it. If Scarlett asked straight out to go to the Tin Hau Temple at five o'clock, she would be more suspicious than ever.

"Is there any news from my dad?" As they cleared the breakfast plates away, Scarlett asked the same question she asked every morning.

"I'm sure he'll call you soon, Scarlett. He's very busy."

"Why can't I call him?"

"It's not possible. China is very difficult." She flicked on the dishwasher. "So where would you like to go today?"

This was the moment Scarlett had been waiting for. She shrugged her shoulders. "I don't know," she said.

"We could go out to Stanley Village. It is on the beach and there are some nice stalls."

Scarlett pretended to consider. "Actually," she said, "I wanted to buy some jade for my friend, Amanda."

Mrs Cheng nodded. "You can find jade in the Hollywood Road. But it's expensive."

"Can't we go to a market?"

"There's a jade market in Kowloon..."

It was exactly what Scarlett wanted her to say. She had read the entire chapter in the guidebook and knew that the most famous jade market in Hong Kong was just round the corner from the temple. If they visited one, they'd be sure to walk over to the other. And that way she would arrive at Tin Hau without even having mentioned it.

She still had to make sure that they got there at the right time, so after they had finished clearing up she announced that she had some school work to do and they didn't leave Wisdom Court until two o'clock. Scarlett would have preferred to have taken the subway that went all the way there but as usual, Mrs Cheng insisted that Karl should drive them. And that meant he would be with them all afternoon. They were certainly keeping her close.

The jade market was in a run-down corner of Kowloon,

227

just off the Nathan Road, which was a long, wide tourist strip known as the "Golden Mile". Not that there was much gold amongst the rather tacky shops which specialized in cheap electronics, fake designer watches and cut-price suits. The market was located in a low-ceilinged warehouse, sheltering under one of the huge flyovers that seemed to be knotted into the city.

The pollution was even worse today. The weather was cold and damp and the mist was thicker than ever. Scarlett could actually feel it clinging to her skin and wondered how the people of Hong Kong put up with it. She noticed that increasing numbers of them had resorted to the white masks on their faces and wondered how long it would be before she joined them.

There were about fifty stalls in the jade market, selling necklaces, bracelets and little figurines. Keeping one eye on her watch, Scarlett made a big deal out of choosing something, haggling with the stallholders, asking Mrs Cheng for advice, before finally settling on a bracelet which cost her all of three pounds. As she handed over the money, it occurred to her that Amanda would actually quite like it – she just hoped that she would be able to give it to her some time soon.

"Do you want to go back down to the Peninsula?" Mrs Cheng suggested as they came back out into the street. Karl was waiting for them, leaning against the car. He never seemed to have any trouble parking in Hong Kong. For some reason, the traffic wardens – if there were any – never came close.

"Not really…" Scarlett looked around her. And she was in luck. There was a signpost pointing to the Tin Hau Temple. They were standing right in front of it. "Can we go there?" she said, trying to make the suggestion sound casual.

"We've already visited a lot of temples."

"Yes. But I'd quite like to see another."

It was true. They'd already been to the Man Mo in Central Hong Kong and to the Kuan Yin only the day before. They were strange places. Chinese temples seemed to mix religion and superstition – with fortune sticks and palm readers sitting comfortably among the altars and the incense. The people who went there didn't pray like an English congregation. They bowed repeatedly, muttering to themselves. They left offerings of food and silk on the tables. They burned sacks of paper in furnaces that were kept going for precisely that purpose. Hong Kong had been Westernized in many ways, but the temples could only belong to the East and provided glimpses of another age.

Tin Hau was just like the others. As Scarlett stepped inside, she found herself facing not one but several altars, surrounded by a collection of life-size statues that could have come out of a bizarre comic book: a cross-legged old man with a beard that was made of real hair, two devil monsters, one bright red, the other blue, both of them more childish than frightening. One of them was crying, wiping its eyes and grimacing at its neighbour. The other stood with a raised hand, trying to calm his friend down. There was a china-doll woman carrying a gift and, in a long row, more than fifty smaller figures, each one a different god, perched on a shelf. The temple was a riot of violent colours, richly patterned curtains, lamps and flowers. The smoke from the incense was so thick that they'd had to install a powerful ventilation system which droned continuously, trying to clear the air.

Scarlett had arrived on time but she had no idea what she

was looking for. There were about a dozen people in the temple, but they were all busy with their devotions and nobody so much as turned her way. Was it possible that she had misunderstood the passage in the guidebook? It had definitely told her to be there at five o'clock and it was already a few minutes past. She waited for someone to approach her, to slip another message into her hand – one of the worshippers, or a tourist perhaps. She even wondered if her father might be there.

Nothing happened. Nobody came close. Scarlett knew that she could only pretend to be interested in the place for so long. Mrs Cheng was watching her with growing suspicion. She certainly hadn't shown much interest in temples the day before – so what was so special about this one?

"Have you had enough, Scarlett?" she demanded.

"Who is that?" Scarlett asked desperately, pointing at one of the statues.

"His name is Kuan Kung, the god of war." Something flickered deep in her eyes. "Maybe you should pray to him."

"Why do you say that, Mrs Cheng?"

"You never know when another war will begin."

In the end, Scarlett had to leave. She had lingered for as long as she could but it seemed clear that nobody was going to come. She was hugely disappointed. Of course, the note had only given her a time. It hadn't told her what day to be there. On the other hand, it was unlikely that she would be able to find an excuse to return, and slipping out of Wisdom Court on her own was out of the question. Nine people had died when the office on the waterfront had burned down. Maybe whoever had sent the message had been among them.

It was beginning to get dark when they emerged into the square. Karl was sitting on a bench with his arms folded, looking about as animated as the statues that they had just seen. A number of stalls were being set up all around. They didn't look particularly interesting – selling socks, hats, reading glasses and useless bits of bric-à-brac – but they were attracting quite a crowd.

"Can we look at them?" Scarlett asked.

It had only struck her there and then. The passage in the guidebook had described the Tin Hau Temple. But it had also gone on about the square outside. Maybe her secret messenger would be waiting there. Mrs Cheng scowled briefly but Scarlett had already set off. She followed.

Scarlett pretended to browse in front of a stand selling cheap alarm clocks and watches. She was determined to spend as much time here as possible. She noticed that the next stall wasn't selling anything. There was a woman with a pack of tarot cards. In fact now that she looked around her she saw that at least half the market was devoted to different methods of fortune-telling.

She walked over to a very old man, a palm reader who was sitting on a plastic stool, close to the ground. His stall was decorated with a banner showing the human hand divided into different segments, each one with a Chinese character. He was examining the palm of a boy of about thirteen, his nose and eyes millimetres away from the skin as if he really could read something there. Scarlett moved on. There was a woman a little further along, also telling the future. But in a very different way.

The woman was small and round with long, grey hair. She

was wearing a red silk jacket, sitting behind a table, arranging half a dozen packets of envelopes which were stacked up in front of her. On one side, there were three cages, each one containing a little yellow bird – a canary or something like it. On the other, she had a mat with a range of different symbols and a jar of seeds. The woman seemed to be completely focused on what she was doing but as Scarlett approached, she suddenly reached out with a single, gnarled finger and, without looking up, tapped one of the symbols on the mat.

It was a five-pointed star.

Scarlett had seen exactly the same thing on the door that had led her to the monastery of the Cry for Mercy. She was careful not to give anything away – Mrs Cheng was standing right next to her – but she felt a rush of excitement. According to Father Gregory, the doors had been built centuries ago to help the Gatekeepers. They were there to help her. Had the woman sent a deliberate signal? Scarlett examined her more closely. She still didn't look up, busying herself with the envelopes and occasionally muttering at the birds.

Scarlett turned to Mrs Cheng. "What's this all about?" she asked.

"She uses the birds to tell fortunes," Mrs Cheng explained.

The old woman had heard the English voices and seemed to notice Scarlett for the first time. She squinted at her and muttered something in Chinese.

"She's offering to tell your fortune," Mrs Cheng translated. "But it will cost you thirty Hong Kong dollars."

"That's about two pounds."

"It's a complete waste of money."

"I don't care." Scarlett dug in her pocket and took out the right amount. She set it down on the mat and then took her place on the plastic seat on her side of the table. The fortune-teller folded the money and transferred it to a little purse that she wore around her neck. Then she reached for a white card and laid it in front of Scarlett. She said something to Mrs Cheng.

"She wants you to make a choice," Mrs Cheng explained.

There were a number of categories set out on the card, written in both Chinese and English. Scarlett could choose which part of her life she wanted to know about: family, love and marriage, health, work, business and wealth or study.

"Maybe I should choose family," she said. "She may be able to tell me what's happened to my dad."

"Your father will be home very soon, Scarlett."

"All right, then. Love and marriage." Scarlett tapped the words on the card and thought briefly of Aidan. She wondered what he was doing right then.

The fortune-teller took the card away and selected one of the piles of envelopes which she had spread out in front of the three cages. Each one had a door in the front and she opened one of them. The little yellow bird hopped out as it had been trained to do, perched on the line of envelopes, then pulled one out with its beak. The old woman rewarded it with a couple of seeds and the bird obediently hopped back in again. It was all over very quickly.

The woman opened the envelope and handed Scarlett the slip of paper which had been inside.

"Do you want me to translate it for you?" Mrs Cheng asked.

Scarlett glanced at the sheet. "No, it's OK," she replied. "It's in English."

"Tell me what it says."

"Good news from Fortune Bird Two." Scarlett read out the words. "You will find your true love in the month of April. Your marriage will be long and happy and you will travel to many countries. When you are old, you will make a great sum of money. Spend it wisely." She folded the page in half. "That's it."

"The note only tells you what you want to hear," Mrs Cheng remarked.

"The bird chose it for me." She held out the page so that Mrs Cheng could see it. "There you are. You can see for yourself. I'm going to be rich."

Mrs Cheng nodded but said nothing. The two of them and Karl walked back to the car. And all the time, Scarlett's heart was racing and she kept the piece of paper close to her. She had folded it quite deliberately. She had only shown Mrs Cheng half of what had been written.

For underneath the printed fortune, there had been another message, written by hand:

Scarlett.

You are in great danger. Do not let the woman read this. Come to The Peak tomorrow afternoon. Follow the path from Lugard Road. We will be waiting.

We are your friends. Trust us if you want to leave Hong Kong alive.

THE PEAK

Scarlett knew something was wrong, the moment she opened her eyes.

A glance at her bedside clock told her that it was eight o'clock in the morning but for some reason the sun wasn't reaching her bedroom. It wasn't just cloudy. It was actually dark. What was going on? She turned over and looked at the window. At first she thought that someone had drawn a black curtain across the glass, but then she realized that it wasn't on the inside. It was outside. How was that possible, twelve storeys up? She propped herself on one elbow, still half-asleep, trying to work it out.

And then the curtain moved. It seemed to fold in on itself and at the same time she heard the beating of tiny wings and understood what she was looking at. It was a great swarm of insects, black flies. They had attached themselves to the window like some single living organism.

She lay where she was, staring at them with complete disgust. She had never seen so many flies, not even in the heat of the summer. And this was a cold day in November! What had brought them here? How had they managed to fly across an entire city to come together on a single pane of glass? She could hear their buzzing and the soft tapping as they threw their bodies against the window. She could make out their legs, thousands of them, sticking to the glass. Their wings were blurring as they held themselves in place. Scarlett felt sick. She was suddenly terrified that they would find their way in. She could imagine them swirling around her head, a great black mass, crawling into her nostrils and mouth. On an impulse, she scooped up her pillow and threw it at the window. It worked. As one, the flies peeled away. For a moment they looked like a long silk scarf, hanging in the breeze. Then they were gone.

For about twenty minutes, Scarlett stayed where she was, almost afraid to get up. She didn't like insects at the best of times but this was something else again. She knew that what she had seen was completely impossible ... just like the door in the church of St Meredith's. And that told her what should have been obvious all along.

She had thought that, at the very least, her sudden departure to Hong Kong would be an escape from what had been happening in London – the monastery, the sense of being followed, the restaurant that had blown up. But of course it wasn't. It was a continuation, part of the same thing. The events that had closed in on her in London had followed her here. She was caught in the same trap. But here it was even worse. She was far from her friends and family, alone in a city that seemed to be hostile in every way.

This was all happening because she was a Gatekeeper. She remembered what Father Gregory had told her. He had talked about an ancient evil ... the Old Ones. Scarlett didn't know exactly what they were but she could imagine the worst. They were here, in Hong Kong. That would explain everything. The Old Ones were toying with her. They were the ones who were controlling the crowd.

What was she going to do?

She could march into the kitchen and tell Mrs Cheng that she didn't want to wait for her father, that she was taking the next flight back to London. She could telephone her mother in Australia or the headmistress at St Genevieve's. They would get her out of here. She could even contact the police.

But she knew that none of it would work. The forces ranged against her were too powerful. She could see it every time she went outside. Hong Kong was sick. There was a sort of cancer that had spread through every alleyway and every street and which had infected everyone who walked there. Did she seriously think that they were just going to let her walk out of here? So far, they hadn't threatened her directly. That hadn't been part of their plan. But if she challenged them, if she tried to assert herself, they would close in on her and she would only make her situation worse.

She had just one hope. The people who were trying to reach her: they had to be on her side. *We are your friends*. That was what they had told her. She just had to behave normally until she reached them. Then, once she knew what was really happening, she would be able to act.

She got up and got dressed. The fortune-teller's note was beside the bed, but now she tucked it away beneath the mattress. Whoever her friends were, they were being very careful. They were contacting her in four separate stages: the guidebook hidden in her luggage, the illuminated sign across the harbour, the bird of fortune at Tin Hau and finally a meeting this afternoon. The question was, how was she going to persuade Mrs Cheng to take her back to The Peak?

They had already been there once. Victoria Peak was the mountain that rose up behind Hong Kong, a must-see for every tourist. Scarlett had gone there on the second day, taking the old wooden tram – it was actually a funicular railway – up the slope to the top, five hundred metres above the city. The views were meant to be spectacular but they hadn't seen very much on account of the pollution. Maybe that was the answer. If the weather cheered up, it would give her an excuse to go back.

Mrs Cheng was in the kitchen, cooking an omelette for Scarlett's breakfast.

"Good morning, Scarlett."

"Good morning, Mrs Cheng."

"Did you sleep well?"

"Very well, thank you."

As Scarlett sat down, it occurred to her that she had never seen the woman eat – not so much as a mouthful. Even when they went to restaurants together, Mrs Cheng only ordered food for Scarlett. In fact she had only ever shown hunger once. That had been at the market when they examined the hideous, sliced-in-half but still-living fish.

"So where would you like to go today, Scarlett?" They were exactly the same words she had used the day before. And she spoke without any real enthusiasm, as if it was simply what she had been programmed to say.

"Why don't we go back to The Peak?" Scarlett suggested. "We didn't see anything very much last time. Maybe we'll get a better view."

Mrs Cheng looked out of the window. "There's a lot of cloud," she remarked.

"But it's going to cheer up this afternoon," Scarlett said. "I saw the forecast on TV." It was grim outside with a non-stop drizzle sweeping across the sky. And the forecast had said it would stay the same for the rest of the week. But somehow Scarlett knew she was right.

"I don't think so." Mrs Cheng shook her head. "Maybe you would like to go to the cinema?"

"Let's see what it's like this afternoon," Scarlett pleaded. "I'm sure it will clear up."

And against all the odds, it did. At around two o'clock, the clouds finally parted and the sun came out, still weak against the ever-present pollution, but definitely there. Even Mrs Cheng had to agree that it was too nice an afternoon to stay indoors and so the two of them set out.

The receptionist was in his usual place as they left Wisdom Court, sitting stiffly behind the desk and wearing the same dark suit and white shirt, watching them with no expression at all. As they went past, Scarlett noticed something. The man had a black spot, a mole, on the side of his face. At least, that was what she thought. Then the spot moved. It crawled over his cheek and began to climb up and she realized that it was actually a fly, one of the fat, black insects that had come to her window that morning. The receptionist didn't move. He didn't try to swat it. He didn't even seem to have noticed it and did nothing as the creature reached the corner of his eye and began to feed.

Scarlett couldn't get out of the building fast enough. Wisdom Court was only a few minutes from the tram station and they could have walked but Karl drove them anyway. But at least he had decided not to come to the top. Mrs Cheng bought tickets for the two of them and she and Scarlett got onto the tram.

Although the station looked new, the tram itself had been built more than a hundred years before. Climbing on board was like stepping back in history. They took their places on the polished, wooden seats and a short while later, with no warning, they set off, trundling up the tracks through thick vegetation with occasional glimpses of the city, ever smaller and more distant as they went. There were about twenty tourists sharing the ride, some of them small children, laughing and pointing. Watching them, Scarlett wished that she could be like them, part of an ordinary family, out here on holiday. She was only a few seats away from them but they could have been inhabiting a different world. Had they really got no inkling about what was happening in Hong Kong? Was she the only one to feel the all-pervading sense of evil?

We will be waiting.

She focused her mind on what lay ahead. Who would be there and why had they chosen The Peak of all places? Maybe it was because it was outside the city, away from the buildings. At the summit there would be no crowds, no surveillance cameras. It was somewhere with room to breathe.

The tram arrived and the passengers poured out, straight into a complex that seemed to have been specially built to make as

240

much money from as many tourists as possible. From the outside it looked like a bizarre observation tower, like something out of *Star Wars*. Inside, it was full of tacky shops and restaurants with a Madame Tussaud's and a Ripley's Believe-it-or-Not with signs inviting visitors to "come and see the world's fattest man". Scarlett couldn't wait to get out.

"Let's go for a walk," she suggested. She was careful to sound as innocent as possible.

Mrs Cheng looked doubtful. She wasn't dressed for a walk – in a short, grey skirt, black stockings and high-heeled shoes. "Maybe a short way…" she muttered.

There was a distinct chill in the air as the two of them made their way down a slope, passing a man who was sweeping leaves. Scarlett knew what she was looking for. A path that led off from the Lugard Road. That was what the fortune-teller's note had said. She saw the sign almost at once. Without even waiting for Mrs Cheng to catch up, she set off.

The path was three miles long, snaking all the way round the mountain, paved all the way. On one side there was The Peak itself, with a tangle of exotic trees and bushes hanging overhead. On the other was an iron railing, to prevent anyone falling down the hill. There weren't many other people around. The changing weather must have dissuaded them, and the other tourists who had come up in the tram had all stayed inside. Soon Scarlett found that she and Mrs Cheng were entirely on their own.

There was a strange atmosphere on The Peak. The mist had returned, hanging in the air, almost blotting out the sun. Everything was washed out, dark green and pale white. There were birds whistling, squawking and rattling in the under-growth, but none of them could be seen. The path was lost in the clouds and it was impossible to see more than twenty metres ahead. As she made her way forward, Scarlett found it easy to imagine that she had somehow travelled back in time, that this was some Eastern version of *Jurassic Park* and that a dinosaur might be waiting for her round the next corner.

But then she arrived at an observation point where the vege-tation had been cut back and Hong Kong appeared, sprawled out below. It was incredible to see so many skyscrapers packed together on both sides of the water. There were hundreds of them, every shape and size, made small and insignificant by the distance – with thousands or even millions of people invisible among them.

Mrs Cheng plodded along behind, saying nothing. Her face was sullen, her hands – loosely curled into fists – hung by her side. Scarlett was quietly amused. Her guardian clearly wasn't enjoying the visit. She wasn't even bothering to glance at the view.

A couple of people walked past them … a woman pushing an old-fashioned pram and a man, jogging. The man was wearing a blue tracksuit and his face was covered by an anti-pollution mask, with only his eyes showing above the white square. Scarlett tensed as each one of them approached. She was waiting for someone to make contact. But neither of them so much as noticed her, both continuing on their way.

They walked for another five minutes, still following the path which curved round the side of The Peak.

"I think we should go back, Scarlett," Mrs Cheng said.

"But it's a circular walk," Scarlett protested. "If we keep going, we'll find ourselves back anyway."

Three more walkers appeared ahead of them: two men and a woman, all Chinese. They were dressed in much the same way with jeans, zip-up jackets and walking shoes. One of the men had a walking stick although he looked young and fit and surely didn't need it. The other man carried a backpack. He was in his thirties, with glasses and a pock-marked face. The two of them were chatting. The woman – she was slim and athletic, her long hair tied back with a pink band – was listening to an iPod. As they drew nearer, they showed no interest in Scarlett at all.

The three of them drew level.

"Scarlett…" Mrs Cheng began.

She never finished the sentence. The man with the back-pack reached behind him and drew out something that was flat and silver. It was a move that he must have rehearsed many times. To Scarlett's eyes, it was as if he had suddenly produced an oversized kitchen knife. Then she realized what it was. A machete. The blade was about half a metre long and razor-sharp. At the same time, the other man twisted the handle of his walking stick, revealing the sword that had been concealed inside. Scarlett saw the glint of metal and heard it slice the air as he pulled it free. The woman wasn't armed. She was looking behind her, checking the path was clear.

Both men plunged their weapons into Audrey Cheng. The Chinese woman screamed – but there was nothing remotely human about the sound. It was a high-pitched howl, almost deaf-ening. Scarlett stared in horror. Her face was unrecognizable, her

243

mouth stretched open in a terrible grimace. Blood was pouring in a torrent over her lower lip. Her eyes had clouded over. She hadn't had time to defend herself or react in any way. Scarlett saw her neck open as if it was hinged and she looked away. She heard the thud as Mrs Cheng's severed head hit the ground. She knew that it was a sound that she would never forget.

The woman ran forward and put an arm around her, comforting her. Some of Mrs Cheng's blood had splattered onto her. There were flecks of it on her jacket. The very air had gone a hazy red.

"I'm sorry you had to see that, Scarlett," she said, in perfect English. "Don't look round. We had to do it. There was no other way."

"You killed her!" Scarlett was in shock. She had never liked Mrs Cheng but she couldn't believe what she had just seen. These people hadn't given her a chance to defend herself. They had murdered her in cold blood.

"Not her. It."

Scarlett stared. "What do you mean?"

"Show her!" one of the men snarled.

"We're your friends," the woman said. "We sent you the message with the fortune-teller. We've come to help you and, believe me, there was no other way." She placed her hands on Scarlett's shoulder. "Turn round and have a look for yourself," she went on. "The woman isn't what you think. She's a shape-changer. We'll show you, but then you have to come with us. They'll know what's happened. They'll have heard her. We don't have much time..."

Scarlett turned round. The man with the sword-stick was

already sheathing it. The other was wiping his machete on a piece of cloth. She swallowed hard, not wanting to do this. There was a lot of blood, spreading across the path.

Mrs Cheng was lying on her back, the legs in their black stockings lying straight out in front of her. There was a dreadful wound in her chest where one of the blades had stabbed her through the heart. The other had decapitated her. Scarlett forced herself to examine the rest of the body. She saw something thick and green coming out of the jacket where Mrs Cheng's neck should have been. It had been severed half way up. But it didn't belong to a human body. It looked like part of a snake.

And the head, lying on the path, wasn't human either. It was the head of an oversized lizard, with yellow and black diamond eyes, scales, a lolling forked tongue. Scarlett glanced back at the body. Mrs Cheng had thrown out one of her arms as she fell. It was also covered in scales.

A shape-changer.

That was what they had said. And in the shock of the moment, all Scarlett could think was – was this the creature she had been living with since she had come to Hong Kong? Audrey Cheng had cooked for her. She had been sleeping in the same flat. And all the time...

She thought she was going to be sick. She couldn't get the hideous images out of her head. But then she heard the sound of an approaching engine, coming down the path towards her. Had they been discovered? The woman and the two men weren't moving. They didn't look alarmed. Scarlett relaxed. Whoever was coming was part of the plan.

A motorbike appeared, speeding round the corner. It was a silver-grey Honda, being ridden by a figure in black leather, gloves and boots. Scarlett guessed that it was a man, but it was hard to be sure as his head was concealed by a helmet with a strip of mirrored plastic across his face. He stopped right in front of them, the wheels tilting underneath him, one leg stretching out to keep the bike upright.

The woman grabbed hold of Scarlett once again. "We need to get you out of here fast," she said. "We don't have time to explain."

"Where are you taking me?"

"Somewhere safe."

They had produced a second helmet. Scarlett hesitated, but only for a few seconds. Audrey Cheng's dead body told her everything she needed to know. She had been living in a nightmare and these people, whoever they were, were rescuing her from it. She grabbed the helmet and put it on, then climbed onto the bike, putting her arms around the driver. At once they were away. She felt the engine roar underneath her as they shot down the path and she tightened her grip, afraid that she would be blown over backwards by the rush of wind.

They shot past a man walking a dog and then a family of local people who had been posing for a photograph but who scattered to get out of the way. They turned another corner. If they went much further, they would surely arrive back at the tram station where Scarlett had begun. On one side there was a small park, on the other a driveway leading up to a house, for there were a few private homes scattered along the upper reaches of The Peak. But that wasn't where they were heading.

Scarlett saw a parked car with two more men waiting. They skidded to a halt.

She got off, quickly removing her helmet. The two men were young, in their twenties, both wearing jeans and sweatshirts. One was Chinese but the other was a foreigner, maybe from Japan or Korea. They both hurried over to her, their faces filled with a mixture of determination and fear.

"You have to come with us," the first one said. He had a thin face and his nose and cheekbones were so sharp-edged that they could almost have been folded out of paper. "We must leave at once."

"Where are we going?"

"Somewhere safe." That was exactly what the woman had said. "Not far. Maybe twenty minutes."

"Wait a minute…"

"No time." He spoke in fractured English, spitting out the words. "You want to die, you stay here. You ask your questions. You want to live, get in the car. Now! They will be coming very soon."

"Who will be coming?"

"Shape-changers. Or worse."

The other man had gone over to the car. But he hadn't opened the door. He had opened the boot.

"You don't expect me to get in there!" Scarlett said.

"It must be this way," the thin-faced man insisted. "You can't be seen. But you'll be all right. We make air-holes…"

"No…" It was too much to ask. Scarlett didn't care how many shape-changers there might be, making their way up The Peak. She wasn't going to be locked in the boot of a car by two

247

people she had never met before and driven off to God knows where. "You can forget it…" she began.

The man had whipped something out of his pocket and he grabbed her before she knew what he was doing. She felt a handkerchief being pressed against her face. She kicked out, trying to knock him off balance, but he was too strong. The fumes of some sort of chemical, sweet and pungent, crept into her nose and mouth. Almost at once, all the strength drained out of her. She felt her legs fold and the world spun. And then she was falling, being guided into the boot, which had become a huge black hole waiting to swallow her up.

The end came very quickly. Darkness. Terror. And then the welcome emptiness of sleep.

LOHAN

She was in a cage, not lying down but standing. And there was something strange. The wall was moving. It seemed to be scrolling downwards in front of her. Or was it she who was moving up?

As consciousness returned, Scarlett realized what was happening. She was in a lift, one of the old-fashioned sort with a folding iron gate instead of a door. What she was looking at was the brickwork between floors in what must be a very tall building. She was pinned between the Japanese man and the one she had decided to call Paper Face. They were supporting her. She could still taste the drug – chloroform or whatever it was – that had knocked her out.

Scarlett groaned and the two men immediately tightened their grip. There was no chance she was going to start a fight in such a confined space, but she had already struggled at the car and they weren't taking any chances.

"You are safe now," Paper Face said.

"Where am I?"

"You will see ... very soon."

The lift slowed down and stopped and the Japanese man

jerked the cage door open. They stepped out into a long, dimly lit corridor with walls that were either grimy or had been deliberately painted the colour of grime. There were doors every few metres. The whole place looked like a cheap hotel.

There was a Chinese man guarding the corridor with a machine gun cradled across his chest. The sight of the weapon struck Scarlett as completely bizarre. It was like something out of a gangster film. But the man didn't look anything like her idea of a gangster. He was dressed in jeans and a loose-hanging shirt. He was skinny, with a wispy beard, a tattoo on his neck and a gold tooth prominent at the front of his mouth. A drug dealer perhaps? Looking at him, it was hard to believe that he was on her side.

The two men took her to the fourth room along the corridor. Paper Face knocked and the door was unlocked from inside. They entered. Machine Gun stayed where he was, opposite the lift.

Scarlett found herself in a large, almost empty flat that looked as if someone had recently moved out ... or in. There were a few pieces of furniture, some of them covered in dust sheets, and no decoration: no carpet, no lampshades, no pictures on the walls. The windows had been blanked out with sheets of papers. Scarlett wondered why. They had to be fairly high up so surely there was no chance of anyone looking in. An archway led into a small kitchen and there was a corridor on the other side, presumably with a bedroom and bathroom at the end.

Another man had been waiting for her to arrive. He was Chinese, more smartly dressed than the others – in a grey suit and grey T-shirt – and everything about him radiated confidence and control. Was he the one in charge? He examined Scarlett briefly. His eyes were very dark, almost black, and gave nothing away. There was a thin scar starting high up on his left cheek and then slanting diagonally across his lips so that the two halves of his face didn't quite meet, like a reflection in a broken mirror. But he was handsome even so. Scarlett guessed that he was barely more than twenty years old.

"How are you?" he said. "You must have been very frightened by your ordeal. I'm sorry that there was no other way."

"Who are you?" Scarlett demanded. "Where am I and who are these people? What do you want with me? And what was that with Mrs Cheng? They said she was a shape-changer. What does that mean?" Once the questions had started, they wouldn't stop.

The man held up a hand. He had long, elegant fingers, like a piano player. "We have a great deal to say to each other," he said. "Would you like a drink?"

"No, thank you."

"But I would." He nodded and Paper Face hurried into the kitchen. He was obviously used to being obeyed. He turned back to Scarlett. "Please, come and sit down."

Scarlett went over to the sofa. She was surprised how quickly the drug had worn off. She sat down. The man followed her and sat opposite. He moved slowly, taking his time. Everything about him was very deliberate.

"My name is Lohan," he said. "Does that answer your first question? I doubt if my colleagues will have very much to say to you but I will tell you their names too. The man in the kitchen

251

is called Draco. And this here…" he nodded in the direction of the Japanese man, "…is Red. Not their real names, you understand. Just the names they use.

"Your next question – what do we want with you? Very simply, we want to get you out of Hong Kong as quickly as possible. Quite frankly, it would have been better for everyone if you had never arrived, but never mind. We couldn't stop you coming, although we tried. It's remarkable how many people you've already managed to get killed."

He certainly wasn't sparing her feelings. But Scarlett wasn't going to let him intimidate her. "I want to see my father," she said. "Do you know where he is?"

"I'm afraid not," Lohan replied. "I have never even met him. For what it's worth, I would imagine that he is dead. A very great many people have died in Hong Kong in the last weeks. He was probably one of them."

"You're telling me my father's dead! Don't you care?"

Lohan shrugged. "I've told you. I've never met him. Why should it matter to me whether he is alive or dead?"

Draco came back in from the kitchen, carrying a tray with a small porcelain bowl and a jar of some sort of spirit – vodka or sake. He set the whole thing down in front of Lohan, bowed and then took his place on a seat beside the front door. Lohan poured himself a drink. He held it briefly between his index finger and his thumb, then threw it back and swallowed. He set the bowl back down.

"You want to know where you are," he continued. "This apartment is in Mong Kok, a couple of blocks north of the Tin Hau Temple, where you had your fortune told. The entire

252

building belongs to us and with a bit of luck, nobody will come up here. While you remain in this room, you are safe. Every minute you spend outside it, you are in more danger than you can possibly imagine."

"You mean shape-changers."

Lohan ignored her. For a moment he gazed past her, as if focusing on something outside the room. Then he began.

"You have to understand the nature of a city," he said. "You live in London and so maybe what I'm about to say will be obvious to you. All cities are the same. They have an atmosphere. More than that. You might call it a flow. The traffic moves in a certain way. The trains pull in and out of the stations. People go to work, they have their lunch, they go shopping, they go home again. Postmen deliver the post. Policemen patrol the streets. Sweepers and refuse collectors come out in the evening. The night bus arrives at the right time and picks up the people who are waiting at the stop and takes them where they expect to go. Everyone is obeying the flow, even if they don't realize it, because if they didn't life would descend into chaos.

"Now ... consider Hong Kong. It is one of the most densely populated cities in the world. There are more than seven million people living here. That works out at around 6,500 people per square kilometre. A few of them are rich. Most of them are very poor. And then there are the millions in between – the doctors and dentists, the shopkeepers, builders, plumbers, teachers..."

"I think I get the point," Scarlett interrupted.

"No, Scarlett. I don't think you do." Lohan hadn't raised his voice. His face was as impassive as ever. But Scarlett realized that she shouldn't have spoken. He wasn't used to being interrupted. "This is the point," he went on. "How many of those people could die, do you think, before you noticed? How many of them could be shot or knifed while they lay in bed before the city seemed any different? Fifty of them? Or five hundred? Or how about five hundred thousand? Can you describe to me, accurately, the man who sold you the ticket when you boarded the tram this morning? Or the driver who took you to The Peak? Or the man who was sweeping the leaves away when you began your walk? Suppose they had all been taken away and replaced with people who looked a little like them but who were not the same? Would you notice? If they and their entire families had been murdered, would you care? We see only what we want to see because that is the way of the city. In a village, in the country, people notice things. But on the streets, we are wilfully blind."

"Are you saying that's what's happened?" Scarlett asked. "Ever since I've been here, I've been seeing weird things. And there's nobody living at Wisdom Court. The whole place is empty. Are you saying they were all killed?"

"In the last three months, Hong Kong has been taken over," Lohan replied. "It happened very quickly, like a virus. It is impossible to know how many people have been killed. Anyone who has noticed what has been going on or who has tried to fight it has been removed. What has happened has been so huge, so terrible that it is almost impossible to understand.

"Of course, some people have guessed, or half-guessed, and they have managed to get out, taking their money and their families with them. Ask them why they have gone and they will lie to you. They will say they wanted a change or they had new business opportunities. But in truth they have gone because they were afraid. Other people are aware that Hong Kong has changed. They have stayed here because they have no choice, because they have nowhere else to go. They are frightened too. But they keep their heads down and they go about their daily business in the hope that, if they ask no questions, they will be left alone. If you are poor, Scarlett, if you run a tiny stall in the street, what does it matter who controls the city? All you care about is your next meal. The city can take care of itself."

"Who has taken over Hong Kong?" Scarlett asked, although she already knew the answer.

"They are called the Old Ones," Lohan said. "At least, that is what you call them. In the East, we talk of *gwei*, evil spirits. We have many names for them."

"I know about all this," Scarlett said. "It's what Father Gregory told me."

"Who is Father Gregory?"

"He's a monk. I went through a door in a wall and I met him…"

"This was at the Church of St Meredith's." Lohan knew the name. Perhaps he had read about it in the newspapers when Scarlett disappeared, but she doubted it. He seemed to know a lot about a lot of things. She wondered how. She still wasn't sure how he fitted in. "You have to understand that we have been interested in you for a long time, Scarlett," Lohan said. It was as if he had been reading her thoughts.

"We?"

"I am referring to the organization to which I have the great honour to belong. In fact, we have been watching you since the day that you were born." He allowed this to sink in, then went on. "Have you ever wondered how you came to find yourself in the Pancoran Kasih Orphanage in Jakarta? Well, I can tell you. We arranged it. Why were you taken to live in Great Britain, thousands of miles away from your true home? We wanted it."

"Why?"

"To keep you safe. To hide you from the enemies that we knew would one day search for you."

"There was an accident in Dulwich. A white van…" Scarlett didn't know why it had come into her mind right then, but she was suddenly sure that it was connected. She had a sense of everything coming together.

Lohan nodded. "It happened when you were thirteen years old," he said. "It was not enough simply to send you far away. My organization had a sacred pledge to protect you, even from your own carelessness. When you stepped in front of the van which was speeding towards you, one of our people was there to push you out of the way. He was able to save you once. Unfortunately, he was less successful a second time."

"He tried to contact me. In London…"

"His task was to give you a message. Under no circumstances were you to come here to Hong Kong. We had hoped to intercept you before you even left for the airport. But by then it was too late. The Old Ones had discovered who you were. They killed him."

"He was waiting for me at the restaurant – the Happy Garden."

Lohan nodded and there was a tiny spark of anger in his

eyes. Perhaps part of him blamed her for the death. "Three people died in the explosion," he said. "And the British authorities didn't even bother to investigate. They just blamed it on us ... Chinese gangs fighting each other. What did it matter to them? A few dead *fei jais*." He used the Cantonese slang for petty criminals. "To the police, it just meant more paperwork."

"This all happened because of the church, didn't it."

It was all making sense. Father Gregory had told her he was going to hand her over to the Old Ones. Scarlett had managed to escape – but not before she had given him her name and address. That had been all he needed. From that moment on she had been in a trap from which there was no escape.

"As soon as you returned, the Old Ones closed in on you," Lohan said. "They knew that they had found one of the Gatekeepers and they weren't going to let you go. From that moment on, they never let you out of their sight."

Scarlett thought back. She had felt all along that she had been under surveillance but it was only now that she realized how true that had been. Every movement she had made had been watched. She had been pushed around like a piece on a board game and the last roll of the dice had brought her here.

"They used my dad to bring me to Hong Kong," she said, and felt a sudden ache of sadness. Lohan had said Paul Adams might be dead. He could well be right.

"We never wanted you to come to this city," Lohan said. "Once you were here, you would be utterly in their power, and you have no idea to what extent that has been true. All day, every day, you have been surrounded by them. Nobody has been allowed to come anywhere near you. Haven't you noticed? Since you have been here, nobody has approached you. Nobody has come near."

"There was a man with a letter…" Scarlett began. "In Queen Street."

Lohan shook his head. "We didn't send him. We knew that it would never have worked." He paused. "The Old Ones control the police, the government and the civil service. They have made deals with the Chinese authorities and anyone who has stood in their way, they have killed. The hospitals, the fire service, the newspapers, the television and radio stations all serve them now. They keep constant watch on us through the surveillance cameras in the streets and know what we buy every time we use a credit card. They have taken over the mobile phone network and the Internet and every call is monitored, every one of the millions of e-mails that are sent every day is read by them. Criticize the government … you die. Even try to tell people what you know … you die. We're back where we started, Scarlett. How many thousands of people can you kill in a city like this without anyone noticing? Only the Old Ones know the answer.

"And they are everywhere. The woman and the driver who pretended to work for your father were both shape-changers. We don't know where they came from or what exactly they are. Many of the crowds that surrounded you were the same. Why do you think Wisdom Court is empty? They wanted to keep you in isolation and every man, woman or child who might come into contact with you was either taken away, killed or replaced."

"Replaced with what?" Scarlett asked.

"With creatures that belong to the Old Ones." Lohan filled the bowl a second time and drank it. The alcohol had no effect on him at all. "The whole city is against you, Scarlett. If you stepped outside now, you would be seen and identified in seconds. That was why you couldn't travel here, sitting in a car. It was also why we had to be so careful reaching you. One of my people added the guidebook to your luggage at the airport. Then we bribed the supervisor of an office building and transmitted a message on the screen. The fortune-teller is part of our organization and she sent you to The Peak. Four different approaches and each time we had to be certain that you alone knew our intentions."

"So what am I going to do?" Scarlett couldn't keep the helplessness out of her voice. This is what it came down to. She was stuck in a room in a dirty block of flats. And outside a whole city was searching for her. She remembered how the day had begun when she had woken up. Even the flies were on their side.

"You must not be weak!" For a moment, Lohan didn't even try to hide his contempt. He spat out the words and his mouth, cut in half by the scar, was twisted into a sneer. "It will not be easy," he said. "The Old Ones chose this city very carefully. You are on an island with only four possible ways out. First, of course, there is the airport, where you arrived. But that is out of the question. Every flight will be watched and even if we disguise you and give you a false passport, the danger is too great.

"The second possibility would be to travel by jet-foil to the island of Macau, which is only an hour away. From there you would be able to fly to Singapore or Taiwan. But again there is too much risk. I don't think that you would even get on board before you were spotted. There is a passport control at the terminal and remember – every single official will be looking for you."

"Can't I go into China?" Scarlett asked.

"It is possible to cross into China at Shenzhen. Many tourists go there to shop because the prices are cheap. But there are police everywhere. The border is well patrolled. And once the Old Ones know you are missing, they will be looking carefully at everyone who crosses."

"So what's the fourth way?"

But she wasn't going to find out. Not then. She hadn't even noticed the telephone in the room but suddenly it rang. The three men froze and she saw at once that it wasn't good news. Lohan didn't answer it himself. He gestured at the Japanese man, Red, who snatched up the phone and listened for a moment in silence. He put it down and muttered a few words in Chinese. Scarlett didn't understand what he'd said but nor did she need to. The call was a warning. The Old Ones were here.

Lohan turned to her, examining her as if for the first time. Even now he seemed undisturbed, refusing to panic.

"Have they found us?" Scarlett blurted out the question.

Lohan nodded slowly. "They're outside. The building is surrounded."

"But how...?"

"We seem to have missed a trick." Lohan's eyes were still fixed on her. For a few seconds, he didn't speak. Then he worked it out. "You have something with you," he said. "The woman –

Mrs Cheng or someone at Nightrise – gave you something to wear."

"No…" Scarlett began. But then she remembered. Her hands went to her throat. "The chairman gave me this…"

She was still wearing the jade pendant. Now, with trembling fingers, she unhooked it and took it off. The little green stone with the carved insect hung at the end of the chain. She handed it over. "It can't be bugged," she said, weakly. "It can't…"

Lohan examined it with cold anger. Then he turned it round and dangled it in front of her face.

Scarlett gasped. The creature inside the pendant – the lizard or the locust or whatever it was – was moving. She saw it blink and shift position. Its legs curled up underneath it. One of its wings fluttered. Scarlett cried out in revulsion. The thing was alive. And all this time it had been around her neck…

Lohan laughed briefly and without humour, then closed his fist over the pendant, winding the chain around his wrist.

"What are we going to do?" Scarlett asked.

Before anyone could reply, there was an explosion in the street. It sounded soft and far away but it was followed at once by screaming and the sound of falling glass. There was the wail of police sirens – not one car but any number of them, closing in from all sides.

Lohan produced an automatic pistol, drawing it out of his back pocket. It was sleek and black and he handled it expertly, loading it with a clip of ammunition, releasing the safety catch and briefly checking the firing mechanism. "You must do whatever we tell you," he said. "No questions. No hesitation. Do you understand?"

Scarlett nodded.

From somewhere in the building came the first burst of machine-gun fire. Lohan threw the door open, signalled and together they began to move.

ACROSS THE ROOF

Lohan was the first out into the corridor, then Draco and Scarlett with Red behind. They were all armed apart from her. The man outside the lift had unhooked his machine gun and was cradling it in his arms. He didn't look scared. In fact, he was completely relaxed, as if this was all in a day's work.

Scarlett was feeling sick with anger. This was her fault. The jade pendant that she had been given was bugged in every sense of the word – and it had told the chairman exactly where she was. Why had she even worn it? She should have left it beside the bed. But it was too late to think about it now. The Hong Kong police had arrived. They were already on their way up.

Her every instinct would have been to get out of there as quickly as they could but they were moving slowly, taking it one step at a time. Lohan was listening out for any sound, his head tilted sideways, his gun level with his shoulder. Scarlett saw him signal to the man at the lift, pointing with two fingers, ordering him to stay where he was, and she knew that it was probably a death sentence. These people had some sort of code among themselves. They did exactly what they were told no matter what it might cost.

For a brief moment, everything was silent. The police cars had turned off their sirens and the gunfire had stopped. The corridor was empty. But then, with a surge of alarm, Scarlett saw a blinking light. There were two arrows next to the lift doors, one pointing up, the other down. One of them was flashing. The lift was on its way up.

Lohan gestured with the gun. "You follow me. This way..."

They set off down the corridor but it seemed to Scarlett that he was leading them the wrong way. It would obviously have been crazy to have tried taking the lift, but wouldn't the emergency stairs be somewhere near by? Lohan was taking them ever further into the building and away from what was surely the only way out.

But nobody argued. Scarlett still had no idea who Lohan was or what authority he had over the others. He had said that he belonged to an organization and one that had been looking out for her from the day she had been born, but he hadn't told her what it was called, who ran it or anything like that. It seemed that he and his people were some sort of resistance, fighting against the Old Ones, the last survivors in a city that had been attacked from within. But they weren't the police. They weren't the army. What did that leave?

It was too late for any more questions now. Lohan was moving a little faster but still on tiptoe, making no noise, as if he expected one of the many doors to spring open and someone to jump out. How high up were they? How long did they have before the lift arrived? The end of the corridor was about thirty metres away with ten doors on either side. A row of light bulbs hanging from the ceiling on wires lit the way ahead. Scarlett heard a loud, metallic click and risked a glance back. The man with the machine gun had released the safety catch. Lohan muttered something under his breath.

There was the ping of a bell.

The lift had arrived.

Scarlett was still watching as the lift doors opened and yellow light flooded out. The man with the machine gun had positioned himself directly opposite with his shoulders planted against the wall. Without any warning, he opened fire, sending a firestorm of bullets into the lift. The noise in the confined space was shocking. She could actually feel it, hammering into her ears. But she couldn't see what the man was shooting at. The entire corridor blazed white and red and she heard a high-pitched scream like nothing she had ever heard before, as whatever was inside the lift was pulverized.

Then something appeared, stretching out of the open doorway. It was impossible to make it out clearly between the gloom of the corridor and the brilliance of the gunfire, the two of them strobing – black, white, black, white – turning everything into slow motion chaos. Some sort of tentacles, extending themselves into the corridor. They reached the man. One slammed into his face. Another curled around his throat. But it was the third that killed him, punching right through his stomach and dragging him up the wall, a great streak of blood following up behind him. The man was screaming, his legs writhing in agony. But his finger was clenched around the trigger and he was still firing. His last bullets went wild, tearing into the ceiling and floor.

Something spilled out of the lift. It seemed to be partly

265

human, but there was smoke everywhere now, adding to the confusion. A second creature followed it. The two had come up together. A huge pincer snapped open and shut. Black eyes on stalks. Straight out of a nightmare. It saw Scarlett and the others and began to move with frightening speed.

"Hurry!"

It was the first time that Lohan had raised his voice. He broke into a run. Scarlett followed him, convinced that they didn't have a chance. There was no emergency exit, nowhere for them to go and the things from the lift must already be closing in on them. Red turned round and fired twice. The bullets had no effect. Then Draco dragged something out of his pocket, brought it to his mouth and threw it. A hand grenade! The pin was still between his teeth. A door to one of the other flats opened and Lohan threw himself in, pulling Scarlett with him, just as there was a deafening explosion and an orange ball of flame in the corridor behind.

Scarlett leant against the wall on the other side of the door. She was choking and there were tears streaming down her face. She wasn't crying. It was the dust and the plaster which had cascaded down, almost blinding her. She wiped a sleeve across her eyes. Red slammed the door shut. It had about half a dozen locks, chains and bolts which he fastened, one after another. Lohan snapped out another command. Draco muttered something in reply.

The flat they were in was very similar to the one they had left but more run down, with even less furniture. There was a woman living here. She had opened the door to let them in and Scarlett recognized her. It took her a moment to work out where they had met but then she realized – it was the fortune-teller from the temple. She was standing by the door, blinking nervously. Her three birds were in their cages on a table, hopping up and down, frightened by all the noise.

Lohan hadn't stopped moving. He was heading towards a second door and the kitchen beyond. "This way, Scarlett," he called out.

Scarlett followed him into a room with a fridge and a cooker and little else. A large hole had been knocked through the wall. The sides were jagged, with old bits of wire and pipework sticking out, but this was the way out. They climbed through the brickwork and into the next-door flat and then into the one after. Each one had been smashed through to provide a passageway that couldn't be seen from the corridor. The last two flats were completely abandoned, with dust and rubble all over the floor. They came to a window with a steel structure on the other side. A fire escape. Lohan jerked the window open. They climbed out.

Scarlett found herself standing on a small, square platform with a series of metal ladders zigzagging all the way down to street level, about twenty floors below. It was very cold up there, the air currents rushing between the buildings, carrying with them a driving rain. She looked down onto the sort of scene she would normally have associated with a major accident. There must have been at least a dozen police cars parked at different angles in the street. They might have turned their sirens off but their lights were flashing, brilliant even in the daylight. Barricades were still being erected around the building and all the traffic had been stopped. Men in black and silver uniforms were holding the crowds back.

They couldn't go down. The fire escape led into the middle of all the chaos and the moment they reached the bottom, they would be seized. Worse still one of the policemen had seen them. He shouted out a warning and pointed. At once a group of armed officers ran forward and began to climb up.

Lohan didn't seem worried. "We don't go down," he muttered. "We go up."

There were just three flights of stairs from the platform to the roof and, aware of the policemen getting nearer all the time, Scarlett made her way up as quickly as she could, keeping close to the wall in case any shots were fired. Draco and Red followed up behind and a minute later they had all reached the roof and were squatting there, catching their breath in the shelter of a rusty water tank. The rain was slicing down. Scarlett was already drenched, her hair clinging to her eyes.

Lohan had taken out a mobile phone. He pressed a direct dial button and spoke urgently into it, then folded it away. The other men hadn't said a word but they seemed to understand what had been agreed. Then Red muttered something and pointed. Scarlett looked up, wondering what he had seen. And shuddered. She had thought their situation couldn't get any worse … but it just had.

There was a cloud of what looked like black smoke in the distance, high above the tower blocks of Kowloon. It was travelling towards them, against the wind. Scarlett knew at once that it couldn't be smoke. It was the swarm of flies. They had come back again. They were heading directly for her.

"Move!"

Lohan set off at once, running across the roof, no longer caring if he was seen or not. He had hung the jade pendant around his neck and Scarlett realized that as long as he was wearing it, the flies would know where he was. But that was his plan. It was the reason he had taken it from her. He was protecting her, making himself the target in her place. He leapt over stacks of cable, moving towards the back of the building. Scarlett followed. She still had no idea where they were heading, how they were going to get down.

They reached the other side and came to a breathless halt. Once again, Scarlett was completely thrown. There was no fire escape, no ladder, no window cleaner's lift. The next apartment block was about twenty metres away and there was no possible means of crossing. Lohan was standing at the very edge of the building. For a moment he looked like a ghost or maybe a scarecrow with his pale skin and his dark clothes, drenched by the falling rain. His black hair had fallen across his face. The scar seemed more prominent than ever.

"Follow me," he instructed. "Don't look down."

And then he stepped into space.

Scarlett waited for him to fall, to be killed twenty storeys below. Instead, impossibly, he seemed to be standing in mid-air, as if he had learned to levitate. More magic? That was her first thought – but then she looked more closely and saw that it was just an incredible trick. There was a bridge constructed between the two buildings, a strip of almost invisible glass or Perspex ... some see-through material strong enough to take his weight. Nobody would have been able to see it from the street or from the air, and even now she might not have been able to make it out but for the rain hitting it and the faint coating of grime that covered the surface. It still looked as if Lohan was suspended between the two buildings. He had walked some distance from the edge of the roof and was standing over the road, the toy cars and people far below.

It would be Scarlett's turn next.

A door burst open on the roof behind her. A staircase led up inside and their pursuers had finally reached them, pouring out onto the roof, nine of them, human from the look of them but with dead eyes and pale, empty faces that might have spent years out of the light. Their hair was ragged, their clothes mouldering away and they wore no shoes. Some of them carried long, jagged knives. Others had lengths of chain hanging down to the ground and wooden clubs spiked with nails. Slowly, they began to fan out.

"You – go!" Red pushed Scarlett forward, propelling her towards the glass bridge. "Draco..." He finished the sentence in Chinese.

There was no time to argue. The creatures were already getting closer. Red moved towards them, away from the safety of the bridge, his own gun raised in front of him. Scarlett looked down. The bridge had no sides, no safety rails. The surface was wet and slippery. Worse still, because it was transparent, it felt completely insubstantial. She could imagine herself falling through it or losing her balance and plunging over the side. And she could see where she would land. The road was there, waiting for her far below.

Red fired a shot and the sound of it propelled her forward. She couldn't look back. She couldn't see what was happening behind her. All her concentration was focused on what she had to do. She took one step, then another. Now she was in mid-air with the wind buffeting her. She felt Draco behind her, urging her on, but fear was paralysing her. Lohan had told her not to look down, but if she didn't, how could she be sure that her foot was coming down in the right place? The rain sliced into her face, half-blinding her. She could feel it running down her cheeks.

There were two more shots but then they stopped and she heard screaming and knew that Red had been caught, that terrible things were being done to him. Scarlett hated herself for doing nothing to help him. He had stayed behind for her, to give her the time to cross, and she was literally walking out on him. All these people were risking their lives for her. The whole apartment block with its knocked-through walls and this incredible glass bridge had been prepared for the time she might need it. And the crazy thing was that she still didn't know who they were or why they had decided to help.

Somehow, she got to the other side, taking the last step with a surge of relief. At that exact moment, Red's screams ended and she turned round to see him being held in the air by a group of the creatures who were standing at the edge of the building she had just left. His body was limp. Blood was pouring from a dozen stab wounds in his arms and chest. Then they let him go. He seemed to glide rather than fall through the air, as if he weighed nothing. Finally, he smashed into one of the parked cars, crumpling the roof and shattering the front windscreen. An alarm went off. With a screech of triumph, the creatures who had killed him lurched themselves onto the bridge.

Lohan was standing, watching them. He let them get about half-way across before he stretched out a hand and closed it around a lever set in a wall. He smiled briefly, malevolently, and pulled. At once, the bridge collapsed. It was like one of those magic wands used by conjurors at children's parties. The different sections folded, then plunged downwards. Five of the creatures went with it, hitting the road in an explosion of bone and blood. The rest were left on the other side, jabbering and shaking their fists, unable to cross.

Behind them, something vague and dark rose up over the side of the rooftop. The swarm of flies had arrived. Lohan signalled and set off across the second apartment block, making for a door on the far side. If he was going to mourn the man who had died, it would have to wait. He went through, waited for Scarlett and Draco, then slammed it shut. There was a flight of stairs on the other side. It led down into a room humming with pipes and banks of machinery. There was a service lift on the other side. Lohan hit the button and the doors opened at once. The three of them piled in. He pressed two buttons: the ground floor and the basement.

Scarlett stood inside the confined space, panting. Her heart was racing at a hundred miles an hour. It felt unnatural to be suddenly standing still, knowing that there was danger all around but there was nothing she could do and nowhere she could go as the lift carried them down. She just hoped there wouldn't be anyone waiting for them at the bottom.

But Lohan was completely relaxed, leaning against the back wall, the pendant hanging around his neck. Water was dripping down his forehead, over his eyes. "You are to go with Draco," he said. "I have made arrangements. There are people waiting. You will be safe with them."

"What about you?" Scarlett asked.

"I will lead them away." He lifted the pendant, glanced at it, let it fall again.

"They'll kill you..."

"If they find me, they will kill me. But my life is not in question here. You are all that matters. You must get away."

"This is my fault." Scarlett felt miserable. She had led the creatures to the apartment. They were only here because of her. "I'm sorry..."

"You are one of the Five!" Lohan stared at her as if he couldn't believe what she had just said. "Do not be sorry. Do not be a little girl. You have the power to destroy them. Use it."

The lift doors opened. They had arrived at the ground floor. Lohan stepped forward and looked outside. Scarlett could hear the wail of police cars, but there was nobody around and she guessed that the police hadn't yet worked out that they had crossed from one building to another. But the jade pendant would bring them soon enough. Lohan gave a last instruction in Chinese to Draco and then he was gone. The doors slid shut behind him.

"You stay with me now," Draco muttered.

Red had been killed. The man with the machine gun was dead. Lohan was probably next. But he didn't seem to care.

The lift continued down to the basement. It opened into an underground car park. There was a shiny black car waiting for them and at first Scarlett couldn't believe what had been arranged for her, what was waiting there beneath the building.

But at the same time, she knew it made complete sense. She remembered what Lohan had told her. The entire city was against her. Every policeman, every surveillance camera, every official was looking out for her. How was she meant to get past them all?

The car was a hearse. There was an open coffin in the back, the inside of it lined with cream-coloured satin with a pillow at one end. Two men were waiting for her. They were dressed in dark suits, like undertakers, but she recognized them from The Peak. They were the ones who had killed Mrs Cheng. One of them made a gesture. Scarlett knew what she had to do.

This time she didn't argue. Without even hesitating, she climbed into the back of the hearse and lay down. It occurred to her that only a few hours ago, when they had tried to lock her in the boot of a car, she had thought it would be like being buried alive. And here it was, happening for real.

She laid her head on the pillow. The two men moved towards her. And then once again darkness claimed her as the lid was bolted into place.

OCEAN TERMINAL

Nobody noticed the hearse as it swung out of the underground car park and began to make its way south towards Victoria Harbour. Everyone's attention was on the building where Lohan and his friends had been found. The hearse emerged on the other side, turned left at a set of traffic lights and set off down the Golden Mile.

It never did more than ten miles an hour. If anyone had been watching it, the fact that it was moving so slowly would only have made it all the more unlikely that it was being used as an escape vehicle. But very soon it had left the crowds and the police cars behind. In the front, the driver and his assistant gazed straight ahead, their grim faces hiding their joint sense of relief.

For Scarlett, it was less easy.

She couldn't see anything. She couldn't do anything. She couldn't even move. She was lying on her back, trapped in a black, airless space with the lid bolted into place only inches above her head. She was completely at the mercy of her own imagination. Every time the car slowed down or stopped, she wondered if they had been discovered. Worse than that, she imagined a nightmare scenario where something had gone horribly wrong and she really was taken to a cemetery and buried alive. Every nerve in her body was screaming. She could hardly breathe.

After what seemed like an hour, she felt the car stop. She heard the doors open and slam shut. A long pause. And then suddenly a crack of daylight appeared, widening as the coffin lid was lifted off. A hand reached out to help her and gratefully she grabbed hold of it. Gently, she was pulled out like a corpse returning to life. She found herself trembling. After all she had been through, she wasn't surprised.

Where was she? The hearse had been parked next to a forklift truck in a warehouse, filled with pallets and crates. There were skylights in the ceiling but it was also lit by neon strips, hanging down in glass cages. One of the men had hit a switch that brought a sliding door rumbling down on castors, but before it reached the floor Scarlett glimpsed water and knew that they were near the harbour. The smell of gunpowder hung in the air. Normally, she might not have recognized it – but there had been plenty of it around in the building she had just left.

The driver was already stripping off his jacket and black tie. The last time Scarlett had seen him, he had been wiping a bloody machete on a cloth up on The Peak. He had been the one with the backpack – long hair and glasses – and he was younger than she had first thought, in his mid-twenties. He was wearing a short-sleeved shirt under the jacket and she noticed a tattoo on his upper arm, a red triangle with a Chinese character inside.

"My name is Jet," he said. Like all the others, he wasn't bothering with surnames. He spoke hesitant English but with a polished accent. "I will be looking after you now. This is Sing."

The other man came over from the door and nodded.

"Where are we?" Scarlett asked.

"Still in Kowloon. This is our warehouse." Jet walked over to one of the crates and pulled off the tarpaulin that half covered it so that she could read the words stencilled underneath. They were written in Chinese and English.

KUNG HING TAO FIREWORK MANUFACTURERS

"Fireworks...?"

"It's good business," Jet explained. "In China, we let off fireworks if someone marries and again when they die. The Bun Festival, the Dragon Boat Festival, the Hungry Ghost Festival and New Year. Everyone wants fireworks! There are one hundred thousand dollars' worth in this warehouse. I suggest you don't smoke."

"You want Coke?" the man called Sing asked. He still had his walking stick with the sword concealed inside. It had been inside the hearse, but he had yanked it out and carried it with him.

"We have a small kitchen and a toilet," Jet said. "We have to stay here for a while."

"How long?"

"Twenty-four hours. But nobody will find you here..."

"What about Lohan?" Scarlett had been worrying about him. She knew that it was her fault that he was in danger.

"He will come. You do not need to be afraid. Very soon you will be on your way out of Hong Kong."

Lohan had spoken of four ways to get out of the city and he had dismissed three of them: the airport, the jet-foil to Macau, the Chinese border. What did that leave? Scarlett had seen the harbour. Perhaps they were going to smuggle her out on a container ship. First a car boot, then a coffin. These people wouldn't think twice about packing her into a crate of fireworks and sending her somewhere in time for Bonfire Night.

Sing had gone into the kitchen but now he came back with three bottles of water and sandwiches on plastic plates. He was still wearing his undertaker's suit but he had taken off the tie. The three of them ate, sitting cross-legged in a circle on the floor. It was only when she took her first bite that Scarlett realized how hungry she was. She'd had little breakfast, no lunch and it was now six o'clock.

"It is not possible to take you out on a container ship." Jet had seen her sizing up the crates and must have guessed what was on her mind. "There's too much security. The ports are all watched day and night, and anyway, it will be the first thing that they expect. We will take you out in public, in front of their eyes."

"How?"

He glanced at the other man who nodded, as if giving him permission to go on.

"Tomorrow morning, a cruise ship arrives in Kowloon. It will dock at the Ocean Terminal on the other side of Harbour City, just ten minutes from here. It spends a day in Hong Kong on its way from Tokyo to the Philippines and then Singapore. That is where it will take you. The ship is called the *Jade Emperor* and it will be full of wealthy tourists. You will be one of them."

"How do I get on board?"

"For their own reasons, the Old Ones do not want the world to know that they have taken over this city. That is good. When the *Jade Emperor* ties up, they will have to be careful. There will be security but it will have to be invisible. They will not want to frighten the tourists. Everything will have to seem normal and that gives us the advantage. We will smuggle you onto the ship with the other passengers. And once you are there, you will be safe."

"What happens when I get to Singapore?"

Jet shrugged. Sing muttered something in Chinese and laughed. "That is the least of your worries," Jet said. "First of all, you have to survive tomorrow. And remember – there are a hundred thousand people who are looking for you. This is a trap and you walked straight into it. Now that you're here, it's not going to be so easy to get you out."

He wasn't being fair. Scarlett hadn't walked into Hong Kong. She had been deliberately drawn in and there had been nothing she could have done to avoid it. But she didn't argue. There was no point.

"We will disguise you," he went on. "We will cut your hair and change its colour and we will dress you as a boy. You must learn to walk in a certain way. We will show you. There is a family joining the boat. Their names are Mr and Mrs Soong and they are part of our organization. Right now, they are travelling with their twelve-year-old son, Eric. You will change places with him and travel on his passport. By midnight tomorrow, you will be in international water and out of danger. Do you understand?"

"How will you make the change?" Scarlett asked.

"We have arranged to meet in a shop in Harbour City. The shop is also owned by us. It pretends to sell tea and Chinese medicine."

"What does it really sell?"

Jet thought for a moment. He was reluctant to answer the question but for some reason he decided to. "Do you really want to know?" He smiled. "Normally it sells opium."

Scarlett spent the night on a mattress behind a row of crates that the two men had arranged to form a private "room". She barely slept at all. It was cold in the warehouse – there wasn't any form of heating – and she had only been given a couple of thin blankets. Every night is trapped between the day before and the day after and she had never been so torn between the two.

She thought about the creatures she had seen coming out of the lift, the flies approaching the tower block and the people – were they actually living people? – who had followed her onto the roof. How could things like that be happening in a modern city – monsters and shape-changers and all the rest of it?

Then she turned her mind to the people she was with. Despite everything that had happened, she still knew almost nothing about them. There were lots of them and they were well organized. Lohan had spoken about them with reverence, almost as if they were a holy order. And yet she had just been told that they sold opium! Opium was a drug that came from the same source as heroin. Could it be that they were some sort of gangsters after all? They carried machine guns and hand grenades. And although they were helping her, none of them was exactly friendly.

Finally, she thought about the next day and the dangers it would bring, walking onto a cruise ship disguised as a boy. Would it really work and what would happen to her if it didn't? As far as she could see, the Old Ones didn't want to kill her. Father Gregory could have done that, and he'd made it clear he had other plans. For some reason, they needed her alive.

Lying on her back, gazing at the skylight, she watched night crawl towards day. In the end, she did manage to sleep – but only fitfully. When she woke up, her neck was aching and she felt even more tired than she had been when she began. Her two bodyguards were already awake. Sing had made breakfast, a plate of noodles, but she hardly ate. Today was her last chance. She knew that if she didn't get out today, she never would.

Nothing happened for the next three hours. Jet and Sing sat silently, waiting, and for some reason Scarlett found herself trying to remember her lines in the school play. She had lost track of the date but guessed that it would be performed – without her – in a couple of weeks' time. All the parents would be there along with some of the boys from The Hall. She thought of Aidan. And as she sat there, trapped in a warehouse full of fireworks, Dulwich seemed a very long way away and she wondered when, if ever, she would see it again.

And then Jet's mobile phone rang. He snapped it open and muttered a few words into it, then nodded at Sing who went and unlocked the door. They opened it just a little bit, enough for Scarlett to see that it had stopped raining outside. Bright sunlight streamed in through the crack, lighting up the dust that hung in the air. Two more people came into the warehouse.

The first of them was Lohan. He went straight over to Scarlett. "Are you OK?" he asked.

Scarlett was relieved to see him. "How about you?" she asked. "What did you do with the pendant?"

"The pendant is on a flight to Australia. Hopefully the Old Ones will follow it there."

"I'm glad you're OK."

"And I will be glad when you have gone."

He gestured at the man who had come with him. He hurried forward, carrying a canvas suitcase about the size of a weekend bag. This man was quite a bit older than the others, wearing a crumpled cardigan and glasses. He placed the suitcase on the floor and opened it to reveal scissors, hair brushes, lots of bottles, pads of cotton wool. There were clothes packed underneath.

It was time for Scarlett to change.

Jet dragged one of the crates over and Scarlett sat down. The older man examined her for a moment, using his fingers to brush her hair back from her face. He nodded as if satisfied, then reached for the scissors.

Scarlett would never forget the way he cut her hair. She wouldn't have said she was particularly vain, but she had always taken care of how she looked. There was something brutal about the way he attacked her, chopping away as if she had no more feelings than a tree. She looked down and saw great locks of her hair hitting the ground. Although she knew that it was necessary and that anyway it would all grow back soon enough, she still felt like a victim, as if she were being assaulted. But the man didn't notice her distress – or if he did, he didn't care.

He kept cutting and soon she felt something she had never felt before: the cold touch of the breeze against her scalp. He finished her hair with a scoop of Brylcreem, then set to work on her face, turning it first one way, then the other, his fingers pressing against her chin. There was absolutely nothing in his eyes. He had done this many times before. It was his business and he did it well. He just wanted to get it over with as quickly as possible.

He painted her skin with a liquid that smelled of vinegar and stung very slightly, then added a few splodges with a thin brush. After that, he set to work on her eyes. Just when Scarlett thought he had finished, he muttered something to Lohan, the first time he had spoken. His voice was completely flat.

"He wants to put in contact lenses," Lohan explained. "They're going to sting."

They did more than that. The man had to clamp Scarlett's head while he pressed them in, the lens balanced on the end of his finger, and when he backed away the entire room was out of focus, hidden behind a blur of tears.

"Now you must get dressed," Lohan said.

They didn't allow her any privacy. The four men stood watching as she stripped down to her underwear and then the man in the cardigan dug a white, padded thing out of his case. Scarlett understood what it was. The boy whose place she was taking must have been quite a bit fatter than her. She slipped the pads over her shoulders and saw at once that she had a completely new body shape and that the slight curve of her breasts had gone. The man handed her a shirt, linen trousers, a blazer and a pair of black leather shoes that added about three centimetres to her height. Finally he gave her a pair of glasses. The disguise was complete.

"Look in the mirror," Lohan said.

They had brought a full length mirror out of the kitchen. Scarlett stood in front of it. She had to admit that the transformation was incredible. She barely recognized herself.

Her hair was now short and spiky, held rigidly in place by the Brylcreem. Her eyes, which were normally green, were now dark brown, the colour magnified by the spectacles which were clumsy and old-fashioned, in plastic frames. There was a touch of acne around her nose. She had become one hundred percent Chinese; a slightly pudgy thirteen-year-old who probably went to an expensive private school and dressed like his dad. She even smelled like a boy. Maybe they had put something in all the chemicals they had used.

"Now you must practise walking," Lohan said. "Walk like a boy, not like a girl."

For the next two hours, Lohan kept her pacing up and down with slouching shoulders, hands in her pockets. Scarlett had never really thought that teenage boys were so different in the way they walked, but she was sensible enough not to argue. Finally, Lohan was satisfied. He crouched next to her. "It is time for you to leave," he said. "But there is something I must tell you before you go."

"What?"

She was alarmed, but he held up a hand, reassuring her. "There is a boy who is coming to meet you," he said. "He is on his way already, travelling from England."

Her first thought was that it was Aidan – but that was ridiculous. Aidan knew nothing about what was happening.

"His name is Matt."

The boy out of her dream! The boy who had led her through the door at the church of St Meredith's. Scarlett felt a surge of hope and excitement. She didn't know why, but if Matt was on his way then she was sure that everything would be all right.

"He is not coming to Hong Kong," Lohan went on. "It is too dangerous here. But he will be in Macau. He is being protected by the Master of the Mountain. He will remain there until he knows that we have been successful and that you have escaped. Then he will follow you and our work will be done."

"Who is the Master of the Mountain?" Scarlett asked.

"He is a very powerful man." That was all Lohan was prepared to say. He straightened up. "Don't speak until you are on the boat. If anyone tries to talk to you, ignore them. When you are with your new parents, hold your mother's hand. She alone will talk to you and you'll smile at her and pretend that you understand. When you are on the *Jade Emperor*, she will take you straight to her cabin. You will remain there until the ship leaves."

"Thank you," Scarlett said. "Thank you for helping me."

Lohan glanced at her and just for a moment she saw the hardness in his eyes and knew that whatever else he was, he would never be her friend. "You do not need to thank us," he said. "Do not imagine that we are helping you because we want to. We are obeying orders from the Master of the Mountain. You are important to him. That is all that matters. Do not let us down."

They opened the warehouse door and, remembering her new walk, Scarlett went out. She found herself in a concrete-lined alleyway. It was after five o'clock and the light was already turning grey. As she stood there, a car drove past and she flinched, afraid of being seen. But she was a boy now: the son of Chinese parents. Nobody was going to look at her twice. Jet and Sing had joined her. The three of them set off together, making their way towards the main road.

The alleyway came out at the very tip of Kowloon, where the Salisbury Road curved round on its way to the ferry terminals. The harbour was in front of them. Scarlett could see all of Hong Kong on the other side of the water with The Nail, the head-quarters of Nightrise, slanting diagonally out of the very centre where it seemed to have been smashed in.

"Walk slowly," Jet whispered. "If you see anyone looking at you, just ignore them. Don't stop…"

They walked down the Salisbury Road, passing the Hong Kong Cultural Centre, a huge, white-tiled building that looked a little bit like a ski slope. The weather had changed again. The sky was clear and the evening sun was dipping down, the water shimmering silver and blood-red. Despite the horror of the last thirty-six hours, everything looked very ordinary. There were several groups of tourists on the promenade, enjoying the view. Crowds of people were pouring out of the terminal for the Star Ferry, on their way home. Young couples holding hands walked together. Newspaper and food sellers stood behind their stalls, waiting for business. A fleet of ships, all different shapes and sizes, were chugging back and forth.

And all the time Scarlett was thinking – what is real and what isn't? Which of these people are shape-changers? How many of them are looking for me? She walked on between Jet and Sing, trying to behave normally but knowing all the time that there were a thousand eyes searching for her. She was already beginning to sweat with all the padding pressing down on her. It made it difficult to breathe.

They passed the Peninsula Hotel. Just a few days before, Scarlett had gone there with Audrey Cheng. They had sat down for tea and sandwiches. It felt like a lifetime ago. They turned into a wide avenue and she found herself walking past a police station. Two men came out, chatting together in dark blue and silver uniforms. Both of them carried guns. Scarlett remembered what Lohan had told her. The Old Ones controlled the police as well as the government and the civil service. These two men would have her description. If they recognized her it would all be over before they got anywhere near the ship.

But they didn't. They continued past and it was only when they had gone that Scarlett realized she had stopped breathing. She felt completely defenceless, waiting for someone to shout her name and for the crowd to close in. A few inches of padding and a handful of make-up was all that stood between her and capture. She was terrified that it wouldn't be enough.

Harbour City lay ahead of them. It was just another shopping centre, though much bigger than any she had visited with Mrs Cheng. They strolled in as if that was what they had always intended to do, as if they were just three friends out for an evening's shopping. The interior was very ugly. It was brightly lit with small, box-like shops standing next to each other in corridors that seemed to go on for ever. They were selling the usual goods: jeans and T-shirts and sunglasses and souvenirs, with fewer famous names than could be found in Hong Kong Central and presumably lower prices.

They continued past a luggage store and there, ahead of them, Scarlett saw a neon sign that read TSIM CHAI KEE HERBAL REMEDIES and knew that they had reached the place where the exchange would happen. The shop was directly in front of them. It was filled with cardboard boxes and glass bottles. Three people were standing with their backs to the front door. A man, a woman and, between them, a boy.

The woman was plump with grey hair, dressed in black. The man was smaller than her, laden down with shopping bags, with a camera around his neck. Their son was dressed exactly the same as Scarlett. They were waiting while the shop assistant wrapped up a packet of tea.

Scarlett walked in. Jet and Sing didn't follow her but continued on their way. At the same time, the boy walked forward, further into the shop and disappeared. The man and the woman stayed exactly where they were so that as Scarlett entered, there was a space between them. And that was it. A moment later she was standing between them. The woman paid for the tea. The shopkeeper handed over some change.

The three of them left together.

A mother, a father and a son had gone into the shop. A mother, a father and a son walked out of it. As they left, Scarlett glanced up and noticed a TV camera in the passageway trained down on them and wondered if there was anybody watching and, if so, whether they could possibly have seen anything that might have aroused their suspicions. But for the first time, she was feeling confident. She was no longer on her own. She was part of a family now. She would be joining hundreds or even thousands of tourists returning to the *Jade Emperor*. Even the Old Ones with all their agents would be unable to spot her.

The family left Harbour City through a set of huge glass doors that brought them straight out onto Ocean Terminal. And there was the ship, tied to the quay by ropes as thick as trees. The *Jade Emperor* was massive, with at least a dozen decks, each one laid out on top of the other with two smoking funnels at the very top. The lower part of the ship was punctuated by a long line of tiny-looking portholes, but further up there were full-sized sliding windows that probably opened onto state rooms for the multi-millionaires on board. The *Jade Emperor* was entirely white, apart from the funnels which were bright green. Crew members, also in spotless white, were hurrying along the corridors, mopping the decks and polishing the brass railings as if it were vital for the ship to look its best before it was allowed to leave.

Scarlett examined her surroundings. The ship was on her left, blocking out the view over to Hong Kong, with a single gangplank, slanting down at its centre. On the right, running the full length of the quay, was a two-storey building lined with flags. This was the back of Harbour City, the shopping centre she had just visited. Between them was a strip of concrete about ten metres across, which they would all have to walk along if they wanted to go on board.

The way was blocked by a series of metal fences that forced passengers to snake round to a control point where half a dozen men in uniforms were checking passports and embarkation slips. The sun was beginning to set now, and although it still sparkled on the water and glinted off the ship's railings, the actual walkway was in shadow. So this was it. Five minutes and maybe fifty paces separated Scarlett from freedom. Once she was on board the *Jade Emperor*, it would be over. Matt was waiting for her. Help had finally arrived. She would set sail and she would never see Hong Kong again.

The woman acting as Scarlett's mother, Mrs Soong, said something and reached out for her hand. Scarlett took it and they began to walk towards the barrier. Nobody stopped them. Nobody even seemed to glance their way. They passed a restaurant with floor-to-ceiling plate glass windows and tables and gas umbrellas outside. It was too late for lunch and too early for dinner so there was hardly anyone there, but as they continued forward Scarlett noticed a man with grey hair and glasses, sipping a glass of beer. He was partly obscured by the window but there was something familiar about him, the way he sat, even the way he held his glass. She stopped dead.

It was Paul Adams.

Maybe if she hadn't stopped so abruptly, he wouldn't have noticed her. But now he looked up and stared at her. Even then he might not have recognized her. But they had made eye contact. That was what did it. Even with the spectacles and the contact lenses, the strange clothes and the short hair, the two of them had made the link.

And Scarlett was glad to see him. For the past week she had been worrying about him, wondering if he was dead or alive. She had hated the thought of skulking out of Hong Kong without letting him know and if there had been any way to warn him what was happening, she would have done so. This was her opportunity. She couldn't just leave him behind.

A second later, he burst out of the restaurant and onto the quay. He still couldn't decide if it was really her. The disguise was that good. But then she smiled at him and he came over to her, his face a mixture of bafflement and relief.

"Scarly... Is that you?"

Scarlett felt Mrs Soong stiffen beside her. Mr Soong stopped, his face filled with alarm. None of the guards at the passport control had noticed them. Tourists were streaming past on both sides, taking out their documents as they approached the fence. Scarlett knew she would have to be quick. She was risking everything even by talking to him but she didn't care. She felt a huge sense of relief. Her father was alive.

"Scarly...?" Paul Adams spoke her name again, peering at her, trying to see through the disguise.

"Dad..." Scarlett whispered. "We can't talk. You have to leave Hong Kong. We're in terrible..."

She didn't finish the sentence.

To her horror, Paul Adams grabbed hold of her, dragging her hand up as if to show her off. His face was flushed with excitement – and something else. He looked demented. There was a sort of terror in his eyes. He was like a man who had just committed murder.

"It's her!" he shouted. "I've got her! She's here!"

"No, Dad…"

But it was already too late. The uniformed policemen had heard. They were already heading towards them. The tourists had stopped moving and in an instant Scarlett saw that half of them weren't tourists at all. They began to close in, their faces blank, their eyes shining with triumph. More people appeared, pouring out of the shopping centre. Matted hair. Dead, white skin. Their mouths hanging open. Dozens of them. And the flies. They burst into the air like a dark geyser and spread out, swarming overhead.

"Dad … what have you done?"

He clung onto her, one hand on her wrist, the other around her neck, strangling her. Mr and Mrs Soong stood there, paralysed, then tried to run. The woman was the first to be brought down. One of the tourists grabbed her. A few seconds earlier he had looked like a grandfather, an Englishman enjoying his retirement. But the mask had slipped. He was grinning and his eyes were ablaze. He was holding her with terrible strength, his hooked fingers gouging into her face, forcing her down to her knees. Then they were all onto her. Mrs Soong disappeared in a crowd that was moving now like a single creature. Mr Soong had taken out a gun. He pointed it at one of the approaching policemen and fired. The bullet hit the policeman in the face, tearing a huge hole in his cheek, but he didn't even flinch. He kept on coming. Mr Soong fired a second time, this time straight into the man's chest. Blood spouted but still the policeman came. Mr Soong was trapped. He had nowhere to run. Scarlett saw him push the barrel of the gun into his own mouth. She closed her eyes a moment before he fired.

It was easy to tell who were the real tourists now. They were screaming, in hysterics, dropping their new purchases and scattering across the quay, unsure what was going on, not wanting to be part of it. A woman in a fur coat slipped and fell. She was immediately trampled underfoot by the rest of the crowd, trying to get past. Two men were knocked over the side into the narrow space between the ship and the quay. Scarlett heard them hit the water and doubted that either of them would ever climb out again.

Her father was still holding her. She couldn't believe what he had done. He had deliberately told them she was there. He had been waiting for her all along. And she had helped him. There had been one final trap and she had fallen into it.

"I'm sorry, Scarly," he was saying. "I had to do it. It was the only way. They've promised that they won't hurt you, and my reward, the reward for both of us – we're going to be rich! You have no idea how much power they have. And we're going to be part of it … their new world."

Of course he had been in it all along. He worked for Nightrise. He had invited her here, made her leave school early with no explanation. He had been skulking somewhere nearby, leaving her in their clutches. And finally he had been positioned here, just in case she tried to get onto the ship…

Scarlett thought of all the people who had tried to help her, all the people who had died because of her. Mr and Mrs Soong had spent just a few minutes with her but it had been enough.

She had killed them.

She listened to this pathetic man – he was still jabbering at her – and she spat in his face.

Then someone grabbed her from behind. It was Karl. She didn't know where he had come from, but the chauffeur was unbelievably strong. He lifted her into the air, then dashed her down. Her head hit the concrete so hard that she thought her skull must have cracked. A bolt of sheer pain ripped across her vision.

In the final moments of consciousness, she saw a whole series of images, flickering across her vision like an out-of-control slide show. There was Matt, the boy she had never met in the real world, on his way to Macau. There were the other three – Scott, Jamie and Pedro – gazing at her helplessly. There was the beach where she had found herself night after night. And there, once again, was the neon sign with a symbol that was shaped like a triangle and two words:

SIGNAL EIGHT

The letters flared in the darkness and looking through them she saw the chairman, Audrey Cheng, Father Gregory and, for one last brief moment, her father.

"It's coming," she managed to whisper to them.

Then the darkness rushed in, slamming into her like an express train and at that moment she felt something unlock inside her. It was like a window being shattered and she knew that she would never be the same again.

And five hundred miles away, in a place called the Strait of

Luzon, between Thailand and the Philippines, the dragon heard her. It was there because she had summoned it. The dragon had been sleeping in the very depths of the ocean but it slowly opened one eye.

SIGNAL NINE

The letters burned in brilliant neon light. There was a symbol beside it, an hour glass and Scarlett almost wanted to laugh because she knew what it was saying. Time's up. The countdown has begun.

The dragon began to move. Nothing could get in its way.

It was heading for Hong Kong.

MATT'S DIARY (3)

I don't think I'm going to be able to write much more of this diary. I don't find it easy, putting all these words together, and anyway, what's the point? Who will ever read it? Richard thought it was a good idea but really it just fills in time.

I can't believe we've finally made it to Macau. Jamie is asleep, worn out with jet lag after another flight across the world, and Richard is in a room next door. In an hour's time, we're going to meet a man called Han Shan-tung who can help us get into Hong Kong. We've waited almost a week for him to turn up and I just hope that we haven't been wasting our time. We have no idea at all what's been happening to Scarlett, whether she is even alive or dead. Harry Foster, the Australian newspaper man who was at the meeting of the Nexus, sent someone to meet her – an assistant from his office. Maybe he managed to track her down but we never heard. The assistant went missing … presumed dead.

The Old Ones are there, waiting for me to arrive. In a way, it's extraordinary that they've managed to keep themselves hidden, but that has always been their way. When I was in Yorkshire, they worked through Jayne Deverill and the villagers who lived at Lesser Malling. In Peru, it was Diego Salamanda. Now it's Nightrise. They like people to do their dirty work for them and when war finally breaks out, as I know it must, my guess is that they won't reveal themselves until the end. And by then it will be too late. They will have won.

Maybe the five days we had in London were worth it after all. Jamie enjoyed himself, seeing all the sights, and in the end I enjoyed being with him. Buckingham Palace, the London Eye, Harrods, the London Dungeon. Richard kept us busy, maybe because he wanted to keep our minds off what lay ahead. We also spoke to Pedro and Scott in Vilcabamba, talking on the satellite phone. Pedro is worried about Scott. He still seems far away, as if he isn't even on our side. I know he's angry that I separated him from Jamie, but I still think it was a good idea. He isn't ready yet.

And then the flight. London to Singapore, followed by Singapore to Macau. I'm too tired to sleep. When I've finished this, I'll have another shower. A cold one, this time. Maybe it will wake me up.

I don't know what to make of Macau. If anyone had asked me about it six months ago, I wouldn't even have been able to point to it on a map. I hadn't heard of it. As it turns out, it's a chunk of land, just ten miles from one end to the other. And it's packed with some of the weirdest buildings I've ever seen. Take the ferry terminal. If you're coming in from Hong Kong on the jet-foil, it's the first building you'll see and you'd have thought they could have made it a bit welcoming. It's not. It's a slab of white concrete, surrounded by flyovers. It's drab and ugly.

But then you come to the casinos and you think you must

have landed on another planet. Macau makes its money out of gambling ... horse racing, greyhound racing, blackjack and roulette. The casinos look like nothing I've ever seen before. One of them is all gold, like a piece of metal bent in the middle. There's another one like a sort of crazy birthday cake. The biggest and the most spectacular reminded me of a giant flower. It was five times taller than anything else in the city. I got a crick in my neck trying to see the top.

The old part of Macau was better. Richard told me that it had once belonged to the Portuguese and he pointed out their influence in some of the palaces with their pillars, arcades and balconies jutting out over the street. But it was still a bit of a dog's dinner. The traffic and the crowds were Chinese. The older buildings seemed to be in better condition than the new ones, which were all dirty and falling down. The Portuguese had built pretty squares and fountains. Then the Chinese had come along and added casinos, shops and blocks of flats, forty or fifty floors high. And now they were all stuck next to each other, like quarrelling neighbours.

Jamie was disappointed too. "I once read a book about China," he told me. "It was in the house when we were in Salt Lake City. I never read very much, but it had dragons and magicians and I thought it must be a really cool place. I guess the book was wrong..."

We were met at the airport by a young Chinese guy who was carrying a big bunch of white flowers. That was a bit weird, but it was the signal we had been given so we would recognize him. He dumped them straight away. There was a Rolls Royce parked outside, numberplate HST 1. I noticed that it had been parked in a NO WAITING zone but nobody had given it a ticket. So that told me something about Han Shan-tung. He likes to show off.

The journey from the airport took about half an hour. It was pouring with rain, which certainly didn't make Macau look any better. Fortunately, it eased off a little by the time we arrived here.

And where are we now?

The driver stopped in front of a wide flight of stairs which climbed up between two old-looking walls that had been painted yellow. The steps were decorated with a black and white mosaic and there were miniature palms growing in neat beds along the side. There were clumps of trees behind the walls. They were still in leaf, filling the sky and blocking out any sight of the shops and apartments. It was like walking through a park. The driver got out of the car and signalled for us to follow him. We grabbed our bags and went about half-way up the stairs until we came to a metal gate that swung open as we approached.

It wasn't a park on the other side. It was a private garden with a courtyard, a marble fountain that had been switched off and, beyond, a really amazing house built in a Spanish style. The house was painted yellow, like the wall, with green shutters on the windows and a balcony on the first floor. It looked a bit like an embassy, somewhere you weren't normally allowed. The house seemed to belong to its own world. It was right in the middle of Macau and yet somehow it was outside it.

"Quite a place," Richard said.

The driver gestured and we went in.

The front door also opened as we walked towards it. A woman was waiting for us on the other side. She was some sort of servant, dressed in a long, black dress with a grey shirt buttoned up to the neck. She bowed and smiled.

"Welcome to the home of Mr Shan-tung. I hope you had a good journey. Please, will you come this way? I will take you to your rooms. Mr Shan-tung invites you to join him for dinner at eight o'clock."

It was one of the most beautiful houses I had ever seen. Everything was very simple but somehow arranged for maximum effect so that a single vase on a shelf, sitting under a spotlight, somehow let you know that it was Ming or something and probably worth a million pounds. The floors were polished wood, the ceilings double height, the walls clean and white. As we went upstairs, we passed paintings by Chinese artists. They were very simple and clean and they probably cost a fortune too.

We all had bedrooms looking out over the garden, on the same floor; Jamie and me sharing, Richard on his own. The beds had already been turned down with sheets that looked brand new. There was a TV and a fridge filled with Coke and fruit juice. It was like being in a five star hotel, but (as Richard said) hopefully without the bill.

We were all dirty and tired after so much travelling and Jamie and I tossed a coin to see who got to shower first. I won and stood naked in a cubicle that would have been big enough to sleep in, with steaming water jetting at me from nine directions. There were towelling robes to put on when we came out. Jamie went next. He was asleep before he was even dry.

I would have liked to have slept.

I've been thinking a lot about the library that I visited. Did I make the right decision? I didn't read the book and I'm beginning to wish I had. Right now I'm just a forty-five minute journey away from Hong Kong and I have no idea what I will find there. The book would have told me. It might have warned me not to go.

But it might also have told me when and how my life will end – and who would want to read that?

It makes me think of a computer game that I used to play when I was living in Ipswich. It was an adventure, a series of puzzles that took you through a whole set of different worlds. Shortly after I met Kelvin, he showed me how to download a cheat. It gave me all the answers. It took away the mystery. Suddenly I knew everything I wanted – but here's the strange thing. I never played the game again. I just wasn't interested.

Why did the Librarian show it to me? What was the point he was trying to make? And for that matter, who was he? He never even told me his name. When I think about it, the dreamworld really annoys me. It's supposed to help us but all it ever gives us is puzzles and clues. I know that it's important to what's going to happen, that it's there for a reason. One day, perhaps, I'll find out what that reason is.

I've written enough. It's twenty to eight. Time to wake Jamie and to meet our host. Han Shan-tung.

Hong Kong is waiting for us. It's out there in the darkness, but I can feel it calling.

Very soon now, I will arrive.

MASTER OF THE MOUNTAIN

Han Shan-tung was one of the most impressive men Matt had ever seen. He was like a bronze Buddha in a Chinese temple. He had the same presence, the same sense of power. He wasn't exactly fat but he was very solid, built like a Sumo wrestler. You could imagine him breaking every one of your fingers when you shook hands.

His hair was black. His face was round, with thick lips and hard, watchful eyes. He was elegantly dressed in a suit that was obviously expensive, possibly silk. His fingers, resting on the table in front of him, were manicured and he wore a slim, silver wedding ring. There was a packet of cigarettes and a gold lighter on the table next to him ... his one vice perhaps. But none of his guests was ever going to give him a lecture on smoking. Everything about the man, even the way he sat there – still and silent – suggested that he wasn't someone to be argued with. He was someone who was used to being obeyed.

And yet his manner was pleasant enough. "Good evening," he said. "Please come and sit down." His English was perfect. Every word was well-modulated and precise.

He was sitting in the dining room, at the head of a long table that could have seated ten people but which had been laid for only four. The room was as elegant as the rest of the house, with floor-to-ceiling windows looking out onto a wooden terrace and views of the garden beyond. Richard, Matt and Jamie took their places. At once, a door at the side slid open and two women appeared, pouring water and shaking out the napkins.

The man waited until they had gone. "My name is Han Shantung," he announced.

"I'm Richard Cole." Richard introduced himself, then the boys. He had already decided he was going to use the names that were on their passports. "This is Martin Hopkins. And Nicholas Helsey."

"I would have said that this was Matthew Freeman and Jamie Tyler," Shan-tung muttered. "And I would add that it is discourteous to lie to a man in his own home – but I will overlook it as I can understand that you are nervous. Let me assure you, Mr Cole. I know everything about all three of you. More, in fact, than you perhaps know about yourselves. Otherwise you would not be here."

"And we know nothing about you," Richard replied. "That's why we have to be careful."

"Very wise. Well, it will be my pleasure to enlighten you. But first we should eat."

As if on cue, the two women returned, carrying plates of food. Silently, they laid out a Chinese dinner. It was a world apart from the sweet and sour, deep-fried grease balls that Matt had once purchased at his local takeaway in Ipswich. The dinner came in about a dozen china bowls – fish, meat, rice, noodles – and it had obviously been cooked by a world-class

303

chef. Matt was glad to see that he had been provided with a spoon and fork. Han Shan-tung ate with chopsticks.

"I must apologize to you," he began. There was no small talk. He didn't ask them about their journey or what they thought of their rooms. "Urgent business took me to America. It was badly timed because it delayed your arrival here. And I'm afraid I have bad news. I had hoped that the object of your journey would have been sitting here with us tonight. I am referring to the girl, Lin Mo." He continued quickly, before Richard could interrupt. "You call her Scarlett Adams. But I refer to her by the name she was given before she was adopted and taken to the West."

"How do you know about Scarlett?" Richard asked.

Shan-tung leaned forward and plucked a prawn off one of the dishes. Despite his large hands, he used the chopsticks very delicately, like a scientist handling a specimen. "I know a great deal about the girl," he replied. "The fact of the matter is that she was with my agents in Hong Kong only yesterday. I have spent a great deal of time and money – not to mention human life – trying to remove her from the city."

Matt played back what Shan-tung had just said and realized that it confirmed exactly what he had thought. "The Old Ones are in Hong Kong," he said.

"The Old Ones have taken over Hong Kong," Shan-tung replied. "They control almost every aspect of the city. From the government and the police to the street cleaners. I do not know how many people they have killed, but the number must run into thousands. My people have been fighting them on your behalf. We are the only remaining resistance."

"Who are your people?" Richard asked.

Shan-tung sighed. "It is unnecessary to keep asking me these things. I am about to tell you anyway."

"I'm sorry." Richard realized his error. "I suppose it's a habit. I used to be a journalist."

"I do not like journalists. It is nothing personal – but they have caused me trouble in the past. I suggest you continue eating. I will tell you everything you need to know."

Han Shan-tung had barely eaten anything. But he laid down his chopsticks and began to speak.

"I have the very considerable honour to be a member of an organization called the Pah Lien. This translates as the White Lotus Society. You might have remarked upon a clue that I sent you at the airport. The man who met you was carrying a bunch of lilies. The lily is part of the lotus family. My society is a very old one. It was founded in the fourth century to resist the foreign invaders known as the Mongols who then ruled over China. The aim of White Lotus remained the same over the next four centuries. It was to help the Chinese people fight against tyranny and oppression.

"But over the years, something very interesting happened. The White Lotus Society changed. It will be difficult for you to understand the nature of this change, so let me explain it to you by referring to a character from your own history. You will, I am sure, know Robin Hood. He stole from the rich and gave to the poor. He was a hero to the peasants in Sherwood Forest. But to the authorities, he was an outlaw, a criminal. They would have hanged him if they could.

"In the early days, the White Lotus Society operated in much the same way. Indeed, it might interest you to know that the society had a motto: *Ta fu – chih p'in*. This translates as 'strike the rich and help the poor'. But here was the crucial difference. As the years passed, White Lotus found that it was enjoying and benefiting from the criminal nature of its activities. It was also remarkably successful in the world of organized crime. It continued to steal from the rich but, as its members became richer themselves, it found itself giving rather less to the poor. It also changed its name. It became known as the Three United Society. There was a reason for this. White Lotus believed that the world was made up of three different parts: heaven, earth and mankind. Its members therefore had a triangle tattooed onto their body. The triangle also appeared on their flags. And in the end, they became known simply as the Triads."

There was a long silence. Matt had heard of the Triads, the criminal gangs that were active all over Asia. They were drug dealers. They were involved in people smuggling, prostitution, extortion and murder. They would torture or kill anyone who got in their way. They were as brutal as they were powerful. And this man was calmly admitting that he was one of them! He glanced at Jamie. The American boy was listening politely. He didn't seem shocked by what he had just heard. Richard, on the other hand, was staring open-mouthed.

"I can see that you are dismayed," Shan-tung remarked. "And before you ask me one of your inane questions, Mr Cole, I will answer you. Yes. I am a criminal. More than that, I am what is known as *Shan Chu*, the Master of the Mountain. This means that I am the supreme leader of my own Triad. I cannot tell you how many people I have murdered to get to where I am today, but at a conservative guess I would say about twenty-five. I do know that I am wanted in exactly nine countries, including the United Kingdom and the United States – and I would have been arrested a long time ago if I hadn't paid the right people a great deal of money to leave me alone.

"You are now wondering if you should be sitting at my table, eating my food. You are asking yourself why I should wish to help you in your struggle against the Old Ones. You are thinking, perhaps, that it would be more natural for me to be on their side. But you would be wrong.

"Until very recently, I controlled all crime in Hong Kong. I have, for example, heroin laboratories in Kowloon and the New Territories. I have illegal casinos and betting shops throughout the island. Immigrants from China were paying $5,000 a time for me to help them cross the border illegally. The arrival of the Old Ones has changed everything. They have no interest in profit. They do not want to do business. They want only to destroy everything around them – and that includes the Triads. They are as much my enemy as anyone's, the only difference being that I have the means to fight back. And that is what I have been doing. There is a certain irony, don't you think? I am undoubtedly a bad man. But a greater evil has come my way and now I am forced to do good.

"And so I have used all my resources within Hong Kong to set up a resistance. I have buildings. I have people. I have weapons ... not that they are of much use against creatures that can form themselves out of flies. Above all, I have determination. I will not be defeated by the Old Ones. They can destroy the world. But they will not destroy me."

"I'm surprised they didn't ask you to work for them," Richard said.

"As it happens, they did indeed ask me to serve them. The Nightrise Corporation approached me exactly a year ago. But the Master of the Mountain does not serve anyone. I mentioned twenty-five victims. The man who put that question to me would have been the twenty-fifth."

"May I ask a question?" Matt asked.

"You have my permission," Han Shan-tung replied. "But I should warn you that soon I have a question to put to you and I very much hope you will be able to provide me with the right answer."

Matt didn't like the sound of that, but he went on anyway. "How do you know about Scarlett?" he asked. "And why did you call her Lin Mo?"

"The White Lotus Society has always known about the Gatekeepers. You must remember that in our early days, almost two thousand years ago, we were to all intents and purposes a religious order. We still are. That means we are the keepers of many secrets ... sacred texts and ancient beliefs. Even when we began to devote ourselves exclusively to crime, we stayed true to ourselves. The secrets were passed on from generation to generation. And I think we always knew that one day we would be called upon to return to our origins, to take up the sword once again.

"As to the second part of your question regarding Lin Mo, that I am not yet prepared to tell you. I need to be persuaded that I can trust you and that is still not the case.

"However, I can say that she was born in a place called

Meizhou. We always knew that the Old Ones would return and look for her ... that she was one of the Gatekeepers. We therefore arranged for her to be adopted and taken to the West. We wanted her to be as far away from here as possible. We hoped that she would be safe."

"It didn't work."

Shan-tung shrugged. "We did everything we could to protect her. It was not our fault that the Old Ones found her. In fact, if anyone is to blame, it is her. Nonetheless, you are right. The Old Ones found her and brought her back."

"You tried to get her out of Hong Kong," Jamie said. He hadn't eaten very much, absorbed in what he was being told.

"Scarlett was kept under guard from the moment she arrived," Shan-tung explained. "With great difficulty, we managed to get a message to her. My most trusted agent in Hong Kong, a man called Lohan, contacted her and arranged for the shape-changer who had been guarding her to be killed. He took her to a safe place where we hoped to keep her hidden, but unfortunately – and again through no fault of our own – she was found again. As I mentioned to you, several of my people died. However, Lohan managed to move her to one of our warehouses and had planned to smuggle her out on a cruise liner. That was yesterday. The plan failed for reasons that are not yet clear. She is now their prisoner."

"So what do we do now?" Jamie asked. "How do we get her back again?"

The Master of the Mountain poured himself a glass of water from a crystal jug and drank it.

"Jamie and I can go into Hong Kong," Matt said. "We can find her..."

"If you go into Hong Kong, you will be doing exactly what they want you to do. They will be waiting for you and although they will not kill you – that is not part of their plan – they will keep you in so much pain that you will wish constantly for death."

"We can't just leave her."

"You may have no choice."

"No, Mr Shan-tung," Matt said. "You don't believe that. Otherwise, why would you have invited me here?" Matt looked him straight in the eyes. "You're going to help us get into Hong Kong," he said. "You've already told us. You've got people over there. You can smuggle us in. We can find Scarlett. And we can be out of there before the Old Ones know what's happened."

Han Shan-tung set his glass down. "I might help you," he said. "But as I mentioned to you earlier, there is still a question you have to answer for me."

"And what is that?"

"I am, by nature, a very careful man. I have told you that I have killed twenty-five times. What I should have added is that there have been as many attempts on my own life. You are here in my house on the recommendation of my friend, Mr Lee. I trust him. He has been useful to me in the past, and he definitely believes that you and the American boy are who you say you are."

"Is that your question?"

"It is exactly that. How can I be sure that you are one of the Five?"

Matt thought for a moment. Then he pointed at the crystal jug. He didn't even need to think about it any more. The jug was swept, instantly, off the table. It fell to the floor and smashed. Shan-tung blinked. It was his only reaction. But then he slowly smiled. "An amusing conjuring trick. But it is still not enough. I do not question your abilities. It is your identity I wish to know."

"I'll read your mind," Jamie said. "You say you know everything about us. In ten seconds I can tell you even more about you."

"I would recommend that you stay out of my mind," Han Shan-tung said. He turned to Matt. "There is a test, a trial you might say, that will prove to me beyond any doubt that you are who you say you are. Only one of you needs to take part in it. But I should warn you though that to fail will cause you great pain and perhaps even death. What do you say?"

Matt shrugged. "We need your help," he said. "We've flown a long way to get it. If there is no other way…"

"There isn't."

"Matt…" Richard muttered.

"Then let's go ahead," Matt said. "What test do you have in mind?"

Han Shan-tung got to his feet. "It is called the sword ladder," he said. He gestured towards a door at the back of the room. "Please … will you come this way."

THE SWORD LADDER

Matt stood up and followed Han Shan-tung. Richard and Jamie came behind. They went through the door into a long corridor, all polished wood but otherwise undecorated. There was a second door at the far end.

It opened into a large, square room that didn't seem to belong to the rest of the house. It reminded Matt of a chapel or perhaps a concert hall that might comfortably seat fifty or sixty people. The walls were plain and wood-panelled, matching the corridor outside, and there were pews arranged around three of the sides. The fourth was concealed by a dark red curtain that had been pulled across, perhaps concealing a stage. There was a gallery above the curtain, but it was high up, arranged in such a way that it was impossible to tell from floor level what it might contain.

"You are inside a Triad lodge," Mr Shan-tung explained. "And you should consider yourselves very privileged. Only Triad members and initiates are allowed in here and normally any outsiders would be instantly killed. We meet in this place on the twenty-fifth day of each Chinese month. There is a separate entrance from the street. You might be interested to know that an initiation ceremony lasts six hours. A new recruit is expected to answer three hundred and thirty-three questions about the society. He learns secret handshakes and recognition signals. A lock of his hair is taken and he signs his name in blood."

"Actually, I wasn't thinking of joining," Richard muttered.

Fortunately, Shan-tung didn't appear to have heard. "I speak of our rituals to remind you that the White Lotus Society is very old," he went on. "Things have, of course, changed with modern times. Nine hundred years ago, initiates would have drunk each other's blood, mixed with wine. And there is another part of the ceremony that has fallen out of use. When China was enslaved by Kublai Khan, it is said, the society searched for a leader, the one man who might liberate them. That man would be known as the Buddhist Messiah and he would show himself by a sign…"

He crossed the room and pulled on a cord that drew back the curtain. Jamie gasped. Matt stepped forward. At first he thought he was looking at a strange ladder leading up to the balcony above but then he realized that it was actually made up of antique swords, each one polished until it shone, lashed together in a wire frame with the edges of the blades facing upwards. Theoretically, it might be possible to climb. But he doubted it. As soon as you rested your body weight on one sword, you would cut your foot in half. Even if you were light enough, the climb to the top would be agony. It was a long way to the balcony. Matt counted nineteen steps. Nineteen chances to slice yourself apart.

"In my time as Master of the Mountain, three initiates have claimed to be the Buddhist Messiah," Shan-tung explained.

"They asked my permission to be allowed to climb the ladder and I was glad to give it. Watching their attempts was a fascinating experience. One of them almost made it to the top before he fainted. Sadly, he broke his neck in the fall."

"What about the other two?" Matt asked.

"One cut off the fingers of his left hand on the first step and chose not to continue. The other bled to death."

"This is insane!" Richard couldn't restrain himself any more. "Matt isn't claiming to be your Buddhist Messiah or whatever you want to call it."

"He is claiming to be one of the Gatekeepers. If he is who he says he is, he has nothing to fear."

"And if we say no? If we refuse to perform your little party trick?"

"Then I will not help you. You will leave Macau. And the girl will die, slowly, on her own."

Richard swore under his breath. Jamie came forward and stood next to Matt. "I don't mind giving it a try," he said, quietly.

"Thanks, Jamie," Matt replied. "But I brought us here. I think this one's down to me..."

He took a step closer but Richard held out a hand. "Forget it, Matt!" he said. "You don't need to do this. There are plenty of ways we can get into Hong Kong without this maniac's help."

"We can't go in on our own," Matt said. "One of us has to try..."

"You're going to cut yourself to pieces."

"After the first finger, I promise I'll stop."

He went over to the ladder. Any hope that it might not be as dangerous as it looked vanished at once. The swords were fixed rigidly in place by the wires. The blades were pointing towards each other so that as he climbed up, the hilts and the points would be on alternate sides. The swords had been sharpened until they were razor-thin. He rested a finger on one and almost cut through the skin just doing that. If he had dropped an envelope onto it, he would have sliced it in two.

Could he do it? Every instinct told him that he couldn't, that it was impossible, that he was being asked to mutilate himself. He closed his eyes. Was there any way out of this? Did they really need this man's help? Hong Kong was only fifty miles away. They could get on a jet-foil and take their chances. Why would they want to involve themselves with gangsters anyway?

But he knew he was fooling himself. Scarlett was in trouble. If he'd wanted to go into Hong Kong on his own, he could have done it a week ago. There was no other way. He opened his eyes. "All right," he said.

"Remove your shoes," Shan-tung commanded.

"Sure," he muttered. "Shame to waste good leather." Right then, he was wondering if he would ever wear shoes again. He took them off, and his socks as well, for good measure. He could feel the wooden floor, cool against the soles of his feet. He flexed his toes.

"Matt..." Richard tried one last time.

"It's OK, Richard."

Matt didn't look at him. He didn't look at any of them. He knew there was only one way this was going to work. He had to focus completely on the task ahead of him. Nineteen steps.

He had once seen people walking on hot coals on television. And in India, fakirs did incredible things with their bodies. Matt remembered what he had done in the Nazca Desert. He had taken a bullet in full flight and turned it back on the person who had fired it. Mind control. That was what this was all about.

He reached out and gently took hold of one of the swords. He felt the blade cut through his skin. It hurt. Blood welled out of the palm of his hand.

"That's enough!" Richard exclaimed. "You can't do this."

"Yes. I can."

Matt gritted his teeth. He knew the mistake he had made. He had been thinking too much about the impossibility of what he was supposed to do. When he moved things without touching them, it never occurred to him that he couldn't do it. That was how the power worked. It was part of him and he could use it any time. This task might seem different but the principle was just the same. Nineteen steps. He wasn't going to hurt himself a second time. He was a Gatekeeper. He had nothing to fear.

He forgot Richard. He forgot where he was. The balcony above him … that was all that mattered. He let the swords blur in front of him. They were no longer there. He reached out with one hand. At the same time, he lifted his left foot and rested his bare sole on the first blade. There was no going back now.

Richard had seen many unforgettable things in his time with Matt, but this was the most incredible of all. He watched Matt begin to climb, one sword at a time, resting his entire weight on edges that were clearly razor sharp. He seemed to be in a self-induced trance, moving steadily upwards as if he were levitating. Already he was half-way up and he hadn't cut himself at all. Next to him, Jamie stared in wonderment. Even Han Shan-tung looked quietly impressed.

He reached the top. He climbed off the ladder and stood on the balcony. Nobody spoke. Shan-tung hurried to the side of the room and took a staircase that also led up. Matt waited for him. There was a single wound on his right palm, the result of his false start, but otherwise he was unharmed.

The Master of the Mountain reached him. He was holding a bandage. He bowed low, then handed it over. "I apologize for questioning you, Matthew," he said – and he sounded completely sincere. "You are indeed one of the Five and it is my honour to be able to help you."

Matt took the bandage and wrapped it round his hand. At the same time, he noticed an altar on the far side of the balcony, hidden from the room below. There were several gold bowls, incense sticks, two crouching Buddhas and, between them, a jade figure of a young girl, slim with long hair falling in waves around her shoulders.

"That is Lin Mo," Han Shan-tung said. "It is the answer to the question that you asked me earlier. Lin Mo is the name of a young girl in Chinese legend. She was born in Meizhou, in the eastern Guangdong province. She had the power to forecast the weather. And she grew up to become the goddess of the sea, very important to the sailors who explored these uncharted waters. She is still worshipped in Macau."

He moved over to the altar and bowed in front of it.

"This figure is very precious to me," he continued. "It is Ming dynasty. From the seventeenth century. It is said to be a true representation of Lin Mo, copied from an earlier work."

Matt recognized the face. He remembered the picture he had seen in the newspaper. "It's Scarlett, isn't it," he said.

"The girl that you know as Scarlett was also born in Meizhou. It was always our belief that she was the reincarnation of Lin Mo. And it is true, yes, that in appearance the two are identical."

"So you're going to help us."

Shan-tung nodded. "You must leave very soon," he said. "Come now with me to my study and we will make the final preparations."

He led Matt over to the staircase and the two of them made their way down. Richard and Jamie were waiting for him.

"That was quite a trick," Richard muttered through clenched teeth.

Jamie said nothing. He rested a hand briefly on Matt's shoulder. He was glad that it hadn't been him.

They followed Shan-tung back down the corridor and into a study that also overlooked the garden. It was an austere room with a large desk, a few shelves of books and little else. His whole manner had changed. He was still in command, a man who was used to being obeyed instantly, but he was being a little quieter about it. Had he really expected Matt to climb the sword ladder? He seemed shaken by what he had seen.

He took out a map and laid it on his desk. Matt glanced at his watch, wondering how long this would take. It was already ten o'clock.

"The Old Ones may control the city," Shan-tung said. "But if they have underestimated the size and extent of the Triads, then they have made a fatal mistake. I have a thousand foot-soldiers that I can place at your service. If called to do so, they will not hesitate to lay down their lives for you. That is our way. The man who commands them is called Lohan. His rank is 438 which we also call Incense Master. He will meet you when you arrive in Hong Kong."

"How do we know we can trust him?" Richard growled.

"Very simply, Mr Cole. He is my eldest son. You will recognize him because his face is scarred." Shan-tung drew a line with his finger, starting on his left cheek and crossing his mouth. "A man was sent to kill me with a *jian*, a Chinese sword. Lohan got in his way. If it were not for him, I would be dead. This is where you will meet…"

His finger stabbed down on the map, at a point close to the waterside.

"I have a legitimate business delivering fireworks to Kowloon. There is a warehouse next to the Salisbury Road and it is there you will be taken. Scarlett was also there before she was captured. You don't need to worry. The location is still secure.

"We are trying to discover where Scarlett is being held prisoner but so far we've had no luck. It is possible that she is here…" He pointed again, this time to a street on the other side of the water. "This is The Nail. It is in Queen Street and it is the headquarters of the Nightrise Corporation. If the girl is there, Lohan will lead an assault on the building. You will be with him.

"The Tai Shan Temple with the door that you were seeking is also in Queen Street." He pointed to a crossroads close to a patch of green with what might be a lake in the middle. "You would be wise not to go there as it is almost certainly being watched. But once you have the girl, the rules will change. It is less than a quarter of a mile away, close to Hong Kong Park. Lohan will help you enter the compound. He will kill anyone who gets in your way. You will enter the temple and the door will take you wherever you want to go."

"But what if Scarlett isn't at The Nail?" Richard asked.

"Then you will have to search for her. Perhaps her father will be able to help you." The finger slid across the page. "Paul Adams has returned to Wisdom Court, the apartment block where he lives. It is here, on Harcourt Road. Be warned. He was with her when she was captured and may have had a hand in what took place. We can't trust him. Even so, he may know where she is."

"And you think he'll tell?"

"We will make him tell us." Han Shan-tung muttered the words casually but there was something about the way he spoke that made the skin crawl.

He seemed to have finished. Matt was exhausted. He was looking forward to getting to bed. But then Han Shan-tung went over to the desk and took a mobile phone out of one of the drawers. He handed it to Richard. "You can use this to contact me at any time of the day or night," he explained. "The speed dial is already set. Just press one and it will connect you directly."

"So when are we leaving?" Jamie asked.

Shan-tung turned and looked at him. There was no expression on his face. "The boat is already waiting for you," he said. "You must enter Hong Kong under cover of darkness. You leave tonight."

INTO HONG KONG

The boat was tied up at Porto Exterior, the outer port of Macau. Han Shan-tung had said a brief goodbye in the hallway of his home and now Matt, Jamie and Richard were being driven across the city through half-empty streets. It was raining again and the pavements, black and glistening, had been deserted by the crowds, many of them sheltering in the casinos, throwing their money after dice and cards in the artificial glare of the chandeliers.

They were all tired. Jamie was half asleep, his head resting on the window, his long hair falling across his face. Richard was sitting next to him. Matt could tell that he was angry – with Shan-tung for arranging the ordeal of the sword ladder and with himself for allowing it. Matt was in the front, beside the driver. The speed of events had taken him by surprise. He had only just arrived in Macau and already he was leaving. He thought about what might lie ahead of him in Hong Kong and wondered if he was doing the right thing. It was obvious now that the whole place was a trap, set up by the Old Ones. And yet, he was walking straight into it.

But they wouldn't be expecting him … not like this. That was what he told himself. And there was no other way. He couldn't leave Scarlett on her own any longer. It had already been too long. It was his responsibility to find her and bring her out. He was a Gatekeeper. It was time to take control.

The ferry terminal was ahead but they didn't drive into it. Instead, the driver took them down a narrow road that led to the water's edge and stopped. They got out, bracing themselves against the cold night air.

For a moment, Matt and Richard found themselves standing next to each other. "Do you really think we should trust these people?" the journalist muttered, putting into words what he had been thinking all along. "They're Triads. Do you know what that means? Drugs and guns. Gambling. Prostitution. They'll chop up anyone who gets in their way – including you and me. Between them and the Old Ones, I wouldn't have said there was a lot to choose."

A few hours ago, Matt might have agreed. But he remembered how Han Shan-tung had looked at the statue of Scarlett, or Lin Mo as he preferred to call her. "I think they're on our side," he said.

"Maybe." Richard reached out for Matt's injured hand and turned it over. There was a dark stain seeping through the bandage. "But he still shouldn't have done that to you."

"I did it to myself," Matt said. "I wasn't concentrating."

Jamie came over to them. "I think he wants us to go with him," he said, glancing at the driver. He yawned. "I just hope this boat has got a decent bed."

There wasn't much to the port: a stretch of white concrete, a couple of gantries and arc lamps spreading a hard, electric glow that only made everything look more unwelcoming. Once again the rain had eased off but a thin drizzle hung in the air. The driver led them over to a boat, moored along the quayside. This was going to take them across.

It was an old, hard-working cargo boat with just two decks. The lower of them had a cargo hold that was open to the elements and looking into it, Matt saw that it was filled with wooden crates, each one marked with a name that had been stencilled in black letters: **KUNG HING TAO**. The cabin was on the upper deck. It was shaped like a greenhouse and not much bigger, with windows all the way round. There were two radio masts jutting into the air, a radar dish and a funnel that was already belching black smoke. The boat was completely ringed with car tyres to stop it colliding with the quay and this, along with the flaking paint and patches of rust, made it look as if it had been rescued from a junkyard. Matt just hoped the sea would be calm.

"We've got company," Richard said.

A man had appeared, climbing down from the cabin, his feet – in wellington boots – clanging against the metal rungs. As he stepped into the light, it became clear that he wasn't Chinese. He was a European, a big man with a beard, dark eyes and curly, black hair. His whole face looked beaten about – cracked lips, broken nose, veins showing through the skin. Either the weather had done it, too many years at sea, or he had once been a boxer … and an unsuccessful one. He was wearing jeans, a thick knitted jersey and a donkey jacket, dark blue, with the rain sparkling on his shoulders. His hands were huge and covered in oil.

"Good evening, my friends," the man said. "You are welcome to *Moon Moth*." He had introduced his ship but not himself. He had a deep voice and a Spanish accent. The words came from somewhere in his chest. "Mr Shan-tung has asked me to look after you. Are you ready to come on board?"

"How long will the journey take us?" Richard asked. He sounded doubtful.

"Three hours … maybe longer. We don't have the power of a jet-foil and the weather's strange. All this rain! It may hold us up, so the sooner we get started, the better." The man took out a pipe and tapped it against his teeth as if checking them for cavities. "I often make the journey at night, if that's what's worrying you," he went on. "Nobody's going to take any notice of us. So let's get out of this weather and be on our way."

He turned and climbed back onto the boat. Richard glanced at Matt. Matt shrugged. The captain hadn't been exactly friendly, but why should they have expected otherwise? These people were criminals. They were only obeying orders. They had no interest in the Gatekeepers or anybody else, so it was pointless to expect first class comfort and smiles.

Richard had brought his backpack with them – it was their only luggage. He picked it up and they followed the man on board. They reached the ladder and Matt was grateful that this one had ordinary rungs instead of swords. As he began to climb, he noticed a Chinese man in filthy jeans and an oil-skin jacket, drawing a tarpaulin over the crates. For a moment their eyes met and Matt found himself being studied with undisguised hostility. The man spat, then went back to work. He seemed to be the only crew.

There wasn't much room in the cabin which looked even older than the ship, with equipment that wouldn't have been out of place in a Second World War film. The captain was sitting on a stool in front of a steering wheel, surrounded by switches and gauges with markings that had largely faded away. The rain had picked up. It was streaming down the windows and the world outside was almost invisible, broken up into beads of water that clung in place, reflecting everything but showing very little. The engines were throbbing sullenly below. The whole cabin was vibrating. It smelled of salt water, diesel fuel and stale tobacco.

There was a low sofa and a couple of chairs for the three passengers. All the furniture was sagging and stained. Richard, Matt and Jamie took their places. The captain sat at the wheel, flicking on a pair of ancient windscreen wipers which began to swing from left to right, clearing the way in front of them. The Chinese crewman cast off and the boat slipped away, unseen, into the night.

A single row of lights shone ahead. There was a road bridge, at least half a mile long, snaking across the entire length of the harbour. But once they had passed underneath it there was nothing. *Moon Moth* had its own spotlights mounted on the bow and the cabin roof, but they barely penetrated the driving rain and showed nothing more than a circle of black water a few metres ahead.

The captain switched on the screens and the cabin glowed green with a soft beeping sound that divided up the silence like commas in a sentence. For about ten minutes nobody said anything but then the crewman appeared, carrying a battered tray with four tin mugs of hot chocolate which he had brought up from a galley somewhere below.

"You haven't told me your names," the captain said. He lit his pipe and blew smoke into the air, making the cabin feel closer and snugger than ever. It was very warm inside, presumably from the heat of the engines below.

Richard introduced them. "I'm Richard. This is Matt and Jamie." They were being smuggled into Hong Kong illegally, and anyway Han Shan-tung already knew who they were. There was no need for false names.

"And I am Hector Machado. But you can call me Captain. That is what everyone calls me – even when I am not on the ship."

"Are you Spanish?" Richard asked.

"Portuguese. I was born in Lisbon. Have you been there?"

Richard shook his head.

"I'm told that it's a beautiful city. I left there when I was three. My father came to Hong Kong to fight against the communists. This was his boat." Machado sucked on his pipe which glowed red. He blew out smoke. "He was shot dead in the very seat where I am sitting now. And the boat is mine."

"How many crew do you have?" Matt was thinking of the man he had seen. Why had he appeared so unfriendly?

"Just Billy. No need for anyone else."

"What's in the crates?"

Machado hesitated, as if afraid of giving too much away. Then he shrugged. "Fireworks. A lot of fireworks. Mr Shan-tung has a business selling them to mainland Hong Kong."

"And what do you carry when you're not delivering

fireworks?" Richard asked. His voice was hostile. It clearly bothered him, being with these people.

"I've carried all sorts of things, Richard. Stuff that maybe it
would be better you didn't know about. I've smuggled people
in, if that's what you want to know. And maybe you should be
grateful. I know the ins and outs. *Moon Moth* may not be much
to look at but she'll outrun the Hong Kong harbour patrols any
time … not that they'll bother themselves about us. Everyone
knows me in these parts. And they leave me alone."

"So how long have you worked for the Triads?"

"You think this is an interview? You want to write about
me?" Machado gestured with the pipe. "I'd get some rest
if I were you. It could be a long night." He slipped the pipe
between his teeth and said no more.

They cruised on into the darkness, guided by the strange,
green light of the radar system. The night was so huge that
it swallowed them completely. There was no moon or stars.
It was impossible to tell if it was still raining as the windows
were being lashed by sea spray. Machado sat where he was,
smoking in silence. Richard, Matt and Jamie sat at the back
of the cabin, out of his way. All three of them were tense and
nervous. They hadn't discussed what they might find in Hong
Kong, but now that they were finally on the way, they could
imagine what they might be up against. A whole city, millions
of people … and the Old Ones infesting everything. They had
to be mad to be going in there. But there seemed to be no
other way to get Scar out.

Jamie finished his hot chocolate and dozed off. Richard
opened his backpack and began to go through his things: he

327

had brought maps, money, a change of clothes. The precious diary – written by Joseph of Cordoba – was also there, sealed in plastic to keep it protected. Matt noticed a glimmer of gold and realized that he was carrying the *tumi* – the Inca knife.

Richard glanced up. "You never know when it may come in handy," he said. "Anyway, I didn't like leaving it behind with that bunch of crooks." He zipped the backpack shut, then lowered his voice. "What do you think?" he asked.

He was referring to Hector Machado, although he didn't need to whisper as the captain would never had heard him above the noise of the engines.

"Shan-tung trusts him," Matt said.

"He doesn't seem to be exactly friendly."

"He doesn't have to be friendly. He just has to get us there."

"Let's hope he does."

The two of them fell silent and soon they were both asleep. But then – it felt like seconds later – Matt found himself being woken by something. It was the boat's engine which had changed tempo, slowing down. He opened his eyes. It was still dark, still raining. But there were lights ahead.

"You can wake up your friends," Captain Machado said. "We're here."

Matt stood up and went over to the window.

And there it was. It was two o'clock in the morning but a city like Hong Kong never really slept. Matt could make out the skyscrapers by the lights that burned all around them, picking out their shapes in brilliant green, blue and pink neon. It was as if someone had drawn the city onto the darkness with a vast, fluorescent crayon. There were advertisements – PHILIPS, SAMSUNG, HITACHI – burning themselves onto the night sky, the colours breaking up in the water, being thrown around by the choppy waves. There were signs in Chinese too, and they reminded him how very different this city would be from London or Miami. This was another world.

It was very misty. Maybe it was an illusion caused by all the neon, but the mist was a strange colour, an ugly, poisonous yellow. It was rolling across the harbour towards them, reaching out to surround them as if it were a living thing and knew who they were. As they continued forward, it pressed itself against the glass of the cabin and the sound of the engines became even more distant.

Richard had joined the captain at the steering wheel. "Why are we going so slowly?" he asked. It was a good question. They were barely moving at all.

"We don't want to draw attention to ourselves," Machado replied.

"I thought you said nobody cared about you anyway."

"There's still no reason to make too much noise."

Another minute passed.

"I thought we were going to Kowloon," Richard said.

"We are."

"But isn't Kowloon on the other side?"

Machado grinned in the half light. He had put the pipe away. "The current will carry us over," he said and at that moment Matt knew that he wasn't telling the truth and felt the familiar tingle of imminent danger. For what seemed like an age, nothing happened. They weren't moving. Machado was standing there, almost daring them to challenge him – to do anything. But there was nothing they could do. They were trapped on board his boat, completely in his power.

And then a searchlight cut through the darkness, pinning *Moon Moth* in its glare. The entire cabin seemed to explode with dazzling light. A second beam swung across. Two boats. They were still some distance away but they were rapidly closing in. They must have been waiting there all the time.

At the same moment, Machado swung his hand, crashing it into the side of Richard's head and then bringing it around on Matt. He was holding a gun. Richard fell. Machado's lips curled in an unpleasant smile. "If you move, I will kill you," he said.

He had betrayed them. He had known the boats were coming. He had led them straight to them.

"The Triads will kill you for this…" Richard muttered. He had pulled himself onto one knee and was cradling his head in his hand. Blood was trickling from a wound just above his eye.

"The Triads are finished," Machado replied. "They're nothing any more."

"So who's paying you?" Matt asked.

"There's a big reward out for you, boy. Two million Hong Kong dollars. More than I've earned with Shan-tung and his friends in ten years. They want you very badly. And they warned me about you. If you even blink, I'll shoot you."

Matt looked out of the window. The boats were getting closer and they had been joined by three more, making five in all, moving in from every side. They were police launches – grey, solid steel with identifying numbers printed on the side. They were coming out of the night like miniature battleships, with bullet-proof windows and bows shaped like knives.

Richard pulled himself to his feet. Machado aimed the gun at him. "Nightrise doesn't want you," he said. "So I hope you don't mind a burial at sea." He was about to fire at point blank range. He licked his lips, enjoying himself. Richard stared at him helplessly.

"Put the gun down," Jamie said.

Machado didn't hesitate. He laid the gun on the floor although his face was filled with puzzlement. He had no idea why he'd done it. But Matt did. In his moment of triumph, the captain had forgotten Jamie. He'd thought he was still asleep … but he'd been wrong. Jamie had seen what was happening and had used his power. If he'd told Machado to stop breathing, the man would have stood there until he died. And, Matt reflected, maybe that was what he deserved.

"This is the Hong Kong police. Heave to…"

The voice echoed out of the water, amplified through a megaphone. There was a man standing on the bow of the nearest boat – except he looked far too tall to be human. He was black and was dressed in the uniform of a senior officer in the Hong Kong police. But it was obvious he was no police-man. He was like something out of a nightmare with his bald head and empty, staring eyes. It was freezing cold out on the water but he wasn't shivering. He showed no feeling or emotion at all.

Richard lunged forward, grabbed hold of the steering wheel and slammed down the throttle. Matt felt the floor tilt beneath him as the cargo boat surged forward. Captain Machado had been standing there, dazed, as if unsure what to do, but now he seized hold of Richard and the two of them began to grapple for the steering wheel.

"Get rid of him, Jamie," Matt said.

"Jump overboard," Jamie commanded.

Machado let go of Richard and lurched out of the cabin, moving as if in a trance. There was shouting, a shot, then a splash as Machado was gunned down even as he hit the sea. The Hong Kong police had assumed he was trying to escape. Or maybe they knew who he was but had decided to kill him anyway. Machado floated face down in the water. He didn't move.

Richard had control of the cargo boat. He spun it round, taking the police by surprise. Seconds later, he burst through them, weaving round one of their boats, heading for the Central side of Hong Kong.

"The gun!" Richard shouted.

Matt snatched it up and handed it to him. Then Jamie shouted and pointed. "Watch out!"

A face had appeared at the window, glaring at them with furious eyes. For a moment Matt thought one of the policemen had somehow boarded *Moon Moth*. Then he remembered the single crewman – Billy – who had sailed with them from Macau. He was holding a gun, bringing it round to aim at the cabin. Richard shot him through the window, a single bullet between the eyes. The boat lurched crazily. The wheel spun. The crewman disappeared.

Then the nearest police launch opened fire. The noise was deafening as the bullets smashed into the metal plates of the cargo boat, cutting a line along the bow and ricocheting back into the water. One of the windows shattered and Richard ducked as tiny fragments of glass showered down onto his shoulders and back. The cold night air rushed into the cabin, carrying with it the spray of water and the foul, decaying smell of the pollution. *Moon Moth* surged forward. Richard was fighting with the wheel, trying not to be shot. Matt looked back. The police launches were regrouping, preparing to come after them. The man at the front suddenly opened his mouth and howled, a sound that split the night, louder than all the boats put together. Matt knew at that moment that he wasn't a man at all.

"We're going to have to jump!" Richard shouted above the roar of the engines and the raging wind. "Jamie, can you swim?"

Jamie nodded.

"I'm going to take us in as close as I can." He turned to Matt. "If we get separated, meet at…"

But Matt didn't hear the rest of the sentence. There was another burst of gunfire, this time strafing the stern and the cargo hold where the fireworks were packed.

"Now!"

Richard abandoned the wheel and the boat began to zigzag. Matt needed to ask him what he had just said, but everything was happening too quickly. Richard snatched up his backpack and forced it over his shoulders. Jamie was right next to him. The five police boats were getting closer, only a few metres behind.

"Go!" Richard shouted.

Jamie hurried out to the deck and without stopping disappeared over the side of the boat. But Richard hadn't followed.

He had climbed down from the cabin and was balancing himself, clinging to a handrail as *Moon Moth*, its engines screaming on full power, swerved drunkenly through the sea. Blood and water streamed down his face and his eyes were wild. Matt had never seen him like this before. Gritting his teeth, he brought the gun up and fired into the crates of fireworks, again and again, emptying the chamber into the same spot.

Nothing happened until the final shot. Then there was a flare of magnesium, burning through the tarpaulin. Richard noticed that Matt was still there, that he hadn't jumped overboard. "Jump!" he pleaded.

Matt jumped.

Even as his feet left the deck, the fireworks went off. There were thousands of pounds worth in the hold. A tonne of gunpowder. But there was nothing beautiful about the explosion. It was just a blinding, burning wheel of fire that seemed to take Richard and hurl him into the air. That was the last thing Matt saw before he hit the water. For a moment everything was panic. The sea was black and freezing. He was still wearing his clothes and trainers. He was being sucked down. He had to fight with all his strength just to get back to the surface.

He emerged, gasping for air, into a brilliant, blazing nightmare. It was as if the whole night was on fire. *Moon Moth* was alight. The fire was burning so intensely that the metal plates would surely melt away. With no one to steer it, the boat had turned a full circle and was ploughing into the police launches, which had been too slow to get out of the way. It was right in the middle of them and Matt could just make out figures in helmets and full riot gear staring at the destruction, knowing that they were too close, that they were part of it. One of their boats was already on fire. The tall man was still howling – but this time in agony. Every part of him was on fire. His suit and the skin beneath it were peeling away. At the very end, his head split open and something began to snake out of it – a second head, but not a human one. Then there was a great rush of white flame as more of the fireworks exploded and he was blown out of sight.

Individual fireworks were going off, one after another and Matt saw cascades of red, blue, white, green and yellow as blazing missiles were shot into the air, reflecting in the water below. About fifty rockets screamed out at once, some of them twisting into the sky, others slamming into the police boats. One of them spluttered across the water and plunged down in front of him, missing his head by inches. He saw a policeman on fire, jumping into the water to save himself. Another was less lucky. He seemed to be holding a spinning Catherine wheel, unable to let go of it even though it was burning into his chest. Fireworks were cracking and buzzing and whining all around him. He didn't make it into the sea. He died where he stood.

Matt was treading water, forcing himself to breathe. He was so cold that his lungs had shut down. He knew that he couldn't stay out here much longer. Two of the police boats were undamaged. Very soon they would be looking for him. But where was Richard? Where was Jamie? The surface of the water was like a black mirror, reflecting the light, but he couldn't see them anywhere. He wanted to shout out for them but he didn't dare. The policemen would have heard him.

There was only one thing he could do. The edge of the water

was about a hundred metres away. He had to get to dry land and hope to find them there. He took one last look and then turned round and began to swim, slowed down by his clothes. The glow from the flames spread out over his shoulders, helping to light the way, and there were more bangs and fizzes as the last fireworks went off. He heard someone shouting an order in Chinese but doubted that they'd seen him. He was wearing dark clothes. His hair was dark. The currents were carrying him away.

He reached land without even realizing it. Suddenly there was a slimy concrete slope under his knees. He crawled onto it and pulled himself out. He was on a building site. That was what it looked like. It was hard to tell as he squatted in the darkness, shivering, filthy water dripping out of his hair.

"Richard? Jamie?"

He didn't dare call too loudly. The whole city – anyone who was awake – must have seen the firework display. The Old Ones knew he was there. They would already be searching.

"Richard? Jamie?"

There was no reply.

He waited ten minutes before he made a decision and set off, moving while he still could. If he stayed still much longer, he would freeze.

It was three o'clock in the morning. He had entered the enemy city. He had no idea where he was going. He was dripping wet. He was unarmed.

And he was alone.

NECROPOLIS

Leaving the water behind him, Matt made for the wall of light that defined the edge of Hong Kong. He came to a main road, empty at this time of the night, with a block of luxury hotels and shopping centres on the far side. The smog was worse than ever. The entire city reeked of it, like a chemical swamp. He had only been there for a few minutes but he already had a nagging headache and his eyes were smarting.

Where were Richard and Jamie? He had to find them. He was lost without them. Jamie had been the first off the boat and although Matt hadn't seen Richard jump, he must surely have followed moments later. Like him, the two of them must have swum ashore – unless the police had managed to find them first. The thought of his friends in captivity sickened him.

He tried to shake off the sense of hopelessness. He had to work out what to do. Get in touch with the Triads. There were a thousand of them, waiting to help him, but the way things had turned out, it wasn't going to be so easy after all. Han Shan-tung had given them a mobile with a direct dial. Richard had been carrying it. But it would have been made useless the moment it hit the water. And then there was Shan-tung's son, Lohan. He would already know that something had gone wrong. Presumably his men would be searching for them all over the city.

But Matt had no way of contacting them. He remembered the address of the place where they were supposed to be going, a warehouse on the Salisbury Road. But that was on the other side of the harbour, in Kowloon. Matt had no map and no money. He was soaking wet. It was the middle of the night. How was he supposed to get there?

He was already finding it hard to walk. Every time his foot came down, his shoes squelched and he felt the water rise over his foot. His shirt and trousers were clinging to him, digging in under his arms and between his legs. As he crossed the road and passed between the first of the buildings, he wondered if it wasn't a little warmer here than it had been in the harbour. But it was only a matter of degrees. He was soaked and shivering and if he didn't want to catch pneumonia he was going to have to find a change of clothes.

He stopped. A man had appeared, coming towards him from round the corner of a building. At first Matt assumed he was drunk, on his way home from a late-night party. The man was wearing a crumpled suit with a tie hanging loosely from his neck, dragged round one side, and he was staggering. Matt thought about hiding but the man obviously had no interest in him. And he wasn't drunk. He was ill. As he drew nearer, Matt saw that his suit was stained with huge sweat patches, and his face was a sickly white. He almost fell, propped himself against a lamppost, then threw up. Matt turned away, but not before he saw that whatever was coming out of his mouth was mixed with blood. The man was dying. He surely wouldn't last the night.

Slowly, the city began to reveal itself. Matt wasn't completely on his own after all. There were street cleaners out, sweeping the pavements, their faces covered by white cloth masks. He saw security men sitting on their own in the neon glare behind the windows, only half awake as they counted the long minutes until dawn. He passed the entrance of a subway station, closed for the night, but there was a woman sitting on the steps, a vagrant, her whole body completely wrapped in old plastic bags. She saw him and laughed, her eyes staring, as if she knew something he didn't. Then she began to cough, a dreadful racking sound. Matt hurried on.

An ambulance raced past, its siren off but its lights flashing, throwing livid blue shadows across the shop windows. It pulled in ahead of him and he saw that a small crowd had gathered round a man lying unconscious on the pavement. The ambulance doors were thrown open and two men climbed out, also wearing white masks. Nobody spoke. The man on the ground wasn't moving. The ambulance men scooped him up like a sack of meat and threw him into the back. He was either dead or dying and they didn't care. There were other bodies in the back, lots of them, piled up on top of one another. The ambulance men slammed the doors then got back in. A moment later, they drove away.

The city was huge, silent, threatening. It seemed to be entirely in the grip of the night, as if the morning would never come. Bald-headed mannequins in furs and diamonds stared out of the shop windows as Matt hurried past. Hundreds of gold and silver watches lay ticking quietly behind armour-plated glass. In the day, in the sunshine, Hong Kong might be a shopper's paradise. But at three o'clock in the morning with the pollution rolling in and the inhabitants sick and dying in the streets, it was something close to hell.

They were looking for him.

He heard the sound of a car approaching and the very speed of it, the angry roar of the engine at this time of the night, told Matt that its journey was urgent and that he should get out of its way. Sure enough, just as he threw himself into a doorway, a police car shot past, immediately followed by a second, both of them heading the way he had just come. He knew that he had to get out of sight before any more arrived. He crossed another wide avenue and began climbing uphill.

And then he heard something coming through the darkness. It was the last thing he would have expected in a modern city and at first he thought he must be mistaken. The clatter of metal against concrete. Horse's hooves…

A man appeared, riding a horse through a set of red traffic lights. The hooves were striking the surface of the road with that strange, unmistakable rhythm, and the echo was being trapped, thrown back and forth between shop windows. The horse paused under a street lamp and in the yellow glare, Matt saw that it was even more horrible than he had imagined. It was skeleton-thin and in an act of dreadful cruelty someone had driven a knife into its head, the blade pointing outwards, so that it looked like a grotesque version of a unicorn.

Matt saw it and remembered Jamie telling him about the fire riders who had taken part in the battle ten thousand years before. Was this one of them? As the man and the beast went past, he ducked behind a parked car, watching them in the wing mirror until they had disappeared from sight.

He was about to stand up, then froze as something huge fluttered through the darkness, high above the skyscrapers. Matt didn't see what it was but guessed that it was some sort of giant bird, maybe even the condor that had been part of the Nazca Lines. It was there, a sweeping shadow, and then it had gone. He knew now that the whole city was possessed: the roads, the water, the very air. It could only be a matter of time before he was seen and captured. Every moment he was on the street he was in terrible danger.

He waited until he was sure there was no one around, then straightened up and hurried on his way, keeping close to the buildings so that he could throw himself into the shadows if anyone approached. He came to a junction. A car had swerved and crashed into a bollard. It was completely smashed up, its horn blaring. Matt could see the driver, half hanging out of the front door, pinned in place by his seat belt, his head and chest covered in blood. No one was coming to help.

A street sign. Matt looked up and read two words directly above him. Harcourt Road. The name meant something.

Paul Adams has returned to Wisdom Court… It is here, on Harcourt Road.

He remembered Han Shan-tung, talking to him in the study, pointing it out on the map. Suddenly he knew what he had to do. Somehow he had stumbled onto the right road. If Paul Adams was at the flat, maybe he would let him in. At the very least he would have somewhere to stay until the break of day.

341

"Help me…"

The man in the car wasn't dead. His eyes, very white, had flicked open. He seemed to be crying, but the tears were blood. There was nothing Matt could do for him. He turned away and began to run.

The road seemed to go on for ever. Matt went past more shopping malls, a hospital, a huge conference centre. He didn't see any more police cars but he heard them in the distance, their sirens slicing through the air. At one point, a taxi rushed past, zigzagging crazily, on the wrong side of the road. He turned a corner and came upon a tram, parked in front of an office building. It was an old-fashioned thing. Apart from the Chinese symbols, it was like something that might have driven through London during the Second World War. And it was full of people. They were just sitting there, slumped in their seats, unmoving. Matt didn't know if they were alive or dead and he didn't hang around to find out. He guessed they were a mix of both.

Somehow he found his way to Wisdom Court. He had only glanced at the map when he was in Macau and he'd got no more than an overview of the city. But there it was, suddenly in front of him, the name on a block of stone and behind it a driveway leading up to a fountain, a wide entrance and, on each side, a statue of a snarling lion. The building was very ordinary, shrouded in darkness, but there was one light burning on the twelfth floor – Matt counted the windows – and he thought he saw a curtain flicker as somebody moved behind.

The driveway hadn't been swept. It was strewn with dead leaves and scraps of paper. The fountain had been turned off.

342

As he walked up to the door, Matt got the feeling that the whole place, apart from that one room on the twelfth floor, might be deserted. There were no cars parked outside. He put his face against the glass door and looked into the reception area. It was empty. The door was locked but there was a panel of buttons next to it, more than a hundred of them, numbered but with no names.

Was this really a good idea? He stood there for a few seconds, cold and wet, and tried to work out his options. Han Shan-tung had suggested that Paul Adams might have been working with the Old Ones. He had been there when Scarlett was taken prisoner. But could he really have sentenced his own daughter to death? Surely not.

At the end of the day it didn't make any difference if Matt trusted him or not. He was freezing. He had to get inside, off the street. He had nowhere else to go.

He began to ring the bells, one after another, beginning with 1200 and moving along, waiting briefly for each one to reply. There was silence until he reached 1213, then a crackle as a voice came over the intercom.

"Yes?"

"Mr Adams?"

"Who is this?"

"I know it's very late, but I'm a friend of Scarlett's. I wonder if I could talk to you."

"Now?"

"Yes. Could you let me in?"

A pause. Then a buzz and the door opened.

As Matt walked into the reception area, he became aware of

a stench – raw sewage. A pipe had burst. He could hear it dripping and the floor was wet underfoot. There was just enough light to make out a staircase leading up, but once he began to climb he had to feel his way in total darkness. He counted twelve floors, sliding his hand along the banister, pressing his shoulder against the wall as he turned each corner. It really was like being blind and he felt smothered, afraid that at any moment something would jump out and grab hold of him. But at last he arrived at a swing door, pushed it open and found himself at the beginning of a long corridor. Light spilled out from an open door about half-way down. Scarlett's father was waiting for him, but Matt couldn't make him out because the light was behind him and he was in silhouette.

"Who are you?" Paul Adams called out.

"My name is Matt."

"You're a friend of Scarly's?"

"I want to help her."

"You can't help her. You're too late."

Matt walked down the corridor, afraid that Paul Adams would go back in and close the door before he could reach him. But Adams waited for him. He reached the door and saw a small, unhappy man with grey hair and glasses. Scarlett's father hadn't shaved for a couple of days, nor had he washed. He was wearing a blue jersey which might have been expensive when he had bought it but now hung off him awkwardly, as if he had been sleeping in it. And he had been drinking. Matt could smell the alcohol on his breath and saw it in the eyes behind the glasses. They were red with exhaustion and self-pity.

"Mr Adams..." Matt began.

"I don't know you." Paul Adams looked at him blankly.

"I told you. My name is Matt."

"You're soaking wet."

"Can I come in?"

Matt didn't wait for an answer. He pushed his way past and entered the flat. The place was a mess. There were dirty plates stacked in the sink and on the kitchen counter. Everything smelled stale and airless with the sewage creeping up from below. It was as if someone had died there … or maybe it was the place itself that had died. Once it had been luxurious. Now it was sordid and sad.

Paul Adams closed the door. "Do you want something to eat?" he asked.

"I'd like some tea," Matt said. The man didn't move so he went into the kitchen and began to make it himself. He looked in the fridge for some food. There were only leftovers but he helped himself anyway. It was only now that he realized how hungry he was. A clock on the oven showed twenty past four. Six hours had passed since he had left Macau.

Paul Adams sat down. He had a glass of whisky and he drank it in one swallow, then refilled it. "You're English…" he said.

"I was at your home in Dulwich," Matt said. He was rummaging through a cupboard for a tea-bag. "I tried to find Scarlett there. But she'd gone."

"They've taken her."

"Do you know where she is?"

"No." He drank again. "I know who you are!" he exclaimed. He had only just worked it out. "You're the boy they're all looking for. You're the reason why they wanted Scarlett."

Matt didn't say anything. The kettle boiled and he made himself the tea, adding two spoons of sugar.

"Matt Freeman. That's who it was. Matt Freeman!" He got up and went over to the kitchen, weaving his way across the carpet. Matt didn't know whether to be saddened or disgusted. He had never seen anyone so utterly lost. Paul Adams leant heavily against the side of the counter and suddenly there were tears in his eyes. "They lied to me," he said. "They told me she'd be all right if I helped them. I was the one who caught her! She'd have got away if it hadn't been for me. But I only did it to protect her. They said they'd kill her if I didn't help them."

"Did they take her to The Nail?" Matt asked.

"She's not there." Paul Adams shook his head.

"Is she still in Hong Kong?"

"Somewhere. They won't tell me." He paused and looked at the window. The first streaks of morning were beginning to bleed through the night sky. "I thought they'd be grateful for what I did, but they said I'd never see her again. They were mocking me. I'd helped them and it was all for nothing. They wanted me to know that." He took off his glasses and wiped his eyes with the back of his hand. "I don't understand what they want, Matt. I don't understand anything any more. This whole city…" His voice trailed away.

"Mr Adams, I can help you," Matt said. "I can find her and get her out of here."

"How? You're just a kid."

"I need to have a shower and get changed." Matt was still dripping water onto the expensive carpet. "Do you have spare clothes?"

346

"I don't know..." He waved vaguely in the direction of the bedroom.

Matt drew on the last of his strength, forcing his mind into gear. He had to find Scarlett. That was the reason he was here. But that wasn't going to be possible, not if she had been taken to some secret location. Was she even still in Hong Kong? He guessed that she would have to be. The Old Ones were using her to get at him. Surely they would keep her there until he arrived.

How to find her? Matt's eyes were desperately heavy. All he wanted to do was go to bed. But somehow he knew that this was his last chance. He had to bring all the pieces together, here in this room. First there was Paul Adams, destroying himself, wracked with guilt and misery. Then there was the man called Lohan, somewhere in Hong Kong with his thousand foot-soldiers. Richard and Jamie. Maybe they had found their way over to them. And the fireworks. What was the name he had seen, stencilled on the crates?

And suddenly he had it.

"Listen to me," he said. "I may be able to find Scarlett, but you're going to have to help me. Will you do that?"

"I'll do anything."

"Does your telephone work here? And do you have a phone book?"

Paul Adams had been expecting something more. How would a simple phone call save his daughter? "It's over there..." He gestured with the hand that was still holding the whisky glass.

Matt went over to the telephone. It was a desperate plan. But he could think of no other way.

He picked it up and began to dial.

* * *

They came for him just after seven o'clock.

Matt was asleep on the sofa, dressed in jeans and a sweater that didn't really fit but were a lot better than the ones he had dumped in the bathroom. He had taken a hot shower, washing the smell of the harbour off his skin and out of his hair. And then he had fallen into a deep, dreamless sleep.

He hadn't heard the police arrive. They had driven down Harcourt Road and turned into Wisdom Court without sirens. He was woken by the sound of the door being smashed open and the shouts of a dozen men as they poured into the flat. Some of them were carrying guns. It was hard to say who was in charge. Suddenly they were everywhere and Matt was surrounded.

He started to get up but something hit him in the chest. It was a dart, fired from what looked like a toy gun, trailing wires behind it. But the next thing he knew, there was an explosion of pain and he was literally thrown off his feet as a bolt of electricity seared through him. He had been hit with a Tasar, a weapon used by police forces all over the world. Despite its appearance, it had fired an electrical charge that had resulted in the total loss of his neuromuscular control. Matt had never felt pain like it. It seemed to shatter every bone in his body. He heard an animal whimper and realized it was him.

Matt collapsed to the ground, unable to move. The policemen weren't taking any chances. They had deliberately neutralized him before he could use his power against them.

A moment later, two of them fell on him. They twisted his arms behind his back and he felt cold steel against his wrists as a pair of handcuffs were locked into place. One of the policemen grabbed him by the hair and twisted him round so that he was in a kneeling position.

Another man appeared at the door.

"So this is Matthew Freeman," he said.

The chairman of the Nightrise Corporation had wanted to make sure that everything was safe before he came in. Now he strutted forward and stood over Matt, looking down at him with a smile on his face. Although he had been hastily summoned out of bed, he was as smartly dressed as always, in a new suit and polished shoes. "What a great pleasure to meet you," he added.

Matt ignored him. He twisted round so that he was facing Paul Adams. His eyes were filled with anger. "What have you done?" he yelled.

"I called them while you were in the shower." Adams went over to the chairman. It was clear he was afraid of him. He stood there, wringing his hands together as if trying to wash them clean. "This is the boy, Mr Chairman," he muttered. "He came to the flat in the middle of the night. I called you the moment I could."

"You've done very well," the chairman muttered. He was still gazing at Matt. "I never thought it would be this easy," he said.

Matt swore at him.

"I knew you were looking for him, Mr Chairman," Paul Adams went on. "And now you have him. So you don't need Scarly. Tell me you'll let Scarly go."

The chairman turned his head slowly and examined Scarlett's

father as if he were a doctor about to break bad news. "I will not let Scarly go," he said. "I will never let Scarly go."

"Then at least let me see her. I've given you the boy. Don't I deserve a reward?"

"You most certainly do," the chairman said.

He nodded at one of the policemen, who shot Paul Adams in the head. Matt saw the spray of blood as the back of his skull was blown off. He was dead instantly. His knees buckled underneath him and he fell to one side.

"A quick death," the chairman remarked. He nodded at Matt. "Soon you'll be wishing you could have had one too."

He turned and walked out of the room. Two of the policemen reached forward and jerked Matt to his feet. Then they dragged him out, along the corridor and down to the city below.

TAI FUNG

SIGNAL ONE

The dragon was moving towards Hong Kong, closing in with deadly precision, gaining strength as it crossed the water. Scarlett had summoned it and it had heard. Even she couldn't turn it back now.

It had begun its life as nothing more than a front of warm air, rising into the sky. But then, very quickly, a swirl of cloud had formed, spinning faster and faster with a dark, unblinking eye at the centre. By the time the weather satellites had transmitted the first pictures from the Strait of Luzon, it was already too late. The dragon was awake. Its appetite was as big as the ocean where it had been born and it would destroy anything that stood in its path.

The dragon was a typhoon.

Tai fung.

The words mean "big wind", but they went nowhere near describing the most powerful force of nature; a storm that contained a hundred storms within it. The typhoon would travel at over two hundred miles an hour. Its eye might be thirty miles wide. The hurricane winds around it would generate as much energy in one second as ten nuclear bombs. To the Chinese, typhoons are also known as "the dragon's breath", as if they come from some terrible monster living deep in the sea.

Since 1884, the Hong Kong Observatory had put out a series of warnings whenever a typhoon had come within five hundred miles and each warning has come with a beacon, or a signal, attached. Signal One was shaped like a letter T and warned the local populace to stand by. Signal Three, an upside down T, was more serious. Now people were told to stay at home, not to travel unless absolutely necessary. Later on came Signal Eight, a triangle, Signal Nine, an hourglass, and finally, most terrifyingly, Signal Ten. Perhaps appropriately, this took the shape of a cross. Signal Ten meant devastation. It would almost certainly bring wholesale loss of life.

And that was what was on its way now.

But there were no warnings. Nobody had been prepared for a typhoon in November, which was months after the storm season should have ended. And anyway, no typhoon could possibly have formed so quickly. It would normally take at least a week. This one had reached its full power in less than a day. The whole thing was impossible.

Nor was there anyone left to send out the signals. Hong Kong Observatory had been abandoned. Many of the scientists had left. The others were too scared to come to work as the city continued its descent into sickness and death.

Unseen, the dragon rushed towards them. The skyscrapers were already in its sight. Suddenly they seemed tiny and insubstantial as, with a great roar, it fell on them. By the time anyone realized what was happening it was already far too late.

SIGNAL TWO

The chairman of the Nightrise Corporation was wondering how many people had died in the last twenty-four hours and how many more would die in the next. He could imagine them, sixty-six floors below, crawling over the pavements, begging for help that would never come, finally losing consciousness in a cloud of misery and pain. He himself would leave Hong Kong very soon. His work here was almost finished. It was time to claim his reward.

The Old Ones were going to give him the whole of Asia to rule over in recognition of what he had achieved. Even Ghengis Khan hadn't been as powerful as that. He would live in a palace, an old-fashioned one with deep, marble baths and banqueting rooms and gardens a mile long. The world leaders who survived would bow in front of him and anyone who had ever offended him, in business or in private life, would die in ingenious ways that he had already designed. He would open a theatre of blood and they would star in it. And anything he wanted he would have. The thought of it made his head spin.

He was behind his desk in his office on the executive floor of

The Nail and he was not alone. There was a man sitting on the same leather sofa that Scarlett Adams had occupied just a week before. The man had travelled a very long way and he was still looking crumpled from his flight. He was elderly, dressed in a shabby, brown suit that didn't quite fit him. It was the right size but it hung awkwardly. The man was bald with two small tufts of white hair around his ears and white eyebrows. He looked ill at ease in this smart office. He was out of place and he knew it. But he was glad to be here. It had been a journey he was determined to make.

His name was Gregor Malenkov. For many years he had been known as Father Gregory, but he planned to put that behind him now. He had left the Monastery of the Cry for Mercy for good. He, too, had come for his reward.

"So how do you like Hong Kong?" the chairman asked.

"It's an extraordinary city," Father Gregory rasped. "Quite extraordinary. I came here as a young man but it was much smaller then. Half the buildings weren't here and the airport was in a different place. All these lights! All the traffic and the noise! I have to say, I hardly recognized it."

"A week from now, it will be completely unrecognizable," the chairman said. "It will have become a necropolis. I'm sure you will understand what that means, a man of your learning."

"A city of the dead."

"Exactly. The entire population has begun to die. In just a matter of days, there will be no one left. The corpses are already piling up in the street. The hospitals are full – not that they would be any use as the doctors and the nurses are dying too. Nobody even bothers to call the cemeteries. There's no room there. And soon things will get much, much worse. It will be interesting to watch."

"How are you killing them?" Father Gregory asked. "Would I be right in thinking it is something to do with the pollution?"

"You would be entirely correct, Father Gregory. Although perhaps I should not call you that, as I understand you are no longer in holy orders." The chairman stood up and went over to the window, but the view had been almost completely obliterated by the mist which swirled around the building, chasing its own tail. There was going to be a storm. He could just make out the water down in the harbour. The water was choppy, rising into angry waves.

"There has always been pollution, blowing in from China," he continued. "And the strange thing is that the people here have tolerated it. Coal-fired power stations. Car exhausts. They have always accepted that it's a price that has to be paid for the comforts of modern life."

"And you have made it worse?"

"The Old Ones have added a few extra chemicals – some very poisonous ones – to the mix. You've seen the results. The elderly and the weak have been the first to go, but the rest of the city will follow if they are exposed to it for very much longer. Which they will be. An unpleasant death. We are safe, of course, inside The Nail. The air is filtered. We just have to be careful not to spend too long in the street."

Father Gregory pressed his fingers together. His sty had got much worse. The eyeball was now jammed, no longer able to move. Only his good eye watched the chairman. "I have to say, I'm disappointed," he said. "I was looking forward to meeting – to actually seeing – the Old Ones."

"The Old Ones have left Hong Kong. They have a great deal of work to do, preparing for a war that will be starting very soon. As soon as they heard that Matthew Freeman had been taken, they went."

"I don't understand why they don't show themselves to the world," Father Gregory said. "You have two of the Gatekeepers. So surely nothing can stop them…"

"It's not the way they work. If the Old Ones told the world that they existed, people would unite against them. That would defeat the point. By keeping themselves hidden, they can let humanity tear itself apart. That is what they enjoy."

There was a moment's silence. Father Gregory licked his lips and something ugly came into his eyes. "I want to see the girl," he said. "I still can't believe that she managed to break free when I had her. I had plans…"

"Yes, that was most unfortunate," the chairman agreed. "Well, right now they are together. The boy came all this way to find her, so I thought it would be amusing to let them spend one day in each other's company."

"Is that safe?"

"The two of them are locked up very securely and nobody knows where they are. The boy has certain abilities which make him dangerous. But as for the girl…"

"What is her power?"

"It seems that she drew the short straw. I'm afraid Scarlett Adams is not quite the superhero one might have imagined." The chairman smiled. "She has the ability to predict the weather. That's all. She can tell if it's going to rain or if the sun is going to shine. As she will never see either of these things again, it will not do her very much good. We are sending her away tonight. To another country."

"You can't kill her of course."

"It's vital that both children are kept alive. In pain, but alive. We are going to bury them in separate rooms, many thousands of miles apart. They will be given limited amounts of food and water, but no human contact. The Old Ones have asked me to blind Matt Freeman and that will be done just before Scarlett leaves. We want her to take the horror of it with her. In the end, she will probably go mad. It will be one of the last memories that she has."

"Excellent. I'd like to be there when it happens."

"That may not be possible."

Father Gregory was disappointed. But he continued anyway. "What about the other boy?" he asked.

"Jamie Tyler?" The chairman was still standing at the window. "He is somewhere here in Hong Kong. We haven't yet been able to find him."

"Have you looked for him?"

The chairman blinked slowly. Far below, two Star Ferries were crossing each other's paths, fighting the storm as they made their way across the harbour. Where had the storm come from? It seemed to be getting stronger. He was surprised the ferries were still operating and looked forward to the time when they finally stopped. It had always annoyed him, watching them go back and forth.

A boat will be the death of you. And it will happen in Hong Kong.

A prophecy that had been made by a fortune-teller. Well, soon there would be no more boats. There would be no more Hong Kong.

"Jamie Tyler can't leave the city," he said. "Unless, of course, he dies in the street and gets thrown into the sea. Either way, he is of no concern to us."

There was another silence.

"But now, my dear Father Gregory," the chairman said. "It is time for you to go."

"I am a little tired," Father Gregory admitted.

"It has been a pleasure meeting you. But – please – let me show you out…"

There was a handle on the edge of one of the windows and the chairman seized hold of it and pulled. The entire window slid aside and the wind rushed in, the mist swirling round and round. Papers fluttered off the desk. The stench of the pollution filled the room.

Father Gregory stared. "I don't understand…" he began.

"It's perfectly simple," the chairman said. "You said it yourself. You let the girl escape. You let her slip through your hands. You don't really think that the Old Ones would let that go unpunished?"

"But … I found her!" Father Gregory was staring at the gap. "If it hadn't been for me, you would never have known who she was!"

"And that is why they have granted you an easy death." The chairman had to shout to make himself heard. "Please don't waste any more of my time, Father Gregory. It's time for you to go!"

Father Gregory stared at the open window, at the clouds rushing past outside. A single tear trickled from his good eye. But he understood. The chairman was right. He had failed.

"I've enjoyed meeting you," he said.

"Goodbye, Father Gregory."

The old man walked across the room and stepped out of the window. The chairman waited a moment, then slid it shut behind him. It was good to be back in the warm again. He wiped some raindrops off his jacket.

The storm was definitely getting worse.

SIGNAL THREE

The Tai Shan Temple was very similar to all the other temples in Hong Kong.

It was perhaps a little larger, with three separate chambers connected by short corridors, but it had the same curving roof made of dark green tiles and it was set back behind a wall, on the edge of a park, in its own private world. Inside, it was filled with smoke, both from the coils of incense that hung from the ceiling and from the oven, which was constantly burning bundles of paper and clothes as sacrifices to the Mountain of the East. There were several altars dedicated to a variety of gods who were represented by standing, sitting and kneeling statues ... a whole crowd of them, brilliantly coloured, staring out with ferocious eyes.

Despite the bad weather, there were about fifteen people at

prayer in the main chamber, bowing with armfuls of incense, muttering quietly to themselves. They were many different ages, men and women, and to all appearances they looked exactly the same as the people who came daily to Man Mo or Tin Hau. And yet there was something about them that suggested that religion was not, in fact, the first thing on their minds. They were too tense, too watchful. Their eyes were fixed on a single entrance at the back of the building – a low, wooden door with a five-pointed star cut into the surface.

The worshippers – who were, in fact, no such thing – had very simple instructions. Any child who passed through that door was to be seized. If they resisted, they could be hurt badly but preferably not killed. The same applied to any young person coming in from the street. They were to be stopped before they got anywhere near the door. The people in the temple were all armed with guns and knives, hidden beneath their clothes. They were in constant touch with The Nail and could call for backup at any time.

This was the ambush that Matt had feared. It was the reason he had refused to take the shortcut to Hong Kong. He had been right from the very start.

The fifteen of them stood there, muttering prayers they didn't believe and bowing to gods they didn't respect. And outside, gusts of wind – growing stronger by the minute – hurled themselves at the temple walls, battering at them as if trying to break through, tearing up the surrounding earth and the grass, whistling around the corners. A tile slid off the roof and smashed on the ground. A shutter came loose and was instantly torn away. The rain, travelling horizontally now, cut into the brickwork. The traffic in the street had completely snarled up. The drivers couldn't see. There was nothing they could do.

The wind rushed in and the flames inside the temple furnace bent, flickered and were suddenly extinguished. Nobody noticed. All their attention was fixed on the doorway. That was what they were there for. Ignoring the storm, they waited for the first of the Gatekeepers to arrive.

SIGNAL FOUR

Scarlett was in a dark place, but someone was nudging her, trying to draw her back into the light. Unwillingly, she opened her eyes to find a boy leaning over her, shaking her awake. She recognized him at once and knew that the fact that he was with her, that he was bruised and dishevelled, could mean only one thing ... and it was the worst news of all. He was here because of her. The Old Ones must have tricked him into coming to Hong Kong and now the two of them were prisoners. Scarlett felt a sense of great anger and bitterness. She had been drawn into this against her will. And it was already over. She had never been given a chance.

"Matt..." she said.

At last the two of them were together. But this wasn't how she had hoped they would meet. She drew herself into a sitting position and rubbed her eyes. They had given her back her own clothes but her hair, cut so short, still felt unfamiliar to her. At least she had lost the contact lenses. She had taken them out the moment she had been left to herself.

"Are you OK?" Matt asked.

"No." She sounded miserable. "How long have I been asleep?"

"I don't know. They only brought me here an hour ago."

"When was that?"

"About eight o'clock."

"Night or day?"

"Day."

Matt examined his surroundings. They were in a bare, windowless room with brick walls and a concrete floor. The only light came from a bulb set in a wire mesh cage. From the moment the door – solid steel – had been closed and locked, he'd had to fight a sense of claustrophobia. They were deep underground. The policemen who had brought him here had forced him down four flights of stairs and then along a corridor that was like a tunnel. Ordinary policemen. The same as the ones who had arrested him. It seemed that the shape-changers, the fly-soldiers and all the other creatures of the Old Ones had decided to leave Hong Kong. He wondered why.

Despite everything, he had been relieved to find Scarlett. She looked very different from the photograph he had seen of her. He couldn't imagine what it must have been like for her, being stuck here on her own.

"Why are you here?" Scarlett asked. She still couldn't keep the disappointment out of her voice.

"I came for you," Matt said. He wanted to tell her more but he didn't dare. There was always a chance that they were being listened to.

"You shouldn't have. I've mucked everything up. I'd have got away if I hadn't..." Scarlett stopped herself. She couldn't bring herself to talk about her last meeting with her father.

Matt sat next to her so that they were shoulder to shoulder with their legs stretched out on the floor. From the way he moved, she could see that he had been hurt. He looked pale and exhausted. "Why don't you tell me everything that happened to you?" he suggested. "You could start by telling me where we are. Do you know?"

She nodded. "The chairman came to see me..."

"Who is the chairman?"

"Just some creep in a suit."

"I think I may have met him."

"He wanted to gloat over me," Scarlett continued. "He told me that you were on your way but I hoped he was lying. This is an old prison. We're right in the middle of Hong Kong. It was left over from Victorian times."

"So when do they serve breakfast?"

"They don't. It's bread and cold soup and they bring it once a day."

Matt lowered his voice. "Hopefully, we won't be here that long," he said. It was as much as he dared tell her, but even so Scarlett felt a glimmer of hope. "You know I went to your home in Dulwich," he said, changing the subject.

"Was that you in the car? There was an accident..."

"It was no accident."

"I knew it had to be you," Scarlett said. "They planned it all very carefully, didn't they? Using me to get you here. Are any of the others with you?"

Matt nodded briefly and Scarlett understood. They both had to be careful what they said. She gazed at him as if seeing him for the first and the last time. "I can't believe you're here. I can't believe I'm really talking to you. Do you know, I've even dreamed about you."

"Don't worry about it," Matt said. "We all dream about each other. It's how it works."

"There's so much I don't understand."

"Join the club."

"It looks like I already have." She took a deep breath. "I don't know where my story even begins, but I suppose I'd better start with St Meredith's…"

She told him – briefly and without fuss – and as she spoke, Matt knew that he was going to like her. She had been through so much, and in a way her experiences reminded him of his own at Lesser Malling, the way she had been reeled into something so completely beyond her understanding. And yet she had coped with it. She had been brought here. She had been locked in this room for three days. But she hadn't cracked. She was ready to fight back.

She finished talking and it seemed to Matt that just for a moment the building trembled as something, a shockwave, travelled through the walls. Scarlett looked up, alarmed. Part of her knew what was happening and had even been expecting it.

"What…?" Matt began.

"It was nothing." She said it so hastily that he could see she didn't want to talk about it, didn't even want to imagine what might be happening outside. "Tell me about yourself," she went on, quickly. "Tell me how you got here. Did you go to the temple? They've got people there waiting for you. They thought you'd come through one of the doors."

"I didn't…"

He told her his own story, or part of it, starting in Peru. It would have taken too long to tell her the whole thing and he was still afraid of being overheard. From Nazca to London to Macau… It had been a long journey and it was only now that they both saw how closely they had been following each other's paths.

Matt finished by explaining how he had found his way to Wisdom Court. This was the difficult part. He had seen Scarlett's father die and he had been at least in part responsible. How was he going to break the news?

But she was already ahead of him. "That jersey you're wearing," she said. She had suddenly realized. "It's his."

"Yes," Matt admitted.

"Where is he now?" Matt didn't answer and she continued. "They've killed him, haven't they?"

Matt nodded. He didn't want to remember what he had seen in the last moments before he had been taken out of Wisdom Court.

Scarlett's face didn't change but suddenly there were tears in her eyes. "It was all his fault," she said. "He thought he could make a deal with these people – the Old Ones – but they would never have got me if it hadn't been for him." She paused. "I don't know, Matt. I suppose that's the way they work. They get ordinary people to do evil things for them. They used him. He really thought he was helping me. And now he's betrayed you too."

The building shivered a second time. It wasn't as strong as

it had been before but they both felt it.

"You know that Hong Kong is dying," Scarlett said. "The chairman told me. They're doing it deliberately. They want to turn it into what they call a necropolis. A city of the dead."

"I saw some of it last night," Matt said. "It was horrible."

"Don't tell me. I lived in it. I can't believe I didn't see what was going on." She sighed. "What will happen to us, Matt? Are we going to be killed?"

"They don't want to kill us," Matt said. "It's complicated. But killing us doesn't really help."

"Then what?"

"They think they've beaten us, but they haven't. The others are still out there. And you and me..."

"What about us?"

"They put us together because they want to crow over us. But that's their mistake. Because..."

He didn't finish the sentence.

There was an explosion. It was loud and immediate – and it came from somewhere inside the building.

"What...?" Scarlett began.

Then the light went out.

SIGNAL FIVE

Lohan had used the storm as cover, closing in on the prison through streets that had quickly emptied as the weather had become more intense. He had only been given one night to prepare the attack, but he had still managed to assemble a small army. He had a hundred men with him, all of them well-armed. The Triads had been smuggling weapons across Asia for many years, supplying anyone from terrorists to mercenaries. Lohan had simply taken what he needed. He had plenty of choice.

Meanwhile, Jet and Sing would be arriving at the Tai Shan Temple. They both had the rank of 426, Red Pole as it was known, making them fighting unit lieutenants. They had another fifty men with them and both operations were to begin at the same moment. There was one door out of Hong Kong. The way there had to be cleared.

Lohan knew where Matt had been taken because he had followed him. This was what Matt had been unable to tell Scarlett. He had played a trick on the chairman. Just for once, he was the one pulling the strings.

Matt had contacted Lohan the night before, the call forwarded through the Kung Hing Tao firework company. The Triad leader already knew what had happened. Richard and Jamie were with him. The two of them had made it out of the water and over to Kowloon. They were standing next to him, worrying desperately about Matt, when the phone rang.

"We have to find Scarlett," Matt had said. "And there's only one way to do it. We have to let the Old Ones capture me."

"How will you do that?"

"Paul Adams – Scarlett's father – will call them and tell them I'm at Wisdom Court. They won't suspect anything. They know that he wants Scarlett back and they'll think he's still trying to help them."

"And then?"

"You have your men outside. You follow me wherever they take me."

"How do you know they'll take you to Scarlett?"

"I don't ... not for sure. But my guess is they'll probably hold us together. I know the way these people think. They'll want to parade us, to boast about how they've beaten us. Having the two of us together will make it more fun for them. Anyway, I haven't got any other ideas so we'll just have to risk it."

Richard had come onto the phone. He had heard what Matt was suggesting. "You can't do this," he pleaded. "It's too dangerous. Please, Matt, think what could go wrong."

"We don't know where she is, Richard. There's no other way we'll find her."

"What about Paul Adams? Once they have no further use for him, you know they'll kill him."

"He's prepared to risk it. He knows what he's done. And he'll do anything to get Scarlett freed."

It had worked out just as Matt had hoped. Six police cars had arrived at Wisdom Court just after seven o'clock. Lohan – with Richard and Jamie crouching next to him – had watched the police go in. They had seen the chairman arrive and leave and they were still there when Matt, semi-conscious and in pain, had been dragged out. Jamie had started forward at that moment, wanting to go to him. But Richard had grabbed hold of him, forcing him to remain still. This was Matt's plan. It was all or nothing.

Matt had been driven across the city, never out of sight of Lohan's men. They had seen him disappear into the prison close to Hollywood Road. So now they knew where he was being held. Hopefully, Scarlett would be there too. As the storm had worsened, Lohan had surrounded the prison, his men closing in from all sides.

The storm.

Lohan was beginning to think that it was getting out of control. In all the years that he had been in Hong Kong, he had never experienced anything like it. When he stood up, he could feel the wind trying to batter him down again. Dust and dead leaves whipped into his face. He could hear the air currents howling as they rushed through the streets. If it got any worse, it would be dangerous out here. But then, of course, it was dangerous anyway. If the storm destroyed the city, it would only be finishing what the Old Ones had already begun.

A crash of thunder. Rain lashing down so hard that he could see it bouncing off the parked cars, turning into miniature rivers that coursed along the side of the road. In seconds, he was soaked. Richard was next to him. "What's going on?" he muttered.

"We must move now," Lohan said.

Victoria Prison was a huge, solid building with barred windows and a single, massive door – the only way in. Six armed guards stood outside it in the rain, dressed in uniforms, with their faces partly obscured by their caps. Lohan, Richard and Jamie were watching from the doorway of an antique shop across the road. Lohan's strategy was simple. There was no time to be clever. He knew he had to break in as quickly, as decisively as possible. Once the enemy knew he was there, they would fight back.

He gave the signal.

There was an explosion – the same explosion that Matt had heard – as a rocket launcher, concealed in a parked van, fired a 40mm shell at the main door. The prison hadn't been built to withstand such an attack. The doors were blown apart in a ball of flame. Half the guards were killed instantly. The rest were cut down by a burst of machine-gun fire as the Triad fighters surged forward, pouring out of alleyways and rising up from behind parked cars. Further down the road, two of Lohan's men, disguised as construction workers, cut off the main power supply, isolating the prison and short circuiting the alarms.

"Move!" Richard and Jamie were unarmed but they ran forward with Lohan and in through the shattered doors.

And then they were inside the prison. Lohan's people were spreading in every direction, through the upper floors, smashing open the doors to reveal the empty cells behind them. Some of them were armed with guns and grenades. Others carried swords and chain-sticks. It was pitch black inside the building now that the electricity had been cut, but they had brought electric torches with them, strapped to their shoulders, the beams slicing through the dark and showing the way ahead. Lohan's orders were clear. Kill anyone who gets in your way. Find Matt and Scarlett. We have only minutes to get them out.

There were more guards on the upper levels. Although the building held only two prisoners, the chairman had taken no chances. Now they opened fire on the invaders. Lohan saw the flash of bullets, heard some of the Triad men cry out. A few bodies fell. Then someone threw a grenade. Another fireball, and one of the guards pitched forward as if diving into a swimming pool, disappearing into the darkness below.

Lohan himself led a group of fighters four floors down into the basement, Richard and Jamie close behind him. Only now was Richard beginning to see the hopelessness of the task. There had to be at least two hundred cells in the prison. Were they really going to blow every one of them open? They came to a corridor with more steel doors set at intervals. A guard ran towards them, bringing his machine gun round to aim.

"Drop the gun!" Jamie said. "Lie on the floor."

The guard did as he was told. A second guard appeared. He was less fortunate. Lohan shot him down. They had been in the prison for less than three minutes but they knew that re-inforcements would already be on the way. There was another explosion upstairs, a scream, the clatter of bullets hitting metal.

Thirty doors stretched out in front of them. There was no point looking for bolts or keys. Lohan rapped out an order and his men blew them open, one at a time, using balls of plastic explosive. Richard and Jamie continued forward as, one after another, the doors were smashed out of their frames, orange flames briefly flaring up. The corridor stank of cordite. Smoke and brick dust filled the air. But every cell was empty. How much more time did they have?

"They're at the end," Jamie said suddenly. "The last door on the left."

Lohan stared at him. But Richard nodded, relief surging through him. Somehow Jamie had managed to connect with them in his own way ... telepathically. Lohan shouted something and his men ran down to the door he had indicated. A final blast. It swung open. Two figures came out into the corridor, choking and covered in dust. It was Matt and Scarlett.

"Matt!" Richard grabbed hold of his friend and embraced him. The night before, when he had pulled himself out of the water, he had been afraid that he would never see him again. "Are you OK?"

Matt nodded. "This is Scarlett."

"I'm delighted to meet you." Richard didn't know what else to say. He examined the girl with the close-cropped hair. She looked worn out.

Jamie said nothing but he went over to her so that the three Gatekeepers were together.

"We have to get to the Tai Shan Temple," Matt said.

Lohan was impressed. The boy was only fifteen but already he had assumed command. The experiences of the past twenty-four hours didn't seem to have had any effect on him. But there was still more trouble to come. Quickly, Lohan took out his mobile phone, pressed a button and spoke a few words. He waited until he had heard what he wanted, then he turned to Matt. "The temple is safe now," he said. "But we have another problem and it may be more serious. There is a storm. In fact my people are saying that it may be something worse…"

But they had all become aware of it. Above the gunfire and the explosions. Beyond the battle that was taking place inside the prison, the wind was screaming. The whole building was shuddering. The full force of the typhoon had fallen on Hong Kong and its total destruction had begun.

SIGNAL SIX

The sun was setting in Cuzco, the ancient city of the Incas, in Peru. There was a band playing and the sound of pan-pipes and the throb of drums rose up into the evening air. The shadows were stretching out over the foothills. The restaurants and cafés were beginning to fill up at the end of another day.

Pedro knew that they shouldn't be here. This wasn't Matt's plan. He wished that they had been able to speak over the satellite telephone, but for the past forty-eight hours there had been only silence. A whole world separated them. They were thousands of miles apart. But he was about to take the single step that would bring them together. He wondered if it was a good idea.

Not that he had been given any choice.

The night before, Pedro had woken up to find Scott leaning over him. The two boys were sharing a stone house in Vilcabamba, high up in the Andes. This was the lost city where Pedro had gone with Matt when they were hiding from Diego Salamanda. It was hidden above the cloud forest in an extraordinary location, a mountain peak that couldn't be seen by anyone. Getting there had involved a helicopter ride and then a one-day hike from Cuzco. The city itself could only be reached by a stone staircase which could vanish in a single moment.

"Scott…? What is it?"

Scott was deathly pale and his eyes were full of worry. Pedro had never seen him like this before. "Jamie's in trouble," he said. "We have to go to Hong Kong."

"We can't..."

"Pedro. You don't understand. We have to go straight away. I have to go to Jamie. I've had a dream."

The dreamworld. All of them had been there. They all knew its significance. They had talked about it often enough. Pedro knew that he couldn't argue. If Scott had been sent a message, they couldn't ignore it, particularly if it involved his brother. And yet the doors were supposed to be too dangerous. It was the whole reason Matt and Jamie had flown to Europe and why the two of them had been left behind.

"Are you sure...?" he began.

Scott wasn't in the mood for an argument. "I'm leaving as soon as it's light," he said. "You can come with me or you can stay behind."

The next morning they left together. One of the Incas escorted them down to the clearing where the helicopter was waiting and then it was a two-hour flight to Cuzco airport. All the time, Scott had been silent and intense. He still hadn't explained what he had seen. He was often reserved but now he seemed miles away, staring ahead with empty eyes. Pedro was trying not to think what they were letting themselves in for. Of all the Gatekeepers, he alone had never been through one of the doors, and the thought of transporting himself half-way round the world filled him with dread.

And here they were now in Cuzco. It was a beautiful evening with hundreds of tourists milling around the brightly coloured stalls that were spread out in front of them. The cathedral would be closing soon. The last visitors were coming out, surrounded by street children, begging for money and sweets. Taxis, like wind-up toys made out of tin, were buzzing around the main square.

Pedro was hungry but he didn't dare suggest that they stop and eat. He knew what the answer would be.

"There it is..." Scott pointed at a great pile of bricks and ornate windows, a Spanish church built on the site of a place of worship that had been there centuries before. The Temple of Coricancha. It was where he and Jamie had found themselves when they first arrived in Peru. Inside was the doorway that had brought them from a cave in Nevada.

Neither of them spoke again. Pedro shook his head and followed as, with grim determination, Scott began to walk across the square.

SIGNAL SEVEN

Matt and Scarlett stood in the shelter of the prison, knowing that they couldn't leave. Hong Kong was being torn apart by a force so devastating it was as if they had arrived at some chapter in the Bible when all the old prophecies happened and Judgement Day finally arrived.

Smashed buildings and debris were being flung along the street as if they weighed nothing. As they looked out of the broken doorway, a huge neon sign spun past like an oversized playing card. It was followed by a table, several crates, a lawn mower, part of a piano... They had somehow been sucked out of the shops and sent on their way as if they were prizes in some insane TV game show. Matt could actually see the air currents. Mixed with the rain, they had become a thousand grey needles that raced along the streets, slamming into cars and tipping them over, flattening everything in their path.

He looked up and saw two clouds rushing together, moving faster than he could have believed. They hit and there was a massive burst of thunder. A bolt of electricity so bright that it hurt his eyes crackled down and smashed into a skyscraper half a mile away, cutting it in two. Shards of glass and broken pieces of metal burst outwards as the top seven storeys of the building leaned over and then fell, trailing wires and pipes. Matt didn't see where they landed or how many people were killed but he heard the massive explosion as they hit the street below. Despite the rain, what remained of the building caught fire. The orange flames licked at the falling water, desperately trying to climb into the air.

"We must wait…" Lohan was right next to him. Matt understood what he meant. If they took so much as one step forward out of the protection of the walls, they would be whisked away. He was having to shout the words to make himself heard.

"We can't wait!" Matt shouted back. "We only have this one chance. We must leave Hong Kong now."

Scarlett was behind him with Richard and Jamie. Matt turned round and their eyes met – and in that moment they both understood what was happening. They could have no secrets from each other. "This is you!" he shouted at her. The wind was still howling. A window on the other side of the road was suddenly torn out, the glass leaping away. "You've done this…"

"No!" Scarlett shook her head, trying to deny it.

"We all have powers. All five of us. This is yours."

And Scarlett knew he was right. In a way, she had known it all along.

Her real name wasn't Scarlett Adams. White Lotus believed that she was a reincarnation of Lin Mo, a figure out of Chinese mythology, a goddess of the sea. And if she had once been a goddess, then she would have a power that went far beyond anything humanly possible. The chairman of Nightrise had made another mistake. He had thought she could only predict the weather. In fact she could control it.

The evidence had always been there. At school in Dulwich, when Scarlett had wanted to go on a history trip, the weather had cleared up against all expectations. The same thing had happened again in Hong Kong when she needed to get to The Peak. Against all the forecasts, the rain had stopped and the sun had suddenly come out.

She had even used the same power at the battle, ten thousand years before. Jamie had once described it to Matt. Just as Pedro had appeared with his reinforcements, a storm had started, the rain coming down so violently that the Old Ones had been unable to see him.

It hadn't been a coincidence.

It had been her.

The chairman had claimed that she was the weakest of the Five. He had been wrong. She was by far the most powerful.

"You can stop it!" Matt shouted.

"I can't!" Scarlett shook her head. She had brought the dragon. She accepted that much. But looking inside herself, after three days in prison, after all she had been through, she knew that she didn't have the strength to turn it back.

"Then you can protect us. You can keep it away."

Scarlett looked out into the road, at the crashing rain, the buildings being scattered like confetti, cars spinning crazily, broken pieces of wood and metal hurtling past. Had she really done this, brought destruction on an entire city? How many people would she have killed? The thought terrified her more than anything else she had seen. Was she really responsible for this?

"I can't do it, Matt…"

"You have to… We have to reach the temple."

Lohan understood. "It's not so far from here," he shouted. "I can show you…"

"Scar…?" Matt looked at her.

And maybe it was simply the fact that he had used that name, a name from ten thousand years ago. Maybe that was the trigger. But in that second, something changed. Scarlett took a deep breath. For too long she had been a victim, pushed around by the chairman, by the Old Ones, even by the Triads. It was time to put that behind her. She was a Gatekeeper. That was what had brought her into all this and suddenly she felt a great anger for everything she had lost – her friends, her home, – even her father. And with the anger came the full knowledge of her own strength. She knew what she had to do.

"Follow me," she said.

They left the prison. First Lohan, then Scarlett and Matt, with Richard and Jamie behind. They stepped outside into the rain, into the wind, into an endless explosion as nature pounded the city with all its strength. They should have been thrown off their feet instantly, or battered senseless to the ground. But the wind spun around them. The rain was lashing everything but they remained dry. They walked into the heart of the typhoon and it swallowed them up without touching them. It was as if they were inside a glass ball that surrounded and protected them. They could barely see. Everything was chaos. But while they stayed together, they were safe.

Lohan led the way but it was Scarlett who made it possible. She seemed to be in a trance, gazing straight ahead, her arms by her side. Matt kept close to her, knowing that his life depended on her protection. All around them, everywhere he looked, brick walls crumbled, buildings fell, windows shattered and, spinning in the rain, lethal shards of broken glass came slashing down. Again and again the thunder sounded. The clouds were a boiling mass.

They didn't hurry. There was no need to. No living thing was going to come out in the typhoon and the five of them were completely invisible. Scarlett was more confident now. She looked almost relaxed. Walking next to her, Matt was amazed by the extent of her power. He could feel it flowing out of her. She was a girl and she was fifteen years old. But she could destroy the entire world.

Another building fell behind them, crumbling in on itself as if it had simply lost the will to live. Bricks showered down, slamming into the pavement, but not near them. The road continued straight ahead. They could see the park. Most of the trees had been uprooted and turned into flying battering rams. The few that remained were bending over, kissing the ground. The Tai Shan Temple was on the other side. Matt was surprised that it was still standing, but perhaps the wall that surrounded it had protected it from the worst of the weather.

Lohan pointed. Scarlett nodded. There was no need for any of them to speak. They had made it. They had crossed Hong Kong in the middle of a Signal Ten typhoon and they had survived.

Moving faster now, they crossed what was left of the park and went in.

SIGNAL EIGHT

The chairman of the Nightrise Corporation was watching the final destruction of his necropolis. He was back in his office on the sixty-sixth floor of The Nail and he could feel the whole building trembling as it was buffeted again and again by the storm. Every now and then there was a grinding sound followed by an explosion of breaking glass as another window burst out of its frame. The lights had long ago flickered and gone out. There was no power in the office. Nor were there any people. The staff had all evacuated, fighting and clawing their way down sixty-six flights of stairs. Some of them might have made it to the basement and would be huddled there now, but he suspected that many more of them would have been killed on the way down – pushed down the stairs or trampled in the general panic. The chairman certainly had no intention of joining them. He was safe here. The Nail could stand up to anything. And it was a spectacular view.

It did trouble him that his plans had somehow gone wrong. The city had been meant to die. That had been the whole idea. But not like this. Indeed, the typhoon might well end up saving many more people than it actually killed because there had been a side-effect. The poisonous gases put in place by the Old Ones had been dispersed. The pollution had been swept away. When the storm finally eased off, the people would be able to breathe again.

He didn't know what had happened at Victoria Prison. All the telephone lines were down and even his mobile didn't work. The whole network must have collapsed. But this devastation couldn't be a coincidence. The girl must have brought it. She was able to predict the weather so at the very least she must have known it was coming. He had put the boy in with her to taunt her, to show her how completely defeated she had been. Perhaps, all in all, it had been a mistake.

He was holding a bottle of Cognac. It had a price tag that made it one of the most expensive in the world and it had always amused him that there were people dying in some countries because they had no water while he could afford to spend five thousand dollars on a drink he didn't even enjoy. Over the years, most of the chairman's taste buds had died. Nothing he ate or drank had any flavour. If he was killed now, it would hardly matter. Most of him was dead anyway.

But he wasn't going to die. Even if Matt and Scarlett had escaped, there was nowhere for them to go. The Tai Shan Temple was protected. They wouldn't be able to reach the door. And soon the typhoon would have passed. He would begin the search through the wreckage immediately, turning it over brick by brick, and next time he would deal with them at once.

He noticed something out of the corner of his eye. It was a speck in the window. At first he thought it was a bird. No. It was extraordinary. As the chairman watched, it grew larger and larger. It was heading towards him.

It was a ship.

Not a huge ship. A wooden *sampan*, one of the Chinese sailing boats that were kept moored up in the harbour, to be photographed by tourists. The wind had grabbed it and torn it free. Even as the chairman watched, it was getting closer, rapidly filling up the window frame. He stood there, transfixed by the sight. He thought about running. Perhaps he could still make it to safety. But what was the point? How could he escape something that had been predicted so many years ago?

He would die in an accident that involved a ship.

He died now.

The *sampan* was thrown at The Nail as if it were a paper dart that had been deliberately aimed. It smashed through the window on the sixty-sixth floor and into the man who stood behind it. At the same time, the wind howled in, scooping up the contents of the room and throwing them out, the files and papers rattling with a sound that was very like applause. The broken body of the chairman went with them, spun once in the air, then plunged down to the pavement below.

Bloodstains on the carpet. A bottle of Cognac with its contents gurgling out. A scattering of broken glass. In the end, that was all that was left.

SIGNAL NINE

There had been a bloody battle inside the Tai Shan Temple. All the bodies had been taken into one of the other chambers, but the evidence was still there in the bullet holes across the walls, rubble and scorch marks from a grenade, a puddle of blood in front of the main altar. One of the porcelain gods was standing with his arms outstretched, but his body now ended at his neck, which was jagged and hollow. His head was in pieces all around him. Another had lost a hand. It was as if they had tried to take part in the fight and had been crippled as a result.

Jet and Sing had been on their own, waiting for Scarlett and the others to arrive. They had no idea how she had managed to cross Hong Kong – it would have been impossible now to leave the building – but they were glad to see her when she walked in. Jet had been wounded. He was holding a dressing against his neck, and his shirt was soaked in blood. Sing was still holding the sword stick that he had used to kill Audrey Cheng. He seemed to be unhurt.

Neither of them had noticed that there was another man in the chamber, hiding underneath the altar. He was one of the chairman's men and he had been shot twice. It was his blood that was pooling out. He knew he didn't have very long. There was a gun inches from his outstretched hand.

Speaking in Chinese, Lohan demanded a report from his two lieutenants. Quickly, they told him what he wanted to know and he translated for Matthew and Richard.

"There were many people waiting here," he said. "They would have killed you if you had tried to reach the door. But they have all been dealt with…"

"Then let's get out of here," Richard said. He turned to Matt. "It's time to go."

Lohan walked forward and shook Scarlett's hand. "Good luck," he said. "The journey that we made together just now is something that I will never forget."

"I'm glad I met you, Lohan," Scarlett said. "Thank you for helping me." She had relaxed a little, but Matt could see that she was still concentrating, keeping the typhoon at bay. She had to stay in control. While she was inside the temple the wind and the rain were barely touching its walls.

The door with the five-pointed star was in front of them. It seemed so small and ordinary that it was hard to believe that it would lead them, not outside and into the storm but to anywhere in the world.

"So where are we going, Matt?" Jamie asked.

The dying man had fumbled for the gun. From where he was lying he could only see the two boys and the Chinese man who had arrived with them. The girl was standing right behind and the other man was somewhere out of sight. He could probably take out at least two of them before he was killed himself. He had decided that was what he would do. After all, it was the reason he was here.

Which one first?

The boy who had just asked the question – the one with the long hair and the American accent – was directly in his sight. Slowly, the man took aim. The boy was only a few steps away.

The man's hand was sticky with his own blood. The gun was covered in it. But he knew exactly what he was doing. There was no way he was going to miss.

Then the door with the five-pointed star opened.

Scott, with Pedro right behind him, burst into the temple. Jamie opened his mouth to speak. Matt was gazing in surprise. What had seemed impossible for so long had finally happened. The Gatekeepers had come together. They were all here, in the same space.

Scott. Jamie. Matt. Pedro. And Scarlett.

The Five.

But Scott hadn't stopped. He ran forward and threw himself at his brother, knocking him aside.

A second later there was a gun shot.

Lohan acted with lightning speed. His own gun was in his hand instantly and he fired five times, the bullets strafing underneath the altar. The man who had been concealed there was killed before he could fire again.

Richard saw that Jamie was all right. Somehow Scott had known and had arrived in time to save him. But then Matt cried out.

The shot had missed Jamie, but Scarlett had been standing right behind him. She had been hit in the head and the wound was a bad one. Blood was pouring down the side of her neck. She toppled sideways. Richard caught her before she hit the floor.

And as she lost consciousness, the whole world exploded.

The typhoon had been kept at bay for too long. Now, as if recognizing what had happened, it fell on the Tai Shan Temple with all its strength. It was like being hit by a bomb, but in slow motion. As the nine of them stood there – the five Gatekeepers with Richard, Lohan, Jet and Sing – the whole building disintegrated around them. The roof was the first to go; torn off as if by a giant hand. Green tiles came crashing down. The wind roared in. Then one of the walls buckled and collapsed, the huge stones toppling forward. For centuries, the gods inside the temple had never seen daylight. Now they were flooded in it as the outside world burst in.

"The door!" Matt shouted.

It was still standing, but it wouldn't be there for long. Once the walls were destroyed it would all be over. The door would go with them. Even now it might be too late. Jamie had joined his brother. The two of them had already turned towards it. Pedro seemed to be confused, frozen to the spot. Matt reached him and spun him round. Richard was hurrying forward, carrying Scarlett who was in his arms, limp, her eyes closed. Lohan followed. One of the spinning tiles had hit him and he was cradling his arm. There was no sign of his two lieutenants, Jet and Sing. They had disappeared beneath the broken wall.

The door had been built for the Gatekeepers, but each of them could take one companion with them. Richard was with Scarlett. Lohan was with Matt. There was still a chance they could all get out alive.

There was another explosion and a great hole suddenly appeared, punched into the wall. Rain and daylight came shafting through. The whole temple was shaking. Scott was the first to reach the door and threw it open. Behind him, the remaining gods were toppling and smashing to pieces on the hard floor.

Pedro was next to him. The others were right behind.

They plunged through just as a last bolt of lightning struck the temple, pulverizing it. The remaining walls were swept away and scattered. Moments later, there was nothing left. Hong Kong Park was empty. And beyond it, Hong Kong itself lay in ruins as the clouds finally parted and the first, small ray of sunlight was allowed through.

SIGNAL TEN

The Necropolis was finished.

Much of it had been destroyed. More than half the skyscrapers had collapsed. Whole streets were buried beneath piles of twisted metal and brickwork that would take years to remove. Scavengers were already hard at work, burrowing into the rubble to find the jewellery – the diamond necklaces and the watches – that must surely lie beneath.

All over the world, people were waking up to the fact that a catastrophe on a massive scale had occurred. Twenty-four hour television news programmes were running the first pictures. There would be thousands dead, but at least the survivors would be able to breathe. The poisonous smog that had been suffocating them for so long had been completely swept aside.

Far away, sitting in the ice palace that he had made his home, the King of the Old Ones saw what had happened. He knew that the chairman had failed him. He knew that the Gatekeepers had escaped.

But it didn't matter.

The Five had entered the door without knowing where they were going, so none of them would have arrived in the same place. They would be as far apart now as they had ever been. Worse than that, the door had been disintegrating even as they had passed through it, and the final blast had played one last trick on them. If the five of them had survived the journey, they would find out very soon.

It would be a very long time before they found each other again.

It was enough.

The King of the Old Ones reached out and gave the order that his disciples had been waiting for. He had made the decision. It was time for the end of the world to begin.

an interview with

ANTHONY
HOROWITZ

When *Necropolis* was first published in hardback, hundreds of readers entered a competition to ask me a question about The Power of Five. You can find many of them with my answers on the website. These were my top 12 favourite questions.

AH

How do you start to write a story? Don't you find it daunting when faced with a blank page?

Patrick, aged 12, Isle of Wight

My favourite part of writing a book is thinking up the ideas and that can start a long time before I actually sit down at my desk. For example, I first visited Hong Kong (the main setting of *Necropolis*) twenty years ago. I got married there – and that was when I first began thinking of a nightmarish adventure taking place there. So I get the idea, I develop it, I keep turning it over in my mind and gradually a story begins to take shape. Then I work out the structure, balancing slow chapters with fast ones, violent moments with more reflective ones. I'm not daunted by a blank page. I look forward to filling it.

What was your inspiration for The Power of Five series?

Archie, aged 9, Peckham

I grew up with the Narnia books and then with Tolkien and *The*

Lord of the Rings ... but I never thought I'd be able to write a full-blooded fantasy sequence. I'm not very good at creating other worlds. I prefer to write about the world as it is. But at the same time, it's often struck me that the real world may not be quite how we imagine it. For example, if I walk past a locked-up church at night, I wonder what's happening inside. Is it actually empty or are there strange creatures coming together in the dark to engage in mysterious rituals? When I was living in North London, I used to pass a shop at the end of the street that always had furniture in the window ... the same things all year round. I became convinced that it was only pretending to sell furniture and I used to like imagining what really happened there. Could it be a meeting place for gangsters? What if it was actually run by visitors from another world?

I love the idea that magic and witchcraft and battles between supernatural creatures could be raging all around us but just out of our sight. This is particularly true of *Necropolis*. Hong Kong is being taken apart piece by piece but nobody has noticed. And I believe it could happen anywhere in the world: as one of the characters says, "We see only what we want to see because that is the way of the city."

Did you base any of the characters on anyone you know?

Dan, aged 11, Hampshire

Scarlett is based on the granddaughter of a friend of mine. The real Scarlett is only nine years old but it was easy to imagine what she might be like when she's fifteen. I also met a boy who was quite like Matt ... he was always in trouble at school and didn't seem to be enjoying himself very much. I went back to Hong Kong recently and many of the characters and locations come from places I visited and people I met. The lady with the birds of fortune, for example, is exactly as I describe her. She told my fortune outside the Tin Hau Temple (although she didn't say if the book would be a success or not).

Did you make up the Old Ones?

Karine, aged 12, Leeds

The Old Ones began life as a bad dream. My wife had bought me a strange piece of pottery – a sort of jug with a horrible creature climbing up the side. Some of her presents are a bit on the weird side! Anyway, that night I had a nightmare in which my house was invaded by beasts and, when I woke up, the pillows were on the floor, the duvet was tied in a knot and I knew I had an idea for a new book.

That said, the Old Ones were named after characters described by a famous horror writer called H.P. Lovecraft which he in turn took from a strange, sixth-century text called the *Necronomicon*. I've read it and I have to admit I didn't quite understand it but it's a cheerful piece of writing that seems to be describing the end of the world. There's even a little snatch of it in *Necropolis*. You'll find it at the end of the prologue.

Did writing from a female point of view feel weird?

Sarah, aged 10, Devon

Not at all. I'm very pleased with the way Scarlett turned out. I was quite nervous at the thought of writing about a girl as the main character – mainly because I'm known as a writer for boys and I was worried that boys might not be interested in following the adventures of a 15-year-old girl. I also wondered if I would be able to do it. Alex Rider, Nick Diamond, Matt Freeman ... all my heroes have been boys and in real life I only have sons. And of course I was once a boy, rather than a girl, myself.

In the end, writing Scarlett was a pleasure. She's a little more emotional than my boy characters. She's not afraid to cry, for example. But she's also just as tough as them ... as you'll discover in chapter 4. I like her because she's so unpredictable. So, in answer to your question – no, it never felt weird. I didn't have to put on a dress or anything like that.

Can you tell us anything about the fifth book in this series?

Tyler, aged 13, Brighton

The fifth book will also be the last book in the series! I don't have a title for it yet but I have a feeling that it's going to be the thickest yet (in terms of how many pages there will be). It's going to be divided into five parts and those parts are going to be called: *Earth*, *Fire*, *Air*, *Water* and *Ice*. I don't want to give too much away but it will finish with a battle, probably in

Antarctica, and although I've often said that I don't like children dying in my books, not all of the five gatekeepers will make it to the end. I should add that if you read the first four episodes carefully, some of the elements that make up the climax are already in place.

Why are all five children orphans?

Callum, aged 10, Nottingham

There are two main reasons. First, it greatly helps the stories if the heroes (and heroine) don't have parents, homes, ordinary lives. It means that they're "out there", on their own, having to depend on their own resources to win the battles. Alex Rider is an orphan for the same reason.

But there's an extra dimension to the five gatekeepers – which is to say, they have more than one identity. If you've read *Nightrise*, you'll know that they all met many thousands of years ago when they defeated the Old Ones for the first time. But they also seem to have an identity that has been drawn from myth and legend. Pedro, for example, is associated with Manco Capac, founder of the Incas (see chapter 15 of *Evil Star*). Scott and Jamie were also known as Flint and Sapling, characters out of Native American mythology. In *Necropolis* you find that Scarlett has another name.

The gatekeepers don't have parents because it is possible that they were never actually born ... even if that thought does rather make my head spin.

Do you believe in telepathy or any other supernatural powers your characters possess?

Tom, aged 10, Kenilworth

I suppose I believe in the possibility of these things – which is to say I have an open mind. I don't think I would be able to write this series if I didn't.

Is it easier to write about faraway settings – e.g. Peru/ Hong Kong rather than England? Have you visited all the places you write about in these books or do you just research in other ways?

Jack, aged 13, Cork

As I've already mentioned, I went back to Hong Kong recently, at the start of 2008, and I visited all the places that you'll read about in the book. I hope I've done the city justice. I was truly amazed by its size and energy, by the millions of people packed together in this relatively tiny space. The street markets that Scarlett visits in chapter 17 are exactly as I describe them … I was particularly struck by the fish sliced in half but still alive and I was as revolted as Scarlett.

I think it is important and I do get inspiration from visiting somewhere different before I write about it. I get so many ideas, just from walking the streets. And you're not going to get the noise and the smells from a guide book!

So, I took the Star Ferry, I walked round The Peak and I even visited Macau where, quite by chance, I stumbled on the extraordinary house that I would turn into the headquarters of the White Lotus Society. But what struck me perhaps more than anything was the pollution in that part of the world. I kept on seeing people wearing white masks to stop themselves breathing in the worst of it and that became a big part of the book. If I hadn't travelled out there, it would never have appeared.

Who/what is the Librarian?

Sacha, aged 11, Bolton

I'm afraid you'll have to wait for the last book to find out – although, in truth, everything you need to know about him is contained in *Necropolis*. I particularly enjoyed writing the chapter when Matt discovers the library. The choice he is given – whether or not to read the book of his life – is at the very heart of what this series is about. Would you read yours?

Is the King of the Old Ones the Devil? Do you believe in the Devil – or a power of evil?

Kieran, aged 12, London

What a very interesting question! The King of the Old Ones is vaguely inspired by the figure of Antichrist which you will find in the Bible. Try reading the Book of Revelations. It's full of monsters and nightmarish things. But for me he's really just a personification of everything that is bad in the world. I call him

Chaos because that's what he represents. An end of law, order and civilization.

Do I believe in the Devil? I don't believe in a figure with horns and a tail. Nor do I think there's a hell we go to if we've been bad (so I'll be more than a bit surprised if I end up there). But I do believe in evil as a force. And when I look at the world, particularly in the past ten years, I think it has been very much in evidence.

If you had to be one of your own characters in The Power of Five, which would it be and why?

Zak, aged 10, Glasgow

Another brilliant question to end with. My first instinct is to say that I would like to be one of the five gatekeepers as I wouldn't mind being a teenager again and I've always fancied having magical powers. On the other hand, I know what happens at the end and I'm not sure I'd want to be in any of their shoes when they reach the last chapter. I'd quite like to be Richard Cole, the journalist, but then again there's something pretty terrible waiting for him in the last book too. It's very difficult – but I think I'm going to choose Lohan. He's young, a great fighter, a leader and a master criminal. I'd quite to be all these things.